THE
INTERNATIONAL SERIES
OF
MONOGRAPHS ON PHYSICS

GENERAL EDITORS

J. BIRMAN   S. F. EDWARDS

C. H. LLEWELLYN SMITH   M. REES

The Ubiquitous Photon
Helicity Method for QED and QCD

R. Gastmans
Institute for Theoretical Physics
University of Leuven, B-3030 Leuven, Belgium

Tai Tsun Wu
Gordon McKay Laboratory
Harvard University, Cambridge MA 02138, USA

CLARENDON PRESS · OXFORD

This book has been printed digitally and produced in a standard specification in order to ensure its continuing availability

OXFORD
UNIVERSITY PRESS

Great Clarendon Street, Oxford OX2 6DP
Oxford University Press is a department of the University of Oxford.
It furthers the University's objective of excellence in research, scholarship,
and education by publishing worldwide in
Oxford New York
Auckland Cape Town Dar es Salaam Hong Kong Karachi
Kuala Lumpur Madrid Melbourne Mexico City Nairobi
New Delhi Shanghai Taipei Toronto
With offices in
Argentina Austria Brazil Chile Czech Republic France Greece
Guatemala Hungary Italy Japan South Korea Poland Portugal
Singapore Switzerland Thailand Turkey Ukraine Vietnam

Oxford is a registered trade mark of Oxford University Press
in the UK and in certain other countries

Published in the United States
by Oxford University Press Inc., New York

© R. Gastmans and Tai Tsun Wu 1990

The moral rights of the author have been asserted

Database right Oxford University Press (maker)

Reprinted 2011

All rights reserved. No part of this publication may be reproduced,
stored in a retrieval system, or transmitted, in any form or by any means,
without the prior permission in writing of Oxford University Press,
or as expressly permitted by law, or under terms agreed with the appropriate
reprographics rights organization. Enquiries concerning reproduction
outside the scope of the above should be sent to the Rights Department,
Oxford University Press, at the address above

You must not circulate this book in any other binding or cover
And you must impose this same condition on any acquirer

ISBN 978-0-19-852043-6

To Ninette
from R. G.

To Professor Ronold W. P. King
from T. T. W.

Preface

Hiking in the mountains with bad shoes can be a very painful experience. On the other hand, it can also lead to some unexpected developments in particle physics... During the 1978 LEP Summer Study in Les Houches, one of us (G) returned from such a hike with a couple of ugly blisters. His subsequent awkward way of walking attracted the attention of the other author (W), who felt the need to deliver some sort of pharmaceutical assistance. And, inevitably, the ensuing conversation turned into a physics discussion about radiative corrections.

At that time, the collider PETRA (Positron-Electron Tandem Ring Accelerator) at DESY, Hamburg, Germany, was just beginning to come into operation, with its energy eventually reaching 46 GeV in the centre-of-mass system. Since this energy is very much larger than the electron mass, it is evident that a systematic treatment of bremsstrahlung processes in the high energy limit is highly desirable. Since the experience of W was then limited mostly to the related but different case of high energy scattering and production processes with fixed momentum transfers, G gave him a thorough lecture on the ways of treating bremsstrahlung processes, where of course both energy and momentum transfers are large.

After the LEP Summer Study, G returned to Belgium while W went to visit CERN, Geneva, Switzerland. A few days later, W decided to carry through carefully what he had learned from G. He sat down early one morning at his table at home and started to do just that: drawing the four Feynman diagrams for the process $e^+e^- \to \mu^+\mu^-\gamma$; writing down the corresponding matrix elements; squaring their sum to get the cross section, which consists of ten terms with different denominators after the evaluation of the traces; and putting these ten terms on a common denominator. In the process, some of the terms involving the electron and muon masses are neglected. Of course, it is not correct to neglect all the mass terms: such a naive approximation fails in certain regions of phase space, the so-called peaking regions.

It took W almost exactly 24 hours to carry through this calculation (with breaks for meals), and he was astonished to find how simple the final answer was. In particular, the numerator for the common denominator was found to factorize, making the answer much shorter than if it had not factorized. Thus he thought that he had understood the instructions of G, and happily sent the results to Belgium. G very promptly wrote back to say that (1) he had found a mistake in one of the terms for the peaking region and corrected it; and (2) he had never seen the answer before in this form. The reason was that G never said that the ten terms should be put on a common denominator, and W had simply misunderstood him! In the same letter, G also added that the simplest way to check whether W's answer was correct was to compare numerically with the previously known, much longer formula. A few day later, he wrote again to say that the numerical comparisons gave agreemen to nine digits.

Encouraged by this numerical agreement, W decided to tackle the muc more intricate case of $e^+e^- \to e^+e^-\gamma$. This involves, to lowest order, eigh Feynman diagrams and, hence, thirty-six terms in the expression for th cross section. Following essentially the same procedure as for the muon case a very similar result, albeit a slightly more complicated one, was obtaine after basically working straight for seventy-two hours. After sending thi result to G, W felt justified to goof off for a few days.

Together with our colleagues Frits Berends and Ronald Kleiss (Universit of Leiden) and Patrick De Causmaecker and Walter Troost (University c Leuven), we then examined several other QED and QCD processes. Success The cross section formulae were again and again much more elegant than w could honestly hope for previously.

Nevertheless, the situation was not satisfying. Simple answers should b obtained in simple ways. But the procedure to get these simple answers fo bremsstrahlung is by no means simple. This unsatisfactory state persiste for a couple of years. Meanwhile, Patrick was working on his doctoral thesis for which he studied the QCD bremsstrahlung processes. Travelling dail between his home and the institute, he disposed of some three hours on th train. These trains being rather crowded, he could not really work wit pencil and paper, but he could do some thinking. Regularly, he would com to the institute with a new idea. One day, he suggested to Walter and (to examine polarization amplitudes rather than calculating cross sections b brute force. G dismissed the idea as being too cumbersome... Patrick an Walter decided to wait, however, until G was absent to work out this new lin of thought on the sly. Using first plane polarization states for the photon and later on photon helicity states, they succeeded in obtaining the cros sections in an alternative way.

Shortly thereafter, W visited Belgium, and he was shown the results c Patrick and Walter. The idea of dealing with matrix elements instead c cross sections pleased him considerably, but it was felt that the calculation could still be simplified. On the train(!) to Marseille, W had plenty of tim to think about it all, and somewhere in the vicinity of Paris, he realize that the helicity states of the photons led to expressions for $\epsilon^{\pm}$ which coul simply be written in terms of the helicity projection operators for the fermion $(1 \pm \gamma_5)/2$. From then on, it was clear that the helicity method was the wa to go about the bremsstrahlung problems.

This book summarizes some ten years of research on the subject. Ou purpose is to give a pedagogical introduction to the helicity method illus trated by many examples. Only some degree of familiarity with Feynma diagrams and Dirac algebra is required from the reader. The book also pre vides an extensive list of helicity amplitudes and cross sections for many c the most important QED and QCD processes at high energies, some being new and not available in the literature previously. Because of the large number of formulae we present in Chapters 9 and 10, it is almost unavoidable that some errors have been made. We would greatly appreciate being promptly informed of such errors. Hopefully, the information contained in this book will be useful for detailed analyses of these processes leading to a better understanding of particle physics at high energies.

We are very grateful to our colleagues in the CALKUL Collaboration named above for bringing this program to a good end. We are especially indebted to Walter Troost for the many enlightening discussions both before and during the writing of this book.

We owe special thanks to Maurice Jacob: not only did he organize the 1978 LEP Summer Study, which led to our collaboration on the helicity program, but also Chapter 12 on beamstrahlung is based mostly on his work. In the past ten years, we often discussed the problems treated in this book with our colleagues, including Jochen Bartels, John Bell, Dirk Danckaert, John Ellis, Alex Grossmann, Francis Halzen, Albert Hofmann, Benedikt Humpert, Harry Lehmann, Alfred Mueller, Per Osland, Roberto Peccei, Paul Söding, Jacques Soffer, Hiroshi Takeda, Sakue Yamada and Da-Hua Zhang.

We also received very valuable advice from Frank Mallezie and Margaret Owens on the use of the wordprocessor for the preparation of the manuscript.

Our research on the helicity method has been supported by the United States Department of Energy and the National Fund for Scientific Research (Belgium), where G is research director. Our collaboration across the Atlantic was facilitated by a NATO Research Grant. We thank the Theory Division of CERN, Geneva, Switzerland, the Gruppe Theorie of DESY, Hamburg, Federal Republic of Germany, and the Centre de Physique Théorique of the CNRS, Luminy, France, for their hospitality during our stays. All this support is gratefully acknowledged.

Leuven and Cambridge, MA   R.G.
1989   T.T.W.

Contents

1 Introduction   **1**
1.1 Elementary particles circa 1989 1
1.2 Radiative processes 2
1.3 The importance of high energies 3

2 Feynman diagrams   **7**
2.1 Feynman rules 7
2.2 Cross sections and decay rates 14
2.3 Example: $Z \to e^+e^-$ 16
2.4 Another example: $e^+e^- \to \gamma\gamma$ 18

3 Helicity states   **21**
3.1 Fermions 21
3.2 Photons and gluons 22
3.3 Example: $e^+e^- \to \gamma\gamma$ again 24
3.4 Ranges of validity 27

4 Single bremsstrahlung in QED   **31**
4.1 The process $e^+e^- \to \gamma\gamma\gamma$ 31
4.2 What if different fermions radiate? 35
4.3 The process $e^+e^- \to \mu^+\mu^-\gamma$ 35
4.4 The process $e^+e^- \to e^+e^-\gamma$ 39
4.5 Inclusion of Z-exchange 44

5 Single bremsstrahlung in QCD   **47**
5.1 Good to know 47
5.2 The process $\bar{q}q \to gg$ 48
5.3 The process $e^+e^- \to 3$ jets 52
5.4 The process $qq' \to qq'g$ 55
5.5 Other QCD processes 64

6 Double bremsstrahlung   **65**
6.1 QED 65
6.2 $e^+e^- \to 4$ jets 70

7 Finite mass effects   **79**
7.1 Their occasional importance 79
7.2 Single bremsstrahlung 80
7.3 An example: $e^+e^- \to \mu^+\mu^-\gamma$ 82
7.4 Mass corrections for amplitudes 84
     7.4.1 General formalism 84

	7.4.2	Single bremsstrahlung 88
	7.4.3	An example: $e^+e^- \to \mu^+\mu^-\gamma$ 91
	7.4.4	Double collinear bremsstrahlung 95
	7.4.5	An example: $e^+e^- \to \mu^+\mu^-\gamma\gamma$ 97

8 The production of quarkonia   101
- 8.1 Framework 101
- 8.2 $^1S_0$ production 106
  - 8.2.1 The amplitude $M(+,+,+)$ 106
  - 8.2.2 The amplitude $M(+,+,-)$ 108
  - 8.2.3 The cross section for $^1S_0$ production 110
- 8.3 The production of other states 111
  - 8.3.1 $^3S_1$ production 112
  - 8.3.2 $^1P_1$ production 112
  - 8.3.3 $^3P_0$ production 114
  - 8.3.4 $^3P_1$ production 115
  - 8.3.5 $^3P_2$ production 116
- 8.4 Conclusions 118

9 Summary of QED formulae   119
- 9.1 $e^+e^- \to \gamma\gamma \, (m=0)$ 119
- 9.2 $e^+e^- \to \mu^+\mu^-$ (no Z-exchange; $m=0$) 120
- 9.3 $e^+e^- \to \mu^+\mu^-$ (with Z-exchange; $m=0$) 121
- 9.4 $e^+e^- \to e^+e^-$ (no Z-exchange; $m=0$) 122
- 9.5 $e^+e^- \to e^+e^-$ (with Z-exchange; $m=0$) 122
- 9.6 $e^+e^- \to \gamma\gamma\gamma \, (m=0)$ 124
- 9.7 $e^+e^- \to \mu^+\mu^-\gamma$ (no Z-exchange; $m=0$) 125
- 9.8 $e^+e^- \to \mu^+\mu^-\gamma$ (with Z-exchange; $m=0$) 127
- 9.9 $e^+e^- \to e^+e^-\gamma$ (no Z-exchange; $m=0$) 130
- 9.10 $e^+e^- \to e^+e^-\gamma$ (with Z-exchange; $m=0$) 133
- 9.11 $e^+e^- \to \gamma\gamma\gamma\gamma \, (m=0)$ 137
- 9.12 $e^+e^- \to \mu^+\mu^-\gamma\gamma$ (no Z-exchange; $m=0$) 139
- 9.13 $e^+e^- \to e^+e^-\gamma\gamma$ (no Z-exchange; $m=0$) 145
- 9.14 $e^+e^- \to \gamma\gamma\gamma \, (m \neq 0)$ 154
  - 9.14.1 $\vec{k}_3$ nearly parallel to $\vec{p}_+$ 154
  - 9.14.2 $\vec{k}_3$ nearly parallel to $\vec{p}_-$ 155
- 9.15 $e^+e^- \to \mu^+\mu^-\gamma$ (no Z-exchange; $m \neq 0$) 157
  - 9.15.1 $\vec{k}$ nearly parallel to $\vec{p}_+$ 157
  - 9.15.2 $\vec{k}$ nearly parallel to $\vec{p}_-$ 159
  - 9.15.3 $\vec{k}$ nearly parallel to $\vec{q}_+$ 161
  - 9.15.4 $\vec{k}$ nearly parallel to $\vec{q}_-$ 164
- 9.16 $e^+e^- \to e^+e^-\gamma$ (no Z-exchange; $m \neq 0$) 166
  - 9.16.1 $\vec{k}$ nearly parallel to $\vec{p}_+$ 166
  - 9.16.2 $\vec{k}$ nearly parallel to $\vec{p}_-$ 169

CONTENTS

    9.16.3 $\vec{k}$ nearly parallel to $\vec{q}_+$ 172
    9.16.4 $\vec{k}$ nearly parallel to $\vec{q}_-$ 175
9.17 $e^+e^- \to \gamma\gamma\gamma\gamma$ $(m \neq 0)$ 179
    9.17.1 $\vec{k}_4$ nearly parallel to $\vec{p}_+$ 180
    9.17.2 $\vec{k}_4$ nearly parallel to $\vec{p}_-$ 182
    9.17.3 $\vec{k}_3$ and $\vec{k}_4$ nearly parallel to $\vec{p}_+$ and $\vec{p}_-$, resp. 185
    9.17.4 $\vec{k}_3$ and $\vec{k}_4$ nearly parallel to $\vec{p}_+$ 187
    9.17.5 $\vec{k}_3$ and $\vec{k}_4$ nearly parallel to $\vec{p}_-$ 190
9.18 $e^+e^- \to \mu^+\mu^-\gamma\gamma$ (no Z-exchange; $m \neq 0$) 193
    9.18.1 $\vec{k}_2$ nearly parallel to $\vec{p}_+$ 193
    9.18.2 $\vec{k}_2$ nearly parallel to $\vec{p}_-$ 197
    9.18.3 $\vec{k}_2$ nearly parallel to $\vec{q}_+$ 201
    9.18.4 $\vec{k}_2$ nearly parallel to $\vec{q}_-$ 206
    9.18.5 $\vec{k}_1$ and $\vec{k}_2$ nearly parallel to $\vec{p}_+$ and $\vec{p}_-$, resp. 210
    9.18.6 $\vec{k}_1$ and $\vec{k}_2$ nearly parallel to $\vec{p}_+$ and $\vec{q}_+$, resp. 215
    9.18.7 $\vec{k}_1$ and $\vec{k}_2$ nearly parallel to $\vec{p}_+$ and $\vec{q}_-$, resp. 221
    9.18.8 $\vec{k}_1$ and $\vec{k}_2$ nearly parallel to $\vec{p}_-$ and $\vec{q}_+$, resp. 227
    9.18.9 $\vec{k}_1$ and $\vec{k}_2$ nearly parallel to $\vec{p}_-$ and $\vec{q}_-$, resp. 233
    9.18.10 $\vec{k}_1$ and $\vec{k}_2$ nearly parallel to $\vec{q}_+$ and $\vec{q}_-$, resp. 239
    9.18.11 $\vec{k}_1$ and $\vec{k}_2$ nearly parallel to $\vec{p}_+$ 244
    9.18.12 $\vec{k}_1$ and $\vec{k}_2$ nearly parallel to $\vec{p}_-$ 249
    9.18.13 $\vec{k}_1$ and $\vec{k}_2$ nearly parallel to $\vec{q}_+$ 254
    9.18.14 $\vec{k}_1$ and $\vec{k}_2$ nearly parallel to $\vec{q}_-$ 260
9.19 $e^+e^- \to e^+e^-\gamma\gamma$ (no Z-exchange; $m \neq 0$) 265
    9.19.1 $\vec{k}_2$ nearly parallel to $\vec{p}_+$ 265
    9.19.2 $\vec{k}_2$ nearly parallel to $\vec{p}_-$ 272
    9.19.3 $\vec{k}_2$ nearly parallel to $\vec{q}_+$ 278
    9.19.4 $\vec{k}_2$ nearly parallel to $\vec{q}_-$ 285
    9.19.5 $\vec{k}_1$ and $\vec{k}_2$ nearly parallel to $\vec{p}_+$ and $\vec{p}_-$, resp. 292
    9.19.6 $\vec{k}_1$ and $\vec{k}_2$ nearly parallel to $\vec{p}_+$ and $\vec{q}_+$, resp. 300
    9.19.7 $\vec{k}_1$ and $\vec{k}_2$ nearly parallel to $\vec{p}_+$ and $\vec{q}_-$, resp. 307
    9.19.8 $\vec{k}_1$ and $\vec{k}_2$ nearly parallel to $\vec{p}_-$ and $\vec{q}_+$, resp. 315
    9.19.9 $\vec{k}_1$ and $\vec{k}_2$ nearly parallel to $\vec{p}_-$ and $\vec{q}_-$, resp. 324
    9.19.10 $\vec{k}_1$ and $\vec{k}_2$ nearly parallel to $\vec{q}_+$ and $\vec{q}_-$, resp. 331
    9.19.11 $\vec{k}_1$ and $\vec{k}_2$ nearly parallel to $\vec{p}_+$ 339
    9.19.12 $\vec{k}_1$ and $\vec{k}_2$ nearly parallel to $\vec{p}_-$ 347
    9.19.13 $\vec{k}_1$ and $\vec{k}_2$ nearly parallel to $\vec{q}_+$ 354
    9.19.14 $\vec{k}_1$ and $\vec{k}_2$ nearly parallel to $\vec{q}_-$ 361

10 Summary of QCD formulae   369
10.1 $e^+e^- \to \bar{q}q$ (no Z-exchange) 369
10.2 $e^+e^- \to \bar{q}qg$ (no Z-exchange) 370
10.3 $e^+e^- \to \bar{q}q\gamma$ (no Z-exchange) 371
10.4 $e^+e^- \to \bar{q}qgg$ (no Z-exchange) 373

10.5	$e^+e^- \to q\bar{q}q'\bar{q}'$ (no Z-exchange)	379
10.6	$e^+e^- \to q\bar{q}q\bar{q}$ (no Z-exchange)	381
10.7	$qq' \to qq'$	385
10.8	$\bar{q}q \to \bar{q}'q'$	386
10.9	$qq \to qq$	387
10.10	$\bar{q}q \to \bar{q}q$	388
10.11	$\gamma q \to \gamma q$	389
10.12	$\gamma q \to gq$	390
10.13	$gq \to \gamma q$	391
10.14	$gq \to gq$	392
10.15	$\gamma\bar{q} \to \gamma\bar{q}$	393
10.16	$\gamma\bar{q} \to g\bar{q}$	394
10.17	$g\bar{q} \to \gamma\bar{q}$	395
10.18	$g\bar{q} \to g\bar{q}$	396
10.19	$\bar{q}q \to \gamma\gamma$	397
10.20	$\bar{q}q \to g\gamma$	398
10.21	$\bar{q}q \to gg$	399
10.22	$\gamma\gamma \to \bar{q}q$	400
10.23	$g\gamma \to \bar{q}q$	401
10.24	$gg \to \bar{q}q$	402
10.25	$gg \to gg$	403
10.26	$qq' \to qq'\gamma$	405
10.27	$qq' \to qq'g$	406
10.28	$\bar{q}q \to \bar{q}'q'\gamma$	409
10.29	$\bar{q}q \to \bar{q}'q'g$	410
10.30	$qq \to qq\gamma$	413
10.31	$qq \to qqg$	415
10.32	$\bar{q}q \to \bar{q}q\gamma$	419
10.33	$\bar{q}q \to \bar{q}qg$	421
10.34	$\gamma q \to \gamma\gamma q$	425
10.35	$\gamma q \to g\gamma q$	427
10.36	$gq \to \gamma\gamma q$	430
10.37	$\gamma q \to ggq$	432
10.38	$gq \to g\gamma q$	435
10.39	$gq \to ggq$	439
10.40	$\gamma\bar{q} \to \gamma\gamma\bar{q}$	445
10.41	$\gamma\bar{q} \to g\gamma\bar{q}$	447
10.42	$g\bar{q} \to \gamma\gamma\bar{q}$	449
10.43	$\gamma\bar{q} \to gg\bar{q}$	452
10.44	$g\bar{q} \to g\gamma\bar{q}$	455
10.45	$g\bar{q} \to gg\bar{q}$	459
10.46	$\bar{q}q \to \gamma\gamma\gamma$	465
10.47	$\bar{q}q \to g\gamma\gamma$	467
10.48	$\bar{q}q \to gg\gamma$	469

CONTENTS xv

10.49 $\bar{q}q \to ggg$ 472
10.50 $\gamma\gamma \to \bar{q}q\gamma$ 477
10.51 $\gamma\gamma \to \bar{q}qg$ 479
10.52 $\gamma g \to \bar{q}q\gamma$ 481
10.53 $\gamma g \to \bar{q}qg$ 484
10.54 $gg \to \bar{q}q\gamma$ 487
10.55 $gg \to \bar{q}qg$ 490
10.56 $gg \to ggg$ 496
10.57 $gg \to {}^1S_0$ 501
10.58 $gg \to {}^3P_0$ 502
10.59 $gg \to {}^3P_2$ 503
10.60 $qg \to q\,{}^1S_0$ 504
10.61 $qg \to q\,{}^3P_0$ 505
10.62 $qg \to q\,{}^3P_1$ 506
10.63 $qg \to q\,{}^3P_2$ 508
10.64 $\bar{q}g \to \bar{q}\,{}^1S_0$ 509
10.65 $\bar{q}g \to \bar{q}\,{}^3P_0$ 510
10.66 $\bar{q}g \to \bar{q}\,{}^3P_1$ 511
10.67 $\bar{q}g \to \bar{q}\,{}^3P_2$ 513
10.68 $\bar{q}q \to g\,{}^1S_0$ 514
10.69 $\bar{q}q \to g\,{}^3P_0$ 515
10.70 $\bar{q}q \to g\,{}^3P_1$ 517
10.71 $\bar{q}q \to g\,{}^3P_2$ 518
10.72 $gg \to g\,{}^1S_0$ 519
10.73 $gg \to g\,{}^3S_1$ 521
10.74 $gg \to g\,{}^1P_1$ 524
10.75 $gg \to g\,{}^3P_0$ 527
10.76 $gg \to g\,{}^3P_1$ 529
10.77 $gg \to g\,{}^3P_2$ 532

11 Polarization   **537**
11.1 Fermions 537
11.2 An example: $e^+e^- \to \mu^+\mu^-$ 539
11.3 Photons and gluons 542
11.4 An example: $gg \to g\,{}^1S_0$ 545

12 Beamstrahlung   **549**
12.1 Electron-positron linear colliders 549
12.2 Nature of approximations 551
    12.2.1 Single-particle approximation 551
    12.2.2 External field approximation 552
    12.2.3 Important length scales 554
    12.2.4 Shape of bunch 555
12.3 Electron wave functions 556

12.4 Cross section for $e + \text{bunch} \to e + \gamma + \text{bunch}$ ... 560
12.4.1 Electrostatic potential ... 560
12.4.2 Initial and final phases ... 561
12.4.3 Integration over transverse momenta ... 564
12.5 Energy distribution of beamstrahlung photon ... 568
12.5.1 Mellin transform ... 569
12.5.2 Residue of $\overline{K}(\zeta)$ at $\zeta = 1$ ... 571
12.5.3 Analytic continuation into the region $1 > \text{Re}\,\zeta > 0$ ... 572
12.5.4 Residue of $\overline{K}(\zeta)$ at $\zeta = 0$ ... 573
12.5.5 Analytic continuation into the region $0 > \text{Re}\,\zeta > -1$ ... 577
12.5.6 Residue of $\overline{K}(\zeta)$ at $\zeta = -1$ ... 577
12.5.7 Hyperbolic secant distribution ... 580
12.5.8 Summary ... 581
12.6 Discussions ... 583

13 Outlook **587**
13.1 The ubiquity of the photon ... 587
13.2 Further developments ... 587
13.2.1 Supersymmetry ... 587
13.2.2 Phase choice of polarization vectors ... 588
13.2.3 Weyl-van der Waerden formalism ... 590
13.2.4 Quantum gravity ... 592
13.2.5 Polarization vectors for massive spin-1 particles ... 593
13.3 Unsolved problems ... 596
13.3.1 Collinear fermions and gluons ... 596
13.3.2 Relation to quaternions ... 597
13.3.3 Loops in Feynman amplitudes ... 598
13.4 Epilogue ... 599

Appendix A Traces: cut-and-paste **601**

Appendix B The process $\gamma\gamma \to \gamma\gamma$ **605**

Appendix C Color traces **613**

Bibliography **619**

Author Index **632**

Subject Index **645**

1
Introduction

> With energy galore,
> one can explore
> proton's core
> more and more.

1.1 Elementary particles circa 1989

Of all fundamental interactions in nature, the electromagnetic interactions between charged particles were the first to be studied extensively. The development of renormalized perturbation theory by Tomonaga, Schwinger, Feynman and Dyson [1] in the late 1940's made quantum electrodynamics (QED) into an extremely successful theory which yielded many very precise predictions which were repeatedly confirmed by experiment, two prominent examples being the Lamb shift [2] and the anomalous magnetic moments of the electron and the muon [3]. QED, a relativistic quantum field theory based on an Abelian gauge symmetry, became the prototype field theory for all other interactions.

Over thirty years ago, in 1954, Yang and Mills [4] showed that non-Abelian gauge symmetries could be introduced for the description of other fundamental interactions. At that time, however, not all ingredients were available to turn this beautiful idea into a realistic theory.

Ten years later, in 1964, Gell-Mann and Zweig [5] proposed quarks as the basic constituents of hadrons. Shortly thereafter, based on the earlier work of Glashow, Weinberg and Salam [6] constructed a Yang-Mills theory of unified weak and electromagnetic interactions. For the construction of this theory, they invoked the idea of the Higgs mechanism [7], discovered in 1964, which allows one to reconcile gauge symmetries with short-range forces. The color degree of freedom, associated with the quarks to preserve their ordinary Fermi-Dirac statistics, led Fritzsch, Gell-Mann and Leutwyler [8] to propose another Yang-Mills theory for the description of strong interactions of the quarks. This theory, called quantum chromodynamics (QCD), together with the Glashow-Weinberg-Salam model of weak and electromagnetic interactions constitutes the standard model.

In both Abelian and non-Abelian gauge theories, one necessarily associates a gauge particle with every local symmetry. For quantum chromodynamics, the gauge particle is the gluon, which interacts with the quarks. For the electroweak theory, the gauge particles are the photon and the intermediate vector bosons, W and Z.

It has taken a number of years for the essential aspects of these theories to be verified experimentally, the major milestones being the following:

i) the discovery in 1979 of the gluon jet at PETRA of the Deutsches Elektronen-Synchrotron by Wu and collaborators [9] and

ii) the discovery in 1983 and 1984 of the intermediate vector bosons at the $p\bar{p}$ Collider of CERN by Rubbia, Van der Meer and collaborators [10].

Presently, the missing link is the Higgs boson [7], which was introduced in the standard model to generate the masses. It will be a most important event when this problem is settled experimentally.

At the present time, the known constituents of matter are the following:

i) neutrinos: there are at least three $(\nu_e, \nu_\mu, \nu_\tau)$;

ii) charged leptons: there are at least three (e, μ, τ);

iii) quarks: there are at least five (u, d, s, c, b); and

iv) gauge particles: there are at least four (γ, W, Z, g).

In this list, the particle and its antiparticle are taken together as one, and internal degrees of freedom, such as spin and color, are not explicitly counted. It is generally believed that there is a sixth quark, the t quark, which however has not been seen experimentally.

1.2 Radiative processes

A radiative process or a bremsstrahlung process is any process in which one or more photons or gluons are emitted. As a rule, bremsstrahlung always accompanies the nonradiative process. Obviously, the cross sections for the radiative processes are smaller by at least one power in the coupling constant, α for QED or α_S for QCD, than for the nonradiative ones.

In many cases, the radiated gauge particle goes undetected. This can occur when its energy is too small to trigger the detector or when its momentum is nearly parallel to another particle's momentum so that their tracks cannot be resolved. Similarly, photons going down the beam pipe will in general not be detected. When this happens, one considers the radiative process as a higher order correction to the nonradiative, lower order process. Because of the smallness of α and α_S, these corrections are typically in the 1-10 % range, but they can be larger in specific situations. For example, initial state radiation in $e^+e^- \to \mu^+\mu^-$ reduces the effective energy available for producing the final state muons, and, at the Z-peak, one expects the lowest order cross section to be reduced by roughly 30% [11].

For precise experiments, the bremsstrahlung corrections must be taken into account. This is the case, for example, in the determination of the beam luminosity for e^+e^- storage rings, which is usually done by measuring the cross section for small angle Bhabha scattering [12]. Bremsstrahlung corrections also play an important role in tests of the electroweak theory, in particular for the study of the angular asymmetry in $e^+e^- \to \mu^+\mu^-$ and other QED tests at high energies. In the case of QCD, the radiative processes are quantitatively even more important because of the larger value of α_S, but the experiments usually do not reach as high a level of accuracy as in the case of QED tests.

Sometimes, the radiative processes are studied for their own sake. Within the framework of perturbative QCD, one associates the production of jets with processes in which quarks and/or gluons are emitted. In lowest order, two-jet production in e^+e^- annihilation is described by the process $e^+e^- \to q\bar{q}$, but, for three-jet production, one has to consider $e^+e^- \to q\bar{q}g$, which is a radiative process. For studying four-jet production, one has to know the cross sections for $e^+e^- \to q\bar{q}q\bar{q}$, $q\bar{q}q'\bar{q}'$ and $q\bar{q}gg$, the last process being a double bremsstrahlung process, etc.

The importance of radiative processes is by no means restricted to e^+e^- collisions. Also for $p\bar{p}$ collider physics, the multi-jet phenomena are described by gluon bremsstrahlung processes. The same is true, for example, for the hadroproduction of heavy quarkonia at large transverse momenta.

It is clear from this nonexhaustive list of examples that radiative processes play an important role in high energy physics and much attention has been devoted to them. Unfortunately, their calculation, although in principle straightforward, is in practice often quite elaborate and cumbersome. This is so because higher order processes are described by a relatively large number of Feynman diagrams. It is the purpose of this book to show the reader how the use of the helicity amplitude method, to be described in the forthcoming chapters, can greatly simplify this task, at least in the high energy limit.

1.3 The importance of high energies

For particle physics, high energy is the name of the game, at least most of the time. The higher the energy, the deeper one can probe the structure of matter. There is a very strong hope among physicists, that ultimately nature will reveal its simplicity at high energies.

At the present moment, the maximum energy range for e^+e^- physics is 100 GeV (LEP at CERN), while for $p\bar{p}$ physics one reaches 2 TeV (Tevatron at Fermilab). This is a very fortunate circumstance for calculations involving leptons or quarks. Indeed, the masses of these particles are so small compared to the energies involved in the collisions that they can safely be neglected in almost all cases.

Putting the masses equal to zero in calculations certainly leads to great simplifications. Yet, at the time when most standard books on Feynman diagrams were written, no special attention was given to the high energy limit. As a result, the techniques for calculating Feynman diagrams in the standard way are not the most efficient ones for obtaining the high energy limit of the cross sections.

In the standard procedure, one adds up all the Feynman amplitudes, one takes the squared absolute value of the sum, which is then summed or averaged over the polarizations or any other degree of freedom. For radiative processes, where one deals with a large number of Feynman diagrams, this is in general a lengthy and cumbersome procedure.

What we propose, alternatively, is to take advantage of the high energy limit. For massless particles, helicity states are Lorentz invariant, and it appears that it is much simpler to calculate first the various helicity amplitudes for a given process. We will show that, by choosing a convenient representation for the polarization vectors of the radiated gauge particles, a sizable fraction of Feynman diagrams gives a vanishing contribution for a specific helicity configuration, thus simplifying the calculation of the corresponding helicity amplitude.

Once the helicity amplitudes are calculated, it suffices to add their squared absolute values to obtain the cross section. This method has the advantage that the calculations are performed at the level of the amplitudes, rather than at the level of their squares, which are in general much more lengthy. Furthermore, the cross sections are obtained as a sum of positive quantities, which for numerical computations circumvents the problem of large cancellations between contributions of opposite signs.

After a summary of the standard Feynman techniques, in Chapter 2, we present the helicity method in detail in Chapters 3 through 6. More specifically, in Chapter 3, we define the helicity states for the fermions, the photons and the gluons. In Chapter 4, we analyse the case of single bremsstrahlung in QED, while, in Chapter 5, we consider single bremsstrahlung in QCD. Finally, in Chapter 6, we present the application of the helicity method to double bremsstrahlung in QED. In all these chapters, we illustrate the procedure by working out several explicit examples.

Even in cases where masses cannot be neglected, the helicity method, or a simple modification of it, can often be used advantageously to obtain the cross section. This somewhat more complicated issue will be treated in Chapters 7 and 8. In Chapter 7, we show how to obtain the finite mass corrections for spin amplitudes. In Chapter 8, the helicity method is applied to the production of heavy quarkonia, i.e., bound $q\bar{q}$ systems where the masses of the quarks are not negligible.

For ease of reference, we list the various helicity (or spin) amplitudes and the cross sections for the simplest QED and QCD processes. Chapter 9 is devoted to single and double bremsstrahlung in QED, the most important processes being $e^+e^- \to \gamma\gamma$, $\mu^+\mu^-$ and e^+e^- with one or two additional photons in the final state. Special attention is given to the spin amplitudes which describe the situations where one or two of the outgoing photons are nearly parallel to one of the fermions. We also treat the case of Z-exchange for the processes $e^+e^- \to \mu^+\mu^-$, $e^+e^- \to e^+e^-$, $e^+e^- \to \mu^+\mu^-\gamma$ and $e^+e^- \to e^+e^-\gamma$. In Chapter 10, we present the formulae for single bremsstrahlung in QCD and for the production of heavy quarkonia. Except for the heavy quarkonia, we assume that all quark masses can be neglected. Taken together, these two chapters may be considered as a handbook of amplitudes and cross sections for QED and QCD processes at high energies. We hope that they are especially useful to experimentalists and theoreticians who are less concerned with their derivation.

In Chapter 11, we explain how the helicity amplitudes can be combined to yield cross section formulae for the case of initial state polarization. In Chapter 12, we discuss the problem of bremsstrahlung in future high-energy electron-positron colliders—known as beamstrahlung. Contrary to the subjects discussed in earlier chapters, the modern treatment of this topic started only about three years ago. Since our knowledge about beamstrahlung and the related process of pair production is at present rapidly evolving, only some of the simplest aspects are treated in this book.

Finally, Chapter 13 presents our summary of the helicity method. We also point out the recent applications, which were not treated in detail in this book, and the further developments, which led to modified versions of the helicity method.

We hope that the helicity method, which we expound, and the list of formulae, which we present, will allow future, more precise investigations of even more processes at high energies. If this is the case, 'The Ubiquitous Photon' will have contributed a little to a deeper understanding of elementary particles and their behaviour.

2
Feynman diagrams

> It's nevertheless
> an endless mess
> before one can assess
> Nature's inventiveness.

2.1 Feynman rules

A Feynman diagram [1] is a graphical representation of the amplitude for a given process. It is composed of three types of elements: external lines, internal lines and vertices. With everyone of these elements one associates a mathematical expression, and the product of these expressions, in an appropriate order (together with some overall factors), yields the amplitude.

The Feynman rules tell us which mathematical expression is to be associated with a given element of the Feynman diagram. They can be derived from first principles, but we simply list the results, using the Minkowski metric

$$g^{\mu\nu} = g_{\mu\nu} = \begin{pmatrix} +1 & 0 & 0 & 0 \\ 0 & -1 & 0 & 0 \\ 0 & 0 & -1 & 0 \\ 0 & 0 & 0 & -1 \end{pmatrix}. \qquad (2.1)$$

In a Feynman diagram for the process

$$a_1 + a_2 + \ldots + a_n \to b_1 + b_2 + \ldots + b_m, \qquad (2.2)$$

the incoming particles, $a_1, a_2, \ldots, a_n$, are shown as lines entering from the left, whereas the outgoing ones, $b_1, b_2, \ldots, b_m$, leave the diagram to the right. For such lines, referred to as external lines, we have the Feynman rules listed in Table 2.1.

In this table, the quantities p or k are the physical four-momenta of the particles, and σ or λ label the polarization degrees of freedom. The spinors $u(p,\sigma)$ and $v(p,\sigma)$ are solutions of the free Dirac equation [13]

$$(\not{p} - m)u(p,\sigma) = 0,$$

$$(\not{p} + m)v(p,\sigma) = 0, \qquad (2.3)$$

where

$$\not{p} \equiv p_\mu \gamma^\mu = p_0 \gamma^0 - \vec{p} \cdot \vec{\gamma}. \qquad (2.4)$$

2. FEYNMAN DIAGRAMS

Table 2.1: Feynman rules for external lines.

external line	expression	particle
—→— p,σ	$u(p,\sigma)$	incoming fermion
—←— p,σ	$\bar{v}(p,\sigma)$	incoming antifermion
—→— p,σ	$\bar{u}(p,\sigma)$	outgoing fermion
—←— p,σ	$v(p,\sigma)$	outgoing antifermion
~~ k,λ	$\epsilon^\mu(k,\lambda)^*$	incoming vector particle
~~ k,λ	$\epsilon^\mu(k,\lambda)$	outgoing vector particle

The γ-matrices in the Dirac equation satisfy the anticommutation relations

$$\gamma^\mu \gamma^\nu + \gamma^\nu \gamma^\mu = 2g^{\mu\nu} . \tag{2.5}$$

There are several representations for the γ-matrices. For our purposes, the following representation is found to be very convenient:

$$\gamma^0 = \begin{pmatrix} 0 & \mathbb{1} \\ \mathbb{1} & 0 \end{pmatrix}, \quad \gamma^i = \begin{pmatrix} 0 & -\sigma^i \\ \sigma^i & 0 \end{pmatrix}, \quad i = 1,2,3, \tag{2.6}$$

where

$$\sigma^1 = \begin{pmatrix} 0 & 1 \\ 1 & 0 \end{pmatrix}, \quad \sigma^2 = \begin{pmatrix} 0 & -i \\ i & 0 \end{pmatrix}, \quad \sigma^3 = \begin{pmatrix} 1 & 0 \\ 0 & -1 \end{pmatrix} \tag{2.7}$$

are the familiar 2×2 Pauli matrices and

$$\mathbb{1} = \begin{pmatrix} 1 & 0 \\ 0 & 1 \end{pmatrix} \tag{2.8}$$

is the 2×2 unit matrix.

The spinors $\bar{u}(p,\sigma)$ and $\bar{v}(p,\sigma)$ are defined by

$$\bar{u}(p,\sigma) = u(p,\sigma)^\dagger \gamma^0, \quad \bar{v}(p,\sigma) = v(p,\sigma)^\dagger \gamma^0, \tag{2.9}$$

where $\dagger$ denotes the adjoint, i.e., the Hermitian conjugate. They satisfy the equations

$$\bar{u}(p,\sigma)(\not{p} - m) = 0,$$

$$\bar{v}(p,\sigma)(\not{p} + m) = 0, \tag{2.10}$$

2.1. FEYNMAN RULES

which follows from eqns (2.3) and the fact that

$$\gamma^0 \gamma^{\mu\dagger} \gamma^0 = \gamma^\mu. \tag{2.11}$$

With external scalar lines, for example, for Higgs bosons [7], one associates the trivial factor 1. Also note that our choice of assigning ϵ (ϵ^*) with outgoing (incoming) vector particles is contrary to the usual conventions of most textbooks. In this book, we deal mostly with outgoing vector particles, and, in order to alleviate the notation, we preferred to omit the asterisks in these cases.

For internal lines, the Feynman rules are listed in Table 2.2, where m denotes the mass of the virtual particle under consideration. In QCD, the quarks and the gluons have an extra degree of freedom, called color. As a result, the quark propagators have an extra factor δ_{ik}, $i, k = 1, 2, 3$, and the gluon propagators an extra factor δ_{ab}, $a, b = 1, \ldots, 8$.

Table 2.2: Feynman rules for internal lines.

internal line	expression	virtual particle
$\overset{k}{\rule[0.5ex]{3em}{0.4pt}}$	$\dfrac{i}{k^2 - m^2 + i\epsilon}$	scalar
$\overset{p}{\rule[0.5ex]{3em}{0.4pt}\!\!\!\to}$	$\dfrac{i(\not{p} + m)}{p^2 - m^2 + i\epsilon}$	lepton
$\underset{j\qquad k}{\overset{p}{\rule[0.5ex]{3em}{0.4pt}\!\!\!\to}}$	$\dfrac{i(\not{p} + m)}{p^2 - m^2 + i\epsilon} \delta_{jk}$	quark
$\underset{\mu\qquad \nu}{\overset{k}{\sim\!\sim\!\sim\!\sim}}$	$\dfrac{-ig_{\mu\nu}}{k^2 + i\epsilon}$	photon
$\underset{\mu, a \qquad \nu, b}{\overset{k}{\sim\!\sim\!\sim\!\sim}}$	$\dfrac{-ig_{\mu\nu}}{k^2 + i\epsilon} \delta_{ab}$	gluon
$\underset{\mu\qquad \nu}{\overset{k}{\sim\!\sim\!\sim\!\sim}}$	$i\dfrac{-g_{\mu\nu} + k_\mu k_\nu/m^2}{k^2 - m^2 + i\epsilon}$	massive spin-1

Table 2.3: Feynman rules for QED and QCD vertices.

vertex	expression	description
μ (photon-lepton diagram)	$ie\gamma_\mu$	photon-lepton vertex
μ; j, k (photon-quark diagram)	$ie_q\gamma_\mu\delta_{jk}$	photon-quark vertex
μ, a; j, k (gluon-quark diagram)	$ig\gamma_\mu T^a_{kj}$	gluon-quark vertex
p,μ,a; q,ν,b; k,ρ,c (three-gluon diagram)	$gf^{abc}[(p-q)_\rho g_{\mu\nu}$ $+(q-k)_\mu g_{\nu\rho}$ $+(k-p)_\nu g_{\rho\mu}]$	three-gluon vertex (all momenta taken to be incoming)
μ,a; σ,d; ν,b; ρ,c (four-gluon diagram)	$-ig^2[f^{abe}f^{cde}(g_{\mu\rho}g_{\nu\sigma}$ $-g_{\mu\sigma}g_{\nu\rho})$ $+f^{ace}f^{bde}(g_{\mu\nu}g_{\rho\sigma}$ $-g_{\mu\sigma}g_{\nu\rho})$ $+f^{ade}f^{cbe}(g_{\mu\rho}g_{\nu\sigma}$ $-g_{\mu\nu}g_{\rho\sigma})]$	four-gluon vertex (summation over e is implied)

2.1. FEYNMAN RULES

For the vertices, we have the Feynman rules of Table 2.3. In this list, the quantity e (e_q) denotes the lepton (quark) charge, while g is the QCD coupling constant. In QED, there is only one type of vertex, i.e., the photon-fermion vertex. However, because of the non-Abelian nature of QCD [4], there are additional vertices involving three or four gluons.

The group structure underlying QCD is SU(3), which was first introduced in particle physics to describe flavor symmetry among the particles [14]. In the context of QCD, one often uses the terminology color SU(3) [8] to distinguish it from the flavor SU(3).

The QCD structure constants f^{abc} are totally antisymmetric, and

$$[T^a, T^b] = i f^{abc} T^c. \tag{2.12}$$

The anticommutation relations

$$\{T^a, T^b\} = d^{abc} T^c + \frac{1}{3}\delta^{ab}, \tag{2.13}$$

define the totally symmetric structure constants d^{abc}. Some useful relations are

$$\sum_a T^a T^a = \frac{4}{3}, \qquad \text{Tr}\,(T^a T^b) = \frac{1}{2}\delta^{ab},$$

$$\text{Tr}\,(T^a T^b T^c) = \frac{1}{4}(d^{abc} + i f^{abc}),$$

$$\sum_{abc} d^2_{abc} = \frac{40}{3}, \qquad \sum_{abc} f^2_{abc} = 24. \tag{2.14}$$

The only nonzero elements (up to permutations) are the following:

$$1 = f_{123} = 2f_{147} = 2f_{246} = 2f_{257} = 2f_{345}$$

$$= -2f_{156} = -2f_{367} = \frac{2}{\sqrt{3}} f_{458} = \frac{2}{\sqrt{3}} f_{678},$$

$$\frac{1}{\sqrt{3}} = d_{118} = d_{228} = d_{338} = -d_{888},$$

$$\frac{-1}{2\sqrt{3}} = d_{448} = d_{558} = d_{668} = d_{778},$$

$$\frac{1}{2} = d_{146} = d_{157} = -d_{247} = d_{256} = d_{344} = d_{355} = -d_{366} = -d_{377}. \tag{2.15}$$

Finally, there are some overall factors which must be included in the Feynman amplitudes (see Table 2.4).

First of all, we note that we have conservation of four-momentum at every vertex. For diagrams without closed loops, i.e., tree diagrams, this prescription determines the momentum associated with internal lines. For closed loop diagrams, some internal momenta will not be fixed, and one must insert $\int d^4k/(2\pi)^4$ to integrate over all values of this internal momentum.

Secondly, there is an extra minus sign for diagrams involving closed fermion loops and a relative minus sign between graphs which differ only by an interchange of two external identical fermion lines. In this context, an incoming electron is identical to an outgoing positron, and similarly for quarks with the same flavor. Of course, for a closed fermion loop, a trace must be taken over the γ-matrices appearing in the loop.

Thirdly, one must include a symmetry factor $1/\mathcal{S}$ to avoid overcounting of identical states in closed loops. More precisely, $\mathcal{S}$ is the symmetry number of the Feynman diagram with the external lines kept fixed [15]. For example, in the tadpole diagram of Fig. 2.1, there is a symmetry factor $1/2$, as the internal boson state with momentum k is identical to the state with momentum $-k$. For the Feynman diagram of Fig. 2.2, the symmetry factor is $1/6$, etc.

Table 2.4: Feynman rules for overall factors.

closed loop	$\frac{1}{(2\pi)^4}\int d^4k$
closed fermion loop	-1
between graphs with interchange of identical fermion lines	relative factor (-1)
symmetry factor to avoid overcounting of identical boson states in closed loops	$1/\mathcal{S}$
conventional overall factor	$-i$

2.1. FEYNMAN RULES

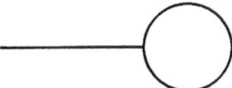

Fig. 2.1: Tadpole diagram with symmetry factor 1/2.

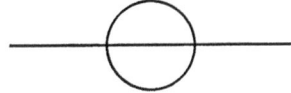

Fig. 2.2: Feynman diagram with symmetry factor 1/6.

Finally, we introduce an overall factor $(-i)$, which is purely conventional, but which gives the optical theorem [16] its familiar form. Let M be the invariant amplitude for elastic scattering in the forward direction constructed according to the above rules, then

$$\text{Im } M = [\lambda(s, m_1^2, m_2^2)]^{\frac{1}{2}} \sigma_{\text{tot}}(s), \qquad (2.16)$$

with, in general,

$$\lambda(a, b, c) = a^2 + b^2 + c^2 - 2bc - 2ca - 2ab. \qquad (2.17)$$

Here, m_1 and m_2 denote the masses of the incoming particles with momenta p_1 and p_2, $s = (p_1 + p_2)^2$, and $\sigma_{\text{tot}}(s)$ denotes the total cross section.

The Feynman rules listed in Tables 2.1 through 2.4 are not complete: they should be supplemented with yet another rule for diagrams involving closed loops with gluons, the so-called ghost contributions [17]. Since we do not discuss such processes in this book, we simply omit this extra rule.

The procedure for writing down an amplitude is then as follows. The type of process one considers determines the number and the nature of the

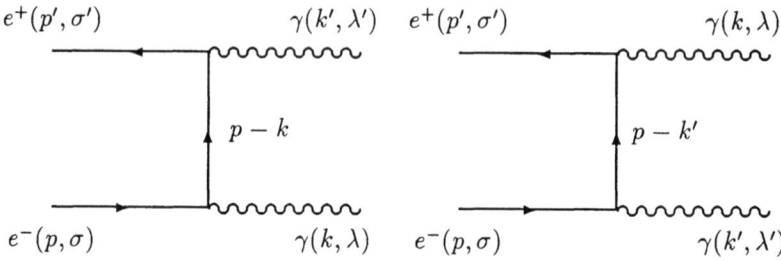

Fig. 2.3: Feynman diagrams for $e^+e^- \to \gamma\gamma$, to order e^2.

external lines. For example, for $e^+e^- \to \gamma\gamma$, one has one incoming electron line, one incoming positron line, and two outgoing photon lines. Using the vertices and the propagators which are at one's disposal, one then draws all the connected Feynman diagrams which link the external lines together. Usually, in perturbation theory, one works to a specific order in the coupling constant, which limits the number of vertices to be included in the diagram. With the help of the Feynman rules which are listed, one writes down the associated amplitude, knowing that a diagram has to be read against the directions of the arrows of its oriented lines.

Thus, for $e^+e^- \to \gamma\gamma$, to order e^2, we have the two Feynman diagrams of Fig. 2.3, and the amplitude reads

$$\begin{aligned} M = -e^2 \bar{v}(p',\sigma') &\left[\slashed{\epsilon}(k',\lambda') \frac{\slashed{p}-\slashed{k}+m}{(p-k)^2-m^2} \slashed{\epsilon}(k,\lambda) \right.\\ &\left. + \slashed{\epsilon}(k,\lambda) \frac{\slashed{p}-\slashed{k}'+m}{(p-k')^2-m^2} \slashed{\epsilon}(k',\lambda') \right] u(p,\sigma), \end{aligned}$$
(2.18)

where, for any four-vector a_μ, we use the notation $\slashed{a} = a_\mu \gamma^\mu$.

Another example: $e^+e^- \to q\bar{q}$ to lowest order. This time, we have only one Feynman diagram which is given in Fig. 2.4. The corresponding amplitude is then simply

$$M = \frac{ee_q}{(p+p')^2} \bar{v}(p',\sigma')\gamma^\mu u(p,\sigma) \bar{u}(q,\tau)\gamma_\mu v(q',\tau'),$$
(2.19)

where $p + p'$ is the four-momentum of the virtual photon.

2.2 Cross sections and decay rates

To obtain the cross section formulae for a given process, one proceeds as follows. First, one determines the amplitude M given by the Feynman rules.

2.2. CROSS SECTIONS AND DECAY RATES

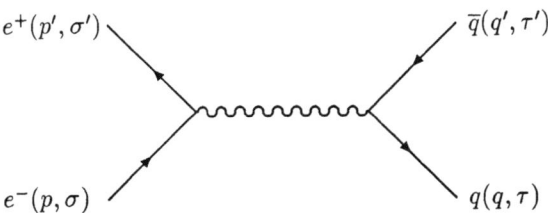

Fig. 2.4: Feynman diagram for $e^+e^- \to q\bar{q}$, in lowest order.

From this, one calculates $\overline{|M|^2}$, i.e., the square of its absolute value summed over the final state polarizations and averaged over the initial state polarizations. Of course, when polarization effects are considered, this summation and averaging procedure must be modified appropriately. The summation over polarization degrees of freedom is performed using the relations:

$$\sum_\sigma u(p,\sigma)\bar{u}(p,\sigma) = \slashed{p}+m, \qquad \text{fermions},$$

$$\sum_\sigma v(p,\sigma)\bar{v}(p,\sigma) = \slashed{p}-m, \qquad \text{antifermions},$$

$$\sum_\lambda \epsilon^\mu(k,\lambda)\epsilon^\nu(k,\lambda)^* = -g^{\mu\nu}, \qquad \text{photons and gluons},$$

$$\sum_\lambda \epsilon^\mu(k,\lambda)\epsilon^\nu(k,\lambda)^* = -g^{\mu\nu}+k^\mu k^\nu/m^2, \quad \text{massive spin-1 particles}.$$

(2.20)

Note that the formula (2.20) for the photons or gluons can only be used when the amplitude is gauge invariant, i.e., when $M = \epsilon^\mu(k,\lambda)M_\mu$ satisfies $k^\mu M_\mu = 0$. If this is not the case, for example, when more than one gluon is present, special care must be taken to perform the gluon polarization sum correctly through the inclusion of so-called ghost contributions. However, the helicity amplitude method to be described in this book nicely circumvents this problem. We, therefore, do not discuss this issue here in more detail.

For a scattering process with two incoming particles and n outgoing ones, $k_1 + k_2 \to p_1 + \ldots + p_n$, one then writes down the multi-differential cross section formula

$$d\sigma = \frac{1}{|\vec{v}_1 - \vec{v}_2|} \frac{1}{4k_{10}k_{20}} \overline{|M|^2} (2\pi)^4 \delta^4(k_1 + k_2 - p_1 - \ldots - p_n)$$

$$\times \frac{d^3\vec{p}_1}{(2\pi)^3 2p_{10}} \cdots \frac{d^3\vec{p}_n}{(2\pi)^3 2p_{n0}}, \qquad (2.21)$$

where $\vec{v}_1$ and $\vec{v}_2$ are the velocities of the incoming particles taken to be collinear. For ultra-relativistic particles, $|\vec{v}| \simeq 1$, and, in the c.m. frame, $|\vec{v}_1 - \vec{v}_2| \simeq 2$.

To obtain the desired differential cross section formula, one must finally integrate over the appropriate phase space variables. For example, for $k_1 + k_2 \to p_1 + p_2$, one finds in the high energy limit in the c.m. frame

$$\left. \frac{d\sigma}{d\Omega} \right|_{c.m.} = \frac{1}{64\pi^2 s} \overline{|M|^2} , \qquad (2.22)$$

with $s = (k_1 + k_2)^2$.

Similarly, the decay rate for a particle with mass m in its rest frame is given by

$$d\Gamma = \frac{1}{2m} \overline{|M|^2} \frac{d^3\vec{p}_1}{(2\pi)^3 2p_{10}} \cdots \frac{d^3\vec{p}_n}{(2\pi)^3 2p_{n0}} (2\pi)^4 \delta^4(k - p_1 - \ldots - p_n) , \qquad (2.23)$$

with $k = (m, \vec{0})$, the four-momentum of the decaying particle.

If r identical particles are present in the final state and one integrates over regions of phase space where they become indistinguishable, one must include a symmetry factor $1/r!$ in the cross section formula to avoid multiple counting of the same kinematical configurations.

2.3 Example: $Z \to e^+ e^-$

The decay of the intermediate vector boson Z, often denoted by Z^0, [6,10] into an e^+e^- pair is given in lowest order by the Feynman diagram of Fig. 2.5, where the four-momenta of the particles are given between parentheses. In the Glashow-Weinberg-Salam model [6], the vertex is given by

$$i e \gamma_\mu (v_e - a_e \gamma_5)/4 \sin\theta_W \cos\theta_W , \qquad (2.24)$$

where $v_e = -1 + 4\sin^2\theta_W$ and $a_e = -1$. The matrix γ_5 is defined by

$$\gamma_5 = i\gamma^0 \gamma^1 \gamma^2 \gamma^3 , \qquad (2.25)$$

and takes the form

$$\gamma_5 = \begin{pmatrix} \mathbb{1} & 0 \\ 0 & -\mathbb{1} \end{pmatrix} \qquad (2.26)$$

in the representation of the γ-matrices given by eqns (2.6).

The corresponding Feynman amplitude is then simply

$$M = \frac{e}{4 \sin\theta_W \cos\theta_W} \epsilon^\mu(k,\lambda)^* \bar{u}(p_-, \sigma_-) \gamma_\mu (v_e - a_e \gamma_5) v(p_+, \sigma_+) . \qquad (2.27)$$

2.3. EXAMPLE: $Z \to e^+ e^-$

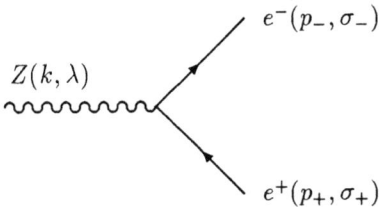

Fig. 2.5: Feynman diagram for $Z \to e^+e^-$, in lowest order.

Hence, neglecting the electron mass and using momentum conservation,

$$\begin{aligned}\overline{|M|^2} &= \frac{1}{3}\sum_{\lambda,\sigma_-,\sigma_+}|M|^2 \\ &= \frac{e^2}{48\sin^2\theta_W\cos^2\theta_W}(-g^{\mu\nu}+k^\mu k^\nu/M_Z^2) \\ &\quad\times \mathrm{Tr}[\gamma_\mu(v_e-a_e\gamma_5)\not{p}_+\gamma_\nu(v_e-a_e\gamma_5)\not{p}_-] \\ &= \frac{e^2(v_e^2+a_e^2)M_Z^2}{12\sin^2\theta_W\cos^2\theta_W}\,.\end{aligned} \quad (2.28)$$

In the Z rest frame, we then have

$$\Gamma = \int \frac{e^2(v_e^2+a_e^2)M_Z}{384\pi^2\sin^2\theta_W\cos^2\theta_W}\delta^4(k-p_+-p_-)\frac{d^3\vec{p}_+\,d^3\vec{p}_-}{p_{+0}\,p_{-0}}. \quad (2.29)$$

With the relation

$$\int\frac{d^3\vec{p}}{2p_0} = \int d^4p\,\delta(p^2-m^2)\theta(p_0)\,, \quad (2.30)$$

we can, for example, integrate over $d^4 p_-$, yielding

$$\Gamma = \frac{e^2(v_e^2+a_e^2)M_Z}{192\pi^2\sin^2\theta_W\cos^2\theta_W}\int\frac{d^3\vec{p}_+}{p_{+0}}\delta(M_Z^2-2M_Zp_{+0})\,\theta(M_Z-p_{+0})\,. \quad (2.31)$$

With $e^2 = 4\pi\alpha$, we obtain

$$\Gamma(Z\to e^+e^-) = \frac{\alpha(v_e^2+a_e^2)M_Z}{48\sin^2\theta_W\cos^2\theta_W}\,. \quad (2.32)$$

With the experimental values for M_Z and $\sin^2\theta_W$ [18], this gives the result $\Gamma(Z\to e^+e^-) = 0.08$ GeV.

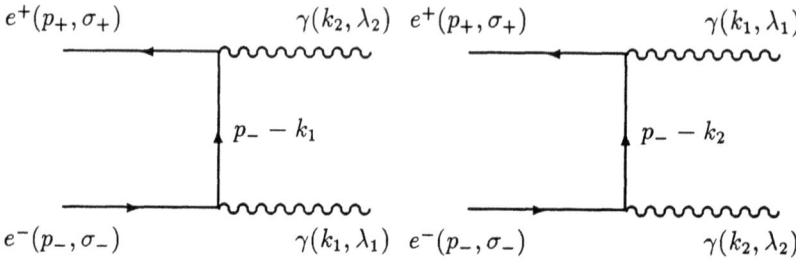

Fig. 2.6: Feynman diagrams for $e^+e^- \to \gamma\gamma$, to order e^2.

2.4 Another example: $e^+e^- \to \gamma\gamma$

For the process of electron-positron annihilation into two photons [19]
$$e^+(p_+) + e^-(p_-) \to \gamma(k_1) + \gamma(k_2), \tag{2.33}$$
we have in lowest order the two Feynman diagrams of Fig. 2.6. In the high energy limit, the corresponding amplitudes are

$$M_1 = \frac{-e^2}{t} \bar{v}(p_+,\sigma_+)\,\not{\epsilon}(k_2,\lambda_2)(\not{p}_- - \not{k}_1)\,\not{\epsilon}(k_1,\lambda_1)u(p_-,\sigma_-),$$

$$M_2 = \frac{-e^2}{u} \bar{v}(p_+,\sigma_+)\,\not{\epsilon}(k_1,\lambda_1)(\not{p}_- - \not{k}_2)\,\not{\epsilon}(k_2,\lambda_2)u(p_-,\sigma_-), \tag{2.34}$$

where we have introduced the Mandelstam variables [20]

$$\begin{aligned} s &= (p_- + p_+)^2 \simeq 2(p_+ \cdot p_-), \\ t &= (p_- - k_1)^2 \simeq -2(p_- \cdot k_1), \\ u &= (p_- - k_2)^2 \simeq -2(p_- \cdot k_2), \\ s + t + u &\simeq 0. \end{aligned} \tag{2.35}$$

It follows that

$$\begin{aligned} \overline{|M|^2} &= \frac{1}{4} \sum_{\sigma_+,\sigma_-,\lambda_1,\lambda_2} |M_1 + M_2|^2 \\ &= \frac{e^4}{4t^2}\mathrm{Tr}\,[\gamma_\mu(\not{p}_- - \not{k}_1)\gamma_\nu \not{p}_- \gamma^\nu(\not{p}_- - \not{k}_1)\gamma^\mu \not{p}_+] \\ &+ \frac{e^4}{4u^2}\mathrm{Tr}\,[\gamma_\nu(\not{p}_- - \not{k}_2)\gamma_\mu \not{p}_- \gamma^\mu(\not{p}_- - \not{k}_2)\gamma^\nu \not{p}_+] \\ &+ \frac{e^4}{2tu}\mathrm{Tr}\,[\gamma_\mu(\not{p}_- - \not{k}_1)\gamma_\nu \not{p}_- \gamma^\mu(\not{p}_- - \not{k}_2)\gamma^\nu \not{p}_+]. \end{aligned} \tag{2.36}$$

2.4. ANOTHER EXAMPLE: $e^+e^- \to \gamma\gamma$

The first term is $\overline{|M_1|^2}$, the second one is $\overline{|M_2|^2}$ and the third term is $2\text{Re}\overline{M_1 M_2^*}$. The evaluation of the traces yields

$$\overline{|M|^2} = \frac{e^4}{t^2}\text{Tr}[\not{k}_1 \not{p}_- \not{k}_1 \not{p}_+] + \frac{e^4}{u^2}\text{Tr}[\not{k}_2 \not{p}_- \not{k}_2 \not{p}_+]$$

$$= 2e^4\left(\frac{u}{t} + \frac{t}{u}\right). \tag{2.37}$$

Note that the interference term vanishes, as it is proportional to the expression $(p_- - k_1 \cdot p_- - k_2) = (s + t + u)/2 = 0$.

In the c.m. frame, the differential cross section becomes

$$d\sigma = \int \frac{e^4}{16\pi^2 s k_{10} k_{20}} \left(\frac{u}{t} + \frac{t}{u}\right) \delta^4(p_+ + p_- - k_1 - k_2)\, d^3\vec{k}_1\, d^3\vec{k}_2$$

$$= \frac{2\alpha^2}{s} \int \left(\frac{u}{t} + \frac{t}{u}\right) \delta[(p_+ + p_- - k_1)^2]\theta(p_{+0} + p_{-0} - k_{10}) \frac{d^3\vec{k}_1}{k_{10}}, \tag{2.38}$$

and

$$\left.\frac{d\sigma}{d\Omega}\right|_{\text{c.m.}} = \frac{\alpha^2}{2s}\left(\frac{u}{t} + \frac{t}{u}\right). \tag{2.39}$$

This formula is divergent for $t = 0$ and $u = 0$, i.e., in the forward and in the backward directions. This divergence, however, is unphysical and originates in the $m = 0$ approximation which was used for the electron and the positron. Eqn (2.39) is therefore not correct for $|t|$ or $|u|$ small as finite mass effects have to be taken into account. A more complete discussion of the ranges of validity of the $m = 0$ approximation will be presented in Section 3.4.

3
Helicity states

*A deep emotion
comes from true devotion
to the right notion
of spin in motion.*

3.1 Fermions

In the following chapters, we shall only work in the high energy limit, which means that we shall neglect the fermion masses. Later on, in Chapter 7, we shall present the modifications to the helicity method which are necessary when the finiteness of the fermion mass has to be considered. For massless fermions, helicity states are Lorentz invariant notions. It is the aim of this book to show how much easier it is to calculate first the helicity amplitudes for a given process and to add their squared absolute values to obtain the cross section rather than following the procedures of Chapter 2. Of course, in Chapter 2, we only treated simple examples, for which the summation of $|M|^2$ over the polarization states was not such a tremendous amount of work. However, when a process is described by n Feynman diagrams, it amounts to writing down $n(n+1)/2$ traces. In the case of $e^+e^- \to e^+e^-\gamma\gamma$, we have 40 Feynman diagrams, and, if we include Z-exchange, this number increases to $n = 80$. Clearly, one does not want to consider 3240 interference terms!

On the other hand, the process $e^+e^- \to e^+e^-\gamma\gamma$ is described by 64 helicity amplitudes, of which one half can be obtained from a parity conjugation on the other 32 amplitudes. Obviously, it is preferable to consider 32 short expressions than to evaluate 3240 long ones.

To proceed with the calculation of the helicity amplitudes, we define the fermion helicities with the equations

$$u_\pm(p) = \tfrac{1}{2}(1 \pm \gamma_5)u_\pm(p),$$

$$v_\pm(p) = \tfrac{1}{2}(1 \mp \gamma_5)v_\pm(p),$$

$$\bar{u}_\pm(p) = \tfrac{1}{2}\bar{u}_\pm(p)(1 \mp \gamma_5),$$

$$\bar{v}_\pm(p) = \tfrac{1}{2}\bar{v}_\pm(p)(1 \pm \gamma_5), \tag{3.1}$$

where the subscripts on the spinors denote the helicities of the corresponding fermions. For this reason, we omit, from now on, the explicit reference to polarization labels, σ, τ, etc., which appeared as arguments of the spinor [in] Chapter 2.

3.2 Photons and gluons

Photons and gluons, being massless, are characterized by two polarization states. In this section, we introduce a convenient representation for [the] helicity states [21].

Let us consider first the case of a photon with four-momentum k. [An] especially convenient choice of the polarization vectors[1] is

$$\epsilon'^{\parallel}_\mu = 2\sqrt{2} N [(q \cdot k) p_\mu - (p \cdot k) q_\mu],$$

$$\epsilon^{\perp}_\mu = 2\sqrt{2} N \epsilon_{\mu\alpha\beta\gamma} q^\alpha p^\beta k^\gamma,$$

where

$$N = \tfrac{1}{4}[(p \cdot q)(p \cdot k)(q \cdot k)]^{-\frac{1}{2}},$$

and $\epsilon_{\alpha\beta\gamma\delta}$ is the totally antisymmetric tensor in four dimensions with $\epsilon_{0123} = +1$.

The prime for $\epsilon'^{\parallel}_\mu$ merely indicates that this is not quite the polarization vector to be used later on. Also, the factor $2\sqrt{2}$ in eqns (3.2) is merely introduced for the sake of simplicity in later formulae. In these expressions we introduced two arbitrary vectors, p and q, which for convenience are taken to be light-like, i.e., $p^2 = q^2 = 0$.

Note that the normalization factor N is the same for both polarization vectors, and that

$$(\epsilon'^{\parallel})^2 = (\epsilon^{\perp})^2 = -1,$$

$$(k \cdot \epsilon'^{\parallel}) = (k \cdot \epsilon^{\perp}) = (\epsilon'^{\parallel} \cdot \epsilon^{\perp}) = 0.$$

These polarization states can alternatively be combined into circularly [po]larized states, characterized by

$$\epsilon'^{\pm}_\mu = \frac{1}{\sqrt{2}}(\epsilon'^{\parallel}_\mu \pm i \epsilon^{\perp}_\mu).$$

Using the identity

$$i\gamma^\mu \epsilon_{\mu\alpha\beta\gamma} = (\gamma_\alpha \gamma_\beta \gamma_\gamma - \gamma_\alpha g_{\beta\gamma} + \gamma_\beta g_{\alpha\gamma} - \gamma_\gamma g_{\alpha\beta})\gamma_5,$$

we can write

$$\not{\epsilon}'^{\pm} = -N[\not{k}\not{p}\not{q}(1 \pm \gamma_5) - \not{p}\not{q}\not{k}(1 \mp \gamma_5) \mp 2(p \cdot q)\not{k}\gamma_5].$$

---
[1]Similar polarization vectors were introduced by Voronov [22] in the context of a quantum gravity calculation.

3.2. PHOTONS AND GLUONS

The last term here is proportional to $\not{k}\gamma_5$. Since we work with massless fermions, it can frequently be omitted because of the conservation of an axial current. This is, for example, the case in pure QED and in some theories involving only vector and/or axial vector couplings. Accordingly, the polarization vectors $\epsilon'^{\pm}_\mu$ are modified to be $\epsilon^{\pm}_\mu$ so that effectively

$$\not{\epsilon}^{\pm} = -N[\not{k}\,\not{p}\,\not{q}(1\pm\gamma_5) - \not{p}\,\not{q}\,\not{k}(1\mp\gamma_5)]. \qquad (3.8)$$

This is the basic formula for the present formalism in the case of photon bremsstrahlung.

There are several reasons why this formula leads to great simplifications in QED calculations.

A. The choice of the four-vectors p and q is still quite arbitrary. When the photon line is next to an external electron or positron line, we can exploit this freedom of choice and take either p or q equal to the fermion momentum. In that case, only one of the terms will give a nonvanishing contribution. The reason is simply that for massless fermions

$$\not{q}\,u(q) = 0, \qquad \bar{u}(p)\,\not{p} = 0, \qquad (3.9)$$

and similarly for v-type spinors.

B. If we fix the fermion helicities, either a factor $1+\gamma_5$ or a factor $1-\gamma_5$ appears on the fermion line in the Feynman diagram. Again, this results in the survival of only one term in the $\not{\epsilon}^{\pm}$ formula.

C. When the photon line is next to the external electron line, there is a cancellation of the denominator. For example,

$$\bar{u}(p)\,\not{\epsilon}^{\pm}\,\frac{\not{p}+\not{k}}{2(p\cdot k)} = -N\bar{u}(p)\,\not{k}\,\not{p}\,\not{q}\,\frac{\not{p}+\not{k}}{2(p\cdot k)}$$

$$= -N\bar{u}(p)\,\not{q}(\not{p}+\not{k}), \qquad (3.10)$$

because $\bar{u}(p)\,\not{p}=0$. Of course, the denominator $(p\cdot k)$ now reappears in the normalization factor N [see eqn (2.42)], but only as an overall factor. This is very important when different amplitudes have to be added. This mechanism of denominator cancellations is to a large extent responsible for the simplicity of the answers obtained previously by brute force [23].

The case of gluon bremsstrahlung can be treated in a very similar way. However, one cannot drop the $\not{k}\gamma_5$ term in $\not{\epsilon}'^{\pm}$. Instead, one can write

$$\not{\epsilon}'^{\pm} = -N[\not{k}\,\not{p}\,\not{q}(1\pm\gamma_5) + \not{q}\,\not{p}\,\not{k}(1\mp\gamma_5) - 2(p\cdot q)\not{k}], \qquad (3.11)$$

but, because of vector current conservation, one can drop the last term proportional to $\not{k}$. Thus, one can use effectively

$$\not{\epsilon}^{\pm} = -N[\not{k}\;\not{p}\;\not{q}(1\pm\gamma_5) + \not{q}\;\not{p}\;\not{k}(1\mp\gamma_5)]\,, \tag{3.12}$$

and most of the advantages A–C remain.

There is an additional difference with the QED case. In QED, the photon polarization vectors ϵ_μ only appear in Feynman diagrams in the combination $\not{\epsilon}$. In QCD, because of the three- and four-gluon vertices (see Table 2.3), the gluon polarization vectors also appear in scalar products with other four-vectors. Nevertheless, the representation (3.12) is still useful as all scalar products can be obtained with the identity

$$(p'\cdot\epsilon) = \tfrac{1}{4}\mathrm{Tr}\,[\not{p}'\;\not{\epsilon}]\,. \tag{3.13}$$

This already shows that, for example, by choosing $q = p'$ in eqn (3.12), the scalar product $(p'\cdot\epsilon)$ can be made to vanish if $p'^2 = 0$.

Also note that no reference is made to a specific choice of frame for the polarization vectors. This means that the calculation of the helicity amplitudes can be naturally performed in a covariant way. Furthermore, we only consider physical polarizations, which means we never have to worry about the so-called ghost contributions, which were mentioned in the preceding chapter.

How the whole procedure of calculating helicity amplitudes is made to work in practice will be shown by a simple example in the coming section.

3.3 Example: $e^+e^- \to \gamma\gamma$ again

We already considered the process of electron-positron annihilation into two photons,

$$e^+(p_+) + e^-(p_-) \to \gamma(k_1) + \gamma(k_2)\,, \tag{3.14}$$

in Section 2.4, where we calculated the cross section using the standard Feynman techniques. Here, we simply want to show how the helicity amplitude method can be applied for the same purpose.

We recall the Feynman diagrams of Fig. 3.1 and the corresponding amplitudes [eqns (2.34)]

$$\begin{aligned}
M_1 &= -\frac{e^2}{t}\,\bar{v}(p_+)\,\not{\epsilon}(k_2)(\not{p}_- - \not{k}_1)\,\not{\epsilon}(k_1)u(p_-)\,,\\[4pt]
M_2 &= -\frac{e^2}{u}\,\bar{v}(p_+)\,\not{\epsilon}(k_1)(\not{p}_- - \not{k}_2)\,\not{\epsilon}(k_2)u(p_-)\,.
\end{aligned} \tag{3.15}$$

Here, as well as in the rest of this book, we omit the photon spin indices λ in the polarization vectors ϵ.

3.3. EXAMPLE: $e^+e^- \to \gamma\gamma$ AGAIN

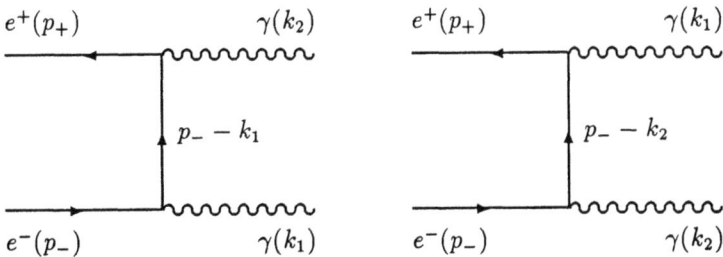

Fig. 3.1: Feynman diagrams for $e^+e^- \to \gamma\gamma$.

There are altogether 16 helicity amplitudes corresponding to the two helicity states for the four particles. Let us denote these amplitudes by $M(\lambda_+, \lambda_-; \lambda_1, \lambda_2)$, where λ_+ (λ_-) is the helicity of the positron (electron), and λ_1 (λ_2) the helicity of photon 1 (2).

When the helicities of the fermions are the same, $\lambda_+ = \lambda_-$, the helicity amplitude vanishes, because of a mismatch in the projection operators $(1 \pm \gamma_5)/2$. Hence,

$$M(+,+;\lambda_1,\lambda_2) = M(-,-;\lambda_1,\lambda_2) = 0, \quad (3.16)$$

for all λ_1 and λ_2.

Let us consider $M(+,-;+,+)$ next. It is especially convenient to choose

$$\not{\epsilon}^+(k_i) = N_i[\not{p}_+ \not{p}_- \not{k}_i(1-\gamma_5) - \not{k}_i \not{p}_+ \not{p}_-(1+\gamma_5)],$$

$$N_i^{-1} = 4[(p_+ \cdot p_-)(p_+ \cdot k_i)(p_- \cdot k_i)]^{\frac{1}{2}}, \quad i = 1,2. \quad (3.17)$$

Because of the projection operator $(1-\gamma_5)/2$ which stands next to $u_-(p_-)$ only the first term in $\not{\epsilon}^+(k_i)$ contributes. But, the substitution of $\not{\epsilon}^+(k_i)$ in M_1 and M_2 then produces the combination $\bar{v}(p_+)\not{p}_+$, which vanishes. Hence,

$$M(+,-;+,+) = 0. \quad (3.18)$$

Similarly, for $M(-,+;+,+)$, the roles of the two terms in $\not{\epsilon}^+(k_i)$ are interchanged, and we find

$$M(-,+;+,+) = 0, \quad (3.19)$$

because of the relation $\not{p}_- u(p_-) = 0$. Because of parity conjugation, which flips all helicities, we also find

$$M(-,+;-,-) = M(+,-;-,-) = 0. \quad (3.20)$$

More precisely, the helicity amplitudes in (3.16) and (3.18)–(3.20) are in fact proportional to some power of the fermion mass, which has been neglected. For a more systematic discussion of these finite mass effects, see Chapter 7.

The helicities of the photons must thus be opposite. Consider next $M(+,-;+,-)$ and let us choose, instead of eqns (3.17),

$$\not{\epsilon}^+(k_1) = N_1[\not{p}_+ \not{p}_- \not{k}_1(1-\gamma_5) - \not{k}_1 \not{p}_+ \not{p}_-(1+\gamma_5)],$$

$$\not{\epsilon}^-(k_2) = N_2[\not{p}_+ \not{p}_- \not{k}_2(1+\gamma_5) - \not{k}_2 \not{p}_+ \not{p}_-(1-\gamma_5)]. \quad (3.21)$$

Substitution of these expressions in M_1 and M_2 [eqns (3.15)] gives

$$M_1 = \frac{e^2}{t} N_1 N_2 \bar{v}(p_+) \frac{1+\gamma_5}{2} \not{k}_2 \not{p}_+ \not{p}_-(1-\gamma_5)$$

$$\times (\not{p}_- - \not{k}_1) \not{p}_+ \not{p}_- \not{k}_1 (1-\gamma_5) \frac{1-\gamma_5}{2} u(p_-)$$

$$= \frac{-e^2}{t} N_1 N_2 \bar{v}(p_+) \not{k}_2 \not{p}_+ \not{p}_- \not{k}_1 \not{p}_+ \not{p}_- \not{k}_1 (1-\gamma_5) u(p_-)$$

$$= 2e^2 N_1 N_2 (p_+ \cdot k_2) \bar{v}(p_+) \not{p}_- \not{k}_1 \not{p}_+ (1-\gamma_5) u(p_-)$$

$$= -4e^2 N_1 N_2 (p_+ \cdot k_2)(p_+ p_-) \bar{v}(p_+) \not{k}_1 (1-\gamma_5) u(p_-)$$

$$= -\frac{e^2}{u} \bar{v}(p_+) \not{k}_1 (1-\gamma_5) u(p_-),$$

$$M_2 = 0. \quad (3.22)$$

We made repeated use of the Dirac equation for the spinors and replaced

$$s = 2(p_+ \cdot p_-) = 2(k_1 \cdot k_2),$$

$$t = -2(p_- \cdot k_1) = -2(p_+ \cdot k_2), \quad (3.23)$$

$$u = -2(p_- \cdot k_2) = -2(p_+ \cdot k_1).$$

We find that M_2 vanishes, thus $|M(+,-;+,-)|^2 = |M_1|^2$, which is easy to calculate:

$$|M(+,-;+,-)|^2 = \frac{e^4}{u^2} \bar{v}(p_+) \not{k}_1 (1-\gamma_5) u(p_-) \bar{u}(p_-) \not{k}_1 (1-\gamma_5) v(p_+)$$

$$= \frac{e^4}{u^2} \bar{v}(p_+) \not{k}_1 (1-\gamma_5) \left(\sum_{\text{pol}} u(p_-) \bar{u}(p_-)\right)$$

$$\times \not{k}_1 (1-\gamma_5) v(p_+)$$

$$= 2\frac{e^4}{u^2} \text{Tr}[\not{p}_+ \not{k}_1 \not{p}_- \not{k}_1 (1-\gamma_5)]$$

$$= 4e^4 t/u. \quad (3.24)$$

The fact that we could freely insert the polarization sum for the spinor $u(p_-)$ [and for $v(p_+)$] is due to the presence of the helicity projection operator $(1-\gamma_5)/2$ in these spinorial expressions. We then used

$$\sum_{\text{pol}} u(p_-)\bar{u}(p_-) = \not{p}_-\,, \qquad \sum_{\text{pol}} v(p_+)\bar{v}(p_+) = \not{p}_+\,, \qquad (3.25)$$

which converted the expression into a simple trace.

Clearly, $M(+,-;-,+)$ can be obtained from $M(+,-;+,-)$ by interchanging $1 \leftrightarrow 2$, which amounts to $t \leftrightarrow u$. We thus have

$$\begin{aligned}\overline{|M|^2} &= \frac{1}{4}\cdot 2\left[|M(+,-;+,-)|^2 + |M(+,-;-,+)|^2\right] \\ &= 2e^4\left(\frac{t}{u} + \frac{u}{t}\right).\end{aligned} \qquad (3.26)$$

The factor $1/4$ is due to the averaging over the fermion helicities, and the factor 2 takes into account the two other helicity amplitudes obtained through parity conjugation. This result coincides of course with the previously obtained formula (2.37).

This simple example already shows many of the features encountered in more complicated cases, the most striking one being that certain Feynman diagrams give vanishing contributions for a given helicity configuration through appropriate choices of $\not{\ell}^{\pm}$.

A comparison between this technique and the standard Feynman technique for this process may fail to impress the reader. This is due to the simplicity of the case which was treated. One should, however, bear in mind that, for more complicated cases with many more Feynman diagrams, the situation becomes quite different. We only made this pedagogical exercise to illustrate the basic techniques of the helicity amplitude method.

3.4 Ranges of validity

In the foregoing presentation of the helicity amplitude method, we put the fermion masses equal to zero and justified this approximation by saying that we only consider the high energy limit. At this point, we want to state more precisely what we mean by this approximation and determine more quantitatively its range of validity.

Every helicity amplitude is a function of the invariants in the process, i.e., of the scalar products of the four-momenta of the particles and, in particular, of the masses. What one then means by the high energy limit is a set of kinematic configurations where all the scalar products involving different four-momenta are much larger than the squared masses of the external particles.

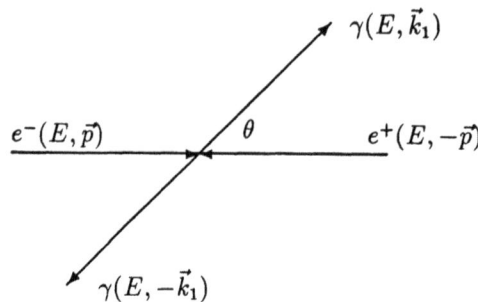

Fig. 3.2: Kinematic configuration for $e^+e^- \to \gamma\gamma$.

Consider, once more, the process

$$e^+(p_+) + e^-(p_-) \to \gamma(k_1) + \gamma(k_2), \qquad (3.27)$$

with $s = (p_+ + p_-)^2$, $t = (p_- - k_1)^2$, $u = (p_- - k_2)^2$, and let m be the lepton mass. The high energy limit is then defined by

$$s \gg m^2, \qquad |t| \gg m^2, \qquad \text{and} \qquad |u| \gg m^2. \qquad (3.28)$$

In practice, this means that not only the incoming energy must be sufficiently large, but also that the photons must be emitted in directions not too close to the directions of the incoming fermions.

In the c.m. frame, we have the kinematic configuration depicted in Fig. 3.2, and the high energy approximation is valid when $E \gg m$, $\theta \gg m/E$, and $\pi - \theta \gg m/E$.

This situation is general. The high energy approximation breaks down under two circumstances:

i) when the incoming energy is too low;

ii) when nearly collinear particles are present provided they have direct couplings to each other.

The second case requires some discussion. For QED bremsstrahlung processes, it implies that the helicity amplitude method cannot be applied without modifications when photons are emitted in directions nearly parallel to charged fermion directions. It also implies that, for example, for Bhabha scattering, $e^+e^- \to e^+e^-$, and radiative Bhabha scattering, $e^+e^- \to e^+e^-\gamma$, the procedure of this chapter cannot be applied directly in the case of very forward scattering, i.e., when the momentum of the outgoing e^+ (e^-) is nearly parallel to that of the incoming e^+ (e^-). Also, for $e^+e^- \to e^+e^-\gamma$ and $e^+e^- \to \mu^+\mu^-\gamma$, the approximation fails when both the outgoing leptons are emitted in nearly parallel directions. On the other hand, nearly collinear

3.4. RANGES OF VALIDITY

photons pose no particular problem in the high energy limit, as photons do not have direct couplings to each other.

In the QCD case, we have to consider finite mass effects when a gluon is nearly collinear to a quark, an antiquark or another gluon. For example, in the process $e^+e^- \to q\bar{q}g$, we cannot neglect the quark mass when the $(q\bar{q})$-pair is nearly collinear or when the gluon is emitted in nearly the same direction as the quark or the antiquark. For the process $e^+e^- \to q\bar{q}gg$, we have to be careful in addition when the two gluons are nearly parallel. There are, however, no difficulties with the finite mass corrections when one of the gluons is collinear with the incoming e^+ or e^-, as gluons do not couple directly to electrons.

As a rule, the effects of a finite fermion mass cannot be neglected when some scalar product becomes of order m^2 instead of being of order E^2. In the coming Chapters, 4 through 6, we shall, however, ignore these complications and continue to work in the $m = 0$ limit. One only has to bear in mind that the formulae are inaccurate in some special configurations. Later on, in Chapter 7, we shall discuss how these finite mass effects can be taken into account and how the helicity amplitude method can be modified accordingly.

4
Single bremsstrahlung in QED

> It's the electron's plight
> to send out light.
> To predict this sight
> is within our might.

4.1 The process $e^+e^- \to \gamma\gamma\gamma$

Consider the process of electron-positron annihilation into three photons,

$$e^+(p_+) + e^-(p_-) \to \gamma(k_1) + \gamma(k_2) + \gamma(k_3), \qquad (4.1)$$

where the four-momenta of the particles are given in the parentheses. In lowest order, there are six Feynman diagrams (see Fig. 4.1), and the corresponding amplitudes are given by

$$M_1 = e^3 \bar{v}(p_+) \slashed{\epsilon}_3 \frac{-\slashed{p}_+ + \slashed{k}_3}{-2(p_+ \cdot k_3)} \slashed{\epsilon}_2 \frac{\slashed{p}_- - \slashed{k}_1}{-2(p_- \cdot k_1)} \slashed{\epsilon}_1 u(p_-),$$

$$M_2 = e^3 \bar{v}(p_+) \slashed{\epsilon}_3 \frac{-\slashed{p}_+ + \slashed{k}_3}{-2(p_+ \cdot k_3)} \slashed{\epsilon}_1 \frac{\slashed{p}_- - \slashed{k}_2}{-2(p_- \cdot k_2)} \slashed{\epsilon}_2 u(p_-),$$

$$M_3 = e^3 \bar{v}(p_+) \slashed{\epsilon}_2 \frac{-\slashed{p}_+ + \slashed{k}_2}{-2(p_+ \cdot k_2)} \slashed{\epsilon}_3 \frac{\slashed{p}_- - \slashed{k}_1}{-2(p_- \cdot k_1)} \slashed{\epsilon}_1 u(p_-),$$

$$M_4 = e^3 \bar{v}(p_+) \slashed{\epsilon}_2 \frac{-\slashed{p}_+ + \slashed{k}_2}{-2(p_+ \cdot k_2)} \slashed{\epsilon}_1 \frac{\slashed{p}_- - \slashed{k}_3}{-2(p_- \cdot k_3)} \slashed{\epsilon}_3 u(p_-),$$

$$M_5 = e^3 \bar{v}(p_+) \slashed{\epsilon}_1 \frac{-\slashed{p}_+ + \slashed{k}_1}{-2(p_+ \cdot k_1)} \slashed{\epsilon}_2 \frac{\slashed{p}_- - \slashed{k}_3}{-2(p_- \cdot k_3)} \slashed{\epsilon}_3 u(p_-),$$

$$M_6 = e^3 \bar{v}(p_+) \slashed{\epsilon}_1 \frac{-\slashed{p}_+ + \slashed{k}_1}{-2(p_+ \cdot k_1)} \slashed{\epsilon}_3 \frac{\slashed{p}_- - \slashed{k}_2}{-2(p_- \cdot k_2)} \slashed{\epsilon}_2 u(p_-). \qquad (4.2)$$

Let us denote the helicity amplitudes by $M(\lambda_+, \lambda_-; \lambda_1, \lambda_2, \lambda_3)$, where λ_+ (λ_-) is the helicity of the positron (electron) and λ_i, $i = 1, 2, 3$, is the helicity of photon i.

Just as in the $e^+e^- \to \gamma\gamma$ case, the fermion helicities must necessarily be opposite. Thus, for all $\lambda_1, \lambda_2, \lambda_3$,

$$M(+, +; \lambda_1, \lambda_2, \lambda_3) = M(-, -; \lambda_1, \lambda_2, \lambda_3) = 0. \qquad (4.3)$$

For the remaining helicity amplitudes, we choose again [24] the photon polarization vectors such that effectively

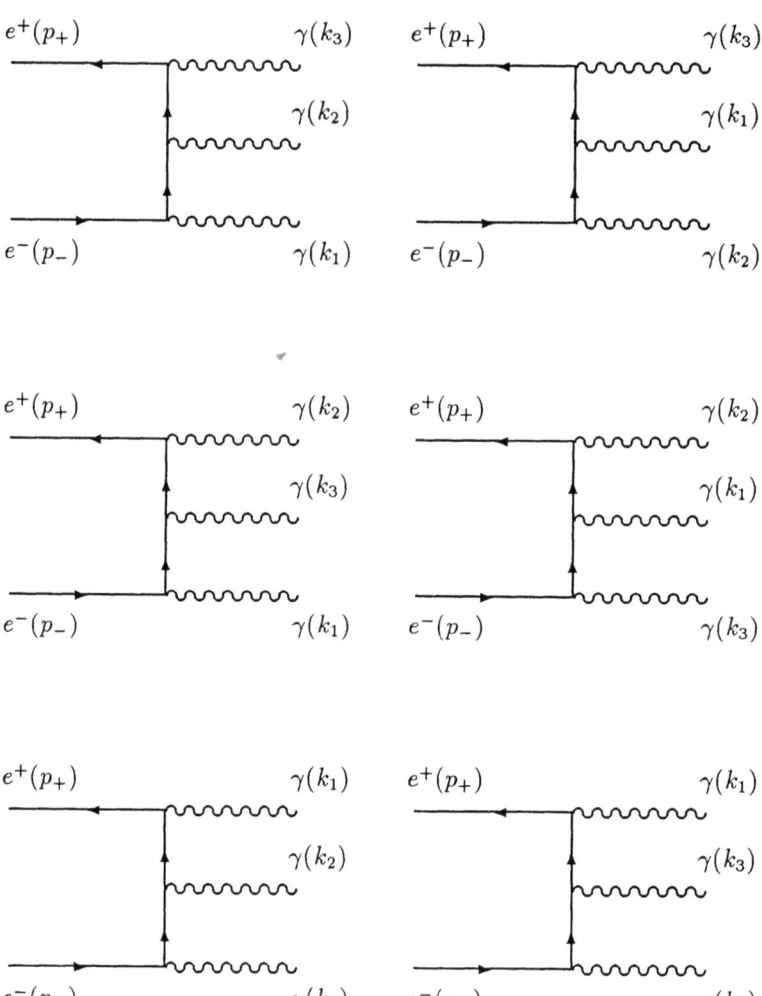

Fig. 4.1: Feynman diagrams for $e^+e^- \to \gamma\gamma\gamma$.

4.1. THE PROCESS $e^+e^- \to \gamma\gamma\gamma$

$$\rlap{/}{k}_i^\pm = N_i[\rlap{/}{p}_+ \rlap{/}{p}_- \rlap{/}{k}_i(1 \mp \gamma_5) - \rlap{/}{k}_i \rlap{/}{p}_+ \rlap{/}{p}_-(1 \pm \gamma_5)],$$

$$N_i^{-1} = 4[(p_+ \cdot p_-)(p_+ \cdot k_i)(p_- \cdot k_i)]^{\frac{1}{2}}, \qquad i = 1,2,3. \qquad (4.4)$$

Consider next $M(+,-;+,+,+)$. The spinor $\bar{v}(p_+)$ comes with the projection operator $(1+\gamma_5)/2$ and, consequently, only the first term in $\rlap{/}{k}_i^+$ contributes. But then, the amplitude vanishes as

$$\bar{v}(p_+) \rlap{/}{p}_+ = 0. \qquad (4.5)$$

More generally, one finds that, for all λ_+ and λ_-,

$$M(\lambda_+, \lambda_-; +, +, +) = M(\lambda_+, \lambda_-; -, -, -) = 0. \qquad (4.6)$$

For the amplitude $M(+,-;+,+,-)$, only the Feynman diagrams M_1 and M_2 contribute:

$$M(+,-;+,+,-)$$

$$= -4e^3 N_1 N_2 N_3 \bar{v}(p_+) \rlap{/}{k}_3 \rlap{/}{p}_+ \rlap{/}{p}_- \frac{\rlap{/}{p}_+ - \rlap{/}{k}_3}{2(p_+ \cdot k_3)} [\rlap{/}{p}_+ \rlap{/}{p}_- \rlap{/}{k}_2 \frac{\rlap{/}{p}_- - \rlap{/}{k}_1}{-2(p_- \cdot k_1)}$$

$$\rlap{/}{p}_+ \rlap{/}{p}_- \rlap{/}{k}_1 + \rlap{/}{p}_+ \rlap{/}{p}_- \rlap{/}{k}_1 \frac{\rlap{/}{p}_- - \rlap{/}{k}_2}{-2(p_- \cdot k_2)} \rlap{/}{p}_+ \rlap{/}{p}_- \rlap{/}{k}_2](1-\gamma_5)u(p_-),$$

$$= -4e^3 N_1 N_2 N_3 \bar{v}(p_+) \rlap{/}{p}_- \rlap{/}{k}_3 \rlap{/}{p}_+ \rlap{/}{p}_-$$

$$\times [\rlap{/}{k}_2(\rlap{/}{p}_- - \rlap{/}{k}_1) + \rlap{/}{k}_1(\rlap{/}{p}_- - \rlap{/}{k}_2)] \rlap{/}{p}_+(1-\gamma_5)u(p_-),$$

$$= -8e^3 N_1 N_2 N_3 [(p_- \cdot k_2) + (p_- \cdot k_1) - (k_1 \cdot k_2)]$$

$$\times \bar{v}(p_+) \rlap{/}{p}_- \rlap{/}{k}_3 \rlap{/}{p}_+ \rlap{/}{p}_- \rlap{/}{p}_+(1-\gamma_5)u(p_-),$$

$$= -16e^3 N_1 N_2 N_3 (p_+ \cdot k_3)(p_+ \cdot p_-)\bar{v}(p_+) \rlap{/}{p}_- \rlap{/}{k}_3 \rlap{/}{p}_+(1-\gamma_5)u(p_-),$$

$$= 32e^3 N_1 N_2 N_3 (p_+ \cdot k_3)(p_+ \cdot p_-)^2 \bar{v}(p_+) \rlap{/}{k}_3(1-\gamma_5)u(p_-). \qquad (4.7)$$

The phase of this amplitude depends on the choice of phases for the spinors. As we only compute unpolarized cross sections here, we can disregard this overall phase and write

$$M(+,-;+,+,-) \doteq 128 e^3 N_1 N_2 N_3 (p_+ \cdot p_-)^2 (p_+ \cdot k_3)[(p_+ \cdot k_3)(p_- \cdot k_3)]^{\frac{1}{2}}. \qquad (4.8)$$

As in the $e^+e^- \to \gamma\gamma$ case, we used the trick of inserting fermion polarization sums to evaluate the absolute value of the spinorial expression. Throughout this book, the symbol '≐' stands for 'an equality sign modulo a phase fac[tor]'. With the definition (4.4) of N_i, we finally obtain

$$M(+,-;+,+,-) \doteq 2e^3 (p_+ \cdot p_-)^{\frac{1}{2}} \frac{(p_+ \cdot k_3)[(p_+ \cdot k_3)(p_- \cdot k_3)]^{\frac{1}{2}}}{\left[\prod_{i=1}^{3}(p_+ \cdot k_i)(p_- \cdot k_i)\right]^{\frac{1}{2}}}.$$

The purpose of this exercise in Diracology was to exhibit the simplicit[y of] the manipulations which lead to this result.

For the helicity amplitude $M(+,-;-,-,+)$, one finds that only the [di]agrams M_4 and M_5 contribute. Similar manipulations then lead to

$$M(+,-;-,-,+) \doteq 2e^3 (p_+ \cdot p_-)^{\frac{1}{2}} \frac{(p_- \cdot k_3)[(p_+ \cdot k_3)(p_- \cdot k_3)]^{\frac{1}{2}}}{\left[\prod_{i=1}^{3}(p_+ \cdot k_i)(p_- \cdot k_i)\right]^{\frac{1}{2}}}. \quad (4.$$

The remaining helicity amplitudes are obtained by permuting the pho[ton] indices and by performing a parity conjugation. Hence, for the case wh[ere] one sums over the final polarizations and averages over the initial ones, [one] finds for the squared matrix element

$$\overline{|M|^2} = 2e^6 (p_+ \cdot p_-) \frac{\sum_{i=1}^{3}[(p_+ \cdot k_i)^2 + (p_- \cdot k_i)^2](p_+ \cdot k_i)(p_- \cdot k_i)}{\prod_{i=1}^{3}(p_+ \cdot k_i)(p_- \cdot k_i)}, \quad (4.$$

and for the cross section

$$d\sigma = \frac{\delta^4(p_+ + p_- - k_1 - k_2 - k_3)}{32(2\pi)^5 (p_+ \cdot p_-)} \overline{|M|^2} \frac{d^3\vec{k}_1 \, d^3\vec{k}_2 \, d^3\vec{k}_3}{k_{10} k_{20} k_{30}}. \quad (4.$$

A few comments should be made about this calculation as it manife[sts] some nice features which we also find in the calculations for the other p[ro]cesses. A first observation is that only a limited number of Feynman diagra[ms] contribute to a given helicity amplitude with our choice of polarization v[ec]tors. Furthermore, their use produces automatically the same denomina[tor] for every amplitude which makes the derivation of the formula for $\overline{|M|^2}$ [a] trivial matter. This should be contrasted with the standard Feynman pro[ce]dure, where most of the work to obtain this result is devoted to writing t[he] various trace results over the same denominator.

When polarization effects are considered, one cannot disregard the phas[e] of the helicity amplitudes. In Chapter 11, we show how the helicity amp[li]tudes must be combined for a description of initial state polarization.

4.2 What if different fermions radiate?

In the two previous cases of electron-positron annihilation, $e^+e^- \to \gamma\gamma$ and $e^+e^- \to \gamma\gamma\gamma$, the choice to be made for the photon polarization vectors was rather obvious. It was indeed very advantageous to express $\epsilon(k_i)$ in terms of the four-vectors p_+ and p_- of the fermions enabling one to use the Dirac equations to simplify the spinor expressions.

When one deals with two different charged fermion lines in the Feynman diagrams, as, for example, in $e^+e^- \to \mu^+\mu^-\gamma$, the choice of ϵ's is slightly more complicated. In order to have the denominator cancellations of the previous sections, it is essential to use the external momenta of the fermion line to which the photon is attached. Thus, when the photon is radiated from the muon line, it is advantageous to express ϵ in terms of the momenta q_+ and q_- of the muons, but, for radiation from the electron line, one would prefer to express ϵ in terms of the electron momenta, p_+ and p_-.

It turns out that it is perfectly possible to use $\epsilon(q_+, q_-)$ for the set of diagrams where the photon is attached to the muon line and to use $\epsilon(p_+, p_-)$ for the remaining diagrams provided one takes into account a simple phase factor. To see this, consider the photon to be moving along the z-axis. As the diagrams with radiation from the muon line and from the electron line form separately gauge invariant sets, we can make gauge transformations such that $\epsilon(p_+, p_-)$ and $\epsilon(q_+, q_-)$ only have components in the xy-plane. But then, as these vectors have the same norm, they can at most differ by a phase factor (and terms proportional to the photon momentum arising from the gauge transformations). Thus,

$$\epsilon_\mu^\pm(q_+, q_-) = e^{\pm i\phi} \epsilon_\mu^\pm(p_+, p_-) + \beta_\pm k_\mu. \tag{4.13}$$

Because of gauge invariance, the constants $\beta_\pm$ are irrelevant. The phase ϕ is given by the scalar product of $\epsilon_\mu^\mp(p_+, p_-)$ and $\epsilon_\mu^\pm(q_+, q_-)$:

$$e^{\pm i\phi} = -\epsilon_\mu^\mp(p_+, p_-) \epsilon^{\pm\mu}(q_+, q_-), \tag{4.14}$$

and must be taken into account.

In Section 4.3, we treat the process $e^+e^- \to \mu^+\mu^-\gamma$, and we shall show that this phase factor, although essential, only leads to a minor complication.

4.3 The process $e^+e^- \to \mu^+\mu^-\gamma$

For the process of radiative μ-pair production,

$$e^+(p_+) + e^-(p_-) \to \mu^+(q_+) + \mu^-(q_-) + \gamma(k), \tag{4.15}$$

we have to consider the four Feynman diagrams of Fig. 4.2, with

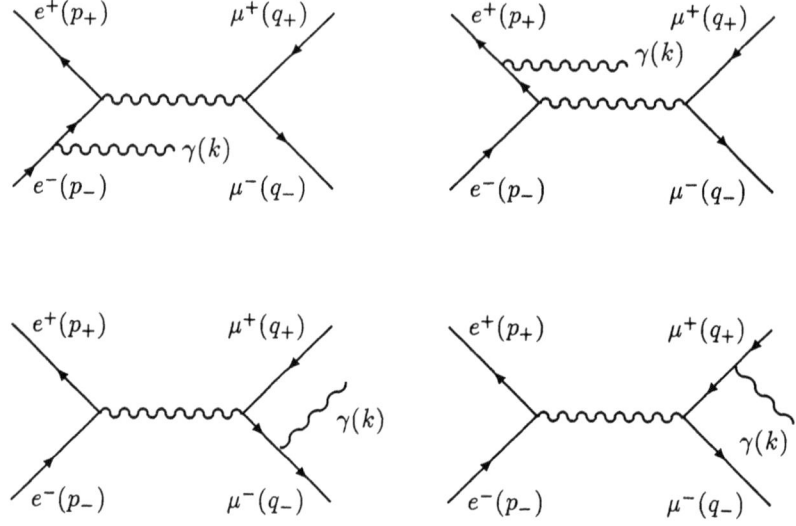

Fig. 4.2: Feynman diagrams for $e^+e^- \to \mu^+\mu^-\gamma$.

$$M_1 = \frac{-e^3}{2(q_+ \cdot q_-)} \bar{v}(p_+)\gamma_\mu \frac{\not{p}_- - \not{k}}{-2(p_- \cdot k)} \not{\epsilon} u(p_-) \bar{u}(q_-)\gamma^\mu v(q_+),$$

$$M_2 = \frac{-e^3}{2(q_+ \cdot q_-)} \bar{v}(p_+) \not{\epsilon} \frac{-\not{p}_+ + \not{k}}{-2(p_+ \cdot k)} \gamma_\mu u(p_-) \bar{u}(q_-)\gamma^\mu v(q_+),$$

$$M_3 = \frac{-e^3}{2(p_+ \cdot p_-)} \bar{v}(p_+)\gamma_\mu u(p_-) \bar{u}(q_-) \not{\epsilon} \frac{\not{q}_- + \not{k}}{2(q_- \cdot k)} \gamma^\mu v(q_+),$$

$$M_4 = \frac{-e^3}{2(p_+ \cdot p_-)} \bar{v}(p_+)\gamma_\mu u(p_-) \bar{u}(q_-)\gamma^\mu \frac{-\not{q}_+ - \not{k}}{2(q_+ \cdot k)} \not{\epsilon} v(q_+). \quad (4.16)$$

As explained in Section 4.2, we introduce two representations for $\not{\epsilon}$. When the photon is radiated from the electron line (M_1 and M_2), we take [24]

$$\not{\epsilon}_p^\pm = N_p[\not{p}_+ \not{p}_- \not{k}(1 \mp \gamma_5) - \not{k} \not{p}_+ \not{p}_-(1 \pm \gamma_5)],$$

$$N_p^{-1} = 4[(p_+ \cdot p_-)(p_+ \cdot k)(p_- \cdot k)]^{\frac{1}{2}}, \quad (4.17)$$

and, for radiation from the muon line (M_3 and M_4), we take

$$\not{\epsilon}_q^\pm = N_q[\not{q}_- \not{q}_+ \not{k}(1 \mp \gamma_5) - \not{k} \not{q}_- \not{q}_+(1 \pm \gamma_5)],$$

$$N_q^{-1} = 4[(q_+ \cdot q_-)(q_+ \cdot k)(q_- \cdot k)]^{\frac{1}{2}}. \quad (4.18)$$

4.3. THE PROCESS $e^+e^- \to \mu^+\mu^-\gamma$

They are, however, related by eqns (4.13) and (4.14):

$$\epsilon_q^\pm = e^{\pm i\phi}\epsilon_p^\pm + \beta_\pm k,$$

$$e^{\pm i\phi} = -(\epsilon_p^\mp \cdot \epsilon_q^\pm)$$

$$= -N_p N_q \text{Tr}[\not{p}_+ \not{p}_- \not{k} \not{q}_- \not{q}_+ \not{k}(1 \mp \gamma_5)]. \quad (4.19)$$

For convenience, let us also introduce the notation

$$s = (p_+ + p_-)^2, \quad t = (p_+ - q_+)^2, \quad u = (p_+ - q_-)^2,$$
$$s' = (q_+ + q_-)^2, \quad t' = (p_- - q_-)^2, \quad u' = (p_- - q_+)^2, \quad (4.20)$$

with

$$s + s' + t + t' + u + u' = 0. \quad (4.21)$$

As usual, we now proceed to calculate the helicity amplitudes. They are denoted by $M(\lambda_1, \lambda_2; \lambda_3, \lambda_4, \lambda_5)$, where λ_1 stands for the positron helicity, λ_2 for the electron, λ_3 for the μ^+, λ_4 for the μ^- and λ_5 for the photon. It is immediately clear that the electron and positron helicities must be opposite, as well as the muon helicities. Hence, for all λ_5,

$$M(+,+;+,+,\lambda_5) = M(+,+;-,-,\lambda_5) = 0,$$
$$M(-,-;+,+,\lambda_5) = M(-,-;-,-,\lambda_5) = 0. \quad (4.22)$$

For the amplitude $M(+,-;+,-,+)$, we find that, with our choice of $\not{k}$, only M_1 and M_4 contribute:

$$M(+,-;+,-,+)$$

$$= \frac{-e^3}{s'}N_p e^{i\phi}\bar{v}(p_+)\gamma_\mu \frac{\not{p}_- - \not{k}}{-2(p_- \cdot k)}\not{p}_+ \not{p}_- \not{k}(1-\gamma_5)u(p_-)$$

$$\times \bar{u}(q_-)\gamma^\mu \frac{1-\gamma_5}{2}v(q_+)$$

$$-\frac{e^3}{s}N_q \bar{v}(p_+)\gamma_\mu \frac{1-\gamma_5}{2}u(p_-)\bar{u}(q_-)\gamma^\mu \frac{-\not{q}_+ - \not{k}}{2(q_+ \cdot k)}$$

$$\times \not{q}_- \not{q}_+ \not{k}(1-\gamma_5)v(q_+)$$

$$= \frac{e^3}{2s'} N_p e^{i\phi} \, \bar{v}(p_+) \gamma_\mu (\slashed{A}_+ + \slashed{A}_-) \, \slashed{p}_+ (1-\gamma_5) u(p_-) \bar{u}(q_-) \gamma^\mu (1-\gamma_5) v(q_+)$$

$$+ \frac{e^3}{2s} N_q \bar{v}(p_+) \gamma_\mu (1-\gamma_5) u(p_-) \bar{u}(q_-) \gamma_\mu (\slashed{p}_+ + \slashed{p}_-) \, \slashed{A}_-(1-\gamma_5) v(q_+).$$
(4.23)

To simplify this result even further, we rewrite, for example, the last term:

$$\frac{e^3}{2s} N_q \, \bar{u}(q_-) \gamma^\mu (\slashed{p}_+ + \slashed{p}_-) \, \slashed{A}_-(1-\gamma_5) v(q_+)$$

$$\times \frac{\bar{v}(q_+) \slashed{A}_-(1-\gamma_5) v(p_+)}{\bar{v}(q_+) \slashed{A}_-(1-\gamma_5) v(p_+)} \bar{v}(p_+) \gamma_\mu (1-\gamma_5) u(p_-)$$

$$= \frac{2e^3}{s} N_q \frac{\bar{u}(q_-) \gamma^\mu (\slashed{p}_+ + \slashed{p}_-) \slashed{A}_- \slashed{A}_+ \slashed{A}_- \slashed{p}_+ \gamma_\mu (1-\gamma_5) u(p_-)}{\bar{v}(q_+) \slashed{A}_-(1-\gamma_5) v(p_+)}$$

$$= -\frac{8e^3}{s} N_q (q_+ \cdot q_-) \frac{\bar{u}(q_-) \slashed{p}_+ \slashed{A}_- \slashed{p}_+ (1-\gamma_5) u(p_-)}{\bar{v}(q_+) \slashed{A}_-(1-\gamma_5) v(p_+)}$$

$$= -\frac{16e^3}{s} N_q (q_+ \cdot q_-)(p_+ \cdot q_-)$$

$$\times \frac{\bar{u}(q_-) \slashed{p}_+ (1-\gamma_5) u(p_-) \bar{v}(p_+) \slashed{A}_-(1-\gamma_5) v(q_+)}{\text{Tr}[\slashed{A}_+ \slashed{A}_-(1-\gamma_5) \slashed{p}_+ \slashed{A}_-(1-\gamma_5)]}$$

$$= -\frac{e^3}{s} N_q \bar{u}(q_-) \slashed{p}_+ (1-\gamma_5) u(p_-) \bar{v}(p_+) \slashed{A}_-(1-\gamma_5) v(q_+).$$
(4.24)

To obtain this result, we repeatedly inserted a summation of fermion helicities to combine products of spinor expressions. This trick was already introduced in Section 3.3 (see also Appendix A).

Similar manipulations can be performed on the first term in the expression for $M(+,-;+,-,+)$, and we find

$$M(+,-;+,-,+) = -e^3 \left[\frac{N_q}{s} + e^{i\phi} \frac{N_p}{s'} \right] \bar{u}(q_-) \slashed{p}_+ (1-\gamma_5) u(p_-)$$

$$\times \bar{v}(p_+) \slashed{A}_-(1-\gamma_5) v(q_+)$$

$$\doteq 4e^3 \left| s' N_q + e^{i\phi} s N_p \right| \frac{u}{(ss')^{\frac{1}{2}}}.$$
(4.25)

With the definitions (4.17), (4.18) and (4.19) of N_p, N_q and $e^{i\phi}$, it is easily verified that

$$\left| s' N_q + e^{i\phi} s N_p \right|^2 = -\frac{1}{8}(v_q - v_p)^2,$$
(4.26)

4.4. THE PROCESS $e^+e^- \to e^+e^-\gamma$

where the four-vectors v_q and v_p are given by

$$v_q = \frac{q_+}{(q_+ \cdot k)} - \frac{q_-}{(q_- \cdot k)}, \quad v_p = \frac{p_+}{(p_+ \cdot k)} - \frac{p_-}{(p_- \cdot k)}. \quad (4.27)$$

Thus,

$$|M(+,-;+,-,+)|^2 = -2e^6(v_q - v_p)^2 \frac{u^2}{ss'}. \quad (4.28)$$

The evaluation of the remaining helicity amplitudes can be done in a completely analogous way. They all have a similar structure, and we refer the reader to Chapter 9 for their explicit expressions.

All nonvanishing helicity amplitudes exhibit a factorization property, one factor being $s'N_q + e^{\pm i\phi}sN_p$, whose squared absolute value is proportional to $(v_q - v_p)^2$. It is amusing to note that formally this factor coincides, in the massless limit, with the 'infrared factor' obtained by Yennie, Frautschi and Suura [25], who considered the limit $k \to 0$ of bremsstrahlung cross sections. In our case, however, the expression $(v_q - v_p)^2$ is not to be evaluated at $k = 0$ since we are dealing with hard photon bremsstrahlung. This factorization property is found to hold for all single bremsstrahlung processes. The case of gluon bremsstrahlung (to be treated in Chapter 5) requires only some simple modifications.

Perhaps more important is the observation that, in the cross section formula, the denominator only contains invariants to the first power. Indeed, all the invariants, like $(p_+ \cdot k)$, s, etc., appear in the Feynman amplitudes in the denominator and one could expect apriori terms like $(p_+ \cdot k)^{-2}$ or s^{-2} in their squares. The fact that this does not happen is largely responsible for the simplicity of the cross section formula. To incorporate this feature in a natural way is inherent to the helicity method.

4.4 The process $e^+e^- \to e^+e^-\gamma$

For the process of radiative Bhabha scattering [12],

$$e^+(p_+) + e^-(p_-) \to e^+(q_+) + e^-(q_-) + \gamma(k), \quad (4.29)$$

we now have eight Feynman diagrams (see Fig. 4.3). The first four are the same as in the muon case, $M_1, ..., M_4$, and the additional t-channel diagrams are given by

$$M_5 = \frac{e^3}{t} \bar{u}(q_-)\gamma_\mu \frac{\not{p}_- - \not{k}}{-2(p_- \cdot k)} \not{\epsilon} u(p_-) \bar{v}(p_+)\gamma^\mu v(q_+),$$

$$M_6 = \frac{e^3}{t'} \bar{u}(q_-)\gamma_\mu u(p_-) \bar{v}(p_+) \not{\epsilon} \frac{-\not{p}_+ + \not{k}}{-2(p_+ \cdot k)} \gamma^\mu v(q_+),$$

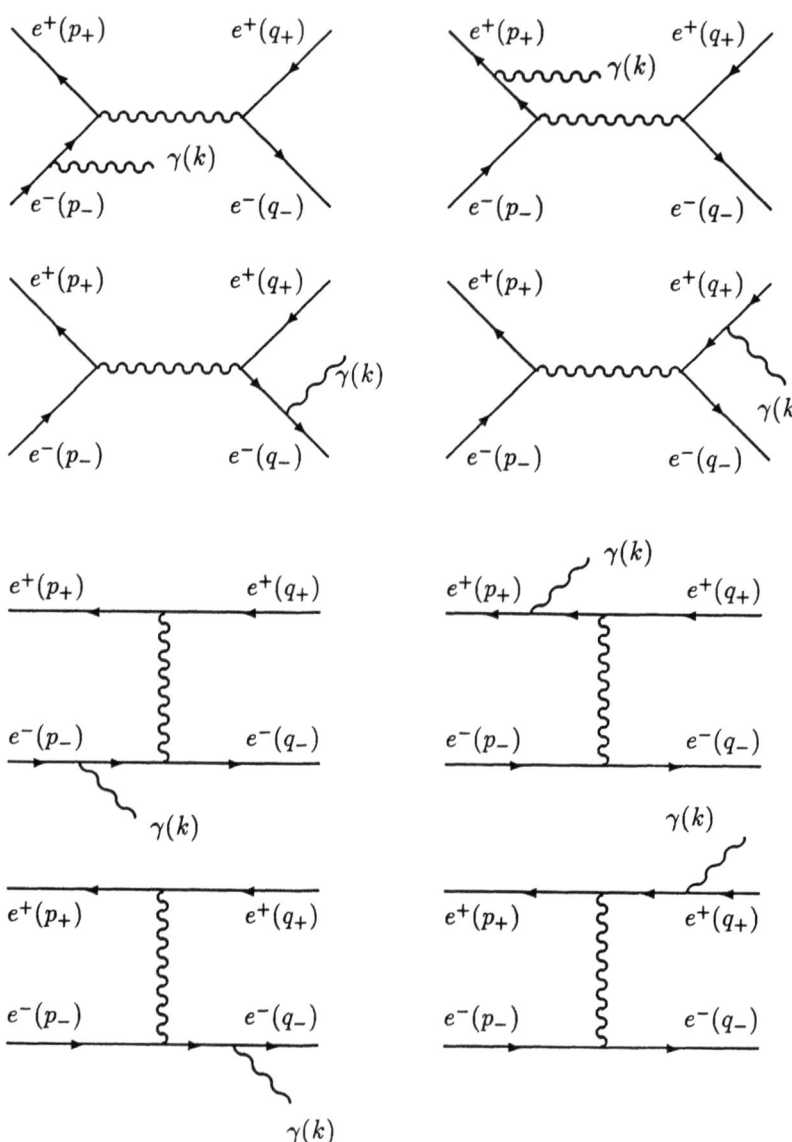

Fig. 4.3: Feynman diagrams for $e^+e^- \to e^+e^-\gamma$.

4.4. THE PROCESS $e^+e^- \to e^+e^-\gamma$

$$M_7 = \frac{e^3}{t} \bar{u}(q_-) \not{\epsilon} \frac{\not{q}_- + \not{k}}{2(q_- \cdot k)} \gamma_\mu u(p_-) \bar{v}(p_+) \gamma^\mu v(q_+),$$

$$M_8 = \frac{e^3}{t'} \bar{u}(q_-) \gamma_\mu u(p_-) \bar{v}(p_+) \gamma^\mu \frac{-\not{q}_+ - \not{k}}{2(q_+ \cdot k)} \not{\epsilon} v(q_+). \quad (4.30)$$

The relative minus sign between the first set of diagrams, $M_1, ..., M_4$, and the second set, $M_5, ..., M_8$, is introduced to satisfy the requirements of Fermi statistics.

In order to deal with these additional diagrams, we introduce two new polarization vectors [24]

$$\not{\epsilon}_+^\pm = N_+ [\not{p}_+ \not{q}_+ \not{k}(1 \mp \gamma_5) - \not{k} \not{p}_+ \not{q}_+ (1 \pm \gamma_5)],$$

$$\not{\epsilon}_-^\pm = N_- [\not{q}_- \not{p}_- \not{k}(1 \mp \gamma_5) - \not{k} \not{q}_- \not{p}_- (1 \pm \gamma_5)],$$

$$N_\pm^{-1} = 4[(p_\pm \cdot q_\pm)(p_\pm \cdot k)(q_\pm \cdot k)]^{\frac{1}{2}}, \quad (4.31)$$

which are again related to one another through a phase factor

$$e^{\pm ix} = -(\epsilon_+^\mp \cdot \epsilon_-^\pm)$$

$$= -N_+ N_- \text{Tr}[\not{p}_+ \not{q}_+ \not{k} \not{q}_- \not{p}_- \not{k}(1 \mp \gamma_5)], \quad (4.32)$$

similar to eqn (4.19).

Let $M_B(\lambda_1, \lambda_2; \lambda_3, \lambda_4, \lambda_5)$ denote the helicity amplitudes for this process, where $\lambda_1, \lambda_2, \lambda_3, \lambda_4$ and λ_5 are respectively the helicities of the incoming e^+, the incoming e^-, the outgoing e^+, the outgoing e^- and the photon. This time, however, $M_B(+,+;+,+,+)$ is no longer zero, but picks up contributions from M_7 and M_8:

$$M_B(+,+;+,+,+)$$

$$= -\frac{e^3}{2t} N_- e^{i\psi} \bar{u}(q_-) \not{p}_-(\not{q}_- + \not{k}) \gamma_\mu (1+\gamma_5) u(p_-) \bar{v}(p_+) \gamma^\mu (1-\gamma_5) v(q_+)$$

$$- \frac{e^3}{2t'} N_+ e^{i(\psi+x)} \bar{u}(q_-) \gamma_\mu (1+\gamma_5) u(p_-) \bar{v}(p_+) \gamma^\mu (\not{q}_+ + \not{k}) \not{p}_+ (1-\gamma_5) v(q_+)$$

$$= -e^3 e^{i\psi} \left[\frac{N_-}{t} + e^{ix} \frac{N_+}{t'} \right] \bar{v}(p_+)(1+\gamma_5) u(p_-) \bar{u}(q_-) \not{p}_- \not{p}_+ (1-\gamma_5) v(q_+)$$

$$\doteq 4e^3 \left| t' N_- + e^{ix} t N_+ \right| \frac{s}{(tt')^{\frac{1}{2}}}, \quad (4.33)$$

where $e^{i\psi}$ is the phase factor relating ϵ_q^+ to ϵ_-^+, i.e.,

$$e^{i\psi} = -(\epsilon_-^- \cdot \epsilon_q^+)$$

$$= -N_q N_- \text{Tr}[\slashed{q}_- \slashed{q}_+ \slashed{k} \slashed{q}_- \slashed{p}_- \slashed{k}(1+\gamma_5)]. \tag{4.34}$$

The calculation for $M_B(+,+;+,+,-)$, with contributions from M_5 and M_6 only, is completely analogous, and $M_B(-,-;-,-,\pm)$ can be obtained by a parity conjugation.

The contribution from the s-channel diagrams to $M_B(+,-;+,-,+)$ has already been calculated in the $\mu^+\mu^-\gamma$ case, and combining with the additional contributions from M_5 and M_8 leads to

$$M_B(+,-;+,-,+)$$

$$= -e^3 \left[\frac{N_q}{s} + e^{i\phi}\frac{N_p}{s'}\right] \bar{u}(q_-)\,\slashed{p}_+(1-\gamma_5)u(p_-)\bar{v}(p_+)\,\slashed{q}_-(1-\gamma_5)v(q_+)$$

$$-e^3 e^{i\psi}\left[\frac{N_-}{t} + e^{ix}\frac{N_+}{t'}\right] \bar{v}(p_+)\,\slashed{q}_-(1-\gamma_5)u(p_-)\bar{u}(q_-)\,\slashed{p}_+(1-\gamma_5)v(q_+). \tag{4.35}$$

We want to rewrite the last spinor expression in $M_B(+,-;+,-,+)$ so that it can be combined with the first term. To this end, note that

$$\bar{v}(p_+)\,\slashed{q}_-(1-\gamma_5)u(p_-) = \frac{1}{2s'}\bar{v}(p_+)\,\slashed{q}_-(1-\gamma_5)\slashed{q}_+ \slashed{q}_-(1-\gamma_5)u(p_-)$$

$$= \frac{1}{2s'}\bar{v}(p_+)\,\slashed{q}_-(1-\gamma_5)v(q_+)$$

$$\times \bar{v}(q_+)\,\slashed{q}_-(1-\gamma_5)u(p_-), \tag{4.36}$$

and, similarly,

$$\bar{u}(q_-)\slashed{p}_+(1-\gamma_5)v(q_+) = \frac{1}{2s}\bar{u}(q_-)\slashed{p}_+(1-\gamma_5)u(p_-)\bar{u}(p_-)\slashed{p}_+(1-\gamma_5)v(q_+). \tag{4.37}$$

Combining these two results, we obtain

$$\bar{v}(p_+)\,\slashed{q}_-(1-\gamma_5)u(p_-)\bar{u}(q_-)\,\slashed{p}_+(1-\gamma_5)v(q_+)$$

$$= \frac{1}{2ss'}\text{Tr}[\slashed{p}_-\,\slashed{p}_+\,\slashed{q}_+\,\slashed{q}_-(1-\gamma_5)]$$

$$\times \bar{v}(p_+)\,\slashed{q}_-(1-\gamma_5)v(q_+)\bar{u}(q_-)\,\slashed{p}_+(1-\gamma_5)u(p_-), \tag{4.38}$$

and

4.4. THE PROCESS $e^+e^- \to e^+e^-\gamma$

$M_B(+,-;+,-,+)$

$$= -e^3\left\{\frac{N_q}{s} + e^{i\phi}\frac{N_p}{s'} + \frac{e^{i\psi}}{2ss'}\text{Tr}[\not{p}_-\not{p}_+\not{A}_+\not{A}_-(1-\gamma_5)]\left[\frac{N_-}{t} + e^{i\chi}\frac{N_+}{t'}\right]\right\}$$

$$\times \bar{u}(q_-)\not{p}_+(1-\gamma_5)u(p_-)\bar{v}(p_+)\not{A}_-(1-\gamma_5)v(q_+). \tag{4.39}$$

Next, we want to simplify the expression within the curly bracket. This is done by showing that

$$s'N_q + e^{i\phi}sN_p + e^{i\psi}(t'N_- + e^{i\chi}tN_+) = 0. \tag{4.40}$$

Indeed, with the definitions of N_p, N_q, $N_\pm$, ϕ, χ and ψ [eqns (4.17)–(4.19), (4.31), (4.32) and (4.34)], we have

$e^{i\psi}(t'N_- + e^{i\chi}tN_+)$

$$= \frac{N_q}{8t'(p_-\cdot k)(q_-\cdot k)}\text{Tr}[\not{A}_-\not{A}_+\not{k}\not{A}_-\not{p}_-\not{k}(1+\gamma_5)]$$

$$\times\left\{t' + \frac{\text{Tr}[\not{p}_+\not{A}_+\not{k}\not{A}_-\not{p}_-\not{k}(1-\gamma_5)]}{8(p_+\cdot k)(q_+\cdot k)}\right\}$$

$$= -\frac{N_q}{4(p_-\cdot k)}\left\{\text{Tr}[\not{A}_-\not{A}_+\not{k}\not{p}_-(1+\gamma_5)]\right.$$

$$\left.+\frac{\text{Tr}[\not{p}_-\not{k}\not{p}_+\not{A}_+\not{k}\not{A}_-\not{p}_-\not{A}_-\not{A}_+\not{k}(1-\gamma_5)]}{4t'(p_+\cdot k)(q_+\cdot k)}\right\}$$

$$= -N_q\left\{s' - \frac{\text{Tr}[\not{p}_+\not{p}_-\not{k}\not{A}_-\not{A}_+\not{k}(1-\gamma_5)]}{8(p_+\cdot k)(p_-\cdot k)}\right\}$$

$$= -[s'N_q + e^{i\phi}sN_p]. \tag{4.41}$$

In deriving this result, we have made use of the 'cut-and-paste' technique, which allows one to write a product of traces involving $(1\pm\gamma_5)$ into a single trace by inserting polarization sums in the appropriate places. This trick was already used in Section 3.3 in the study of $e^+e^- \to \gamma\gamma$, but its applicability is quite general. For more details about this technique, we refer the reader to Appendix A.

It then follows that

$$M_B(+,-;+,-,+) = -e^3[s'N_q + e^{i\phi}sN_p]\left\{1 - \frac{\text{Tr}[\not{p}_-\not{p}_+\not{A}_+\not{A}_-(1-\gamma_5)]}{2tt'}\right\}$$

$$\times \bar{u}(q_-)\not{p}_+(1-\gamma_5)u(p_-)\bar{v}(p_+)\not{A}_-(1-\gamma_5)v(q_+)/ss', \tag{4.42}$$

and we recover the factorization property for the helicity amplitude, one factor being $s'N_q + e^{i\phi}sN_p$, just as in the $\mu^+\mu^-\gamma$ case. To obtain the absolute value of this amplitude, we only have to observe that

$$|\text{Tr}[\slashed{p}_- \slashed{p}_+ \slashed{A}_+ \slashed{A}_-(1-\gamma_5)]|^2$$

$$= \text{Tr}[\slashed{p}_- \slashed{p}_+ \slashed{A}_+ \slashed{A}_-(1-\gamma_5)] \, \text{Tr}[\slashed{A}_- \slashed{A}_+ \slashed{p}_+ \slashed{p}_-(1+\gamma_5)]$$

$$= 2\,\text{Tr}[\slashed{A}_- \slashed{p}_- \slashed{p}_+ \slashed{A}_+ \slashed{A}_- \slashed{A}_+ \slashed{p}_+ \slashed{p}_-(1+\gamma_5)]$$

$$= 4ss'tt', \qquad (4.43)$$

hence,

$$M_B(+,-;+,-,+) \doteq 4e^3 \left|s'N_q + e^{i\phi}sN_p\right| u \left(\frac{uu'}{ss'tt'}\right)^{\frac{1}{2}}, \qquad (4.44)$$

and

$$|M_B(+,-;+,-,+)|^2 = -2e^6(v_q - v_p)^2 u^2 \frac{uu'}{ss'tt'}. \qquad (4.45)$$

This result is written in terms of the four-vectors v_q and v_p introduced in eqns (4.27).

All the remaining helicity amplitudes can be computed in the same way and are listed in Chapter 9. Of course, this calculation was somewhat more involved than in the $e^+e^- \to \mu^+\mu^-\gamma$ case, but one should not forget that we had to treat twice as many Feynman diagrams. Still, because of the fact that all helicity amplitudes factorize with a common factor, we are able to obtain a simple expression for the squared matrix element

$$\overline{|M|^2} = -e^6(v_q - v_p)^2[ss'(s^2+s'^2) + tt'(t^2+t'^2) + uu'(u^2+u'^2)]/ss'tt'. \qquad (4.46)$$

Again, we note the absence of double pole terms of the type s^{-2}, t^{-2}, etc., in this expression.

4.5 Inclusion of Z-exchange

One of the features of the standard model is that, whenever a photon can be exchanged between charged fermions, the exchange of the neutral heavy boson Z also takes place. The importance of Z-exchange was first observed at PETRA through the charge asymmetry measurement in $e^+e^- \to \mu^+\mu^-$ [26]. It can also be observed in high-precision measurements of Bhabha scattering ($e^+e^- \to e^+e^-$). With increasing energy, these effects are expected to become even more important. We thus want to know the cross section formulae for $e^+e^- \to \mu^+\mu^-\gamma$ and $e^+e^- \to e^+e^-\gamma$ including Z-exchange.

4.5. INCLUSION OF Z-EXCHANGE

Let the coupling of the Z to the leptons be of the vector and axial vector type, i.e., a vertex given by

$$ie[a\gamma_\mu(1-\gamma_5) + b\gamma_\mu(1+\gamma_5)]. \tag{4.47}$$

In terms of the coupling constants v_e and a_e of eqns (2.24), we have

$$a = (v_e + a_e)/8\sin\theta_W \cos\theta_W,$$

$$b = (v_e - a_e)/8\sin\theta_W \cos\theta_W. \tag{4.48}$$

The propagator for a massive spin-1 particle is given by (see Table 2.2)

$$-i\frac{g_{\mu\nu} - k_\mu k_\nu/m^2}{k^2 - m^2 + i\epsilon}, \tag{4.49}$$

but, for our applications, we can drop the $k_\mu k_\nu$ term. This is due to the fact that it is always contracted with a γ^μ-matrix on a fermion line, where it gives a vanishing contribution in the massless fermion case. To incorporate the effects of the finite width of the Z, we modify the Z-propagator to read

$$\frac{-ig_{\mu\nu}}{k^2 - M_Z^2 + iM_Z\Gamma}, \tag{4.50}$$

where Γ is the full width of the Z.

For the case of $e^+e^- \to \mu^+\mu^-\gamma$, we have the four QED Feynman amplitudes M_1, M_2, M_3 and M_4, already listed in eqns (4.16) and four additional ones due to Z-exchange. They are

$$Z_1 = \frac{-e^3}{s' - M_Z^2 + iM_Z\Gamma} \bar{u}(q_-)\gamma_\mu[a(1-\gamma_5) + b(1+\gamma_5)]v(q_+)$$

$$\times \bar{v}(p_+)\gamma^\mu[a(1-\gamma_5) + b(1+\gamma_5)]\frac{\not{p}_- - \not{k}}{-2(p_- \cdot k)}\not{\epsilon}u(p_-),$$

$$Z_2 = \frac{-e^3}{s' - M_Z^2 + iM_Z\Gamma} \bar{u}(q_-)\gamma_\mu[a(1-\gamma_5) + b(1+\gamma_5)]v(q_+)$$

$$\times \bar{v}(p_+)\not{\epsilon}\frac{-\not{p}_+ + \not{k}}{-2(p_+ \cdot k)}\gamma^\mu[a(1-\gamma_5) + b(1+\gamma_5)]u(p_-),$$

$$Z_3 = \frac{-e^3}{s - M_Z^2 + iM_Z\Gamma} \bar{v}(p_+)\gamma^\mu[a(1-\gamma_5) + b(1+\gamma_5)]u(p_-)$$

$$\times \bar{u}(q_-)\not{\epsilon}\frac{\not{q}_- + \not{k}}{2(q_- \cdot k)}\gamma_\mu[a(1-\gamma_5) + b(1+\gamma_5)]v(q_+),$$

$$Z_4 = \frac{-e^3}{s - M_Z^2 + iM_Z\Gamma} \bar{v}(p_+)\gamma^\mu[a(1-\gamma_5) + b(1+\gamma_5)]u(p_-)$$

$$\times \bar{u}(q_-)\gamma_\mu[a(1-\gamma_5) + b(1+\gamma_5)]\frac{-\not{q}_+ - \not{k}}{2(q_+ \cdot k)}\not{k}v(q_+). \quad ($$

It is clear that, if we make a specific choice for the fermion helicities, one of the two terms in the Z-fermion vertex can contribute, either the a- or the b-term. But then, the application of the helicity amplitude me can be carried through just as in the pure QED case, the only modific being an overall factor in the matrix element. For example,

$$M(+,-;+,-,+) \doteq 4e^3 \left| s' N_q A(a^2,s) + e^{i\phi} s N_p A(a^2, s') \right| u/(ss')^{\frac{1}{2}}, \quad ($$

where N_p, N_q and ϕ are defined by eqns (4.17), (4.18) and (4.19) and

$$A(x,y) = 1 + \frac{4xy}{y - M_Z^2 + iM_Z\Gamma}. \quad ($$

Clearly, when $M_Z \to \infty$, $A(x,y) \to 1$, and we recover the pure QED re

As the quantity $A(x,y)$ is complex, we need to know $\text{Im}[ss'N_pN_qe^{-i\phi}]$ well as $\text{Re}[ss'N_pN_qe^{-i\phi}]$ for the calculation of the cross section. It is give

$$\text{Im}[ss'N_pN_qe^{-i\phi}] = -ss'N_p^2N_q^2 \text{Tr}[\not{p}_+ \not{p}_- \not{k} \not{q}_- \not{q}_+ \not{k}(1+\gamma_5)]$$

$$= \frac{(s-s')\,\epsilon(p_+, p_-, q_+, q_-)}{16(p_+ \cdot k)(p_- \cdot k)(q_+ \cdot k)(q_- \cdot k)}, \quad ($$

where we introduced the notation

$$\epsilon(p_+, p_-, q_+, q_-) = \epsilon_{\alpha\beta\gamma\delta}\, p_+^\alpha p_-^\beta q_+^\gamma q_-^\delta, \quad (\epsilon_{0123} = +1). \quad ($$

In exactly the same way, one can treat the inclusion of Z-exchange i: process $e^+e^- \to e^+e^-\gamma$. For both cases, the helicity amplitudes and the sections can be found in Chapter 9. The cross section formulae, howevei rather lengthy, whereas the helicity amplitudes are rather simple. If wants to evaluate numerically the cross section in a given point of p space, it might therefore be more advantageous to evaluate separatel: various helicity amplitudes as complex numbers and to add their squ amplitudes, rather than computing $\overline{|M|^2}$ directly.

5
Single bremsstrahlung in QCD

> Color is a complication
> for gluon radiation:
> Its nicest implication
> is jet fragmentation.

5.1 Good to know

We now want to apply the helicity amplitude method for the calculation of QCD processes. Compared to QED calculations, there are three major sources of complications:

i) There is a complication due to the appearance in the Feynman diagrams of color matrices or SU(3) structure constants. We shall see that a good choice for the polarization vectors of the gluons has the net result that one or more Feynman diagrams can be made to vanish. In this way, the helicity amplitude method reduces the amount of color algebra one has to perform to obtain cross section formulae.

ii) As stated in Section 3.2, we have to use polarization vectors of the type

$$\not{\epsilon}^\pm = N[\not{k}\,\not{p}\,\not{q}(1\pm\gamma_5) + \not{q}\,\not{p}\,\not{k}(1\mp\gamma_5)], \tag{5.1}$$

for gluons with four-momentum k. If p and q are four-momenta of external fermions, this implies that at most one of the terms can give a vanishing contribution by virtue of the Dirac equation. This is because in this example $\not{p}$ always appears in the middle. As a consequence, we have fewer simplifications compared to the QED case.

iii) Three- and four-gluon vertices can be present in the Feynman diagrams, generating longer expressions. When this is the case, it is often useful to choose the four-momenta of other gluons in the process to play the role of $\not{p}$ and $\not{q}$ in $\not{\epsilon}$. Suppose, for example, that we have three external gluons with momenta k_1, k_2 and k_3, with helicities $+, -$ and $-$. Suppose, furthermore, that we choose

$$\not{\epsilon}^+(k_1) = N[\not{k}_1\,\not{k}_2\,\not{k}_3(1+\gamma_5) + \not{k}_3\,\not{k}_2\,\not{k}_1(1-\gamma_5)],$$

$$\not{\epsilon}^-(k_2) = N[\not{k}_1\,\not{k}_3\,\not{k}_2(1+\gamma_5) + \not{k}_2\,\not{k}_3\,\not{k}_1(1-\gamma_5)],$$

$$\not{\epsilon}^-(k_3) = N[\not{k}_1\,\not{k}_2\,\not{k}_3(1+\gamma_5) + \not{k}_3\,\not{k}_2\,\not{k}_1(1-\gamma_5)]. \tag{5.2}$$

As gluons 1 and 3 have opposite helicities, we can indeed take $\not{\epsilon}^+(k_1) = \not{\epsilon}^-(k_3)$. It follows that

$$\left(\epsilon^+(k_1) \cdot \epsilon^-(k_3)\right) = 0, \tag{5.3}$$

which can also be derived using the explicit expressions, viz.,

$$\left(\epsilon^+(k_1) \cdot \epsilon^-(k_3)\right) = N^2 \, \text{Tr}[\not{k}_1 \not{k}_2 \not{k}_3 \not{k}_3 \not{k}_2 \not{k}_1 (1 - \gamma_5)] = 0. \tag{5.4}$$

At the same time, we have

$$\left(\epsilon^+(k_1) \cdot \epsilon^-(k_2)\right) = \left(\epsilon^-(k_2) \cdot \epsilon^-(k_3)\right) = 0. \tag{5.5}$$

As a result, we can omit any Feynman diagram with a four-gluon vertex where these three gluons appear as external ones, provided of course we are interested in the above mentioned helicity configuration. Every term in the four-gluon vertex contains a scalar product of these gluon polarization vectors and thus the whole four-gluon vertex vanishes.

In the following sections, we shall illustrate the use of the helicity amplitude method for QCD processes for two simple examples: $q\bar{q} \to gg$ and $e^+e^- \to 3$ jets. We hope to show that, at least for the simpler processes, QCD is not much more difficult than QED.

5.2 The process $\bar{q}q \to gg$

For the process of quark-antiquark annihilation into two gluons,

$$\bar{q}(p_+, i) + q(p_-, j) \to g(k_1, a) + g(k_2, b), \tag{5.6}$$

we have to consider, in lowest order, the three Feynman diagrams of Fig. 5.1. Here, i and j denote the color degrees of freedom for the antiquark and the quark, while a and b denote those for the gluons. As already discussed in Section 2.1, i and j take the values 1,2,3, while a and b run from 1 to 8.

With the Feynman rules of Section 2.1, we can write down the corresponding amplitudes

$$\begin{aligned} M_1 &= -g^2 \, \bar{v}(p_+) \, \not{\epsilon}(k_2) T^b_{ik} \frac{\not{p}_- - \not{k}_1}{-2(p_- \cdot k_1)} \, \not{\epsilon}(k_1) T^a_{kj} u(p_-), \\ M_2 &= -g^2 \, \bar{v}(p_+) \, \not{\epsilon}(k_1) T^a_{ik} \frac{\not{p}_- - \not{k}_2}{-2(p_- \cdot k_2)} \, \not{\epsilon}(k_2) T^b_{kj} u(p_-), \\ M_3 &= -ig^2 \, \bar{v}(p_+) T^c_{ij} \gamma^\mu u(p_-) \frac{1}{(k_1 + k_2)^2} f^{cab} \Big[2\epsilon_\mu(k_1)\big(k_1 \cdot \epsilon(k_2)\big) \\ &\quad + (k_2 - k_1)_\mu \big(\epsilon(k_1) \cdot \epsilon(k_2)\big) - 2\epsilon_\mu(k_2)\big(k_2 \cdot \epsilon(k_1)\big) \Big], \end{aligned} \tag{5.7}$$

5.2. THE PROCESS $\bar{q}q \to gg$

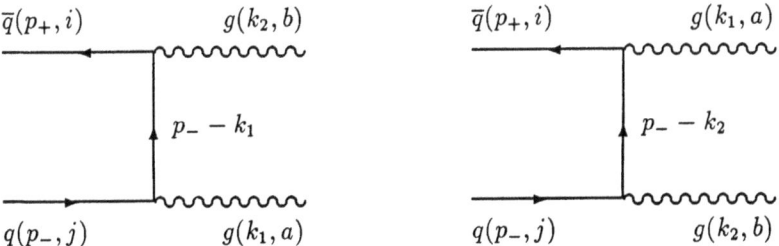

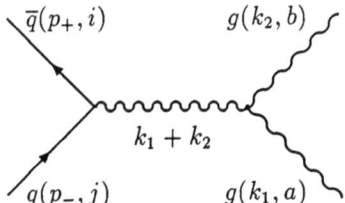

Fig. 5.1: Feynman diagrams for $\bar{q}q \to gg$.

where we used the relations

$$\bigl(k_1 \cdot \epsilon(k_1)\bigr) = \bigl(k_2 \cdot \epsilon(k_2)\bigr) = 0. \tag{5.8}$$

We denote the helicity amplitudes by $M(\lambda_+, \lambda_-; \lambda_1, \lambda_2)$ with λ_+, λ_-, λ_1 and λ_2 the helicities of $\bar{q}$, q, $g(k_1)$ and $g(k_2)$. We immediately find the familiar result that, for all λ_1 and λ_2,

$$M(+,+;\lambda_1,\lambda_2) = 0, \qquad M(-,-;\lambda_1,\lambda_2) = 0. \tag{5.9}$$

Consider next $M(+,-;+,+)$, and let us choose

$$\not{\epsilon}^+(k_i) = N_i[\not{k}_i \not{p}_- \not{p}_+(1+\gamma_5) + \not{p}_+ \not{p}_- \not{k}_i(1-\gamma_5)],$$

$$N_i^{-1} = 4[(p_+ \cdot p_-)(p_+ \cdot k_i)(p_- \cdot k_i)]^{\frac{1}{2}}, \qquad i=1,2. \tag{5.10}$$

Because of the fermion helicities, only the second term in $\not{\epsilon}^+(k_i)$ contributes, and one sees directly that M_1 and M_2 give vanishing contributions. With this choice, we also have

$$\bigl(\epsilon^+(k_1) \cdot \epsilon^+(k_2)\bigr) = 0, \tag{5.11}$$

hence, M_3 does not contribute either. Consequently,

$$M(+,-;+,+) = 0. \tag{5.12}$$

By interchanging p_+ and p_- in $\not{\epsilon}^+(k_i)$, the same considerations lead to

$$M(-,+;+,+) = 0. \tag{5.13}$$

A parity conjugation then tells us that also

$$M(+,-;-,-) = 0, \qquad M(-,+;-,-) = 0. \tag{5.14}$$

Just as in the $e^+e^- \to \gamma\gamma$ case, we find that the helicities of the gauge particles must be opposite.

For $M(+,-;+,-)$, we choose

$$\begin{aligned}
\not{\epsilon}^+(k_1) &= N_1[\not{k}_1\not{p}_-\not{p}_+(1+\gamma_5) + \not{p}_+\not{p}_-\not{k}_1(1-\gamma_5)], \\
\not{\epsilon}^-(k_2) &= N_2[\not{p}_+\not{p}_-\not{k}_2(1+\gamma_5) + \not{k}_2\not{p}_-\not{p}_+(1-\gamma_5)].
\end{aligned} \tag{5.15}$$

With the usual notation

$$s = (p_- + p_+)^2, \quad t = (p_- - k_1)^2, \quad u = (p_- - k_2)^2, \quad s+t+u = 0, \tag{5.16}$$

we have

$$\begin{aligned}
\left(\epsilon^+(k_1)\cdot\epsilon^-(k_2)\right) &= N_1N_2\text{Tr}[\not{k}_1\not{p}_-\not{p}_+\not{k}_2\not{p}_-\not{p}_+(1-\gamma_5)] = 1, \\
\left(k_2\cdot\epsilon^+(k_1)\right) &= \tfrac{1}{2}N_1\text{Tr}[\not{k}_2\not{k}_1\not{p}_-\not{p}_+(1+\gamma_5)] = -N_1st, \\
\left(k_1\cdot\epsilon^-(k_2)\right) &= \tfrac{1}{2}N_2\text{Tr}[\not{k}_1\not{p}_+\not{p}_-\not{k}_2(1+\gamma_5)] = -N_2su, \\
N_1 &= N_2 = (2stu)^{-\frac{1}{2}}.
\end{aligned} \tag{5.17}$$

With this choice of $\not{\epsilon}(k_i)$, we find that M_2 does not contribute, and

$$\begin{aligned}
M_1 &= -2g^2\frac{N_1N_2}{t}(T^bT^a)_{ij}\,\bar{v}(p_+)\,\not{k}_2\not{p}_-\not{p}_+(\not{p}_- - \not{k}_1) \\
&\qquad\qquad\qquad\qquad\qquad\qquad \times \not{p}_+\not{p}_-\not{k}_1(1-\gamma_5)u(p_-) \\
&= 2g^2N_1N_2(T^bT^a)_{ij}\,\bar{v}(p_+)\,\not{k}_2\not{p}_-\not{p}_+\not{k}_2\not{p}_+(1-\gamma_5)u(p_-) \\
&= -\frac{g^2}{su}(T^bT^a)_{ij}\,\bar{v}(p_+)\,\not{k}_2\not{p}_-\not{p}_+(1-\gamma_5)u(p_-)
\end{aligned}$$

5.2. THE PROCESS $\bar{q}q \to gg$

$$= -\frac{g^2}{u}(T^bT^a)_{ij}\,\bar{v}(p_+)\,\not{k}_2(1-\gamma_5)u(p_-),$$

$$M_3 = -i\frac{g^2}{s}f^{cab}T^c_{ij}[\bar{v}(p_+)\,\not{k}_2(1-\gamma_5)u(p_-)$$

$$+2N_1N_2st\bar{v}(p_+)\,\not{k}_2\,\not{p}_-\,\not{p}_+(1-\gamma_5)u(p_-)]$$

$$= \frac{ig^2t}{su}f^{abc}T^c_{ij}\,\bar{v}(p_+)\,\not{k}_2(1-\gamma_5)u(p_-). \tag{5.18}$$

We find that M_1 and M_3 are proportional to the same spinor expression, and

$$M(+,-;+,-) \doteq 2g^2\left(\frac{t}{u}\right)^{\frac{1}{2}}\left[(T^bT^a)_{ij} - i\frac{t}{s}f^{abc}T^c_{ij}\right]$$

$$\doteq \frac{2g^2}{s}\left(\frac{t}{u}\right)^{\frac{1}{2}}\left[t(T^aT^b)_{ij} + u(T^bT^a)_{ij}\right]. \tag{5.19}$$

For the summation over the color degrees of freedom, we have to know that, using eqns (2.14),

$$\sum_{ab}\mathrm{Tr}\,(T^bT^aT^aT^b) = \frac{4}{3}\sum_b\mathrm{Tr}\,(T^bT^b) = \frac{16}{3},$$

$$\sum_{abc}\mathrm{Tr}\,(T^bT^aT^c)f^{abc} = -\frac{i}{4}\sum_{abc}(f^{abc})^2 = -6i, \tag{5.20}$$

$$\sum_{abcd}f^{abc}\mathrm{Tr}\,(T^cT^d)f^{abd} = \frac{1}{2}\sum_{abc}(f^{abc})^2 = 12.$$

More relations for traces of products of color matrices can be found in Appendix C. Substitution into eqn (5.19) gives

$$|M(+,-;+,-)|^2 = 48g^4\frac{t}{u}\left(\frac{4}{9} + \frac{t}{s} + \frac{t^2}{s^2}\right)$$

$$= 48g^4\left(\frac{4t}{9u} - \frac{t^2}{s^2}\right). \tag{5.21}$$

The expression for $|M(+,-;-,+)|^2$ is derived from the above one by interchanging $1 \leftrightarrow 2$, which amounts to $t \leftrightarrow u$. Thus,

$$|M(+,-;-,+)|^2 = 48g^4\left(\frac{4u}{9t} - \frac{u^2}{s^2}\right). \tag{5.22}$$

Adding the parity conjugate contributions and averaging over the initial degrees of freedom, we finally obtain

$$\overline{|M|^2} = \frac{8}{3}g^4(t^2+u^2)\left(\frac{4}{9tu} - \frac{1}{s^2}\right). \tag{5.23}$$

The conclusion from this exercise is that, apart from the complications due to the color algebra, the entire calculation for $\bar{q}q \to gg$ is very similar to the calculation for the $e^+e^- \to \gamma\gamma$ case. One only has to keep in mind that the chosen representation of $\not{\epsilon}$ in the gluon case is different from the one in the photon case, where we could drop the $\not{k}\gamma_5$ term.

5.3 The process $e^+e^- \to$ 3 jets

Within the framework of perturbative QCD, the process $e^+e^- \to$ 3 jets is described by the subprocess

$$e^+(p_+) + e^-(p_-) \to q(q_-, i) + \bar{q}(q_+, j) + g(k, a). \tag{5.24}$$

In lowest order, the Feynman diagrams are given by Fig. 5.2, and, correspondingly,

$$M_1 = \frac{-ee_q g}{2(p_+ \cdot p_-)} \bar{v}(p_+)\gamma^\mu u(p_-)\bar{u}(q_-)\not{\epsilon}(k)T^a_{ij}\frac{\not{q}_- + \not{k}}{2(q_- \cdot k)}\gamma_\mu v(q_+),$$

$$M_2 = \frac{-ee_q g}{2(p_+ \cdot p_-)} \bar{v}(p_+)\gamma^\mu u(p_-)\bar{u}(q_-)\gamma_\mu \frac{-\not{q}_+ - \not{k}}{2(q_+ \cdot k)} \not{\epsilon}(k)T^a_{ij}v(q_+).$$
(5.25)

These expressions are very similar to the Feynman amplitudes M_3 and M_4 for the QED process $e^+e^- \to \mu^+\mu^-\gamma$ [see eqns (4.16)].

We have again that the e^+ and e^- helicities must be opposite as well as the q and $\bar{q}$ helicities. Let us therefore proceed with the calculation of $M(+,-;+,-,+)$, where the order of the helicities is given by e^+, e^-, $\bar{q}$, q, g. In analogy with the QED case, we take

$$\not{\epsilon}^+(k) = N[\not{k}\not{q}_+\not{q}_-(1+\gamma_5) + \not{q}_-\not{q}_+\not{k}(1-\gamma_5)],$$

$$N^{-1} = 4[(q_+ \cdot q_-)(q_+ \cdot k)(q_- \cdot k)]^{\frac{1}{2}}. \tag{5.26}$$

Because of the quark helicity projection operators, only the second term in $\not{\epsilon}^+(k)$ can contribute. This leads immediately to a vanishing contribution from M_1. Thus,

5.3. THE PROCESS $e^+e^- \to 3$ JETS

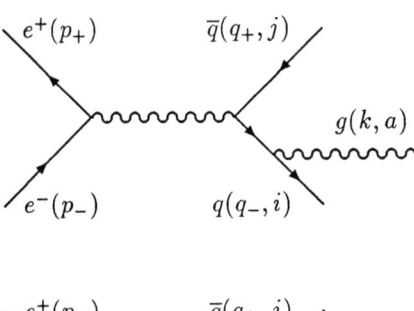

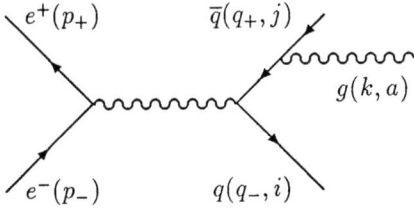

Fig. 5.2: Feynman diagrams for $e^+e^- \to q\bar{q}g$.

$$M(+,-;+,-,+) = \frac{ee_q gN}{8(p_+ \cdot p_-)(q_+ \cdot k)} T^a_{ij} \bar{v}(p_+)\gamma^\mu(1-\gamma_5)u(p_-)$$

$$\times \bar{u}(q_-)\gamma_\mu(\slashed{q}_+ + \slashed{k})\,\slashed{A}_-\,\slashed{A}_+\,\slashed{k}(1-\gamma_5)v(q_+)$$

$$= \frac{ee_q gN}{4(p_+ \cdot p_-)} T^a_{ij} \bar{v}(p_+)\gamma^\mu(1-\gamma_5)u(p_-)\bar{u}(q_-)\gamma_\mu(\slashed{q}_+ + \slashed{k})\,\slashed{A}_-(1-\gamma_5)v(q_+). \quad (5.27)$$

With the cut-and-paste technique of Appendix A, we eliminate the repeated index μ as follows:

$$A = \bar{v}(p_+)\gamma^\mu(1-\gamma_5)u(p_-)\,\bar{u}(q_-)\gamma_\mu(\slashed{q}_+ + \slashed{k})\,\slashed{A}_-(1-\gamma_5)v(q_+)$$

$$= \bar{v}(p_+)\gamma^\mu(1-\gamma_5)u(p_-)\,\frac{\bar{u}(p_-)\,\slashed{A}(1-\gamma_5)u(q_-)}{\bar{u}(p_-)\,\slashed{A}(1-\gamma_5)u(q_-)}$$

$$\times \bar{u}(q_-)\gamma_\mu(\slashed{q}_+ + \slashed{k})\,\slashed{A}_-(1-\gamma_5)v(q_+)$$

$$= 4\,\frac{\bar{v}(p_+)\gamma^\mu\,\slashed{p}_-\,\slashed{A}\,\slashed{A}_-\gamma_\mu(\slashed{q}_+ + \slashed{k})\,\slashed{A}_-(1-\gamma_5)v(q_+)}{\bar{u}(p_-)\,\slashed{A}(1-\gamma_5)u(q_-)}$$

$$= -8\,\frac{\bar{v}(p_+)\,\slashed{A}_-\,\slashed{A}\,\slashed{p}_-(\slashed{q}_+ + \slashed{k})\,\slashed{A}_-(1-\gamma_5)v(q_+)}{\bar{u}(p_-)\,\slashed{A}(1-\gamma_5)u(q_-)}. \quad (5.28)$$

Choosing $a = p_+$ and using momentum conservation, we have

$$A = -16(p_+ \cdot q_-) \frac{\bar{v}(p_+) \not{p}_- \not{p}_+ \not{A}_-(1-\gamma_5)v(q_+)}{\bar{u}(p_-) \not{p}_+(1-\gamma_5)u(q_-)}$$

$$= -32(p_+ \cdot q_-)(p_+ \cdot p_-) \frac{\bar{v}(p_+) \not{A}_-(1-\gamma_5)v(q_+)}{\bar{u}(p_-) \not{p}_+(1-\gamma_5)u(q_-)}$$

$$\doteq 32(p_+ \cdot q_-)[(p_+ \cdot p_-)(q_+ \cdot q_-)]^{\frac{1}{2}}.$$

Thus,

$$M(+,-;+,-,+) \doteq 8ee_q g N T_{ij}^a (p_+ \cdot q_-) \left[\frac{(q_+ \cdot q_-)}{(p_+ \cdot p_-)}\right]^{\frac{1}{2}}$$

$$\doteq 2ee_q g T_{ij}^a (p_+ \cdot q_-)[(p_+ \cdot p_-)(q_+ \cdot k)(q_- \cdot k)]^{-\frac{1}{2}}$$

and, summing over the color degrees of freedom,

$$|M(+,-;+,-,+)|^2 = 8(ee_q g)^2 \frac{u^2}{s(q_+ \cdot k)(q_- \cdot k)},$$

where, as usual,

$$s = (p_+ + p_-)^2, \qquad t = (p_+ - q_+)^2, \qquad u = (p_+ - q_-)^2$$

$$s' = (q_+ + q_-)^2, \qquad t' = (p_- - q_-)^2, \qquad u' = (p_- - q_+)^2$$

For the amplitude $M(-,+;+,-,+)$, we have with the same choice $\not{\epsilon}^+$:

$$M(-,+;+,-,+) = \frac{ee_q g N}{2s} T_{ij}^a \bar{v}(p_+)\gamma^\mu(1+\gamma_5)u(p_-)$$

$$\times \bar{u}(q_-)\gamma_\mu(\not{A}_+ + \not{k}) \not{A}_-(1-\gamma_5)v(q_+).$$

This time, we have to simplify the expression

$$A' = \bar{v}(p_+)\gamma^\mu(1+\gamma_5)u(p_-)\,\bar{u}(q_-)\gamma_\mu(\not{A}_+ + \not{k}) \not{A}_-(1-\gamma_5)v(q_+)$$

$$= \bar{v}(p_+)\gamma^\mu(1+\gamma_5)u(p_-) \frac{\bar{u}(p_-)(1-\gamma_5)u(q_-)}{\bar{u}(p_-)(1-\gamma_5)u(q_-)}$$

$$\times \bar{u}(q_-)\gamma_\mu(\not{A}_+ + \not{k}) \not{A}_-(1-\gamma_5)v(q_+)$$

5.4. THE PROCESS $qq' \to qq'g$

$$= 4\frac{\bar{v}(p_+)\gamma^\mu \not{p}_- \not{A}_-\gamma_\mu(\not{A}_+ + \not{k}) \not{A}_-(1-\gamma_5)v(q_+)}{\bar{u}(p_-)(1-\gamma_5)u(q_-)}$$

$$= 16(p_- \cdot q_-)\frac{\bar{v}(p_+) \not{p}_- \not{A}_-(1-\gamma_5)v(q_+)}{\bar{u}(p_-)(1-\gamma_5)u(q_-)}$$

$$\doteq 8t'(ss')^{\frac{1}{2}}, \qquad (5.34)$$

yielding

$$M(-,+;+,-,+) \doteq \sqrt{2}ee_q g\, T^a_{ij}\, t' \left[s(q_+ \cdot k)(q_- \cdot k)\right]^{-\frac{1}{2}}. \qquad (5.35)$$

Performing the color summation, using eqns (2.14), we find

$$|M(-,+;+,-,+)|^2 = 8(ee_q g)^2 \frac{t'^2}{s(q_+ \cdot k)(q_- \cdot k)}, \qquad (5.36)$$

which corresponds to the substitution $p_+ \leftrightarrow p_-$ in $|M(+,-;+,-,+)|^2$.

For the amplitudes with opposite quark helicities, it is better to take

$$\not{\epsilon}^+(k) = N[\not{k}\,\not{A}_-\,\not{A}_+(1+\gamma_5) + \not{A}_+\,\not{A}_-\,\not{k}(1-\gamma_5)]. \qquad (5.37)$$

Now, only the first term in $\not{\epsilon}^+$ can contribute, and, because of the Dirac equation, M_2 is made to vanish. The explicit calculation of the helicity amplitudes is again straightforward and is left as an exercise to the reader. Reversing the quark helicities amounts to the interchange $q_+ \leftrightarrow q_-$. The results are to be found in Section 10.2.

Combining the contributions from the different helicity amplitudes then leads to the simple formula

$$\overline{|M|^2} = 4(ee_q g)^2 \frac{t^2 + t'^2 + u^2 + u'^2}{s(q_+ \cdot k)(q_- \cdot k)} \qquad (5.38)$$

for the spin averaged squared matrix element. This formula is very similar to the one for $e^+e^- \to \mu^+\mu^-\gamma$, if we consider radiation from the muon lines only.

5.4 The process $qq' \to qq'g$

In this section, we present another QCD bremsstrahlung process [27], viz.,

$$q(p,j) + q'(q,n) \to q(p',i) + q'(q',m) + g(k,a), \qquad (5.39)$$

as this case presents a complication which was not encountered in the previous examples. Here, q and q' denote quarks with different flavors.

The Feynman diagrams are shown in Fig. 5.3, and the amplitudes read

$$M_1 = -\frac{g^3}{t'}\overline{u}(p')\not{\epsilon}T^a_{ik}\frac{\not{p}'+\not{k}}{2(p'\cdot k)}\gamma^\mu T^b_{kj}u(p)\,\overline{u}(q')\gamma_\mu T^b_{mn}u(q),$$

$$M_2 = -\frac{g^3}{t'}\overline{u}(p')\gamma^\mu T^b_{ik}\frac{\not{p}-\not{k}}{-2(p\cdot k)}\not{\epsilon}T^a_{kj}u(p)\,\overline{u}(q')\gamma_\mu T^b_{mn}u(q),$$

$$M_3 = -\frac{g^3}{t}\overline{u}(p')\gamma^\mu T^b_{ij}u(p)\,\overline{u}(q')\not{\epsilon}T^a_{mk}\frac{\not{q}'+\not{k}}{2(q'\cdot k)}\gamma_\mu T^b_{kn}u(q),$$

$$M_4 = -\frac{g^3}{t}\overline{u}(p')\gamma^\mu T^b_{ij}u(p)\,\overline{u}(q')\gamma_\mu T^b_{mk}\frac{\not{q}-\not{k}}{-2(q\cdot k)}\not{\epsilon}T^a_{kn}u(q),$$

$$M_5 = -\frac{ig^3}{tt'}\overline{u}(p')\gamma^\mu T^c_{ij}u(p)\,\overline{u}(q')\gamma^\nu T^b_{mn}u(q)f^{cba}$$

$$\times[(p-p'-q+q'\cdot\epsilon)g_{\mu\nu}+(q-q'+k)_\mu\epsilon_\nu+(p'-k-p)_\nu\epsilon_\mu],$$

(5.40)

where, this time, we defined

$$s = (p+q)^2, \quad t = (p-p')^2, \quad u = (p-q')^2,$$
$$s' = (p'+q')^2, \quad t' = (q-q')^2, \quad u' = (p'-q)^2.$$

(5.41)

As all the lines in the Feynman diagrams carry color, there is no gauge invariant subset of diagrams. Yet, for the application of the helicity amplitude formalism, we would like to use a polarization vector for the gluon expressed in terms of the momenta p and p' for M_1 and M_2, and a similar expression in terms of q and q' for M_3 and M_4. But how should one treat M_5 in this case? We shall show that there exists a natural way of separating M_5 into two contributions which can be added to $M_1 + M_2$ and to $M_3 + M_4$, yielding gauge invariant combinations, for which we can simply apply our technique.

To this end, observe that the general three-gluon vertex (see Table 2.3)

$$V_{\mu\nu\rho}(p,q,k) = (p-q)_\rho g_{\mu\nu} + (q-k)_\mu g_{\nu\rho} + (k-p)_\nu g_{\rho\mu},$$ (5.42)

can be rewritten as

$$V_{\mu\nu\rho}(p,q,k) = \tfrac{1}{8}\{\text{Tr}(\not{p}\gamma_\mu[\gamma_\nu,\gamma_\rho]) + \text{Tr}(\not{q}\gamma_\nu[\gamma_\rho,\gamma_\mu]) + \text{Tr}(\not{k}\gamma_\rho[\gamma_\mu,\gamma_\nu])\}.$$

(5.43)

Because of the antisymmetry in (μ,ν,ρ,σ) of $\text{Tr}[\gamma_\mu\gamma_\nu\gamma_\rho\gamma_\sigma\gamma_5]$ and because of momentum conservation at the vertex, we have

$$\text{Tr}(\not{p}\gamma_\mu[\gamma_\nu,\gamma_\rho]\gamma_5) + \text{Tr}(\not{q}\gamma_\nu[\gamma_\rho,\gamma_\mu]\gamma_5) + \text{Tr}(\not{k}\gamma_\rho[\gamma_\mu,\gamma_\nu]\gamma_5)$$

$$= 2\text{Tr}[(\not{p}+\not{q}+\not{k})\gamma_\mu\gamma_\nu\gamma_\rho\gamma_5] = 0.$$ (5.44)

5.4. THE PROCESS $qq' \to qq'g$

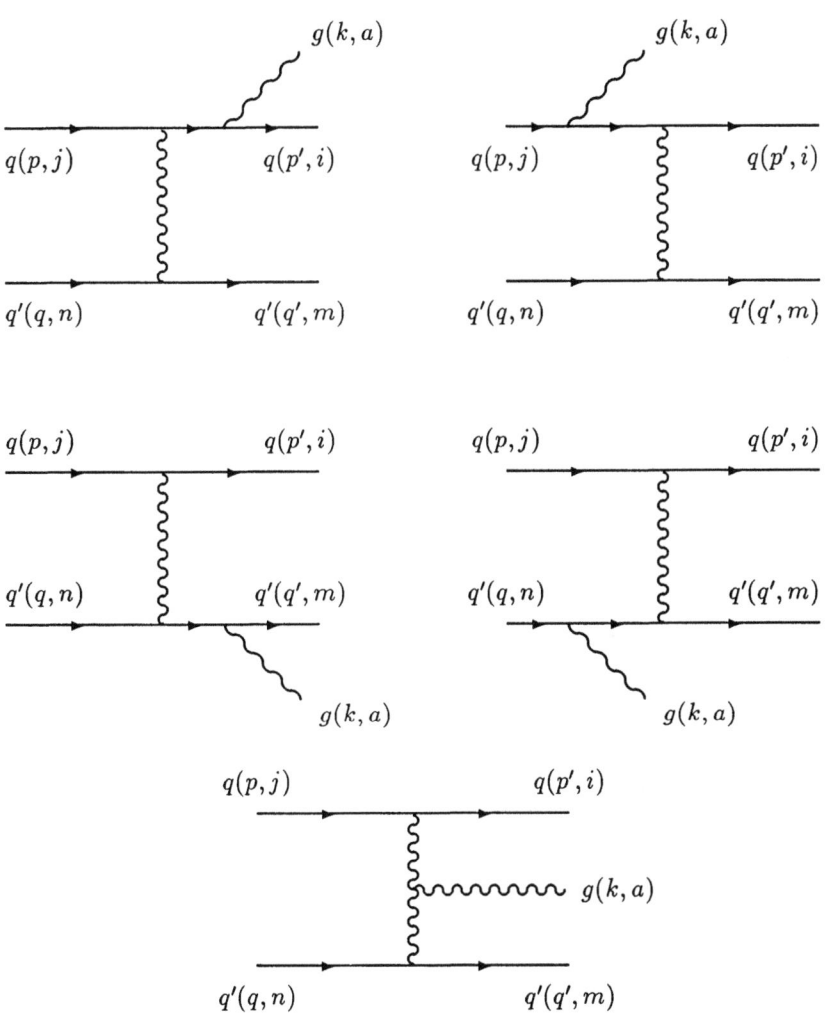

Fig. 5.3: Feynman diagrams for $qq' \to qq'g$.

Consequently, we can rewrite the three-gluon vertex in the form

$$V_{\mu\nu\rho}(p,q,k) = \tfrac{1}{8}\{\text{Tr}[\slashed{p}\gamma_\mu[\gamma_\nu,\gamma_\rho](1\pm\gamma_5)]$$

$$+\text{Tr}[\slashed{q}\gamma_\nu[\gamma_\rho,\gamma_\mu](1\pm\gamma_5)] + \text{Tr}[\slashed{k}\gamma_\rho[\gamma_\mu,\gamma_\nu](1\pm\gamma_5)]\}. \quad (5.45)$$

and we are free to choose the sign in front of the γ_5-matrix. The amplitude M_5 then becomes

$$M_5 = -\frac{ig^3}{8tt'}\,\bar{u}(p')\gamma^\mu T^c_{ij} u(p)\,\bar{u}(q')\gamma^\nu T^b_{mn} u(q)\,f^{cba}$$

$$\times\{\text{Tr}[(\slashed{p}-\slashed{p}')\gamma_\mu[\gamma_\nu,\slashed{\epsilon}](1\pm\gamma_5)] + \text{Tr}[(\slashed{q}-\slashed{q}')\gamma_\nu[\slashed{\epsilon},\gamma_\mu](1\pm\gamma_5)]$$

$$-\text{Tr}[\slashed{k}\,\slashed{\epsilon}[\gamma_\mu,\gamma_\nu](1\pm\gamma_5)]\}. \quad (5.46)$$

We denote by $M(\lambda_1,\lambda_2;\lambda_3,\lambda_4,\lambda_5)$ the helicity amplitudes for $qq' \to qq'g$, where the helicities correspond respectively to those of $q(p)$, $q(p')$, $q'(q)$, $q'(q')$ and $g(k)$. Conservation of helicity along the quark lines tells us that the helicity amplitudes vanish, unless $\lambda_1 = \lambda_2$ and $\lambda_3 = \lambda_4$. Suppose we calculate first $M(-,-;-,-,+)$. For $\slashed{\epsilon}^+(k)$, we take an expression of the type

$$\slashed{\epsilon}^+(k) = N[\slashed{k}\,\slashed{p}''\,\slashed{q}''(1+\gamma_5) + \slashed{q}''\,\slashed{p}''\,\slashed{k}(1-\gamma_5)], \quad (5.47)$$

where N is a normalization factor and p'' and q'' are two distinct arbitrary four-vectors. Clearly,

$$\slashed{k}\,\slashed{\epsilon}^+(k)(1+\gamma_5) = 0. \quad (5.48)$$

Therefore, if we choose the + sign in front of the γ_5-matrix in the expression for M_5, we find that its last term vanishes and that M_5 naturally separates into two terms. Its contribution to $M(-,-;-,-,+)$ then becomes

$$M_5 = M_5^{(1)} + M_5^{(2)},$$

$$M_5^{(1)} = -\frac{ig^3}{32tt'}\,\bar{u}(p')\gamma^\mu T^c_{ij}(1-\gamma_5)u(p)\,\bar{u}(q')\gamma^\nu T^b_{mn}(1-\gamma_5)u(q)\,f^{cba}$$

$$\times\text{Tr}\{(\slashed{p}-\slashed{p}')\gamma_\mu[\gamma_\nu,\slashed{\epsilon}^+](1+\gamma_5)\},$$

$$M_5^{(2)} = -\frac{ig^3}{32tt'}\,\bar{u}(p')\gamma^\mu T^c_{ij}(1-\gamma_5)u(p)\,\bar{u}(q')\gamma^\nu T^b_{mn}(1-\gamma_5)u(q)\,f^{cba}$$

$$\times\text{Tr}\{(\slashed{q}-\slashed{q}')\gamma_\nu[\slashed{\epsilon}^+,\gamma_\mu](1+\gamma_5)\}. \quad (5.49)$$

Clearly, $M_5^{(1)}$ goes together with M_1 and M_2, whereas $M_5^{(2)}$ should be combined with M_3 and M_4.

5.4. THE PROCESS $qq' \to qq'g$

Using formula (A.11) of Appendix A

$$\gamma_\mu \text{Tr}[\gamma^\mu \slashed{A} \slashed{B} \slashed{C}(1 \pm \gamma_5)] = 2[\slashed{A} \slashed{B} \slashed{C}(1 \pm \gamma_5) + \slashed{C} \slashed{B} \slashed{A}(1 \mp \gamma_5)], \qquad (5.50)$$

we can rewrite the $M_5^{(1)}$ contribution in a form suited for its combination with M_1 and M_2:

$$\begin{aligned}
M_5^{(1)} &= \frac{ig^3}{8tt'} \overline{u}(p')[\gamma_\nu, \slashed{\epsilon}^+] \slashed{p}' T_{ij}^c (1-\gamma_5) u(p) \, \overline{u}(q') \gamma^\nu T_{mn}^b (1-\gamma_5) u(q) f^{cba} \\
&= \frac{-g^3}{8tt'} [T^a, T^b]_{ij} T_{mn}^b \overline{u}(p')[\gamma_\nu, \slashed{\epsilon}^+] \slashed{p}'(1-\gamma_5) u(p) \, \overline{u}(q') \gamma^\nu (1-\gamma_5) u(q) \\
&= -\frac{g^3}{4tt'} [T^a, T^b]_{ij} T_{mn}^b \overline{u}(p') \gamma_\nu \slashed{\epsilon}^+ \slashed{p}'(1-\gamma_5) u(p) \, \overline{u}(q') \gamma^\nu (1-\gamma_5) u(q) \\
&= -\frac{g^3}{4tt'} (T^a T^b)_{ij} T_{mn}^b \overline{u}(p') \gamma_\nu \slashed{\epsilon}^+ \slashed{p}'(1-\gamma_5) u(p) \, \overline{u}(q') \gamma^\nu (1-\gamma_5) u(q) \\
&\quad + \frac{g^3}{4tt'} (T^b T^a)_{ij} T_{mn}^b \overline{u}(p') \gamma_\nu \slashed{\epsilon}^+ \slashed{p}'(1-\gamma_5) u(p) \, \overline{u}(q') \gamma^\nu (1-\gamma_5) u(q) \\
&= -\frac{g^3}{8tt'(p' \cdot k)} (T^a T^b)_{ij} T_{mn}^b \overline{u}(p') \gamma_\nu \slashed{k} \slashed{p}' \slashed{\epsilon}^+ \slashed{p}'(1-\gamma_5) u(p) \\
&\qquad \times \overline{u}(q') \gamma^\nu (1-\gamma_5) u(q) \\
&\quad + \frac{g^3}{4tt'} (T^b T^a)_{ij} T_{mn}^b \overline{u}(p') \gamma_\nu \slashed{\epsilon}^+ \slashed{p}'(1-\gamma_5) u(p) \, \overline{u}(q') \gamma^\nu (1-\gamma_5) u(q) \\
&= -\frac{g^3}{4tt'} \frac{(p' \cdot \epsilon^+)}{(p' \cdot k)} (T^a T^b)_{ij} T_{mn}^b \overline{u}(p') \gamma_\nu \slashed{k} \slashed{p}'(1-\gamma_5) u(p) \\
&\qquad \times \overline{u}(q') \gamma^\nu (1-\gamma_5) u(q) \\
&\quad + \frac{g^3}{4tt'} (T^b T^a)_{ij} T_{mn}^b \overline{u}(p') \gamma_\nu \slashed{\epsilon}^+ \slashed{p}'(1-\gamma_5) u(p) \, \overline{u}(q') \gamma^\nu (1-\gamma_5) u(q).
\end{aligned} \qquad (5.51)$$

We now rewrite the contribution of M_1 to $M(-,-;-,-,+)$. In order to combine it with $M_5^{(1)}$, we introduce a factor t in the numerator and the denominator. We also make use of the relation $\slashed{k} \slashed{\epsilon}^+(1+\gamma_5) = 0$ and obtain

$$M_1 = \frac{g^3}{4tt'}(T^aT^b)_{ij}T^b_{mn}\bar{u}(p')\not{\epsilon}^+ \frac{\not{p}'}{2(p'\cdot k)}\gamma^\mu \not{p}\not{p}'(1-\gamma_5)u(p)$$

$$\times \bar{u}(q')\gamma_\mu(1-\gamma_5)u(q)$$

$$= \frac{g^3}{4tt'}(T^aT^b)_{ij}T^b_{mn}\frac{(p'\cdot\epsilon^+)}{(p'\cdot k)}\bar{u}(p')\gamma^\mu \not{p}\not{p}'(1-\gamma_5)u(p)\bar{u}(q')\gamma_\mu(1-\gamma_5)u(q). \quad (5.52)$$

It follows that

$$M_1 + M_5^{(1)} = \frac{g^3}{4tt'}(T^aT^b)_{ij}T^b_{mn}\frac{(p'\cdot\epsilon^+)}{(p'\cdot k)}\bar{u}(p')\gamma^\mu(\not{p}-\not{k})\not{p}'(1-\gamma_5)u(p)$$

$$\times \bar{u}(q')\gamma_\mu(1-\gamma_5)u(q)$$

$$+ \frac{g^3}{4tt'}(T^bT^a)_{ij}T^b_{mn}\bar{u}(p')\gamma^\mu \not{\epsilon}^+ \not{p}'(1-\gamma_5)u(p)$$

$$\times \bar{u}(q')\gamma_\mu(1-\gamma_5)u(q). \quad (5.53)$$

Next, we rewrite the contribution from M_2 to $M(-,-;-,-,+)$ in analogy with the manipulations on M_1. We find

$$M_2 = -\frac{g^3}{4tt'}(T^bT^a)_{ij}T^b_{mn}\bar{u}(p')\gamma^\mu \frac{\not{p}-\not{k}}{-2(p\cdot k)}\not{\epsilon}^+(1-\gamma_5)u(p)$$

$$\times \bar{u}(q')\gamma_\mu(1-\gamma_5)u(q)$$

$$= \frac{g^3}{4tt'}(T^bT^a)_{ij}T^b_{mn}\bar{u}(p')\gamma^\mu \frac{\not{p}-\not{k}}{-2(p\cdot k)}(\not{p}'\not{p}+\not{p}\not{p}')\not{\epsilon}^+(1-\gamma_5)u(p)$$

$$\times \bar{u}(q')\gamma_\mu(1-\gamma_5)u(q)$$

$$= -\frac{g^3}{4tt'}(T^bT^a)_{ij}T^b_{mn}\frac{(p\cdot\epsilon^+)}{(p\cdot k)}\bar{u}(p')\gamma^\mu(\not{p}-\not{k})\not{p}'(1-\gamma_5)u(p)$$

$$\times \bar{u}(q')\gamma_\mu(1-\gamma_5)u(q)$$

$$-\frac{g^3}{4tt'}(T^bT^a)_{ij}T^b_{mn}\bar{u}(p')\gamma^\mu \frac{\not{k}}{2(p\cdot k)}\not{p}\not{\epsilon}^+ \not{p}'(1-\gamma_5)u(p)$$

$$\times \bar{u}(q')\gamma_\mu(1-\gamma_5)u(q)$$

5.4. THE PROCESS $qq' \to qq'g$

$$= -\frac{g^3}{4tt'}(T^bT^a)_{ij}T^b_{mn}\frac{(p \cdot \epsilon^+)}{(p \cdot k)}\overline{u}(p')\gamma^\mu(\slashed{p}-\slashed{k})\slashed{p}'(1-\gamma_5)u(p)$$

$$\times \overline{u}(q')\gamma_\mu(1-\gamma_5)u(q)$$

$$-\frac{g^3}{4tt'}(T^bT^a)_{ij}T^b_{mn}\overline{u}(p')\gamma^\mu \slashed{\epsilon}^+ \slashed{p}'(1-\gamma_5)u(p)$$

$$\times \overline{u}(q')\gamma_\mu(1-\gamma_5)u(q). \qquad (5.54)$$

Combining M_1, M_2 and $M_5^{(1)}$, we obtain

$$M_1 + M_2 + M_5^{(1)} = \frac{g^3}{4tt'}\left[\frac{(p' \cdot \epsilon^+)}{(p' \cdot k)}(T^aT^b)_{ij}T^b_{mn} - \frac{(p \cdot \epsilon^+)}{(p \cdot k)}(T^bT^a)_{ij}T^b_{mn}\right]$$

$$\times \overline{u}(p')\gamma^\mu(\slashed{p}-\slashed{k})\slashed{p}'(1-\gamma_5)u(p)\,\overline{u}(q')\gamma_\mu(1-\gamma_5)u(q). \qquad (5.55)$$

With the cut-and-paste technique of Appendix A, we can rewrite the spinor expression as follows:

$$A = \overline{u}(p')\gamma^\mu(\slashed{p}-\slashed{k})\slashed{p}'(1-\gamma_5)u(p)\,\overline{u}(q')\gamma_\mu(1-\gamma_5)u(q)$$

$$= \overline{u}(p')\gamma^\mu(\slashed{p}-\slashed{k})\slashed{p}'(1-\gamma_5)u(p)\frac{\overline{u}(p)\,\slashed{a}(1-\gamma_5)u(q')}{\overline{u}(p)\,\slashed{a}(1-\gamma_5)u(q')}$$

$$\times \overline{u}(q')\gamma_\mu(1-\gamma_5)u(q)$$

$$= 4\frac{\overline{u}(p')\gamma^\mu(\slashed{p}-\slashed{k})\slashed{p}'\,\slashed{p}\,\slashed{a}\,\slashed{a}'\gamma_\mu(1-\gamma_5)u(q)\overline{u}(q')\,\slashed{a}(1-\gamma_5)u(p)}{\text{Tr}[\slashed{a}(1-\gamma_5)\,\slashed{a}'\,\slashed{a}(1-\gamma_5)\,\slashed{p}]}$$

$$= \frac{-1}{2(p \cdot a)(q' \cdot a)}\overline{u}(p')\,\slashed{a}'\,\slashed{a}\,\slashed{p}\,\slashed{p}'(\slashed{p}-\slashed{k})(1-\gamma_5)u(q)$$

$$\times \overline{u}(q')\,\slashed{a}(1-\gamma_5)u(p). \qquad (5.56)$$

Choosing $a = p'$ and using momentum conservation, we have

$$A = -2\,\overline{u}(p')\,\slashed{k}'(1-\gamma_5)u(q)\,\overline{u}(q')\,\slashed{p}'(1-\gamma_5)u(p). \qquad (5.57)$$

Thus,

$$M_1 + M_2 + M_5^{(1)} = \frac{g^3}{2tt'}\left[\frac{(p \cdot \epsilon^+)}{(p \cdot k)}(T^bT^a)_{ij}T^b_{mn} - \frac{(p' \cdot \epsilon^+)}{(p' \cdot k)}(T^aT^b)_{ij}T^b_{mn}\right]$$

$$\times \overline{u}(p')\,\slashed{k}'(1-\gamma_5)u(q)\,\overline{u}(q')\,\slashed{p}'(1-\gamma_5)u(p). \qquad (5.58)$$

Similar manipulations on $M_3 + M_4 + M_5^{(2)}$ give an analogous result, and, for the helicity amplitude itself, we find the rather nice result

$$M(-,-;-,-,+) = \frac{g^3}{2tt'}\left[\frac{(p\cdot\epsilon^+)}{(p\cdot k)}(T^bT^a)_{ij}T^b_{mn} - \frac{(p'\cdot\epsilon^+)}{(p'\cdot k)}(T^aT^b)_{ij}T^b_{mn}\right.$$

$$\left. + \frac{(q\cdot\epsilon^+)}{(q\cdot k)}T^b_{ij}(T^bT^a)_{mn} - \frac{(q'\cdot\epsilon^+)}{(q'\cdot k)}T^b_{ij}(T^aT^b)_{mn}\right]$$

$$\times \overline{u}(p')\,\slashed{A}'(1-\gamma_5)u(q)\,\overline{u}(q')\,\slashed{p}'(1-\gamma_5)u(p)$$

$$\doteq 2g^3\left[\frac{(p\cdot\epsilon^+)}{(p\cdot k)}(T^bT^a)_{ij}T^b_{mn} - \frac{(p'\cdot\epsilon^+)}{(p'\cdot k)}(T^aT^b)_{ij}T^b_{mn}\right.$$

$$\left. + \frac{(q\cdot\epsilon^+)}{(q\cdot k)}T^b_{ij}(T^bT^a)_{mn} - \frac{(q'\cdot\epsilon^+)}{(q'\cdot k)}T^b_{ij}(T^aT^b)_{mn}\right]\frac{s'}{(tt')^{\frac{1}{2}}}.$$

(5.59)

We find that the amplitude factorizes again. The expression within the brackets is a generalization for this QCD process of the 'infrared factor', which we encountered in Chapter 4 on single bremsstrahlung in QED.

To proceed with the calculation of the cross section, we have to square this amplitude and sum over the color degrees of freedom. Using eqns (2.14), this leads among other things to formulae of the type

$$\sum_{abc}\text{Tr}(T^bT^aT^cT^a)\,\text{Tr}(T^bT^c) = \tfrac{1}{2}\sum_{ab}\text{Tr}(T^bT^aT^bT^a)$$

$$= \tfrac{1}{2}\left[\sum_{ab}\text{Tr}(T^bT^bT^aT^a) + i\sum_{abc}f^{abc}\text{Tr}(T^bT^cT^a)\right]$$

$$= -\tfrac{1}{3}. \quad (5.60)$$

Thus,

$$\sum_{\text{color}}|M(-,-;-,-,+)|^2$$

$$= \frac{4g^6}{3}\left[7\left|\frac{(p'\cdot\epsilon^+)}{(p'\cdot k)} - \frac{(q\cdot\epsilon^+)}{(q\cdot k)}\right|^2 + 7\left|\frac{(q'\cdot\epsilon^+)}{(q'\cdot k)} - \frac{(p\cdot\epsilon^+)}{(p\cdot k)}\right|^2\right.$$

$$- \left|\frac{(p'\cdot\epsilon^+)}{(p'\cdot k)} - \frac{(p\cdot\epsilon^+)}{(p\cdot k)}\right|^2 - \left|\frac{(q'\cdot\epsilon^+)}{(q'\cdot k)} - \frac{(q\cdot\epsilon^+)}{(q\cdot k)}\right|^2$$

$$\left. + 2\left|\frac{(p'\cdot\epsilon^+)}{(p'\cdot k)} - \frac{(q'\cdot\epsilon^+)}{(q'\cdot k)}\right|^2 + 2\left|\frac{(q\cdot\epsilon^+)}{(q\cdot k)} - \frac{(p\cdot\epsilon^+)}{(p\cdot k)}\right|^2\right]\frac{s'^2}{tt'}. \quad (5.61)$$

5.4. THE PROCESS $qq' \to qq'g$

The six combinations in the bracket are separately gauge invariant. To evaluate them, one can thus use a different representation for the gluon polarization vector in each combination separately. For example, for the first combination,

$$B = \left| \frac{(p' \cdot \epsilon^+)}{(p' \cdot k)} - \frac{(q \cdot \epsilon^+)}{(q \cdot k)} \right|^2, \tag{5.62}$$

we take

$$\not{\epsilon}^+ = N[\not{k} \not{p}' \not{q}(1+\gamma_5) + \not{q} \not{p}' \not{k}(1-\gamma_5)],$$

$$N^{-1} = 4[(p' \cdot q)(p' \cdot k)(q \cdot k)]^{\frac{1}{2}}, \tag{5.63}$$

which yields

$$(q \cdot \epsilon^+) = 0,$$

$$(p' \cdot \epsilon^+) = \frac{N}{2}\text{Tr}[\not{p}' \not{k} \not{p}' \not{q}(1+\gamma_5)] = \left[\frac{(p' \cdot k)(p' \cdot q)}{(q \cdot k)}\right]^{\frac{1}{2}}. \tag{5.64}$$

Thus,

$$B = \frac{(p' \cdot q)}{(p' \cdot k)(q \cdot k)}. \tag{5.65}$$

For the second combination, it is obviously preferable to express $\not{\epsilon}^+$ in terms of p and q', etc. This way, all six combinations are readily evaluated, yielding

$$\sum_{\text{color}} |M(-,-;-,-,+)|^2$$

$$= \frac{4g^6}{3}\left[7\frac{(p' \cdot q)}{(p' \cdot k)(q \cdot k)} + 7\frac{(p \cdot q')}{(p \cdot k)(q' \cdot k)} - \frac{(p \cdot p')}{(p \cdot k)(p' \cdot k)}\right.$$

$$\left. - \frac{(q \cdot q')}{(q \cdot k)(q' \cdot k)} + 2\frac{(p' \cdot q')}{(p' \cdot k)(q' \cdot k)} + 2\frac{(p \cdot q)}{(p \cdot k)(q \cdot k)}\right]\frac{s'^2}{tt'}. \tag{5.66}$$

All the other helicity amplitudes can be calculated in a completely analogous way. The result is simply a replacement in the above formula of the quantity s' by s, u or u'. To obtain the cross section, we add the squared absolute values of all the helicity amplitudes, we average over the initial spin states, which produces a factor $\frac{1}{4}$, and we average over the initial color degrees of freedom, which gives a factor $\frac{1}{9}$. Hence, for $qq' \to qq'g$,

$$\overline{|M|^2} = \frac{2g^6}{27}\left[\frac{7(p' \cdot q)}{(p' \cdot k)(q \cdot k)} + \frac{7(p \cdot q')}{(p \cdot k)(q' \cdot k)} - \frac{(p \cdot p')}{(p \cdot k)(p' \cdot k)} - \frac{(q \cdot q')}{(q \cdot k)(q' \cdot k)}\right.$$

$$\left. + \frac{2(p' \cdot q')}{(p' \cdot k)(q' \cdot k)} + \frac{2(p \cdot q)}{(p \cdot k)(q \cdot k)}\right]\frac{s^2 + s'^2 + u^2 + u'^2}{tt'}, \tag{5.67}$$

which is again a quite simple formula.

5.5 Other QCD processes

The other pure QCD processes of the type $2 \to 3$ particles are $q\bar{q} \to q\bar{q}g$, $q\bar{q} \to ggg$, $gg \to ggg$, and the processes obtained by crossing. They can all be treated in the way the previous examples were handled. The general strategy is to choose a representation for the gluon polarizations which annihilates as many Feynman diagrams as possible.

In the case $gg \to ggg$, where there are 25 Feynman diagrams, it is always possible to annihilate the 10 Feynman diagrams involving the four-gluon vertex, thus reducing the algebra considerably. It is amusing to note that, although no fermions are involved in this process, it is computationally advantageous [27] to introduce spinor algebra by rewriting the three-gluon vertex as a trace, as we did in Section 5.4. If one expresses the gluon polarization vectors in terms of the four-momenta of the gluons, one finds that many cancellations occur, ultimately leading to a simple cross section formula.

For the explicit formulae for these QCD processes, we refer to Chapter 10.

6
Double bremsstrahlung

> When electrons radiate twice,
> one must realize:
> The spectrum looks so nice
> it becomes fun to analyse.

6.1 QED

Double bremsstrahlung processes in QED, such as $e^+e^- \to \gamma\gamma\gamma\gamma$, $e^+e^- \to \mu^+\mu^-\gamma\gamma$ and $e^+e^- \to e^+e^-\gamma\gamma$, are of interest for several reasons. They have all been observed in experiments and provide high energy QED tests of order α^4 [28]. The agreement between theory and experiment for the process $e^+e^- \to \mu^+\mu^-\gamma\gamma$ has been widely used to set lower limits on the mass of the excited muon. This was done both at PEP [29] at $\sqrt{s} = 29$ GeV and at PETRA [30] at higher energies. It is also not inconceivable that the effects of this process have to be taken into account if the accuracy of the charge asymmetry measurements in $e^+e^- \to \mu^+\mu^-$ continues to increase. Finally, the process $e^+e^- \to e^+e^-\gamma\gamma$ gives an α^4 contribution to small angle Bhabha scattering, which is generally used to determine the beam luminosity of e^+e^- colliders [28].

From the theoretical point of view, these processes are not so easy to evaluate because of the large number of Feynman diagrams which are involved: 24 for $e^+e^- \to \gamma\gamma\gamma\gamma$, 20 for $e^+e^- \to \mu^+\mu^-\gamma\gamma$ and 40 for $e^+e^- \to e^+e^-\gamma\gamma$, not counting the Z-exchange diagrams for the latter two processes. Even with symbolic manipulation programs, the use of the standard Feynman techniques is cumbersome, and, eventually, the resulting formulae are inconvenient for the analysis of the experiments due to their lengths. In this section, we shall illustrate with the example $e^+e^- \to \gamma\gamma\gamma\gamma$ how the helicity amplitude method can be applied to yield more practical formulae.

The process of electron-positron annihilation into four photons [31],

$$e^+(p_+) + e^-(p_-) \to \gamma(k_1) + \gamma(k_2) + \gamma(k_3) + \gamma(k_4), \qquad (6.1)$$

is described by the Feynman diagrams of Fig. 6.1. The Feynman amplitude is

$$M = -e^4 \bar{v}(p_+) \not{k}_4 \frac{\not{k}_4 - \not{p}_+}{-2(p_+ \cdot k_4)} \not{k}_3 \frac{\not{p}_- - \not{k}_1 - \not{k}_2}{(p_- - k_1 - k_2)^2} \not{k}_2 \frac{\not{p}_- - \not{k}_1}{-2(p_- \cdot k_1)} \not{k}_1 u(p_-)$$

$$+ \text{ 23 other permutations of } (1,2,3,4). \qquad (6.2)$$

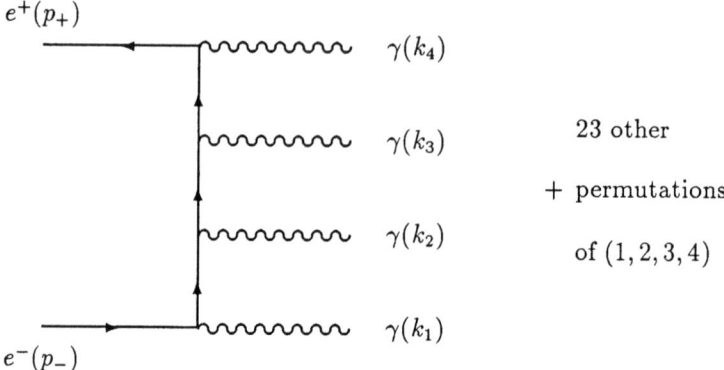

Fig. 6.1: Feynman diagrams for $e^+e^- \to \gamma\gamma\gamma\gamma$.

The obvious choice for the photon polarization vectors in this case is that of eqns (4.4),

$$\not{\epsilon}_i^\pm = N_i[\not{p}_+ \not{p}_- \not{k}_i(1 \mp \gamma_5) - \not{k}_i \not{p}_+ \not{p}_-(1 \pm \gamma_5)],$$

$$N_i^{-1} = 4[(p_+ \cdot p_-)(p_+ \cdot k_i)(p_- \cdot k_i)]^{\frac{1}{2}}, \qquad i = 1,\ldots,4. \qquad (6.3)$$

Let $M(\lambda_+, \lambda_-; \lambda_1, \lambda_2, \lambda_3, \lambda_4)$ be the helicity amplitudes, with λ_+ (λ_-) the helicity of the positron (electron) and λ_i, $i = 1,\ldots,4$, the helicity of photon i. Clearly, for $\lambda_+ = \lambda_-$, the helicity amplitudes vanish. Moreover, for $\lambda_+ = -\lambda_-$, only one of the two terms in $\not{\epsilon}_i^{\lambda_i}$ gives a nonvanishing contribution. But, if all photon helicities are equal, either $\not{p}_+$ or $\not{p}_-$ necessarily appears next to its corresponding spinor in the expression for the helicity amplitude. By virtue of the Dirac equation for massless spinors, this leads again to a vanishing contribution. We thus have to consider only the case where at least one photon helicity is opposite to another photon helicity.

For $M(+, -; +, -, -, -)$, we find

$$M(+,-;+,-,-,-) = 8e^4 N_1 N_2 N_3 N_4 \bar{v}(p_+) \not{k}_4 \not{p}_+ \not{p}_- \frac{\not{k}_4 - \not{p}_+}{-2(p_+ \cdot k_4)} \not{k}_3 \not{p}_+$$

$$\times \not{p}_- \frac{\not{p}_- - \not{k}_1 - \not{k}_2}{(p_- - k_1 - k_2)^2} \not{k}_2 \not{p}_+ \not{p}_- \frac{\not{p}_- - \not{k}_1}{-2(p_- \cdot k_1)} \not{p}_+ \not{p}_- \not{k}_1 (1 - \gamma_5) u(p_-)$$

$$+ \text{ 5 other permutations of } (2,3,4)$$

6.1. QED

$$= -8e^4 N_1 N_2 N_3 N_4 \bar{v}(p_+) \not{p}_-(\not{k}_4 - \not{p}_+) \not{k}_3 \not{p}_+ \not{p}_- \frac{\not{p}_- - \not{k}_1 - \not{k}_2}{(p_- - k_1 - k_2)^2}$$

$$\times \not{k}_2 \not{p}_+ \not{p}_- \not{k}_1 \not{p}_+ (1-\gamma_5) u(p_-)$$

$$+ \text{5 other permutations of } (2,3,4)$$

$$= -\tfrac{1}{2} e^4 A (p_+ \cdot p_-)(p_- \cdot k_1) \bar{v}(p_+) \not{k}_1 (1-\gamma_5) u(p_-)$$

$$\doteq 2 e^4 A (p_+ \cdot p_-)(p_- \cdot k_1) \sqrt{(p_+ \cdot k_1)(p_- \cdot k_1)}, \tag{6.4}$$

with

$$A = \left[\prod_{i=1}^{4} (p_+ \cdot k_i)(p_- \cdot k_i) \right]^{-\frac{1}{2}}. \tag{6.5}$$

Note that all factors of $(p_+ \cdot p_-)$ in the denominator have been cancelled, as well as the factors of the type $(p_- - k_1 - k_2)^2$. By a parity conjugation and/or by permuting the photon indices, one can obtain all the remaining helicity amplitudes for which one photon has the opposite helicity of the three other photons.

For the amplitudes of the type $M(+,-;+,+,-,-)$, the formulae are somewhat more lengthy due to the fact that the pole terms of the type $(p_- - k_1 - k_3)^2$ do not cancel. For example,

$$M(+,-;+,+,-,-)$$

$$= \tfrac{1}{4} e^4 A (p_+ \cdot p_-)(p_- - k_1 - k_2)^2 \bar{v}(p_+)(\not{k}_1 + \not{k}_2)(1-\gamma_5) u(p_-)$$

$$+ \tfrac{1}{8} e^4 A \bar{v}(p_+) \Big\{ \frac{1}{\Delta_{13}} \not{k}_4 \not{p}_- \not{k}_2 (\not{p}_- - \not{k}_1) \not{k}_3 \not{p}_+ \not{k}_1$$

$$+ \frac{1}{\Delta_{14}} \not{k}_3 \not{p}_- \not{k}_2 (\not{p}_- - \not{k}_1) \not{k}_4 \not{p}_+ \not{k}_1$$

$$+ \frac{1}{\Delta_{23}} \not{k}_4 \not{p}_- \not{k}_1 (\not{p}_- - \not{k}_2) \not{k}_3 \not{p}_+ \not{k}_2$$

$$+ \frac{1}{\Delta_{24}} \not{k}_3 \not{p}_- \not{k}_1 (\not{p}_- - \not{k}_2) \not{k}_4 \not{p}_+ \not{k}_2 \Big\} (1-\gamma_5) u(p_-), \tag{6.6}$$

where

$$\Delta_{ij} = -2(p_- \cdot k_i) - 2(p_- \cdot k_j) + 2(k_i \cdot k_j), \qquad i,j = 1,\ldots,4. \tag{6.7}$$

To obtain the cross section for $e^+ e^- \to \gamma\gamma\gamma\gamma$, we must know the squared absolute value of the helicity amplitudes. For the amplitudes of the type

$M(+,-;+,+,-,-)$, this leads, however, to very lengthy traces and hardly any simplifications occur in the result. A much more convenient method consists in evaluating the helicity amplitudes directly for any given point in phase space as complex functions of the components of the four-vectors in the process. To this end, we introduce explicit representations for the γ-matrices and the spinors [32].

Suppose we go to the e^+e^- c.m. frame, with the z-direction along $\vec{p}_+$, and that we introduce the notation

$$k_\pm = k_0 \pm k_z, \qquad k_\perp = k_x + ik_y = |k_\perp|e^{i\phi_k}, \qquad (6.8)$$

for any vector k_μ. With the representation (2.6) for the γ-matrices, we have

$$\gamma_5 = \begin{pmatrix} \mathbb{1} & 0 \\ 0 & -\mathbb{1} \end{pmatrix}. \qquad (6.9)$$

Then, we can take for any light-like vector k,

$$u_+(k) = v_-(k) = \begin{pmatrix} \sqrt{k_+} \\ \sqrt{k_-}e^{i\phi_k} \\ 0 \\ 0 \end{pmatrix}, \qquad (6.10)$$

$$u_-(k) = v_+(k) = \begin{pmatrix} 0 \\ 0 \\ -\sqrt{k_-}e^{-i\phi_k} \\ \sqrt{k_+} \end{pmatrix}. \qquad (6.11)$$

Clearly, the first spinor is an eigenstate of $1+\gamma_5$, and the second one of $1-\gamma_5$.

With these formulae, it is easy to see that

$$\overline{u}(k_i)(1+\gamma_5)u(k_j) = [\overline{u}(k_j)(1-\gamma_5)u(k_i)]^* = \frac{2k_{j\perp}Z^*_{ij}}{k_{j-}\sqrt{k_{i+}k_{j+}}}, \qquad (6.12)$$

where

$$Z_{ij} = k_{i+}k_{j-} - k^*_{i\perp}k_{j\perp}. \qquad (6.13)$$

Similarly,

$$\begin{aligned}
\overline{u}(k)(1-\gamma_5)u(p_-) &= \overline{u}(p_-)(1+\gamma_5)u(k) &= -2\sqrt{2Ek_+}, \\
\overline{u}(k)(1+\gamma_5)u(p_-) &= \overline{u}(p_-)(1-\gamma_5)u(k) &= 2\sqrt{2Ek_+}, \\
\overline{u}(k)(1-\gamma_5)u(p_+) &= [\overline{u}(p_+)(1+\gamma_5)u(k)]^* &= 2\sqrt{\frac{2E}{k_+}}k^*_\perp, \\
\overline{u}(k)(1+\gamma_5)u(p_+) &= [\overline{u}(p_+)(1-\gamma_5)u(k)]^* &= -2\sqrt{\frac{2E}{k_+}}k_\perp,
\end{aligned} \qquad (6.14)$$

6.1. QED

where $E = p_{+0} = p_{-0}$ is the beam energy.

With the cut-and-paste technique described in Appendix A, we can reduce any spinor expression to a product of simple expressions of the above type. All helicity amplitudes are thus expressed as complex functions of the components of k_i, $i = 1, \ldots, 4$.

In terms of these variables, the helicity amplitudes become

$$M(+,-;+,-,-,-) = 4e^4 B k_{1+} k_{1\perp},$$

$$M(+,-;+,+,-,-) = -2e^4 B E^{-1} F(1,2,3,4), \quad (6.15)$$

with

$$B = [k_{1+} k_{1-} k_{2+} k_{2-} k_{3+} k_{3-} k_{4+} k_{4-}]^{-\frac{1}{2}}, \quad (6.16)$$

and

$$F(1,2,3,4) = (k_{1\perp} + k_{2\perp})\Delta_{12}$$

$$+ \frac{k_{4\perp}}{\Delta_{13}}(2Ek_{2+}k_{3\perp}^* k_{1\perp} + Z_{21} Z_{13}^*) + \frac{k_{3\perp}}{\Delta_{14}}(2Ek_{2+}k_{4\perp}^* k_{1\perp} + Z_{21} Z_{14}^*)$$

$$+ \frac{k_{4\perp}}{\Delta_{23}}(2Ek_{1+}k_{3\perp}^* k_{2\perp} + Z_{12} Z_{23}^*) + \frac{k_{3\perp}}{\Delta_{24}}(2Ek_{1+}k_{4\perp}^* k_{2\perp} + Z_{12} Z_{24}^*).$$

(6.17)

By adding the squared absolute values of all helicity amplitudes and by averaging over the initial spins, we obtain the unpolarized squared matrix element:

$$\overline{|M|^2} = 2e^8 B^2 \left\{ 4 \sum_{i=1}^{4} (k_{i+}^2 + k_{i-}^2) k_{i+} k_{i-} \right.$$

$$+ E^{-2} \Big[|F(1,2,3,4)|^2 + |F(1,3,2,4)|^2 + |F(1,4,2,3)|^2$$

$$\left. + |F(2,3,1,4)|^2 + |F(2,4,1,3)|^2 + |F(3,4,1,2)|^2 \Big] \right\}.$$

(6.18)

With this equation, it becomes straightforward to evaluate the cross section numerically for any given point in phase space provided one is willing to introduce complex arithmetic.

In this way, one can also obtain the helicity amplitudes and the cross section formulae for $e^+e^- \to \mu^+\mu^-\gamma\gamma$ and $e^+e^- \to e^+e^-\gamma\gamma$, which are listed in Chapter 9.

6.2 $e^+e^- \to 4$ jets

The study of four-jet events in e^+e^--annihilation [33] allows one to make additional comparisons with perturbative QCD. According to this model, one has to know the cross sections for the subprocesses $e^+e^- \to q\bar{q}gg$ and $e^+e^- \to q\bar{q}q\bar{q}$ in lowest order. They were calculated by various authors [34] using the standard Feynman techniques, but, in this section, we want to show that the helicity amplitude method can be used to obtain more practical formulae for these processes.

Consider first the double bremsstrahlung process [32]

$$e^+(p_+) + e^-(p_-) \to \bar{q}(k_3,j) + q(k_4,i) + g(k_1,a) + g(k_2,b), \qquad (6.19)$$

described by the Feynman diagrams of Fig. 6.2, which yield the Feynman amplitudes

$$M_1 = \frac{ee_q g^2}{s}(T^a T^b)_{ij}\bar{u}(k_4)\,\slashed{\epsilon}_1 \frac{\slashed{k}_1 + \slashed{k}_4}{2(k_1\cdot k_4)}\,\slashed{\epsilon}_2 \frac{\slashed{k}_1+\slashed{k}_2+\slashed{k}_4}{(k_1+k_2+k_4)^2}\gamma^\mu v(k_3)$$

$$\times \bar{v}(p_+)\gamma_\mu u(p_-),$$

$$M_2 = \frac{ee_q g^2}{s}(T^b T^a)_{ij}\bar{u}(k_4)\,\slashed{\epsilon}_2 \frac{\slashed{k}_2 + \slashed{k}_4}{2(k_2\cdot k_4)}\,\slashed{\epsilon}_1 \frac{\slashed{k}_1+\slashed{k}_2+\slashed{k}_4}{(k_1+k_2+k_4)^2}\gamma^\mu v(k_3)$$

$$\times \bar{v}(p_+)\gamma_\mu u(p_-),$$

$$M_3 = \frac{ee_q g^2}{s}(T^a T^b)_{ij}\bar{u}(k_4)\gamma^\mu \frac{-\slashed{k}_1-\slashed{k}_2-\slashed{k}_3}{(k_1+k_2+k_3)^2}\,\slashed{\epsilon}_1 \frac{-\slashed{k}_2-\slashed{k}_3}{2(k_2\cdot k_3)}\,\slashed{\epsilon}_2 v(k_3)$$

$$\times \bar{v}(p_+)\gamma_\mu u(p_-),$$

$$M_4 = \frac{ee_q g^2}{s}(T^b T^a)_{ij}\bar{u}(k_4)\gamma^\mu \frac{-\slashed{k}_1-\slashed{k}_2-\slashed{k}_3}{(k_1+k_2+k_3)^2}\,\slashed{\epsilon}_2 \frac{-\slashed{k}_1-\slashed{k}_3}{2(k_1\cdot k_3)}\,\slashed{\epsilon}_1 v(k_3)$$

$$\times \bar{v}(p_+)\gamma_\mu u(p_-),$$

$$M_5 = \frac{ee_q g^2}{s}(T^b T^a)_{ij}\bar{u}(k_4)\,\slashed{\epsilon}_2 \frac{\slashed{k}_2+\slashed{k}_4}{2(k_2\cdot k_4)}\gamma^\mu \frac{-\slashed{k}_1-\slashed{k}_3}{2(k_1\cdot k_3)}\,\slashed{\epsilon}_1 v(k_3)$$

$$\times \bar{v}(p_+)\gamma_\mu u(p_-),$$

$$M_6 = \frac{ee_q g^2}{s}(T^a T^b)_{ij}\bar{u}(k_4)\,\slashed{\epsilon}_1 \frac{\slashed{k}_1+\slashed{k}_4}{2(k_1\cdot k_4)}\gamma^\mu \frac{-\slashed{k}_2-\slashed{k}_3}{2(k_2\cdot k_3)}\,\slashed{\epsilon}_2 v(k_3)$$

$$\times \bar{v}(p_+)\gamma_\mu u(p_-),$$

6.2. $e^+e^- \rightarrow$ 4 JETS

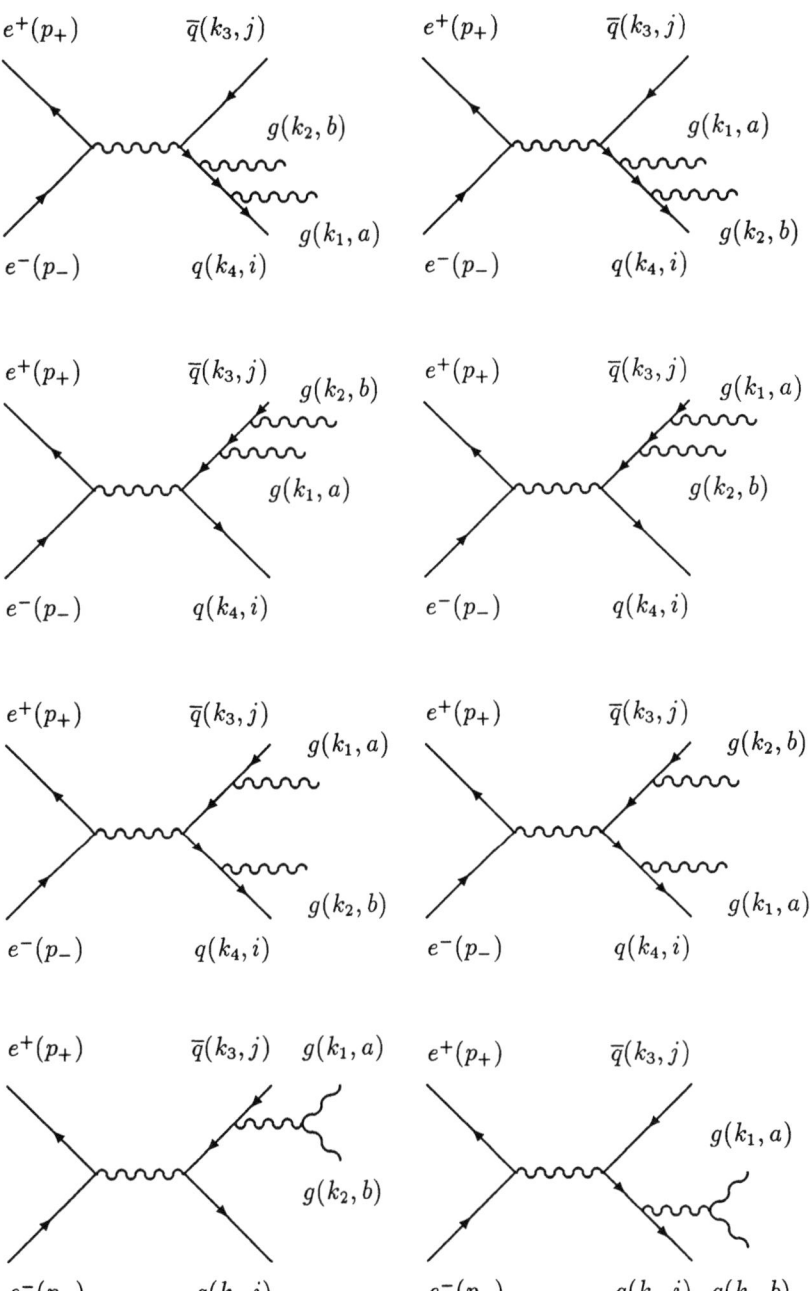

Fig. 6.2: Feynman diagrams for $e^+e^- \rightarrow q\bar{q}gg$.

$$M_7 = \frac{iee_q g^2}{s} T_{ij}^c \bar{v}(p_+)\gamma_\mu u(p_-)\bar{u}(k_4)\gamma^\mu \frac{-\slashed{k}_1 - \slashed{k}_2 - \slashed{k}_3}{(k_1+k_2+k_3)^2}\gamma^\nu v(k_3)$$

$$\times \frac{f^{cba}}{2(k_1 \cdot k_2)}[2(k_2 \cdot \epsilon_1)\epsilon_{2\nu} + (k_1-k_2)_\nu(\epsilon_1 \cdot \epsilon_2) - 2(k_1 \cdot \epsilon_2)\epsilon_{1\nu}],$$

$$M_8 = \frac{iee_q g^2}{s} T_{ij}^c \bar{v}(p_+)\gamma_\mu u(p_-)\bar{u}(k_4)\gamma^\nu \frac{\slashed{k}_1 + \slashed{k}_2 + \slashed{k}_4}{(k_1+k_2+k_4)^2}\gamma^\mu v(k_3)$$

$$\times \frac{f^{cba}}{2(k_1 \cdot k_2)}[2(k_2 \cdot \epsilon_1)\epsilon_{2\nu} + (k_1-k_2)_\nu(\epsilon_1 \cdot \epsilon_2) - 2(k_1 \cdot \epsilon_2)\epsilon_{1\nu}],$$

(6.20)

with $s = (p_+ + p_-)^2 = 4E^2$, E being the beam energy.

Let $M(\lambda_1, \lambda_2; \lambda_3, \lambda_4, \lambda_5, \lambda_6)$ denote the helicity amplitude with λ_1 the helicity for the e^+, λ_2 the one for e^-, λ_3 for $\bar{q}$, λ_4 for q, λ_5 the one for $g(k_1)$ and λ_6 for $g(k_2)$. As usual, we introduce explicit polarization vectors for the radiated gluons of the form

$$\slashed{\epsilon}_i^\pm = N_i[\slashed{q} \slashed{q}' \slashed{k}_i(1 \mp \gamma_5) + \slashed{k}_i \slashed{q}' \slashed{q}(1 \pm \gamma_5)],$$

$$N_i^{-1} = 4[(q \cdot q')(q \cdot k_i)(q' \cdot k_i)]^{\frac{1}{2}}, \qquad i=1,2. \qquad (6.21)$$

A very simple nonzero helicity amplitude is then $M(-,+;-,+,-,-)$ for which we choose $q = k_4$, $q' = k_3$ in the above expressions for $\slashed{\epsilon}_i^\pm$ and N_i. Of the eight Feynman amplitudes, only three contribute and we find, after the elimination of the repeated index,

$$M(-,+;-,+,-,-)$$

$$= -\frac{ee_q g^2 N_1 N_2}{8E^2(k_1 \cdot k_2)}\bar{v}(p_+) \slashed{k}_4(1+\gamma_5)v(k_3)\bar{u}(k_4) \slashed{p}_+(1+\gamma_5)u(p_-)$$

$$\times \{(T^a T^b)_{ij}\text{Tr}[\slashed{k}_4 \slashed{k}_3 \slashed{k}_1 \slashed{k}_2(1-\gamma_5)]$$

$$+ (T^b T^a)_{ij}\text{Tr}[\slashed{k}_4 \slashed{k}_3 \slashed{k}_2 \slashed{k}_1(1-\gamma_5)]\}. \qquad (6.22)$$

Summing over the color degrees of freedom, it readily follows that

$$|M(-,+;-,+,-,-)|^2 = \frac{4e^2 e_q^2 g^4}{3A} k_{4-}^2 (k_3 \cdot k_4) \qquad (6.23)$$

$$\times [9(k_1 \cdot k_3)(k_2 \cdot k_4) + 9(k_1 \cdot k_4)(k_2 \cdot k_3) - (k_1 \cdot k_2)(k_3 \cdot k_4)],$$

where

$$A = \prod_{i>j}(k_i \cdot k_j). \qquad (6.24)$$

We have gone to the e^+e^- c.m. frame, with the z-direction along $\vec{p}_+$, and we have introduced the notation

$$k_\pm = k_0 \pm k_z, \qquad k_\perp = k_x + ik_y = |k_\perp|e^{i\phi_k}, \qquad (6.25)$$

for all vectors k_i.

All other nonzero helicity amplitudes for which the gluon helicities are equal have the same structure. They only differ in the appearance of k_{3+}, k_{3-} or k_{4+} instead of k_{4-}.

Somewhat more complicated are the helicity amplitudes with opposite helicities for the gluons. This time, it is more convenient to choose $q = k_4$, $q' = k_3$ for $\not{\epsilon}_1$ and $q = k_3$, $q' = k_4$ for $\not{\epsilon}_2$ in the calculation of $M(-,+;-,+,-,+)$. Five Feynman diagrams are found to contribute, yielding

$$M(-,+;-,+,-,+) \doteq \frac{ee_q g^2 N_1 N_2}{4E^2(k_1 \cdot k_2)}$$

$$\times \Big\{ -(T^aT^b)_{ij}\bar{u}(k_4)\not{k}_2\not{k}_4(1-\gamma_5)v(p_+)\bar{u}(p_-)\not{k}_3\not{k}_1(1+\gamma_5)v(k_3)$$

$$-(T^bT^a)_{ij}\bar{u}(k_4)\not{k}_1(\not{k}_2+\not{k}_4)(1-\gamma_5)v(p_+)$$

$$\times \bar{u}(p_-)(\not{k}_1+\not{k}_3)\not{k}_2(1+\gamma_5)v(k_3)$$

$$+2\frac{(T^aT^b)_{ij}(k_2 \cdot k_4)+(T^bT^a)_{ij}(k_1 \cdot k_4)}{(k_1+k_2+k_4)^2}\bar{u}(k_4)\not{k}_2(1+\gamma_5)v(k_3)$$

$$\times \bar{u}(p_-)\not{k}_3\not{k}_1(\not{k}_2+\not{k}_4)(1-\gamma_5)v(p_+)$$

$$+2\frac{(T^aT^b)_{ij}(k_1 \cdot k_3)+(T^bT^a)_{ij}(k_2 \cdot k_3)}{(k_1+k_2+k_3)^2}\bar{u}(k_4)\not{k}_1(1+\gamma_5)v(k_3)$$

$$\times \bar{u}(p_-)(\not{k}_1+\not{k}_3)\not{k}_2\not{k}_4(1-\gamma_5)v(p_+)\Big\}. \qquad (6.26)$$

We now introduce the same explicit representation for the γ-matrices and the spinors as in Eqs(6.9), (6.10) and (6.11). This allows us to rewrite the helicity amplitude in the following form:

$$M(-,+;-,+,-,+) \doteq \frac{4ee_q g^2 N_1 N_2 (k_3 \cdot k_4)}{E(k_1 \cdot k_2)k_{1+}k_{2-}(k_{3+}k_{4-})^{\frac{1}{2}}}[c_1 T^a T^b + c_2 T^b T^a], \qquad (6.27)$$

with

$$c_1 = |\alpha|^2 + \frac{(k_2 \cdot k_4)}{2E(E-k_{30})}\alpha\beta + \frac{(k_1 \cdot k_3)}{2E(E-k_{40})}\alpha^*\gamma,$$

$$c_2 = \beta\gamma + \frac{(k_1 \cdot k_4)}{2E(E - k_{30})}\alpha\beta + \frac{(k_2 \cdot k_3)}{2E(E - k_{40})}\alpha^*\gamma,$$

$$\alpha = Z_{12}Z_{34} - Z_{14}Z_{32},,$$

$$\beta = Z_{32}^*(Z_{12}^* + Z_{14}^*),$$

$$\gamma = Z_{14}(Z_{12} + Z_{32}),$$

$$Z_{ij} = k_{i+}k_{j-} - k_{i\perp}^*k_{j\perp}, \qquad i,j = 1,\ldots,4. \tag{6.28}$$

The summation of the color degrees of freedom then gives

$$|M(-,+;-,+,-,+)|^2 = \frac{e^2 e_q^2 g^4}{48A} \frac{(k_3 \cdot k_4)}{E^2(k_1 \cdot k_2)k_{1+}^2 k_{2-}^2 k_{3+}k_{4-}}$$

$$\times [7\,|c_1 + c_2|^2 + 9\,|c_1 - c_2|^2]. \tag{6.29}$$

The remaining nonzero helicity amplitudes, with opposite gluon helicities, can all be obtained from this equation by interchanging $k_1 \leftrightarrow k_2$, and/or $k_3 \leftrightarrow k_4$, or by simply applying parity conjugation.

To obtain the four-jet cross section, one must sum all the squared absolute values of the helicity amplitudes, perform the color sum, average over the initial lepton helicities, symmetrize appropriately in the four-momenta of the outgoing particles, and sum over the quark flavors. For the process $e^+e^- \to q\bar{q}gg$, we find

$$\overline{|M_g|^2} = \frac{2e^4 g^4}{3A}\left(\sum_f Q_f^2\right) \sum_P [G_1(1,2,3,4) + G_2(1,2,3,4)], \tag{6.30}$$

where Q_f is the fractional charge of the quark with flavor f and

$$G_1(1,2,3,4) = (k_{4+}^2 + k_{4-}^2)(k_3 \cdot k_4)[18(k_1 \cdot k_3)(k_2 \cdot k_4) - (k_1 \cdot k_2)(k_3 \cdot k_4)],$$

$$G_2(1,2,3,4) = \frac{(k_3 \cdot k_4)[7\,|c_1 + c_2|^2 + 9\,|c_1 - c_2|^2]}{32E^2(k_1 \cdot k_2)k_{1+}^2 k_{2-}^2 k_{3+}k_{4-}}. \tag{6.31}$$

In eqn (6.30), the first summation runs over all relevant quark flavors, while the second summation runs over all 24 permutations of k_1, k_2, k_3 and k_4.

Consider next the process

$$e^+(p_+) + e^-(p_-) \to q(k_1,i) + \bar{q}(k_2,j) + q'(k_3,m) + \bar{q}'(k_4,n), \tag{6.32}$$

with $q \neq q'$, i.e., different flavors. In lowest order, we have the four Feynman diagrams of Fig. 6.3, with the Feynman amplitudes

6.2. $e^+e^- \to 4$ JETS

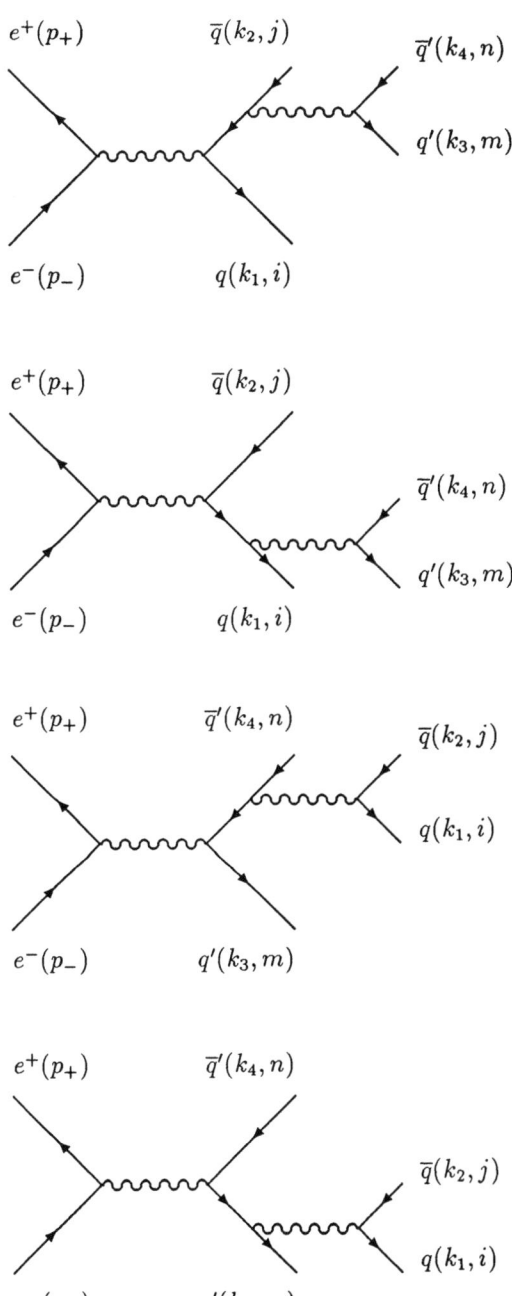

Fig. 6.3: Feynman diagrams for $e^+e^- \to q\bar{q}q'\bar{q}'$.

$$M_1 = -\frac{ee_q g^2}{2s(k_3 \cdot k_4)} T^a_{ij} T^a_{mn} \bar{u}(k_1)\gamma_\mu \frac{-\not{k}_2 - \not{k}_3 - \not{k}_4}{(k_2 + k_3 + k_4)^2} \gamma_\nu v(k_2)$$

$$\times \bar{u}(k_3)\gamma^\nu v(k_4) \bar{v}(p_+)\gamma^\mu u(p_-),$$

$$M_2 = -\frac{ee_q g^2}{2s(k_3 \cdot k_4)} T^a_{ij} T^a_{mn} \bar{u}(k_1)\gamma_\nu \frac{\not{k}_1 + \not{k}_3 + \not{k}_4}{(k_1 + k_3 + k_4)^2} \gamma_\mu v(k_2)$$

$$\times \bar{u}(k_3)\gamma^\nu v(k_4) \bar{v}(p_+)\gamma^\mu u(p_-),$$

$$M_3 = -\frac{ee'_q g^2}{2s(k_1 \cdot k_2)} T^a_{ij} T^a_{mn} \bar{u}(k_3)\gamma_\mu \frac{-\not{k}_1 - \not{k}_2 - \not{k}_4}{(k_1 + k_2 + k_4)^2} \gamma_\nu v(k_4)$$

$$\times \bar{u}(k_1)\gamma^\nu v(k_2) \bar{v}(p_+)\gamma^\mu u(p_-),$$

$$M_4 = -\frac{ee'_q g^2}{2s(k_1 \cdot k_2)} T^a_{ij} T^a_{mn} \bar{u}(k_3)\gamma_\nu \frac{\not{k}_1 + \not{k}_2 + \not{k}_3}{(k_1 + k_2 + k_3)^2} \gamma_\mu v(k_4)$$

$$\times \bar{u}(k_1)\gamma^\nu v(k_2) \bar{v}(p_+)\gamma^\mu u(p_-). \quad (6.33)$$

By inserting the helicity projection operators in the Feynman amplitudes, one finds again that the lepton and quark helicities must be opposite to yield nonzero helicity amplitudes. One then proceeds to eliminate the repeated indices and to introduce explicit spinors, as was done in Section 6.1. For the helicity amplitudes $M(\lambda_1, \lambda_2; \lambda_3, \lambda_4, \lambda_5, \lambda_6)$, where λ_i, $i = 1, \ldots, 6$, refers to the helicities of the $e^+, e^-, q, \bar{q}, q', \bar{q}'$, we find

$$M(+,-;+,-,-,+) \doteq BT^a_{ij}T^a_{mn}[Q_f F(1,2,3,4) - Q'_f F(4,3,2,1)],$$

$$M(+,-;-,+,+,-) \doteq BT^a_{ij}T^a_{mn}[-Q_f F(2,1,4,3) + Q'_f F(3,4,1,2)],$$

$$M(+,-;+,-,+,-) \doteq BT^a_{ij}T^a_{mn}[Q_f F(1,2,4,3) + Q'_f F(3,4,2,1)],$$

$$M(+,-;-,+,-,+) \doteq BT^a_{ij}T^a_{mn}[-Q_f F(2,1,3,4) - Q'_f F(4,3,2,1)],$$
$$(6.34)$$

with

$$F(1,2,3,4) = \frac{k_{4\perp}}{(k_3 \cdot k_4)k_{4-}} \left[\frac{k_{2+}Z^*_{14}(Z_{31} + Z_{34})}{E - k_{20}} + \frac{k_{1\perp}k^*_{3\perp}Z_{23}(Z^*_{24} + Z^*_{34})}{k_{3-}(E - k_{10})} \right],$$
$$(6.35)$$

and

$$B = \frac{e^2 g^2}{4E^2 [k_{1+}k_{2+}k_{3+}k_{4+}]^{\frac{1}{2}}}. \tag{6.36}$$

The quantities F and B here are of course different from those of eqns (6.16) and (6.17). In the above formulae, Q_f (Q'_f) represent the fractional charge of quark q (q'). The remaining helicity amplitudes can again be obtained by applying a parity conjugation.

The case of identical quark flavors can be treated in the same way. We only have to take into account that there are more Feynman diagrams and that a relative minus sign has to be introduced between the Feynman amplitudes which differ by the exchange of identical quarks. We merely list the results:

$$M(+,-;+,-,-,+) \doteq BQ_f T^a_{ij} T^a_{mn}[F(1,2,3,4) - F(4,3,2,1)],$$

$$M(+,-;-,+,+,-) \doteq BQ_f T^a_{ij} T^a_{mn}[-F(2,1,4,3) + F(3,4,1,2)],$$

$$M(+,-;+,-,+,-) \doteq BQ_f\{T^a_{ij} T^a_{mn}[F(1,2,4,3) + F(3,4,2,1)]$$

$$-T^a_{mj} T^a_{in}[F(3,2,4,1) + F(1,4,2,3)]\},$$

$$M(+,-;-,+,-,+) \doteq BQ_f\{T^a_{ij} T^a_{mn}[-F(2,1,3,4) - F(4,3,1,2)]$$

$$-T^a_{mj} T^a_{in}[-F(2,3,1,4) - F(4,1,3,2)]\},$$

$$M(+,-;+,+,-,-) \doteq -BQ_f T^a_{mj} T^a_{in}[-F(2,3,4,1) + F(1,4,3,2)],$$

$$M(+,-;-,-,+,+) \doteq -BQ_f T^a_{mj} T^a_{in}[F(3,2,1,4) - F(4,1,2,3)]. \tag{6.37}$$

To obtain the four-jet cross section, one must sum all the squared absolute values of the helicity amplitudes, perform the color sum, average over the initial lepton helicities, symmetrize appropriately in the four-momenta of the outgoing particles, and sum over the quark flavors. This yields

$$\overline{|M_q|^2} = \frac{2B^2}{3}\left(\sum_f Q_f^2\right)\sum_P\{[6N_f F(1,2,3,4) + F(1,3,2,4)$$

$$+F(4,2,3,1)]F^*(1,2,3,4)\}, \tag{6.38}$$

where N_f is the number of relevant quark flavors. Note that the terms proportional to $(\sum Q_f)^2$ have dropped out after momentum symmetrization.

Including the phase space factors, we thus obtain for the four-jet cross section:

$$d\sigma(4\text{-jet}) = \frac{\delta^4(p_+ + p_- - k_1 - k_2 - k_3 - k_4)}{128E^2(2\pi)^8}$$

$$\times \left[\overline{|M_g|^2} + \overline{|M_q|^2}\right] \frac{d^3\vec{k}_1\, d^3\vec{k}_2\, d^3\vec{k}_3\, d^3\vec{k}_4}{k_{10}\, k_{20}\, k_{30}\, k_{40}}, \qquad (6.39)$$

with $\overline{|M_g|^2}$ and $\overline{|M_q|^2}$ given by eqns (6.30) and (6.38).

This example shows again that, for double bremsstrahlung processes, it is in general much more efficient to evaluate separately the various helicity amplitudes as complex numbers in terms of the components of the four-vectors in the process, rather than having to consider lengthy formulae for the cross section.

One last comment should be made concerning these formulae. Because of the appearance of factors like $(k_3 \cdot k_4)$, k_{1+}, k_{2-}, etc. in the denominators, the expressions look quite singular in collinear configurations, for example, $\vec{k}_3 \| \vec{k}_4$, $\vec{k}_1 \| \vec{p}_-$, $\vec{k}_2 \| \vec{p}_+$. It turns out, however, that also the numerators vanish in these limits, but a numerical evaluation of these expressions without sufficient care could quite possibly result in a loss of numerical precision. Of course, when all jets point in different directions, away from the beam axis, no special care is needed. This draws the attention to the need for a systematic treatment of collinear configurations within the framework of the helicity amplitude method, a topic which is analysed in detail in the coming chapter.

7
Finite mass effects

> The mere neglect
> of a mass effect
> is quite suspect
> and often not correct.

7.1 Their occasional importance

In the previous chapters, we developed a general formalism for calculating multiple bremsstrahlung in gauge theories at high energies. By considering the limit of vanishing fermion masses, we were able to obtain simple expressions for the various helicity amplitudes. For high energy processes, the massless fermion limit is a good approximation unless nearly collinear particle configurations are encountered. In these kinematical situations, finite mass corrections have to be introduced for a correct description of the process.

This can be understood as follows. Consider a photon or a gluon with momentum $\vec{k}$ and a fermion with mass m and momentum $\vec{p}$. If $\vec{k}$ is nearly parallel to $\vec{p}$ such that the scalar product $(p \cdot k)$ is of order m^2, then, when $(p \cdot k)$ appears in the denominator of a cross section formula (see Chapter 4), one must also include terms proportional to $m^2 (p \cdot k)^{-2}$ as they are of the same order of magnitude.

In the preceding chapters, such terms proportional to $m^2 (p \cdot k)^{-2}$ were neglected. Consequently, the formulae could not be used in the case of collinear photon or gluon bremsstrahlung. This point was already emphasized in Section 3.4. In Section 7.2, we first show how the finite mass effects can be treated at the level of the cross section formulae in the case of single bremsstrahlung, and, in Section 7.3, we present an explicit example: $e^+e^- \to \mu^+\mu^-\gamma$. For multiple bremsstrahlung, we find that it is essential to consider the mass corrections to the various spin amplitudes of the process, rather than calculating these effects for the cross section. How this can be done is explained in Section 7.4.

We have also explained in Section 3.4 why the zero fermion mass limit breaks down when collinear fermions are present, for example, for Bhabha scattering in the very forward direction or for collinear muons in $e^+e^- \to \mu^+\mu^-\gamma$. In principle, it should be possible to calculate the finite mass effects for these cases using the techniques of Section 7.4, but, to the best of our knowledge, this analysis has not been carried out systematically. In this chapter, the fermions are assumed to be not collinear.

7.2 Single bremsstrahlung

Following the method of Berends et al. [24], we show how it is possible, in general, to calculate the finite fermion mass effects for single bremsstrahlung at the level of the cross section. Consider a process with an incoming fermion with momentum p. The generalization to incoming antifermions or outgoing (anti)fermions is easy to make. The amplitude for the lower order, nonradiative process can be written in the form

$$M_0 = A(q_i)u(p), \qquad (7.1)$$

where q_i stands for all occurring momenta except p. Because of momentum conservation, we can always write an amplitude in this form.

The nonradiative cross section, summed over all polarizations, is then given by

$$\begin{aligned}|M_0|^2 &= \sum A(q_i)u(p)\,\bar{u}(p)\overline{A}(q_i) \\ &= \sum A(q_i)(\not{p}+m)\overline{A}(q_i),\end{aligned} \qquad (7.2)$$

where m is the mass of the fermion and $\overline{A}(q_i) = \gamma^0 A^\dagger(q_i)$. Again, because of momentum conservation, this expression depends only on q_i:

$$|M_0|^2 = f_0(q_i). \qquad (7.3)$$

We now let the particle corresponding to $u(p)$ radiate a photon. The bremsstrahlung amplitude becomes

$$M_B = \frac{e}{2(p \cdot k)} A(q_i)(\not{p} - \not{k} + m)\,\not{\epsilon}(k) u(p). \qquad (7.4)$$

Note that the momentum p in eqn (7.4) is different from the momentum p in eqn (7.1); in fact, it corresponds to $p+k$. Taking the square of this amplitude and summing over the photon polarizations, we find, after applying some Dirac algebra,

$$\begin{aligned}|M_B|^2 &= -\frac{e^2}{4(p\cdot k)^2} \sum A(q_i)(\not{p}-\not{k}+m)\gamma_\alpha(\not{p}+m)\gamma^\alpha(\not{p}-\not{k}+m)\overline{A}(q_i) \\ &= -\frac{e^2 m^2}{(p\cdot k)^2}\sum A(q_i)(\not{p}-\not{k}+m)\overline{A}(q_i) \\ &\quad + \frac{e^2}{(p\cdot k)}\sum A(q_i)(\not{k}+m)\overline{A}(q_i).\end{aligned} \qquad (7.5)$$

We are only interested in the leading contributions when $m \to 0$ with $(p\cdot k) = \mathcal{O}(m^2)$. Thus, to leading order,

7.2. SINGLE BREMSSTRAHLUNG

$$|M_B|^2 = -\frac{e^2 m^2}{(p \cdot k)^2} \sum A(q_i)(\not{p} - \not{k})\overline{A}(q_i)$$

$$+\frac{e^2}{(p \cdot k)} \sum A(q_i) \not{k}\overline{A}(q_i). \tag{7.6}$$

The first term, the finite mass correction, has a double pole in $(p \cdot k)$ and, being proportional to m^2, it is only relevant when $\vec{k}$ is nearly parallel to $\vec{p}$. The second term, however, is the only one which survives in the limit $m = 0$ and is thus precisely the contribution to the cross section we calculated previously.

Comparing $|M_B|^2$ with $|M_0|^2$, we see that the mass correction is just

$$-\frac{e^2 m^2}{(p \cdot k)^2} f_0(q_i). \tag{7.7}$$

This means that, in general, the double pole terms relevant for collinear photon bremsstrahlung can be obtained from the lower order, nonradiative cross section formula by eliminating the momentum p of the radiating fermion, i.e., using the variables q_i only.

For the case where more charged fermions take part in the process, one has to supplement the zero-mass cross section formula with more finite mass correction terms, one for each external fermion. Every time, one should express f_0 in terms of the momenta not associated with the fermion under consideration.

This procedure for single photon bremsstrahlung is easily generalized to the QCD case [27], the only complication being the color structure of the amplitudes. Summing over polarizations and color, we can write the cross section for the lower order, nonradiative process in the form

$$|M_0|^2 = \sum A(q_i)(\not{p} + m)\overline{A}(q_i) \equiv f_0(q_i), \tag{7.8}$$

where m this time is the mass of the quark with momentum p. If this quark now emits a gluon with momentum k, the bremsstrahlung amplitude becomes

$$M_B = \frac{g}{2(p \cdot k)} A(q_i)(\not{p} - \not{k} + m) \not{\epsilon}(k) T^a_{mn} u(p). \tag{7.9}$$

Using the identity

$$\sum_{a,n} T^a_{mn} T^a_{nk} = \tfrac{4}{3}\delta_{mk}, \tag{7.10}$$

we can simply repeat the above argument. Hence, summing over all polarizations, we find

$$|M_B|^2 = -\frac{4g^2 m^2}{3(p \cdot k)^2} \sum A(q_i)(\not{p} - \not{k} + m)\overline{A}(q_i)$$

$$+\frac{4g^2}{3(p \cdot k)} \sum A(q_i)(\not{k} + m)\overline{A}(q_i), \tag{7.11}$$

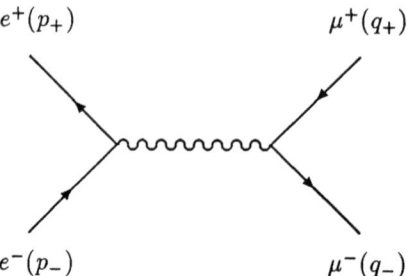

Fig. 7.1: Feynman diagram for $e^+e^- \to \mu^+\mu^-$.

and the finite mass correction term, for $\vec{k}$ nearly parallel to $\vec{p}$, is given by

$$-\frac{4g^2 m^2}{3(p\cdot k)^2} f_0(q_i). \qquad (7.12)$$

In the coming section, we illustrate this procedure with a simple QED example.

7.3 An example: $e^+e^- \to \mu^+\mu^-\gamma$

From the preceding section, we know that in order to calculate the finite fermion mass corrections for $e^+e^- \to \mu^+\mu^-\gamma$, we must first calculate the lower order cross section for the nonradiative process, i.e., $e^+(p_+) + e^-(p_-) \to \mu^+(q_+) + \mu^-(q_-)$. This process is described by the Feynman diagram of Fig. 7.1, and the Feynman amplitude reads

$$M_0 = \frac{e^2}{(p_+ + p_-)^2} \bar{u}(q_-)\gamma^\mu v(q_+) \bar{v}(p_+)\gamma_\mu u(p_-). \qquad (7.13)$$

The only nonzero helicity amplitudes are

$$\begin{aligned}
M(+,-;+,-) &= \frac{e^2}{4(p_+ + p_-)^2} \bar{u}(q_-)\gamma^\mu(1-\gamma_5)v(q_+) \bar{v}(p_+)\gamma_\mu(1-\gamma_5)u(p_-) \\
&\doteq 2e^2 \frac{(p_+ - q_-)^2}{(p_+ + p_-)^2}, \\
M(+,-;-,+) &= \frac{e^2}{4(p_+ + p_-)^2} \bar{u}(q_-)\gamma^\mu(1+\gamma_5)v(q_+) \bar{v}(p_+)\gamma_\mu(1-\gamma_5)u(p_-) \\
&\doteq 2e^2 \frac{(p_+ - q_+)^2}{(p_+ + p_-)^2}, \qquad (7.14)
\end{aligned}$$

7.3. AN EXAMPLE: $e^+e^- \to \mu^+\mu^-\gamma$

and their parity conjugates. Averaging over the initial state polarizations and summing over the final state degrees of freedom, we obtain

$$\overline{|M_0|^2} = 2e^4 \frac{(p_+ - q_+)^4 + (p_+ - q_-)^4}{(p_+ + p_-)^4}. \tag{7.15}$$

For the process $e^+(p_+) + e^-(p_-) \to \mu^+(q_+) + \mu^-(q_-) + \gamma(k)$, we can now consider photon emission in a direction close to $\vec{p}_+$. From Section 7.2, we know that, in order to obtain the finite mass correction in this case, we have to express $\overline{|M_0|^2}$ in terms of p_-, q_+ and q_-:

$$2e^4 \frac{(p_- - q_-)^4 + (p_- - q_+)^4}{(q_+ + q_-)^4}, \tag{7.16}$$

and the mass correction term in the cross section becomes

$$-\frac{2e^6 m^2}{(p_+ \cdot k)^2} \frac{(p_- - q_-)^4 + (p_- - q_+)^4}{(q_+ + q_-)^4}, \tag{7.17}$$

where m is the electron mass.

Repeating this procedure for photon emission close to the $\vec{p}_-$, $\vec{q}_+$ and $\vec{q}_-$ directions, we obtain all the finite mass correction terms:

$$-\frac{2e^6 m^2}{(q_+ + q_-)^4} \left[\frac{(p_- - q_-)^4 + (p_- - q_+)^4}{(p_+ \cdot k)^2} + \frac{(p_+ - q_+)^4 + (p_+ - q_-)^4}{(p_- \cdot k)^2} \right]$$

$$-\frac{2e^6 \mu^2}{(p_+ + p_-)^4} \left[\frac{(p_- - q_-)^4 + (p_+ - q_-)^4}{(q_+ \cdot k)^2} + \frac{(p_+ - q_+)^4 + (p_- - q_+)^4}{(q_- \cdot k)^2} \right]. \tag{7.18}$$

Here, μ denotes the muon mass. Expressing this result in the usual variables

$$s = (p_+ + p_-)^2, \quad t = (p_+ - q_+)^2, \quad u = (p_+ - q_-)^2,$$
$$s' = (q_+ + q_-)^2, \quad t' = (p_- - q_-)^2, \quad u' = (p_- - q_+)^2, \tag{7.19}$$

and adding the zero-mass formula for this process, we obtain the unpolarized squared amplitude to leading order in m^2 and μ^2:

$$\overline{|M_B|^2} = -2e^6 \left\{ \frac{m^2}{s'^2} \left[\frac{t'^2 + u'^2}{(p_+ \cdot k)^2} + \frac{t^2 + u^2}{(p_- \cdot k)^2} \right] + \frac{\mu^2}{s^2} \left[\frac{t'^2 + u^2}{(q_+ \cdot k)^2} + \frac{t^2 + u'^2}{(q_- \cdot k)^2} \right] \right\}$$

$$- e^6 (v_q - v_p)^2 \frac{t^2 + t'^2 + u^2 + u'^2}{ss'}. \tag{7.20}$$

We used the same notation as in Section 4.3, with the four-vectors v_q and v_p defined by eqns (4.27),

$$v_q = \frac{q_+}{(q_+ \cdot k)} - \frac{q_-}{(q_- \cdot k)}, \quad v_p = \frac{p_+}{(p_+ \cdot k)} - \frac{p_-}{(p_- \cdot k)}. \tag{7.21}$$

Eqn (7.20) is valid for all kinematical configurations without nearly collinear fermions.

This procedure can easily be applied to all other QED and QCD single bremsstrahlung processes. All one needs to do is to express the nonradiative cross section formula in the appropriate variables to obtain the mass correction terms and to add the zero mass expression of the radiative process. We refer again to Chapter 9, where one can find a list of cross section formulae including the finite mass corrections.

7.4 Mass corrections for amplitudes

7.4.1 General formalism

The technique for obtaining the finite mass corrections, which was presented in Section 7.2, works nicely for the case of single bremsstrahlung. It is, however, not powerful enough for the case of multiple bremsstrahlung. The reason is simply that, for multiple bremsstrahlung, not just one denominator like $(p \cdot k)$ can be of order m^2, but also denominators of the type $(p+k+k')^2$, where k and k' are photon or gluon momenta. In this section, we present a systematic treatment [35] of the finite mass effects for multiple bremsstrahlung in which an arbitrary number of gauge particles are radiated nearly parallel to fermion directions. We now find that it is essential to consider the mass corrections to the various spin amplitudes of the process, rather than calculating them for the cross section as was done previously. In the zero mass limit, these spin amplitudes reduce to the helicity amplitudes we considered earlier, except for some additional amplitudes which are directly proportional to a fermion mass. For simplicity, we shall present the technique in the case of QED only.

Let us consider any QED process in the tree approximation, in which n photons are emitted in directions nearly parallel to a fermion direction described by the four-momentum p (later we shall consider the more general case where additional photons are emitted in directions parallel to other fermion directions). Let k_i, $i = 1, 2, \ldots, n$, be the four-momenta of these collinear photons, and let m be the fermion mass, i.e., $p^2 = m^2$. There will then be Feynman diagrams describing the process in which fermion propagators will have small denominators:

$$\Delta_i = (p - k_i)^2 - m^2,$$

$$\Delta_{ij} = (p - k_i - k_j)^2 - m^2,$$

$$\vdots$$

$$\Delta_{12\ldots n} = (p - k_1 - k_2 - \cdots - k_n)^2 - m^2. \quad (7.22)$$

We consider the case of nearly collinear photons such that the quantities

7.4. MASS CORRECTIONS FOR AMPLITUDES

$(p \cdot k_i)$ and $(k_i \cdot k_j)$ are of order m^2. In this case, all the quantities Δ are also of order m^2. It is clear, therefore, that even in the high energy limit the fermion mass must be taken into account.

By introducing a generally positioned light-like vector q, we can write down the following representation for the polarization vectors of the photons:

$$\rlap{/}{\epsilon}_i^{\lambda_i} = N_i[\rlap{/}{k}_i \rlap{/}{p} \rlap{/}{q}(1 + \lambda_i \gamma_5) + \rlap{/}{q} \rlap{/}{p} \rlap{/}{k}_i(1 - \lambda_i \gamma_5)],$$

$$N_i^{-2} = 16(p \cdot q)(p \cdot k_i)(q \cdot k_i) - 8m^2(q \cdot k_i)^2,$$

$$\lambda_i = \pm 1, \qquad\qquad i = 1, 2, \ldots, n. \qquad (7.23)$$

Note that all components of $\epsilon_i^{\lambda_i}$ are of order 1, even in the collinear limit. Also, the representation for $\rlap{/}{\epsilon}_i^{\lambda_i}$ differs from the effective one, eqn (3.8), which was used earlier. This is due to the fact that, for massive fermions, we do not have conservation of axial current, and, consequently, we cannot omit the $\rlap{/}{k}_i \gamma_5$ terms in $\rlap{/}{\epsilon}_i^{\lambda_i}$. However, for $m = 0$, the two representations are gauge equivalent.

With the present choice for $\rlap{/}{\epsilon}_i^{\lambda_i}$, we now show that, in the collinear situation, the amplitudes are at most of order m^{-n}. To this order in m, the contributions to the amplitude will come only from the Feynman diagrams for which the collinear photons are attached directly next to the external fermion with momentum p. This follows from the fact that only these diagrams have all the denominators Δ.

Consider the case that the photons are collinear with an incoming fermion with momentum p (the remaining cases of an incoming antifermion or an outgoing (anti)fermion can be treated in the same way). The diagrams with the collinear photons close to the spinor $u(p)$ contain the expression

$$A_n = \frac{\rlap{/}{p} - \rlap{/}{k}_1 - \ldots - \rlap{/}{k}_n + m}{\Delta_{12\ldots n}} \rlap{/}{\epsilon}_n^{\lambda_n} \ldots$$

$$\times \rlap{/}{\epsilon}_3^{\lambda_3} \frac{\rlap{/}{p} - \rlap{/}{k}_1 - \rlap{/}{k}_2 + m}{\Delta_{12}} \rlap{/}{\epsilon}_2^{\lambda_2} \frac{\rlap{/}{p} - \rlap{/}{k}_1 + m}{\Delta_1} \rlap{/}{\epsilon}_1^{\lambda_1} u(p)$$

$$+(n! - 1) \text{ other permutations of } (1, 2, \ldots, n). \qquad (7.24)$$

But,

$$A_1 = \frac{\rlap{/}{p} - \rlap{/}{k}_1 + m}{\Delta_1} \rlap{/}{\epsilon}_1^{\lambda_1} u(p)$$

$$= \frac{1}{\Delta_1} \left[2(p \cdot \epsilon_1^{\lambda_1}) - \rlap{/}{k}_1 \rlap{/}{\epsilon}_1^{\lambda_1} \right] u(p), \qquad (7.25)$$

and, with our choice, eqns (7.23), for $\epsilon_i^{\lambda_i}$, we see that $(p \cdot \epsilon_i^{\lambda_i}) = \mathcal{O}(m)$. Similarly, one can show that $\rlap{/}{k}_i u(p) = \mathcal{O}(m)$. To this end, we introduce two additional four-vectors t_1 and t_2, defined by

$$(p \cdot t_1) = (q \cdot t_1) = 0, \qquad t_1^2 = -1,$$

$$t_{2\mu} = \epsilon_{\alpha\beta\gamma\mu} q^\alpha p^\beta t_1^\gamma /(p \cdot q), \qquad (7.26)$$

where the sign of $\epsilon_{\alpha\beta\gamma\mu}$ is defined by $\epsilon_{0123} = +1$ as in eqn (4.55). It follows that

$$(p \cdot t_2) = (q \cdot t_2) = (t_1 \cdot t_2) = 0, \qquad t_2^2 = -1. \qquad (7.27)$$

As the vectors p, q, t_1 and t_2 are linearly independent, we can decompose k_i, $i = 1, \ldots, n$, as follows:

$$k_i^\mu = \frac{(q \cdot k_i)}{(q \cdot p)} p^\mu + \frac{1}{(q \cdot p)}\left[(p \cdot k_i) - m^2 \frac{(q \cdot k_i)}{(q \cdot p)}\right] q^\mu - (k_i \cdot t_1) t_1^\mu - (k_i \cdot t_2) t_2^\mu. \qquad (7.28)$$

This relation, together with the fact that $k_i^2 = 0$, leads immediately to the result

$$(k_i \cdot t_1) = \mathcal{O}(m), \qquad (k_i \cdot t_2) = \mathcal{O}(m). \qquad (7.29)$$

From the decomposition of k_i^μ, it then follows that $\not{k}_i u(p) = \mathcal{O}(m)$. Hence, the entire expression A_1 is of order m^{-1}. It also follows that

$$A_2 = \frac{1}{\Delta_{12}\Delta_1}(\not{p} - \not{k}_1 - \not{k}_2 + m)\not{\epsilon}_2^{\lambda_2}(\not{p} - \not{k}_1 + m)\not{\epsilon}_1^{\lambda_1} u(p)$$

$$= \frac{1}{\Delta_{12}}\left\{[2(p - k_1 \cdot \epsilon_2^{\lambda_2}) - \not{k}_2 \not{\epsilon}_2^{\lambda_2}] F_1(k_1, \lambda_1) - \not{\epsilon}_2^{\lambda_2} \not{\epsilon}_1^{\lambda_1}\right\} u(p), \qquad (7.30)$$

where we introduced the notation

$$F_1(k_1, \lambda_1) = \frac{1}{\Delta_1}[2(p \cdot \epsilon_1^{\lambda_1}) - \not{k}_1 \not{\epsilon}_1^{\lambda_1}]. \qquad (7.31)$$

Because of the symmetrization between photons 1 and 2 contained in the expression for A, we can effectively replace $\not{\epsilon}_2^{\lambda_2} \not{\epsilon}_1^{\lambda_1}$ by $(\epsilon_2^{\lambda_2} \cdot \epsilon_1^{\lambda_1})$ in A_2. Furthermore,

$$\not{k}_2 \not{\epsilon}_2^{\lambda_2} \not{k}_1 \not{\epsilon}_1^{\lambda_1} u(p)$$

$$= 2 N_1 N_2 \delta_{\lambda_1 \lambda_2} \not{k}_2 \not{A} \not{p} \not{k}_2 \not{k}_1 \not{A} \not{p} \not{k}_1 (1 - \lambda_1 \gamma_5) u(p)$$

$$= N_1 N_2 \delta_{\lambda_1 \lambda_2} \mathrm{Tr}[\not{A} \not{p} \not{k}_2 \not{k}_1 (1 + \lambda_2 \gamma_5)] \not{k}_2 \not{A} \not{p} \not{k}_1 (1 - \lambda_1 \gamma_5) u(p)$$

$$= 2 N_1 \delta_{\lambda_1 \lambda_2} (k_1 \cdot \epsilon_2^{\lambda_2}) \not{k}_2 \not{A} \not{p} \not{k}_1 (1 - \lambda_1 \gamma_5) u(p). \qquad (7.32)$$

7.4. MASS CORRECTIONS FOR AMPLITUDES

This time, one easily verifies that $(k_i \cdot \epsilon_j^{\lambda_j}) = \mathcal{O}(m)$, and that $\not{p}\not{k_i}u(p) = \mathcal{O}(m^2)$. Hence, the whole preceding expression is of order m^2. It follows that A_2 is of order m^{-2} and that it can effectively be replaced by

$$F_2(k_1, \lambda_1; k_2, \lambda_2)u(p)$$

$$= \frac{1}{\Delta_{12}}\left\{[2(p - k_1 \cdot \epsilon_2^{\lambda_2}) - \not{k_2}\not{\epsilon_2^{\lambda_2}}]F_1(k_1, \lambda_1) - (\epsilon_1^{\lambda_1} \cdot \epsilon_2^{\lambda_2})\right\}u(p). \quad (7.33)$$

This procedure can now be continued for the remaining photons. We obtain the result that

$$A_n = F_n(k_1, \lambda_1; \ldots; k_n, \lambda_n)u(p) \quad (7.34)$$

is of order m^{-n} and that F_n is given by the following recursion relation:

$$F_n(k_1, \lambda_1; \ldots; k_n, \lambda_n)$$

$$= \frac{1}{\Delta_{1\ldots n}}\left\{[2(p - k_1 - \ldots - k_{n-1} \cdot \epsilon_n^{\lambda_n}) - \not{k_n}\not{\epsilon_n^{\lambda_n}}]\right.$$

$$\times F_{n-1}(k_1, \lambda_1; \ldots; k_{n-1}, \lambda_{n-1})$$

$$\left. - (\epsilon_n^{\lambda_n} \cdot \epsilon_{n-1}^{\lambda_{n-1}})F_{n-2}(k_1, \lambda_1; \ldots; k_{n-2}, \lambda_{n-2})\right\}, \quad (7.35)$$

with $F_0 = 1$. Once F_1 and F_2 are known, it is thus possible to evaluate the higher order functions F_n. A very useful property for this purpose is the relation

$$\not{k_n}\not{\epsilon_n^{\lambda_n}}\not{k_{n-1}}\not{\epsilon_{n-1}^{\lambda_{n-1}}}\ldots\not{k_1}\not{\epsilon_1^{\lambda_1}}u(p)$$

$$= 2^{n-1}N_1\delta_{\lambda_n\lambda_1}\delta_{\lambda_{n-1}\lambda_1}\ldots\delta_{\lambda_2\lambda_1}(k_{n-1} \cdot \epsilon_n^{\lambda_n})$$

$$\times (k_{n-2} \cdot \epsilon_{n-1}^{\lambda_{n-1}})\ldots(k_1 \cdot \epsilon_2^{\lambda_2})\not{k_n}\not{A}\not{p}\not{k_1}(1 - \lambda_1\gamma_5)u(p). \quad (7.36)$$

The remaining part of the diagram, involving the spinor structure A can now be treated in the massless limit. This means that if we fix the helicities for the remaining fermions, the terms of the above type will give zero whenever the fermion helicity operators $\frac{1}{2}(1 \pm \gamma_5)$ kill the $(1 - \lambda_1\gamma_5)$ factor. Of course, this term already vanishes when some of the collinear photons have different helicities. Furthermore, the relation

$$(\epsilon_i^\lambda \cdot \epsilon_j^\lambda) = 0, \quad (7.37)$$

valid for all i and j, can often be used to eliminate the last term in the expression for F_n.

The other diagrams, which do not have all the collinear photons next to $u(p)$, are necessarily smaller by at least one power of m. This is due to the fact that as soon as an acollinear photon is inserted in the string of collinear photons all the fermion propagators from that point on are of order 1 instead of being of order m^{-1}.

To summarize, we can say that, with our choice of representation for $\not{\epsilon}_i^{\lambda_i}$, the amplitudes in the collinear situation are at most of order m^{-n} if there are n collinear photons. To this order in m, the relevant Feynman diagrams are those in which the collinear photons are attached immediately next to the external fermion which determines the collinear direction. These Feynman diagrams can easily be evaluated using the function F_n, while the rest of the diagram can be treated in the massless fermion limit. The evaluation of F_n is further simplified using the last two relations (7.36) and (7.37).

This analysis also shows that when all $(p \cdot k_i)$ are small, but much larger than m^2, the same set of diagrams is the only relevant one. In this case, all manipulations in the numerator of the Feynman diagrams can be done in the massless limit, all finite mass effects being of subleading order.

We shall now work out in detail the special cases of single and double collinear bremsstrahlung.

7.4.2 Single bremsstrahlung

In this section, we present the relevant formulae for the case of one photon being emitted in a collinear configuration. Suppose that this photon has a momentum $\vec{k}$ nearly parallel to the momentum $\vec{p}$ of an incoming fermion. Let λ and λ_p be their helicities. We already know that the relevant Feynman diagrams are those where $\not{\epsilon}^\lambda$ stands next to $u(p)$. On the other end of the fermion line stands another fermion spinor, which can be taken to be massless. If we calculate the helicity amplitudes of the process, this spinor will produce a helicity projection operator $\frac{1}{2}(1 + \lambda' \gamma_5)$, with $\lambda' = \pm 1$, multiplying from the left the expression A_n given in eqn (7.24). Since one photon is radiated, we need to calculate $F_1(k, \lambda)$. However, $F_1(k, \lambda)$ appears only in the combination

$$\tfrac{1}{2}(1 + \lambda'\gamma_5) F_1(k, \lambda) u_{\lambda_p}(p), \qquad (7.38)$$

where $u_{\lambda_p}(p)$ is the fermion helicity state with helicity λ_p. It is therefore convenient to define the 'collinearity factor' $F_1(\lambda; \lambda', \lambda_p)$ by

$$(1 + \lambda'\gamma_5) F_1(k, \lambda) u_{\lambda_p}(p) = (1 + \lambda'\gamma_5) F_1(\lambda; \lambda', \lambda_p) u_{\lambda'}(p) + \mathcal{O}(1). \quad (7.39)$$

Two comments should be made about this formula. Firstly, it does not define $F_1(\lambda; \lambda', \lambda_p)$ uniquely. We shall see that $F_1(\lambda; \lambda', \lambda_p)$ is of order m^{-1}. As we neglect higher order terms in m, we can choose the simplest form for $F_1(\lambda; \lambda', \lambda_p)$. Secondly, we shall also see that, to leading order in m, the effect

7.4. MASS CORRECTIONS FOR AMPLITUDES

of $(1 + \lambda' \gamma_5) F_1(k, \lambda)$ operating on $u_{\lambda_p}(p)$ is to retain the helicity assignment of $u_{\lambda_p}(p)$ for the case $\lambda' = \lambda_p$ and to reverse the helicity of $u_{\lambda_p}(p)$ in the case $\lambda' = -\lambda_p$. In both cases, the radiation of the photon is described by the 'collinearity factor' times a lower order amplitude for which the helicity of the fermion is taken to be λ'.

Suppose we consider first the case $\lambda' = \lambda$. As

$$\not{k} \not{\epsilon}^\lambda = N \not{k} \not{q} \not{p} \not{k} (1 - \lambda \gamma_5),$$

$$N^{-2} = 16(p \cdot q)(p \cdot k)(q \cdot k) - 8m^2 (q \cdot k)^2, \tag{7.40}$$

and

$$F_1(k, \lambda) = \frac{-1}{2(p \cdot k)} \left[2(p \cdot \epsilon^\lambda) - \not{k} \not{\epsilon}^\lambda \right], \tag{7.41}$$

we find that the last term in F_1 does not contribute as it is multiplied from the left by $(1 + \lambda' \gamma_5)$ and $\lambda' = \lambda$. Hence, for all λ,

$$F_1(\lambda; \lambda, \lambda_p) = -\frac{(p \cdot \epsilon^\lambda)}{(p \cdot k)} \delta_{\lambda \lambda_p}$$

$$= -[4N(p \cdot k)(q \cdot k)]^{-1} \delta_{\lambda \lambda_p}. \tag{7.42}$$

To deal with the other helicity combinations, we first establish the following result:

$$[(k \cdot t_1) \not{t}_1 + (k \cdot t_2) \not{t}_2](1 \pm \gamma_5) u(p)$$

$$= \pm e^{\pm i \alpha} (k \cdot t_1 \mp i t_2)(1 \mp \gamma_5) \tilde{u}(p) + \mathcal{O}(m^2), \tag{7.43}$$

where

$$e^{i\alpha} = \frac{t_{1+}}{p_+} p_\perp - t_{1\perp}, \tag{7.44}$$

and where $\tilde{u}(p)$ is the spinor which has its spin flipped compared to $u(p)$. What we mean more precisely with this statement will be made clear after eqn (7.47).

Indeed, from the definition (7.26) of t_2 and the identity (3.6), it follows that

$$\not{t}_2 = -i \gamma_5 \not{t}_1 (\not{q} \not{p} - \not{p} \not{q})/2(p \cdot q), \tag{7.45}$$

and

$$i \not{t}_2 (1 \pm \gamma_5) u(p) = \mp \frac{1}{2(p \cdot q)} (\not{q} \not{p} \not{t}_1 - \not{t}_1 \not{p} \not{q})(1 \pm \gamma_5) u(p)$$

$$= \pm \not{t}_1 (1 \pm \gamma_5) u(p) + \mathcal{O}(m). \tag{7.46}$$

But then,

$$[(k \cdot t_1)\,\rlap{/}t_1 + (k \cdot t_2)\,\rlap{/}t_2](1 \pm \gamma_5)u(p)$$

$$= (k \cdot t_1 \mp it_2)\,\rlap{/}t_1(1 \pm \gamma_5)u(p) + \mathcal{O}(m^2). \tag{7.47}$$

As $(k \cdot t_1 \mp it_2) = \mathcal{O}(m)$ and terms of order $\mathcal{O}(m^2)$ are neglected, we can also neglect terms of order m in $\rlap{/}t_1(1 \pm \gamma_5)u(p)$, which means that it can be taken in the $m = 0$ limit. On the other hand, the form $\rlap{/}t_1(1 \pm \gamma_5)u(p)$ is a solution of the Dirac equation and an eigenvector of $(1 \mp \gamma_5)$. It must therefore be proportional to $(1 \mp \gamma_5)\tilde{u}(p)$, where $\tilde{u}(p)$ is obtained from the spinor $u(p)$ by flipping the helicity. Taking into account our conventions (6.10) and (6.11) for the fermion helicity states, the phase factor $e^{i\alpha}$ is then completely determined and given by eqn (7.44). This establishes the quoted result, eqn (7.43).

Next, we take $\lambda' = -\lambda$ and $\lambda_p = -\lambda$. The expression $F_1(k, \lambda)$ is now sandwiched between the projection operators $\frac{1}{2}(1 - \lambda\gamma_5)$. Using the decomposition of $\rlap{/}k$, eqn (7.28), we see that the terms proportional to $\rlap{/}t_1$ and $\rlap{/}t_2$ do not contribute. Hence, to leading order,

$$F_1(\lambda; -\lambda, -\lambda)$$

$$\simeq \frac{-1}{2(p \cdot k)}\left\{2(p \cdot \epsilon^\lambda) - 2N\,\rlap{/}k\,\rlap{/}A\,\rlap{/}p\left[\frac{(q \cdot k)}{(q \cdot p)}\,\rlap{/}p\right.\right.$$

$$\left.\left. + \frac{1}{(q \cdot p)}\left((p \cdot k) - m^2\frac{(q \cdot k)}{(q \cdot p)}\right)\rlap{/}A\right]\right\}$$

$$\simeq \frac{-1}{(p \cdot k)}\left\{(p \cdot \epsilon^\lambda) - N\left[m^2\frac{(q \cdot k)}{(q \cdot p)}\,\rlap{/}k\,\rlap{/}A + 2\,\rlap{/}k\,\rlap{/}A\bigl((p \cdot k)\right.\right.$$

$$\left.\left. - m^2\frac{(q \cdot k)}{(q \cdot p)}\bigr)\right]\right\}$$

$$\simeq \frac{-1}{(p \cdot k)}\left\{(p \cdot \epsilon^\lambda) - 2N(q \cdot k)\left[2(p \cdot k) - m^2\frac{(q \cdot k)}{(q \cdot p)}\right]\right\}$$

$$\simeq -\frac{(q \cdot p - k)}{4N(p \cdot q)(p \cdot k)(q \cdot k)}. \tag{7.48}$$

Finally, for $\lambda' = -\lambda$ and $\lambda_p = \lambda$, the $(p \cdot \epsilon^\lambda)$ term does not contribute, and, in the decomposition of $\rlap{/}k$, only the terms proportional to $\rlap{/}t_1$ and $\rlap{/}t_2$ give nonvanishing contributions. They have, however, the effect of flipping the helicity of the spinor $u(p)$. Thus, by eqn (7.43),

7.4. MASS CORRECTIONS FOR AMPLITUDES

$$F_1(+;-,+) = -\frac{Ne^{i\alpha}(k \cdot t_1 - it_2)}{(p \cdot k)} \not{k} \not{A} \not{p}$$
$$= -\frac{2mNe^{i\alpha}(k \cdot t_1 - it_2)(q \cdot k)}{(p \cdot k)}, \tag{7.49}$$

and

$$F_1(-;+,-) = \frac{2mNe^{-i\alpha}(k \cdot t_1 + it_2)(q \cdot k)}{(p \cdot k)}. \tag{7.50}$$

In all these derivations, one should keep in mind that the quantities F_1 stand next to the spinor $u(p)$ and that $\not{k}u(p) = \mathcal{O}(m)$. Also note the relation

$$F_1(-\lambda; -\lambda', -\lambda_p) = [F_1(\lambda; \lambda', \lambda_p)]^\star, \tag{7.51}$$

where the superscript $\star$ denotes complex conjugation and the replacement $m \to -m$.

To summarize the single collinear bremsstrahlung case, we find that all helicity amplitudes reduce to a product of a collinearity factor F_1 times an amplitude for the process in which the collinear photon is removed. For the 'allowed' amplitudes, $\lambda_p = \lambda'$, this lower order amplitude retains the helicities of the fermions, but for the 'forbidden' amplitudes, $\lambda_p = -\lambda'$, the helicity of the spinor $u(p)$ is flipped.

7.4.3 An example: $e^+e^- \to \mu^+\mu^-\gamma$

We now illustrate the general techniques of the preceding section for the case of

$$e^+(p_+) + e^-(p_-) \to \mu^+(q_+) + \mu^-(q_-) + \gamma(k), \tag{7.52}$$

where $\vec{k}$ is nearly collinear to $\vec{p}_-$, i.e.,

$$(p_- \cdot k) = \mathcal{O}(m^2), \qquad k_+ = \mathcal{O}(m^2), \qquad k_\perp = \mathcal{O}(m), \tag{7.53}$$

where, as usual, we have gone to the e^+e^- c.m. frame with the z-axis along $\vec{p}_+$ and $k_\pm = k_0 \pm k_z$, $k_\perp = k_x + ik_y$.

We know that in this limit we only have to consider one Feynman amplitude, viz.,

$$M = -\frac{e^3}{s'} \bar{v}(p_+)\gamma_\mu \frac{\not{p}_- - \not{k} + m}{-2(p_- \cdot k)} \not{k} u(p_-) \bar{u}(q_-)\gamma^\mu v(q_+), \tag{7.54}$$

with

$$s = (p_+ + p_-)^2, \qquad t = (p_+ - q_+)^2, \qquad u = (p_+ - q_-)^2,$$
$$s' = (q_+ + q_-)^2, \qquad t' = (p_- - q_-)^2, \qquad u' = (p_- - q_+)^2. \tag{7.55}$$

We also know that the spin amplitudes in the collinear limit are given by

$$M(+,-;+,-,+) = -eF_1(+;-,-)M_0(+,-;+,-),$$

$$M(+,-;-,+,+) = -eF_1(+;-,-)M_0(+,-;-,+),$$

$$M(-,+;+,-,+) = -eF_1(+;+,+)M_0(-,+;+,-),$$

$$M(-,+;-,+,+) = -eF_1(+;+,+)M_0(-,+;-,+),$$

$$M(+,+;+,-,+) = -eF_1(+;-,+)M_0(+,-;+,-),$$

$$M(+,+;-,+,+) = -eF_1(+;-,+)M_0(+,-;-,+),$$

$$M(-,-;-,+,+) = -eF_1(+;+,-)M_0(-,+;-,+),$$

$$M(-,-;+,-,+) = -eF_1(+;+,-)M_0(-,+;+,-), \qquad (7.56)$$

where the quantities M_0 are obtained from the spin amplitudes M by removing the photon emission parts. Note that in this case the first argument of F_1 is the last argument of M, the second one of F_1 is the second one of M_0, while the third one of F_1 is the second one of M.

These quantities M_0 are the massless helicity amplitudes for $e^+e^- \to \mu^+\mu^-$. Here,

$$M_0(-\lambda,\lambda;-\lambda',\lambda') = \frac{e^2}{s'}\bar{v}(p_+)\gamma_\mu\frac{1+\lambda\gamma_5}{2}u(p_-)\bar{u}(q_-)\gamma^\mu\frac{1+\lambda'\gamma_5}{2}v(q_+), \qquad (7.57)$$

as all other helicity combinations vanish. They are easily evaluated, yielding

$$M_0(+,-;+,-) \doteq 2e^2(uu')^{\frac{1}{2}}/s',$$

$$M_0(+,-;-,+) \doteq 2e^2(tt')^{\frac{1}{2}}/s',$$

$$M_0(-,+;+,-) \doteq 2e^2(tt')^{\frac{1}{2}}/s',$$

$$M_0(-,+;-,+) \doteq 2e^2(uu')^{\frac{1}{2}}/s'. \qquad (7.58)$$

The amplitudes for negative helicity photon emission can then be obtained by applying a parity conjugation, together with eqn (7.51).

For the evaluation of the collinearity factors F_1, we have $p = p_-$ and we can take $q = p_+$. If E denotes the beam energy, we find that the normaliza-

7.4. MASS CORRECTIONS FOR AMPLITUDES

tion factor for ℓ^+, to leading order in m^2, is given by

$$N^{-2} = 32E^4 k_+ k_-, \tag{7.59}$$

as

$$(p_+ \cdot k) = Ek_0 - k_z\sqrt{E^2 - m^2} = Ek_- + \mathcal{O}(m^2),$$

$$(p_- \cdot k) = Ek_0 + k_z\sqrt{E^2 - m^2} = Ek_+ + \frac{m^2 k_-}{4E} + \mathcal{O}(m^4). \tag{7.60}$$

As a result, we have effectively

$$F_1(+;+,+) = -\frac{8NE^3 k_+}{(p_- \cdot k)},$$

$$F_1(+;-,-) = -\frac{4NE^2 k_+(2E - k_-)}{(p_- \cdot k)},$$

$$F_1(+;-,+) = -\frac{2NmEk_- k_\perp^*}{(p_- \cdot k)},$$

$$F_1(+;+,-) = 0, \tag{7.61}$$

and

$$|F_1(+;+,+)|^2 + |F_1(+;-,-)|^2 + |F_1(+;-,+)|^2 + |F_1(+;+,-)|^2$$

$$= \frac{k_+}{2k_-(p_- \cdot k)^2}\left[(2E)^2 + (2E - k_-)^2 + \frac{m^2 k_-^3}{4E^2 k_+}\right]. \tag{7.62}$$

This formula is useful for calculating the unpolarized cross section in the case $\vec{k}$ nearly parallel to $\vec{p}_-$. Using the fact that, in this collinear limit, $2Es' \simeq s(2E - k_-)$, we have

$$\overline{|M|^2} = 2e^6 F(k, p_-)\frac{tt' + uu'}{ss'}, \tag{7.63}$$

with

$$F(k, p_-) = \frac{Ek_+}{(2E - k_-)k_-(p_- \cdot k)^2}\left[(2E)^2 + (2E - k_-)^2 + \frac{m^2 k_-^3}{4E^2 k_+}\right]. \tag{7.64}$$

To obtain this result, we added the contributions for negative helicity photon emission and averaged over the initial state polarizations. We also used the fact that, for $k^2 = 0$,

$$|k_\perp|^2 = k_+ k_-. \tag{7.65}$$

If one wants the analogous formula for $\vec{k}$ along $\vec{p}_+$, one can proceed in exactly the same way. All one has to do is to interchange the roles of p_+ and p_-, which implies $k_+ \leftrightarrow k_-$. Hence, in this limit,

$$\overline{|M|^2} = 2e^6 F(k, p_+) \frac{tt' + uu'}{ss'}, \tag{7.66}$$

with

$$F(k, p_+) = \frac{Ek_-}{(2E - k_+)k_+(p_+ \cdot k)^2} \left[(2E)^2 + (2E - k_+)^2 + \frac{m^2 k_+^3}{4E^2 k_-} \right]. \tag{7.67}$$

For $\vec{k}$ along $\vec{q}_+$, the collinearity factor F is most easily evaluated in the frame for which $\vec{q}_+$ determines the positive z-axis. To distinguish the components of the four-vectors in this frame from the ones in the e^+e^- c.m. frame, we add primes to them. Thus, $k'_\pm = k_0 - k'_z$, etc., where k'_z is the component of $\vec{k}$ in the direction of $\vec{q}_+$. Denoting the muon mass by μ, we then have

$$F(k, q_+) = \frac{q_{+0} k'_-}{(2q_{+0} + k'_+)k'_+(q_+ \cdot k)^2} \left[(2q_{+0})^2 + (2q_{+0} + k'_+)^2 + \frac{\mu^2 k'^3_+}{4q_{+0}^2 k'_-} \right], \tag{7.68}$$

and

$$\overline{|M|^2} = 2e^6 F(k, q_+) \frac{tt' + uu'}{ss'}. \tag{7.69}$$

Similarly, for $\vec{k}$ along $\vec{q}_-$, we have

$$\overline{|M|^2} = 2e^6 F(k, q_-) \frac{tt' + uu'}{ss'}, \tag{7.70}$$

with

$$F(k, q_-) = \frac{q_{-0} k''_-}{(2q_{-0} + k''_+)k''_+(q_- \cdot k)^2} \left[(2q_{-0})^2 + (2q_{-0} + k''_+)^2 + \frac{\mu^2 k''^3_+}{4q_{-0}^2 k''_-} \right], \tag{7.71}$$

where q_{-0} denotes the μ^- energy, and the double primes refer to a frame in which $\vec{q}_-$ determines the positive z-axis.

The choice of these rotated frames to describe collinear bremsstrahlung along $\vec{q}_+$ and $\vec{q}_-$ may seem somewhat awkward. Their main advantage is that they clearly exhibit the appropriate orders of μ^2 which have to be considered. There is, however, another reason why they are useful in practice. Here, we treated the case of $e^+e^- \to \mu^+\mu^-\gamma$ for the sake of simplicity, but one should not forget that, in experimental studies of jet phenomena, one cannot always separate a nearly collinear gluon-quark pair into two jets. Hence, one may want to integrate over the gluon variables in these collinear regions. For that purpose, the primed variables are much more convenient as the experimental cuts are readily translated in these variables.

7.4. MASS CORRECTIONS FOR AMPLITUDES

All other single bremsstrahlung processes can be treated in the same way. The same collinearity factors occur in the expressions for the cross sections and only the lower order, nonradiative part has to be adjusted for each process. This is a sufficiently straightforward matter and we simply refer to Chapter 9 for the explicit expressions of the cross sections in the various collinear configurations.

7.4.4 Double collinear bremsstrahlung

When two collinear photons are emitted, it may happen that they are both nearly parallel to the same fermion direction, or that each of them is nearly parallel to a different direction.

In the first case, we assume that $\vec{k}_1$ and $\vec{k}_2$ are close to the direction of $\vec{p}$, the momentum of the incoming fermion. We now have to evaluate $F_2(k_1, \lambda_1; k_2, \lambda_2)$ of Section 7.4.1 for the various helicity configurations. The calculation proceeds exactly as in the single collinear bremsstrahlung case. In complete analogy with eqn (7.39), we can define a collinearity factor $F_2(\lambda_1, \lambda_2; \lambda', \lambda_p)$ by

$$(1 + \lambda'\gamma_5)F_2(k_1, \lambda_1; k_2, \lambda_2)u_{\lambda_p}(p)$$

$$= (1 + \lambda'\gamma_5)F_2(\lambda_1, \lambda_2; \lambda', \lambda_p)u_{\lambda'}(p) + \mathcal{O}(m^{-1}), \quad (7.72)$$

where λ' is the signature of the helicity projection operator which is associated with the fermion spinor on the other end of the fermion line which connects with $u_{\lambda_p}(p)$.

First of all, note that the relation

$$F_2(\lambda_1, \lambda_2; \lambda', \lambda_p) = [F_2(-\lambda_1, -\lambda_2; -\lambda', -\lambda_p)]^\star \quad (7.73)$$

holds, where as before the symbol $\star$ denotes complex conjugation and the replacement $m \to -m$. This relation is proven by replacing γ_5 by $-\gamma_5$.

We proceed to list the results. For the allowed amplitudes, $\lambda' = \lambda_p$, we have

$$F_2(+,+;+,+) = \frac{1}{\Delta_{12}\Delta_1} \frac{N_2}{2N_1(q \cdot k_1)} \text{Tr}[(\not{p} - \not{k}_1) \not{k}_2 \not{p} \not{q}(1 + \gamma_5)],$$

$$F_2(+,+;-,-) = \frac{1}{\Delta_{12}\Delta_1} \frac{N_2(p - k_1 - k_2 \cdot q)}{2N_1(q \cdot k_1)(p \cdot q)} \text{Tr}[(\not{p} - \not{k}_1) \not{k}_2 \not{p} \not{q}(1 + \gamma_5)],$$

$$F_2(+,-;+,+) = \frac{-1}{\Delta_{12}\Delta_1} \frac{1}{2N_1(q \cdot k_1)} \left\{ \frac{1}{2N_2(p \cdot q)} \right.$$

$$\left. - N_2 \text{Tr}[(\not{p} - \not{k}_1) \not{k}_2 \not{p} \not{q}(1 - \gamma_5)] \right\},$$

$$F_2(+,-;-,-) = \frac{1}{\Delta_{12}\Delta_1} \frac{N_2(p-k_1\cdot q)}{2N_1(q\cdot k_1)(p\cdot q)} \text{Tr}[(\not{p}-\not{k}_1)\not{k}_2\not{p}\not{A}(1-\gamma_5)]$$

$$-\frac{N_2}{\Delta_{12}}\left\{2N_1(p\cdot q)\text{Tr}[\not{k}_1\not{p}\not{A}\not{k}_2(1-\gamma_5)] - \frac{(q\cdot k_2)}{N_1(q\cdot k_1)}\right\}.$$
(7.74)

In these formulae, N_1 and N_2 are the normalization coefficients (7.23) for the photon polarization vectors, expressed in terms of k_i, p and the generally positioned four-vector q. Furthermore,

$$\Delta_1 = (p-k_1)^2 - m^2, \qquad \Delta_{12} = (p-k_1-k_2)^2 - m^2. \qquad (7.75)$$

The allowed helicity amplitudes are now given by a sum of expressions $F_2(\lambda_1,\lambda_2;\lambda',\lambda) + F_2(\lambda_2,\lambda_1;\lambda',\lambda)$ times the helicity amplitude for the massless lower order amplitude in which the two photons are removed, while all other particles retain their helicity assignments.

For the forbidden amplitudes, $\lambda' = -\lambda_p$, we have

$$F_2(+,+;+,-) = 0,$$

$$F_2(+,+;-,+) = \frac{2me^{i\alpha}}{\Delta_{12}\Delta_1}\Big\{2N_1N_2(q\cdot k_1+k_2)(k_1\cdot t_1-it_2)$$

$$\times \text{Tr}[\not{k}_1\not{k}_2\not{p}\not{A}(1+\gamma_5)]$$

$$-\frac{N_1(q\cdot k_1)}{N_2(q\cdot k_2)}(k_1\cdot t_1-it_2) - \frac{N_2(q\cdot k_2)}{N_1(q\cdot k_1)}(k_2\cdot t_1-it_2)\Big\},$$

$$F_2(+,-;+,-) = \frac{2me^{-i\alpha}}{\Delta_{12}\Delta_1}\frac{N_2(q\cdot k_2)}{N_1(q\cdot k_1)}(k_2\cdot t_1+it_2),$$

$$F_2(+,-;-,+) = -\frac{4me^{i\alpha}}{\Delta_{12}\Delta_1}N_1N_2(q\cdot k_1)(k_1\cdot t_1-it_2)$$

$$\times \text{Tr}[(\not{p}-\not{k}_1)\not{k}_2\not{p}\not{A}(1-\gamma_5)], \qquad (7.76)$$

where $e^{i\alpha}$ is given by eqn (7.44).

The forbidden helicity amplitudes are given by the sum of expressions $F_2(\lambda_1,\lambda_2;\lambda',-\lambda') + F_2(\lambda_2,\lambda_1;\lambda',-\lambda')$ times the lower order amplitude without the collinear photons, but with the flipped helicity for $u(p)$.

For the case where two photons are nearly parallel to two different fermion directions, it becomes cumbersome to give general formulae analogous to the above ones. The reason is that a given spin amplitude can become a linear combination of two different lower order amplitudes, one for which the fermion helicities are unchanged and one for which the two fermions,

7.4. MASS CORRECTIONS FOR AMPLITUDES

that specify the parallel directions, have their helicities flipped. Using the techniques of the foregoing sections, it is, however, straightforward to work out the amplitudes case by case.

7.4.5 An example: $e^+e^- \to \mu^+\mu^-\gamma\gamma$

Consider once more the process

$$e^+(p_+) + e^-(p_-) \to \mu^+(q_+) + \mu^-(q_-) + \gamma(k_1) + \gamma(k_2), \qquad (7.77)$$

and suppose that we are interested in the double collinear limit where $\vec{k}_1$ and $\vec{k}_2$ are both nearly parallel to $\vec{p}_-$, i.e.,

$$(p_- \cdot k_i) = \mathcal{O}(m^2), \quad k_{i+} = \mathcal{O}(m^2), \quad k_{i\perp} = \mathcal{O}(m), \quad i = 1, 2. \qquad (7.78)$$

We know that all spin amplitudes can be written as a product of two factors: a helicity amplitude for $e^+e^- \to \mu^+\mu^-$ and a collinearity factor which was called F_2 in eqn (7.33). These quantities F_2 depend on the momenta of the collinear photons, whereas the helicity amplitude for $e^+e^- \to \mu^+\mu^-$ does not.

It is a simple matter to evaluate the functions F_2 in the e^+e^- c.m. frame with the z-direction along $\vec{p}_+$. They contain an arbitrary four-vector q, which can conveniently be chosen equal to p_+. We then find

$$F_2(+,+;+,+) = 128CE^5 k_{1+}(2Ek_{2+} - k_{1-}k_{2+} + k_{1\perp}k_{2\perp}^*),$$

$$F_2(+,+;+,-) = 0,$$

$$F_2(+,+;-,+) = 32CmE^3[2Ek_{1-}k_{2+}k_{1\perp}^* + 2Ek_{1+}k_{2-}k_{2\perp}^*$$
$$+ k_{1-}(k_{1-} + k_{2-})(k_{1+}k_{2\perp}^* - k_{2+}k_{1\perp}^*)],$$

$$F_2(+,-;+,+) = 32CE^4[4Ek_{1+}(2Ek_{2+} - k_{1-}k_{2+} - k_{2+}k_{2-} + k_{1\perp}^*k_{2\perp})$$
$$+ \Delta_1 k_{1\perp}^* k_{2\perp}],$$

$$F_2(+,+;-,-) = 64CE^4 k_{1+}(2E - k_{1-} - k_{2-})(2Ek_{2+} - k_{1-}k_{2+} + k_{1\perp}k_{2\perp}^*),$$

$$F_2(+,-;+,-) = -64CmE^4 k_{1+}k_{2-}k_{2\perp},$$

$$F_2(+,-;-,+) = 32CmE^3 k_{1-}k_{1\perp}^*(2Ek_{2+} - k_{1-}k_{2+} + k_{1\perp}^*k_{2\perp}),$$

$$F_2(+,-;-,-) = 32CE^4[2k_{1+}(2E-k_{1-})(2Ek_{2+} - k_{1-}k_{2+} + k_{1\perp}^*k_{2\perp})$$

$$+ \Delta_1 k_{1\perp}^* k_{2\perp}], \quad (7.79)$$

with

$$C = \frac{N_1 N_2}{4(p_- \cdot k_1)[(p_- \cdot k_1) + (p_- \cdot k_2) - (k_1 \cdot k_2)]}, \quad (7.80)$$

where, as usual, N_1 and N_2 are the normalization factors (7.23) for the photon polarization vectors and E is the beam energy. The first two labels of F_2 denote the helicities of photons 1 and 2, photon 1 being closest to $u(p_-)$ in the Feynman diagrams. The third label is minus the helicity of the e^+, and the last one denotes the e^- helicity.

In this collinear limit, we now have

$$M(-\lambda, \lambda; \lambda_+, \lambda_-, \lambda_1, \lambda_2) = e^2[F_2(\lambda_1, \lambda_2; \lambda, \lambda) + (1 \leftrightarrow 2)]M_0(-\lambda, \lambda; \lambda_+, \lambda_-),$$

$$M(\lambda, \lambda; \lambda_+, \lambda_-, \lambda_1, \lambda_2) = e^2[F_2(\lambda_1, \lambda_2; -\lambda, \lambda) + (1 \leftrightarrow 2)]M_0(\lambda, -\lambda; \lambda_+, \lambda_-). \quad (7.81)$$

Here, the labels of M denote the helicities of the e^+, e^-, μ^+, μ^-, $\gamma(k_1)$ and $\gamma(k_2)$. Also, M_0 denotes the nonradiative helicity amplitudes for $e^+e^- \to \mu^+\mu^-$, which we already encountered in eqns (7.57) and (7.58).

With the explicit expressions for F_2 and M_0, it is now a simple matter to evaluate the unpolarized cross section. For $\vec{k}_1$ and $\vec{k}_2$ both nearly parallel to $\vec{p}_-$, we have

$$\overline{|M|^2} = 2e^8 \frac{tt' + uu'}{ss'} \frac{1}{k_{1+}k_{1-}k_{2+}k_{2-}} \frac{2E}{2E - k_{1-} - k_{2-}}$$

$$\times \left\{ \left[1 + \left(\frac{2E - k_{1-} - k_{2-}}{2E}\right)^2\right] |A_1|^2 + |A_2(1,2)|^2 + |A_2(2,1)|^2 \right.$$

$$\left. + |A_3|^2 + |A_4(1,2)|^2 + |A_4(2,1)|^2 \right\}, \quad (7.82)$$

with

$$A_1 = \frac{2E^2 k_{1+} k_{2+}}{(p_- \cdot k_1)(p_- \cdot k_2)} \left[1 + \frac{m^2 Z_{12} Z_{21}}{4E^2 k_{1+} k_{2+} \Delta_{12}}\right],$$

7.4. MASS CORRECTIONS FOR AMPLITUDES

$$A_2(1,2) = \frac{1}{\Delta_{12}}\left\{\frac{k_{1+}(2E-k_{1-})(2Ek_{2+}-Z_{21})}{(p_-\cdot k_1)}\right.$$
$$\left.+\frac{1}{(p_-\cdot k_2)}\Big[2Ek_{1+}k_{2+}(2E-k_{1-}-k_{2-}) - m^2 k_{2-}k_{1\perp}k_{2\perp}^*/2E\Big]\right\},$$

$$A_3 = \frac{m}{2(p_-\cdot k_1)(p_-\cdot k_2)}\left[k_{1-}k_{2+}k_{1\perp}^* + k_{1+}k_{2-}k_{2\perp}^*\right.$$
$$\left.-\frac{(k_{1\perp}^* + k_{2\perp}^*)Z_{12}Z_{21}}{\Delta_{12}}\right],$$

$$A_4(1,2) = \frac{mk_{1-}k_{1\perp}}{2E\Delta_{12}}\left[\frac{2Ek_{2+}-Z_{21}}{(p_-\cdot k_1)} + \frac{2Ek_{2+}}{(p_-\cdot k_2)}\right], \qquad (7.83)$$

and

$$\Delta_{12} = -2(p_-\cdot k_1) - 2(p_-\cdot k_2) + 2(k_1\cdot k_2),$$

$$Z_{ij} = k_{i+}k_{j-} - k_{i\perp}^* k_{j\perp}, \qquad i,j=1,2. \qquad (7.84)$$

To obtain the unpolarized squared matrix element for the collinear limit when $\vec{k}_1$ and $\vec{k}_2$ are both nearly parallel to $\vec{p}_+$, it suffices to interchange p_+ and p_- in the above equations. This amounts to interchanging the subscripts $+$ and $-$ in these formulae. As a result, Z_{ij} becomes Z_{ji}^*.

The limit $\vec{k}_1$ and $\vec{k}_2$ both nearly parallel to $\vec{q}_+$ is worked out in the same way. As in the single bremsstrahlung case, the quantities F_2 are most easily evaluated in the frame where $\vec{q}_+$ determines the positive z-axis. The arbitrary four-vector q in F_2 is best chosen to be q_+^R, the four-vector obtained by applying a space reflection to q_+. Then,

$$\overline{|M|^2} = 2e^8 \frac{tt'+uu'}{ss'} \frac{1}{k'_{1+}k'_{1-}k'_{2+}k'_{2-}} \frac{2q_{+0}}{2q_{+0}+k'_{1+}+k'_{2+}}$$
$$\times \left\{\left[1 + \left(\frac{2q_{+0}+k'_{1+}+k'_{2+}}{2q_{+0}}\right)^2\right]|A'_1|^2 + |A'_2(1,2)|^2 + |A'_2(2,1)|^2\right.$$
$$\left. + |A'_3|^2 + |A'_4(1,2)|^2 + |A'_4(2,1)|^2\right\}, \qquad (7.85)$$

with

$$A'_1 = \frac{2q_{+0}^2 k'_{1-}k'_{2-}}{(q_+\cdot k_1)(q_+\cdot k_2)}\left[1 + \frac{\mu^2 Z'_{12}Z'_{21}}{4q_{+0}^2 k'_{1-}k'_{2-}\Delta'_{12}}\right],$$

$$A'_2(1,2) = \frac{1}{\Delta'_{12}} \left\{ \frac{k'_{1-}(2q_{+0} + k'_{1+})(2q_{+0}k'_{2-} + Z'_{12})}{(q_+ \cdot k_1)} \right.$$

$$+ \frac{1}{(q_+ \cdot k_2)} \left[2q_{+0}k'_{1-}k'_{2-}(2q_{+0} + k'_{1+} + k'_{2+}) \right.$$

$$\left. \left. + \frac{\mu^2 k'_{2+} k'^*_{1\perp} k'_{2\perp}}{2q_{+0}} \right] \right\},$$

$$A'_3 = \frac{\mu}{2(q_+ \cdot k_1)(q_+ \cdot k_2)} \left[k'_{1+}k'_{2-}k'^*_{1\perp} + k'_{1-}k'_{2+}k'_{2\perp} \right.$$

$$\left. - \frac{(k'^*_{1\perp} + k'_{2\perp})Z'_{12}Z'_{21}}{\Delta'_{12}} \right],$$

$$A'_4(1,2) = \frac{\mu k'_{1+}k'^*_{1\perp}}{2q_{+0}\Delta'_{12}} \left[\frac{2q_{+0}k'_{2-} + Z'^*_{12}}{(q_+ \cdot k_1)} + \frac{2q_{+0}k'_{2-}}{(q_+ \cdot k_2)} \right], \quad (7.86)$$

and

$$\Delta'_{12} = 2(q_+ \cdot k_1) + 2(q_+ \cdot k_2) + 2(k_1 \cdot k_2). \quad (7.87)$$

Analogously, for $\vec{k}_1$ and $\vec{k}_2$ both nearly parallel to $\vec{q}_-$, it suffices to replace in the last set of equations q_+ by q_-, and the primed quantities k'_i and Z'_{ij} by k''_i and Z''_{ij}, i.e., to evaluate the components of the four-vectors k_i in the rotated frame where $\vec{q}_-$ determines the positive z-axis.

For the mixed double collinear limit, where the photons have directions nearly parallel to two different fermion directions, the situation is somewhat simpler. We find that the cross section is composed of two single 'collinearity factors' and a lowest order cross section written in the appropriate variables.

Let p and q denote any two vectors of the set of fermion momenta, p_+, p_-, q_+ and q_-, with $p \neq q$. For $\vec{k}_1$ and $\vec{k}_2$ nearly parallel to $\vec{p}$ and $\vec{q}$ respectively, we have the following simple expression for the unpolarized squared matrix element in these limits:

$$\overline{|M|^2} = 2e^8 \frac{tt' + uu'}{ss'} F(k_1, p) F(k_2, q). \quad (7.88)$$

The collinearity factors F [see eqns (7.64), (7.67), (7.68) and (7.71)] are explicitly given in Section 7.4.3. This formula can then be used in the 12 different cases depending on the values for p and q.

An analysis of the other QED double bremsstrahlung processes, $e^+e^- \to \gamma\gamma\gamma\gamma$ and $e^+e^- \to e^+e^-\gamma\gamma$, shows the universality of the collinearity factors in both the single and the double collinear situations. Each time, only the expression for the lower order cross section has to be adjusted. We therefore find it unnecessary to present the derivation of the formulae in detail and refer, once more, to Chapter 9, where they are listed.

8
The production of quarkonia

8.1 Framework

The applications of the helicity method, which were presented up to now, dealt with processes in which either the masses were completely neglected or treated to leading order. In Chapters 4 through 6, we considered QED and QCD processes, in which the quark and lepton masses were put equal to zero, while, in Chapter 7, we studied the finite mass effects to leading order in m^2. This is completely justified in the high energy limit, where these masses are small compared to the energies in the process. The technical reasons for the simplifications we encountered in the high energy limit is the concurrence of the facts that helicity states are Lorentz invariant concepts for massless particles and that the polarization vectors for the gauge particles which we introduced naturally combined with the helicity states of the fermions, because of the appearance of the combinations $1 \pm \gamma_5$ in $\not{k}$.

The usefulness of the helicity method is, however, by no means restricted to processes involving only massless (or nearly massless) particles. We have already shown in Section 4.5 that the inclusion of the Z-exchange can be taken into account without much effort. In this chapter, we want to show that, also for processes in which heavy external particles are present, the helicity method can be used advantageously, provided that at least some of the other external particles are massless. For this purpose, we present the calculation of the cross sections for $g + g \to {}^{2S+1}L_J + g$, where ${}^{2S+1}L_J$ is the spectroscopic notation for a heavy quarkonium state, for example, a $(c\bar{c})$ state. In this case, the mass of the quarkonium cannot be neglected and its effects must be included to all orders.

For a detailed comparison between theory and experiment in the case of J/ψ hadroproduction for example, it is not sufficient to calculate direct J/ψ production only. Indeed, some of the $(c\bar{c})$ excited states are known to have large branching ratios into the J/ψ [36]. It thus becomes necessary to know, for example, the production cross sections for the ${}^3P$ states.

Within the framework of perturbative QCD, the hadroproduction of heavy quarkonia is described by several subprocesses, such as $gg \to {}^{2S+1}L_J$, $q\bar{q} \to {}^{2S+1}L_J$, $gq \to {}^{2S+1}L_J q$, $g\bar{q} \to {}^{2S+1}L_J \bar{q}$, $q\bar{q} \to {}^{2S+1}L_J g$, $gg \to {}^{2S+1}L_J g$, etc. Using the standard Feynman techniques, one readily evaluates the cross sections for these processes [37], one exception being the case $gg \to {}^{2S+1}L_J g$, described by the Feynman diagrams of Fig. 8.1. The problem in this case is the relatively large number of Feynman diagrams and the long expressions which are generated by the covariant summation over the polarization degrees of freedom. As a result, one had to resort to symbolic manipulation programs and the cross section formulae one obtained were quite cumbersome.

Of these subprocesses, the last one, $gg \to {}^{2S+1}L_J\, g$, is the most complicated. We thus present here the calculation of this process with the helicity method. For reasons of symmetry, it is useful to think of this process as $ggg \to {}^{2S+1}L_J$. We shall explain how the crossing of a gluon can be applied to yield the desired formulae for $gg \to {}^{2S+1}L_J\, g$. We shall also limit ourselves to the most important cases $L = S$ or P, i.e., S- or P-wave quarkonia. The corresponding problems of electroproduction, which involve basically the replacement of a gluon by a photon, have been studied by Berger and Jones and by Kunszt [38].

Consider first the process

$$g(k_1, a) + g(k_2, b) + g(k_3, c) \to q(\tfrac{1}{2}p + q) + \bar{q}(\tfrac{1}{2}p - q), \qquad (8.1)$$

where k_1, k_2 and k_3 are the four-momenta of the gluons, while $\tfrac{1}{2}p + q$ and $\tfrac{1}{2}p - q$ denote those of the quark and the antiquark. As we are only interested in color singlet $q\bar{q}$ states, we can omit all the Feynman diagrams in which only one gluon couples to the $q\bar{q}$ system. This is because the quark belongs to a color octet. This reduces the number of Feynman diagrams to twelve, in lowest order. They are given in Fig. 8.1, where each diagram, in fact, stands for several diagrams. The first three diagrams represent each two diagrams: the $q\bar{q}$ lines must also be interchanged. In the last diagram, the six permutations of the gluons are implicit.

We introduce the standard notation

$$s = (k_1 + k_2)^2, \qquad t = (k_2 + k_3)^2, \qquad u = (k_3 + k_1)^2. \qquad (8.2)$$

For the first diagrams (s-channel), we have the amplitude

$$M_s = \frac{ig^3 f^{abd}}{s}[(k_1 - k_2)^\alpha (\epsilon_1^* \cdot \epsilon_2^*) + 2(k_2 \cdot \epsilon_1^*)\epsilon_2^{*\alpha} - 2(k_1 \cdot \epsilon_2^*)\epsilon_1^{*\alpha}]$$

$$\times \bar{u}(\tfrac{1}{2}p + q)\left[\frac{T_{ij}^d \gamma_\alpha(-\tfrac{1}{2}\slashed{p} + \slashed{q} + \slashed{k}_3 + m)\slashed{\epsilon}_3^* T_{jk}^c}{-(p \cdot k_3) + 2(q \cdot k_3)}\right.$$

$$\left. + \frac{T_{ij}^c \slashed{\epsilon}_3^*(\tfrac{1}{2}\slashed{p} + \slashed{q} - \slashed{k}_3 + m)\gamma_\alpha T_{jk}^d}{-(p \cdot k_3) - 2(q \cdot k_3)}\right] v(\tfrac{1}{2}p - q) \frac{\delta_{ik}}{\sqrt{3}}, \qquad (8.3)$$

with m the quark mass. The last term in this equation is obtained from the previous one by interchanging the quark lines, and the factor $\delta_{ik}/\sqrt{3}$ combines the quark and the antiquark into a color singlet state. Defining the quantity $\mathcal{O}_s(p, q)$ by the equation

$$M_s = \bar{u}(\tfrac{1}{2}p + q)\,\mathcal{O}_s(p, q)\,v(\tfrac{1}{2}p - q), \qquad (8.4)$$

and using the formula

$$\mathrm{Tr}(T^d T^c) = \tfrac{1}{2}\delta_{dc}, \qquad (8.5)$$

8.1. FRAMEWORK

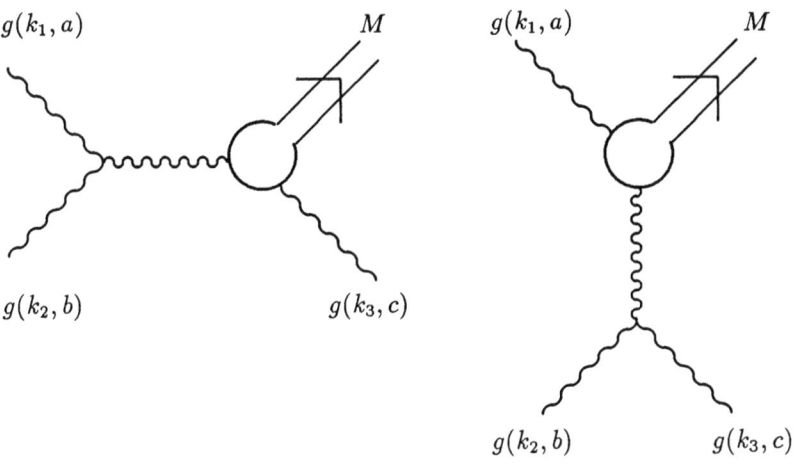

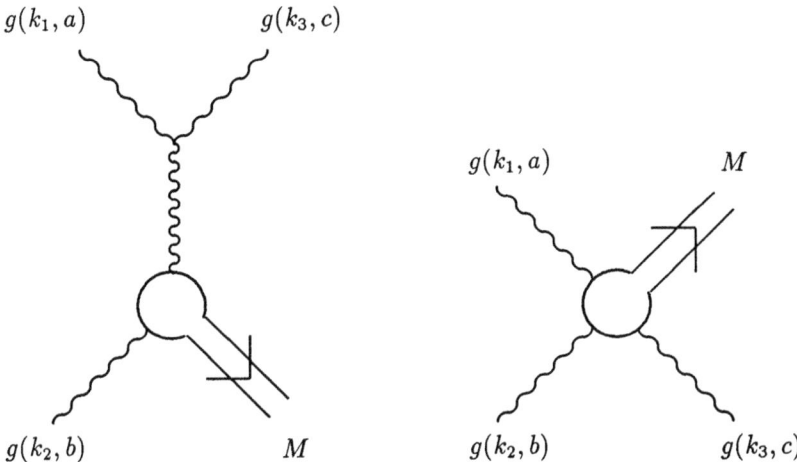

+ permutations

Fig. 8.1: Feynman diagrams for $g + g \to {}^{2S+1}L_J + g$.

we have

$$\mathcal{O}_s(p,q) = \frac{-ig^3 f^{abc}}{2s\sqrt{3}}[(k_2-k_1)^\alpha(\epsilon_1^*\cdot\epsilon_2^*) + 2(k_1\cdot\epsilon_2^*)\epsilon_1^{*\alpha} - 2(k_2\cdot\epsilon_1^*)\epsilon_2^{*c}$$

$$\times \left[\frac{\gamma_\alpha(\slashed{p}-2\slashed{q}-2\slashed{k}_3-2m)\slashed{\epsilon}_3^*}{2(p\cdot k_3)-4(q\cdot k_3)}\right.$$

$$\left.-\frac{\slashed{\epsilon}_3^*(\slashed{p}+2\slashed{q}-2\slashed{k}_3+2m)\gamma_\alpha}{2(p\cdot k_3)+4(q\cdot k_3)}\right].$$

For the t- and u-channel amplitudes, one readily writes down similar expressions.

For the last diagrams, not involving the three-gluon vertex, one finds amplitudes

$$\mathcal{O}_D(p,q) = \frac{-ig^3}{\sqrt{3}}(f^{abc}+id^{abc})$$

$$\times \frac{[(p\cdot\epsilon_1^*)+2(q\cdot\epsilon_1^*)+\slashed{k}_1\slashed{\epsilon}_1^*]\slashed{\epsilon}_2^*[(p\cdot\epsilon_3^*)-2(q\cdot\epsilon_3^*)+\slashed{\epsilon}_3^*\slashed{k}_3]}{[2(p\cdot k_1)+4(q\cdot k_1)][2(p\cdot k_3)-4(q\cdot k_3)]}$$

$+ 5$ other permutations of $(1,2,3$ and $a,b,c)$.

We now combine the free quarks into a quarkonium state. This is using the nonrelativistic bound state approximation

$$M \simeq 2m,$$

where M is the quarkonium mass. It also means that for S-waves we put $q = 0$, while for P-waves we have to retain the terms linear in q in amplitudes, q being the relative momentum of the quarks.

Following the method of Guberina, Kühn, Peccei and Rückl [39], we introduce spin projection operators, which combine the quark and antiquark spins to the appropriate singlet ($S = 0$) or triplet ($S = 1$) states, and into account the nonrelativistic bound state wave function of the quarks. refer to Guberina et al. [39] for the details of the method and simply list different amplitudes for quarkonium production:

$$A(^1S_0) = \frac{R_0}{\sqrt{16\pi M}}\text{Tr}[\mathcal{O}(\slashed{p}-M)\gamma_5],$$

$$A(^3S_1) = \frac{R_0}{\sqrt{16\pi M}}\text{Tr}[\mathcal{O}(\slashed{p}-M)\slashed{\epsilon}],$$

$$A(^1P_1) = -iR_1'\left(\frac{3}{16\pi M}\right)^{\frac{1}{2}}\text{Tr}[\epsilon^\alpha\mathcal{O}_\alpha(\slashed{p}-M)\gamma_5 + 2\mathcal{O}\slashed{\epsilon}\slashed{p}\gamma_5/M],$$

8.1. FRAMEWORK

$$A(^3P_0) = \frac{iR_1'}{\sqrt{16\pi M}} \text{Tr}[\mathcal{O}_\alpha(\gamma^\alpha - p^\alpha/M)(\slashed{p} + M) + 6\mathcal{O}],$$

$$A(^3P_1) = -R_1'\left(\frac{3}{32\pi M^3}\right)^{\frac{1}{2}} \epsilon^{\alpha\beta\gamma\delta} p_\gamma \epsilon_\delta$$

$$\times \text{Tr}[\mathcal{O}_\alpha \gamma_\beta (\slashed{p} + M) - 2\mathcal{O}\gamma_\alpha \slashed{p}\gamma_\beta/M],$$

$$A(^3P_2) = iR_1'\left(\frac{3}{16\pi M}\right)^{\frac{1}{2}} \epsilon^{\alpha\beta} \text{Tr}[\mathcal{O}_\alpha \gamma_\beta (\slashed{p} + M)]. \quad (8.9)$$

In these formulae, we have introduced

$$\mathcal{O} = \mathcal{O}_s(p,0) + \mathcal{O}_t(p,0) + \mathcal{O}_u(p,0) + \mathcal{O}_D(p,0),$$

$$\mathcal{O}_\alpha = \frac{\partial}{\partial q^\alpha}[\mathcal{O}_s(p,q) + \mathcal{O}_t(p,q) + \mathcal{O}_u(p,q) + \mathcal{O}_D(p,q)]\bigg|_{q=0}, \quad (8.10)$$

and the quantity R_0, which is the S-wave wave function evaluated at the origin, and R_1', the derivative of the P-wave wave function also evaluated at the origin. The quantity R_0 is simply related to the leptonic width of the $^3S_1$ state through the formula

$$R_0^2 = M^2 \Gamma(^3S_1 \to e^+e^-)/4\alpha^2 Q_f^2, \quad (8.11)$$

with $\alpha \simeq 1/137$, the fine structure constant, and Q_f the fractional charge of the quarks. Using the quarkonium potential of Hagiwara et al. [40] with $\Lambda = 0.2$ GeV, one finds for the $(c\bar{c})$ system

$$R_1'^2/M_\chi^2 = 0.006(\text{GeV})^3, \quad (8.12)$$

where M_χ is the mass of a ^3P state.

Finally, the quantities ϵ^α and $\epsilon^{\alpha\beta}$ in the expressions for the amplitudes describe the polarization states for the massive spin-1 and spin-2 quarkonia.

Eqns (8.9) are obtained by applying several approximations to a relativistic Bethe-Salpeter equation describing the production of quarkonia. Because of these approximations, it is not clear how accurate these equations are. Nevertheless, in the absence of more accurate ones, the calculations in this chapter will be based on eqns (8.9).

We now apply the helicity method to the calculation of the production cross sections of these quarkonia states [41]. Let us denote by $M(\lambda_1, \lambda_2, \lambda_3)$ any helicity amplitude for $ggg \to {}^{2S+1}L_J$, where the labels λ_i, $i = 1, 2, 3$, refer to the helicities of the three gluons. It is sufficient to consider $M(+,+,+)$ and $M(+,+,-)$ only, as all other helicity amplitudes can be obtained by permuting the gluons, or by a parity conjugation, which flips all helicities. We present the calculation of the $^1S_0$ production in some detail, as it allows us to illustrate, through a sufficiently simple example, the basic features of the helicity technique when massive particles are present. For the other states, we shall merely present the results of the calculations as the computational techniques are almost identical.

8.2 $^1S_0$ production

8.2.1 The amplitude $M(+,+,+)$

For incoming gluons with positive helicities, we can choose the following polarization vectors:

$$\epsilon_1^*(k_1) = N[\slashed{k}_1 \slashed{k}_2 \slashed{k}_3(1-\gamma_5) + \slashed{k}_3 \slashed{k}_2 \slashed{k}_1(1+\gamma_5)],$$

$$\epsilon_2^*(k_2) = N[\slashed{k}_2 \slashed{k}_3 \slashed{k}_1(1-\gamma_5) + \slashed{k}_1 \slashed{k}_3 \slashed{k}_2(1+\gamma_5)],$$

$$\epsilon_3^*(k_3) = N[\slashed{k}_3 \slashed{k}_1 \slashed{k}_2(1-\gamma_5) + \slashed{k}_2 \slashed{k}_1 \slashed{k}_3(1+\gamma_5)], \qquad (8.13)$$

with the normalization factor

$$N = (2stu)^{-\frac{1}{2}}. \qquad (8.14)$$

This choice of polarization vectors has the advantage of being invariant under cyclic permutations of (1,2,3).

With the identity

$$4(a \cdot b) = \text{Tr}[\slashed{a} \slashed{b}], \qquad (8.15)$$

one can now express all the scalar products of four-vectors which appear in a given amplitude in terms of s, t and u. For example,

$$\begin{aligned}
(p \cdot \epsilon_1^*) &= (k_2 \cdot \epsilon_1^*) = Nst, \\
(p \cdot \epsilon_2^*) &= (k_3 \cdot \epsilon_2^*) = Ntu, \\
(p \cdot \epsilon_3^*) &= (k_1 \cdot \epsilon_3^*) = Nus, \\
(\epsilon_1^* \cdot \epsilon_2^*) &= (\epsilon_2^* \cdot \epsilon_3^*) = (\epsilon_3^* \cdot \epsilon_1^*) = 1,
\end{aligned} \qquad (8.16)$$

while many other scalar products, like $(k_1 \cdot \epsilon_1^*)$, $(k_3 \cdot \epsilon_1^*)$, $(k_1 \cdot \epsilon_2^*)$, etc., vanish. Also note that

$$p = k_1 + k_2 + k_3, \qquad s+t+u = M^2, \qquad (8.17)$$

and

$$2(p \cdot k_1) = s+u, \qquad 2(p \cdot k_2) = s+t, \qquad 2(p \cdot k_3) = t+u. \qquad (8.18)$$

8.2. $^1S_0$ PRODUCTION

For the case of $^1S_0$ production, we first calculate

$$M(+,+,+) = \frac{R_0}{\sqrt{16\pi M}} \text{Tr}[\mathcal{O}(\slashed{p}-M)\gamma_5], \tag{8.19}$$

with our choice for the quantities ϵ_i^*, $i=1,2,3$, eqns (8.13). The s-channel diagrams give

$$\mathcal{O}_s(p,0) = \frac{ig^3 f^{abc}}{2\sqrt{3}s(s-M^2)}$$

$$\times [(\slashed{k}_2-\slashed{k}_1)(\slashed{p}-2\slashed{k}_3-M)\slashed{\epsilon}_3^* - 2Nst\,\slashed{\epsilon}_2^*(\slashed{p}-2\slashed{k}_3-M)\slashed{\epsilon}_3^*$$

$$- \slashed{\epsilon}_3^*(\slashed{p}-2\slashed{k}_3+M)(\slashed{k}_2-\slashed{k}_1) + 2Nst\,\slashed{\epsilon}_3^*(\slashed{p}-2\slashed{k}_3+M)\slashed{\epsilon}_2^*]. \tag{8.20}$$

and denoting their contribution to $M(+,+,+)$ by A_s, we have

$$A_s = \frac{2ig^3 f^{abc} R_0}{2\sqrt{3\pi M}s(s-M^2)} \text{Tr}[(\slashed{k}_1-\slashed{k}_2)\slashed{k}_3\slashed{\epsilon}_3^*\slashed{p}\gamma_5 + 2Nst\,\slashed{\epsilon}_2^*\slashed{k}_3\slashed{\epsilon}_3^*\slashed{p}\gamma_5]. \tag{8.21}$$

Because of the γ_5 appearing in the trace, the last two terms in $\mathcal{O}_s$ were found to give the same contributions as the first two terms. We can now substitute the expressions for $\slashed{\epsilon}_2^*$ and $\slashed{\epsilon}_3^*$ in A_s. As

$$\slashed{k}_3\slashed{\epsilon}_3^* = N\slashed{k}_3\slashed{k}_2\slashed{k}_1\slashed{k}_3(1+\gamma_5), \tag{8.22}$$

we find that only one term in $\slashed{\epsilon}_2^*$ contributes, and

$$A_s = \frac{ig^3 f^{abc} R_0 N}{2\sqrt{3\pi M}s(s-M^2)} \text{Tr}[-(\slashed{k}_1-\slashed{k}_2)\slashed{k}_3\slashed{k}_2\slashed{k}_1\slashed{k}_3\slashed{p}(1-\gamma_5)$$

$$-4N^2 st\,\slashed{k}_1\slashed{k}_3\slashed{k}_2\slashed{k}_3\slashed{k}_2\slashed{k}_1\slashed{k}_3\slashed{p}(1-\gamma_5)]. \tag{8.23}$$

Using momentum conservation, the traces can now be written with $\slashed{k}_1$, $\slashed{k}_2$ and $\slashed{k}_3$ only. Such traces are easily evaluated as $k_i^2 = 0$, $i=1,2,3$. The result is

$$A_s = -\frac{2ig^3 f^{abc} R_0 N}{\sqrt{3\pi M}} t. \tag{8.24}$$

The analogous contributions from the t- and u-channel diagrams can be obtained by cyclic permutation, and

$$A_s + A_t + A_u = -\frac{2ig^3 f^{abc} R_0 N}{\sqrt{3\pi M}} M^2. \tag{8.25}$$

Next, we have

$$A_D = -\frac{ig^3 f^{abc} R_0}{2\sqrt{3\pi M}(s-M^2)(t-M^2)} \text{Tr}[\slashed{k}_1\slashed{\epsilon}_1^*\slashed{\epsilon}_2^*\slashed{\epsilon}_3^*\slashed{k}_3\slashed{p}\gamma_5$$

$$+(p\cdot\epsilon_1^*)\slashed{\epsilon}_2^*\slashed{\epsilon}_3^*\slashed{k}_3\slashed{p}\gamma_5 + (p\cdot\epsilon_3^*)\slashed{k}_1\slashed{\epsilon}_1^*\slashed{\epsilon}_2^*\slashed{p}\gamma_5]$$

$$+ 2 \text{ cyclic permutations of } (1,2,3). \tag{8.26}$$

Indeed, because of the γ_5 in the trace, we find that the d^{abc} contribution vanishes and that each nonvanishing term appears twice. Substituting [eqn (8.13)] then shows that the first term in A_D vanishes, and

$$A_D = -\frac{ig^3 f^{abc} R_0 N^3}{\sqrt{3\pi} M(s-M^2)(t-M^2)}$$

$$\times \mathrm{Tr}[-st \not k_1 \not k_3 \not k_2 \not k_3 \not k_1 \not k_2 \not k_3 \not p(1-\gamma_5)$$

$$+su \not k_1 \not k_3 \not k_2 \not k_1 \not k_2 \not k_3 \not k_1 \not p(1+\gamma_5)]$$

$$+ 2 \text{ cyclic permutations of } (1,2,3). \qquad (8.2?)$$

Similar manipulations to those in the A_s case then lead to the result

$$A_D = \frac{2ig^3 f^{abc} R_0 N}{\sqrt{3\pi} M} \frac{M^2 stu}{(s-M^2)(t-M^2)(u-M^2)}, \qquad (8.2?)$$

and

$$M(+,+,+) = -\frac{2ig^3 f^{abc} R_0 N M^2}{\sqrt{3\pi} M} \left[1 - \frac{stu}{(s-M^2)(t-M^2)(u-M^2)} \right]$$

$$= \frac{2ig^3 f^{abc} R_0 N}{\sqrt{3\pi} M} \frac{M^4(st+tu+us)}{(s-M^2)(t-M^2)(u-M^2)}. \qquad (8.2?)$$

8.2.2 The amplitude $M(+,+,-)$

This time, the third gluon has an opposite helicity, and we find it convenient to choose

$$\not\epsilon_1^*(k_1) = N[\not k_1 \not k_2 \not k_3 (1-\gamma_5) + \not k_3 \not k_2 \not k_1 (1+\gamma_5)],$$

$$\not\epsilon_2^*(k_2) = -N[\not k_2 \not k_1 \not k_3 (1-\gamma_5) + \not k_3 \not k_1 \not k_2 (1+\gamma_5)],$$

$$\not\epsilon_3^*(k_3) = N[\not k_3 \not k_1 \not k_2 (1+\gamma_5) + \not k_2 \not k_1 \not k_3 (1-\gamma_5)], \qquad (8.3?)$$

where the normalization factor N is defined by eqn (8.14). Note that, because of the opposite helicities of gluons 2 and 3, it is perfectly possible to take $\epsilon_2^* - \epsilon_3^*$. The minus sign in $\epsilon_2^*(k_2)$ is introduced to ensure gauge equivalence with the choice of $\epsilon_2^*(k_2)$ in eqn (8.13). Indeed, it suffices to anticommute $\not k_3$ and $\not k_1$ in eqn (8.13) to find that the two expressions differ by a term proportional to $\not k_2$, which can be omitted because of vector current conservation.

8.2. $^1S_0$ PRODUCTION

With the choice (8.30), all scalar products among the ϵ^*'s are made to vanish, and

$$(p \cdot \epsilon_1^*) = (k_2 \cdot \epsilon_1^*) = Nst,$$

$$(p \cdot \epsilon_2^*) = (k_1 \cdot \epsilon_2^*) = -Nsu, \qquad (8.31)$$

$$(p \cdot \epsilon_3^*) = (k_1 \cdot \epsilon_3^*) = Nsu.$$

Again, many other scalar products vanish, and the evaluation of the helicity amplitude is greatly simplified.

The s-channel diagrams now give

$$\mathcal{O}_s(p,0) = \frac{ig^3 f^{abc} N}{\sqrt{3}(s - M^2)}$$

$$\times [u \,\not{\epsilon}_1^*(\not{p} - 2\not{k}_3 + M)\, \not{\epsilon}_3^* - u \,\not{\epsilon}_3^*(\not{p} - 2\not{k}_3 + M)\, \not{\epsilon}_1^*$$

$$- t \,\not{\epsilon}_2^*(\not{p} - 2\not{k}_3 + M)\, \not{\epsilon}_3^* + t \,\not{\epsilon}_3^*(\not{p} - 2\not{k}_3 + M)\, \not{\epsilon}_2^*], \qquad (8.32)$$

and

$$A_s = \frac{2ig^3 f^{abc} R_0 N}{\sqrt{3\pi} M} \operatorname{Tr}[u \,\not{\epsilon}_1^* \not{k}_3 \not{\epsilon}_3^* \not{p}\gamma_5 - t \,\not{\epsilon}_2^* \not{k}_3 \not{\epsilon}_3^* \not{p}\gamma_5]. \qquad (8.33)$$

But, with the choice of ϵ_i^* in eqns (8.30), we have

$$\not{\epsilon}_1^* \not{k}_3 \not{\epsilon}_3^* = \not{\epsilon}_2^* \not{k}_3 \not{\epsilon}_3^* = 0, \qquad (8.34)$$

hence $A_s = 0$. Similarly, $A_t = 0$. However,

$$\mathcal{O}_u(p,0) = -\frac{ig^3 f^{abc} N s}{\sqrt{3}(u - M^2)}[\not{\epsilon}_1^*(\not{p} - 2\not{k}_2 + M)\, \not{\epsilon}_2^* - \not{\epsilon}_2^*(\not{p} - 2\not{k}_2 + M)\, \not{\epsilon}_1^*], \qquad (8.35)$$

and

$$A_u = \frac{ig^3 f^{abc} R_0 N s}{\sqrt{3\pi} M (u - M^2)} \operatorname{Tr}[\not{\epsilon}_1^* \not{k}_2 \not{\epsilon}_2^* \not{p}\gamma_5]$$

$$= \frac{2ig^3 f^{abc} R_0 N^3 s}{\sqrt{3\pi} M (u - M^2)} \operatorname{Tr}[\not{k}_3 \not{k}_2 \not{k}_1 \not{k}_2 \not{k}_3 \not{k}_1 \not{k}_2 \not{p}(1 - \gamma_5)]$$

$$= \frac{2ig^3 f^{abc} R_0 N}{\sqrt{3\pi} M} \frac{s^2}{u - M^2}. \qquad (8.36)$$

The calculation of A_D for this helicity amplitude is somewhat more involved:

$$A_D = -\frac{ig^3 f^{abc} R_0}{2\sqrt{3\pi} M(s-M^2)(t-M^2)} \text{Tr}[\slashed{k}_1 \slashed{\epsilon}_1^* \slashed{\epsilon}_2^* \slashed{\epsilon}_3^* \slashed{k}_3 \slashed{p}\gamma_5$$

$$+(p\cdot\epsilon_1^*)\slashed{\epsilon}_2^* \slashed{\epsilon}_3^* \slashed{k}_3 \slashed{p}\gamma_5 + (p\cdot\epsilon_3^*)\slashed{k}_1 \slashed{\epsilon}_1^* \slashed{\epsilon}_2^* \slashed{p}\gamma_5]$$

$$+ \text{ 2 cyclic permutations of (1,2,3)}. \tag{8.37}$$

The fact that

$$\slashed{\epsilon}_1^* \slashed{k}_3 \slashed{\epsilon}_3^* = \slashed{\epsilon}_2^* \slashed{k}_3 \slashed{\epsilon}_3^* = \slashed{k}_2 \slashed{\epsilon}_2^* \slashed{\epsilon}_3^* = 0 \tag{8.38}$$

simplifies this expression considerably. It follows that

$$A_D = -\frac{ig^3 f^{abc} R_0}{2\sqrt{3\pi} M} \text{Tr}\left[\frac{(p\cdot\epsilon_3^*)\slashed{k}_1 \slashed{\epsilon}_1^* \slashed{\epsilon}_2^* \slashed{p}\gamma_5}{(s-M^2)(t-M^2)}\right.$$

$$+\frac{(p\cdot\epsilon_2^*)\slashed{\epsilon}_3^* \slashed{\epsilon}_1^* \slashed{k}_1 \slashed{p}\gamma_5}{(t-M^2)(u-M^2)} + \left.\frac{(p\cdot\epsilon_3^*)\slashed{\epsilon}_1^* \slashed{\epsilon}_2^* \slashed{k}_2 \slashed{p}\gamma_5}{(u-M^2)(s-M^2)}\right]$$

$$= \frac{ig^3 f^{abc} R_0 N^3 su}{\sqrt{3\pi} M} \text{Tr}\left[\frac{\slashed{k}_1 \slashed{k}_3 \slashed{k}_2 \slashed{k}_1 \slashed{k}_2 \slashed{k}_1 \slashed{k}_3 \slashed{p}(1+\gamma_5)}{(s-M^2)(t-M^2)}\right.$$

$$-\frac{\slashed{k}_3 \slashed{k}_1 \slashed{k}_2 \slashed{k}_1 \slashed{k}_2 \slashed{k}_3 \slashed{k}_1 \slashed{p}(1-\gamma_5)}{(t-M^2)(u-M^2)}$$

$$\left.-\frac{\slashed{k}_3 \slashed{k}_2 \slashed{k}_1 \slashed{k}_2 \slashed{k}_1 \slashed{k}_3 \slashed{k}_2 \slashed{p}(1-\gamma_5)}{(u-M^2)(s-M^2)}\right]$$

$$= -\frac{2ig^3 f^{abc} R_0 N}{\sqrt{3\pi} M} \frac{s^2 u^2}{(s-M^2)(t-M^2)(u-M^2)}. \tag{8.39}$$

Adding A_u and A_D, we obtain the helicity amplitude

$$M(+,+,-) = \frac{2ig^3 f^{abc} R_0 N}{\sqrt{3\pi} M} \frac{s^2(st+tu+us)}{(s-M^2)(t-M^2)(u-M^2)}. \tag{8.40}$$

Obviously, the helicity amplitude $M(+,-,+)$ is obtained from the above one by replacing $s \leftrightarrow u$, and $M(-,+,+)$ by $s \leftrightarrow t$.

8.2.3 The cross section for $^1S_0$ production

In Sections 8.2.1 and 8.2.2, we derived the helicity amplitudes for the process

$$g(k_1) + g(k_2) + g(k_3) \to {}^1S_0(p). \tag{8.41}$$

Ultimately, we are interested in the process

$$g(k_1) + g(k_2) \to {}^1S_0(p) + g(k_3). \tag{8.42}$$

To obtain the helicity amplitudes of the latter process, it suffices to replace k_3 by $-k_3$ in the helicity amplitudes of the former process and to flip the helicity of gluon 3. Hence, for $gg \to {}^1S_0 g$,

$$M(+,+;-) = \frac{2ig^3 f^{abc} R_0 N}{\sqrt{3\pi} M} \frac{M^4(st+tu+us)}{(s-M^2)(t-M^2)(u-M^2)},$$

$$M(+,+;+) = \frac{2ig^3 f^{abc} R_0 N}{\sqrt{3\pi} M} \frac{s^2(st+tu+us)}{(s-M^2)(t-M^2)(u-M^2)},$$

$$M(+,-;-) = \frac{2ig^3 f^{abc} R_0 N}{\sqrt{3\pi} M} \frac{u^2(st+tu+us)}{(s-M^2)(t-M^2)(u-M^2)},$$

$$M(-,+;-) = \frac{2ig^3 f^{abc} R_0 N}{\sqrt{3\pi} M} \frac{t^2(st+tu+us)}{(s-M^2)(t-M^2)(u-M^2)}, \quad (8.43)$$

where

$$s = (k_1 + k_2)^2, \qquad t = (k_2 - k_3)^2, \qquad u = (k_1 - k_3)^2. \quad (8.44)$$

All the remaining helicity amplitudes can be obtained by applying a parity conjugation to the above ones.

The cross section, averaged over the initial polarizations and color degrees of freedom, then reads

$$\frac{d\sigma}{dt} = \frac{\pi \alpha_S^3 R_0^2}{2Ms^2} \left[\frac{st+tu+us}{(s-M^2)(t-M^2)(u-M^2)} \right]^2 \frac{M^8 + s^4 + t^4 + u^4}{stu}, \quad (8.45)$$

with $\alpha_S = g^2/4\pi$. Using the variables

$$P = st + tu + us, \qquad Q = stu, \quad (8.46)$$

which are symmetrical under all permutations of s, t and u, one can rewrite this cross section formula in the following form:

$$\frac{d\sigma}{dt} = \frac{\pi \alpha_S^3 R_0^2}{Ms^2} \frac{P^2(M^8 - 2M^4 P + P^2 + 2M^2 Q)}{Q(Q - M^2 P)^2}. \quad (8.47)$$

8.3 The production of other states

In this section, we present the results of the calculation for the production of the other quarkonium states ${}^3S_1$, ${}^1P_1$, ${}^3P_0$, ${}^3P_1$ and ${}^3P_2$. The given helicity amplitudes always refer to the process $ggg \to {}^{2S+1}L_J$, but we have explained in Section 8.2.3 how the crossing of the third gluon should be performed.

8.3.1 $^3S_1$ production

For this process, only the last diagrams of Fig. 8.1, not involving the three-gluon vertex, contribute. We use the same polarization vectors for the gluons as in the $^1S_0$ case and we make use of the relation $(p \cdot \epsilon) = 0$, where ϵ is the polarization vector for the spin-1 quarkonium. We then find

$$M(+,+,+) = 0,$$

$$M(+,+,-) = -\frac{2g^3 d^{abc} R_0 N}{\sqrt{3\pi M}} \frac{Ms}{(s-M^2)(t-M^2)(u-M^2)}$$

$$\times [t(t+u)(k_1 \cdot \epsilon) - u(t+u)(k_2 \cdot \epsilon) - s(t-u)(k_3 \cdot \epsilon)$$

$$+ 2i(t+u)\epsilon^{\alpha\beta\gamma\delta} k_{1\alpha} k_{2\beta} k_{3\gamma} \epsilon_\delta]. \qquad (8.48)$$

Using the relation

$$\sum_{\text{pol}} \epsilon_\mu \epsilon_\nu^* = -g_{\mu\nu} + p_\mu p_\nu / M^2, \qquad (8.49)$$

and summing over the gluon color degrees of freedom, we find

$$\sum_{\text{pol}} |M(+,+,-)|^2 = \frac{10240 \pi^2 \alpha_S^3 R_0^2 M s^2}{9(t-M^2)(u-M^2)^2}. \qquad (8.50)$$

Symmetrizing this equation with respect to s, t and u, and averaging over the initial state degrees of freedom then yields the cross section formula:

$$\frac{d\sigma}{dt} = \frac{5\pi \alpha_S^3 R_0^2 M}{9s^2} \left[\frac{s^2}{(t-M^2)^2 (u-M^2)^2} \right.$$

$$\left. + \frac{t^2}{(u-M^2)^2 (s-M^2)^2} + \frac{u^2}{(s-M^2)^2 (t-M^2)^2} \right]. \qquad (8.51)$$

In terms of the variables P and Q, eqns (8.46), this reads

$$\frac{d\sigma}{dt} = \frac{10\pi \alpha_S^3 R_0^2}{9s^2} \frac{M(P^2 - M^2 Q)}{(Q - M^2 P)^2}. \qquad (8.52)$$

8.3.2 $^1P_1$ production

As in the $^3S_1$ case, the s-, t- and u-channel diagrams do not contribute, and the amplitude $A(^1P_1)$ is again proportional to d^{abc}. We obtain

$$M(+,+,+) = -\frac{ig^3 d^{abc} R_1'}{\sqrt{\pi M}} \left[a_1(\epsilon \cdot \epsilon_1^*) + a_2(\epsilon \cdot \epsilon_2^*) + a_3(\epsilon \cdot \epsilon_3^*) \right.$$

$$\left. + \frac{2b_1}{t-M^2}(k_1 \cdot \epsilon) + \frac{2b_2}{u-M^2}(k_2 \cdot \epsilon) + \frac{2b_3}{s-M^2}(k_3 \cdot \epsilon) \right], \qquad (8.53)$$

8.3. THE PRODUCTION OF OTHER STATES

with

$$a_1 = -\frac{2(2t+u)}{(s-M^2)(t-M^2)},$$

$$a_2 = -\frac{2(2u+s)}{(t-M^2)(u-M^2)},$$

$$a_3 = -\frac{2(2s+t)}{(u-M^2)(s-M^2)},$$

$$b_1 = \frac{2N}{t-M^2}\left[\frac{-s(t^2+ut-uM^2)}{s-M^2}+\frac{t(u^2+su-sM^2)}{u-M^2}\right],$$

$$b_2 = \frac{2N}{u-M^2}\left[\frac{-t(u^2+su-sM^2)}{t-M^2}+\frac{u(s^2+st-tM^2)}{s-M^2}\right],$$

$$b_3 = \frac{2N}{s-M^2}\left[\frac{-u(s^2+ts-tM^2)}{u-M^2}+\frac{s(t^2+tu-uM^2)}{t-M^2}\right]. \quad (8.54)$$

After summation over the polarization states of $^1P_1$ and the color states of the gluons, we obtain

$$\sum_{\text{pol}}|M(+,+,+)|^2 = \frac{40960\pi^2\alpha_s^3 R_1'^2}{3M}\frac{M^8(-M^2P+5Q)}{(Q-M^2P)^3}. \quad (8.55)$$

Similar manipulations give for the amplitude $M(+,+,-)$ a formula as in the $M(+,+,+)$ case, but now,

$$a_1 = 0,$$

$$a_2 = \frac{2s}{(t-M^2)(u-M^2)},$$

$$a_3 = \frac{2s(t+u-2M^2)}{(s-M^2)(t-M^2)(u-M^2)},$$

$$b_1 = -\frac{2Ns^2u}{t-M^2}\left[\frac{1}{s-M^2}+\frac{1}{u-M^2}\right],$$

$$b_2 = \frac{2Ns^2u}{u-M^2}\left[\frac{1}{t-M^2}-\frac{1}{s-M^2}\right],$$

$$b_3 = \frac{2Ns^2u}{s-M^2}\left[\frac{1}{u-M^2}+\frac{1}{t-M^2}\right], \quad (8.56)$$

and, consequently,

$$\sum_{\text{pol}} |M(+,+,-)|^2 = \frac{40960\pi^2\alpha_S^3 R_1'^2}{3M}$$

$$\times \left[\frac{M^4 s^2}{(s-M^2)^2(t-M^2)^2(u-M^2)^2} \right.$$

$$\left. + \frac{2stu(s^4 + s^2 M^4)}{(s-M^2)^3(t-M^2)^3(u-M^2)^3} \right]. \quad (8.57)$$

The cross section for $^1P_1$ production then becomes:

$$\frac{d\sigma}{dt} = \frac{20\pi\alpha_S^3 R_1'^2}{3Ms^2} \frac{1}{[(s-M^2)(t-M^2)(u-M^2)]^2}$$

$$\times \left\{ M^4(s^2 + t^2 + u^2 + M^4) \right.$$

$$\left. + \frac{2stu[s^4 + t^4 + u^4 + M^4(s^2 + t^2 + u^2) + 2M^8]}{(s-M^2)(t-M^2)(u-M^2)} \right\}$$

$$= \frac{40\pi\alpha_S^3 R_1'^2}{3Ms^2} \frac{-M^{10}P + M^6 P^2 + Q(5M^8 - 7M^4 P + 2P^2) + 4M^2 Q^2}{(Q - M^2 P)^3}.$$

$$(8.58)$$

8.3.3 $^3P_0$ production

The evaluation of the helicity amplitudes for $^3P_0$ production is straightforward. We find

$$M(+,+,+) = -\frac{12g^3 f^{abc} R_1' N}{\sqrt{3\pi M^3}} \frac{M^4 P}{(s-M^2)(t-M^2)(u-M^2)},$$

$$M(+,+,-) = -\frac{4g^3 f^{abc} R_1' N}{\sqrt{3\pi M^3}} \frac{s^2[t^2 u^2 - stu(s - 6M^2) - 3M^2 s^2(s - M^2)]}{(s-M^2)(t-M^2)^2(u-M^2)^2}.$$

$$(8.59)$$

The resulting cross section reads:

8.3. THE PRODUCTION OF OTHER STATES

$$\frac{d\sigma}{dt} = \frac{2\pi\alpha_S^3 R_1'^2}{M^3 s^2} \frac{1}{[(s-M^2)(t-M^2)(u-M^2)]^2}$$

$$\times \left\{ 8M^2 \left[\frac{tu(t^4 - t^2 u^2 + u^4)}{(s-M^2)^2} + \frac{us(u^4 - u^2 s^2 + s^4)}{(t-M^2)^2} \right. \right.$$

$$\left. + \frac{st(s^4 - s^2 t^2 + t^4)}{(u-M^2)^2} \right] + 4M^2 [M^2(st + tu + us) - 5stu]$$

$$+ (st + tu + us)^2 \left[\frac{9M^8}{stu} + \frac{1}{(s-M^2)(t-M^2)(u-M^2)} \right.$$

$$\times \left(-16M^2 stu + 8M^4(s^2 + t^2 + u^2) \right.$$

$$\left. \left. + \left(1 - 9M^2\left(\frac{1}{s} + \frac{1}{t} + \frac{1}{u}\right)\right)(s^4 + t^4 + u^4)\right)\right]\right\}$$

$$= \frac{4\pi\alpha_S^3 R_1'^2}{M^3 s^2} \frac{1}{Q(Q-M^2 P)^4} \left[9M^4 P^4 (M^8 - 2M^4 P + P^2) \right.$$

$$- 6M^2 P^3 Q(2M^8 - 5M^4 P + P^2) - P^2 Q^2 (M^8 + 2M^4 P - P^2)$$

$$\left. + 2M^2 PQ^3 (M^4 - P) + 6M^4 Q^4 \right], \quad (8.60)$$

with the quantities P and Q defined in eqns (8.46).

8.3.4 $^3P_1$ production

For the $^3P_1$ case, a somewhat lengthy, but straightforward calculation shows that

$$M(+,+,+) = 0. \quad (8.61)$$

Also,

$$M(+,+,-) = -\frac{4ig^3 f^{abc} R_1' N}{\sqrt{2\pi} M} \frac{s}{(s-M^2)(t-M^2)^2(u-M^2)^2}$$

$$\times \{ 2\epsilon_{\alpha\beta\gamma\delta} k_1^\alpha k_2^\beta k_3^\gamma \epsilon^\delta (t-u)(st + tu + us - s^2)$$

$$- i(k_1 \cdot \epsilon)(s+t)[st(t-s) + tu(t-u) + us(u-s)]$$

$$- i(k_2 \cdot \epsilon)(s+u)[su(u-s) + ut(u-t) + ts(t-s)] \},$$

$$(8.62)$$

where ϵ is the polarization vector of the $^3P_1$ quarkonium. Summing over its polarization states and over the color states of the gluons, we obtain

$$\sum_{\text{pol}} |M(+,+,-)|^2 = \frac{6144\,\pi^2 \alpha_S^3 R_1'^2}{M^3} \frac{s^2}{(s-M^2)^2(t-M^2)^4(u-M^2)^4}$$

$$\times \{(s+t)(t+u)(u+s)[st(s-t)^2 + tu(t-u)^2 + us(u-s)^2]$$

$$+ M^2(t-u)^2(st+tu+us-s^2)^2\}. \tag{8.63}$$

Symmetrizing this equation in s, t and u, and multiplying with the appropriate factors, we obtain the cross section formula

$$\frac{d\sigma}{dt} = \frac{12\pi^2 \alpha_S^3 R_1'^2}{M^3 s^2} \frac{1}{[(s-M^2)(t-M^2)(u-M^2)]^2}$$

$$\times \left\{ M^2 \left[\frac{t^2 u^2 (t^2+u^2)}{(s-M^2)^2} + \frac{u^2 s^2 (u^2+s^2)}{(t-M^2)^2} + \frac{s^2 t^2 (s^2+t^2)}{(u-M^2)^2} \right] \right.$$

$$\left. + \frac{2(s^2 t^2 + t^2 u^2 + u^2 s^2)(s^2 t^2 + t^2 u^2 + u^2 s^2 + M^2 stu)}{(s-M^2)(t-M^2)(u-M^2)} \right\}$$

$$= \frac{12\pi \alpha_S^3 R_1'^2}{M^3 s^2}$$

$$\times \frac{P^2[M^2 P^2(M^4 - 4P) - 2Q(M^8 - 5M^4 P - P^2) - 15M^2 Q^2]}{(Q-M^2 P)^4}.$$

$$\tag{8.64}$$

8.3.5 $^3P_2$ production

For the case of $^3P_2$ production, we find again that

$$M(+,+,+) = 0. \tag{8.65}$$

The calculation of the $M(+,+,-)$ amplitude is somewhat more involved. We find

8.3. THE PRODUCTION OF OTHER STATES

$$M(+,+,-) = -\frac{8g^3 f^{abc} R_1' N}{\sqrt{\pi}} \sqrt{M}\, \epsilon^{\alpha\beta}$$

$$\times \left\{ s^2 \left[\frac{k_{1\alpha}}{t(t-M^2)} - \frac{k_{2\alpha}}{u(u-M^2)} \right] \left[t(t+u)k_{1\beta} - u(t+u)k_{2\beta} \right. \right.$$

$$\left. - s(t-u)k_{3\beta} \right] + \frac{st+tu+us}{stu} \left[(s-t-u)(tk_{1\alpha} - uk_{2\alpha})(tk_{1\beta} - uk_{2\beta}) \right.$$

$$\left. + sk_{3\alpha} \left(t(2t-s)k_{1\beta} + u(2u-s)k_{2\beta} \right) + \frac{s^2(t^2+u^2)}{s-M^2} k_{3\alpha} k_{3\beta} \right]$$

$$+ 2i\epsilon_{\mu\nu\rho\alpha} k_1^\mu k_2^\nu k_3^\rho \left[k_{1\beta} \left(\frac{u^2(t-s)}{t(t-M^2)} - t + u - \frac{tu}{s} - \frac{t(st+tu+us)}{su} \right) \right.$$

$$\left. \left. - k_{2\beta} \left(\frac{t^2(u-s)}{u(u-M^2)} - u + t - \frac{ut}{s} - \frac{u(st+tu+us)}{st} \right) \right] \right\}. \tag{8.66}$$

In this equation, $\epsilon^{\alpha\beta}$ is the polarization tensor of the $^3P_2$ quarkonium, which satisfies the relations

$$\epsilon^{\alpha\beta} = \epsilon^{\beta\alpha}, \qquad p_\alpha \epsilon^{\alpha\beta} = 0, \qquad \epsilon^\alpha_\alpha = 0. \tag{8.67}$$

To obtain the cross section, we have to sum over the polarization states of the $^3P_2$ system using

$$\sum_{\text{pol}} \epsilon_{\alpha\beta} \epsilon^*_{\mu\nu} = \tfrac{1}{2}(P_{\alpha\mu} P_{\beta\nu} + P_{\alpha\nu} P_{\beta\mu}) - \tfrac{1}{3} P_{\alpha\beta} P_{\mu\nu}, \tag{8.68}$$

with

$$P_{\mu\nu} = -g_{\mu\nu} + p_\mu p_\nu / M^2. \tag{8.69}$$

Also summing over the color degrees of freedom, we find

$$\sum_{\text{pol}} |M(+,+,-)|^2 = \frac{4096 \pi^2 \alpha_S^3 R_1'^2}{M^3 Q(Q-M^2 P)^4}$$

$$\times \left[12 M^8 P^4 (3s^2 - 4M^2 s + M^4) - 12 M^4 P^5 s(s - 3M^2) \right.$$

$$- 3 M^6 P^3 Q (25 s^2 - 33 M^2 s + 8 M^4) + 12 M^2 P^4 Q (s^2 - 4 M^2 s - 3 M^4)$$

$$+ M^4 P^2 Q^2 (8 s^2 + 9 M^2 s - 15 M^4) - 2 P^3 Q^2 (s^2 - 5 M^2 s - 30 M^4)$$

$$+ M^2 P Q^3 (29 s^2 - 51 M^2 s + 18 M^4) + 2 P^2 Q^3 (s - 11 M^2)$$

$$\left. - M^2 Q^4 (9s - 11 M^2) \right]. \tag{8.70}$$

It follows that, for $^3P_2$ production,

$$\frac{d\sigma}{dt} = \frac{4\pi\alpha_S^3 R_1'^2}{M^3 s^2} \frac{1}{Q(Q-M^2P)^4}$$

$$\times \Big[12M^4P^4(M^8 - 2M^4P + P^2) - 3M^2P^3Q(8M^8 - M^4P + P^2)$$

$$-2P^2Q^2(7M^8 - 43M^4P - P^2) + M^2PQ^3(16M^4 - 61P) + 12M^4Q^4\Big]$$

$$= \frac{4\pi\alpha_S^3 R_1'^2}{M^3 s^2} \frac{1}{[(s-M^2)(t-M^2)(u-M^2)]^2}$$

$$\times \bigg\{ M^2 \bigg[\frac{t^2u^2(t^2 + 4tu + u^2)}{(s-M^2)^2} + \frac{u^2s^2(u^2 + 4us + s^2)}{(t-M^2)^2}$$

$$+ \frac{s^2t^2(s^2 + 4st + t^2)}{(u-M^2)^2} \bigg]$$

$$+ 12M^2\Big[3(s^3t + t^3u + u^3s + st^3 + tu^3 + us^3) + 4M^2stu\Big]$$

$$+ \frac{2(st + tu + us - M^4)(st + tu + us)^2}{(s-M^2)(t-M^2)(u-M^2)} \bigg[st + tu + us - 24M^4$$

$$-6M^2(st + tu + us - M^4)\Big(\frac{1}{s} + \frac{1}{t} + \frac{1}{u}\Big)\bigg]\bigg\}. \quad (8.71)$$

8.4 Conclusions

We have shown that the application of the helicity formalism leads to simple and useful formulae for the process $g + g \to {}^{2S+1}L_J + g$, with $L = S$ or P. For a detailed list of formulae, see Chapter 10. The simplicity of our formulae, especially in the ^3P-cases, allows one to make more refined studies because of the savings in computer time.

The fact that we obtained the separate helicity amplitudes for these processes, enables one to consider the more complicated case of quarkonium production from polarized gluons. We explain in Chapter 11 how the various helicity amplitudes should be combined in this case.

Although the helicity method was originally designed for massless fermion processes, it is seen from these examples that the formalism can also be used advantageously in the case where massive quark pairs are produced.

9
Summary of QED formulae

All the following formulae are presented in the centre-of-mass system, with total energy denoted by $2E$. Unless otherwise stated, we take the positive z-axis along $\vec{p}_+$. For an arbitrary four-vector k, we frequently use the notation [see eqns (6.8)]

$$k_\pm = k_0 \pm k_z, \qquad k_\perp = k_x + ik_y. \tag{9.1}$$

The indication $m = 0$ in the section headings implies that, throughout the section, we neglect the effects of a finite electron and/or muon mass. Consequently, the formulae are not valid in certain kinematical configurations. For a discussion of the ranges of validity of the approximation, we refer to Section 3.4. Alternatively, the indication $m \neq 0$ implies that the formulae in the section are valid to leading order in the fermion masses.

The labels of the listed helicity amplitudes always refer to the particles in the order in which they appear in the given process. For example, for $e^+e^- \to \mu^+\mu^-\gamma$, the helicity amplitude $M(+,-;+,-,+)$ describes the annihilation of a positive helicity e^+ and a negative helicity e^- into a positive helicity μ^+, a negative helicity μ^- and a positive helicity γ. As usual, the symbol '$\dot{=}$' is to be read as 'an equality sign modulo a phase factor'. In this chapter, the symbol $\overline{|M|^2}$ is used to denote the square of the absolute value of M summed over the final state polarizations and averaged over the initial state polarizations.

9.1 $e^+e^- \to \gamma\gamma$ $(m = 0)$

Process:
$$e^+(p_+) + e^-(p_-) \to \gamma(k_1) + \gamma(k_2). \tag{9.2}$$

Invariants:
$$s = 2(p_+ \cdot p_-), \qquad t = -2(p_+ \cdot k_1), \qquad u = -2(p_+ \cdot k_2). \tag{9.3}$$

Photon polarizations:
$$\slashed{\epsilon}_i^\pm = N_i[\slashed{p}_+ \slashed{p}_- \slashed{k}_i(1 \mp \gamma_5) - \slashed{k}_i \slashed{p}_+ \slashed{p}_-(1 \pm \gamma_5)],$$
$$N_i^{-1} = E^2[32k_{i+}k_{i-}]^{\frac{1}{2}}, \qquad i = 1,2. \tag{9.4}$$

Nonvanishing helicity amplitudes:

$$M(+,-;+,-) = 2e^2 \frac{k_{1\perp}}{k_{1-}} \doteq 2e^2 \left(\frac{k_{1+}}{k_{1-}}\right)^{\frac{1}{2}},$$

$$M(+,-;-,+) = 2e^2 \frac{k_{2\perp}}{k_{2-}} \doteq 2e^2 \left(\frac{k_{2+}}{k_{2-}}\right)^{\frac{1}{2}},$$

$$M(-,+;+,-) = 2e^2 \frac{k_{2\perp}^*}{k_{2-}} \doteq 2e^2 \left(\frac{k_{2+}}{k_{2-}}\right)^{\frac{1}{2}},$$

$$M(-,+;-,+) = 2e^2 \frac{k_{1\perp}^*}{k_{1-}} \doteq 2e^2 \left(\frac{k_{1+}}{k_{1-}}\right)^{\frac{1}{2}}.$$

(9.5)

Unpolarized squared matrix element:

$$\overline{|M|^2} = 2e^4 \left(\frac{t}{u} + \frac{u}{t}\right). \tag{9.6}$$

Unpolarized cross section:

$$\frac{d\sigma}{dt} = \frac{2\pi\alpha^2}{s^2} \left(\frac{t}{u} + \frac{u}{t}\right). \tag{9.7}$$

9.2 $e^+e^- \to \mu^+\mu^-$ (no Z-exchange; $m = 0$)

Process:
$$e^+(p_+) + e^-(p_-) \to \mu^+(q_+) + \mu^-(q_-). \tag{9.8}$$

Invariants:
$$s = 2(p_+ \cdot p_-), \qquad t = -2(p_+ \cdot q_+), \qquad u = -2(p_+ \cdot q_-). \tag{9.9}$$

Nonvanishing helicity amplitudes:

$$M(+,-;+,-) = e^2 \frac{q_{+\perp}}{E} \left(\frac{q_{++}}{q_{-+}}\right)^{\frac{1}{2}} \doteq e^2 \frac{q_{++}}{E},$$

$$M(+,-;-,+) = e^2 \frac{q_{-\perp}}{E} \left(\frac{q_{-+}}{q_{++}}\right)^{\frac{1}{2}} \doteq e^2 \frac{q_{-+}}{E},$$

$$M(-,+;+,-) = e^2 \frac{q_{-\perp}^*}{E} \left(\frac{q_{-+}}{q_{++}}\right)^{\frac{1}{2}} \doteq e^2 \frac{q_{-+}}{E},$$

$$M(-,+;-,+) = e^2 \frac{q_{+\perp}^*}{E} \left(\frac{q_{++}}{q_{-+}}\right)^{\frac{1}{2}} \doteq e^2 \frac{q_{++}}{E}.$$

(9.10)

Unpolarized squared matrix element:

$$\overline{|M|^2} = 2e^4 \frac{t^2 + u^2}{s^2}. \tag{9.11}$$

Unpolarized cross section:

$$\frac{d\sigma}{dt} = 2\pi\alpha^2 \frac{t^2 + u^2}{s^4}. \tag{9.12}$$

9.3 $e^+e^- \to \mu^+\mu^-$ (with Z-exchange; $m = 0$)

Process:
$$e^+(p_+) + e^-(p_-) \to \mu^+(q_+) + \mu^-(q_-). \tag{9.13}$$

Invariants and definitions:

$$s = 2(p_+ \cdot p_-), \qquad t = -2(p_+ \cdot q_+), \qquad u = -2(p_+ \cdot q_-),$$

$$A(x,y) = 1 + \frac{4xy}{y - M_Z^2 + iM_Z\Gamma}. \tag{9.14}$$

Nonvanishing helicity amplitudes [for the definitions of a and b, the coupling constants of the Z to the leptons, see eqns (4.47) and (4.48)]:

$$M(+,-;+,-) = e^2 A(a^2,s) \frac{q_{+\perp}}{E} \left(\frac{q_{++}}{q_{-+}}\right)^{\frac{1}{2}} \doteq e^2|A(a^2,s)| \frac{q_{++}}{E},$$

$$M(+,-;-,+) = e^2 A(ab,s) \frac{q_{-\perp}}{E} \left(\frac{q_{-+}}{q_{++}}\right)^{\frac{1}{2}} \doteq e^2|A(ab,s)| \frac{q_{-+}}{E},$$

$$M(-,+;+,-) = e^2 A(ab,s) \frac{q^*_{-\perp}}{E} \left(\frac{q_{-+}}{q_{++}}\right)^{\frac{1}{2}} \doteq e^2|A(ab,s)| \frac{q_{-+}}{E}, \tag{9.15}$$

$$M(-,+;-,+) = e^2 A(b^2,s) \frac{q^*_{+\perp}}{E} \left(\frac{q_{++}}{q_{-+}}\right)^{\frac{1}{2}} \doteq e^2|A(b^2,s)| \frac{q_{++}}{E}.$$

Unpolarized squared matrix element:

$$\overline{|M|^2} = \frac{e^4}{s^2} \left[2t^2|A(ab,s)|^2 + u^2\left(|A(a^2,s)|^2 + |A(b^2,s)|^2\right)\right]. \tag{9.16}$$

Unpolarized cross section:

$$\frac{d\sigma}{dt} = \frac{\pi\alpha^2}{s^4} \left[2t^2|A(ab,s)|^2 + u^2\left(|A(a^2,s)|^2 + |A(b^2,s)|^2\right)\right]. \tag{9.17}$$

9.4 $e^+e^- \to e^+e^-$ (no Z-exchange; $m = 0$)

Process:
$$e^+(p_+) + e^-(p_-) \to e^+(q_+) + e^-(q_-). \qquad (9.18)$$

Invariants:
$$s = 2(p_+ \cdot p_-), \qquad t = -2(p_+ \cdot q_+), \qquad u = -2(p_+ \cdot q_-). \qquad (9.19)$$

Nonvanishing helicity amplitudes:

$$
\begin{aligned}
M(+,+;+,+) &= 4e^2 \frac{Eq_{-\perp}^*}{(q_{++}q_{+-}^3)^{\frac{1}{2}}} \doteq 4e^2 \frac{E}{q_{+-}}, \\
M(+,-;+,-) &= e^2 \frac{q_{-\perp}}{E} \left(\frac{q_{++}}{q_{+-}}\right)^{\frac{3}{2}} \doteq e^2 \frac{q_{++}^2}{Eq_{+-}}, \\
M(+,-;-,+) &= e^2 \frac{q_{-\perp}}{E} \left(\frac{q_{-+}}{q_{++}}\right)^{\frac{1}{2}} \doteq e^2 \frac{q_{-+}}{E}, \\
M(-,+;+,-) &= e^2 \frac{q_{-\perp}^*}{E} \left(\frac{q_{-+}}{q_{++}}\right)^{\frac{1}{2}} \doteq e^2 \frac{q_{-+}}{E}, \\
M(-,+;-,+) &= e^2 \frac{q_{-\perp}^*}{E} \left(\frac{q_{++}}{q_{+-}}\right)^{\frac{3}{2}} \doteq e^2 \frac{q_{++}^2}{Eq_{+-}}, \\
M(-,-;-,-) &= 4e^2 \frac{Eq_{-\perp}}{(q_{++}q_{+-}^3)^{\frac{1}{2}}} \doteq 4e^2 \frac{E}{q_{+-}}.
\end{aligned}
\qquad (9.20)
$$

Unpolarized squared matrix element:
$$\overline{|M|^2} = 2e^4 \frac{s^4 + t^4 + u^4}{s^2 t^2}. \qquad (9.21)$$

Unpolarized cross section:
$$\frac{d\sigma}{dt} = 2\pi\alpha^2 \frac{s^4 + t^4 + u^4}{s^4 t^2}. \qquad (9.22)$$

9.5 $e^+e^- \to e^+e^-$ (with Z-exchange; $m = 0$)

Process:
$$e^+(p_+) + e^-(p_-) \to e^+(q_+) + e^-(q_-). \qquad (9.23)$$

Invariants and definitions:
$$s = 2(p_+ \cdot p_-), \qquad t = -2(p_+ \cdot q_+), \qquad u = -2(p_+ \cdot q_-),$$

9.5. $e^+e^- \to e^+e^-$ (WITH Z-EXCHANGE; $m = 0$)

$$A(x,y) = 1 + \frac{4xy}{y - M_Z^2 + iM_Z\Gamma}. \tag{9.24}$$

Nonvanishing helicity amplitudes [for the definitions of a and b, the coupling constants of the Z to the leptons, see eqns (4.47) and (4.48)]:

$$M(+,+;+,+) = 4e^2 A(ab,t) \frac{E q_{-\perp}^*}{(q_{++}q_{+-}^3)^{\frac{1}{2}}}$$

$$\doteq 4e^2 |A(ab,t)| \frac{E}{q_{+-}},$$

$$M(+,-;+,-) = e^2 \left[A(a^2,s) - \frac{2E}{q_{+-}} A(a^2,t) \right] \frac{q_{+\perp}}{E} \left(\frac{q_{++}}{q_{+-}} \right)^{\frac{1}{2}}$$

$$\doteq e^2 \left| A(a^2,s) - \frac{2E}{q_{+-}} A(a^2,t) \right| \frac{q_{++}}{E},$$

$$M(+,-;-,+) = e^2 A(ab,s) \frac{q_{-\perp}}{E} \left(\frac{q_{-+}}{q_{++}} \right)^{\frac{1}{2}}$$

$$\doteq e^2 |A(ab,s)| \frac{q_{+-}}{E},$$

$$M(-,+;+,-) = e^2 A(ab,s) \frac{q_{-\perp}^*}{E} \left(\frac{q_{-+}}{q_{++}} \right)^{\frac{1}{2}}$$

$$\doteq e^2 |A(ab,s)| \frac{q_{+-}}{E},$$

$$M(-,+;-,+) = e^2 \left[A(b^2,s) - \frac{2E}{q_{+-}} A(b^2,t) \right] \frac{q_{+\perp}^*}{E} \left(\frac{q_{++}}{q_{-+}} \right)^{\frac{1}{2}}$$

$$\doteq e^2 \left| A(b^2,s) - \frac{2E}{q_{+-}} A(b^2,t) \right| \frac{q_{++}}{E},$$

$$M(-,-;-,-) = 4e^2 A(ab,t) \frac{E q_{-\perp}}{(q_{++}q_{+-}^3)^{\frac{1}{2}}}$$

$$\doteq 4e^2 |A(ab,t)| \frac{E}{q_{+-}}. \tag{9.25}$$

Unpolarized squared matrix element:

$$\overline{|M|^2} = \frac{e^4}{s^2 t^2}\left\{t^2 u^2\left[|A(a^2,s)|^2 + \frac{s^2}{t^2}|A(a^2,t)|^2 + \frac{2s}{t}\text{Re}\big(A(a^2,s)A^*(a^2,t)\big)\right.\right.$$

$$\left.+|A(b^2,s)|^2 + \frac{s^2}{t^2}|A(b^2,t)|^2 + \frac{2s}{t}\text{Re}\big(A(b^2,s)A^*(b^2,t)\big)\right]$$

$$\left.+2t^4|A(ab,s)|^2 + 2s^4|A(ab,t)|^2\right\}. \qquad (9.26)$$

Unpolarized cross section:

$$\frac{d\sigma}{dt} = \frac{\pi\alpha^2}{s^4 t^2}\left\{t^2 u^2\left[|A(a^2,s)|^2 + \frac{s^2}{t^2}|A(a^2,t)|^2 + \frac{2s}{t}\text{Re}\big(A(a^2,s)A^*(a^2,t)\big)\right.\right.$$

$$\left.+|A(b^2,s)|^2 + \frac{s^2}{t^2}|A(b^2,t)|^2 + \frac{2s}{t}\text{Re}\big(A(b^2,s)A^*(b^2,t)\big)\right]$$

$$\left.+2t^4|A(ab,s)|^2 + 2s^4|A(ab,t)|^2\right\}. \qquad (9.27)$$

9.6 $e^+ e^- \to \gamma\gamma\gamma$ $(m=0)$

Process:
$$e^+(p_+) + e^-(p_-) \to \gamma(k_1) + \gamma(k_2) + \gamma(k_3). \qquad (9.28)$$

Invariants and definitions:

$$s = 2(p_+ \cdot p_-) \qquad\qquad C = (k_{1+}k_{1-}k_{2+}k_{2-}k_{3+}k_{3-})^{-\frac{1}{2}}. \qquad (9.29)$$

Photon polarizations:

$$\rlap{/}{\epsilon}_i^\pm = N_i[\rlap{/}{p}_+ \rlap{/}{p}_- \rlap{/}{k}_i(1 \mp \gamma_5) - \rlap{/}{k}_i \rlap{/}{p}_+ \rlap{/}{p}_-(1 \pm \gamma_5)],$$

$$N_i^{-1} = E^2[32 k_{i+} k_{i-}]^{\frac{1}{2}}, \qquad\qquad i = 1,2,3. \qquad (9.30)$$

Nonvanishing helicity amplitudes:

$$M(+,-;+,+,-) = -e^3\sqrt{8}C k_{3-}k_{3\perp} \doteq e^3\sqrt{8}C(k_{3+}k_{3-}^3)^{\frac{1}{2}},$$

$$M(+,-;+,-,+) = -e^3\sqrt{8}C k_{2-}k_{2\perp} \doteq e^3\sqrt{8}C(k_{2+}k_{2-}^3)^{\frac{1}{2}},$$

$$M(+,-;-,+,+) = -e^3\sqrt{8}C k_{1-}k_{1\perp} \doteq e^3\sqrt{8}C(k_{1+}k_{1-}^3)^{\frac{1}{2}},$$

$$M(-,+;+,+,-) = e^3\sqrt{8}C k_{3+}k_{3\perp}^* \doteq e^3\sqrt{8}C(k_{3+}^3 k_{3-})^{\frac{1}{2}},$$

9.7. $e^+e^- \to \mu^+\mu^-\gamma$ (NO Z-EXCHANGE; m = 0)

$$\begin{aligned}
M(-,+;+,-,+) &= e^3\sqrt{8}Ck_{2+}k_{2\perp}^* \doteq e^3\sqrt{8}C(k_{2+}^3 k_{2-})^{\frac{1}{2}}, \\
M(-,+;-,+,+) &= e^3\sqrt{8}Ck_{1+}k_{1\perp}^* \doteq e^3\sqrt{8}C(k_{1+}^3 k_{1-})^{\frac{1}{2}}, \\
M(+,-;+,-,-) &= e^3\sqrt{8}Ck_{1+}k_{1\perp} \doteq e^3\sqrt{8}C(k_{1+}^3 k_{1-})^{\frac{1}{2}}, \\
M(+,-;-,+,-) &= e^3\sqrt{8}Ck_{2+}k_{2\perp} \doteq e^3\sqrt{8}C(k_{2+}^3 k_{2-})^{\frac{1}{2}}, \\
M(+,-;-,-,+) &= e^3\sqrt{8}Ck_{3+}k_{3\perp} \doteq e^3\sqrt{8}C(k_{3+}^3 k_{3-})^{\frac{1}{2}}, \\
M(-,+;+,-,-) &= -e^3\sqrt{8}Ck_{1-}k_{1\perp}^* \doteq e^3\sqrt{8}C(k_{1+}k_{1-}^3)^{\frac{1}{2}}, \\
M(-,+;-,+,-) &= -e^3\sqrt{8}Ck_{2-}k_{2\perp}^* \doteq e^3\sqrt{8}C(k_{2+}k_{2-}^3)^{\frac{1}{2}}, \\
M(-,+;-,-,+) &= -e^3\sqrt{8}Ck_{3-}k_{3\perp}^* \doteq e^3\sqrt{8}C(k_{3+}k_{3-}^3)^{\frac{1}{2}}.
\end{aligned} \quad (9.31)$$

Unpolarized squared matrix element:

$$\overline{|M|^2} = 2e^6 s \frac{\sum_{i=1}^{3}(p_+ \cdot k_i)(p_- \cdot k_i)[(p_+ \cdot k_i)^2 + (p_- \cdot k_i)^2]}{\prod_{i=1}^{3}(p_+ \cdot k_i)(p_- \cdot k_i)}. \quad (9.32)$$

Unpolarized cross section:

$$\begin{aligned}
d\sigma &= \frac{\alpha^3}{4\pi^2} \frac{\sum_{i=1}^{3}(p_+ \cdot k_i)(p_- \cdot k_i)[(p_+ \cdot k_i)^2 + (p_- \cdot k_i)^2]}{\prod_{i=1}^{3}(p_+ \cdot k_i)(p_- \cdot k_i)} \\
&\times \delta^4(p_+ + p_- - k_1 - k_2 - k_3)\frac{d^3\vec{k}_1\, d^3\vec{k}_2\, d^3\vec{k}_3}{k_{10}\, k_{20}\, k_{30}}.
\end{aligned} \quad (9.33)$$

9.7 $e^+e^- \to \mu^+\mu^-\gamma$ (no Z-exchange; $m = 0$)

Process:
$$e^+(p_+) + e^-(p_-) \to \mu^+(q_+) + \mu^-(q_-) + \gamma(k). \quad (9.34)$$

Invariants and definitions:

$$\begin{aligned}
s &= 2(p_+ \cdot p_-), & t &= -2(p_+ \cdot q_+), & u &= -2(p_+ \cdot q_-), \\
s' &= 2(q_+ \cdot q_-), & t' &= -2(p_- \cdot q_-), & u' &= -2(p_- \cdot q_+),
\end{aligned}$$

$$B = 1 + \frac{2E}{s'}\left[2q_{-z} + \frac{k_\perp q^*_{-\perp}}{k_+} - \frac{k^*_\perp q_{-\perp}}{k_-}\right],$$

$$Z_{+-} = q_{++}q_{--} - q^*_{+\perp}q_{-\perp},$$

$$Z_{-+} = q_{-+}q_{+-} - q^*_{-\perp}q_{+\perp}. \tag{9.35}$$

Photon polarizations:

$$\not{k}^\pm = N[\not{p}_+ \not{p}_- \not{k}(1 \mp \gamma_5) - \not{k} \not{p}_+ \not{p}_-(1 \pm \gamma_5)],$$

$$N^{-1} = E^2[32k_+k_-]^{\frac{1}{2}}. \tag{9.36}$$

Nonvanishing helicity amplitudes:

$$M(+,-;+,-,+) = -\frac{e^3 q_{-\perp} Z_{+-} B}{[q_{++}q_{-+}(q_+ \cdot q_-)(q_+ \cdot k)(q_- \cdot k)]^{\frac{1}{2}}}$$

$$\doteq e^3 q_{--}|B|\left[\frac{2}{(q_+ \cdot k)(q_- \cdot k)}\right]^{\frac{1}{2}},$$

$$M(+,-;-,+,+) = -\frac{e^3 q_{+\perp} Z_{-+} B}{[q_{++}q_{-+}(q_+ \cdot q_-)(q_+ \cdot k)(q_- \cdot k)]^{\frac{1}{2}}}$$

$$\doteq e^3 q_{+-}|B|\left[\frac{2}{(q_+ \cdot k)(q_- \cdot k)}\right]^{\frac{1}{2}},$$

$$M(-,+;+,-,+) = -\frac{e^3 q^*_{+\perp} Z_{-+} B}{q_{+-}}\left[\frac{q_{-+}}{q_{++}(q_+ \cdot q_-)(q_+ \cdot k)(q_- \cdot k)}\right]^{\frac{1}{2}}$$

$$\doteq e^3 q_{-+}|B|\left[\frac{2}{(q_+ \cdot k)(q_- \cdot k)}\right]^{\frac{1}{2}},$$

$$M(-,+;-,+,+) = -\frac{e^3 q^*_{-\perp} Z_{+-} B}{q_{--}}\left[\frac{q_{++}}{q_{-+}(q_+ \cdot q_-)(q_+ \cdot k)(q_- \cdot k)}\right]^{\frac{1}{2}}$$

$$\doteq e^3 q_{++}|B|\left[\frac{2}{(q_+ \cdot k)(q_- \cdot k)}\right]^{\frac{1}{2}},$$

$$M(+,-;+,-,-) = -\frac{e^3 q_{-\perp} Z^*_{+-} B^*}{q_{--}}\left[\frac{q_{++}}{q_{-+}(q_+ \cdot q_-)(q_+ \cdot k)(q_- \cdot k)}\right]^{\frac{1}{2}}$$

$$\doteq e^3 q_{++}|B|\left[\frac{2}{(q_+ \cdot k)(q_- \cdot k)}\right]^{\frac{1}{2}},$$

$$M(+,-;-,+,-) = -\frac{e^3 q_{+\perp} Z^*_{-+} B^*}{q_{+-}} \left[\frac{q_{-+}}{q_{++}(q_+ \cdot q_-)(q_+ \cdot k)(q_- \cdot k)} \right]^{\frac{1}{2}}$$

$$\doteq e^3 q_{-+} |B| \left[\frac{2}{(q_+ \cdot k)(q_- \cdot k)} \right]^{\frac{1}{2}},$$

$$M(-,+;+,-,-) = -\frac{e^3 q^*_{+\perp} Z^*_{-+} B^*}{[q_{++} q_{-+}(q_+ \cdot q_-)(q_+ \cdot k)(q_- \cdot k)]^{\frac{1}{2}}}$$

$$\doteq e^3 q_{+-} |B| \left[\frac{2}{(q_+ \cdot k)(q_- \cdot k)} \right]^{\frac{1}{2}},$$

$$M(-,+;-,+,-) = -\frac{e^3 q^*_{-\perp} Z^*_{+-} B^*}{[q_{++} q_{-+}(q_+ \cdot q_-)(q_+ \cdot k)(q_- \cdot k)]^{\frac{1}{2}}}$$

$$\doteq e^3 q_{--} |B| \left[\frac{2}{(q_+ \cdot k)(q_- \cdot k)} \right]^{\frac{1}{2}}. \tag{9.37}$$

Unpolarized squared matrix element:

$$\overline{|M|^2} = -e^6 (v_p - v_q)^2 \frac{t^2 + t'^2 + u^2 + u'^2}{ss'}, \tag{9.38}$$

where [see eqns (4.27)]

$$v_p = \frac{p_+}{(p_+ \cdot k)} - \frac{p_-}{(p_- \cdot k)}, \qquad v_q = \frac{q_+}{(q_+ \cdot k)} - \frac{q_-}{(q_- \cdot k)}. \tag{9.39}$$

Unpolarized cross section:

$$d\sigma = -\frac{\alpha^3}{8\pi^2}(v_p - v_q)^2 \frac{t^2 + t'^2 + u^2 + u'^2}{s^2 s'}$$

$$\times \delta^4(p_+ + p_- - q_+ - q_- - k) \frac{d^3\vec{q}_+ \, d^3\vec{q}_- \, d^3\vec{k}}{q_{+0} \, q_{-0} \, k_0}. \tag{9.40}$$

9.8 $e^+ e^- \to \mu^+ \mu^- \gamma$ (with Z-exchange; $m = 0$)

Process:
$$e^+(p_+) + e^-(p_-) \to \mu^+(q_+) + \mu^-(q_-) + \gamma(k). \tag{9.41}$$

Invariants and definitions:

$$s = 2(p_+ \cdot p_-), \qquad t = -2(p_+ \cdot q_+), \qquad u = -2(p_+ \cdot q_-),$$

$$s' = 2(q_+ \cdot q_-), \qquad t' = -2(p_- \cdot q_-), \qquad u' = -2(p_- \cdot q_+),$$

$$B_1 = \frac{E}{(q_+ \cdot q_-)}\left[2q_{-z} + \frac{k_\perp q^*_{-\perp}}{k_+} - \frac{k^*_\perp q_{-\perp}}{k_-}\right],$$

$$Z_{+-} = q_{++}q_{--} - q^*_{+\perp}q_{-\perp},$$

$$Z_{-+} = q_{-+}q_{+-} - q^*_{-\perp}q_{+\perp},$$

$$A(x,y) = 1 + \frac{4xy}{y - M_Z^2 + iM_Z\Gamma}. \tag{9.42}$$

Photon polarizations:

$$\not{\epsilon}^{\pm} = N[\not{p}_+ \not{p}_- \not{k}(1 \mp \gamma_5) - \not{k}\not{p}_+\not{p}_-(1 \pm \gamma_5)],$$

$$N^{-1} = E^2[32k_+k_-]^{\frac{1}{2}}. \tag{9.43}$$

Nonvanishing helicity amplitudes [for the definitions of a and b, the coupling constants of the Z to the leptons, see eqns (4.47) and (4.48)]:

$$M(+,-;+,-,+) = -\frac{e^3 q_{-\perp} Z_{+-}[A(a^2,s) + B_1 A(a^2,s')]}{[q_{++}q_{-+}(q_+ \cdot q_-)(q_+ \cdot k)(q_- \cdot k)]^{\frac{1}{2}}}$$

$$\doteq e^3 q_{--}\left[\frac{2}{(q_+ \cdot k)(q_- \cdot k)}\right]^{\frac{1}{2}} |A(a^2,s) + B_1 A(a^2,s')|,$$

$$M(+,-;-,+,+) = -\frac{e^3 q_{+\perp} Z_{-+}[A(ab,s) + B_1 A(ab,s')]}{[q_{++}q_{-+}(q_+ \cdot q_-)(q_+ \cdot k)(q_- \cdot k)]^{\frac{1}{2}}}$$

$$\doteq e^3 q_{+-}\left[\frac{2}{(q_+ \cdot k)(q_- \cdot k)}\right]^{\frac{1}{2}} |A(ab,s) + B_1 A(ab,s')|,$$

$$M(-,+;+,-,+) = -\frac{e^3 q^*_{+\perp} Z_{-+}}{q_{+-}}\left[\frac{q_{-+}}{q_{++}(q_+ \cdot q_-)(q_+ \cdot k)(q_- \cdot k)}\right]^{\frac{1}{2}}$$

$$\times [A(ab,s) + B_1 A(ab,s')]$$

$$\doteq e^3 q_{-+}\left[\frac{2}{(q_+ \cdot k)(q_- \cdot k)}\right]^{\frac{1}{2}} |A(ab,s) + B_1 A(ab,s')|,$$

9.8. $e^+e^- \to \mu^+\mu^-\gamma$ (WITH Z-EXCHANGE; $m = 0$)

$$M(-,+;-,+,+) = -\frac{e^3 q_{-\perp}^* Z_{+-}}{q_{--}} \left[\frac{q_{++}}{q_{-+}(q_+ \cdot q_-)(q_+ \cdot k)(q_- \cdot k)}\right]^{\frac{1}{2}}$$

$$\times [A(b^2, s) + B_1 A(b^2, s')]$$

$$\doteq e^3 q_{++} \left[\frac{2}{(q_+ \cdot k)(q_- \cdot k)}\right]^{\frac{1}{2}} |A(b^2, s) + B_1 A(b^2, s')|,$$

$$M(+,-;+,-,-) = -\frac{e^3 q_{-\perp} Z_{+-}^* \sqrt{q_{++}} [A(a^2, s) + B_1^* A(a^2, s')]}{q_{--}[q_{-+}(q_+ \cdot q_-)(q_+ \cdot k)(q_- \cdot k)]^{\frac{1}{2}}}$$

$$\doteq e^3 q_{++} \left[\frac{2}{(q_+ \cdot k)(q_- \cdot k)}\right]^{\frac{1}{2}} |A(a^2, s) + B_1^* A(a^2, s')|,$$

$$M(+,-;-,+,-) = -\frac{e^3 q_{+\perp} Z_{-+}^* \sqrt{q_{-+}} [A(ab, s) + B_1^* A(ab, s')]}{q_{+-}[q_{++}(q_+ \cdot q_-)(q_+ \cdot k)(q_- \cdot k)]^{\frac{1}{2}}}$$

$$\doteq e^3 q_{-+} \left[\frac{2}{(q_+ \cdot k)(q_- \cdot k)}\right]^{\frac{1}{2}} |A(ab, s) + B_1^* A(ab, s')|,$$

$$M(-,+;+,-,-) = -\frac{e^3 q_{+\perp}^* Z_{-+}^* [A(ab, s) + B_1^* A(ab, s')]}{[q_{++}q_{-+}(q_+ \cdot q_-)(q_+ \cdot k)(q_- \cdot k)]^{\frac{1}{2}}}$$

$$\doteq e^3 q_{+-} \left[\frac{2}{(q_+ \cdot k)(q_- \cdot k)}\right]^{\frac{1}{2}} |A(ab, s) + B_1^* A(ab, s')|,$$

$$M(-,+;-,+,-) = -\frac{e^3 q_{-\perp}^* Z_{+-}^* [A(b^2, s) + B_1^* A(b^2, s')]}{[q_{++}q_{-+}(q_+ \cdot q_-)(q_+ \cdot k)(q_- \cdot k)]^{\frac{1}{2}}}$$

$$\doteq e^3 q_{--} \left[\frac{2}{(q_+ \cdot k)(q_- \cdot k)}\right]^{\frac{1}{2}} |A(b^2, s) + B_1^* A(b^2, s')|. \quad (9.44)$$

Unpolarized squared matrix element:

$$\overline{|M|^2} = \frac{e^6}{2(q_+ \cdot k)(q_- \cdot k)}$$

$$\times \left\{ q_{++}^2 \left[|A(b^2, s) + B_1 A(b^2, s')|^2 + |A(a^2, s) + B_1^* A(a^2, s')|^2\right] \right.$$

$$+ (q_{+-}^2 + q_{-+}^2)\left[|A(ab, s) + B_1 A(ab, s')|^2 + |A(ab, s) + B_1^* A(ab, s')|^2\right]$$

$$\left. + q_{--}^2 \left[|A(a^2, s) + B_1 A(a^2, s')|^2 + |A(a^2, s) + B_1^* A(a^2, s')|^2\right] \right\}.$$

$$(9.45)$$

Unpolarized cross section:

$$d\sigma = \frac{\alpha^3}{16\pi^2 s(q_+ \cdot k)(q_- \cdot k)}$$

$$\times \left\{ q_{++}^2 \left[|A(b^2,s) + B_1 A(b^2,s')|^2 + |A(a^2,s) + B_1^* A(a^2,s')|^2 \right] \right.$$

$$+ (q_{+-}^2 + q_{-+}^2) \left[|A(ab,s) + B_1 A(ab,s')|^2 + |A(ab,s) + B_1^* A(ab,s')|^2 \right]$$

$$\left. + q_{--}^2 \left[|A(a^2,s) + B_1 A(a^2,s')|^2 + |A(a^2,s) + B_1^* A(a^2,s')|^2 \right] \right\}$$

$$\times \delta^4(p_+ + p_- - q_+ - q_- - k) \frac{d^3\vec{q}_+\, d^3\vec{q}_-\, d^3\vec{k}}{q_{+0}\, q_{-0}\, k_0}. \quad (9.46)$$

9.9 $e^+e^- \to e^+e^-\gamma$ (no Z-exchange; $m = 0$)

Process:
$$e^+(p_+) + e^-(p_-) \to e^+(q_+) + e^-(q_-) + \gamma(k). \quad (9.47)$$

Invariants and definitions:

$$s = 2(p_+ \cdot p_-), \quad t = -2(p_+ \cdot q_+), \quad u = -2(p_+ \cdot q_-),$$

$$s' = 2(q_+ \cdot q_-), \quad t' = -2(p_- \cdot q_-), \quad u' = -2(p_- \cdot q_+),$$

$$B = 1 + \frac{E}{(q_+ \cdot q_-)} \left[2q_{-z} + \frac{k_\perp q_{-\perp}^*}{k_+} - \frac{k_\perp^* q_{-\perp}}{k_-} \right],$$

$$B_0 = 2 - \frac{k_\perp q_{+\perp}^*}{k_+ q_{+-}} - \frac{k_\perp^* q_{-\perp}}{k_- q_{-+}},$$

$$Z_{+-} = q_{++} q_{--} - q_{+\perp}^* q_{-\perp},$$

$$Z_{-+} = q_{-+} q_{+-} - q_{-\perp}^* q_{+\perp}. \quad (9.48)$$

Photon polarizations:

$$\not{e}^\pm = N[\not{p}_+ \not{p}_- \not{k}(1 \mp \gamma_5) - \not{k} \not{p}_+ \not{p}_-(1 \pm \gamma_5)],$$

$$N^{-1} = E^2 [32 k_+ k_-]^{\frac{1}{2}}. \quad (9.49)$$

9.9. $e^+e^- \to e^+e^-\gamma$ (NO Z-EXCHANGE; $m = 0$)

Nonvanishing helicity amplitudes:

$$M(+,+;+,+,+) = -\frac{4e^3 E q_{+\perp}^* B}{q_{+-}} \left[\frac{(q_+ \cdot q_-)}{q_{++}q_{-+}(q_+ \cdot k)(q_- \cdot k)}\right]^{\frac{1}{2}}$$

$$\doteq 2e^3 E|B_0| \left[\frac{2}{(q_+ \cdot k)(q_- \cdot k)}\right]^{\frac{1}{2}},$$

$$M(+,+;+,+,-) = -\frac{e^3 q_{+\perp}^* Z_{-+} B_0^*}{E q_{+-}} \left[\frac{(q_+ \cdot q_-)}{q_{++}q_{-+}(q_+ \cdot k)(q_- \cdot k)}\right]^{\frac{1}{2}}$$

$$\doteq \frac{e^3 (q_+ \cdot q_-)|B_0|}{E} \left[\frac{2}{(q_+ \cdot k)(q_- \cdot k)}\right]^{\frac{1}{2}},$$

$$M(+,-;+,-,+) = -\frac{e^3 q_{-\perp}^2 q_{+\perp}^* Z_{+-} B}{q_{+-}[q_{++}q_{-+}^3(q_+ \cdot q_-)(q_+ \cdot k)(q_- \cdot k)]^{\frac{1}{2}}}$$

$$\doteq e^3 |B_0| \left[\frac{q_{++}q_{--}^3}{(q_+ \cdot q_-)(q_+ \cdot k)(q_- \cdot k)}\right]^{\frac{1}{2}},$$

$$M(+,-;-,+,+) = -\frac{e^3 q_{+\perp} Z_{-+} B}{[q_{++}q_{-+}(q_+ \cdot q_-)(q_+ \cdot k)(q_- \cdot k)]^{\frac{1}{2}}}$$

$$\doteq e^3 q_{+-}|B| \left[\frac{2}{(q_+ \cdot k)(q_- \cdot k)}\right]^{\frac{1}{2}},$$

$$M(-,+;+,-,+) = -\frac{e^3 q_{+\perp}^* Z_{-+} B}{q_{+-}} \left[\frac{q_{-+}}{q_{++}(q_+ \cdot q_-)(q_+ \cdot k)(q_- \cdot k)}\right]^{\frac{1}{2}}$$

$$\doteq e^3 q_{-+}|B| \left[\frac{2}{(q_+ \cdot k)(q_- \cdot k)}\right]^{\frac{1}{2}},$$

$$M(-,+;-,+,+) = -\frac{e^3 q_{+\perp}^* Z_{+-} B}{q_{+-}} \left[\frac{q_{++}}{q_{-+}(q_+ \cdot q_-)(q_+ \cdot k)(q_- \cdot k)}\right]^{\frac{1}{2}}$$

$$\doteq e^3 |B_0| \left[\frac{q_{++}^3 q_{--}}{(q_+ \cdot q_-)(q_+ \cdot k)(q_- \cdot k)}\right]^{\frac{1}{2}},$$

$$M(+,-;+,-,-) = -\frac{e^3 q_{+\perp} Z_{+-}^* B^*}{q_{+-}} \left[\frac{q_{++}}{q_{-+}(q_+ \cdot q_-)(q_+ \cdot k)(q_- \cdot k)}\right]^{\frac{1}{2}}$$

$$\doteq e^3 |B_0| \left[\frac{q_{++}^3 q_{--}}{(q_+ \cdot q_-)(q_+ \cdot k)(q_- \cdot k)}\right]^{\frac{1}{2}},$$

$$M(+,-;-,+,-) = -\frac{e^3 q_{+\perp} Z^*_{-+} B^*}{q_{+-}} \left[\frac{q_{-+}}{q_{++}(q_+ \cdot q_-)(q_+ \cdot k)(q_- \cdot k)}\right]^{\frac{1}{2}}$$

$$\doteq e^3 q_{-+}|B| \left[\frac{2}{(q_+ \cdot k)(q_- \cdot k)}\right]^{\frac{1}{2}},$$

$$M(-,+;+,-,-) = -\frac{e^3 q^*_{+\perp} Z^*_{-+} B^*}{[q_{++} q_{-+}(q_+ \cdot q_-)(q_+ \cdot k)(q_- \cdot k)]^{\frac{1}{2}}}$$

$$\doteq e^3 q_{+-}|B| \left[\frac{2}{(q_+ \cdot k)(q_- \cdot k)}\right]^{\frac{1}{2}},$$

$$M(-,+;-,+,-) = -\frac{e^3 q^{*2}_{-\perp} q_{+\perp} Z^*_{+-} B^*}{q_{+-}[q_{++} q^3_{-+}(q_+ \cdot q_-)(q_+ \cdot k)(q_- \cdot k)]^{\frac{1}{2}}}$$

$$\doteq e^3 |B_0| \left[\frac{q_{++} q^3_{--}}{(q_+ \cdot q_-)(q_+ \cdot k)(q_- \cdot k)}\right]^{\frac{1}{2}},$$

$$M(-,-;-,-,+) = -\frac{e^3 q_{+\perp} Z^*_{-+} B_0}{E q_{+-}} \left[\frac{(q_+ \cdot q_-)}{q_{++} q_{-+}(q_+ \cdot k)(q_- \cdot k)}\right]^{\frac{1}{2}}$$

$$\doteq \frac{e^3(q_+ \cdot q_-)|B_0|}{E} \left[\frac{2}{(q_+ \cdot k)(q_- \cdot k)}\right]^{\frac{1}{2}},$$

$$M(-,-;-,-,-) = -\frac{4e^3 E q_{+\perp} B^*}{q_{+-}} \left[\frac{(q_+ \cdot q_-)}{q_{++} q_{-+}(q_+ \cdot k)(q_- \cdot k)}\right]^{\frac{1}{2}}$$

$$\doteq 2e^3 E|B_0| \left[\frac{2}{(q_+ \cdot k)(q_- \cdot k)}\right]^{\frac{1}{2}}. \qquad (9.50)$$

Unpolarized squared matrix element:

$$\overline{|M|^2} = -e^6(v_p - v_q)^2 \frac{ss'(s^2 + s'^2) + tt'(t^2 + t'^2) + uu'(u^2 + u'^2)}{ss'tt'}, \qquad (9.51)$$

where [see eqns (4.27)]

$$v_p = \frac{p_+}{(p_+ \cdot k)} - \frac{p_-}{(p_- \cdot k)}, \qquad v_q = \frac{q_+}{(q_+ \cdot k)} - \frac{q_-}{(q_- \cdot k)}. \qquad (9.52)$$

Unpolarized cross section:

$$d\sigma = -\frac{\alpha^3}{8\pi^2}(v_p - v_q)^2 \frac{ss'(s^2 + s'^2) + tt'(t^2 + t'^2) + uu'(u^2 + u'^2)}{s^2 s' tt'}$$

$$\times \delta^4(p_+ + p_- - q_+ - q_- - k) \frac{d^3\vec{q}_+ \, d^3\vec{q}_- \, d^3\vec{k}}{q_{+0}\, q_{-0}\, k_0}. \qquad (9.53)$$

9.10 $e^+e^- \to e^+e^-\gamma$ (with Z-exchange; $m = 0$)

Process:
$$e^+(p_+) + e^-(p_-) \to e^+(q_+) + e^-(q_-) + \gamma(k). \qquad (9.54)$$

Invariants and definitions:

$$s = 2(p_+ \cdot p_-), \quad t = -2(p_+ \cdot q_+), \quad u = -2(p_+ \cdot q_-),$$

$$s' = 2(q_+ \cdot q_-), \quad t' = -2(p_- \cdot q_-), \quad u' = -2(p_- \cdot q_+),$$

$$B_1 = \frac{E}{(q_+ \cdot q_-)}\left[2q_{-z} + \frac{k_\perp q^*_{-\perp}}{k_+} - \frac{k^*_\perp q_{-\perp}}{k_-}\right],$$

$$B_2 = 1 - \frac{k_\perp q^*_{+\perp}}{k_+ q_{+-}}, \qquad B_3 = 1 - \frac{k^*_\perp q_{-\perp}}{k_- q_{-+}},$$

$$Z_{+-} = q_{++}q_{--} - q^*_{+\perp}q_{-\perp}, \qquad Z_{-+} = q_{-+}q_{+-} - q^*_{-\perp}q_{+\perp},$$

$$A(x,y) = 1 + \frac{4xy}{y - M_Z^2 + iM_Z\Gamma}. \qquad (9.55)$$

Photon polarizations:

$$\not{e}^\pm = N[\not{p}_+ \not{p}_- \not{k}(1 \mp \gamma_5) - \not{k} \not{p}_+ \not{p}_-(1 \pm \gamma_5)],$$

$$N^{-1} = E^2[32k_+k_-]^{\frac{1}{2}}. \qquad (9.56)$$

Nonvanishing helicity amplitudes [for the definitions of a and b, the coupling constants of the Z to the leptons, see eqns (4.47) and (4.48)]:

$$M(+,+;+,+,+) = -\frac{2e^3 E q^*_{+\perp} Z_{-+}[B_2 A(ab,t) + B_3 A(ab,t')]}{q_{+-}[q_{++}q_{-+}(q_+ \cdot q_-)(q_+ \cdot k)(q_- \cdot k)]^{\frac{1}{2}}}$$

$$\doteq 2e^3 E \left[\frac{2}{(q_+ \cdot k)(q_- \cdot k)}\right]^{\frac{1}{2}} |B_2 A(ab,t) + B_3 A(ab,t')|,$$

$$M(+,+;+,+,-) = -\frac{e^3 q^*_{+\perp} Z_{-+}}{E q_{+-}} \left[\frac{(q_+ \cdot q_-)}{q_{++}q_{-+}(q_+ \cdot k)(q_- \cdot k)}\right]^{\frac{1}{2}}$$

$$\times [B_2^* A(ab,t) + B_3^* A(ab,t')]$$

$$\doteq \frac{e^3(q_+ \cdot q_-)}{E}\left[\frac{2}{(q_+ \cdot k)(q_- \cdot k)}\right]^{\frac{1}{2}}|B_2^* A(ab,t) + B_3^* A(ab,t')|,$$

$$M(+,-;+,-,+) = -\frac{e^3 q_{-\perp} Z_{+-}}{[q_{++}q_{-+}(q_+ \cdot q_-)(q_+ \cdot k)(q_- \cdot k)]^{\frac{1}{2}}}$$
$$\times [A(a^2,s) + B_1 A(a^2,s') - B_2 A(a^2,t) - B_3 A(a^2,t')]$$
$$\doteq e^3 q_{--} \left[\frac{2}{(q_+ \cdot k)(q_- \cdot k)}\right]^{\frac{1}{2}}$$
$$\times |A(a^2,s) + B_1 A(a^2,s') - B_2 A(a^2,t) - B_3 A(a^2,t')|,$$

$$M(+,-;-,+,+) = -\frac{e^3 q_{+\perp} Z_{-+}[A(ab,s) + B_1 A(ab,s')]}{[q_{++}q_{-+}(q_+ \cdot q_-)(q_+ \cdot k)(q_- \cdot k)]^{\frac{1}{2}}}$$
$$\doteq e^3 q_{+-} \left[\frac{2}{(q_+ \cdot k)(q_- \cdot k)}\right]^{\frac{1}{2}} |A(ab,s) + B_1 A(ab,s')|,$$

$$M(-,+;+,-,+) = -\frac{e^3 q_{+\perp}^* Z_{-+} \sqrt{q_{-+}} [A(ab,s) + B_1 A(ab,s')]}{q_{+-}[q_{++}(q_+ \cdot q_-)(q_+ \cdot k)(q_- \cdot k)]^{\frac{1}{2}}}$$
$$\doteq e^3 q_{-+} \left[\frac{2}{(q_+ \cdot k)(q_- \cdot k)}\right]^{\frac{1}{2}} |A(ab,s) + B_1 A(ab,s')|,$$

$$M(-,+;-,+,+) = -\frac{e^3 q_{-\perp}^* Z_{+-}}{q_{--}} \left[\frac{q_{++}}{q_{-+}(q_+ \cdot q_-)(q_+ \cdot k)(q_- \cdot k)}\right]^{\frac{1}{2}}$$
$$\times [A(b^2,s) + B_1 A(b^2,s') - B_2 A(b^2,t) - B_3 A(b^2,t')]$$
$$\doteq e^3 q_{++} \left[\frac{2}{(q_+ \cdot k)(q_- \cdot k)}\right]^{\frac{1}{2}}$$
$$\times |A(b^2,s) + B_1 A(b^2,s') - B_2 A(b^2,t) - B_3 A(b^2,t')|,$$

$$M(+,-;+,-,-) = -\frac{e^3 q_{-\perp} Z_{+-}^*}{q_{--}} \left[\frac{q_{++}}{q_{-+}(q_+ \cdot q_-)(q_+ \cdot k)(q_- \cdot k)}\right]^{\frac{1}{2}}$$
$$\times [A(b^2,s) + B_1^* A(b^2,s') - B_2^* A(b^2,t) - B_3^* A(b^2,t')]$$
$$\doteq e^3 q_{++} \left[\frac{2}{(q_+ \cdot k)(q_- \cdot k)}\right]^{\frac{1}{2}}$$
$$\times |A(b^2,s) + B_1^* A(b^2,s') - B_2^* A(b^2,t) - B_3^* A(b^2,t')|,$$

9.10. $e^+e^- \to e^+e^-\gamma$ (WITH Z-EXCHANGE; m = 0)

$$M(+,-;-,+,-) = -\frac{e^3 q_{+\perp} Z^*_{-+}}{q_{+-}} \left[\frac{q_{-+}}{q_{++}(q_+ \cdot q_-)(q_+ \cdot k)(q_- \cdot k)}\right]^{\frac{1}{2}}$$

$$\times [A(ab,s) + B_1^* A(ab,s')]$$

$$\doteq e^3 q_{-+} \left[\frac{2}{(q_+ \cdot k)(q_- \cdot k)}\right]^{\frac{1}{2}} |A(ab,s) + B_1^* A(ab,s')|,$$

$$M(-,+;+,-,-) = -\frac{e^3 q^*_{+\perp} Z^*_{-+}[A(ab,s) + B_1^* A(ab,s')]}{[q_{++}q_{-+}(q_+ \cdot q_-)(q_+ \cdot k)(q_- \cdot k)]^{\frac{1}{2}}}$$

$$\doteq e^3 q_{+-} \left[\frac{2}{(q_+ \cdot k)(q_- \cdot k)}\right]^{\frac{1}{2}} |A(ab,s) + B_1^* A(ab,s')|,$$

$$M(-,+;-,+,-) = -\frac{e^3 q^*_{-\perp} Z^*_{+-}}{[q_{++}q_{-+}(q_+ \cdot q_-)(q_+ \cdot k)(q_- \cdot k)]^{\frac{1}{2}}}$$

$$\times [A(a^2,s) + B_1^* A(a^2,s') - B_2^* A(a^2,t) - B_3^* A(a^2,t')]$$

$$\doteq e^3 q_{--} \left[\frac{2}{(q_+ \cdot k)(q_- \cdot k)}\right]^{\frac{1}{2}}$$

$$\times |A(a^2,s) + B_1^* A(a^2,s') - B_2^* A(a^2,t) - B_3^* A(a^2,t')|,$$

$$M(-,-;-,-,+) = -\frac{e^3 q_{+\perp} Z^*_{-+}}{E q_{+-}} \left[\frac{(q_+ \cdot q_-)}{q_{++}q_{-+}(q_+ \cdot k)(q_- \cdot k)}\right]^{\frac{1}{2}}$$

$$\times [B_2 A(ab,t) + B_3 A(ab,t')]$$

$$\doteq \frac{e^3(q_+ \cdot q_-)}{E} \left[\frac{2}{(q_+ \cdot k)(q_- \cdot k)}\right]^{\frac{1}{2}} |B_2 A(ab,t) + B_3 A(ab,t')|,$$

$$M(-,-;-,-,-) = -\frac{2e^3 E q_{+\perp} Z^*_{-+}[B_2^* A(ab,t) + B_3^* A(ab,t')]}{q_{+-}[q_{++}q_{-+}(q_+ \cdot q_-)(q_+ \cdot k)(q_- \cdot k)]^{\frac{1}{2}}}$$

$$\doteq 2e^3 E \left[\frac{2}{(q_+ \cdot k)(q_- \cdot k)}\right]^{\frac{1}{2}} |B_2^* A(ab,t) + B_3^* A(ab,t')|. \quad (9.57)$$

Unpolarized squared matrix element:

$$\overline{|M|^2} = \frac{e^6}{2(q_+ \cdot k)(q_- \cdot k)}$$

$$\times \left\{ \left(s + \frac{s'^2}{s} \right) \left[|B_2 A(ab,t) + B_3 A(ab,t')|^2 + |B_2^* A(ab,t) + B_3^* A(ab,t')|^2 \right] \right.$$

$$+ q_{--}^2 \left[|A(a^2,s) + B_1 A(a^2,s') - B_2 A(a^2,t) - B_3 A(a^2,t')|^2 \right.$$

$$\left. + |A(b^2,s) + B_1^* A(b^2,s') - B_2^* A(b^2,t) - B_3^* A(b^2,t')|^2 \right]$$

$$+ q_{++}^2 \left[|A(b^2,s) + B_1 A(b^2,s') - B_2 A(b^2,t) - B_3 A(b^2,t')|^2 \right.$$

$$\left. + |A(a^2,s) + B_1^* A(a^2,s') - B_2^* A(a^2,t) - B_3^* A(a^2,t')|^2 \right]$$

$$\left. + (q_{+-}^2 + q_{-+}^2) \left[|A(ab,s) + B_1 A(ab,s')|^2 + |A(ab,s) + B_1^* A(ab,s')|^2 \right] \right\}.$$

(9.58)

Unpolarized cross section:

$$d\sigma = \frac{\alpha^3}{16\pi^2 s(q_+ \cdot k)(q_- \cdot k)}$$

$$\times \left\{ \left(s + \frac{s'^2}{s} \right) \left[|B_2 A(ab,t) + B_3 A(ab,t')|^2 + |B_2^* A(ab,t) + B_3^* A(ab,t')|^2 \right] \right.$$

$$+ q_{--}^2 \left[|A(a^2,s) + B_1 A(a^2,s') - B_2 A(a^2,t) - B_3 A(a^2,t')|^2 \right.$$

$$\left. + |A(b^2,s) + B_1^* A(b^2,s') - B_2^* A(b^2,t) - B_3^* A(b^2,t')|^2 \right]$$

$$+ q_{++}^2 \left[|A(b^2,s) + B_1 A(b^2,s') - B_2 A(b^2,t) - B_3 A(b^2,t')|^2 \right.$$

$$\left. + |A(a^2,s) + B_1^* A(a^2,s') - B_2^* A(a^2,t) - B_3^* A(a^2,t')|^2 \right]$$

$$\left. + (q_{+-}^2 + q_{-+}^2) \left[|A(ab,s) + B_1 A(ab,s')|^2 + |A(ab,s) + B_1^* A(ab,s')|^2 \right] \right\}$$

$$\times \delta^4(p_+ + p_- - q_+ - q_- - k) \frac{d^3\vec{q}_+ \, d^3\vec{q}_- \, d^3\vec{k}}{q_{+0}\, q_{-0}\, k_0}.$$

(9.59)

9.11 $e^+e^- \to \gamma\gamma\gamma\gamma$ $(m=0)$

Process:
$$e^+(p_+) + e^-(p_-) \to \gamma(k_1) + \gamma(k_2) + \gamma(k_3) + \gamma(k_4). \quad (9.60)$$

Invariants and definitions:
$$B = (k_{1+}k_{1-}k_{2+}k_{2-}k_{3+}k_{3-}k_{4+}k_{4-})^{-\frac{1}{2}},$$

$$Z_{ij} = k_{i+}k_{j-} - k_{i\perp}^* k_{j\perp},$$

$$\Delta_{ij} = -2(p_- \cdot k_i) - 2(p_- \cdot k_j) + 2(k_i \cdot k_j), \quad i,j = 1,\ldots,4,$$

$$F(1,2,3,4) = (k_{1\perp} + k_{2\perp})\Delta_{12} + \frac{k_{4\perp}}{\Delta_{13}}(2Ek_{2+}k_{3\perp}^* k_{1\perp} + Z_{21}Z_{13}^*)$$

$$+ \frac{k_{3\perp}}{\Delta_{14}}(2Ek_{2+}k_{4\perp}^* k_{1\perp} + Z_{21}Z_{14}^*)$$

$$+ \frac{k_{4\perp}}{\Delta_{23}}(2Ek_{1+}k_{3\perp}^* k_{2\perp} + Z_{12}Z_{23}^*)$$

$$+ \frac{k_{3\perp}}{\Delta_{24}}(2Ek_{1+}k_{4\perp}^* k_{2\perp} + Z_{12}Z_{24}^*). \quad (9.61)$$

Photon polarizations:
$$\not{\epsilon}_i^{\pm} = N_i[\not{p}_+ \not{p}_- \not{k}_i(1 \mp \gamma_5) - \not{k}_i \not{p}_+ \not{p}_-(1 \pm \gamma_5)],$$

$$N_i^{-1} = E^2[32k_{i+}k_{i-}]^{\frac{1}{2}}, \qquad i = 1,2,3,4. \quad (9.62)$$

Nonvanishing helicity amplitudes:
$$M(+,-;+,+,+,-) = -4e^4 B k_{4-} k_{4\perp} \doteq 4e^4 B (k_{4+} k_{4-}^3)^{\frac{1}{2}},$$

$$M(+,-;+,+,-,+) = -4e^4 B k_{3-} k_{3\perp} \doteq 4e^4 B (k_{3+} k_{3-}^3)^{\frac{1}{2}},$$

$$M(+,-;+,-,+,+) = -4e^4 B k_{2-} k_{2\perp} \doteq 4e^4 B (k_{2+} k_{2-}^3)^{\frac{1}{2}},$$

$$M(+,-;-,+,+,+) = -4e^4 B k_{1-} k_{1\perp} \doteq 4e^4 B (k_{1+} k_{1-}^3)^{\frac{1}{2}},$$

$$M(-,+;+,+,+,-) = 4e^4 B k_{4+} k_{4\perp}^* \doteq 4e^4 B (k_{4+}^3 k_{4-})^{\frac{1}{2}},$$

9. SUMMARY OF QED FORMULAE

$$M(-,+;+,+,-,+) = 4e^4 B k_{3+} k_{3\perp}^* \doteq 4e^4 B (k_{3+}^3 k_{3-})^{\frac{1}{2}},$$

$$M(-,+;+,-,+,+) = 4e^4 B k_{2+} k_{2\perp}^* \doteq 4e^4 B (k_{2+}^3 k_{2-})^{\frac{1}{2}},$$

$$M(-,+;-,+,+,+) = 4e^4 B k_{1+} k_{1\perp}^* \doteq 4e^4 B (k_{1+}^3 k_{1-})^{\frac{1}{2}},$$

$$M(+,-;+,-,-,-) = 4e^4 B k_{1+} k_{1\perp} \doteq 4e^4 B (k_{1+}^3 k_{1-})^{\frac{1}{2}},$$

$$M(+,-;-,+,-,-) = 4e^4 B k_{2+} k_{2\perp} \doteq 4e^4 B (k_{2+}^3 k_{2-})^{\frac{1}{2}},$$

$$M(+,-;-,-,+,-) = 4e^4 B k_{3+} k_{3\perp} \doteq 4e^4 B (k_{3+}^3 k_{3-})^{\frac{1}{2}},$$

$$M(+,-;-,-,-,+) = 4e^4 B k_{4+} k_{4\perp} \doteq 4e^4 B (k_{4+}^3 k_{4-})^{\frac{1}{2}},$$

$$M(-,+;+,-,-,-) = -4e^4 B k_{1-} k_{1\perp}^* \doteq 4e^4 B (k_{1+} k_{1-}^3)^{\frac{1}{2}},$$

$$M(-,+;-,+,-,-) = -4e^4 B k_{2-} k_{2\perp}^* \doteq 4e^4 B (k_{2+} k_{2-}^3)^{\frac{1}{2}},$$

$$M(-,+;-,-,+,-) = -4e^4 B k_{3-} k_{3\perp}^* \doteq 4e^4 B (k_{3+} k_{3-}^3)^{\frac{1}{2}},$$

$$M(-,+;-,-,-,+) = -4e^4 B k_{4-} k_{4\perp}^* \doteq 4e^4 B (k_{4+} k_{4-}^3)^{\frac{1}{2}},$$

$$M(+,-;+,+,-,-) = -2e^4 \frac{BF(1,2,3,4)}{E} \doteq 2e^4 \frac{B|F(1,2,3,4)|}{E},$$

$$M(+,-;+,-,+,-) = -2e^4 \frac{BF(1,3,2,4)}{E} \doteq 2e^4 \frac{B|F(1,3,2,4)|}{E},$$

$$M(+,-;+,-,-,+) = -2e^4 \frac{BF(1,4,2,3)}{E} \doteq 2e^4 \frac{B|F(1,4,2,3)|}{E},$$

$$M(+,-;-,+,+,-) = -2e^4 \frac{BF(2,3,1,4)}{E} \doteq 2e^4 \frac{B|F(2,3,1,4)|}{E},$$

$$M(+,-;-,+,-,+) = -2e^4 \frac{BF(2,4,1,3)}{E} \doteq 2e^4 \frac{B|F(2,4,1,3)|}{E},$$

$$M(+,-;-,-,+,+) = -2e^4 \frac{BF(3,4,1,2)}{E} \doteq 2e^4 \frac{B|F(3,4,1,2)|}{E},$$

$$M(-,+;+,+,-,-) = -2e^4 \frac{BF^*(3,4,1,2)}{E} \doteq 2e^4 \frac{B|F(3,4,1,2)|}{E},$$

$$M(-,+;+,-,+,-) = -2e^4 \frac{BF^*(2,4,1,3)}{E} \doteq 2e^4 \frac{B|F(2,4,1,3)|}{E},$$

$$M(-,+;+,-,-,+) = -2e^4 \frac{BF^*(2,3,1,4)}{E} \doteq 2e^4 \frac{B|F(2,3,1,4)|}{E},$$

$$M(-,+;-,+,+,-) = -2e^4 \frac{BF^*(1,4,2,3)}{E} \doteq 2e^4 \frac{B|F(1,4,2,3)|}{E},$$

$$M(-,+;-,+,-,+) = -2e^4 \frac{BF^*(1,3,2,4)}{E} \doteq 2e^4 \frac{B|F(1,3,2,4)|}{E},$$

$$M(-,+;-,-,+,+) = -2e^4 \frac{BF^*(1,2,3,4)}{E} \doteq 2e^4 \frac{B|F(1,2,3,4)|}{E}.$$
(9.63)

Unpolarized squared matrix element:

$$\overline{|M|^2} = 2e^8 B^2 \Biggl\{ \sum_{i=1}^{4} (k_{i+}^2 + k_{i-}^2) k_{i+} k_{i-}$$

$$+ E^{-2} \Bigl[|F(1,2,3,4)|^2 + |F(1,3,2,4)|^2 + |F(1,4,2,3)|^2$$

$$+ |F(2,3,1,4)|^2 + |F(2,4,1,3)|^2 + |F(3,4,1,2)|^2 \Bigr] \Biggr\}.$$
(9.64)

Unpolarized cross section:

$$d\sigma = \frac{\alpha^4 B^2}{64\pi^4 E^2} \Biggl\{ \sum_{i=1}^{4} (k_{i+}^2 + k_{i-}^2) k_{i+} k_{i-}$$

$$+ E^{-2} \Bigl[|F(1,2,3,4)|^2 + |F(1,3,2,4)|^2 + |F(1,4,2,3)|^2$$

$$+ |F(2,3,1,4)|^2 + |F(2,4,1,3)|^2 + |F(3,4,1,2)|^2 \Bigr] \Biggr\}$$

$$\times \delta^4(p_+ + p_- - k_1 - k_2 - k_3 - k_4) \frac{d^3\vec{k}_1 \, d^3\vec{k}_2 \, d^3\vec{k}_3 \, d^3\vec{k}_4}{k_{10} \, k_{20} \, k_{30} \, k_{40}}.$$
(9.65)

9.12 $e^+e^- \to \mu^+\mu^-\gamma\gamma$ (no Z-exchange; $m = 0$)

Process:

$$e^+(p_+) + e^-(p_-) \to \mu^+(q_+) + \mu^-(q_-) + \gamma(k_1) + \gamma(k_2).$$
(9.66)

Invariants and definitions:

$$s = 2(p_+ \cdot p_-), \quad t = -2(p_+ \cdot q_+), \quad u = -2(p_+ \cdot q_-),$$

$$s' = 2(q_+ \cdot q_-), \quad t' = -2(p_- \cdot q_-), \quad u' = -2(p_- \cdot q_+),$$

$$s_1 = (p_+ + p_- - k_1)^2, \quad s_2 = (p_+ + p_- - k_2)^2,$$

$$B_i = 1 + \frac{2E}{s'}\left[2q_{-z} + \frac{k_{i\perp}q^*_{-\perp}}{k_{i+}} - \frac{k^*_{i\perp}q_{-\perp}}{k_{i-}}\right], \quad i = 1,2,$$

$$Q = (q_+ \cdot k_1)(q_- \cdot k_1)(q_+ \cdot k_2)(q_- \cdot k_2),$$

$$Z_{+-} = q_{++}q_{--} - q^*_{+\perp}q_{-\perp}, \qquad Z_{-+} = q_{-+}q_{+-} - q^*_{-\perp}q_{+\perp},$$

$$w_+ = q_{+\perp}/q_{++}, \qquad w_- = q_{-\perp}/q_{-+},$$

$$w_i = k_{i\perp}/k_{i+}, \quad i = 1,2,$$

$$A = -\frac{(w_+ - w_1)(w_- - w_1)(w_+ - w_2)^*(w_- - w_2)^*}{|w_+ - w_1|\,|w_- - w_1|\,|w_+ - w_2|\,|w_- - w_2|},$$

$$a = q_{+\perp} + k_{1\perp},$$

$$b = 2E - k_{1-} + k_{1\perp}w^*_+ + k^*_{1\perp}w_- - k_{1+}w^*_+w_-,$$

$$c = 2E - k_{1-} + k_{1\perp}w^*_+,$$

$$d = (2E - k_{2+})w_- + k_{2\perp},$$

$$a_1 = q_{-\perp} + k_{1\perp},$$

$$b_1 = 2E - k_{1-} + k_{1\perp}w^*_- + k^*_{1\perp}w_+ - k_{1+}w_+w^*_- = b^*,$$

$$c_1 = 2E - k_{1-} + k_{1\perp}w^*_-,$$

$$d_1 = (2E - k_{2+})w_+ + k_{2\perp},$$

$$a_2 = q^*_{+\perp} + k^*_{2\perp},$$

$$b_2 = 2E - k_{2-} + k^*_{2\perp}w_+ + k_{2\perp}w^*_- - k_{2+}w_+w^*_-,$$

$$c_2 = 2E - k_{2-} + k^*_{2\perp}w_+,$$

9.12. $e^+e^- \to \mu^+\mu^-\gamma\gamma$ (NO Z-EXCHANGE; m = 0)

$$d_2 = (2E - k_{1+})w_-^* + k_{1\perp}^*,$$

$$a_3 = q_{-\perp}^* + k_{2\perp}^*,$$

$$b_3 = 2E - k_{2-} + k_{2\perp}^*w_- + k_{2\perp}w_+^* - k_{2+}w_+^*w_- = b_2^*,$$

$$c_3 = 2E - k_{2-} + k_{2\perp}^*w_-,$$

$$d_3 = (2E - k_{1+})w_+^* + k_{1\perp}^*. \tag{9.67}$$

Photon polarizations:

$$\not{k}_i^{\pm} = N_i[\not{p}_+ \not{p}_- \not{k}_i(1\mp\gamma_5) - \not{k}_i \not{p}_+ \not{p}_-(1\pm\gamma_5)],$$

$$N_i^{-1} = E^2[32k_{i+}k_{i-}]^{\frac{1}{2}}, \qquad i=1,2. \tag{9.68}$$

Nonvanishing helicity amplitudes:

$$M(+,-;+,-,+,+) = -\frac{e^4 q_{-\perp} Z_{+-} B_1 B_2}{[q_{++}q_{-+}Q]^{\frac{1}{2}}}$$

$$\doteq e^4 q_{--}|B_1 B_2|\left[\frac{2(q_+ \cdot q_-)}{Q}\right]^{\frac{1}{2}},$$

$$M(+,-;-,+,+,+) = -\frac{e^4 q_{+\perp} Z_{-+} B_1 B_2}{[q_{++}q_{-+}Q]^{\frac{1}{2}}}$$

$$\doteq e^4 q_{+-}|B_1 B_2|\left[\frac{2(q_+ \cdot q_-)}{Q}\right]^{\frac{1}{2}},$$

$$M(-,+;+,-,+,+) = -\frac{e^4 q_{+\perp}^* Z_{-+} B_1 B_2 \sqrt{q_{-+}}}{q_{+-}[q_{++}Q]^{\frac{1}{2}}}$$

$$\doteq e^4 q_{-+}|B_1 B_2|\left[\frac{2(q_+ \cdot q_-)}{Q}\right]^{\frac{1}{2}},$$

$$M(-,+;-,+,+,+) = -\frac{e^4 q_{-\perp}^* Z_{+-} B_1 B_2 \sqrt{q_{++}}}{q_{--}[q_{-+}Q]^{\frac{1}{2}}}$$

$$\doteq e^4 q_{++}|B_1 B_2|\left[\frac{2(q_+ \cdot q_-)}{Q}\right]^{\frac{1}{2}},$$

$$M(+,-;+,-,-,-) = -\frac{e^4 q_{-\perp} Z^*_{+-} B^*_1 B^*_2 \sqrt{q_{++}}}{q_{--}[q_{-+}Q]^{\frac{1}{2}}}$$

$$\doteq e^4 q_{++} |B_1 B_2| \left[\frac{2(q_+ \cdot q_-)}{Q}\right]^{\frac{1}{2}},$$

$$M(+,-;-,+,-,-) = -\frac{e^4 q_{+\perp} Z^*_{-+} B^*_1 B^*_2 \sqrt{q_{-+}}}{q_{+-}[q_{++}Q]^{\frac{1}{2}}}$$

$$\doteq e^4 q_{-+} |B_1 B_2| \left[\frac{2(q_+ \cdot q_-)}{Q}\right]^{\frac{1}{2}},$$

$$M(-,+;+,-,-,-) = -\frac{e^4 q^*_{+\perp} Z^*_{-+} B^*_1 B^*_2}{[q_{++}q_{-+}Q]^{\frac{1}{2}}}$$

$$\doteq e^4 q_{+-} |B_1 B_2| \left[\frac{2(q_+ \cdot q_-)}{Q}\right]^{\frac{1}{2}},$$

$$M(-,+;-,+,-,-) = -\frac{e^4 q^*_{-\perp} Z^*_{+-} B^*_1 B^*_2}{[q_{++}q_{-+}Q]^{\frac{1}{2}}}$$

$$\doteq e^4 q_{--} |B_1 B_2| \left[\frac{2(q_+ \cdot q_-)}{Q}\right]^{\frac{1}{2}},$$

$$M(+,-;+,-,+,-) = 4e^4 A \left\{ \frac{(q_{++}q_{-+})^{\frac{1}{2}}[w_2 b + (w_- - w_2)c]}{(p_- - k_1 - k_2)^2 s'} \right.$$

$$+ \frac{2Ecd}{(q_{++}q_{-+})^{\frac{1}{2}} s k_{1+} k_{2+} (w_+ - w_2)^*(w_- - w_2)^*(w_+ - w_1)(w_- - w_1)}$$

$$+ \frac{(q_{++}q_{-+})^{\frac{1}{2}} cd}{2Es' k_{1+} k_{2+} w_1 w_2^*} - \frac{(q_{++}q_{-+})^{\frac{1}{2}} w_2[b + (w_+ - w_1)^*d]}{(p_+ - k_1 - k_2)^2 s' w_1 w_2^*}$$

$$+ \left(\frac{q_{++}}{q_{-+}}\right)^{\frac{1}{2}} \frac{c^2}{s_1 k_{1+} k_{2+} w_1 (w_+ - w_2)^*(w_- - w_2)^*}$$

$$+ \left(\frac{q_{-+}}{q_{++}}\right)^{\frac{1}{2}} \frac{d^2}{s_2 k_{1+} k_{2+} w_2^* (w_+ - w_1)(w_- - w_1)}$$

$$- \left(\frac{q_{-+}}{q_{++}}\right)^{\frac{1}{2}} \frac{2E(w_+ - w_1)^*[w_2 d + (w_2 - w_-)a]}{(q_+ + k_1 + k_2)^2 s(w_+ - w_1)(w_+ - w_2)^*}$$

$$\left. + \left(\frac{q_{++}}{q_{-+}}\right)^{\frac{1}{2}} \frac{2E(w_- - w_2)[c + (w_1 - w_+)^*a]}{(q_- + k_1 + k_2)^2 s(w_- - w_1)(w_- - w_2)^*} \right\},$$

9.12. $e^+e^- \to \mu^+\mu^-\gamma\gamma$ (NO Z-EXCHANGE; $m=0$)

$$M(+,-;-,+,+,-) = 4e^4 A \left\{ \frac{(q_{++}q_{-+})^{\frac{1}{2}}[w_2 b_1 + (w_- - w_2)c_1]}{(p_- - k_1 - k_2)^2 s'} \right.$$

$$+ \frac{2E c_1 d_1}{(q_{++}q_{-+})^{\frac{1}{2}} s k_{1+} k_{2+}(w_+ - w_1)(w_- - w_1)(w_+ - w_2)^*(w_- - w_2)^*}$$

$$+ \frac{(q_{++}q_{-+})^{\frac{1}{2}} c_1 d_1}{2E s' k_{1+} k_{2+} w_1 w_2^*} - \frac{(q_{++}q_{-+})^{\frac{1}{2}} w_2 [b_1 + (w_+ - w_1)^* d_1]}{(p_+ - k_1 - k_2)^2 s' w_1 w_2^*}$$

$$- \left(\frac{q_{-+}}{q_{++}}\right)^{\frac{1}{2}} \frac{c_1^2}{s_1 k_{1+} k_{2+} w_1 (w_+ - w_2)^*(w_- - w_2)^*}$$

$$- \left(\frac{q_{++}}{q_{-+}}\right)^{\frac{1}{2}} \frac{d_1^2}{s_2 k_{1+} k_{2+} w_2^* (w_+ - w_1)(w_- - w_1)}$$

$$- \left(\frac{q_{++}}{q_{-+}}\right)^{\frac{1}{2}} \frac{2E(w_- - w_1)^*[w_2 d_1 + (w_2 - w_+)a_1]}{(q_- + k_1 + k_2)^2 s(w_- - w_1)(w_- - w_2)^*}$$

$$+ \left. \left(\frac{q_{-+}}{q_{++}}\right)^{\frac{1}{2}} \frac{2E(w_+ - w_2)[c_1 + (w_1 - w_-)^* a_1]}{(q_+ + k_1 + k_2)^2 s(w_+ - w_1)(w_+ - w_2)^*} \right\},$$

$$M(-,+;+,-,+,-) = 4e^4 A \left\{ \frac{(q_{++}q_{-+})^{\frac{1}{2}}[w_1^* b_3 + (w_- - w_1)^* c_3]}{(p_- - k_1 - k_2)^2 s'} \right.$$

$$+ \frac{2E c_3 d_3}{(q_{++}q_{-+})^{\frac{1}{2}} s k_{1+} k_{2+}(w_+ - w_1)(w_- - w_1)(w_+ - w_2)^*(w_- - w_2)^*}$$

$$+ \frac{(q_{++}q_{-+})^{\frac{1}{2}} c_3 d_3}{2E s' k_{1+} k_{2+} w_1 w_2^*} - \frac{(q_{++}q_{-+})^{\frac{1}{2}} w_1^* [b_3 + (w_+ - w_2) d_3]}{(p_+ - k_1 - k_2)^2 s' w_1 w_2^*}$$

$$- \left(\frac{q_{-+}}{q_{++}}\right)^{\frac{1}{2}} \frac{c_3^2}{s_2 k_{1+} k_{2+} w_2^* (w_+ - w_1)(w_- - w_1)}$$

$$- \left(\frac{q_{++}}{q_{-+}}\right)^{\frac{1}{2}} \frac{d_3^2}{s_1 k_{1+} k_{2+} w_1 (w_+ - w_2)^*(w_- - w_2)^*}$$

$$- \left(\frac{q_{++}}{q_{-+}}\right)^{\frac{1}{2}} \frac{2E(w_- - w_2)[w_1^* d_3 + (w_1 - w_+)a_3]}{(q_- + k_1 + k_2)^2 s(w_- - w_1)(w_- - w_2)^*}$$

$$+ \left. \left(\frac{q_{-+}}{q_{++}}\right)^{\frac{1}{2}} \frac{2E(w_+ - w_1)^*[c_3 + (w_2 - w_-)^* a_3]}{(q_+ + k_1 + k_2)^2 s(w_+ - w_1)(w_+ - w_2)^*} \right\},$$

$$M(-,+;-,+,+,-) = 4e^4 A \left\{ \frac{(q_{++}q_{-+})^{\frac{1}{2}}[w_1^* b_2 + (w_- - w_1)^* c_2]}{(p_- - k_1 - k_2)^2 s'} \right.$$

$$+ \frac{2Ec_2 d_2}{(q_{++}q_{-+})^{\frac{1}{2}} sk_{1+}k_{2+}(w_+ - w_1)(w_- - w_1)(w_+ - w_2)^*(w_- - w_2)^*}$$

$$+ \frac{(q_{++}q_{-+})^{\frac{1}{2}} c_2 d_2}{2Es' k_{1+}k_{2+} w_1 w_2^*} - \frac{(q_{++}q_{-+})^{\frac{1}{2}} w_1^*[b_2 + (w_+ - w_2)d_2]}{(p_+ - k_1 - k_2)^2 s' w_1 w_2^*}$$

$$+ \left(\frac{q_{++}}{q_{-+}}\right)^{\frac{1}{2}} \frac{c_2^2}{s_2 k_{1+}k_{2+} w_2^*(w_+ - w_1)(w_- - w_1)}$$

$$+ \left(\frac{q_{-+}}{q_{++}}\right)^{\frac{1}{2}} \frac{d_2^2}{s_1 k_{1+}k_{2+} w_1 (w_+ - w_2)^*(w_- - w_2)^*}$$

$$- \left(\frac{q_{-+}}{q_{++}}\right)^{\frac{1}{2}} \frac{2E(w_+ - w_2)[w_1^* d_2 + (w_1 - w_-)^* a_2]}{(q_+ + k_1 + k_2)^2 s(w_+ - w_1)(w_+ - w_2)^*}$$

$$+ \left.\left(\frac{q_{++}}{q_{-+}}\right)^{\frac{1}{2}} \frac{2E(w_- - w_1)^*[c_2 + (w_2 - w_+)a_2]}{(q_- + k_1 + k_2)^2 s(w_- - w_1)(w_- - w_2)^*} \right\},$$

$$M(+,-;+,-,-,+) = [M(-,+;-,+,+,-)]^*,$$

$$M(+,-;-,+,-,+) = [M(-,+;+,-,+,-)]^*,$$

$$M(-,+;+,-,-,+) = [M(+,-;-,+,+,-)]^*,$$

$$M(-,+;-,+,-,+) = [M(+,-;+,-,+,-)]^*. \quad (9.69)$$

Unpolarized squared matrix element:

$$\overline{|M|^2} = \tfrac{1}{2} e^8 (v_{1p} - v_{1q})^2 (v_{2p} - v_{2q})^2 \frac{t^2 + t'^2 + u^2 + u'^2}{ss'}$$

$$+ \tfrac{1}{2} \Big[|M(+,-;+,-,+,-)|^2 + |M(+,-;-,+,+,-)|^2$$

$$+ |M(-,+;-,+,+,-)|^2 + |M(-,+;+,-,+,-)|^2 \Big], \quad (9.70)$$

where [see eqns (4.27)]

$$v_{ip} = \frac{p_+}{(p_+ \cdot k_i)} - \frac{p_-}{(p_- \cdot k_i)}, \quad v_{iq} = \frac{q_+}{(q_+ \cdot k_i)} - \frac{q_-}{(q_- \cdot k_i)}, \quad i = 1, 2. \quad (9.71)$$

9.13. $e^+e^- \to e^+e^-\gamma\gamma$ (NO Z-EXCHANGE; $m=0$)

Unpolarized cross section:

$$d\sigma = \frac{\alpha^4}{64\pi^4 s}$$

$$\times \left\{ (v_{1p} - v_{1q})^2 (v_{2p} - v_{2q})^2 \frac{t^2 + t'^2 + u^2 + u'^2}{ss'} \right.$$

$$+ |M(+,-;+,-,+,-)|^2 + |M(+,-;-,+,+,-)|^2$$

$$\left. + |M(-,+;-,+,+,-)|^2 + |M(-,+;+,-,+,-)|^2 \right\}$$

$$\times \delta^4(p_+ + p_- - q_+ - q_- - k_1 - k_2) \frac{d^3\vec{q}_+ \, d^3\vec{q}_- \, d^3\vec{k}_1 \, d^3\vec{k}_2}{q_{+0} \, q_{-0} \, k_{10} \, k_{20}}. \tag{9.72}$$

9.13 $e^+e^- \to e^+e^-\gamma\gamma$ (no Z-exchange; $m=0$)

Process:

$$e^+(p_+) + e^-(p_-) \to e^+(q_+) + e^-(q_-) + \gamma(k_1) + \gamma(k_2). \tag{9.73}$$

Invariants and definitions:

$$s = 2(p_+ \cdot p_-), \quad t = -2(p_+ \cdot q_+), \quad u = -2(p_+ \cdot q_-),$$

$$s' = 2(q_+ \cdot q_-), \quad t' = -2(p_- \cdot q_-), \quad u' = -2(p_- \cdot q_+),$$

$$s_1 = (p_+ + p_- - k_1)^2, \quad s_2 = (p_+ + p_- - k_2)^2,$$

$$t_1 = (p_+ - q_+ - k_1)^2, \quad t_2 = (p_+ - q_+ - k_2)^2,$$

$$u_1 = (p_+ - q_- - k_1)^2, \quad u_2 = (p_+ - q_- - k_2)^2,$$

$$B_i = 1 + \frac{2E}{s'}\left[2q_{-z} + \frac{k_{i\perp}q^*_{-\perp}}{k_{i+}} - \frac{k^*_{i\perp}q_{-\perp}}{k_{i-}}\right], \quad i = 1,2,$$

$$Q = (q_+ \cdot k_1)(q_- \cdot k_1)(q_+ \cdot k_2)(q_- \cdot k_2),$$

$$Z_{+-} = q_{++}q_{--} - q^*_{+\perp}q_{-\perp}, \quad Z_{-+} = q_{-+}q_{+-} - q^*_{-\perp}q_{+\perp},$$

$$w_+ = q_{+\perp}/q_{++}, \quad w_- = q_{-\perp}/q_{-+},$$

$$w_i = k_{i\perp}/k_{i+}, \qquad i = 1, 2,$$

$$A = -\frac{(w_+ - w_1)(w_- - w_1)(w_+ - w_2)^*(w_- - w_2)^*}{|w_+ - w_1|\,|w_- - w_1|\,|w_+ - w_2|\,|w_- - w_2|},$$

$$a = q_{+\perp} + k_{1\perp},$$

$$b = 2E - k_{1-} + k_{1\perp}w_+^* + k_{1\perp}^*w_- - k_{1+}w_+^*w_-,$$

$$c = 2E - k_{1-} + k_{1\perp}w_+^*, \qquad d = (2E - k_{2+})w_- + k_{2\perp},$$

$$a_1 = q_{-\perp} + k_{1\perp},$$

$$b_1 = 2E - k_{1-} + k_{1\perp}w_-^* + k_{1\perp}^*w_+ - k_{1+}w_+w_-^* = b^*,$$

$$c_1 = 2E - k_{1-} + k_{1\perp}w_-^*, \qquad d_1 = (2E - k_{2+})w_+ + k_{2\perp},$$

$$a_2 = q_{+\perp}^* + k_{2\perp}^*,$$

$$b_2 = 2E - k_{2-} + k_{2\perp}^*w_+ + k_{2\perp}w_-^* - k_{2+}w_+w_-^*,$$

$$c_2 = 2E - k_{2-} + k_{2\perp}^*w_+, \qquad d_2 = (2E - k_{1+})w_-^* + k_{1\perp}^*,$$

$$a_3 = q_{-\perp}^* + k_{2\perp}^*,$$

$$b_3 = 2E - k_{2-} + k_{2\perp}^*w_- + k_{2\perp}w_+^* - k_{2+}w_+^*w_- = b_2^*,$$

$$c_3 = 2E - k_{2-} + k_{2\perp}^*w_-,$$

$$d_3 = (2E - k_{1+})w_+^* + k_{1\perp}^*. \qquad (9.74)$$

Photon polarizations:

$$\rlap{/}k_i^{\pm} = N_i[\rlap{/}p_+ \rlap{/}p_- \rlap{/}k_i(1 \mp \gamma_5) - \rlap{/}k_i \rlap{/}p_+ \rlap{/}p_-(1 \pm \gamma_5)],$$

$$N_i^{-1} = E^2[32k_{i+}k_{i-}]^{\frac{1}{2}}, \qquad\qquad i = 1, 2. \qquad (9.75)$$

9.13. $e^+e^- \to e^+e^-\gamma\gamma$ (NO Z-EXCHANGE; $m=0$)

Nonvanishing helicity amplitudes:

$$M(+,+;+,+,+,+) = -\frac{4e^4 E(q_+ \cdot q_-) q^*_{+\perp} B_1 B_2}{q_{+-}[q_{++}q_{-+}Q]^{\frac{1}{2}}}$$

$$\doteq \frac{4e^4 E(q_+ \cdot q_-)|B_1 B_2|}{[q_{+-}q_{-+}Q]^{\frac{1}{2}}},$$

$$M(+,+;+,+,-,-) = -\frac{e^4(q_+ \cdot q_-)q^*_{+\perp} Z^2_{-+} B^*_1 B^*_2}{Eq^2_{+-}[q_{++}q^3_{-+}Q]^{\frac{1}{2}}}$$

$$\doteq \frac{2e^4(q_+ \cdot q_-)^2 |B_1 B_2|}{E[q_{+-}q_{-+}Q]^{\frac{1}{2}}},$$

$$M(+,-;+,-,+,+) = -\frac{e^4 q^2_{-\perp} q^*_{+\perp} Z_{+-} B_1 B_2}{q_{+-}[q_{++}q^3_{-+}Q]^{\frac{1}{2}}}$$

$$\doteq e^4 |B_1 B_2| \left[\frac{2q_{++}q^3_{--}(q_+ \cdot q_-)}{q_{+-}q_{-+}Q} \right]^{\frac{1}{2}},$$

$$M(+,-;-,+,+,+) = -\frac{e^4 q_{+\perp} Z_{-+} B_1 B_2}{[q_{++}q_{-+}Q]^{\frac{1}{2}}}$$

$$\doteq e^4 q_{+-} |B_1 B_2| \left[\frac{2(q_+ \cdot q_-)}{Q} \right]^{\frac{1}{2}},$$

$$M(-,+;+,-,+,+) = -\frac{e^4 q^*_{+\perp} Z_{-+} B_1 B_2 \sqrt{q_{-+}}}{q_{+-}[q_{++}Q]^{\frac{1}{2}}}$$

$$\doteq e^4 q_{-+} |B_1 B_2| \left[\frac{2(q_+ \cdot q_-)}{Q} \right]^{\frac{1}{2}},$$

$$M(-,+;-,+,+,+) = -\frac{e^4 q^*_{+\perp} Z_{+-} B_1 B_2 \sqrt{q_{++}}}{q_{+-}[q_{-+}Q]^{\frac{1}{2}}}$$

$$\doteq e^4 |B_1 B_2| \left[\frac{2q^3_{++}q_{--}(q_+ \cdot q_-)}{q_{+-}q_{-+}Q} \right]^{\frac{1}{2}},$$

$$M(+,-;+,-,-,-) = -\frac{e^4 q_{+\perp} Z^*_{+-} B^*_1 B^*_2 \sqrt{q_{++}}}{q_{+-}[q_{-+}Q]^{\frac{1}{2}}}$$

$$\doteq e^4 |B_1 B_2| \left[\frac{2q^3_{++}q_{--}(q_+ \cdot q_-)}{q_{+-}q_{-+}Q} \right]^{\frac{1}{2}},$$

$$M(+,-;-,+,-,-) = -\frac{e^4 q_{+\perp} Z^*_{-+} B^*_1 B^*_2 \sqrt{q_{-+}}}{q_{+-}[q_{++}Q]^{\frac{1}{2}}}$$

$$\doteq e^4 q_{-+} |B_1 B_2| \left[\frac{2(q_+ \cdot q_-)}{Q}\right]^{\frac{1}{2}},$$

$$M(-,+;+,-,-,-) = -\frac{e^4 q^*_{+\perp} Z^*_{-+} B^*_1 B^*_2}{[q_{++}q_{-+}Q]^{\frac{1}{2}}}$$

$$\doteq e^4 q_{+-} |B_1 B_2| \left[\frac{2(q_+ \cdot q_-)}{Q}\right]^{\frac{1}{2}},$$

$$M(-,+;-,+,-,-) = -\frac{e^4 q^{*2}_{-\perp} q_{+\perp} Z^*_{+-} B^*_1 B^*_2}{q_{+-}[q_{++}q^3_{-+}Q]^{\frac{1}{2}}}$$

$$\doteq e^4 |B_1 B_2| \left[\frac{2q_{++}q^3_{--}(q_+ \cdot q_-)}{q_{+-}q_{-+}Q}\right]^{\frac{1}{2}},$$

$$M(-,-;-,-,+,+) = -\frac{e^4 (q_+ \cdot q_-) q_{+\perp} Z^{*2}_{-+} B_1 B_2}{E q^2_{+-}[q_{++}q^3_{-+}Q]^{\frac{1}{2}}}$$

$$\doteq \frac{2e^4 (q_+ \cdot q_-)^2 |B_1 B_2|}{E[q_{+-}q_{-+}Q]^{\frac{1}{2}}},$$

$$M(-,-;-,-,-,-) = -\frac{4e^4 E(q_+ \cdot q_-) q_{+\perp} B^*_1 B^*_2}{q_{+-}[q_{++}q_{-+}Q]^{\frac{1}{2}}}$$

$$\doteq \frac{4e^4 E(q_+ \cdot q_-) |B_1 B_2|}{[q_{+-}q_{-+}Q]^{\frac{1}{2}}},$$

$$M(+,+;+,+,+,-) = 4e^4 A \left\{ \left(\frac{q_{++}}{q_{-+}}\right)^{\frac{1}{2}} \frac{d^*_1 c^*_3}{tk_{1+}k_{2+}(w_- - w_1)(w_- - w_2)^*} \right.$$

$$+ \left(\frac{q_{-+}}{q_{++}}\right)^{\frac{1}{2}} \frac{d^*_1 c^*_3}{t' k_{1+} k_{2+} w_1 w^*_2 (w_+ - w_1)(w_+ - w_2)^*}$$

$$- \left(\frac{q_{++}}{q_{-+}}\right)^{\frac{1}{2}} \frac{d^{*2}_1}{t_2 k_{1+} k_{2+} w^*_2 (w_- - w_1)(w_+ - w_2)^*}$$

$$- \left(\frac{q_{-+}}{q_{++}}\right)^{\frac{1}{2}} \frac{c^{*2}_3}{t_1 k_{1+} k_{2+} w_1 (w_+ - w_1)(w_- - w_2)^*}$$

$$+ \frac{(q_{++}q_{-+})^{\frac{1}{2}} w_2 [(w_+ - w_1)^* d^* + (w_1 - w_-)^* d^*_1]}{(p_+ - k_1 - k_2)^2 t' w_1 w^*_2}$$

$$- \frac{(q_{++}q_{-+})^{\frac{1}{2}} [(w_1 - w_+)^* c^*_3 + (w_- - w_1)^* c^*_2]}{(p_- - k_1 - k_2)^2 t}$$

9.13. $e^+e^- \to e^+e^- \gamma\gamma$ *(NO Z-EXCHANGE; m = 0)*

$$+ \left(\frac{q_{-+}}{q_{++}}\right)^{\frac{1}{2}} \frac{2E(w_+ - w_1)^*[c_3^* + w_2 d^*]}{(q_+ + k_1 + k_2)^2 t'(w_+ - w_1)(w_+ - w_2)^*}$$

$$- \left(\frac{q_{++}}{q_{-+}}\right)^{\frac{1}{2}} \frac{2E(w_- - w_1)^*[c_2^* + w_2 d_1^*]}{(q_- + k_1 + k_2)^2 t(w_- - w_1)(w_- - w_2)^*} \bigg\} ,$$

$$M(+,-;+,-,+,-) = 4e^4 A \bigg\{ \frac{(q_{++}q_{-+})^{\frac{1}{2}} b^2}{2Et_2 k_{1+} k_{2+} w_2^*(w_- - w_1)(w_+ - w_2)^*}$$

$$\left(\frac{q_{++}}{q_{-+}}\right)^{\frac{1}{2}} \frac{c^2}{s_1 k_{1+} k_{2+} w_1 (w_+ - w_2)^*(w_- - w_2)^*}$$

$$+ \left(\frac{q_{-+}}{q_{++}}\right)^{\frac{1}{2}} \frac{d^2}{s_2 k_{1+} k_{2+} w_2^*(w_+ - w_1)(w_- - w_1)}$$

$$+ \frac{2Ecd}{(q_{++}q_{-+})^{\frac{1}{2}} s k_{1+} k_{2+}(w_+ - w_2)^*(w_- - w_2)^*(w_+ - w_1)(w_- - w_1)}$$

$$+ \frac{(q_{++}q_{-+})^{\frac{1}{2}} cd}{2Es' k_{1+} k_{2+} w_1 w_2^*} + \frac{2Ea^2}{(q_{++}q_{-+})^{\frac{1}{2}} t_1 k_{1+} k_{2+} w_1 (w_+ - w_1)^*(w_- - w_2)}$$

$$- \left(\frac{q_{++}}{q_{-+}}\right)^{\frac{1}{2}} \frac{ab}{t k_{1+} k_{2+}(w_- - w_1)(w_- - w_2)^*}$$

$$- \left(\frac{q_{-+}}{q_{++}}\right)^{\frac{1}{2}} \frac{ab}{t' k_{1+} k_{2+} w_1 w_2^*(w_+ - w_1)(w_+ - w_2)^*}$$

$$- \frac{(q_{++}q_{-+})^{\frac{1}{2}} w_2 [b + (w_+ - w_1)^* d]}{(p_+ - k_1 - k_2)^2 w_1 w_2^*} \left(\frac{1}{s'} + \frac{1}{t'}\right)$$

$$+ \frac{(q_{++}q_{-+})^{\frac{1}{2}} [w_2 b + (w_- - w_2)c]}{(p_- - k_1 - k_2)^2} \left(\frac{1}{s'} + \frac{1}{t}\right)$$

$$- \left(\frac{q_{-+}}{q_{++}}\right)^{\frac{1}{2}} \frac{2E(w_+ - w_1)^*[w_2 d + (w_2 - w_-)a]}{(q_+ + k_1 + k_2)^2 (w_+ - w_1)(w_+ - w_2)^*} \left(\frac{1}{s} + \frac{1}{t'}\right)$$

$$+ \left(\frac{q_{++}}{q_{-+}}\right)^{\frac{1}{2}} \frac{2E(w_- - w_2)[c + (w_1 - w_+)^* a]}{(q_- + k_1 + k_2)^2 (w_- - w_1)(w_- - w_2)^*} \left(\frac{1}{s} + \frac{1}{t}\right) \bigg\} ,$$

$$M(+,-;-,+,+,-) = 4e^4 A \left\{ \frac{(q_{++}q_{-+})^{\frac{1}{2}}[w_2 b_1 + (w_- - w_2)c_1]}{(p_- - k_1 - k_2)^2 s'} \right.$$

$$+ \frac{2Ec_1 d_1}{(q_{++}q_{-+})^{\frac{1}{2}} s k_{1+} k_{2+} (w_+ - w_1)(w_- - w_1)(w_+ - w_2)^*(w_- - w_2)^*}$$

$$+ \frac{(q_{++}q_{-+})^{\frac{1}{2}} c_1 d_1}{2Es' k_{1+} k_{2+} w_1 w_2^*} - \frac{(q_{++}q_{-+})^{\frac{1}{2}} w_2 [b_1 + (w_+ - w_1)^* d_1]}{(p_+ - k_1 - k_2)^2 s' w_1 w_2^*}$$

$$- \left(\frac{q_{-+}}{q_{++}} \right)^{\frac{1}{2}} \frac{c_1^2}{s_1 k_{1+} k_{2+} w_1 (w_+ - w_2)^*(w_- - w_2)^*}$$

$$- \left(\frac{q_{++}}{q_{-+}} \right)^{\frac{1}{2}} \frac{d_1^2}{s_2 k_{1+} k_{2+} w_2^* (w_+ - w_1)(w_- - w_1)}$$

$$- \left(\frac{q_{++}}{q_{-+}} \right)^{\frac{1}{2}} \frac{2E(w_- - w_1)^*[w_2 d_1 + (w_2 - w_+)a_1]}{(q_- + k_1 + k_2)^2 s(w_- - w_1)(w_- - w_2)^*}$$

$$+ \left. \left(\frac{q_{-+}}{q_{++}} \right)^{\frac{1}{2}} \frac{2E(w_+ - w_2)[c_1 + (w_1 - w_-)^* a_1]}{(q_+ + k_1 + k_2)^2 s(w_+ - w_1)(w_+ - w_2)^*} \right\},$$

$$M(-,+;+,-,+,-) = 4e^4 A \left\{ \frac{(q_{++}q_{-+})^{\frac{1}{2}}[w_1^* b_3 + (w_- - w_1)^* c_3]}{(p_- - k_1 - k_2)^2 s'} \right.$$

$$+ \frac{2Ec_3 d_3}{(q_{++}q_{-+})^{\frac{1}{2}} s k_{1+} k_{2+} (w_+ - w_1)(w_- - w_1)(w_+ - w_2)^*(w_- - w_2)^*}$$

$$+ \frac{(q_{++}q_{-+})^{\frac{1}{2}} c_3 d_3}{2Es' k_{1+} k_{2+} w_1 w_2^*} - \frac{(q_{++}q_{-+})^{\frac{1}{2}} w_1^* [b_3 + (w_+ - w_2) d_3]}{(p_+ - k_1 - k_2)^2 s' w_1 w_2^*}$$

$$- \left(\frac{q_{-+}}{q_{++}} \right)^{\frac{1}{2}} \frac{c_3^2}{s_2 k_{1+} k_{2+} w_2^* (w_+ - w_1)(w_- - w_1)}$$

$$- \left(\frac{q_{++}}{q_{-+}} \right)^{\frac{1}{2}} \frac{d_3^2}{s_1 k_{1+} k_{2+} w_1 (w_+ - w_2)^*(w_- - w_2)^*}$$

$$- \left(\frac{q_{++}}{q_{-+}} \right)^{\frac{1}{2}} \frac{2E(w_- - w_2)[w_1^* d_3 + (w_1 - w_+)a_3]}{(q_- + k_1 + k_2)^2 s(w_- - w_1)(w_- - w_2)^*}$$

$$+ \left. \left(\frac{q_{-+}}{q_{++}} \right)^{\frac{1}{2}} \frac{2E(w_+ - w_1)^*[c_3 + (w_2 - w_-)^* a_3]}{(q_+ + k_1 + k_2)^2 s(w_+ - w_1)(w_+ - w_2)^*} \right\},$$

9.13. $e^+e^- \to e^+e^- \gamma\gamma$ (NO Z-EXCHANGE; $m = 0$)

$$M(-,+;-,+,+,-) = 4e^4 A$$

$$\times \left\{ \left(\frac{q_{++}}{q_{-+}}\right)^{\frac{1}{2}} \frac{c_2^2}{s_2 k_{1+} k_{2+} w_2^*(w_+ - w_1)(w_- - w_1)} \right.$$

$$+ \left(\frac{q_{-+}}{q_{++}}\right)^{\frac{1}{2}} \frac{d_2^2}{s_1 k_{1+} k_{2+} w_1 (w_+ - w_2)^*(w_- - w_2)^*}$$

$$+ \frac{2E c_2 d_2}{(q_{++}q_{-+})^{\frac{1}{2}} s k_{1+} k_{2+} (w_+ - w_1)(w_- - w_1)(w_+ - w_2)^*(w_- - w_2)^*}$$

$$+ \frac{(q_{++}q_{-+})^{\frac{1}{2}} c_2 d_2}{2E s' k_{1+} k_{2+} w_1 w_2^*} + \frac{2E a_2^2}{(q_{++}q_{-+})^{\frac{1}{2}} t_2 k_{1+} k_{2+} w_2^*(w_+ - w_1)(w_+ - w_2)^*}$$

$$+ \frac{(q_{++}q_{-+})^{\frac{1}{2}} b_2^2}{2E t_1 k_{1+} k_{2+} w_1 (w_+ - w_1)(w_- - w_2)^*}$$

$$- \left(\frac{q_{++}}{q_{-+}}\right)^{\frac{1}{2}} \frac{a_2 b_2}{t k_{1+} k_{2+} (w_- - w_1)(w_- - w_2)^*}$$

$$- \left(\frac{q_{-+}}{q_{++}}\right)^{\frac{1}{2}} \frac{a_2 b_2}{t' k_{1+} k_{2+} w_1 w_2^* (w_+ - w_1)(w_+ - w_2)^*}$$

$$- \frac{(q_{++}q_{-+})^{\frac{1}{2}} w_1^* [b_2 + (w_+ - w_2) d_2]}{(p_+ - k_1 - k_2)^2 w_1 w_2^*} \left(\frac{1}{s'} + \frac{1}{t'}\right)$$

$$+ \frac{(q_{++}q_{-+})^{\frac{1}{2}} [w_1^* b_2 + (w_- - w_1)^* c_2]}{(p_- - k_1 - k_2)^2} \left(\frac{1}{s'} + \frac{1}{t}\right)$$

$$- \left(\frac{q_{-+}}{q_{++}}\right)^{\frac{1}{2}} \frac{2E(w_+ - w_2)[w_1^* d_2 + (w_1 - w_-)^* a_2]}{(q_+ + k_1 + k_2)^2 (w_+ - w_1)(w_+ - w_2)^*} \left(\frac{1}{s} + \frac{1}{t'}\right)$$

$$+ \left.\left(\frac{q_{++}}{q_{-+}}\right)^{\frac{1}{2}} \frac{2E(w_- - w_1)^*[c_2 + (w_2 - w_+) a_2]}{(q_- + k_1 + k_2)^2 (w_- - w_1)(w_- - w_2)^*} \left(\frac{1}{s} + \frac{1}{t}\right) \right\},$$

$$M(-,-;-,-,+,-) = 4e^4 A$$

$$\times \left\{ -\left(\frac{q_{++}}{q_{-+}}\right)^{\frac{1}{2}} \frac{d_3^{*2}}{t_2 k_{1+} k_{2+} w_2 (w_- - w_1)^* (w_+ - w_2)} \right.$$

$$-\left(\frac{q_{-+}}{q_{++}}\right)^{\frac{1}{2}} \frac{c_2^{*2}}{t_1 k_{1+} k_{2+} w_1^* (w_+ - w_1)^* (w_- - w_2)}$$

$$+\left(\frac{q_{++}}{q_{-+}}\right)^{\frac{1}{2}} \frac{d_3^* c_2^*}{t k_{1+} k_{2+} (w_- - w_1)^* (w_- - w_2)}$$

$$+\left(\frac{q_{-+}}{q_{++}}\right)^{\frac{1}{2}} \frac{d_3^* c_2^*}{t' k_{1+} k_{2+} w_1^* w^* (w_+ - w_1)^* (w_+ - w_2)}$$

$$+\frac{(q_{++} q_{-+})^{\frac{1}{2}} w_2^* [(w_+ - w_1) d_2^* + (w_1 - w_-) d_3^*]}{(p_+ - k_1 - k_2)^2 t' w_1^* w_2}$$

$$-\frac{(q_{++} q_{-+})^{\frac{1}{2}} [(w_1 - w_+) c_2^* + (w_- - w_1) c^*]}{(p_- - k_1 - k_2)^2 t}$$

$$+\left(\frac{q_{-+}}{q_{++}}\right)^{\frac{1}{2}} \frac{2E(w_+ - w_1)[c_2^* + w_2^* d_2^*]}{(q_+ + k_1 + k_2)^2 t' (w_+ - w_1)^* (w_+ - w_2)}$$

$$\left. -\left(\frac{q_{++}}{q_{-+}}\right)^{\frac{1}{2}} \frac{2E(w_- - w_1)[c^* + w_2^* d_3^*]}{(q_- + k_1 + k_2)^2 t (w_- - w_1)^* (w_- - w_2)} \right\},$$

$$M(+,+;+,+,-,+) = [M(-,-;-,-,+,-)]^*,$$

$$M(+,-;+,-,-,+) = [M(-,+;-,+,+,-)]^*,$$

$$M(+,-;-,+,-,+) = [M(-,+;+,-,+,-)]^*,$$

$$M(-,+;+,-,-,+) = [M(+,-;-,+,+,-)]^*,$$

$$M(-,+;-,+,-,+) = [M(+,-;+,-,+,-)]^*,$$

$$M(-,-;-,-,-,+) = [M(+,+;+,+,+,-)]^*. \tag{9.76}$$

9.13. $e^+e^- \to e^+e^-\gamma\gamma$ (NO Z-EXCHANGE; $m = 0$)

Unpolarized squared matrix element:

$$\overline{|M|^2} = \tfrac{1}{2}e^8(v_{1p} - v_{1q})^2(v_{2p} - v_{2q})^2$$

$$\times \frac{ss'(s^2 + s'^2) + tt'(t^2 + t'^2) + uu'(u^2 + u'^2)}{ss'tt'}$$

$$+ \tfrac{1}{2}\Big[|M(+,-;+,-,+,-)|^2 + |M(+,-;-,+,+,-)|^2$$

$$+ |M(-,+;-,+,+,-)|^2 + |M(-,+;+,-,+,-)|^2$$

$$+ |M(+,+;+,+,+,-)|^2 + |M(-,-;-,-,+,-)|^2\Big], \qquad (9.77)$$

where [see eqns (4.27)]

$$v_{ip} = \frac{p_+}{(p_+ \cdot k_i)} - \frac{p_-}{(p_- \cdot k_i)}, \qquad v_{iq} = \frac{q_+}{(q_+ \cdot k_i)} - \frac{q_-}{(q_- \cdot k_i)}, \qquad i = 1,2. \tag{9.78}$$

Unpolarized cross section:

$$d\sigma = \frac{\alpha^4}{64\pi^4 s}\Big\{(v_{1p} - v_{1q})^2(v_{2p} - v_{2q})^2$$

$$\times \frac{ss'(s^2 + s'^2) + tt'(t^2 + t'^2) + uu'(u^2 + u'^2)}{ss'tt'}$$

$$+ |M(+,-;+,-,+,-)|^2 + |M(+,-;-,+,+,-)|^2$$

$$+ |M(-,+;-,+,+,-)|^2 + |M(-,+;+,-,+,-)|^2$$

$$+ |M(+,+;+,+,+,-)|^2 + |M(-,-;-,-,+,-)|^2\Big\}$$

$$\times \delta^4(p_+ + p_- - q_+ - q_- - k_1 - k_2)\frac{d^3\vec{q}_+\, d^3\vec{q}_-\, d^3\vec{k}_1\, d^3\vec{k}_2}{q_{+0}\, q_{-0}\, k_{10}\, k_{20}}.$$

$$\tag{9.79}$$

9.14 $e^+e^- \to \gamma\gamma\gamma$ $(m \neq 0)$

Process:
$$e^+(p_+) + e^-(p_-) \to \gamma(k_1) + \gamma(k_2) + \gamma(k_3). \qquad (9.80)$$

9.14.1 $\vec{k}_3$ nearly parallel to $\vec{p}_+$

Photon polarizations:

$$\not{\epsilon}_1^{\pm} = N_1[\not{p}_+ \not{p}_- \not{k}_1(1 \mp \gamma_5) - \not{k}_1 \not{p}_+ \not{p}_-(1 \pm \gamma_5)],$$

$$\not{\epsilon}_2^{\pm} = N_2[\not{p}_+ \not{p}_- \not{k}_2(1 \mp \gamma_5) - \not{k}_2 \not{p}_+ \not{p}_-(1 \pm \gamma_5)],$$

$$\not{\epsilon}_3^{\pm} = N_3[\not{k}_3 \not{p}_+ \not{p}_-(1 \pm \gamma_5) + \not{p}_- \not{p}_+ \not{k}_3(1 \mp \gamma_5)],$$

$$N_i^{-1} = E^2[32 k_{i+} k_{i-}]^{\frac{1}{2}}, \qquad i = 1, 2, 3. \qquad (9.81)$$

Nonvanishing helicity amplitudes:

$$M(+,+;+,-,+) = \frac{e^3 m k_{2\perp}^* k_{3\perp}}{E k_{1+}(p_+ \cdot k_3)} \left[\frac{k_{3+}}{2 k_{3-}}\right]^{\frac{1}{2}} \doteq \frac{e^3 m k_{3+}}{E(p_+ \cdot k_3)} \left[\frac{k_{1-}}{2 k_{1+}}\right]^{\frac{1}{2}},$$

$$M(+,+;-,+,+) = \frac{e^3 m k_{1\perp}^* k_{3\perp}}{E k_{2+}(p_+ \cdot k_3)} \left[\frac{k_{3+}}{2 k_{3-}}\right]^{\frac{1}{2}} \doteq \frac{e^3 m k_{3+}}{E(p_+ \cdot k_3)} \left[\frac{k_{2-}}{2 k_{2+}}\right]^{\frac{1}{2}},$$

$$M(+,-;+,-,+) = \frac{e^3 E k_{2\perp}}{k_{2+}(p_+ \cdot k_3)} \left[\frac{8 k_{3-}}{k_{3+}}\right]^{\frac{1}{2}} \doteq \frac{e^3 E}{(p_+ \cdot k_3)} \left[\frac{8 k_{2-} k_{3-}}{k_{2+} k_{3+}}\right]^{\frac{1}{2}},$$

$$M(+,-;-,+,+) = \frac{e^3 E k_{1\perp}}{k_{1+}(p_+ \cdot k_3)} \left[\frac{8 k_{3-}}{k_{3+}}\right]^{\frac{1}{2}} \doteq \frac{e^3 E}{(p_+ \cdot k_3)} \left[\frac{8 k_{1-} k_{3-}}{k_{1+} k_{3+}}\right]^{\frac{1}{2}},$$

$$M(-,+;+,-,+) = \frac{e^3 E k_{1\perp}^*}{k_{2-}(p_+ \cdot k_3)} \left[\frac{8 k_{3-}}{k_{3+}}\right]^{\frac{1}{2}} \doteq \frac{e^3 E}{(p_+ \cdot k_3)} \left[\frac{8 k_{2+} k_{3-}}{k_{2-} k_{3+}}\right]^{\frac{1}{2}},$$

$$M(-,+;-,+,+) = \frac{e^3 E k_{2\perp}^*}{k_{1-}(p_+ \cdot k_3)} \left[\frac{8 k_{3-}}{k_{3+}}\right]^{\frac{1}{2}} \doteq \frac{e^3 E}{(p_+ \cdot k_3)} \left[\frac{8 k_{1+} k_{3-}}{k_{1-} k_{3+}}\right]^{\frac{1}{2}},$$

$$M(+,-;+,-,-) = \frac{e^3 E k_{2\perp}}{k_{1-}(p_+ \cdot k_3)} \left[\frac{8 k_{3-}}{k_{3+}}\right]^{\frac{1}{2}} \doteq \frac{e^3 E}{(p_+ \cdot k_3)} \left[\frac{8 k_{2+} k_{3-}}{k_{2-} k_{3+}}\right]^{\frac{1}{2}},$$

$$M(+,-;-,+,-) = \frac{e^3 E k_{1\perp}}{k_{2-}(p_+ \cdot k_3)} \left[\frac{8 k_{3-}}{k_{3+}}\right]^{\frac{1}{2}} \doteq \frac{e^3 E}{(p_+ \cdot k_3)} \left[\frac{8 k_{1+} k_{3-}}{k_{1-} k_{3+}}\right]^{\frac{1}{2}},$$

9.14. $e^+e^- \to \gamma\gamma\gamma$ $(m \neq 0)$

$$M(-,+;+,-,-) = \frac{e^3 E k_{1\perp}^*}{k_{1+}(p_+ \cdot k_3)} \left[\frac{8k_{3-}}{k_{3+}}\right]^{\frac{1}{2}} \doteq \frac{e^3 E}{(p_+ \cdot k_3)} \left[\frac{8k_{1-}k_{3-}}{k_{1+}k_{3+}}\right]^{\frac{1}{2}},$$

$$M(-,+;-,+,-) = \frac{e^3 E k_{2\perp}^*}{k_{2+}(p_+ \cdot k_3)} \left[\frac{8k_{3-}}{k_{3+}}\right]^{\frac{1}{2}} \doteq \frac{e^3 E}{(p_+ \cdot k_3)} \left[\frac{8k_{2-}k_{3-}}{k_{2+}k_{3+}}\right]^{\frac{1}{2}},$$

$$M(-,-;+,-,-) = \frac{e^3 m k_{2\perp} k_{3\perp}^*}{E k_{2+}(p_+ \cdot k_3)} \left[\frac{k_{3+}}{2k_{3-}}\right]^{\frac{1}{2}} \doteq \frac{e^3 m k_{3+}}{E(p_+ \cdot k_3)} \left[\frac{k_{2-}}{2k_{2+}}\right]^{\frac{1}{2}},$$

$$M(-,-;-,+,-) = \frac{e^3 m k_{1\perp} k_{3\perp}^*}{E k_{1+}(p_+ \cdot k_3)} \left[\frac{k_{3+}}{2k_{3-}}\right]^{\frac{1}{2}} \doteq \frac{e^3 m k_{3+}}{E(p_+ \cdot k_3)} \left[\frac{k_{1-}}{2k_{1+}}\right]^{\frac{1}{2}}.$$
(9.82)

Unpolarized squared matrix element:

$$\overline{|M|^2} = \frac{2e^6 E k_{3-}}{(2E - k_{3+})k_{3+}(p_+ \cdot k_3)^2} \left[4E^2 + (2E - k_{3+})^2 + \frac{m^2 k_{3+}^3}{4E^2 k_{3-}}\right]$$

$$\times \left(\frac{k_{1+}}{k_{2+}} + \frac{k_{2+}}{k_{1+}}\right).$$
(9.83)

Unpolarized cross section:

$$d\sigma = \frac{\alpha^3 k_{3-}}{16\pi^2 E k_{3+}(2E - k_{3+})(p_+ \cdot k_3)^2} \left[4E^2 + (2E - k_{3+})^2 + \frac{m^2 k_{3+}^3}{4E^2 k_{3-}}\right]$$

$$\times \left(\frac{k_{1+}}{k_{2+}} + \frac{k_{2+}}{k_{1+}}\right) \delta^4(p_+ + p_- - k_1 - k_2 - k_3) \frac{d^3\vec{k}_1 \, d^3\vec{k}_2 \, d^3\vec{k}_3}{k_{10} \, k_{20} \, k_{30}}.$$
(9.84)

9.14.2 $\vec{k}_3$ nearly parallel to $\vec{p}_-$

Photon polarizations:

$$\not{\epsilon}_1^{\pm} = N_1[\not{p}_+ \not{p}_- \not{k}_1(1 \mp \gamma_5) - \not{k}_1 \not{p}_+ \not{p}_-(1 \pm \gamma_5)],$$

$$\not{\epsilon}_2^{\pm} = N_2[\not{p}_+ \not{p}_- \not{k}_2(1 \mp \gamma_5) - \not{k}_2 \not{p}_+ \not{p}_-(1 \pm \gamma_5)],$$

$$\not{\epsilon}_3^{\pm} = N_3[\not{k}_3 \not{p}_- \not{p}_+(1 \pm \gamma_5) + \not{p}_+ \not{p}_- \not{k}_3(1 \mp \gamma_5)],$$

$$N_i^{-1} = E^2[32 k_{i+} k_{i-}]^{\frac{1}{2}}, \qquad i = 1, 2, 3. \quad (9.85)$$

Nonvanishing helicity amplitudes:

$$M(+,+;+,-,+) = \frac{e^3 m k_{1\perp} k_{3\perp}^*}{E k_{1-}(p_- \cdot k_3)} \left[\frac{k_{3-}}{2k_{3+}}\right]^{\frac{1}{2}} \doteq \frac{e^3 m k_{3-}}{E(p_- \cdot k_3)} \left[\frac{k_{1+}}{2k_{1-}}\right]^{\frac{1}{2}},$$

$$M(+,+;-,+,+) = \frac{e^3 m k_{2\perp} k_{3\perp}^*}{Ek_{2-}(p_- \cdot k_3)} \left[\frac{k_{3-}}{2k_{3+}}\right]^{\frac{1}{2}} \doteq \frac{e^3 m k_{3-}}{E(p_- \cdot k_3)} \left[\frac{k_{2+}}{2k_{2-}}\right]^{\frac{1}{2}},$$

$$M(+,-;+,-,+) = \frac{e^3 E k_{1\perp}}{k_{2+}(p_- \cdot k_3)} \left[\frac{8k_{3+}}{k_{3-}}\right]^{\frac{1}{2}} \doteq \frac{e^3 E}{(p_- \cdot k_3)} \left[\frac{8k_{2-}k_{3+}}{k_{2+}k_{3-}}\right]^{\frac{1}{2}},$$

$$M(+,-;-,+,+) = \frac{e^3 E k_{2\perp}}{k_{1+}(p_- \cdot k_3)} \left[\frac{8k_{3+}}{k_{3-}}\right]^{\frac{1}{2}} \doteq \frac{e^3 E}{(p_- \cdot k_3)} \left[\frac{8k_{1-}k_{3+}}{k_{1+}k_{3-}}\right]^{\frac{1}{2}},$$

$$M(-,+;+,-,+) = \frac{e^3 E k_{2\perp}^*}{k_{2-}(p_- \cdot k_3)} \left[\frac{8k_{3+}}{k_{3-}}\right]^{\frac{1}{2}} \doteq \frac{e^3 E}{(p_- \cdot k_3)} \left[\frac{8k_{2+}k_{3+}}{k_{2-}k_{3-}}\right]^{\frac{1}{2}},$$

$$M(-,+;-,+,+) = \frac{e^3 E k_{1\perp}^*}{k_{1-}(p_- \cdot k_3)} \left[\frac{8k_{3+}}{k_{3-}}\right]^{\frac{1}{2}} \doteq \frac{e^3 E}{(p_- \cdot k_3)} \left[\frac{8k_{1+}k_{3+}}{k_{1-}k_{3-}}\right]^{\frac{1}{2}},$$

$$M(+,-;+,-,-) = \frac{e^3 E k_{1\perp}}{k_{1-}(p_- \cdot k_3)} \left[\frac{8k_{3+}}{k_{3-}}\right]^{\frac{1}{2}} \doteq \frac{e^3 E}{(p_- \cdot k_3)} \left[\frac{8k_{1+}k_{3+}}{k_{1-}k_{3-}}\right]^{\frac{1}{2}},$$

$$M(+,-;-,+,-) = \frac{e^3 E k_{2\perp}}{k_{2-}(p_- \cdot k_3)} \left[\frac{8k_{3+}}{k_{3-}}\right]^{\frac{1}{2}} \doteq \frac{e^3 E}{(p_- \cdot k_3)} \left[\frac{8k_{2+}k_{3+}}{k_{2-}k_{3-}}\right]^{\frac{1}{2}},$$

$$M(-,+;+,-,-) = \frac{e^3 E k_{2\perp}^*}{k_{1+}(p_- \cdot k_3)} \left[\frac{8k_{3+}}{k_{3-}}\right]^{\frac{1}{2}} \doteq \frac{e^3 E}{(p_- \cdot k_3)} \left[\frac{8k_{1-}k_{3+}}{k_{1+}k_{3-}}\right]^{\frac{1}{2}},$$

$$M(-,+;-,+,-) = \frac{e^3 E k_{1\perp}^*}{k_{2+}(p_- \cdot k_3)} \left[\frac{8k_{3+}}{k_{3-}}\right]^{\frac{1}{2}} \doteq \frac{e^3 E}{(p_- \cdot k_3)} \left[\frac{8k_{2-}k_{3+}}{k_{2+}k_{3-}}\right]^{\frac{1}{2}},$$

$$M(-,-;+,-,-) = \frac{e^3 m k_{1\perp}^* k_{3\perp}}{Ek_{2-}(p_- \cdot k_3)} \left[\frac{k_{3-}}{2k_{3+}}\right]^{\frac{1}{2}} \doteq \frac{e^3 m k_{3-}}{E(p_- \cdot k_3)} \left[\frac{k_{2+}}{2k_{2-}}\right]^{\frac{1}{2}},$$

$$M(-,-;-,+,-) = \frac{e^3 m k_{2\perp}^* k_{3\perp}}{Ek_{1-}(p_- \cdot k_3)} \left[\frac{k_{3-}}{2k_{3+}}\right]^{\frac{1}{2}} \doteq \frac{e^3 m k_{3-}}{E(p_- \cdot k_3)} \left[\frac{k_{1+}}{2k_{1-}}\right]^{\frac{1}{2}}.$$

(9.86)

Unpolarized squared matrix element:

$$\overline{|M|^2} = \frac{2e^6 E k_{3+}}{(2E - k_{3-})k_{3-}(p_- \cdot k_3)^2} \left[4E^2 + (2E - k_{3-})^2 + \frac{m^2 k_{3-}^3}{4E^2 k_{3+}}\right]$$
$$\times \left(\frac{k_{1-}}{k_{2-}} + \frac{k_{2-}}{k_{1-}}\right).$$

(9.87)

9.15. $e^+e^- \to \mu^+\mu^-\gamma$ (NO Z-EXCHANGE; $m \neq 0$)

Unpolarized cross section:

$$d\sigma = \frac{\alpha^3 k_{3+}}{16\pi^2 E k_{3-}(2E-k_{3-})(p_-\cdot k_3)^2}\left[4E^2 + (2E-k_{3-})^2 + \frac{m^2 k_{3-}^3}{4E^2 k_{3+}}\right]$$
$$\times \left(\frac{k_{1-}}{k_{2-}} + \frac{k_{2-}}{k_{1-}}\right) \delta^4(p_+ + p_- - k_1 - k_2 - k_3) \frac{d^3\vec{k}_1\, d^3\vec{k}_2\, d^3\vec{k}_3}{k_{10}\, k_{20}\, k_{30}}.$$
(9.88)

9.15 $e^+e^- \to \mu^+\mu^-\gamma$ (no Z-exchange; $m \neq 0$)

Process:
$$e^+(p_+) + e^-(p_-) \to \mu^+(q_+) + \mu^-(q_-) + \gamma(k).$$
(9.89)

Invariants:
$$s = 2(p_+ \cdot p_-), \quad t = -2(p_+ \cdot q_+), \quad u = -2(p_+ \cdot q_-),$$
$$s' = 2(q_+ \cdot q_-), \quad t' = -2(p_- \cdot q_-), \quad u' = -2(p_- \cdot q_+).$$
(9.90)

9.15.1 $\vec{k}$ nearly parallel to $\vec{p}_+$

Photon polarizations:

$$\not{\epsilon}^\pm = N[\not{k}\not{p}_+\not{p}_-(1\pm\gamma_5) + \not{p}_-\not{p}_+\not{k}(1\mp\gamma_5)],$$

$$N^{-1} = E^2[32k_+k_-]^{\frac{1}{2}}.$$
(9.91)

Nonvanishing helicity amplitudes:

$$M(+,+;+,-,+) = \frac{e^3 m k_\perp q^*_{-\perp}}{sq_{++}(p_+\cdot k)}\left[\frac{2k_+ q_{+-} q_{--}}{k_-}\right]^{\frac{1}{2}}$$

$$\doteq \frac{e^3 m k_+ q_{+-}}{s(p_+\cdot k)}\left[\frac{2q_{--}}{q_{++}}\right]^{\frac{1}{2}},$$

$$M(+,+;-,+,+) = \frac{e^3 m k_\perp q^*_{+\perp}}{sq_{-+}(p_+\cdot k)}\left[\frac{2k_+ q_{+-} q_{--}}{k_-}\right]^{\frac{1}{2}}$$

$$\doteq \frac{e^3 m k_+ q_{--}}{s(p_+\cdot k)}\left[\frac{2q_{+-}}{q_{-+}}\right]^{\frac{1}{2}},$$

$$M(+,-;+,-,+) = \frac{e^3 q_{--} q_{-\perp}}{(p_+ \cdot k)} \left[\frac{2k_-}{k_+ q_{++} q_{-+}}\right]^{\frac{1}{2}}$$

$$\doteq \frac{e^3}{(p_+ \cdot k)} \left[\frac{2k_- q_{--}^3}{k_+ q_{++}}\right]^{\frac{1}{2}},$$

$$M(+,-;-,+,+) = \frac{e^3 q_{+\perp}}{q_{++}(p_+ \cdot k)} \left[\frac{2k_- q_{+-} q_{--}}{k_+}\right]^{\frac{1}{2}}$$

$$\doteq \frac{e^3 q_{+-}}{(p_+ \cdot k)} \left[\frac{2k_- q_{--}}{k_+ q_{++}}\right]^{\frac{1}{2}},$$

$$M(-,+;+,-,+) = \frac{e^3 q_{+\perp}^*}{(p_+ \cdot k)} \left[\frac{2k_- q_{-+}}{k_+ q_{++}}\right]^{\frac{1}{2}}$$

$$\doteq \frac{e^3}{(p_+ \cdot k)} \left[\frac{2k_- q_{+-} q_{-+}}{k_+}\right]^{\frac{1}{2}},$$

$$M(-,+;-,+,+) = \frac{e^3 q_{-\perp}^*}{(p_+ \cdot k)} \left[\frac{2k_- q_{++}}{k_+ q_{-+}}\right]^{\frac{1}{2}}$$

$$\doteq \frac{e^3}{(p_+ \cdot k)} \left[\frac{2k_- q_{++} q_{--}}{k_+}\right]^{\frac{1}{2}},$$

$$M(+,-;+,-,-) = \frac{e^3 q_{-\perp}}{(p_+ \cdot k)} \left[\frac{2k_- q_{++}}{k_+ q_{-+}}\right]^{\frac{1}{2}}$$

$$\doteq \frac{e^3}{(p_+ \cdot k)} \left[\frac{2k_- q_{++} q_{--}}{k_+}\right]^{\frac{1}{2}},$$

$$M(+,-;-,+,-) = \frac{e^3 q_{+\perp}}{(p_+ \cdot k)} \left[\frac{2k_- q_{-+}}{k_+ q_{++}}\right]^{\frac{1}{2}}$$

$$\doteq \frac{e^3}{(p_+ \cdot k)} \left[\frac{2k_- q_{+-} q_{-+}}{k_+}\right]^{\frac{1}{2}},$$

$$M(-,+;+,-,-) = \frac{e^3 q_{+\perp}^*}{q_{++}(p_+ \cdot k)} \left[\frac{2k_- q_{+-} q_{--}}{k_+}\right]^{\frac{1}{2}}$$

$$\doteq \frac{e^3 q_{+-}}{(p_+ \cdot k)} \left[\frac{2k_- q_{--}}{k_+ q_{++}}\right]^{\frac{1}{2}},$$

9.15. $e^+e^- \to \mu^+\mu^-\gamma$ (NO Z-EXCHANGE; $m \neq 0$)

$$M(-,+;-,+,-) = \frac{e^3 q_{--} q^*_{-\perp}}{(p_+ \cdot k)} \left[\frac{2k_-}{k_+ q_{++} q_{-+}}\right]^{\frac{1}{2}}$$

$$\doteq \frac{e^3}{(p_+ \cdot k)} \left[\frac{2k_- q^3_{--}}{k_+ q_{++}}\right]^{\frac{1}{2}},$$

$$M(-,-;+,-,-) = \frac{e^3 m k^*_\perp q_{-\perp}}{s q_{-+} (p_+ \cdot k)} \left[\frac{2k_+ q_{+-} q_{--}}{k_-}\right]^{\frac{1}{2}}$$

$$\doteq \frac{e^3 m k_+ q_{--}}{s(p_+ \cdot k)} \left[\frac{2q_{+-}}{q_{-+}}\right]^{\frac{1}{2}},$$

$$M(-,-;-,+,-) = \frac{e^3 m k^*_\perp q_{+\perp}}{s q_{++} (p_+ \cdot k)} \left[\frac{2k_+ q_{+-} q_{--}}{k_-}\right]^{\frac{1}{2}}$$

$$\doteq \frac{e^3 m k_+}{s(p_+ \cdot k)} \left[\frac{2q^3_{+-}}{q_{-+}}\right]^{\frac{1}{2}}. \tag{9.92}$$

Unpolarized squared matrix element:

$$\overline{|M|^2} = \frac{2e^6 E k_-}{(2E - k_+) k_+ (p_+ \cdot k)^2} \left[s + (2E - k_+)^2 + \frac{m^2 k^3_+}{s k_-}\right] \frac{t'^2 + u'^2}{s'^2}. \tag{9.93}$$

Unpolarized cross section:

$$d\sigma = \frac{\alpha^3 E k_-}{4\pi^2 (2E - k_+) k_+ (p_+ \cdot k)^2} \left[s + (2E - k_+)^2 + \frac{m^2 k^3_+}{s k_-}\right]$$

$$\times \frac{t'^2 + u'^2}{ss'^2} \delta^4(p_+ + p_- - q_+ - q_- - k) \frac{d^3 \vec{q}_+ \, d^3 \vec{q}_- \, d^3 \vec{k}}{q_{+0} \, q_{-0} \, k_0}. \tag{9.94}$$

9.15.2 $\vec{k}$ nearly parallel to $\vec{p}_-$

Photon polarizations:

$$\rlap{/}{\epsilon}^\pm = N[\rlap{/}{k}\, \rlap{/}{p}_- \rlap{/}{p}_+ (1 \pm \gamma_5) + \rlap{/}{p}_+ \rlap{/}{p}_- \rlap{/}{k}(1 \mp \gamma_5)],$$

$$N^{-1} = E^2 [32 k_+ k_-]^{\frac{1}{2}}. \tag{9.95}$$

Nonvanishing helicity amplitudes:

$$M(+,+;+,-,+) = \frac{e^3 m k^*_\perp q_{+\perp}}{s q_{+-} (p_- \cdot k)} \left[\frac{2k_- q_{++} q_{-+}}{k_+}\right]^{\frac{1}{2}}$$

$$\doteq \frac{e^3 m k_- q_{++}}{s(p_- \cdot k)} \left[\frac{2q_{-+}}{q_{+-}}\right]^{\frac{1}{2}},$$

$$M(+,+;-,+,+) = \frac{e^3 m k_\perp^* q_{-\perp}}{s q_{--}(p_- \cdot k)} \left[\frac{2k_- q_{++} q_{-+}}{k_+}\right]^{\frac{1}{2}}$$

$$\doteq \frac{e^3 m k_- q_{-+}}{s(p_- \cdot k)} \left[\frac{2q_{++}}{q_{--}}\right]^{\frac{1}{2}},$$

$$M(+,-;+,-,+) = \frac{e^3 q_{+\perp}}{(p_- \cdot k)} \left[\frac{2k_+ q_{++}}{k_- q_{-+}}\right]^{\frac{1}{2}}$$

$$\doteq \frac{e^3}{(p_- \cdot k)} \left[\frac{2k_+ q_{++} q_{--}}{k_-}\right]^{\frac{1}{2}},$$

$$M(+,-;-,+,+) = \frac{e^3 q_{-\perp}}{(p_- \cdot k)} \left[\frac{2k_+ q_{-+}}{k_- q_{++}}\right]^{\frac{1}{2}}$$

$$\doteq \frac{e^3}{(p_- \cdot k)} \left[\frac{2k_+ q_{+-} q_{-+}}{k_-}\right]^{\frac{1}{2}},$$

$$M(-,+;+,-,+) = \frac{e^3 q_{-\perp}^*}{q_{--}(p_- \cdot k)} \left[\frac{2k_+ q_{++} q_{-+}}{k_-}\right]^{\frac{1}{2}}$$

$$\doteq \frac{e^3 q_{-+}}{(p_- \cdot k)} \left[\frac{2k_+ q_{++}}{k_- q_{--}}\right]^{\frac{1}{2}},$$

$$M(-,+;-,+,+) = \frac{e^3 q_{+\perp}^*}{q_{+-}(p_- \cdot k)} \left[\frac{2k_+ q_{++} q_{-+}}{k_-}\right]^{\frac{1}{2}}$$

$$\doteq \frac{e^3 q_{++}}{(p_- \cdot k)} \left[\frac{2k_+ q_{-+}}{k_- q_{+-}}\right]^{\frac{1}{2}},$$

$$M(+,-;+,-,-) = \frac{e^3 q_{+\perp}}{q_{+-}(p_- \cdot k)} \left[\frac{2k_+ q_{++} q_{-+}}{k_-}\right]^{\frac{1}{2}}$$

$$\doteq \frac{e^3 q_{++}}{(p_- \cdot k)} \left[\frac{2k_+ q_{-+}}{k_- q_{+-}}\right]^{\frac{1}{2}},$$

$$M(+,-;-,+,-) = \frac{e^3 q_{-\perp}}{q_{--}(p_- \cdot k)} \left[\frac{2k_+ q_{++} q_{-+}}{k_-}\right]^{\frac{1}{2}}$$

$$\doteq \frac{e^3 q_{-+}}{(p_- \cdot k)} \left[\frac{2k_+ q_{++}}{k_- q_{--}}\right]^{\frac{1}{2}},$$

9.15. $e^+e^- \to \mu^+\mu^-\gamma$ (NO Z-EXCHANGE; $m \neq 0$)

$$M(-,+;+,-,-) = \frac{e^3 q^*_{-\perp}}{(p_- \cdot k)} \left[\frac{2k_+ q_{-+}}{k_- q_{++}}\right]^{\frac{1}{2}}$$

$$\doteq \frac{e^3}{(p_- \cdot k)} \left[\frac{2k_+ q_{+-} q_{-+}}{k_-}\right]^{\frac{1}{2}},$$

$$M(-,+;-,+,-) = \frac{e^3 q^*_{+\perp}}{(p_- \cdot k)} \left[\frac{2k_+ q_{++}}{k_- q_{-+}}\right]^{\frac{1}{2}}$$

$$\doteq \frac{e^3}{(p_+ \cdot k)} \left[\frac{2k_+ q_{++} q_{--}}{k_-}\right]^{\frac{1}{2}},$$

$$M(-,-;+,-,-) = \frac{e^3 m_\perp q^*_{+\perp}}{s q_{--}(p_- \cdot k)} \left[\frac{2k_- q_{++} q_{-+}}{k_+}\right]^{\frac{1}{2}}$$

$$\doteq \frac{e^3 m k_- q_{-+}}{s(p_- \cdot k)} \left[\frac{2q_{++}}{q_{--}}\right]^{\frac{1}{2}},$$

$$M(-,-;-,+,-) = \frac{e^3 m_\perp q^*_{-\perp}}{s q_{+-}(p_- \cdot k)} \left[\frac{2k_- q_{++} q_{-+}}{k_+}\right]^{\frac{1}{2}}$$

$$\doteq \frac{e^3 m k_- q_{++}}{s(p_- \cdot k)} \left[\frac{2q_{-+}}{q_{+-}}\right]^{\frac{1}{2}}. \quad (9.96)$$

Unpolarized squared matrix element:

$$\overline{|M|^2} = \frac{2e^6 E k_+}{(2E - k_-)k_-(p_- \cdot k)^2} \left[s + (2E - k_-)^2 + \frac{m^2 k_-^3}{s k_+}\right]$$

$$\times \frac{t^2 + u^2}{s'^2}. \quad (9.97)$$

Unpolarized cross section:

$$d\sigma = \frac{\alpha^3 E k_+}{4\pi^2 (2E - k_-)k_-(p_- \cdot k)^2} \left[s + (2E - k_-)^2 + \frac{m^2 k_-^3}{s k_+}\right]$$

$$\times \frac{t^2 + u^2}{ss'^2} \delta^4(p_+ + p_- - q_+ - q_- - k) \frac{d^3\vec{q}_+ \, d^3\vec{q}_- \, d^3\vec{k}}{q_{+0} \, q_{-0} \, k_0}. \quad (9.98)$$

9.15.3 $\vec{k}$ nearly parallel to $\vec{q}_+$

The primed quantities $k'_\pm$ and $k'_\perp$ are evaluated in the rotated frame where $\vec{q}_+$ determines the positive z-axis (see also Section 7.4.3). The vector q in eqn (9.99) is obtained by applying a space reflection to q_+. The quantity μ denotes the muon mass.

Photon polarizations:

$$\rlap{/}{\epsilon}^{\pm} = N[\rlap{/}{k}\,\rlap{/}{A}_+\,\rlap{/}{A}(1\pm\gamma_5) + \rlap{/}{A}\,\rlap{/}{A}_+\,\rlap{/}{k}(1\mp\gamma_5)],$$

$$N^{-1} = q_{+0}^2[32k'_+ k'_-]^{\frac{1}{2}}.\tag{9}$$

Nonvanishing helicity amplitudes:

$$M(+,-;+,+,-) = -\frac{e^3\mu k'^*_\perp q_{+\perp}}{Eq_{+0}(q_+\cdot k)}\left[\frac{k'_+ q_{-+}}{8k'_- q_{++}}\right]^{\frac{1}{2}}$$

$$\doteq \frac{e^3\mu k'_+}{Eq_{+0}(q_+\cdot k)}\left[\frac{q_{+-}q_{-+}}{8}\right]^{\frac{1}{2}},$$

$$M(+,-;+,-,+) = -\frac{e^3 q_{+0} q_{-\perp}}{Eq_{+-}(q_+\cdot k)}\left[\frac{2k'_- q_{++} q_{-+}}{k'_+}\right]^{\frac{1}{2}}$$

$$\doteq \frac{e^3 q_{+0}}{E(q_+\cdot k)}\left[\frac{2k'_- q^3_{--}}{k'_+ q_{++}}\right]^{\frac{1}{2}},$$

$$M(+,-;-,+,+) = -\frac{e^3 q_{+0} q_{+\perp}}{E(q_+\cdot k)}\left[\frac{2k'_- q_{-+}}{k'_+ q_{++}}\right]^{\frac{1}{2}}$$

$$\doteq \frac{e^3 q_{+0}}{E(q_+\cdot k)}\left[\frac{2k'_- q_{+-} q_{-+}}{k'_+}\right]^{\frac{1}{2}},$$

$$M(-,+;+,+,-) = -\frac{e^3\mu k'^*_\perp q^*_{-\perp}}{Eq_{+0}(q_+\cdot k)}\left[\frac{k'_+ q_{++}}{8k'_- q_{-+}}\right]^{\frac{1}{2}}$$

$$\doteq \frac{e^3\mu k'_+}{Eq_{+0}(q_+\cdot k)}\left[\frac{q_{++} q_{--}}{8}\right]^{\frac{1}{2}},$$

$$M(-,+;+,-,+) = -\frac{e^3 q_{+0} q^*_{+\perp}}{Eq_{+-}(q_+\cdot k)}\left[\frac{2k'_- q^3_{-+}}{k'_+ q_{++}}\right]^{\frac{1}{2}}$$

$$\doteq \frac{e^3 q_{+0}}{E(q_+\cdot k)}\left[\frac{2k'_- q^3_{-+}}{k'_+ q_{+-}}\right]^{\frac{1}{2}},$$

$$M(-,+;-,+,+) = -\frac{e^3 q_{+0} q^*_{-\perp}}{E(q_+\cdot k)}\left[\frac{2k'_- q_{++}}{k'_+ q_{-+}}\right]^{\frac{1}{2}}$$

$$\doteq \frac{e^3 q_{+0}}{E(q_+\cdot k)}\left[\frac{2k'_- q_{++} q_{--}}{k'_+}\right]^{\frac{1}{2}},$$

9.15. $e^+e^- \to \mu^+\mu^-\gamma$ (NO Z-EXCHANGE; $m \neq 0$)

$$M(+,-;+,-,-) = -\frac{e^3 q_{+0} q_{-\perp}}{E(q_+ \cdot k)} \left[\frac{2k'_- q_{++}}{k'_+ q_{-+}}\right]^{\frac{1}{2}}$$

$$\doteq \frac{e^3 q_{+0}}{E(q_+ \cdot k)} \left[\frac{2k'_- q_{++} q_{--}}{k'_+}\right]^{\frac{1}{2}},$$

$$M(+,-;-,+,-) = -\frac{e^3 q_{+0} q_{+\perp}}{E q_{+-}(q_+ \cdot k)} \left[\frac{2k'_- q^3_{-+}}{k'_+ q_{++}}\right]^{\frac{1}{2}}$$

$$\doteq \frac{e^3 q_{+0}}{E(q_+ \cdot k)} \left[\frac{2k'_- q^3_{-+}}{k'_+ q_{+-}}\right]^{\frac{1}{2}},$$

$$M(+,-;-,-,+) = \frac{e^3 \mu k'_\perp q_{-\perp}}{E q_{+0}(q_+ \cdot k)} \left[\frac{k'_+ q_{++}}{8k'_- q_{-+}}\right]^{\frac{1}{2}}$$

$$\doteq \frac{e^3 \mu k'_+}{E q_{+0}(q_+ \cdot k)} \left[\frac{q_{++} q_{--}}{8}\right]^{\frac{1}{2}},$$

$$M(-,+;+,-,-) = -\frac{e^3 q_{+0} q^*_{+\perp}}{E(q_+ \cdot k)} \left[\frac{2k'_- q_{-+}}{k'_+ q_{++}}\right]^{\frac{1}{2}}$$

$$\doteq \frac{e^3 q_{+0}}{E(q_+ \cdot k)} \left[\frac{2k'_- q_{+-} q_{-+}}{k'_+}\right]^{\frac{1}{2}},$$

$$M(-,+;-,+,-) = -\frac{e^3 q_{+0} q^*_{-\perp}}{E q_{+-}(q_+ \cdot k)} \left[\frac{2k'_- q_{++} q_{-+}}{k'_+}\right]^{\frac{1}{2}}$$

$$\doteq \frac{e^3 q_{+0}}{E(q_+ \cdot k)} \left[\frac{2k'_- q^3_{--}}{k'_+ q_{++}}\right]^{\frac{1}{2}},$$

$$M(-,+;-,-,+) = \frac{e^3 \mu k'_\perp q^*_{+\perp}}{E q_{+0}(q_+ \cdot k)} \left[\frac{k'_+ q_{-+}}{8k'_- q_{++}}\right]^{\frac{1}{2}}$$

$$\doteq \frac{e^3 \mu k'_+}{E q_{+0}(q_+ \cdot k)} \left[\frac{q_{+-} q_{-+}}{8}\right]^{\frac{1}{2}}. \tag{9.100}$$

Unpolarized squared matrix element:

$$\overline{|M|^2} = \frac{2e^6 q_{+0} k'_-}{(2q_{+0} + k'_+) k'_+ (q_+ \cdot k)^2} \left[4q^2_{+0} + (2q_{+0} + k'_+)^2 + \frac{\mu^2 k'^3_+}{4q^2_{+0} k'_-}\right]$$

$$\times \frac{t'^2 + u^2}{s^2}. \tag{9.101}$$

Unpolarized cross section:

$$d\sigma = \frac{\alpha^3 k'_-}{4\pi^2(2q_{+0}+k'_+)k'_+(q_+\cdot k)^2}\left[4q_{+0}^2+(2q_{+0}+k'_+)^2+\frac{\mu^2 k'^3_+}{4q_{+0}^2 k'_-}\right]$$

$$\times \frac{t'^2+u^2}{s^3}\delta^4(p_++p_--q_+-q_--k)\frac{d^3\vec{q}_+\,d^3\vec{q}_-\,d^3\vec{k}}{q_{-0}k_0}. \quad (9.102)$$

9.15.4 $\vec{k}$ nearly parallel to $\vec{q}_-$

The doubly primed quantities k''_+ and k''_- are evaluated in the rotated frame where $\vec{q}_-$ determines the positive z-axis (see also Section 7.4.3). The vector q in eqn (9.103) is obtained by applying a space reflection to q_-. The quantity μ denotes the muon mass.

Photon polarizations:

$$\not{\epsilon}^{\pm} = N[\not{k}\,\not{q}_-\,\not{q}(1\pm\gamma_5)+\not{q}\,\not{q}_-\,\not{k}(1\mp\gamma_5)],$$

$$N^{-1} = q_{-0}^2[32k''_+k''_-]^{\frac{1}{2}}. \quad (9.103)$$

Nonvanishing helicity amplitudes:

$$M(+,-;+,+,-) = \frac{e^3\mu k''^*_\perp q_{-\perp}}{Eq_{-0}(q_-\cdot k)}\left[\frac{k''_+q_{++}}{8k''_-q_{-+}}\right]^{\frac{1}{2}}$$

$$\doteq \frac{e^3\mu k''_+}{Eq_{-0}(q_-\cdot k)}\left[\frac{q_{++}q_{--}}{8}\right]^{\frac{1}{2}},$$

$$M(+,-;+,-,+) = \frac{e^3 q_{-0}q_{-\perp}}{E(q_-\cdot k)}\left[\frac{2k''_-q_{++}}{k''_+q_{-+}}\right]^{\frac{1}{2}}$$

$$\doteq \frac{e^3 q_{-0}}{E(q_-\cdot k)}\left[\frac{2k''_-q_{++}q_{--}}{k''_+}\right]^{\frac{1}{2}},$$

$$M(+,-;-,+,+) = \frac{e^3 q_{-0}q_{+\perp}}{E(q_-\cdot k)}\left[\frac{2k''_-q_{+-}}{k''_+q_{--}}\right]^{\frac{1}{2}}$$

$$\doteq \frac{e^3 q_{-0}}{E(q_-\cdot k)}\left[\frac{2k''_-q^3_{+-}}{k''_+q_{-+}}\right]^{\frac{1}{2}},$$

$$M(-,+;+,+,-) = \frac{e^3\mu k''^*_\perp q^*_{+\perp}}{Eq_{-0}(q_-\cdot k)}\left[\frac{k''_+q_{-+}}{8k''_-q_{++}}\right]^{\frac{1}{2}}$$

$$\doteq \frac{e^3\mu k''_+}{Eq_{-0}(q_-\cdot k)}\left[\frac{q_{+-}q_{-+}}{8}\right]^{\frac{1}{2}},$$

9.15. $e^+e^- \to \mu^+\mu^-\gamma$ (NO Z-EXCHANGE; $m \neq 0$)

$$M(-,+;+,-,+) = \frac{e^3 q_{-0} q_{+\perp}^*}{E(q_- \cdot k)} \left[\frac{2k''_- q_{-+}}{k''_+ q_{++}}\right]^{\frac{1}{2}}$$

$$\doteq \frac{e^3 q_{-0}}{E(q_- \cdot k)} \left[\frac{2k''_- q_{+-} q_{-+}}{k''_+}\right]^{\frac{1}{2}},$$

$$M(-,+;-,+,+) = \frac{e^3 q_{-0} q_{-\perp}^*}{E q_{--}(q_- \cdot k)} \left[\frac{2k''_- q_{++}^3}{k''_+ q_{-+}}\right]^{\frac{1}{2}}$$

$$\doteq \frac{e^3 q_{-0}}{E(q_- \cdot k)} \left[\frac{2k''_- q_{++}^3}{k''_+ q_{--}}\right]^{\frac{1}{2}},$$

$$M(+,-;+,-,-) = \frac{e^3 q_{-0} q_{-\perp}}{E q_{--}(q_- \cdot k)} \left[\frac{2k''_- q_{++}^3}{k''_+ q_{-+}}\right]^{\frac{1}{2}}$$

$$\doteq \frac{e^3 q_{-0}}{E(q_- \cdot k)} \left[\frac{2k''_- q_{++}^3}{k''_+ q_{--}}\right]^{\frac{1}{2}},$$

$$M(+,-;-,+,-) = \frac{e^3 q_{-0} q_{+\perp}}{E(q_- \cdot k)} \left[\frac{2k''_- q_{-+}}{k''_+ q_{++}}\right]^{\frac{1}{2}}$$

$$\doteq \frac{e^3 q_{-0}}{E(q_- \cdot k)} \left[\frac{2k''_- q_{+-} q_{-+}}{k''_+}\right]^{\frac{1}{2}},$$

$$M(+,-;-,-,+) = -\frac{e^3 \mu k''_\perp q_{+\perp}}{E q_{-0}(q_- \cdot k)} \left[\frac{k''_+ q_{-+}}{8 k''_- q_{++}}\right]^{\frac{1}{2}}$$

$$\doteq \frac{e^3 \mu k''_+}{E q_{-0}(q_- \cdot k)} \left[\frac{q_{+-} q_{-+}}{8}\right]^{\frac{1}{2}},$$

$$M(-,+;+,-,-) = \frac{e^3 q_{-0} q_{+\perp}^*}{E(q_- \cdot k)} \left[\frac{2k''_- q_{+-}}{k''_+ q_{--}}\right]^{\frac{1}{2}}$$

$$\doteq \frac{e^3 q_{-0}}{E(q_- \cdot k)} \left[\frac{2k''_- q_{+-}^3}{k''_+ q_{-+}}\right]^{\frac{1}{2}},$$

$$M(-,+;-,+,-) = \frac{e^3 q_{-0} q_{-\perp}^*}{E(q_- \cdot k)} \left[\frac{2k''_- q_{++}}{k''_+ q_{-+}}\right]^{\frac{1}{2}}$$

$$\doteq \frac{e^3 q_{-0}}{E(q_- \cdot k)} \left[\frac{2k''_- q_{++} q_{--}}{k''_+}\right]^{\frac{1}{2}},$$

$$M(-,+;-,-,+) = -\frac{e^3\mu k''_\perp q^*_{-\perp}}{Eq_{-0}(q_- \cdot k)}\left[\frac{k''_+ q_{++}}{8k''_- q_{-+}}\right]^{\frac{1}{2}}$$

$$\doteq \frac{e^3\mu k''_+}{Eq_{-0}(q_- \cdot k)}\left[\frac{q_{++}q_{--}}{8}\right]^{\frac{1}{2}}. \quad (9.104)$$

Unpolarized squared matrix element:

$$\overline{|M|^2} = \frac{2e^6 q_{-0} k''_-}{(2q_{-0}+k''_+)k''_+(q_- \cdot k)^2}\left[4q^2_{-0}+(2q_{-0}+k''_+)^2+\frac{\mu^2 k''^3_+}{4q^2_{-0}k''_-}\right]\frac{t^2+u'^2}{s^2}. \quad (9.105)$$

Unpolarized cross section:

$$d\sigma = \frac{\alpha^3 k''_-}{4\pi^2(2q_{-0}+k''_+)k''_+(q_- \cdot k)^2}\left[4q^2_{-0}+(2q_{-0}+k''_+)^2+\frac{\mu^2 k''^3_+}{4q^2_{-0}k''_-}\right]$$

$$\times \frac{t^2+u'^2}{s^3}\delta^4(p_++p_--q_+-q_--k)\frac{d^3\vec{q}_+\,d^3\vec{q}_-\,d^3\vec{k}}{q_{+0}k_0}. \quad (9.106)$$

9.16 $e^+e^- \to e^+e^-\gamma$ (no Z-exchange; $m \neq 0$)

Process:
$$e^+(p_+) + e^-(p_-) \to e^+(q_+) + e^-(q_-) + \gamma(k). \quad (9.107)$$

Invariants:

$$s = 2(p_+ \cdot p_-), \quad t = -2(p_+ \cdot q_+), \quad u = -2(p_+ \cdot q_-),$$

$$s' = 2(q_+ \cdot q_-), \quad t' = -2(p_- \cdot q_-), \quad u' = -2(p_- \cdot q_+). \quad (9.108)$$

9.16.1 $\vec{k}$ nearly parallel to $\vec{p}_+$

Photon polarizations:

$$\rlap{/}{t}^\pm = N[\rlap{/}{k}\,\rlap{/}{p}_+\,\rlap{/}{p}_-(1\pm\gamma_5) + \rlap{/}{p}_-\,\rlap{/}{p}_+\,\rlap{/}{k}(1\mp\gamma_5)],$$

$$N^{-1} = E^2[32k_+k_-]^{\frac{1}{2}}. \quad (9.109)$$

Nonvanishing helicity amplitudes:

$$M(+,+;+,+,+) = \frac{e^3 s q^*_{+\perp}}{q_{+-}(p_+ \cdot k)}\left[\frac{2k_-}{k_+q_{++}q_{-+}}\right]^{\frac{1}{2}}$$

$$\doteq \frac{e^3 s}{(p_+ \cdot k)}\left[\frac{2k_-}{k_+ q_{+-} q_{-+}}\right]^{\frac{1}{2}},$$

9.16. $e^+e^- \to e^+e^-\gamma$ (NO Z-EXCHANGE; $m \neq 0$)

$$M(+,+;+,+,-) = \frac{e^3 s q_{+\perp}^*}{(p_+ \cdot k)} \left[\frac{2k_-}{k_+ q_{+-}^3}\right]^{\frac{1}{2}}$$

$$\doteq \frac{e^3 s}{(p_+ \cdot k)} \left[\frac{2k_- q_{-+}}{k_+ q_{+-}^3}\right]^{\frac{1}{2}},$$

$$M(+,+;+,-,+) = \frac{e^3 m k_\perp q_{-\perp}^*}{s q_{++}(p_+ \cdot k)} \left[\frac{2k_+ q_{--} q_{+-}}{k_-}\right]^{\frac{1}{2}}$$

$$\doteq \frac{e^3 m k_+}{s(p_+ \cdot k)} \left[\frac{2q_{+-}^3}{q_{-+}}\right]^{\frac{1}{2}},$$

$$M(+,+;-,+,+) = \frac{e^3 m q_{--} k_\perp q_{-\perp}^*}{s(p_+ \cdot k)} \left[\frac{2k_+ q_{++}}{k_- q_{-+}^3}\right]^{\frac{1}{2}}$$

$$\doteq \frac{e^3 m k_+ \left[2 q_{++} q_{--}^3\right]^{\frac{1}{2}}}{s q_{-+}(p_+ \cdot k)},$$

$$M(+,-;+,-,+) = \frac{e^3 q_{--} q_{+\perp}}{(p_+ \cdot k)} \left[\frac{2k_- q_{++}}{k_+ q_{-+}^3}\right]^{\frac{1}{2}}$$

$$\doteq \frac{e^3}{q_{-+}(p_+ \cdot k)} \left[\frac{2k_- q_{++} q_{--}^3}{k_+}\right]^{\frac{1}{2}},$$

$$M(+,-;-,+,+) = \frac{e^3 q_{+\perp}}{q_{++}(p_+ \cdot k)} \left[\frac{2k_- q_{+-} q_{--}}{k_+}\right]^{\frac{1}{2}}$$

$$\doteq \frac{e^3 q_{+-}}{(p_+ \cdot k)} \left[\frac{2k_- q_{--}}{k_+ q_{++}}\right]^{\frac{1}{2}},$$

$$M(-,+;+,+,-) = \frac{e^3 m k_\perp^* q_{+\perp}^*}{q_{+-}(p_+ \cdot k)} \left[\frac{2k_+}{k_- q_{++} q_{-+}}\right]^{\frac{1}{2}}$$

$$\doteq \frac{e^3 m k_+}{(p_+ \cdot k)} \left[\frac{2}{q_{+-} q_{-+}}\right]^{\frac{1}{2}},$$

$$M(-,+;+,-,+) = \frac{e^3 q_{+\perp}^*}{(p_+ \cdot k)} \left[\frac{2k_- q_{-+}}{k_+ q_{++}}\right]^{\frac{1}{2}}$$

$$\doteq \frac{e^3}{(p_+ \cdot k)} \left[\frac{2k_- q_{+-} q_{-+}}{k_+}\right]^{\frac{1}{2}},$$

$$M(-,+;-,+,+) = \frac{e^3 q_{+\perp}^*}{(p_+ \cdot k)} \left[\frac{2k_- q_{++}^3}{k_+ q_{-+}^3} \right]^{\frac{1}{2}}$$

$$\doteq \frac{e^3 q_{++}^2}{(p_+ \cdot k)} \left[\frac{2k_- q_{+-}}{k_+ q_{-+}^3} \right]^{\frac{1}{2}},$$

$$M(+,-;+,-,-) = \frac{e^3 q_{+\perp}}{(p_+ \cdot k)} \left[\frac{2k_- q_{++}^3}{k_+ q_{-+}^3} \right]^{\frac{1}{2}}$$

$$\doteq \frac{e^3 q_{++}^2}{(p_+ \cdot k)} \left[\frac{2k_- q_{+-}}{k_+ q_{-+}^3} \right]^{\frac{1}{2}},$$

$$M(+,-;-,+,-) = \frac{e^3 q_{+\perp}}{(p_+ \cdot k)} \left[\frac{2k_- q_{-+}}{k_+ q_{++}} \right]^{\frac{1}{2}}$$

$$\doteq \frac{e^3}{(p_+ \cdot k)} \left[\frac{2k_- q_{+-} q_{-+}}{k_+} \right]^{\frac{1}{2}},$$

$$M(+,-;-,-,+) = \frac{e^3 m k_\perp q_{-\perp}}{q_{+-}(p_+ \cdot k)} \left[\frac{2k_+}{k_- q_{++} q_{-+}} \right]^{\frac{1}{2}}$$

$$\doteq \frac{e^3 m k_+}{(p_+ \cdot k)} \left[\frac{2}{q_{+-} q_{-+}} \right]^{\frac{1}{2}},$$

$$M(-,+;+,-,-) = \frac{e^3 q_{+\perp}^*}{q_{++}(p_+ \cdot k)} \left[\frac{2k_- q_{+-} q_{--}}{k_+} \right]^{\frac{1}{2}}$$

$$\doteq \frac{e^3}{(p_+ \cdot k)} \left[\frac{2k_- q_{--}}{k_+ q_{++}} \right]^{\frac{1}{2}},$$

$$M(-,+;-,+,-) = \frac{e^3 q_{--} q_{+\perp}^*}{(p_+ \cdot k)} \left[\frac{2k_- q_{++}}{k_+ q_{-+}^3} \right]^{\frac{1}{2}}$$

$$\doteq \frac{e^3}{q_{-+}(p_+ \cdot k)} \left[\frac{2k_- q_{++} q_{--}^3}{k_+} \right]^{\frac{1}{2}},$$

$$M(-,-;+,-,-) = \frac{e^3 m q_{--} k_\perp^* q_{+\perp}}{s(p_+ \cdot k)} \left[\frac{2k_+ q_{++}}{k_- q_{-+}^3} \right]^{\frac{1}{2}}$$

$$\doteq \frac{e^3 m k_+ \left[2 q_{++} q_{--}^3 \right]^{\frac{1}{2}}}{s q_{-+}(p_+ \cdot k)},$$

9.16. $e^+e^- \to e^+e^-\gamma$ (NO Z-EXCHANGE; $m \neq 0$)

$$M(-,-;-,+,-) = \frac{e^3 m k_\perp^* q_{+\perp}}{s q_{++}(p_+ \cdot k)} \left[\frac{2k_+ q_{+-} q_{--}}{k_-}\right]^{\frac{1}{2}}$$

$$\doteq \frac{e^3 m k_+}{s(p_+ \cdot k)} \left[\frac{2q_{+-}^3}{q_{-+}}\right]^{\frac{1}{2}},$$

$$M(-,-;-,-,+) = \frac{e^3 s q_{+\perp}}{(p_+ \cdot k)} \left[\frac{2k_-}{k_+ q_{+-}^3 q_{--}}\right]^{\frac{1}{2}}$$

$$\doteq \frac{e^3 s}{(p_+ \cdot k)} \left[\frac{2k_- q_{-+}}{k_+ q_{+-}^3}\right]^{\frac{1}{2}},$$

$$M(-,-;-,-,-) = \frac{e^3 s q_{+\perp}}{q_{+-}(p_+ \cdot k)} \left[\frac{2k_-}{k_+ q_{++} q_{-+}}\right]^{\frac{1}{2}}$$

$$\doteq \frac{e^3 s}{(p_+ \cdot k)} \left[\frac{2k_-}{k_+ q_{+-} q_{-+}}\right]^{\frac{1}{2}}. \tag{9.110}$$

Unpolarized squared matrix element:

$$\overline{|M|^2} = \frac{2e^6 E k_-}{(2E - k_+) k_+ (p_+ \cdot k)^2} \left[s + (2E - k_+)^2 + \frac{m^2 k_+^3}{s k_-}\right] \frac{s'^4 + t'^4 + u'^4}{s'^2 t'^2}. \tag{9.111}$$

Unpolarized cross section:

$$d\sigma = \frac{\alpha^3 k_-}{16\pi^2 E k_+ (2E - k_+)(p_+ \cdot k)^2} \left[s + (2E - k_+)^2 + \frac{m^2 k_+^3}{s k_-}\right]$$

$$\times \frac{s'^4 + t'^4 + u'^4}{s'^2 t'^2} \delta^4(p_+ + p_- - q_+ - q_- - k) \frac{d^3\vec{q}_+ \, d^3\vec{q}_- \, d^3\vec{k}}{q_{+0} \, q_{-0} \, k_0}. \tag{9.112}$$

9.16.2 $\vec{k}$ nearly parallel to $\vec{p}_-$

Photon polarizations:

$$\not{\epsilon}^{\pm} = N[\not{k}\, \not{p}_-\, \not{p}_+(1 \pm \gamma_5) + \not{p}_+\, \not{p}_-\, \not{k}(1 \mp \gamma_5)],$$

$$N^{-1} = E^2[32 k_+ k_-]^{\frac{1}{2}}. \tag{9.113}$$

Nonvanishing helicity amplitudes:

$$M(+,+;+,+,+) = \frac{e^3 s q_{-\perp}^*}{q_{+-}(p_- \cdot k)} \left[\frac{2k_+}{k_- q_{++} q_{-+}}\right]^{\frac{1}{2}}$$

$$\doteq \frac{e^3 s}{(p_- \cdot k)} \left[\frac{2k_+}{k_- q_{+-} q_{-+}}\right]^{\frac{1}{2}},$$

$$M(+,+;+,+,-) = \frac{e^3 s q^*_{-\perp}}{(p_- \cdot k)} \left[\frac{2k_+}{k_- q_{++} q^3_{-+}}\right]^{\frac{1}{2}}$$

$$\doteq \frac{e^3 s}{(p_- \cdot k)} \left[\frac{2k_+ q_{+-}}{k_- q^3_{-+}}\right]^{\frac{1}{2}},$$

$$M(+,+;+,-,+) = \frac{e^3 m k^*_\perp q_{-\perp}}{s q_{+-} (p_- \cdot k)} \left[\frac{2k_- q^3_{++}}{k_+ q_{-+}}\right]^{\frac{1}{2}}$$

$$\doteq \frac{e^3 m k_- \left[2 q^3_{++} q_{--}\right]^{\frac{1}{2}}}{s q_{+-} (p_- \cdot k)},$$

$$M(+,+;-,+,+) = \frac{e^3 m k^*_\perp q_{-\perp}}{s q_{--} (p_- \cdot k)} \left[\frac{2k_- q_{++} q_{-+}}{k_+}\right]^{\frac{1}{2}}$$

$$\doteq \frac{e^3 m k_- q_{-+}}{s (p_- \cdot k)} \left[\frac{2 q_{++}}{q_{--}}\right]^{\frac{1}{2}},$$

$$M(+,-;+,+,-) = \frac{e^3 m k_\perp q^*_{+\perp}}{q_{+-} (p_- \cdot k)} \left[\frac{2k_-}{k_+ q_{++} q_{-+}}\right]^{\frac{1}{2}}$$

$$\doteq \frac{e^3 m k_-}{(p_- \cdot k)} \left[\frac{2}{q_{+-} q_{-+}}\right]^{\frac{1}{2}},$$

$$M(+,-;+,-,+) = \frac{e^3 q_{-\perp}}{(p_- \cdot k)} \left[\frac{2k_+ q^3_{++}}{k_- q^3_{-+}}\right]^{\frac{1}{2}}$$

$$\doteq \frac{e^3}{q_{-+} (p_- \cdot k)} \left[\frac{2k_+ q^3_{++} q_{--}}{k_-}\right]^{\frac{1}{2}},$$

$$M(+,-;-,+,+) = \frac{e^3 q_{-\perp}}{(p_- \cdot k)} \left[\frac{2k_+ q_{-+}}{k_- q_{++}}\right]^{\frac{1}{2}}$$

$$\doteq \frac{e^3}{(p_- \cdot k)} \left[\frac{2k_+ q_{+-} q_{-+}}{k_-}\right]^{\frac{1}{2}},$$

$$M(-,+;+,-,+) = \frac{e^3 q^*_{-\perp}}{q_{--} (p_- \cdot k)} \left[\frac{2k_+ q_{++} q_{-+}}{k_-}\right]^{\frac{1}{2}}$$

$$\doteq \frac{e^3 q_{-+}}{(p_- \cdot k)} \left[\frac{2k_+ q_{++}}{k_- q_{--}}\right]^{\frac{1}{2}},$$

9.16. $e^+e^- \to e^+e^-\gamma$ (NO Z-EXCHANGE; $m \neq 0$)

$$M(-,+;-,+,+) = \frac{e^3 q^*_{-\perp}}{q_{+-}(p_- \cdot k)} \left[\frac{2k_+ q^3_{++}}{k_- q_{-+}}\right]^{\frac{1}{2}}$$

$$\doteq \frac{e^3}{q_{+-}(p_- \cdot k)} \left[\frac{2k_+ q^3_{++} q_{--}}{k_-}\right]^{\frac{1}{2}},$$

$$M(+,-;+,-,-) = \frac{e^3 q_{-\perp}}{q_{+-}(p_- \cdot k)} \left[\frac{2k_+ q^3_{++}}{k_- q_{-+}}\right]^{\frac{1}{2}}$$

$$\doteq \frac{e^3}{q_{+-}(p_- \cdot k)} \left[\frac{2k_+ q^3_{++} q_{--}}{k_-}\right]^{\frac{1}{2}},$$

$$M(+,-;-,+,-) = \frac{e^3 q_{-\perp}}{q_{--}(p_- \cdot k)} \left[\frac{2k_+ q_{++} q_{-+}}{k_-}\right]^{\frac{1}{2}}$$

$$\doteq \frac{e^3 q_{-+}}{(p_- \cdot k)} \left[\frac{2k_+ q_{++}}{k_- q_{--}}\right]^{\frac{1}{2}},$$

$$M(-,+;+,-,-) = \frac{e^3 q^*_{-\perp}}{(p_- \cdot k)} \left[\frac{2k_+ q_{-+}}{k_- q_{++}}\right]^{\frac{1}{2}}$$

$$\doteq \frac{e^3}{(p_- \cdot k)} \left[\frac{2k_+ q_{+-} q_{-+}}{k_-}\right]^{\frac{1}{2}},$$

$$M(-,+;-,+,-) = \frac{e^3 q^*_{-\perp}}{(p_- \cdot k)} \left[\frac{2k_+ q^3_{++}}{k_- q^3_{-+}}\right]^{\frac{1}{2}}$$

$$\doteq \frac{e^3}{q_{-+}(p_- \cdot k)} \left[\frac{2k_+ q^3_{++} q_{--}}{k_-}\right]^{\frac{1}{2}},$$

$$M(-,+;-,-,+) = \frac{e^3 m k^*_\perp q_{-\perp}}{q_{+-}(p_- \cdot k)} \left[\frac{2k_-}{k_+ q_{++} q_{-+}}\right]^{\frac{1}{2}}$$

$$\doteq \frac{e^3 m k_-}{q_{+-}(p_- \cdot k)} \left[\frac{2q_{--}}{q_{++}}\right]^{\frac{1}{2}},$$

$$M(-,-;+,-,-) = \frac{e^3 m k_\perp q^*_{+\perp}}{s q_{--}(p_- \cdot k)} \left[\frac{2k_- q_{++} q_{-+}}{k_+}\right]^{\frac{1}{2}}$$

$$\doteq \frac{e^3 m k_- q_{-+}}{s(p_- \cdot k)} \left[\frac{2q_{++}}{q_{--}}\right]^{\frac{1}{2}},$$

$$M(-,-;-,+,-) = \frac{e^3 m k_\perp q_{+\perp}^*}{s q_{+-}(p_- \cdot k)} \left[\frac{2k_- q_{++}^3}{k_+ q_{-+}}\right]^{\frac{1}{2}}$$

$$\doteq \frac{e^3 m k_- \left[2q_{++}^3 q_{--}\right]^{\frac{1}{2}}}{s q_{+-}(p_- \cdot k)},$$

$$M(-,-;-,-,+) = \frac{e^3 s q_{-\perp}}{(p_- \cdot k)} \left[\frac{2k_+}{k_- q_{++} q_{-+}^3}\right]^{\frac{1}{2}}$$

$$\doteq \frac{e^3 s}{(p_- \cdot k)} \left[\frac{2k_+ q_{+-}}{k_- q_{-+}^3}\right]^{\frac{1}{2}},$$

$$M(-,-;-,-,-) = \frac{e^3 s q_{-\perp}}{q_{+-}(p_- \cdot k)} \left[\frac{2k_+}{k_- q_{++} q_{-+}}\right]^{\frac{1}{2}}$$

$$\doteq \frac{e^3 s}{(p_- \cdot k)} \left[\frac{2k_+}{k_- q_{+-} q_{-+}}\right]^{\frac{1}{2}}. \tag{9.114}$$

Unpolarized squared matrix element:

$$\overline{|M|^2} = \frac{2e^6 E k_+}{(2E - k_-) k_-(p_- \cdot k)^2} \left[s + (2E - k_-)^2 + \frac{m^2 k_-^3}{s k_+}\right] \frac{s'^4 + t^4 + u^4}{s'^2 t^2}. \tag{9.115}$$

Unpolarized cross section:

$$d\sigma = \frac{\alpha^3 k_+}{16\pi^2 E(2E - k_-) k_-(p_- \cdot k)^2} \left[s + (2E - k_-)^2 + \frac{m^2 k_-^3}{s k_+}\right]$$

$$\times \frac{s'^4 + t^4 + u^4}{s'^2 t^2} \delta^4(p_+ + p_- - q_+ - q_- - k) \frac{d^3 \vec{q}_+ \, d^3 \vec{q}_- \, d^3 \vec{k}}{q_{+0} \, q_{-0} \, k_0}. \tag{9.116}$$

9.16.3 $\vec{k}$ nearly parallel to $\vec{q}_+$

The primed quantities $k'_\pm$ and $k'_\perp$ are evaluated in the rotated frame where $\vec{q}_+$ determines the positive z-axis (see also Section 7.4.3). The vector q in eqn (9.118) is obtained by applying a space reflection to q_+.

Definition:
$$Z_{+-} = q_{++} q_{--} - q_{+\perp}^* q_{-\perp}. \tag{9.117}$$

Photon polarizations:

$$\not{\epsilon}^\pm = N[\not{k} \not{A}_+ \not{A}(1 \pm \gamma_5) + \not{A} \not{A}_+ \not{k}(1 \mp \gamma_5)],$$

$$N^{-1} = q_{+0}^2 [32 k'_+ k'_-]^{\frac{1}{2}}. \tag{9.118}$$

9.16. $e^+e^- \to e^+e^-\gamma$ (NO Z-EXCHANGE; $m \neq 0$)

Nonvanishing helicity amplitudes:

$$M(+,+;+,+,+) = \frac{e^3 q_{+0} q_{-\perp}^* Z_{+-}}{(q_+ \cdot k)} \left[\frac{8k'_-}{k'_+ q^3_{++} q^3_{-+}}\right]^{\frac{1}{2}}$$

$$\doteq \frac{e^3 E q_{+0}}{q_{-+}(q_+ \cdot k)} \left[\frac{32 k'_- q_{--}}{k'_+ q_{++}}\right]^{\frac{1}{2}},$$

$$M(+,+;+,+,-) = \frac{e^3 q_{+0} q_{-\perp}^* Z_{+-}}{q_{--}(q_+ \cdot k)} \left[\frac{8k'_-}{k'_+ q_{++} q^3_{-+}}\right]^{\frac{1}{2}}$$

$$\doteq \frac{e^3 E q_{+0}}{q_{-+}(q_+ \cdot k)} \left[\frac{32 k'_- q_{++}}{k'_+ q_{--}}\right]^{\frac{1}{2}},$$

$$M(+,+;-,+,+) = -\frac{e^3 m k'_\perp q_{-\perp}^* Z_{+-}}{q_{+0} q_{--}(q_+ \cdot k)} \left[\frac{k'_+}{2k'_- q_{++} q^3_{-+}}\right]^{\frac{1}{2}}$$

$$\doteq \frac{e^3 m E k'_+}{q_{+0} q_{-+}(q_+ \cdot k)} \left[\frac{2q_{++}}{q_{--}}\right]^{\frac{1}{2}},$$

$$M(+,-;+,+,-) = -\frac{e^3 m k'^*_\perp q_{+\perp}}{E q_{+0}(q_+ \cdot k)} \left[\frac{k'_+ q_{-+}}{8 k'_- q_{++}}\right]^{\frac{1}{2}}$$

$$\doteq \frac{e^3 m k'_+}{E q_{+0}(q_+ \cdot k)} \left[\frac{q_{+-} q_{-+}}{8}\right]^{\frac{1}{2}},$$

$$M(+,-;+,-,+) = \frac{e^3 q_{+0} q_{-\perp}}{E q^2_{+-}(q_+ \cdot k)} \left[\frac{2k'_- q^3_{++} q_{-+}}{k'_+}\right]^{\frac{1}{2}}$$

$$\doteq \frac{e^3 q_{+0}}{E q_{+-}(q_+ \cdot k)} \left[\frac{2k'_- q_{++} q^3_{--}}{k'_+}\right]^{\frac{1}{2}},$$

$$M(+,-;-,+,+) = -\frac{e^3 q_{+0} q_{+\perp}}{E(q_+ \cdot k)} \left[\frac{2k'_- q_{-+}}{k'_+ q_{++}}\right]^{\frac{1}{2}}$$

$$\doteq \frac{e^3 q_{+0}}{E(q_+ \cdot k)} \left[\frac{2k'_- q_{+-} q_{-+}}{k'_+}\right]^{\frac{1}{2}},$$

$$M(-,+;+,+,-) = \frac{e^3 m q_{--} k'^*_\perp q^*_\perp}{E q_{+0}(q_+ \cdot k)} \left[\frac{k'_+ q_{++}}{8 k'_- q^3_{-+}}\right]^{\frac{1}{2}}$$

$$\doteq \frac{e^3 m k'_+}{E q_{+0} q_{-+}(q_+ \cdot k)} \left[\frac{q_{++} q^3_{--}}{8}\right]^{\frac{1}{2}},$$

$$M(-,+;+,-,+) = -\frac{e^3 q_{+0} q_{+\perp}^*}{E q_{+-}(q_+\cdot k)}\left[\frac{2k'_- q^3_{-+}}{k'_+ q_{++}}\right]^{\frac{1}{2}}$$

$$\doteq \frac{e^3 q_{+0}}{E(q_+\cdot k)}\left[\frac{2k'_- q^3_{-+}}{k'_+ q_{+-}}\right]^{\frac{1}{2}},$$

$$M(-,+;-,+,+) = \frac{e^3 q_{+0} q_{--} q^*_{-\perp}}{E(q_+\cdot k)}\left[\frac{2k'_- q_{++}}{k'_+ q^3_{-+}}\right]^{\frac{1}{2}}$$

$$\doteq \frac{e^3 q_{+0}}{E q_{-+}(q_+\cdot k)}\left[\frac{2k'_- q_{++} q^3_{--}}{k'_+}\right]^{\frac{1}{2}},$$

$$M(+,-;+,-,-) = \frac{e^3 q_{+0} q_{--} q_{-\perp}}{E(q_+\cdot k)}\left[\frac{2k'_- q_{++}}{k'_+ q^3_{-+}}\right]^{\frac{1}{2}}$$

$$\doteq \frac{e^3 q_{+0}}{E q_{-+}(q_+\cdot k)}\left[\frac{2k'_- q_{++} q^3_{--}}{k'_+}\right]^{\frac{1}{2}},$$

$$M(+,-;-,+,-) = -\frac{e^3 q_{+0} q_{+\perp}}{E q_{+-}(q_+\cdot k)}\left[\frac{2k'_- q^3_{-+}}{k'_+ q_{++}}\right]^{\frac{1}{2}}$$

$$\doteq \frac{e^3 q_{+0}}{E(q_+\cdot k)}\left[\frac{2k'_- q^3_{-+}}{k'_+ q_{+-}}\right]^{\frac{1}{2}},$$

$$M(+,-;-,-,+) = -\frac{e^3 m q_{--} k'_\perp q_{-\perp}}{E q_{+0}(q_+\cdot k)}\left[\frac{k'_+ q_{++}}{8 k'_- q^3_{-+}}\right]^{\frac{1}{2}}$$

$$\doteq \frac{e^3 m k'_+}{E q_{+0} q_{-+}(q_+\cdot k)}\left[\frac{q_{++} q^3_{--}}{8}\right]^{\frac{1}{2}},$$

$$M(-,+;+,-,-) = -\frac{e^3 q_{+0} q^*_{+\perp}}{E(q_+\cdot k)}\left[\frac{2k'_- q_{-+}}{k'_+ q_{++}}\right]^{\frac{1}{2}}$$

$$\doteq \frac{e^3 q_{+0}}{E(q_+\cdot k)}\left[\frac{2k'_- q_{+-} q_{-+}}{k'_+}\right]^{\frac{1}{2}},$$

$$M(-,+;-,+,-) = \frac{e^3 q_{+0} q^*_{-\perp}}{E q^2_{+-}(q_+\cdot k)}\left[\frac{2k'_- q^3_{++} q_{-+}}{k'_+}\right]^{\frac{1}{2}}$$

$$\doteq \frac{e^3 q_{+0}}{E q_{+-}(q_+\cdot k)}\left[\frac{2k'_- q_{++} q^3_{--}}{k'_+}\right]^{\frac{1}{2}},$$

$$M(-,+;-,-,+) = \frac{e^3 m k'_\perp q^*_{+\perp}}{E q_{+0}(q_+ \cdot k)} \left[\frac{k'_+ q_{-+}}{8 k'_- q_{++}}\right]^{\frac{1}{2}}$$

$$\doteq \frac{e^3 m k'_+}{E q_{+0}(q_+ \cdot k)} \left[\frac{q_{+-} q_{-+}}{8}\right]^{\frac{1}{2}},$$

$$M(-,-;+,-,-) = \frac{e^3 m k'^*_\perp q_{-\perp} Z^*_{+-}}{q_{+0} q_{--}(q_+ \cdot k)} \left[\frac{k'_+}{2 k'_- q_{++} q^3_{-+}}\right]^{\frac{1}{2}}$$

$$\doteq \frac{e^3 m E k'_+}{q_{+0} q_{-+}(q_+ \cdot k)} \left[\frac{2 q_{++}}{q_{--}}\right]^{\frac{1}{2}},$$

$$M(-,-;-,-,+) = \frac{e^3 q_{+0} q_{-\perp} Z^*_{+-}}{q_{--}(q_+ \cdot k)} \left[\frac{8 k'_-}{k'_+ q_{++} q^3_{-+}}\right]^{\frac{1}{2}}$$

$$\doteq \frac{e^3 E q_{+0}}{q_{-+}(q_+ \cdot k)} \left[\frac{32 k'_- q_{++}}{k'_+ q_{--}}\right]^{\frac{1}{2}},$$

$$M(-,-;-,-,-) = \frac{e^3 q_{+0} q_{-\perp} Z^*_{+-}}{(q_+ \cdot k)} \left[\frac{8 k'_-}{k'_+ q^3_{++} q^3_{-+}}\right]^{\frac{1}{2}}$$

$$\doteq \frac{e^3 E q_{+0}}{q_{-+}(q_+ \cdot k)} \left[\frac{32 k'_- q_{--}}{k'_+ q_{++}}\right]^{\frac{1}{2}}. \qquad (9.119)$$

Unpolarized squared matrix element:

$$\overline{|M|^2} = \frac{2 e^6 q_{+0} k'_-}{(2 q_{+0} + k'_+) k'_+ (q_+ \cdot k)^2} \left[4 q^2_{+0} + (2 q_{+0} + k'_+)^2 + \frac{\mu^2 k'^3_+}{4 q^2_{+0} k'_-}\right]$$

$$\times \frac{s^4 + t'^4 + u^4}{s^2 t'^2}. \qquad (9.120)$$

Unpolarized cross section:

$$d\sigma = \frac{\alpha^3 k'_-}{4 \pi^2 (2 q_{+0} + k'_+) k'_+ (q_+ \cdot k)^2} \left[4 q^2_{+0} + (2 q_{+0} + k'_+)^2 + \frac{\mu^2 k'^3_+}{4 q^2_{+0} k'_-}\right]$$

$$\times \frac{s^4 + t'^4 + u^4}{s^3 t'^2} \delta^4(p_+ + p_- - q_+ - q_- - k) \frac{d^3 \vec{q}_+ \, d^3 \vec{q}_- \, d^3 \vec{k}}{q_{-0} \, k_0}. \qquad (9.121)$$

9.16.4 $\vec{k}$ nearly parallel to $\vec{q}_-$

The doubly primed quantities $k''_\pm$ and $k''_\perp$ are evaluated in the rotated frame where $\vec{q}_-$ determines the positive z-axis (see also Section 7.4.3). The vector q in eqn (9.123) is obtained by applying a space reflection to q_-.

Definition:
$$Z_{+-} = q_{++}q_{--} - q_{+\perp}^* q_{-\perp}. \tag{9.122}$$

Photon polarizations:
$$\not{e}^\pm = N[\not{k}\,\not{k}_-\,\not{A}(1\pm\gamma_5) + \not{A}\,\not{k}_-\,\not{k}(1\mp\gamma_5)],$$

$$N^{-1} = q_{-0}^2[32k_+''k_-'']^{\frac{1}{2}}. \tag{9.123}$$

Nonvanishing helicity amplitudes:

$$M(+,+;+,+,+) = -\frac{e^3 q_{-0} q_{-\perp}^* Z_{+-}}{q_{--}(q_-\cdot k)}\left[\frac{8k_-''}{k_+'' q_{++} q_{-+}^3}\right]^{\frac{1}{2}}$$

$$\doteq \frac{e^3 E q_{-0}}{q_{-+}(q_-\cdot k)}\left[\frac{32 k_-'' q_{--}}{k_+'' q_{++}}\right]^{\frac{1}{2}},$$

$$M(+,+;+,+,-) = -\frac{e^3 q_{-0} q_{-\perp}^* Z_{+-}}{(q_-\cdot k)}\left[\frac{8 k_-''}{k_+'' q_{+-}^3 q_{--}^3}\right]^{\frac{1}{2}}$$

$$\doteq \frac{e^3 E q_{-0}}{(q_-\cdot k)}\left[\frac{32 k_-'' q_{-+}}{k_+'' q_{+-}^3}\right]^{\frac{1}{2}},$$

$$M(+,+;+,-,+) = \frac{e^3 m k_\perp'' q_{-\perp}^* Z_{+-}}{q_{-0}(q_-\cdot k)}\left[\frac{k_+''}{2k_-'' q_{+-}^3 q_{--}^3}\right]^{\frac{1}{2}}$$

$$\doteq \frac{e^3 m E k_+''}{q_{-0}(q_-\cdot k)}\left[\frac{2 q_{-+}}{q_{+-}^3}\right]^{\frac{1}{2}},$$

$$M(+,-;+,+,-) = -\frac{e^3 m k_\perp''^* q_{-\perp}}{E q_{-0} q_{+-}(q_-\cdot k)}\left[\frac{k_+'' q_{++}^3}{8 k_-'' q_{-+}}\right]^{\frac{1}{2}}$$

$$\doteq \frac{e^3 m k_+''}{E q_{-0} q_{+-}(q_-\cdot k)}\left[\frac{q_{++}^3 q_{--}}{8}\right]^{\frac{1}{2}},$$

$$M(+,-;+,-,+) = -\frac{e^3 q_{-0} q_{-\perp}}{E q_{+-}(q_-\cdot k)}\left[\frac{2 k_-'' q_{++}^3}{k_+'' q_{-+}}\right]^{\frac{1}{2}}$$

$$\doteq \frac{e^3 q_{-0}}{E q_{+-}(q_-\cdot k)}\left[\frac{2 k_-'' q_{++}^3 q_{--}}{k_+''}\right]^{\frac{1}{2}},$$

$$M(+,-;-,+,+) = \frac{e^3 q_{-0} q_{+\perp}}{E(q_-\cdot k)}\left[\frac{2 k_-'' q_{+-}}{k_+'' q_{--}}\right]^{\frac{1}{2}}$$

$$\doteq \frac{e^3 q_{-0}}{E(q_-\cdot k)}\left[\frac{2 k_-'' q_{+-}^3}{k_+'' q_{-+}}\right]^{\frac{1}{2}},$$

9.16. $e^+e^- \to e^+e^-\gamma$ (NO Z-EXCHANGE; $m \neq 0$)

$$M(-,+;+,+,-) = \frac{e^3 m k_\perp'''^* q_{+\perp}^*}{Eq_{-0}(q_-\cdot k)}\left[\frac{k_+'' q_{-+}}{8k_-'' q_{++}}\right]^{\frac{1}{2}}$$

$$\doteq \frac{e^3 m k_+''}{Eq_{-0}(q_-\cdot k)}\left[\frac{q_{+-} q_{-+}}{8}\right]^{\frac{1}{2}},$$

$$M(-,+;+,-,+) = \frac{e^3 q_{-0} q_{+\perp}^*}{E(q_-\cdot k)}\left[\frac{2k_-'' q_{-+}}{k_+'' q_{++}}\right]^{\frac{1}{2}}$$

$$\doteq \frac{e^3 q_{-0}}{E(q_-\cdot k)}\left[\frac{2k_-'' q_{+-} q_{-+}}{k_+''}\right]^{\frac{1}{2}},$$

$$M(-,+;-,+,+) = -\frac{e^3 q_{-0} q_{-\perp}^*}{E(q_-\cdot k)}\left[\frac{2k_-'' q_{++}^3}{k_+'' q_{-+}^3}\right]^{\frac{1}{2}}$$

$$\doteq \frac{e^3 q_{-0}}{Eq_{-+}(q_-\cdot k)}\left[\frac{2k_-'' q_{++}^3 q_{--}}{k_+''}\right]^{\frac{1}{2}},$$

$$M(+,-;+,-,-) = -\frac{e^3 q_{-0} q_{-\perp}}{E(q_-\cdot k)}\left[\frac{2k_-'' q_{++}^3}{k_+'' q_{-+}^3}\right]^{\frac{1}{2}}$$

$$\doteq \frac{e^3 q_{-0}}{Eq_{-+}(q_-\cdot k)}\left[\frac{2k_-'' q_{++}^3 q_{--}}{k_+''}\right]^{\frac{1}{2}},$$

$$M(+,-;-,+,-) = \frac{e^3 q_{-0} q_{+\perp}}{E(q_-\cdot k)}\left[\frac{2k_-'' q_{-+}}{k_+'' q_{++}}\right]^{\frac{1}{2}}$$

$$\doteq \frac{e^3 q_{-0}}{E(q_-\cdot k)}\left[\frac{2k_-'' q_{+-} q_{-+}}{k_+''}\right]^{\frac{1}{2}},$$

$$M(+,-;-,-,+) = -\frac{e^3 m k_\perp'' q_{+\perp}}{Eq_{-0}(q_-\cdot k)}\left[\frac{k_+'' q_{-+}}{8k_-'' q_{++}}\right]^{\frac{1}{2}}$$

$$\doteq \frac{e^3 m k_+''}{Eq_{-0}(q_-\cdot k)}\left[\frac{q_{+-} q_{-+}}{8}\right]^{\frac{1}{2}},$$

$$M(-,+;+,-,-) = \frac{e^3 q_{-0} q_{+\perp}^*}{E(q_-\cdot k)}\left[\frac{2k_-'' q_{+-}}{k_+'' q_{--}}\right]^{\frac{1}{2}}$$

$$\doteq \frac{e^3 q_{-0}}{E(q_-\cdot k)}\left[\frac{2k_-'' q_{+-}^3}{k_+'' q_{-+}}\right]^{\frac{1}{2}},$$

$$M(-,+;-,+,-) = -\frac{e^3 q_{-0} q^*_{-\perp}}{Eq_{+-}(q_- \cdot k)} \left[\frac{2k''_- q^3_{++}}{k''_+ q_{-+}}\right]^{\frac{1}{2}}$$

$$\doteq \frac{e^3 q_{-0}}{Eq_{+-}(q_- \cdot k)} \left[\frac{2k''_- q^3_{++} q_{--}}{k''_+}\right]^{\frac{1}{2}},$$

$$M(-,+;-,-,+) = \frac{e^3 m k''_\perp q^*_{-\perp}}{Eq_{-0}q_{+-}(q_- \cdot k)} \left[\frac{k''_+ q^3_{++}}{8k''_- q_{-+}}\right]^{\frac{1}{2}}$$

$$\doteq \frac{e^3 m k''_+}{Eq_{-0}q_{+-}(q_- \cdot k)} \left[\frac{q^3_{++} q_{--}}{8}\right]^{\frac{1}{2}},$$

$$M(-,-;-,+,-) = -\frac{e^3 m k''^*_\perp q_{-\perp} Z^*_{+-}}{q_{-0}(q_- \cdot k)} \left[\frac{k''_+}{2k''_- q^3_{+-} q^3_{--}}\right]^{\frac{1}{2}}$$

$$\doteq \frac{e^3 m E k''_+}{q_{-0}(q_- \cdot k)} \left[\frac{2q_{-+}}{q^3_{+-}}\right]^{\frac{1}{2}},$$

$$M(-,-;-,-,+) = -\frac{e^3 q_{-0} q_{-\perp} Z^*_{+-}}{(q_- \cdot k)} \left[\frac{8k''_-}{k''_+ q^3_{+-} q^3_{--}}\right]^{\frac{1}{2}}$$

$$\doteq \frac{e^3 E q_{-0}}{(q_- \cdot k)} \left[\frac{32 k''_- q_{-+}}{k''_+ q^3_{+-}}\right]^{\frac{1}{2}},$$

$$M(-,-;-,-,-) = -\frac{e^3 q_{-0} q_{-\perp} Z^*_{+-}}{q_{--}(q_- \cdot k)} \left[\frac{8k''_-}{k''_+ q_{++} q^3_{-+}}\right]^{\frac{1}{2}}$$

$$\doteq \frac{e^3 E q_{-0}}{(q_- \cdot k)} \left[\frac{32 k''_-}{k''_+ q_{+-} q_{-+}}\right]^{\frac{1}{2}}. \quad (9.124)$$

Unpolarized squared matrix element:

$$\overline{|M|^2} = \frac{2e^6 q_{-0} k''_-}{(2q_{-0} + k''_+) k''_+ (q_- \cdot k)^2} \left[4q^2_{-0} + (2q_{-0} + k''_+)^2 + \frac{m^2 k''^3_+}{4q^2_{-0} k''_-}\right]$$

$$\times \frac{s^4 + t^4 + u'^4}{s^2 t^2}. \quad (9.125)$$

Unpolarized cross section:

$$d\sigma = \frac{\alpha^3 k''_-}{4\pi^2 (2q_{-0} + k''_+) k''_+ (q_- \cdot k)^2} \left[4q^2_{-0} + (2q_{-0} + k''_+)^2 + \frac{m^2 k''^3_+}{4q^2_{-0} k''_-}\right]$$

$$\times \frac{s^4 + t^4 + u'^4}{s^3 t^2} \delta^4(p_+ + p_- - q_+ - q_- - k) \frac{d^3\vec{q}_+ d^3\vec{q}_- d^3\vec{k}}{q_{+0} k_0}.$$

$$(9.126)$$

9.17 $e^+e^- \to \gamma\gamma\gamma\gamma$ $(m \neq 0)$

Process:
$$e^+(p_+) + e^-(p_-) \to \gamma(k_1) + \gamma(k_2) + \gamma(k_3) + \gamma(k_4). \qquad (9.127)$$

Definitions:
$$C = [k_{1+}k_{1-}k_{2+}k_{2-}k_{3+}k_{3-}]^{-1},$$

$$Z_{ij} = k_{i+}k_{j-} - k_{i\perp}^* k_{j\perp},$$

$$\Delta_{ij} = -2(p_- \cdot k_i) - 2(p_- \cdot k_j) + 2(k_i \cdot k_j),$$

$$\Delta'_{ij} = -2(p_+ \cdot k_i) - 2(p_+ \cdot k_j) + 2(k_i \cdot k_j), \qquad i,j = 3,4,$$

$$A_1 = \frac{2E^2}{(p_- \cdot k_3)(p_- \cdot k_4)} \left[k_{3+}k_{4+} + \frac{m^2 Z_{34} Z_{43}}{4E^2 \Delta_{34}} \right],$$

$$A_2(3,4) = \frac{1}{\Delta_{34}} \left[\frac{k_{3+}(2E - k_{3-})(2Ek_{4+} - Z_{43})}{(p_- \cdot k_3)} \right.$$
$$\left. + \frac{2Ek_{3+}k_{4+}(2E - k_{3-} - k_{4-}) - m^2 k_{4-} k_{3\perp} k_{4\perp}^*/2E}{(p_- \cdot k_4)} \right],$$

$$A_3 = \frac{m}{2(p_- \cdot k_3)(p_- \cdot k_4)} \left[k_{3-}k_{4+}k_{3\perp}^* + k_{3+}k_{4-}k_{4\perp}^* \right.$$
$$\left. - \frac{(k_{3\perp}^* + k_{4\perp}^*) Z_{34} Z_{43}}{\Delta_{34}} \right],$$

$$A_4(3,4) = \frac{mk_{3-}k_{3\perp}}{2E\Delta_{34}} \left[\frac{2Ek_{4+} - Z_{43}}{(p_- \cdot k_3)} + \frac{2Ek_{4+}}{(p_- \cdot k_4)} \right],$$

$$B_1 = \frac{2E^2}{(p_+ \cdot k_3)(p_+ \cdot k_4)} \left[k_{3-}k_{4-} + \frac{m^2 Z_{34} Z_{43}}{4E^2 \Delta'_{34}} \right],$$

$$B_2(3,4) = \frac{1}{\Delta'_{34}} \left[\frac{k_{3-}(2E - k_{3+})(2Ek_{4-} - Z_{34})}{(p_+ \cdot k_3)} \right.$$
$$\left. + \frac{2Ek_{3-}k_{4-}(2E - k_{3+} - k_{4+}) - m^2 k_{4+} k_{3\perp}^* k_{4\perp}/2E}{(p_+ \cdot k_4)} \right],$$

$$B_3 = \frac{m}{2(p_+ \cdot k_3)(p_+ \cdot k_4)} \left[k_{3+}k_{4-}k_{3\perp} + k_{3-}k_{4+}k_{4\perp} \right.$$
$$\left. - \frac{(k_{3\perp} + k_{4\perp}) Z_{34} Z_{43}}{\Delta'_{34}} \right],$$

$$B_4(3,4) = \frac{mk_{3+}k_{3\perp}}{2E\Delta'_{34}} \left[\frac{2Ek_{4-} - Z^*_{34}}{(p_+ \cdot k_3)} + \frac{2Ek_{4-}}{(p_+ \cdot k_4)} \right]. \tag{9.128}$$

9.17.1 $\vec{k}_4$ nearly parallel to $\vec{p}_+$

Photon polarizations:

$$\not{\epsilon}_i^{\pm} = N_i[\not{p}_+ \not{p}_- \not{k}_i(1 \mp \gamma_5) - \not{k}_i \not{p}_+ \not{p}_-(1 \pm \gamma_5)], \qquad i = 1, 2, 3,$$

$$\not{\epsilon}_4^{\pm} = N_4[\not{k}_4 \not{p}_+ \not{p}_-(1 \pm \gamma_5) + \not{p}_- \not{p}_+ \not{k}_4(1 \mp \gamma_5)],$$

$$N_i^{-1} = E^2[32k_{i+}k_{i-}]^{\frac{1}{2}}, \qquad i = 1, \ldots, 4. \tag{9.129}$$

Nonvanishing helicity amplitudes, up to permutations of photons 1,2 and 3 [for example, the helicity amplitude $M(+,-;-,+,+,+)$ is obtained from $M(+,-;+,+,-,+)$ through the interchange $k_1 \leftrightarrow k_3$]:

$$M(+,+;+,+,-,+) = \frac{2e^4 mk^*_{3\perp}k_{4\perp}}{(2E - k_{4+})(p_+ \cdot k_4)} \left[\frac{k_{3+}k_{4+}}{k_{1+}k_{1-}k_{2+}k_{2-}k_{3-}k_{4-}} \right]^{\frac{1}{2}}$$

$$\doteq \frac{2e^4 mk_{3+}k_{4+}}{(2E - k_{4+})(p_+ \cdot k_4)[k_{1+}k_{1-}k_{2+}k_{2-}]^{\frac{1}{2}}},$$

$$M(+,+;-,-,+,+) = -\frac{e^4 mk^*_{3\perp}k_{4\perp}}{E(p_+ \cdot k_4)} \left[\frac{k_{3-}k_{4+}}{k_{1+}k_{1-}k_{2+}k_{2-}k_{3+}k_{4-}} \right]^{\frac{1}{2}}$$

$$\doteq \frac{e^4 mk_{3-}k_{4+}}{E(p_+ \cdot k_4)[k_{1+}k_{1-}k_{2+}k_{2-}]^{\frac{1}{2}}},$$

$$M(+,-;+,+,-,+) = \frac{4e^4 Ek_{3\perp}}{(p_+ \cdot k_4)} \left[\frac{k_{3-}k_{4-}}{k_{1+}k_{1-}k_{2+}k_{2-}k_{3+}k_{4+}} \right]^{\frac{1}{2}}$$

$$\doteq \frac{4e^4 Ek_{3-}}{(p_+ \cdot k_4)} \left[\frac{k_{4-}}{k_{1+}k_{1-}k_{2+}k_{2-}k_{4+}} \right]^{\frac{1}{2}},$$

$$M(-,+;+,+,-,+) = -\frac{4e^4 Ek^*_{3\perp}}{(p_+ \cdot k_4)} \left[\frac{k_{3+}k_{4-}}{k_{1+}k_{1-}k_{2+}k_{2-}k_{3-}k_{4+}} \right]^{\frac{1}{2}}$$

$$\doteq \frac{4e^4 Ek_{3+}}{(p_+ \cdot k_4)} \left[\frac{k_{4-}}{k_{1+}k_{1-}k_{2+}k_{2-}k_{4+}} \right]^{\frac{1}{2}},$$

9.17. $e^+e^- \to \gamma\gamma\gamma\gamma$ ($m \neq 0$)

$$M(+,-;+,+,-,-) = \frac{2e^4(2E-k_{4+})k_{3\perp}}{(p_+ \cdot k_4)} \left[\frac{k_{3-}k_{4-}}{k_{1+}k_{1-}k_{2+}k_{2-}k_{3+}k_{4+}}\right]^{\frac{1}{2}}$$

$$\doteq \frac{2e^4(2E-k_{4+})k_{3-}}{(p_+ \cdot k_4)} \left[\frac{k_{4-}}{k_{1+}k_{1-}k_{2+}k_{2-}k_{4+}}\right]^{\frac{1}{2}},$$

$$M(+,-;-,-,+,+) = \frac{-8e^4 E^2 k_{3\perp}}{(2E-k_{4+})(p_+ \cdot k_4)} \left[\frac{k_{3+}k_{4-}}{k_{1+}k_{1-}k_{2+}k_{2-}k_{3-}k_{4+}}\right]^{\frac{1}{2}}$$

$$\doteq \frac{8e^4 E^2 k_{3+}}{(2E-k_{4+})(p_+ \cdot k_4)} \left[\frac{k_{4-}}{k_{1+}k_{1-}k_{2+}k_{2-}k_{4+}}\right]^{\frac{1}{2}},$$

$$M(-,+;+,+,-,-) = \frac{-8e^4 E^2 k_{3\perp}^*}{(2E-k_{4+})(p_+ \cdot k_4)} \left[\frac{k_{3+}k_{4-}}{k_{1+}k_{1-}k_{2+}k_{2-}k_{3-}k_{4+}}\right]^{\frac{1}{2}}$$

$$\doteq \frac{8e^4 E^2 k_{3+}}{(2E-k_{4+})(p_+ \cdot k_4)} \left[\frac{k_{4-}}{k_{1+}k_{1-}k_{2+}k_{2-}k_{4+}}\right]^{\frac{1}{2}},$$

$$M(-,+;-,-,+,+) = \frac{2e^4(2E-k_{4+})k_{3\perp}^*}{(p_+ \cdot k_4)} \left[\frac{k_{3-}k_{4-}}{k_{1+}k_{1-}k_{2+}k_{2-}k_{3+}k_{4+}}\right]^{\frac{1}{2}}$$

$$\doteq \frac{2e^4(2E-k_{4+})k_{3-}}{(p_+ \cdot k_4)} \left[\frac{k_{4-}}{k_{1+}k_{1-}k_{2+}k_{2-}k_{4+}}\right]^{\frac{1}{2}},$$

$$M(+,-;-,-,+,-) = -\frac{4e^4 E k_{3\perp}}{(p_+ \cdot k_4)} \left[\frac{k_{3+}k_{4-}}{k_{1+}k_{1-}k_{2+}k_{2-}k_{3-}k_{4+}}\right]^{\frac{1}{2}}$$

$$\doteq \frac{4e^4 E k_{3+}}{(p_+ \cdot k_4)} \left[\frac{k_{4-}}{k_{1+}k_{1-}k_{2+}k_{2-}k_{4+}}\right]^{\frac{1}{2}},$$

$$M(-,+;-,-,+,-) = \frac{4e^4 E k_{3\perp}^*}{(p_+ \cdot k_4)} \left[\frac{k_{3-}k_{4-}}{k_{1+}k_{1-}k_{2+}k_{2-}k_{3+}k_{4+}}\right]^{\frac{1}{2}}$$

$$\doteq \frac{4e^4 E k_{3-}}{(p_+ \cdot k_4)} \left[\frac{k_{4-}}{k_{1+}k_{1-}k_{2+}k_{2-}k_{4+}}\right]^{\frac{1}{2}},$$

$$M(-,-;+,+,-,-) = \frac{e^4 m k_{3\perp} k_{4\perp}^*}{E(p_+ \cdot k_4)} \left[\frac{k_{3-}k_{4+}}{k_{1+}k_{1-}k_{2+}k_{2-}k_{3+}k_{4-}}\right]^{\frac{1}{2}}$$

$$\doteq \frac{e^4 m k_{3-}k_{4+}}{E(p_+ \cdot k_4)[k_{1+}k_{1-}k_{2+}k_{2-}]^{\frac{1}{2}}},$$

$$M(-,-;-,-,+,-) = \frac{-2e^4 m k_{3\perp} k_{4\perp}^*}{(2E-k_{4+})(p_+ \cdot k_4)} \left[\frac{k_{3+}k_{4+}}{k_{1+}k_{1-}k_{2+}k_{2-}k_{3-}k_{4-}}\right]^{\frac{1}{2}}$$

$$\doteq \frac{2e^4 m k_{3+} k_{4+}}{(2E-k_{4+})(p_+ \cdot k_4)[k_{1+}k_{1-}k_{2+}k_{2-}]^{\frac{1}{2}}} \cdot$$
(9.130)

Unpolarized squared matrix element:

$$\overline{|M|^2} = \frac{2e^8 C k_{4-}}{k_{4+}(p_+ \cdot k_4)^2}\left[4E^2 + (2E-k_{4+})^2 + \frac{m^2 k_{4+}^3}{4E^2 k_{4-}}\right]$$

$$\times \sum_{i=1}^{3} k_{i+}k_{i-}\left[k_{i-}^2 + \left(\frac{2E k_{i+}}{2E-k_{4+}}\right)^2\right].$$
(9.131)

Unpolarized cross section:

$$d\sigma = \frac{\alpha^4 C k_{4-}}{64\pi^2 E^2 k_{4+}(p_+ \cdot k_4)^2}\left[4E^2 + (2E-k_{4+})^2 + \frac{m^2 k_{4+}^3}{4E^2 k_{4-}}\right]$$

$$\times \sum_{i=1}^{3} k_{i+}k_{i-}\left[k_{i-}^2 + \left(\frac{2E k_{i+}}{2E-k_{4+}}\right)^2\right]$$

$$\times \delta^4(p_+ + p_- - k_1 - k_2 - k_3 - k_4)\frac{d^3\vec{k}_1 \, d^3\vec{k}_2 \, d^3\vec{k}_3 \, d^3\vec{k}_4}{k_{10}\, k_{20}\, k_{30}\, k_{40}}.$$
(9.132)

9.17.2 $\vec{k}_4$ nearly parallel to $\vec{p}_-$

Photon polarizations:

$$\rlap{/}{k}_i^{\pm} = N_i[\rlap{/}{p}_+ \rlap{/}{p}_- \rlap{/}{k}_i(1\mp\gamma_5) - \rlap{/}{k}_i \rlap{/}{p}_+ \rlap{/}{p}_-(1\pm\gamma_5)], \qquad i=1,2,3,$$

$$\rlap{/}{k}_4^{\pm} = N_4[\rlap{/}{k}_4 \rlap{/}{p}_- \rlap{/}{p}_+(1\pm\gamma_5) + \rlap{/}{p}_+ \rlap{/}{p}_- \rlap{/}{k}_4(1\mp\gamma_5)],$$

$$N_i^{-1} = E^2[32 k_{i+} k_{i-}]^{\frac{1}{2}}, \qquad i=1,\ldots,4.$$
(9.133)

Nonvanishing helicity amplitudes, up to permutations of photons 1,2 and 3 [for example, the helicity amplitude $M(-,+;-,+,+,+)$ is obtained from $M(-,+;+,+,-,+)$ through the interchange $k_1 \leftrightarrow k_3$]:

$$M(+,+;+,+,-,+) = \frac{-2e^4 m k_{3\perp} k_{4\perp}^*}{(2E-k_{4-})(p_- \cdot k_4)}\left[\frac{k_{3-}k_{4-}}{k_{1+}k_{1-}k_{2+}k_{2-}k_{3+}k_{4+}}\right]^{\frac{1}{2}}$$

$$\doteq \frac{2e^4 m k_{3-} k_{4-}}{(2E-k_{4-})(p_- \cdot k_4)[k_{1+}k_{1-}k_{2+}k_{2-}]^{\frac{1}{2}}},$$

9.17. $e^+e^- \to \gamma\gamma\gamma$ ($m \neq 0$)

$$M(+,+;-,-,+,+) = \frac{e^4 m k_{3\perp} k_{4\perp}^*}{E(p_- \cdot k_4)} \left[\frac{k_{3+} k_{4-}}{k_{1+} k_{1-} k_{2+} k_{2-} k_{3-} k_{4+}} \right]^{\frac{1}{2}}$$

$$\doteq \frac{e^4 m k_{3+} k_{4-}}{E(p_- \cdot k_4) [k_{1+} k_{1-} k_{2+} k_{2-}]^{\frac{1}{2}}},$$

$$M(+,-;+,+,-,+) = -\frac{4e^4 E k_{3\perp}}{(p_- \cdot k_4)} \left[\frac{k_{3-} k_{4+}}{k_{1+} k_{1-} k_{2+} k_{2-} k_{3+} k_{4-}} \right]^{\frac{1}{2}}$$

$$\doteq \frac{4e^4 E k_{3-}}{(p_- \cdot k_4)} \left[\frac{k_{4+}}{k_{1+} k_{1-} k_{2+} k_{2-} k_{4-}} \right]^{\frac{1}{2}},$$

$$M(-,+;+,+,-,+) = \frac{4e^4 E k_{3\perp}^*}{(p_- \cdot k_4)} \left[\frac{k_{3+} k_{4+}}{k_{1+} k_{1-} k_{2+} k_{2-} k_{3-} k_{4-}} \right]^{\frac{1}{2}}$$

$$\doteq \frac{4e^4 E k_{3+}}{(p_- \cdot k_4)} \left[\frac{k_{4+}}{k_{1+} k_{1-} k_{2+} k_{2-} k_{4-}} \right]^{\frac{1}{2}},$$

$$M(+,-;+,+,-,-) = \frac{-8e^4 E^2 k_{3\perp}}{(2E - k_{4-})(p_- \cdot k_4)} \left[\frac{k_{3-} k_{4+}}{k_{1+} k_{1-} k_{2+} k_{2-} k_{3+} k_{4-}} \right]^{\frac{1}{2}}$$

$$\doteq \frac{8e^4 E^2 k_{3-}}{(2E - k_{4-})(p_- \cdot k_4)} \left[\frac{k_{4+}}{k_{1+} k_{1-} k_{2+} k_{2-} k_{4-}} \right]^{\frac{1}{2}},$$

$$M(+,-;-,-,+,+) = \frac{2e^4 (2E - k_{4-}) k_{3\perp}}{(p_- \cdot k_4)} \left[\frac{k_{3+} k_{4+}}{k_{1+} k_{1-} k_{2+} k_{2-} k_{3-} k_{4-}} \right]^{\frac{1}{2}}$$

$$\doteq \frac{2e^4 (2E - k_{4-}) k_{3+}}{(p_- \cdot k_4)} \left[\frac{k_{4+}}{k_{1+} k_{1-} k_{2+} k_{2-} k_{4-}} \right]^{\frac{1}{2}},$$

$$M(-,+;+,+,-,-) = \frac{2e^4 (2E - k_{4-}) k_{3\perp}^*}{(p_- \cdot k_4)} \left[\frac{k_{3+} k_{4+}}{k_{1+} k_{1-} k_{2+} k_{2-} k_{3-} k_{4-}} \right]^{\frac{1}{2}}$$

$$\doteq \frac{2e^4 (2E - k_{4-}) k_{3+}}{(p_- \cdot k_4)} \left[\frac{k_{4+}}{k_{1+} k_{1-} k_{2+} k_{2-} k_{4-}} \right]^{\frac{1}{2}},$$

$$M(-,+;-,-,+,+) = \frac{-8e^4 E^2 k_{3\perp}^*}{(2E - k_{4-})(p_- \cdot k_4)} \left[\frac{k_{3-} k_{4+}}{k_{1+} k_{1-} k_{2+} k_{2-} k_{3+} k_{4-}} \right]^{\frac{1}{2}}$$

$$\doteq \frac{8e^4 E^2 k_{3-}}{(2E - k_{4-})(p_- \cdot k_4)} \left[\frac{k_{4+}}{k_{1+} k_{1-} k_{2+} k_{2-} k_{4-}} \right]^{\frac{1}{2}},$$

$$M(+,-;-,-,+,-) = \frac{4e^4 E k_{3\perp}}{(p_- \cdot k_4)} \left[\frac{k_{3+}k_{4+}}{k_{1+}k_{1-}k_{2+}k_{2-}k_{3-}k_{4-}}\right]^{\frac{1}{2}}$$

$$\doteq \frac{4e^4 E k_{3+}}{(p_- \cdot k_4)} \left[\frac{k_{4+}}{k_{1+}k_{1-}k_{2+}k_{2-}k_{4-}}\right]^{\frac{1}{2}},$$

$$M(-,+;-,-,+,-) = -\frac{4e^4 E k_{3\perp}^*}{(p_- \cdot k_4)} \left[\frac{k_{3-}k_{4+}}{k_{1+}k_{1-}k_{2+}k_{2-}k_{3+}k_{4-}}\right]^{\frac{1}{2}}$$

$$\doteq \frac{4e^4 E k_{3-}}{(p_- \cdot k_4)} \left[\frac{k_{4+}}{k_{1+}k_{1-}k_{2+}k_{2-}k_{4-}}\right]^{\frac{1}{2}},$$

$$M(-,-;+,+,-,-) = -\frac{e^4 m k_{3\perp}^* k_{4\perp}}{E(p_- \cdot k_4)} \left[\frac{k_{3+}k_{4-}}{k_{1+}k_{1-}k_{2+}k_{2-}k_{3-}k_{4+}}\right]^{\frac{1}{2}}$$

$$\doteq \frac{e^4 m k_{3+}k_{4-}}{E(p_- \cdot k_4)[k_{1+}k_{1-}k_{2+}k_{2-}]^{\frac{1}{2}}},$$

$$M(-,-;-,-,+,-) = \frac{2e^4 m k_{3\perp}^* k_{4\perp}}{(2E-k_{4-})(p_- \cdot k_4)} \left[\frac{k_{3-}k_{4-}}{k_{1+}k_{1-}k_{2+}k_{2-}k_{3+}k_{4+}}\right]^{\frac{1}{2}}$$

$$\doteq \frac{2e^4 m k_{3-}k_{4-}}{(2E-k_{4-})(p_- \cdot k_4)[k_{1+}k_{1-}k_{2+}k_{2-}]^{\frac{1}{2}}}.$$

(9.134)

Unpolarized squared matrix element:

$$\overline{|M|^2} = \frac{2e^8 C k_{4+}}{k_{4-}(p_- \cdot k_4)^2} \left[4E^2 + (2E-k_{4-})^2 + \frac{m^2 k_{4-}^3}{4E^2 k_{4+}}\right]$$

$$\times \sum_{i=1}^{3} k_{i+}k_{i-} \left[k_{i+}^2 + \left(\frac{2Ek_{i-}}{2E-k_{4-}}\right)^2\right].$$

(9.135)

Unpolarized cross section:

$$d\sigma = \frac{\alpha^4 C k_{4+}}{64\pi^2 E^2 k_{4-}(p_- \cdot k_4)^2} \left[4E^2 + (2E-k_{4-})^2 + \frac{m^2 k_{4-}^3}{4E^2 k_{4+}}\right]$$

$$\times \sum_{i=1}^{3} k_{i+}k_{i-} \left[k_{i+}^2 + \left(\frac{2Ek_{i-}}{2E-k_{4-}}\right)^2\right]$$

$$\times \delta^4(p_+ + p_- - k_1 - k_2 - k_3 - k_4) \frac{d^3\vec{k}_1 d^3\vec{k}_2 d^3\vec{k}_3 d^3\vec{k}_4}{k_{10} k_{20} k_{30} k_{40}}.$$

(9.136)

9.17.3 $\vec{k}_3$ and $\vec{k}_4$ nearly parallel to $\vec{p}_+$ and $\vec{p}_-$, resp.

Photon polarizations:

$$\not{\epsilon}_i^\pm = N_i[\not{p}_+ \not{p}_- \not{k}_i(1 \mp \gamma_5) - \not{k}_i \not{p}_+ \not{p}_-(1 \pm \gamma_5)], \qquad i = 1, 2,$$

$$\not{\epsilon}_3^\pm = N_3[\not{k}_3 \not{p}_+ \not{p}_-(1 \pm \gamma_5) + \not{p}_- \not{p}_+ \not{k}_3(1 \mp \gamma_5)],$$

$$\not{\epsilon}_4^\pm = N_4[\not{k}_4 \not{p}_- \not{p}_+(1 \pm \gamma_5) + \not{p}_+ \not{p}_- \not{k}_4(1 \mp \gamma_5)],$$

$$N_i^{-1} = E^2[32 k_{i+} k_{i-}]^{\frac{1}{2}}, \qquad i = 1, \ldots, 4. \qquad (9.137)$$

Nonvanishing helicity amplitudes, up to permutations of photons 1 and 2 [for example, the helicity amplitude $M(+, +; -, +, +, +)$ is obtained from $M(+, +; +, -, +, +)$ through the interchange $k_1 \leftrightarrow k_2$]:

$$M(+, +; +, -, +, +) = \frac{2e^4 mE}{(2E - k_{4-})(p_+ \cdot k_3)(p_- \cdot k_4)}$$

$$\times \left\{ \frac{k_{2\perp} k_{4\perp}^*}{k_{2+}} \left[\frac{k_{3-} k_{4-}}{k_{3+} k_{4+}}\right]^{\frac{1}{2}} + \frac{k_{2\perp}^* k_{3\perp}}{k_{1+}} \left[\frac{k_{3+} k_{4+}}{k_{3-} k_{4-}}\right]^{\frac{1}{2}} \right\},$$

$$M(+, +; +, -, +, -) = \frac{e^4 m k_{2\perp}^* k_{3\perp}}{k_{1+}(p_+ \cdot k_3)(p_- \cdot k_4)} \left[\frac{k_{3+} k_{4+}}{k_{3-} k_{4-}}\right]^{\frac{1}{2}}$$

$$\doteq \frac{e^4 m k_{3+}}{(p_+ \cdot k_3)(p_- \cdot k_4)} \left[\frac{k_{1-} k_{4+}}{k_{1+} k_{4-}}\right]^{\frac{1}{2}},$$

$$M(+, +; +, -, -, +) = \frac{e^4 m k_{2\perp} k_{4\perp}^*}{k_{1-}(p_+ \cdot k_3)(p_- \cdot k_4)} \left[\frac{k_{3-} k_{4-}}{k_{3+} k_{4+}}\right]^{\frac{1}{2}}$$

$$\doteq \frac{e^4 m k_{4-}}{(p_+ \cdot k_3)(p_- \cdot k_4)} \left[\frac{k_{1+} k_{3-} k_{4-}}{k_{1-} k_{3+} k_{4+}}\right]^{\frac{1}{2}},$$

$$M(+, -; +, -, +, +) = \frac{4e^4 E^2 k_{2\perp}}{k_{2+}(p_+ \cdot k_3)(p_- \cdot k_4)} \left[\frac{k_{3-} k_{4+}}{k_{3+} k_{4-}}\right]^{\frac{1}{2}}$$

$$\doteq \frac{4e^4 E^2}{(p_+ \cdot k_3)(p_- \cdot k_4)} \left[\frac{k_{2-} k_{3-} k_{4+}}{k_{2+} k_{3+} k_{4-}}\right]^{\frac{1}{2}},$$

$$M(-, +; +, -, +, +) = \frac{4e^4 E^2 k_{1\perp}^*}{k_{2-}(p_+ \cdot k_3)(p_- \cdot k_4)} \left[\frac{k_{3-} k_{4+}}{k_{3+} k_{4-}}\right]^{\frac{1}{2}}$$

$$\doteq \frac{4e^4 E^2}{(p_+ \cdot k_3)(p_- \cdot k_4)} \left[\frac{k_{2+} k_{3-} k_{4+}}{k_{2-} k_{3+} k_{4-}}\right]^{\frac{1}{2}},$$

$$M(+,-;+,-,+,-) = \frac{e^4}{(2E-k_{4-})(p_+\cdot k_3)(p_-\cdot k_4)} \left\{ \frac{8E^3 k_{2\perp}}{k_{2+}} \left[\frac{k_{3-}k_{4+}}{k_{3+}k_{4-}}\right]^{\frac{1}{2}} \right.$$
$$\left. - \frac{m^2 k_{2\perp}^* k_{3\perp} k_{4\perp}}{2E k_{1+}} \left[\frac{k_{3+}k_{4-}}{k_{3-}k_{4+}}\right]^{\frac{1}{2}} \right\},$$

$$M(+,-;+,-,-,+) = \frac{2e^4 E(2E-k_{3+})k_{2\perp}}{k_{2+}(p_+\cdot k_3)(p_-\cdot k_4)} \left[\frac{k_{3-}k_{4+}}{k_{3+}k_{4-}}\right]^{\frac{1}{2}}$$
$$\doteq \frac{2e^4 E(2E-k_{3+})}{(p_+\cdot k_3)(p_-\cdot k_4)} \left[\frac{k_{2-}k_{3-}k_{4+}}{k_{2+}k_{3+}k_{4-}}\right]^{\frac{1}{2}},$$

$$M(-,+;+,-,+,-) = \frac{2e^4 E(2E-k_{3+})k_{1\perp}^*}{k_{1+}(p_+\cdot k_3)(p_-\cdot k_4)} \left[\frac{k_{3-}k_{4+}}{k_{3+}k_{4-}}\right]^{\frac{1}{2}}$$
$$\doteq \frac{2e^4 E(2E-k_{3+})}{(p_+\cdot k_3)(p_-\cdot k_4)} \left[\frac{k_{1-}k_{3-}k_{4+}}{k_{1+}k_{3+}k_{4-}}\right]^{\frac{1}{2}},$$

$$M(-,+;+,-,-,+) = \frac{e^4}{(2E-k_{4-})(p_+\cdot k_3)(p_-\cdot k_4)} \left\{ \frac{8E^3 k_{1\perp}^*}{k_{1+}} \left[\frac{k_{3-}k_{4+}}{k_{3+}k_{4-}}\right]^{\frac{1}{2}} \right.$$
$$\left. - \frac{m^2 k_{1\perp} k_{3\perp}^* k_{4\perp}^*}{2E k_{2+}} \left[\frac{k_{3+}k_{4-}}{k_{3-}k_{4+}}\right]^{\frac{1}{2}} \right\},$$

$$M(+,-;+,-,-,-) = \frac{4e^4 E^2 k_{2\perp}}{k_{1-}(p_+\cdot k_3)(p_-\cdot k_4)} \left[\frac{k_{3-}k_{4+}}{k_{3+}k_{4-}}\right]^{\frac{1}{2}}$$
$$\doteq \frac{4e^4 E^2}{(p_+\cdot k_3)(p_-\cdot k_4)} \left[\frac{k_{1+}k_{3-}k_{4+}}{k_{1-}k_{3+}k_{4-}}\right]^{\frac{1}{2}},$$

$$M(-,+;+,-,-,-) = \frac{4e^4 E^2 k_{1\perp}^*}{k_{1+}(p_+\cdot k_3)(p_-\cdot k_4)} \left[\frac{k_{3-}k_{4+}}{k_{3+}k_{4-}}\right]^{\frac{1}{2}}$$
$$\doteq \frac{4e^4 E^2}{(p_+\cdot k_3)(p_-\cdot k_4)} \left[\frac{k_{1-}k_{3-}k_{4+}}{k_{1+}k_{3+}k_{4-}}\right]^{\frac{1}{2}},$$

$$M(-,-;+,-,+,-) = \frac{e^4 m k_{2\perp}^* k_{4\perp}}{k_{2-}(p_+\cdot k_3)(p_-\cdot k_4)} \left[\frac{k_{3-}k_{4-}}{k_{3+}k_{4+}}\right]^{\frac{1}{2}}$$
$$\doteq \frac{e^4 m k_{4-}}{(p_+\cdot k_3)(p_-\cdot k_4)} \left[\frac{k_{2+}k_{3-}}{k_{2-}k_{3+}}\right]^{\frac{1}{2}},$$

$$M(-,-;+,-,-,+) = \frac{e^4 m k_{2\perp} k_{3\perp}^*}{k_{2+}(p_+ \cdot k_3)(p_- \cdot k_4)} \left[\frac{k_{3+}k_{4+}}{k_{3-}k_{4-}}\right]^{\frac{1}{2}}$$

$$\doteq \frac{e^4 m k_{3+}}{(p_+ \cdot k_3)(p_- \cdot k_4)} \left[\frac{k_{2-}k_{4+}}{k_{2+}k_{4-}}\right]^{\frac{1}{2}},$$

$$M(-,-;+,-,-,-) = \frac{2e^4 m E}{(2E - k_{4-})(p_+ \cdot k_3)(p_- \cdot k_4)}$$

$$\times \left\{ \frac{k_{2\perp}k_{3\perp}^*}{k_{2+}} \left[\frac{k_{3+}k_{4+}}{k_{3-}k_{4-}}\right]^{\frac{1}{2}} + \frac{k_{2\perp}^* k_{4\perp}}{k_{1+}} \left[\frac{k_{3-}k_{4-}}{k_{3+}k_{4+}}\right]^{\frac{1}{2}} \right\}.$$

(9.138)

Unpolarized squared matrix element:

$$\overline{|M|^2} = \frac{2e^8 E^2 k_{3-} k_{4+}}{k_{3+} k_{4-} (p_+ \cdot k_3)^2 (p_- \cdot k_4)^2} \left[4E^2 + (2E - k_{3+})^2 + \frac{m^2 k_{3+}^3}{4E^2 k_{3-}}\right]$$

$$\times \left[4E^2 + (2E - k_{4-})^2 + \frac{m^2 k_{4-}^3}{4E^2 k_{4+}}\right] \frac{k_{1+}^2 + k_{2+}^2}{(2E - k_{3+})^2 k_{1+} k_{1-}}.$$

(9.139)

Unpolarized cross section:

$$d\sigma = \frac{\alpha^4 k_{3-} k_{4+}}{64\pi^2 k_{3+} k_{4-} (p_+ \cdot k_3)^2 (p_- \cdot k_4)^2} \left[4E^2 + (2E - k_{3+})^2 + \frac{m^2 k_{3+}^3}{4E^2 k_{3-}}\right]$$

$$\times \left[4E^2 + (2E - k_{4-})^2 + \frac{m^2 k_{4-}^3}{4E^2 k_{4+}}\right] \frac{k_{1+}^2 + k_{2+}^2}{(2E - k_{3+})^2 k_{1+} k_{1-}}$$

$$\times \delta^4(p_+ + p_- - k_1 - k_2 - k_3 - k_4) \frac{d^3\vec{k}_1 \, d^3\vec{k}_2 \, d^3\vec{k}_3 \, d^3\vec{k}_4}{k_{10} k_{20} k_{30} k_{40}}.$$

(9.140)

9.17.4 $\vec{k}_3$ and $\vec{k}_4$ nearly parallel to $\vec{p}_+$

Photon polarizations:

$$\not{\epsilon}_i^\pm = N_i [\not{p}_+ \not{p}_- \not{k}_i (1 \mp \gamma_5) - \not{k}_i \not{p}_+ \not{p}_- (1 \pm \gamma_5)], \qquad i = 1, 2,$$

$$\not{\epsilon}_j^\pm = N_j [\not{k}_j \not{p}_+ \not{p}_- (1 \pm \gamma_5) + \not{p}_- \not{p}_+ \not{k}_j (1 \mp \gamma_5)], \qquad j = 3, 4,$$

$$N_i^{-1} = E^2 [32 k_{i+} k_{i-}]^{\frac{1}{2}}, \qquad i = 1, \ldots, 4.$$

(9.141)

Nonvanishing helicity amplitudes, up to permutations of photons 1 and 2 [for example, the helicity amplitude $M(+,+;-,+,+,+)$ is obtained from $M(+,+;+,-,+,+)$ through the interchange $k_1 \leftrightarrow k_2$]:

$$M(+,+;+,-,+,+) = \frac{2e^4 B_3 k_{1\perp}^*}{k_{1+}[k_{3+}k_{3-}k_{4+}k_{4-}]^{\frac{1}{2}}}$$

$$\doteq 2e^4 |B_3| \left[\frac{k_{1-}}{k_{1+}k_{3+}k_{3-}k_{4+}k_{4-}}\right]^{\frac{1}{2}},$$

$$M(+,+;+,-,+,-) = \frac{2e^4 B_4(3,4) k_{2\perp}^*}{k_{1+}[k_{3+}k_{3-}k_{4+}k_{4-}]^{\frac{1}{2}}}$$

$$\doteq 2e^4 |B_4(3,4)| \left[\frac{k_{1-}}{k_{1+}k_{3+}k_{3-}k_{4+}k_{4-}}\right]^{\frac{1}{2}},$$

$$M(+,+;+,-,-,+) = \frac{2e^4 B_4(4,3) k_{2\perp}^*}{k_{1+}[k_{3+}k_{3-}k_{4+}k_{4-}]^{\frac{1}{2}}}$$

$$\doteq 2e^4 |B_4(4,3)| \left[\frac{k_{1-}}{k_{1+}k_{3+}k_{3-}k_{4+}k_{4-}}\right]^{\frac{1}{2}},$$

$$M(+,-;+,-,+,+) = \frac{2e^4 B_1 k_{1\perp}}{k_{2+}[k_{3+}k_{3-}k_{4+}k_{4-}]^{\frac{1}{2}}}$$

$$\doteq 2e^4 |B_1| \left[\frac{k_{2-}}{k_{2+}k_{3+}k_{3-}k_{4+}k_{4-}}\right]^{\frac{1}{2}},$$

$$M(-,+;+,-,+,+) = \frac{e^4 (2E - k_{3+} - k_{4+}) B_1 k_{2\perp}^*}{E k_{1+}[k_{3+}k_{3-}k_{4+}k_{4-}]^{\frac{1}{2}}}$$

$$\doteq \frac{e^4 (2E - k_{3+} - k_{4+})|B_1|}{E} \left[\frac{k_{1-}}{k_{1+}k_{3+}k_{3-}k_{4+}k_{4-}}\right]^{\frac{1}{2}},$$

$$M(+,-;+,-,+,-) = \frac{2e^4 B_2(4,3) k_{2\perp}}{k_{2+}[k_{3+}k_{3-}k_{4+}k_{4-}]^{\frac{1}{2}}}$$

$$\doteq 2e^4 |B_2(4,3)| \left[\frac{k_{2-}}{k_{2+}k_{3+}k_{3-}k_{4+}k_{4-}}\right]^{\frac{1}{2}},$$

$$M(+,-;+,-,-,+) = \frac{2e^4 B_2(3,4) k_{2\perp}}{k_{2+}[k_{3+}k_{3-}k_{4+}k_{4-}]^{\frac{1}{2}}}$$

$$\doteq 2e^4 |B_2(3,4)| \left[\frac{k_{2-}}{k_{2+}k_{3+}k_{3-}k_{4+}k_{4-}}\right]^{\frac{1}{2}},$$

9.17. $e^+e^- \to \gamma\gamma\gamma$ $(m \neq 0)$

$$\begin{aligned}
M(-,+;+,-,+,-) &= \frac{2e^4 B_2^*(3,4) k_{1\perp}^*}{k_{1+} [k_{3+}k_{3-}k_{4+}k_{4-}]^{\frac{1}{2}}} \\
&\doteq 2e^4 |B_2(3,4)| \left[\frac{k_{1-}}{k_{1+}k_{3+}k_{3-}k_{4+}k_{4-}} \right]^{\frac{1}{2}}, \\
M(-,+;+,-,-,+) &= \frac{2e^4 B_2^*(4,3) k_{1\perp}^*}{k_{1+} [k_{3+}k_{3-}k_{4+}k_{4-}]^{\frac{1}{2}}} \\
&\doteq 2e^4 |B_2(4,3)| \left[\frac{k_{1-}}{k_{1+}k_{3+}k_{3-}k_{4+}k_{4-}} \right]^{\frac{1}{2}}, \\
M(+,-;+,-,-,-) &= \frac{e^4 (2E - k_{3+} - k_{4+}) B_1^* k_{1\perp}}{E k_{2+} [k_{3+}k_{3-}k_{4+}k_{4-}]^{\frac{1}{2}}} \\
&\doteq \frac{e^4 (2E - k_{3+} - k_{4+}) |B_1|}{E} \left[\frac{k_{2-}}{k_{2+}k_{3+}k_{3-}k_{4+}k_{4-}} \right]^{\frac{1}{2}}, \\
M(-,+;+,-,-,-) &= \frac{2e^4 B_1^* k_{2\perp}^*}{k_{1+} [k_{3+}k_{3-}k_{4+}k_{4-}]^{\frac{1}{2}}} \\
&\doteq 2e^4 |B_1| \left[\frac{k_{1-}}{k_{1+}k_{3+}k_{3-}k_{4+}k_{4-}} \right]^{\frac{1}{2}}, \\
M(-,-;+,-,+,-) &= \frac{2e^4 B_4^*(4,3) k_{2\perp}}{k_{2+} [k_{3+}k_{3-}k_{4+}k_{4-}]^{\frac{1}{2}}} \\
&\doteq 2e^4 |B_4(4,3)| \left[\frac{k_{2-}}{k_{2+}k_{3+}k_{3-}k_{4+}k_{4-}} \right]^{\frac{1}{2}}, \\
M(-,-;+,-,-,+) &= \frac{2e^4 B_4^*(3,4) k_{2\perp}}{k_{2+} [k_{3+}k_{3-}k_{4+}k_{4-}]^{\frac{1}{2}}} \\
&\doteq 2e^4 |B_4(3,4)| \left[\frac{k_{2-}}{k_{2+}k_{3+}k_{3-}k_{4+}k_{4-}} \right]^{\frac{1}{2}}, \\
M(-,-;+,-,-,-) &= \frac{2e^4 B_3^* k_{1\perp}}{k_{2+} [k_{3+}k_{3-}k_{4+}k_{4-}]^{\frac{1}{2}}} \\
&\doteq 2e^4 |B_3| \left[\frac{k_{2-}}{k_{2+}k_{3+}k_{3-}k_{4+}k_{4-}} \right]^{\frac{1}{2}}. \quad (9.142)
\end{aligned}$$

Unpolarized squared matrix element:

$$\overline{|M|^2} = \frac{2e^8}{k_{3+}k_{3-}k_{4+}k_{4-}} \left(\frac{k_{1-}}{k_{1+}} + \frac{k_{2-}}{k_{2+}}\right) \left\{ \left[1 + \left(\frac{2E - k_{3+} - k_{4+}}{2E}\right)^2\right] |B_1|^2 \right.$$

$$\left. + |B_2(3,4)|^2 + |B_2(4,3)|^2 + |B_3|^2 + |B_4(3,4)|^2 + |B_4(4,3)|^2 \right\}.$$
(9.143)

Unpolarized cross section:

$$d\sigma = \frac{\alpha^4}{64\pi^2 E^2 k_{3+}k_{3-}k_{4+}k_{4-}} \left(\frac{k_{1-}}{k_{1+}} + \frac{k_{2-}}{k_{2+}}\right) \left\{ \left[1 + \left(\frac{2E - k_{3+} - k_{4+}}{2E}\right)^2\right] |B_1|^2 \right.$$

$$\left. + |B_2(3,4)|^2 + |B_2(4,3)|^2 + |B_3|^2 + |B_4(3,4)|^2 + |B_4(4,3)|^2 \right\}$$

$$\times \delta^4(p_+ + p_- - k_1 - k_2 - k_3 - k_4) \frac{d^3\vec{k}_1\, d^3\vec{k}_2\, d^3\vec{k}_3\, d^3\vec{k}_4}{k_{10}\, k_{20}\, k_{30}\, k_{40}}.$$
(9.144)

9.17.5 $\vec{k}_3$ and $\vec{k}_4$ nearly parallel to $\vec{p}_-$

Photon polarizations:

$$\not{e}_i^{\pm} = N_i[\not{p}_+ \not{p}_- \not{k}_i(1 \mp \gamma_5) - \not{k}_i \not{p}_+ \not{p}_-(1 \pm \gamma_5)], \qquad i = 1, 2,$$

$$\not{e}_j^{\pm} = N_j[\not{k}_j \not{p}_- \not{p}_+(1 \pm \gamma_5) + \not{p}_+ \not{p}_- \not{k}_j(1 \mp \gamma_5)], \qquad j = 3, 4,$$

$$N_i^{-1} = E^2[32 k_{i+} k_{i-}]^{\frac{1}{2}}, \qquad i = 1, \ldots, 4.$$
(9.145)

Nonvanishing helicity amplitudes, up to permutations of photons 1 and 2 [for example, the helicity amplitude $M(+, +; -, +, +, +)$ is obtained from $M(+, +; +, -, +, +)$ through the interchange $k_1 \leftrightarrow k_2$]:

$$M(+, +; +, -, +, +) = \frac{2e^4 A_3 k_{2\perp}}{k_{1-}[k_{3+}k_{3-}k_{4+}k_{4-}]^{\frac{1}{2}}}$$

$$\doteq 2e^4 |A_3| \left[\frac{k_{1+}}{k_{1-}k_{3+}k_{3-}k_{4+}k_{4-}}\right]^{\frac{1}{2}},$$

$$M(+, +; +, -, +, -) = \frac{2e^4 A_4^*(3,4) k_{2\perp}}{k_{1-}[k_{3+}k_{3-}k_{4+}k_{4-}]^{\frac{1}{2}}}$$

$$\doteq 2e^4 |A_4(3,4)| \left[\frac{k_{1+}}{k_{1-}k_{3+}k_{3-}k_{4+}k_{4-}}\right]^{\frac{1}{2}},$$

9.17. $e^+e^- \to \gamma\gamma\gamma$ $(m \neq 0)$

$$M(+,+;+,-,-,+) = \frac{2e^4 A_4^*(4,3)k_{2\perp}}{k_{1-}[k_{3+}k_{3-}k_{4+}k_{4-}]^{\frac{1}{2}}}$$

$$\doteq 2e^4|A_4(4,3)|\left[\frac{k_{1+}}{k_{1-}k_{3+}k_{3-}k_{4+}k_{4-}}\right]^{\frac{1}{2}},$$

$$M(+,-;+,-,+,+) = \frac{e^4(2E - k_{3-} - k_{4-})A_1 k_{1\perp}}{Ek_{1-}[k_{3+}k_{3-}k_{4+}k_{4-}]^{\frac{1}{2}}}$$

$$\doteq \frac{e^4(2E - k_{3-} - k_{4-})|A_1|}{E}\left[\frac{k_{1+}}{k_{1-}k_{3+}k_{3-}k_{4+}k_{4-}}\right]^{\frac{1}{2}},$$

$$M(-,+;+,-,+,+) = \frac{2e^4 A_1 k_{2\perp}^*}{k_{2-}[k_{3+}k_{3-}k_{4+}k_{4-}]^{\frac{1}{2}}}$$

$$\doteq 2e^4|A_1|\left[\frac{k_{2+}}{k_{2-}k_{3+}k_{3-}k_{4+}k_{4-}}\right]^{\frac{1}{2}},$$

$$M(+,-;+,-,+,-) = \frac{2e^4 A_2^*(3,4)k_{2\perp}}{k_{1-}[k_{3+}k_{3-}k_{4+}k_{4-}]^{\frac{1}{2}}}$$

$$\doteq 2e^4|A_2(3,4)|\left[\frac{k_{1+}}{k_{1-}k_{3+}k_{3-}k_{4+}k_{4-}}\right]^{\frac{1}{2}},$$

$$M(+,-;+,-,-,+) = \frac{2e^4 A_2^*(4,3)k_{2\perp}}{k_{1-}[k_{3+}k_{3-}k_{4+}k_{4-}]^{\frac{1}{2}}}$$

$$\doteq 2e^4|A_2(4,3)|\left[\frac{k_{1+}}{k_{1-}k_{3+}k_{3-}k_{4+}k_{4-}}\right]^{\frac{1}{2}},$$

$$M(-,+;+,-,+,-) = \frac{2e^4 A_2(4,3)k_{1\perp}^*}{k_{2-}[k_{3+}k_{3-}k_{4+}k_{4-}]^{\frac{1}{2}}}$$

$$\doteq 2e^4|A_2(4,3)|\left[\frac{k_{2+}}{k_{2-}k_{3+}k_{3-}k_{4+}k_{4-}}\right]^{\frac{1}{2}},$$

$$M(-,+;+,-,-,+) = \frac{2e^4 A_2(3,4)k_{1\perp}^*}{k_{2-}[k_{3+}k_{3-}k_{4+}k_{4-}]^{\frac{1}{2}}}$$

$$\doteq 2e^4|A_2(3,4)|\left[\frac{k_{2+}}{k_{2-}k_{3+}k_{3-}k_{4+}k_{4-}}\right]^{\frac{1}{2}},$$

$$M(+,-;+,-,-,-) = \frac{2e^4 A_1^* k_{1\perp}}{k_{1-}[k_{3+}k_{3-}k_{4+}k_{4-}]^{\frac{1}{2}}}$$

$$\doteq 2e^4 |A_1| \left[\frac{k_{1+}}{k_{1-}k_{3+}k_{3-}k_{4+}k_{4-}}\right]^{\frac{1}{2}},$$

$$M(-,+;+,-,-,-) = \frac{e^4(2E-k_{3-}-k_{4-})A_1^* k_{2\perp}^*}{Ek_{2-}[k_{3+}k_{3-}k_{4+}k_{4-}]^{\frac{1}{2}}}$$

$$\doteq \frac{e^4(2E-k_{3-}-k_{4-})|A_1|}{E} \left[\frac{k_{2+}}{k_{2-}k_{3+}k_{3-}k_{4+}k_{4-}}\right]^{\frac{1}{2}},$$

$$M(-,-;+,-,+,-) = \frac{2e^4 A_4(4,3) k_{2\perp}^*}{k_{2-}[k_{3+}k_{3-}k_{4+}k_{4-}]^{\frac{1}{2}}}$$

$$\doteq 2e^4 |A_4(4,3)| \left[\frac{k_{2+}}{k_{2-}k_{3+}k_{3-}k_{4+}k_{4-}}\right]^{\frac{1}{2}},$$

$$M(-,-;+,-,-,+) = \frac{2e^4 A_4(3,4) k_{2\perp}^*}{k_{2-}[k_{3+}k_{3-}k_{4+}k_{4-}]^{\frac{1}{2}}}$$

$$\doteq 2e^4 |A_4(3,4)| \left[\frac{k_{2+}}{k_{2-}k_{3+}k_{3-}k_{4+}k_{4-}}\right]^{\frac{1}{2}},$$

$$M(-,-;+,-,-,-) = \frac{2e^4 A_3^* k_{2\perp}^*}{k_{2-}[k_{3+}k_{3-}k_{4+}k_{4-}]^{\frac{1}{2}}}$$

$$\doteq 2e^4 |A_3| \left[\frac{k_{2+}}{k_{2-}k_{3+}k_{3-}k_{4+}k_{4-}}\right]^{\frac{1}{2}}.$$

(9.146)

Unpolarized squared matrix element:

$$\overline{|M|^2} = \frac{2e^8}{k_{3+}k_{3-}k_{4+}k_{4-}} \left(\frac{k_{1+}}{k_{1-}} + \frac{k_{2+}}{k_{2-}}\right)$$

$$\times \left\{\left[1 + \left(\frac{2E-k_{3-}-k_{4-}}{2E}\right)^2\right] |A_1|^2 + |A_2(3,4)|^2 \right.$$

$$\left. + |A_2(4,3)|^2 + |A_3|^2 + |A_4(3,4)|^2 + |A_4(4,3)|^2 \right\}.$$

(9.147)

9.18. $e^+e^- \to \mu^+\mu^-\gamma\gamma$ (NO Z-EXCHANGE; $m \neq 0$)

Unpolarized cross section:

$$d\sigma = \frac{\alpha^4}{64\pi^2 E^2 k_{3+}k_{3-}k_{4+}k_{4-}} \left(\frac{k_{1+}}{k_{1-}} + \frac{k_{2+}}{k_{2-}}\right)$$

$$\times \left\{ \left[1 + \left(\frac{2E - k_{3-} - k_{4-}}{2E}\right)^2\right] |A_1|^2 \right.$$

$$\left. + |A_2(3,4)|^2 + |A_2(4,3)|^2 + |A_3|^2 + |A_4(3,4)|^2 + |A_4(4,3)|^2 \right\}$$

$$\times \delta^4(p_+ + p_- - k_1 - k_2 - k_3 - k_4) \frac{d^3\vec{k}_1 \, d^3\vec{k}_2 \, d^3\vec{k}_3 \, d^3\vec{k}_4}{k_{10} \, k_{20} \, k_{30} \, k_{40}}. \quad (9.148)$$

9.18 $e^+e^- \to \mu^+\mu^-\gamma\gamma$ (no Z-exchange; $m \neq 0$)

Process:

$$e^+(p_+) + e^-(p_-) \to \mu^+(q_+) + \mu^-(q_-) + \gamma(k_1) + \gamma(k_2). \quad (9.149)$$

Definitions:

$$Z_{+-} = q_{++}q_{--} - q^*_{+\perp}q_{-\perp}, \qquad Z_{-+} = q_{-+}q_{+-} - q^*_{-\perp}q_{+\perp},$$

$$B = 1 + \frac{E}{(q_+ \cdot q_-)} \left[2q_{-z} + \frac{k_{2\perp}q^*_{-\perp}}{k_{2+}} - \frac{k^*_{2\perp}q_{-\perp}}{k_{2-}}\right],$$

$$C = (q_+ \cdot q_-)(q_+ \cdot k_1)(q_- \cdot k_1). \quad (9.150)$$

9.18.1 $\vec{k}_2$ nearly parallel to $\vec{p}_+$

Photon polarizations:

$$\not{\epsilon}_1^{\pm} = N_1[\not{p}_+ \not{p}_- \not{k}_1(1 \mp \gamma_5) - \not{k}_1 \not{p}_+ \not{p}_-(1 \pm \gamma_5)],$$

$$\not{\epsilon}_2^{\pm} = N_2[\not{k}_2 \not{p}_+ \not{p}_-(1 \pm \gamma_5) + \not{p}_- \not{p}_+ \not{k}_2(1 \mp \gamma_5)],$$

$$N_i^{-1} = E^2[32k_{i+}k_{i-}]^{\frac{1}{2}}, \qquad i = 1,2. \quad (9.151)$$

Nonvanishing helicity amplitudes:

$$M(+,+;+,-,+,+) = -\frac{e^4 m k_{2\perp} q^*_{+\perp} Z_{-+} B}{(2E - k_{2+})q_{+-}(p_+ \cdot k_2)} \left[\frac{k_{2+}q_{-+}}{2C k_{2-}q_{++}}\right]^{\frac{1}{2}}$$

$$\doteq \frac{e^4 m k_{2+} q_{-+} |B|}{(2E - k_{2+})(p_+ \cdot k_2)[(q_+ \cdot k_1)(q_- \cdot k_1)]^{\frac{1}{2}}},$$

$$M(+,+;-,+,+,+) = -\frac{e^4 m q_{-\perp}^* k_{2\perp} Z_{+-} B}{(2E-k_{2+})q_{--}(p_+\cdot k_2)}\left[\frac{k_{2+}q_{++}}{2Ck_{2-}q_{-+}}\right]^{\frac{1}{2}}$$

$$\doteq \frac{e^4 m k_{2+} q_{++}|B|}{(2E-k_{2+})(p_+\cdot k_2)[(q_+\cdot k_1)(q_-\cdot k_1)]^{\frac{1}{2}}},$$

$$M(+,+;+,-,-,+) = -\frac{e^4 m k_{2\perp} q_{+\perp}^* Z_{-+}^* B^*}{E(p_+\cdot k_2)}\left[\frac{k_{2+}}{8Ck_{2-}q_{++}q_{-+}}\right]^{\frac{1}{2}}$$

$$\doteq \frac{e^4 m k_{2+} q_{+-}|B|}{2E(p_+\cdot k_2)[(q_+\cdot k_1)(q_-\cdot k_1)]^{\frac{1}{2}}},$$

$$M(+,+;-,+,-,+) = -\frac{e^4 m k_{2\perp} q_{-\perp}^* Z_{+-}^* B^*}{E(p_+\cdot k_2)}\left[\frac{k_{2+}}{8Ck_{2-}q_{++}q_{-+}}\right]^{\frac{1}{2}}$$

$$\doteq \frac{e^4 m k_{2+} q_{--}|B|}{2E(p_+\cdot k_2)[(q_+\cdot k_1)(q_-\cdot k_1)]^{\frac{1}{2}}},$$

$$M(+,-;+,-,+,+) = \frac{e^4 E q_{-\perp} Z_{+-} B}{(p_+\cdot k_2)}\left[\frac{2k_{2-}}{Ck_{2+}q_{++}q_{-+}}\right]^{\frac{1}{2}}$$

$$\doteq \frac{2e^4 E q_{--}|B|}{(p_+\cdot k_2)}\left[\frac{k_{2-}}{k_{2+}(q_+\cdot k_1)(q_-\cdot k_1)}\right]^{\frac{1}{2}},$$

$$M(+,-;-,+,+,+) = \frac{e^4 E q_{+\perp} Z_{-+} B}{(p_+\cdot k_2)}\left[\frac{2k_{2-}}{Ck_{2+}q_{++}q_{--+}}\right]^{\frac{1}{2}}$$

$$\doteq \frac{2e^4 E q_{+-}|B|}{(p_+\cdot k_2)}\left[\frac{k_{2-}}{k_{2+}(q_+\cdot k_1)(q_-\cdot k_1)}\right]^{\frac{1}{2}},$$

$$M(-,+;+,-,+,+) = \frac{e^4 E q_{+\perp}^* Z_{-+} B}{q_{+-}(p_+\cdot k_2)}\left[\frac{2k_{2-}q_{-+}}{Ck_{2+}q_{++}}\right]^{\frac{1}{2}}$$

$$\doteq \frac{2e^4 E q_{-+}|B|}{(p_+\cdot k_2)}\left[\frac{k_{2-}}{k_{2+}(q_+\cdot k_1)(q_-\cdot k_1)}\right]^{\frac{1}{2}},$$

$$M(-,+;-,+,+,+) = \frac{e^4 E q_{-\perp}^* Z_{+-} B}{q_{--}(p_+\cdot k_2)}\left[\frac{2k_{2-}q_{++}}{Ck_{2+}q_{-+}}\right]^{\frac{1}{2}}$$

$$\doteq \frac{2e^4 E q_{++}|B|}{(p_+\cdot k_2)}\left[\frac{k_{2-}}{k_{2+}(q_+\cdot k_1)(q_-\cdot k_1)}\right]^{\frac{1}{2}},$$

9.18. $e^+e^- \to \mu^+\mu^-\gamma\gamma$ (NO Z-EXCHANGE; $m \neq 0$)

$$M(+,-;+,-,+,-) = \frac{e^4(2E-k_{2+})q_{-\perp}Z_{+-}B}{(p_+ \cdot k_2)}\left[\frac{k_{2-}}{2Ck_{2+}q_{++}q_{-+}}\right]^{\frac{1}{2}}$$

$$\doteq \frac{e^4(2E-k_{2+})q_{--}|B|}{(p_+ \cdot k_2)}\left[\frac{k_{2-}}{k_{2+}(q_+ \cdot k_1)(q_- \cdot k_1)}\right]^{\frac{1}{2}},$$

$$M(+,-;+,-,-,+) = \frac{e^4 E^2 q_{-\perp}Z_{+-}^*B^*}{(2E-k_{2+})q_{--}(p_+ \cdot k_2)}\left[\frac{8k_{2-}q_{++}}{Ck_{2+}q_{-+}}\right]^{\frac{1}{2}}$$

$$\doteq \frac{4e^4 E^2 q_{++}|B|}{(2E-k_{2+})(p_+ \cdot k_2)}\left[\frac{k_{2-}}{k_{2+}(q_+ \cdot k_1)(q_- \cdot k_1)}\right]^{\frac{1}{2}},$$

$$M(+,-;-,+,+,-) = \frac{e^4(2E-k_{2+})q_{+\perp}Z_{-+}B}{(p_+ \cdot k_2)}\left[\frac{k_{2-}}{2Ck_{2+}q_{++}q_{-+}}\right]^{\frac{1}{2}}$$

$$\doteq \frac{e^4(2E-k_{2+})q_{+-}|B|}{(p_+ \cdot k_2)}\left[\frac{k_{2-}}{k_{2+}(q_+ \cdot k_1)(q_- \cdot k_1)}\right]^{\frac{1}{2}},$$

$$M(+,-;-,+,-,+) = \frac{e^4 E^2 q_{+\perp}Z_{-+}^*B^*}{(2E-k_{2+})q_{+-}(p_+ \cdot k_2)}\left[\frac{8k_{2-}q_{-+}}{Ck_{2+}q_{++}}\right]^{\frac{1}{2}}$$

$$\doteq \frac{4e^4 E^2 q_{-+}|B|}{(2E-k_{2+})(p_+ \cdot k_2)}\left[\frac{k_{2-}}{k_{2+}(q_+ \cdot k_1)(q_- \cdot k_1)}\right]^{\frac{1}{2}},$$

$$M(-,+;+,-,+,-) = \frac{e^4 E^2 q_{+\perp}^*Z_{-+}B}{(2E-k_{2+})q_{+-}(p_+ \cdot k_2)}\left[\frac{8k_{2-}q_{-+}}{Ck_{2+}q_{++}}\right]^{\frac{1}{2}}$$

$$\doteq \frac{4e^4 E^2 q_{-+}|B|}{(2E-k_{2+})(p_+ \cdot k_2)}\left[\frac{k_{2-}}{k_{2+}(q_+ \cdot k_1)(q_- \cdot k_1)}\right]^{\frac{1}{2}},$$

$$M(-,+;+,-,-,+) = \frac{e^4(2E-k_{2+})q_{+\perp}^*Z_{-+}^*B^*}{(p_+ \cdot k_2)}\left[\frac{k_{2-}}{2Ck_{2+}q_{++}q_{-+}}\right]^{\frac{1}{2}}$$

$$\doteq \frac{e^4(2E-k_{2+})q_{+-}|B|}{(p_+ \cdot k_2)}\left[\frac{k_{2-}}{k_{2+}(q_+ \cdot k_1)(q_- \cdot k_1)}\right]^{\frac{1}{2}},$$

$$M(-,+;-,+,+,-) = \frac{e^4 E^2 q_{-\perp}^*Z_{+-}B}{(2E-k_{2+})q_{--}(p_+ \cdot k_2)}\left[\frac{8k_{2-}q_{++}}{Ck_{2+}q_{-+}}\right]^{\frac{1}{2}}$$

$$\doteq \frac{4e^4 E^2 q_{++}|B|}{(2E-k_{2+})(p_+ \cdot k_2)}\left[\frac{k_{2-}}{k_{2+}(q_+ \cdot k_1)(q_- \cdot k_1)}\right]^{\frac{1}{2}},$$

$$M(-,+;-,+,-,+) = \frac{e^4(2E-k_{2+})q^*_{-\perp}Z^*_{+-}B^*}{(p_+\cdot k_2)}\left[\frac{k_{2-}}{2Ck_{2+}q_{++}q_{-+}}\right]^{\frac{1}{2}}$$

$$\doteq \frac{e^4(2E-k_{2+})q_{--}|B|}{(p_+\cdot k_2)}\left[\frac{k_{2-}}{k_{2+}(q_+\cdot k_1)(q_-\cdot k_1)}\right]$$

$$M(+,-;+,-,-,-) = \frac{e^4 Eq_{-\perp}Z^*_{+-}B^*}{q_{--}(p_+\cdot k_2)}\left[\frac{2k_{2-}q_{++}}{Ck_{2+}q_{-+}}\right]^{\frac{1}{2}}$$

$$\doteq \frac{2e^4 Eq_{++}|B|}{(p_+\cdot k_2)}\left[\frac{k_{2-}}{k_{2+}(q_+\cdot k_1)(q_-\cdot k_1)}\right]^{\frac{1}{2}},$$

$$M(+,-;-,+,-,-) = \frac{e^4 Eq_{+\perp}Z^*_{-+}B^*}{q_{+-}(p_+\cdot k_2)}\left[\frac{2k_{2-}q_{-+}}{Ck_{2+}q_{++}}\right]^{\frac{1}{2}}$$

$$\doteq \frac{2e^4 Eq_{-+}|B|}{(p_+\cdot k_2)}\left[\frac{k_{2-}}{k_{2+}(q_+\cdot k_1)(q_-\cdot k_1)}\right]^{\frac{1}{2}},$$

$$M(-,+;+,-,-,-) = \frac{e^4 Eq^*_{+\perp}Z^*_{-+}B^*}{(p_+\cdot k_2)}\left[\frac{2k_{2-}}{Ck_{2+}q_{++}q_{-+}}\right]^{\frac{1}{2}}$$

$$\doteq \frac{2e^4 Eq_{+-}|B|}{(p_+\cdot k_2)}\left[\frac{k_{2-}}{k_{2+}(q_+\cdot k_1)(q_-\cdot k_1)}\right]^{\frac{1}{2}},$$

$$M(-,+;-,+,-,-) = \frac{e^4 Eq^*_{-\perp}Z^*_{+-}B^*}{(p_+\cdot k_2)}\left[\frac{2k_{2-}}{Ck_{2+}q_{++}q_{-+}}\right]^{\frac{1}{2}}$$

$$\doteq \frac{2e^4 Eq_{--}|B|}{(p_+\cdot k_2)}\left[\frac{k_{2-}}{k_{2+}(q_+\cdot k_1)(q_-\cdot k_1)}\right]^{\frac{1}{2}},$$

$$M(-,-;+,-,+,-) = \frac{e^4 mq_{-\perp}k^*_{2\perp}Z_{+-}B}{E(p_+\cdot k_2)}\left[\frac{k_{2+}}{8Ck_{2-}q_{++}q_{-+}}\right]^{\frac{1}{2}}$$

$$\doteq \frac{e^4 mk_{2+}q_{--}|B|}{2E(p_+\cdot k_2)[(q_+\cdot k_1)(q_-\cdot k_1)]^{\frac{1}{2}}},$$

$$M(-,-;-,+,+,-) = \frac{e^4 mq_{+\perp}k^*_{2\perp}Z_{-+}B}{E(p_+\cdot k_2)}\left[\frac{k_{2+}}{8Ck_{2-}q_{++}q_{-+}}\right]^{\frac{1}{2}}$$

$$\doteq \frac{e^4 mk_{2+}q_{+-}|B|}{2E(p_+\cdot k_2)[(q_+\cdot k_1)(q_-\cdot k_1)]^{\frac{1}{2}}},$$

9.18. $e^+e^- \to \mu^+\mu^-\gamma\gamma$ (NO Z-EXCHANGE; $m \neq 0$)

$$M(-,-;+,-,-,-) = \frac{e^4 m q_{-\perp} k_{2\perp}^* Z_{+-}^* B^*}{(2E - k_{2+})q_{--}(p_+ \cdot k_2)} \left[\frac{k_{2+}q_{++}}{2Ck_{2-}q_{-+}}\right]^{\frac{1}{2}}$$

$$\doteq \frac{e^4 m k_{2+} q_{++} |B|}{(2E - k_{2+})(p_+ \cdot k_2)[(q_+ \cdot k_1)(q_- \cdot k_1)]^{\frac{1}{2}}},$$

$$M(-,-;-,+,-,-) = \frac{e^4 m q_{+\perp} k_{2\perp}^* Z_{-+}^* B^*}{(2E - k_{2+})q_{+-}(p_+ \cdot k_2)} \left[\frac{k_{2+}q_{-+}}{2Ck_{2-}q_{++}}\right]^{\frac{1}{2}}$$

$$\doteq \frac{e^4 m k_{2+} q_{-+} |B|}{(2E - k_{2+})(p_+ \cdot k_2)[(q_+ \cdot k_1)(q_- \cdot k_1)]^{\frac{1}{2}}}.$$

(9.152)

Unpolarized squared matrix element:

$$\overline{|M|^2} = \frac{e^8|B|^2 k_{2-}}{2k_{2+}(p_+ \cdot k_2)^2(q_+ \cdot k_1)(q_- \cdot k_1)}$$

$$\times \left[4E^2 + (2E - k_{2+})^2 + \frac{m^2 k_{2+}^3}{4E^2 k_{2-}}\right]$$

$$\times \left[q_{+-}^2 + q_{--}^2 + \left(\frac{2Eq_{++}}{2E - k_{2+}}\right)^2 + \left(\frac{2Eq_{-+}}{2E - k_{2+}}\right)^2\right]. \quad (9.153)$$

Unpolarized cross section:

$$d\sigma = \frac{\alpha^4 |B|^2 k_{2-}}{256\pi^2 E^2 k_{2+}(p_+ \cdot k_2)^2 (q_+ \cdot k_1)(q_- \cdot k_1)}$$

$$\times \left[4E^2 + (2E - k_{2+})^2 + \frac{m^2 k_{2+}^3}{4E^2 k_{2-}}\right]$$

$$\times \left[q_{+-}^2 + q_{--}^2 + \left(\frac{2Eq_{++}}{2E - k_{2+}}\right)^2 + \left(\frac{2Eq_{-+}}{2E - k_{2+}}\right)^2\right]$$

$$\times \delta^4(p_+ + p_- - q_+ - q_- - k_1 - k_2) \frac{d^3\vec{q}_+ \, d^3\vec{q}_- \, d^3\vec{k}_1 \, d^3\vec{k}_2}{q_{+0} \, q_{-0} \, k_{10} \, k_{20}}. \quad (9.154)$$

9.18.2 $\vec{k}_2$ nearly parallel to $\vec{p}_-$

Photon polarizations:

$$\not{e}_1^{\pm} = N_1[\not{p}_+ \not{p}_- \not{k}_1(1 \mp \gamma_5) - \not{k}_1 \not{p}_+ \not{p}_-(1 \pm \gamma_5)],$$

$$\not{e}_2^{\pm} = N_2[\not{k}_2 \not{p}_- \not{p}_+(1 \pm \gamma_5) + \not{p}_+ \not{p}_- \not{k}_2(1 \mp \gamma_5)],$$

$$N_i^{-1} = E^2[32k_{i+}k_{i-}]^{\frac{1}{2}}, \qquad i = 1,2. \quad (9.155)$$

9. SUMMARY OF QED FORMULAE

Nonvanishing helicity amplitudes:

$$M(+,+;+,-,+,+) = \frac{e^4 m k_{2\perp}^* q_{-\perp} Z_{+-} B}{(2E - k_{2-})(p_- \cdot k_2)} \left[\frac{k_{2-}}{2C k_{2+} q_{++} q_{-+}}\right]^{\frac{1}{2}}$$

$$\doteq \frac{e^4 m k_{2-} q_{--} |B|}{(2E - k_{2-})(p_- \cdot k_2)[(q_+ \cdot k_1)(q_- \cdot k_1)]^{\frac{1}{2}}},$$

$$M(+,+;-,+,+,+) = \frac{e^4 m q_{+\perp} k_{2\perp}^* Z_{-+} B}{(2E - k_{2-})(p_- \cdot k_2)} \left[\frac{k_{2-}}{2C k_{2+} q_{++} q_{-+}}\right]^{\frac{1}{2}}$$

$$\doteq \frac{e^4 m k_{2-} q_{+-} |B|}{(2E - k_{2-})(p_- \cdot k_2)[(q_+ \cdot k_1)(q_- \cdot k_1)]^{\frac{1}{2}}},$$

$$M(+,+;+,-,-,+) = \frac{e^4 m k_{2\perp}^* q_{-\perp} Z_{+-}^* B^*}{E q_{--}(p_- \cdot k_2)} \left[\frac{k_{2-} q_{++}}{8C k_{2+} q_{-+}}\right]^{\frac{1}{2}}$$

$$\doteq \frac{e^4 m k_{2-} q_{++} |B|}{2E(p_- \cdot k_2)[(q_+ \cdot k_1)(q_- \cdot k_1)]^{\frac{1}{2}}},$$

$$M(+,+;-,+,-,+) = \frac{e^4 m k_{2\perp}^* q_{+\perp} Z_{-+}^* B^*}{E q_{+-}(p_- \cdot k_2)} \left[\frac{k_{2-} q_{-+}}{8C k_{2+} q_{++}}\right]^{\frac{1}{2}}$$

$$\doteq \frac{e^4 m k_{2-} q_{-+} |B|}{2E(p_- \cdot k_2)[(q_+ \cdot k_1)(q_- \cdot k_1)]^{\frac{1}{2}}},$$

$$M(+,-;+,-,+,+) = \frac{e^4 E q_{-\perp} Z_{+-} B}{(p_- \cdot k_2)} \left[\frac{2k_{2+}}{C k_{2-} q_{++} q_{-+}}\right]^{\frac{1}{2}}$$

$$\doteq \frac{2e^4 E q_{--} |B|}{(p_- \cdot k_2)} \left[\frac{k_{2+}}{k_{2-}(q_+ \cdot k_1)(q_- \cdot k_1)}\right]^{\frac{1}{2}},$$

$$M(+,-;-,+,+,+) = \frac{e^4 E q_{+\perp} Z_{-+} B}{(p_- \cdot k_2)} \left[\frac{2k_{2+}}{C k_{2-} q_{++} q_{-+}}\right]^{\frac{1}{2}}$$

$$\doteq \frac{2e^4 E q_{+-} |B|}{(p_- \cdot k_2)} \left[\frac{k_{2+}}{k_{2-}(q_+ \cdot k_1)(q_- \cdot k_1)}\right]^{\frac{1}{2}},$$

$$M(-,+;+,-,+,+) = \frac{e^4 E q_{+\perp}^* Z_{-+} B}{q_{+-}(p_- \cdot k_2)} \left[\frac{2k_{2+} q_{-+}}{C k_{2-} q_{++}}\right]^{\frac{1}{2}}$$

$$\doteq \frac{2e^4 E q_{-+} |B|}{(p_- \cdot k_2)} \left[\frac{k_{2+}}{k_{2-}(q_+ \cdot k_1)(q_- \cdot k_1)}\right]^{\frac{1}{2}},$$

9.18. $e^+e^- \to \mu^+\mu^-\gamma\gamma$ (NO Z-EXCHANGE; $m \neq 0$)

$$M(-,+;-,+,+,+) = \frac{e^4 E q_{-\perp}^* Z_{+-} B}{q_{--}(p_- \cdot k_2)} \left[\frac{2k_{2+}q_{++}}{Ck_{2-}q_{-+}}\right]^{\frac{1}{2}}$$

$$\doteq \frac{2e^4 E q_{++} |B|}{(p_- \cdot k_2)} \left[\frac{k_{2+}}{k_{2-}(q_+ \cdot k_1)(q_- \cdot k_1)}\right]^{\frac{1}{2}},$$

$$M(+,-;+,-,+,-) = \frac{e^4 E^2 q_{-\perp} Z_{+-} B}{(2E-k_{2-})(p_- \cdot k_2)} \left[\frac{8k_{2+}}{Ck_{2-}q_{++}q_{-+}}\right]^{\frac{1}{2}}$$

$$\doteq \frac{4e^4 E^2 q_{--} |B|}{(2E-k_{2-})(p_- \cdot k_2)} \left[\frac{k_{2+}}{k_{2-}(q_+ \cdot k_1)(q_- \cdot k_1)}\right]^{\frac{1}{2}},$$

$$M(+,-;+,-,-,+) = \frac{e^4 (2E-k_{2-}) q_{-\perp} Z_{+-}^* B^*}{q_{--}(p_- \cdot k_2)} \left[\frac{k_{2+}q_{++}}{2Ck_{2-}q_{-+}}\right]^{\frac{1}{2}}$$

$$\doteq \frac{e^4 (2E-k_{2-}) q_{++} |B|}{(p_- \cdot k_2)} \left[\frac{k_{2+}}{k_{2-}(q_+ \cdot k_1)(q_- \cdot k_1)}\right]^{\frac{1}{2}},$$

$$M(+,-;-,+,+,-) = \frac{e^4 E^2 q_{+\perp} Z_{-+} B}{(2E-k_{2-})(p_- \cdot k_2)} \left[\frac{8k_{2+}}{Ck_{2-}q_{++}q_{-+}}\right]^{\frac{1}{2}}$$

$$\doteq \frac{4e^4 E^2 q_{+-} |B|}{(2E-k_{2-})(p_- \cdot k_2)} \left[\frac{k_{2+}}{k_{2-}(q_+ \cdot k_1)(q_- \cdot k_1)}\right]^{\frac{1}{2}},$$

$$M(+,-;-,+,-,+) = \frac{e^4 (2E-k_{2-}) q_{+\perp} Z_{-+}^* B^*}{q_{+-}(p_- \cdot k_2)} \left[\frac{k_{2+}q_{-+}}{2Ck_{2-}q_{++}}\right]^{\frac{1}{2}}$$

$$\doteq \frac{e^4 (2E-k_{2-}) q_{-+} |B|}{(p_- \cdot k_2)} \left[\frac{k_{2+}}{k_{2-}(q_+ \cdot k_1)(q_- \cdot k_1)}\right]^{\frac{1}{2}},$$

$$M(-,+;+,-,+,-) = \frac{e^4 (2E-k_{2-}) q_{+\perp}^* Z_{-+} B}{q_{+-}(p_- \cdot k_2)} \left[\frac{k_{2+}q_{-+}}{2Ck_{2-}q_{++}}\right]^{\frac{1}{2}}$$

$$\doteq \frac{e^4 (2E-k_{2-}) q_{-+} |B|}{(p_- \cdot k_2)} \left[\frac{k_{2+}}{k_{2-}(q_+ \cdot k_1)(q_- \cdot k_1)}\right]^{\frac{1}{2}},$$

$$M(-,+;+,-,-,+) = \frac{e^4 E^2 q_{+\perp}^* Z_{-+}^* B^*}{(2E-k_{2-})(p_- \cdot k_2)} \left[\frac{8k_{2+}}{Ck_{2-}q_{++}q_{-+}}\right]^{\frac{1}{2}}$$

$$\doteq \frac{4e^4 E^2 q_{+-} |B|}{(2E-k_{2-})(p_- \cdot k_2)} \left[\frac{k_{2+}}{k_{2-}(q_+ \cdot k_1)(q_- \cdot k_1)}\right]^{\frac{1}{2}},$$

$$M(-,+;-,+,+,-) = \frac{e^4(2E-k_{2-})q_{-\perp}^* Z_{+-}B}{q_{--}(p_- \cdot k_2)}\left[\frac{k_{2+}q_{++}}{2Ck_{2-}q_{-+}}\right]^{\frac{1}{2}}$$

$$\doteq \frac{e^4(2E-k_{2-})q_{++}|B|}{(p_- \cdot k_2)}\left[\frac{k_{2+}}{k_{2-}(q_+ \cdot k_1)(q_- \cdot k_1)}\right]^{\frac{1}{2}},$$

$$M(-,+;-,+,-,+) = \frac{e^4 E^2 q_{-\perp}^* Z_{+-}^* B^*}{(2E-k_{2-})(p_- \cdot k_2)}\left[\frac{8k_{2+}}{Ck_{2-}q_{++}q_{-+}}\right]^{\frac{1}{2}}$$

$$\doteq \frac{4e^4 E^2 q_{--}|B|}{(2E-k_{2-})(p_- \cdot k_2)}\left[\frac{k_{2+}}{k_{2-}(q_+ \cdot k_1)(q_- \cdot k_1)}\right]^{\frac{1}{2}},$$

$$M(+,-;+,-,-,-) = \frac{e^4 E q_{-\perp} Z_{+-}^* B^*}{q_{--}(p_- \cdot k_2)}\left[\frac{2k_{2+}q_{++}}{Ck_{2-}q_{-+}}\right]^{\frac{1}{2}}$$

$$\doteq \frac{2e^4 E q_{++}|B|}{(p_- \cdot k_2)}\left[\frac{k_{2+}}{k_{2-}(q_+ \cdot k_1)(q_- \cdot k_1)}\right]^{\frac{1}{2}},$$

$$M(+,-;-,+,-,-) = \frac{e^4 E q_{+\perp} Z_{-+}^* B^*}{q_{+-}(p_- \cdot k_2)}\left[\frac{2k_{2+}q_{-+}}{Ck_{2-}q_{++}}\right]^{\frac{1}{2}}$$

$$\doteq \frac{2e^4 E q_{-+}|B|}{(p_- \cdot k_2)}\left[\frac{k_{2+}}{k_{2-}(q_+ \cdot k_1)(q_- \cdot k_1)}\right]^{\frac{1}{2}},$$

$$M(-,+;+,-,-,-) = \frac{e^4 E q_{+\perp}^* Z_{-+}^* B^*}{(p_- \cdot k_2)}\left[\frac{2k_{2+}}{Ck_{2-}q_{++}q_{-+}}\right]^{\frac{1}{2}}$$

$$\doteq \frac{2e^4 E q_{+-}|B|}{(p_- \cdot k_2)}\left[\frac{k_{2+}}{k_{2-}(q_+ \cdot k_1)(q_- \cdot k_1)}\right]^{\frac{1}{2}},$$

$$M(-,+;-,+,-,-) = \frac{e^4 E q_{-\perp}^* Z_{+-}^* B^*}{(p_- \cdot k_2)}\left[\frac{2k_{2+}}{Ck_{2-}q_{++}q_{-+}}\right]^{\frac{1}{2}}$$

$$\doteq \frac{2e^4 E q_{--}|B|}{(p_- \cdot k_2)}\left[\frac{k_{2+}}{k_{2-}(q_+ \cdot k_1)(q_- \cdot k_1)}\right]^{\frac{1}{2}},$$

$$M(-,-;+,-,+,-) = -\frac{e^4 m q_{+\perp}^* k_{2\perp} Z_{-+} B}{E q_{+-}(p_- \cdot k_2)}\left[\frac{k_{2-}q_{-+}}{8Ck_{2+}q_{++}}\right]^{\frac{1}{2}}$$

$$\doteq \frac{e^4 m k_{2-}q_{-+}|B|}{2E(p_- \cdot k_2)\left[(q_+ \cdot k_1)(q_- \cdot k_1)\right]^{\frac{1}{2}}},$$

9.18. $e^+e^- \to \mu^+\mu^-\gamma\gamma$ (NO Z-EXCHANGE; $m \neq 0$)

$$M(-,-;-,+,+,-) = -\frac{e^4 m q_{-\perp}^* k_{2\perp} Z_{+-} B}{E q_{--}(p_- \cdot k_2)} \left[\frac{k_{2-}q_{++}}{8C k_{2+}q_{-+}}\right]^{\frac{1}{2}}$$

$$\doteq \frac{e^4 m k_{2-} q_{++} |B|}{2E(p_- \cdot k_2)[(q_+ \cdot k_1)(q_- \cdot k_1)]^{\frac{1}{2}}},$$

$$M(-,-;+,-,-,-) = -\frac{e^4 m q_{+\perp}^* k_{2\perp} Z_{-+}^* B^*}{(2E-k_{2-})(p_- \cdot k_2)} \left[\frac{k_{2-}}{2C k_{2+}q_{++}q_{-+}}\right]^{\frac{1}{2}}$$

$$\doteq \frac{e^4 m k_{2-} q_{+-} |B|}{(2E-k_{2-})(p_- \cdot k_2)[(q_+ \cdot k_1)(q_- \cdot k_1)]^{\frac{1}{2}}},$$

$$M(-,-;-,+,-,-) = -\frac{e^4 m q_{-\perp}^* k_{2\perp} Z_{+-}^* B^*}{(2E-k_{2-})(p_- \cdot k_2)} \left[\frac{k_{2-}}{2C k_{2+}q_{++}q_{-+}}\right]^{\frac{1}{2}}$$

$$\doteq \frac{e^4 m k_{2-} q_{--} |B|}{(2E-k_{2-})(p_- \cdot k_2)[(q_+ \cdot k_1)(q_- \cdot k_1)]^{\frac{1}{2}}}.$$
(9.156)

Unpolarized squared matrix element:

$$\overline{|M|^2} = \frac{e^8 |B|^2 k_{2+}}{2k_{2-}(p_- \cdot k_2)^2 (q_+ \cdot k_1)(q_- \cdot k_1)}$$

$$\times \left[4E^2 + (2E-k_{2-})^2 + \frac{m^2 k_{2-}^3}{4E^2 k_{2+}}\right]$$

$$\times \left[q_{++}^2 + q_{-+}^2 + \left(\frac{2Eq_{+-}}{2E-k_{2-}}\right)^2 + \left(\frac{2Eq_{--}}{2E-k_{2-}}\right)^2\right]. \quad (9.157)$$

Unpolarized cross section:

$$d\sigma = \frac{\alpha^4 |B|^2 k_{2+}}{256\pi^2 E^2 k_{2-}(p_- \cdot k_2)^2 (q_+ \cdot k_1)(q_- \cdot k_1)}$$

$$\times \left[4E^2 + (2E-k_{2-})^2 + \frac{m^2 k_{2-}^3}{4E^2 k_{2+}}\right]$$

$$\times \left[q_{++}^2 + q_{-+}^2 + \left(\frac{2Eq_{+-}}{2E-k_{2-}}\right)^2 + \left(\frac{2Eq_{--}}{2E-k_{2-}}\right)^2\right]$$

$$\times \delta^4(p_+ + p_- - q_+ - q_- - k_1 - k_2) \frac{d^3\vec{q}_+ \, d^3\vec{q}_- \, d^3\vec{k}_1 \, d^3\vec{k}_2}{q_{+0} \, q_{-0} \, k_{10} \, k_{20}}. \quad (9.158)$$

9.18.3 $\vec{k}_2$ nearly parallel to $\vec{q}_+$

The primed quantities $k'_{2\pm}$ and $k'_{2\perp}$ are evaluated in the rotated frame where $\vec{q}_+$ determines the positive z-axis (see also Section 7.4.3). The four-vector q

in eqn (9.159) is obtained by applying a space reflection to q_+. The quantity μ denotes the muon mass.

Photon polarizations:

$$\not{e}_1^\pm = N_1[\not{p}_+ \not{p}_- \not{k}_1(1 \mp \gamma_5) - \not{k}_1 \not{p}_+ \not{p}_-(1 \pm \gamma_5)],$$

$$N_1^{-1} = E^2[32k_{1+}k_{1-}]^{\frac{1}{2}},$$

$$\not{e}_2^\pm = N_2[\not{k}_2 \not{q}_+ \not{q}(1 \pm \gamma_5) + \not{q} \not{q}_+ \not{k}_2(1 \mp \gamma_5)],$$

$$N_2^{-1} = q_{+0}^2[32k'_{2+}k'_{2-}]^{\frac{1}{2}}. \tag{9.159}$$

Nonvanishing helicity amplitudes:

$$M(+,-;+,+,+,-) = -\frac{e^4 \mu k'^{*}_{2\perp} q_{+\perp} Z_{-+} B}{q_{+0}(q_+ \cdot k_2)} \left[\frac{k'_{2+}}{8Ck'_{2-}q_{++}q_{-+}}\right]^{\frac{1}{2}}$$

$$\doteq \frac{e^4 \mu k'_{2+} q_{+-}|B|}{2q_{+0}(q_+ \cdot k_2)[(q_+ \cdot k_1)(q_- \cdot k_1)]^{\frac{1}{2}}},$$

$$M(+,-;+,-,+,+) = -\frac{e^4 q_{+0} q_{-\perp} Z_{+-} B}{(q_+ \cdot k_2)} \left[\frac{2k'_{2-}}{Ck'_{2+}q_{++}q_{-+}}\right]^{\frac{1}{2}}$$

$$\doteq \frac{2e^4 q_{+0} q_{--}|B|}{(q_+ \cdot k_2)} \left[\frac{k'_{2-}}{k'_{2+}(q_+ \cdot k_1)(q_- \cdot k_1)}\right]^{\frac{1}{2}},$$

$$M(+,-;-,+,+,+) = -\frac{e^4 q_{+0} q_{+\perp} Z_{-+} B}{(q_+ \cdot k_2)} \left[\frac{2k'_{2-}}{Ck'_{2+}q_{++}q_{-+}}\right]^{\frac{1}{2}}$$

$$\doteq \frac{2e^4 q_{+0} q_{+-}|B|}{(q_+ \cdot k_2)} \left[\frac{k'_{2-}}{k'_{2+}(q_+ \cdot k_1)(q_- \cdot k_1)}\right]^{\frac{1}{2}},$$

$$M(-,+;+,+,+,-) = -\frac{e^4 \mu q^*_{-\perp} k'^{*}_{2\perp} Z_{+-} B}{q_{+0} q_{--}(q_+ \cdot k_2)} \left[\frac{k'_{2+} q_{++}}{8Ck'_{2-} q_{-+}}\right]^{\frac{1}{2}}$$

$$\doteq \frac{e^4 \mu k'_{2+} q_{++}|B|}{2q_{+0}(q_+ \cdot k_2)[(q_+ \cdot k_1)(q_- \cdot k_1)]^{\frac{1}{2}}},$$

$$M(-,+;+,-,+,+) = -\frac{e^4 q_{+0} q^*_{+\perp} Z_{-+} B}{q_{+-}(q_+ \cdot k_2)} \left[\frac{2k'_{2-} q_{-+}}{Ck'_{2+} q_{++}}\right]^{\frac{1}{2}}$$

$$\doteq \frac{2e^4 q_{+0} q_{-+}|B|}{(q_+ \cdot k_2)} \left[\frac{k'_{2-}}{k'_{2+}(q_+ \cdot k_1)(q_- \cdot k_1)}\right]^{\frac{1}{2}},$$

9.18. $e^+e^- \to \mu^+\mu^-\gamma\gamma$ (NO Z-EXCHANGE; $m \neq 0$)

$$M(-,+;-,+,+,+) = -\frac{e^4 q_{+0} q_{-\perp}^* Z_{+-} B}{q_{--}(q_+ \cdot k_2)} \left[\frac{2k'_{2-}q_{++}}{Ck'_{2+}q_{-+}}\right]^{\frac{1}{2}}$$

$$\doteq \frac{2e^4 q_{+0} q_{++} |B|}{(q_+ \cdot k_2)} \left[\frac{k'_{2-}}{k'_{2+}(q_+ \cdot k_1)(q_- \cdot k_1)}\right]^{\frac{1}{2}},$$

$$M(+,-;+,+,-,-) = -\frac{e^4 \mu k'^*_{2\perp} q_{+\perp} Z^*_{-+} B^*}{(2q_{+0}+k'_{2+})q_{+-}(q_+ \cdot k_2)} \left[\frac{k'_{2+}q_{-+}}{2Ck'_{2-}q_{++}}\right]^{\frac{1}{2}}$$

$$\doteq \frac{e^4 \mu k'_{2+} q_{-+} |B|}{(2q_{+0}+k'_{2+})(q_+ \cdot k_2)[(q_+ \cdot k_1)(q_- \cdot k_1)]^{\frac{1}{2}}},$$

$$M(+,-;+,-,+,-) = -\frac{e^4 q_{+0}^2 q_{-\perp} Z_{+-} B}{(2q_{+0}+k'_{2+})(q_+ \cdot k_2)} \left[\frac{8k'_{2-}}{Ck'_{2+}q_{++}q_{-+}}\right]^{\frac{1}{2}}$$

$$\doteq \frac{4e^4 q_{+0}^2 q_{--} |B|}{(2q_{+0}+k'_{2+})(q_+ \cdot k_2)} \left[\frac{k'_{2-}}{k'_{2+}(q_+ \cdot k_1)(q_- \cdot k_1)}\right]^{\frac{1}{2}},$$

$$M(+,-;+,-,-,+) = -\frac{e^4 (2q_{+0}+k'_{2+})q_{-\perp} Z^*_{+-} B^*}{q_{--}(q_+ \cdot k_2)} \left[\frac{k'_{2-}q_{++}}{2Ck'_{2+}q_{-+}}\right]^{\frac{1}{2}}$$

$$\doteq \frac{e^4 (2q_{+0}+k'_{2+})q_{++} |B|}{(q_+ \cdot k_2)} \left[\frac{k'_{2-}}{k'_{2+}(q_+ \cdot k_1)(q_- \cdot k_1)}\right]^{\frac{1}{2}},$$

$$M(+,-;-,+,+,-) = -\frac{e^4 (2q_{+0}+k'_{2+})q_{+\perp} Z_{-+} B}{(q_+ \cdot k_2)} \left[\frac{k'_{2-}}{Ck'_{2+}q_{++}q_{-+}}\right]^{\frac{1}{2}}$$

$$\doteq \frac{e^4 (2q_{+0}+k'_{2+})q_{+-} |B|}{(q_+ \cdot k_2)} \left[\frac{k'_{2-}}{k'_{2+}(q_+ \cdot k_1)(q_- \cdot k_1)}\right]^{\frac{1}{2}},$$

$$M(+,-;-,+,-,+) = -\frac{e^4 q_{+0}^2 q_{+\perp} Z^*_{-+} B^*}{(2q_{+0}+k'_{2+})q_{+-}(q_+ \cdot k_2)} \left[\frac{8k'_{2-}q_{-+}}{Ck'_{2+}q_{++}}\right]^{\frac{1}{2}}$$

$$\doteq \frac{4e^4 q_{+0}^2 q_{-+} |B|}{(2q_{+0}+k'_{2+})(q_+ \cdot k_2)} \left[\frac{k'_{2-}}{k'_{2+}(q_+ \cdot k_1)(q_- \cdot k_1)}\right]^{\frac{1}{2}},$$

$$M(+,-;-,-,+,+) = \frac{e^4 \mu q_{-\perp} k'_{2\perp} Z_{+-} B}{(2q_{+0}+k'_{2+})(q_+ \cdot k_2)} \left[\frac{k'_{2+}}{2Ck'_{2-}q_{++}q_{-+}}\right]^{\frac{1}{2}}$$

$$\doteq \frac{e^4 \mu k'_{2+} q_{--} |B|}{(2q_{+0}+k'_{2+})(q_+ \cdot k_2)[(q_+ \cdot k_1)(q_- \cdot k_1)]^{\frac{1}{2}}},$$

$$M(-,+;+,+,-,-) = -\frac{e^4 \mu k_{21}'^* q_{-\perp}^* Z_{+-}^* B^*}{(2q_{+0} + k_{2+}')(q_+ \cdot k_2)} \left[\frac{k_{2+}'}{2Ck_{2-}' q_{++} q_{-+}}\right]^{\frac{1}{2}}$$

$$\doteq \frac{e^4 \mu k_{2+}' q_{--}|B|}{(2q_{+0} + k_{2+}')(q_+ \cdot k_2)[(q_+ \cdot k_1)(q_- \cdot k_1)]^{\frac{1}{2}}},$$

$$M(-,+;+,-,+,-) = -\frac{e^4 q_{+0}^2 q_{+\perp}^* Z_{-+} B}{(2q_{+0} + k_{2+}')q_{+-}(q_+ \cdot k_2)} \left[\frac{8k_{2-}' q_{-+}}{Ck_{2+}' q_{++}}\right]^{\frac{1}{2}}$$

$$\doteq \frac{4e^4 q_{+0}^2 q_{-+}|B|}{(2q_{+0} + k_{2+}')(q_+ \cdot k_2)} \left[\frac{k_{2-}'}{k_{2+}'(q_+ \cdot k_1)(q_- \cdot k_1)}\right]^{\frac{1}{2}},$$

$$M(-,+;+,-,-,+) = -\frac{e^4(2q_{+0} + k_{2+}')q_{+\perp}^* Z_{-+}^* B^*}{(q_+ \cdot k_2)} \left[\frac{k_{2-}'}{2Ck_{2+}' q_{++} q_{-+}}\right]^{\frac{1}{2}}$$

$$\doteq \frac{e^4(2q_{+0} + k_{2+}')q_{+-}|B|}{(q_+ \cdot k_2)} \left[\frac{k_{2-}'}{k_{2+}'(q_+ \cdot k_1)(q_- \cdot k_1)}\right]^{\frac{1}{2}},$$

$$M(-,+;-,+,+,-) = -\frac{e^4(2q_{+0} + k_{2+}')q_{-\perp}^* Z_{+-} B}{q_{--}(q_+ \cdot k_2)} \left[\frac{k_{2-}' q_{++}}{2Ck_{2+}' q_{-+}}\right]^{\frac{1}{2}}$$

$$\doteq \frac{e^4(2q_{+0} + k_{2+}')q_{++}|B|}{(q_+ \cdot k_2)} \left[\frac{k_{2-}'}{k_{2+}'(q_+ \cdot k_1)(q_- \cdot k_1)}\right]^{\frac{1}{2}},$$

$$M(-,+;-,+,-,+) = -\frac{e^4 q_{+0}^2 q_{-\perp}^* Z_{+-}^* B^*}{(2q_{+0} + k_{2+}')(q_+ \cdot k_2)} \left[\frac{8k_{2-}'}{Ck_{2+}' q_{++} q_{-+}}\right]^{\frac{1}{2}}$$

$$\doteq \frac{4e^4 q_{+0}^2 q_{--}|B|}{(2q_{+0} + k_{2+}')(q_+ \cdot k_2)} \left[\frac{k_{2-}'}{k_{2+}'(q_+ \cdot k_1)(q_- \cdot k_1)}\right]^{\frac{1}{2}},$$

$$M(-,+;-,-,+,+) = \frac{e^4 \mu q_{+\perp}^* k_{21}' Z_{-+} B}{(2q_{+0} + k_{2+}')q_{+-}(q_+ \cdot k_2)} \left[\frac{k_{2+}' q_{-+}}{2Ck_{2-}' q_{++}}\right]^{\frac{1}{2}}$$

$$\doteq \frac{e^4 \mu k_{2+}' q_{-+}|B|}{(2q_{+0} + k_{2+}')(q_+ \cdot k_2)[(q_+ \cdot k_1)(q_- \cdot k_1)]^{\frac{1}{2}}},$$

$$M(+,-;+,-,-,-) = -\frac{e^4 q_{+0} q_{-\perp} Z_{+-}^* B^*}{q_{--}(q_+ \cdot k_2)} \left[\frac{2k_{2-}' q_{++}}{Ck_{2+}' q_{-+}}\right]^{\frac{1}{2}}$$

$$\doteq \frac{2e^4 q_{+0} q_{++}|B|}{(q_+ \cdot k_2)} \left[\frac{k_{2-}'}{k_{2+}'(q_+ \cdot k_1)(q_- \cdot k_1)}\right]^{\frac{1}{2}},$$

9.18. $e^+e^- \to \mu^+\mu^-\gamma\gamma$ (NO Z-EXCHANGE; $m \neq 0$)

$$M(+,-;-,+,-,-) = -\frac{e^4 q_{+0} q_{+\perp} Z^*_{-+} B^*}{q_{+-}(q_+ \cdot k_2)} \left[\frac{2k'_{2-}q_{-+}}{Ck'_{2+}q_{++}}\right]^{\frac{1}{2}}$$

$$\doteq \frac{2e^4 q_{+0} q_{-+} |B|}{(q_+ \cdot k_2)} \left[\frac{k'_{2-}}{k'_{2+}(q_+ \cdot k_1)(q_- \cdot k_1)}\right]^{\frac{1}{2}},$$

$$M(+,-;-,-,-,+) = \frac{e^4 \mu q_{-\perp} k'_{2\perp} Z^*_{+-} B^*}{q_{+0} q_{--}(q_+ \cdot k_2)} \left[\frac{k'_{2+}q_{++}}{8Ck'_{2-}q_{-+}}\right]^{\frac{1}{2}}$$

$$\doteq \frac{e^4 \mu k'_{2+} q_{++} |B|}{2q_{+0}(q_+ \cdot k_2)\left[(q_+ \cdot k_1)(q_- \cdot k_1)\right]^{\frac{1}{2}}},$$

$$M(-,+;+,-,-,-) = -\frac{e^4 q_{+0} q^*_{+\perp} Z^*_{-+} B^*}{(q_+ \cdot k_2)} \left[\frac{2k'_{2-}}{Ck'_{2+}q_{++}q_{-+}}\right]^{\frac{1}{2}}$$

$$\doteq \frac{2e^4 q_{+0} q_{+-} |B|}{(q_+ \cdot k_2)} \left[\frac{k'_{2-}}{k'_{2+}(q_+ \cdot k_1)(q_- \cdot k_1)}\right]^{\frac{1}{2}},$$

$$M(-,+;-,+,-,-) = -\frac{e^4 q_{+0} q^*_{-\perp} Z^*_{+-} B^*}{(q_+ \cdot k_2)} \left[\frac{2k'_{2-}}{Ck'_{2+}q_{++}q_{-+}}\right]^{\frac{1}{2}}$$

$$\doteq \frac{2e^4 q_{+0} q_{--} |B|}{(q_+ \cdot k_2)} \left[\frac{k'_{2-}}{k'_{2+}(q_+ \cdot k_1)(q_- \cdot k_1)}\right]^{\frac{1}{2}},$$

$$M(-,+;-,-,-,+) = \frac{e^4 \mu q^*_{+\perp} k'_{2\perp} Z^*_{-+} B^*}{q_{+0}(q_+ \cdot k_2)} \left[\frac{k'_{2+}}{8Ck'_{2-}q_{++}q_{-+}}\right]^{\frac{1}{2}}$$

$$\doteq \frac{e^4 \mu k'_{2+} q_{+-} |B|}{2q_{+0}(q_+ \cdot k_2)\left[(q_+ \cdot k_1)(q_- \cdot k_1)\right]^{\frac{1}{2}}}. \quad (9.160)$$

Unpolarized squared matrix element:

$$\overline{|M|^2} = \frac{e^8 |B|^2 k'_{2-}}{2k'_{2+}(q_+ \cdot k_2)^2(q_+ \cdot k_1)(q_- \cdot k_1)}$$

$$\times \left[4q^2_{+0} + (2q_{+0} + k'_{2+})^2 + \frac{\mu^2 k'^3_{2+}}{4q^2_{+0} k'_{2-}}\right]$$

$$\times \left[q^2_{++} + q^2_{+-} + \left(\frac{2q_{+0} q_{-+}}{2q_{+0} + k'_{2+}}\right)^2 + \left(\frac{2q_{+0} q_{--}}{2q_{+0} + k'_{2+}}\right)^2\right]. \quad (9.161)$$

Unpolarized cross section:

$$d\sigma = \frac{\alpha^4 |B|^2 k'_{2-}}{256\pi^2 E^2 k'_{2+}(q_+ \cdot k_2)^2 (q_+ \cdot k_1)(q_- \cdot k_1)}$$

$$\times \left[4q_{+0}^2 + (2q_{+0} + k'_{2+})^2 + \frac{\mu^2 k'^3_{2+}}{4q_{+0}^2 k'_{2-}} \right]$$

$$\times \left[q_{++}^2 + q_{+-}^2 + \left(\frac{2q_{+0} q_{-+}}{2q_{+0} + k'_{2+}} \right)^2 + \left(\frac{2q_{+0} q_{--}}{2q_{+0} + k'_{2+}} \right)^2 \right]$$

$$\times \delta^4(p_+ + p_- - q_+ - q_- - k_1 - k_2) \frac{d^3\vec{q}_+ \, d^3\vec{q}_- \, d^3\vec{k}_1 \, d^3\vec{k}_2}{q_{+0} \, q_{-0} \, k_{10} \, k_{20}}. \quad (9.162)$$

9.18.4 $\vec{k}_2$ nearly parallel to $\vec{q}_-$

The doubly primed quantities $k''_{2\pm}$ and k''_{21} are evaluated in the rotated frame where $\vec{q}_-$ determines the positive z-axis (see also Section 7.4.3). The four-vector q in eqn (9.163) is obtained by applying a space reflection to q_-. The quantity μ denotes the muon mass.

Photon polarizations:

$$\rlap{/}{\epsilon}_1^{\pm} = N_1[\rlap{/}{p}_+ \, \rlap{/}{p}_- \, \rlap{/}{k}_1(1 \mp \gamma_5) - \rlap{/}{k}_1 \, \rlap{/}{p}_+ \, \rlap{/}{p}_-(1 \pm \gamma_5)],$$

$$N_1^{-1} = E^2 [32 k_{1+} k_{1-}]^{\frac{1}{2}},$$

$$\rlap{/}{\epsilon}_2^{\pm} = N_2[\rlap{/}{k}_2 \, \rlap{/}{q}_- \, \rlap{/}{q}(1 \pm \gamma_5) + \rlap{/}{q} \, \rlap{/}{q}_- \, \rlap{/}{k}_2(1 \mp \gamma_5)],$$

$$N_2^{-1} = q_{-0}^2 [32 k''_{2+} k''_{2-}]^{\frac{1}{2}}. \quad (9.163)$$

Nonvanishing helicity amplitudes:

$$M(+,-;+,+,+,-) = \frac{e^4 \mu k''^*_{21} q_{-\perp} Z_{+-} B}{q_{-0}(q_- \cdot k_2)} \left[\frac{k''_{2+}}{8 C k''_{2-} q_{++} q_{-+}} \right]^{\frac{1}{2}}$$

$$\doteq \frac{e^4 \mu k''_{2+} q_{--} |B|}{2 q_{-0}(q_- \cdot k_2) [(q_+ \cdot k_1)(q_- \cdot k_1)]^{\frac{1}{2}}},$$

$$M(+,-;+,-,+,+) = \frac{e^4 q_{-0} q_{-\perp} Z_{+-} B}{(q_- \cdot k_2)} \left[\frac{2 k''_{2-}}{C k''_{2+} q_{++} q_{-+}} \right]^{\frac{1}{2}}$$

$$\doteq \frac{2 e^4 q_{-0} q_{--} |B|}{(q_- \cdot k_2)} \left[\frac{k''_{2-}}{k''_{2+}(q_+ \cdot k_1)(q_- \cdot k_1)} \right]^{\frac{1}{2}},$$

9.18. $e^+e^- \to \mu^+\mu^-\gamma\gamma$ (NO Z-EXCHANGE; $m \neq 0$)

$$M(+,-;-,+,+,+) = \frac{e^4 q_{-0} q_{+\perp} Z_{-+} B}{(q_- \cdot k_2)} \left[\frac{2k''_{2-}}{Ck''_{2+}q_{++}q_{-+}} \right]^{\frac{1}{2}}$$

$$\doteq \frac{2e^4 q_{-0} q_{+-} |B|}{(q_- \cdot k_2)} \left[\frac{k''_{2-}}{k''_{2+}(q_+ \cdot k_1)(q_- \cdot k_1)} \right]^{\frac{1}{2}},$$

$$M(-,+;+,+,+,-) = \frac{e^4 \mu q^*_{+\perp} k''^*_{21} Z_{-+} B}{q_{-0} q_{+-}(q_- \cdot k_2)} \left[\frac{k''_{2+} q_{-+}}{8Ck''_{2-} q_{++}} \right]^{\frac{1}{2}}$$

$$\doteq \frac{e^4 \mu k''_{2+} q_{-+} |B|}{2q_{-0}(q_- \cdot k_2)[(q_+ \cdot k_1)(q_- \cdot k_1)]^{\frac{1}{2}}},$$

$$M(-,+;+,-,+,+) = \frac{e^4 q_{-0} q^*_{+\perp} Z_{-+} B}{q_{+-}(q_- \cdot k_2)} \left[\frac{2k''_{2-} q_{-+}}{Ck''_{2+} q_{++}} \right]^{\frac{1}{2}}$$

$$\doteq \frac{2e^4 q_{-0} q_{-+} |B|}{(q_- \cdot k_2)} \left[\frac{k''_{2-}}{k''_{2+}(q_+ \cdot k_1)(q_- \cdot k_1)} \right]^{\frac{1}{2}},$$

$$M(-,+;-,+,+,+) = \frac{e^4 q_{-0} q^*_{-\perp} Z_{+-} B}{q_{--}(q_- \cdot k_2)} \left[\frac{2k''_{2-} q_{++}}{Ck''_{2+} q_{-+}} \right]^{\frac{1}{2}}$$

$$\doteq \frac{2e^4 q_{-0} q_{++} |B|}{(q_- \cdot k_2)} \left[\frac{k''_{2-}}{k''_{2+}(q_+ \cdot k_1)(q_- \cdot k_1)} \right]^{\frac{1}{2}},$$

$$M(+,-;+,+,-,-) = \frac{e^4 \mu k''^*_{21} q_{-\perp} Z^*_{+-} B^*}{(2q_{-0} + k''_{2+})q_{--}(q_- \cdot k_2)} \left[\frac{k''_{2+} q_{++}}{2Ck''_{2-} q_{-+}} \right]^{\frac{1}{2}}$$

$$\doteq \frac{e^4 \mu k''_{2+} q_{++} |B|}{(2q_{-0} + k''_{2+})(q_- \cdot k_2)[(q_+ \cdot k_1)(q_- \cdot k_1)]^{\frac{1}{2}}},$$

$$M(+,-;+,-,+,-) = \frac{e^4 (2q_{-0} + k''_{2+}) q_{-\perp} Z_{+-} B}{(q_- \cdot k_2)} \left[\frac{k''_{2-}}{2Ck''_{2+} q_{++} q_{-+}} \right]^{\frac{1}{2}}$$

$$\doteq \frac{e^4 (2q_{-0} + k''_{2+}) q_{--} |B|}{(q_- \cdot k_2)} \left[\frac{k''_{2-}}{k''_{2+}(q_+ \cdot k_1)(q_- \cdot k_1)} \right]^{\frac{1}{2}},$$

$$M(+,-;+,-,-,+) = \frac{e^4 q^2_{-0} q_{-\perp} Z^*_{+-} B^*}{(2q_{-0} + k''_{2+})q_{--}(q_- \cdot k_2)} \left[\frac{8k''_{2-} q_{++}}{Ck''_{2+} q_{-+}} \right]^{\frac{1}{2}}$$

$$\doteq \frac{4e^4 q^2_{-0} q_{++} |B|}{(2q_{-0} + k''_{2+})(q_- \cdot k_2)} \left[\frac{k''_{2-}}{k''_{2+}(q_+ \cdot k_1)(q_- \cdot k_1)} \right]^{\frac{1}{2}},$$

$$M(+,-;-,+,+,-) = \frac{e^4 q_{-0}^2 q_{+\perp} Z_{-+} B}{(2q_{-0} + k_{2+}'')(q_- \cdot k_2)} \left[\frac{8k_{2-}''}{Ck_{2+}'' q_{++} q_{-+}}\right]^{\frac{1}{2}}$$

$$\doteq \frac{4e^4 q_{-0}^2 q_{+-}|B|}{(2q_{-0} + k_{2+}'')(q_- \cdot k_2)} \left[\frac{k_{2-}''}{k_{2+}''(q_+ \cdot k_1)(q_- \cdot k_1)}\right]^{\frac{1}{2}},$$

$$M(+,-;-,+,-,+) = \frac{e^4 (2q_{-0} + k_{2+}'') q_{+\perp} Z_{-+}^* B^*}{q_{+-}(q_- \cdot k_2)} \left[\frac{k_{2-}'' q_{-+}}{2Ck_{2+}'' q_{++}}\right]^{\frac{1}{2}}$$

$$\doteq \frac{e^4 (2q_{-0} + k_{2+}'') q_{-+}|B|}{(q_- \cdot k_2)} \left[\frac{k_{2-}''}{k_{2+}''(q_+ \cdot k_1)(q_- \cdot k_1)}\right]^{\frac{1}{2}},$$

$$M(+,-;-,-,+,+) = -\frac{e^4 \mu q_{+\perp} k_{2\perp}'' Z_{-+} B}{(2q_{-0} + k_{2+}'')(q_- \cdot k_2)} \left[\frac{k_{2+}''}{2Ck_{2-}'' q_{++} q_{-+}}\right]^{\frac{1}{2}}$$

$$\doteq \frac{e^4 \mu k_{2+}'' q_{+-}|B|}{(2q_{-0} + k_{2+}'')(q_- \cdot k_2)[(q_+ \cdot k_1)(q_- \cdot k_1)]^{\frac{1}{2}}},$$

$$M(-,+;+,+,-,-) = \frac{e^4 \mu k_{2\perp}''^* q_{+\perp}^* Z_{-+}^* B^*}{(2q_{-0} + k_{2+}'')(q_- \cdot k_2)} \left[\frac{k_{2+}''}{2Ck_{2-}'' q_{++} q_{-+}}\right]^{\frac{1}{2}}$$

$$\doteq \frac{e^4 \mu k_{2+}'' q_{+-}|B|}{(2q_{-0} + k_{2+}'')(q_- \cdot k_2)[(q_+ \cdot k_1)(q_- \cdot k_1)]^{\frac{1}{2}}},$$

$$M(-,+;+,-,+,-) = \frac{e^4 (2q_{-0} + k_{2+}'') q_{+\perp}^* Z_{-+} B}{q_{+-}(q_- \cdot k_2)} \left[\frac{k_{2-}'' q_{-+}}{2Ck_{2+}'' q_{++}}\right]^{\frac{1}{2}}$$

$$\doteq \frac{e^4 (2q_{-0} + k_{2+}'') q_{-+}|B|}{(q_- \cdot k_2)} \left[\frac{k_{2-}''}{k_{2+}''(q_+ \cdot k_1)(q_- \cdot k_1)}\right]^{\frac{1}{2}},$$

$$M(-,+;+,-,-,+) = \frac{e^4 q_{-0}^2 q_{+\perp}^* Z_{-+}^* B^*}{(2q_{-0} + k_{2+}'')(q_- \cdot k_2)} \left[\frac{8k_{2-}''}{Ck_{2+}'' q_{++} q_{-+}}\right]^{\frac{1}{2}}$$

$$\doteq \frac{4e^4 q_{-0}^2 q_{+-}|B|}{(2q_{-0} + k_{2+}'')(q_- \cdot k_2)} \left[\frac{k_{2-}''}{k_{2+}''(q_+ \cdot k_1)(q_- \cdot k_1)}\right]^{\frac{1}{2}},$$

$$M(-,+;-,+,+,-) = \frac{e^4 q_{-0}^2 q_{-\perp}^* Z_{+-} B}{(2q_{-0} + k_{2+}'') q_{--}(q_- \cdot k_2)} \left[\frac{8k_{2-}'' q_{++}}{Ck_{2+}'' q_{-+}}\right]^{\frac{1}{2}}$$

$$\doteq \frac{4e^4 q_{-0}^2 q_{++}|B|}{(2q_{-0} + k_{2+}'')(q_- \cdot k_2)} \left[\frac{k_{2-}''}{k_{2+}''(q_+ \cdot k_1)(q_- \cdot k_1)}\right]^{\frac{1}{2}},$$

9.18. $e^+e^- \to \mu^+\mu^-\gamma\gamma$ (NO Z-EXCHANGE; $m \neq 0$)

$$M(-,+;-,+,-,+) = \frac{e^4(2q_{-0} + k''_{2+})q^*_{-\perp}Z^*_{+-}B^*}{(q_- \cdot k_2)}\left[\frac{k''_{2-}}{2Ck''_{2+}q_{++}q_{-+}}\right]^{\frac{1}{2}}$$

$$\doteq \frac{e^4(2q_{-0} + k''_{2+})q_{--}|B|}{(q_- \cdot k_2)}\left[\frac{k''_{2-}}{k''_{2+}(q_+ \cdot k_1)(q_- \cdot k_1)}\right]^{\frac{1}{2}},$$

$$M(-,+;-,-,+,+) = -\frac{e^4\mu q^*_{-\perp}k''_{2\perp}Z_{+-}B}{(2q_{-0} + k''_{2+})q_{--}(q_- \cdot k_2)}\left[\frac{k''_{2+}q_{++}}{2Ck''_{2-}q_{-+}}\right]^{\frac{1}{2}}$$

$$\doteq \frac{e^4\mu k''_{2+}q_{++}|B|}{(2q_{-0} + k''_{2+})(q_- \cdot k_2)[(q_+ \cdot k_1)(q_- \cdot k_1)]^{\frac{1}{2}}},$$

$$M(+,-;+,-,-,-) = \frac{e^4 q_{-0} q_{-\perp} Z^*_{+-}B^*}{q_{--}(q_- \cdot k_2)}\left[\frac{2k''_{2-}q_{++}}{Ck''_{2+}q_{-+}}\right]^{\frac{1}{2}}$$

$$\doteq \frac{2e^4 q_{-0} q_{++}|B|}{(q_- \cdot k_2)}\left[\frac{k''_{2-}}{k''_{2+}(q_+ \cdot k_1)(q_- \cdot k_1)}\right]^{\frac{1}{2}},$$

$$M(+,-;-,+,-,-) = \frac{e^4 q_{-0} q_{+\perp} Z^*_{-+}B^*}{q_{+-}(q_- \cdot k_2)}\left[\frac{2k''_{2-}q_{-+}}{Ck''_{2+}q_{++}}\right]^{\frac{1}{2}}$$

$$\doteq \frac{2e^4 q_{-0} q_{-+}|B|}{(q_- \cdot k_2)}\left[\frac{k''_{2-}}{k''_{2+}(q_+ \cdot k_1)(q_- \cdot k_1)}\right]^{\frac{1}{2}},$$

$$M(+,-;-,-,-,+) = -\frac{e^4\mu q_{+\perp}k''_{2\perp}Z^*_{-+}B^*}{q_{-0}q_{+-}(q_- \cdot k_2)}\left[\frac{k''_{2+}q_{-+}}{8Ck''_{2-}q_{++}}\right]^{\frac{1}{2}}$$

$$\doteq \frac{e^4\mu k''_{2+}q_{-+}|B|}{2q_{-0}(q_- \cdot k_2)[(q_+ \cdot k_1)(q_- \cdot k_1)]^{\frac{1}{2}}},$$

$$M(-,+;+,-,-,-) = \frac{e^4 q_{-0} q^*_{+\perp} Z^*_{-+}B^*}{(q_- \cdot k_2)}\left[\frac{2k''_{2-}}{Ck''_{2+}q_{++}q_{-+}}\right]^{\frac{1}{2}}$$

$$\doteq \frac{2e^4 q_{-0} q_{+-}|B|}{(q_- \cdot k_2)}\left[\frac{k''_{2-}}{k''_{2+}(q_+ \cdot k_1)(q_- \cdot k_1)}\right]^{\frac{1}{2}},$$

$$M(-,+;-,+,-,-) = \frac{e^4 q_{-0} q^*_{-\perp} Z^*_{+-}B^*}{(q_- \cdot k_2)}\left[\frac{2k''_{2-}}{Ck''_{2+}q_{++}q_{-+}}\right]^{\frac{1}{2}}$$

$$\doteq \frac{2e^4 q_{-0} q_{--}|B|}{(q_- \cdot k_2)}\left[\frac{k''_{2-}}{k''_{2+}(q_+ \cdot k_1)(q_- \cdot k_1)}\right]^{\frac{1}{2}},$$

$$M(-,+;-,-,-,+) = -\frac{e^4 \mu q_{-\perp}^* k_{2\perp}'' Z_{+-}^* B^*}{q_{-0}(q_- \cdot k_2)} \left[\frac{k_{2+}''}{8C k_{2-}'' q_{++} q_{-+}}\right]^{\frac{1}{2}}$$

$$\doteq \frac{e^4 \mu k_{2+}'' q_{--} |B|}{2 q_{-0}(q_- \cdot k_2) \left[(q_+ \cdot k_1)(q_- \cdot k_1)\right]^{\frac{1}{2}}}. \quad (9.164)$$

Unpolarized squared matrix element:

$$\overline{|M|^2} = \frac{e^8 |B|^2 k_{2-}''}{2 k_{2+}''(q_- \cdot k_2)^2 (q_+ \cdot k_1)(q_- \cdot k_1)}$$

$$\times \left[4 q_{-0}^2 + (2q_{-0} + k_{2+}'')^2 + \frac{\mu^2 k_{2+}''^3}{4 q_{-0}^2 k_{2-}''}\right]$$

$$\times \left[q_{-+}^2 + q_{--}^2 + \left(\frac{2 q_{-0} q_{++}}{2 q_{-0} + k_{2+}''}\right)^2 + \left(\frac{2 q_{-0} q_{+-}}{2 q_{-0} + k_{2+}''}\right)^2\right]. \quad (9.165)$$

Unpolarized cross section:

$$d\sigma = \frac{\alpha^4 |B|^2 k_{2-}''}{256 \pi^2 E^2 k_{2+}''(q_- \cdot k_2)^2(q_+ \cdot k_1)(q_- \cdot k_1)}$$

$$\times \left[4 q_{-0}^2 + (2q_{-0} + k_{2+}'')^2 + \frac{\mu^2 k_{2+}''^3}{4 q_{-0}^2 k_{2-}''}\right]$$

$$\times \left[q_{-+}^2 + q_{--}^2 + \left(\frac{2 q_{-0} q_{++}}{2 q_{-0} + k_{2+}''}\right)^2 + \left(\frac{2 q_{-0} q_{+-}}{2 q_{-0} + k_{2+}''}\right)^2\right]$$

$$\times \delta^4(p_+ + p_- - q_+ - q_- - k_1 - k_2) \frac{d^3 \vec{q}_+ \, d^3 \vec{q}_- \, d^3 \vec{k}_1 \, d^3 \vec{k}_2}{q_{+0} \, q_{-0} \, k_{10} \, k_{20}}. \quad (9.166)$$

9.18.5 $\vec{k}_1$ and $\vec{k}_2$ nearly parallel to $\vec{p}_+$ and $\vec{p}_-$, resp.

Photon polarizations:

$$\not{e}_1^{\pm} = N_1[\not{k}_1 \not{p}_+ \not{p}_-(1 \pm \gamma_5) + \not{p}_- \not{p}_+ \not{k}_1(1 \mp \gamma_5)],$$

$$\not{e}_2^{\pm} = N_2[\not{k}_2 \not{p}_- \not{p}_+(1 \pm \gamma_5) + \not{p}_+ \not{p}_- \not{k}_2(1 \mp \gamma_5)],$$

$$N_i^{-1} = E^2[32 k_{i+} k_{i-}]^{\frac{1}{2}}, \qquad i = 1, 2. \quad (9.167)$$

9.18. $e^+e^- \to \mu^+\mu^-\gamma\gamma$ (NO Z-EXCHANGE; $m \neq 0$)

Nonvanishing helicity amplitudes:

$$M(+,+;+,-,+,+) = \frac{e^4 mE}{(q_+ \cdot q_-)(p_+ \cdot k_1)(p_- \cdot k_2)} \left\{ k_{1\perp} q_{-\perp}^* \left[\frac{k_{1+}k_{2+}q_{-+}}{k_{1-}k_{2-}q_{++}} \right]^{\frac{1}{2}} \right.$$
$$\left. + k_{2\perp}^* q_{-\perp} \left[\frac{k_{1-}k_{2-}q_{++}}{k_{1+}k_{2+}q_{-+}} \right]^{\frac{1}{2}} \right\},$$

$$M(+,+;-,+,+,+) = \frac{e^4 mE}{(q_+ \cdot q_-)(p_+ \cdot k_1)(p_- \cdot k_2)} \left\{ k_{1\perp} q_{+\perp}^* \left[\frac{k_{1+}k_{2+}q_{++}}{k_{1-}k_{2-}q_{-+}} \right]^{\frac{1}{2}} \right.$$
$$\left. + k_{2\perp}^* q_{+\perp} \left[\frac{k_{1-}k_{2-}q_{-+}}{k_{1+}k_{2+}q_{++}} \right]^{\frac{1}{2}} \right\},$$

$$M(+,+;+,-,+,-) = \frac{e^4 m k_{1\perp} q_{-\perp}^*}{(2E-k_{1+})(p_+ \cdot k_1)(p_- \cdot k_2)} \left[\frac{k_{1+}k_{2+}q_{-+}}{k_{1-}k_{2-}q_{++}} \right]^{\frac{1}{2}}$$
$$\doteq \frac{e^4 m k_{1+} q_{-+}}{(2E-k_{1+})(p_+ \cdot k_1)(p_- \cdot k_2)} \left[\frac{k_{2+}q_{--}}{k_{2-}q_{++}} \right]^{\frac{1}{2}},$$

$$M(+,+;+,-,-,+) = \frac{e^4 m k_{2\perp}^* q_{-\perp}}{(2E-k_{2-})(p_+ \cdot k_1)(p_- \cdot k_2)} \left[\frac{k_{1-}k_{2-}q_{++}}{k_{1+}k_{2+}q_{-+}} \right]^{\frac{1}{2}}$$
$$\doteq \frac{e^4 m k_{2-} q_{++}}{(2E-k_{2-})(p_+ \cdot k_1)(p_- \cdot k_2)} \left[\frac{k_{1-}q_{+-}}{k_{1+}q_{-+}} \right]^{\frac{1}{2}},$$

$$M(+,+;-,+,+,-) = \frac{e^4 m k_{1\perp} q_{+\perp}^*}{(2E-k_{1+})(p_+ \cdot k_1)(p_- \cdot k_2)} \left[\frac{k_{1+}k_{2+}q_{++}}{k_{1-}k_{2-}q_{-+}} \right]^{\frac{1}{2}}$$
$$\doteq \frac{e^4 m k_{1+} q_{++}}{(2E-k_{1+})(p_+ \cdot k_1)(p_- \cdot k_2)} \left[\frac{k_{2+}q_{+-}}{k_{2-}q_{-+}} \right]^{\frac{1}{2}},$$

$$M(+,+;-,+,-,+) = \frac{e^4 m k_{2\perp}^* q_{+\perp}}{(2E-k_{2-})(p_+ \cdot k_1)(p_- \cdot k_2)} \left[\frac{k_{1-}k_{2-}q_{-+}}{k_{1+}k_{2+}q_{++}} \right]^{\frac{1}{2}}$$
$$\doteq \frac{e^4 m k_{2-} q_{-+}}{(2E-k_{2-})(p_+ \cdot k_1)(p_- \cdot k_2)} \left[\frac{k_{1-}q_{--}}{k_{1+}q_{++}} \right]^{\frac{1}{2}},$$

$$M(+,-;+,-,+,+) = \frac{4e^4 E^2 q_{-\perp}}{(2E-k_{1+})(p_+ \cdot k_1)(p_- \cdot k_2)} \left[\frac{k_{1-}k_{2+}q_{++}}{k_{1+}k_{2-}q_{-+}} \right]^{\frac{1}{2}}$$
$$\doteq \frac{4e^4 E^2 q_{++}}{(2E-k_{1+})(p_+ \cdot k_1)(p_- \cdot k_2)} \left[\frac{k_{1-}k_{2+}q_{+-}}{k_{1+}k_{2-}q_{-+}} \right]^{\frac{1}{2}},$$

$$M(+,-;-,+,+,+) = \frac{4e^4 E^2 q_{+\perp}}{(2E-k_{1+})(p_+ \cdot k_1)(p_- \cdot k_2)} \left[\frac{k_{1-}k_{2+}q_{-+}}{k_{1+}k_{2-}q_{++}}\right]^{\frac{1}{2}}$$

$$\doteq \frac{4e^4 E^2 q_{-+}}{(2E-k_{1+})(p_+ \cdot k_1)(p_- \cdot k_2)} \left[\frac{k_{1-}k_{2+}q_{--}}{k_{1+}k_{2-}q_{++}}\right]^{\frac{1}{2}},$$

$$M(-,+;+,-,+,+) = \frac{4e^4 E^2 q^*_{+\perp}}{(2E-k_{2-})(p_+ \cdot k_1)(p_- \cdot k_2)} \left[\frac{k_{1-}k_{2+}q_{-+}}{k_{1+}k_{2-}q_{++}}\right]^{\frac{1}{2}}$$

$$\doteq \frac{4e^4 E^2 q_{-+}}{(2E-k_{2-})(p_+ \cdot k_1)(p_- \cdot k_2)} \left[\frac{k_{1-}k_{2+}q_{--}}{k_{1+}k_{2-}q_{++}}\right]^{\frac{1}{2}},$$

$$M(-,+;-,+,+,+) = \frac{4e^4 E^2 q^*_{-\perp}}{(2E-k_{2-})(p_+ \cdot k_1)(p_- \cdot k_2)} \left[\frac{k_{1-}k_{2+}q_{++}}{k_{1+}k_{2-}q_{-+}}\right]^{\frac{1}{2}}$$

$$\doteq \frac{4e^4 E^2 q_{++}}{(2E-k_{2-})(p_+ \cdot k_1)(p_- \cdot k_2)} \left[\frac{k_{1-}k_{2+}q_{+-}}{k_{1+}k_{2-}q_{-+}}\right]^{\frac{1}{2}},$$

$$M(+,-;+,-,+,-) = \frac{e^4}{(q_+ \cdot q_-)(p_+ \cdot k_1)(p_- \cdot k_2)} \left\{4E^3 q_{-\perp} \left[\frac{k_{1-}k_{2+}q_{++}}{k_{1+}k_{2-}q_{-+}}\right]^{\frac{1}{2}} \right.$$

$$\left. + \frac{m^2 k_{1\perp}k_{2\perp}q^*_{+\perp}}{4E} \left[\frac{k_{1+}k_{2-}q_{-+}}{k_{1-}k_{2+}q_{++}}\right]^{\frac{1}{2}}\right\},$$

$$M(+,-;+,-,-,+) = \frac{2e^4 E q_{-\perp}}{(p_+ \cdot k_1)(p_- \cdot k_2)} \left[\frac{k_{1-}k_{2+}q_{++}}{k_{1+}k_{2-}q_{-+}}\right]^{\frac{1}{2}}$$

$$\doteq \frac{2e^4 E q_{++}}{(p_+ \cdot k_1)(p_- \cdot k_2)} \left[\frac{k_{1-}k_{2+}q_{+-}}{k_{1+}k_{2-}q_{-+}}\right]^{\frac{1}{2}},$$

$$M(+,-;-,+,+,-) = \frac{e^4}{(q_+ \cdot q_-)(p_+ \cdot k_1)(p_- \cdot k_2)} \left\{4E^3 q_{+\perp} \left[\frac{k_{1-}k_{2+}q_{-+}}{k_{1+}k_{2-}q_{++}}\right]^{\frac{1}{2}} \right.$$

$$\left. + \frac{m^2 k_{1\perp}k_{2\perp}q^*_{-\perp}}{4E} \left[\frac{k_{1+}k_{2-}q_{++}}{k_{1-}k_{2+}q_{-+}}\right]^{\frac{1}{2}}\right\},$$

$$M(+,-;-,+,-,+) = \frac{2e^4 E q_{+\perp}}{(p_+ \cdot k_1)(p_- \cdot k_2)} \left[\frac{k_{1-}k_{2+}q_{-+}}{k_{1+}k_{2-}q_{++}}\right]^{\frac{1}{2}}$$

$$\doteq \frac{2e^4 E q_{-+}}{(p_+ \cdot k_1)(p_- \cdot k_2)} \left[\frac{k_{1-}k_{2+}q_{--}}{k_{1+}k_{2-}q_{++}}\right]^{\frac{1}{2}},$$

9.18. $e^+e^- \to \mu^+\mu^-\gamma\gamma$ (NO Z-EXCHANGE; $m \neq 0$)

$$M(-,+;+,-,+,-) = \frac{2e^4 E q^*_{+\perp}}{(p_+ \cdot k_1)(p_- \cdot k_2)} \left[\frac{k_{1-}k_{2+}q_{-+}}{k_{1+}k_{2-}q_{++}}\right]^{\frac{1}{2}}$$

$$\doteq \frac{2e^4 E q_{-+}}{(p_+ \cdot k_1)(p_- \cdot k_2)} \left[\frac{k_{1-}k_{2+}q_{--}}{k_{1+}k_{2-}q_{++}}\right]^{\frac{1}{2}},$$

$$M(-,+;+,-,-,+) = \frac{e^4}{(q_+ \cdot q_-)(p_+ \cdot k_1)(p_- \cdot k_2)} \left\{ 4E^3 q^*_{+\perp} \left[\frac{k_{1-}k_{2+}q_{-+}}{k_{1+}k_{2-}q_{++}}\right]^{\frac{1}{2}} \right.$$

$$\left. + \frac{m^2 k^*_{1\perp} k^*_{2\perp} q_{-\perp}}{4E} \left[\frac{k_{1+}k_{2-}q_{++}}{k_{1-}k_{2+}q_{-+}}\right]^{\frac{1}{2}} \right\},$$

$$M(-,+;-,+,+,-) = \frac{2e^4 E q^*_{-\perp}}{(p_+ \cdot k_1)(p_- \cdot k_2)} \left[\frac{k_{1-}k_{2+}q_{++}}{k_{1+}k_{2-}q_{-+}}\right]^{\frac{1}{2}}$$

$$\doteq \frac{2e^4 E q_{++}}{(p_+ \cdot k_1)(p_- \cdot k_2)} \left[\frac{k_{1-}k_{2+}q_{+-}}{k_{1+}k_{2-}q_{-+}}\right]^{\frac{1}{2}},$$

$$M(-,+;-,+,-,+) = \frac{e^4}{(q_+ \cdot q_-)(p_+ \cdot k_1)(p_- \cdot k_2)} \left\{ 4E^3 q^*_{-\perp} \left[\frac{k_{1-}k_{2+}q_{++}}{k_{1+}k_{2-}q_{-+}}\right]^{\frac{1}{2}} \right.$$

$$\left. + \frac{m^2 k^*_{1\perp} k^*_{2\perp} q_{+\perp}}{4E} \left[\frac{k_{1+}k_{2-}q_{-+}}{k_{1-}k_{2+}q_{++}}\right]^{\frac{1}{2}} \right\},$$

$$M(+,-;+,-,-,-) = \frac{4e^4 E^2 q_{-\perp}}{(2E - k_{2-})(p_+ \cdot k_1)(p_- \cdot k_2)} \left[\frac{k_{1-}k_{2+}q_{++}}{k_{1+}k_{2-}q_{-+}}\right]^{\frac{1}{2}}$$

$$\doteq \frac{4e^4 E^2 q_{++}}{(2E - k_{2-})(p_+ \cdot k_1)(p_- \cdot k_2)} \left[\frac{k_{1-}k_{2+}q_{+-}}{k_{1+}k_{2-}q_{-+}}\right]^{\frac{1}{2}},$$

$$M(+,-;-,+,-,-) = \frac{4e^4 E^2 q_{+\perp}}{(2E - k_{2-})(p_+ \cdot k_1)(p_- \cdot k_2)} \left[\frac{k_{1-}k_{2+}q_{-+}}{k_{1+}k_{2-}q_{++}}\right]^{\frac{1}{2}}$$

$$\doteq \frac{4e^4 E^2 q_{-+}}{(2E - k_{2-})(p_+ \cdot k_1)(p_- \cdot k_2)} \left[\frac{k_{1-}k_{2+}q_{--}}{k_{1+}k_{2-}q_{++}}\right]^{\frac{1}{2}},$$

$$M(-,+;+,-,-,-) = \frac{4e^4 E^2 q^*_{+\perp}}{(2E - k_{1+})(p_+ \cdot k_1)(p_- \cdot k_2)} \left[\frac{k_{1-}k_{2+}q_{-+}}{k_{1+}k_{2-}q_{++}}\right]^{\frac{1}{2}}$$

$$\doteq \frac{4e^4 E^2 q_{-+}}{(2E - k_{1+})(p_+ \cdot k_1)(p_- \cdot k_2)} \left[\frac{k_{1-}k_{2+}q_{--}}{k_{1+}k_{2-}q_{++}}\right]^{\frac{1}{2}},$$

$$M(-,+;-,+,-,-) = \frac{4e^4 E^2 q_{-\perp}^*}{(2E - k_{1+})(p_+ \cdot k_1)(p_- \cdot k_2)} \left[\frac{k_{1-}k_{2+}q_{++}}{k_{1+}k_{2-}q_{-+}} \right]^{\frac{1}{2}}$$

$$\doteq \frac{4e^4 E^2 q_{++}}{(2E - k_{1+})(p_+ \cdot k_1)(p_- \cdot k_2)} \left[\frac{k_{1-}k_{2+}q_{+-}}{k_{1+}k_{2-}q_{-+}} \right]^{\frac{1}{2}},$$

$$M(-,-;+,-,+,-) = \frac{e^4 m k_{2\perp} q_{-\perp}^*}{(2E - k_{2-})(p_+ \cdot k_1)(p_- \cdot k_2)} \left[\frac{k_{1-}k_{2-}q_{-+}}{k_{1+}k_{2+}q_{++}} \right]^{\frac{1}{2}}$$

$$\doteq \frac{e^4 m k_{2-} q_{-+}}{(2E - k_{2-})(p_+ \cdot k_1)(p_- \cdot k_2)} \left[\frac{k_{1-}q_{--}}{k_{1+}q_{++}} \right]^{\frac{1}{2}},$$

$$M(-,-;+,-,-,+) = \frac{e^4 m k_{1\perp}^* q_{-\perp}}{(2E - k_{1+})(p_+ \cdot k_1)(p_- \cdot k_2)} \left[\frac{k_{1+}k_{2+}q_{++}}{k_{1-}k_{2-}q_{-+}} \right]^{\frac{1}{2}}$$

$$\doteq \frac{e^4 m k_{1+} q_{++}}{(2E - k_{1+})(p_+ \cdot k_1)(p_- \cdot k_2)} \left[\frac{k_{2+}q_{+-}}{k_{2-}q_{-+}} \right]^{\frac{1}{2}},$$

$$M(-,-;-,+,+,-) = \frac{e^4 m k_{2\perp} q_{+\perp}^*}{(2E - k_{2-})(p_+ \cdot k_1)(p_- \cdot k_2)} \left[\frac{k_{1-}k_{2-}q_{++}}{k_{1+}k_{2+}q_{-+}} \right]^{\frac{1}{2}}$$

$$\doteq \frac{e^4 m k_{2-} q_{++}}{(2E - k_{2-})(p_+ \cdot k_1)(p_- \cdot k_2)} \left[\frac{k_{1-}q_{+-}}{k_{1+}q_{-+}} \right]^{\frac{1}{2}},$$

$$M(-,-;-,+,-,+) = \frac{e^4 m k_{1\perp}^* q_{+\perp}}{(2E - k_{1+})(p_+ \cdot k_1)(p_- \cdot k_2)} \left[\frac{k_{1+}k_{2+}q_{-+}}{k_{1-}k_{2-}q_{++}} \right]^{\frac{1}{2}}$$

$$\doteq \frac{e^4 m k_{1+} q_{-+}}{(2E - k_{1+})(p_+ \cdot k_1)(p_- \cdot k_2)} \left[\frac{k_{2+}q_{--}}{k_{2-}q_{++}} \right]^{\frac{1}{2}},$$

$$M(-,-;+,-,-,-) = \frac{e^4 m E}{(q_+ \cdot q_-)(p_+ \cdot k_1)(p_- \cdot k_2)} \left\{ k_{2\perp} q_{-\perp}^* \left[\frac{k_{1-}k_{2-}q_{-+}}{k_{1+}k_{2+}q_{++}} \right]^{\frac{1}{2}} \right.$$

$$\left. + k_{1\perp}^* q_{-\perp} \left[\frac{k_{1+}k_{2+}q_{++}}{k_{1-}k_{2-}q_{-+}} \right]^{\frac{1}{2}} \right\},$$

$$M(-,-;-,+,-,-) = \frac{e^4 m E}{(q_+ \cdot q_-)(p_+ \cdot k_1)(p_- \cdot k_2)} \left\{ k_{2\perp} q_{+\perp}^* \left[\frac{k_{1-}k_{2-}q_{++}}{k_{1+}k_{2+}q_{-+}} \right]^{\frac{1}{2}} \right.$$

$$\left. + k_{1\perp}^* q_{+\perp} \left[\frac{k_{1+}k_{2+}q_{-+}}{k_{1-}k_{2-}q_{++}} \right]^{\frac{1}{2}} \right\}. \quad (9.168)$$

9.18. $e^+e^- \to \mu^+\mu^-\gamma\gamma$ (NO Z-EXCHANGE; $m \neq 0$)

Unpolarized squared matrix element:

$$\overline{|M|^2} = \frac{e^8 E^2 k_{1-} k_{2+}}{2 k_{1+} k_{2-} [(q_+ \cdot q_-)(p_+ \cdot k_1)(p_- \cdot k_2)]^2}$$

$$\times \left[4E^2 + (2E - k_{1+})^2 + \frac{m^2 k_{1+}^3}{4 E^2 k_{1-}} \right]$$

$$\times \left[4E^2 + (2E - k_{2-})^2 + \frac{m^2 k_{2-}^3}{4 E^2 k_{2+}} \right] (q_{++} q_{--} + q_{+-} q_{-+}). \quad (9.169)$$

Unpolarized cross section:

$$d\sigma = \frac{\alpha^4 k_{1-} k_{2+}}{256 \pi^2 k_{1+} k_{2-} [(q_+ \cdot q_-)(p_+ \cdot k_1)(p_- \cdot k_2)]^2}$$

$$\times \left[4E^2 + (2E - k_{1+})^2 + \frac{m^2 k_{1+}^3}{4 E^2 k_{1-}} \right]$$

$$\times \left[4E^2 + (2E - k_{2-})^2 + \frac{m^2 k_{2-}^3}{4 E^2 k_{2+}} \right] (q_{++} q_{--} + q_{+-} q_{-+})$$

$$\times \delta^4(p_+ + p_- - q_+ - q_- - k_1 - k_2) \frac{d^3\vec{q}_+ \, d^3\vec{q}_- \, d^3\vec{k}_1 \, d^3\vec{k}_2}{q_{+0} \, q_{-0} \, k_{10} \, k_{20}}. \quad (9.170)$$

9.18.6 $\vec{k}_1$ and $\vec{k}_2$ nearly parallel to $\vec{p}_+$ and $\vec{q}_+$, resp.

The primed quantities $k'_{2\pm}$ and $k'_{2\perp}$ are evaluated in the rotated frame where $\vec{q}_+$ determines the positive z-axis (see also Section 7.4.3). The four-vector q in eqn (9.171) is obtained by applying a space reflection to q_+. The quantity μ denotes the muon mass.
Photon polarizations:

$$\slashed{\epsilon}_1^\pm = N_1 [\slashed{k}_1 \slashed{p}_+ \slashed{p}_-(1 \pm \gamma_5) + \slashed{p}_- \slashed{p}_+ \slashed{k}_1 (1 \mp \gamma_5)],$$

$$N_1^{-1} = E^2 [32 k_{1+} k_{1-}]^{\frac{1}{2}},$$

$$\slashed{\epsilon}_2^\pm = N_2 [\slashed{k}_2 \slashed{q}_+ \slashed{q}(1 \pm \gamma_5) + \slashed{q} \slashed{q}_+ \slashed{k}_2 (1 \mp \gamma_5)],$$

$$N_2^{-1} = q_{+0}^2 [32 k'_{2+} k'_{2-}]^{\frac{1}{2}}. \quad (9.171)$$

Nonvanishing helicity amplitudes:

$$M(+,+;+,+,+,-) = -\frac{e^4 m\mu k_{1\perp} k'^{*}_{2\perp} q^{*}_{-\perp}}{4Eq_{+0}(2E-k_{1+})(p_+ \cdot k_1)(q_+ \cdot k_2)} \left[\frac{k_{1+}k'_{2+}q_{++}}{k_{1-}k'_{2-}q_{-+}}\right]^{\frac{1}{2}}$$

$$\doteq \frac{e^4 m\mu k_{1+}k'_{2+}[q_{++}q_{--}]^{\frac{1}{2}}}{4Eq_{+0}(2E-k_{1+})(p_+ \cdot k_1)(q_+ \cdot k_2)},$$

$$M(+,+;+,-,+,+) = -\frac{e^4 mq_{+0}q_{--}k_{1\perp}q^{*}_{+\perp}}{2E^2(p_+ \cdot k_1)(q_+ \cdot k_2)} \left[\frac{k_{1+}k'_{2-}q_{-+}}{k_{1-}k'_{2+}q^{3}_{++}}\right]^{\frac{1}{2}}$$

$$\doteq \frac{e^4 mq_{+0}q_{--}k_{1+}}{2E^2 q_{++}(p_+ \cdot k_1)(q_+ \cdot k_2)} \left[\frac{k'_{2-}q_{+-}q_{-+}}{k'_{2+}}\right]^{\frac{1}{2}},$$

$$M(+,+;-,+,+,+) = -\frac{e^4 mq_{+0}k_{1\perp}q^{*}_{-\perp}}{E(2E-k_{1+})(p_+ \cdot k_1)(q_+ \cdot k_2)} \left[\frac{k_{1+}k'_{2-}q_{++}}{k_{1-}k'_{2+}q_{-+}}\right]^{\frac{1}{2}}$$

$$\doteq \frac{e^4 mq_{+0}k_{1+}}{E(2E-k_{1+})(p_+ \cdot k_1)(q_+ \cdot k_2)} \left[\frac{k'_{2-}q_{++}q_{--}}{k'_{2+}}\right]^{\frac{1}{2}},$$

$$M(+,+;+,-,+,-) = -\frac{e^4 mq_{+0}k_{1\perp}q^{*}_{+\perp}}{E(2E-k_{1+})(p_+ \cdot k_1)(q_+ \cdot k_2)} \left[\frac{k_{1+}k'_{2-}q_{-+}}{k_{1-}k'_{2+}q_{++}}\right]^{\frac{1}{2}}$$

$$\doteq \frac{e^4 mq_{+0}k_{1+}}{E(2E-k_{1+})(p_+ \cdot k_1)(q_+ \cdot k_2)} \left[\frac{k'_{2-}q_{+-}q_{-+}}{k'_{2+}}\right]^{\frac{1}{2}},$$

$$M(+,+;-,+,+,-) = -\frac{e^4 mq_{+0}q_{--}k_{1\perp}q^{*}_{-\perp}}{2E^2(p_+ \cdot k_1)(q_+ \cdot k_2)} \left[\frac{k_{1+}k'_{2-}}{k_{1-}k'_{2+}q_{++}q_{-+}}\right]^{\frac{1}{2}}$$

$$\doteq \frac{e^4 mq_{+0}k_{1+}}{2E^2(p_+ \cdot k_1)(q_+ \cdot k_2)} \left[\frac{k'_{2-}q^{3}_{--}}{k'_{2+}q_{++}}\right]^{\frac{1}{2}},$$

$$M(+,+;-,-,+,+) = \frac{e^4 m\mu k_{1\perp} k'_{2\perp} q^{*}_{+\perp}}{4Eq_{+0}(2E-k_{1+})(p_+ \cdot k_1)(q_+ \cdot k_2)} \left[\frac{k_{1+}k'_{2+}q_{-+}}{k_{1-}k'_{2-}q_{++}}\right]^{\frac{1}{2}}$$

$$\doteq \frac{e^4 m\mu k_{1+}k'_{2+}[q_{+-}q_{-+}]^{\frac{1}{2}}}{4Eq_{+0}(2E-k_{1+})(p_+ \cdot k_1)(q_+ \cdot k_2)},$$

$$M(+,-;+,+,+,-) = \frac{e^4 \mu E k'^{*}_{2\perp} q_{+\perp}}{q_{+0}(2E-k_{1+})(p_+ \cdot k_1)(q_+ \cdot k_2)} \left[\frac{k_{1-}k'_{2+}q_{-+}}{k_{1+}k'_{2-}q_{++}}\right]^{\frac{1}{2}}$$

$$\doteq \frac{e^4 \mu E k'_{2+}}{q_{+0}(2E-k_{1+})(p_+ \cdot k_1)(q_+ \cdot k_2)} \left[\frac{k_{1-}q_{+-}q_{-+}}{k_{1+}}\right]^{\frac{1}{2}},$$

9.18. $e^+e^- \to \mu^+\mu^-\gamma\gamma$ (NO Z-EXCHANGE; $m \neq 0$)

$$M(+,-;+,-,+,+) = \frac{2e^4 q_{+0} q_{--} q_{-\perp}}{(p_+ \cdot k_1)(q_+ \cdot k_2)} \left[\frac{k_{1-}k'_{2-}}{k_{1+}k'_{2+}q_{++}q_{-+}}\right]^{\frac{1}{2}}$$

$$\doteq \frac{2e^4 q_{+0}}{(p_+ \cdot k_1)(q_+ \cdot k_2)} \left[\frac{k_{1-}k'_{2-}q^3_{--}}{k_{1+}k'_{2+}q_{++}}\right]^{\frac{1}{2}},$$

$$M(+,-;-,+,+,+) = \frac{4e^4 E q_{+0} q_{+\perp}}{(2E-k_{1+})(p_+ \cdot k_1)(q_+ \cdot k_2)} \left[\frac{k_{1-}k'_{2-}q_{-+}}{k_{1+}k'_{2+}q_{++}}\right]^{\frac{1}{2}}$$

$$\doteq \frac{4e^4 E q_{+0}}{(2E-k_{1+})(p_+ \cdot k_1)(q_+ \cdot k_2)} \left[\frac{k_{1-}k'_{2-}q_{+-}q_{-+}}{k_{1+}k'_{2+}}\right]^{\frac{1}{2}},$$

$$M(-,+;+,+,+,-) = \frac{e^4 \mu k'^*_{2\perp} q^*_{-\perp}}{2q_{+0}(p_+ \cdot k_1)(q_+ \cdot k_2)} \left[\frac{k_{1-}k'_{2+}q_{++}}{k_{1+}k'_{2-}q_{-+}}\right]^{\frac{1}{2}}$$

$$\doteq \frac{e^4 \mu k'_{2+}}{2q_{+0}(p_+ \cdot k_1)(q_+ \cdot k_2)} \left[\frac{k_{1-}q_{++}q_{--}}{k_{1+}}\right]^{\frac{1}{2}},$$

$$M(-,+;+,-,+,+) = \frac{e^4 (2q_{+0}+k'_{2+}) q^*_{+\perp}}{(p_+ \cdot k_1)(q_+ \cdot k_2)} \left[\frac{k_{1-}k'_{2-}q_{-+}}{k_{1+}k'_{2+}q_{++}}\right]^{\frac{1}{2}}$$

$$\doteq \frac{e^4 (2q_{+0}+k'_{2+})}{(p_+ \cdot k_1)(q_+ \cdot k_2)} \left[\frac{k_{1-}k'_{2-}q_{+-}q_{-+}}{k_{1+}k'_{2+}}\right]^{\frac{1}{2}},$$

$$M(-,+;-,+,+,+) = \frac{2e^4 q_{+0} q^*_{-\perp}}{(p_+ \cdot k_1)(q_+ \cdot k_2)} \left[\frac{k_{1-}k'_{2-}q_{++}}{k_{1+}k'_{2+}q_{-+}}\right]^{\frac{1}{2}}$$

$$\doteq \frac{2e^4 q_{+0}}{(p_+ \cdot k_1)(q_+ \cdot k_2)} \left[\frac{k_{1-}k'_{2-}q_{++}q_{--}}{k_{1+}k'_{2+}}\right]^{\frac{1}{2}},$$

$$M(+,-;+,+,-,-) = \frac{e^4 \mu k'^*_{2\perp} q_{+\perp}}{2q_{+0}(p_+ \cdot k_1)(q_+ \cdot k_2)} \left[\frac{k_{1-}k'_{2+}q_{-+}}{k_{1+}k'_{2-}q_{++}}\right]^{\frac{1}{2}}$$

$$\doteq \frac{e^4 \mu k'_{2+}}{2q_{+0}(p_+ \cdot k_1)(q_+ \cdot k_2)} \left[\frac{k_{1-}q_{+-}q_{-+}}{k_{1+}}\right]^{\frac{1}{2}},$$

$$M(+,-;+,-,+,-) = \frac{4e^4 E q_{+0} q_{-\perp}}{(2E-k_{1+})(p_+ \cdot k_1)(q_+ \cdot k_2)} \left[\frac{k_{1-}k'_{2-}q_{++}}{k_{1+}k'_{2+}q_{-+}}\right]^{\frac{1}{2}}$$

$$\doteq \frac{4e^4 E q_{+0}}{(2E-k_{1+})(p_+ \cdot k_1)(q_+ \cdot k_2)} \left[\frac{k_{1-}k'_{2-}q_{++}q_{--}}{k_{1+}k'_{2+}}\right]^{\frac{1}{2}},$$

$$M(+,-;+,-,-,+) = \frac{e^4(2q_{+0}+k'_{2+})q_{-\perp}}{(p_+\cdot k_1)(q_+\cdot k_2)}\left[\frac{k_{1-}k'_{2-}q_{++}}{k_{1+}k'_{2+}q_{-+}}\right]^{\frac{1}{2}}$$

$$\doteq \frac{e^4(2q_{+0}+k'_{2+})}{(p_+\cdot k_1)(q_+\cdot k_2)}\left[\frac{k_{1-}k'_{2-}q_{++}q_{--}}{k_{1+}k'_{2+}}\right]^{\frac{1}{2}},$$

$$M(+,-;-,+,+,-) = \frac{2e^4 q_{+0}q_{--}q_{+\perp}}{(p_+\cdot k_1)(q_+\cdot k_2)}\left[\frac{k_{1-}k'_{2-}q_{-+}}{k_{1+}k'_{2+}q_{++}^3}\right]^{\frac{1}{2}}$$

$$\doteq \frac{2e^4 q_{+0}q_{--}}{q_{++}(p_+\cdot k_1)(q_+\cdot k_2)}\left[\frac{k_{1-}k'_{2-}q_{+-}q_{-+}}{k_{1+}k'_{2+}}\right]^{\frac{1}{2}},$$

$$M(+,-;-,+,-,+) = \frac{2e^4 q_{+0}q_{+\perp}}{(p_+\cdot k_1)(q_+\cdot k_2)}\left[\frac{k_{1-}k'_{2-}q_{-+}}{k_{1+}k'_{2+}q_{++}}\right]^{\frac{1}{2}}$$

$$\doteq \frac{2e^4 q_{+0}}{(p_+\cdot k_1)(q_+\cdot k_2)}\left[\frac{k_{1-}k'_{2-}q_{+-}q_{-+}}{k_{1+}k'_{2+}}\right]^{\frac{1}{2}},$$

$$M(+,-;-,-,+,+) = -\frac{e^4\mu E k'_{2\perp}q_{-\perp}}{q_{+0}(2E-k_{1+})(p_+\cdot k_1)(q_+\cdot k_2)}\left[\frac{k_{1-}k'_{2+}q_{++}}{k_{1+}k'_{2-}q_{-+}}\right]^{\frac{1}{2}}$$

$$\doteq \frac{e^4\mu E k'_{2+}}{q_{+0}(2E-k_{1+})(p_+\cdot k_1)(q_+\cdot k_2)}\left[\frac{k_{1-}q_{++}q_{--}}{k_{1+}}\right]^{\frac{1}{2}},$$

$$M(-,+;+,+,-,-) = \frac{e^4\mu E k'^{*}_{2\perp}q^*_{-\perp}}{q_{+0}(2E-k_{1+})(p_+\cdot k_1)(q_+\cdot k_2)}\left[\frac{k_{1-}k'_{2+}q_{++}}{k_{1+}k'_{2-}q_{-+}}\right]^{\frac{1}{2}}$$

$$\doteq \frac{e^4\mu E k'_{2+}}{q_{+0}(2E-k_{1+})(p_+\cdot k_1)(q_+\cdot k_2)}\left[\frac{k_{1-}q_{++}q_{--}}{k_{1+}}\right]^{\frac{1}{2}},$$

$$M(-,+;+,-,+,-) = \frac{2e^4 q_{+0}q^*_{+\perp}}{(p_+\cdot k_1)(q_+\cdot k_2)}\left[\frac{k_{1-}k'_{2-}q_{-+}}{k_{1+}k'_{2+}q_{++}}\right]^{\frac{1}{2}}$$

$$\doteq \frac{2e^4 q_{+0}}{(p_+\cdot k_1)(q_+\cdot k_2)}\left[\frac{k_{1-}k'_{2-}q_{+-}q_{-+}}{k_{1+}k'_{2+}}\right]^{\frac{1}{2}},$$

$$M(-,+;+,-,-,+) = \frac{2e^4 q_{+0}q_{--}q^*_{+\perp}}{(p_+\cdot k_1)(q_+\cdot k_2)}\left[\frac{k_{1-}k'_{2-}q_{-+}}{k_{1+}k'_{2+}q_{++}^3}\right]^{\frac{1}{2}}$$

$$\doteq \frac{2e^4 q_{+0}q_{--}}{q_{++}(p_+\cdot k_1)(q_+\cdot k_2)}\left[\frac{k_{1-}k'_{2-}q_{+-}q_{-+}}{k_{1+}k'_{2+}}\right]^{\frac{1}{2}},$$

9.18. $e^+e^- \to \mu^+\mu^-\gamma\gamma$ (NO Z-EXCHANGE; $m \neq 0$)

$$M(-,+;-,+,+,-) = \frac{e^4(2q_{+0}+k'_{2+})q^*_{-\perp}}{(p_+ \cdot k_1)(q_+ \cdot k_2)}\left[\frac{k_{1-}k'_{2-}q_{++}}{k_{1+}k'_{2+}q_{-+}}\right]^{\frac{1}{2}}$$

$$\doteq \frac{e^4(2q_{+0}+k'_{2+})}{(p_+ \cdot k_1)(q_+ \cdot k_2)}\left[\frac{k_{1-}k'_{2-}q_{++}q_{--}}{k_{1+}k'_{2+}}\right]^{\frac{1}{2}},$$

$$M(-,+;-,+,-,+) = \frac{4e^4 E q_{+0} q^*_{-\perp}}{(2E-k_{1+})(p_+ \cdot k_1)(q_+ \cdot k_2)}\left[\frac{k_{1-}k'_{2-}q_{++}}{k_{1+}k'_{2+}q_{-+}}\right]^{\frac{1}{2}}$$

$$\doteq \frac{4e^4 E q_{+0}}{(2E-k_{1+})(p_+ \cdot k_1)(q_+ \cdot k_2)}\left[\frac{k_{1-}k'_{2-}q_{++}q_{--}}{k_{1+}k'_{2+}}\right]^{\frac{1}{2}},$$

$$M(-,+;-,-,+,+) = -\frac{e^4 \mu k'_{2\perp} q^*_{+\perp}}{2q_{+0}(p_+ \cdot k_1)(q_+ \cdot k_2)}\left[\frac{k_{1-}k'_{2+}q_{-+}}{k_{1+}k'_{2-}q_{++}}\right]^{\frac{1}{2}}$$

$$\doteq \frac{e^4 \mu k'_{2+}}{2q_{+0}(p_+ \cdot k_1)(q_+ \cdot k_2)}\left[\frac{k_{1-}q_{+-}q_{-+}}{k_{1+}}\right]^{\frac{1}{2}},$$

$$M(+,-;+,-,-,-) = \frac{2e^4 q_{+0} q_{-\perp}}{(p_+ \cdot k_1)(q_+ \cdot k_2)}\left[\frac{k_{1-}k'_{2-}q_{++}}{k_{1+}k'_{2+}q_{-+}}\right]^{\frac{1}{2}}$$

$$\doteq \frac{2e^4 q_{+0}}{(p_+ \cdot k_1)(q_+ \cdot k_2)}\left[\frac{k_{1-}k'_{2-}q_{++}q_{--}}{k_{1+}k'_{2+}}\right]^{\frac{1}{2}},$$

$$M(+,-;-,+,-,-) = \frac{e^4(2q_{+0}+k'_{2+})q_{+\perp}}{(p_+ \cdot k_1)(q_+ \cdot k_2)}\left[\frac{k_{1-}k'_{2-}q_{-+}}{k_{1+}k'_{2+}q_{++}}\right]^{\frac{1}{2}}$$

$$\doteq \frac{e^4(2q_{+0}+k'_{2+})}{(p_+ \cdot k_1)(q_+ \cdot k_2)}\left[\frac{k_{1-}k'_{2-}q_{+-}q_{-+}}{k_{1+}k'_{2+}}\right]^{\frac{1}{2}},$$

$$M(+,-;-,-,-,+) = -\frac{e^4 \mu k'_{2\perp} q_{-\perp}}{2q_{+0}(p_+ \cdot k_1)(q_+ \cdot k_2)}\left[\frac{k_{1-}k'_{2+}q_{++}}{k_{1+}k'_{2-}q_{-+}}\right]^{\frac{1}{2}}$$

$$\doteq \frac{e^4 \mu k'_{2+}}{2q_{+0}(p_+ \cdot k_1)(q_+ \cdot k_2)}\left[\frac{k_{1-}q_{++}q_{--}}{k_{1+}}\right]^{\frac{1}{2}},$$

$$M(-,+;+,-,-,-) = \frac{4e^4 E q_{+0} q^*_{+\perp}}{(2E-k_{1+})(p_+ \cdot k_1)(q_+ \cdot k_2)}\left[\frac{k_{1-}k'_{2-}q_{-+}}{k_{1+}k'_{2+}q_{++}}\right]^{\frac{1}{2}}$$

$$\doteq \frac{4e^4 E q_{+0}}{(2E-k_{1+})(p_+ \cdot k_1)(q_+ \cdot k_2)}\left[\frac{k_{1-}k'_{2-}q_{+-}q_{-+}}{k_{1+}k'_{2+}}\right]^{\frac{1}{2}},$$

$$M(-,+;-,+,-,-) = \frac{2e^4 q_{+0} q_{--} q_{-\perp}^*}{(p_+ \cdot k_1)(q_+ \cdot k_2)} \left[\frac{k_{1-} k'_{2-}}{k_{1+} k'_{2+} q_{++} q_{-+}} \right]^{\frac{1}{2}}$$

$$\doteq \frac{2e^4 q_{+0}}{(p_+ \cdot k_1)(q_+ \cdot k_2)} \left[\frac{k_{1-} k'_{2-} q^3_{--}}{k_{1+} k'_{2+} q_{++}} \right]^{\frac{1}{2}},$$

$$M(-,+;-,-,-,+) = -\frac{e^4 \mu E k'_{2\perp} q^*_{+\perp}}{q_{+0}(2E - k_{1+})(p_+ \cdot k_1)(q_+ \cdot k_2)} \left[\frac{k_{1-} k'_{2+} q_{-+}}{k_{1+} k'_{2-} q_{++}} \right]$$

$$\doteq \frac{e^4 \mu E k'_{2+}}{q_{+0}(2E - k_{1+})(p_+ \cdot k_1)(q_+ \cdot k_2)} \left[\frac{k_{1-} q_{+-} q_{-+}}{k_{1+}} \right]^{\frac{1}{2}}$$

$$M(-,-;+,+,-,-) = \frac{e^4 m \mu k^*_{1\perp} k'^*_{2\perp} q_{+\perp}}{4E q_{+0}(2E - k_{1+})(p_+ \cdot k_1)(q_+ \cdot k_2)} \left[\frac{k_{1+} k'_{2+} q_{-+}}{k_{1-} k'_{2-} q_{++}} \right]$$

$$\doteq \frac{e^4 m \mu k_{1+} k'_{2+} [q_{+-} q_{-+}]^{\frac{1}{2}}}{4E q_{+0}(2E - k_{1+})(p_+ \cdot k_1)(q_+ \cdot k_2)},$$

$$M(-,-;+,-,-,+) = \frac{e^4 m q_{+0} q_{--} k^*_{1\perp} q_{-\perp}}{2E^2 (p_+ \cdot k_1)(q_+ \cdot k_2)} \left[\frac{k_{1+} k'_{2-}}{k_{1-} k'_{2+} q_{++} q_{-+}} \right]^{\frac{1}{2}}$$

$$\doteq \frac{e^4 m q_{+0} k_{1+}}{2E^2 (p_+ \cdot k_1)(q_+ \cdot k_2)} \left[\frac{k'_{2-} q^3_{--}}{k'_{2+} q_{++}} \right]^{\frac{1}{2}},$$

$$M(-,-;-,+,-,+) = \frac{e^4 m q_{+0} k^*_{1\perp} q_{+\perp}}{E(2E - k_{1+})(p_+ \cdot k_1)(q_+ \cdot k_2)} \left[\frac{k_{1+} k'_{2-} q_{-+}}{k_{1-} k'_{2+} q_{++}} \right]^{\frac{1}{2}}$$

$$\doteq \frac{e^4 m q_{+0} k_{1+}}{E(2E - k_{1+})(p_+ \cdot k_1)(q_+ \cdot k_2)} \left[\frac{k'_{2-} q_{+-} q_{-+}}{k'_{2+}} \right]^{\frac{1}{2}},$$

$$M(-,-;+,-,-,-) = \frac{e^4 m q_{+0} k^*_{1\perp} q_{-\perp}}{E(2E - k_{1+})(p_+ \cdot k_1)(q_+ \cdot k_2)} \left[\frac{k_{1+} k'_{2-} q_{++}}{k_{1-} k'_{2+} q_{-+}} \right]^{\frac{1}{2}}$$

$$\doteq \frac{e^4 m q_{+0} k_{1+}}{E(2E - k_{1+})(p_+ \cdot k_1)(q_+ \cdot k_2)} \left[\frac{k'_{2-} q_{++} q_{--}}{k'_{2+}} \right]^{\frac{1}{2}},$$

$$M(-,-;-,+,-,-) = \frac{e^4 m q_{+0} q_{--} k^*_{1\perp} q_{+\perp}}{2E^2 (p_+ \cdot k_1)(q_+ \cdot k_2)} \left[\frac{k_{1+} k'_{2-} q_{-+}}{k_{1-} k'_{2+} q^3_{++}} \right]^{\frac{1}{2}}$$

$$\doteq \frac{e^4 m q_{+0} q_{--} k_{1+}}{2E^2 q_{++}(p_+ \cdot k_1)(q_+ \cdot k_2)} \left[\frac{k'_{2-} q_{+-} q_{-+}}{k'_{2+}} \right]^{\frac{1}{2}},$$

9.18. $e^+e^- \to \mu^+\mu^-\gamma\gamma$ (NO Z-EXCHANGE; $m \neq 0$)

$$M(-,-;-,-,-,+) = -\frac{e^4 m\mu k_{1\perp}^* k_{2\perp}' q_{-\perp}}{4Eq_{+0}(2E-k_{1+})(p_+ \cdot k_1)(q_+ \cdot k_2)} \left[\frac{k_{1+}k_{2+}'q_{++}}{k_{1-}k_{2-}'q_{-+}}\right]^{\frac{1}{2}}$$

$$\doteq \frac{e^4 m\mu k_{1+} k_{2+}' [q_{++}q_{--}]^{\frac{1}{2}}}{4Eq_{+0}(2E-k_{1+})(p_+ \cdot k_1)(q_+ \cdot k_2)}. \quad (9.172)$$

Unpolarized squared matrix element:

$$\overline{|M|^2} = \frac{e^8 k_{1-} k_{2-}'}{2k_{1+} k_{2+}' [(2E-k_{1+})(p_+ \cdot k_1)(q_+ \cdot k_2)]^2}$$

$$\times \left[4E^2 + (2E-k_{1+})^2 + \frac{m^2 k_{1+}^3}{4E^2 k_{1-}}\right]$$

$$\times \left[4q_{+0}^2 + (2q_{+0}+k_{2+}')^2 + \frac{\mu^2 k_{2+}'^3}{4q_{+0}^2 k_{2-}'}\right] (q_{++}q_{--} + q_{+-}q_{-+}). \quad (9.173)$$

Unpolarized cross section:

$$d\sigma = \frac{\alpha^4 k_{1-} k_{2-}'}{256\pi^2 k_{1+} k_{2+}' [E(2E-k_{1+})(p_+ \cdot k_1)(q_+ \cdot k_2)]^2}$$

$$\times \left[4E^2 + (2E-k_{1+})^2 + \frac{m^2 k_{1+}^3}{4E^2 k_{1-}}\right]$$

$$\times \left[4q_{+0}^2 + (2q_{+0}+k_{2+}')^2 + \frac{\mu^2 k_{2+}'^3}{4q_{+0}^2 k_{2-}'}\right] (q_{++}q_{--} + q_{+-}q_{-+})$$

$$\times \delta^4(p_+ + p_- - q_+ - q_- - k_1 - k_2) \frac{d^3\vec{q}_+ \, d^3\vec{q}_- \, d^3\vec{k}_1 \, d^3\vec{k}_2}{q_{+0} \, q_{-0} \, k_{10} \, k_{20}}. \quad (9.174)$$

9.18.7 $\vec{k}_1$ and $\vec{k}_2$ nearly parallel to $\vec{p}_+$ and $\vec{q}_-$, resp.

The doubly primed quantities $k_{2\pm}''$ and $k_{2\perp}''$ are evaluated in the rotated frame where $\vec{q}_-$ determines the positive z-axis (see also Section 7.4.3). The four-vector q in eqn (9.175) is obtained by applying a space reflection to q_-. The quantity μ denotes the muon mass.

Photon polarizations:

$$\not{\epsilon}_1^\pm = N_1[\not{k}_1 \not{p}_+ \not{p}_-(1 \pm \gamma_5) + \not{p}_- \not{p}_+ \not{k}_1(1 \mp \gamma_5)],$$

$$N_1^{-1} = E^2[32k_{1+}k_{1-}]^{\frac{1}{2}},$$

$$\not{\epsilon}_2^\pm = N_2[\not{k}_2 \not{q}_- \not{q}(1 \pm \gamma_5) + \not{q} \not{q}_- \not{k}_2(1 \mp \gamma_5)],$$

$$N_2^{-1} = q_{-0}^2[32k_{2+}'' k_{2-}'']^{\frac{1}{2}}. \quad (9.175)$$

Nonvanishing helicity amplitudes:

$$M(+,+;+,+,+,-) = \frac{e^4 m \mu k_{1\perp} k_{21}^{\prime\prime*} q_{+\perp}^*}{4Eq_{-0}(2E-k_{1+})(p_+ \cdot k_1)(q_- \cdot k_2)} \left[\frac{k_{1+} k_{2+}^{\prime\prime} q_{--}}{k_{1-} k_{2-}^{\prime\prime} q_{+-}}\right]$$

$$\doteq \frac{e^4 m \mu k_{1+} k_{2+}^{\prime\prime} [q_{+-} q_{-+}]^{\frac{1}{2}}}{4Eq_{-0}(2E-k_{1+})(p_+ \cdot k_1)(q_- \cdot k_2)},$$

$$M(+,+;+,-,+,+) = \frac{e^4 m q_{-0} k_{1\perp} q_{+\perp}^*}{E(2E-k_{1+})(p_+ \cdot k_1)(q_- \cdot k_2)} \left[\frac{k_{1+} k_{2-}^{\prime\prime} q_{-+}}{k_{1-} k_{2+}^{\prime\prime} q_{++}}\right]^{\frac{1}{2}}$$

$$\doteq \frac{e^4 m q_{-0} k_{1+}}{E(2E-k_{1+})(p_+ \cdot k_1)(q_- \cdot k_2)} \left[\frac{k_{2-}^{\prime\prime} q_{+-} q_{-+}}{k_{2+}^{\prime\prime}}\right]^{\frac{1}{2}},$$

$$M(+,+;-,+,+,+) = \frac{e^4 m q_{-0} q_{+-} k_{1\perp} q_{-\perp}^*}{2E^2 (p_+ \cdot k_1)(q_- \cdot k_2)} \left[\frac{k_{1+} k_{2-}^{\prime\prime} q_{++}}{k_{1-} k_{2+}^{\prime\prime} q_{-+}^3}\right]^{\frac{1}{2}}$$

$$\doteq \frac{e^4 m q_{-0} q_{+-} k_{1+}}{2E^2 q_{-+}(p_+ \cdot k_1)(q_- \cdot k_2)} \left[\frac{k_{2-}^{\prime\prime} q_{++} q_{--}}{k_{2+}^{\prime\prime}}\right]^{\frac{1}{2}},$$

$$M(+,+;+,-,+,-) = \frac{e^4 m q_{-0} q_{+-} k_{1\perp} q_{+\perp}^*}{2E^2 (p_+ \cdot k_1)(q_- \cdot k_2)} \left[\frac{k_{1+} k_{2-}^{\prime\prime}}{k_{1-} k_{2+}^{\prime\prime} q_{++} q_{-+}}\right]^{\frac{1}{2}}$$

$$\doteq \frac{e^4 m q_{-0} k_{1+}}{2E^2 (p_+ \cdot k_1)(q_- \cdot k_2)} \left[\frac{k_{2-}^{\prime\prime} q_{+-}^3}{k_{2+}^{\prime\prime} q_{-+}}\right]^{\frac{1}{2}},$$

$$M(+,+;-,+,+,-) = \frac{e^4 m q_{-0} k_{1\perp} q_{-\perp}^*}{E(2E-k_{1+})(p_+ \cdot k_1)(q_- \cdot k_2)} \left[\frac{k_{1+} k_{2-}^{\prime\prime} q_{++}}{k_{1-} k_{2+}^{\prime\prime} q_{-+}}\right]^{\frac{1}{2}}$$

$$\doteq \frac{e^4 m q_{-0} k_{1+}}{E(2E-k_{1+})(p_+ \cdot k_1)(q_- \cdot k_2)} \left[\frac{k_{2-}^{\prime\prime} q_{++} q_{--}}{k_{2+}^{\prime\prime}}\right]^{\frac{1}{2}},$$

$$M(+,+;-,-,+,+) = \frac{-e^4 m \mu k_{1\perp} k_{21}^{\prime\prime} q_{-\perp}^*}{4Eq_{-0}(2E-k_{1+})(p_+ \cdot k_1)(q_- \cdot k_2)} \left[\frac{k_{1+} k_{2+}^{\prime\prime} q_{++}}{k_{1-} k_{2-}^{\prime\prime} q_{-+}}\right]$$

$$\doteq \frac{e^4 m \mu k_{1+} k_{2+}^{\prime\prime} [q_{++} q_{--}]^{\frac{1}{2}}}{4Eq_{-0}(2E-k_{1+})(p_+ \cdot k_1)(q_- \cdot k_2)},$$

$$M(+,-;+,+,+,-) = \frac{-e^4 \mu E k_{21}^{\prime\prime*} q_{-\perp}}{q_{-0}(2E-k_{1+})(p_+ \cdot k_1)(q_- \cdot k_2)} \left[\frac{k_{1-} k_{2+}^{\prime\prime} q_{++}}{k_{1+} k_{2-}^{\prime\prime} q_{-+}}\right]^{\frac{1}{2}}$$

$$\doteq \frac{e^4 \mu E k_{2+}^{\prime\prime}}{q_{-0}(2E-k_{1+})(p_+ \cdot k_1)(q_- \cdot k_2)} \left[\frac{k_{1-} q_{++} q_{--}}{k_{1+}}\right]^{\frac{1}{2}}$$

9.18. $e^+e^- \to \mu^+\mu^-\gamma\gamma$ (NO Z-EXCHANGE; $m \neq 0$)

$$M(+,-;+,-,+,+) = -\frac{4e^4 E q_{-0} q_{-\perp}}{(2E-k_{1+})(p_+ \cdot k_1)(q_- \cdot k_2)} \left[\frac{k_{1-}k''_{2-}q_{++}}{k_{1+}k''_{2+}q_{-+}}\right]^{\frac{1}{2}}$$

$$\doteq \frac{4e^4 E q_{-0}}{(2E-k_{1+})(p_+ \cdot k_1)(q_- \cdot k_2)} \left[\frac{k_{1-}k''_{2-}q_{++}q_{--}}{k_{1+}k''_{2+}}\right]^{\frac{1}{2}},$$

$$M(+,-;-,+,+,+) = -\frac{2e^4 q_{-0} q_{+-} q_{+\perp}}{(p_+ \cdot k_1)(q_- \cdot k_2)} \left[\frac{k_{1-}k''_{2-}}{k_{1+}k''_{2+}q_{++}q_{-+}}\right]^{\frac{1}{2}}$$

$$\doteq \frac{2e^4 q_{-0}}{(p_+ \cdot k_1)(q_- \cdot k_2)} \left[\frac{k_{1-}k''_{2-}q_{+-}^3}{k_{1+}k''_{2+}q_{-+}}\right]^{\frac{1}{2}},$$

$$M(-,+;+,+,+,-) = -\frac{e^4 \mu k''^*_{2\perp} q^*_{+\perp}}{2q_{-0}(p_+ \cdot k_1)(q_- \cdot k_2)} \left[\frac{k_{1-}k''_{2+}q_{-+}}{k_{1+}k''_{2-}q_{++}}\right]^{\frac{1}{2}}$$

$$\doteq \frac{e^4 \mu k''_{2+}}{2q_{-0}(p_+ \cdot k_1)(q_- \cdot k_2)} \left[\frac{k_{1-}q_{+-}q_{-+}}{k_{1+}}\right]^{\frac{1}{2}},$$

$$M(-,+;+,-,+,+) = -\frac{2e^4 q_{-0} q^*_{+\perp}}{(p_+ \cdot k_1)(q_- \cdot k_2)} \left[\frac{k_{1-}k''_{2-}q_{-+}}{k_{1+}k''_{2+}q_{++}}\right]^{\frac{1}{2}}$$

$$\doteq \frac{2e^4 q_{-0}}{(p_+ \cdot k_1)(q_- \cdot k_2)} \left[\frac{k_{1-}k''_{2-}q_{+-}q_{-+}}{k_{1+}k''_{2+}}\right]^{\frac{1}{2}},$$

$$M(-,+;-,+,+,+) = -\frac{e^4(2q_{-0}+k''_{2+})q^*_{-\perp}}{(p_+ \cdot k_1)(q_- \cdot k_2)} \left[\frac{k_{1-}k''_{2-}q_{++}}{k_{1+}k''_{2+}q_{-+}}\right]^{\frac{1}{2}}$$

$$\doteq \frac{e^4(2q_{-0}+k''_{2+})}{(p_+ \cdot k_1)(q_- \cdot k_2)} \left[\frac{k_{1-}k''_{2-}q_{++}q_{--}}{k_{1+}k''_{2+}}\right]^{\frac{1}{2}},$$

$$M(+,-;+,+,-,-) = -\frac{e^4 \mu k''^*_{2\perp} q_{-\perp}}{2q_{-0}(p_+ \cdot k_1)(q_- \cdot k_2)} \left[\frac{k_{1-}k''_{2+}q_{++}}{k_{1+}k''_{2-}q_{-+}}\right]^{\frac{1}{2}}$$

$$\doteq \frac{e^4 \mu k''_{2+}}{2q_{-0}(p_+ \cdot k_1)(q_- \cdot k_2)} \left[\frac{k_{1-}q_{++}q_{--}}{k_{1+}}\right]^{\frac{1}{2}},$$

$$M(+,-;+,-,+,-) = -\frac{2e^4 q_{-0} q_{+-} q_{-\perp}}{(p_+ \cdot k_1)(q_- \cdot k_2)} \left[\frac{k_{1-}k''_{2-}q_{++}}{k_{1+}k''_{2+}q_{-+}^3}\right]^{\frac{1}{2}}$$

$$\doteq \frac{2e^4 q_{-0} q_{+-}}{q_{-+}(p_+ \cdot k_1)(q_- \cdot k_2)} \left[\frac{k_{1-}k''_{2-}q_{++}q_{--}}{k_{1+}k''_{2+}}\right]^{\frac{1}{2}},$$

$$M(+,-;+,-,-,+) = -\frac{2e^4 q_{-0} q_{-\perp}}{(p_+ \cdot k_1)(q_- \cdot k_2)} \left[\frac{k_{1-} k''_{2-} q_{++}}{k_{1+} k''_{2+} q_{-+}}\right]^{\frac{1}{2}}$$

$$\doteq \frac{2e^4 q_{-0}}{(p_+ \cdot k_1)(q_- \cdot k_2)} \left[\frac{k_{1-} k''_{2-} q_{++} q_{--}}{k_{1+} k''_{2+}}\right]^{\frac{1}{2}},$$

$$M(+,-;-,+,-,+) = -\frac{e^4(2q_{-0} + k''_{2+}) q_{+\perp}}{(p_+ \cdot k_1)(q_- \cdot k_2)} \left[\frac{k_{1-} k''_{2-} q_{-+}}{k_{1+} k''_{2+} q_{++}}\right]^{\frac{1}{2}}$$

$$\doteq \frac{e^4(2q_{-0} + k''_{2+})}{(p_+ \cdot k_1)(q_- \cdot k_2)} \left[\frac{k_{1-} k''_{2-} q_{+-} q_{-+}}{k_{1+} k''_{2+}}\right]^{\frac{1}{2}},$$

$$M(+,-;-,+,+,-) = -\frac{4e^4 E q_{-0} q_{+\perp}}{(2E - k_{1+})(p_+ \cdot k_1)(q_- \cdot k_2)} \left[\frac{k_{1-} k''_{2-} q_{-+}}{k_{1+} k''_{2+} q_{++}}\right]^{\frac{1}{2}}$$

$$\doteq \frac{4e^4 E q_{-0}}{(2E - k_{1+})(p_+ \cdot k_1)(q_- \cdot k_2)} \left[\frac{k_{1-} k''_{2-} q_{+-} q_{-+}}{k_{1+} k''_{2+}}\right]^{\frac{1}{2}},$$

$$M(+,-;-,-,+,+) = \frac{e^4 \mu E k''_{2\perp} q_{+\perp}}{q_{-0}(2E - k_{1+})(p_+ \cdot k_1)(q_- \cdot k_2)} \left[\frac{k_{1-} k''_{2+} q_{-+}}{k_{1+} k''_{2-} q_{++}}\right]^{\frac{1}{2}}$$

$$\doteq \frac{e^4 \mu k''_{2+}}{q_{-0}(2E - k_{1+})(p_+ \cdot k_1)(q_- \cdot k_2)} \left[\frac{k_{1-} q_{+-} q_{-+}}{k_{1+}}\right]^{\frac{1}{2}},$$

$$M(-,+;+,+,-,-) = -\frac{e^4 \mu E k''^*_{2\perp} q^*_{+\perp}}{q_{-0}(2E - k_{1+})(p_+ \cdot k_1)(q_- \cdot k_2)} \left[\frac{k_{1-} k''_{2+} q_{-+}}{k_{1+} k''_{2-} q_{++}}\right]^{\frac{1}{2}}$$

$$\doteq \frac{e^4 \mu E k''_{2+}}{q_{-0}(2E - k_{1+})(p_+ \cdot k_1)(q_- \cdot k_2)} \left[\frac{k_{1-} q_{+-} q_{-+}}{k_{1+}}\right]^{\frac{1}{2}},$$

$$M(-,+;+,-,+,-) = -\frac{e^4(2q_{-0} + k''_{2+}) q^*_{+\perp}}{(p_+ \cdot k_1)(q_- \cdot k_2)} \left[\frac{k_{1-} k''_{2-} q_{-+}}{k_{1+} k''_{2+} q_{++}}\right]^{\frac{1}{2}}$$

$$\doteq \frac{e^4(2q_{-0} + k''_{2+})}{(p_+ \cdot k_1)(q_- \cdot k_2)} \left[\frac{k_{1-} k''_{2-} q_{+-} q_{-+}}{k_{1+} k''_{2+}}\right]^{\frac{1}{2}},$$

$$M(-,+;+,-,-,+) = -\frac{4e^4 E q_{-0} q^*_{+\perp}}{(2E - k_{1+})(p_+ \cdot k_1)(q_- \cdot k_2)} \left[\frac{k_{1-} k''_{2-} q_{-+}}{k_{1+} k''_{2+} q_{++}}\right]^{\frac{1}{2}}$$

$$\doteq \frac{4e^4 E q_{-0}}{(2E - k_{1+})(p_+ \cdot k_1)(q_- \cdot k_2)} \left[\frac{k_{1-} k''_{2-} q_{+-} q_{-+}}{k_{1+} k''_{2+}}\right]^{\frac{1}{2}},$$

9.18. $e^+e^- \to \mu^+\mu^-\gamma\gamma$ (NO Z-EXCHANGE; $m \neq 0$)

$$M(-,+;-,+,+,-) = -\frac{2e^4 q_{-0} q_{-\perp}^*}{(p_+ \cdot k_1)(q_- \cdot k_2)} \left[\frac{k_{1-} k_{2-}'' q_{++}}{k_{1+} k_{2+}'' q_{-+}}\right]^{\frac{1}{2}}$$

$$\doteq \frac{2e^4 q_{-0}}{(p_+ \cdot k_1)(q_- \cdot k_2)} \left[\frac{k_{1-} k_{2-}'' q_{++} q_{--}}{k_{1+} k_{2+}''}\right]^{\frac{1}{2}},$$

$$M(-,+;-,+,-,+) = -\frac{2e^4 q_{-0} q_{+-} q_{-\perp}^*}{(p_+ \cdot k_1)(q_- \cdot k_2)} \left[\frac{k_{1-} k_{2-}'' q_{++}}{k_{1+} k_{2+}'' q_{-+}^3}\right]^{\frac{1}{2}}$$

$$\doteq \frac{2e^4 q_{-0} q_{+-}}{q_{-+}(p_+ \cdot k_1)(q_- \cdot k_2)} \left[\frac{k_{1-} k_{2-}'' q_{++} q_{--}}{k_{1+} k_{2+}''}\right]^{\frac{1}{2}},$$

$$M(-,+;-,-,+,+) = \frac{e^4 \mu k_{2\perp}'' q_{-\perp}^*}{2q_{-0}(p_+ \cdot k_1)(q_- \cdot k_2)} \left[\frac{k_{1-} k_{2+}'' q_{++}}{k_{1+} k_{2-}'' q_{-+}}\right]^{\frac{1}{2}}$$

$$\doteq \frac{e^4 \mu k_{2+}''}{2q_{-0}(p_+ \cdot k_1)(q_- \cdot k_2)} \left[\frac{k_{1-} q_{++} q_{--}}{k_{1+}}\right]^{\frac{1}{2}},$$

$$M(+,-;+,-,-,-) = -\frac{e^4(2q_{-0}+k_{2+}'') q_{-\perp}}{(p_+ \cdot k_1)(q_- \cdot k_2)} \left[\frac{k_{1-} k_{2-}'' q_{++}}{k_{1+} k_{2+}'' q_{-+}}\right]^{\frac{1}{2}}$$

$$\doteq \frac{e^4(2q_{-0}+k_{2+}'')}{(p_+ \cdot k_1)(q_- \cdot k_2)} \left[\frac{k_{1-} k_{2-}'' q_{++} q_{--}}{k_{1+} k_{2+}''}\right]^{\frac{1}{2}},$$

$$M(+,-;-,+,-,-) = -\frac{2e^4 q_{-0} q_{+\perp}}{(p_+ \cdot k_1)(q_- \cdot k_2)} \left[\frac{k_{1-} k_{2-}'' q_{-+}}{k_{1+} k_{2+}'' q_{++}}\right]^{\frac{1}{2}}$$

$$\doteq \frac{2e^4 q_{-0}}{(p_+ \cdot k_1)(q_- \cdot k_2)} \left[\frac{k_{1-} k_{2-}'' q_{+-} q_{-+}}{k_{1+} k_{2+}''}\right]^{\frac{1}{2}},$$

$$M(+,-;-,-,-,+) = \frac{e^4 \mu k_{2\perp}'' q_{+\perp}}{2q_{-0}(p_+ \cdot k_1)(q_- \cdot k_2)} \left[\frac{k_{1-} k_{2+}'' q_{-+}}{k_{1+} k_{2-}'' q_{++}}\right]^{\frac{1}{2}}$$

$$\doteq \frac{e^4 \mu k_{2+}''}{2q_{-0}(p_+ \cdot k_1)(q_- \cdot k_2)} \left[\frac{k_{1-} q_{+-} q_{-+}}{k_{1+}}\right]^{\frac{1}{2}},$$

$$M(-,+;+,-,-,-) = -\frac{2e^4 q_{-0} q_{+-} q_{+\perp}^*}{(p_+ \cdot k_1)(q_- \cdot k_2)} \left[\frac{k_{1-} k_{2-}''}{k_{1+} k_{2+}'' q_{++} q_{-+}}\right]^{\frac{1}{2}}$$

$$\doteq \frac{2e^4 q_{-0}}{(p_+ \cdot k_1)(q_- \cdot k_2)} \left[\frac{k_{1-} k_{2-}'' q_{+-}^3}{k_{1+} k_{2+}'' q_{-+}}\right]^{\frac{1}{2}},$$

$$M(-,+;-,+,-,-) = -\frac{4e^4 E q_{-0} q^*_{-\perp}}{(2E-k_{1+})(p_+ \cdot k_1)(q_- \cdot k_2)} \left[\frac{k_{1-} k''_{2-} q_{++}}{k_{1+} k''_{2+} q_{-+}}\right]^{\frac{1}{2}}$$

$$\doteq \frac{4e^4 E q_{-0}}{(2E-k_{1+})(p_+ \cdot k_1)(q_- \cdot k_2)} \left[\frac{k_{1-} k''_{2-} q_{++} q_{--}}{k_{1+} k''_{2+}}\right]^{\frac{1}{2}},$$

$$M(-,+;-,-,-,+) = \frac{e^4 \mu E k''_{2\perp} q^*_{-\perp}}{q_{-0}(2E-k_{1+})(p_+ \cdot k_1)(q_- \cdot k_2)} \left[\frac{k_{1-} k''_{2+} q_{++}}{k_{1+} k''_{2-} q_{-+}}\right]^{\frac{1}{2}}$$

$$\doteq \frac{e^4 \mu k''_{2+}}{q_{-0}(2E-k_{1+})(p_+ \cdot k_1)(q_- \cdot k_2)} \left[\frac{k_{1-} q_{++} q_{--}}{k_{1+}}\right]^{\frac{1}{2}},$$

$$M(-,-;+,+,-,-) = \frac{-e^4 m \mu k^*_{1\perp} k'''^*_{2\perp} q_{-\perp}}{4E q_{-0}(2E-k_{1+})(p_+ \cdot k_1)(q_- \cdot k_2)} \left[\frac{k_{1+} k''_{2+} q_{++}}{k_{1-} k''_{2-} q_{-+}}\right]^{\frac{1}{2}}$$

$$\doteq \frac{e^4 m \mu k_{1+} k''_{2+} [q_{++} q_{--}]^{\frac{1}{2}}}{4E q_{-0}(2E-k_{1+})(p_+ \cdot k_1)(q_- \cdot k_2)},$$

$$M(-,-;+,-,-,+) = -\frac{e^4 m q_{-0} k^*_{1\perp} q_{-\perp}}{E(2E-k_{1+})(p_+ \cdot k_1)(q_- \cdot k_2)} \left[\frac{k_{1+} k''_{2-} q_{++}}{k_{1-} k''_{2+} q_{-+}}\right]^{\frac{1}{2}}$$

$$\doteq \frac{e^4 m q_{-0} k_{1+}}{E(2E-k_{1+})(p_+ \cdot k_1)(q_- \cdot k_2)} \left[\frac{k''_{2-} q_{++} q_{--}}{k''_{2+}}\right]^{\frac{1}{2}},$$

$$M(-,-;-,+,-,+) = -\frac{e^4 m q_{-0} q_{+-} k^*_{1\perp} q_{+\perp}}{2E^2 (p_+ \cdot k_1)(q_- \cdot k_2)} \left[\frac{k_{1+} k''_{2-}}{k_{1-} k''_{2+} q_{++} q_{-+}}\right]^{\frac{1}{2}}$$

$$\doteq \frac{e^4 m q_{-0} k_{1+}}{2E^2 (p_+ \cdot k_1)(q_- \cdot k_2)} \left[\frac{k''_{2-} q^3_{+-}}{k''_{2+} q_{-+}}\right]^{\frac{1}{2}},$$

$$M(-,-;+,-,-,-) = -\frac{e^4 m q_{-0} q_{+-} k^*_{1\perp} q_{-\perp}}{2E^2 (p_+ \cdot k_1)(q_- \cdot k_2)} \left[\frac{k_{1+} k''_{2-} q_{++}}{k_{1-} k''_{2+} q^3_{-+}}\right]^{\frac{1}{2}}$$

$$\doteq \frac{e^4 m q_{-0} q_{+-} k_{1+}}{2E^2 q_{-+}(p_+ \cdot k_1)(q_- \cdot k_2)} \left[\frac{k''_{2-} q_{++} q_{--}}{k''_{2+}}\right]^{\frac{1}{2}},$$

$$M(-,-;-,+,-,-) = -\frac{e^4 m q_{-0} k^*_{1\perp} q_{+\perp}}{E(2E-k_{1+})(p_+ \cdot k_1)(q_- \cdot k_2)} \left[\frac{k_{1+} k''_{2-} q_{-+}}{k_{1-} k''_{2+} q_{++}}\right]^{\frac{1}{2}}$$

$$\doteq \frac{e^4 m q_{-0} k_{1+}}{E(2E-k_{1+})(p_+ \cdot k_1)(q_- \cdot k_2)} \left[\frac{k''_{2-} q_{+-} q_{-+}}{k''_{2+}}\right]^{\frac{1}{2}},$$

9.18. $e^+e^- \to \mu^+\mu^-\gamma\gamma$ (NO Z-EXCHANGE; $m \neq 0$)

$$M(-,-;-,-,-,+) = \frac{e^4 m\mu k_{1\perp}^* k_{2\perp}'' q_{+\perp}}{4Eq_{-0}(2E-k_{1+})(p_+ \cdot k_1)(q_- \cdot k_2)} \left[\frac{k_{1+}k_{2+}''q_{-+}}{k_{1-}k_{2-}''q_{++}}\right]^{\frac{1}{2}}$$

$$\doteq \frac{e^4 m\mu k_{1+}k_{2+}''[q_{+-}q_{-+}]^{\frac{1}{2}}}{4Eq_{-0}(2E-k_{1+})(p_+ \cdot k_1)(q_- \cdot k_2)}. \quad (9.176)$$

Unpolarized squared matrix element:

$$\overline{|M|^2} = \frac{e^8 k_{1-}k_{2-}''}{2k_{1+}k_{2+}''[(2E-k_{1+})(p_+ \cdot k_1)(q_- \cdot k_2)]^2}$$

$$\times \left[4E^2 + (2E-k_{1+})^2 + \frac{m^2 k_{1+}^3}{4E^2 k_{1-}}\right]$$

$$\times \left[4q_{-0}^2 + (2q_{-0}+k_{2+}'')^2 + \frac{\mu^2 k_{2+}''^3}{4q_{-0}^2 k_{2-}''}\right](q_{++}q_{--}+q_{+-}q_{-+}). \quad (9.177)$$

Unpolarized cross section:

$$d\sigma = \frac{\alpha^4 k_{1-}k_{2-}''}{256\pi^2 k_{1+}k_{2+}''[E(2E-k_{1+})(p_+ \cdot k_1)(q_- \cdot k_2)]^2}$$

$$\times \left[4E^2 + (2E-k_{1+})^2 + \frac{m^2 k_{1+}^3}{4E^2 k_{1-}}\right]$$

$$\times \left[4q_{-0}^2 + (2q_{-0}+k_{2+}'')^2 + \frac{\mu^2 k_{2+}''^3}{4q_{-0}^2 k_{2-}''}\right](q_{++}q_{--}+q_{+-}q_{-+})$$

$$\times \delta^4(p_+ + p_- - q_+ - q_- - k_1 - k_2)\frac{d^3\vec{q}_+ d^3\vec{q}_- d^3\vec{k}_1 d^3\vec{k}_2}{q_{+0}\,q_{-0}\,k_{10}\,k_{20}}. \quad (9.178)$$

9.18.8 $\vec{k}_1$ and $\vec{k}_2$ nearly parallel to $\vec{p}_-$ and $\vec{q}_+$, resp.

The primed quantities $k_{2\pm}'$ and $k_{2\perp}'$ are evaluated in the rotated frame where $\vec{q}_+$ determines the positive z-axis (see also Section 7.4.3). The four-vector q in eqn (9.179) is obtained by applying a space reflection to q_+. The quantity μ denotes the muon mass. Photon polarizations:

$$\not{e}_1^{\pm} = N_1[\not{k}_1\not{p}_-\not{p}_+(1\pm\gamma_5) + \not{p}_+\not{p}_-\not{k}_1(1\mp\gamma_5)],$$

$$N_1^{-1} = E^2[32k_{1+}k_{1-}]^{\frac{1}{2}},$$

$$\not{e}_2^{\pm} = N_2[\not{k}_2\not{q}_+\not{q}(1\pm\gamma_5) + \not{q}\not{q}_+\not{k}_2(1\mp\gamma_5)],$$

$$N_2^{-1} = q_{+0}^2[32k_{2+}'k_{2-}']^{\frac{1}{2}}. \quad (9.179)$$

Nonvanishing helicity amplitudes:

$$M(+,+;+,+,+,-) = -\frac{e^4 m\mu k_{1\perp}^* k_{2\perp}'^* q_{+\perp}}{4Eq_{+0}(2E-k_{1-})(p_-\cdot k_1)(q_+\cdot k_2)} \left[\frac{k_{1-}k_{2+}'q_{-+}}{k_{1+}k_{2-}'q_{++}}\right]^{\frac{1}{2}}$$

$$\doteq \frac{e^4 m\mu k_{1-}k_{2+}'[q_{+-}q_{-+}]^{\frac{1}{2}}}{4Eq_{+0}(2E-k_{1-})(p_-\cdot k_1)(q_+\cdot k_2)},$$

$$M(+,+;+,-,+,+) = -\frac{e^4 mq_{+0}k_{1\perp}^* q_{-\perp}}{2E^2 q_{+-}(p_-\cdot k_1)(q_+\cdot k_2)} \left[\frac{k_{1-}k_{2+}'q_{++}q_{-+}}{k_{1+}k_{2+}'}\right]^{\frac{1}{2}}$$

$$\doteq \frac{e^4 mq_{+0}q_{-+}k_{1-}}{2E^2 q_{+-}(p_-\cdot k_1)(q_+\cdot k_2)} \left[\frac{k_{2-}'q_{++}q_{--}}{k_{2+}'}\right]^{\frac{1}{2}},$$

$$M(+,+;-,+,+,+) = -\frac{e^4 mq_{+0}k_{1\perp}^* q_{+\perp}}{E(2E-k_{1-})(p_-\cdot k_1)(q_+\cdot k_2)} \left[\frac{k_{1-}k_{2-}'q_{-+}}{k_{1+}k_{2+}'q_{++}}\right]^{\frac{1}{2}}$$

$$\doteq \frac{e^4 mq_{+0}k_{1-}}{E(2E-k_{1-})(p_-\cdot k_1)(q_+\cdot k_2)} \left[\frac{k_{2-}'q_{+-}q_{-+}}{k_{2+}'}\right]^{\frac{1}{2}},$$

$$M(+,+;+,-,+,-) = -\frac{e^4 mq_{+0}k_{1\perp}^* q_{-\perp}}{E(2E-k_{1-})(p_-\cdot k_1)(q_+\cdot k_2)} \left[\frac{k_{1-}k_{2-}'q_{++}}{k_{1+}k_{2+}'q_{-+}}\right]^{\frac{1}{2}}$$

$$\doteq \frac{e^4 mq_{+0}k_{1-}}{E(2E-k_{1-})(p_-\cdot k_1)(q_+\cdot k_2)} \left[\frac{k_{2-}'q_{++}q_{--}}{k_{2+}'}\right]^{\frac{1}{2}},$$

$$M(+,+;-,+,+,-) = -\frac{e^4 mq_{+0}k_{1\perp}^* q_{+\perp}}{2E^2 q_{+-}(p_-\cdot k_1)(q_+\cdot k_2)} \left[\frac{k_{1-}k_{2-}'q_{-+}^3}{k_{1+}k_{2+}'q_{++}}\right]^{\frac{1}{2}}$$

$$\doteq \frac{e^4 mq_{+0}k_{1-}}{2E^2 (p_-\cdot k_1)(q_+\cdot k_2)} \left[\frac{k_{2-}'q_{-+}^3}{k_{2+}'q_{+-}}\right]^{\frac{1}{2}},$$

$$M(+,+;-,-,+,+) = \frac{e^4 m\mu k_{1\perp}^* k_{2\perp}' q_{-\perp}}{4Eq_{+0}(2E-k_{1-})(p_-\cdot k_1)(q_+\cdot k_2)} \left[\frac{k_{1-}k_{2+}'q_{++}}{k_{1+}k_{2-}'q_{-+}}\right]^{\frac{1}{2}}$$

$$\doteq \frac{e^4 m\mu k_{1-}k_{2+}'[q_{++}q_{--}]^{\frac{1}{2}}}{4Eq_{+0}(2E-k_{1-})(p_-\cdot k_1)(q_+\cdot k_2)},$$

$$M(+,-;+,+,+,-) = -\frac{e^4 \mu k_{2\perp}'^* q_{+\perp}}{2q_{+0}(p_-\cdot k_1)(q_+\cdot k_2)} \left[\frac{k_{1+}k_{2+}'q_{-+}}{k_{1-}k_{2-}'q_{++}}\right]^{\frac{1}{2}}$$

$$\doteq \frac{e^4 \mu k_{2+}'}{2q_{+0}(p_-\cdot k_1)(q_+\cdot k_2)} \left[\frac{k_{1+}q_{+-}q_{-+}}{k_{1-}}\right]^{\frac{1}{2}},$$

9.18. $e^+e^- \to \mu^+\mu^-\gamma\gamma$ (NO Z-EXCHANGE; $m \neq 0$)

$$M(+,-;+,-,+,+) = -\frac{e^4(2q_{+0}+k'_{2+})q_{-\perp}}{(p_-\cdot k_1)(q_+\cdot k_2)}\left[\frac{k_{1+}k'_{2-}q_{++}}{k_{1-}k'_{2+}q_{-+}}\right]^{\frac{1}{2}}$$

$$\doteq \frac{e^4(2q_{+0}+k'_{2+})}{(p_-\cdot k_1)(q_+\cdot k_2)}\left[\frac{k_{1+}k'_{2-}q_{++}q_{--}}{k_{1-}k'_{2+}}\right]^{\frac{1}{2}},$$

$$M(+,-;-,+,+,+) = -\frac{2e^4 q_{+0}q_{+\perp}}{(p_-\cdot k_1)(q_+\cdot k_2)}\left[\frac{k_{1+}k'_{2-}q_{-+}}{k_{1-}k'_{2+}q_{++}}\right]^{\frac{1}{2}}$$

$$\doteq \frac{2e^4 q_{+0}}{(p_-\cdot k_1)(q_+\cdot k_2)}\left[\frac{k_{1+}k'_{2-}q_{+-}q_{-+}}{k_{1-}k'_{2+}}\right]^{\frac{1}{2}},$$

$$M(-,+;+,+,+,-) = -\frac{e^4\mu E k'^*_{2\perp}q^*_{-\perp}}{q_{+0}(2E-k_{1-})(p_-\cdot k_1)(q_+\cdot k_2)}\left[\frac{k_{1+}k'_{2+}q_{++}}{k_{1-}k'_{2-}q_{-+}}\right]^{\frac{1}{2}}$$

$$\doteq \frac{e^4\mu E k'_{2+}}{q_{+0}(2E-k_{1-})(p_-\cdot k_1)(q_+\cdot k_2)}\left[\frac{k_{1+}q_{++}q_{--}}{k_{1-}}\right]^{\frac{1}{2}},$$

$$M(-,+;+,-,+,+) = -\frac{2e^4 q_{+0}q^*_{+\perp}}{q_{+-}(p_-\cdot k_1)(q_+\cdot k_2)}\left[\frac{k_{1+}k'_{2-}q^3_{-+}}{k_{1-}k'_{2+}q_{++}}\right]^{\frac{1}{2}}$$

$$\doteq \frac{2e^4 q_{+0}}{(p_-\cdot k_1)(q_+\cdot k_2)}\left[\frac{k_{1+}k'_{2-}q^3_{-+}}{k_{1-}k'_{2+}q_{+-}}\right]^{\frac{1}{2}},$$

$$M(-,+;-,+,+,+) = -\frac{4e^4 E q_{+0}q^*_{-\perp}}{(2E-k_{1-})(p_-\cdot k_1)(q_+\cdot k_2)}\left[\frac{k_{1+}k'_{2-}q_{++}}{k_{1-}k'_{2+}q_{-+}}\right]^{\frac{1}{2}}$$

$$\doteq \frac{4e^4 E q_{+0}}{(2E-k_{1-})(p_-\cdot k_1)(q_+\cdot k_2)}\left[\frac{k_{1+}k'_{2-}q_{++}q_{--}}{k_{1-}k'_{2+}}\right]^{\frac{1}{2}},$$

$$M(+,-;+,+,-,-) = -\frac{e^4\mu E k'^*_{2\perp}q_{+\perp}}{q_{+0}(2E-k_{1-})(p_-\cdot k_1)(q_+\cdot k_2)}\left[\frac{k_{1+}k'_{2+}q_{-+}}{k_{1-}k'_{2-}q_{++}}\right]^{\frac{1}{2}}$$

$$\doteq \frac{e^4\mu E k'_{2+}}{q_{+0}(2E-k_{1-})(p_-\cdot k_1)(q_+\cdot k_2)}\left[\frac{k_{1+}q_{+-}q_{-+}}{k_{1-}}\right]^{\frac{1}{2}},$$

$$M(+,-;+,-,+,-) = -\frac{2e^4 q_{+0}q_{-\perp}}{(p_-\cdot k_1)(q_+\cdot k_2)}\left[\frac{k_{1+}k'_{2-}q_{++}}{k_{1-}k'_{2+}q_{-+}}\right]^{\frac{1}{2}}$$

$$\doteq \frac{2e^4 q_{+0}}{(p_-\cdot k_1)(q_+\cdot k_2)}\left[\frac{k_{1+}k'_{2-}q_{++}q_{--}}{k_{1-}k'_{2+}}\right]^{\frac{1}{2}},$$

$$M(+,-;+,-,-,+) = -\frac{2e^4 q_{+0} q_{-\perp}}{q_{+-}(p_- \cdot k_1)(q_+ \cdot k_2)} \left[\frac{k_{1+} k'_{2-} q_{++} q_{-+}}{k_{1-} k'_{2+}}\right]^{\frac{1}{2}}$$

$$\doteq \frac{2e^4 q_{+0} q_{-+}}{q_{+-}(p_- \cdot k_1)(q_+ \cdot k_2)} \left[\frac{k_{1+} k'_{2-} q_{++} q_{--}}{k_{1-} k'_{2+}}\right]^{\frac{1}{2}},$$

$$M(+,-;-,+,+,-) = -\frac{e^4(2q_{+0} + k'_{2+}) q_{+\perp}}{(p_- \cdot k_1)(q_+ \cdot k_2)} \left[\frac{k_{1+} k'_{2-} q_{-+}}{k_{1-} k'_{2+} q_{++}}\right]^{\frac{1}{2}}$$

$$\doteq \frac{e^4(2q_{+0} + k'_{2+})}{(p_- \cdot k_1)(q_+ \cdot k_2)} \left[\frac{k_{1+} k'_{2-} q_{+-} q_{-+}}{k_{1-} k'_{2+}}\right]^{\frac{1}{2}},$$

$$M(+,-;-,+,-,+) = -\frac{4e^4 E q_{+0} q_{+\perp}}{(2E - k_{1-})(p_- \cdot k_1)(q_+ \cdot k_2)} \left[\frac{k_{1+} k'_{2-} q_{-+}}{k_{1-} k'_{2+} q_{++}}\right]^{\frac{1}{2}}$$

$$\doteq \frac{4e^4 E q_{+0}}{(2E - k_{1-})(p_- \cdot k_1)(q_+ \cdot k_2)} \left[\frac{k_{1+} k'_{2-} q_{+-} q_{-+}}{k_{1-} k'_{2+}}\right]^{\frac{1}{2}}$$

$$M(+,-;-,-,+,+) = \frac{e^4 \mu k'_{2\perp} q_{-\perp}}{2q_{+0}(p_- \cdot k_1)(q_+ \cdot k_2)} \left[\frac{k_{1+} k'_{2+} q_{++}}{k_{1-} k'_{2-} q_{-+}}\right]^{\frac{1}{2}}$$

$$\doteq \frac{e^4 \mu k'_{2+}}{2q_{+0}(p_- \cdot k_1)(q_+ \cdot k_2)} \left[\frac{k_{1+} q_{++} q_{--}}{k_{1-}}\right]^{\frac{1}{2}},$$

$$M(-,+;+,+,-,-) = -\frac{e^4 \mu k'^{*}_{2\perp} q^{*}_{-\perp}}{2q_{+0}(p_- \cdot k_1)(q_+ \cdot k_2)} \left[\frac{k_{1+} k'_{2+} q_{++}}{k_{1-} k'_{2-} q_{-+}}\right]^{\frac{1}{2}}$$

$$\doteq \frac{e^4 \mu k'_{2+}}{2q_{+0}(p_- \cdot k_1)(q_+ \cdot k_2)} \left[\frac{k_{1+} q_{++} q_{--}}{k_{1-}}\right]^{\frac{1}{2}},$$

$$M(-,+;+,-,+,-) = -\frac{4e^4 E q_{+0} q^{*}_{+\perp}}{(2E - k_{1-})(p_- \cdot k_1)(q_+ \cdot k_2)} \left[\frac{k_{1+} k'_{2-} q_{-+}}{k_{1-} k'_{2+} q_{++}}\right]^{\frac{1}{2}}$$

$$\doteq \frac{4e^4 E q_{+0}}{(2E - k_{1-})(p_- \cdot k_1)(q_+ \cdot k_2)} \left[\frac{k_{1+} k'_{2-} q_{+-} q_{-+}}{k_{1-} k'_{2+}}\right]^{\frac{1}{2}}$$

$$M(-,+;+,-,-,+) = -\frac{e^4(2q_{+0} + k'_{2+}) q^{*}_{+\perp}}{(p_- \cdot k_1)(q_+ \cdot k_2)} \left[\frac{k_{1+} k'_{2-} q_{-+}}{k_{1-} k'_{2+} q_{++}}\right]^{\frac{1}{2}}$$

$$\doteq \frac{e^4(2q_{+0} + k'_{2+})}{(p_- \cdot k_1)(q_+ \cdot k_2)} \left[\frac{k_{1+} k'_{2-} q_{+-} q_{-+}}{k_{1-} k'_{2+}}\right]^{\frac{1}{2}},$$

9.18. $e^+e^- \to \mu^+\mu^-\gamma\gamma$ (NO Z-EXCHANGE; $m \neq 0$)

$$M(-,+;-,+,+,-) = -\frac{2e^4 q_{+0} q_{-\perp}^*}{q_{+-}(p_- \cdot k_1)(q_+ \cdot k_2)} \left[\frac{k_{1+}k'_{2-}q_{++}q_{-+}}{k_{1-}k'_{2+}}\right]^{\frac{1}{2}}$$

$$\doteq \frac{2e^4 q_{+0} q_{-+}}{q_{+-}(p_- \cdot k_1)(q_+ \cdot k_2)} \left[\frac{k_{1+}k'_{2-}q_{++}q_{--}}{k_{1-}k'_{2+}}\right]^{\frac{1}{2}},$$

$$M(-,+;-,+,-,+) = -\frac{2e^4 q_{+0} q_{-\perp}^*}{(p_- \cdot k_1)(q_+ \cdot k_2)} \left[\frac{k_{1+}k'_{2-}q_{++}}{k_{1-}k'_{2+}q_{-+}}\right]^{\frac{1}{2}}$$

$$\doteq \frac{2e^4 q_{+0}}{(p_- \cdot k_1)(q_+ \cdot k_2)} \left[\frac{k_{1+}k'_{2-}q_{++}q_{--}}{k_{1-}k'_{2+}}\right]^{\frac{1}{2}},$$

$$M(-,+;-,-,+,+) = \frac{e^4 \mu E k'_{2\perp} q_{+\perp}^*}{q_{+0}(2E - k_{1-})(p_- \cdot k_1)(q_+ \cdot k_2)} \left[\frac{k_{1+}k'_{2+}q_{-+}}{k_{1-}k'_{2-}q_{++}}\right]^{\frac{1}{2}}$$

$$\doteq \frac{e^4 \mu E k'_{2+}}{q_{+0}(2E - k_{1-})(p_- \cdot k_1)(q_+ \cdot k_2)} \left[\frac{k_{1+}q_{+-}q_{-+}}{k_{1-}}\right]^{\frac{1}{2}},$$

$$M(+,-;+,-,-,-) = -\frac{4e^4 E q_{+0} q_{-\perp}}{(2E - k_{1-})(p_- \cdot k_1)(q_+ \cdot k_2)} \left[\frac{k_{1+}k'_{2-}q_{++}}{k_{1-}k'_{2+}q_{-+}}\right]^{\frac{1}{2}}$$

$$\doteq \frac{4e^4 E q_{+0}}{(2E - k_{1-})(p_- \cdot k_1)(q_+ \cdot k_2)} \left[\frac{k_{1+}k'_{2-}q_{++}q_{--}}{k_{1-}k'_{2+}}\right]^{\frac{1}{2}},$$

$$M(+,-;-,+,-,-) = -\frac{2e^4 q_{+0} q_{+\perp}}{q_{+-}(p_- \cdot k_1)(q_+ \cdot k_2)} \left[\frac{k_{1+}k'_{2-}q_{-+}^3}{k_{1-}k'_{2+}q_{++}}\right]^{\frac{1}{2}}$$

$$\doteq \frac{2e^4 q_{+0}}{(p_- \cdot k_1)(q_+ \cdot k_2)} \left[\frac{k_{1+}k'_{2-}q_{-+}^3}{k_{1-}k'_{2+}q_{+-}}\right]^{\frac{1}{2}},$$

$$M(+,-;-,-,-,+) = \frac{e^4 \mu E k'_{2\perp} q_{-\perp}}{q_{+0}(2E - k_{1-})(p_- \cdot k_1)(q_+ \cdot k_2)} \left[\frac{k_{1+}k'_{2+}q_{++}}{k_{1-}k'_{2-}q_{-+}}\right]^{\frac{1}{2}}$$

$$\doteq \frac{e^4 \mu E k'_{2+}}{q_{+0}(2E - k_{1-})(p_- \cdot k_1)(q_+ \cdot k_2)} \left[\frac{k_{1+}q_{++}q_{--}}{k_{1-}}\right]^{\frac{1}{2}},$$

$$M(-,+;+,-,-,-) = -\frac{2e^4 q_{+0} q_{+\perp}^*}{(p_- \cdot k_1)(q_+ \cdot k_2)} \left[\frac{k_{1+}k'_{2-}q_{-+}}{k_{1-}k'_{2+}q_{++}}\right]^{\frac{1}{2}}$$

$$\doteq \frac{2e^4 q_{+0}}{(p_- \cdot k_1)(q_+ \cdot k_2)} \left[\frac{k_{1+}k'_{2-}q_{+-}q_{-+}}{k_{1-}k'_{2+}}\right]^{\frac{1}{2}},$$

$$M(-,+;-,+,-,-) = -\frac{e^4(2q_{+0} + k'_{2+})q^*_{-\perp}}{(p_- \cdot k_1)(q_+ \cdot k_2)}\left[\frac{k_{1+}k'_{2-}q_{++}}{k_{1-}k'_{2+}q_{-+}}\right]^{\frac{1}{2}}$$

$$\doteq \frac{e^4(2q_{+0} + k'_{2+})}{(p_- \cdot k_1)(q_+ \cdot k_2)}\left[\frac{k_{1+}k'_{2-}q_{++}q_{--}}{k_{1-}k'_{2+}}\right]^{\frac{1}{2}},$$

$$M(-,+;-,-,-,+) = \frac{e^4\mu k'_{2\perp}q^*_{+\perp}}{2q_{+0}(p_- \cdot k_1)(q_+ \cdot k_2)}\left[\frac{k_{1+}k'_{2+}q_{-+}}{k_{1-}k'_{2-}q_{++}}\right]^{\frac{1}{2}}$$

$$\doteq \frac{e^4\mu k'_{2+}}{2q_{+0}(p_- \cdot k_1)(q_+ \cdot k_2)}\left[\frac{k_{1+}q_{+-}q_{-+}}{k_{1-}}\right]^{\frac{1}{2}},$$

$$M(-,-;+,+,-,-) = -\frac{e^4 m\mu k_{1\perp}k'^*_{2\perp}q^*_{-\perp}}{4Eq_{+0}(2E-k_{1-})(p_- \cdot k_1)(q_+ \cdot k_2)}\left[\frac{k_{1-}k'_{2+}q_{++}}{k_{1+}k'_{2-}q_{-+}}\right]^{\frac{1}{2}}$$

$$\doteq \frac{e^4 m\mu k_{1-}k'_{2+}[q_{++}q_{--}]^{\frac{1}{2}}}{4Eq_{+0}(2E-k_{1-})(p_- \cdot k_1)(q_+ \cdot k_2)},$$

$$M(-,-;+,-,-,+) = \frac{e^4 mq_{+0}k_{1\perp}q^*_{+\perp}}{2E^2 q_{+-}(p_- \cdot k_1)(q_+ \cdot k_2)}\left[\frac{k_{1-}k'_{2-}q^3_{-+}}{k_{1+}k'_{2+}q_{++}}\right]^{\frac{1}{2}}$$

$$\doteq \frac{e^4 mq_{+0}k_{1-}}{2E^2(p_- \cdot k_1)(q_+ \cdot k_2)}\left[\frac{k'_{2-}q^3_{-+}}{k'_{2+}q_{+-}}\right]^{\frac{1}{2}},$$

$$M(-,-;-,+,-,+) = \frac{e^4 mq_{+0}k_{1\perp}q^*_{-\perp}}{E(2E-k_{1-})(p_- \cdot k_1)(q_+ \cdot k_2)}\left[\frac{k_{1-}k'_{2-}q_{++}}{k_{1+}k'_{2+}q_{-+}}\right]^{\frac{1}{2}}$$

$$\doteq \frac{e^4 mq_{+0}k_{1-}}{E(2E-k_{1-})(p_- \cdot k_1)(q_+ \cdot k_2)}\left[\frac{k'_{2-}q_{++}q_{--}}{k'_{2+}}\right]^{\frac{1}{2}},$$

$$M(-,-;+,-,-,-) = \frac{e^4 mq_{+0}k_{1\perp}q^*_{+\perp}}{E(2E-k_{1-})(p_- \cdot k_1)(q_+ \cdot k_2)}\left[\frac{k_{1-}k'_{2-}q_{-+}}{k_{1+}k'_{2+}q_{++}}\right]^{\frac{1}{2}}$$

$$\doteq \frac{e^4 mq_{+0}k_{1-}}{E(2E-k_{1-})(p_- \cdot k_1)(q_+ \cdot k_2)}\left[\frac{k'_{2-}q_{+-}q_{-+}}{k'_{2+}}\right]^{\frac{1}{2}},$$

$$M(-,-;-,+,-,-) = \frac{e^4 mq_{+0}k_{1\perp}q^*_{-\perp}}{2E^2 q_{+-}(p_- \cdot k_1)(q_+ \cdot k_2)}\left[\frac{k_{1-}k'_{2-}q_{++}q_{-+}}{k_{1+}k'_{2+}}\right]^{\frac{1}{2}}$$

$$\doteq \frac{e^4 mq_{+0}q_{-+}k_{1-}}{2E^2 q_{+-}(p_- \cdot k_1)(q_+ \cdot k_2)}\left[\frac{k'_{2-}q_{++}q_{--}}{k'_{2+}}\right]^{\frac{1}{2}},$$

9.18. $e^+e^- \to \mu^+\mu^-\gamma\gamma$ (NO Z-EXCHANGE; $m \neq 0$)

$$M(-,-;-,-,-,+) = -\frac{e^4 m\mu k_{1\perp} k'_{2\perp} q^*_{+\perp}}{4Eq_{+0}(2E-k_{1-})(p_-\cdot k_1)(q_+\cdot k_2)} \left[\frac{k_{1-}k'_{2+}q_{-+}}{k_{1+}k'_{2-}q_{++}}\right]^{\frac{1}{2}}$$

$$\doteq \frac{e^4 m\mu k_{1-} k'_{2+} [q_{+-}q_{-+}]^{\frac{1}{2}}}{4Eq_{+0}(2E-k_{1-})(p_-\cdot k_1)(q_+\cdot k_2)}. \qquad (9.180)$$

Unpolarized squared matrix element:

$$\overline{|M|^2} = \frac{e^8 k_{1+} k'_{2-}}{2k_{1-} k'_{2+}[(2E-k_{1-})(p_-\cdot k_1)(q_+\cdot k_2)]^2}$$

$$\times \left[4E^2 + (2E-k_{1-})^2 + \frac{m^2 k_{1-}^3}{4E^2 k_{1+}}\right]$$

$$\times \left[4q_{+0}^2 + (2q_{+0}+k'_{2+})^2 + \frac{\mu^2 k'^3_{2+}}{4q_{+0}^2 k'_{2-}}\right](q_{++}q_{--}+q_{+-}q_{-+}). \qquad (9.181)$$

Unpolarized cross section:

$$d\sigma = \frac{\alpha^4 k_{1+} k'_{2-}}{256\pi^2 k_{1-} k'_{2+}[E(2E-k_{1-})(p_-\cdot k_1)(q_+\cdot k_2)]^2}$$

$$\times \left[4E^2 + (2E-k_{1-})^2 + \frac{m^2 k_{1-}^3}{4E^2 k_{1+}}\right]$$

$$\times \left[4q_{+0}^2 + (2q_{+0}+k'_{2+})^2 + \frac{\mu^2 k'^3_{2+}}{4q_{+0}^2 k'_{2-}}\right](q_{++}q_{--}+q_{+-}q_{-+})$$

$$\times \delta^4(p_+ + p_- - q_+ - q_- - k_1 - k_2) \frac{d^3\vec{q}_+ \, d^3\vec{q}_- \, d^3\vec{k}_1 \, d^3\vec{k}_2}{q_{+0} \, q_{-0} \, k_{10} \, k_{20}}. \qquad (9.182)$$

9.18.9 $\vec{k}_1$ and $\vec{k}_2$ nearly parallel to $\vec{p}_-$ and $\vec{q}_-$, resp.

The doubly primed quantities $k''_{2\pm}$ and $k''_{2\perp}$ are evaluated in the rotated frame where $\vec{q}_-$ determines the positive z-axis (see also Section 7.4.3). The four-vector q in eqn (9.183) is obtained by applying a space reflection to q_-. The quantity μ denotes the muon mass.

Photon polarizations:

$$\not{\epsilon}_1^{\pm} = N_1[\not{k}_1 \not{p}_- \not{p}_+(1\pm\gamma_5) + \not{p}_+ \not{p}_- \not{k}_1(1\mp\gamma_5)],$$

$$N_1^{-1} = E^2[32k_{1+}k_{1-}]^{\frac{1}{2}},$$

$$\not{\epsilon}_2^{\pm} = N_2[\not{k}_2 \not{q}_- \not{q}(1\pm\gamma_5) + \not{q} \not{q}_- \not{k}_2(1\mp\gamma_5)],$$

$$N_2^{-1} = q_{-0}^2[32k''_{2+}k''_{2-}]^{\frac{1}{2}}. \qquad (9.183)$$

Nonvanishing helicity amplitudes:

$$M(+,+;+,+,+,-) = \frac{e^4 m\mu k_{1\perp}^* k_{21}''^* q_{-\perp}}{4Eq_{-0}(2E-k_{1-})(p_-\cdot k_1)(q_-\cdot k_2)}\left[\frac{k_{1-}k_{2+}''q_{++}}{k_{1+}k_{2-}''q_{-+}}\right]^{\frac{1}{2}}$$

$$\doteq \frac{e^4 m\mu k_{1-}k_{2+}''[q_{++}q_{--}]^{\frac{1}{2}}}{4Eq_{-0}(2E-k_{1-})(p_-\cdot k_1)(q_-\cdot k_2)},$$

$$M(+,+;+,-,+,+) = \frac{e^4 mq_{-0}k_{1\perp}^* q_{-\perp}}{E(2E-k_{1-})(p_-\cdot k_1)(q_-\cdot k_2)}\left[\frac{k_{1-}k_{2-}''q_{++}}{k_{1+}k_{2+}''q_{-+}}\right]^{\frac{1}{2}}$$

$$\doteq \frac{e^4 mq_{-0}k_{1-}}{E(2E-k_{1-})(p_-\cdot k_1)(q_-\cdot k_2)}\left[\frac{k_{2-}''q_{++}q_{--}}{k_{2+}''}\right]^{\frac{1}{2}},$$

$$M(+,+;-,+,+,+) = \frac{e^4 mq_{-0}k_{1\perp}^* q_{+\perp}}{2E^2 q_{--}(p_-\cdot k_1)(q_-\cdot k_2)}\left[\frac{k_{1-}k_{2-}''q_{++}q_{-+}}{k_{1+}k_{2+}''}\right]^{\frac{1}{2}}$$

$$\doteq \frac{e^4 mq_{-0}q_{++}k_{1-}}{2E^2 q_{--}(p_-\cdot k_1)(q_-\cdot k_2)}\left[\frac{k_{2-}''q_{+-}q_{-+}}{k_{2+}''}\right]^{\frac{1}{2}},$$

$$M(+,+;+,-,+,-) = \frac{e^4 mq_{-0}k_{1\perp}^* q_{-\perp}}{2E^2 q_{--}(p_-\cdot k_1)(q_-\cdot k_2)}\left[\frac{k_{1-}k_{2-}''q_{++}^3}{k_{1+}k_{2+}''q_{-+}}\right]^{\frac{1}{2}}$$

$$\doteq \frac{e^4 mq_{-0}k_{1-}}{2E^2 (p_-\cdot k_1)(q_-\cdot k_2)}\left[\frac{k_{2-}''q_{++}^3}{k_{2+}''q_{--}}\right]^{\frac{1}{2}},$$

$$M(+,+;-,+,+,-) = \frac{e^4 mq_{-0}k_{1\perp}^* q_{+\perp}}{E(2E-k_{1-})(p_-\cdot k_1)(q_-\cdot k_2)}\left[\frac{k_{1-}k_{2-}''q_{-+}}{k_{1+}k_{2+}''q_{++}}\right]^{\frac{1}{2}}$$

$$\doteq \frac{e^4 mq_{-0}k_{1-}}{E(2E-k_{1-})(p_-\cdot k_1)(q_-\cdot k_2)}\left[\frac{k_{2-}''q_{+-}q_{-+}}{k_{2+}''}\right]^{\frac{1}{2}},$$

$$M(+,+;-,-,+,+) = -\frac{e^4 m\mu k_{1\perp}^* k_{21}'' q_{+\perp}}{4Eq_{-0}(2E-k_{1-})(p_-\cdot k_1)(q_-\cdot k_2)}\left[\frac{k_{1-}k_{2+}''q_{-+}}{k_{1+}k_{2-}''q_{++}}\right]$$

$$\doteq \frac{e^4 m\mu k_{1-}k_{2+}''[q_{+-}q_{-+}]^{\frac{1}{2}}}{4Eq_{-0}(2E-k_{1-})(p_-\cdot k_1)(q_-\cdot k_2)},$$

$$M(+,-;+,+,+,-) = \frac{e^4 \mu k_{21}''^* q_{-\perp}}{2q_{-0}(p_-\cdot k_1)(q_-\cdot k_2)}\left[\frac{k_{1+}k_{2+}''q_{++}}{k_{1-}k_{2-}''q_{-+}}\right]^{\frac{1}{2}}$$

$$\doteq \frac{e^4 \mu k_{2+}''}{2q_{-0}(p_-\cdot k_1)(q_-\cdot k_2)}\left[\frac{k_{1+}q_{++}q_{--}}{k_{1-}}\right]^{\frac{1}{2}},$$

9.18. $e^+e^- \to \mu^+\mu^-\gamma\gamma$ (NO Z-EXCHANGE; $m \neq 0$)

$$M(+,-;+,-,+,+) = \frac{2e^4 q_{-0} q_{-\perp}}{(p_- \cdot k_1)(q_- \cdot k_2)} \left[\frac{k_{1+}k''_{2-}q_{++}}{k_{1-}k''_{2+}q_{-+}}\right]^{\frac{1}{2}}$$

$$\doteq \frac{2e^4 q_{-0}}{(p_- \cdot k_1)(q_- \cdot k_2)} \left[\frac{k_{1+}k''_{2-}q_{++}q_{--}}{k_{1-}k''_{2+}}\right]^{\frac{1}{2}},$$

$$M(+,-;-,+,+,+) = \frac{e^4(2q_{-0}+k''_{2+})q_{+\perp}}{(p_- \cdot k_1)(q_- \cdot k_2)} \left[\frac{k_{1+}k''_{2-}q_{-+}}{k_{1-}k''_{2+}q_{++}}\right]^{\frac{1}{2}}$$

$$\doteq \frac{e^4(2q_{-0}+k''_{2+})}{(p_- \cdot k_1)(q_- \cdot k_2)} \left[\frac{k_{1+}k''_{2-}q_{+-}q_{-+}}{k_{1-}k''_{2+}}\right]^{\frac{1}{2}},$$

$$M(-,+;+,+,+,-) = \frac{e^4 \mu E k''^*_{21} q^*_{+\perp}}{q_{-0}(2E-k_{1-})(p_- \cdot k_1)(q_- \cdot k_2)} \left[\frac{k_{1+}k''_{2+}q_{-+}}{k_{1-}k''_{2-}q_{++}}\right]^{\frac{1}{2}}$$

$$\doteq \frac{e^4 \mu E k''_{2+}}{q_{-0}(2E-k_{1-})(p_- \cdot k_1)(q_- \cdot k_2)} \left[\frac{k_{1+}q_{+-}q_{-+}}{k_{1-}}\right]^{\frac{1}{2}},$$

$$M(-,+;+,-,+,+) = \frac{4e^4 E q_{-0} q^*_{+\perp}}{(2E-k_{1-})(p_- \cdot k_1)(q_- \cdot k_2)} \left[\frac{k_{1+}k''_{2-}q_{-+}}{k_{1-}k''_{2+}q_{++}}\right]^{\frac{1}{2}}$$

$$\doteq \frac{4e^4 E q_{-0}}{(2E-k_{1-})(p_- \cdot k_1)(q_- \cdot k_2)} \left[\frac{k_{1+}k''_{2-}q_{+-}q_{-+}}{k_{1-}k''_{2+}}\right]^{\frac{1}{2}},$$

$$M(-,+;-,+,+,+) = \frac{2e^4 q_{-0} q^*_{-\perp}}{q_{--}(p_- \cdot k_1)(q_- \cdot k_2)} \left[\frac{k_{1+}k''_{2-}q^3_{++}}{k_{1-}k''_{2+}q_{-+}}\right]^{\frac{1}{2}}$$

$$\doteq \frac{2e^4 q_{-0}}{(p_- \cdot k_1)(q_- \cdot k_2)} \left[\frac{k_{1+}k''_{2-}q^3_{++}}{k_{1-}k''_{2+}q_{--}}\right]^{\frac{1}{2}},$$

$$M(+,-;+,+,-,-) = \frac{e^4 \mu E k''^*_{21} q_{-\perp}}{q_{-0}(2E-k_{1-})(p_- \cdot k_1)(q_- \cdot k_2)} \left[\frac{k_{1+}k''_{2+}q_{++}}{k_{1-}k''_{2-}q_{-+}}\right]^{\frac{1}{2}}$$

$$\doteq \frac{e^4 \mu E k''_{2+}}{q_{-0}(2E-k_{1-})(p_- \cdot k_1)(q_- \cdot k_2)} \left[\frac{k_{1+}q_{++}q_{--}}{k_{1-}}\right]^{\frac{1}{2}},$$

$$M(+,-;+,-,+,-) = \frac{e^4(2q_{-0}+k''_{2+})q_{-\perp}}{(p_- \cdot k_1)(q_- \cdot k_2)} \left[\frac{k_{1+}k''_{2-}q_{++}}{k_{1-}k''_{2+}q_{-+}}\right]^{\frac{1}{2}}$$

$$\doteq \frac{e^4(2q_{-0}+k''_{2+})}{(p_- \cdot k_1)(q_- \cdot k_2)} \left[\frac{k_{1+}k''_{2-}q_{++}q_{--}}{k_{1-}k''_{2+}}\right]^{\frac{1}{2}},$$

$$M(+,-;+,-,-,+) = \frac{4e^4 E q_{-0} q_{-\perp}}{(2E - k_{1-})(p_- \cdot k_1)(q_- \cdot k_2)} \left[\frac{k_{1+} k''_{2-} q_{++}}{k_{1-} k''_{2+} q_{-+}} \right]^{\frac{1}{2}}$$

$$\doteq \frac{4e^4 E q_{-0}}{(2E - k_{1-})(p_- \cdot k_1)(q_- \cdot k_2)} \left[\frac{k_{1+} k''_{2-} q_{++} q_{--}}{k_{1-} k''_{2+}} \right]^{\frac{1}{2}},$$

$$M(+,-;-,+,+,-) = \frac{2e^4 q_{-0} q_{+\perp}}{(p_- \cdot k_1)(q_- \cdot k_2)} \left[\frac{k_{1+} k''_{2-} q_{-+}}{k_{1-} k''_{2+} q_{++}} \right]^{\frac{1}{2}}$$

$$\doteq \frac{2e^4 q_{-0}}{(p_- \cdot k_1)(q_- \cdot k_2)} \left[\frac{k_{1+} k''_{2-} q_{+-} q_{-+}}{k_{1-} k''_{2+}} \right]^{\frac{1}{2}},$$

$$M(+,-;-,+,-,+) = \frac{2e^4 q_{-0} q_{+\perp}}{q_{--}(p_- \cdot k_1)(q_- \cdot k_2)} \left[\frac{k_{1+} k''_{2-} q_{++} q_{-+}}{k_{1-} k''_{2+}} \right]^{\frac{1}{2}}$$

$$\doteq \frac{2e^4 q_{-0} q_{++}}{q_{--}(p_- \cdot k_1)(q_- \cdot k_2)} \left[\frac{k_{1+} k''_{2-} q_{+-} q_{-+}}{k_{1-} k''_{2+}} \right]^{\frac{1}{2}},$$

$$M(+,-;-,-,+,+) = -\frac{e^4 \mu k''_{2\perp} q_{+\perp}}{2 q_{-0}(p_- \cdot k_1)(q_- \cdot k_2)} \left[\frac{k_{1+} k''_{2+} q_{-+}}{k_{1-} k''_{2-} q_{++}} \right]^{\frac{1}{2}}$$

$$\doteq \frac{e^4 \mu k''_{2+}}{2 q_{-0}(p_- \cdot k_1)(q_- \cdot k_2)} \left[\frac{k_{1+} q_{+-} q_{-+}}{k_{1-}} \right]^{\frac{1}{2}},$$

$$M(-,+;+,+,-,-) = \frac{e^4 \mu k'''^*_{2\perp} q^*_{+\perp}}{2 q_{-0}(p_- \cdot k_1)(q_- \cdot k_2)} \left[\frac{k_{1+} k''_{2+} q_{-+}}{k_{1-} k''_{2-} q_{++}} \right]^{\frac{1}{2}}$$

$$\doteq \frac{e^4 \mu k''_{2+}}{2 q_{-0}(p_- \cdot k_1)(q_- \cdot k_2)} \left[\frac{k_{1+} q_{+-} q_{-+}}{k_{1-}} \right]^{\frac{1}{2}},$$

$$M(-,+;+,-,+,-) = \frac{2e^4 q_{-0} q^*_{+\perp}}{q_{--}(p_- \cdot k_1)(q_- \cdot k_2)} \left[\frac{k_{1+} k''_{2-} q_{++} q_{-+}}{k_{1-} k''_{2+}} \right]^{\frac{1}{2}}$$

$$\doteq \frac{2e^4 q_{-0} q_{++}}{q_{--}(p_- \cdot k_1)(q_- \cdot k_2)} \left[\frac{k_{1+} k''_{2-} q_{+-} q_{-+}}{k_{1-} k''_{2+}} \right]^{\frac{1}{2}},$$

$$M(-,+;+,-,-,+) = \frac{2e^4 q_{-0} q^*_{+\perp}}{(p_- \cdot k_1)(q_- \cdot k_2)} \left[\frac{k_{1+} k''_{2-} q_{-+}}{k_{1-} k''_{2+} q_{++}} \right]^{\frac{1}{2}}$$

$$\doteq \frac{2e^4 q_{-0}}{(p_- \cdot k_1)(q_- \cdot k_2)} \left[\frac{k_{1+} k''_{2-} q_{+-} q_{-+}}{k_{1-} k''_{2+}} \right]^{\frac{1}{2}},$$

9.18. $e^+e^- \to \mu^+\mu^-\gamma\gamma$ (NO Z-EXCHANGE; $m \neq 0$)

$$M(-,+;-,+,+,-) = \frac{4e^4 E q_{-0} q_{-\perp}^*}{(2E - k_{1-})(p_- \cdot k_1)(q_- \cdot k_2)} \left[\frac{k_{1+}k_{2-}'' q_{++}}{k_{1-}k_{2+}'' q_{-+}} \right]^{\frac{1}{2}}$$

$$\doteq \frac{4e^4 E q_{-0}}{(2E - k_{1-})(p_- \cdot k_1)(q_- \cdot k_2)} \left[\frac{k_{1+}k_{2-}'' q_{++} q_{--}}{k_{1-}k_{2+}''} \right]^{\frac{1}{2}},$$

$$M(-,+;-,+,-,+) = \frac{e^4(2q_{-0} + k_{2+}'') q_{-\perp}^*}{(p_- \cdot k_1)(q_- \cdot k_2)} \left[\frac{k_{1+}k_{2-}'' q_{++}}{k_{1-}k_{2+}'' q_{-+}} \right]^{\frac{1}{2}}$$

$$\doteq \frac{e^4(2q_{-0} + k_{2+}'')}{(p_- \cdot k_1)(q_- \cdot k_2)} \left[\frac{k_{1+}k_{2-}'' q_{++} q_{--}}{k_{1-}k_{2+}''} \right]^{\frac{1}{2}},$$

$$M(-,+;-,-,+,+) = -\frac{e^4 \mu E k_{2\perp}'' q_{-\perp}^*}{q_{-0}(2E - k_{1-})(p_- \cdot k_1)(q_- \cdot k_2)} \left[\frac{k_{1+}k_{2+}'' q_{++}}{k_{1-}k_{2-}'' q_{-+}} \right]^{\frac{1}{2}}$$

$$\doteq \frac{e^4 \mu E k_{2+}''}{q_{-0}(2E - k_{1-})(p_- \cdot k_1)(q_- \cdot k_2)} \left[\frac{k_{1+} q_{++} q_{--}}{k_{1-}} \right]^{\frac{1}{2}},$$

$$M(+,-;+,-,-,-) = \frac{2e^4 q_{-0} q_{-\perp}}{q_{--}(p_- \cdot k_1)(q_- \cdot k_2)} \left[\frac{k_{1+}k_{2-}'' q_{++}^3}{k_{1-}k_{2+}'' q_{-+}} \right]^{\frac{1}{2}}$$

$$\doteq \frac{2e^4 q_{-0}}{(p_- \cdot k_1)(q_- \cdot k_2)} \left[\frac{k_{1+}k_{2-}'' q_{++}^3}{k_{1-}k_{2+}'' q_{--}} \right]^{\frac{1}{2}},$$

$$M(+,-;-,+,-,-) = \frac{4e^4 E q_{-0} q_{+\perp}}{(2E - k_{1-})(p_- \cdot k_1)(q_- \cdot k_2)} \left[\frac{k_{1+}k_{2-}'' q_{-+}}{k_{1-}k_{2+}'' q_{++}} \right]^{\frac{1}{2}}$$

$$\doteq \frac{4e^4 E q_{-0}}{(2E - k_{1-})(p_- \cdot k_1)(q_- \cdot k_2)} \left[\frac{k_{1+}k_{2-}'' q_{+-} q_{-+}}{k_{1-}k_{2+}''} \right]^{\frac{1}{2}},$$

$$M(+,-;-,-,-,+) = -\frac{e^4 \mu E k_{2\perp}'' q_{+\perp}}{q_{-0}(2E - k_{1-})(p_- \cdot k_1)(q_- \cdot k_2)} \left[\frac{k_{1+}k_{2+}'' q_{-+}}{k_{1-}k_{2-}'' q_{++}} \right]^{\frac{1}{2}}$$

$$\doteq \frac{e^4 \mu E k_{2+}''}{q_{-0}(2E - k_{1-})(p_- \cdot k_1)(q_- \cdot k_2)} \left[\frac{k_{1+} q_{+-} q_{-+}}{k_{1-}} \right]^{\frac{1}{2}},$$

$$M(-,+;+,-,-,-) = \frac{e^4(2q_{-0} + k_{2+}'') q_{+\perp}^*}{(p_- \cdot k_1)(q_- \cdot k_2)} \left[\frac{k_{1+}k_{2-}'' q_{-+}}{k_{1-}k_{2+}'' q_{++}} \right]^{\frac{1}{2}}$$

$$\doteq \frac{e^4(2q_{-0} + k_{2+}'')}{(p_- \cdot k_1)(q_- \cdot k_2)} \left[\frac{k_{1+}k_{2-}'' q_{+-} q_{-+}}{k_{1-}k_{2+}''} \right]^{\frac{1}{2}},$$

9. SUMMARY OF QED FORMULAE

$$M(-,+;-,+,-,-) = \frac{2e^4 q_{-0} q_{-\perp}^*}{(p_- \cdot k_1)(q_- \cdot k_2)} \left[\frac{k_{1+} k_{2-}'' q_{++}}{k_{1-} k_{2+}'' q_{-+}}\right]^{\frac{1}{2}}$$

$$\doteq \frac{2e^4 q_{-0}}{(p_- \cdot k_1)(q_- \cdot k_2)} \left[\frac{k_{1+} k_{2-}'' q_{++} q_{--}}{k_{1-} k_{2+}''}\right]^{\frac{1}{2}},$$

$$M(-,+;-,-,-,+) = -\frac{e^4 \mu k_{2\perp}'' q_{-\perp}^*}{2q_{-0}(p_- \cdot k_1)(q_- \cdot k_2)} \left[\frac{k_{1+} k_{2+}'' q_{++}}{k_{1-} k_{2-}'' q_{-+}}\right]^{\frac{1}{2}}$$

$$\doteq \frac{e^4 \mu k_{2+}''}{2q_{-0}(p_- \cdot k_1)(q_- \cdot k_2)} \left[\frac{k_{1+} q_{++} q_{--}}{k_{1-}}\right]^{\frac{1}{2}},$$

$$M(-,-;+,+,-,-) = -\frac{e^4 m \mu k_{1\perp} k_{2\perp}''^* q_{+\perp}^*}{4E q_{-0}(2E-k_{1-})(p_- \cdot k_1)(q_- \cdot k_2)} \left[\frac{k_{1-} k_{2+}'' q_{-+}}{k_{1+} k_{2-}'' q_{++}}\right]^{\frac{1}{2}}$$

$$\doteq \frac{e^4 m \mu k_{1-} k_{2+}'' [q_{+-} q_{-+}]^{\frac{1}{2}}}{4E q_{-0}(2E-k_{1-})(p_- \cdot k_1)(q_- \cdot k_2)},$$

$$M(-,-;+,-,-,+) = -\frac{e^4 m q_{-0} k_{1\perp} q_{+\perp}^*}{E(2E-k_{1-})(p_- \cdot k_1)(q_- \cdot k_2)} \left[\frac{k_{1-} k_{2-}'' q_{-+}}{k_{1+} k_{2+}'' q_{++}}\right]^{\frac{1}{2}}$$

$$\doteq \frac{e^4 m q_{-0} k_{1-}}{E(2E-k_{1-})(p_- \cdot k_1)(q_- \cdot k_2)} \left[\frac{k_{2-}'' q_{+-} q_{-+}}{k_{2+}''}\right]^{\frac{1}{2}},$$

$$M(-,-;-,+,-,+) = -\frac{e^4 m q_{-0} k_{1\perp} q_{-\perp}^*}{2E^2 q_{--}(p_- \cdot k_1)(q_- \cdot k_2)} \left[\frac{k_{1+} k_{2-}'' q_{++}^3}{k_{1+} k_{2+}'' q_{-+}}\right]^{\frac{1}{2}}$$

$$\doteq \frac{e^4 m q_{-0} k_{1-}}{2E^2 (p_- \cdot k_1)(q_- \cdot k_2)} \left[\frac{k_{2-}'' q_{++}^3}{k_{2+}'' q_{--}}\right]^{\frac{1}{2}},$$

$$M(-,-;+,-,-,-) = -\frac{e^4 m q_{-0} k_{1\perp} q_{+\perp}^*}{2E^2 q_{--}(p_- \cdot k_1)(q_- \cdot k_2)} \left[\frac{k_{1-} k_{2-}'' q_{++} q_{-+}}{k_{1+} k_{2+}''}\right]^{\frac{1}{2}}$$

$$\doteq \frac{e^4 m q_{-0} q_{++} k_{1-}}{2E^2 q_{--}(p_- \cdot k_1)(q_- \cdot k_2)} \left[\frac{k_{2-}'' q_{+-} q_{-+}}{k_{2+}''}\right]^{\frac{1}{2}},$$

$$M(-,-;-,+,-,-) = -\frac{e^4 m q_{-0} k_{1\perp} q_{-\perp}^*}{E(2E-k_{1-})(p_- \cdot k_1)(q_- \cdot k_2)} \left[\frac{k_{1-} k_{2-}'' q_{++}}{k_{1+} k_{2+}'' q_{-+}}\right]^{\frac{1}{2}}$$

$$\doteq \frac{e^4 m q_{-0} k_{1-}}{E(2E-k_{1-})(p_- \cdot k_1)(q_- \cdot k_2)} \left[\frac{k_{2-}'' q_{++} q_{--}}{k_{2+}''}\right]^{\frac{1}{2}},$$

9.18. $e^+e^- \to \mu^+\mu^-\gamma\gamma$ (NO Z-EXCHANGE; $m \neq 0$)

$$M(-,-;-,-,-,+) = \frac{e^4 m\mu k_{1\perp} k_{2\perp}'' q_{-\perp}^*}{4Eq_{-0}(2E-k_{1-})(p_-\cdot k_1)(q_-\cdot k_2)} \left[\frac{k_{1-}k_{2+}''q_{++}}{k_{1+}k_{2-}''q_{-+}}\right]^{\frac{1}{2}}$$

$$\doteq \frac{e^4 m\mu k_{1-} k_{2+}'' [q_{++}q_{--}]^{\frac{1}{2}}}{4Eq_{-0}(2E-k_{1-})(p_-\cdot k_1)(q_-\cdot k_2)}. \quad (9.184)$$

Unpolarized squared matrix element:

$$\overline{|M|^2} = \frac{e^8 k_{1+} k_{2-}''}{2 k_{1-} k_{2+}''[(2E-k_{1-})(p_-\cdot k_1)(q_-\cdot k_2)]^2}$$

$$\times \left[4E^2 + (2E-k_{1-})^2 + \frac{m^2 k_{1-}^3}{4E^2 k_{1+}}\right]$$

$$\times \left[4q_{-0}^2 + (2q_{-0}+k_{2+}'')^2 + \frac{\mu^2 k_{2+}''^3}{4q_{-0}^2 k_{2-}''}\right] (q_{++}q_{--}+q_{+-}q_{-+}). \quad (9.185)$$

Unpolarized cross section:

$$d\sigma = \frac{\alpha^4 k_{1+} k_{2-}''}{256\pi^2 k_{1-} k_{2+}''[E(2E-k_{1-})(p_-\cdot k_1)(q_-\cdot k_2)]^2}$$

$$\times \left[4E^2 + (2E-k_{1-})^2 + \frac{m^2 k_{1-}^3}{4E^2 k_{1+}}\right]$$

$$\times \left[4q_{-0}^2 + (2q_{-0}+k_{2+}'')^2 + \frac{\mu^2 k_{2+}''^3}{4q_{-0}^2 k_{2-}''}\right] (q_{++}q_{--}+q_{+-}q_{-+})$$

$$\times \delta^4(p_+ + p_- - q_+ - q_- - k_1 - k_2) \frac{d^3\vec{q}_+ \, d^3\vec{q}_- \, d^3\vec{k}_1 \, d^3\vec{k}_2}{q_{+0}\,q_{-0}\,k_{10}\,k_{20}}. \quad (9.186)$$

9.18.10 $\vec{k}_1$ and $\vec{k}_2$ nearly parallel to $\vec{q}_+$ and $\vec{q}_-$, resp.

The primed (doubly primed) quantities $k_{1\pm}'$ and $k_{1\perp}'$ ($k_{2\pm}''$ and $k_{2\perp}''$) are evaluated in the rotated frame where $\vec{q}_+$ ($\vec{q}_-$) determines the positive z-axis (see also Section 7.4.3). The four-vector q (q') in eqns (9.187) is obtained by applying a space reflection to q_+ (q_-). The quantity μ denotes the muon mass. Photon polarizations:

$$\not{\epsilon}_1^{\pm} = N_1[\not{k}_1 \not{q}_+ \not{q}(1 \pm \gamma_5) + \not{q} \not{q}_+ \not{k}_1(1 \mp \gamma_5)],$$

$$N_1^{-1} = q_{+0}^2 [32 k_{1+}' k_{1-}']^{\frac{1}{2}},$$

$$\not{\epsilon}_2^{\pm} = N_2[\not{k}_2 \not{q}_- \not{q}'(1 \pm \gamma_5) + \not{q}' \not{q}_- \not{k}_2(1 \mp \gamma_5)],$$

$$N_2^{-1} = q_{-0}^2 [32 k_{2+}'' k_{2-}'']^{\frac{1}{2}}. \quad (9.187)$$

Nonvanishing helicity amplitudes:

$$M(+,-;+,+,+,-) = \frac{e^4\mu(2q_{+0}+k'_{1+})k''^*_{2\perp}q_{-\perp}}{4Eq_{-0}(q_+\cdot k_1)(q_-\cdot k_2)}\left[\frac{k'_{1-}k''_{2+}q_{++}}{k'_{1+}k''_{2-}q_{-+}}\right]^{\frac{1}{2}}$$

$$\doteq \frac{e^4\mu(2q_{+0}+k'_{1+})k''_{2+}}{4Eq_{-0}(q_+\cdot k_1)(q_-\cdot k_2)}\left[\frac{k'_{1-}q_{++}q_{--}}{k'_{1+}}\right]^{\frac{1}{2}},$$

$$M(+,-;+,+,-,+) = \frac{e^4\mu(2q_{-0}+k''_{2+})k'^*_{1\perp}q_{+\perp}}{4Eq_{+0}(q_+\cdot k_1)(q_-\cdot k_2)}\left[\frac{k'_{1+}k''_{2-}q_{-+}}{k'_{1-}k''_{2+}q_{++}}\right]^{\frac{1}{2}}$$

$$\doteq \frac{e^4\mu(2q_{-0}+k''_{2+})k'_{1+}}{4Eq_{+0}(q_+\cdot k_1)(q_-\cdot k_2)}\left[\frac{k''_{2-}q_{+-}q_{-+}}{k''_{2+}}\right]^{\frac{1}{2}},$$

$$M(+,-;+,-,+,+) = \frac{e^4q_{-0}(2q_{+0}+k'_{1+})q_{-\perp}}{E(q_+\cdot k_1)(q_-\cdot k_2)}\left[\frac{k'_{1-}k''_{2-}q_{++}}{k'_{1+}k''_{2+}q_{-+}}\right]^{\frac{1}{2}}$$

$$\doteq \frac{e^4q_{-0}(2q_{+0}+k'_{1+})}{E(q_+\cdot k_1)(q_-\cdot k_2)}\left[\frac{k'_{1-}k''_{2-}q_{++}q_{--}}{k'_{1+}k''_{2+}}\right]^{\frac{1}{2}},$$

$$M(+,-;-,+,+,+) = \frac{e^4q_{+0}(2q_{-0}+k''_{2+})q_{+\perp}}{E(q_+\cdot k_1)(q_-\cdot k_2)}\left[\frac{k'_{1-}k''_{2-}q_{-+}}{k'_{1+}k''_{2+}q_{++}}\right]^{\frac{1}{2}}$$

$$\doteq \frac{e^4q_{+0}(2q_{-0}+k''_{2+})}{E(q_+\cdot k_1)(q_-\cdot k_2)}\left[\frac{k'_{1-}k''_{2-}q_{+-}q_{-+}}{k'_{1+}k''_{2+}}\right]^{\frac{1}{2}},$$

$$M(-,+;+,+,+,-) = \frac{e^4\mu(2q_{+0}+k'_{1+})k''^*_{2\perp}q^*_{+\perp}}{4Eq_{-0}(q_+\cdot k_1)(q_-\cdot k_2)}\left[\frac{k'_{1-}k''_{2+}q_{-+}}{k'_{1+}k''_{2-}q_{++}}\right]^{\frac{1}{2}}$$

$$\doteq \frac{e^4\mu(2q_{+0}+k'_{1+})k''_{2+}}{4Eq_{-0}(q_+\cdot k_1)(q_-\cdot k_2)}\left[\frac{k'_{1-}q_{+-}q_{-+}}{k'_{1+}}\right]^{\frac{1}{2}},$$

$$M(-,+;+,+,-,+) = \frac{e^4\mu(2q_{-0}+k''_{2+})k'^*_{1\perp}q^*_{-\perp}}{4Eq_{+0}(q_+\cdot k_1)(q_-\cdot k_2)}\left[\frac{k'_{1+}k''_{2-}q_{++}}{k'_{1-}k''_{2+}q_{-+}}\right]^{\frac{1}{2}}$$

$$\doteq \frac{e^4\mu(2q_{-0}+k''_{2+})k'_{1+}}{4Eq_{+0}(q_+\cdot k_1)(q_-\cdot k_2)}\left[\frac{k''_{2-}q_{++}q_{--}}{k''_{2+}}\right]^{\frac{1}{2}},$$

$$M(-,+;+,-,+,+) = \frac{e^4q_{-0}(2q_{+0}+k'_{1+})q^*_{+\perp}}{E(q_+\cdot k_1)(q_-\cdot k_2)}\left[\frac{k'_{1-}k''_{2-}q_{-+}}{k'_{1+}k''_{2+}q_{++}}\right]^{\frac{1}{2}}$$

$$\doteq \frac{e^4q_{-0}(2q_{+0}+k'_{1+})}{E(q_+\cdot k_1)(q_-\cdot k_2)}\left[\frac{k'_{1-}k''_{2-}q_{+-}q_{-+}}{k'_{1+}k''_{2+}}\right]^{\frac{1}{2}},$$

9.18. $e^+e^- \to \mu^+\mu^-\gamma\gamma$ (NO Z-EXCHANGE; $m \neq 0$)

$$M(-,+;-,+,+,+) = \frac{e^4 q_{+0}(2q_{-0} + k''_{2+})q^*_{-\perp}}{E(q_+ \cdot k_1)(q_- \cdot k_2)} \left[\frac{k'_{1-}k''_{2-}q_{++}}{k'_{1+}k''_{2+}q_{-+}}\right]^{\frac{1}{2}}$$

$$\doteq \frac{e^4 q_{+0}(2q_{-0} + k''_{2+})}{E(q_+ \cdot k_1)(q_- \cdot k_2)} \left[\frac{k'_{1-}k''_{2-}q_{++}q_{--}}{k'_{1+}k''_{2+}}\right]^{\frac{1}{2}},$$

$$M(+,-;+,+,-,-) = \frac{e^4 \mu}{2E(q_+ \cdot k_1)(q_- \cdot k_2)} \left\{\frac{q_{-0}k'^{*}_{1\perp}q_{+\perp}}{q_{+0}} \left[\frac{k'_{1+}k''_{2-}q_{-+}}{k'_{1-}k''_{2+}q_{++}}\right]^{\frac{1}{2}}\right.$$

$$\left. + \frac{q_{+0}k''^{*}_{2\perp}q_{-\perp}}{q_{-0}} \left[\frac{k'_{1-}k''_{2+}q_{++}}{k'_{1+}k''_{2-}q_{-+}}\right]^{\frac{1}{2}}\right\},$$

$$M(+,-;+,-,+,-) = \frac{e^4(2q_{+0} + k'_{1+})(2q_{-0} + k''_{2+})q_{-\perp}}{2E(q_+ \cdot k_1)(q_- \cdot k_2)} \left[\frac{k'_{1-}k''_{2-}q_{++}}{k'_{1+}k''_{2+}q_{-+}}\right]^{\frac{1}{2}}$$

$$\doteq \frac{e^4(2q_{+0} + k'_{1+})(2q_{-0} + k''_{2+})}{2E(q_+ \cdot k_1)(q_- \cdot k_2)} \left[\frac{k'_{1-}k''_{2-}q_{++}q_{--}}{k'_{1+}k''_{2+}}\right]^{\frac{1}{2}},$$

$$M(+,-;+,-,-,+) = \frac{e^4}{E(q_+ \cdot k_1)(q_- \cdot k_2)} \left\{2q_{+0}q_{-0}q_{-\perp}\left[\frac{k'_{1-}k''_{2-}q_{++}}{k'_{1+}k''_{2+}q_{-+}}\right]^{\frac{1}{2}}\right.$$

$$\left. - \frac{\mu^2 k'^{*}_{1\perp}k''_{2\perp}q_{+\perp}}{q_{+0}q_{-0}}\left[\frac{k'_{1+}k''_{2+}q_{-+}}{k'_{1-}k''_{2-}q_{++}}\right]^{\frac{1}{2}}\right\},$$

$$M(+,-;-,+,+,-) = \frac{e^4}{E(q_+ \cdot k_1)(q_- \cdot k_2)} \left\{2q_{+0}q_{-0}q_{+\perp}\left[\frac{k'_{1-}k''_{2-}q_{-+}}{k'_{1+}k''_{2+}q_{++}}\right]^{\frac{1}{2}}\right.$$

$$\left. - \frac{\mu^2 k'_{1\perp}k''^{*}_{2\perp}q_{-\perp}}{8q_{+0}q_{-0}}\left[\frac{k'_{1+}k''_{2+}q_{++}}{k'_{1-}k''_{2-}q_{-+}}\right]^{\frac{1}{2}}\right\},$$

$$M(+,-;-,+,-,+) = \frac{4e^4 E q_{+0}q_{-0}q_{+\perp}}{(q_+\cdot q_-)(q_+\cdot k_1)(q_-\cdot k_2)} \left[\frac{k'_{1-}k''_{2-}q_{-+}}{k'_{1+}k''_{2+}q_{++}}\right]^{\frac{1}{2}}$$

$$\doteq \frac{4e^4 E q_{+0}q_{-0}}{(q_+\cdot q_-)(q_+\cdot k_1)(q_-\cdot k_2)} \left[\frac{k'_{1-}k''_{2-}q_{+-}q_{-+}}{k'_{1+}k''_{2+}}\right]^{\frac{1}{2}},$$

$$M(+,-;-,-,+,+) = -\frac{e^4 \mu}{2E(q_+ \cdot k_1)(q_- \cdot k_2)} \left\{\frac{q_{-0}k'_{1\perp}q_{-\perp}}{q_{+0}}\left[\frac{k'_{1+}k''_{2-}q_{++}}{k'_{1-}k''_{2+}q_{-+}}\right]^{\frac{1}{2}}\right.$$

$$\left. + \frac{q_{+0}k''_{2\perp}q_{+\perp}}{q_{-0}}\left[\frac{k'_{1-}k''_{2+}q_{-+}}{k'_{1+}k''_{2-}q_{++}}\right]^{\frac{1}{2}}\right\},$$

$$M(-,+;+,+,-,-) = \frac{e^4\mu}{2E(q_+\cdot k_1)(q_-\cdot k_2)} \left\{ \frac{q_{+0}k_{21}''^* q_{+\perp}^*}{q_{-0}} \left[\frac{k_{1-}'k_{2+}''q_{-+}}{k_{1+}'k_{2-}''q_{++}}\right]^{\frac{1}{2}} \right.$$

$$\left. + \frac{q_{-0}k_{11}'^* q_{-\perp}^*}{q_{+0}} \left[\frac{k_{1+}'k_{2-}''q_{++}}{k_{1-}'k_{2+}''q_{-+}}\right]^{\frac{1}{2}} \right\},$$

$$M(-,+;+,-,+,-) = \frac{4e^4 E q_{+0} 2q_{-0} q_{+\perp}^*}{(q_+\cdot q_-)(q_+\cdot k_1)(q_-\cdot k_2)} \left[\frac{k_{1-}'k_{2-}''q_{-+}}{k_{1+}'k_{2+}''q_{++}}\right]^{\frac{1}{2}}$$

$$\doteq \frac{4e^4 E q_{+0} q_{-0}}{(q_+\cdot q_-)(q_+\cdot k_1)(q_-\cdot k_2)} \left[\frac{k_{1-}'k_{2-}''q_{+-}q_{-+}}{k_{1+}'k_{2+}''}\right]^{\frac{1}{2}},$$

$$M(-,+;+,-,-,+) = \frac{e^4}{E(q_+\cdot k_1)(q_-\cdot k_2)} \left\{ 2q_{+0}q_{-0}q_{+\perp}^* \left[\frac{k_{1-}'k_{2-}''q_{-+}}{k_{1+}'k_{2+}''q_{++}}\right]^{\frac{1}{2}} \right.$$

$$\left. - \frac{\mu^2 k_{11}'^* k_{21}'' q_{-\perp}^*}{q_{+0}q_{-0}} \left[\frac{k_{1+}'k_{2+}''q_{++}}{k_{1-}'k_{2-}''q_{-+}}\right]^{\frac{1}{2}} \right\},$$

$$M(-,+;-,+,+,-) = \frac{e^4}{E(q_+\cdot k_1)(q_-\cdot k_2)} \left\{ 2q_{+0}q_{-0}q_{-\perp}^* \left[\frac{k_{1-}'k_{2-}''q_{++}}{k_{1+}'k_{2+}''q_{-+}}\right]^{\frac{1}{2}} \right.$$

$$\left. - \frac{\mu^2 k_{11}' k_{21}''^* q_{+\perp}^*}{8q_{+0}q_{-0}} \left[\frac{k_{1+}'k_{2+}''q_{-+}}{k_{1-}'k_{2-}''q_{++}}\right]^{\frac{1}{2}} \right\},$$

$$M(-,+;-,+,-,+) = \frac{e^4(2q_{+0}+k_{1+}')(2q_{-0}+k_{2+}'')q_{-\perp}^*}{2E(q_+\cdot k_1)(q_-\cdot k_2)} \left[\frac{k_{1-}'k_{2-}''q_{++}}{k_{1+}'k_{2+}''q_{-+}}\right]^{\frac{1}{2}}$$

$$\doteq \frac{2e^4(2q_{+0}+k_{1+}')(2q_{-0}+k_{2+}'')}{2E(q_+\cdot k_1)(q_-\cdot k_2)} \left[\frac{k_{1-}'k_{2-}''q_{++}q_{--}}{k_{1+}'k_{2+}''}\right]^{\frac{1}{2}},$$

$$M(-,+;-,-,+,+) = -\frac{e^4\mu}{2E(q_+\cdot k_1)(q_-\cdot k_2)} \left\{ \frac{q_{-0}k_{11}'q_{+\perp}^*}{q_{+0}} \left[\frac{k_{1+}'k_{2-}''q_{-+}}{k_{1-}'k_{2+}''q_{++}}\right]^{\frac{1}{2}} \right.$$

$$\left. + \frac{q_{+0}k_{21}''q_{-\perp}^*}{q_{-0}} \left[\frac{k_{1-}'k_{2+}''q_{++}}{k_{1+}'k_{2-}''q_{-+}}\right]^{\frac{1}{2}} \right\},$$

$$M(+,-;+,-,-,-) = \frac{e^4 q_{+0}(2q_{-0}+k_{2+}'')q_{-\perp}}{E(q_+\cdot k_1)(q_-\cdot k_2)} \left[\frac{k_{1-}'k_{2-}''q_{++}}{k_{1+}'k_{2+}''q_{-+}}\right]^{\frac{1}{2}}$$

$$\doteq \frac{e^4 q_{+0}(2q_{-0}+k_{2+}'')}{E(q_+\cdot k_1)(q_-\cdot k_2)} \left[\frac{k_{1-}'k_{2-}''q_{++}q_{--}}{k_{1+}'k_{2+}''}\right]^{\frac{1}{2}},$$

9.18. $e^+e^- \to \mu^+\mu^-\gamma\gamma$ (NO Z-EXCHANGE; $m \neq 0$)

$$M(+,-;-,+,-,-) = \frac{e^4 q_{-0}(2q_{+0} + k'_{1+})q_{+\perp}}{E(q_+ \cdot k_1)(q_- \cdot k_2)} \left[\frac{k'_{1-}k''_{2-}q_{-+}}{k'_{1+}k''_{2+}q_{++}}\right]^{\frac{1}{2}}$$

$$\doteq \frac{e^4 q_{-0}(2q_{+0} + k'_{1+})}{E(q_+ \cdot k_1)(q_- \cdot k_2)} \left[\frac{k'_{1-}k''_{2-}q_{+-}q_{-+}}{k'_{1+}k''_{2+}}\right]^{\frac{1}{2}},$$

$$M(+,-;-,-,+,-) = -\frac{e^4\mu(2q_{-0} + k''_{2+})k'_{1\perp}q_{-\perp}}{4Eq_{+0}(q_+ \cdot k_1)(q_- \cdot k_2)} \left[\frac{k'_{1+}k''_{2-}q_{++}}{k'_{1-}k''_{2+}q_{-+}}\right]^{\frac{1}{2}}$$

$$\doteq \frac{e^4\mu(2q_{-0} + k''_{2+})k'_{1+}}{4Eq_{+0}(q_+ \cdot k_1)(q_- \cdot k_2)} \left[\frac{k''_{2-}q_{++}q_{--}}{k''_{2+}}\right]^{\frac{1}{2}},$$

$$M(+,-;-,-,-,+) = -\frac{e^4\mu(2q_{+0} + k'_{1+})k''_{2\perp}q_{+\perp}}{4Eq_{-0}(q_+ \cdot k_1)(q_- \cdot k_2)} \left[\frac{k'_{1-}k''_{2+}q_{-+}}{k'_{1+}k''_{2-}q_{++}}\right]^{\frac{1}{2}}$$

$$\doteq \frac{e^4\mu(2q_{+0} + k'_{1+})k''_{2+}}{4Eq_{-0}(q_+ \cdot k_1)(q_- \cdot k_2)} \left[\frac{k'_{1-}q_{+-}q_{-+}}{k'_{1+}}\right]^{\frac{1}{2}},$$

$$M(-,+;+,-,-,-) = \frac{e^4 q_{+0}(2q_{-0} + k''_{2+})q^*_{+\perp}}{E(q_+ \cdot k_1)(q_- \cdot k_2)} \left[\frac{k'_{1-}k''_{2-}q_{-+}}{k'_{1+}k''_{2+}q_{++}}\right]^{\frac{1}{2}}$$

$$\doteq \frac{e^4 q_{+0}(2q_{-0} + k''_{2+})}{E(q_+ \cdot k_1)(q_- \cdot k_2)} \left[\frac{k'_{1-}k''_{2-}q_{+-}q_{-+}}{k'_{1+}k''_{2+}}\right]^{\frac{1}{2}},$$

$$M(-,+;-,+,-,-) = \frac{e^4 q_{-0}(2q_{+0} + k'_{1+})q^*_{-\perp}}{E(q_+ \cdot k_1)(q_- \cdot k_2)} \left[\frac{k'_{1-}k''_{2-}q_{++}}{k'_{1+}k''_{2+}q_{-+}}\right]^{\frac{1}{2}}$$

$$\doteq \frac{e^4 q_{-0}(2q_{+0} + k'_{1+})}{E(q_+ \cdot k_1)(q_- \cdot k_2)} \left[\frac{k'_{1-}k''_{2-}q_{++}q_{--}}{k'_{1+}k''_{2+}}\right]^{\frac{1}{2}},$$

$$M(-,+;-,-,+,-) = -\frac{e^4\mu(2q_{-0} + k''_{2+})k'_{1\perp}q^*_{+\perp}}{4Eq_{+0}(q_+ \cdot k_1)(q_- \cdot k_2)} \left[\frac{k'_{1+}k''_{2-}q_{-+}}{k'_{1-}k''_{2+}q_{++}}\right]^{\frac{1}{2}}$$

$$\doteq \frac{e^4\mu(2q_{-0} + k''_{2+})k'_{1+}}{4Eq_{+0}(q_+ \cdot k_1)(q_- \cdot k_2)} \left[\frac{k''_{2-}q_{+-}q_{-+}}{k''_{2+}}\right]^{\frac{1}{2}},$$

$$M(-,+;-,-,-,+) = -\frac{e^4\mu(2q_{+0} + k'_{1+})k''_{2\perp}q^*_{-\perp}}{4Eq_{-0}(q_+ \cdot k_1)(q_- \cdot k_2)} \left[\frac{k'_{1-}k''_{2+}q_{++}}{k'_{1+}k''_{2-}q_{-+}}\right]^{\frac{1}{2}}$$

$$\doteq \frac{e^4\mu(2q_{+0} + k'_{1+})k''_{2+}}{4Eq_{-0}(q_+ \cdot k_1)(q_- \cdot k_2)} \left[\frac{k'_{1-}q_{++}q_{--}}{k'_{1+}}\right]^{\frac{1}{2}}.$$

(9.188)

Unpolarized squared matrix element:

$$\overline{|M|^2} = \frac{e^8 k'_{1-} k''_{2-} (q_{++}q_{--} + q_{+-}q_{-+})}{8 k'_{1+} k''_{2+} [E(q_+ \cdot k_1)(q_- \cdot k_2)]^2} \left[4q_{+0}^2 + (2q_{+0} + k'_{1+})^2 + \frac{\mu^2 k'^3_{1+}}{4 q_{+0}^2 k'_{1-}} \right]$$

$$\times \left[4q_{-0}^2 + (2q_{-0} + k''_{2+})^2 + \frac{\mu^2 k''^3_{2+}}{4 q_{-0}^2 k''_{2-}} \right]. \quad (9.189)$$

Unpolarized cross section:

$$d\sigma = \frac{\alpha^4 k'_{1-} k''_{2-} (q_{++}q_{--} + q_{+-}q_{-+})}{1024 \pi^2 E^4 k'_{1+} k''_{2+} [(q_+ \cdot k_1)(q_- \cdot k_2)]^2}$$

$$\times \left[4q_{+0}^2 + (2q_{+0} + k'_{1+})^2 + \frac{\mu^2 k'^3_{1+}}{4 q_{+0}^2 k'_{1-}} \right] \left[4q_{-0}^2 + (2q_{-0} + k''_{2+})^2 + \frac{\mu^2 k''^3_{2+}}{4 q_{-0}^2 k''_{2-}} \right]$$

$$\times \delta^4(p_+ + p_- - q_+ - q_- - k_1 - k_2) \frac{d^3 \vec{q}_+ \, d^3 \vec{q}_- \, d^3 \vec{k}_1 \, d^3 \vec{k}_2}{q_{+0} \, q_{-0} \, k_{10} \, k_{20}}. \quad (9.190)$$

9.18.11 $\vec{k}_1$ and $\vec{k}_2$ nearly parallel to $\vec{p}_+$

Definitions:

$$\Delta'_{12} = -2(p_+ \cdot k_1) - 2(p_+ \cdot k_2) + 2(k_1 \cdot k_2),$$

$$Z_{ij} = k_{i+} k_{j-} - k^*_{i\perp} k_{j\perp}, \qquad i,j = 1, 2,$$

$$B_1 = \frac{2E^2}{(p_+ \cdot k_1)(p_+ \cdot k_2)} \left[k_{1-} k_{2-} + \frac{m^2 Z_{12} Z_{21}}{4 E^2 \Delta'_{12}} \right],$$

$$B_2(1,2) = \frac{1}{\Delta'_{12}} \left[\frac{k_{1-} (2E - k_{1+})(2E k_{2-} - Z_{12})}{(p_+ \cdot k_1)} \right.$$

$$\left. + \frac{2E k_{1-} k_{2-} (2E - k_{1+} - k_{2+}) - m^2 k_{2+} k^*_{1\perp} k_{2\perp}/2E}{(p_+ \cdot k_2)} \right],$$

$$B_3 = \frac{m}{2(p_+ \cdot k_1)(p_+ \cdot k_2)} \left[k_{1+} k_{2-} k_{1\perp} + k_{1-} k_{2+} k_{2\perp} \right.$$

$$\left. - \frac{(k_{1\perp} + k_{2\perp}) Z_{12} Z_{21}}{\Delta'_{12}} \right],$$

$$B_4(1,2) = \frac{m k_{1+} k_{1\perp}}{2E \Delta'_{12}} \left[\frac{2E k_{2-} - Z^*_{12}}{(p_+ \cdot k_1)} + \frac{2E k_{2-}}{(p_+ \cdot k_2)} \right]. \quad (9.191)$$

9.18. $e^+e^- \rightarrow \mu^+\mu^-\gamma\gamma$ (NO Z-EXCHANGE; $m \neq 0$)

Photon polarizations:

$$\not{k}_i^\pm = N_i[\not{k}_i \not{p}_+ \not{p}_-(1\pm\gamma_5) + \not{p}_- \not{p}_+ \not{k}_i(1\mp\gamma_5)],$$

$$N_i^{-1} = E^2[32k_{i+}k_{i-}]^{\frac{1}{2}}, \qquad i=1,2. \qquad (9.192)$$

Nonvanishing helicity amplitudes:

$$M(+,+;+,-,+,+) = \frac{2e^4 E B_3 q_{+\perp}^*}{(q_+ \cdot q_-)} \left[\frac{q_{-+}}{q_{++}k_{1+}k_{1-}k_{2+}k_{2-}}\right]^{\frac{1}{2}}$$

$$\doteq \frac{2e^4 E |B_3|}{(q_+ \cdot q_-)} \left[\frac{q_{+-}q_{-+}}{k_{1+}k_{1-}k_{2+}k_{2-}}\right]^{\frac{1}{2}},$$

$$M(+,+;-,+,+,+) = \frac{2e^4 E B_3 q_{-\perp}^*}{(q_+ \cdot q_-)} \left[\frac{q_{++}}{q_{-+}k_{1+}k_{1-}k_{2+}k_{2-}}\right]^{\frac{1}{2}}$$

$$\doteq \frac{2e^4 E |B_3|}{(q_+ \cdot q_-)} \left[\frac{q_{++}q_{--}}{k_{1+}k_{1-}k_{2+}k_{2-}}\right]^{\frac{1}{2}},$$

$$M(+,+;+,-,+,-) = \frac{2e^4 E B_4(1,2) q_{-\perp}^*}{(q_+ \cdot q_-)} \left[\frac{q_{-+}}{q_{++}k_{1+}k_{1-}k_{2+}k_{2-}}\right]^{\frac{1}{2}}$$

$$\doteq \frac{2e^4 E |B_4(1,2)|}{(q_+ \cdot q_-)} \left[\frac{q_{+-}q_{-+}}{k_{1+}k_{1-}k_{2+}k_{2-}}\right]^{\frac{1}{2}},$$

$$M(+,+;+,-,-,+) = \frac{2e^4 E B_4(2,1) q_{-\perp}^*}{(q_+ \cdot q_-)} \left[\frac{q_{-+}}{q_{++}k_{1+}k_{1-}k_{2+}k_{2-}}\right]^{\frac{1}{2}}$$

$$\doteq \frac{2e^4 E |B_4(2,1)|}{(q_+ \cdot q_-)} \left[\frac{q_{+-}q_{-+}}{k_{1+}k_{1-}k_{2+}k_{2-}}\right]^{\frac{1}{2}},$$

$$M(+,+;-,+,+,-) = \frac{2e^4 E B_4(1,2) q_{+\perp}^*}{(q_+ \cdot q_-)} \left[\frac{q_{++}}{q_{-+}k_{1+}k_{1-}k_{2+}k_{2-}}\right]^{\frac{1}{2}}$$

$$\doteq \frac{2e^4 E |B_4(1,2)|}{(q_+ \cdot q_-)} \left[\frac{q_{++}q_{--}}{k_{1+}k_{1-}k_{2+}k_{2-}}\right]^{\frac{1}{2}},$$

$$M(+,+;-,+,-,+) = \frac{2e^4 E B_4(2,1) q_{+\perp}^*}{(q_+ \cdot q_-)} \left[\frac{q_{++}}{q_{-+}k_{1+}k_{1-}k_{2+}k_{2-}}\right]^{\frac{1}{2}}$$

$$\doteq \frac{2e^4 E |B_4(2,1)|}{(q_+ \cdot q_-)} \left[\frac{q_{++}q_{--}}{k_{1+}k_{1-}k_{2+}k_{2-}}\right]^{\frac{1}{2}},$$

$$M(+,-;+,-,+,+) = \frac{2e^4 E B_1 q_{+\perp}}{(q_+ \cdot q_-)} \left[\frac{q_{++}}{q_{-+} k_{1+} k_{1-} k_{2+} k_{2-}} \right]^{\frac{1}{2}}$$

$$\doteq \frac{2e^4 E |B_1|}{(q_+ \cdot q_-)} \left[\frac{q_{++} q_{--}}{k_{1+} k_{1-} k_{2+} k_{2-}} \right]^{\frac{1}{2}},$$

$$M(+,-;-,+,+,+) = \frac{2e^4 E B_1 q_{-\perp}}{(q_+ \cdot q_-)} \left[\frac{q_{-+}}{q_{++} k_{1+} k_{1-} k_{2+} k_{2-}} \right]^{\frac{1}{2}}$$

$$\doteq \frac{2e^4 E |B_1|}{(q_+ \cdot q_-)} \left[\frac{q_{+-} q_{-+}}{k_{1+} k_{1-} k_{2+} k_{2-}} \right]^{\frac{1}{2}},$$

$$M(-,+;+,-,+,+) = \frac{e^4 (2E - k_{1+} - k_{2+}) B_1 q_{-\perp}^*}{(q_+ \cdot q_-)} \left[\frac{q_{-+}}{q_{++} k_{1+} k_{1-} k_{2+} k_{2-}} \right]^{\frac{1}{2}}$$

$$\doteq \frac{e^4 (2E - k_{1+} - k_{2+}) |B_1|}{(q_+ \cdot q_-)} \left[\frac{q_{+-} q_{-+}}{k_{1+} k_{1-} k_{2+} k_{2-}} \right]^{\frac{1}{2}},$$

$$M(-,+;-,+,+,+) = \frac{e^4 (2E - k_{1+} - k_{2+}) B_1 q_{+\perp}^*}{(q_+ \cdot q_-)} \left[\frac{q_{++}}{q_{-+} k_{1+} k_{1-} k_{2+} k_{2-}} \right]^{\frac{1}{2}}$$

$$\doteq \frac{e^4 (2E - k_{1+} - k_{2+}) |B_1|}{(q_+ \cdot q_-)} \left[\frac{q_{++} q_{--}}{k_{1+} k_{1-} k_{2+} k_{2-}} \right]^{\frac{1}{2}},$$

$$M(+,-;+,-,+,-) = \frac{2e^4 E B_2(2,1) q_{-\perp}}{(q_+ \cdot q_-)} \left[\frac{q_{++}}{q_{-+} k_{1+} k_{1-} k_{2+} k_{2-}} \right]^{\frac{1}{2}}$$

$$\doteq \frac{2e^4 E |B_2(2,1)|}{(q_+ \cdot q_-)} \left[\frac{q_{++} q_{--}}{k_{1+} k_{1-} k_{2+} k_{2-}} \right]^{\frac{1}{2}},$$

$$M(+,-;+,-,-,+) = \frac{2e^4 E B_2(1,2) q_{-\perp}}{(q_+ \cdot q_-)} \left[\frac{q_{++}}{q_{-+} k_{1+} k_{1-} k_{2+} k_{2-}} \right]^{\frac{1}{2}}$$

$$\doteq \frac{2e^4 E |B_2(1,2)|}{(q_+ \cdot q_-)} \left[\frac{q_{++} q_{--}}{k_{1+} k_{1-} k_{2+} k_{2-}} \right]^{\frac{1}{2}},$$

$$M(+,-;-,+,+,-) = \frac{2e^4 E B_2(2,1) q_{+\perp}}{(q_+ \cdot q_-)} \left[\frac{q_{-+}}{q_{++} k_{1+} k_{1-} k_{2+} k_{2-}} \right]^{\frac{1}{2}}$$

$$\doteq \frac{2e^4 E |B_2(2,1)|}{(q_+ \cdot q_-)} \left[\frac{q_{+-} q_{-+}}{k_{1+} k_{1-} k_{2+} k_{2-}} \right]^{\frac{1}{2}},$$

9.18. $e^+e^- \to \mu^+\mu^-\gamma\gamma$ (NO Z-EXCHANGE; $m \neq 0$)

$$M(+,-;-,+,-,+) = \frac{2e^4 E B_2(1,2) q_{+\perp}}{(q_+ \cdot q_-)} \left[\frac{q_{-+}}{q_{++}k_{1+}k_{1-}k_{2+}k_{2-}}\right]^{\frac{1}{2}}$$

$$\doteq \frac{2e^4 E |B_2(1,2)|}{(q_+ \cdot q_-)} \left[\frac{q_{+-}q_{-+}}{k_{1+}k_{1-}k_{2+}k_{2-}}\right]^{\frac{1}{2}},$$

$$M(-,+;+,-,+,-) = \frac{2e^4 E B_2^*(1,2) q_{+\perp}^*}{(q_+ \cdot q_-)} \left[\frac{q_{-+}}{q_{++}k_{1+}k_{1-}k_{2+}k_{2-}}\right]^{\frac{1}{2}}$$

$$\doteq \frac{2e^4 E |B_2(1,2)|}{(q_+ \cdot q_-)} \left[\frac{q_{+-}q_{-+}}{k_{1+}k_{1-}k_{2+}k_{2-}}\right]^{\frac{1}{2}},$$

$$M(-,+;+,-,-,+) = \frac{2e^4 E B_2^*(2,1) q_{+\perp}^*}{(q_+ \cdot q_-)} \left[\frac{q_{-+}}{q_{++}k_{1+}k_{1-}k_{2+}k_{2-}}\right]^{\frac{1}{2}}$$

$$\doteq \frac{2e^4 E |B_2(2,1)|}{(q_+ \cdot q_-)} \left[\frac{q_{+-}q_{-+}}{k_{1+}k_{1-}k_{2+}k_{2-}}\right]^{\frac{1}{2}},$$

$$M(-,+;-,+,+,-) = \frac{2e^4 E B_2^*(1,2) q_{-\perp}}{(q_+ \cdot q_-)} \left[\frac{q_{++}}{q_{-+}k_{1+}k_{1-}k_{2+}k_{2-}}\right]^{\frac{1}{2}}$$

$$\doteq \frac{2e^4 E |B_2(1,2)|}{(q_+ \cdot q_-)} \left[\frac{q_{++}q_{--}}{k_{1+}k_{1-}k_{2+}k_{2-}}\right]^{\frac{1}{2}},$$

$$M(-,+;-,+,-,+) = \frac{2e^4 E B_2^*(2,1) q_{-\perp}^*}{(q_+ \cdot q_-)} \left[\frac{q_{++}}{q_{-+}k_{1+}k_{1-}k_{2+}k_{2-}}\right]^{\frac{1}{2}}$$

$$\doteq \frac{2e^4 E |B_2(2,1)|}{(q_+ \cdot q_-)} \left[\frac{q_{++}q_{--}}{k_{1+}k_{1-}k_{2+}k_{2-}}\right]^{\frac{1}{2}},$$

$$M(+,-;+,-,-,-) = \frac{e^4 (2E - k_{1+} - k_{2+}) B_1^* q_{+\perp}}{(q_+ \cdot q_-)} \left[\frac{q_{++}}{q_{-+}k_{1+}k_{1-}k_{2+}k_{2-}}\right]^{\frac{1}{2}}$$

$$\doteq \frac{e^4 (2E - k_{1+} - k_{2+}) |B_1|}{(q_+ \cdot q_-)} \left[\frac{q_{++}q_{--}}{k_{1+}k_{1-}k_{2+}k_{2-}}\right]^{\frac{1}{2}},$$

$$M(+,-;-,+,-,-) = \frac{e^4 (2E - k_{1+} - k_{2+}) B_1^* q_{-\perp}}{(q_+ \cdot q_-)} \left[\frac{q_{-+}}{q_{++}k_{1+}k_{1-}k_{2+}k_{2-}}\right]^{\frac{1}{2}}$$

$$\doteq \frac{e^4 (2E - k_{1+} - k_{2+}) |B_1|}{(q_+ \cdot q_-)} \left[\frac{q_{+-}q_{-+}}{k_{1+}k_{1-}k_{2+}k_{2-}}\right]^{\frac{1}{2}},$$

$$M(-,+;+,-,-,-) = \frac{2e^4 E B_1^* q_{-\perp}^*}{(q_+ \cdot q_-)} \left[\frac{q_{-+}}{q_{++}k_{1+}k_{1-}k_{2+}k_{2-}} \right]^{\frac{1}{2}}$$

$$\doteq \frac{2e^4 E |B_1|}{(q_+ \cdot q_-)} \left[\frac{q_{+-}q_{-+}}{k_{1+}k_{1-}k_{2+}k_{2-}} \right]^{\frac{1}{2}},$$

$$M(-,+;-,+,-,-) = \frac{2e^4 E B_1^* q_{+\perp}^*}{(q_+ \cdot q_-)} \left[\frac{q_{++}}{q_{-+}k_{1+}k_{1-}k_{2+}k_{2-}} \right]^{\frac{1}{2}}$$

$$\doteq \frac{2e^4 E |B_1|}{(q_+ \cdot q_-)} \left[\frac{q_{++}q_{--}}{k_{1+}k_{1-}k_{2+}k_{2-}} \right]^{\frac{1}{2}},$$

$$M(-,-;+,-,+,-) = \frac{2e^4 E B_4^*(2,1) q_{-\perp}}{(q_+ \cdot q_-)} \left[\frac{q_{++}}{q_{-+}k_{1+}k_{1-}k_{2+}k_{2-}} \right]^{\frac{1}{2}}$$

$$\doteq \frac{2e^4 E |B_4(2,1)|}{(q_+ \cdot q_-)} \left[\frac{q_{++}q_{--}}{k_{1+}k_{1-}k_{2+}k_{2-}} \right]^{\frac{1}{2}},$$

$$M(-,-;+,-,-,+) = \frac{2e^4 E B_4^*(1,2) q_{-\perp}}{(q_+ \cdot q_-)} \left[\frac{q_{++}}{q_{-+}k_{1+}k_{1-}k_{2+}k_{2-}} \right]^{\frac{1}{2}}$$

$$\doteq \frac{2e^4 E |B_4(1,2)|}{(q_+ \cdot q_-)} \left[\frac{q_{++}q_{--}}{k_{1+}k_{1-}k_{2+}k_{2-}} \right]^{\frac{1}{2}},$$

$$M(-,-;-,+,+,-) = \frac{2e^4 E B_4^*(2,1) q_{+\perp}}{(q_+ \cdot q_-)} \left[\frac{q_{-+}}{q_{++}k_{1+}k_{1-}k_{2+}k_{2-}} \right]^{\frac{1}{2}}$$

$$\doteq \frac{2e^4 E |B_4(2,1)|}{(q_+ \cdot q_-)} \left[\frac{q_{+-}q_{-+}}{k_{1+}k_{1-}k_{2+}k_{2-}} \right]^{\frac{1}{2}},$$

$$M(-,-;-,+,-,+) = \frac{2e^4 E B_4^*(1,2) q_{+\perp}}{(q_+ \cdot q_-)} \left[\frac{q_{-+}}{q_{++}k_{1+}k_{1-}k_{2+}k_{2-}} \right]^{\frac{1}{2}}$$

$$\doteq \frac{2e^4 E |B_4(1,2)|}{(q_+ \cdot q_-)} \left[\frac{q_{+-}q_{-+}}{k_{1+}k_{1-}k_{2+}k_{2-}} \right]^{\frac{1}{2}},$$

$$M(-,-;+,-,-,-) = \frac{2e^4 E B_3^* q_{+\perp}}{(q_+ \cdot q_-)} \left[\frac{q_{++}}{q_{-+}k_{1+}k_{1-}k_{2+}k_{2-}} \right]^{\frac{1}{2}}$$

$$\doteq \frac{2e^4 E |B_3|}{(q_+ \cdot q_-)} \left[\frac{q_{++}q_{--}}{k_{1+}k_{1-}k_{2+}k_{2-}} \right]^{\frac{1}{2}},$$

9.18. $e^+e^- \rightarrow \mu^+\mu^-\gamma\gamma$ (NO Z-EXCHANGE; $m \neq 0$)

$$M(-,-;-,+,-,-) = \frac{2e^4 E B_3^* q_{-\perp}}{(q_+ \cdot q_-)} \left[\frac{q_{-+}}{q_{++}k_{1+}k_{1-}k_{2+}k_{2-}} \right]^{\frac{1}{2}}$$

$$\doteq \frac{2e^4 E |B_3|}{(q_+ \cdot q_-)} \left[\frac{q_{+-}q_{-+}}{k_{1+}k_{1-}k_{2+}k_{2-}} \right]^{\frac{1}{2}}. \quad (9.193)$$

Unpolarized squared matrix element:

$$\overline{|M|^2} = \frac{2e^8 E^2 (q_{++}q_{--} + q_{+-}q_{-+})}{k_{1+}k_{1-}k_{2+}k_{2-}(q_+ \cdot q_-)^2} \left\{ \left[1 + \left(\frac{2E - k_{1+} - k_{2+}}{2E} \right)^2 \right] |B_1|^2 \right.$$

$$\left. + |B_2(1,2)|^2 + |B_2(2,1)|^2 + |B_3|^2 + |B_4(1,2)|^2 + |B_4(2,1)|^2 \right\}.$$

$$(9.194)$$

Unpolarized cross section:

$$d\sigma = \frac{\alpha^4 (q_{++}q_{--} + q_{+-}q_{-+})}{64\pi^2 k_{1+}k_{1-}k_{2+}k_{2-}(q_+ \cdot q_-)^2} \left\{ \left[1 + \left(\frac{2E - k_{1+} - k_{2+}}{2E} \right)^2 \right] |B_1|^2 \right.$$

$$\left. + |B_2(1,2)|^2 + |B_2(2,1)|^2 + |B_3|^2 + |B_4(1,2)|^2 + |B_4(2,1)|^2 \right\}$$

$$\times \delta^4(p_+ + p_- - q_+ - q_- - k_1 - k_2) \frac{d^3\vec{q}_+ \, d^3\vec{q}_- \, d^3\vec{k}_1 \, d^3\vec{k}_2}{q_{+0} q_{-0} k_{10} k_{20}}. \quad (9.195)$$

9.18.12 $\vec{k}_1$ and $\vec{k}_2$ nearly parallel to $\vec{p}_-$

Definitions:

$$\Delta_{12} = -2(p_- \cdot k_1) - 2(p_- \cdot k_2) + 2(k_1 \cdot k_2),$$

$$Z_{ij} = k_{i+}k_{j-} - k_{i\perp}^* k_{j\perp}, \qquad i,j = 1,2,$$

$$A_1 = \frac{2E^2}{(p_- \cdot k_1)(p_- \cdot k_2)} \left[k_{1+}k_{2+} + \frac{m^2 Z_{12}Z_{21}}{4E^2 \Delta_{12}} \right],$$

$$A_2(1,2) = \frac{1}{\Delta_{12}} \left[\frac{k_{1+}(2E - k_{1-})(2Ek_{2+} - Z_{21})}{(p_- \cdot k_1)} \right.$$

$$\left. + \frac{2Ek_{1+}k_{2+}(2E - k_{1-} - k_{2-}) - m^2 k_{2-}k_{1\perp}k_{2\perp}^*/2E}{(p_- \cdot k_2)} \right],$$

$$A_3 = \frac{m}{2(p_- \cdot k_1)(p_- \cdot k_2)} \left[k_{1-}k_{2+}k_{1\perp}^* + k_{1+}k_{2-}k_{2\perp}^* \right.$$

$$\left. - \frac{(k_{1\perp}^* + k_{2\perp}^*)Z_{12}Z_{21}}{\Delta_{12}} \right],$$

$$A_4(1,2) = \frac{mk_{1-}k_{1\perp}}{2E\Delta_{12}} \left[\frac{2Ek_{2+} - Z_{21}}{(p_- \cdot k_1)} + \frac{2Ek_{2+}}{(p_- \cdot k_2)} \right]. \qquad (9.196)$$

Photon polarizations:

$$\slashed{k}_i^\pm = N_i[\slashed{k}_i \slashed{p}_- \slashed{p}_+(1 \pm \gamma_5) + \slashed{p}_+ \slashed{p}_- \slashed{k}_i(1 \mp \gamma_5)],$$

$$N_i^{-1} = E^2[32k_{i+}k_{i-}]^{\frac{1}{2}}, \qquad i = 1, 2. \qquad (9.197)$$

Nonvanishing helicity amplitudes:

$$M(+,+;+,-,+,+) = \frac{2e^4 E A_3 q_{-\perp}}{(q_+ \cdot q_-)} \left[\frac{q_{++}}{q_{-+}k_{1+}k_{1-}k_{2+}k_{2-}} \right]^{\frac{1}{2}}$$

$$\doteq \frac{2e^4 E |A_3|}{(q_+ \cdot q_-)} \left[\frac{q_{++}q_{--}}{k_{1+}k_{1-}k_{2+}k_{2-}} \right]^{\frac{1}{2}},$$

$$M(+,+;-,+,+,+) = \frac{2e^4 E A_3 q_{+\perp}}{(q_+ \cdot q_-)} \left[\frac{q_{-+}}{q_{++}k_{1+}k_{1-}k_{2+}k_{2-}} \right]^{\frac{1}{2}}$$

$$\doteq \frac{2e^4 E |A_3|}{(q_+ \cdot q_-)} \left[\frac{q_{+-}q_{-+}}{k_{1+}k_{1-}k_{2+}k_{2-}} \right]^{\frac{1}{2}},$$

$$M(+,+;+,-,+,-) = \frac{2e^4 E A_4^*(1,2) q_{-\perp}}{(q_+ \cdot q_-)} \left[\frac{q_{++}}{q_{-+}k_{1+}k_{1-}k_{2+}k_{2-}} \right]^{\frac{1}{2}}$$

$$\doteq \frac{2e^4 E |A_4(1,2)|}{(q_+ \cdot q_-)} \left[\frac{q_{++}q_{--}}{k_{1+}k_{1-}k_{2+}k_{2-}} \right]^{\frac{1}{2}},$$

$$M(+,+;+,-,-,+) = \frac{2e^4 E A_4^*(2,1) q_{-\perp}}{(q_+ \cdot q_-)} \left[\frac{q_{++}}{q_{-+}k_{1+}k_{1-}k_{2+}k_{2-}} \right]^{\frac{1}{2}}$$

$$\doteq \frac{2e^4 E |A_4(2,1)|}{(q_+ \cdot q_-)} \left[\frac{q_{++}q_{--}}{k_{1+}k_{1-}k_{2+}k_{2-}^*} \right]^{\frac{1}{2}},$$

$$M(+,+;-,+,+,-) = \frac{2e^4 E A_4^*(1,2) q_{+\perp}}{(q_+ \cdot q_-)} \left[\frac{q_{-+}}{q_{++}k_{1+}k_{1-}k_{2+}k_{2-}} \right]^{\frac{1}{2}}$$

$$\doteq \frac{2e^4 E |A_4(1,2)|}{(q_+ \cdot q_-)} \left[\frac{q_{+-}q_{-+}}{k_{1+}k_{1-}k_{2+}k_{2-}} \right]^{\frac{1}{2}},$$

9.18. $e^+e^- \to \mu^+\mu^-\gamma\gamma$ (NO Z-EXCHANGE; $m \neq 0$)

$$M(+,+;-,+,-,+) = \frac{2e^4 E A_4^*(2,1) q_{+\perp}}{(q_+ \cdot q_-)} \left[\frac{q_{-+}}{q_{++}k_{1+}k_{1-}k_{2+}k_{2-}}\right]^{\frac{1}{2}}$$

$$\doteq \frac{2e^4 E |A_4(2,1)|}{(q_+ \cdot q_-)} \left[\frac{q_{+-}q_{-+}}{k_{1+}k_{1-}k_{2+}k_{2-}}\right]^{\frac{1}{2}},$$

$$M(+,-;+,-,+,+) = \frac{e^4(2E - k_{1-} - k_{2-}) A_1 q_{+\perp}}{(q_+ \cdot q_-)} \left[\frac{q_{++}}{q_{-+}k_{1+}k_{1-}k_{2+}k_{2-}}\right]^{\frac{1}{2}}$$

$$\doteq \frac{e^4(2E - k_{1-} - k_{2-})|A_1|}{(q_+ \cdot q_-)} \left[\frac{q_{++}q_{--}}{k_{1+}k_{1-}k_{2+}k_{2-}}\right]^{\frac{1}{2}},$$

$$M(+,-;-,+,+,+) = \frac{e^4(2E - k_{1-} - k_{2-}) A_1 q_{-\perp}}{(q_+ \cdot q_-)} \left[\frac{q_{-+}}{q_{++}k_{1+}k_{1-}k_{2+}k_{2-}}\right]^{\frac{1}{2}}$$

$$\doteq \frac{e^4(2E - k_{1-} - k_{2-})|A_1|}{(q_+ \cdot q_-)} \left[\frac{q_{+-}q_{-+}}{k_{1+}k_{1-}k_{2+}k_{2-}}\right]^{\frac{1}{2}},$$

$$M(-,+;+,-,+,+) = \frac{2e^4 E A_1 q_{-\perp}^*}{(q_+ \cdot q_-)} \left[\frac{q_{-+}}{q_{++}k_{1+}k_{1-}k_{2+}k_{2-}}\right]^{\frac{1}{2}}$$

$$\doteq \frac{2e^4 E |A_1|}{(q_+ \cdot q_-)} \left[\frac{q_{+-}q_{-+}}{k_{1+}k_{1-}k_{2+}k_{2-}}\right]^{\frac{1}{2}},$$

$$M(-,+;-,+,+,+) = \frac{2e^4 E A_1 q_{+\perp}^*}{(q_+ \cdot q_-)} \left[\frac{q_{++}}{q_{-+}k_{1+}k_{1-}k_{2+}k_{2-}}\right]^{\frac{1}{2}}$$

$$\doteq \frac{2e^4 E |A_1|}{(q_+ \cdot q_-)} \left[\frac{q_{++}q_{--}}{k_{1+}k_{1-}k_{2+}k_{2-}}\right]^{\frac{1}{2}},$$

$$M(+,-;+,-,+,-) = \frac{2e^4 E A_2^*(1,2) q_{-\perp}}{(q_+ \cdot q_-)} \left[\frac{q_{++}}{q_{-+}k_{1+}k_{1-}k_{2+}k_{2-}}\right]^{\frac{1}{2}}$$

$$\doteq \frac{2e^4 E |A_2(1,2)|}{(q_+ \cdot q_-)} \left[\frac{q_{++}q_{--}}{k_{1+}k_{1-}k_{2+}k_{2-}}\right]^{\frac{1}{2}},$$

$$M(+,-;+,-,-,+) = \frac{2e^4 E A_2^*(2,1) q_{-\perp}}{(q_+ \cdot q_-)} \left[\frac{q_{++}}{q_{-+}k_{1+}k_{1-}k_{2+}k_{2-}}\right]^{\frac{1}{2}}$$

$$\doteq \frac{2e^4 E |A_2(2,1)|}{(q_+ \cdot q_-)} \left[\frac{q_{++}q_{--}}{k_{1+}k_{1-}k_{2+}k_{2-}}\right]^{\frac{1}{2}},$$

$$M(+,-;-,+,+,-) = \frac{2e^4 E A_2^*(1,2) q_{+\perp}}{(q_+ \cdot q_-)} \left[\frac{q_{-+}}{q_{++} k_{1+} k_{1-} k_{2+} k_{2-}} \right]^{\frac{1}{2}}$$

$$\doteq \frac{2e^4 E |A_2(1,2)|}{(q_+ \cdot q_-)} \left[\frac{q_{+-} q_{-+}}{k_{1+} k_{1-} k_{2+} k_{2-}} \right]^{\frac{1}{2}},$$

$$M(+,-;-,+,-,+) = \frac{2e^4 E A_2^*(2,1) q_{+\perp}}{(q_+ \cdot q_-)} \left[\frac{q_{-+}}{q_{++} k_{1+} k_{1-} k_{2+} k_{2-}} \right]^{\frac{1}{2}}$$

$$\doteq \frac{2e^4 E |A_2(2,1)|}{(q_+ \cdot q_-)} \left[\frac{q_{+-} q_{-+}}{k_{1+} k_{1-} k_{2+} k_{2-}} \right]^{\frac{1}{2}},$$

$$M(-,+;+,-,+,-) = \frac{2e^4 E A_2(2,1) q_{+\perp}^*}{(q_+ \cdot q_-)} \left[\frac{q_{-+}}{q_{++} k_{1+} k_{1-} k_{2+} k_{2-}} \right]^{\frac{1}{2}}$$

$$\doteq \frac{2e^4 E |A_2(1,2)|}{(q_+ \cdot q_-)} \left[\frac{q_{+-} q_{-+}}{k_{1+} k_{1-} k_{2+} k_{2-}} \right]^{\frac{1}{2}},$$

$$M(-,+;+,-,-,+) = \frac{2e^4 E A_2(1,2) q_{+\perp}^*}{(q_+ \cdot q_-)} \left[\frac{q_{-+}}{q_{++} k_{1+} k_{1-} k_{2+} k_{2-}} \right]^{\frac{1}{2}}$$

$$\doteq \frac{2e^4 E |A_2(1,2)|}{(q_+ \cdot q_-)} \left[\frac{q_{+-} q_{-+}}{k_{1+} k_{1-} k_{2+} k_{2-}} \right]^{\frac{1}{2}},$$

$$M(-,+;-,+,+,-) = \frac{2e^4 E A_2(2,1) q_{-\perp}^*}{(q_+ \cdot q_-)} \left[\frac{q_{++}}{q_{-+} k_{1+} k_{1-} k_{2+} k_{2-}} \right]^{\frac{1}{2}}$$

$$\doteq \frac{2e^4 E |A_2(1,2)|}{(q_+ \cdot q_-)} \left[\frac{q_{++} q_{--}}{k_{1+} k_{1-} k_{2+} k_{2-}} \right]^{\frac{1}{2}},$$

$$M(-,+;-,+,-,+) = \frac{2e^4 E A_2(2,1) q_{-\perp}^*}{(q_+ \cdot q_-)} \left[\frac{q_{++}}{q_{-+} k_{1+} k_{1-} k_{2+} k_{2-}} \right]^{\frac{1}{2}}$$

$$\doteq \frac{2e^4 E |A_2(2,1)|}{(q_+ \cdot q_-)} \left[\frac{q_{++} q_{--}}{k_{1+} k_{1-} k_{2+} k_{2-}} \right]^{\frac{1}{2}},$$

$$M(+,-;+,-,-,-) = \frac{2e^4 E A_1^* q_{+\perp}}{(q_+ \cdot q_-)} \left[\frac{q_{++}}{q_{-+} k_{1+} k_{1-} k_{2+} k_{2-}} \right]^{\frac{1}{2}}$$

$$\doteq \frac{2e^4 E |A_1|}{(q_+ \cdot q_-)} \left[\frac{q_{++} q_{--}}{k_{1+} k_{1-} k_{2+} k_{2-}} \right]^{\frac{1}{2}},$$

9.18. $e^+e^- \to \mu^+\mu^-\gamma\gamma$ (NO Z-EXCHANGE; $m \neq 0$)

$$M(+,-;-,+,-,-) = \frac{2e^4 E A_1^* q_{-\perp}}{(q_+ \cdot q_-)} \left[\frac{q_{-+}}{q_{++}k_{1+}k_{1-}k_{2+}k_{2-}}\right]^{\frac{1}{2}}$$

$$\doteq \frac{2e^4 E |A_1|}{(q_+ \cdot q_-)} \left[\frac{q_{+-}q_{-+}}{k_{1+}k_{1-}k_{2+}k_{2-}}\right]^{\frac{1}{2}},$$

$$M(-,+;+,-,-,-) = \frac{e^4(2E - k_{1-} - k_{2-}) A_1^* q_{-\perp}^*}{(q_+ \cdot q_-)} \left[\frac{q_{-+}}{q_{++}k_{1+}k_{1-}k_{2+}k_{2-}}\right]^{\frac{1}{2}}$$

$$\doteq \frac{e^4(2E - k_{1-} - k_{2-})|A_1|}{(q_+ \cdot q_-)} \left[\frac{q_{+-}q_{-+}}{k_{1+}k_{1-}k_{2+}k_{2-}}\right]^{\frac{1}{2}},$$

$$M(-,+;-,+,-,-) = \frac{e^4(2E - k_{1-} - k_{2-}) A_1^* q_{+\perp}^*}{(q_+ \cdot q_-)} \left[\frac{q_{++}}{q_{-+}k_{1+}k_{1-}k_{2+}k_{2-}}\right]^{\frac{1}{2}}$$

$$\doteq \frac{e^4(2E - k_{1-} - k_{2-})|A_1|}{(q_+ \cdot q_-)} \left[\frac{q_{++}q_{--}}{k_{1+}k_{1-}k_{2+}k_{2-}}\right]^{\frac{1}{2}},$$

$$M(-,-;+,-,+,-) = \frac{2e^4 E A_4(2,1) q_{-\perp}^*}{(q_+ \cdot q_-)} \left[\frac{q_{-+}}{q_{++}k_{1+}k_{1-}k_{2+}k_{2-}}\right]^{\frac{1}{2}}$$

$$\doteq \frac{2e^4 E |A_4(2,1)|}{(q_+ \cdot q_-)} \left[\frac{q_{+-}q_{-+}}{k_{1+}k_{1-}k_{2+}k_{2-}}\right]^{\frac{1}{2}},$$

$$M(-,-;+,-,-,+) = \frac{2e^4 E A_4(1,2) q_{-\perp}^*}{(q_+ \cdot q_-)} \left[\frac{q_{-+}}{q_{++}k_{1+}k_{1-}k_{2+}k_{2-}}\right]^{\frac{1}{2}}$$

$$\doteq \frac{2e^4 E |A_4(1,2)|}{(q_+ \cdot q_-)} \left[\frac{q_{+-}q_{-+}}{k_{1+}k_{1-}k_{2+}k_{2-}}\right]^{\frac{1}{2}},$$

$$M(-,-;-,+,+,-) = \frac{2e^4 E A_4(2,1) q_{+\perp}^*}{(q_+ \cdot q_-)} \left[\frac{q_{++}}{q_{-+}k_{1+}k_{1-}k_{2+}k_{2-}}\right]^{\frac{1}{2}}$$

$$\doteq \frac{2e^4 E |A_4(2,1)|}{(q_+ \cdot q_-)} \left[\frac{q_{++}q_{--}}{k_{1+}k_{1-}k_{2+}k_{2-}}\right]^{\frac{1}{2}},$$

$$M(-,-;-,+,-,+) = \frac{2e^4 E A_4(1,2) q_{+\perp}^*}{(q_+ \cdot q_-)} \left[\frac{q_{++}}{q_{-+}k_{1+}k_{1-}k_{2+}k_{2-}}\right]^{\frac{1}{2}}$$

$$\doteq \frac{2e^4 E |A_4(1,2)|}{(q_+ \cdot q_-)} \left[\frac{q_{++}q_{--}}{k_{1+}k_{1-}k_{2+}k_{2-}}\right]^{\frac{1}{2}},$$

$$M(-,-;+,-,-,-) = \frac{2e^4 E A_3^* q_{-\perp}^*}{(q_+ \cdot q_-)} \left[\frac{q_{-+}}{q_{++}k_{1+}k_{1-}k_{2+}k_{2-}} \right]^{\frac{1}{2}}$$

$$\doteq \frac{2e^4 E|A_3|}{(q_+ \cdot q_-)} \left[\frac{q_{+-}q_{-+}}{k_{1+}k_{1-}k_{2+}k_{2-}} \right]^{\frac{1}{2}},$$

$$M(-,-;-,+,-,-) = \frac{2e^4 E A_3^* q_{+\perp}^*}{(q_+ \cdot q_-)} \left[\frac{q_{++}}{q_{-+}k_{1+}k_{1-}k_{2+}k_{2-}} \right]^{\frac{1}{2}}$$

$$\doteq \frac{2e^4 E|A_3|}{(q_+ \cdot q_-)} \left[\frac{q_{++}q_{--}}{k_{1+}k_{1-}k_{2+}k_{2-}} \right]^{\frac{1}{2}}. \qquad (9.198)$$

Unpolarized squared matrix element:

$$\overline{|M|^2} = \frac{2e^8 E^2 (q_{++}q_{--} + q_{+-}q_{-+})}{(q_+ \cdot q_-)^2 k_{1+}k_{1-}k_{2+}k_{2-}} \left\{ \left[1 + \left(\frac{2E - k_{1-} - k_{2-}}{2E} \right)^2 \right] |A_1|^2 \right.$$

$$\left. + |A_2(1,2)|^2 + |A_2(2,1)|^2 + |A_3|^2 + |A_4(1,2)|^2 + |A_4(2,1)|^2 \right\}.$$

$$(9.199)$$

Unpolarized cross section:

$$d\sigma = \frac{\alpha^4 (q_{++}q_{--} + q_{+-}q_{-+})}{64\pi^2 (q_+ \cdot q_-)^2 k_{1+}k_{1-}k_{2+}k_{2-}} \left\{ \left[1 + \left(\frac{2E - k_{1-} - k_{2-}}{2E} \right)^2 \right] |A_1|^2 \right.$$

$$\left. + |A_2(1,2)|^2 + |A_2(2,1)|^2 + |A_3|^2 + |A_4(1,2)|^2 + |A_4(2,1)|^2 \right\}$$

$$\times \delta^4 (p_+ + p_- - q_+ - q_- - k_1 - k_2) \frac{d^3 \vec{q}_+ \, d^3 \vec{q}_- \, d^3 \vec{k}_1 \, d^3 \vec{k}_2}{q_{+0} \, q_{-0} \, k_{10} \, k_{20}}. \qquad (9.200)$$

9.18.13 $\vec{k}_1$ and $\vec{k}_2$ nearly parallel to $\vec{q}_+$

The primed quantities $k'_{i\pm}$ and $k'_{i\perp}$, $i = 1,2$, are evaluated in the rotated frame where $\vec{q}_+$ determines the positive z-axis (see also Section 7.4.3). The four-vector q in eqns (9.202) is obtained by applying a space reflection to q_+. The quantity μ denotes the muon mass.
Definitions:

$$\Delta'_{12} = 2(q_+ \cdot k_1) + 2(q_+ \cdot k_2) + 2(k_1 \cdot k_2),$$

$$Z'_{ij} = k'_{i+} k'_{j-} - k'^*_{i\perp} k'_{j\perp}, \qquad i,j = 1,2,$$

9.18. $e^+e^- \to \mu^+\mu^-\gamma\gamma$ (NO Z-EXCHANGE; $m \neq 0$)

$$A'_1 = \frac{2q_{+0}^2}{(q_+ \cdot k_1)(q_+ \cdot k_2)} \left[k'_{1-}k'_{2-} + \frac{\mu^2 Z'_{12} Z'_{21}}{4q_{+0}^2 \Delta'_{12}} \right],$$

$$A'_2(1,2) = \frac{1}{\Delta'_{12}} \left[\frac{k'_{1-}(2q_{+0} + k'_{1+})(2q_{+0}k'_{2-} + Z'_{12})}{(q_+ \cdot k_1)} \right.$$

$$\left. + \frac{2q_{+0}k'_{1-}k'_{2-}(2q_{+0} + k'_{1+} + k'_{2+}) + \mu^2 k'_{2+}k'^{*}_{1\perp}k'_{2\perp}/2q_{+0}}{(q_+ \cdot k_2)} \right],$$

$$A'_3 = \frac{\mu}{(q_+ \cdot k_1)(q_+ \cdot k_2)} \left[k'_{1+}k'_{2-}k'_{1\perp} + k'_{1-}k'_{2+}k'_{2\perp} \right.$$

$$\left. - \frac{(k'_{1\perp} + k'_{2\perp})Z'_{12}Z'_{21}}{\Delta'_{12}} \right],$$

$$A'_4(1,2) = \frac{\mu k'_{1+}k'_{1\perp}}{2q_{+0}\Delta'_{12}} \left[\frac{2q_{+0}k'_{2-} + Z'^{*}_{12}}{(q_+ \cdot k_1)} + \frac{2q_{+0}k'_{2-}}{(q_+ \cdot k_2)} \right]. \qquad (9.201)$$

Photon polarizations:

$$\not{\epsilon}_i^{\pm} = N_i[\not{k}_i \not{q}_+ \not{A}(1 \pm \gamma_5) + \not{A} \not{q}_+ \not{k}_i(1 \mp \gamma_5)],$$

$$N_i^{-1} = q_{+0}^2 [32 k'_{i+} k'_{i-}]^{\frac{1}{2}}, \qquad i = 1,2. \qquad (9.202)$$

Nonvanishing helicity amplitudes:

$$M(+,-;+,+,+,-) = -\frac{e^4 A'^{*}_4(2,1) q_{+\perp}}{E} \left[\frac{q_{-+}}{q_{++}k'_{1+}k'_{1-}k'_{2+}k'_{2-}} \right]^{\frac{1}{2}}$$

$$\doteq \frac{e^4 |A'_4(2,1)|}{E} \left[\frac{q_{+-}q_{-+}}{k'_{1+}k'_{1-}k'_{2+}k'_{2-}} \right]^{\frac{1}{2}},$$

$$M(+,-;+,+,-,+) = -\frac{e^4 A'^{*}_4(1,2) q_{+\perp}}{E} \left[\frac{q_{-+}}{q_{++}k'_{1+}k'_{1-}k'_{2+}k'_{2-}} \right]^{\frac{1}{2}}$$

$$\doteq \frac{e^4 |A'_4(1,2)|}{E} \left[\frac{q_{+-}q_{-+}}{k'_{1+}k'_{1-}k'_{2+}k'_{2-}} \right]^{\frac{1}{2}},$$

$$M(+,-;+,-,+,+) = -\frac{e^4 A'_1 q_{-\perp}}{E} \left[\frac{q_{--}}{q_{+-}k'_{1+}k'_{1-}k'_{2+}k'_{2-}} \right]^{\frac{1}{2}}$$

$$\doteq \frac{e^4 |A'_1|}{E} \left[\frac{q_{--}^3}{q_{++}k'_{1+}k'_{1-}k'_{2+}k'_{2-}} \right]^{\frac{1}{2}},$$

$$M(+,-;-,+,+,+) = -\frac{e^4 A_1' q_{+\perp}}{E}\left[\frac{q_{-+}}{q_{++}k_{1+}'k_{1-}'k_{2+}'k_{2-}'}\right]^{\frac{1}{2}}$$

$$\doteq \frac{e^4|A_1'|}{E}\left[\frac{q_{+-}q_{-+}}{k_{1+}'k_{1-}'k_{2+}'k_{2-}'}\right]^{\frac{1}{2}},$$

$$M(-,+;+,+,-,+) = -\frac{e^4 A_4'^*(1,2) q_{-\perp}^*}{E}\left[\frac{q_{++}}{q_{-+}k_{1+}'k_{1-}'k_{2+}'k_{2-}'}\right]^{\frac{1}{2}}$$

$$\doteq \frac{e^4|A_4'(1,2)|}{E}\left[\frac{q_{++}q_{--}}{k_{1+}'k_{1-}'k_{2+}'k_{2-}'}\right]^{\frac{1}{2}},$$

$$M(-,+;+,-,+,+) = -\frac{2E e^4 A_1' q_{+\perp}^*}{(q_+\cdot q_-)}\left[\frac{q_{-+}}{q_{++}k_{1+}'k_{1-}'k_{2+}'k_{2-}'}\right]^{\frac{1}{2}}$$

$$\doteq \frac{2e^4 E|A_1'|}{(q_+\cdot q_-)}\left[\frac{q_{+-}q_{-+}}{k_{1+}'k_{1-}'k_{2+}'k_{2-}'}\right]^{\frac{1}{2}},$$

$$M(-,+;-,+,+,+) = -\frac{e^4 A_1' q_{-\perp}^*}{E}\left[\frac{q_{++}}{q_{-+}k_{1+}'k_{1-}'k_{2+}'k_{2-}'}\right]^{\frac{1}{2}}$$

$$\doteq \frac{e^4|A_1'|}{E}\left[\frac{q_{++}q_{--}}{k_{1+}'k_{1-}'k_{2+}'k_{2-}'}\right]^{\frac{1}{2}},$$

$$M(+,-;+,+,-,-) = -\frac{e^4 A_3'^* q_{+\perp}}{E}\left[\frac{q_{-+}}{q_{++}k_{1+}'k_{1-}'k_{2+}'k_{2-}'}\right]^{\frac{1}{2}}$$

$$\doteq \frac{e^4|A_3'|}{E}\left[\frac{q_{+-}q_{-+}}{k_{1+}'k_{1-}'k_{2+}'k_{2-}'}\right]^{\frac{1}{2}},$$

$$M(+,-;+,-,+,-) = -\frac{e^4 A_2'^*(1,2) q_{-\perp}}{E}\left[\frac{q_{++}}{q_{-+}k_{1+}'k_{1-}'k_{2+}'k_{2-}'}\right]^{\frac{1}{2}}$$

$$\doteq \frac{e^4|A_2'(1,2)|}{E}\left[\frac{q_{++}q_{--}}{k_{1+}'k_{1-}'k_{2+}'k_{2-}'}\right]^{\frac{1}{2}},$$

$$M(+,-;+,-,-,+) = -\frac{e^4 A_2'^*(2,1) q_{-\perp}}{E}\left[\frac{q_{++}}{q_{-+}k_{1+}'k_{1-}'k_{2+}'k_{2-}'}\right]^{\frac{1}{2}}$$

$$\doteq \frac{e^4|A_2'(2,1)|}{E}\left[\frac{q_{++}q_{--}}{k_{1+}'k_{1-}'k_{2+}'k_{2-}'}\right]^{\frac{1}{2}},$$

9.18. $e^+e^- \to \mu^+\mu^-\gamma\gamma$ (NO Z-EXCHANGE; $m \neq 0$)

$$M(+,-;-,+,+,-) = -\frac{e^4 A_2'(2,1) q_{+\perp}}{E} \left[\frac{q_{-+}}{q_{++} k_{1+}' k_{1-}' k_{2+}' k_{2-}'}\right]^{\frac{1}{2}}$$

$$\doteq \frac{e^4 |A_2'(2,1)|}{E} \left[\frac{q_{+-} q_{-+}}{k_{1+}' k_{1-}' k_{2+}' k_{2-}'}\right]^{\frac{1}{2}},$$

$$M(+,-;-,+,-,+) = -\frac{e^4 A_2'(1,2) q_{+\perp}}{E} \left[\frac{q_{-+}}{q_{++} k_{1+}' k_{1-}' k_{2+}' k_{2-}'}\right]^{\frac{1}{2}}$$

$$\doteq \frac{e^4 |A_2'(1,2)|}{E} \left[\frac{q_{+-} q_{-+}}{k_{1+}' k_{1-}' k_{2+}' k_{2-}'}\right]^{\frac{1}{2}},$$

$$M(+,-;-,-,+,+) = \frac{e^4 A_3' q_{-\perp}}{E} \left[\frac{q_{++}}{q_{-+} k_{1+}' k_{1-}' k_{2+}' k_{2-}'}\right]^{\frac{1}{2}}$$

$$\doteq \frac{e^4 |A_3'|}{E} \left[\frac{q_{++} q_{--}}{k_{1+}' k_{1-}' k_{2+}' k_{2-}'}\right]^{\frac{1}{2}},$$

$$M(-,+;+,+,+,-) = -\frac{e^4 A_4'^*(2,1) q_{-\perp}^*}{E} \left[\frac{q_{++}}{q_{-+} k_{1+}' k_{1-}' k_{2+}' k_{2-}'}\right]^{\frac{1}{2}}$$

$$\doteq \frac{e^4 |A_4'(2,1)|}{E} \left[\frac{q_{++} q_{--}}{k_{1+}' k_{1-}' k_{2+}' k_{2-}'}\right]^{\frac{1}{2}},$$

$$M(-,+;+,+,-,-) = -\frac{e^4 A_3'^* q_{-\perp}^*}{E} \left[\frac{q_{++}}{q_{-+} k_{1+}' k_{1-}' k_{2+}' k_{2-}'}\right]^{\frac{1}{2}}$$

$$\doteq \frac{e^4 |A_3'|}{E} \left[\frac{q_{++} q_{--}}{k_{1+}' k_{1-}' k_{2+}' k_{2-}'}\right]^{\frac{1}{2}},$$

$$M(-,+;+,-,+,-) = -\frac{e^4 A_2'^*(1,2) q_{+\perp}^*}{E} \left[\frac{q_{-+}}{q_{++} k_{1+}' k_{1-}' k_{2+}' k_{2-}'}\right]^{\frac{1}{2}}$$

$$\doteq \frac{e^4 |A_2'(1,2)|}{E} \left[\frac{q_{+-} q_{-+}}{k_{1+}' k_{1-}' k_{2+}' k_{2-}'}\right]^{\frac{1}{2}},$$

$$M(-,+;+,-,-,+) = -\frac{e^4 A_2'^*(2,1) q_{+\perp}^*}{E} \left[\frac{q_{-+}}{q_{++} k_{1+}' k_{1-}' k_{2+}' k_{2-}'}\right]^{\frac{1}{2}}$$

$$\doteq \frac{e^4 |A_2'(2,1)|}{E} \left[\frac{q_{+-} q_{-+}}{k_{1+}' k_{1-}' k_{2+}' k_{2-}'}\right]^{\frac{1}{2}},$$

$$M(-,+;-,+,+,-) = -\frac{e^4 A_2'(2,1)q_{-\perp}^*}{E}\left[\frac{q_{++}}{q_{-+}k_{1+}'k_{1-}'k_{2+}'k_{2-}'}\right]^{\frac{1}{2}}$$

$$\doteq \frac{e^4 |A_2'(2,1)|}{E}\left[\frac{q_{++}q_{--}}{k_{1+}'k_{1-}'k_{2+}'k_{2-}'}\right]^{\frac{1}{2}},$$

$$M(-,+;-,+,-,+) = -\frac{e^4 A_2'(1,2)q_{-\perp}^*}{E}\left[\frac{q_{++}}{q_{-+}k_{1+}'k_{1-}'k_{2+}'k_{2-}'}\right]^{\frac{1}{2}}$$

$$\doteq \frac{e^4 |A_2'(1,2)|}{E}\left[\frac{q_{++}q_{--}}{k_{1+}'k_{1-}'k_{2+}'k_{2-}'}\right]^{\frac{1}{2}},$$

$$M(-,+;-,-,+,+) = \frac{e^4 A_3' q_{+\perp}^*}{E}\left[\frac{q_{-+}}{q_{++}k_{1+}'k_{1-}'k_{2+}'k_{2-}'}\right]^{\frac{1}{2}}$$

$$\doteq \frac{e^4 |A_3'|}{E}\left[\frac{q_{+-}q_{-+}}{k_{1+}'k_{1-}'k_{2+}'k_{2-}'}\right]^{\frac{1}{2}},$$

$$M(+,-;+,-,-,-) = -\frac{e^4 A_1'^* q_{-\perp}}{E}\left[\frac{q_{++}}{q_{-+}k_{1+}'k_{1-}'k_{2+}'k_{2-}'}\right]^{\frac{1}{2}}$$

$$\doteq \frac{e^4 |A_1'|}{E}\left[\frac{q_{++}q_{--}}{k_{1+}'k_{1-}'k_{2+}'k_{2-}'}\right]^{\frac{1}{2}},$$

$$M(+,-;-,+,-,-) = -\frac{2e^4 E A_1'^* q_{+\perp}}{(q_+ \cdot q_-)}\left[\frac{q_{-+}}{q_{++}k_{1+}'k_{1-}'k_{2+}'k_{2-}'}\right]^{\frac{1}{2}}$$

$$\doteq \frac{2e^4 E |A_1'|}{(q_+ \cdot q_-)}\left[\frac{q_{+-}q_{-+}}{k_{1+}'k_{1-}'k_{2+}'k_{2-}'}\right]^{\frac{1}{2}},$$

$$M(+,-;-,-,+,-) = \frac{e^4 A_4'(1,2)q_{-\perp}}{E}\left[\frac{q_{++}}{q_{-+}k_{1+}'k_{1-}'k_{2+}'k_{2-}'}\right]^{\frac{1}{2}}$$

$$\doteq \frac{e^4 |A_4'(1,2)|}{E}\left[\frac{q_{++}q_{--}}{k_{1+}'k_{1-}'k_{2+}'k_{2-}'}\right]^{\frac{1}{2}},$$

$$M(+,-;-,-,-,+) = \frac{e^4 A_4'(2,1)q_{-\perp}}{E}\left[\frac{q_{++}}{q_{-+}k_{1+}'k_{1-}'k_{2+}'k_{2-}'}\right]^{\frac{1}{2}}$$

$$\doteq \frac{e^4 |A_4'(2,1)|}{E}\left[\frac{q_{++}q_{--}}{k_{1+}'k_{1-}'k_{2+}'k_{2-}'}\right]^{\frac{1}{2}},$$

9.18. $e^+e^- \rightarrow \mu^+\mu^-\gamma\gamma$ (NO Z-EXCHANGE; $m \neq 0$)

$$M(-,+;+,-,-,-) = -\frac{e^4 A_1'^* q_{+\perp}^*}{E}\left[\frac{q_{-+}}{q_{++}k_{1+}'k_{1-}'k_{2+}'k_{2-}'}\right]^{\frac{1}{2}}$$

$$\doteq \frac{e^4|A_1'|}{E}\left[\frac{q_{+-}q_{-+}}{k_{1+}'k_{1-}'k_{2+}'k_{2-}'}\right]^{\frac{1}{2}},$$

$$M(-,+;-,+,-,-) = -\frac{e^4 A_1'^* q_{-\perp}^*}{E}\left[\frac{q_{--}}{q_{+-}k_{1+}'k_{1-}'k_{2+}'k_{2-}'}\right]^{\frac{1}{2}}$$

$$\doteq \frac{e^4|A_1'|}{E}\left[\frac{q_{--}^3}{q_{++}k_{1+}'k_{1-}'k_{2+}'k_{2-}'}\right]^{\frac{1}{2}},$$

$$M(-,+;-,-,+,-) = \frac{e^4 A_4'(1,2)q_{+\perp}^*}{E}\left[\frac{q_{-+}}{q_{++}k_{1+}'k_{1-}'k_{2+}'k_{2-}'}\right]^{\frac{1}{2}}$$

$$\doteq \frac{e^4|A_4'(1,2)|}{E}\left[\frac{q_{+-}q_{-+}}{k_{1+}'k_{1-}'k_{2+}'k_{2-}'}\right]^{\frac{1}{2}},$$

$$M(-,+;-,-,-,+) = \frac{e^4 A_4'(2,1)q_{+\perp}^*}{E}\left[\frac{q_{-+}}{q_{++}k_{1+}'k_{1-}'k_{2+}'k_{2-}'}\right]^{\frac{1}{2}}$$

$$\doteq \frac{e^4|A_4'(2,1)|}{E}\left[\frac{q_{+-}q_{-+}}{k_{1+}'k_{1-}'k_{2+}'k_{2-}'}\right]^{\frac{1}{2}}. \quad (9.203)$$

Unpolarized squared matrix element:

$$\overline{|M|^2} = \frac{e^8(q_{++}q_{--}+q_{+-}q_{-+})}{2E^2 k_{1+}'k_{1-}'k_{2+}'k_{2-}'}\left\{\left[1+\left(\frac{2q_{+0}+k_{1+}'+k_{2+}'}{2q_{+0}}\right)^2\right]|A_1'|^2\right.$$

$$\left. + |A_2'(1,2)|^2 + |A_2'(2,1)|^2 + |A_3'|^2 + |A_4'(1,2)|^2 + |A_4'(2,1)|^2\right\}. \quad (9.204)$$

Unpolarized cross section:

$$d\sigma = \frac{\alpha^4(q_{++}q_{--}+q_{+-}q_{-+})}{256\pi^2 E^4 k_{1+}'k_{1-}'k_{2+}'k_{2-}'}\left\{\left[1+\left(\frac{2q_{+0}+k_{1+}'+k_{2+}'}{2q_{+0}}\right)^2\right]|A_1'|^2\right.$$

$$\left. + |A_2'(1,2)|^2 + |A_2'(2,1)|^2 + |A_3'|^2 + |A_4'(1,2)|^2 + |A_4'(2,1)|^2\right\}$$

$$\times \delta^4(p_+ + p_- - q_+ - q_- - k_1 - k_2)\frac{d^3\vec{q}_+\, d^3\vec{q}_-\, d^3\vec{k}_1\, d^3\vec{k}_2}{q_{+0}\, q_{-0}\, k_{10}\, k_{20}}. \quad (9.205)$$

9.18.14 $\vec{k}_1$ and $\vec{k}_2$ nearly parallel to $\vec{q}_-$

The doubly primed quantities $k''_{i\pm}$ and $k''_{i\perp}$, $i = 1, 2$, are evaluated rotated frame where $\vec{q}_-$ determines the positive z-axis (see also Section The four-vector q in eqns (9.207) is obtained by applying a space refl to q_-. The quantity μ denotes the muon mass.

Definitions:

$$\Delta''_{12} = 2(q_- \cdot k_1) + 2(q_- \cdot k_2) + 2(k_1 \cdot k_2),$$

$$Z''_{ij} = k''_{i+} k''_{j-} - k''^*_{i\perp} k''_{j\perp}, \qquad i, j = 1, 2,$$

$$A''_1 = \frac{2q^2_{-0}}{(q_- \cdot k_1)(q_- \cdot k_2)} \left[k''_{1-} k''_{2-} + \frac{\mu^2 Z''_{12} Z''_{21}}{4q^2_{-0} \Delta''_{12}} \right],$$

$$A''_2(1,2) = \frac{1}{\Delta''_{12}} \left[\frac{k''_{1-}(2q_{-0} + k''_{1+})(2q_{-0} k''_{2-} + Z''_{12})}{(q_- \cdot k_1)} \right.$$

$$\left. + \frac{2q_{-0} k''_{1-} k''_{2-}(2q_{-0} + k''_{1+} + k''_{2+}) + \mu^2 k''_{2+} k''^*_{1\perp} k''_{2\perp}/2q_{-0}}{(q_- \cdot k_2)} \right]$$

$$A''_3 = \frac{\mu}{(q_- \cdot k_1)(q_- \cdot k_2)} \left[k''_{1+} k''_{2-} k''_{1\perp} + k''_{1-} k''_{2+} k''_{2\perp} \right.$$

$$\left. - \frac{(k''_{1\perp} + k''_{2\perp}) Z''_{12} Z''_{21}}{\Delta''_{12}} \right]$$

$$A''_4(1,2) = \frac{\mu k''_{1+} k''_{1\perp}}{2q_{-0} \Delta''_{12}} \left[\frac{2q_{-0} k''_{2-} + Z''^*_{12}}{(q_- \cdot k_1)} + \frac{2q_{-0} k''_{2-}}{(q_- \cdot k_2)} \right]. \qquad ($$

Photon polarizations:

$$\rlap{/}{k}^{\pm}_i = N_i [\rlap{/}{k}_i \rlap{/}{k}_- \rlap{/}{A}(1 \pm \gamma_5) + \rlap{/}{A} \rlap{/}{k}_- \rlap{/}{k}_i (1 \mp \gamma_5)],$$

$$N_i^{-1} = q^2_{-0} [32 k''_{i+} k''_{i-}]^{\frac{1}{2}}, \qquad i = 1, 2. \qquad ($$

Nonvanishing helicity amplitudes:

$$M(+, -; +, +, +, -) = -\frac{e^4 A''^*_4(2,1) q_{-\perp}}{E} \left[\frac{q_{++}}{q_{-+} k''_{1+} k''_{1-} k''_{2+} k''_{2-}} \right]^{\frac{1}{2}}$$

$$\doteq \frac{e^4 |A''_4(2,1)|}{E} \left[\frac{q_{++} q_{--}}{k''_{1+} k''_{1-} k''_{2+} k''_{2-}} \right]^{\frac{1}{2}},$$

9.18. $e^+e^- \to \mu^+\mu^-\gamma\gamma$ (NO Z-EXCHANGE; $m \neq 0$)

$$M(+,-;+,+,-,+) = -\frac{e^4 A_4''^*(1,2) q_{-\perp}}{E}\left[\frac{q_{++}}{q_{-+}k_{1+}''k_{1-}''k_{2+}''k_{2-}''}\right]^{\frac{1}{2}}$$

$$\doteq \frac{e^4|A_4''(1,2)|}{E}\left[\frac{q_{++}q_{--}}{k_{1+}''k_{1-}''k_{2+}''k_{2-}''}\right]^{\frac{1}{2}},$$

$$M(+,-;+,-,+,+) = -\frac{e^4 A_1'' q_{-\perp}}{E}\left[\frac{q_{++}}{q_{-+}k_{1+}''k_{1-}''k_{2+}''k_{2-}''}\right]^{\frac{1}{2}}$$

$$\doteq \frac{e^4|A_1''|}{E}\left[\frac{q_{++}q_{--}}{k_{1+}''k_{1-}''k_{2+}''k_{2-}''}\right]^{\frac{1}{2}},$$

$$M(+,-;-,+,+,+) = -\frac{e^4 A_1'' q_{+\perp}}{E}\left[\frac{q_{+-}}{q_{--}k_{1+}''k_{1-}''k_{2+}''k_{2-}''}\right]^{\frac{1}{2}}$$

$$\doteq \frac{e^4 q_{+-}|A_1''|}{E}\left[\frac{q_{++}}{q_{--}k_{1+}''k_{1-}''k_{2+}''k_{2-}''}\right]^{\frac{1}{2}},$$

$$M(-,+;+,+,+,-) = -\frac{e^4 A_4''^*(2,1) q_{+\perp}^*}{E}\left[\frac{q_{-+}}{q_{++}k_{1+}''k_{1-}''k_{2+}''k_{2-}''}\right]^{\frac{1}{2}}$$

$$\doteq \frac{e^4|A_4''(2,1)|}{E}\left[\frac{q_{+-}q_{-+}}{k_{1+}''k_{1-}''k_{2+}''k_{2-}''}\right]^{\frac{1}{2}},$$

$$M(-,+;+,+,-,+) = -\frac{e^4 A_4''^*(1,2) q_{+\perp}^*}{E}\left[\frac{q_{-+}}{q_{++}k_{1+}''k_{1-}''k_{2+}''k_{2-}''}\right]^{\frac{1}{2}}$$

$$\doteq \frac{e^4|A_4''(1,2)|}{E}\left[\frac{q_{+-}q_{-+}}{k_{1+}''k_{1-}''k_{2+}''k_{2-}''}\right]^{\frac{1}{2}},$$

$$M(-,+;+,-,+,+) = -\frac{e^4 A_1'' q_{+\perp}^*}{E}\left[\frac{q_{-+}}{q_{++}k_{1+}''k_{1-}''k_{2+}''k_{2-}''}\right]^{\frac{1}{2}}$$

$$\doteq \frac{e^4|A_1''|}{E}\left[\frac{q_{+-}q_{-+}}{k_{1+}''k_{1-}''k_{2+}''k_{2-}''}\right]^{\frac{1}{2}},$$

$$M(-,+;-,+,+,+) = -\frac{2e^4 E A_1'' q_{-\perp}^*}{q_+ \cdot q_-}\left[\frac{q_{++}}{q_{-+}k_{1+}''k_{1-}''k_{2+}''k_{2-}''}\right]^{\frac{1}{2}}$$

$$\doteq \frac{2e^4 E|A_1''|}{q_+ \cdot q_-}\left[\frac{q_{++}q_{--}}{k_{1+}''k_{1-}''k_{2+}''k_{2-}''}\right]^{\frac{1}{2}},$$

$$M(+,-;+,+,-,-) = -\frac{e^4 A_3''^* q_{-\perp}}{E} \left[\frac{q_{++}}{q_{-+}k_{1+}''k_{1-}''k_{2+}''k_{2-}''}\right]^{\frac{1}{2}}$$

$$\doteq \frac{e^4 |A_3''|}{E} \left[\frac{q_{++}q_{--}}{k_{1+}''k_{1-}''k_{2+}''k_{2-}''}\right]^{\frac{1}{2}},$$

$$M(+,-;+,-,+,-) = -\frac{e^4 A_2''(2,1) q_{-\perp}}{E} \left[\frac{q_{++}}{q_{-+}k_{1+}''k_{1-}''k_{2+}''k_{2-}''}\right]^{\frac{1}{2}}$$

$$\doteq \frac{e^4 |A_2''(2,1)|}{E} \left[\frac{q_{++}q_{--}}{k_{1+}''k_{1-}''k_{2+}''k_{2-}''}\right]^{\frac{1}{2}},$$

$$M(+,-;+,-,-,+) = -\frac{e^4 A_2''(1,2) q_{-\perp}}{E} \left[\frac{q_{++}}{q_{-+}k_{1+}''k_{1-}''k_{2+}''k_{2-}''}\right]^{\frac{1}{2}}$$

$$\doteq \frac{e^4 |A_2''(1,2)|}{E} \left[\frac{q_{++}q_{--}}{k_{1+}''k_{1-}''k_{2+}''k_{2-}''}\right]^{\frac{1}{2}},$$

$$M(+,-;-,+,+,-) = -\frac{e^4 A_2''^*(1,2) q_{+\perp}}{E} \left[\frac{q_{-+}}{q_{++}k_{1+}''k_{1-}''k_{2+}''k_{2-}''}\right]^{\frac{1}{2}}$$

$$\doteq \frac{e^4 |A_2''(1,2)|}{E} \left[\frac{q_{+-}q_{-+}}{k_{1+}''k_{1-}''k_{2+}''k_{2-}''}\right]^{\frac{1}{2}},$$

$$M(+,-;-,+,-,+) = -\frac{e^4 A_2''^*(2,1) q_{+\perp}}{E} \left[\frac{q_{-+}}{q_{++}k_{1+}''k_{1-}''k_{2+}''k_{2-}''}\right]^{\frac{1}{2}}$$

$$\doteq \frac{e^4 |A_2''(2,1)|}{E} \left[\frac{q_{+-}q_{-+}}{k_{1+}''k_{1-}''k_{2+}''k_{2-}''}\right]^{\frac{1}{2}},$$

$$M(+,-;-,-,+,+) = \frac{e^4 A_3'' q_{+\perp}}{E} \left[\frac{q_{-+}}{q_{++}k_{1+}''k_{1-}''k_{2+}''k_{2-}''}\right]^{\frac{1}{2}}$$

$$\doteq \frac{e^4 |A_3''|}{E} \left[\frac{q_{+-}q_{-+}}{k_{1+}''k_{1-}''k_{2+}''k_{2-}''}\right]^{\frac{1}{2}},$$

$$M(-,+;+,+,-,-) = -\frac{e^4 A_3''^* q_{+\perp}^*}{E} \left[\frac{q_{-+}}{q_{++}k_{1+}''k_{1-}''k_{2+}''k_{2-}''}\right]^{\frac{1}{2}}$$

$$\doteq \frac{e^4 |A_3''|}{E} \left[\frac{q_{+-}q_{-+}}{k_{1+}''k_{1-}''k_{2+}''k_{2-}''}\right]^{\frac{1}{2}},$$

9.18. $e^+e^- \to \mu^+\mu^-\gamma\gamma$ (NO Z-EXCHANGE; $m \neq 0$)

$$M(-,+;+,-,+,-) = -\frac{e^4 A_2''(2,1)q_{+\perp}^*}{E}\left[\frac{q_{-+}}{q_{++}k_{1+}''k_{1-}''k_{2+}''k_{2-}''}\right]^{\frac{1}{2}}$$

$$\doteq \frac{e^4|A_2''(2,1)|}{E}\left[\frac{q_{+-}q_{-+}}{k_{1+}''k_{1-}''k_{2+}''k_{2-}''}\right]^{\frac{1}{2}},$$

$$M(-,+;+,-,-,+) = -\frac{e^4 A_2''(1,2)q_{+\perp}^*}{E}\left[\frac{q_{-+}}{q_{++}k_{1+}''k_{1-}''k_{2+}''k_{2-}''}\right]^{\frac{1}{2}}$$

$$\doteq \frac{e^4|A_2''(1,2)|}{E}\left[\frac{q_{+-}q_{-+}}{k_{1+}''k_{1-}''k_{2+}''k_{2-}''}\right]^{\frac{1}{2}},$$

$$M(-,+;-,+,+,-) = -\frac{e^4 A_2''^*(1,2)q_{-\perp}^*}{E}\left[\frac{q_{++}}{q_{-+}k_{1+}''k_{1-}''k_{2+}''k_{2-}''}\right]^{\frac{1}{2}}$$

$$\doteq \frac{e^4|A_2''(1,2)|}{E}\left[\frac{q_{++}q_{--}}{k_{1+}''k_{1-}''k_{2+}''k_{2-}''}\right]^{\frac{1}{2}},$$

$$M(-,+;-,+,-,+) = -\frac{e^4 A_2''^*(2,1)q_{-\perp}^*}{E}\left[\frac{q_{++}}{q_{-+}k_{1+}''k_{1-}''k_{2+}''k_{2-}''}\right]^{\frac{1}{2}}$$

$$\doteq \frac{e^4|A_2''(2,1)|}{E}\left[\frac{q_{++}q_{--}}{k_{1+}''k_{1-}''k_{2+}''k_{2-}''}\right]^{\frac{1}{2}},$$

$$M(-,+;-,-,+,+) = \frac{e^4 A_3'' q_{-\perp}^*}{E}\left[\frac{q_{++}}{q_{-+}k_{1+}''k_{1-}''k_{2+}''k_{2-}''}\right]^{\frac{1}{2}}$$

$$\doteq \frac{e^4|A_3''|}{E}\left[\frac{q_{++}q_{--}}{k_{1+}''k_{1-}''k_{2+}''k_{2-}''}\right]^{\frac{1}{2}},$$

$$M(+,-;+,-,-,-) = -\frac{2e^4 A_1''^* q_{-\perp}}{(q_+\cdot q_-)}\left[\frac{q_{++}}{q_{-+}k_{1+}''k_{1-}''k_{2+}''k_{2-}''}\right]^{\frac{1}{2}}$$

$$\doteq \frac{2e^4 E|A_1''|}{q_+\cdot q_-)}\left[\frac{q_{++}q_{--}}{k_{1+}''k_{1-}''k_{2+}''k_{2-}''}\right]^{\frac{1}{2}},$$

$$M(+,-;-,+,-,-) = -\frac{e^4 A_1''^* q_{+\perp}}{E}\left[\frac{q_{-+}}{q_{++}k_{1+}''k_{1-}''k_{2+}''k_{2-}''}\right]^{\frac{1}{2}}$$

$$\doteq \frac{e^4|A_1''|}{E}\left[\frac{q_{+-}q_{-+}}{k_{1+}''k_{1-}''k_{2+}''k_{2-}''}\right]^{\frac{1}{2}},$$

$$M(+,-;-,-,+,-) = \frac{e^4 A_4''(1,2) q_{+\perp}}{E} \left[\frac{q_{-+}}{q_{++} k_{1+}'' k_{1-}'' k_{2+}'' k_{2-}''}\right]^{\frac{1}{2}}$$

$$\doteq \frac{e^4 |A_4''(1,2)|}{E} \left[\frac{q_{+-} q_{-+}}{k_{1+}'' k_{1-}'' k_{2+}'' k_{2-}''}\right]^{\frac{1}{2}},$$

$$M(+,-;-,-,-,+) = \frac{e^4 A_4''(2,1) q_{+\perp}}{E} \left[\frac{q_{-+}}{q_{++} k_{1+}'' k_{1-}'' k_{2+}'' k_{2-}''}\right]^{\frac{1}{2}}$$

$$\doteq \frac{e^4 |A_4''(2,1)|}{E} \left[\frac{q_{+-} q_{-+}}{k_{1+}'' k_{1-}'' k_{2+}'' k_{2-}''}\right]^{\frac{1}{2}},$$

$$M(-,+;+,-,-,-) = -\frac{e^4 A_1''^* q_{+\perp}^*}{E} \left[\frac{q_{+-}}{q_{--} k_{1+}'' k_{1-}'' k_{2+}'' k_{2-}''}\right]^{\frac{1}{2}}$$

$$\doteq \frac{e^4 q_{+-} |A_1''|}{E} \left[\frac{q_{++}}{q_{--} k_{1+}'' k_{1-}'' k_{2+}'' k_{2-}''}\right]^{\frac{1}{2}},$$

$$M(-,+;-,+,-,-) = -\frac{e^4 A_1''^* q_{-\perp}^*}{E} \left[\frac{q_{++}}{q_{-+} k_{1+}'' k_{1-}'' k_{2+}'' k_{2-}''}\right]^{\frac{1}{2}}$$

$$\doteq \frac{e^4 |A_1''|}{E} \left[\frac{q_{++} q_{--}}{k_{1+}'' k_{1-}'' k_{2+}'' k_{2-}''}\right]^{\frac{1}{2}},$$

$$M(-,+;-,-,+,-) = \frac{e^4 A_4''(1,2) q_{-\perp}^*}{E} \left[\frac{q_{++}}{q_{-+} k_{1+}'' k_{1-}'' k_{2+}'' k_{2-}''}\right]^{\frac{1}{2}}$$

$$\doteq \frac{e^4 |A_4''(1,2)|}{E} \left[\frac{q_{++} q_{--}}{k_{1+}'' k_{1-}'' k_{2+}'' k_{2-}''}\right]^{\frac{1}{2}},$$

$$M(-,+;-,-,-,+) = \frac{e^4 A_4''(2,1) q_{-\perp}^*}{E} \left[\frac{q_{++}}{q_{-+} k_{1+}'' k_{1-}'' k_{2+}'' k_{2-}''}\right]^{\frac{1}{2}}$$

$$\doteq \frac{e^4 |A_4''(2,1)|}{E} \left[\frac{q_{++} q_{--}}{k_{1+}'' k_{1-}'' k_{2+}'' k_{2-}''}\right]^{\frac{1}{2}}. \quad (9.208)$$

Unpolarized squared matrix element:

$$\overline{|M|^2} = \frac{e^8 (q_{++} q_{--} + q_{+-} q_{-+})}{2 E^2 k_{1+}'' k_{1-}'' k_{2+}'' k_{2-}''} \left\{ \left[1 + \left(\frac{2 q_{-0} + k_{1+}'' + k_{2+}''}{2 q_{-0}}\right)^2\right] |A_1''|^2 \right.$$

$$\left. + |A_2''(1,2)|^2 + |A_2''(2,1)|^2 + |A_3''|^2 + |A_4''(1,2)|^2 + |A_4''(2,1)|^2 \right\}.$$

$$(9.209)$$

9.19. $e^+e^- \to e^+e^- \gamma\gamma$ (NO Z-EXCHANGE; $m \neq 0$)

Unpolarized cross section:

$$d\sigma = \frac{\alpha^4(q_{++}q_{--} + q_{+-}q_{-+})}{256\pi^2 E^4 k''_{1+} k''_{1-} k''_{2+} k''_{2-}} \left\{ \left[1 + \left(\frac{2q_{-0} + k''_{1+} + k''_{2+}}{2q_{-0}} \right)^2 \right] |A''_1|^2 \right.$$

$$+ |A''_2(1,2)|^2 + |A''_2(2,1)|^2 + |A''_3|^2 + |A''_4(1,2)|^2 + |A''_4(2,1)|^2 \Big\}$$

$$\times \delta^4(p_+ + p_- - q_+ - q_- - k_1 - k_2) \frac{d^3\vec{q}_+ \, d^3\vec{q}_- \, d^3\vec{k}_1 \, d^3\vec{k}_2}{q_{+0} \, q_{-0} \, k_{10} \, k_{20}}. \quad (9.210)$$

9.19 $e^+e^- \to e^+e^- \gamma\gamma$ (no Z-exchange; $m \neq 0$)

Process:

$$e^+(p_+) + e^-(p_-) \to e^+(q_+) + e^-(q_-) + \gamma(k_1) + \gamma(k_2). \quad (9.211)$$

Definitions:

$$B = 1 + \frac{E}{(q_+ \cdot q_-)} \left[2q_{-z} + \frac{k_{2\perp} q^*_{-\perp}}{k_{2+}} - \frac{k^*_{2\perp} q_{-\perp}}{k_{2-}} \right],$$

$$B_0 = 2 - \frac{k_{2\perp} q^*_{+\perp}}{k_{2+} q_{+-}} - \frac{k^*_{2\perp} q_{-\perp}}{k_{2-} q_{-+}},$$

$$C = (q_+ \cdot q_-)(q_+ \cdot k_1)(q_- \cdot k_1),$$

$$Z_{+-} = q_{++}q_{--} - q^*_{+\perp}q_{-\perp}, \qquad Z_{-+} = q_{-+}q_{+-} - q^*_{-\perp}q_{+\perp}. \quad (9.212)$$

9.19.1 $\vec{k}_2$ nearly parallel to $\vec{p}_+$

Photon polarizations:

$$\rlap{/}{\epsilon}_1^{\pm} = N_1[\rlap{/}{p}_+ \rlap{/}{p}_- \rlap{/}{k}_1(1 \mp \gamma_5) - \rlap{/}{k}_1 \rlap{/}{p}_+ \rlap{/}{p}_-(1 \pm \gamma_5)],$$

$$\rlap{/}{\epsilon}_2^{\pm} = N_2[\rlap{/}{k}_2 \rlap{/}{p}_+ \rlap{/}{p}_-(1 \pm \gamma_5) + \rlap{/}{p}_- \rlap{/}{p}_+ \rlap{/}{k}_2(1 \mp \gamma_5)],$$

$$N_i^{-1} = E^2[32k_{i+}k_{i-}]^{\frac{1}{2}}, \qquad i = 1,2. \quad (9.213)$$

Nonvanishing helicity amplitudes:

$$M(+,+;+,+,+,+) = \frac{e^4 E^2 q^*_{+\perp} B}{q_{+-}(p_+ \cdot k_2)} \left[\frac{32k_{2-}(q_+ \cdot q_-)}{k_{2+}q_{++}q_{-+}(q_+ \cdot k_1)(q_- \cdot k_1)} \right]^{\frac{1}{2}}$$

$$\doteq \frac{e^4 E^2 |B|}{(p_+ \cdot k_2)} \left[\frac{32k_{2-}(q_+ \cdot q_-)}{k_{2+}q_{+-}q_{-+}(q_+ \cdot k_1)(q_- \cdot k_1)} \right]^{\frac{1}{2}},$$

$$M(+,+;+,+,+,-) = \frac{e^4 E(2E-k_{2+})q^*_{+\perp}B}{q_{+-}(p_+\cdot k_2)}$$

$$\times \left[\frac{8k_{2-}(q_+\cdot q_-)}{k_{2+}q_{++}q_{-+}(q_+\cdot k_1)(q_-\cdot k_1)}\right]^{\frac{1}{2}}$$

$$\doteq \frac{e^4 E(2E-k_{2+})|B|}{(p_+\cdot k_2)}$$

$$\times \left[\frac{8k_{2-}(q_+\cdot q_-)}{k_{2+}q_{+-}q_{-+}(q_+\cdot k_1)(q_-\cdot k_1)}\right]^{\frac{1}{2}},$$

$$M(+,+;+,+,-,+) = \frac{e^4 E q^*_{+\perp} Z_{-+} B^*_0}{(2E-k_{2+})q_{+-}(p_+\cdot k_2)}$$

$$\times \left[\frac{8k_{2-}(q_+\cdot q_-)}{k_{2+}q_{++}q_{-+}(q_+\cdot k_1)(q_-\cdot k_1)}\right]^{\frac{1}{2}}$$

$$\doteq \frac{4e^4 E(q_+\cdot q_-)|B_0|}{(2E-k_{2+})(p_+\cdot k_2)}\left[\frac{k_{2-}}{k_{2+}(q_+\cdot k_1)(q_-\cdot k_1)}\right]^{\frac{1}{2}},$$

$$M(+,+;+,-,+,+) = -\frac{e^4 m k_{2\perp} q^*_{+\perp} Z_{-+} B}{(2E-k_{2+})q_{+-}(p_+\cdot k_2)}\left[\frac{k_{2+}q_{-+}}{2Ck_{2-}q_{++}}\right]^{\frac{1}{2}}$$

$$\doteq \frac{e^4 m k_{2+} q_{-+}|B|}{(2E-k_{2+})(p_+\cdot k_2)[(q_+\cdot k_1)(q_-\cdot k_1)]^{\frac{1}{2}}},$$

$$M(+,+;-,+,+,+) = -\frac{e^4 m k_{2\perp} q^*_{+\perp} Z_{+-} B}{(2E-k_{2+})q_{+-}(p_+\cdot k_2)}\left[\frac{k_{2+}q_{++}}{2Ck_{2-}q_{-+}}\right]^{\frac{1}{2}}$$

$$\doteq \frac{e^4 m k_{2+}|B|}{(2E-k_{2+})(p_+\cdot k_2)}\left[\frac{q^3_{++}q_{--}}{q_{+-}q_{-+}(q_+\cdot k_1)(q_-\cdot k_1)}\right]^{\frac{1}{2}},$$

$$M(+,+;+,+,-,-) = \frac{e^4 q^*_{+\perp} Z_{-+} B^*_0}{q_{+-}(p_+\cdot k_2)}\left[\frac{2k_{2-}(q_+\cdot q_-)}{k_{2+}q_{++}q_{-+}(q_+\cdot k_1)(q_-\cdot k_1)}\right]^{\frac{1}{2}}$$

$$\doteq \frac{2e^4(q_+\cdot q_-)|B_0|}{(p_+\cdot k_2)}\left[\frac{k_{2-}}{k_{2+}(q_+\cdot k_1)(q_-\cdot k_1)}\right]^{\frac{1}{2}},$$

$$M(+,+;+,-,-,+) = -\frac{e^4 m k_{2\perp} q^*_{+\perp} Z^*_{-+} B^*}{E(p_+\cdot k_2)}\left[\frac{k_{2+}}{8Ck_{2-}q_{++}q_{-+}}\right]^{\frac{1}{2}}$$

$$\doteq \frac{e^4 m k_{2+} q_{+-}|B|}{2E(p_+\cdot k_2)[(q_+\cdot k_1)(q_-\cdot k_1)]^{\frac{1}{2}}},$$

9.19. $e^+e^- \to e^+e^-\gamma\gamma$ (NO Z-EXCHANGE; $m \neq 0$)   267

$$M(+,+;-,+,-,+) = -\frac{e^4 m k_{2\perp} q_{-\perp}^{*2} q_{+\perp} Z_{+-}^* B^*}{E q_{+-}(p_+ \cdot k_2)}\left[\frac{k_{2+}}{8Ck_{2-}q_{++}q_{-+}^3}\right]^{\frac{1}{2}}$$

$$\doteq \frac{e^4 m k_{2+}|B|}{2E(p_+ \cdot k_2)}\left[\frac{q_{++}q_{--}^3}{q_{+-}q_{-+}(q_+ \cdot k_1)(q_- \cdot k_1)}\right]^{\frac{1}{2}},$$

$$M(+,-;+,-,+,+) = \frac{e^4 E q_{-\perp}^2 q_{+\perp}^* Z_{+-} B}{q_{+-}(p_+ \cdot k_2)}\left[\frac{2k_{2-}}{Ck_{2+}q_{++}q_{-+}^3}\right]^{\frac{1}{2}}$$

$$\doteq \frac{2e^4 E|B|}{(p_+ \cdot k_2)}\left[\frac{k_{2-}q_{++}q_{--}^3}{k_{2+}q_{+-}q_{-+}(q_+ \cdot k_1)(q_- \cdot k_1)}\right]^{\frac{1}{2}},$$

$$M(+,-;-,+,+,+) = \frac{e^4 E q_{+\perp} Z_{-+} B}{(p_+ \cdot k_2)}\left[\frac{2k_{2-}}{Ck_{2+}q_{++}q_{-+}}\right]^{\frac{1}{2}}$$

$$\doteq \frac{2e^4 E q_{+-}|B|}{(p_+ \cdot k_2)}\left[\frac{k_{2-}}{k_{2+}(q_+ \cdot k_1)(q_- \cdot k_1)}\right]^{\frac{1}{2}},$$

$$M(-,+;+,+,+,-) = \frac{e^4 m k_{2\perp}^* q_{+\perp}^* B}{q_{+-}(p_+ \cdot k_2)}\left[\frac{2k_{2+}(q_+ \cdot q_-)}{k_{2-}q_{++}q_{-+}(q_+ \cdot k_1)(q_- \cdot k_1)}\right]^{\frac{1}{2}}$$

$$\doteq \frac{e^4 m k_{2+}|B|}{(p_+ \cdot k_2)}\left[\frac{2(q_+ \cdot q_-)}{q_{+-}q_{-+}(q_+ \cdot k_1)(q_- \cdot k_1)}\right]^{\frac{1}{2}},$$

$$M(-,+;+,-,+,+) = \frac{e^4 E q_{+\perp}^* Z_{-+} B}{q_{+-}(p_+ \cdot k_2)}\left[\frac{2k_{2-}q_{-+}}{Ck_{2+}q_{++}}\right]^{\frac{1}{2}}$$

$$\doteq \frac{2e^4 E q_{-+}|B|}{(p_+ \cdot k_2)}\left[\frac{k_{2-}}{k_{2+}(q_+ \cdot k_1)(q_- \cdot k_1)}\right]^{\frac{1}{2}},$$

$$M(-,+;-,+,+,+) = \frac{e^4 E q_{+\perp}^* Z_{+-} B}{q_{+-}(p_+ \cdot k_2)}\left[\frac{2k_{2-}q_{++}}{Ck_{2+}q_{-+}}\right]^{\frac{1}{2}}$$

$$\doteq \frac{2e^4 E|B|}{(p_+ \cdot k_2)}\left[\frac{k_{2-}q_{++}^3 q_{--}}{k_{2+}q_{+-}q_{-+}(q_+ \cdot k_1)(q_- \cdot k_1)}\right]^{\frac{1}{2}},$$

$$M(+,-;+,-,+,-) = \frac{e^4(2E-k_{2+})q_{-\perp}^2 q_{+\perp}^* Z_{+-} B}{q_{+-}(p_+ \cdot k_2)}\left[\frac{k_{2-}}{2Ck_{2+}q_{++}q_{-+}^3}\right]^{\frac{1}{2}}$$

$$\doteq \frac{e^4(2E-k_{2+})|B|}{(p_+ \cdot k_2)}$$

$$\times \left[\frac{k_{2-}q_{++}q_{--}^3}{k_{2+}q_{+-}q_{-+}(q_+ \cdot k_1)(q_- \cdot k_1)}\right]^{\frac{1}{2}},$$

$$M(+,-;+,-,-,+) = \frac{e^4 E^2 q_{+\perp} Z^*_{+-} B^*}{(2E-k_{2+})q_{+-}(p_+ \cdot k_2)} \left[\frac{8k_{2-}q_{++}}{Ck_{2+}q_{-+}}\right]^{\frac{1}{2}}$$

$$\doteq \frac{4e^4 E^2 |B|}{(2E-k_{2+})(p_+ \cdot k_2)}$$

$$\times \left[\frac{k_{2-}q_{++}^3 q_{--}}{k_{2+}q_{+-}q_{-+}(q_+ \cdot k_1)(q_- \cdot k_1)}\right]^{\frac{1}{2}},$$

$$M(+,-;-,+,+,-) = \frac{e^4(2E-k_{2+})q_{+\perp} Z_{-+} B}{(p_+ \cdot k_2)} \left[\frac{k_{2-}}{2Ck_{2+}q_{++}q_{-+}}\right]^{\frac{1}{2}}$$

$$\doteq \frac{e^4(2E-k_{2+})q_{+-}|B|}{(p_+ \cdot k_2)} \left[\frac{k_{2-}}{k_{2+}(q_+ \cdot k_1)(q_- \cdot k_1)}\right]^{\frac{1}{2}},$$

$$M(+,-;-,+,-,+) = \frac{e^4 E^2 q_{+\perp} Z^*_{-+} B^*}{(2E-k_{2+})q_{+-}(p_+ \cdot k_2)} \left[\frac{8k_{2-}q_{-+}}{Ck_{2+}q_{++}}\right]^{\frac{1}{2}}$$

$$\doteq \frac{4e^4 E^2 q_{-+}|B|}{(2E-k_{2+})(p_+ \cdot k_2)} \left[\frac{k_{2-}}{k_{2+}(q_+ \cdot k_1)(q_- \cdot k_1)}\right]^{\frac{1}{2}},$$

$$M(+,-;-,-,+,+) = -\frac{e^4 m k_{2\perp} q_{+\perp} Z^*_{-+} B_0}{E(2E-k_{2+})q_{+-}(p_+ \cdot k_2)}$$

$$\times \left[\frac{k_{2+}(q_+ \cdot q_-)}{2k_{2-}q_{++}q_{-+}(q_+ \cdot k_1)(q_- \cdot k_1)}\right]^{\frac{1}{2}}$$

$$\doteq \frac{e^4 m k_{2+}(q_+ \cdot q_-)|B_0|}{E(2E-k_{2+})(p_+ \cdot k_2)[(q_+ \cdot k_1)(q_- \cdot k_1)]^{\frac{1}{2}}},$$

$$M(-,+;+,+,-,-) = \frac{e^4 m k^*_{2\perp} q^*_{+\perp} Z_{-+} B^*_0}{E(2E-k_{2+})q_{+-}(p_+ \cdot k_2)}$$

$$\times \left[\frac{k_{2+}(q_+ \cdot q_-)}{2k_{2-}q_{++}q_{-+}(q_+ \cdot k_1)(q_- \cdot k_1)}\right]^{\frac{1}{2}}$$

$$\doteq \frac{e^4 m k_{2+}(q_+ \cdot q_-)|B_0|}{E(2E-k_{2+})(p_+ \cdot k_2)[(q_+ \cdot k_1)(q_- \cdot k_1)]^{\frac{1}{2}}},$$

$$M(-,+;+,-,+,-) = \frac{e^4 E^2 q^*_{+\perp} Z_{-+} B}{(2E-k_{2+})q_{+-}(p_+ \cdot k_2)} \left[\frac{8k_{2-}q_{-+}}{Ck_{2+}q_{++}}\right]^{\frac{1}{2}}$$

$$\doteq \frac{4e^4 E^2 q_{-+}|B|}{(2E-k_{2+})(p_+ \cdot k_2)} \left[\frac{k_{2-}}{k_{2+}(q_+ \cdot k_1)(q_- \cdot k_1)}\right]^{\frac{1}{2}},$$

9.19. $e^+e^- \to e^+e^- \gamma\gamma$ (NO Z-EXCHANGE; $m \neq 0$)

$$M(-,+;+,-,-,+) = \frac{e^4(2E - k_{2+})q_{+\perp}^* Z_{-+}^* B^*}{(p_+ \cdot k_2)} \left[\frac{k_{2-}}{2Ck_{2+}q_{++}q_{-+}}\right]^{\frac{1}{2}}$$

$$\doteq \frac{e^4(2E - k_{2+})q_{+-}|B|}{(p_+ \cdot k_2)} \left[\frac{k_{2-}}{k_{2+}(q_+ \cdot k_1)(q_- \cdot k_1)}\right]^{\frac{1}{2}},$$

$$M(-,+;-,+,+,-) = \frac{e^4 E^2 q_{+\perp}^* Z_{+-} B}{(2E - k_{2+})q_{+-}(p_+ \cdot k_2)} \left[\frac{8k_{2-}q_{++}}{Ck_{2+}q_{-+}}\right]^{\frac{1}{2}}$$

$$\doteq \frac{4e^4 E^2 |B|}{(2E - k_{2+})(p_+ \cdot k_2)}$$

$$\times \left[\frac{k_{2-}q_{++}^3 q_{--}}{k_{2+}q_{+-}q_{-+}(q_+ \cdot k_1)(q_- \cdot k_1)}\right]^{\frac{1}{2}},$$

$$M(-,+;-,+,-,+) = \frac{e^4(2E - k_{2+})q_{-\perp}^{*2} q_{+\perp} Z_{+-}^* B^*}{q_{+-}(p_+ \cdot k_2)} \left[\frac{k_{2-}}{2Ck_{2+}q_{++}q_{-+}^3}\right]^{\frac{1}{2}}$$

$$\doteq \frac{e^4(2E - k_{2+})|B|}{(p_+ \cdot k_2)} \left[\frac{k_{2-}q_{++}q_{--}^3}{k_{2+}q_{+-}q_{-+}(q_+ \cdot k_1)(q_- \cdot k_1)}\right]^{\frac{1}{2}},$$

$$M(+,-;+,-,-,-) = \frac{e^4 E q_{+\perp} Z_{+-}^* B^*}{q_{+-}(p_+ \cdot k_2)} \left[\frac{2k_{2-}q_{++}}{Ck_{2+}q_{-+}}\right]^{\frac{1}{2}}$$

$$\doteq \frac{2e^4 E |B|}{(p_+ \cdot k_2)} \left[\frac{k_{2-}q_{++}^3 q_{--}}{k_{2+}q_{+-}q_{-+}(q_+ \cdot k_1)(q_- \cdot k_1)}\right]^{\frac{1}{2}},$$

$$M(+,-;-,+,-,-) = \frac{e^4 E q_{+\perp} Z_{-+}^* B^*}{q_{+-}(p_+ \cdot k_2)} \left[\frac{2k_{2-}q_{-+}}{Ck_{2+}q_{++}}\right]^{\frac{1}{2}}$$

$$\doteq \frac{2e^4 E q_{-+}|B|}{(p_+ \cdot k_2)} \left[\frac{k_{2-}}{k_{2+}(q_+ \cdot k_1)(q_- \cdot k_1)}\right]^{\frac{1}{2}},$$

$$M(+,-;-,-,-,+) = -\frac{e^4 m k_{2\perp} q_{+\perp} B^*}{q_{+-}(p_+ \cdot k_2)} \left[\frac{2k_{2+}(q_+ \cdot q_-)}{k_{2-}q_{++}q_{-+}(q_+ \cdot k_1)(q_- \cdot k_1)}\right]^{\frac{1}{2}}$$

$$\doteq \frac{e^4 m k_{2+}|B|}{(p_+ \cdot k_2)} \left[\frac{2(q_+ \cdot q_-)}{q_{+-}q_{-+}(q_+ \cdot k_1)(q_- \cdot k_1)}\right]^{\frac{1}{2}},$$

$$M(-,+;+,-,-,-) = \frac{e^4 E q_{+\perp}^* Z_{-+}^* B^*}{(p_+ \cdot k_2)} \left[\frac{2k_{2-}}{Ck_{2+}q_{++}q_{-+}}\right]^{\frac{1}{2}}$$

$$\doteq \frac{2e^4 E q_{+-}|B|}{(p_+ \cdot k_2)} \left[\frac{k_{2-}}{k_{2+}(q_+ \cdot k_1)(q_- \cdot k_1)}\right]^{\frac{1}{2}},$$

$$M(-,+;-,+,-,-) = \frac{e^4 E q_{-\perp}^{*2} q_{+\perp} Z_{+-}^* B^*}{q_{+-}(p_+ \cdot k_2)} \left[\frac{2k_{2-}}{Ck_{2+}q_{++}q_{-+}^3}\right]^{\frac{1}{2}}$$

$$\doteq \frac{2e^4 E|B|}{(p_+ \cdot k_2)} \left[\frac{k_{2-}q_{++}q_{--}^3}{k_{2+}q_{+-}q_{-+}(q_+ \cdot k_1)(q_- \cdot k_1)}\right]^{\frac{1}{2}},$$

$$M(-,-;+,-,+,-) = \frac{e^4 m k_{2\perp}^* q_{-\perp}^2 q_{+\perp}^* Z_{+-} B}{E q_{+-}(p_+ \cdot k_2)} \left[\frac{k_{2+}}{8Ck_{2-}q_{++}q_{-+}^3}\right]^{\frac{1}{2}}$$

$$\doteq \frac{e^4 m k_{2+}|B|}{2E(p_+ \cdot k_2)} \left[\frac{q_{++}q_{--}^3}{q_{+-}q_{-+}(q_+ \cdot k_1)(q_- \cdot k_1)}\right]^{\frac{1}{2}},$$

$$M(-,-;-,+,+,-) = \frac{e^4 m k_{2\perp}^* q_{+\perp} Z_{-+} B}{E(p_+ \cdot k_2)} \left[\frac{k_{2+}}{8Ck_{2-}q_{++}q_{-+}}\right]^{\frac{1}{2}}$$

$$\doteq \frac{e^4 m k_{2+} q_{+-}|B|}{2E(p_+ \cdot k_2)[(q_+ \cdot k_1)(q_- \cdot k_1)]^{\frac{1}{2}}},$$

$$M(-,-;-,-,+,+) = \frac{e^4 q_{+\perp} Z_{-+}^* B_0}{q_{+-}(p_+ \cdot k_2)} \left[\frac{2k_{2-}(q_+ \cdot q_-)}{k_{2+}q_{++}q_{-+}(q_+ \cdot k_1)(q_- \cdot k_1)}\right]^{\frac{1}{2}}$$

$$\doteq \frac{2e^4(q_+ \cdot q_-)|B_0|}{(p_+ \cdot k_2)} \left[\frac{k_{2-}}{k_{2+}(q_+ \cdot k_1)(q_- \cdot k_1)}\right]^{\frac{1}{2}},$$

$$M(-,-;+,-,-,-) = \frac{e^4 m k_{2\perp}^* q_{+\perp} Z_{+-}^* B^*}{(2E - k_{2+})q_{+-}(p_+ \cdot k_2)} \left[\frac{k_{2+}q_{++}}{2Ck_{2-}q_{-+}}\right]^{\frac{1}{2}}$$

$$\doteq \frac{e^4 m k_{2+}|B|}{(2E - k_{2+})(p_+ \cdot k_2)} \left[\frac{q_{++}^3 q_{--}}{q_{+-}q_{-+}(q_+ \cdot k_1)(q_- \cdot k_1)}\right]^{\frac{1}{2}},$$

$$M(-,-;-,+,-,-) = \frac{e^4 m k_{2\perp}^* q_{+\perp} Z_{-+}^* B^*}{(2E - k_{2+})q_{+-}(p_+ \cdot k_2)} \left[\frac{k_{2+}q_{-+}}{2Ck_{2-}q_{++}}\right]^{\frac{1}{2}}$$

$$\doteq \frac{e^4 m k_{2+} q_{-+}|B|}{(2E - k_{2+})(p_+ \cdot k_2)[(q_+ \cdot k_1)(q_- \cdot k_1)]^{\frac{1}{2}}},$$

$$M(-,-;-,-,+,-) = \frac{e^4 E q_{+\perp} Z_{-+}^* B_0}{(2E - k_{2+})q_{+-}(p_+ \cdot k_2)}$$

$$\times \left[\frac{8k_{2-}(q_+ \cdot q_-)}{k_{2+}q_{++}q_{-+}(q_+ \cdot k_1)(q_- \cdot k_1)}\right]^{\frac{1}{2}}$$

$$\doteq \frac{4e^4 E(q_+ \cdot q_-)|B_0|}{(2E - k_{2+})(p_+ \cdot k_2)} \left[\frac{k_{2-}}{k_{2+}(q_+ \cdot k_1)(q_- \cdot k_1)}\right]^{\frac{1}{2}},$$

9.19. $e^+e^- \to e^+e^-\gamma\gamma$ (NO Z-EXCHANGE; $m \neq 0$)

$$M(-,-;-,-,-,+) = \frac{e^4 E(2E - k_{2+})q_{+\perp}B^*}{q_{+-}(p_+ \cdot k_2)}$$

$$\times \left[\frac{8k_{2-}(q_+ \cdot q_-)}{k_{2+}q_{++}q_{-+}(q_+ \cdot k_1)(q_- \cdot k_1)}\right]^{\frac{1}{2}}$$

$$\doteq \frac{e^4 E(2E - k_{2+})|B|}{(p_+ \cdot k_2)}$$

$$\times \left[\frac{8k_{2-}(q_+ \cdot q_-)}{k_{2+}q_{+-}q_{-+}(q_+ \cdot k_1)(q_- \cdot k_1)}\right]^{\frac{1}{2}},$$

$$M(-,-;-,-,-,-) = \frac{e^4 E^2 q_{+\perp}B^*}{q_{+-}(p_+ \cdot k_2)}\left[\frac{32k_{2-}(q_+ \cdot q_-)}{k_{2+}q_{++}q_{-+}(q_+ \cdot k_1)(q_- \cdot k_1)}\right]^{\frac{1}{2}}$$

$$\doteq \frac{e^4 E^2 |B|}{(p_+ \cdot k_2)}\left[\frac{32k_{2-}(q_+ \cdot q_-)}{k_{2+}q_{+-}q_{-+}(q_+ \cdot k_1)(q_- \cdot k_1)}\right]^{\frac{1}{2}}.$$

(9.214)

Unpolarized squared matrix element:

$$\overline{|M|^2} = \frac{e^8 |B|^2 k_{2-}}{2k_{2+}(p_+ \cdot k_2)^2 (q_+ \cdot k_1)(q_- \cdot k_1)q_{+-}q_{-+}}$$

$$\times \left[4E^2 + (2E - k_{2+})^2 + \frac{m^2 k_{2+}^3}{4E^2 k_{2-}}\right]\left[q_{++}q_{--}^3 + q_{+-}^3 q_{-+} + 8E^2(q_+ \cdot q_-)\right.$$

$$\left. + \left(\frac{2E}{2E - k_{2+}}\right)^2\left(q_{++}^3 q_{--} + q_{+-}q_{-+}^3 + \frac{2(q_+ \cdot q_-)^3}{E^2}\right)\right].$$

(9.215)

Unpolarized cross section:

$$d\sigma = \frac{\alpha^4 |B|^2 k_{2-}}{256\pi^2 E^2 k_{2+}(p_+ \cdot k_2)^2 (q_+ \cdot k_1)(q_- \cdot k_1)q_{+-}q_{-+}}$$

$$\times \left[4E^2 + (2E - k_{2+})^2 + \frac{m^2 k_{2+}^3}{4E^2 k_{2-}}\right]\left[q_{++}q_{--}^3 + q_{+-}^3 q_{-+} + 8E^2(q_+ \cdot q_-)\right.$$

$$\left. + \left(\frac{2E}{2E - k_{2+}}\right)^2\left(q_{++}^3 q_{--} + q_{+-}q_{-+}^3 + \frac{2(q_+ \cdot q_-)^3}{E^2}\right)\right]$$

$$\times \delta^4(p_+ + p_- - q_+ - q_- - k_1 - k_2)\frac{d^3\vec{q}_+ d^3\vec{q}_- d^3\vec{k}_1 d^3\vec{k}_2}{q_{+0}q_{-0}k_{10}k_{20}}.$$

(9.216)

9.19.2 $\vec{k}_2$ nearly parallel to $\vec{p}_-$

Photon polarizations:

$$\not{k}_1^{\pm} = N_1[\not{p}_+ \not{p}_- \not{k}_1(1 \mp \gamma_5) - \not{k}_1 \not{p}_+ \not{p}_-(1 \pm \gamma_5)],$$

$$\not{k}_2^{\pm} = N_2[\not{k}_2 \not{p}_- \not{p}_+(1 \pm \gamma_5) + \not{p}_+ \not{p}_- \not{k}_2(1 \mp \gamma_5)],$$

$$N_i^{-1} = E^2[32 k_{i+} k_{i-}]^{\frac{1}{2}}, \qquad\qquad i = 1, 2. \qquad (9.217)$$

Nonvanishing helicity amplitudes:

$$M(+,+;+,+,+,+) = \frac{e^4 E^2 q_{+\perp}^* B}{q_{+-}(p_- \cdot k_2)} \left[\frac{32 k_{2+}(q_+ \cdot q_-)}{k_{2-} q_{++} q_{-+}(q_+ \cdot k_1)(q_- \cdot k_1)}\right]^{\frac{1}{2}}$$

$$\doteq \frac{e^4 E^2 |B|}{(p_- \cdot k_2)} \left[\frac{32 k_{2+}(q_+ \cdot q_-)}{k_{2-} q_{+-} q_{-+}(q_+ \cdot k_1)(q_- \cdot k_1)}\right]^{\frac{1}{2}},$$

$$M(+,+;+,+,+,-) = \frac{e^4 E(2E - k_{2-}) q_{+\perp}^* B}{q_{+-}(p_- \cdot k_2)}$$

$$\times \left[\frac{8 k_{2+}(q_+ \cdot q_-)}{k_{2-} q_{++} q_{-+}(q_+ \cdot k_1)(q_- \cdot k_1)}\right]^{\frac{1}{2}}$$

$$\doteq \frac{e^4 E(2E - k_{2-})|B|}{(p_- \cdot k_2)}$$

$$\times \left[\frac{8 k_{2+}(q_+ \cdot q_-)}{k_{2-} q_{+-} q_{-+}(q_+ \cdot k_1)(q_- \cdot k_1)}\right]^{\frac{1}{2}},$$

$$M(+,+;+,+,-,+) = \frac{e^4 E q_{+\perp}^* Z_{-+} B_0^*}{(2E - k_{2-}) q_{+-}(p_- \cdot k_2)}$$

$$\times \left[\frac{8 k_{2+}(q_+ \cdot q_-)}{k_{2-} q_{++} q_{-+}(q_+ \cdot k_1)(q_- \cdot k_1)}\right]^{\frac{1}{2}}$$

$$\doteq \frac{4 e^4 E(q_+ \cdot q_-)|B_0|}{(2E - k_{2-})(p_- \cdot k_2)} \left[\frac{k_{2+}}{k_{2-}(q_+ \cdot k_1)(q_- \cdot k_1)}\right]^{\frac{1}{2}},$$

$$M(+,+;+,-,+,+) = \frac{e^4 m k_{2\perp}^* q_{-\perp}^2 q_{+\perp}^* Z_{+-} B}{(2E - k_{2-}) q_{+-}(p_- \cdot k_2)} \left[\frac{k_{2-}}{2C k_{2+} q_{++} q_{-+}^3}\right]^{\frac{1}{2}}$$

$$\doteq \frac{e^4 m k_{2-}|B|}{(2E - k_{2-})(p_- \cdot k_2)} \left[\frac{q_{++} q_{--}^3}{q_{+-} q_{-+}(q_+ \cdot k_1)(q_- \cdot k_1)}\right]^{\frac{1}{2}},$$

9.19. $e^+e^- \to e^+e^-\gamma\gamma$ (NO Z-EXCHANGE; $m \neq 0$)

$$M(+,+;-,+,+,+) = \frac{e^4 m k_{2\perp}^* q_{+\perp} Z_{-+} B}{(2E - k_{2-})(p_- \cdot k_2)} \left[\frac{k_{2-}}{2C k_{2+} q_{++} q_{-+}} \right]^{\frac{1}{2}}$$

$$\doteq \frac{e^4 m k_{2-} q_{+-} |B|}{(2E - k_{2-})(p_- \cdot k_2) \left[(q_+ \cdot k_1)(q_- \cdot k_1) \right]^{\frac{1}{2}}},$$

$$M(+,+;+,+,-,-) = \frac{e^4 q_{+\perp}^* Z_{-+} B_0^*}{q_{+-}(p_- \cdot k_2)} \left[\frac{2k_{2+}(q_+ \cdot q_-)}{k_{2-} q_{++} q_{-+} (q_+ \cdot k_1)(q_- \cdot k_1)} \right]^{\frac{1}{2}}$$

$$\doteq \frac{2e^4 (q_+ \cdot q_-) |B_0|}{(p_- \cdot k_2)} \left[\frac{k_{2+}}{k_{2-}(q_+ \cdot k_1)(q_- \cdot k_1)} \right]^{\frac{1}{2}},$$

$$M(+,+;+,-,-,+) = \frac{e^4 m k_{2\perp}^* q_{+\perp} Z_{+-}^* B^*}{E q_{+-}(p_- \cdot k_2)} \left[\frac{k_{2-} q_{++}}{8C k_{2+} q_{-+}} \right]^{\frac{1}{2}}$$

$$\doteq \frac{e^4 m k_{2-} |B|}{2E(p_- \cdot k_2)} \left[\frac{q_{++}^3 q_{--}}{q_{+-} q_{-+} (q_+ \cdot k_1)(q_- \cdot k_1)} \right]^{\frac{1}{2}},$$

$$M(+,+;-,+,-,+) = \frac{e^4 m k_{2\perp}^* q_{+\perp} Z_{-+}^* B^*}{E q_{+-}(p_- \cdot k_2)} \left[\frac{k_{2-} q_{-+}}{8C k_{2+} q_{++}} \right]^{\frac{1}{2}}$$

$$\doteq \frac{e^4 m k_{2-} q_{-+} |B|}{2E(p_- \cdot k_2) \left[(q_+ \cdot k_1)(q_- \cdot k_1) \right]^{\frac{1}{2}}},$$

$$M(+,-;+,+,+,-) = -\frac{e^4 m k_{2\perp} q_{+\perp}^* B}{q_{+-}(p_- \cdot k_2)} \left[\frac{2k_{2-}(q_+ \cdot q_-)}{k_{2+} q_{++} q_{-+} (q_+ \cdot k_1)(q_- \cdot k_1)} \right]^{\frac{1}{2}}$$

$$\doteq \frac{e^4 m k_{2-} |B|}{(p_- \cdot k_2)} \left[\frac{2(q_+ \cdot q_-)}{q_{+-} q_{-+} (q_+ \cdot k_1)(q_- \cdot k_1)} \right]^{\frac{1}{2}},$$

$$M(+,-;+,-,+,+) = \frac{e^4 E q_{\perp}^2 q_{+\perp}^* Z_{+-} B}{q_{+-}(p_- \cdot k_2)} \left[\frac{2k_{2+}}{C k_{2-} q_{++} q_{-+}^3} \right]^{\frac{1}{2}}$$

$$\doteq \frac{2e^4 E |B|}{(p_- \cdot k_2)} \left[\frac{k_{2+} q_{++} q_{--}^3}{k_{2-} q_{+-} q_{-+} (q_+ \cdot k_1)(q_- \cdot k_1)} \right]^{\frac{1}{2}},$$

$$M(+,-;-,+,+,+) = \frac{e^4 E q_{+\perp} Z_{-+} B}{(p_- \cdot k_2)} \left[\frac{2k_{2+}}{C k_{2-} q_{++} q_{-+}} \right]^{\frac{1}{2}}$$

$$\doteq \frac{2e^4 E q_{+-} |B|}{(p_- \cdot k_2)} \left[\frac{k_{2+}}{k_{2-}(q_+ \cdot k_1)(q_- \cdot k_1)} \right]^{\frac{1}{2}},$$

$$M(-,+;+,-,+,+) = \frac{e^4 E q_{+\perp}^* Z_{-+} B}{q_{+-}(p_- \cdot k_2)} \left[\frac{2k_{2+}q_{-+}}{Ck_{2-}q_{++}}\right]^{\frac{1}{2}}$$

$$\doteq \frac{2e^4 E q_{-+}|B|}{(p_- \cdot k_2)} \left[\frac{k_{2+}}{k_{2-}(q_+ \cdot k_1)(q_- \cdot k_1)}\right]^{\frac{1}{2}},$$

$$M(-,+;-,+,+,+) = \frac{e^4 E q_{+\perp}^* Z_{+-} B}{q_{+-}(p_- \cdot k_2)} \left[\frac{2k_{2+}q_{++}}{Ck_{2-}q_{-+}}\right]^{\frac{1}{2}}$$

$$\doteq \frac{2e^4 E|B|}{(p_- \cdot k_2)} \left[\frac{k_{2+}q_{++}^3 q_{--}}{k_{2-}q_{+-}q_{-+}(q_+ \cdot k_1)(q_- \cdot k_1)}\right]^{\frac{1}{2}},$$

$$M(+,-;+,+,-,-) = -\frac{e^4 m k_{2\perp} q_{+\perp}^* Z_{-+} B_0^*}{E(2E - k_{2-})q_{+-}(p_- \cdot k_2)}$$

$$\times \left[\frac{k_{2-}(q_+ \cdot q_-)}{2k_{2+}q_{++}q_{-+}(q_+ \cdot k_1)(q_- \cdot k_1)}\right]^{\frac{1}{2}}$$

$$\doteq \frac{e^4 m k_{2-}(q_+ \cdot q_-)|B_0|}{E(2E - k_{2-})(p_- \cdot k_2)[(q_+ \cdot k_1)(q_- \cdot k_1)]^{\frac{1}{2}}},$$

$$M(+,-;+,-,+,-) = \frac{e^4 E^2 q_{-\perp}^2 q_{+\perp}^* Z_{+-} B}{(2E - k_{2-})(p_- \cdot k_2)} \left[\frac{8k_{2+}}{Ck_{2-}q_{++}q_{-+}^3}\right]^{\frac{1}{2}}$$

$$\doteq \frac{4e^4 E^2 |B|}{(2E - k_{2-})(p_- \cdot k_2)}$$

$$\times \left[\frac{k_{2+}q_{++}q_{--}^3}{k_{2-}q_{+-}q_{-+}(q_+ \cdot k_1)(q_- \cdot k_1)}\right]^{\frac{1}{2}},$$

$$M(+,-;+,-,-,+) = \frac{e^4 (2E - k_{2-})q_{+\perp} Z_{+-}^* B^*}{q_{+-}(p_- \cdot k_2)} \left[\frac{k_{2+}q_{++}}{2Ck_{2-}q_{-+}}\right]^{\frac{1}{2}}$$

$$\doteq \frac{e^4 (2E - k_{2-})|B|}{(p_- \cdot k_2)}$$

$$\times \left[\frac{k_{2+}q_{++}^3 q_{--}}{k_{2-}q_{+-}q_{-+}(q_+ \cdot k_1)(q_- \cdot k_1)}\right]^{\frac{1}{2}},$$

$$M(+,-;-,+,+,-) = \frac{e^4 E^2 q_{+\perp} Z_{-+} B}{(2E - k_{2-})(p_- \cdot k_2)} \left[\frac{8k_{2+}}{Ck_{2-}q_{++}q_{-+}}\right]^{\frac{1}{2}}$$

$$\doteq \frac{4e^4 E^2 q_{+-}|B|}{(2E - k_{2-})(p_- \cdot k_2)} \left[\frac{k_{2+}}{k_{2-}(q_+ \cdot k_1)(q_- \cdot k_1)}\right]^{\frac{1}{2}},$$

9.19. $e^+e^- \to e^+e^-\gamma\gamma$ (NO Z-EXCHANGE; $m \neq 0$) 275

$$M(+,-;-,+,-,+) = \frac{e^4(2E-k_{2-})q_{+\perp}Z^*_{-+}B^*}{q_{+-}(p_-\cdot k_2)}\left[\frac{k_{2+}q_{-+}}{2Ck_{2-}q_{++}}\right]^{\frac{1}{2}}$$

$$\doteq \frac{e^4(2E-k_{2-})q_{-+}|B|}{(p_-\cdot k_2)}\left[\frac{k_{2+}}{k_{2-}(q_+\cdot k_1)(q_-\cdot k_1)}\right]^{\frac{1}{2}},$$

$$M(-,+;+,-,+,-) = \frac{e^4(2E-k_{2-})q^*_{+\perp}Z_{-+}B}{q_{+-}(p_-\cdot k_2)}\left[\frac{k_{2+}q_{-+}}{2Ck_{2-}q_{++}}\right]^{\frac{1}{2}}$$

$$\doteq \frac{e^4(2E-k_{2-})q_{-+}|B|}{(p_-\cdot k_2)}\left[\frac{k_{2+}}{k_{2-}(q_+\cdot k_1)(q_-\cdot k_1)}\right]^{\frac{1}{2}},$$

$$M(-,+;+,-,-,+) = \frac{e^4 E^2 q^*_{+\perp}Z^*_{-+}B^*}{(2E-k_{2-})(p_-\cdot k_2)}\left[\frac{8k_{2+}}{Ck_{2-}q_{++}q_{-+}}\right]^{\frac{1}{2}}$$

$$\doteq \frac{4e^4 E^2 q_{+-}|B|}{(2E-k_{2-})(p_-\cdot k_2)}\left[\frac{k_{2+}}{k_{2-}(q_+\cdot k_1)(q_-\cdot k_1)}\right]^{\frac{1}{2}},$$

$$M(-,+;-,+,+,-) = \frac{e^4(2E-k_{2-})q^*_{+\perp}Z_{+-}B}{q_{+-}(p_-\cdot k_2)}\left[\frac{k_{2+}q_{++}}{2Ck_{2-}q_{-+}}\right]^{\frac{1}{2}}$$

$$\doteq \frac{e^4(2E-k_{2-})|B|}{(p_-\cdot k_2)}$$

$$\times \left[\frac{k_{2+}q^3_{++}q_{--}}{k_{2-}q_{+-}q_{-+}(q_+\cdot k_1)(q_-\cdot k_1)}\right]^{\frac{1}{2}},$$

$$M(-,+;-,+,-,+) = \frac{e^4 E q^{*2}_{-\perp}q_{+\perp}Z^*_{+-}B^*}{q_{+-}(p_-\cdot k_2)}\left[\frac{2k_{2+}}{Ck_{2-}q_{++}q^3_{-+}}\right]^{\frac{1}{2}}$$

$$\doteq \frac{2e^4 E|B|}{(p_-\cdot k_2)}\left[\frac{k_{2+}q_{++}q^3_{--}}{k_{2-}q_{+-}q_{-+}(q_+\cdot k_1)(q_-\cdot k_1)}\right]^{\frac{1}{2}},$$

$$M(-,+;-,-,+,+) = \frac{e^4 m k^*_{2\perp}q_{+\perp}Z^*_{-+}B_0}{E(2E-k_{2-+})q_{+-}(p_-\cdot k_2)}$$

$$\times \left[\frac{k_{2-}(q_+\cdot q_-)}{2k_{2+}q_{++}q_{-+}(q_+\cdot k_1)(q_-\cdot k_1)}\right]^{\frac{1}{2}}$$

$$\doteq \frac{e^4 m k_{2-}(q_+\cdot q_-)|B_0|}{E(2E-k_{2-})(p_-\cdot k_2)[(q_+\cdot k_1)(q_-\cdot k_1)]^{\frac{1}{2}}},$$

9. SUMMARY OF QED FORMULA1

$$M(+,-;+,-,-,-) = \frac{e^4 E q_{+\perp} Z^*_{+-} B^*}{q_{+-}(p_- \cdot k_2)} \left[\frac{2k_{2+} q_{++}}{C k_{2-} q_{-+}} \right]^{\frac{1}{2}}$$

$$\doteq \frac{2e^4 E |B|}{(p_- \cdot k_2)} \left[\frac{k_{2+} q^3_{++} q_{--}}{k_{2-} q_{+-} q_{-+} (q_+ \cdot k_1)(q_- \cdot k_1)} \right]^{\frac{1}{2}},$$

$$M(+,-;-,+,-,-) = \frac{e^4 E q_{+\perp} Z^*_{-+} B^*}{q_{+-}(p_- \cdot k_2)} \left[\frac{2k_{2+} q_{-+}}{C k_{2-} q_{++}} \right]^{\frac{1}{2}}$$

$$\doteq \frac{2e^4 E q_{-+} |B|}{(p_- \cdot k_2)} \left[\frac{k_{2+}}{k_{2-}(q_+ \cdot k_1)(q_- \cdot k_1)} \right]^{\frac{1}{2}},$$

$$M(-,+;+,-,-,-) = \frac{e^4 E q^*_{+\perp} Z^*_{-+} B^*}{(p_- \cdot k_2)} \left[\frac{2k_{2+}}{C k_{2-} q_{++} q_{-+}} \right]^{\frac{1}{2}}$$

$$\doteq \frac{2e^4 E q_{+-} |B|}{(p_- \cdot k_2)} \left[\frac{k_{2+}}{k_{2-}(q_+ \cdot k_1)(q_- \cdot k_1)} \right]^{\frac{1}{2}},$$

$$M(-,+;-,+,-,-) = \frac{e^4 E q^{*2}_{-\perp} q_{+\perp} Z^*_{+-} B^*}{q_{+-}(p_- \cdot k_2)} \left[\frac{2k_{2+}}{C k_{2-} q_{++} q^3_{-+}} \right]^{\frac{1}{2}}$$

$$\doteq \frac{2e^4 E |B|}{(p_- \cdot k_2)} \left[\frac{k_{2+} q_{++} q^3_{--}}{k_{2-} q_{+-} q_{-+} (q_+ \cdot k_1)(q_- \cdot k_1)} \right]^{\frac{1}{2}},$$

$$M(-,+;-,-,-,+) = \frac{e^4 m k^*_{2\perp} q_{+\perp} B^*}{q_{+-}(p_- \cdot k_2)} \left[\frac{2k_{2-}(q_+ \cdot q_-)}{k_{2+} q_{++} q_{-+}(q_+ \cdot k_1)(q_- \cdot k_1)} \right]^{\frac{1}{2}}$$

$$\doteq \frac{e^4 m k_{2-} |B|}{(p_- \cdot k_2)} \left[\frac{2(q_+ \cdot q_-)}{q_{+-} q_{-+}(q_+ \cdot k_1)(q_- \cdot k_1)} \right]^{\frac{1}{2}},$$

$$M(-,-;+,-,+,-) = -\frac{e^4 m k_{2\perp} q^*_{+\perp} Z_{-+} B}{E q_{+-}(p_- \cdot k_2)} \left[\frac{k_{2-} q_{-+}}{8C k_{2+} q_{++}} \right]^{\frac{1}{2}}$$

$$\doteq \frac{e^4 m k_{2-} q_{-+} |B|}{2E(p_- \cdot k_2) [(q_+ \cdot k_1)(q_- \cdot k_1)]^{\frac{1}{2}}},$$

$$M(-,-;-,+,+,-) = -\frac{e^4 m k_{2\perp} q^*_{+\perp} Z_{+-} B}{E q_{+-}(p_- \cdot k_2)} \left[\frac{k_{2-} q_{++}}{8C k_{2+} q_{-+}} \right]^{\frac{1}{2}}$$

$$\doteq \frac{e^4 m k_{2-} |B|}{2E(p_- \cdot k_2)} \left[\frac{q^3_{++} q_{--}}{q_{+-} q_{-+}(q_+ \cdot k_1)(q_- \cdot k_1)} \right]^{\frac{1}{2}},$$

9.19. $e^+e^- \to e^+e^-\gamma\gamma$ (NO Z-EXCHANGE; $m \neq 0$)

$$\begin{aligned}
M(-,-;-,-,+,+) &= \frac{e^4 q_{+\perp} Z_{-+}^* B_0}{q_{+-}(p_- \cdot k_2)} \left[\frac{2k_{2+}(q_+ \cdot q_-)}{k_{2-}q_{++}q_{-+}(q_+ \cdot k_1)(q_- \cdot k_1)}\right]^{\frac{1}{2}} \\
&\doteq \frac{2e^4(q_+ \cdot q_-)|B_0|}{(p_- \cdot k_2)} \left[\frac{k_{2+}}{k_{2-}(q_+ \cdot k_1)(q_- \cdot k_1)}\right]^{\frac{1}{2}}, \\
M(-,-;+,-,-,-) &= -\frac{e^4 m k_{2\perp} q_{+\perp}^* Z_{-+}^* B^*}{(2E - k_{2-})(p_- \cdot k_2)} \left[\frac{k_{2-}}{2Ck_{2+}q_{++}q_{-+}}\right]^{\frac{1}{2}} \\
&\doteq \frac{e^4 m k_{2-} q_{+-}|B|}{(2E - k_{2-})(p_- \cdot k_2)[(q_+ \cdot k_1)(q_- \cdot k_1)]^{\frac{1}{2}}}, \\
M(-,-;-,+,-,-) &= -\frac{e^4 m k_{2\perp} q_{-\perp}^{*2} q_{+\perp} Z_{+-}^* B^*}{(2E - k_{2-}) q_{+-}(p_- \cdot k_2)} \left[\frac{k_{2-}}{2Ck_{2+}q_{++}q_{-+}^3}\right]^{\frac{1}{2}} \\
&\doteq \frac{e^4 m k_{2-}|B|}{(2E - k_{2-})(p_- \cdot k_2)} \left[\frac{q_{++} q_{--}^3}{q_{+-}q_{-+}(q_+ \cdot k_1)(q_- \cdot k_1)}\right]^{\frac{1}{2}}, \\
M(-,-;-,-,+,-) &= \frac{e^4 E q_{+\perp} Z_{-+}^* B_0}{(2E - k_{2-}) q_{+-}(p_- \cdot k_2)} \\
&\quad \times \left[\frac{8k_{2+}(q_+ \cdot q_-)}{k_{2-}q_{++}q_{-+}(q_+ \cdot k_1)(q_- \cdot k_1)}\right]^{\frac{1}{2}} \\
&\doteq \frac{4e^4 E(q_+ \cdot q_-)|B_0|}{(2E - k_{2-})(p_- \cdot k_2)} \left[\frac{k_{2+}}{k_{2-}(q_+ \cdot k_1)(q_- \cdot k_1)}\right]^{\frac{1}{2}}, \\
M(-,-;-,-,-,+) &= \frac{e^4 E(2E - k_{2-}) q_{+\perp} B^*}{q_{+-}(p_- \cdot k_2)} \\
&\quad \times \left[\frac{8k_{2+}(q_+ \cdot q_-)}{k_{2-}q_{++}q_{-+}(q_+ \cdot k_1)(q_- \cdot k_1)}\right]^{\frac{1}{2}} \\
&\doteq \frac{e^4 E(2E - k_{2-})|B|}{(p_- \cdot k_2)} \\
&\quad \times \left[\frac{8k_{2+}(q_+ \cdot q_-)}{k_{2-}q_{+-}q_{-+}(q_+ \cdot k_1)(q_- \cdot k_1)}\right]^{\frac{1}{2}}, \\
M(-,-;-,-,-,-) &= \frac{e^4 E^2 q_{+\perp} B^*}{q_{+-}(p_- \cdot k_2)} \left[\frac{32k_{2+}(q_+ \cdot q_-)}{k_{2-}q_{++}q_{-+}(q_+ \cdot k_1)(q_- \cdot k_1)}\right]^{\frac{1}{2}} \\
&\doteq \frac{e^4 E^2 |B|}{(p_- \cdot k_2)} \left[\frac{32k_{2+}(q_+ \cdot q_-)}{k_{2-}q_{+-}q_{-+}(q_+ \cdot k_1)(q_- \cdot k_1)}\right]^{\frac{1}{2}}.
\end{aligned}$$

(9.218)

Unpolarized squared matrix element:

$$\overline{|M|^2} = \frac{e^8|B|^2 k_{2+}}{2k_{2-}(p_- \cdot k_2)^2 (q_+ \cdot k_1)(q_- \cdot k_1) q_{+-} q_{-+}}$$

$$\times \left[4E^2 + (2E - k_{2-})^2 + \frac{m^2 k_{2-}^3}{4E^2 k_{2+}} \right] \left[q_{++}^3 q_{--} + q_{+-} q_{-+}^3 + 8E^2 (q_+ \cdot q_-) \right.$$

$$\left. + \left(\frac{2E}{2E - k_{2-}} \right)^2 \left(q_{++} q_{--}^3 + q_{+-}^3 q_{-+} + \frac{2(q_+ \cdot q_-)^3}{E^2} \right) \right]. \quad (9.219)$$

Unpolarized cross section:

$$d\sigma = \frac{\alpha^4 |B|^2 k_{2+}}{256\pi^2 E^2 k_{2-}(p_- \cdot k_2)^2 (q_+ \cdot k_1)(q_- \cdot k_1) q_{+-} q_{-+}}$$

$$\times \left[4E^2 + (2E - k_{2-})^2 + \frac{m^2 k_{2-}^3}{4E^2 k_{2+}} \right] \left[q_{++}^3 q_{--} + q_{+-} q_{-+}^3 + +8E^2 (q_+ \cdot q_-) \right.$$

$$\left. + \left(\frac{2E}{2E - k_{2-}} \right)^2 \left(q_{++} q_{--}^3 + q_{+-}^3 q_{-+} + \frac{2(q_+ \cdot q_-)^3}{E^2} \right) \right]$$

$$\times \delta^4 (p_+ + p_- - q_+ - q_- - k_1 - k_2) \frac{d^3 \vec{q}_+ \, d^3 \vec{q}_- \, d^3 \vec{k}_1 \, d^3 \vec{k}_2}{q_{+0} \, q_{-0} \, k_{10} \, k_{20}}. \quad (9.220)$$

9.19.3 $\vec{k}_2$ nearly parallel to $\vec{q}_+$

The primed quantities $k'_{2\pm}$ and $k'_{2\perp}$ are evaluated in the rotated frame where $\vec{q}_+$ determines the positive z-axis (see also Section 7.4.3). The four-vector q in eqn (9.221) is obtained by applying a space reflection to q_+.
Photon polarizations:

$$\not{e}_1^\pm = N_1 [\not{p}_+ \not{p}_- \not{k}_1 (1 \mp \gamma_5) - \not{k}_1 \not{p}_+ \not{p}_- (1 \pm \gamma_5)],$$

$$N_1^{-1} = E^2 [32 k_{1+} k_{1-}]^{\frac{1}{2}},$$

$$\not{e}_2^\pm = N_2 [\not{k}_2 \not{q}_+ \not{q} (1 \pm \gamma_5) + \not{q} \not{q}_+ \not{k}_2 (1 \mp \gamma_5)],$$

$$N_2^{-1} = q_{+0}^2 [32 k'_{2+} k'_{2-}]^{\frac{1}{2}}. \quad (9.221)$$

9.19. $e^+e^- \to e^+e^- \gamma\gamma$ (NO Z-EXCHANGE; $m \neq 0$)

Nonvanishing helicity amplitudes:

$$M(+,+;+,+,+,+) = -\frac{e^4 E q_{+0} q_{+\perp}^* B}{q_{+-}(q_+ \cdot k_2)} \left[\frac{32 k'_{2-}(q_+ \cdot q_-)}{k'_{2+} q_{++} q_{-+}(q_+ \cdot k_1)(q_- \cdot k_1)} \right]^{\frac{1}{2}}$$

$$\doteq \frac{e^4 E q_{+0}|B|}{(q_+ \cdot k_2)} \left[\frac{32 k'_{2-}(q_+ \cdot q_-)}{k'_{2+} q_{+-} q_{-+}(q_+ \cdot k_1)(q_- \cdot k_1)} \right]^{\frac{1}{2}},$$

$$M(+,+;+,+,+,-) = -\frac{e^4 E q_{+0}^2 q_{+\perp} B}{(2q_{+0} + k'_{2+}) q_{+-}(q_+ \cdot k_2)}$$

$$\times \left[\frac{128 k'_{2-}(q_+ \cdot q_-)}{k'_{2+} q_{++} q_{-+}(q_+ \cdot k_1)(q_- \cdot k_1)} \right]^{\frac{1}{2}}$$

$$\doteq \frac{e^4 E q_{+0}^2 |B|}{(2q_{+0} + k'_{2+})(q_+ \cdot k_2)}$$

$$\times \left[\frac{128 k'_{2-}(q_+ \cdot q_-)}{k'_{2+} q_{+-} q_{-+}(q_+ \cdot k_1)(q_- \cdot k_1)} \right]^{\frac{1}{2}},$$

$$M(+,+;+,+,-,+) = -\frac{e^4(2q_{+0} + k'_{2+}) q_{+\perp}^* Z_{+-} B_0^*}{E q_{+-}(q_+ \cdot k_2)}$$

$$\times \left[\frac{k'_{2-}(q_+ \cdot q_-)}{2 k'_{2+}(q_+ \cdot k_1)(q_- \cdot k_1)} \right]^{\frac{1}{2}}$$

$$\doteq \frac{e^4(2q_{+0} + k'_{2+})(q_+ \cdot q_-)|B_0|}{E(q_+ \cdot k_2)}$$

$$\times \left[\frac{k'_{2-}}{k'_{2+}(q_+ \cdot k_1)(q_- \cdot k_1)} \right]^{\frac{1}{2}},$$

$$M(+,+;-,+,+,+) = \frac{e^4 m E k'_{2\perp} q_{+\perp}^* B}{(2q_{+0} + k'_{2+}) q_{+-}(q_+ \cdot k_2)}$$

$$\times \left[\frac{8 k'_{2+}(q_+ \cdot q_-)}{k'_{2-} q_{++} q_{-+}(q_+ \cdot k_1)(q_- \cdot k_1)} \right]^{\frac{1}{2}}$$

$$\doteq \frac{e^4 m E k'_{2+}|B|}{(2q_{+0} + k'_{2+})(q_+ \cdot k_2)} \left[\frac{8(q_+ \cdot q_-)}{q_{+-} q_{-+}(q_+ \cdot k_1)(q_- \cdot k_1)} \right]^{\frac{1}{2}},$$

$$M(+,+;+,+,-,-) = -\frac{e^4 q_{+0} q_{+\perp}^* Z_{-+} B_0^*}{E q_{+-}(q_+ \cdot k_2)} \left[\frac{2 k'_{2-}(q_+ \cdot q_-)}{k'_{2+} q_{++} q_{-+}(q_+ \cdot k_1)(q_- \cdot k_1)} \right]^{\frac{1}{2}}$$

$$\doteq \frac{2 e^4 q_{+0}(q_+ \cdot q_-)|B_0|}{E(q_+ \cdot k_2)} \left[\frac{k'_{2-}}{k'_{2+}(q_+ \cdot k_1)(q_- \cdot k_1)} \right]^{\frac{1}{2}},$$

$$M(+,+;-,+,-,+) = \frac{e^4 m k'_{2\perp} q^*_{+\perp} Z_{-+} B^*_0}{E q_{+0} q_{+-} (q_+ \cdot k_2)}$$

$$\times \left[\frac{k'_{2+}(q_+ \cdot q_-)}{8 k'_{2+} q_{++} q_{-+} (q_+ \cdot k_1)(q_- \cdot k_1)} \right]$$

$$\doteq \frac{e^4 m k'_{2+}(q_+ \cdot q_-)|B_0|}{2 E q_{+0}(q_+ \cdot k_2)[(q_+ \cdot k_1)(q_- \cdot k_1)]^{\frac{1}{2}}},$$

$$M(+,-;+,+,+,-) = -\frac{e^4 m k'^*_{2\perp} q_{+\perp} Z_{-+} B}{q_{+0}(q_+ \cdot k_2)} \left[\frac{k'_{2+}}{8C k'_{2-} q_{++} q_{-+}} \right]^{\frac{1}{2}}$$

$$\doteq \frac{e^4 m k'_{2+} q_{+-}|B|}{2 q_{+0}(q_+ \cdot k_2)[(q_+ \cdot k_1)(q_- \cdot k_1)]^{\frac{1}{2}}},$$

$$M(+,-;+,-,+,+) = -\frac{e^4 q_{+0} q^2_{-\perp} q^*_{+\perp} Z_{+-} B}{q_{+-}(q_+ \cdot k_2)} \left[\frac{2 k'_{2-}}{C k'_{2+} q_{++} q^3_{-+}} \right]^{\frac{1}{2}}$$

$$\doteq \frac{2 e^4 q_{+0}|B|}{(q_+ \cdot k_2)} \left[\frac{k'_{2-} q_{++} q^3_{--}}{k'_{2+} q_{+-} q_{-+}(q_+ \cdot k_1)(q_- \cdot k_1)} \right]^{\frac{1}{2}},$$

$$M(+,-;-,+,+,+) = -\frac{e^4 q_{+0} q_{+\perp} Z_{-+} B}{(q_+ \cdot k_2)} \left[\frac{2 k'_{2-}}{C k'_{2+} q_{++} q_{-+}} \right]^{\frac{1}{2}}$$

$$\doteq \frac{2 e^4 q_{+0} q_{+-}|B|}{(q_+ \cdot k_2)} \left[\frac{k'_{2-}}{k'_{2+}(q_+ \cdot k_1)(q_- \cdot k_1)} \right]^{\frac{1}{2}},$$

$$M(-,+;+,+,+,-) = -\frac{e^4 m k'^*_{2\perp} q^*_{+\perp} Z_{+-} B}{q_{+0} q_{+-}(q_+ \cdot k_2)} \left[\frac{k'_{2+} q_{++}}{8C k'_{2-} q_{-+}} \right]^{\frac{1}{2}}$$

$$\doteq \frac{e^4 m k'_{2+}|B|}{2 q_{+0}(q_+ \cdot k_2)} \left[\frac{q^3_{++} q_{--}}{q_{+-} q_{-+}(q_+ \cdot k_1)(q_- \cdot k_1)} \right]^{\frac{1}{2}},$$

$$M(-,+;+,-,+,+) = -\frac{e^4 q_{+0} q^*_{+\perp} Z_{-+} B}{q_{+-}(q_+ \cdot k_2)} \left[\frac{2 k'_{2-} q_{-+}}{C k'_{2+} q_{++}} \right]^{\frac{1}{2}}$$

$$\doteq \frac{2 e^4 q_{+0} q_{-+}|B|}{(q_+ \cdot k_2)} \left[\frac{k'_{2-}}{k'_{2+}(q_+ \cdot k_1)(q_- \cdot k_1)} \right]^{\frac{1}{2}},$$

$$M(-,+;-,+,+,+) = -\frac{e^4 q_{+0} q^*_{+\perp} Z_{+-} B}{q_{+-}(q_+ \cdot k_2)} \left[\frac{2 k'_{2-} q_{++}}{C k'_{2+} q_{-+}} \right]^{\frac{1}{2}}$$

$$\doteq \frac{2 e^4 q_{+0}|B|}{(q_+ \cdot k_2)} \left[\frac{k'_{2-} q^3_{++} q_{--}}{k'_{2+} q_{+-} q_{-+}(q_+ \cdot k_1)(q_- \cdot k_1)} \right]^{\frac{1}{2}},$$

9.19. $e^+e^- \to e^+e^- \gamma\gamma$ (NO Z-EXCHANGE; $m \neq 0$)

$$M(+,-;+,+,-,-) = -\frac{e^4 m k_{2\perp}'^* q_{+\perp} Z_{-+}^* B^*}{(2q_{+0}+k_{2+}')q_{+-}(q_+ \cdot k_2)} \left[\frac{k_{2+}' q_{-+}}{2Ck_{2-}' q_{++}}\right]^{\frac{1}{2}}$$

$$\doteq \frac{e^4 m k_{2+}' q_{-+} |B|}{(2q_{+0}+k_{2+}')(q_+ \cdot k_2)[(q_+ \cdot k_1)(q_- \cdot k_1)]^{\frac{1}{2}}},$$

$$M(+,-;+,-,+,-) = -\frac{e^4 q_{+0}^2 q_{-\perp}^2 q_{+\perp}^* Z_{+-} B}{(2q_{+0}+k_{2+}')q_{+-}(q_+ \cdot k_2)} \left[\frac{8k_{2-}'}{Ck_{2+}' q_{++} q_{-+}^3}\right]^{\frac{1}{2}}$$

$$\doteq \frac{4e^4 q_{+0}^2 |B|}{(2q_{+0}+k_{2+}')(q_+ \cdot k_2)}$$

$$\times \left[\frac{k_{2-}' q_{++} q_{--}^3}{k_{2+}' q_{+-} q_{-+}(q_+ \cdot k_1)(q_- \cdot k_1)}\right]^{\frac{1}{2}},$$

$$M(+,-;+,-,-,+) = -\frac{e^4 (2q_{+0}+k_{2+}') q_{+\perp} Z_{+-}^* B^*}{q_{+-}(q_+ \cdot k_2)} \left[\frac{k_{2-}' q_{++}}{2Ck_{2+}' q_{-+}}\right]^{\frac{1}{2}}$$

$$\doteq \frac{e^4 (2q_{+0}+k_{2+}') |B|}{(q_+ \cdot k_2)}$$

$$\times \left[\frac{k_{2-}' q_{++}^3 q_{--}}{k_{2+}' q_{+-} q_{-+}(q_+ \cdot k_1)(q_- \cdot k_1)}\right]^{\frac{1}{2}},$$

$$M(+,-;-,+,+,-) = -\frac{e^4 (2q_{+0}+k_{2+}') q_{+\perp} Z_{-+} B}{(q_+ \cdot k_2)} \left[\frac{k_{2-}'}{2Ck_{2+}' q_{++} q_{-+}}\right]^{\frac{1}{2}}$$

$$\doteq \frac{e^4 (2q_{+0}+k_{2+}') q_{+-} |B|}{(q_+ \cdot k_2)} \left[\frac{k_{2-}'}{k_{2+}'(q_+ \cdot k_1)(q_- \cdot k_1)}\right]^{\frac{1}{2}},$$

$$M(+,-;-,+,-,+) = -\frac{e^4 q_{+0}^2 q_{+\perp} Z_{-+}^* B^*}{(2q_{+0}+k_{2+}')q_{+-}(q_+ \cdot k_2)} \left[\frac{8k_{2-}' q_{-+}}{Ck_{2+}' q_{++}}\right]^{\frac{1}{2}}$$

$$\doteq \frac{4e^4 q_{+0}^2 q_{-+} |B|}{(2q_{+0}+k_{2+}')(q_+ \cdot k_2)} \left[\frac{k_{2-}'}{k_{2+}'(q_+ \cdot k_1)(q_- \cdot k_1)}\right]^{\frac{1}{2}},$$

$$M(+,-;-,-,+,+) = \frac{e^4 m k_{2\perp}' q_{-\perp}^2 q_{+\perp}^* Z_{+-} B}{(2q_{+0}+k_{2+}')q_{+-}(q_+ \cdot k_2)} \left[\frac{k_{2+}'}{2Ck_{2-}' q_{++} q_{-+}^3}\right]^{\frac{1}{2}}$$

$$\doteq \frac{e^4 m k_{2+}' |B|}{(2q_{+0}+k_{2+}')(q_+ \cdot k_2)} \left[\frac{q_{++} q_{--}^3}{q_{+-} q_{-+}(q_+ \cdot k_1)(q_- \cdot k_1)}\right]^{\frac{1}{2}},$$

$$M(-,+;+,+,-,-) = -\frac{e^4 m k_{2\perp}'^* q_{-\perp}^{*2} q_{+\perp} Z_{+-}^* B^*}{(2q_{+0}+k_{2+}')q_{+-}(q_+\cdot k_2)}\left[\frac{k_{2+}'}{2Ck_{2-}'q_{++}q_{-+}^3}\right]^{\frac{1}{2}}$$

$$\doteq \frac{e^4 m k_{2+}'|B|}{(2q_{+0}+k_{2+}')(q_+\cdot k_2)}\left[\frac{q_{++}q_{--}^3}{q_{+-}q_{-+}(q_+\cdot k_1)(q_-\cdot k_1)}\right]$$

$$M(-,+;+,-,+,-) = -\frac{e^4 q_{+0}^2 q_{+\perp}^* Z_{-+} B}{(2q_{+0}+k_{2+}')q_{+-}(q_+\cdot k_2)}\left[\frac{8k_{2-}'q_{-+}}{Ck_{2+}'q_{++}}\right]^{\frac{1}{2}}$$

$$\doteq \frac{4e^4 q_{+0}^2 q_{-+}|B|}{(2q_{+0}+k_{2+}')(q_+\cdot k_2)}\left[\frac{k_{2-}'}{k_{2+}'(q_+\cdot k_1)(q_-\cdot k_1)}\right]^{\frac{1}{2}},$$

$$M(-,+;+,-,-,+) = -\frac{e^4 (2q_{+0}+k_{2+}')q_{+\perp}^* Z_{-+}^* B^*}{(q_+\cdot k_2)}\left[\frac{k_{2-}'}{2Ck_{2+}'q_{++}q_{-+}}\right]^{\frac{1}{2}}$$

$$\doteq \frac{e^4 (2q_{+0}+k_{2+}')q_{+-}|B|}{(q_+\cdot k_2)}\left[\frac{k_{2-}'}{k_{2+}'(q_+\cdot k_1)(q_-\cdot k_1)}\right]^{\frac{1}{2}},$$

$$M(-,+;-,+,+,-) = -\frac{e^4 (2q_{+0}+k_{2+}')q_{+\perp}^* Z_{+-} B}{q_{+-}(q_+\cdot k_2)}\left[\frac{k_{2-}'q_{++}}{2Ck_{2+}'q_{-+}}\right]^{\frac{1}{2}}$$

$$\doteq \frac{e^4 (2q_{+0}+k_{2+}')|B|}{(q_+\cdot k_2)}\left[\frac{k_{2-}'q_{++}^3 q_{--}}{k_{2+}'q_{+-}q_{-+}(q_+\cdot k_1)(q_-\cdot k_1)}\right]$$

$$M(-,+;-,+,-,+) = -\frac{e^4 q_{+0}^2 q_{-\perp}^{*2} q_{+\perp} Z_{+-}^* B^*}{(2q_{+0}+k_{2+}')q_{+-}(q_+\cdot k_2)}\left[\frac{8k_{2-}'}{Ck_{2+}'q_{++}q_{-+}^3}\right]^{\frac{1}{2}}$$

$$\doteq \frac{4e^4 q_{+0}^2|B|}{(2q_{+0}+k_{2+}')(q_+\cdot k_2)}$$

$$\times\left[\frac{k_{2-}'q_{++}q_{--}^3}{k_{2+}'q_{+-}q_{-+}(q_+\cdot k_1)(q_-\cdot k_1)}\right]^{\frac{1}{2}},$$

$$M(-,+;-,-,+,+) = \frac{e^4 m k_{2\perp}' q_{+\perp}^* Z_{-+} B}{(2q_{+0}+k_{2+}')q_{+-}(q_+\cdot k_2)}\left[\frac{k_{2+}'q_{-+}}{2Ck_{2-}'q_{++}}\right]^{\frac{1}{2}}$$

$$\doteq \frac{e^4 m k_{2+}'q_{-+}|B|}{(2q_{+0}+k_{2+}')(q_+\cdot k_2)[(q_+\cdot k_1)(q_-\cdot k_1)]^{\frac{1}{2}}},$$

$$M(+,-;+,-,-,-) = -\frac{e^4 q_{+0} q_{+\perp} Z_{+-}^* B^*}{q_{+-}(q_+\cdot k_2)}\left[\frac{2k_{2-}'q_{++}}{Ck_{2+}'q_{-+}}\right]^{\frac{1}{2}}$$

$$\doteq \frac{2e^4 q_{+0}|B|}{(q_+\cdot k_2)}\left[\frac{k_{2-}'q_{++}^3 q_{--}}{k_{2+}'q_{+-}q_{-+}(q_+\cdot k_1)(q_-\cdot k_1)}\right]^{\frac{1}{2}},$$

9.19. $e^+e^- \to e^+e^-\gamma\gamma$ (NO Z-EXCHANGE; $m \neq 0$)

$$M(+,-;-,+,-,-) = -\frac{e^4 q_{+0} q_{+\perp} Z^*_{-+} B^*}{q_{+-}(q_+ \cdot k_2)} \left[\frac{2k'_{2-} q_{-+}}{C k'_{2+} q_{++}}\right]^{\frac{1}{2}}$$

$$\doteq \frac{2e^4 q_{+0} q_{-+} |B|}{(q_+ \cdot k_2)} \left[\frac{k'_{2-}}{k'_{2+}(q_+ \cdot k_1)(q_- \cdot k_1)}\right]^{\frac{1}{2}},$$

$$M(+,-;-,-,-,+) = \frac{e^4 m k'_{2\perp} q_{+\perp} Z^*_{+-} B^*}{q_{+0} q_{+-}(q_+ \cdot k_2)} \left[\frac{k'_{2+} q_{++}}{8 C k'_{2-} q_{-+}}\right]^{\frac{1}{2}}$$

$$\doteq \frac{e^4 m k'_{2+} |B|}{2 q_{+0}(q_+ \cdot k_2)} \left[\frac{q^3_{++} q_{--}}{q_{+-} q_{-+}(q_+ \cdot k_1)(q_- \cdot k_1)}\right]^{\frac{1}{2}},$$

$$M(-,+;+,-,-,-) = -\frac{e^4 q_{+0} q^*_{+\perp} Z^*_{-+} B^*}{(q_+ \cdot k_2)} \left[\frac{2k'_{2-}}{C k'_{2+} q_{++} q_{-+}}\right]^{\frac{1}{2}}$$

$$\doteq \frac{2e^4 q_{+0} q_{+-} |B|}{(q_+ \cdot k_2)} \left[\frac{k'_{2-}}{k'_{2+}(q_+ \cdot k_1)(q_- \cdot k_1)}\right]^{\frac{1}{2}},$$

$$M(-,+;-,+,-,-) = -\frac{e^4 q_{+0} q^{*2}_{-\perp} q_{+\perp} Z^*_{+-} B^*}{q_{+-}(q_+ \cdot k_2)} \left[\frac{2k'_{2-}}{C k'_{2+} q_{++} q^3_{-+}}\right]^{\frac{1}{2}}$$

$$\doteq \frac{2e^4 q_{+0} |B|}{(q_+ \cdot k_2)} \left[\frac{k'_{2-} q_{++} q^3_{--}}{k'_{2+} q_{+-} q_{-+}(q_+ \cdot k_1)(q_- \cdot k_1)}\right]^{\frac{1}{2}},$$

$$M(-,+;-,-,-,+) = \frac{e^4 m k'_{2\perp} q^*_{+\perp} Z^*_{-+} B^*}{q_{+0}(q_+ \cdot k_2)} \left[\frac{k'_{2+}}{8 C k'_{2-} q_{++} q_{-+}}\right]^{\frac{1}{2}}$$

$$\doteq \frac{e^4 m k'_{2+} q_{+-} |B|}{2 q_{+0}(q_+ \cdot k_2)[(q_+ \cdot k_1)(q_- \cdot k_1)]^{\frac{1}{2}}},$$

$$M(-,-;+,-,+,-) = -\frac{e^4 m k'^*_{2\perp} q_{+\perp} Z^*_{-+} B_0}{E q_{+0} q_{+-}(q_+ \cdot k_2)}$$

$$\times \left[\frac{k'_{2+}(q_+ \cdot q_-)}{8 k'_{2+} q_{++} q_{-+}(q_+ \cdot k_1)(q_- \cdot k_1)}\right]^{\frac{1}{2}}$$

$$\doteq \frac{e^4 m k'_{2+}(q_+ \cdot q_-) |B_0|}{2 E q_{+0}(q_+ \cdot k_2)[(q_+ \cdot k_1)(q_- \cdot k_1)]^{\frac{1}{2}}},$$

$$M(-,-;-,-,+,+) = -\frac{e^4 q_{+0} q_{+\perp} Z^*_{-+} B_0}{E q_{+-}(q_+ \cdot k_2)} \left[\frac{2 k'_{2-}(q_+ \cdot q_-)}{k'_{2+} q_{++} q_{-+}(q_+ \cdot k_1)(q_- \cdot k_1)}\right]^{\frac{1}{2}}$$

$$\doteq \frac{2e^4 q_{+0}(q_+ \cdot q_-) |B_0|}{E(q_+ \cdot k_2)} \left[\frac{k'_{2-}}{k'_{2+}(q_+ \cdot k_1)(q_- \cdot k_1)}\right]^{\frac{1}{2}},$$

$$M(-,-;+,-,-,-) = -\frac{e^4 mE k'^*_{2\perp} q_{+\perp} B^*}{(2q_{+0} + k'_{2+})q_{+-}(q_+ \cdot k_2)}$$

$$\times \left[\frac{8k'_{2+}(q_+ \cdot q_-)}{k'_{2-}q_{++}q_{-+}(q_+ \cdot k_1)(q_- \cdot k_1)}\right]^{\frac{1}{2}}$$

$$\doteq \frac{e^4 mE k'_{2+}|B|}{(2q_{+0} + k'_{2+})(q_+ \cdot k_2)}\left[\frac{8(q_+ \cdot q_-)}{q_{+-}q_{-+}(q_+ \cdot k_1)(q_- \cdot k_1)}\right]^{\frac{1}{2}}$$

$$M(-,-;-,-,+,-) = -\frac{e^4(2q_{+0} + k'_{2+})q_{+\perp} Z^*_{+-} B_0}{Eq_{+-}(q_+ \cdot k_2)}$$

$$\times \left[\frac{k'_{2-}(q_+ \cdot q_-)}{2k'_{2+}(q_+ \cdot k_1)(q_- \cdot k_1)}\right]^{\frac{1}{2}}$$

$$\doteq \frac{e^4(2q_{+0} + k'_{2+})(q_+ \cdot q_-)|B_0|}{E(q_+ \cdot k_2)}$$

$$\times \left[\frac{k'_{2-}}{k'_{2+}(q_+ \cdot k_1)(q_- \cdot k_1)}\right]^{\frac{1}{2}}$$

$$M(-,-;-,-,-,+) = -\frac{e^4 E q^2_{+0} q^*_{+\perp} B^*}{(2q_{+0} + k'_{2+})q_{+-}(q_+ \cdot k_2)}$$

$$\times \left[\frac{128 k'_{2-}(q_+ \cdot q_-)}{k'_{2+}q_{++}q_{-+}(q_+ \cdot k_1)(q_- \cdot k_1)}\right]^{\frac{1}{2}}$$

$$\doteq \frac{e^4 E q^2_{+0}|B|}{(2q_{+0} + k'_{2+})(q_+ \cdot k_2)}$$

$$\times \left[\frac{128 k'_{2-}(q_+ \cdot q_-)}{k'_{2+}q_{+-}q_{-+}(q_+ \cdot k_1)(q_- \cdot k_1)}\right]^{\frac{1}{2}},$$

$$M(-,-;-,-,-,-) = -\frac{e^4 E q_{+0} q_{+\perp} B^*}{q_{+-}(q_+ \cdot k_2)}\left[\frac{32 k'_{2-}(q_+ \cdot q_-)}{k'_{2+}q_{++}q_{-+}(q_+ \cdot k_1)(q_- \cdot k_1)}\right]^{\frac{1}{2}}$$

$$\doteq \frac{e^4 E q_{+0}|B|}{(q_+ \cdot k_2)}\left[\frac{32 k'_{2-}(q_+ \cdot q_-)}{k'_{2+}q_{+-}q_{-+}(q_+ \cdot k_1)(q_- \cdot k_1)}\right]^{\frac{1}{2}}.$$

(9.22)

9.19. $e^+e^- \to e^+e^-\gamma\gamma$ (NO Z-EXCHANGE; $m \neq 0$)

Unpolarized squared matrix element:

$$\overline{|M|^2} = \frac{e^8|B|^2 k'_{2-}}{2k'_{2+}(q_+ \cdot k_2)^2(q_+ \cdot k_1)(q_- \cdot k_1)q_{+-}q_{-+}}$$

$$\times \left[4q_{+0}^2 + (2q_{+0} + k'_{2+})^2 + \frac{m^2 k'^3_{2+}}{4q_{+0}^2 k'_{2-}} \right] \left[q_{++}^3 q_{--} + q_{+-}^3 q_{-+} + \frac{2(q_+ \cdot q_-)^3}{E^2} \right.$$

$$\left. + \left(\frac{2q_{+0}}{2q_{+0} + k'_{2+}} \right)^2 \left(q_{++} q_{--}^3 + q_{+-} q_{-+}^3 + 8E^2(q_+ \cdot q_-) \right) \right] . \qquad (9.223)$$

Unpolarized cross section:

$$d\sigma = \frac{\alpha^4 |B|^2 k'_{2-}}{256\pi^2 E^2 k'_{2+}(q_+ \cdot k_2)^2(q_+ \cdot k_1)(q_- \cdot k_1)q_{+-}q_{-+}}$$

$$\times \left[4q_{+0}^2 + (2q_{+0} + k'_{2+})^2 + \frac{m^2 k'^3_{2+}}{4q_{+0}^2 k'_{2-}} \right] \left[q_{++}^3 q_{--} + q_{+-}^3 q_{-+} + \frac{2(q_+ \cdot q_-)^3}{E^2} \right.$$

$$\left. + \left(\frac{2q_{+0}}{2q_{+0} + k'_{2+}} \right)^2 \left(q_{++} q_{--}^3 + q_{+-} q_{-+}^3 + 8E^2(q_+ \cdot q_-) \right) \right]$$

$$\times \delta^4(p_+ + p_- - q_+ - q_- - k_1 - k_2) \frac{d^3\vec{q}_+ \, d^3\vec{q}_- \, d^3\vec{k}_1 \, d^3\vec{k}_2}{q_{+0} \, q_{-0} \, k_{10} \, k_{20}} . \qquad (9.224)$$

9.19.4 $\vec{k}_2$ nearly parallel to $\vec{q}_-$

The doubly primed quantities $k''_{2\pm}$ and $k''_{2\perp}$ are evaluated in the rotated frame where $\vec{q}_-$ determines the positive z-axis (see also Section 7.4.3). The four-vector q in eqn (9.225) is obtained by applying a space reflection to q_-.
Photon polarizations:

$$\rlap{/}{\epsilon}_1^\pm = N_1[\rlap{/}{p}_+ \rlap{/}{p}_- \rlap{/}{k}_1(1 \mp \gamma_5) - \rlap{/}{k}_1 \rlap{/}{p}_+ \rlap{/}{p}_-(1 \pm \gamma_5)],$$

$$N_1^{-1} = E^2[32k_{1+}k_{1-}]^{\frac{1}{2}},$$

$$\rlap{/}{\epsilon}_2^\pm = N_2[\rlap{/}{k}_2 \rlap{/}{q}_- \rlap{/}{q}(1 \pm \gamma_5) + \rlap{/}{q} \rlap{/}{q}_- \rlap{/}{k}_2(1 \mp \gamma_5)],$$

$$N_2^{-1} = q_{-0}^2[32k''_{2+}k''_{2-}]^{\frac{1}{2}} . \qquad (9.225)$$

Nonvanishing helicity amplitudes:

$$M(+,+;+,+,+,+) = \frac{e^4 E q_{-0} q_{+\perp}^* B}{q_{+-}(q_- \cdot k_2)} \left[\frac{32 k''_{2-}(q_+ \cdot q_-)}{k''_{2+} q_{++} q_{-+}(q_+ \cdot k_1)(q_- \cdot k_1)} \right]^{\frac{1}{2}}$$

$$\doteq \frac{e^4 E q_{-0} |B|}{(q_- \cdot k_2)} \left[\frac{32 k''_{2-}(q_+ \cdot q_-)}{k''_{2+} q_{+-} q_{-+}(q_+ \cdot k_1)(q_- \cdot k_1)} \right]^{\frac{1}{2}} ,$$

$$M(+,+;+,+,+,-) = \frac{e^4 E q_{-0}^2 q_{+\perp}^* B}{(2q_{-0} + k_{2+}'')q_{+-}(q_- \cdot k_2)}$$

$$\times \left[\frac{128 k_{2-}''(q_+ \cdot q_-)}{k_{2+}'' q_{++} q_{-+}(q_+ \cdot k_1)(q_- \cdot k_1)} \right]^{\frac{1}{2}}$$

$$\doteq \frac{e^4 E q_{-0}^2 |B|}{(q_- \cdot k_2)} \left[\frac{128 k_{2-}''(q_+ \cdot q_-)}{k_{2+}'' q_{++} q_{-+}(q_+ \cdot k_1)(q_- \cdot k_1)} \right]^{\frac{1}{2}},$$

$$M(+,+;+,+,-,+) = \frac{e^4 (2q_{-0} + k_{2+}'') q_{+\perp}^* Z_{-+} B_0^*}{E q_{+-}(q_- \cdot k_2)}$$

$$\times \left[\frac{k_{2-}''(q_+ \cdot q_-)}{2 k_{2+}'' q_{++} q_{-+}(q_+ \cdot k_1)(q_- \cdot k_1)} \right]^{\frac{1}{2}}$$

$$\doteq \frac{e^4 (2q_{-0} + k_{2+}'')(q_+ \cdot q_-)|B_0|}{E(q_- \cdot k_2)}$$

$$\times \left[\frac{k_{2+}''}{k_{2+}''(q_+ \cdot k_1)(q_- \cdot k_1)} \right]^{\frac{1}{2}},$$

$$M(+,+;+,-,+,+) = -\frac{e^4 m E k_{2\perp}'' q_{+\perp}^* B}{(2q_{-0} + k_{2+}'') q_{+-}(q_- \cdot k_2)}$$

$$\times \left[\frac{8 k_{2+}''(q_+ \cdot q_-)}{k_{2-}'' q_{++} q_{-+}(q_+ \cdot k_1)(q_- \cdot k_1)} \right]^{\frac{1}{2}}$$

$$\doteq \frac{e^4 m E k_{2+}'' |B|}{(2q_{-0} + k_{2+}'')(q_- \cdot k_2)} \left[\frac{8(q_+ \cdot q_-)}{q_{+-} q_{-+}(q_+ \cdot k_1)(q_- \cdot k_1)} \right]^{\frac{1}{2}},$$

$$M(+,+;+,+,-,-) = \frac{e^4 q_{-0} q_{+\perp}^* Z_{-+} B_0^*}{E q_{+-}(q_- \cdot k_2)} \left[\frac{2 k_{2-}''(q_+ \cdot q_-)}{k_{2+}'' q_{++} q_{-+}(q_+ \cdot k_1)(q_- \cdot k_1)} \right]^{\frac{1}{2}}$$

$$\doteq \frac{2 e^4 q_{-0}(q_+ \cdot q_-)|B_0|}{E(q_- \cdot k_2)} \left[\frac{k_{2-}''}{k_{2+}''(q_+ \cdot k_1)(q_- \cdot k_1)} \right]^{\frac{1}{2}},$$

$$M(+,+;+,-,-,+) = -\frac{e^4 m k_{2\perp}'' q_{+\perp}^* Z_{-+} B_0^*}{E q_{-0} q_{+-}(q_- \cdot k_2)}$$

$$\times \left[\frac{k_{2+}''(q_+ \cdot q_-)}{8 k_{2-}'' q_{++} q_{-+}(q_+ \cdot k_1)(q_- \cdot k_1)} \right]^{\frac{1}{2}}$$

$$\doteq \frac{e^4 m k_{2+}''(q_+ \cdot q_-)|B_0|}{2 E q_{-0}(q_- \cdot k_2)[(q_+ \cdot k_1)(q_- \cdot k_1)]^{\frac{1}{2}}},$$

9.19. $e^+e^- \to e^+e^-\gamma\gamma$ (NO Z-EXCHANGE; $m \neq 0$)

$$M(+,-;+,+,+,-) = \frac{e^4 m k_{21}^{'''*} q_{-\perp}^2 q_{+\perp}^* Z_{+-} B}{q_{-0} q_{+-} (q_- \cdot k_2)} \left[\frac{k_{2+}''}{8C k_{2-}'' q_{++} q_{-+}^3} \right]^{\frac{1}{2}}$$

$$\doteq \frac{e^4 m k_{2+}'' |B|}{2 q_{-0} (q_- \cdot k_2)} \left[\frac{q_{++} q_{--}^3}{q_{+-} q_{-+} (q_+ \cdot k_1)(q_- \cdot k_1)} \right]^{\frac{1}{2}},$$

$$M(+,-;+,-,+,+) = \frac{e^4 q_{-0} q_{-\perp}^2 q_{+\perp}^* Z_{+-} B}{q_{+-}(q_- \cdot k_2)} \left[\frac{2 k_{2-}''}{C k_{2+}'' q_{++} q_{-+}^3} \right]^{\frac{1}{2}}$$

$$\doteq \frac{2 e^4 q_{-0} |B|}{(q_- \cdot k_2)} \left[\frac{k_{2-}'' q_{++} q_{--}^3}{k_{2+}'' q_{+-} q_{-+} (q_+ \cdot k_1)(q_- \cdot k_1)} \right]^{\frac{1}{2}},$$

$$M(+,-;-,+,+,+) = \frac{e^4 q_{-0} q_{+\perp} Z_{-+} B}{(q_- \cdot k_2)} \left[\frac{2 k_{2-}''}{C k_{2+}'' q_{++} q_{-+}} \right]^{\frac{1}{2}}$$

$$\doteq \frac{2 e^4 q_{-0} q_{+-} |B|}{(q_- \cdot k_2)} \left[\frac{k_{2-}''}{k_{2+}'' (q_+ \cdot k_1)(q_- \cdot k_1)} \right]^{\frac{1}{2}},$$

$$M(-,+;+,+,+,-) = \frac{e^4 m k_{21}^{'''*} q_{+\perp}^* Z_{-+} B}{q_{-0} q_{+-}(q_- \cdot k_2)} \left[\frac{k_{2+}'' q_{-+}}{8 C k_{2-}'' q_{++}} \right]^{\frac{1}{2}}$$

$$\doteq \frac{e^4 m k_{2+}'' q_{-+} |B|}{2 q_{-0}(q_- \cdot k_2) [(q_+ \cdot k_1)(q_- \cdot k_1)]^{\frac{1}{2}}},$$

$$M(-,+;+,-,+,+) = \frac{e^4 q_{-0} q_{+\perp}^* Z_{-+} B}{q_{+-}(q_- \cdot k_2)} \left[\frac{2 k_{2-}'' q_{-+}}{C k_{2+}'' q_{++}} \right]^{\frac{1}{2}}$$

$$\doteq \frac{2 e^4 q_{-0} q_{-+} |B|}{(q_- \cdot k_2)} \left[\frac{k_{2-}''}{k_{2+}''(q_+ \cdot k_1)(q_- \cdot k_1)} \right]^{\frac{1}{2}},$$

$$M(-,+;-,+,+,+) = \frac{e^4 q_{-0} q_{+\perp}^* Z_{+-} B}{q_{+-}(q_- \cdot k_2)} \left[\frac{2 k_{2-}'' q_{++}}{C k_{2+}'' q_{-+}} \right]^{\frac{1}{2}}$$

$$\doteq \frac{2 e^4 q_{-0} |B|}{(q_- \cdot k_2)} \left[\frac{k_{2-}'' q_{++}^3 q_{--}}{k_{2+}'' q_{+-} q_{-+}(q_+ \cdot k_1)(q_- \cdot k_1)} \right]^{\frac{1}{2}},$$

$$M(+,-;+,+,-,-) = \frac{e^4 m k_{21}^{'''*} q_{+\perp} Z_{+-}^* B^*}{(2q_{-0} + k_{2+}'') q_{+-}(q_- \cdot k_2)} \left[\frac{k_{2+}'' q_{++}}{2 C k_{2-}'' q_{-+}} \right]^{\frac{1}{2}}$$

$$\doteq \frac{e^4 m k_{2+}'' |B|}{(2q_{-0} + k_{2+}'')(q_- \cdot k_2)} \left[\frac{q_{++}^3 q_{--}}{q_{+-} q_{-+}(q_+ \cdot k_1)(q_- \cdot k_1)} \right]^{\frac{1}{2}},$$

$$M(+,-;+,-,+,-) = \frac{e^4(2q_{-0}+k_{2+}'')q_{-\perp}^2 q_{+\perp}^* Z_{+-}B}{q_{+-}(q_{-}\cdot k_2)}\left[\frac{k_{2-}''}{2Ck_{2+}''q_{++}q_{-+}^3}\right]^{\frac{1}{2}}$$

$$\doteq \frac{e^4(2q_{-0}+k_{2+}'')|B|}{(q_{-}\cdot k_2)}$$

$$\times \left[\frac{k_{2-}''q_{++}q_{--}^3}{k_{2+}''q_{+-}q_{-+}(q_{+}\cdot k_1)(q_{-}\cdot k_1)}\right]^{\frac{1}{2}},$$

$$M(+,-;+,-,-,+) = \frac{e^4 q_{-0}^2 q_{+\perp} Z_{+-}^* B^*}{(2q_{-0}+k_{2+}'')q_{+-}(q_{-}\cdot k_2)}\left[\frac{8k_{2-}''q_{++}}{Ck_{2+}''q_{-+}}\right]^{\frac{1}{2}}$$

$$\doteq \frac{4e^4 q_{-0}^2|B|}{(2q_{-0}+k_{2+}'')(q_{-}\cdot k_2)}$$

$$\times \left[\frac{k_{2-}''q_{++}^3 q_{--}}{k_{2+}''q_{+-}q_{-+}(q_{+}\cdot k_1)(q_{-}\cdot k_1)}\right]^{\frac{1}{2}},$$

$$M(+,-;-,+,+,-) = \frac{e^4 q_{-0}^2 q_{+\perp} Z_{-+}B}{(2q_{-0}+k_{2+}'')(q_{-}\cdot k_2)}\left[\frac{8k_{2-}''}{Ck_{2+}''q_{++}q_{-+}}\right]^{\frac{1}{2}}$$

$$\doteq \frac{4e^4 q_{-0}^2 q_{+-}|B|}{(2q_{-0}+k_{2+}'')(q_{-}\cdot k_2)}\left[\frac{k_{2-}''}{k_{2+}''(q_{+}\cdot k_1)(q_{-}\cdot k_1)}\right]^{\frac{1}{2}},$$

$$M(+,-;-,+,-,+) = \frac{e^4(2q_{-0}+k_{2+}'')q_{+\perp}Z_{-+}^* B^*}{q_{+-}(q_{-}\cdot k_2)}\left[\frac{k_{2-}''q_{-+}}{2Ck_{2+}''q_{++}}\right]^{\frac{1}{2}}$$

$$\doteq \frac{e^4(2q_{-0}+k_{2+}'')q_{-+}|B|}{(q_{-}\cdot k_2)}\left[\frac{k_{2-}''}{k_{2+}''(q_{+}\cdot k_1)(q_{-}\cdot k_1)}\right]^{\frac{1}{2}},$$

$$M(+,-;-,-,+,+) = -\frac{e^4 mk_{2\perp}'' q_{+\perp} Z_{-+}B}{(2q_{-0}+k_{2+}'')(q_{-}\cdot k_2)}\left[\frac{k_{2+}''}{2Ck_{2-}''q_{++}q_{-+}}\right]^{\frac{1}{2}}$$

$$\doteq \frac{e^4 mk_{2+}'' q_{+-}|B|}{(2q_{-0}+k_{2+}'')(q_{-}\cdot k_2)[(q_{+}\cdot k_1)(q_{-}\cdot k_1)]^{\frac{1}{2}}},$$

$$M(-,+;+,+,-,-) = \frac{e^4 mk_{2\perp}''^* q_{+\perp}^* Z_{-+}^* B^*}{(2q_{-0}+k_{2+}'')(q_{-}\cdot k_2)}\left[\frac{k_{2+}''}{2Ck_{2-}''q_{++}q_{-+}}\right]^{\frac{1}{2}}$$

$$\doteq \frac{e^4 mk_{2+}'' q_{+-}|B|}{(2q_{-0}+k_{2+}'')(q_{-}\cdot k_2)[(q_{+}\cdot k_1)(q_{-}\cdot k_1)]^{\frac{1}{2}}},$$

9.19. $e^+e^- \to e^+e^-\gamma\gamma$ (NO Z-EXCHANGE; $m \neq 0$)

$$M(-,+;+,-,+,-) = \frac{e^4(2q_{-0}+k_{2+}'')q_{+\perp}^* Z_{-+} B}{q_{+-}(q_- \cdot k_2)} \left[\frac{k_{2-}''q_{-+}}{2Ck_{2+}''q_{++}}\right]^{\frac{1}{2}}$$

$$\doteq \frac{e^4(2q_{-0}+k_{2+}'')q_{-+}|B|}{(q_- \cdot k_2)} \left[\frac{k_{2-}''}{k_{2+}''(q_+ \cdot k_1)(q_- \cdot k_1)}\right]^{\frac{1}{2}},$$

$$M(-,+;+,-,-,+) = \frac{e^4 q_{-0}^2 q_{+\perp}^* Z_{-+}^* B^*}{(2q_{-0}+k_{2+}'')(q_- \cdot k_2)} \left[\frac{8k_{2-}''}{Ck_{2+}''q_{++}q_{-+}}\right]^{\frac{1}{2}}$$

$$\doteq \frac{4e^4 q_{-0}^2 q_{+-}|B|}{(2q_{-0}+k_{2+}'')(q_- \cdot k_2)} \left[\frac{k_{2-}''}{k_{2+}''(q_+ \cdot k_1)(q_- \cdot k_1)}\right]^{\frac{1}{2}},$$

$$M(-,+;-,+,+,-) = \frac{e^4 q_{-0}^2 q_{+\perp}^* Z_{+-} B}{(2q_{-0}+k_{2+}'')q_{+-}(q_- \cdot k_2)} \left[\frac{8k_{2-}''q_{++}}{Ck_{2+}''q_{-+}}\right]^{\frac{1}{2}}$$

$$\doteq \frac{4e^4 q_{-0}^2 |B|}{(2q_{-0}+k_{2+}'')(q_- \cdot k_2)}$$

$$\times \left[\frac{k_{2-}'' q_{++}^3 q_{--}}{k_{2+}'' q_{+-} q_{-+}(q_+ \cdot k_1)(q_- \cdot k_1)}\right]^{\frac{1}{2}},$$

$$M(-,+;-,+,-,+) = \frac{e^4(2q_{-0}+k_{2+}'')q_{-\perp}^{*2} q_{+\perp} Z_{+-}^* B^*}{q_{+-}(q_- \cdot k_2)} \left[\frac{k_{2-}''}{2Ck_{2+}'' q_{++} q_{-+}^3}\right]^{\frac{1}{2}}$$

$$\doteq \frac{e^4(2q_{-0}+k_{2+}'')|B|}{(q_- \cdot k_2)}$$

$$\times \left[\frac{k_{2-}'' q_{++} q_{--}^3}{k_{2+}'' q_{+-} q_{-+}(q_+ \cdot k_1)(q_- \cdot k_1)}\right]^{\frac{1}{2}},$$

$$M(-,+;-,-,+,+) = -\frac{e^4 m k_{2\perp}'' q_{+\perp}^* Z_{+-} B}{(2q_{-0}+k_{2+}'')q_{+-}(q_- \cdot k_2)} \left[\frac{k_{2+}'' q_{++}}{2Ck_{2-}'' q_{-+}}\right]^{\frac{1}{2}}$$

$$\doteq \frac{e^4 m k_{2+}'' |B|}{(2q_{-0}+k_{2+}'')(q_- \cdot k_2)}$$

$$\times \left[\frac{q_{++}^3 q_{--}}{q_{+-} q_{-+}(q_+ \cdot k_1)(q_- \cdot k_1)}\right]^{\frac{1}{2}},$$

$$M(+,-;+,-,-,-) = \frac{e^4 q_{-0} q_{+\perp} Z_{+-}^* B^*}{q_{+-}(q_- \cdot k_2)} \left[\frac{2k_{2-}'' q_{++}}{Ck_{2+}'' q_{-+}}\right]^{\frac{1}{2}}$$

$$\doteq \frac{2e^4 q_{-0}|B|}{(q_- \cdot k_2)} \left[\frac{k_{2-}'' q_{++}^3 q_{--}}{k_{2+}'' q_{+-} q_{-+}(q_+ \cdot k_1)(q_- \cdot k_1)}\right]^{\frac{1}{2}},$$

$$M(+,-;-,+,-,-) = \frac{e^4 q_{-0} q_{+\perp} Z^*_{-+} B^*}{q_{+-}(q_- \cdot k_2)} \left[\frac{2k''_{2-} q_{-+}}{C k''_{2+} q_{++}} \right]^{\frac{1}{2}}$$

$$\doteq \frac{2e^4 q_{-0} q_{-+} |B|}{(q_- \cdot k_2)} \left[\frac{k''_{2-}}{k''_{2+}(q_+ \cdot k_1)(q_- \cdot k_1)} \right]^{\frac{1}{2}},$$

$$M(+,-;-,-,-,+) = -\frac{e^4 m k''_{2\perp} q_{+\perp} Z^*_{-+} B^*}{q_{-0} q_{+-} (q_- \cdot k_2)} \left[\frac{k''_{2+} q_{-+}}{8 C k''_{2-} q_{++}} \right]^{\frac{1}{2}}$$

$$\doteq \frac{e^4 m k''_{2+} q_{-+} |B|}{2 q_{-0} (q_- \cdot k_2) [(q_+ \cdot k_1)(q_- \cdot k_1)]^{\frac{1}{2}}},$$

$$M(-,+;+,-,-,-) = \frac{e^4 q_{-0} q^*_{+\perp} Z^*_{-+} B^*}{(q_- \cdot k_2)} \left[\frac{2k''_{2-}}{C k''_{2+} q_{++} q_{-+}} \right]^{\frac{1}{2}}$$

$$\doteq \frac{2e^4 q_{-0} q_{+-} |B|}{(q_- \cdot k_2)} \left[\frac{k''_{2-}}{k''_{2+}(q_+ \cdot k_1)(q_- \cdot k_1)} \right]^{\frac{1}{2}},$$

$$M(-,+;-,+,-,-) = \frac{e^4 q_{-0} q^{*2}_{-\perp} q_{+\perp} Z^*_{+-} B^*}{q_{+-}(q_- \cdot k_2)} \left[\frac{2k''_{2-}}{C k''_{2+} q_{++} q^3_{-+}} \right]^{\frac{1}{2}}$$

$$\doteq \frac{2e^4 q_{-0} |B|}{(q_- \cdot k_2)} \left[\frac{k''_{2-} q_{++} q^3_{--}}{k''_{2+} q_{+-} q_{-+}(q_+ \cdot k_1)(q_- \cdot k_1)} \right]^{\frac{1}{2}},$$

$$M(-,+;-,-,-,+) = -\frac{e^4 m k''_{2\perp} q^{*2}_{-\perp} q_{+\perp} Z^*_{+-} B^*}{q_{-0} q_{+-} (q_- \cdot k_2)} \left[\frac{k''_{2+}}{8 C k''_{2-} q_{++} q^3_{-+}} \right]^{\frac{1}{2}}$$

$$\doteq \frac{e^4 m k''_{2+} |B|}{2 q_{-0} (q_- \cdot k_2)} \left[\frac{q_{++} q^3_{--}}{q_{+-} q_{-+}(q_+ \cdot k_1)(q_- \cdot k_1)} \right]^{\frac{1}{2}},$$

$$M(-,-;-,+,+,-) = \frac{e^4 m k'''^*_{2\perp} q_{+\perp} Z^*_{-+} B_0}{E q_{-0} q_{+-} (q_- \cdot k_2)}$$

$$\times \left[\frac{k''_{2+}(q_+ \cdot q_-)}{8 k''_{2-} q_{++} q_{-+}(q_+ \cdot k_1)(q_- \cdot k_1)} \right]$$

$$\doteq \frac{e^4 m k''_{2+}(q_+ \cdot q_-) |B_0|}{2 E q_{-0} (q_- \cdot k_2) [(q_+ \cdot k_1)(q_- \cdot k_1)]^{\frac{1}{2}}},$$

$$M(-,-;-,-,+,+) = \frac{e^4 q_{-0} q_{+\perp} Z^*_{-+} B_0}{E q_{+-}(q_- \cdot k_2)} \left[\frac{2k''_{2-}(q_+ \cdot q_-)}{k''_{2+} q_{++} q_{-+}(q_+ \cdot k_1)(q_- \cdot k_1)} \right]$$

$$\doteq \frac{2 e^4 q_{-0}(q_+ \cdot q_-) |B_0|}{E(q_- \cdot k_2)} \left[\frac{k''_{2-}}{k''_{2+}(q_+ \cdot k_1)(q_- \cdot k_1)} \right]^{\frac{1}{2}},$$

9.19. $e^+e^- \to e^+e^-\gamma\gamma$ (NO Z-EXCHANGE; $m \neq 0$)

$$M(-,-;-,+,-,-) = \frac{e^4 m E k_{21}''^* q_{+\perp} B^*}{(2q_{-0} + k_{2+}'')q_{+-}(q_- \cdot k_2)}$$

$$\times \left[\frac{8k_{2+}''(q_+ \cdot q_-)}{k_{2-}'' q_{++} q_{-+}(q_+ \cdot k_1)(q_- \cdot k_1)}\right]^{\frac{1}{2}}$$

$$\doteq \frac{e^4 m E k_{2+}''|B|}{(2q_{-0} + k_{2+}'')(q_- \cdot k_2)} \left[\frac{8(q_+ \cdot q_-)}{q_{+-}q_{-+}(q_+ \cdot k_1)(q_- \cdot k_1)}\right]^{\frac{1}{2}},$$

$$M(-,-;-,-,+,-) = \frac{e^4(2q_{-0} + k_{2+}'')q_{+\perp} Z_{-+}^* B_0}{Eq_{+-}(q_- \cdot k_2)}$$

$$\times \left[\frac{k_{2-}''(q_+ \cdot q_-)}{2k_{2+}'' q_{++} q_{-+}(q_+ \cdot k_1)(q_- \cdot k_1)}\right]^{\frac{1}{2}}$$

$$\doteq \frac{e^4(2q_{-0} + k_{2+}'')(q_+ \cdot q_-)|B_0|}{E(q_- \cdot k_2)}$$

$$\times \left[\frac{k_{2+}''}{k_{2+}''(q_+ \cdot k_1)(q_- \cdot k_1)}\right]^{\frac{1}{2}},$$

$$M(-,-;-,-,-,+) = \frac{e^4 E q_{-0}^2 q_{+\perp} B^*}{(2q_{-0} + k_{2+}'')q_{+-}(q_- \cdot k_2)}$$

$$\times \left[\frac{128 k_{2-}''(q_+ \cdot q_-)}{k_{2+}'' q_{++} q_{-+}(q_+ \cdot k_1)(q_- \cdot k_1)}\right]^{\frac{1}{2}}$$

$$\doteq \frac{e^4 E q_{-0}^2 |B|}{(q_- \cdot k_2)} \left[\frac{128 k_{2-}''(q_+ \cdot q_-)}{k_{2+}'' q_{+-} q_{-+}(q_+ \cdot k_1)(q_- \cdot k_1)}\right]^{\frac{1}{2}},$$

$$M(-,-;-,-,-,-) = \frac{e^4 E q_{-0} q_{+\perp} B^*}{q_{+-}(q_- \cdot k_2)} \left[\frac{32 k_{2-}''(q_+ \cdot q_-)}{k_{2+}'' q_{++} q_{-+}(q_+ \cdot k_1)(q_- \cdot k_1)}\right]^{\frac{1}{2}}$$

$$\doteq \frac{e^4 E q_{-0}|B|}{(q_- \cdot k_2)} \left[\frac{32 k_{2-}''(q_+ \cdot q_-)}{k_{2+}'' q_{+-} q_{-+}(q_+ \cdot k_1)(q_- \cdot k_1)}\right]^{\frac{1}{2}}. \quad (9.226)$$

Unpolarized squared matrix element:

$$\overline{|M|^2} = \frac{e^8 |B|^2 k_{2-}''}{2k_{2+}''(q_- \cdot k_2)^2 (q_+ \cdot k_1)(q_- \cdot k_1) q_{+-} q_{-+}}$$

$$\times \left[4q_{-0}^2 + (2q_{-0} + k_{2+}'')^2 + \frac{m^2 k_{2+}''^3}{4q_{-0}^2 k_{2-}''}\right] \left[q_{++} q_{--}^3 + q_{+-} q_{-+}^3 + \frac{2(q_+ \cdot q_-)^3}{E^2}\right]$$

$$+ \left(\frac{2q_{-0}}{2q_{-0} + k_{2+}''}\right)^2 \left(q_{++}^3 q_{--} + q_{+-}^3 q_{-+} + 8E^2 (q_+ \cdot q_-)\right)\right]. \quad (9.227)$$

Unpolarized cross section:

$$d\sigma = \frac{\alpha^4 |B|^2 k''_{2-}}{256\pi^2 E^2 k''_{2+}(q_- \cdot k_2)^2 (q_+ \cdot k_1)(q_- \cdot k_1) q_{+-} q_{-+}}$$

$$\times \left[4q_{-0}^2 + (2q_{-0} + k''_{2+})^2 + \frac{m^2 k''^3_{2+}}{4 q_{-0}^2 k''_{2-}} \right] \left[q_{++} q_{--}^3 + q_{+-} q_{-+}^3 + \frac{2(q_+ \cdot q_-)^3}{E^2} \right]$$

$$+ \left(\frac{2q_{-0}}{2q_{-0} + k''_{2+}} \right)^2 \left(q_{++}^3 q_{--} + q_{+-}^3 q_{-+} + 8E^2 (q_+ \cdot q_-) \right) \Bigg]$$

$$\times \delta^4(p_+ + p_- - q_+ - q_- - k_1 - k_2) \frac{d^3\vec{q}_+ \, d^3\vec{q}_- \, d^3\vec{k}_1 \, d^3\vec{k}_2}{q_{+0} \, q_{-0} \, k_{10} \, k_{20}}. \qquad (9.228)$$

9.19.5 $\vec{k}_1$ and $\vec{k}_2$ nearly parallel to $\vec{p}_+$ and $\vec{p}_-$, resp.

Photon polarizations:

$$\not{\epsilon}_1^{\pm} = N_1 [\not{k}_1 \not{p}_+ \not{p}_-(1 \pm \gamma_5) + \not{p}_- \not{p}_+ \not{k}_1 (1 \mp \gamma_5)],$$

$$\not{\epsilon}_2^{\pm} = N_2 [\not{k}_2 \not{p}_- \not{p}_+(1 \pm \gamma_5) + \not{p}_+ \not{p}_- \not{k}_2 (1 \mp \gamma_5)],$$

$$N_i^{-1} = E^2 [32 k_{i+} k_{i-}]^{\frac{1}{2}}, \qquad i = 1, 2. \qquad (9.229)$$

Nonvanishing helicity amplitudes:

$$M(+,+;+,+,+,+) = \frac{8e^4 E^3 q^*_{+\perp}}{q_{+-}(p_+ \cdot k_1)(p_- \cdot k_2)} \left[\frac{k_{1-} k_{2+}}{k_{1+} k_{2-} q_{++} q_{-+}} \right]^{\frac{1}{2}}$$

$$\doteq \frac{8e^4 E^3}{(p_+ \cdot k_1)(p_- \cdot k_2)} \left[\frac{k_{1-} k_{2+}}{k_{1+} k_{2-} q_{+-} q_{-+}} \right]^{\frac{1}{2}},$$

$$M(+,+;+,+,+,-) = \frac{4e^4 E^2 (2E - k_{2-}) q^*_{+\perp}}{q_{+-}(p_+ \cdot k_1)(p_- \cdot k_2)} \left[\frac{k_{1-} k_{2+}}{k_{1+} k_{2-} q_{++} q_{-+}} \right]^{\frac{1}{2}}$$

$$\doteq \frac{4e^4 E^2 (2E - k_{2-})}{(p_+ \cdot k_1)(p_- \cdot k_2)} \left[\frac{k_{1-} k_{2+}}{k_{1+} k_{2-} q_{+-} q_{-+}} \right]^{\frac{1}{2}},$$

9.19. $e^+e^- \to e^+e^- \gamma\gamma$ (NO Z-EXCHANGE; $m \neq 0$) 293

$$M(+,+;+,+,-,+) = \frac{4e^4 E^2 (2E - k_{1+}) q^*_{+\perp}}{q_{+-}(p_+ \cdot k_1)(p_- \cdot k_2)} \left[\frac{k_{1-}k_{2+}}{k_{1+}k_{2-}q_{++}q_{-+}}\right]^{\frac{1}{2}}$$

$$\doteq \frac{4e^4 E^2 (2E - k_{1+})}{(p_+ \cdot k_1)(p_- \cdot k_2)} \left[\frac{k_{1-}k_{2+}}{k_{1+}k_{2-}q_{+-}q_{-+}}\right]^{\frac{1}{2}},$$

$$M(+,+;+,-,+,+) = \frac{e^4 mE}{(q_+ \cdot q_-)(p_+ \cdot k_1)(p_- \cdot k_2)} \left\{ k_{1\perp} q^*_{-\perp} \left[\frac{k_{1+}k_{2+}q_{-+}}{k_{1-}k_{2-}q_{++}}\right]^{\frac{1}{2}} \right.$$

$$\left. + k^*_{2\perp} q_{+\perp} \left[\frac{k_{1-}k_{2-}q^3_{++}}{k_{1+}k_{2+}q^3_{-+}}\right]^{\frac{1}{2}} \right\},$$

$$M(+,+;-,+,+,+) = \frac{e^4 mE}{(q_+ \cdot q_-)(p_+ \cdot k_1)(p_- \cdot k_2)} \left\{ k_{1\perp} q^*_{-\perp} \left[\frac{k_{1+}k_{2+}q^3_{++}}{k_{1-}k_{2-}q^3_{-+}}\right]^{\frac{1}{2}} \right.$$

$$\left. + k^*_{2\perp} q_{+\perp} \left[\frac{k_{1-}k_{2-}q_{-+}}{k_{1+}k_{2+}q_{++}}\right]^{\frac{1}{2}} \right\},$$

$$M(+,+;+,+,-,-) = \frac{4e^4 E (q_+ \cdot q_-) q^*_{+\perp}}{q_{+-}(p_+ \cdot k_1)(p_- \cdot k_2)} \left[\frac{k_{1-}k_{2+}}{k_{1+}k_{2-}q_{++}q_{-+}}\right]^{\frac{1}{2}}$$

$$\doteq \frac{4e^4 E (q_+ \cdot q_-)}{(p_+ \cdot k_1)(p_- \cdot k_2)} \left[\frac{k_{1-}k_{2+}}{k_{1+}k_{2-}q_{+-}q_{-+}}\right]^{\frac{1}{2}},$$

$$M(+,+;+,-,+,-) = \frac{e^4 m k_{1\perp} q^*_{-\perp}}{(2E - k_{1+})(p_+ \cdot k_1)(p_- \cdot k_2)} \left[\frac{k_{1+}k_{2+}q_{-+}}{k_{1-}k_{2-}q_{++}}\right]^{\frac{1}{2}}$$

$$\doteq \frac{e^4 m k_{1+}}{(2E - k_{1+})(p_+ \cdot k_1)(p_- \cdot k_2)} \left[\frac{k_{2+}q_{+-}q_{-+}}{k_{2-}}\right]^{\frac{1}{2}},$$

$$M(+,+;+,-,-,+) = \frac{e^4 m k^*_{2\perp} q_{+\perp}}{(2E - k_{2-})(p_+ \cdot k_1)(p_- \cdot k_2)} \left[\frac{k_{1-}k_{2-}q^3_{++}}{k_{1+}k_{2+}q^3_{-+}}\right]^{\frac{1}{2}}$$

$$\doteq \frac{e^4 m k_{2-} q^2_{++}}{(2E - k_{2-})(p_+ \cdot k_1)(p_- \cdot k_2)} \left[\frac{k_{1-}q_{+-}}{k_{1+}q^3_{-+}}\right]^{\frac{1}{2}},$$

$$M(+,+;-,+,+,-) = \frac{e^4 m k_{1\perp} q^*_{-\perp}}{(2E - k_{1+})(p_+ \cdot k_1)(p_- \cdot k_2)} \left[\frac{k_{1+}k_{2+}q^3_{++}}{k_{1-}k_{2-}q^3_{-+}}\right]^{\frac{1}{2}}$$

$$\doteq \frac{e^4 m k_{1+} q^2_{++}}{(2E - k_{1+})(p_+ \cdot k_1)(p_- \cdot k_2)} \left[\frac{k_{2+}q_{+-}}{k_{2-}q^3_{-+}}\right]^{\frac{1}{2}},$$

$$M(+,+;-,+,-,+) = \frac{e^4 m k_{2\perp}^* q_{+\perp}}{(2E-k_{2-})(p_+\cdot k_1)(p_-\cdot k_2)} \left[\frac{k_{1-}k_{2-}q_{-+}}{k_{1+}k_{2+}q_{++}}\right]^{\frac{1}{2}}$$

$$\doteq \frac{e^4 m k_{2-}}{(2E-k_{2-})(p_+\cdot k_1)(p_-\cdot k_2)} \left[\frac{k_{1-}q_{+-}q_{-+}}{k_{1+}}\right]^{\frac{1}{2}},$$

$$M(+,+;-,-,+,+) = \frac{e^4 m^2 k_{1\perp} k_{2\perp}^* q_{-\perp}^*}{2Eq_{+-}(p_+\cdot k_1)(p_-\cdot k_2)} \left[\frac{k_{1+}k_{2-}}{k_{1-}k_{2+}q_{++}q_{-+}}\right]^{\frac{1}{2}}$$

$$\doteq \frac{e^4 m^2 k_{1+}k_{2-}}{2E(p_+\cdot k_1)(p_-\cdot k_2)\left[q_{+-}q_{-+}\right]^{\frac{1}{2}}},$$

$$M(+,-;+,+,+,-) = \frac{2e^4 mE k_{2\perp}q_{-\perp}^*}{q_{+-}(p_+\cdot k_1)(p_-\cdot k_2)} \left[\frac{k_{1-}k_{2-}}{k_{1+}k_{2+}q_{++}q_{-+}}\right]^{\frac{1}{2}}$$

$$\doteq \frac{2e^4 mE k_{2-}}{(p_+\cdot k_1)(p_-\cdot k_2)} \left[\frac{k_{1-}}{k_{1+}q_{+-}q_{-+}}\right]^{\frac{1}{2}},$$

$$M(+,-;+,-,+,+) = \frac{4e^4 E^2 q_{+\perp}}{(2E-k_{1+})(p_+\cdot k_1)(p_-\cdot k_2)} \left[\frac{k_{1-}k_{2+}q_{++}^3}{k_{1+}k_{2-}q_{-+}^3}\right]^{\frac{1}{2}}$$

$$\doteq \frac{4e^4 E^2 q_{++}^2}{(2E-k_{1+})(p_+\cdot k_1)(p_-\cdot k_2)} \left[\frac{k_{1-}k_{2+}q_{+-}}{k_{1+}k_{2-}q_{-+}^3}\right]^{\frac{1}{2}},$$

$$M(+,-;-,+,+,+) = \frac{4e^4 E^2 q_{+\perp}}{(2E-k_{1+})(p_+\cdot k_1)(p_-\cdot k_2)} \left[\frac{k_{1-}k_{2+}q_{-+}}{k_{1+}k_{2-}q_{++}}\right]^{\frac{1}{2}}$$

$$\doteq \frac{4e^4 E^2}{(2E-k_{1+})(p_+\cdot k_1)(p_-\cdot k_2)} \left[\frac{k_{1-}k_{2+}q_{+-}q_{-+}}{k_{1+}k_{2-}}\right]^{\frac{1}{2}},$$

$$M(-,+;+,+,-,+) = \frac{2e^4 mE k_{1\perp}^* q_{+\perp}^*}{q_{+-}(p_+\cdot k_1)(p_-\cdot k_2)} \left[\frac{k_{1+}k_{2+}}{k_{1-}k_{2-}q_{++}q_{-+}}\right]^{\frac{1}{2}}$$

$$\doteq \frac{2e^4 mE k_{1+}}{(p_+\cdot k_1)(p_-\cdot k_2)} \left[\frac{k_{2+}}{k_{2-}q_{+-}q_{-+}}\right]^{\frac{1}{2}},$$

$$M(-,+;+,-,+,+) = \frac{4e^4 E^2 q_{+\perp}^*}{(2E-k_{2-})(p_+\cdot k_1)(p_-\cdot k_2)} \left[\frac{k_{1-}k_{2+}q_{-+}}{k_{1+}k_{2-}q_{++}}\right]^{\frac{1}{2}}$$

$$\doteq \frac{4e^4 E^2}{(2E-k_{2-})(p_+\cdot k_1)(p_-\cdot k_2)} \left[\frac{k_{1-}k_{2+}q_{+-}q_{-+}}{k_{1+}k_{2-}}\right]^{\frac{1}{2}},$$

9.19. $e^+e^- \to e^+e^- \gamma\gamma$ (NO Z-EXCHANGE; $m \neq 0$)

$$M(-,+;-,+,+,+) = \frac{4e^4 E^2 q_{+\perp}^*}{(2E-k_{2-})(p_+\cdot k_1)(p_-\cdot k_2)} \left[\frac{k_{1-}k_{2+}q_{++}^3}{k_{1+}k_{2-}q_{-+}^3}\right]^{\frac{1}{2}}$$

$$\doteq \frac{4e^4 E^2 q_{++}^2}{(2E-k_{2-})(p_+\cdot k_1)(p_-\cdot k_2)} \left[\frac{k_{1-}k_{2+}q_{+-}}{k_{1+}k_{2-}q_{-+}^3}\right]^{\frac{1}{2}},$$

$$M(+,-;+,+,-,-) = \frac{e^4 m(2E-k_{1+})k_{2\perp}q_{-\perp}^*}{q_{+-}(p_+\cdot k_1)(p_-\cdot k_2)} \left[\frac{k_{1-}k_{2-}}{k_{1+}k_{2+}q_{++}q_{-+}}\right]^{\frac{1}{2}}$$

$$\doteq \frac{e^4 m(2E-k_{1+})k_{2-}}{(p_+\cdot k_1)(p_-\cdot k_2)} \left[\frac{k_{1-}}{k_{1+}q_{+-}q_{-+}}\right]^{\frac{1}{2}},$$

$$M(+,-;+,-,+,-) = \frac{e^4}{(q_+\cdot q_-)(p_+\cdot k_1)(p_-\cdot k_2)} \left\{4E^3 q_{+\perp}\left[\frac{k_{1-}k_{2+}q_{++}^3}{k_{1+}k_{2-}q_{-+}^3}\right]^{\frac{1}{2}}\right.$$

$$\left. + \frac{m^2 k_{1\perp}k_{2\perp}q_{+\perp}^*}{4E}\left[\frac{k_{1+}k_{2-}q_{-+}}{k_{1-}k_{2+}q_{++}}\right]^{\frac{1}{2}}\right\},$$

$$M(+,-;+,-,-,+) = \frac{2e^4 E q_{+\perp}}{(p_+\cdot k_1)(p_-\cdot k_2)}\left[\frac{k_{1-}k_{2+}q_{++}^3}{k_{1+}k_{2-}q_{-+}^3}\right]^{\frac{1}{2}}$$

$$\doteq \frac{2e^4 E q_{++}^2}{(p_+\cdot k_1)(p_-\cdot k_2)}\left[\frac{k_{1-}k_{2+}q_{+-}}{k_{1+}k_{2-}q_{-+}^3}\right]^{\frac{1}{2}},$$

$$M(+,-;-,+,+,-) = \frac{e^4}{(q_+\cdot q_-)(p_+\cdot k_1)(p_-\cdot k_2)}\left\{4E^3 q_{+\perp}\left[\frac{k_{1-}k_{2+}q_{-+}}{k_{1+}k_{2-}q_{++}}\right]^{\frac{1}{2}}\right.$$

$$\left. + \frac{m^2 k_{1\perp}k_{2\perp}q_{+\perp}^*}{4E}\left[\frac{k_{1+}k_{2-}q_{++}^3}{k_{1-}k_{2+}q_{-+}^3}\right]^{\frac{1}{2}}\right\},$$

$$M(+,-;-,+,-,+) = \frac{2e^4 E q_{+\perp}}{(p_+\cdot k_1)(p_-\cdot k_2)}\left[\frac{k_{1-}k_{2+}q_{-+}}{k_{1+}k_{2-}q_{++}}\right]^{\frac{1}{2}}$$

$$\doteq \frac{2e^4 E}{(p_+\cdot k_1)(p_-\cdot k_2)}\left[\frac{k_{1-}k_{2+}q_{+-}q_{-+}}{k_{1+}k_{2-}}\right]^{\frac{1}{2}},$$

$$M(+,-;-,-,+,+) = \frac{e^4 m(2E-k_{2-})k_{1\perp}q_{-\perp}}{q_{+-}(p_+\cdot k_1)(p_-\cdot k_2)}\left[\frac{k_{1+}k_{2+}}{k_{1-}k_{2-}q_{++}q_{-+}}\right]^{\frac{1}{2}}$$

$$\doteq \frac{e^4 m(2E-k_{2-})k_{1+}}{(p_+\cdot k_1)(p_-\cdot k_2)}\left[\frac{k_{2+}}{k_{2-}q_{+-}q_{-+}}\right]^{\frac{1}{2}},$$

$$M(-,+;+,+,-,-) = \frac{e^4 m(2E-k_{2-})k_{1\perp}^* q_{+\perp}^*}{q_{+-}(p_+\cdot k_1)(p_-\cdot k_2)}\left[\frac{k_{1+}k_{2+}}{k_{1-}k_{2-}q_{++}q_{-+}}\right]^{\frac{1}{2}}$$

$$\doteq \frac{e^4 m(2E-k_{2-})k_{1+}}{(p_+\cdot k_1)(p_-\cdot k_2)}\left[\frac{k_{2+}}{k_{2-}q_{+-}q_{-+}}\right]^{\frac{1}{2}},$$

$$M(-,+;+,-,+,-) = \frac{2e^4 E q_{+\perp}^*}{(p_+\cdot k_1)(p_-\cdot k_2)}\left[\frac{k_{1-}k_{2+}q_{-+}}{k_{1+}k_{2-}q_{++}}\right]^{\frac{1}{2}}$$

$$\doteq \frac{2e^4 E}{(p_+\cdot k_1)(p_-\cdot k_2)}\left[\frac{k_{1-}k_{2+}q_{+-}q_{-+}}{k_{1+}k_{2-}}\right]^{\frac{1}{2}},$$

$$M(-,+;+,-,-,+) = \frac{e^4}{(q_+\cdot q_-)(p_+\cdot k_1)(p_-\cdot k_2)}\left\{4E^3 q_{+\perp}^*\left[\frac{k_{1-}k_{2+}q_{-+}}{k_{1+}k_{2-}q_{++}}\right]^{\frac{1}{2}}\right.$$

$$\left.+\frac{m^2 k_{1\perp}^* k_{2\perp}^* q_{+\perp}}{4E}\left[\frac{k_{1+}k_{2-}q_{++}^3}{k_{1-}k_{2+}q_{-+}^3}\right]^{\frac{1}{2}}\right\},$$

$$M(-,+;-,+,+,-) = \frac{2e^4 E q_{+\perp}^*}{(p_+\cdot k_1)(p_-\cdot k_2)}\left[\frac{k_{1-}k_{2+}q_{++}^3}{k_{1+}k_{2-}q_{-+}^3}\right]^{\frac{1}{2}}$$

$$\doteq \frac{2e^4 E q_{++}^2}{(p_+\cdot k_1)(p_-\cdot k_2)}\left[\frac{k_{1-}k_{2+}q_{+-}}{k_{1+}k_{2-}q_{-+}^3}\right]^{\frac{1}{2}},$$

$$M(-,+;-,+,-,+) = \frac{e^4}{(q_+\cdot q_-)(p_+\cdot k_1)(p_-\cdot k_2)}\left\{4E^3 q_{+\perp}^*\left[\frac{k_{1-}k_{2+}q_{++}^3}{k_{1+}k_{2-}q_{-+}^3}\right]^{\frac{1}{2}}\right.$$

$$\left.+\frac{m^2 k_{1\perp}^* k_{2\perp}^* q_{+\perp}}{4E}\left[\frac{k_{1+}k_{2-}q_{-+}}{k_{1-}k_{2+}q_{++}}\right]^{\frac{1}{2}}\right\},$$

$$M(-,+;-,-,+,+) = \frac{e^4 m(2E-k_{1+})k_{2\perp}^* q_{+\perp}}{q_{+-}(p_+\cdot k_1)(p_-\cdot k_2)}\left[\frac{k_{1-}k_{2-}}{k_{1+}k_{2+}q_{++}q_{-+}}\right]^{\frac{1}{2}}$$

$$\doteq \frac{e^4 m(2E-k_{1+})k_{2-}}{(p_+\cdot k_1)(p_-\cdot k_2)}\left[\frac{k_{1-}}{k_{1+}q_{+-}q_{-+}}\right]^{\frac{1}{2}},$$

$$M(+,-;+,-,-,-) = \frac{4e^4 E^2 q_{+\perp}}{(2E-k_{2-})(p_+\cdot k_1)(p_-\cdot k_2)}\left[\frac{k_{1-}k_{2+}q_{++}^3}{k_{1+}k_{2-}q_{-+}^3}\right]^{\frac{1}{2}}$$

$$\doteq \frac{4e^4 E^2 q_{++}^2}{(2E-k_{2-})(p_+\cdot k_1)(p_-\cdot k_2)}\left[\frac{k_{1-}k_{2+}q_{+-}}{k_{1+}k_{2-}q_{-+}^3}\right]^{\frac{1}{2}},$$

9.19. $e^+e^- \to e^+e^-\gamma\gamma$ (NO Z-EXCHANGE; $m \neq 0$)

$$M(+,-;-,+,-,-) = \frac{4e^4 E^2 q_{+\perp}}{(2E - k_{2-})(p_+ \cdot k_1)(p_- \cdot k_2)} \left[\frac{k_{1-}k_{2+}q_{-+}}{k_{1+}k_{2-}q_{++}}\right]^{\frac{1}{2}}$$

$$\doteq \frac{4e^4 E^2}{(2E - k_{2-})(p_+ \cdot k_1)(p_- \cdot k_2)} \left[\frac{k_{1-}k_{2+}q_{+-}q_{-+}}{k_{1+}k_{2-}}\right]^{\frac{1}{2}},$$

$$M(+,-;-,-,+,-) = \frac{2e^4 m E k_{1\perp} q_{-\perp}}{q_{+-}(p_+ \cdot k_1)(p_- \cdot k_2)} \left[\frac{k_{1+}k_{2+}}{k_{1-}k_{2-}q_{++}q_{-+}}\right]^{\frac{1}{2}}$$

$$\doteq \frac{2e^4 m E k_{1+}}{(p_+ \cdot k_1)(p_- \cdot k_2)} \left[\frac{k_{2+}}{k_{2-}q_{+-}q_{-+}}\right]^{\frac{1}{2}},$$

$$M(-,+;+,-,-,-) = \frac{4e^4 E^2 q^*_{+\perp}}{(2E - k_{1+})(p_+ \cdot k_1)(p_- \cdot k_2)} \left[\frac{k_{1-}k_{2+}q_{-+}}{k_{1+}k_{2-}q_{++}}\right]^{\frac{1}{2}}$$

$$\doteq \frac{4e^4 E^2}{(2E - k_{1+})(p_+ \cdot k_1)(p_- \cdot k_2)} \left[\frac{k_{1-}k_{2+}q_{+-}q_{-+}}{k_{1+}k_{2-}}\right]^{\frac{1}{2}},$$

$$M(-,+;-,+,-,-) = \frac{4e^4 E^2 q^*_{+\perp}}{(2E - k_{1+})(p_+ \cdot k_1)(p_- \cdot k_2)} \left[\frac{k_{1-}k_{2+}q^3_{++}}{k_{1+}k_{2-}q^3_{-+}}\right]^{\frac{1}{2}}$$

$$\doteq \frac{4e^4 E^2 q^2_{++}}{(2E - k_{1+})(p_+ \cdot k_1)(p_- \cdot k_2)} \left[\frac{k_{1-}k_{2+}q_{+-}}{k_{1+}k_{2-}q^3_{-+}}\right]^{\frac{1}{2}},$$

$$M(-,+;-,-,-,+) = \frac{2e^4 m E k^*_{2\perp} q_{+\perp}}{q_{+-}(p_+ \cdot k_1)(p_- \cdot k_2)} \left[\frac{k_{1-}k_{2-}}{k_{1+}k_{2+}q_{++}q_{-+}}\right]^{\frac{1}{2}}$$

$$\doteq \frac{2e^4 m E k_{2-}}{(p_+ \cdot k_1)(p_- \cdot k_2)} \left[\frac{k_{1-}}{k_{1+}q_{+-}q_{-+}}\right]^{\frac{1}{2}},$$

$$M(-,-;+,+,-,-) = \frac{e^4 m^2 k^*_{1\perp} k_{2\perp} q^*_{-\perp}}{2E q_{+-}(p_+ \cdot k_1)(p_- \cdot k_2)} \left[\frac{k_{1+}k_{2-}}{k_{1-}k_{2+}q_{++}q_{-+}}\right]^{\frac{1}{2}}$$

$$\doteq \frac{e^4 m^2 k_{1+}k_{2-}}{2E(p_+ \cdot k_1)(p_- \cdot k_2)\left[q_{+-}q_{-+}\right]^{\frac{1}{2}}},$$

$$M(-,-;+,-,+,-) = \frac{e^4 m k_{2\perp} q^*_{-\perp}}{(2E - k_{2-})(p_+ \cdot k_1)(p_- \cdot k_2)} \left[\frac{k_{1-}k_{2-}q_{-+}}{k_{1+}k_{2+}q_{++}}\right]^{\frac{1}{2}}$$

$$\doteq \frac{e^4 m k_{2-}}{(2E - k_{2-})(p_+ \cdot k_1)(p_- \cdot k_2)} \left[\frac{k_{1-}q_{+-}q_{-+}}{k_{1+}}\right]^{\frac{1}{2}},$$

$$M(-,-;+,-,-,+) = \frac{e^4 m k_{1\perp}^* q_{+\perp}}{(2E - k_{1+})(p_+ \cdot k_1)(p_- \cdot k_2)} \left[\frac{k_{1+}k_{2+}q_{++}^3}{k_{1-}k_{2-}q_{-+}^3}\right]^{\frac{1}{2}}$$

$$\doteq \frac{e^4 m k_{1+} q_{++}^2}{(2E - k_{1+})(p_+ \cdot k_1)(p_- \cdot k_2)} \left[\frac{k_{2+}q_{+-}}{k_{2-}q_{-+}^3}\right]^{\frac{1}{2}},$$

$$M(-,-;-,+,+,-) = \frac{e^4 m k_{2\perp} q_{-\perp}^*}{(2E - k_{2-})(p_+ \cdot k_1)(p_- \cdot k_2)} \left[\frac{k_{1-}k_{2-}q_{++}^3}{k_{1+}k_{2+}q_{-+}^3}\right]^{\frac{1}{2}}$$

$$\doteq \frac{e^4 m k_{2-} q_{++}^2}{(2E - k_{2-})(p_+ \cdot k_1)(p_- \cdot k_2)} \left[\frac{k_{1-}q_{+-}}{k_{1+}q_{-+}^3}\right]^{\frac{1}{2}},$$

$$M(-,-;-,+,-,+) = \frac{e^4 m k_{1\perp}^* q_{+\perp}}{(2E - k_{1+})(p_+ \cdot k_1)(p_- \cdot k_2)} \left[\frac{k_{1+}k_{2+}q_{-+}}{k_{1-}k_{2-}q_{++}}\right]^{\frac{1}{2}}$$

$$\doteq \frac{e^4 m k_{1+}}{(2E - k_{1+})(p_+ \cdot k_1)(p_- \cdot k_2)} \left[\frac{k_{2+}q_{+-}q_{-+}}{k_{2-}}\right]^{\frac{1}{2}},$$

$$M(-,-;-,-,+,+) = \frac{4e^4 E(q_+ \cdot q_-) q_{+\perp}}{q_{+-}(p_+ \cdot k_1)(p_- \cdot k_2)} \left[\frac{k_{1-}k_{2+}}{k_{1+}k_{2-}q_{++}q_{-+}}\right]^{\frac{1}{2}}$$

$$\doteq \frac{4e^4 E(q_+ \cdot q_-)}{(p_+ \cdot k_1)(p_- \cdot k_2)} \left[\frac{k_{1-}k_{2+}}{k_{1+}k_{2-}q_{+-}q_{-+}}\right]^{\frac{1}{2}},$$

$$M(-,-;+,-,-,-) = \frac{e^4 m E}{(q_+ \cdot q_-)(p_+ \cdot k_1)(p_- \cdot k_2)} \left\{ k_{2\perp} q_{-\perp}^* \left[\frac{k_{1-}k_{2-}q_{-+}}{k_{1+}k_{2+}q_{++}}\right] \right.$$

$$\left. + k_{1\perp}^* q_{+\perp} \left[\frac{k_{1+}k_{2+}q_{++}^3}{k_{1-}k_{2-}q_{-+}^3}\right]^{\frac{1}{2}} \right\},$$

$$M(-,-;-,+,-,-) = \frac{e^4 m E}{(q_+ \cdot q_-)(p_+ \cdot k_1)(p_- \cdot k_2)} \left\{ k_{2\perp} q_{-\perp}^* \left[\frac{k_{1-}k_{2-}q_{++}^3}{k_{1+}k_{2+}q_{-+}^3}\right]^{\frac{1}{2}} \right.$$

$$\left. + k_{1\perp}^* q_{+\perp} \left[\frac{k_{1+}k_{2+}q_{-+}}{k_{1-}k_{2-}q_{++}}\right]^{\frac{1}{2}} \right\},$$

$$M(-,-;-,-,+,-) = \frac{4e^4 E^2 (2E - k_{1+}) q_{+\perp}}{q_{+-}(p_+ \cdot k_1)(p_- \cdot k_2)} \left[\frac{k_{1-}k_{2+}}{k_{1+}k_{2-}q_{++}q_{-+}}\right]^{\frac{1}{2}}$$

$$\doteq \frac{4e^4 E^2 (2E - k_{1+})}{(p_+ \cdot k_1)(p_- \cdot k_2)} \left[\frac{k_{1-}k_{2+}}{k_{1+}k_{2-}q_{+-}q_{-+}}\right]^{\frac{1}{2}},$$

9.19. $e^+e^- \to e^+e^- \gamma\gamma$ (NO Z-EXCHANGE; $m \neq 0$)

$$M(-,-;-,-,-,+) = \frac{4e^4 E^2 (2E - k_{2-}) q_{+\perp}}{q_{+-}(p_+ \cdot k_1)(p_- \cdot k_2)} \left[\frac{k_{1-} k_{2+}}{k_{1+} k_{2-} q_{++} q_{-+}}\right]^{\frac{1}{2}}$$

$$\doteq \frac{4e^4 E^2 (2E - k_{2-})}{(p_+ \cdot k_1)(p_- \cdot k_2)} \left[\frac{k_{1-} k_{2+}}{k_{1+} k_{2-} q_{+-} q_{-+}}\right]^{\frac{1}{2}},$$

$$M(-,-;-,-,-,-) = \frac{8e^4 E^3 q_{+\perp}}{q_{+-}(p_+ \cdot k_1)(p_- \cdot k_2)} \left[\frac{k_{1-} k_{2+}}{k_{1+} k_{2-} q_{++} q_{-+}}\right]^{\frac{1}{2}}$$

$$\doteq \frac{8e^4 E^3}{(p_+ \cdot k_1)(p_- \cdot k_2)} \left[\frac{k_{1-} k_{2+}}{k_{1+} k_{2-} q_{+-} q_{-+}}\right]^{\frac{1}{2}}. \quad (9.230)$$

Unpolarized squared matrix element:

$$\overline{|M|^2} = \frac{e^8 E^2 k_{1-} k_{2+}}{2 k_{1+} k_{2-} q_{+-} q_{-+} [(q_+ \cdot q_-)(p_+ \cdot k_1)(p_- \cdot k_2)]^2}$$

$$\times \left[4E^2 + (2E - k_{1+})^2 + \frac{m^2 k_{1+}^3}{4E^2 k_{1-}}\right]$$

$$\times \left[4E^2 + (2E - k_{2-})^2 + \frac{m^2 k_{2-}^3}{4E^2 k_{2+}}\right]$$

$$\times \left[4(q_+ \cdot q_-)^2 + q_{++}^2 q_{--}^2 + q_{+-}^2 q_{-+}^2\right]. \quad (9.231)$$

Unpolarized cross section:

$$d\sigma = \frac{\alpha^4 k_{1-} k_{2+}}{256\pi^2 k_{1+} k_{2-} q_{+-} q_{-+} [(q_+ \cdot q_-)(p_+ \cdot k_1)(p_- \cdot k_2)]^2}$$

$$\times \left[4E^2 + (2E - k_{1+})^2 + \frac{m^2 k_{1+}^3}{4E^2 k_{1-}}\right]$$

$$\times \left[4E^2 + (2E - k_{2-})^2 + \frac{m^2 k_{2-}^3}{4E^2 k_{2+}}\right]$$

$$\times \left[4(q_+ \cdot q_-)^2 + q_{++}^2 q_{--}^2 + q_{+-}^2 q_{-+}^2\right]$$

$$\times \delta^4(p_+ + p_- - q_+ - q_- - k_1 - k_2) \frac{d^3\vec{q}_+ \, d^3\vec{q}_- \, d^3\vec{k}_1 \, d^3\vec{k}_2}{q_{+0} \, q_{-0} \, k_{10} \, k_{20}}.$$

$$(9.232)$$

9.19.6 $\vec{k}_1$ and $\vec{k}_2$ nearly parallel to $\vec{p}_+$ and $\vec{q}_+$, resp.

The primed quantities $k'_{2\pm}$ and $k'_{2\perp}$ are evaluated in the rotated frame where $\vec{q}_+$ determines the positive z-axis (see also Section 7.4.3). The four-vector q in eqn (9.234) is obtained by applying a space reflection to q_+.
Definition:
$$Z_{+-} = q_{++}q_{--} - q^*_{+\perp}q_{-\perp}. \tag{9.233}$$

Photon polarizations:

$$\not{\epsilon}_1^{\pm} = N_1[\not{k}_1 \not{p}_+ \not{p}_-(1 \pm \gamma_5) + \not{p}_- \not{p}_+ \not{k}_1(1 \mp \gamma_5)],$$

$$N_1^{-1} = E^2[32k_{1+}k_{1-}]^{\frac{1}{2}},$$

$$\not{\epsilon}_2^{\pm} = N_2[\not{k}_2 \not{q}_+ \not{q}(1 \pm \gamma_5) + \not{q} \not{q}_+ \not{k}_2(1 \mp \gamma_5)],$$

$$N_2^{-1} = q_{+0}^2[32k'_{2+}k'_{2-}]^{\frac{1}{2}}. \tag{9.234}$$

Nonvanishing helicity amplitudes:

$$M(+,+;+,+,+,+) = -\frac{2e^4 E(2q_{+0} + k'_{2+})q^*_{-\perp}Z_{+-}}{q_{--}(p_+ \cdot k_1)(q_+ \cdot k_2)}\left[\frac{k_{1-}k'_{2-}}{k_{1+}k'_{2+}q_{++}q^3_{-+}}\right]^{\frac{1}{2}}$$

$$\doteq \frac{e^4 E(2q_{+0} + k'_{2+})}{q_{-+}(p_+ \cdot k_1)(q_+ \cdot k_2)}\left[\frac{8(q_+ \cdot q_-)k_{1-}k'_{2-}}{k_{1+}k'_{2+}}\right]^{\frac{1}{2}},$$

$$M(+,+;+,+,+,-) = \frac{e^4 q^*_{-\perp}}{(p_+ \cdot k_1)(q_+ \cdot k_2)}\left\{-\frac{4Eq_{+0}Z_{+-}}{q_{--}}\left[\frac{k_{1-}k'_{2-}}{k_{1+}k'_{2+}q_{++}q^3_{-+}}\right]^{\frac{1}{2}}\right.$$

$$\left.+\frac{m^2 q_{--}k_{1\perp}k'^*_{2\perp}}{8E^2 q_{+0}}\left[\frac{k_{1+}k'_{2+}q_{++}}{k_{1-}k'_{2-}q^3_{-+}}\right]^{\frac{1}{2}}\right\},$$

$$M(+,+;+,+,-,+) = -\frac{e^4(2E - k_{1+})(2q_{+0} + k'_{2+})q^*_{-\perp}Z_{+-}}{q_{--}(p_+ \cdot k_1)(q_+ \cdot k_2)}$$

$$\times \left[\frac{k_{1-}k'_{2-}}{k_{1+}k'_{2+}q_{++}q^3_{-+}}\right]^{\frac{1}{2}}$$

$$\doteq \frac{e^4(2E - k_{1+})(2q_{+0} + k'_{2+})}{q_{-+}(p_+ \cdot k_1)(q_+ \cdot k_2)}\left[\frac{2(q_+ \cdot q_-)k_{1-}k'_{2-}}{k_{1+}k'_{2+}}\right]^{\frac{1}{2}},$$

9.19. $e^+e^- \to e^+e^- \gamma\gamma$ (NO Z-EXCHANGE; $m \neq 0$)

$$M(+,+;+,-,+,+) = -\frac{e^4 m q_{+0} q_{--} k_{1\perp} q^*_{+\perp}}{2E^2 (p_+ \cdot k_1)(q_+ \cdot k_2)} \left[\frac{k_{1+} k'_{2-} q_{-+}}{k_{1-} k'_{2+} q^3_{++}}\right]^{\frac{1}{2}}$$

$$\doteq \frac{e^4 m q_{+0} q_{--} k_{1+}}{2E^2 q_{++} (p_+ \cdot k_1)(q_+ \cdot k_2)} \left[\frac{k'_{2-} q_{+-} q_{-+}}{k'_{2+}}\right]^{\frac{1}{2}},$$

$$M(+,+;-,+,+,+) = \frac{e^4 m q^*_{-\perp}}{(p_+ \cdot k_1)(q_+ \cdot k_2)} \left\{ \frac{q_{+0} q_{--} k_{1\perp}}{2E^2} \left[\frac{k_{1+} k'_{2-} q_{++}}{k_{1-} k'_{2+} q^3_{-+}}\right]^{\frac{1}{2}} \right.$$

$$\left. + \frac{E k'_{2\perp} Z_{+-}}{q_{+0} q_{--}} \left[\frac{k_{1-} k'_{2+}}{k_{1+} k'_{2-} q_{++} q^3_{-+}}\right]^{\frac{1}{2}} \right\},$$

$$M(+,+;+,+,-,-) = -\frac{2 e^4 q_{+0} (2E - k_{1+}) q^*_{-\perp} Z_{+-}}{q_{--} (p_+ \cdot k_1)(q_+ \cdot k_2)} \left[\frac{k_{1-} k'_{2-}}{k_{1+} k'_{2+} q_{++} q^3_{-+}}\right]^{\frac{1}{2}}$$

$$\doteq \frac{e^4 q_{+0} (2E - k_{1+})}{q_{-+} (p_+ \cdot k_1)(q_+ \cdot k_2)} \left[\frac{8(q_+ \cdot q_-) k_{1-} k'_{2-}}{k_{1+} k'_{2+}}\right]^{\frac{1}{2}},$$

$$M(+,+;+,-,+,-) = -\frac{e^4 m q_{+0} k_{1\perp} q^*_{+\perp}}{E(2E - k_{1+})(p_+ \cdot k_1)(q_+ \cdot k_2)} \left[\frac{k_{1+} k'_{2-} q_{-+}}{k_{1-} k'_{2+} q_{++}}\right]^{\frac{1}{2}}$$

$$\doteq \frac{e^4 m q_{+0} k_{1+}}{E(2E - k_{1+})(p_+ \cdot k_1)(q_+ \cdot k_2)} \left[\frac{k'_{2+}}{k'_{2-} q_{+-} q_{-+}}\right]^{\frac{1}{2}},$$

$$M(+,+;-,+,+,-) = \frac{e^4 m q_{--} (2 q_{+0} + k'_{2+}) k_{1\perp} q^*_{-\perp}}{4 E^2 (p_+ \cdot k_1)(q_+ \cdot k_2)} \left[\frac{k_{1+} k'_{2-} q_{++}}{k_{1-} k'_{2+} q^3_{-+}}\right]^{\frac{1}{2}}$$

$$\doteq \frac{e^4 m (2 q_{+0} + k'_{2+}) k_{1+}}{4 E^2 q_{-+} (p_+ \cdot k_1)(q_+ \cdot k_2)} \left[\frac{k'_{2-} q_{++} q^3_{--}}{k'_{2+}}\right]^{\frac{1}{2}},$$

$$M(+,+;-,+,-,+) = \frac{e^4 m (2E - k_{1+}) k'_{2\perp} q^*_{-\perp} Z_{+-}}{2 q_{+0} q_{--} (p_+ \cdot k_1)(q_+ \cdot k_2)} \left[\frac{k_{1-} k'_{2+}}{k_{1+} k'_{2-} q_{++} q^3_{-+}}\right]^{\frac{1}{2}}$$

$$\doteq \frac{e^4 m (2E - k_{1+}) k'_{2+}}{q_{+0} q_{-+} (p_+ \cdot k_1)(q_+ \cdot k_2)} \left[\frac{(q_+ \cdot q_-) k_{1-}}{2 k_{1+}}\right]^{\frac{1}{2}},$$

$$M(+,+;-,-,+,+) = \frac{e^4 m^2 k_{1\perp} k'_{2\perp} q^*_{+\perp}}{4 E q_{+0} (2E - k_{1+})(p_+ \cdot k_1)(q_+ \cdot k_2)} \left[\frac{k_{1+} k'_{2+} q_{-+}}{k_{1-} k'_{2-} q_{++}}\right]^{\frac{1}{2}}$$

$$\doteq \frac{e^4 m^2 k_{1+} k'_{2+} [q_{+-} q_{-+}]^{\frac{1}{2}}}{4 E q_{+0} (2E - k_{1+})(p_+ \cdot k_1)(q_+ \cdot k_2)},$$

$$M(+,-;+,+,+,-) = \frac{e^4 m E k'^*_{2\perp} q_{+\perp}}{q_{+0}(2E - k_{1+})(p_+ \cdot k_1)(q_+ \cdot k_2)} \left[\frac{k_{1-}k'_{2+}q_{-+}}{k_{1+}k'_{2-}q_{++}}\right]^{\frac{1}{2}}$$

$$\doteq \frac{e^4 m E k'_{2+}}{q_{+0}(2E - k_{1+})(p_+ \cdot k_1)(q_+ \cdot k_2)} \left[\frac{k_{1-}q_{+-}q_{-+}}{k_{1+}}\right]^{\frac{1}{2}},$$

$$M(+,-;+,-,+,+) = -\frac{e^4(2q_{+0} + k'_{2+})q_{--}q_{-\perp}}{(p_+ \cdot k_1)(q_+ \cdot k_2)} \left[\frac{k_{1-}k'_{2-}q_{++}}{k_{1+}k'_{2+}q^3_{-+}}\right]^{\frac{1}{2}}$$

$$\doteq \frac{e^4(2q_{+0} + k'_{2+})}{q_{-+}(p_+ \cdot k_1)(q_+ \cdot k_2)} \left[\frac{k_{1-}k'_{2-}q_{++}q^3_{--}}{k_{1+}k'_{2+}}\right]^{\frac{1}{2}},$$

$$M(+,-;-,+,+,+) = \frac{4e^4 E q_{+0} q_{+\perp}}{(2E - k_{1+})(p_+ \cdot k_1)(q_+ \cdot k_2)} \left[\frac{k_{1-}k'_{2-}q_{-+}}{k_{1+}k'_{2+}q_{++}}\right]^{\frac{1}{2}}$$

$$\doteq \frac{4e^4 E q_{+0}}{(2E - k_{1+})(p_+ \cdot k_1)(q_+ \cdot k_2)} \left[\frac{k_{1-}k'_{2-}q_{+-}q_{-+}}{k_{1+}k'_{2+}}\right]^{\frac{1}{2}},$$

$$M(-,+;+,+,+,-) = -\frac{e^4 m q_{--}(2E - k_{1+})k'^*_{2\perp}q^*_{-\perp}}{4E q_{+0}(p_+ \cdot k_1)(q_+ \cdot k_2)} \left[\frac{k_{1-}k'_{2+}q_{++}}{k_{1+}k'_{2-}q^3_{-+}}\right]^{\frac{1}{2}}$$

$$\doteq \frac{e^4 m(2E - k_{1+})k'_{2+}}{4E q_{+0} q_{-+}(p_+ \cdot k_1)(q_+ \cdot k_2)} \left[\frac{k_{1-}q_{++}q^3_{--}}{k_{1+}}\right]^{\frac{1}{2}},$$

$$M(-,+;+,+,-,+) = -\frac{e^4 m(2q_{+0} + k'_{2+})k^*_{1\perp}q^*_{-\perp}Z_{+-}}{2Eq_{--}(p_+ \cdot k_1)(q_+ \cdot k_2)} \left[\frac{k_{1+}k'_{2-}}{k_{1-}k'_{2+}q_{++}q^3_{-+}}\right]^{\frac{1}{2}}$$

$$\doteq \frac{e^4 m(2q_{+0} + k'_{2+})k_{1+}}{Eq_{-+}(p_+ \cdot k_1)(q_+ \cdot k_2)} \left[\frac{(q_+ \cdot q_-)k'_{2-}}{2k'_{2+}}\right]^{\frac{1}{2}},$$

$$M(-,+;+,-,+,+) = \frac{e^4(2q_{+0} + k'_{2+})q^*_{+\perp}}{(p_+ \cdot k_1)(q_+ \cdot k_2)} \left[\frac{k_{1-}k'_{2-}q_{-+}}{k_{1+}k'_{2+}q_{++}}\right]^{\frac{1}{2}}$$

$$\doteq \frac{e^4(2q_{+0} + k'_{2+})}{(p_+ \cdot k_1)(q_+ \cdot k_2)} \left[\frac{k_{1-}k'_{2-}q_{+-}q_{-+}}{k_{1+}k'_{2+}}\right]^{\frac{1}{2}},$$

$$M(-,+;-,+,+,+) = -\frac{e^4 q_{+0} q_{--}(2E - k_{1+})q^*_{-\perp}}{E(p_+ \cdot k_1)(q_+ \cdot k_2)} \left[\frac{k_{1-}k'_{2-}q_{++}}{k_{1+}k'_{2+}q^3_{-+}}\right]^{\frac{1}{2}}$$

$$\doteq \frac{e^4 q_{+0}(2E - k_{1+})}{Eq_{-+}(p_+ \cdot k_1)(q_+ \cdot k_2)} \left[\frac{k_{1-}k'_{2-}q_{++}q^3_{--}}{k_{1+}k'_{2+}}\right]^{\frac{1}{2}},$$

9.19. $e^+e^- \to e^+e^-\gamma\gamma$ (NO Z-EXCHANGE; $m \neq 0$)

$$M(+,-;+,+,-,-) = \frac{e^4 m k_{2\perp}'^* q_{+\perp}}{2q_{+0}(p_+ \cdot k_1)(q_+ \cdot k_2)} \left[\frac{k_{1-}k_{2+}'q_{-+}}{k_{1+}k_{2-}'q_{++}}\right]^{\frac{1}{2}}$$

$$\doteq \frac{e^4 m k_{2+}'}{2q_{+0}(p_+ \cdot k_1)(q_+ \cdot k_2)} \left[\frac{k_{1-}q_{+-}q_{-+}}{k_{1+}}\right]^{\frac{1}{2}},$$

$$M(+,-;+,-,+,-) = \frac{e^4 q_{-\perp}}{(p_+ \cdot k_1)(q_+ \cdot k_2)} \left\{-2q_{+0}q_{--}\left[\frac{k_{1-}k_{2-}'q_{++}}{k_{1+}k_{2+}'q_{-+}^3}\right]^{\frac{1}{2}}\right.$$

$$\left. + \frac{m^2 k_{1\perp}k_{2\perp}'^* Z_{+-}^*}{4E q_{+0} q_{--}}\left[\frac{k_{1+}k_{2+}'}{k_{1-}k_{2-}'q_{++}q_{-+}^3}\right]^{\frac{1}{2}}\right\},$$

$$M(+,-;+,-,-,+) = -\frac{e^4 q_{--}(2E - k_{1+})(2q_{+0} + k_{2+}')q_{-\perp}}{2E(p_+ \cdot k_1)(q_+ \cdot k_2)} \left[\frac{k_{1-}k_{2-}'q_{++}}{k_{1+}k_{2+}'q_{-+}^3}\right]^{\frac{1}{2}}$$

$$\doteq \frac{e^4(2E - k_{1+})(2q_{+0} + k_{2+}')}{2E q_{-+}(p_+ \cdot k_1)(q_+ \cdot k_2)} \left[\frac{k_{1-}k_{2-}'q_{++}q_{--}^3}{k_{1+}k_{2+}'}\right]^{\frac{1}{2}},$$

$$M(+,-;-,+,+,-) = \frac{2e^4 q_{+0}q_{--}q_{+\perp}}{(p_+ \cdot k_1)(q_+ \cdot k_2)} \left[\frac{k_{1-}k_{2-}'q_{-+}}{k_{1+}k_{2+}'q_{++}^3}\right]^{\frac{1}{2}}$$

$$\doteq \frac{2e^4 q_{+0}q_{--}}{q_{++}(p_+ \cdot k_1)(q_+ \cdot k_2)} \left[\frac{k_{1-}k_{2-}'q_{+-}q_{-+}}{k_{1+}k_{2+}'}\right]^{\frac{1}{2}},$$

$$M(+,-;-,+,-,+) = \frac{2e^4 q_{+0}q_{+\perp}}{(p_+ \cdot k_1)(q_+ \cdot k_2)} \left[\frac{k_{1-}k_{2-}'q_{-+}}{k_{1+}k_{2+}'q_{++}}\right]^{\frac{1}{2}}$$

$$\doteq \frac{2e^4 q_{+0}}{(p_+ \cdot k_1)(q_+ \cdot k_2)} \left[\frac{k_{1-}k_{2-}'q_{+-}q_{-+}}{k_{1+}k_{2+}'}\right]^{\frac{1}{2}},$$

$$M(+,-;-,-,+,+) = \frac{e^4 m q_{-\perp}}{(p_+ \cdot k_1)(q_+ \cdot k_2)} \left\{\frac{q_{+0}k_{1\perp}Z_{+-}^*}{E q_{--}}\left[\frac{k_{1+}k_{2-}'}{k_{1-}k_{2+}'q_{++}q_{-+}^3}\right]^{\frac{1}{2}}\right.$$

$$\left. + \frac{q_{--}k_{2\perp}'}{2q_{+0}}\left[\frac{k_{1-}k_{2+}'q_{++}}{k_{1+}k_{2-}'q_{-+}^3}\right]^{\frac{1}{2}}\right\},$$

$$M(-,+;+,+,-,-) = -\frac{e^4 m q_{-\perp}^*}{(p_+ \cdot k_1)(q_+ \cdot k_2)} \left\{\frac{q_{+0}k_{1\perp}^* Z_{+-}}{E q_{--}}\left[\frac{k_{1+}k_{2-}'}{k_{1-}k_{2+}'q_{++}q_{-+}^3}\right]^{\frac{1}{2}}\right.$$

$$\left. + \frac{q_{--}k_{2\perp}'^*}{2q_{+0}}\left[\frac{k_{1-}k_{2+}'q_{++}}{k_{1+}k_{2-}'q_{-+}^3}\right]^{\frac{1}{2}}\right\},$$

$$M(-,+;+,-,+,-) = \frac{2e^4 q_{+0} q_{+\perp}^*}{(p_+\cdot k_1)(q_+\cdot k_2)}\left[\frac{k_{1-}k_{2-}'q_{-+}}{k_{1+}k_{2+}'q_{++}}\right]^{\frac{1}{2}}$$

$$\doteq \frac{2e^4 q_{+0}}{(p_+\cdot k_1)(q_+\cdot k_2)}\left[\frac{k_{1-}k_{2-}'q_{+-}q_{-+}}{k_{1+}k_{2+}'}\right]^{\frac{1}{2}},$$

$$M(-,+;+,-,-,+) = \frac{2e^4 q_{+0}q_{--}q_{+\perp}^*}{(p_+\cdot k_1)(q_+\cdot k_2)}\left[\frac{k_{1-}k_{2-}'q_{-+}}{k_{1+}k_{2+}'q_{++}^3}\right]^{\frac{1}{2}}$$

$$\doteq \frac{2e^4 q_{+0}q_{--}}{q_{++}(p_+\cdot k_1)(q_+\cdot k_2)}\left[\frac{k_{1-}k_{2-}'q_{+-}q_{-+}}{k_{1+}k_{2+}'}\right]^{\frac{1}{2}},$$

$$M(-,+;-,+,+,-) = -\frac{e^4 q_{--}(2E-k_{1+})(2q_{+0}+k_{2+}')q_{-\perp}^*}{2E(p_+\cdot k_1)(q_+\cdot k_2)}\left[\frac{k_{1-}k_{2-}'q}{k_{1+}k_{2+}'q}\right.$$

$$\doteq \frac{e^4(2E-k_{1+})(2q_{+0}+k_{2+}')}{2Eq_{-+}(p_+\cdot k_1)(q_+\cdot k_2)}\left[\frac{k_{1-}k_{2-}'q_{++}q_{--}^3}{k_{1+}k_{2+}'}\right]^{\frac{1}{2}}$$

$$M(-,+;-,+,-,+) = \frac{e^4 q_{-\perp}^*}{(p_+\cdot k_1)(q_+\cdot k_2)}\left\{-2q_{+0}q_{--}\left[\frac{k_{1-}k_{2-}'q_{++}}{k_{1+}k_{2+}'q_{-+}^3}\right]^{\frac{1}{2}}\right.$$

$$\left.+\frac{m^2 k_{1\perp}^* k_{2\perp}' Z_{+-}}{4Eq_{+0}q_{--}}\left[\frac{k_{1+}k_{2+}'}{k_{1-}k_{2-}'q_{++}q_{-+}^3}\right]\right\}$$

$$M(-,+;-,-,+,+) = -\frac{e^4 m k_{2\perp}' q_{+\perp}^*}{2q_{+0}(p_+\cdot k_1)(q_+\cdot k_2)}\left[\frac{k_{1-}k_{2+}'q_{-+}}{k_{1+}k_{2-}'q_{++}}\right]^{\frac{1}{2}}$$

$$\doteq \frac{e^4 m k_{2+}'}{2q_{+0}(p_+\cdot k_1)(q_+\cdot k_2)}\left[\frac{k_{1-}q_{+-}q_{-+}}{k_{1+}}\right]^{\frac{1}{2}},$$

$$M(+,-;+,-,-,-) = -\frac{e^4 q_{+0}q_{--}(2E-k_{1+})q_{-\perp}}{E(p_+\cdot k_1)(q_+\cdot k_2)}\left[\frac{k_{1-}k_{2-}'q_{++}}{k_{1+}k_{2+}'q_{-+}^3}\right]^{\frac{1}{2}}$$

$$\doteq \frac{e^4 q_{+0}(2E-k_{1+})}{Eq_{-+}(p_+\cdot k_1)(q_+\cdot k_2)}\left[\frac{k_{1-}k_{2-}'q_{++}q_{--}^3}{k_{1+}k_{2+}'}\right]^{\frac{1}{2}},$$

$$M(+,-;-,+,-,-) = \frac{e^4(2q_{+0}+k_{2+}')q_{+\perp}}{(p_+\cdot k_1)(q_+\cdot k_2)}\left[\frac{k_{1-}k_{2-}'q_{-+}}{k_{1+}k_{2+}'q_{++}}\right]^{\frac{1}{2}}$$

$$\doteq \frac{e^4(2q_{+0}+k_{2+}')}{(p_+\cdot k_1)(q_+\cdot k_2)}\left[\frac{k_{1-}k_{2-}'q_{+-}q_{-+}}{k_{1+}k_{2+}'}\right]^{\frac{1}{2}},$$

9.19. $e^+e^- \to e^+e^- \gamma\gamma$ (NO Z-EXCHANGE; $m \neq 0$)

$$M(+,-;-,-,+,-) = \frac{e^4 m(2q_{+0} + k'_{2+})k_{1\perp}q_{-\perp}Z^*_{+-}}{2Eq_{--}(p_+ \cdot k_1)(q_+ \cdot k_2)} \left[\frac{k_{1+}k'_{2-}}{k_{1-}k'_{2+}q_{++}q^3_{-+}}\right]^{\frac{1}{2}}$$

$$\doteq \frac{e^4 m(2q_{+0} + k'_{2+})k_{1+}}{Eq_{-+}(p_+ \cdot k_1)(q_+ \cdot k_2)} \left[\frac{(q_+ \cdot q_-)k'_{2-}}{2k'_{2+}}\right]^{\frac{1}{2}},$$

$$M(+,-;-,-,-,+) = \frac{e^4 m q_{--}(2E - k_{1+})k'_{2\perp}q_{-\perp}}{4Eq_{+0}(p_+ \cdot k_1)(q_+ \cdot k_2)} \left[\frac{k_{1-}k'_{2+}q_{++}}{k_{1+}k'_{2-}q^3_{-+}}\right]^{\frac{1}{2}}$$

$$\doteq \frac{e^4 m(2E - k_{1+})k'_{2+}}{4Eq_{+0}q_{-+}(p_+ \cdot k_1)(q_+ \cdot k_2)} \left[\frac{k_{1-}q_{++}q^3_{--}}{k_{1+}}\right]^{\frac{1}{2}},$$

$$M(-,+;+,-,-,-) = \frac{4e^4 E q_{+0}q^*_{+\perp}}{(2E - k_{1+})(p_+ \cdot k_1)(q_+ \cdot k_2)} \left[\frac{k_{1-}k'_{2-}q_{-+}}{k_{1+}k'_{2+}q_{++}}\right]^{\frac{1}{2}}$$

$$\doteq \frac{4e^4 E q_{+0}}{(2E - k_{1+})(p_+ \cdot k_1)(q_+ \cdot k_2)} \left[\frac{k_{1-}k'_{2-}q_{+-}q_{-+}}{k_{1+}k'_{2+}}\right]^{\frac{1}{2}},$$

$$M(-,+;-,+,-,-) = -\frac{e^4(2q_{+0} + k'_{2+})q_{--}q^*_{-\perp}}{(p_+ \cdot k_1)(q_+ \cdot k_2)} \left[\frac{k_{1-}k'_{2-}q_{++}}{k_{1+}k'_{2+}q^3_{-+}}\right]^{\frac{1}{2}}$$

$$\doteq \frac{e^4(2q_{+0} + k'_{2+})}{q_{-+}(p_+ \cdot k_1)(q_+ \cdot k_2)} \left[\frac{k_{1-}k'_{2-}q_{++}q^3_{--}}{k_{1+}k'_{2+}}\right]^{\frac{1}{2}},$$

$$M(-,+;-,-,-,+) = -\frac{e^4 m E k'_{2\perp}q^*_{+\perp}}{q_{+0}(2E - k_{1+})(p_+ \cdot k_1)(q_+ \cdot k_2)} \left[\frac{k_{1-}k'_{2+}q_{-+}}{k_{1+}k'_{2-}q_{++}}\right]^{\frac{1}{2}}$$

$$\doteq \frac{e^4 m E k'_{2+}}{q_{+0}(2E - k_{1+})(p_+ \cdot k_1)(q_+ \cdot k_2)} \left[\frac{k_{1-}q_{+-}q_{-+}}{k_{1+}}\right]^{\frac{1}{2}},$$

$$M(-,-;+,+,-,-) = \frac{e^4 m^2 k^*_{1\perp}k'^*_{2\perp}q_{+\perp}}{4Eq_{+0}(2E - k_{1+})(p_+ \cdot k_1)(q_+ \cdot k_2)} \left[\frac{k_{1+}k'_{2+}q_{-+}}{k_{1-}k'_{2-}q_{++}}\right]^{\frac{1}{2}}$$

$$\doteq \frac{e^4 m^2 k_{1+}k'_{2+}[q_{+-}q_{-+}]^{\frac{1}{2}}}{4Eq_{+0}(2E - k_{1+})(p_+ \cdot k_1)(q_+ \cdot k_2)},$$

$$M(-,-;+,-,+,-) = -\frac{e^4 m(2E - k_{1+})k'^*_{2\perp}q_{-\perp}Z^*_{+-}}{2q_{+0}q_{--}(p_+ \cdot k_1)(q_+ \cdot k_2)} \left[\frac{k_{1-}k'_{2+}}{k_{1+}k'_{2-}q_{++}q^3_{-+}}\right]^{\frac{1}{2}}$$

$$\doteq \frac{e^4 m(2E - k_{1+})k'_{2+}}{q_{+0}q_{-+}(p_+ \cdot k_1)(q_+ \cdot k_2)} \left[\frac{(q_+ \cdot q_-)k_{1-}}{2k_{1+}}\right]^{\frac{1}{2}},$$

$$M(-,-;+,-,-,+) = -\frac{e^4 m q_{--}(2q_{+0} + k'_{2+})k^*_{1\perp}q_{-\perp}}{4E^2(p_+ \cdot k_1)(q_+ \cdot k_2)}\left[\frac{k_{1+}k'_{2-}q_{++}}{k_{1-}k'_{2+}q^3_{-+}}\right]^{\frac{1}{2}}$$

$$\doteq \frac{e^4 m k_{1+}(2q_{+0} + k'_{2+})}{4E^2 q_{-+}(p_+ \cdot k_1)(q_+ \cdot k_2)}\left[\frac{k'_{2-}q_{++}q^3_{--}}{k'_{2+}}\right]^{\frac{1}{2}},$$

$$M(-,-;-,+,-,+) = \frac{e^4 m q_{+0} k^*_{1\perp}q_{+\perp}}{E(2E - k_{1+})(p_+ \cdot k_1)(q_+ \cdot k_2)}\left[\frac{k_{1+}k'_{2-}q_{-+}}{k_{1-}k'_{2+}q_{++}}\right]^{\frac{1}{2}}$$

$$\doteq \frac{e^4 m q_{+0} k_{1+}}{E(2E - k_{1+})(p_+ \cdot k_1)(q_+ \cdot k_2)}\left[\frac{k'_{2-}q_{+-}q_{-+}}{k'_{2+}}\right]^{\frac{1}{2}},$$

$$M(-,-;-,-,+,+) = -\frac{2e^4 q_{+0}(2E - k_{1+})q_{-\perp}Z^*_{+-}}{q_{--}(p_+ \cdot k_1)(q_+ \cdot k_2)}\left[\frac{k_{1-}k'_{2-}}{k_{1+}k'_{2+}q_{++}q^3_{-+}}\right]^{\frac{1}{2}}$$

$$\doteq \frac{e^4 q_{+0}(2E - k_{1+})}{q_{-+}(p_+ \cdot k_1)(q_+ \cdot k_2)}\left[\frac{8(q_+ \cdot q_-)k_{1-}k'_{2-}}{k_{1+}k'_{2+}}\right]^{\frac{1}{2}},$$

$$M(-,-;+,-,-,-) = -\frac{e^4 m q_{-\perp}}{(p_+ \cdot k_1)(q_+ \cdot k_2)}\left\{\frac{q_{+0}q_{--}k^*_{1\perp}}{2E^2}\left[\frac{k_{1+}k'_{2-}q_{++}}{k_{1-}k'_{2+}q^3_{-+}}\right]^{\frac{1}{2}}\right.$$

$$\left. + \frac{E k'^*_{2\perp}Z^*_{+-}}{q_{+0}q_{--}}\left[\frac{k_{1-}k'_{2+}}{k_{1+}k'_{2-}q_{++}q^3_{-+}}\right]^{\frac{1}{2}}\right\},$$

$$M(-,-;-,+,-,-) = \frac{e^4 m q_{+0}q_{--}k^*_{1\perp}q_{+\perp}}{2E^2(p_+ \cdot k_1)(q_+ \cdot k_2)}\left[\frac{k_{1+}k'_{2-}q_{-+}}{k_{1-}k'_{2+}q^3_{++}}\right]^{\frac{1}{2}}$$

$$\doteq \frac{e^4 m q_{+0}q_{--}k_{1+}}{2E^2 q_{++}(p_+ \cdot k_1)(q_+ \cdot k_2)}\left[\frac{k'_{2-}q_{+-}q_{-+}}{k'_{2+}}\right]^{\frac{1}{2}},$$

$$M(-,-;-,-,+,-) = -\frac{e^4(2E - k_{1+})(2q_{+0} + k'_{2+})q_{-\perp}Z^*_{+-}}{q_{--}(p_+ \cdot k_1)(q_+ \cdot k_2)}$$

$$\times \left[\frac{k_{1-}k'_{2-}}{k_{1+}k'_{2+}q_{++}q^3_{-+}}\right]^{\frac{1}{2}}$$

$$\doteq \frac{e^4(2E - k_{1+})(2q_{+0} + k'_{2+})}{q_{-+}(p_+ \cdot k_1)(q_+ \cdot k_2)}\left[\frac{2(q_+ \cdot q_-)k_{1-}k'_{2-}}{k_{1+}k'_{2+}}\right]^{\frac{1}{2}},$$

9.19. $e^+e^- \to e^+e^-\gamma\gamma$ (NO Z-EXCHANGE; $m \neq 0$)

$$M(-,-;-,-,-,+) = \frac{e^4 q_{-\perp}}{(p_+ \cdot k_1)(q_+ \cdot k_2)} \left\{ -\frac{4Eq_{+0}Z^*_{+-}}{q_{--}} \left[\frac{k_{1-}k'_{2-}}{k_{1+}k'_{2+}q_{++}q^3_{-+}} \right]^{\frac{1}{2}} \right.$$

$$\left. + \frac{m^2 q_{--} k^*_{1\perp} k'_{2\perp}}{8E^2 q_{+0}} \left[\frac{k_{1+}k'_{2+}q_{++}}{k_{1-}k'_{2-}q^3_{-+}} \right]^{\frac{1}{2}} \right\},$$

$$M(-,-;-,-,-,-) = -\frac{2e^4 E(2q_{+0} + k'_{2+})q_{-\perp}Z^*_{+-}}{q_{--}(p_+ \cdot k_1)(q_+ \cdot k_2)} \left[\frac{k_{1-}k'_{2-}}{k_{1+}k'_{2+}q_{++}q^3_{-+}} \right]^{\frac{1}{2}}$$

$$\doteq \frac{e^4 E(2q_{+0} + k'_{2+})}{q_{-+}(p_+ \cdot k_1)(q_+ \cdot k_2)} \left[\frac{8(q_+ \cdot q_-)k_{1-}k'_{2-}}{k_{1+}k'_{2+}} \right]^{\frac{1}{2}}.$$

(9.235)

Unpolarized squared matrix element:

$$\overline{|M|^2} = \frac{e^8 k_{1-}k'_{2-} \left[4(q_+ \cdot q_-)^2 + q^2_{++}q^2_{--} + q^2_{+-}q^2_{-+} \right]}{2k_{1+}k'_{2+}q_{+-}q_{-+}[(2E - k_{1+})(p_+ \cdot k_1)(q_+ \cdot k_2)]^2}$$

$$\times \left[4E^2 + (2E - k_{1+})^2 + \frac{m^2 k^3_{1+}}{4E^2 k_{1-}} \right]$$

$$\times \left[4q^2_{+0} + (2q_{+0} + k'_{2+})^2 + \frac{m^2 k'^3_{2+}}{4q^2_{+0}k'_{2-}} \right]. \quad (9.236)$$

Unpolarized cross section:

$$d\sigma = \frac{\alpha^4 k_{1-}k'_{2-} \left[4(q_+ \cdot q_-)^2 + q^2_{++}q^2_{--} + q^2_{+-}q^2_{-+} \right]}{256\pi^2 k_{1+}k'_{2+}q_{+-}q_{-+}[E(2E - k_{1+})(p_+ \cdot k_1)(q_+ \cdot k_2)]^2}$$

$$\times \left[4E^2 + (2E - k_{1+})^2 + \frac{m^2 k^3_{1+}}{4E^2 k_{1-}} \right]$$

$$\times \left[4q^2_{+0} + (2q_{+0} + k'_{2+})^2 + \frac{m^2 k'^3_{2+}}{4q^2_{+0}k'_{2-}} \right]$$

$$\times \delta^4(p_+ + p_- - q_+ - q_- - k_1 - k_2) \frac{d^3\vec{q}_+ \, d^3\vec{q}_- \, d^3\vec{k}_1 \, d^3\vec{k}_2}{q_{+0} q_{-0} k_{10} k_{20}}. \quad (9.237)$$

9.19.7 $\vec{k}_1$ and $\vec{k}_2$ nearly parallel to $\vec{p}_+$ and $\vec{q}_-$, resp.

The doubly primed quantities $k''_{2\pm}$ and $k''_{2\perp}$ are evaluated in the rotated frame where $\vec{q}_-$ determines the positive z-axis (see also Section 7.4.3). The four-vector q in eqn (9.239) is obtained by applying a space reflection to q_-.
Definition:

$$Z_{-+} = q_{-+}q_{+-} - q^*_{-\perp}q_{+\perp}. \quad (9.238)$$

Photon polarizations:

$$\epsilon_1^\pm = N_1[\slashed{k}_1 \slashed{p}_+ \slashed{p}_-(1 \pm \gamma_5) + \slashed{p}_- \slashed{p}_+ \slashed{k}_1(1 \mp \gamma_5)],$$

$$N_1^{-1} = E^2[32 k_{1+} k_{1-}]^{\frac{1}{2}},$$

$$\epsilon_2^\pm = N_2[\slashed{k}_2 \slashed{q}_- \slashed{q}(1 \pm \gamma_5) + \slashed{q} \slashed{q}_- \slashed{k}_2(1 \mp \gamma_5)],$$

$$N_2^{-1} = q_{-0}^2 [32 k_{2+}'' k_{2-}'']^{\frac{1}{2}}. \tag{9.239}$$

Nonvanishing helicity amplitudes:

$$M(+,+;+,+,+,+) = -\frac{4e^4 E q_{-0} q_{+\perp}^* Z_{-+}}{q_{+-}(p_+ \cdot k_1)(q_- \cdot k_2)} \left[\frac{k_{1-} k_{2-}''}{k_{1+} k_{2+}'' q_{++} q_{-+}^3}\right]^{\frac{1}{2}}$$

$$\doteq \frac{e^4 E q_{-0}}{q_{-+}(p_+ \cdot k_1)(q_- \cdot k_2)} \left[\frac{32(q_+ \cdot q_-) k_{1-} k_{2-}''}{k_{1+} k_{2+}''}\right]^{\frac{1}{2}},$$

$$M(+,+;+,+,+,-) = \frac{e^4 q_{+\perp}^*}{(2E - k_{1+})(p_+ \cdot k_1)(q_- \cdot k_2)}$$

$$\times \left\{ -\frac{8E^2 q_{-0} Z_{-+}}{q_{+-}^2} \left[\frac{k_{1-} k_{2-}''}{k_{1+} k_{2+}'' q_{++} q_{-+}}\right]^{\frac{1}{2}} \right.$$

$$\left. +\frac{m^2 k_{1\perp} k_{2\perp}''^*}{4 E q_{-0}} \left[\frac{k_{1+} k_{2+}'' q_{-+}}{k_{1-} k_{2-}'' q_{++}}\right]^{\frac{1}{2}} \right\},$$

$$M(+,+;+,+,-,+) = -\frac{2e^4 q_{-0}(2E - k_{1+}) q_{+\perp}^* Z_{-+}}{q_{+-}(p_+ \cdot k_1)(q_- \cdot k_2)} \left[\frac{k_{1-} k_{2-}''}{k_{1+} k_{2+}'' q_{++} q_{-+}^3}\right]^{\frac{1}{2}}$$

$$\doteq \frac{e^4 q_{-0}(2E - k_{1+})}{q_{-+}(p_+ \cdot k_1)(q_- \cdot k_2)} \left[\frac{8(q_+ \cdot q_-) k_{1-} k_{2-}''}{k_{1+} k_{2+}''}\right]^{\frac{1}{2}},$$

$$M(+,+;+,-,+,+) = \frac{e^4 m q_{+\perp}^*}{(2E - k_{1+})(p_+ \cdot k_1)(q_- \cdot k_2)}$$

$$\times \left\{ \frac{2E^2 k_{2\perp}'' Z_{-+}}{q_{-0} q_{+-}^2} \left[\frac{k_{1-} k_{2+}''}{k_{1+} k_{2-}'' q_{++} q_{-+}}\right]^{\frac{1}{2}} \right.$$

$$\left. +\frac{q_{-0} k_{1\perp}}{E} \left[\frac{k_{1+} k_{2-}'' q_{-+}}{k_{1-} k_{2+}'' q_{++}}\right]^{\frac{1}{2}} \right\},$$

9.19. $e^+e^- \to e^+e^-\gamma\gamma$ (NO Z-EXCHANGE; $m \neq 0$)

$$M(+,+;-,+,+,+) = -\frac{e^4 m(2q_{-0}+k''_{2+})q_{--}k_{1\perp}q^*_{-\perp}}{4E^2(p_+\cdot k_1)(q_-\cdot k_2)}\left[\frac{k_{1+}k''_{2-}q_{++}}{k_{1-}k''_{2+}q^3_{-+}}\right]^{\frac{1}{2}}$$

$$\doteq \frac{e^4 m(2q_{-0}+k''_{2+})k_{1+}}{4E^2 q_{-+}(p_+\cdot k_1)(q_-\cdot k_2)}\left[\frac{k''_{2-}q_{++}q^3_{--}}{k''_{2+}}\right]^{\frac{1}{2}},$$

$$M(+,+;+,+,-,-) = -\frac{4e^4 E q_{-0}q^*_{+\perp}Z_{-+}}{q^2_{+-}(p_+\cdot k_1)(q_-\cdot k_2)}\left[\frac{k_{1-}k''_{2-}}{k_{1+}k''_{2+}q_{++}q_{-+}}\right]^{\frac{1}{2}}$$

$$\doteq \frac{e^4 E q_{-0}}{q_{+-}(p_+\cdot k_1)(q_-\cdot k_2)}\left[\frac{32(q_+\cdot q_-)k_{1-}k''_{2-}}{k_{1+}k''_{2+}}\right]^{\frac{1}{2}},$$

$$M(+,+;+,-,+,-) = \frac{e^4 m q_{-0}q_{+-}k_{1\perp}q^*_{+\perp}}{2E^2(p_+\cdot k_1)(q_-\cdot k_2)}\left[\frac{k_{1+}k''_{2-}}{k_{1-}k''_{2+}q_{++}q_{-+}}\right]^{\frac{1}{2}}$$

$$\doteq \frac{e^4 m q_{-0}k_{1+}}{2E^2(p_+\cdot k_1)(q_-\cdot k_2)}\left[\frac{k''_{2-}q^3_{+-}}{k''_{2+}q_{-+}}\right]^{\frac{1}{2}},$$

$$M(+,+;+,-,-,+) = \frac{e^4 m E k''_{2\perp}q^*_{+\perp}Z_{-+}}{q_{-0}q^2_{+-}(p_+\cdot k_1)(q_-\cdot k_2)}\left[\frac{k_{1-}k''_{2+}}{k_{1+}k''_{2-}q_{++}q_{-+}}\right]^{\frac{1}{2}}$$

$$\doteq \frac{e^4 m E k''_{2+}}{q_{-0}q_{+-}(p_+\cdot k_1)(q_-\cdot k_2)}\left[\frac{2(q_+\cdot q_-)k_{1-}}{k_{1+}}\right]^{\frac{1}{2}},$$

$$M(+,+;-,+,+,-) = -\frac{e^4 m q_{-0}q_{--}k_{1\perp}q^*_{-\perp}}{2E^2(p_+\cdot k_1)(q_-\cdot k_2)}\left[\frac{k_{1+}k''_{2-}q_{++}}{k_{1-}k''_{2+}q^3_{-+}}\right]^{\frac{1}{2}}$$

$$\doteq \frac{e^4 m q_{-0}k_{1+}}{2E^2 q_{-+}(p_+\cdot k_1)(q_-\cdot k_2)}\left[\frac{k''_{2-}q_{++}q^3_{--}}{k''_{2+}}\right]^{\frac{1}{2}},$$

$$M(+,+;-,-,+,+) = \frac{e^4 m^2 q_{--}k_{1\perp}k''_{2\perp}q^*_{-\perp}}{8E^2 q_{-0}(p_+\cdot k_1)(q_-\cdot k_2)}\left[\frac{k_{1+}k''_{2+}q_{++}}{k_{1-}k''_{2-}q^3_{-+}}\right]^{\frac{1}{2}}$$

$$\doteq \frac{e^4 m^2 k_{1+}k''_{2+}\left[q_{++}q^3_{--}\right]^{\frac{1}{2}}}{8E^2 q_{-0}q_{-+}(p_+\cdot k_1)(q_-\cdot k_2)},$$

$$M(+,-;+,+,+,-) = \frac{e^4 m q_{--}k''^*_{2\perp}q_{-\perp}}{2q_{-0}(p_+\cdot k_1)(q_-\cdot k_2)}\left[\frac{k_{1-}k''_{2+}q_{++}}{k_{1+}k''_{2-}q^3_{-+}}\right]^{\frac{1}{2}}$$

$$\doteq \frac{e^4 m k''_{2+}}{2q_{-0}q_{-+}(p_+\cdot k_1)(q_-\cdot k_2)}\left[\frac{k_{1-}q_{++}q^3_{--}}{k_{1+}}\right]^{\frac{1}{2}},$$

$$M(+,-;+,-,+,+) = \frac{2e^4 q_{-0} q_{--} q_{-\perp}}{(p_+ \cdot k_1)(q_- \cdot k_2)} \left[\frac{k_{1-} k''_{2-} q_{++}}{k_{1+} k''_{2+} q^3_{-+}}\right]^{\frac{1}{2}}$$

$$\doteq \frac{2e^4 q_{-0}}{q_{-+}(p_+ \cdot k_1)(q_- \cdot k_2)} \left[\frac{k_{1-} k''_{2-} q_{++} q^3_{--}}{k_{1+} k''_{2+}}\right]^{\frac{1}{2}},$$

$$M(+,-;-,+,+,+) = -\frac{2e^4 q_{-0} q_{+-} q_{+\perp}}{(p_+ \cdot k_1)(q_- \cdot k_2)} \left[\frac{k_{1-} k''_{2-}}{k_{1+} k''_{2+} q_{++} q_{-+}}\right]^{\frac{1}{2}}$$

$$\doteq \frac{2e^4 q_{-0}}{(p_+ \cdot k_1)(q_- \cdot k_2)} \left[\frac{k_{1-} k''_{2-} q^3_{+-}}{k_{1+} k''_{2+} q_{-+}}\right]^{\frac{1}{2}},$$

$$M(-,+;+,+,+,-) = -\frac{e^4 m k'''_{2\perp} q^*_{+\perp}}{2 q_{-0}(p_+ \cdot k_1)(q_- \cdot k_2)} \left[\frac{k_{1-} k''_{2+} q_{-+}}{k_{1+} k''_{2-} q_{++}}\right]^{\frac{1}{2}}$$

$$\doteq \frac{e^4 m k''_{2+}}{2 q_{-0}(p_+ \cdot k_1)(q_- \cdot k_2)} \left[\frac{k_{1-} q_{+-} q_{-+}}{k_{1+}}\right]^{\frac{1}{2}},$$

$$M(-,+;+,+,-,+) = -\frac{e^4 m q_{-0} k^*_{1\perp} q_{+\perp} Z_{-+}}{E q_{+-}(p_+ \cdot k_1)(q_- \cdot k_2)} \left[\frac{k_{1+} k''_{2-}}{k_{1-} k''_{2+} q_{++} q^3_{-+}}\right]^{\frac{1}{2}}$$

$$\doteq \frac{e^4 m q_{-0} k_{1+}}{E q_{-+}(p_+ \cdot k_1)(q_- \cdot k_2)} \left[\frac{2(q_+ \cdot q_-) k''_{2-}}{k''_{2+}}\right]^{\frac{1}{2}},$$

$$M(-,+;+,-,+,+) = -\frac{2e^4 q_{-0} q^*_{+\perp}}{(p_+ \cdot k_1)(q_- \cdot k_2)} \left[\frac{k_{1-} k''_{2-} q_{-+}}{k_{1+} k''_{2+} q_{++}}\right]^{\frac{1}{2}}$$

$$\doteq \frac{2e^4 q_{-0}}{(p_+ \cdot k_1)(q_- \cdot k_2)} \left[\frac{k_{1-} k''_{2-} q_{+-} q_{-+}}{k_{1+} k''_{2+}}\right]^{\frac{1}{2}},$$

$$M(-,+;-,+,+,+) = \frac{2e^4 q_{-0} q^*_{-\perp}}{(p_+ \cdot k_1)(q_- \cdot k_2)} \left[\frac{k_{1-} k''_{2-} q^3_{++}}{k_{1+} k''_{2+} q^3_{-+}}\right]^{\frac{1}{2}}$$

$$\doteq \frac{2e^4 q_{-0}}{q_{-+}(p_+ \cdot k_1)(q_- \cdot k_2)} \left[\frac{k_{1-} k''_{2-} q^3_{++} q_{--}}{k_{1+} k''_{2+}}\right]^{\frac{1}{2}},$$

$$M(+,-;+,+,-,-) = \frac{e^4 m (2E - k_{1+}) q_{--} k'''^*_{2\perp} q_{-\perp}}{4 E q_{-0}(p_+ \cdot k_1)(q_- \cdot k_2)} \left[\frac{k_{1-} k''_{2+} q_{++}}{k_{1+} k''_{2-} q^3_{-+}}\right]^{\frac{1}{2}}$$

$$\doteq \frac{e^4 m (2E - k_{1+}) k''_{2+}}{4 E q_{-0} q_{-+}(p_+ \cdot k_1)(q_- \cdot k_2)} \left[\frac{k_{1-} q_{++} q^3_{--}}{k_{1+}}\right]^{\frac{1}{2}},$$

9.19. $e^+e^- \to e^+e^-\gamma\gamma$ (NO Z-EXCHANGE; $m \neq 0$)

$$M(+,-;+,-,+,-) = \frac{e^4(2q_{-0}+k''_{2+})q_{--}q_{-\perp}}{(p_+\cdot k_1)(q_-\cdot k_2)}\left[\frac{k_{1-}k''_{2-}q_{++}}{k_{1+}k''_{2+}q^3_{-+}}\right]^{\frac{1}{2}}$$

$$\doteq \frac{e^4(2q_{-0}+k''_{2+})}{q_{-+}(p_+\cdot k_1)(q_-\cdot k_2)}\left[\frac{k_{1-}k''_{2-}q_{++}q^3_{--}}{k_{1+}k''_{2+}}\right]^{\frac{1}{2}},$$

$$M(+,-;+,-,-,+) = \frac{e^4 q_{-0}(2E-k_{1+})q_{--}q_{-\perp}}{E(p_+\cdot k_1)(q_-\cdot k_2)}\left[\frac{k_{1-}k''_{2-}q_{++}}{k_{1+}k''_{2+}q^3_{-+}}\right]^{\frac{1}{2}}$$

$$\doteq \frac{e^4 q_{-0}(2E-k_{1+})}{Eq_{-+}(p_+\cdot k_1)(q_-\cdot k_2)}\left[\frac{k_{1-}k''_{2-}q_{++}q^3_{--}}{k_{1+}k''_{2+}}\right]^{\frac{1}{2}},$$

$$M(+,-;-,+,+,-) = \frac{e^4 q_{+\perp}}{(2E-k_{1+})(p_+\cdot k_1)(q_-\cdot k_2)}$$

$$\times\left\{-4Eq_{-0}\left[\frac{k_{1-}k''_{2-}q_{-+}}{k_{1+}k''_{2+}q_{++}}\right]^{\frac{1}{2}}\right.$$

$$\left.+\frac{m^2 k_{1\perp}k''^*_{2\perp}Z^*_{-+}}{2q_{-0}q^2_{+-}}\left[\frac{k_{1+}k''_{2+}}{k_{1-}k''_{2-}q_{++}q_{-+}}\right]^{\frac{1}{2}}\right\},$$

$$M(+,-;-,+,-,+) = -\frac{e^4(2q_{-0}+k''_{2+})q_{+\perp}}{(p_+\cdot k_1)(q_-\cdot k_2)}\left[\frac{k_{1-}k''_{2-}q_{-+}}{k_{1+}k''_{2+}q_{++}}\right]^{\frac{1}{2}}$$

$$\doteq \frac{e^4(2q_{-0}+k''_{2+})}{(p_+\cdot k_1)(q_-\cdot k_2)}\left[\frac{k_{1-}k''_{2-}q_{+-}q_{-+}}{k_{1+}k''_{2+}}\right]^{\frac{1}{2}},$$

$$M(+,-;-,-,+,+) = \frac{e^4 m q_{+\perp}}{(2E-k_{1+})(p_+\cdot k_1)(q_-\cdot k_2)}$$

$$\times\left\{\frac{2q_{-0}k_{1\perp}Z^*_{-+}}{q^2_{+-}}\left[\frac{k_{1+}k''_{2-}}{k_{1-}k''_{2+}q_{++}q_{-+}}\right]^{\frac{1}{2}}\right.$$

$$\left.+\frac{Ek''_{2\perp}}{q_{-0}}\left[\frac{k_{1-}k''_{2+}q_{-+}}{k_{1+}k''_{2-}q_{++}}\right]^{\frac{1}{2}}\right\},$$

$$M(-,+;+,+,-,-) = \frac{-e^4 m q^*_{+\perp}}{(2E-k_{1+})(p_+\cdot k_1)(q_-\cdot k_2)}\left\{\frac{Ek''^*_{2\perp}}{q_{-0}}\left[\frac{k_{1-}k''_{2+}q_{-+}}{k_{1+}k''_{2-}q_{++}}\right]^{\frac{1}{2}}\right.$$

$$\left.+\frac{2q_{-0}k^*_{1\perp}Z_{-+}}{q^2_{+-}}\left[\frac{k_{1+}k''_{2-}}{k_{1-}k''_{2+}q_{++}q_{-+}}\right]^{\frac{1}{2}}\right\},$$

$$M(-,+;+,-,+,-) = -\frac{e^4(2q_{-0}+k''_{2+})q^*_{+\perp}}{(p_+\cdot k_1)(q_-\cdot k_2)}\left[\frac{k_{1-}k''_{2-}q_{-+}}{k_{1+}k''_{2+}q_{++}}\right]^{\frac{1}{2}}$$

$$\doteq \frac{e^4(2q_{-0}+k''_{2+})}{(p_+\cdot k_1)(q_-\cdot k_2)}\left[\frac{k_{1-}k''_{2-}q_{+-}q_{-+}}{k_{1+}k''_{2+}}\right]^{\frac{1}{2}},$$

$$M(-,+;+,-,-,+) = \frac{e^4 q^*_{+\perp}}{(2E-k_{1+})(p_+\cdot k_1)(q_-\cdot k_2)}$$

$$\times\left\{-4Eq_{-0}\left[\frac{k_{1-}k''_{2-}q_{-+}}{k_{1+}k''_{2+}q_{++}}\right]^{\frac{1}{2}}\right.$$

$$\left.+\frac{m^2 k^*_{1\perp}k''_{2\perp}Z_{-+}}{2q_{-0}q^2_{+-}}\left[\frac{k_{1+}k''_{2+}}{k_{1-}k''_{2-}q_{++}q_{-+}}\right]^{\frac{1}{2}}\right\},$$

$$M(-,+;-,+,+,-) = \frac{e^4 q_{-0}(2E-k_{1+})q_{--}q^*_{-\perp}}{E(p_+\cdot k_1)(q_-\cdot k_2)}\left[\frac{k_{1-}k''_{2-}q_{++}}{k_{1+}k''_{2+}q^3_{-+}}\right]^{\frac{1}{2}}$$

$$\doteq \frac{e^4 q_{-0}(2E-k_{1+})}{Eq_{-+}(p_+\cdot k_1)(q_-\cdot k_2)}\left[\frac{k_{1-}k''_{2-}q_{++}q^3_{--}}{k_{1+}k''_{2+}}\right]^{\frac{1}{2}},$$

$$M(-,+;-,+,-,+) = \frac{e^4(2q_{-0}+k''_{2+})q_{--}q^*_{-\perp}}{(p_+\cdot k_1)(q_-\cdot k_2)}\left[\frac{k_{1-}k''_{2-}q_{++}}{k_{1+}k''_{2+}q^3_{-+}}\right]^{\frac{1}{2}}$$

$$\doteq \frac{e^4(2q_{-0}+k''_{2+})}{q_{-+}(p_+\cdot k_1)(q_-\cdot k_2)}\left[\frac{k_{1-}k''_{2-}q_{++}q^3_{--}}{k_{1+}k''_{2+}}\right]^{\frac{1}{2}},$$

$$M(-,+;-,-,+,+) = -\frac{e^4 m(2E-k_{1+})q_{--}k''_{2\perp}q^*_{-\perp}}{4Eq_{-0}(p_+\cdot k_1)(q_-\cdot k_2)}\left[\frac{k''_{2+}q_{++}}{k_{1+}k''_{2-}q^3_{-+}}\right]^{\frac{1}{2}}$$

$$\doteq \frac{e^4 m(2E-k_{1+})k''_{2+}}{4Eq_{-0}q_{-+}(p_+\cdot k_1)(q_-\cdot k_2)}\left[\frac{k_{1-}q_{++}q^3_{--}}{k_{1+}}\right]^{\frac{1}{2}},$$

$$M(+,-;+,-,-,-) = \frac{2e^4 q_{-0}q_{-\perp}}{(p_+\cdot k_1)(q_-\cdot k_2)}\left[\frac{k_{1-}k''_{2-}q^3_{++}}{k_{1+}k''_{2+}q^3_{-+}}\right]^{\frac{1}{2}}$$

$$\doteq \frac{2e^4 q_{-0}}{q_{-+}(p_+\cdot k_1)(q_-\cdot k_2)}\left[\frac{k_{1-}k''_{2-}q^3_{++}q_{--}}{k_{1+}k''_{2+}}\right]^{\frac{1}{2}},$$

$$M(+,-;-,+,-,-) = -\frac{2e^4 q_{-0}q_{+\perp}}{(p_+\cdot k_1)(q_-\cdot k_2)}\left[\frac{k_{1-}k''_{2-}q_{-+}}{k_{1+}k''_{2+}q_{++}}\right]^{\frac{1}{2}}$$

$$\doteq \frac{2e^4 q_{-0}}{(p_+\cdot k_1)(q_-\cdot k_2)}\left[\frac{k_{1-}k''_{2-}q_{+-}q_{-+}}{k_{1+}k''_{2+}}\right]^{\frac{1}{2}},$$

9.19. $e^+e^- \to e^+e^- \gamma\gamma$ (NO Z-EXCHANGE; $m \neq 0$)

$$M(+,-;-,-,+,-) = \frac{e^4 m q_{-0} k_{1\perp} q_{+\perp} Z^*_{-+}}{E q_{+-}(p_+ \cdot k_1)(q_- \cdot k_2)} \left[\frac{k_{1+} k''_{2-}}{k_{1-} k''_{2+} q_{++} q^3_{-+}}\right]^{\frac{1}{2}}$$

$$\doteq \frac{e^4 m q_{-0} k_{1+}}{E q_{-+}(p_+ \cdot k_1)(q_- \cdot k_2)} \left[\frac{2(q_+ \cdot q_-) k''_{2-}}{k''_{2+}}\right]^{\frac{1}{2}},$$

$$M(+,-;-,-,-,+) = \frac{e^4 m k''_{2\perp} q_{+\perp}}{2 q_{-0}(p_+ \cdot k_1)(q_- \cdot k_2)} \left[\frac{k_{1-} k''_{2+} q_{-+}}{k_{1+} k''_{2-} q_{++}}\right]^{\frac{1}{2}}$$

$$\doteq \frac{e^4 m k''_{2+}}{2 q_{-0}(p_+ \cdot k_1)(q_- \cdot k_2)} \left[\frac{k_{1-} q_{+-} q_{-+}}{k_{1+}}\right]^{\frac{1}{2}},$$

$$M(-,+;+,-,-,-) = -\frac{2 e^4 q_{-0} q_{+-} q^*_{+\perp}}{(p_+ \cdot k_1)(q_- \cdot k_2)} \left[\frac{k_{1-} k''_{2-}}{k_{1+} k''_{2+} q_{++} q_{-+}}\right]^{\frac{1}{2}}$$

$$\doteq \frac{2 e^4 q_{-0}}{(p_+ \cdot k_1)(q_- \cdot k_2)} \left[\frac{k_{1-} k''_{2-} q^3_{+-}}{k_{1+} k''_{2+} q_{-+}}\right]^{\frac{1}{2}},$$

$$M(-,+;-,+,-,-) = \frac{2 e^4 q_{-0} q_{--} q^*_{-\perp}}{(p_+ \cdot k_1)(q_- \cdot k_2)} \left[\frac{k_{1-} k''_{2-} q_{++}}{k_{1+} k''_{2+} q^3_{-+}}\right]^{\frac{1}{2}}$$

$$\doteq \frac{2 e^4 q_{-0}}{q_{-+}(p_+ \cdot k_1)(q_- \cdot k_2)} \left[\frac{k_{1-} k''_{2-} q_{++} q^3_{--}}{k_{1+} k''_{2+}}\right]^{\frac{1}{2}},$$

$$M(-,+;-,-,-,+) = -\frac{e^4 m q_{--} k''_{2\perp} q^*_{-\perp}}{2 q_{-0}(p_+ \cdot k_1)(q_- \cdot k_2)} \left[\frac{k_{1-} k''_{2+} q_{++}}{k_{1+} k''_{2-} q^3_{-+}}\right]^{\frac{1}{2}}$$

$$\doteq \frac{e^4 m k''_{2+}}{2 q_{-0} q_{-+}(p_+ \cdot k_1)(q_- \cdot k_2)} \left[\frac{k_{1-} q_{++} q^3_{--}}{k_{1+}}\right]^{\frac{1}{2}},$$

$$M(-,-;+,+,-,-) = \frac{e^4 m^2 q_{--} k^*_{1\perp} k'''_{2\perp} q_{-\perp}}{8 E^2 q_{-0}(p_+ \cdot k_1)(q_- \cdot k_2)} \left[\frac{k_{1+} k''_{2+} q_{++}}{k_{1-} k''_{2-} q^3_{-+}}\right]^{\frac{1}{2}}$$

$$\doteq \frac{e^4 m^2 k_{1+} k''_{2+} \left[q_{++} q^3_{--}\right]^{\frac{1}{2}}}{8 E^2 q_{-0} q_{-+}(p_+ \cdot k_1)(q_- \cdot k_2)},$$

$$M(-,-;+,-,-,+) = \frac{e^4 m q_{-0} q_{--} k^*_{1\perp} q_{-\perp}}{2 E^2 (p_+ \cdot k_1)(q_- \cdot k_2)} \left[\frac{k_{1+} k''_{2-} q_{++}}{k_{1-} k''_{2+} q^3_{-+}}\right]^{\frac{1}{2}}$$

$$\doteq \frac{e^4 m q_{-0} k_{1+}}{2 E^2 q_{-+}(p_+ \cdot k_1)(q_- \cdot k_2)} \left[\frac{k''_{2-} q_{++} q^3_{--}}{k''_{2+}}\right]^{\frac{1}{2}},$$

$$M(-,-;-,+,+,-) = -\frac{e^4 m E k_{21}^{\prime\prime*} q_{+\perp} Z_{-+}^*}{q_{-0} q_{+-}^2 (p_+ \cdot k_1)(q_- \cdot k_2)} \left[\frac{k_{1-} k_{2+}^{\prime\prime}}{k_{1+} k_{2-}^{\prime\prime} q_{++} q_{-+}}\right]^{\frac{1}{2}}$$

$$\doteq \frac{e^4 m E k_{2+}^{\prime\prime}}{q_{-0} q_{+-} (p_+ \cdot k_1)(q_- \cdot k_2)} \left[\frac{2(q_+ \cdot q_-) k_{1-}}{k_{1+}}\right]^{\frac{1}{2}},$$

$$M(-,-;-,+,-,+) = -\frac{e^4 m q_{-0} q_{+-} k_{1\perp}^* q_{+\perp}}{2E^2 (p_+ \cdot k_1)(q_- \cdot k_2)} \left[\frac{k_{1+} k_{2-}^{\prime\prime}}{k_{1-} k_{2+}^{\prime\prime} q_{++} q_{-+}}\right]^{\frac{1}{2}}$$

$$\doteq \frac{e^4 m q_{-0} k_{1+}}{2E^2 (p_+ \cdot k_1)(q_- \cdot k_2)} \left[\frac{k_{2-}^{\prime\prime} q_{+-}^3}{k_{2+}^{\prime\prime} q_{-+}}\right]^{\frac{1}{2}},$$

$$M(-,-;-,-,+,+) = -\frac{4e^4 E q_{-0} q_{+\perp} Z_{-+}^*}{q_{+-}^2 (p_+ \cdot k_1)(q_- \cdot k_2)} \left[\frac{k_{1-} k_{2-}^{\prime\prime}}{k_{1+} k_{2+}^{\prime\prime} q_{++} q_{-+}}\right]^{\frac{1}{2}}$$

$$\doteq \frac{e^4 E q_{-0}}{q_{+-} (p_+ \cdot k_1)(q_- \cdot k_2)} \left[\frac{32(q_+ \cdot q_-) k_{1-} k_{2-}^{\prime\prime}}{k_{1+} k_{2+}^{\prime\prime}}\right]^{\frac{1}{2}},$$

$$M(-,-;+,-,-,-) = \frac{e^4 m (2q_{-0} + k_{2+}^{\prime\prime}) q_{--} k_{1\perp}^* q_{-\perp}}{4E^2 (p_+ \cdot k_1)(q_- \cdot k_2)} \left[\frac{k_{1+} k_{2-}^{\prime\prime} q_{++}}{k_{1-} k_{2+}^{\prime\prime} q_{-+}^3}\right]^{\frac{1}{2}}$$

$$\doteq \frac{e^4 m (2q_{-0} + k_{2+}^{\prime\prime}) k_{1+}}{4E^2 q_{-+} (p_+ \cdot k_1)(q_- \cdot k_2)} \left[\frac{k_{2-}^{\prime\prime} q_{++} q_{--}^3}{k_{2+}^{\prime\prime}}\right]^{\frac{1}{2}},$$

$$M(-,-;-,+,-,-) = -\frac{e^4 m q_{+\perp}}{(2E - k_{1+})(p_+ \cdot k_1)(q_- \cdot k_2)}$$

$$\times \left\{ \frac{2E^2 k_{21}^{\prime\prime*} Z_{-+}^*}{q_{-0} q_{+-}^2} \left[\frac{k_{1-} k_{2+}^{\prime\prime}}{k_{1+} k_{2-}^{\prime\prime} q_{++} q_{-+}}\right]^{\frac{1}{2}} \right.$$

$$\left. + \frac{q_{-0} k_{1\perp}^*}{E} \left[\frac{k_{1+} k_{2-}^{\prime\prime} q_{-+}}{k_{1-} k_{2+}^{\prime\prime} q_{++}}\right]^{\frac{1}{2}} \right\},$$

$$M(-,-;-,-,+,-) = -\frac{2e^4 q_{-0} (2E - k_{1+}) q_{+\perp} Z_{-+}^*}{q_{+-} (p_+ \cdot k_1)(q_- \cdot k_2)} \left[\frac{k_{1-} k_{2-}^{\prime\prime}}{k_{1+} k_{2+}^{\prime\prime} q_{++} q_{-+}^3}\right]^{\frac{1}{2}}$$

$$\doteq \frac{e^4 q_{-0} (2E - k_{1+})}{q_{-+} (p_+ \cdot k_1)(q_- \cdot k_2)} \left[\frac{8(q_+ \cdot q_-) k_{1-} k_{2-}^{\prime\prime}}{k_{1+} k_{2+}^{\prime\prime}}\right]^{\frac{1}{2}},$$

9.19. $e^+e^- \to e^+e^-\gamma\gamma$ (NO Z-EXCHANGE; $m \neq 0$)

$$M(-,-;-,-,-,+) = \frac{e^4 q_{+\perp}}{(2E-k_{1+})(p_+ \cdot k_1)(q_- \cdot k_2)}$$

$$\times \left\{ -\frac{8E^2 q_{-0} Z^*_{-+}}{q^2_{+-}} \left[\frac{k_{1-}k''_{2-}}{k_{1+}k''_{2+}q_{++}q_{-+}} \right]^{\frac{1}{2}} \right.$$

$$\left. + \frac{m^2 k^*_{1\perp} k''_{2\perp}}{4Eq_{-0}} \left[\frac{k_{1+}k''_{2+}q_{-+}}{k_{1-}k''_{2-}q_{++}} \right]^{\frac{1}{2}} \right\},$$

$$M(-,-;-,-,-,-) = -\frac{4e^4 E q_{-0} q_{+\perp} Z^*_{-+}}{q_{+-}(p_+ \cdot k_1)(q_- \cdot k_2)} \left[\frac{k_{1-}k''_{2-}}{k_{1+}k''_{2+}q_{++}q^3_{-+}} \right]^{\frac{1}{2}}$$

$$\doteq \frac{e^4 E q_{-0}}{q_{-+}(p_+ \cdot k_1)(q_- \cdot k_2)} \left[\frac{32(q_+ \cdot q_-) k_{1-} k''_{2-}}{k_{1+} k''_{2+}} \right]^{\frac{1}{2}}.$$

(9.240)

Unpolarized squared matrix element:

$$\overline{|M|^2} = \frac{e^8 k_{1-} k''_{2-} \left[4(q_+ \cdot q_-)^2 + q^2_{++} q^2_{--} + q^2_{+-} q^2_{-+} \right]}{2 k_{1+} k''_{2+} q_{+-} q_{-+} [(2E-k_{1+})(p_+ \cdot k_1)(q_- \cdot k_2)]^2}$$

$$\times \left[4E^2 + (2E-k_{1+})^2 + \frac{m^2 k^3_{1+}}{4E^2 k_{1-}} \right]$$

$$\times \left[4q^2_{-0} + (2q_{-0} + k''_{2+})^2 + \frac{m^2 k''^3_{2+}}{4q^2_{-0} k''_{2-}} \right].$$

(9.241)

Unpolarized cross section:

$$d\sigma = \frac{\alpha^4 k_{1-} k''_{2-} \left[4(q_+ \cdot q_-)^2 + q^2_{++} q^2_{--} + q^2_{+-} q^2_{-+} \right]}{256\pi^2 k_{1+} k''_{2+} q_{+-} q_{-+} [E(2E-k_{1+})(p_+ \cdot k_1)(q_- \cdot k_2)]^2}$$

$$\times \left[4E^2 + (2E-k_{1+})^2 + \frac{m^2 k^3_{1+}}{4E^2 k_{1-}} \right]$$

$$\times \left[4q^2_{-0} + (2q_{-0} + k''_{2+})^2 + \frac{m^2 k''^3_{2+}}{4q^2_{-0} k''_{2-}} \right]$$

$$\times \delta^4(p_+ + p_- - q_+ - q_- - k_1 - k_2) \frac{d^3\vec{q}_+ \, d^3\vec{q}_- \, d^3\vec{k}_1 \, d^3\vec{k}_2}{q_{+0} \, q_{-0} \, k_{10} \, k_{20}}.$$ (9.242)

9.19.8 $\vec{k}_1$ and $\vec{k}_2$ nearly parallel to $\vec{p}_-$ and $\vec{q}_+$, resp.

The primed quantities $k'_{2\pm}$ and $k'_{2\perp}$ are evaluated in the rotated frame where $\vec{q}_+$ determines the positive z-axis (see also Section 7.4.3). The four-vector q

in eqn (9.244) is obtained by applying a space reflection to q_+.
Definition:

$$Z_{+-} = q_{++}q_{--} - q_{+\perp}^* q_{-\perp}. \tag{9.243}$$

Photon polarizations:

$$\begin{aligned}
\not{\epsilon}_1^{\pm} &= N_1[\not{k}_1 \not{p}_- \not{p}_+(1 \pm \gamma_5) + \not{p}_+ \not{p}_- \not{k}_1(1 \mp \gamma_5)], \\
N_1^{-1} &= E^2[32k_{1+}k_{1-}]^{\frac{1}{2}}, \\
\not{\epsilon}_2^{\pm} &= N_2[\not{k}_2 \not{A}_+ \not{A}(1 \pm \gamma_5) + \not{A} \not{A}_+ \not{k}_2(1 \mp \gamma_5)], \\
N_2^{-1} &= q_{+0}^2[32k'_{2+}k'_{2-}]^{\frac{1}{2}}. \tag{9.244}
\end{aligned}$$

Nonvanishing helicity amplitudes:

$$M(+,+;+,+,+,+) = \frac{4e^4 E q_{+0} q_{-\perp}^* Z_{+-}}{q_{+-}q_{--}(p_- \cdot k_1)(q_+ \cdot k_2)} \left[\frac{k_{1+}k'_{2-}}{k_{1-}k'_{2+}q_{++}q_{-+}} \right]^{\frac{1}{2}}$$

$$\doteq \frac{e^4 E q_{+0}}{q_{+-}(p_- \cdot k_1)(q_+ \cdot k_2)} \left[\frac{32(q_+ \cdot q_-)k_{1+}k'_{2-}}{k_{1-}k'_{2+}} \right]^{\frac{1}{2}},$$

$$M(+,+;+,+,+,-) = \frac{e^4}{(2E - k_{1-})(p_- \cdot k_1)(q_+ \cdot k_2)}$$

$$\times \left\{ \frac{8E^2 q_{+0} q_{-\perp}^* Z_{+-}}{q_{--}} \left[\frac{k_{1+}k'_{2-}}{k_{1-}k'_{2+}q_{++}q_{-+}^3} \right]^{\frac{1}{2}} \right.$$

$$\left. - \frac{m^2 k_{1\perp}^* k_{2\perp}'^* q_{+\perp}}{4E q_{+0}} \left[\frac{k_{1-}k'_{2+}q_{-+}}{k_{1+}k'_{2-}q_{++}} \right]^{\frac{1}{2}} \right\},$$

$$M(+,+;+,+,-,+) = \frac{2e^4 E(2q_{+0} + k'_{2+}) q_{-\perp}^* Z_{+-}}{q_{--}(p_- \cdot k_1)(q_+ \cdot k_2)} \left[\frac{k_{1+}k'_{2-}}{k_{1-}k'_{2+}q_{++}q_{-+}^3} \right]^{\frac{1}{2}}$$

$$\doteq \frac{e^4 E(2q_{+0} + k'_{2+})}{q_{-+}(p_- \cdot k_1)(q_+ \cdot k_2)} \left[\frac{8(q_+ \cdot q_-)k_{1+}k'_{2-}}{k_{1-}k'_{2+}} \right]^{\frac{1}{2}},$$

9.19. $e^+e^- \to e^+e^-\gamma\gamma$ (NO Z-EXCHANGE; $m \neq 0$)

$$M(+,+;+,-,+,+) = \frac{e^4 m(2q_{+0} + k'_{2+})k^*_{1\perp}q_{-\perp}}{4E^2 q_{+-}(p_- \cdot k_1)(q_+ \cdot k_2)} \left[\frac{k_{1-}k'_{2-}q^3_{++}}{k_{1+}k'_{2+}q_{-+}}\right]^{\frac{1}{2}}$$

$$\doteq \frac{e^4 m(2q_{+0} + k'_{2+})k_{1-}}{4E^2 q_{+-}(p_- \cdot k_1)(q_+ \cdot k_2)} \left[\frac{k'_{2-}q^3_{++}q_{--}}{k'_{2+}}\right]^{\frac{1}{2}},$$

$$M(+,+;-,+,+,+) = -\frac{e^4 m}{(2E - k_{1-})(p_- \cdot k_1)(q_+ \cdot k_2)}$$

$$\times \left\{\frac{2E^2 k'_{2\perp}q^*_{-\perp}Z_{+-}}{q_{+0}q_{--}} \left[\frac{k_{1+}k'_{2+}}{k_{1-}k'_{2-}q_{++}q^3_{-+}}\right]^{\frac{1}{2}}\right.$$

$$\left. + \frac{q_{+0}k^*_{1\perp}q_{+\perp}}{E}\left[\frac{k_{1-}k'_{2-}q_{-+}}{k_{1+}k'_{2+}q_{++}}\right]^{\frac{1}{2}}\right\},$$

$$M(+,+;+,+,-,-) = \frac{4e^4 E q_{+0}q^*_{-\perp}Z_{+-}}{q_{--}(p_- \cdot k_1)(q_+ \cdot k_2)} \left[\frac{k_{1+}k'_{2-}}{k_{1-}k'_{2+}q_{++}q^3_{-+}}\right]^{\frac{1}{2}}$$

$$\doteq \frac{e^4 E q_{+0}}{q_{-+}(p_- \cdot k_1)(q_+ \cdot k_2)} \left[\frac{32(q_+ \cdot q_-)k_{1+}k'_{2-}}{k_{1-}k'_{2+}}\right]^{\frac{1}{2}},$$

$$M(+,+;+,-,+,-) = \frac{e^4 m q_{+0}k^*_{1\perp}q_{-\perp}}{2E^2 q_{+-}(p_- \cdot k_1)(q_+ \cdot k_2)} \left[\frac{k_{1-}k'_{2-}q^3_{++}}{k_{1+}k'_{2+}q_{-+}}\right]^{\frac{1}{2}}$$

$$\doteq \frac{e^4 m q_{+0}k_{1-}}{2E^2 q_{+-}(p_- \cdot k_1)(q_+ \cdot k_2)} \left[\frac{k'_{2-}q^3_{++}q_{--}}{k'_{2+}}\right]^{\frac{1}{2}},$$

$$M(+,+;-,+,+,-) = -\frac{e^4 m q_{+0}k^*_{1\perp}q_{+\perp}}{2E^2 q_{+-}(p_- \cdot k_1)(q_+ \cdot k_2)} \left[\frac{k_{1-}k'_{2-}q^3_{-+}}{k_{1+}k'_{2+}q_{++}}\right]^{\frac{1}{2}}$$

$$\doteq \frac{e^4 m q_{+0}k_{1-}}{2E^2(p_- \cdot k_1)(q_+ \cdot k_2)} \left[\frac{k'_{2-}q^3_{-+}}{k'_{2+}q_{+-}}\right]^{\frac{1}{2}},$$

$$M(+,+;-,+,-,+) = -\frac{e^4 m E k'_{2\perp}q^*_{-\perp}Z_{+-}}{q_{+0}q_{--}(p_- \cdot k_1)(q_+ \cdot k_2)} \left[\frac{k_{1+}k'_{2+}}{k_{1-}k'_{2-}q_{++}q^3_{-+}}\right]^{\frac{1}{2}}$$

$$\doteq \frac{e^4 m E k'_{2+}}{q_{+0}q_{-+}(p_- \cdot k_1)(q_+ \cdot k_2)} \left[\frac{2(q_+ \cdot q_-)k_{1+}}{k_{1-}}\right]^{\frac{1}{2}},$$

$$M(+,+;-,-,+,+) = -\frac{e^4 m^2 k_{1\perp}^* k_{2\perp}' q_{-\perp}}{8E^2 q_{+0} q_{+-}(p_- \cdot k_1)(q_+ \cdot k_2)} \left[\frac{k_{1-} k_{2+}' q_{++}^3}{k_{1+} k_{2-}' q_{-+}}\right]^{\frac{1}{2}}$$

$$\doteq \frac{e^4 m^2 k_{1-} k_{2+}' \left[q_{++}^3 q_{--}\right]^{\frac{1}{2}}}{8E^2 q_{+0} q_{+-}(p_- \cdot k_1)(q_+ \cdot k_2)},$$

$$M(+,-;+,+,+,-) = -\frac{e^4 m k_{2\perp}'^* q_{+\perp}}{2 q_{+0}(p_- \cdot k_1)(q_+ \cdot k_2)} \left[\frac{k_{1+} k_{2+}' q_{-+}}{k_{1-} k_{2-}' q_{++}}\right]^{\frac{1}{2}}$$

$$\doteq \frac{e^4 m k_{2+}'}{2 q_{+0}(p_- \cdot k_1)(q_+ \cdot k_2)} \left[\frac{k_{1+} q_{+-} q_{-+}}{k_{1-}}\right]^{\frac{1}{2}},$$

$$M(+,-;+,+,-,+) = -\frac{e^4 m q_{+0} k_{1\perp} q_{-\perp}^* Z_{+-}}{E q_{+-} q_{--}(p_- \cdot k_1)(q_+ \cdot k_2)} \left[\frac{k_{1-} k_{2-}'}{k_{1+} k_{2+}' q_{++} q_{-+}}\right]^{\frac{1}{2}}$$

$$\doteq \frac{e^4 m q_{+0} k_{1-}}{E q_{+-}(p_- \cdot k_1)(q_+ \cdot k_2)} \left[\frac{2(q_+ \cdot q_-) k_{2-}'}{k_{2+}'}\right]^{\frac{1}{2}},$$

$$M(+,-;+,-,+,+) = \frac{2 e^4 q_{+0} q_{--} q_{-\perp}}{q_{+-}(p_- \cdot k_1)(q_+ \cdot k_2)} \left[\frac{k_{1+} k_{2-}' q_{++}}{k_{1-} k_{2+}' q_{-+}}\right]^{\frac{1}{2}}$$

$$\doteq \frac{2 e^4 q_{+0}}{q_{+-}(p_- \cdot k_1)(q_+ \cdot k_2)} \left[\frac{k_{1+} k_{2-}' q_{++} q_{--}^3}{k_{1-} k_{2+}'}\right]^{\frac{1}{2}},$$

$$M(+,-;-,+,+,+) = -\frac{2 e^4 q_{+0} q_{+\perp}}{(p_- \cdot k_1)(q_+ \cdot k_2)} \left[\frac{k_{1+} k_{2-}' q_{-+}}{k_{1-} k_{2+}' q_{++}}\right]^{\frac{1}{2}}$$

$$\doteq \frac{2 e^4 q_{+0}}{(p_- \cdot k_1)(q_+ \cdot k_2)} \left[\frac{k_{1+} k_{2-}' q_{+-} q_{-+}}{k_{1-} k_{2+}'}\right]^{\frac{1}{2}},$$

$$M(-,+;+,+,+,-) = \frac{e^4 m k_{2\perp}'^* q_{-\perp}^*}{2 q_{+0} q_{+-}(p_- \cdot k_1)(q_+ \cdot k_2)} \left[\frac{k_{1+} k_{2+}' q_{++}^3}{k_{1-} k_{2-}' q_{-+}}\right]^{\frac{1}{2}}$$

$$\doteq \frac{e^4 m k_{2+}'}{2 q_{+0} q_{+-}(p_- \cdot k_1)(q_+ \cdot k_2)} \left[\frac{k_{1+} q_{++}^3 q_{--}}{k_{1-}}\right]^{\frac{1}{2}},$$

$$M(-,+;+,-,+,+) = -\frac{2 e^4 q_{+0} q_{+\perp}^*}{q_{+-}(p_- \cdot k_1)(q_+ \cdot k_2)} \left[\frac{k_{1+} k_{2-}' q_{-+}^3}{k_{1-} k_{2+}' q_{++}}\right]^{\frac{1}{2}}$$

$$\doteq \frac{2 e^4 q_{+0}}{(p_- \cdot k_1)(q_+ \cdot k_2)} \left[\frac{k_{1+} k_{2-}' q_{-+}^3}{k_{1-} k_{2+}' q_{+-}}\right]^{\frac{1}{2}},$$

9.19. $e^+e^- \to e^+e^-\gamma\gamma$ (NO Z-EXCHANGE; $m \neq 0$)

$$M(-,+;-,+,+,+) = \frac{2e^4 q_{+0} q_{-\perp}^*}{q_{+-}(p_-\cdot k_1)(q_+\cdot k_2)} \left[\frac{k_{1+}k'_{2-}q^3_{++}}{k_{1-}k'_{2+}q_{-+}}\right]^{\frac{1}{2}}$$

$$\doteq \frac{2e^4 q_{+0}}{q_{+-}(p_-\cdot k_1)(q_+\cdot k_2)} \left[\frac{k_{1+}k'_{2-}q^3_{++}q_{--}}{k_{1-}k'_{2+}}\right]^{\frac{1}{2}},$$

$$M(+,-;+,+,-,-) = -\frac{e^4 m}{(2E-k_{1-})(p_-\cdot k_1)(q_+\cdot k_2)}$$

$$\times \left\{ \frac{2q_{+0}k_{1\perp}q^*_{-\perp}Z_{+-}}{q_{--}} \left[\frac{k_{1-}k'_{2-}}{k_{1+}k'_{2+}q_{++}q^3_{-+}}\right]^{\frac{1}{2}} \right.$$

$$\left. + \frac{Ek'^*_{2\perp}q_{+\perp}}{q_{+0}} \left[\frac{k_{1+}k'_{2+}q_{-+}}{k_{1-}k'_{2-}q_{++}}\right]^{\frac{1}{2}} \right\},$$

$$M(+,-;+,-,+,-) = \frac{e^4 q_{+0}(2E-k_{1-})q_{-\perp}}{Eq_{+-}(p_-\cdot k_1)(q_+\cdot k_2)} \left[\frac{k_{1+}k'_{2-}q^3_{++}}{k_{1-}k'_{2+}q_{-+}}\right]^{\frac{1}{2}}$$

$$\doteq \frac{e^4 q_{+0}(2E-k_{1-})}{Eq_{+-}(p_-\cdot k_1)(q_+\cdot k_2)} \left[\frac{k_{1+}k'_{2-}q^3_{++}q_{--}}{k_{1-}k'_{2+}}\right]^{\frac{1}{2}},$$

$$M(+,-;+,-,-,+) = \frac{e^4(2q_{+0}+k'_{2+})q_{-\perp}}{q_{+-}(p_-\cdot k_1)(q_+\cdot k_2)} \left[\frac{k_{1+}k'_{2-}q^3_{++}}{k_{1-}k'_{2+}q_{-+}}\right]^{\frac{1}{2}}$$

$$\doteq \frac{e^4(2q_{+0}+k'_{2+})}{q_{+-}(p_-\cdot k_1)(q_+\cdot k_2)} \left[\frac{k_{1+}k'_{2-}q^3_{++}q_{--}}{k_{1-}k'_{2+}}\right]^{\frac{1}{2}},$$

$$M(+,-;-,+,+,-) = -\frac{e^4(2q_{+0}+k'_{2+})q_{+\perp}}{(p_-\cdot k_1)(q_+\cdot k_2)} \left[\frac{k_{1+}k'_{2-}q_{-+}}{k_{1-}k'_{2+}q_{++}}\right]^{\frac{1}{2}}$$

$$\doteq \frac{e^4(2q_{+0}+k'_{2+})}{(p_-\cdot k_1)(q_+\cdot k_2)} \left[\frac{k_{1+}k'_{2-}q_{+-}q_{-+}}{k_{1-}k'_{2+}}\right]^{\frac{1}{2}},$$

$$M(+,-;-,+,-,+) = \frac{e^4}{(2E-k_{1-})(p_-\cdot k_1)(q_+\cdot k_2)}$$

$$\times \left\{ -4Eq_{+0}q_{+\perp} \left[\frac{k_{1+}k'_{2-}q_{-+}}{k_{1-}k'_{2+}q_{++}}\right]^{\frac{1}{2}} \right.$$

$$\left. + \frac{m^2 k_{1\perp}k'_{2\perp}q^*_{-\perp}Z_{+-}}{2q_{+0}q_{--}} \left[\frac{k_{1-}k'_{2+}}{k_{1+}k'_{2-}q_{++}q^3_{-+}}\right]^{\frac{1}{2}} \right\},$$

$$M(+,-;-,-,+,+) = -\frac{e^4 m(2E-k_{1-})k'_{2\perp}q_{-\perp}}{4Eq_{+0}q_{+-}(p_-\cdot k_1)(q_+\cdot k_2)}\left[\frac{k_{1+}k'_{2+}q^3_{++}}{k_{1-}k'_{2-}q_{-+}}\right]^{\frac{1}{2}}$$

$$\doteq \frac{e^4 m(2E-k_{1-})k'_{2+}}{4Eq_{+0}q_{+-}(p_-\cdot k_1)(q_+\cdot k_2)}\left[\frac{k_{1+}q^3_{++}q_{--}}{k_{1-}}\right]^{\frac{1}{2}},$$

$$M(-,+;+,+,-,-) = \frac{e^4 m(2E-k_{1-})k'^*_{2\perp}q^*_{-\perp}}{4Eq_{+0}q_{+-}(p_-\cdot k_1)(q_+\cdot k_2)}\left[\frac{k_{1+}k'_{2+}q^3_{++}}{k_{1-}k'_{2-}q_{-+}}\right]^{\frac{1}{2}}$$

$$\doteq \frac{e^4 m(2E-k_{1-})k'_{2+}}{4Eq_{+0}q_{+-}(p_-\cdot k_1)(q_+\cdot k_2)}\left[\frac{k_{1+}q^3_{++}q_{--}}{k_{1-}}\right]^{\frac{1}{2}},$$

$$M(-,+;+,-,+,-) = \frac{e^4}{(2E-k_{1-})(p_-\cdot k_1)(q_+\cdot k_2)}$$

$$\times\left\{-4Eq_{+0}q^*_{+\perp}\left[\frac{k_{1+}k'_{2-}q_{-+}}{k_{1-}k'_{2+}q_{++}}\right]^{\frac{1}{2}}\right.$$

$$\left.+\frac{m^2 k^*_{1\perp}k'^*_{2\perp}q_{-\perp}Z^*_{+-}}{2q_{+0}q_{--}}\left[\frac{k_{1-}k'_{2+}}{k_{1+}k'_{2-}q_{++}q^3_{-+}}\right]^{\frac{1}{2}}\right\},$$

$$M(-,+;+,-,-,+) = -\frac{e^4(2q_{+0}+k'_{2+})q^*_{+\perp}}{(p_-\cdot k_1)(q_+\cdot k_2)}\left[\frac{k_{1+}k'_{2-}q_{-+}}{k_{1-}k'_{2+}q_{++}}\right]^{\frac{1}{2}}$$

$$\doteq \frac{e^4(2q_{+0}+k'_{2+})}{(p_-\cdot k_1)(q_+\cdot k_2)}\left[\frac{k_{1+}k'_{2-}q_{+-}q_{-+}}{k_{1-}k'_{2+}}\right]^{\frac{1}{2}},$$

$$M(-,+;-,+,+,-) = \frac{e^4(2q_{+0}+k'_{2+})q^*_{-\perp}}{q_{+-}(p_-\cdot k_1)(q_+\cdot k_2)}\left[\frac{k_{1+}k'_{2-}q^3_{++}}{k_{1-}k'_{2+}q_{-+}}\right]^{\frac{1}{2}}$$

$$\doteq \frac{e^4(2q_{+0}+k'_{2+})}{q_{+-}(p_-\cdot k_1)(q_+\cdot k_2)}\left[\frac{k_{1+}k'_{2-}q^3_{++}q_{--}}{k_{1-}k'_{2+}}\right]^{\frac{1}{2}},$$

$$M(-,+;-,+,-,+) = \frac{e^4 q_{+0}(2E-k_{1-})q^*_{-\perp}}{Eq_{+-}(p_-\cdot k_1)(q_+\cdot k_2)}\left[\frac{k_{1+}k'_{2-}q^3_{++}}{k_{1-}k'_{2+}q_{-+}}\right]^{\frac{1}{2}}$$

$$\doteq \frac{e^4 q_{+0}(2E-k_{1-})}{Eq_{+-}(p_-\cdot k_1)(q_+\cdot k_2)}\left[\frac{k_{1+}k'_{2-}q^3_{++}q_{--}}{k_{1-}k'_{2+}}\right]^{\frac{1}{2}},$$

$$M(+,-;+,-,-,-) = \frac{2e^4 q_{+0}q_{-\perp}}{q_{+-}(p_-\cdot k_1)(q_+\cdot k_2)}\left[\frac{k_{1+}k'_{2-}q^3_{++}}{k_{1-}k'_{2+}q_{-+}}\right]^{\frac{1}{2}}$$

$$\doteq \frac{2e^4 q_{+0}}{q_{+-}(p_-\cdot k_1)(q_+\cdot k_2)}\left[\frac{k_{1+}k'_{2-}q^3_{++}q_{--}}{k_{1-}k'_{2+}}\right]^{\frac{1}{2}},$$

9.19. $e^+e^- \to e^+e^-\gamma\gamma$ (NO Z-EXCHANGE; $m \neq 0$)

$$M(-,+;-,-,+,+) = \frac{e^4 m}{(2E - k_{1-})(p_- \cdot k_1)(q_+ \cdot k_2)}$$

$$\times \left\{ \frac{E k'_{2\perp} q^*_{+\perp}}{q_{+0}} \left[\frac{k_{1+} k'_{2+} q_{-+}}{k_{1-} k'_{2-} q_{++}} \right]^{\frac{1}{2}} \right.$$

$$\left. + \frac{2 q_{+0} k^*_{1\perp} q_{-\perp} Z^*_{+-}}{q_{--}} \left[\frac{k_{1-} k'_{2-}}{k_{1+} k'_{2+} q_{++} q^3_{-+}} \right]^{\frac{1}{2}} \right\},$$

$$M(+,-;-,+,-,-) = -\frac{2e^4 q_{+0} q_{+\perp}}{q_{+-}(p_- \cdot k_1)(q_+ \cdot k_2)} \left[\frac{k_{1+} k'_{2-} q^3_{-+}}{k_{1-} k'_{2+} q_{++}} \right]^{\frac{1}{2}}$$

$$\doteq \frac{2e^4 q_{+0}}{(p_- \cdot k_1)(q_+ \cdot k_2)} \left[\frac{k_{1+} k'_{2-} q^3_{-+}}{k_{1-} k'_{2+} q_{+-}} \right]^{\frac{1}{2}},$$

$$M(+,-;-,-,-,+) = -\frac{e^4 m k'_{2\perp} q_{-\perp}}{2 q_{+0} q_{+-}(p_- \cdot k_1)(q_+ \cdot k_2)} \left[\frac{k_{1+} k'_{2+} q^3_{++}}{k_{1-} k'_{2-} q_{-+}} \right]^{\frac{1}{2}}$$

$$\doteq \frac{e^4 m k'_{2+}}{2 q_{+0} q_{+-}(p_- \cdot k_1)(q_+ \cdot k_2)} \left[\frac{k_{1+} q^3_{++} q_{--}}{k_{1-}} \right]^{\frac{1}{2}},$$

$$M(-,+;+,-,-,-) = -\frac{2e^4 q_{+0} q^*_{+\perp}}{(p_- \cdot k_1)(q_+ \cdot k_2)} \left[\frac{k_{1+} k'_{2-} q_{-+}}{k_{1-} k'_{2+} q_{++}} \right]^{\frac{1}{2}}$$

$$\doteq \frac{2e^4 q_{+0}}{(p_- \cdot k_1)(q_+ \cdot k_2)} \left[\frac{k_{1+} k'_{2-} q_{+-} q_{-+}}{k_{1-} k'_{2+}} \right]^{\frac{1}{2}},$$

$$M(-,+;-,+,-,-) = \frac{2e^4 q_{+0} q_{--} q^*_{-\perp}}{q_{+-}(p_- \cdot k_1)(q_+ \cdot k_2)} \left[\frac{k_{1+} k'_{2-} q_{++}}{k_{1-} k'_{2+} q_{-+}} \right]^{\frac{1}{2}}$$

$$\doteq \frac{2e^4 q_{+0}}{q_{+-}(p_- \cdot k_1)(q_+ \cdot k_2)} \left[\frac{k_{1+} k'_{2-} q_{++} q^3_{--}}{k_{1-} k'_{2+}} \right]^{\frac{1}{2}},$$

$$M(-,+;-,-,+,-) = \frac{e^4 m q_{+0} k^*_{1\perp} q_{-\perp} Z^*_{+-}}{E q_{+-} q_{--}(p_- \cdot k_1)(q_+ \cdot k_2)} \left[\frac{k_{1-} k'_{2-}}{k_{1+} k'_{2+} q_{++} q_{-+}} \right]^{\frac{1}{2}}$$

$$\doteq \frac{e^4 m q_{+0} k_{1-}}{E q_{+-}(p_- \cdot k_1)(q_+ \cdot k_2)} \left[\frac{2(q_+ \cdot q_-) k'_{2-}}{k'_{2+}} \right]^{\frac{1}{2}},$$

$$M(-,+;-,-,-,+) = \frac{e^4 m k'_{2\perp} q^*_{+\perp}}{2 q_{+0}(p_- \cdot k_1)(q_+ \cdot k_2)} \left[\frac{k_{1+} k'_{2+} q_{-+}}{k_{1-} k'_{2-} q_{++}} \right]^{\frac{1}{2}}$$

$$\doteq \frac{e^4 m k'_{2+}}{2 q_{+0}(p_- \cdot k_1)(q_+ \cdot k_2)} \left[\frac{k_{1+} q_{+-} q_{-+}}{k_{1-}} \right]^{\frac{1}{2}},$$

$$M(-,-;+,+,-,-) = -\frac{e^4 m^2 k_{1\perp} k'^{*}_{2\perp} q^{*}_{-\perp}}{8E^2 q_{+0} q_{+-} (p_- \cdot k_1)(q_+ \cdot k_2)} \left[\frac{k_{1-} k'_{2+} q^3_{++}}{k_{1+} k'_{2-} q_{-+}}\right]^{\frac{1}{2}}$$

$$\doteq \frac{e^4 m^2 k_{1-} k'_{2+} \left[q^3_{++} q_{--}\right]^{\frac{1}{2}}}{8E^2 q_{+0} q_{+-} (p_- \cdot k_1)(q_+ \cdot k_2)},$$

$$M(-,-;+,-,+,-) = \frac{e^4 m E k'^{*}_{2\perp} q_{-\perp} Z^{*}_{+-}}{q_{+0} q_{--} (p_- \cdot k_1)(q_+ \cdot k_2)} \left[\frac{k_{1+} k'_{2+}}{k_{1-} k'_{2-} q_{++} q^3_{-+}}\right]^{\frac{1}{2}}$$

$$\doteq \frac{e^4 m E k'_{2+}}{q_{+0} q_{-+} (p_- \cdot k_1)(q_+ \cdot k_2)} \left[\frac{2(q_+ \cdot q_-) k_{1+}}{k_{1-}}\right]^{\frac{1}{2}},$$

$$M(-,-;+,-,-,+) = \frac{e^4 m q_{+0} k_{1\perp} q^{*}_{+\perp}}{2E^2 q_{+-} (p_- \cdot k_1)(q_+ \cdot k_2)} \left[\frac{k_{1-} k'_{2-} q^3_{-+}}{k_{1+} k'_{2+} q_{++}}\right]^{\frac{1}{2}}$$

$$\doteq \frac{e^4 m q_{+0} k_{1-}}{2E^2 (p_- \cdot k_1)(q_+ \cdot k_2)} \left[\frac{k'_{2-} q^3_{-+}}{k'_{2+} q_{+-}}\right]^{\frac{1}{2}},$$

$$M(-,-;-,+,-,+) = -\frac{e^4 m q_{+0} k_{1\perp} q^{*}_{-\perp}}{2E^2 q_{+-} (p_- \cdot k_1)(q_+ \cdot k_2)} \left[\frac{k_{1-} k'_{2-} q^3_{++}}{k_{1+} k'_{2+} q_{-+}}\right]^{\frac{1}{2}}$$

$$\doteq \frac{e^4 m q_{+0} k_{1-}}{2E^2 q_{+-} (p_- \cdot k_1)(q_+ \cdot k_2)} \left[\frac{k'_{2-} q^3_{++} q_{--}}{k'_{2+}}\right]^{\frac{1}{2}},$$

$$M(-,-;-,-,+,+) = \frac{4 e^4 q_{+0} q_{-\perp} Z^{*}_{+-}}{q_{--} (p_- \cdot k_1)(q_+ \cdot k_2)} \left[\frac{k_{1+} k'_{2-}}{k_{1-} k'_{2+} q_{++} q^3_{-+}}\right]^{\frac{1}{2}}$$

$$\doteq \frac{e^4 E q_{+0}}{q_{-+} (p_- \cdot k_1)(q_+ \cdot k_2)} \left[\frac{32(q_+ \cdot q_-) k_{1+} k'_{2-}}{k_{1-} k'_{2+}}\right]^{\frac{1}{2}},$$

$$M(-,-;-,+,-,-) = -\frac{e^4 m (2q_{+0} + k'_{2+}) k_{1\perp} q^{*}_{-\perp}}{4E^2 q_{+-} (p_- \cdot k_1)(q_+ \cdot k_2)} \left[\frac{k_{1-} k'_{2-} q^3_{++}}{k_{1+} k'_{2+} q_{-+}}\right]^{\frac{1}{2}}$$

$$\doteq \frac{e^4 m (2q_{+0} + k'_{2+}) k_{1-}}{4E^2 q_{+-} (p_- \cdot k_1)(q_+ \cdot k_2)} \left[\frac{k'_{2-} q^3_{++} q_{--}}{k'_{2+}}\right]^{\frac{1}{2}},$$

$$M(-,-;-,-,+,-) = \frac{2 e^4 E (2q_{+0} + k'_{2+}) q_{-\perp} Z^{*}_{+-}}{q_{--} (p_- \cdot k_1)(q_+ \cdot k_2)} \left[\frac{k_{1+} k'_{2-}}{k_{1-} k'_{2+} q_{++} q^3_{-+}}\right]^{\frac{1}{2}}$$

$$\doteq \frac{e^4 E (2q_{+0} + k'_{2+})}{q_{-+} (p_- \cdot k_1)(q_+ \cdot k_2)} \left[\frac{8(q_+ \cdot q_-) k_{1+} k'_{2-}}{k_{1-} k'_{2+}}\right]^{\frac{1}{2}},$$

9.19. $e^+e^- \to e^+e^- \gamma\gamma$ (NO Z-EXCHANGE; $m \neq 0$)

$$M(-,-;+,-,-,-) = \frac{e^4 m}{(2E - k_{1-})(p_- \cdot k_1)(q_+ \cdot k_2)}$$

$$\times \left\{ \frac{q_{+0} k_{1\perp} q_{+\perp}^*}{E} \left[\frac{k_{1-} k_{2-}' q_{-+}}{k_{1+} k_{2+}' q_{++}} \right]^{\frac{1}{2}} \right.$$

$$\left. + \frac{2E^2 k_{2\perp}'^* q_{-\perp} Z_{+-}^*}{q_{+0} q_{--}} \left[\frac{k_{1+} k_{2+}'}{k_{1-} k_{2-}' q_{++} q_{-+}^3} \right]^{\frac{1}{2}} \right\},$$

$$M(-,-;-,-,-,+) = \frac{e^4}{(2E - k_{1-})(p_- \cdot k_1)(q_+ \cdot k_2)}$$

$$\times \left\{ \frac{8E^2 q_{+0} q_{-\perp} Z_{+-}^*}{q_{--}} \left[\frac{k_{1+} k_{2-}'}{k_{1-} k_{2+}' q_{++} q_{-+}^3} \right]^{\frac{1}{2}} \right.$$

$$\left. - \frac{m^2 k_{1\perp} k_{2\perp}' q_{+\perp}^*}{4E q_{+0}} \left[\frac{k_{1-} k_{2+}' q_{-+}}{k_{1+} k_{2-}' q_{++}} \right]^{\frac{1}{2}} \right\},$$

$$M(-,-;-,-,-,-) = \frac{4e^4 E q_{+0} q_{-\perp} Z_{+-}^*}{q_{+-} q_{--} (p_- \cdot k_1)(q_+ \cdot k_2)} \left[\frac{k_{1+} k_{2-}'}{k_{1-} k_{2+}' q_{++} q_{-+}} \right]^{\frac{1}{2}}$$

$$\doteq \frac{e^4 E q_{+0}}{q_{+-} (p_- \cdot k_1)(q_+ \cdot k_2)} \left[\frac{32(q_+ \cdot q_-) k_{1+} k_{2-}'}{k_{1-} k_{2+}'} \right]^{\frac{1}{2}}.$$

(9.245)

Unpolarized squared matrix element:

$$\overline{|M|^2} = \frac{e^8 k_{1+} k_{2-}' \left[4(q_+ \cdot q_-)^2 + q_{++}^2 q_{--}^2 + q_{+-}^2 q_{-+}^2 \right]}{2 k_{1-} k_{2+}' q_{+-} q_{-+} [(2E - k_{1-})(p_- \cdot k_1)(q_+ \cdot k_2)]^2}$$

$$\times \left[4E^2 + (2E - k_{1-})^2 + \frac{m^2 k_{1-}^3}{4E^2 k_{1+}} \right] \left[4q_{+0}^2 + (2q_{+0} + k_{2+}')^2 + \frac{m^2 k_{2+}'^3}{4 q_{+0}^2 k_{2-}'} \right].$$

(9.246)

Unpolarized cross section:

$$d\sigma = \frac{\alpha^4 k_{1+} k_{2-}' \left[4(q_+ \cdot q_-)^2 + q_{++}^2 q_{--}^2 + q_{+-}^2 q_{-+}^2 \right]}{256 \pi^2 k_{1-} k_{2+}' q_{+-} q_{-+} [E(2E - k_{1-})(p_- \cdot k_1)(q_+ \cdot k_2)]^2}$$

$$\times \left[4E^2 + (2E - k_{1-})^2 + \frac{m^2 k_{1-}^3}{4E^2 k_{1+}} \right] \left[4q_{+0}^2 + (2q_{+0} + k_{2+}')^2 + \frac{m^2 k_{2+}'^3}{4 q_{+0}^2 k_{2-}'} \right]$$

$$\times \delta^4(p_+ + p_- - q_+ - q_- - k_1 - k_2) \frac{d^3\vec{q}_+ \, d^3\vec{q}_- \, d^3\vec{k}_1 \, d^3\vec{k}_2}{q_{+0} \, q_{-0} \, k_{10} \, k_{20}}.$$

(9.247)

9.19.9 $\vec{k}_1$ and $\vec{k}_2$ nearly parallel to $\vec{p}_-$ and $\vec{q}_-$, resp.

The doubly primed quantities $k''_{2\pm}$ and $k''_{2\perp}$ are evaluated in the rotated frame where $\vec{q}_-$ determines the positive z-axis (see also Section 7.4.3). The four-vector q in eqn (9.249) is obtained by applying a space reflection to q_-.
Definition:

$$Z_{+-} = q_{++}q_{--} - q^*_{+\perp}q_{-\perp}. \tag{9.248}$$

Photon polarizations:

$$\not{\epsilon}_1^\pm = N_1[\not{k}_1 \not{p}_- \not{p}_+(1\pm\gamma_5) + \not{p}_+ \not{p}_- \not{k}_1(1\mp\gamma_5)],$$

$$N_1^{-1} = E^2[32k_{1+}k_{1-}]^{\frac{1}{2}},$$

$$\not{\epsilon}_2^\pm = N_2[\not{k}_2 \not{q}_- \not{q}(1\pm\gamma_5) + \not{q} \not{q}_- \not{k}_2(1\mp\gamma_5)],$$

$$N_2^{-1} = q^2_{-0}[32k''_{2+}k''_{2-}]^{\frac{1}{2}}. \tag{9.249}$$

Nonvanishing helicity amplitudes:

$$M(+,+;+,+,+,+) = -\frac{2e^4 E(2q_{-0}+k''_{2+})q^*_{-\perp}Z_{+-}}{q_{+-}q_{--}(p_-\cdot k_1)(q_-\cdot k_2)}\left[\frac{k_{1+}k''_{2-}}{k_{1-}k''_{2+}q_{++}q_{-+}}\right]^{\frac{1}{2}}$$

$$\doteq \frac{e^4 E(2q_{-0}+k''_{2+})}{q_{+-}(p_-\cdot k_1)(q_-\cdot k_2)}\left[\frac{8(q_+\cdot q_-)k_{1+}k''_{2-}}{k_{1-}k''_{2+}}\right]^{\frac{1}{2}},$$

$$M(+,+;+,+,+,-) = -\frac{e^4}{q_{+-}(p_-\cdot k_1)(q_-\cdot k_2)}$$

$$\times \left\{\frac{4Eq_{-0}q^*_{-\perp}Z_{+-}}{q_{--}}\left[\frac{k_{1+}k''_{2-}}{k_{1-}k''_{2+}q_{++}q_{-+}}\right]^{\frac{1}{2}}\right.$$

$$\left.+\frac{m^2 k^*_{1\perp}k''^*_{2\perp}q_{-\perp}}{8E^2 q_{-0}}\left[\frac{k_{1-}k''_{2+}q^3_{++}}{k_{1+}k''_{2-}q_{-+}}\right]^{\frac{1}{2}}\right\},$$

$$M(+,+;+,+,-,+) = -\frac{4e^4 Eq_{-0}q^*_{-\perp}Z_{+-}}{q_{--}(p_-\cdot k_1)(q_-\cdot k_2)}\left[\frac{k_{1+}k''_{2-}}{k_{1-}k''_{2+}q_{++}q^3_{-+}}\right]^{\frac{1}{2}}$$

$$\doteq \frac{e^4 Eq_{-0}}{q_{-+}(p_-\cdot k_1)(q_-\cdot k_2)}\left[\frac{32(q_+\cdot q_-)k_{1+}k''_{2-}}{k_{1-}k''_{2+}}\right]^{\frac{1}{2}},$$

9.19. $e^+e^- \to e^+e^- \gamma\gamma$ (NO Z-EXCHANGE; $m \neq 0$)

$$M(+,+;+,-,+,+) = \frac{e^4 m}{q_{+-}(p_- \cdot k_1)(q_- \cdot k_2)}$$

$$\times \left\{ \frac{E k_{2\perp}'' q_{-\perp}^* Z_{+-}}{q_{-0} q_{--}} \left[\frac{k_{1+} k_{2+}''}{k_{1-} k_{2-}'' q_{++} q_{-+}} \right]^{\frac{1}{2}} \right.$$

$$\left. - \frac{q_{-0} k_{1\perp}^* q_{-\perp}}{2E^2} \left[\frac{k_{1-} k_{2-}'' q_{++}^3}{k_{1+} k_{2+}'' q_{-+}} \right]^{\frac{1}{2}} \right\},$$

$$M(+,+;-,+,+,+) = \frac{e^4 m q_{-0} k_{1\perp}^* q_{+\perp}}{2E^2 q_{--}(p_- \cdot k_1)(q_- \cdot k_2)} \left[\frac{k_{1-} k_{2-}'' q_{++} q_{-+}}{k_{1+} k_{2+}''} \right]^{\frac{1}{2}}$$

$$\doteq \frac{e^4 m q_{-0} q_{++} k_{1-}}{2E^2 q_{--}(p_- \cdot k_1)(q_- \cdot k_2)} \left[\frac{k_{2-}'' q_{+-} q_{-+}}{k_{2+}''} \right]^{\frac{1}{2}},$$

$$M(+,+;+,+,-,-) = -\frac{2 e^4 q_{-0}(2E - k_{1-}) q_{-\perp}^* Z_{+-}}{q_{+-} q_{--}(p_- \cdot k_1)(q_- \cdot k_2)} \left[\frac{k_{1+} k_{2-}''}{k_{1-} k_{2+}'' q_{++} q_{-+}} \right]^{\frac{1}{2}}$$

$$\doteq \frac{e^4 q_{-0}(2E - k_{1-})}{q_{+-}(p_- \cdot k_1)(q_- \cdot k_2)} \left[\frac{8(q_+ \cdot q_-) k_{1+} k_{2-}''}{k_{1-} k_{2+}''} \right]^{\frac{1}{2}},$$

$$M(+,+;+,-,+,-) = -\frac{e^4 m (2q_{-0} + k_{2+}'') k_{1\perp}^* q_{-\perp}}{4 E^2 q_{+-}(p_- \cdot k_1)(q_- \cdot k_2)} \left[\frac{k_{1-} k_{2-}'' q_{++}^3}{k_{1+} k_{2+}'' q_{-+}} \right]^{\frac{1}{2}}$$

$$\doteq \frac{e^4 m (2q_{-0} + k_{2+}'') k_{1-}}{4 E^2 q_{+-}(p_- \cdot k_1)(q_- \cdot k_2)} \left[\frac{k_{2-}'' q_{++}^3 q_{--}}{k_{2+}''} \right]^{\frac{1}{2}},$$

$$M(+,+;+,-,-,+) = \frac{e^4 m (2E - k_{1-}) k_{2\perp}'' q_{-\perp}^* Z_{+-}}{2 q_{-0} q_{+-} q_{--}(p_- \cdot k_1)(q_- \cdot k_2)} \left[\frac{k_{1+} k_{2+}''}{k_{1-} k_{2-}'' q_{++} q_{-+}} \right]^{\frac{1}{2}}$$

$$\doteq \frac{e^4 m (2E - k_{1-}) k_{2+}''}{q_{-0} q_{+-}(p_- \cdot k_1)(q_- \cdot k_2)} \left[\frac{(q_+ \cdot q_-) k_{1+}}{2 k_{1-}} \right]^{\frac{1}{2}},$$

$$M(+,+;-,+,+,-) = \frac{e^4 m q_{-0} k_{1\perp}^* q_{+\perp}}{E(2E - k_{1-})(p_- \cdot k_1)(q_- \cdot k_2)} \left[\frac{k_{1-} k_{2-}'' q_{-+}}{k_{1+} k_{2+}'' q_{++}} \right]^{\frac{1}{2}}$$

$$\doteq \frac{e^4 m q_{-0} k_{1-}}{E(2E - k_{1-})(p_- \cdot k_1)(q_- \cdot k_2)} \left[\frac{k_{2-}'' q_{+-} q_{-+}}{k_{2+}''} \right]^{\frac{1}{2}},$$

$$M(+,+;-,-,+,+) = \frac{-e^4 m^2 k_{1\perp}^* k_{2\perp}'' q_{+\perp}}{4 E q_{-0}(2E - k_{1-})(p_- \cdot k_1)(q_- \cdot k_2)} \left[\frac{k_{1-} k_{2+}'' q_{-+}}{k_{1+} k_{2-}'' q_{++}} \right]^{\frac{1}{2}}$$

$$\doteq \frac{e^4 m^2 k_{1-} k_{2+}'' [q_{+-} q_{-+}]^{\frac{1}{2}}}{4 E q_{-0}(2E - k_{1-})(p_- \cdot k_1)(q_- \cdot k_2)},$$

$$M(+,-;+,+,+,-) = -\frac{e^4 m(2E - k_{1-})k_{2\perp}^{\prime\prime*}q_{-\perp}}{4Eq_{-0}q_{+-}(p_- \cdot k_1)(q_- \cdot k_2)}\left[\frac{k_{1+}k_{2+}^{\prime\prime}q_{++}^3}{k_{1-}k_{2-}^{\prime\prime}q_{-+}}\right]^{\frac{1}{2}}$$

$$\doteq \frac{e^4 m(2E - k_{1-})k_{2+}^{\prime\prime}}{4Eq_{-0}q_{+-}(p_- \cdot k_1)(q_- \cdot k_2)}\left[\frac{k_{1+}q_{++}^3 q_{--}}{k_{1-}}\right]^{\frac{1}{2}},$$

$$M(+,-;+,+,-,+) = \frac{e^4 m(2q_{-0} + k_{2+}^{\prime\prime})k_{1\perp}q_{-\perp}^* Z_{+-}}{2Eq_{+-}q_{--}(p_- \cdot k_1)(q_- \cdot k_2)}\left[\frac{k_{1-}k_{2-}^{\prime\prime}}{k_{1+}k_{2+}^{\prime\prime}q_{++}q_{-+}}\right]^{\frac{1}{2}}$$

$$\doteq \frac{e^4 m(2q_{-0} + k_{2+}^{\prime\prime})k_{1-}}{Eq_{+-}(p_- \cdot k_1)(q_- \cdot k_2)}\left[\frac{(q_+ \cdot q_-)k_{2-}^{\prime\prime}}{2k_{2+}^{\prime\prime}}\right]^{\frac{1}{2}},$$

$$M(+,-;+,-,+,+) = -\frac{e^4 q_{-0}(2E - k_{1-})q_{-\perp}}{Eq_{+-}(p_- \cdot k_1)(q_- \cdot k_2)}\left[\frac{k_{1+}k_{2-}^{\prime\prime}q_{++}^3}{k_{1-}k_{2+}^{\prime\prime}q_{-+}}\right]^{\frac{1}{2}}$$

$$\doteq \frac{e^4 q_{-0}(2E - k_{1-})}{Eq_{+-}(p_- \cdot k_1)(q_- \cdot k_2)}\left[\frac{k_{1+}k_{2-}^{\prime\prime}q_{++}^3 q_{--}}{k_{1-}k_{2+}^{\prime\prime}}\right]^{\frac{1}{2}},$$

$$M(+,-;-,+,+,+) = \frac{e^4(2q_{-0} + k_{2+}^{\prime\prime})q_{+\perp}}{(p_- \cdot k_1)(q_- \cdot k_2)}\left[\frac{k_{1+}k_{2-}^{\prime\prime}q_{-+}}{k_{1-}k_{2+}^{\prime\prime}q_{++}}\right]^{\frac{1}{2}}$$

$$\doteq \frac{e^4(2q_{-0} + k_{2+}^{\prime\prime})}{(p_- \cdot k_1)(q_- \cdot k_2)}\left[\frac{k_{1+}k_{2-}^{\prime\prime}q_{+-}q_{-+}}{k_{1-}k_{2+}^{\prime\prime}}\right]^{\frac{1}{2}},$$

$$M(-,+;+,+,+,-) = \frac{e^4 m E k_{2\perp}^{\prime\prime*}q_{+\perp}^*}{q_{-0}(2E - k_{1-})(p_- \cdot k_1)(q_- \cdot k_2)}\left[\frac{k_{1+}k_{2+}^{\prime\prime}q_{-+}}{k_{1-}k_{2-}^{\prime\prime}q_{++}}\right]^{\frac{1}{2}}$$

$$\doteq \frac{e^4 m E k_{2+}^{\prime\prime}}{q_{-0}(2E - k_{1-})(p_- \cdot k_1)(q_- \cdot k_2)}\left[\frac{k_{1+}q_{+-}q_{-+}}{k_{1-}}\right]^{\frac{1}{2}},$$

$$M(-,+;+,-,+,+) = \frac{4e^4 E q_{-0}q_{+\perp}^*}{(2E - k_{1-})(p_- \cdot k_1)(q_- \cdot k_2)}\left[\frac{k_{1+}k_{2-}^{\prime\prime}q_{-+}}{k_{1-}k_{2+}^{\prime\prime}q_{++}}\right]^{\frac{1}{2}}$$

$$\doteq \frac{4e^4 E q_{-0}}{(2E - k_{1-})(p_- \cdot k_1)(q_- \cdot k_2)}\left[\frac{k_{1+}k_{2-}^{\prime\prime}q_{+-}q_{-+}}{k_{1-}k_{2+}^{\prime\prime}}\right]^{\frac{1}{2}},$$

$$M(-,+;-,+,+,+) = -\frac{e^4(2q_{-0} + k_{2+}^{\prime\prime})q_{-\perp}^*}{q_{+-}(p_- \cdot k_1)(q_- \cdot k_2)}\left[\frac{k_{1+}k_{2-}^{\prime\prime}q_{++}^3}{k_{1-}k_{2+}^{\prime\prime}q_{-+}}\right]^{\frac{1}{2}}$$

$$\doteq \frac{e^4(2q_{-0} + k_{2+}^{\prime\prime})}{q_{+-}(p_- \cdot k_1)(q_- \cdot k_2)}\left[\frac{k_{1+}k_{2-}^{\prime\prime}q_{++}^3 q_{--}}{k_{1-}k_{2+}^{\prime\prime}}\right]^{\frac{1}{2}},$$

9.19. $e^+e^- \to e^+e^-\gamma\gamma$ (NO Z-EXCHANGE; $m \neq 0$)

$$M(+,-;+,+,-,-) = \frac{e^4 m}{q_{+-}(p_- \cdot k_1)(q_- \cdot k_2)}$$

$$\times \left\{ \frac{q_{-0} k_{1\perp} q_{-\perp}^* Z_{+-}}{E q_{--}} \left[\frac{k_{1-} k_{2-}''}{k_{1+} k_{2+}'' q_{++} q_{-+}} \right]^{\frac{1}{2}} \right.$$

$$\left. - \frac{k_{2\perp}''^* q_{-\perp}}{2 q_{-0}} \left[\frac{k_{1+} k_{2+}'' q_{++}^3}{k_{1-} k_{2-}'' q_{--}} \right]^{\frac{1}{2}} \right\},$$

$$M(+,-;+,-,+,-) = -\frac{2 e^4 q_{-0} q_{-\perp}}{(p_- \cdot k_1)(q_- \cdot k_2)} \left[\frac{k_{1+} k_{2-}'' q_{++}^3}{k_{1-} k_{2+}'' q_{-+}^3} \right]^{\frac{1}{2}}$$

$$\doteq \frac{2 e^4 q_{-0}}{q_{-+}(p_- \cdot k_1)(q_- \cdot k_2)} \left[\frac{k_{1+} k_{2-}'' q_{++}^3 q_{--}}{k_{1-} k_{2+}''} \right]^{\frac{1}{2}},$$

$$M(+,-;+,-,-,+) = -\frac{e^4}{q_{+-}(p_- \cdot k_1)(q_- \cdot k_2)}$$

$$\times \left\{ 2 q_{-0} q_{-\perp} \left[\frac{k_{1+} k_{2-}'' q_{++}^3}{k_{1-} k_{2+}'' q_{-+}} \right]^{\frac{1}{2}} \right.$$

$$\left. + \frac{m^2 k_{1\perp} k_{2\perp}'' q_{-\perp}^* Z_{+-}}{4 E q_{-0} q_{--}} \left[\frac{k_{1-} k_{2+}''}{k_{1+} k_{2-}'' q_{++} q_{-+}} \right]^{\frac{1}{2}} \right\},$$

$$M(+,-;-,+,+,-) = \frac{2 e^4 q_{-0} q_{+\perp}}{(p_- \cdot k_1)(q_- \cdot k_2)} \left[\frac{k_{1+} k_{2-}'' q_{-+}}{k_{1-} k_{2+}'' q_{++}} \right]^{\frac{1}{2}}$$

$$\doteq \frac{2 e^4 q_{-0}}{(p_- \cdot k_1)(q_- \cdot k_2)} \left[\frac{k_{1+} k_{2-}'' q_{+-} q_{-+}}{k_{1-} k_{2+}''} \right]^{\frac{1}{2}},$$

$$M(+,-;-,+,-,+) = \frac{2 e^4 q_{-0} q_{+\perp}}{q_{--}(p_- \cdot k_1)(q_- \cdot k_2)} \left[\frac{k_{1+} k_{2-}'' q_{++} q_{-+}}{k_{1-} k_{2+}''} \right]^{\frac{1}{2}}$$

$$\doteq \frac{2 e^4 q_{-0} q_{++}}{q_{--}(p_- \cdot k_1)(q_- \cdot k_2)} \left[\frac{k_{1+} k_{2-}'' q_{+-} q_{-+}}{k_{1-} k_{2+}''} \right]^{\frac{1}{2}},$$

$$M(+,-;-,-,+,+) = -\frac{e^4 m k_{2\perp}'' q_{+\perp}}{2 q_{-0}(p_- \cdot k_1)(q_- \cdot k_2)} \left[\frac{k_{1+} k_{2+}'' q_{-+}}{k_{1-} k_{2-}'' q_{++}} \right]^{\frac{1}{2}}$$

$$\doteq \frac{e^4 m k_{2+}''}{2 q_{-0}(p_- \cdot k_1)(q_- \cdot k_2)} \left[\frac{k_{1+} q_{+-} q_{-+}}{k_{1-}} \right]^{\frac{1}{2}},$$

$$M(-,+;+,+,-,-) = \frac{e^4 m k_{2\perp}^{\prime\prime\prime*} q_{+\perp}^*}{2q_{-0}(p_- \cdot k_1)(q_- \cdot k_2)} \left[\frac{k_{1+}k_{2+}^{\prime\prime}q_{-+}}{k_{1-}k_{2-}^{\prime\prime}q_{++}} \right]^{\frac{1}{2}}$$

$$\doteq \frac{e^4 m k_{2+}^{\prime\prime}}{2q_{-0}(p_- \cdot k_1)(q_- \cdot k_2)} \left[\frac{k_{1+}q_{+-}q_{-+}}{k_{1-}} \right]^{\frac{1}{2}},$$

$$M(-,+;+,-,+,-) = \frac{2e^4 q_{-0} q_{+\perp}^*}{q_{--}(p_- \cdot k_1)(q_- \cdot k_2)} \left[\frac{k_{1+}k_{2-}^{\prime\prime}q_{++}q_{-+}}{k_{1-}k_{2+}^{\prime\prime}} \right]^{\frac{1}{2}}$$

$$\doteq \frac{2e^4 q_{-0} q_{++}}{q_{--}(p_- \cdot k_1)(q_- \cdot k_2)} \left[\frac{k_{1+}k_{2-}^{\prime\prime}q_{+-}q_{-+}}{k_{1-}k_{2+}^{\prime\prime}} \right]^{\frac{1}{2}},$$

$$M(-,+;+,-,-,+) = \frac{2e^4 q_{-0} q_{+\perp}^*}{(p_- \cdot k_1)(q_- \cdot k_2)} \left[\frac{k_{1+}k_{2-}^{\prime\prime}q_{-+}}{k_{1-}k_{2+}^{\prime\prime}q_{++}} \right]^{\frac{1}{2}}$$

$$\doteq \frac{2e^4 q_{-0}}{(p_- \cdot k_1)(q_- \cdot k_2)} \left[\frac{k_{1+}k_{2-}^{\prime\prime}q_{+-}q_{-+}}{k_{1-}k_{2+}^{\prime\prime}} \right]^{\frac{1}{2}},$$

$$M(-,+;-,+,+,-) = -\frac{e^4}{q_{+-}(p_- \cdot k_1)(q_- \cdot k_2)} \left\{ 2q_{-0} q_{-\perp}^* \left[\frac{k_{1+}k_{2-}^{\prime\prime}q_{++}^3}{k_{1-}k_{2+}^{\prime\prime}q_{-+}} \right]^{\frac{1}{2}} \right.$$

$$\left. + \frac{m^2 k_{1\perp}^* k_{2\perp}^{\prime\prime\prime*} q_{-\perp} Z_{+-}^*}{4Eq_{-0}q_{--}} \left[\frac{k_{1-}k_{2+}^{\prime\prime}}{k_{1+}k_{2-}^{\prime\prime}q_{++}q_{-+}} \right]^{\frac{1}{2}} \right\},$$

$$M(-,+;-,+,-,+) = -\frac{2e^4 q_{-0} q_{-\perp}^*}{(p_- \cdot k_1)(q_- \cdot k_2)} \left[\frac{k_{1+}k_{2-}^{\prime\prime}q_{++}^3}{k_{1-}k_{2+}^{\prime\prime}q_{-+}^3} \right]^{\frac{1}{2}}$$

$$\doteq \frac{2e^4 q_{-0}}{q_{-+}(p_- \cdot k_1)(q_- \cdot k_2)} \left[\frac{k_{1+}k_{2-}^{\prime\prime}q_{++}^3 q_{--}}{k_{1-}k_{2+}^{\prime\prime}} \right]^{\frac{1}{2}},$$

$$M(-,+;-,-,+,+) = \frac{e^4 m}{q_{+-}(p_- \cdot k_1)(q_- \cdot k_2)} \left\{ \frac{k_{2\perp}^{\prime\prime} q_{-\perp}^*}{2q_{-0}} \left[\frac{k_{1+}k_{2+}^{\prime\prime}q_{++}^3}{k_{1-}k_{2-}^{\prime\prime}q_{--}} \right]^{\frac{1}{2}} \right.$$

$$\left. - \frac{q_{-0} k_{1\perp}^* q_{-\perp} Z_{+-}^*}{Eq_{--}} \left[\frac{k_{1-}k_{2-}^{\prime\prime}}{k_{1+}k_{2+}^{\prime\prime}q_{++}q_{-+}} \right]^{\frac{1}{2}} \right\},$$

$$M(+,-;+,-,-,-) = -\frac{e^4 (2q_{-0} + k_{2+}^{\prime\prime}) q_{-\perp}}{q_{+-}(p_- \cdot k_1)(q_- \cdot k_2)} \left[\frac{k_{1+}k_{2-}^{\prime\prime}q_{++}^3}{k_{1-}k_{2+}^{\prime\prime}q_{-+}} \right]^{\frac{1}{2}}$$

$$\doteq \frac{e^4 (2q_{-0} + k_{2+}^{\prime\prime})}{q_{+-}(p_- \cdot k_1)(q_- \cdot k_2)} \left[\frac{k_{1+}k_{2-}^{\prime\prime}q_{++}^3 q_{--}}{k_{1-}k_{2+}^{\prime\prime}} \right]^{\frac{1}{2}},$$

9.19. $e^+e^- \to e^+e^-\gamma\gamma$ (NO Z-EXCHANGE; $m \neq 0$)

$$M(+,-;-,+,-,-) = \frac{4e^4 E q_{-0} q_{+\perp}}{(2E-k_{1-})(p_-\cdot k_1)(q_-\cdot k_2)} \left[\frac{k_{1+}k''_{2-}q_{-+}}{k_{1-}k''_{2+}q_{++}}\right]^{\frac{1}{2}}$$

$$\doteq \frac{4e^4 E q_{-0}}{(2E-k_{1-})(p_-\cdot k_1)(q_-\cdot k_2)} \left[\frac{k_{1+}k''_{2-}q_{+-}q_{-+}}{k_{1-}k''_{2+}}\right]^{\frac{1}{2}},$$

$$M(+,-;-,-,-,+) = -\frac{e^4 m E k''_{2\perp} q_{+\perp}}{q_{-0}(2E-k_{1-})(p_-\cdot k_1)(q_-\cdot k_2)} \left[\frac{k_{1+}k''_{2+}q_{-+}}{k_{1-}k''_{2-}q_{++}}\right]^{\frac{1}{2}}$$

$$\doteq \frac{e^4 m E k''_{2+}}{q_{-0}(2E-k_{1-})(p_-\cdot k_1)(q_-\cdot k_2)} \left[\frac{k_{1+}q_{+-}q_{-+}}{k_{1-}}\right]^{\frac{1}{2}},$$

$$M(-,+;+,-,-,-) = \frac{e^4(2q_{-0}+k''_{2+})q^*_{+\perp}}{(p_-\cdot k_1)(q_-\cdot k_2)} \left[\frac{k_{1+}k''_{2-}q_{-+}}{k_{1-}k''_{2+}q_{++}}\right]^{\frac{1}{2}}$$

$$\doteq \frac{e^4(2q_{-0}+k''_{2+})}{(p_-\cdot k_1)(q_-\cdot k_2)} \left[\frac{k_{1+}k''_{2-}q_{+-}q_{-+}}{k_{1-}k''_{2+}}\right]^{\frac{1}{2}},$$

$$M(-,+;-,+,-,-) = -\frac{e^4 q_{-0}(2E-k_{1-})q^*_{-\perp}}{E q_{+-}(p_-\cdot k_1)(q_-\cdot k_2)} \left[\frac{k_{1+}k''_{2-}q^3_{++}}{k_{1-}k''_{2+}q_{-+}}\right]^{\frac{1}{2}}$$

$$\doteq \frac{e^4 q_{-0}(2E-k_{1-})}{E q_{+-}(p_-\cdot k_1)(q_-\cdot k_2)} \left[\frac{k_{1+}k''_{2-}q^3_{++}q_{--}}{k_{1-}k''_{2+}}\right]^{\frac{1}{2}},$$

$$M(-,+;-,-,+,-) = -\frac{e^4 m(2q_{-0}+k''_{2+})k^*_{1\perp}q_{-\perp}Z^*_{+-}}{2E q_{+-}q_{--}(p_-\cdot k_1)(q_-\cdot k_2)} \left[\frac{k_{1-}k''_{2-}}{k_{1+}k''_{2+}q_{++}q_{-+}}\right]^{\frac{1}{2}}$$

$$\doteq \frac{e^4 m(2q_{-0}+k''_{2+})k_{1-}}{E q_{+-}(p_-\cdot k_1)(q_-\cdot k_2)} \left[\frac{(q_+\cdot q_-)k''_{2-}}{2k''_{2+}}\right]^{\frac{1}{2}},$$

$$M(-,+;-,-,-,+) = \frac{e^4 m(2E-k_{1-})k''_{2\perp}q^*_{-\perp}}{4E q_{-0}q_{+-}(p_-\cdot k_1)(q_-\cdot k_2)} \left[\frac{k_{1+}k''_{2+}q^3_{++}}{k_{1-}k''_{2-}q_{-+}}\right]^{\frac{1}{2}}$$

$$\doteq \frac{e^4 m(2E-k_{1-})k''_{2+}}{4E q_{-0}q_{+-}(p_-\cdot k_1)(q_-\cdot k_2)} \left[\frac{k_{1+}q^3_{++}q_{--}}{k_{1-}}\right]^{\frac{1}{2}},$$

$$M(-,-;+,+,-,-) = \frac{-e^4 m^2 k_{1\perp}k'''_{2\perp}q^*_{+\perp}}{4E q_{-0}(2E-k_{1-})(p_-\cdot k_1)(q_-\cdot k_2)} \left[\frac{k_{1-}k''_{2+}q_{-+}}{k_{1+}k''_{2-}q_{++}}\right]^{\frac{1}{2}}$$

$$\doteq \frac{e^4 m^2 k_{1-}k''_{2+}[q_{+-}q_{-+}]^{\frac{1}{2}}}{4E q_{-0}(2E-k_{1-})(p_-\cdot k_1)(q_-\cdot k_2)},$$

$$M(-,-;+,-,-,+) = -\frac{e^4 m q_{-0} k_{1\perp} q^*_{+\perp}}{E(2E-k_{1-})(p_-\cdot k_1)(q_-\cdot k_2)} \left[\frac{k_{1-} k''_{2-} q_{-+}}{k_{1+} k''_{2+} q_{++}}\right]^{\frac{1}{2}}$$

$$\doteq \frac{e^4 m q_{-0} k_{1-}}{E(2E-k_{1-})(p_-\cdot k_1)(q_-\cdot k_2)} \left[\frac{k''_{2-} q_{+-} q_{-+}}{k''_{2+}}\right]^{\frac{1}{2}},$$

$$M(-,-;-,+,+,-) = -\frac{e^4 m(2E-k_{1-}) k''^*_{2\perp} q_{-\perp} Z^*_{+-}}{2 q_{-0} q_{+-} q_{--}(p_-\cdot k_1)(q_-\cdot k_2)} \left[\frac{k_{1+} k''_{2+}}{k_{1-} k''_{2-} q_{++} q_{-+}}\right]^{\frac{1}{2}}$$

$$\doteq \frac{e^4 m(2E-k_{1-}) k''_{2+}}{q_{-0} q_{+-}(p_-\cdot k_1)(q_-\cdot k_2)} \left[\frac{(q_+\cdot q_-) k_{1+}}{2 k_{1-}}\right]^{\frac{1}{2}},$$

$$M(-,-;-,+,-,+) = \frac{e^4 m(2q_{-0}+k''_{2+}) k_{1\perp} q^*_{-\perp}}{4E^2 q_{+-}(p_-\cdot k_1)(q_-\cdot k_2)} \left[\frac{k_{1+} k''_{2-} q^3_{++}}{k_{1+} k''_{2+} q_{-+}}\right]^{\frac{1}{2}}$$

$$\doteq \frac{e^4 m(2q_{-0}+k''_{2+}) k_{1-}}{4E^2 q_{+-}(p_-\cdot k_1)(q_-\cdot k_2)} \left[\frac{k''_{2-} q^3_{++} q_{--}}{k''_{2+}}\right]^{\frac{1}{2}},$$

$$M(-,-;-,-,+,+) = -\frac{2e^4 q_{-0}(2E-k_{1-}) q_{-\perp} Z^*_{+-}}{q_{+-} q_{--}(p_-\cdot k_1)(q_-\cdot k_2)} \left[\frac{k_{1+} k''_{2-}}{k_{1-} k''_{2+} q_{++} q_{-+}}\right]^{\frac{1}{2}}$$

$$\doteq \frac{e^4 q_{-0}(2E-k_{1-})}{q_{+-}(p_-\cdot k_1)(q_-\cdot k_2)} \left[\frac{8(q_+\cdot q_-) k_{1+} k''_{2-}}{k_{1-} k''_{2+}}\right]^{\frac{1}{2}},$$

$$M(-,-;+,-,-,-) = -\frac{e^4 m q_{-0} k_{1\perp} q^*_{+\perp}}{2E^2 q_{--}(p_-\cdot k_1)(q_-\cdot k_2)} \left[\frac{k_{1-} k''_{2-} q_{++} q_{-+}}{k_{1+} k''_{2+}}\right]^{\frac{1}{2}}$$

$$\doteq \frac{e^4 m q_{-0} q_{++} k_{1-}}{2E^2 q_{--}(p_-\cdot k_1)(q_-\cdot k_2)} \left[\frac{k''_{2-} q_{+-} q_{-+}}{k''_{2+}}\right]^{\frac{1}{2}},$$

$$M(-,-;-,+,-,-) = \frac{e^4 m}{q_{+-}(p_-\cdot k_1)(q_-\cdot k_2)} \left\{\frac{q_{-0} k_{1\perp} q^*_{-\perp}}{2E^2} \left[\frac{k_{1-} k''_{2-} q^3_{++}}{k_{1+} k''_{2+} q_{-+}}\right]^{\frac{1}{2}}\right.$$

$$\left. -\frac{E k''^*_{2\perp} q_{-\perp} Z^*_{+-}}{q_{-0} q_{--}} \left[\frac{k_{1+} k''_{2+}}{k_{1-} k''_{2-} q_{++} q_{-+}}\right]^{\frac{1}{2}}\right\},$$

$$M(-,-;-,-,+,-) = -\frac{4e^4 E q_{-0} q_{-\perp} Z^*_{+-}}{q_{--}(p_-\cdot k_1)(q_-\cdot k_2)} \left[\frac{k_{1+} k''_{2-}}{k_{1-} k''_{2+} q_{++} q^3_{-+}}\right]^{\frac{1}{2}}$$

$$\doteq \frac{e^4 E q_{-0}}{q_{-+}(p_-\cdot k_1)(q_-\cdot k_2)} \left[\frac{32(q_+\cdot q_-) k_{1+} k''_{2-}}{k_{1-} k''_{2+}}\right]^{\frac{1}{2}},$$

9.19. $e^+e^- \to e^+e^-\gamma\gamma$ (NO Z-EXCHANGE; $m \neq 0$)

$$M(-,-;-,-,-,+) = -\frac{e^4}{q_{+-}(p_- \cdot k_1)(q_- \cdot k_2)}$$

$$\times \left\{ \frac{4Eq_{-0}q_{-\perp}Z^*_{+-}}{q_{--}} \left[\frac{k_{1+}k''_{2-}}{k_{1-}k''_{2+}q_{++}q_{-+}} \right]^{\frac{1}{2}} \right.$$

$$\left. + \frac{m^2 k_{1\perp} k''_{2\perp} q^*_{-\perp}}{8E^2 q_{-0}} \left[\frac{k_{1-}k''_{2+}q^3_{++}}{k_{1+}k''_{2-}q_{-+}} \right]^{\frac{1}{2}} \right\},$$

$$M(-,-;-,-,-,-) = -\frac{e^4 E(2q_{-0} + k''_{2+})q_{-\perp}Z^*_{+-}}{q_{+-}(p_- \cdot k_1)(q_- \cdot k_2)} \left[\frac{k_{1+}k''_{2-}}{k_{1-}k''_{2+}q_{++}q_{-+}} \right]^{\frac{1}{2}}$$

$$\doteq \frac{e^4 E(2q_{-0} + k''_{2+})}{q_{+-}(p_- \cdot k_1)(q_- \cdot k_2)} \left[\frac{8(q_+ \cdot q_-)k_{1+}k''_{2-}}{k_{1-}k''_{2+}} \right]^{\frac{1}{2}}. \tag{9.250}$$

Unpolarized squared matrix element:

$$\overline{|M|^2} = \frac{e^8 k_{1+}k''_{2-} \left[4(q_+ \cdot q_-)^2 + q^2_{++}q^2_{--} + q^2_{+-}q^2_{-+} \right]}{2k_{1-}k''_{2+}q_{+-}q_{-+}[(2E - k_{1-})(p_- \cdot k_1)(q_- \cdot k_2)]^2}$$

$$\times \left[4E^2 + (2E - k_{1-})^2 + \frac{m^2 k^3_{1-}}{4E^2 k_{1+}} \right] \left[4q^2_{-0} + (2q_{-0} + k''_{2+})^2 + \frac{m^2 k''^3_{2+}}{4q^2_{-0}k''_{2-}} \right]. \tag{9.251}$$

Unpolarized cross section:

$$d\sigma = \frac{\alpha^4 k_{1+}k''_{2-} \left[4(q_+ \cdot q_-)^2 + q^2_{++}q^2_{--} + q^2_{+-}q^2_{-+} \right]}{256\pi^2 k_{1-}k''_{2+}q_{+-}q_{-+}[E(2E - k_{1-})(p_- \cdot k_1)(q_- \cdot k_2)]^2}$$

$$\times \left[4E^2 + (2E - k_{1-})^2 + \frac{m^2 k^3_{1-}}{4E^2 k_{1+}} \right] \left[4q^2_{-0} + (2q_{-0} + k''_{2+})^2 + \frac{m^2 k''^3_{2+}}{4q^2_{-0}k''_{2-}} \right]$$

$$\times \delta^4(p_+ + p_- - q_+ - q_- - k_1 - k_2) \frac{d^3\vec{q}_+ d^3\vec{q}_- d^3\vec{k}_1 d^3\vec{k}_2}{q_{+0}q_{-0}k_{10}k_{20}}. \tag{9.252}$$

9.19.10 $\vec{k}_1$ and $\vec{k}_2$ nearly parallel to $\vec{q}_+$ and $\vec{q}_-$, resp.

The primed (doubly primed) quantities $k'_{1\pm}$ and $k''_{1\pm}$ ($k''_{2\pm}$ and $k'_{2\pm}$) are evaluated in the rotated frame where $\vec{q}_+$ ($\vec{q}_-$) determines the positive z-axis (see also Section 7.4.3). The four-vector q (q') in eqns (9.254) is obtained by applying a space reflection to q_+ (q_-).
Definition:

$$Z_{+-} = q_{++}q_{--} - q^*_{+\perp}q_{-\perp}. \tag{9.253}$$

Photon polarizations:

$$\rlap{/}{\epsilon}_1^{\pm} = N_1[\rlap{/}{k}_1 \rlap{/}{A}_+ \rlap{/}{A}(1 \pm \gamma_5) + \rlap{/}{A} \rlap{/}{A}_+ \rlap{/}{k}_1(1 \mp \gamma_5)],$$

$$N_1^{-1} = q_{+0}^2[32k'_{1+}k'_{1-}]^{\frac{1}{2}},$$

$$\rlap{/}{\epsilon}_2^{\pm} = N_2[\rlap{/}{k}_2 \rlap{/}{A}_- \rlap{/}{A}'(1 \pm \gamma_5) + \rlap{/}{A}' \rlap{/}{A}_- \rlap{/}{k}_2(1 \mp \gamma_5)],$$

$$N_2^{-1} = q_{-0}^2[32k''_{2+}k''_{2-}]^{\frac{1}{2}}. \qquad (9.28)$$

Nonvanishing helicity amplitudes:

$$M(+,+;+,+,+,+) = -\frac{2e^4 q_{-0}(2q_{+0} + k'_{1+})q^*_{-\perp}Z_{+-}}{q_{--}(q_+ \cdot k_1)(q_- \cdot k_2)} \left[\frac{k'_{1-}k''_{2-}}{k'_{1+}k''_{2+}q_{++}q^3_{-+}}\right]$$

$$\doteq \frac{e^4 q_{-0}(2q_{+0} + k'_{1+})}{q_{-+}(q_+ \cdot k_1)(q_- \cdot k_2)} \left[\frac{8(q_+ \cdot q_-)k'_{1-}k''_{2-}}{k'_{1+}k''_{2+}}\right]^{\frac{1}{2}},$$

$$M(+,+;+,+,+,-) = -\frac{4e^4 q_{+0} q_{-0} q^*_{-\perp} Z_{+-}}{(q_+ \cdot k_1)(q_- \cdot k_2)} \left[\frac{k'_{1-}k''_{2-}}{k'_{1+}k''_{2+}q^3_{++}q^3_{-+}}\right]^{\frac{1}{2}}$$

$$\doteq \frac{e^4 q_{+0} q_{-0} q_{--}}{q_{++} q_{+-}(q_+ \cdot k_1)(q_- \cdot k_2)} \left[\frac{32(q_+ \cdot q_-)k'_{1-}k''_{2-}}{k'_{1+}k''_{2+}}\right]^{\frac{1}{2}},$$

$$M(+,+;+,+,-,+) = -\frac{4e^4 q_{+0} q_{-0} q^*_{-\perp} Z_{+-}}{q_{--}(q_+ \cdot k_1)(q_- \cdot k_2)} \left[\frac{k'_{1-}k''_{2-}}{k'_{1+}k''_{2+}q_{++}q^3_{-+}}\right]^{\frac{1}{2}}$$

$$\doteq \frac{e^4 q_{+0} q_{-0}}{q_{-+}(q_+ \cdot k_1)(q_- \cdot k_2)} \left[\frac{32(q_+ \cdot q_-)k'_{1-}k''_{2-}}{k'_{1+}k''_{2+}}\right]^{\frac{1}{2}},$$

$$M(+,+;+,-,+,+) = \frac{e^4 m q_{+0} k''_{2\perp} q^*_{-\perp} Z_{+-}}{q_{-0}(q_+ \cdot k_1)(q_- \cdot k_2)} \left[\frac{k'_{1-}k''_{2+}}{k'_{1+}k''_{2-}q^3_{++}q^3_{-+}}\right]^{\frac{1}{2}}$$

$$\doteq \frac{e^4 m q_{+0} q_{--} k''_{2+}}{q_{+0} q_{++} q_{-+}(q_+ \cdot k_1)(q_- \cdot k_2)} \left[\frac{2(q_+ \cdot q_-)k'_{1-}}{k'_{1+}}\right]^{\frac{1}{2}},$$

$$M(+,+;-,+,+,+) = \frac{e^4 m q_{-0} k'_{1\perp} q^*_{-\perp} Z_{+-}}{q_{+0} q_{--}(q_+ \cdot k_1)(q_- \cdot k_2)} \left[\frac{k'_{1+}k''_{2-}}{k'_{1-}k''_{2+}q_{++}q^3_{-+}}\right]^{\frac{1}{2}}$$

$$\doteq \frac{e^4 m q_{-0} k'_{1+}}{q_{+0} q_{-+}(q_+ \cdot k_1)(q_- \cdot k_2)} \left[\frac{2(q_+ \cdot q_-)k''_{2-}}{k''_{2+}}\right]^{\frac{1}{2}},$$

9.19. $e^+e^- \to e^+e^-\gamma\gamma$ (NO Z-EXCHANGE; $m \neq 0$)

$$M(+,+;+,+,-,-) = -\frac{8e^4 q_{+0} q_{-0}^2 q_{-\perp}^* Z_{+-}}{q_{--}(2q_{-0}+k_{2+}'')(q_+\cdot k_1)(q_-\cdot k_2)}$$

$$\times \left[\frac{k_{1-}' k_{2-}''}{k_{1+}' k_{2+}'' q_{++} q_{-+}^3}\right]^{\frac{1}{2}}$$

$$\doteq \frac{e^4 q_{+0} q_{-0}^2}{q_{-+}(2q_{-0}+k_{2+}'')(q_+\cdot k_1)(q_-\cdot k_2)}$$

$$\times \left[\frac{128(q_+\cdot q_-) k_{1-}' k_{2-}''}{k_{1+}' k_{2+}''}\right]^{\frac{1}{2}},$$

$$M(+,+;+,-,-,+) = \frac{-2e^4 m q_{+0} k_{2\perp}'' q_{-\perp}^* Z_{+-}}{q_{--}(2q_{-0}+k_{2+}'')(q_+\cdot k_1)(q_-\cdot k_2)} \left[\frac{k_{1-}' k_{2+}''}{k_{1+}' k_{2-}'' q_{++} q_{-+}^3}\right]^{\frac{1}{2}}$$

$$\doteq \frac{e^4 m q_{+0} k_{2+}''}{q_{-+}(2q_{-0}+k_{2+}'')(q_+\cdot k_1)(q_-\cdot k_2)} \left[\frac{8(q_+\cdot q_-) k_{1-}'}{k_{1+}'}\right]^{\frac{1}{2}},$$

$$M(+,+;-,+,+,-) = \frac{2e^4 m q_{-0}^2 k_{1\perp}' q_{-\perp}^* Z_{+-}}{q_{+0} q_{--}(2q_{-0}+k_{2+}'')(q_+\cdot k_1)(q_-\cdot k_2)}$$

$$\times \left[\frac{k_{1+}' k_{2-}''}{k_{1-}' k_{2+}'' q_{++} q_{-+}^3}\right]^{\frac{1}{2}}$$

$$\doteq \frac{e^4 m q_{-0}^2 k_{1+}'}{q_{+0} q_{-+}(2q_{-0}+k_{2+}'')(q_+\cdot k_1)(q_-\cdot k_2)}$$

$$\times \left[\frac{8(q_+\cdot q_-) k_{2-}''}{k_{2+}''}\right]^{\frac{1}{2}},$$

$$M(+,+;-,-,+,+) = -\frac{e^4 m^2 k_{1\perp}' k_{2\perp}'' q_{-\perp}^* Z_{+-}}{2 q_{+0} q_{--}(2q_{-0}+k_{2+}'')(q_+\cdot k_1)(q_-\cdot k_2)}$$

$$\times \left[\frac{k_{1+}' k_{2+}''}{k_{1-}' k_{2-}'' q_{++} q_{-+}^3}\right]^{\frac{1}{2}}$$

$$\doteq \frac{e^4 m^2 k_{1+}' k_{2+}''}{q_{+0} q_{-+}(2q_{-0}+k_{2+}'')(q_+\cdot k_1)(q_-\cdot k_2)} \left[\frac{(q_+\cdot q_-)}{2}\right]^{\frac{1}{2}},$$

$$M(+,-;+,+,+,-) = \frac{e^4 m(2q_{+0}+k_{1+}') k_{2\perp}''^* q_{-\perp}}{4E q_{-0} q_{+-}(q_+\cdot k_1)(q_-\cdot k_2)} \left[\frac{k_{1-}' k_{2+}'' q_{++}^3}{k_{1+}' k_{2-}'' q_{-+}}\right]^{\frac{1}{2}}$$

$$\doteq \frac{e^4 m(2q_{+0}+k_{1+}') k_{2+}''}{4E q_{-0} q_{+-}(q_+\cdot k_1)(q_-\cdot k_2)} \left[\frac{k_{1-}' q_{++}^3 q_{--}}{k_{1+}'}\right]^{\frac{1}{2}},$$

$$M(+,-;+,+,-,+) = \frac{e^4 m(2q_{-0}+k''_{2+})k'^*_{1\perp}q_{+\perp}}{4Eq_{+0}(q_+\cdot k_1)(q_-\cdot k_2)}\left[\frac{k'_{1+}k''_{2-}q_{-+}}{k'_{1-}k''_{2+}q_{++}}\right]^{\frac{1}{2}}$$

$$\doteq \frac{e^4 m(2q_{-0}+k''_{2+})k'_{1+}}{4Eq_{+0}(q_+\cdot k_1)(q_-\cdot k_2)}\left[\frac{k''_{2-}q_{+-}q_{-+}}{k''_{2+}}\right]^{\frac{1}{2}},$$

$$M(+,-;+,-,+,+) = \frac{e^4 q_{-0}(2q_{+0}+k'_{1+})q_{-\perp}}{Eq_{+-}(q_+\cdot k_1)(q_-\cdot k_2)}\left[\frac{k'_{1-}k''_{2-}q^3_{++}}{k'_{1+}k''_{2+}q_{-+}}\right]^{\frac{1}{2}}$$

$$\doteq \frac{e^4 q_{-0}(2q_{+0}+k'_{1+})}{Eq_{+-}(q_+\cdot k_1)(q_-\cdot k_2)}\left[\frac{k'_{1-}k''_{2-}q^3_{++}q_{--}}{k'_{1+}k''_{2+}}\right]^{\frac{1}{2}},$$

$$M(+,-;-,+,+,+) = \frac{e^4 q_{+0}(2q_{-0}+k''_{2+})q_{+\perp}}{E(q_+\cdot k_1)(q_-\cdot k_2)}\left[\frac{k'_{1-}k''_{2-}q_{-+}}{k'_{1+}k''_{2+}q_{++}}\right]^{\frac{1}{2}}$$

$$\doteq \frac{e^4 q_{+0}(2q_{-0}+k''_{2+})}{E(q_+\cdot k_1)(q_-\cdot k_2)}\left[\frac{k'_{1-}k''_{2-}q_{+-}q_{-+}}{k'_{1+}k''_{2+}}\right]^{\frac{1}{2}},$$

$$M(-,+;+,+,+,-) = \frac{e^4 m(2q_{+0}+k'_{1+})k''^*_{2\perp}q^*_{+\perp}}{4Eq_{-0}(q_+\cdot k_1)(q_-\cdot k_2)}\left[\frac{k'_{1-}k''_{2+}q_{-+}}{k'_{1+}k''_{2-}q_{++}}\right]^{\frac{1}{2}}$$

$$\doteq \frac{e^4 m(2q_{+0}+k'_{1+})k''_{2+}}{4Eq_{-0}(q_+\cdot k_1)(q_-\cdot k_2)}\left[\frac{k'_{1-}q_{+-}q_{-+}}{k'_{1+}}\right]^{\frac{1}{2}},$$

$$M(-,+;+,+,-,+) = \frac{e^4 m(2q_{-0}+k''_{2+})k'^*_{1\perp}q^*_{-\perp}}{4Eq_{+0}q_{+-}(q_+\cdot k_1)(q_-\cdot k_2)}\left[\frac{k'_{1+}k''_{2-}q^3_{++}}{k'_{1-}k''_{2+}q_{-+}}\right]^{\frac{1}{2}}$$

$$\doteq \frac{e^4 m(2q_{-0}+k''_{2+})k'_{1+}}{4Eq_{+0}q_{+-}(q_+\cdot k_1)(q_-\cdot k_2)}\left[\frac{k''_{2-}q^3_{++}q_{--}}{k''_{2+}}\right]^{\frac{1}{2}},$$

$$M(-,+;+,-,+,+) = \frac{e^4 q_{-0}(2q_{+0}+k'_{1+})q^*_{+\perp}}{E(q_+\cdot k_1)(q_-\cdot k_2)}\left[\frac{k'_{1-}k''_{2-}q_{-+}}{k'_{1+}k''_{2+}q_{++}}\right]^{\frac{1}{2}}$$

$$\doteq \frac{e^4 q_{-0}(2q_{+0}+k'_{1+})}{E(q_+\cdot k_1)(q_-\cdot k_2)}\left[\frac{k'_{1-}k''_{2-}q_{+-}q_{-+}}{k'_{1+}k''_{2+}}\right]^{\frac{1}{2}},$$

$$M(-,+;-,+,+,+) = \frac{e^4 q_{+0}(2q_{-0}+k''_{2+})q^*_{-\perp}}{Eq_{+-}(q_+\cdot k_1)(q_-\cdot k_2)}\left[\frac{k'_{1-}k''_{2-}q^3_{++}}{k'_{1+}k''_{2+}q_{-+}}\right]^{\frac{1}{2}}$$

$$\doteq \frac{e^4 q_{+0}(2q_{-0}+k''_{2+})}{Eq_{+-}(q_+\cdot k_1)(q_-\cdot k_2)}\left[\frac{k'_{1-}k''_{2-}q^3_{++}q_{--}}{k'_{1+}k''_{2+}}\right]^{\frac{1}{2}},$$

9.19. $e^+e^- \to e^+e^- \gamma\gamma$ (NO Z-EXCHANGE; $m \neq 0$) 335

$$M(+,-;+,+,-,-) = \frac{e^4 m}{2E(q_+ \cdot k_1)(q_- \cdot k_2)} \left\{ \frac{q_{-0} k'^*_{1\perp} q_{+\perp}}{q_{+0}} \left[\frac{k'_{1-} k''_{2-} q_{-+}}{k'_{1-} k''_{2+} q_{++}} \right]^{\frac{1}{2}} \right.$$

$$\left. + \frac{q_{+0} k''^*_{2\perp} q_{-\perp}}{q_{-0} q_{+-}} \left[\frac{k'_{1-} k''_{2+} q^3_{++}}{k'_{1+} k''_{2-} q_{-+}} \right]^{\frac{1}{2}} \right\},$$

$$M(+,-;+,-,+,-) = \frac{4e^4 E q_{+0} q_{-0} q_{-\perp}}{q_{+-}(q_+ \cdot q_-)(q_+ \cdot k_1)(q_- \cdot k_2)} \left[\frac{k'_{1-} k''_{2-} q^3_{++}}{k'_{1+} k''_{2+} q_{-+}} \right]^{\frac{1}{2}}$$

$$\doteq \frac{4e^4 E q_{+0} q_{-0}}{q_{+-}(q_+ \cdot q_-)(q_+ \cdot k_1)(q_- \cdot k_2)} \left[\frac{k'_{1-} k''_{2-} q^3_{++} q_{--}}{k'_{1+} k''_{2+}} \right]^{\frac{1}{2}},$$

$$M(+,-;+,-,-,+) = \frac{e^4}{E(q_+ \cdot k_1)(q_- \cdot k_2)} \left\{ \frac{2 q_{+0} q_{-0} q_{-\perp}}{q_{+-}} \left[\frac{k'_{1-} k''_{2-} q^3_{++}}{k'_{1+} k''_{2+} q_{-+}} \right]^{\frac{1}{2}} \right.$$

$$\left. - \frac{m^2 k'^*_{1\perp} k''_{2\perp} q_{+\perp}}{8 q_{+0} q_{-0}} \left[\frac{k'_{1+} k''_{2+} q_{-+}}{k'_{1-} k''_{2-} q_{++}} \right]^{\frac{1}{2}} \right\},$$

$$M(+,-;-,+,+,-) = \frac{e^4}{E(q_+ \cdot k_1)(q_- \cdot k_2)} \left\{ 2 q_{+0} q_{-0} q_{+\perp} \left[\frac{k'_{1-} k''_{2-} q_{-+}}{k'_{1+} k''_{2+} q_{++}} \right]^{\frac{1}{2}} \right.$$

$$\left. - \frac{m^2 k'_{1\perp} k''^*_{2\perp} q_{-\perp}}{8 q_{+0} q_{-0} q_{+-}} \left[\frac{k'_{1+} k''_{2+} q^3_{++}}{k'_{1-} k''_{2-} q_{-+}} \right]^{\frac{1}{2}} \right\},$$

$$M(+,-;-,+,-,+) = \frac{4e^4 E q_{+0} q_{-0} q_{+\perp}}{(q_+ \cdot q_-)(q_+ \cdot k_1)(q_- \cdot k_2)} \left[\frac{k'_{1-} k''_{2-} q_{-+}}{k'_{1+} k''_{2+} q_{++}} \right]^{\frac{1}{2}}$$

$$\doteq \frac{4e^4 E q_{+0} q_{-0}}{(q_+ \cdot q_-)(q_+ \cdot k_1)(q_- \cdot k_2)} \left[\frac{k'_{1-} k''_{2-} q_{+-} q_{-+}}{k'_{1+} k''_{2+}} \right]^{\frac{1}{2}},$$

$$M(+,-;-,-,+,+) = -\frac{e^4 m}{2E(q_+ \cdot k_1)(q_- \cdot k_2)} \left\{ \frac{q_{-0} k'_{1\perp} q_{-\perp}}{q_{+0} q_{+-}} \left[\frac{k'_{1+} k''_{2-} q^3_{++}}{k'_{1-} k''_{2+} q_{-+}} \right]^{\frac{1}{2}} \right.$$

$$\left. + \frac{q_{+0} k''_{2\perp} q_{+\perp}}{q_{-0}} \left[\frac{k'_{1-} k''_{2+} q_{-+}}{k'_{1+} k''_{2-} q_{++}} \right]^{\frac{1}{2}} \right\},$$

$$M(-,+;+,+,-,-) = \frac{e^4 m}{2E(q_+ \cdot k_1)(q_- \cdot k_2)} \left\{ \frac{q_{+0} k''^*_{2\perp} q^*_{+\perp}}{q_{-0}} \left[\frac{k'_{1-} k''_{2+} q_{-+}}{k'_{1+} k''_{2-} q_{++}} \right]^{\frac{1}{2}} \right.$$

$$\left. + \frac{q_{-0} k'^*_{1\perp} q^*_{-\perp}}{q_{+0} q_{+-}} \left[\frac{k'_{1+} k''_{2-} q^3_{++}}{k'_{1-} k''_{2+} q_{-+}} \right]^{\frac{1}{2}} \right\},$$

$$M(-,+;+,-,+,-) = \frac{4e^4 E q_{+0} q_{-0} q_{+\perp}^*}{(q_+ \cdot q_-)(q_+ \cdot k_1)(q_- \cdot k_2)} \left[\frac{k'_{1-} k''_{2-} q_{-+}}{k'_{1+} k''_{2+} q_{++}} \right]^{\frac{1}{2}}$$

$$\doteq \frac{4e^4 E q_{+0} q_{-0}}{(q_+ \cdot q_-)(q_+ \cdot k_1)(q_- \cdot k_2)} \left[\frac{k'_{1-} k''_{2-} q_{+-} q_{-+}}{k'_{1+} k''_{2+}} \right]^{\frac{1}{2}},$$

$$M(-,+;+,-,-,+) = \frac{e^4}{E(q_+ \cdot k_1)(q_- \cdot k_2)} \left\{ 2 q_{+0} q_{-0} q_{+\perp}^* \left[\frac{k'_{1-} k''_{2-} q_{-+}}{k'_{1+} k''_{2+} q_{++}} \right]^{\frac{1}{2}} \right.$$

$$\left. - \frac{m^2 k'^{*}_{1\perp} k''_{2\perp} q_{-\perp}^*}{8 q_{+0} q_{-0} q_{+-}} \left[\frac{k'_{1+} k''_{2+} q^3_{++}}{k'_{1-} k''_{2-} q_{-+}} \right]^{\frac{1}{2}} \right\},$$

$$M(-,+;-,+,+,-) = \frac{e^4}{E(q_+ \cdot k_1)(q_- \cdot k_2)} \left\{ \frac{2 q_{+0} q_{-0} q_{-\perp}^*}{q_{+-}} \left[\frac{k'_{1-} k''_{2-} q^3_{++}}{k'_{1+} k''_{2+} q_{-+}} \right]^{\frac{1}{2}} \right.$$

$$\left. - \frac{m^2 k'_{1\perp} k'''^{*}_{2\perp} q_{+\perp}^*}{8 q_{+0} q_{-0}} \left[\frac{k'_{1+} k''_{2+} q_{-+}}{k'_{1-} k''_{2-} q_{++}} \right]^{\frac{1}{2}} \right\},$$

$$M(-,+;-,+,-,+) = \frac{4e^4 E q_{+0} q_{-0} q_{-\perp}^*}{q_{+-}(q_+ \cdot q_-)(q_+ \cdot k_1)(q_- \cdot k_2)} \left[\frac{k'_{1-} k''_{2-} q^3_{++}}{k'_{1+} k''_{2+} q_{-+}} \right]^{\frac{1}{2}}$$

$$\doteq \frac{4e^4 E q_{+0} q_{-0}}{q_{+-}(q_+ \cdot q_-)(q_+ \cdot k_1)(q_- \cdot k_2)} \left[\frac{k'_{1-} k''_{2-} q^3_{++} q_{--}}{k'_{1+} k''_{2+}} \right]^{\frac{1}{2}},$$

$$M(-,+;-,-,+,+) = -\frac{e^4 m}{2E(q_+ \cdot k_1)(q_- \cdot k_2)} \left\{ \frac{q_{-0} k'_{1\perp} q_{+\perp}^*}{q_{+0}} \left[\frac{k'_{1+} k''_{2-} q_{-+}}{k'_{1-} k''_{2+} q_{++}} \right]^{\frac{1}{2}} \right.$$

$$\left. + \frac{q_{+0} k''_{2\perp} q_{-\perp}^*}{q_{-0} q_{+-}} \left[\frac{k'_{1-} k''_{2+} q^3_{++}}{k'_{1+} k''_{2-} q_{-+}} \right]^{\frac{1}{2}} \right\},$$

$$M(+,-;+,-,-,-) = \frac{e^4 q_{+0}(2 q_{-0} + k''_{2+}) q_{-\perp}}{E q_{+-}(q_+ \cdot k_1)(q_- \cdot k_2)} \left[\frac{k'_{1-} k''_{2-} q^3_{++}}{k'_{1+} k''_{2+} q_{-+}} \right]^{\frac{1}{2}}$$

$$\doteq \frac{e^4 q_{+0}(2 q_{-0} + k''_{2+})}{E q_{+-}(q_+ \cdot k_1)(q_- \cdot k_2)} \left[\frac{k'_{1-} k''_{2-} q^3_{++} q_{--}}{k'_{1+} k''_{2+}} \right]^{\frac{1}{2}},$$

$$M(+,-;-,+,-,-) = \frac{e^4 q_{-0}(2 q_{+0} + k'_{1+}) q_{+\perp}}{E(q_+ \cdot k_1)(q_- \cdot k_2)} \left[\frac{k'_{1-} k''_{2-} q_{-+}}{k'_{1+} k''_{2+} q_{++}} \right]^{\frac{1}{2}}$$

$$\doteq \frac{e^4 q_{-0}(2 q_{+0} + k'_{1+})}{E(q_+ \cdot k_1)(q_- \cdot k_2)} \left[\frac{k'_{1-} k''_{2-} q_{+-} q_{-+}}{k'_{1+} k''_{2+}} \right]^{\frac{1}{2}},$$

9.19. $e^+e^- \to e^+e^- \gamma\gamma$ (NO Z-EXCHANGE; $m \neq 0$)

$$M(+,-;-,-,+,-) = -\frac{e^4 m(2q_{-0} + k''_{2+})k'_{1\perp}q_{-\perp}}{4Eq_{+0}q_{+-}(q_+ \cdot k_1)(q_- \cdot k_2)}\left[\frac{k'_{1+}k''_{2-}q^3_{++}}{k'_{1-}k''_{2+}q_{-+}}\right]^{\frac{1}{2}}$$

$$\doteq \frac{e^4 m(2q_{-0} + k''_{2+})k'_{1+}}{4Eq_{+0}q_{+-}(q_+ \cdot k_1)(q_- \cdot k_2)}\left[\frac{k''_2 - q^3_{++}q_{--}}{k''_{2+}}\right]^{\frac{1}{2}},$$

$$M(+,-;-,-,-,+) = -\frac{e^4 m(2q_{+0} + k'_{1+})k''_{2\perp}q_{+\perp}}{4Eq_{-0}(q_+ \cdot k_1)(q_- \cdot k_2)}\left[\frac{k'_{1-}k''_{2+}q_{-+}}{k'_{1+}k''_{2-}q_{++}}\right]^{\frac{1}{2}}$$

$$\doteq \frac{e^4 m(2q_{+0} + k'_{1+})k''_{2+}}{4Eq_{-0}(q_+ \cdot k_1)(q_- \cdot k_2)}\left[\frac{k'_{1-}q_{+-}q_{-+}}{k'_{1+}}\right]^{\frac{1}{2}},$$

$$M(-,+;+,-,-,-) = \frac{e^4 q_{+0}(2q_{-0} + k''_{2+})q^*_{+\perp}}{E(q_+ \cdot k_1)(q_- \cdot k_2)}\left[\frac{k'_{1-}k''_{2-}q_{-+}}{k'_{1+}k''_{2+}q_{++}}\right]^{\frac{1}{2}}$$

$$\doteq \frac{e^4 q_{+0}(2q_{-0} + k''_{2+})}{E(q_+ \cdot k_1)(q_- \cdot k_2)}\left[\frac{k'_{1-}k''_{2-}q_{+-}q_{-+}}{k'_{1+}k''_{2+}}\right]^{\frac{1}{2}},$$

$$M(-,+;-,+,-,-) = \frac{e^4 q_{-0}(2q_{+0} + k'_{1+})q^*_{-\perp}}{Eq_{+-}(q_+ \cdot k_1)(q_- \cdot k_2)}\left[\frac{k'_{1-}k''_{2-}q^3_{++}}{k'_{1+}k''_{2+}q_{-+}}\right]^{\frac{1}{2}}$$

$$\doteq \frac{e^4 q_{-0}(2q_{+0} + k'_{1+})}{Eq_{+-}(q_+ \cdot k_1)(q_- \cdot k_2)}\left[\frac{k'_{1-}k''_{2-}q^3_{++}q_{--}}{k'_{1+}k''_{2+}}\right]^{\frac{1}{2}},$$

$$M(-,+;-,-,+,-) = -\frac{e^4 m(2q_{-0} + k''_{2+})k'_{1\perp}q^*_{+\perp}}{4Eq_{+0}(q_+ \cdot k_1)(q_- \cdot k_2)}\left[\frac{k'_{1+}k''_{2-}q_{-+}}{k'_{1-}k''_{2+}q_{++}}\right]^{\frac{1}{2}}$$

$$\doteq \frac{e^4 m(2q_{-0} + k''_{2+})k'_{1+}}{4Eq_{+0}(q_+ \cdot k_1)(q_- \cdot k_2)}\left[\frac{k''_2 - q_{+-}q_{-+}}{k''_{2+}}\right]^{\frac{1}{2}},$$

$$M(-,+;-,-,-,+) = -\frac{e^4 m(2q_{+0} + k'_{1+})k''_{2\perp}q^*_{-\perp}}{4Eq_{-0}q_{+-}(q_+ \cdot k_1)(q_- \cdot k_2)}\left[\frac{k'_{1-}k''_{2+}q^3_{++}}{k'_{1+}k''_{2-}q_{-+}}\right]^{\frac{1}{2}}$$

$$\doteq \frac{e^4 m(2q_{+0} + k'_{1+})k''_{2+}}{4Eq_{-0}q_{+-}(q_+ \cdot k_1)(q_- \cdot k_2)}\left[\frac{k'_{1-}q^3_{++}q_{--}}{k'_{1+}}\right]^{\frac{1}{2}},$$

$$M(-,-;+,+,-,-) = -\frac{e^4 m^2 k'^*_{1\perp}k''^*_{2\perp}q_{-\perp}Z^*_{+-}}{2q_{+0}q_{--}(2q_{-0} + k''_{2+})(q_+ \cdot k_1)(q_- \cdot k_2)}$$

$$\times \left[\frac{k'_{1+}k''_{2+}}{k'_{1-}k''_{2-}q_{++}q^3_{-+}}\right]^{\frac{1}{2}}$$

$$\doteq \frac{e^4 m^2 k'_{1+}k''_{2+}}{q_{+0}q_{-+}(2q_{-0} + k''_{2+})(q_+ \cdot k_1)(q_- \cdot k_2)}\left[\frac{(q_+ \cdot q_-)}{2}\right]^{\frac{1}{2}},$$

$$M(-,-;+,-,-,+) = -\frac{2e^4mq_{-0}^2k_{1\perp}'^*q_{-\perp}Z_{+-}^*}{q_{+0}q_{--}(2q_{-0}+k_{2+}'')(q_+\cdot k_1)(q_-\cdot k_2)}$$

$$\times \left[\frac{k_{1+}'k_{2-}''}{k_{1-}'k_{2+}''q_{++}q_{-+}^3}\right]^{\frac{1}{2}}$$

$$\doteq \frac{e^4mq_{-0}^2k_{1+}'}{q_{+0}q_{-+}(2q_{-0}+k_{2+}'')(q_+\cdot k_1)(q_-\cdot k_2)}$$

$$\times \left[\frac{8(q_+\cdot q_-)k_{2-}''}{k_{2+}''}\right]^{\frac{1}{2}},$$

$$M(-,-;-,+,+,-) = \frac{-2e^4mq_{+0}k_{2\perp}'''q_{-\perp}Z_{+-}^*}{q_{--}(2q_{-0}+k_{2+}'')(q_+\cdot k_1)(q_-\cdot k_2)}\left[\frac{k_{1-}'k_{2+}''}{k_{1+}'k_{2-}''q_{++}q_{-+}^3}\right]^{\frac{1}{2}}$$

$$\doteq \frac{e^4mq_{+0}k_{2+}''}{q_{-+}(2q_{-0}+k_{2+}'')(q_+\cdot k_1)(q_-\cdot k_2)}\left[\frac{8(q_+\cdot q_-)k_{1-}'}{k_{1+}'}\right]^{\frac{1}{2}},$$

$$M(-,-;-,-,+,+) = \frac{-8e^4q_{+0}q_{-0}^2q_{-\perp}Z_{+-}^*}{q_{--}(2q_{-0}+k_{2+}'')(q_+\cdot k_1)(q_-\cdot k_2)}\left[\frac{k_{1-}'k_{2-}''}{k_{1+}'k_{2+}''q_{++}q_{-+}^3}\right]^{\frac{1}{2}}$$

$$\doteq \frac{e^4q_{+0}q_{-0}^2}{q_{-+}(2q_{-0}+k_{2+}'')(q_+\cdot k_1)(q_-\cdot k_2)}$$

$$\times \left[\frac{128(q_+\cdot q_-)k_{1-}'k_{2-}''}{k_{1+}'k_{2+}''}\right]^{\frac{1}{2}},$$

$$M(-,-;+,-,-,-) = -\frac{e^4mq_{-0}k_{1\perp}'^*q_{-\perp}Z_{+-}^*}{q_{+0}q_{--}(q_+\cdot k_1)(q_-\cdot k_2)}\left[\frac{k_{1+}'k_{2-}''}{k_{1-}'k_{2+}''q_{++}q_{-+}^3}\right]^{\frac{1}{2}}$$

$$\doteq \frac{e^4mq_{-0}k_{1+}'}{q_{+0}q_{-+}(q_+\cdot k_1)(q_-\cdot k_2)}\left[\frac{2(q_+\cdot q_-)k_{2-}''}{k_{2+}''}\right]^{\frac{1}{2}},$$

$$M(-,-;-,+,-,-) = -\frac{e^4mq_{+0}k_{2\perp}'''^*q_{-\perp}Z_{+-}^*}{q_{-0}(q_+\cdot k_1)(q_-\cdot k_2)}\left[\frac{k_{1-}'k_{2+}''}{k_{1+}'k_{2-}''q_{++}^3q_{-+}^3}\right]^{\frac{1}{2}}$$

$$\doteq \frac{e^4mq_{+0}q_{--}k_{2+}''}{q_{-0}q_{++}q_{-+}(q_+\cdot k_1)(q_-\cdot k_2)}\left[\frac{2(q_+\cdot q_-)k_{1-}'}{k_{1+}'}\right]^{\frac{1}{2}},$$

$$M(-,-;-,-,+,-) = -\frac{4e^4q_{+0}q_{-0}q_{-\perp}Z_{+-}^*}{q_{--}(q_+\cdot k_1)(q_-\cdot k_2)}\left[\frac{k_{1-}'k_{2-}''}{k_{1+}'k_{2+}''q_{++}q_{-+}^3}\right]^{\frac{1}{2}}$$

$$\doteq \frac{e^4q_{+0}q_{-0}}{q_{-+}(q_+\cdot k_1)(q_-\cdot k_2)}\left[\frac{32(q_+\cdot q_-)k_{1-}'k_{2-}''}{k_{1+}'k_{2+}''}\right]^{\frac{1}{2}},$$

9.19. $e^+e^- \to e^+e^-\gamma\gamma$ (NO Z-EXCHANGE; $m \neq 0$)

$$M(-,-;-,-,-,+) = -\frac{4e^4 q_{+0} q_{-0} q_{-\perp} Z^*_{+-}}{(q_+ \cdot k_1)(q_- \cdot k_2)} \left[\frac{k'_{1-}k''_{2-}}{k'_{1+}k''_{2+}q^3_{++}q^3_{-+}}\right]^{\frac{1}{2}}$$

$$\doteq \frac{e^4 q_{+0} q_{-0} q_{--}}{q_{++}q_{+-}(q_+ \cdot k_1)(q_- \cdot k_2)} \left[\frac{32(q_+ \cdot q_-)k'_{1-}k''_{2-}}{k'_{1+}k''_{2+}}\right]^{\frac{1}{2}},$$

$$M(-,-;-,-,-,-) = -\frac{2e^4 q_{-0}(2q_{+0} + k'_{1+})q_{-\perp}Z^*_{+-}}{q_{--}(q_+ \cdot k_1)(q_- \cdot k_2)} \left[\frac{k'_{1-}k''_{2-}}{k'_{1+}k''_{2+}q_{++}q^3_{-+}}\right]^{\frac{1}{2}}$$

$$\doteq \frac{e^4 q_{-0}(2q_{+0} + k'_{1+})}{q_{-+}(q_+ \cdot k_1)(q_- \cdot k_2)} \left[\frac{8(q_+ \cdot q_-)k'_{1-}k''_{2-}}{k'_{1+}k''_{2+}}\right]^{\frac{1}{2}}.$$

(9.255)

Unpolarized squared matrix element:

$$\overline{|M|^2} = \frac{e^8 k'_{1-} k''_{2-} \left[4(q_+ \cdot q_-)^2 + q^2_{++}q^2_{--} + q^2_{+-}q^2_{-+}\right]}{8k'_{1+}k''_{2+}q_{+-}q_{-+}[E(q_+ \cdot k_1)(q_- \cdot k_2)]^2}$$

$$\times \left[4q^2_{+0} + (2q_{+0} + k'_{1+})^2 + \frac{m^2 k'^3_{1+}}{4q^2_{+0}k'_{1-}}\right] \left[4q^2_{-0} + (2q_{-0} + k''_{2+})^2 + \frac{m^2 k''^3_{2+}}{4q^2_{-0}k''_{2-}}\right].$$

(9.256)

Unpolarized cross section:

$$d\sigma = \frac{\alpha^4 k'_{1-} k''_{2-} \left[4(q_+ \cdot q_-)^2 + q^2_{++}q^2_{--} + q^2_{+-}q^2_{-+}\right]}{1024\pi^2 E^4 k'_{1+}k''_{2+}q_{+-}q_{-+}[(q_+ \cdot k_1)(q_- \cdot k_2)]^2}$$

$$\times \left[4q^2_{+0} + (2q_{+0} + k'_{1+})^2 + \frac{m^2 k'^3_{1+}}{4q^2_{+0}k'_{1-}}\right] \left[4q^2_{-0} + (2q_{-0} + k''_{2+})^2 + \frac{m^2 k''^3_{2+}}{4q^2_{-0}k''_{2-}}\right]$$

$$\times \delta^4(p_+ + p_- - q_+ - q_- - k_1 - k_2) \frac{d^3\vec{q}_+ \, d^3\vec{q}_- \, d^3\vec{k}_1 \, d^3\vec{k}_2}{q_{+0} \, q_{-0} \, k_{10} \, k_{20}}.$$

(9.257)

9.19.11 $\vec{k}_1$ and $\vec{k}_2$ nearly parallel to $\vec{p}_+$

Definitions:

$$\Delta'_{12} = -2(p_+ \cdot k_1) - 2(p_+ \cdot k_2) + 2(k_1 \cdot k_2),$$

$$Z_{ij} = k_{i+}k_{j-} - k^*_{i\perp}k_{j\perp}, \qquad i,j = 1,2,$$

$$B_1 = \frac{2E^2}{(p_+ \cdot k_1)(p_+ \cdot k_2)} \left[k_{1-}k_{2-} + \frac{m^2 Z_{12}Z_{21}}{4E^2 \Delta'_{12}}\right],$$

$$B_2(1,2) = \frac{1}{\Delta'_{12}}\left[\frac{k_{1-}(2E - k_{1+})(2Ek_{2-} - Z_{12})}{(p_+ \cdot k_1)}\right.$$

$$\left.+\frac{2Ek_{1-}k_{2-}(2E - k_{1+} - k_{2+}) - m^2 k_{2+}k^*_{1\perp}k_{2\perp}/2E}{(p_+ \cdot k_2)}\right],$$

$$B_3 = \frac{m}{2(p_+ \cdot k_1)(p_+ \cdot k_2)}\left[k_{1+}k_{2-}k_{1\perp} + k_{1-}k_{2+}k_{2\perp}\right.$$

$$\left.-\frac{(k_{1\perp} + k_{2\perp})Z_{12}Z_{21}}{\Delta'_{12}}\right],$$

$$B_4(1,2) = \frac{mk_{1+}k_{1\perp}}{2E\Delta'_{12}}\left[\frac{2Ek_{2-} - Z^*_{12}}{(p_+ \cdot k_1)} + \frac{2Ek_{2-}}{(p_+ \cdot k_2)}\right]. \qquad (9.258)$$

Photon polarizations:

$$\not{k}^{\pm}_i = N_i[\not{k}_i \not{p}_+ \not{p}_-(1 \pm \gamma_5) + \not{p}_- \not{p}_+ \not{k}_i(1 \mp \gamma_5)],$$

$$N_i^{-1} = E^2[32k_{i+}k_{i-}]^{\frac{1}{2}}, \qquad\qquad i = 1, 2. \quad (9.259)$$

Nonvanishing helicity amplitudes:

$$M(+,+;+,+,+,+) = \frac{4e^4 E B_1 q^*_{-\perp}}{q_{--}}\left[\frac{q_{++}}{q^3_{-+}k_{1+}k_{1-}k_{2+}k_{2-}}\right]^{\frac{1}{2}}$$

$$\doteq \frac{4e^4 E|B_1|}{[q_{+-}q_{-+}k_{1+}k_{1-}k_{2+}k_{2-}]^{\frac{1}{2}}},$$

$$M(+,+;+,+,+,-) = \frac{4e^4 E B_2(2,1) q^*_{+\perp}}{q_{--}}\left[\frac{q_{++}}{q^3_{-+}k_{1+}k_{1-}k_{2+}k_{2-}}\right]^{\frac{1}{2}}$$

$$\doteq \frac{4e^4 E|B_2(2,1)|}{[q_{+-}q_{-+}k_{1+}k_{1-}k_{2+}k_{2-}]^{\frac{1}{2}}},$$

$$M(+,+;+,+,-,+) = \frac{4e^4 E B_2(1,2) q^*_{+\perp}}{q_{--}}\left[\frac{q_{++}}{q^3_{-+}k_{1+}k_{1-}k_{2+}k_{2-}}\right]^{\frac{1}{2}}$$

$$\doteq \frac{4e^4 E|B_2(1,2)|}{[q_{+-}q_{-+}k_{1+}k_{1-}k_{2+}k_{2-}]^{\frac{1}{2}}},$$

$$M(+,+;+,-,+,+) = \frac{2e^4 E B_3 q^*_{+\perp}}{(q_+ \cdot q_-)}\left[\frac{q_{-+}}{q_{++}k_{1+}k_{1-}k_{2+}k_{2-}}\right]^{\frac{1}{2}}$$

$$\doteq \frac{2e^4 E|B_3|}{(q_+ \cdot q_-)}\left[\frac{q_{+-}q_{-+}}{k_{1+}k_{1-}k_{2+}k_{2-}}\right]^{\frac{1}{2}},$$

9.19. $e^+e^- \to e^+e^- \gamma\gamma$ (NO Z-EXCHANGE; $m \neq 0$)

$$M(+,+;-,+,+,+) = \frac{2e^4 E B_3 q^*_{+\perp}}{(q_+ \cdot q_-)} \left[\frac{q^3_{++}}{q^3_{-+} k_{1+} k_{1-} k_{2+} k_{2-}} \right]^{\frac{1}{2}}$$

$$\doteq \frac{2e^4 E |B_3| q_{++} q_{--}}{(q_+ \cdot q_-)[q_{+-} q_{-+} k_{1+} k_{1-} k_{2+} k_{2-}]^{\frac{1}{2}}},$$

$$M(+,+;+,+,-,-) = \frac{2e^4(2E - k_{1+} - k_{2+}) B^*_1 q^*_{-\perp}}{q_{--}} \left[\frac{q_{++}}{q^3_{-+} k_{1+} k_{1-} k_{2+} k_{2-}} \right]^{\frac{1}{2}}$$

$$\doteq \frac{2e^4(2E - k_{1+} - k_{2+})|B_1|}{[q_{+-} q_{-+} k_{1+} k_{1-} k_{2+} k_{2-}]^{\frac{1}{2}}},$$

$$M(+,+;+,-,+,-) = \frac{2e^4 E B_4(1,2) q^*_{-\perp}}{(q_+ \cdot q_-)} \left[\frac{q_{-+}}{q_{++} k_{1+} k_{1-} k_{2+} k_{2-}} \right]^{\frac{1}{2}}$$

$$\doteq \frac{2e^4 E |B_4(1,2)|}{(q_+ \cdot q_-)} \left[\frac{q_{+-} q_{-+}}{k_{1+} k_{1-} k_{2+} k_{2-}} \right]^{\frac{1}{2}},$$

$$M(+,+;+,-,-,+) = \frac{2e^4 E B_4(2,1) q^*_{-\perp}}{(q_+ \cdot q_-)} \left[\frac{q_{+-}}{q_{++} k_{1+} k_{1-} k_{2+} k_{2-}} \right]^{\frac{1}{2}}$$

$$\doteq \frac{2e^4 E |B_4(2,1)|}{(q_+ \cdot q_-)} \left[\frac{q_{+-} q_{-+}}{k_{1+} k_{1-} k_{2+} k_{2-}} \right]^{\frac{1}{2}},$$

$$M(+,+;-,+,+,-) = \frac{2e^4 E B_4(1,2) q^*_{-\perp}}{(q_+ \cdot q_-)} \left[\frac{q^3_{++}}{q^3_{-+} k_{1+} k_{1-} k_{2+} k_{2-}} \right]^{\frac{1}{2}}$$

$$\doteq \frac{2e^4 E |B_4(1,2)| q_{++} q_{--}}{(q_+ \cdot q_-)[q_{+-} q_{-+} k_{1+} k_{1-} k_{2+} k_{2-}]^{\frac{1}{2}}},$$

$$M(+,+;-,+,-,+) = \frac{2e^4 E B_4(2,1) q^*_{-\perp}}{(q_+ \cdot q_-)} \left[\frac{q^3_{++}}{q^3_{-+} k_{1+} k_{1-} k_{2+} k_{2-}} \right]^{\frac{1}{2}}$$

$$\doteq \frac{2e^4 E |B_4(2,1)| q_{++} q_{--}}{(q_+ \cdot q_-)[q_{+-} q_{-+} k_{1+} k_{1-} k_{2+} k_{2-}]^{\frac{1}{2}}},$$

$$M(+,-;+,-,+,+) = \frac{2e^4 E B_1 q_{-\perp}}{(q_+ \cdot q_-)} \left[\frac{q^3_{++}}{q^3_{-+} k_{1+} k_{1-} k_{2+} k_{2-}} \right]^{\frac{1}{2}}$$

$$\doteq \frac{2e^4 E |B_1| q_{++} q_{--}}{(q_+ \cdot q_-)[q_{+-} q_{-+} k_{1+} k_{1-} k_{2+} k_{2-}]^{\frac{1}{2}}},$$

$$M(+,-;-,+,+,+) = \frac{2e^4 E B_1 q_{-\perp}}{(q_+ \cdot q_-)} \left[\frac{q_{-+}}{q_{++} k_{1+} k_{1-} k_{2+} k_{2-}} \right]^{\frac{1}{2}}$$

$$\doteq \frac{2e^4 E |B_1|}{(q_+ \cdot q_-)} \left[\frac{q_{+-} q_{-+}}{k_{1+} k_{1-} k_{2+} k_{2-}} \right]^{\frac{1}{2}},$$

$$M(-,+;+,+,+,-) = \frac{4e^4 E B_4^*(2,1) q_{+\perp}^*}{q_{--}} \left[\frac{q_{++}}{q_{-+}^3 k_{1+} k_{1-} k_{2+} k_{2-}} \right]^{\frac{1}{2}}$$

$$\doteq \frac{4e^4 E |B_4(2,1)|}{[q_{+-} q_{-+} k_{1+} k_{1-} k_{2+} k_{2-}]^{\frac{1}{2}}},$$

$$M(-,+;+,+,-,+) = \frac{4e^4 E B_4^*(1,2) q_{+\perp}^*}{q_{--}} \left[\frac{q_{++}}{q_{-+}^3 k_{1+} k_{1-} k_{2+} k_{2-}} \right]^{\frac{1}{2}}$$

$$\doteq \frac{4e^4 E |B_4(1,2)|}{[q_{+-} q_{-+} k_{1+} k_{1-} k_{2+} k_{2-}]^{\frac{1}{2}}},$$

$$M(-,+;+,-,+,+) = \frac{e^4 (2E - k_{1+} - k_{2+}) B_1 q_{-\perp}^*}{(q_+ \cdot q_-)} \left[\frac{q_{-+}}{q_{++} k_{1+} k_{1-} k_{2+} k_{2-}} \right]^{\frac{1}{2}}$$

$$\doteq \frac{e^4 (2E - k_{1+} - k_{2+}) |B_1|}{(q_+ \cdot q_-)} \left[\frac{q_{+-} q_{-+}}{k_{1+} k_{1-} k_{2+} k_{2-}} \right]^{\frac{1}{2}},$$

$$M(-,+;-,+,+,+) = \frac{e^4 (2E - k_{1+} - k_{2+}) B_1 q_{-\perp}^*}{(q_+ \cdot q_-)} \left[\frac{q_{++}^3}{q_{-+}^3 k_{1+} k_{1-} k_{2+} k_{2-}} \right]^{\frac{1}{2}}$$

$$\doteq \frac{e^4 (2E - k_{1+} - k_{2+}) |B_1| q_{++} q_{--}}{(q_+ \cdot q_-) [q_{+-} q_{-+} k_{1+} k_{1-} k_{2+} k_{2-}]^{\frac{1}{2}}},$$

$$M(+,-;+,-,+,-) = \frac{2e^4 E B_2(2,1) q_{+\perp}}{(q_+ \cdot q_-)} \left[\frac{q_{++}^3}{q_{-+}^3 k_{1+} k_{1-} k_{2+} k_{2-}} \right]^{\frac{1}{2}}$$

$$\doteq \frac{2e^4 E |B_2(2,1)| q_{++} q_{--}}{(q_+ \cdot q_-) [q_{+-} q_{-+} k_{1+} k_{1-} k_{2+} k_{2-}]^{\frac{1}{2}}},$$

$$M(+,-;+,-,-,+) = \frac{2e^4 E B_2(1,2) q_{+\perp}}{(q_+ \cdot q_-)} \left[\frac{q_{++}^3}{q_{-+}^3 k_{1+} k_{1-} k_{2+} k_{2-}} \right]^{\frac{1}{2}}$$

$$\doteq \frac{2e^4 E |B_2(1,2)| q_{++} q_{--}}{(q_+ \cdot q_-) [q_{+-} q_{-+} k_{1+} k_{1-} k_{2+} k_{2-}]^{\frac{1}{2}}},$$

9.19. $e^+e^- \to e^+e^-\gamma\gamma$ (NO Z-EXCHANGE; $m \neq 0$)

$$M(+,-;-,+,+,-) = \frac{2e^4 E B_2(2,1) q_{+\perp}}{(q_+ \cdot q_-)} \left[\frac{q_{-+}}{q_{++} k_{1+} k_{1-} k_{2+} k_{2-}}\right]^{\frac{1}{2}}$$

$$\doteq \frac{2e^4 E |B_2(2,1)|}{(q_+ \cdot q_-)} \left[\frac{q_{+-} q_{-+}}{k_{1+} k_{1-} k_{2+} k_{2-}}\right]^{\frac{1}{2}},$$

$$M(+,-;-,+,-,+) = \frac{2e^4 E B_2(1,2) q_{+\perp}}{(q_+ \cdot q_-)} \left[\frac{q_{-+}}{q_{++} k_{1+} k_{1-} k_{2+} k_{2-}}\right]^{\frac{1}{2}}$$

$$\doteq \frac{2e^4 E |B_2(1,2)|}{(q_+ \cdot q_-)} \left[\frac{q_{+-} q_{-+}}{k_{1+} k_{1-} k_{2+} k_{2-}}\right]^{\frac{1}{2}},$$

$$M(+,-;-,-,+,+) = \frac{4e^4 E B_3 q_{+\perp}}{q_{--}} \left[\frac{q_{++}}{q_{-+}^3 k_{1+} k_{1-} k_{2+} k_{2-}}\right]^{\frac{1}{2}}$$

$$\doteq \frac{4e^4 E |B_3|}{[q_{+-} q_{-+} k_{1+} k_{1-} k_{2+} k_{2-}]^{\frac{1}{2}}},$$

$$M(-,+;+,+,-,-) = \frac{4e^4 E B_3^* q_{-\perp}^*}{q_{--}} \left[\frac{q_{++}}{q_{-+}^3 k_{1+} k_{1-} k_{2+} k_{2-}}\right]^{\frac{1}{2}}$$

$$\doteq \frac{4e^4 E |B_3|}{[q_{+-} q_{-+} k_{1+} k_{1-} k_{2+} k_{2-}]^{\frac{1}{2}}},$$

$$M(-,+;+,-,+,-) = \frac{2e^4 E B_2^*(1,2) q_{+\perp}^*}{(q_+ \cdot q_-)} \left[\frac{q_{-+}}{q_{++} k_{1+} k_{1-} k_{2+} k_{2-}}\right]^{\frac{1}{2}}$$

$$\doteq \frac{2e^4 E |B_2(1,2)|}{(q_+ \cdot q_-)} \left[\frac{q_{+-} q_{-+}}{k_{1+} k_{1-} k_{2+} k_{2-}}\right]^{\frac{1}{2}},$$

$$M(-,+;+,-,-,+) = \frac{2e^4 E B_2^*(2,1) q_{+\perp}^*}{(q_+ \cdot q_-)} \left[\frac{q_{-+}}{q_{++} k_{1+} k_{1-} k_{2+} k_{2-}}\right]^{\frac{1}{2}}$$

$$\doteq \frac{2e^4 E |B_2(2,1)|}{(q_+ \cdot q_-)} \left[\frac{q_{+-} q_{-+}}{k_{1+} k_{1-} k_{2+} k_{2-}}\right]^{\frac{1}{2}},$$

$$M(-,+;-,+,+,-) = \frac{2e^4 E B_2^*(1,2) q_{+\perp}^*}{(q_+ \cdot q_-)} \left[\frac{q_{++}^3}{q_{-+}^3 k_{1+} k_{1-} k_{2+} k_{2-}}\right]^{\frac{1}{2}}$$

$$\doteq \frac{2e^4 E |B_2(1,2)| q_{++} q_{--}}{(q_+ \cdot q_-)[q_{+-} q_{-+} k_{1+} k_{1-} k_{2+} k_{2-}]^{\frac{1}{2}}},$$

$$M(-,+;-,+,-,+) = \frac{2e^4 E B_2^*(2,1) q_{+\perp}^*}{(q_+ \cdot q_-)} \left[\frac{q_{++}^3}{q_{-+}^3 k_{1+} k_{1-} k_{2+} k_{2-}} \right]^{\frac{1}{2}}$$

$$\doteq \frac{2e^4 E |B_2(2,1)| q_{++} q_{--}}{(q_+ \cdot q_-) [q_{+-} q_{-+} k_{1+} k_{1-} k_{2+} k_{2-}]^{\frac{1}{2}}},$$

$$M(+,-;+,-,-,-) = \frac{e^4 (2E - k_{1+} - k_{2+}) B_1^* q_{-\perp}}{(q_+ \cdot q_-)} \left[\frac{q_{++}^3}{q_{-+}^3 k_{1+} k_{1-} k_{2+} k_{2-}} \right]^{\frac{1}{2}}$$

$$\doteq \frac{e^4 (2E - k_{1+} - k_{2+}) |B_1| q_{++} q_{--}}{(q_+ \cdot q_-) [q_{+-} q_{-+} k_{1+} k_{1-} k_{2+} k_{2-}]^{\frac{1}{2}}},$$

$$M(+,-;-,+,-,-) = \frac{e^4 (2E - k_{1+} - k_{2+}) B_1^* q_{-\perp}}{(q_+ \cdot q_-)} \left[\frac{q_{-+}}{q_{++} k_{1+} k_{1-} k_{2+} k_{2-}} \right]^{\frac{1}{2}}$$

$$\doteq \frac{e^4 (2E - k_{1+} - k_{2+}) |B_1|}{(q_+ \cdot q_-)} \left[\frac{q_{+-} q_{-+}}{k_{1+} k_{1-} k_{2+} k_{2-}} \right]^{\frac{1}{2}},$$

$$M(+,-;-,-,+,-) = \frac{4e^4 E B_4(1,2) q_{-\perp}}{q_{--}} \left[\frac{q_{++}}{q_{-+}^3 k_{1+} k_{1-} k_{2+} k_{2-}} \right]^{\frac{1}{2}}$$

$$\doteq \frac{4e^4 E |B_4(1,2)|}{[q_{+-} q_{-+} k_{1+} k_{1-} k_{2+} k_{2-}]^{\frac{1}{2}}},$$

$$M(+,-;-,-,-,+) = \frac{4e^4 E B_4(2,1) q_{-\perp}}{q_{--}} \left[\frac{q_{++}}{q_{-+}^3 k_{1+} k_{1-} k_{2+} k_{2-}} \right]^{\frac{1}{2}}$$

$$\doteq \frac{4e^4 E |B_4(2,1)|}{[q_{+-} q_{-+} k_{1+} k_{1-} k_{2+} k_{2-}]^{\frac{1}{2}}},$$

$$M(-,+;+,-,-,-) = \frac{2e^4 E B_1^* q_{-\perp}^*}{(q_+ \cdot q_-)} \left[\frac{q_{-+}}{q_{++} k_{1+} k_{1-} k_{2+} k_{2-}} \right]^{\frac{1}{2}}$$

$$\doteq \frac{2e^4 E |B_1|}{(q_+ \cdot q_-)} \left[\frac{q_{+-} q_{-+}}{k_{1+} k_{1-} k_{2+} k_{2-}} \right]^{\frac{1}{2}},$$

$$M(-,+;-,+,-,-) = \frac{2e^4 E B_1^* q_{-\perp}^*}{(q_+ \cdot q_-)} \left[\frac{q_{++}^3}{q_{-+}^3 k_{1+} k_{1-} k_{2+} k_{2-}} \right]^{\frac{1}{2}}$$

$$\doteq \frac{2e^4 E |B_1| q_{++} q_{--}}{(q_+ \cdot q_-) [q_{+-} q_{-+} k_{1+} k_{1-} k_{2+} k_{2-}]^{\frac{1}{2}}},$$

9.19. $e^+e^- \to e^+e^- \gamma\gamma$ (NO Z-EXCHANGE; $m \neq 0$)

$$M(-,-;+,-,+,-) = \frac{2e^4 E B_4^*(2,1) q_{+\perp}}{(q_+ \cdot q_-)} \left[\frac{q_{++}^3}{q_{-+}^3 k_{1+} k_{1-} k_{2+} k_{2-}}\right]^{\frac{1}{2}}$$

$$\doteq \frac{2e^4 E |B_4(2,1)| q_{++} q_{--}}{(q_+ \cdot q_-) [q_{+-} q_{-+} k_{1+} k_{1-} k_{2+} k_{2-}]^{\frac{1}{2}}},$$

$$M(-,-;+,-,-,+) = \frac{2e^4 E B_4^*(1,2) q_{+\perp}}{(q_+ \cdot q_-)} \left[\frac{q_{++}^3}{q_{-+}^3 k_{1+} k_{1-} k_{2+} k_{2-}}\right]^{\frac{1}{2}}$$

$$\doteq \frac{2e^4 E |B_4(1,2)| q_{++} q_{--}}{(q_+ \cdot q_-) [q_{+-} q_{-+} k_{1+} k_{1-} k_{2+} k_{2-}]^{\frac{1}{2}}},$$

$$M(-,-;-,+,+,-) = \frac{2e^4 E B_4^*(2,1) q_{+\perp}}{(q_+ \cdot q_-)} \left[\frac{q_{-+}}{q_{++} k_{1+} k_{1-} k_{2+} k_{2-}}\right]^{\frac{1}{2}}$$

$$\doteq \frac{2e^4 E |B_4(2,1)|}{(q_+ \cdot q_-)} \left[\frac{q_{+-} q_{-+}}{k_{1+} k_{1-} k_{2+} k_{2-}}\right]^{\frac{1}{2}},$$

$$M(-,-;-,+,-,+) = \frac{2e^4 E B_4^*(1,2) q_{+\perp}}{(q_+ \cdot q_-)} \left[\frac{q_{-+}}{q_{++} k_{1+} k_{1-} k_{2+} k_{2-}}\right]^{\frac{1}{2}}$$

$$\doteq \frac{2e^4 E |B_4(1,2)|}{(q_+ \cdot q_-)} \left[\frac{q_{+-} q_{-+}}{k_{1+} k_{1-} k_{2+} k_{2-}}\right]^{\frac{1}{2}},$$

$$M(-,-;-,-,+,+) = \frac{2e^4 (2E - k_{1+} - k_{2+}) B_1 q_{-\perp}}{q_{--}} \left[\frac{q_{++}}{q_{-+}^3 k_{1+} k_{1-} k_{2+} k_{2-}}\right]^{\frac{1}{2}}$$

$$\doteq \frac{2e^4 (2E - k_{1+} - k_{2+}) |B_1|}{[q_{+-} q_{-+} k_{1+} k_{1-} k_{2+} k_{2-}]^{\frac{1}{2}}},$$

$$M(-,-;+,-,-,-) = \frac{2e^4 E B_3^* q_{-\perp}}{(q_+ \cdot q_-)} \left[\frac{q_{++}^3}{q_{-+}^3 k_{1+} k_{1-} k_{2+} k_{2-}}\right]^{\frac{1}{2}}$$

$$\doteq \frac{2e^4 E |B_3| q_{++} q_{--}}{(q_+ \cdot q_-) [q_{+-} q_{-+} k_{1+} k_{1-} k_{2+} k_{2-}]^{\frac{1}{2}}},$$

$$M(-,-;-,+,-,-) = \frac{2e^4 E B_3^* q_{-\perp}}{(q_+ \cdot q_-)} \left[\frac{q_{-+}}{q_{++} k_{1+} k_{1-} k_{2+} k_{2-}}\right]^{\frac{1}{2}}$$

$$\doteq \frac{2e^4 E |B_3|}{(q_+ \cdot q_-)} \left[\frac{q_{+-} q_{-+}}{k_{1+} k_{1-} k_{2+} k_{2-}}\right]^{\frac{1}{2}},$$

$$M(-,-;-,-,+,-) = \frac{4e^4 E B_2^*(1,2) q_{+\perp}}{q_{--}} \left[\frac{q_{++}}{q_{-+}^3 k_{1+} k_{1-} k_{2+} k_{2-}}\right]^{\frac{1}{2}}$$

$$\doteq \frac{4e^4 E |B_2(1,2)|}{[q_{+-}q_{-+}k_{1+}k_{1-}k_{2+}k_{2-}]^{\frac{1}{2}}},$$

$$M(-,-;-,-,-,+) = \frac{4e^4 E B_2^*(2,1) q_{+\perp}}{q_{--}} \left[\frac{q_{++}}{q_{-+}^3 k_{1+} k_{1-} k_{2+} k_{2-}}\right]^{\frac{1}{2}}$$

$$\doteq \frac{4e^4 E |B_2(2,1)|}{[q_{+-}q_{-+}k_{1+}k_{1-}k_{2+}k_{2-}]^{\frac{1}{2}}},$$

$$M(-,-;-,-,-,-) = \frac{4e^4 E B_1^* q_{-\perp}}{q_{--}} \left[\frac{q_{++}}{q_{-+}^3 k_{1+} k_{1-} k_{2+} k_{2-}}\right]^{\frac{1}{2}}$$

$$\doteq \frac{4e^4 E |B_1|}{[q_{+-}q_{-+}k_{1+}k_{1-}k_{2+}k_{2-}]^{\frac{1}{2}}}. \quad (9.260)$$

Unpolarized squared matrix element:

$$\overline{|M|^2} = \frac{2e^8 E^2 \left[4(q_+ \cdot q_-)^2 + q_{++}^2 q_{--}^2 + q_{+-}^2 q_{-+}^2\right]}{k_{1+}k_{1-}k_{2+}k_{2-}q_{+-}q_{-+}(q_+ \cdot q_-)^2}$$

$$\times \left\{\left[1 + \left(\frac{2E - k_{1+} - k_{2+}}{2E}\right)^2\right] |B_1|^2 + |B_2(1,2)|^2 \right.$$

$$\left. + |B_2(2,1)|^2 + |B_3|^2 + |B_4(1,2)|^2 + |B_4(2,1)|^2 \right\}.$$

$$(9.261)$$

Unpolarized cross section:

$$d\sigma = \frac{\alpha^4 \left[4(q_+ \cdot q_-)^2 + q_{++}^2 q_{--}^2 + q_{+-}^2 q_{-+}^2\right]}{64\pi^2 k_{1+}k_{1-}k_{2+}k_{2-}q_{+-}q_{-+}(q_+ \cdot q_-)^2}$$

$$\times \left\{\left[1 + \left(\frac{2E - k_{1+} - k_{2+}}{2E}\right)^2\right] |B_1|^2 \right.$$

$$\left. + |B_2(1,2)|^2 + |B_2(2,1)|^2 + |B_3|^2 + |B_4(1,2)|^2 + |B_4(2,1)|^2 \right\}$$

$$\times \delta^4(p_+ + p_- - q_+ - q_- - k_1 - k_2) \frac{d^3\vec{q}_+ \, d^3\vec{q}_- \, d^3\vec{k}_1 \, d^3\vec{k}_2}{q_{+0} \, q_{-0} \, k_{10} \, k_{20}}. \quad (9.262)$$

9.19.12 $\vec{k}_1$ and $\vec{k}_2$ nearly parallel to $\vec{p}_-$

Definitions:

$$\Delta_{12} = -2(p_- \cdot k_1) - 2(p_- \cdot k_2) + 2(k_1 \cdot k_2),$$

$$Z_{ij} = k_{i+}k_{j-} - k^*_{i\perp}k_{j\perp}, \qquad i,j = 1,2,$$

$$A_1 = \frac{2E^2}{(p_- \cdot k_1)(p_- \cdot k_2)}\left[k_{1+}k_{2+} + \frac{m^2 Z_{12}Z_{21}}{4E^2\Delta_{12}}\right],$$

$$A_2(1,2) = \frac{1}{\Delta_{12}}\left[\frac{k_{1+}(2E - k_{1-})(2Ek_{2+} - Z_{21})}{(p_- \cdot k_1)}\right.$$
$$\left. + \frac{2Ek_{1+}k_{2+}(2E - k_{1-} - k_{2-}) - m^2 k_{2-}k_{1\perp}k^*_{2\perp}/2E}{(p_- \cdot k_2)}\right],$$

$$A_3 = \frac{m}{2(p_- \cdot k_1)(p_- \cdot k_2)}\left[k_{1-}k_{2+}k^*_{1\perp} + k_{1+}k_{2-}k^*_{2\perp}\right.$$
$$\left. - \frac{(k^*_{1\perp} + k^*_{2\perp})Z_{12}Z_{21}}{\Delta_{12}}\right],$$

$$A_4(1,2) = \frac{mk_{1-}k_{1\perp}}{2E\Delta_{12}}\left[\frac{2Ek_{2+} - Z_{21}}{(p_- \cdot k_1)} + \frac{2Ek_{2+}}{(p_- \cdot k_2)}\right]. \qquad (9.263)$$

Photon polarizations:

$$\not{\!\epsilon}^{\pm}_i = N_i[\not{\!k}_i \not{\!p}_- \not{\!p}_+(1 \pm \gamma_5) + \not{\!p}_+ \not{\!p}_- \not{\!k}_i(1 \mp \gamma_5)],$$

$$N_i^{-1} = E^2[32k_{i+}k_{i-}]^{\frac{1}{2}}, \qquad i = 1,2. \qquad (9.264)$$

Nonvanishing helicity amplitudes:

$$M(+,+;+,+,+,+) = \frac{4e^4 E A_1 q^*_{-\perp}}{q_{+-}[q_{++}q_{-+}k_{1+}k_{1-}k_{2+}k_{2-}]^{\frac{1}{2}}}$$

$$\doteq \frac{4e^4 E |A_1|}{[q_{+-}q_{-+}k_{1+}k_{1-}k_{2+}k_{2-}]^{\frac{1}{2}}},$$

$$M(+,+;+,+,+,-) = \frac{4e^4 E A_2(2,1) q^*_{+\perp}}{q_{+-}[q_{++}q_{-+}k_{1+}k_{1-}k_{2+}k_{2-}]^{\frac{1}{2}}}$$

$$\doteq \frac{4e^4 E |A_2(2,1)|}{[q_{+-}q_{-+}k_{1+}k_{1-}k_{2+}k_{2-}]^{\frac{1}{2}}},$$

$$M(+,+;+,+,-,+) = \frac{4e^4 E A_2(1,2) q_{+\perp}^*}{q_{+-}[q_{++}q_{-+}k_{1+}k_{1-}k_{2+}k_{2-}]^{\frac{1}{2}}}$$

$$\doteq \frac{4e^4 E |A_2(1,2)|}{[q_{+-}q_{-+}k_{1+}k_{1-}k_{2+}k_{2-}]^{\frac{1}{2}}},$$

$$M(+,+;+,-,+,+) = \frac{2e^4 E A_3 q_{+\perp}}{(q_+ \cdot q_-)} \left[\frac{q_{++}^3}{q_{-+}^3 k_{1+}k_{1-}k_{2+}k_{2-}}\right]^{\frac{1}{2}}$$

$$\doteq \frac{2e^4 E |A_3| q_{++}q_{--}}{(q_+ \cdot q_-)[q_{+-}q_{-+}k_{1+}k_{1-}k_{2+}k_{2-}]^{\frac{1}{2}}},$$

$$M(+,+;-,+,+,+) = \frac{2e^4 E A_3 q_{+\perp}}{(q_+ \cdot q_-)} \left[\frac{q_{-+}}{q_{++}k_{1+}k_{1-}k_{2+}k_{2-}}\right]^{\frac{1}{2}}$$

$$\doteq \frac{2e^4 E |A_3|}{(q_+ \cdot q_-)} \left[\frac{q_{+-}q_{-+}}{k_{1+}k_{1-}k_{2+}k_{2-}}\right]^{\frac{1}{2}},$$

$$M(+,+;+,+,-,-) = \frac{2e^4 (2E - k_{1-} - k_{2-}) A_1^* q_{-\perp}^*}{q_{+-}[q_{++}q_{-+}k_{1+}k_{1-}k_{2+}k_{2-}]^{\frac{1}{2}}}$$

$$\doteq \frac{2e^4 (2E - k_{1-} - k_{2-}) |A_1|}{[q_{+-}q_{-+}k_{1+}k_{1-}k_{2+}k_{2-}]^{\frac{1}{2}}},$$

$$M(+,+;+,-,+,-) = \frac{2e^4 E A_4^*(1,2) q_{+\perp}}{(q_+ \cdot q_-)} \left[\frac{q_{++}^3}{q_{-+}^3 k_{1+}k_{1-}k_{2+}k_{2-}}\right]^{\frac{1}{2}}$$

$$\doteq \frac{2e^4 E |A_4(1,2)| q_{++}q_{--}}{(q_+ \cdot q_-)[q_{+-}q_{-+}k_{1+}k_{1-}k_{2+}k_{2-}]^{\frac{1}{2}}},$$

$$M(+,+;+,-,-,+) = \frac{2e^4 E A_4^*(2,1) q_{+\perp}}{(q_+ \cdot q_-)} \left[\frac{q_{++}^3}{q_{-+}^3 k_{1+}k_{1-}k_{2+}k_{2-}}\right]^{\frac{1}{2}}$$

$$\doteq \frac{2e^4 E |A_4(2,1)| q_{++}q_{--}}{(q_+ \cdot q_-)[q_{+-}q_{-+}k_{1+}k_{1-}k_{2+}k_{2-}]^{\frac{1}{2}}},$$

$$M(+,+;-,+,+,-) = \frac{2e^4 E A_4^*(1,2) q_{+\perp}}{(q_+ \cdot q_-)} \left[\frac{q_{-+}}{q_{++}k_{1+}k_{1-}k_{2+}k_{2-}}\right]^{\frac{1}{2}}$$

$$\doteq \frac{2e^4 E |A_4(1,2)|}{(q_+ \cdot q_-)} \left[\frac{q_{+-}q_{-+}}{k_{1+}k_{1-}k_{2+}k_{2-}}\right]^{\frac{1}{2}},$$

9.19. $e^+e^- \to e^+e^-\gamma\gamma$ *(NO Z-EXCHANGE; $m \neq 0$)*   349

$$M(+,+;-,+,-,+) = \frac{2e^4 E A_4^*(2,1) q_{+\perp}}{(q_+ \cdot q_-)} \left[\frac{q_{-+}}{q_{++}k_{1+}k_{1-}k_{2+}k_{2-}} \right]^{\frac{1}{2}}$$

$$\doteq \frac{2e^4 E |A_4(2,1)|}{(q_+ \cdot q_-)} \left[\frac{q_{+-}q_{-+}}{k_{1+}k_{1-}k_{2+}k_{2-}} \right]^{\frac{1}{2}},$$

$$M(+,-;+,+,+,-) = \frac{4e^4 E A_4(2,1) q_{-\perp}^*}{q_{+-} [q_{++}q_{-+}k_{1+}k_{1-}k_{2+}k_{2-}]^{\frac{1}{2}}}$$

$$\doteq \frac{4e^4 E |A_4(2,1)|}{[q_{+-}q_{-+}k_{1+}k_{1-}k_{2+}k_{2-}]^{\frac{1}{2}}},$$

$$M(+,-;+,+,-,+) = \frac{4e^4 E A_4(1,2) q_{-\perp}^*}{q_{+-} [q_{++}q_{-+}k_{1+}k_{1-}k_{2+}k_{2-}]^{\frac{1}{2}}}$$

$$\doteq \frac{4e^4 E |A_4(1,2)|}{[q_{+-}q_{-+}k_{1+}k_{1-}k_{2+}k_{2-}]^{\frac{1}{2}}},$$

$$M(+,-;+,-,+,+) = \frac{e^4 (2E - k_{1-} - k_{2-}) A_1 q_{-\perp}}{(q_+ \cdot q_-)} \left[\frac{q_{++}^3}{q_{-+}^3 k_{1+}k_{1-}k_{2+}k_{2-}} \right]^{\frac{1}{2}}$$

$$\doteq \frac{e^4 (2E - k_{1-} - k_{2-}) |A_1| q_{++} q_{--}}{(q_+ \cdot q_-) [q_{+-} q_{-+} k_{1+}k_{1-}k_{2+}k_{2-}]^{\frac{1}{2}}},$$

$$M(+,-;-,+,+,+) = \frac{e^4 (2E - k_{1-} - k_{2-}) A_1 q_{-\perp}}{(q_+ \cdot q_-)} \left[\frac{q_{-+}}{q_{++} k_{1+} k_{1-} k_{2+} k_{2-}} \right]^{\frac{1}{2}}$$

$$\doteq \frac{e^4 (2E - k_{1-} - k_{2-}) |A_1|}{(q_+ \cdot q_-)} \left[\frac{q_{+-} q_{-+}}{k_{1+} k_{1-} k_{2+} k_{2-}} \right]^{\frac{1}{2}},$$

$$M(-,+;+,-,+,+) = \frac{2e^4 E A_1 q_{-\perp}^*}{(q_+ \cdot q_-)} \left[\frac{q_{-+}}{q_{++} k_{1+} k_{1-} k_{2+} k_{2-}} \right]^{\frac{1}{2}}$$

$$\doteq \frac{2e^4 E |A_1|}{(q_+ \cdot q_-)} \left[\frac{q_{+-} q_{-+}}{k_{1+} k_{1-} k_{2+} k_{2-}} \right]^{\frac{1}{2}},$$

$$M(-,+;-,+,+,+) = \frac{2e^4 E A_1 q_{-\perp}^*}{(q_+ \cdot q_-)} \left[\frac{q_{++}^3}{q_{-+}^3 k_{1+} k_{1-} k_{2+} k_{2-}} \right]^{\frac{1}{2}}$$

$$\doteq \frac{2e^4 E |A_1| q_{++} q_{--}}{(q_+ \cdot q_-) [q_{+-} q_{-+} k_{1+} k_{1-} k_{2+} k_{2-}]^{\frac{1}{2}}},$$

$$M(+,-;+,+,-,-) = \frac{4e^4 E A_3^* q_{-\perp}^*}{q_{+-}[q_{++}q_{-+}k_{1+}k_{1-}k_{2+}k_{2-}]^{\frac{1}{2}}}$$

$$\doteq \frac{4e^4 E |A_3|}{[q_{+-}q_{-+}k_{1+}k_{1-}k_{2+}k_{2-}]^{\frac{1}{2}}},$$

$$M(+,-;+,-,+,-) = \frac{2e^4 E A_2^*(1,2) q_{+\perp}}{(q_+ \cdot q_-)} \left[\frac{q_{++}^3}{q_{-+}^3 k_{1+}k_{1-}k_{2+}k_{2-}}\right]^{\frac{1}{2}}$$

$$\doteq \frac{2e^4 E |A_2(1,2)| q_{++}q_{--}}{(q_+ \cdot q_-)[q_{+-}q_{-+}k_{1+}k_{1-}k_{2+}k_{2-}]^{\frac{1}{2}}},$$

$$M(+,-;+,-,-,+) = \frac{2e^4 E A_2^*(2,1) q_{+\perp}}{(q_+ \cdot q_-)} \left[\frac{q_{++}^3}{q_{-+}^3 k_{1+}k_{1-}k_{2+}k_{2-}}\right]^{\frac{1}{2}}$$

$$\doteq \frac{2e^4 E |A_2(2,1)| q_{++}q_{--}}{(q_+ \cdot q_-)[q_{+-}q_{-+}k_{1+}k_{1-}k_{2+}k_{2-}]^{\frac{1}{2}}},$$

$$M(+,-;-,+,+,-) = \frac{2e^4 E A_2^*(1,2) q_{+\perp}}{(q_+ \cdot q_-)} \left[\frac{q_{-+}}{q_{++}k_{1+}k_{1-}k_{2+}k_{2-}}\right]^{\frac{1}{2}}$$

$$\doteq \frac{2e^4 E |A_2(1,2)|}{(q_+ \cdot q_-)} \left[\frac{q_{+-}q_{-+}}{k_{1+}k_{1-}k_{2+}k_{2-}}\right]^{\frac{1}{2}},$$

$$M(+,-;-,+,-,+) = \frac{2e^4 E A_2^*(2,1) q_{+\perp}}{(q_+ \cdot q_-)} \left[\frac{q_{-+}}{q_{++}k_{1+}k_{1-}k_{2+}k_{2-}}\right]^{\frac{1}{2}}$$

$$\doteq \frac{2e^4 E |A_2(2,1)|}{(q_+ \cdot q_-)} \left[\frac{q_{+-}q_{-+}}{k_{1+}k_{1-}k_{2+}k_{2-}}\right]^{\frac{1}{2}},$$

$$M(-,+;+,-,+,-) = \frac{2e^4 E A_2(2,1) q_{+\perp}^*}{(q_+ \cdot q_-)} \left[\frac{q_{-+}}{q_{++}k_{1+}k_{1-}k_{2+}k_{2-}}\right]^{\frac{1}{2}}$$

$$\doteq \frac{2e^4 E |A_2(2,1)|}{(q_+ \cdot q_-)} \left[\frac{q_{+-}q_{-+}}{k_{1+}k_{1-}k_{2+}k_{2-}}\right]^{\frac{1}{2}},$$

$$M(-,+;+,-,-,+) = \frac{2e^4 E A_2(1,2) q_{+\perp}^*}{(q_+ \cdot q_-)} \left[\frac{q_{-+}}{q_{++}k_{1+}k_{1-}k_{2+}k_{2-}}\right]^{\frac{1}{2}}$$

$$\doteq \frac{2e^4 E |A_2(1,2)|}{(q_+ \cdot q_-)} \left[\frac{q_{+-}q_{-+}}{k_{1+}k_{1-}k_{2+}k_{2-}}\right]^{\frac{1}{2}},$$

9.19. $e^+e^- \to e^+e^-\gamma\gamma$ (NO Z-EXCHANGE; $m \neq 0$)

$$M(-,+;-,+,+,-) = \frac{2e^4 E A_2(2,1) q^*_{+\perp}}{(q_+ \cdot q_-)} \left[\frac{q^3_{++}}{q^3_{-+} k_{1+} k_{1-} k_{2+} k_{2-}} \right]^{\frac{1}{2}}$$

$$\doteq \frac{2e^4 E |A_2(2,1)| q_{++} q_{--}}{(q_+ \cdot q_-)[q_{+-} q_{-+} k_{1+} k_{1-} k_{2+} k_{2-}]^{\frac{1}{2}}},$$

$$M(-,+;-,+,-,+) = \frac{2e^4 E A_2(1,2) q^*_{+\perp}}{(q_+ \cdot q_-)} \left[\frac{q^3_{++}}{q^3_{-+} k_{1+} k_{1-} k_{2+} k_{2-}} \right]^{\frac{1}{2}}$$

$$\doteq \frac{2e^4 E |A_2(1,2)| q_{++} q_{--}}{(q_+ \cdot q_-)[q_{+-} q_{-+} k_{1+} k_{1-} k_{2+} k_{2-}]^{\frac{1}{2}}},$$

$$M(-,+;-,-,+,+) = \frac{4e^4 E A_3 q_{+\perp}}{q_{+-}[q_{++} q_{-+} k_{1+} k_{1-} k_{2+} k_{2-}]^{\frac{1}{2}}}$$

$$\doteq \frac{4e^4 E |A_3|}{[q_{+-} q_{-+} k_{1+} k_{1-} k_{2+} k_{2-}]^{\frac{1}{2}}},$$

$$M(+,-;+,-,-,-) = \frac{2e^4 E A^*_1 q_{-\perp}}{(q_+ \cdot q_-)} \left[\frac{q^3_{++}}{q^3_{-+} k_{1+} k_{1-} k_{2+} k_{2-}} \right]^{\frac{1}{2}}$$

$$\doteq \frac{2e^4 E |A_1| q_{++} q_{--}}{(q_+ \cdot q_-)[q_{+-} q_{-+} k_{1+} k_{1-} k_{2+} k_{2-}]^{\frac{1}{2}}},$$

$$M(+,-;-,+,-,-) = \frac{2e^4 E A^*_1 q_{-\perp}}{(q_+ \cdot q_-)} \left[\frac{q_{-+}}{q_{++} k_{1+} k_{1-} k_{2+} k_{2-}} \right]^{\frac{1}{2}}$$

$$\doteq \frac{2e^4 E |A_1|}{(q_+ \cdot q_-)} \left[\frac{q_{+-} q_{-+}}{k_{1+} k_{1-} k_{2+} k_{2-}} \right]^{\frac{1}{2}},$$

$$M(-,+;+,-,-,-) = \frac{e^4(2E - k_{1-} - k_{2-}) A^*_1 q^*_{-\perp}}{(q_+ \cdot q_-)} \left[\frac{q_{-+}}{q_{++} k_{1+} k_{1-} k_{2+} k_{2-}} \right]^{\frac{1}{2}}$$

$$\doteq \frac{e^4(2E - k_{1-} - k_{2-})|A_1|}{(q_+ \cdot q_-)} \left[\frac{q_{+-} q_{-+}}{k_{1+} k_{1-} k_{2+} k_{2-}} \right]^{\frac{1}{2}},$$

$$M(-,+;-,+,-,-) = \frac{e^4(2E - k_{1-} - k_{2-}) A^*_1 q^*_{-\perp}}{(q_+ \cdot q_-)} \left[\frac{q^3_{++}}{q^3_{-+} k_{1+} k_{1-} k_{2+} k_{2-}} \right]^{\frac{1}{2}}$$

$$\doteq \frac{e^4(2E - k_{1-} - k_{2-})|A_1| q_{++} q_{--}}{(q_+ \cdot q_-)[q_{+-} q_{-+} k_{1+} k_{1-} k_{2+} k_{2-}]^{\frac{1}{2}}},$$

$$M(-,+;-,-,+,-) = \frac{4e^4 E A_4^*(1,2) q_{+\perp}}{q_{+-}[q_{++}q_{-+}k_{1+}k_{1-}k_{2+}k_{2-}]^{\frac{1}{2}}}$$

$$\doteq \frac{4e^4 E |A_4(1,2)|}{[q_{+-}q_{-+}k_{1+}k_{1-}k_{2+}k_{2-}]^{\frac{1}{2}}},$$

$$M(-,+;-,-,-,+) = \frac{4e^4 E A_4^*(2,1) q_{+\perp}}{q_{+-}[q_{++}q_{-+}k_{1+}k_{1-}k_{2+}k_{2-}]^{\frac{1}{2}}}$$

$$\doteq \frac{4e^4 E |A_4(2,1)|}{[q_{+-}q_{-+}k_{1+}k_{1-}k_{2+}k_{2-}]^{\frac{1}{2}}},$$

$$M(-,-;+,-,+,-) = \frac{2e^4 E A_4(2,1) q_{-\perp}^*}{(q_+ \cdot q_-)} \left[\frac{q_{-+}}{q_{++}k_{1+}k_{1-}k_{2+}k_{2-}} \right]^{\frac{1}{2}}$$

$$\doteq \frac{2e^4 E |A_4(2,1)|}{(q_+ \cdot q_-)} \left[\frac{q_{+-}q_{-+}}{k_{1+}k_{1-}k_{2+}k_{2-}} \right]^{\frac{1}{2}},$$

$$M(-,-;+,-,-,+) = \frac{2e^4 E A_4(1,2) q_{-\perp}^*}{(q_+ \cdot q_-)} \left[\frac{q_{-+}}{q_{++}k_{1+}k_{1-}k_{2+}k_{2-}} \right]^{\frac{1}{2}}$$

$$\doteq \frac{4e^4 E |A_4(1,2)|}{(q_+ \cdot q_-)} \left[\frac{q_{+-}q_{-+}}{k_{1+}k_{1-}k_{2+}k_{2-}} \right]^{\frac{1}{2}},$$

$$M(-,-;-,+,+,-) = \frac{2e^4 E A_4(2,1) q_{-\perp}^*}{(q_+ \cdot q_-)} \left[\frac{q_{++}^3}{q_{-+}^3 k_{1+}k_{1-}k_{2+}k_{2-}} \right]^{\frac{1}{2}}$$

$$\doteq \frac{2e^4 E |A_4(2,1)| q_{++}q_{--}}{(q_+ \cdot q_-)[q_{+-}q_{-+}k_{1+}k_{1-}k_{2+}k_{2-}]^{\frac{1}{2}}},$$

$$M(-,-;-,+,-,+) = \frac{2e^4 E A_4(1,2) q_{-\perp}^*}{(q_+ \cdot q_-)} \left[\frac{q_{++}^3}{q_{-+}^3 k_{1+}k_{1-}k_{2+}k_{2-}} \right]^{\frac{1}{2}}$$

$$\doteq \frac{2e^4 E |A_4(1,2)| q_{++}q_{--}}{(q_+ \cdot q_-)[q_{+-}q_{-+}k_{1+}k_{1-}k_{2+}k_{2-}]^{\frac{1}{2}}},$$

$$M(-,-;-,-,+,+) = \frac{2e^4 (2E - k_{1-} - k_{2-}) A_1 q_{-\perp}}{q_{+-}[q_{++}q_{-+}k_{1+}k_{1-}k_{2+}k_{2-}]^{\frac{1}{2}}}$$

$$\doteq \frac{2e^4 (2E - k_{1-} - k_{2-}) |A_1|}{[q_{+-}q_{-+}k_{1+}k_{1-}k_{2+}k_{2-}]^{\frac{1}{2}}},$$

9.19. $e^+e^- \to e^+e^-\gamma\gamma$ (NO Z-EXCHANGE; $m \neq 0$)

$$M(-,-;+,-,-,-) = \frac{2e^4 E A_3^* q_{-\perp}^*}{(q_+ \cdot q_-)} \left[\frac{q_{-+}}{q_{++}k_{1+}k_{1-}k_{2+}k_{2-}}\right]^{\frac{1}{2}}$$

$$\doteq \frac{2e^4 E |A_3|}{(q_+ \cdot q_-)} \left[\frac{q_{+-}q_{-+}}{k_{1+}k_{1-}k_{2+}k_{2-}}\right]^{\frac{1}{2}},$$

$$M(-,-;-,+,-,-) = \frac{2e^4 E A_3^* q_{-\perp}^*}{(q_+ \cdot q_-)} \left[\frac{q_{++}^3}{q_{-+}^3 k_{1+}k_{1-}k_{2+}k_{2-}}\right]^{\frac{1}{2}}$$

$$\doteq \frac{2e^4 E |A_3| q_{++} q_{--}}{(q_+ \cdot q_-)[q_{+-}q_{-+}k_{1+}k_{1-}k_{2+}k_{2-}]^{\frac{1}{2}}},$$

$$M(-,-;-,-,+,-) = \frac{4e^4 E A_2^*(1,2) q_{+\perp}}{q_{+-}[q_{++}q_{-+}k_{1+}k_{1-}k_{2+}k_{2-}]^{\frac{1}{2}}}$$

$$\doteq \frac{4e^4 E |A_2(1,2)|}{[q_{+-}q_{-+}k_{1+}k_{1-}k_{2+}k_{2-}]^{\frac{1}{2}}},$$

$$M(-,-;-,-,-,+) = \frac{4e^4 E A_2^*(2,1) q_{+\perp}}{q_{+-}[q_{++}q_{-+}k_{1+}k_{1-}k_{2+}k_{2-}]^{\frac{1}{2}}}$$

$$\doteq \frac{4e^4 E |A_2(2,1)|}{[q_{+-}q_{-+}k_{1+}k_{1-}k_{2+}k_{2-}]^{\frac{1}{2}}},$$

$$M(-,-;-,-,-,-) = \frac{4e^4 E A_1^* q_{-\perp}}{q_{+-}[q_{++}q_{-+}k_{1+}k_{1-}k_{2+}k_{2-}]^{\frac{1}{2}}}$$

$$\doteq \frac{4e^4 E |A_1|}{[q_{+-}q_{-+}k_{1+}k_{1-}k_{2+}k_{2-}]^{\frac{1}{2}}}. \quad (9.265)$$

Unpolarized squared matrix element:

$$\overline{|M|^2} = \frac{2e^8 E^2 \left[4(q_+ \cdot q_-)^4 + q_{++}^2 q_{--}^2 + q_{+-}^2 q_{-+}^2\right]}{k_{1+}k_{1-}k_{2+}k_{2-}q_{+-}q_{-+}(q_+ \cdot q_-)^2}$$

$$\times \left\{\left[1 + \left(\frac{2E - k_{1-} - k_{2-}}{2E}\right)^2\right]|A_1|^2 + |A_2(1,2)|^2 \right.$$

$$\left. + |A_2(2,1)|^2 + |A_3|^2 + |A_4(1,2)|^2 + |A_4(2,1)|^2\right\}.$$

$$(9.266)$$

Unpolarized cross section:

$$d\sigma = \frac{\alpha^4 \left[4(q_+ \cdot q_-)^2 + q_{++}^2 q_{--}^2 + q_{+-}^2 q_{-+}^2\right]}{64\pi^2 k_{1+} k_{1-} k_{2+} k_{2-} q_{+-} q_{-+} (q_+ \cdot q_-)^2}$$

$$\times \left\{ \left[1 + \left(\frac{2E - k_{1-} - k_{2-}}{2E}\right)^2\right] |A_1|^2 \right.$$

$$\left. + |A_2(1,2)|^2 + |A_2(2,1)|^2 + |A_3|^2 + |A_4(1,2)|^2 + |A_4(2,1)|^2 \right\}$$

$$\times \delta^4(p_+ + p_- - q_+ - q_- - k_1 - k_2) \frac{d^3\vec{q}_+ \, d^3\vec{q}_- \, d^3\vec{k}_1 \, d^3\vec{k}_2}{q_{+0} \, q_{-0} \, k_{10} \, k_{20}}. \quad (9.267)$$

9.19.13 $\vec{k}_1$ and $\vec{k}_2$ nearly parallel to $\vec{q}_+$

The primed quantities $k'_{i\pm}$ and $k'_{i\perp}$, $i = 1, 2$, are evaluated in the rotated frame where $\vec{q}_+$ determines the positive z-axis (see also Section 7.4.3). The four-vector q in eqns (9.269) is obtained by applying a space reflection to q_+. Definitions:

$$Z_{+-} = q_{++} q_{--} - q^*_{+\perp} q_{-\perp},$$

$$Z'_{ij} = k'_{i+} k'_{j-} - k'^*_{i\perp} k'_{j\perp}, \qquad i,j = 1,2,$$

$$\Delta'_{12} = 2(q_+ \cdot k_1) + 2(q_+ \cdot k_2) + 2(k_1 \cdot k_2),$$

$$A'_1 = \frac{2q_{+0}^2}{(q_+ \cdot k_1)(q_+ \cdot k_2)} \left[k'_{1-} k'_{2-} + \frac{m^2 Z'_{12} Z'_{21}}{4q_{+0}^2 \Delta'_{12}}\right],$$

$$A'_2(1,2) = \frac{1}{\Delta'_{12}} \left[\frac{k'_{1-}(2q_{+0} + k'_{1+})(2q_{+0} k'_{2-} + Z'_{12})}{(q_+ \cdot k_1)} \right.$$

$$\left. + \frac{2q_{+0} k'_{1-} k'_{2-}(2q_{+0} + k'_{1+} + k'_{2+}) + m^2 k'_{2+} k'^*_{1\perp} k'_{2\perp}/2q_{+0}}{(q_+ \cdot k_2)} \right],$$

$$A'_3 = \frac{m}{(q_+ \cdot k_1)(q_+ \cdot k_2)} \left[k'_{1+} k'_{2-} k'_{1\perp} + k'_{1-} k'_{2+} k'_{2\perp} \right.$$

$$\left. - \frac{(k'_{1\perp} + k'_{2\perp}) Z'_{12} Z'_{21}}{\Delta'_{12}} \right],$$

$$A'_4(1,2) = \frac{m k'_{1+} k'_{1\perp}}{2q_{+0} \Delta'_{12}} \left[\frac{2q_{+0} k'_{2-} + Z'^*_{12}}{(q_+ \cdot k_1)} + \frac{2q_{+0} k'_{2-}}{(q_+ \cdot k_2)} \right]. \quad (9.268)$$

9.19. $e^+e^- \to e^+e^-\gamma\gamma$ (NO Z-EXCHANGE; $m \neq 0$)

Photon polarizations:

$$\not{\epsilon}_i^{\pm} = N_i[\not{k}_i \not{q}_+ \not{q}(1 \pm \gamma_5) + \not{q} \not{q}_+ \not{k}_i(1 \mp \gamma_5)],$$

$$N_i^{-1} = q_{+0}^2[32k'_{i+}k'_{i-}]^{\frac{1}{2}}, \qquad i = 1, 2. \qquad (9.269)$$

Nonvanishing helicity amplitudes:

$$M(+,+;+,+,+,+) = \frac{2e^4 A'_1 q^*_{-\perp} Z_{+-}}{q_{--}[q_{++}^3 q_{-+}^3 k'_{1+}k'_{1-}k'_{2+}k'_{2-}]^{\frac{1}{2}}}$$

$$\doteq \frac{e^4|A'_1|}{q_{+-}}\left[\frac{8(q_+ \cdot q_-)}{k'_{1+}k'_{1-}k'_{2+}k'_{2-}}\right]^{\frac{1}{2}},$$

$$M(+,+;+,+,+,-) = \frac{2e^4 A_2^{'*}(1,2)q^*_{-\perp} Z_{+-}}{q_{--}[q_{++}q_{-+}^3 k'_{1+}k'_{1-}k'_{2+}k'_{2-}]^{\frac{1}{2}}}$$

$$\doteq \frac{e^4|A'_2(1,2)|}{q_{-+}}\left[\frac{8(q_+ \cdot q_-)}{k'_{1+}k'_{1-}k'_{2+}k'_{2-}}\right]^{\frac{1}{2}},$$

$$M(+,+;+,+,-,+) = \frac{2e^4 A_2^{'*}(2,1)q^*_{-\perp} Z_{+-}}{q_{--}[q_{++}q_{-+}^3 k'_{1+}k'_{1-}k'_{2+}k'_{2-}]^{\frac{1}{2}}}$$

$$\doteq \frac{e^4|A'_2(2,1)|}{q_{-+}}\left[\frac{8(q_+ \cdot q_-)}{k'_{1+}k'_{1-}k'_{2+}k'_{2-}}\right]^{\frac{1}{2}},$$

$$M(+,+;-,+,+,+) = -\frac{2e^4 A'_3 q^*_{-\perp} Z_{+-}}{q_{--}[q_{++}q_{-+}^3 k'_{1+}k'_{1-}k'_{2+}k'_{2-}]^{\frac{1}{2}}}$$

$$\doteq \frac{e^4|A'_3|}{q_{-+}}\left[\frac{8(q_+ \cdot q_-)}{k'_{1+}k'_{1-}k'_{2+}k'_{2-}}\right]^{\frac{1}{2}},$$

$$M(+,+;+,+,-,-) = \frac{2e^4 A_1^{'*} q^*_{-\perp} Z_{+-}}{q_{--}[q_{++}q_{-+}^3 k'_{1+}k'_{1-}k'_{2+}k'_{2-}]^{\frac{1}{2}}}$$

$$\doteq \frac{e^4|A'_1|}{q_{-+}}\left[\frac{8(q_+ \cdot q_-)}{k'_{1+}k'_{1-}k'_{2+}k'_{2-}}\right]^{\frac{1}{2}},$$

$$M(+,+;-,+,+,-) = -\frac{2e^4 A'_4(1,2)q^*_{-\perp} Z_{+-}}{q_{--}[q_{++}q_{-+}^3 k'_{1+}k'_{1-}k'_{2+}k'_{2-}]^{\frac{1}{2}}}$$

$$\doteq \frac{e^4|A'_4(1,2)|}{q_{-+}}\left[\frac{8(q_+ \cdot q_-)}{k'_{1+}k'_{1-}k'_{2+}k'_{2-}}\right]^{\frac{1}{2}},$$

$$M(+,+;-,+,-,+) = -\frac{2e^4 A'_4(2,1)q^*_{-\perp}Z_{+-}}{q_{--}[q_{++}q^3_{-+}k'_{1+}k'_{1-}k'_{2+}k'_{2-}]^{\frac{1}{2}}}$$

$$\doteq \frac{e^4|A'_4(2,1)|}{q_{-+}}\left[\frac{8(q_+\cdot q_-)}{k'_{1+}k'_{1-}k'_{2+}k'_{2-}}\right]^{\frac{1}{2}},$$

$$M(+,-;+,+,+,-) = -\frac{e^4 A'^*_4(2,1)q_{+\perp}}{E}\left[\frac{q_{-+}}{q_{++}k'_{1+}k'_{1-}k'_{2+}k'_{2-}}\right]^{\frac{1}{2}}$$

$$\doteq \frac{e^4|A'_4(2,1)|}{E}\left[\frac{q_{+-}q_{-+}}{k'_{1+}k'_{1-}k'_{2+}k'_{2-}}\right]^{\frac{1}{2}},$$

$$M(+,-;+,+,-,+) = -\frac{e^4 A'^*_4(1,2)q_{+\perp}}{E}\left[\frac{q_{-+}}{q_{++}k'_{1+}k'_{1-}k'_{2+}k'_{2-}}\right]^{\frac{1}{2}}$$

$$\doteq \frac{e^4|A'_4(1,2)|}{E}\left[\frac{q_{+-}q_{-+}}{k'_{1+}k'_{1-}k'_{2+}k'_{2-}}\right]^{\frac{1}{2}},$$

$$M(+,-;+,-,+,+) = \frac{e^4 A'_1 q_{--}q_{-\perp}}{Eq_{+-}}\left[\frac{q_{++}}{q_{-+}k'_{1+}k'_{1-}k'_{2+}k'_{2-}}\right]^{\frac{1}{2}}$$

$$\doteq \frac{e^4|A'_1|}{Eq_{+-}}\left[\frac{q_{++}q^3_{--}}{k'_{1+}k'_{1-}k'_{2+}k'_{2-}}\right]^{\frac{1}{2}},$$

$$M(+,-;-,+,+,+) = -\frac{e^4 A'_1 q_{+\perp}}{E}\left[\frac{q_{-+}}{q_{++}k'_{1+}k'_{1-}k'_{2+}k'_{2-}}\right]^{\frac{1}{2}}$$

$$\doteq \frac{e^4|A'_1|}{E}\left[\frac{q_{+-}q_{-+}}{k'_{1+}k'_{1-}k'_{2+}k'_{2-}}\right]^{\frac{1}{2}},$$

$$M(-,+;+,+,+,-) = \frac{2e^4 E A'^*_4(2,1)q^*_{-\perp}}{(q_+\cdot q_-)}\left[\frac{q^3_{++}}{q^3_{-+}k'_{1+}k'_{1-}k'_{2+}k'_{2-}}\right]^{\frac{1}{2}}$$

$$\doteq \frac{e^4|A'_4(2,1)|q_{++}q_{--}}{E[q_{+-}q_{-+}k'_{1+}k'_{1-}k'_{2+}k'_{2-}]^{\frac{1}{2}}},$$

$$M(-,+;+,+,-,+) = \frac{2e^4 E A'^*_4(1,2)q^*_{-\perp}}{(q_+\cdot q_-)}\left[\frac{q^3_{++}}{q^3_{-+}k'_{1+}k'_{1-}k'_{2+}k'_{2-}}\right]^{\frac{1}{2}}$$

$$\doteq \frac{e^4|A'_4(1,2)|q_{++}q_{--}}{E[q_{+-}q_{-+}k'_{1+}k'_{1-}k'_{2+}k'_{2-}]^{\frac{1}{2}}},$$

9.19. $e^+e^- \to e^+e^- \gamma\gamma$ (NO Z-EXCHANGE; $m \neq 0$)

$$M(-,+;+,-,+,+) = -\frac{2e^4 E A_1' q_{+\perp}^*}{(q_+ \cdot q_-)} \left[\frac{q_{-+}}{q_{++}k_{1+}'k_{1-}'k_{2+}'k_{2-}'}\right]^{\frac{1}{2}}$$

$$\doteq \frac{2e^4 E |A_1'|}{(q_+ \cdot q_-)} \left[\frac{q_{+-}q_{-+}}{k_{1+}'k_{1-}'k_{2+}'k_{2-}'}\right]^{\frac{1}{2}},$$

$$M(-,+;-,+,+,+) = \frac{2e^4 E A_1' q_{-\perp}^*}{(q_+ \cdot q_-)} \left[\frac{q_{++}^3}{q_{-+}^3 k_{1+}'k_{1-}'k_{2+}'k_{2-}'}\right]^{\frac{1}{2}}$$

$$\doteq \frac{e^4 |A_1'| q_{++}q_{--}}{E [q_{+-}q_{-+}k_{1+}'k_{1-}'k_{2+}'k_{2-}']^{\frac{1}{2}}},$$

$$M(+,-;+,+,-,-) = -\frac{e^4 A_3'^* q_{+\perp}}{E} \left[\frac{q_{-+}}{q_{++}k_{1+}'k_{1-}'k_{2+}'k_{2-}'}\right]^{\frac{1}{2}}$$

$$\doteq \frac{e^4 |A_3'|}{E} \left[\frac{q_{+-}q_{-+}}{k_{1+}'k_{1-}'k_{2+}'k_{2-}'}\right]^{\frac{1}{2}},$$

$$M(+,-;+,-,+,-) = \frac{2e^4 E A_2'^*(1,2) q_{-\perp}}{(q_+ \cdot q_-)} \left[\frac{q_{++}^3}{q_{-+}^3 k_{1+}'k_{1-}'k_{2+}'k_{2-}'}\right]^{\frac{1}{2}}$$

$$\doteq \frac{e^4 |A_2'(1,2)| q_{++}q_{--}}{E [q_{+-}q_{-+}k_{1+}'k_{1-}'k_{2+}'k_{2-}']^{\frac{1}{2}}},$$

$$M(+,-;+,-,-,+) = \frac{2e^4 E A_2'^*(2,1) q_{-\perp}}{(q_+ \cdot q_-)} \left[\frac{q_{++}^3}{q_{-+}^3 k_{1+}'k_{1-}'k_{2+}'k_{2-}'}\right]^{\frac{1}{2}}$$

$$\doteq \frac{e^4 |A_2'(2,1)| q_{++}q_{--}}{E [q_{+-}q_{-+}k_{1+}'k_{1-}'k_{2+}'k_{2-}']^{\frac{1}{2}}},$$

$$M(+,-;-,+,+,-) = -\frac{e^4 A_2'(2,1) q_{+\perp}}{E} \left[\frac{q_{-+}}{q_{++}k_{1+}'k_{1-}'k_{2+}'k_{2-}'}\right]^{\frac{1}{2}}$$

$$\doteq \frac{e^4 |A_2'(2,1)|}{E} \left[\frac{q_{+-}q_{-+}}{k_{1+}'k_{1-}'k_{2+}'k_{2-}'}\right]^{\frac{1}{2}},$$

$$M(+,-;-,+,-,+) = -\frac{e^4 A_2'(1,2) q_{+\perp}}{E} \left[\frac{q_{-+}}{q_{++}k_{1+}'k_{1-}'k_{2+}'k_{2-}'}\right]^{\frac{1}{2}}$$

$$\doteq \frac{e^4 |A_2'(1,2)|}{E} \left[\frac{q_{+-}q_{-+}}{k_{1+}'k_{1-}'k_{2+}'k_{2-}'}\right]^{\frac{1}{2}},$$

$$M(+,-;-,-,+,+) = -\frac{2e^4 E A_3' q_{-\perp}}{(q_+ \cdot q_-)} \left[\frac{q_{++}^3}{q_{-+}^3 k_{1+}' k_{1-}' k_{2+}' k_{2-}'}\right]^{\frac{1}{2}}$$

$$\doteq \frac{e^4 |A_3'| q_{++} q_{--}}{E \left[q_{+-} q_{-+} k_{1+}' k_{1-}' k_{2+}' k_{2-}'\right]^{\frac{1}{2}}},$$

$$M(-,+;+,+,-,-) = \frac{2e^4 E A_3'^* q_{-\perp}^*}{(q_+ \cdot q_-)} \left[\frac{q_{++}^3}{q_{-+}^3 k_{1+}' k_{1-}' k_{2+}' k_{2-}'}\right]^{\frac{1}{2}}$$

$$\doteq \frac{e^4 |A_3'| q_{++} q_{--}}{E \left[q_{+-} q_{-+} k_{1+}' k_{1-}' k_{2+}' k_{2-}'\right]^{\frac{1}{2}}},$$

$$M(-,+;+,-,+,-) = -\frac{e^4 A_2'^*(1,2) q_{+\perp}^*}{E} \left[\frac{q_{-+}}{q_{++} k_{1+}' k_{1-}' k_{2+}' k_{2-}'}\right]^{\frac{1}{2}}$$

$$\doteq \frac{e^4 |A_2'(1,2)|}{E} \left[\frac{q_{+-} q_{-+}}{k_{1+}' k_{1-}' k_{2+}' k_{2-}'}\right]^{\frac{1}{2}},$$

$$M(-,+;+,-,-,+) = -\frac{e^4 A_2'^*(2,1) q_{+\perp}^*}{E} \left[\frac{q_{-+}}{q_{++} k_{1+}' k_{1-}' k_{2+}' k_{2-}'}\right]^{\frac{1}{2}}$$

$$\doteq \frac{e^4 |A_2'(2,1)|}{E} \left[\frac{q_{+-} q_{-+}}{k_{1+}' k_{1-}' k_{2+}' k_{2-}'}\right]^{\frac{1}{2}},$$

$$M(-,+;-,+,+,-) = \frac{2e^4 E A_2'(2,1) q_{-\perp}^*}{(q_+ \cdot q_-)} \left[\frac{q_{++}^3}{q_{-+}^3 k_{1+}' k_{1-}' k_{2+}' k_{2-}'}\right]^{\frac{1}{2}}$$

$$\doteq \frac{e^4 |A_2'(2,1)| q_{++} q_{--}}{E \left[q_{+-} q_{-+} k_{1+}' k_{1-}' k_{2+}' k_{2-}'\right]^{\frac{1}{2}}},$$

$$M(-,+;-,+,-,+) = \frac{2e^4 A_2'(1,2) q_{-\perp}^*}{(q_+ \cdot q_-)} \left[\frac{q_{++}^3}{q_{-+}^3 k_{1+}' k_{1-}' k_{2+}' k_{2-}'}\right]^{\frac{1}{2}}$$

$$\doteq \frac{e^4 |A_2'(1,2)| q_{++} q_{--}}{E \left[q_{+-} q_{-+} k_{1+}' k_{1-}' k_{2+}' k_{2-}'\right]^{\frac{1}{2}}},$$

$$M(-,+;-,-,+,+) = \frac{e^4 A_3' q_{+\perp}^*}{E} \left[\frac{q_{-+}}{q_{++} k_{1+}' k_{1-}' k_{2+}' k_{2-}'}\right]^{\frac{1}{2}}$$

$$\doteq \frac{e^4 |A_3'|}{E} \left[\frac{q_{+-} q_{-+}}{k_{1+}' k_{1-}' k_{2+}' k_{2-}'}\right]^{\frac{1}{2}},$$

9.19. $e^+e^- \to e^+e^-\gamma\gamma$ (NO Z-EXCHANGE; $m \neq 0$)

$$M(+,-;+,-,-,-) = \frac{2e^4 E A_1'^* q_{-\perp}}{(q_+ \cdot q_-)} \left[\frac{q_{++}^3}{q_{-+}^3 k_{1+}' k_{1-}' k_{2+}' k_{2-}'}\right]^{\frac{1}{2}}$$

$$\doteq \frac{e^4 |A_1'| q_{++} q_{--}}{E \left[q_{+-} q_{-+} k_{1+}' k_{1-}' k_{2+}' k_{2-}'\right]^{\frac{1}{2}}},$$

$$M(+,-;-,+,-,-) = -\frac{2e^4 E A_1'^* q_{+\perp}}{(q_+ \cdot q_-)} \left[\frac{q_{-+}}{q_{++} k_{1+}' k_{1-}' k_{2+}' k_{2-}'}\right]^{\frac{1}{2}}$$

$$\doteq \frac{2e^4 E |A_1'|}{(q_+ \cdot q_-)} \left[\frac{q_{+-} q_{-+}}{k_{1+}' k_{1-}' k_{2+}' k_{2-}'}\right]^{\frac{1}{2}},$$

$$M(+,-;-,-,+,-) = -\frac{2e^4 E A_4'(1,2) q_{-\perp}}{(q_+ \cdot q_-)} \left[\frac{q_{++}^3}{q_{-+}^3 k_{1+}' k_{1-}' k_{2+}' k_{2-}'}\right]^{\frac{1}{2}}$$

$$\doteq \frac{e^4 |A_4'(1,2)| q_{++} q_{--}}{E \left[q_{+-} q_{-+} k_{1+}' k_{1-}' k_{2+}' k_{2-}'\right]^{\frac{1}{2}}},$$

$$M(+,-;-,-,-,+) = -\frac{2e^4 E A_4'(2,1) q_{-\perp}}{(q_+ \cdot q_-)} \left[\frac{q_{++}^3}{q_{-+}^3 k_{1+}' k_{1-}' k_{2+}' k_{2-}'}\right]^{\frac{1}{2}}$$

$$\doteq \frac{e^4 |A_4'(2,1)| q_{++} q_{--}}{E \left[q_{+-} q_{-+} k_{1+}' k_{1-}' k_{2+}' k_{2-}'\right]^{\frac{1}{2}}},$$

$$M(-,+;+,-,-,-) = -\frac{e^4 A_1'^* q_{+\perp}^*}{E} \left[\frac{q_{-+}}{q_{++} k_{1+}' k_{1-}' k_{2+}' k_{2-}'}\right]^{\frac{1}{2}}$$

$$\doteq \frac{e^4 |A_1'|}{E} \left[\frac{q_{+-} q_{-+}}{k_{1+}' k_{1-}' k_{2+}' k_{2-}'}\right]^{\frac{1}{2}},$$

$$M(-,+;-,+,-,-) = \frac{e^4 A_1'^* q_{--} q_{-\perp}^*}{E q_{+-}} \left[\frac{q_{++}}{q_{-+} k_{1+}' k_{1-}' k_{2+}' k_{2-}'}\right]^{\frac{1}{2}}$$

$$\doteq \frac{e^4 |A_1'|}{E q_{+-}} \left[\frac{q_{++} q_{--}^3}{k_{1+}' k_{1-}' k_{2+}' k_{2-}'}\right]^{\frac{1}{2}},$$

$$M(-,+;-,-,+,-) = \frac{e^4 A_4'(1,2) q_{+\perp}^*}{E} \left[\frac{q_{-+}}{q_{++} k_{1+}' k_{1-}' k_{2+}' k_{2-}'}\right]^{\frac{1}{2}}$$

$$\doteq \frac{e^4 |A_4'(1,2)|}{E} \left[\frac{q_{+-} q_{-+}}{k_{1+}' k_{1-}' k_{2+}' k_{2-}'}\right]^{\frac{1}{2}},$$

$$M(-,+;-,-,-,+) = \frac{e^4 A_4'(2,1) q_{+\perp}^*}{E} \left[\frac{q_{-+}}{q_{++} k_{1+}' k_{1-}' k_{2+}' k_{2-}'} \right]^{\frac{1}{2}}$$

$$\doteq \frac{e^4 |A_4'(2,1)|}{E} \left[\frac{q_{+-} q_{-+}}{k_{1+}' k_{1-}' k_{2+}' k_{2-}'} \right]^{\frac{1}{2}},$$

$$M(-,-;+,-,+,-) = \frac{2e^4 A_4'^*(2,1) q_{-\perp} Z_{+-}^*}{q_{--} [q_{++} q_{-+}^3 k_{1+}' k_{1-}' k_{2+}' k_{2-}']^{\frac{1}{2}}}$$

$$\doteq \frac{e^4 |A_4'(2,1)|}{q_{-+}} \left[\frac{8(q_+ \cdot q_-)}{k_{1+}' k_{1-}' k_{2+}' k_{2-}'} \right]^{\frac{1}{2}},$$

$$M(-,-;+,-,-,+) = \frac{2e^4 A_4'^*(1,2) q_{-\perp} Z_{+-}^*}{q_{--} [q_{++} q_{-+}^3 k_{1+}' k_{1-}' k_{2+}' k_{2-}']^{\frac{1}{2}}}$$

$$\doteq \frac{e^4 |A_4'(1,2)|}{q_{-+}} \left[\frac{8(q_+ \cdot q_-)}{k_{1+}' k_{1-}' k_{2+}' k_{2-}'} \right]^{\frac{1}{2}},$$

$$M(-,-;-,-,+,+) = \frac{2e^4 A_1' q_{-\perp} Z_{+-}^*}{q_{--} [q_{++} q_{-+}^3 k_{1+}' k_{1-}' k_{2+}' k_{2-}']^{\frac{1}{2}}}$$

$$\doteq \frac{e^4 |A_1'|}{q_{-+}} \left[\frac{8(q_+ \cdot q_-)}{k_{1+}' k_{1-}' k_{2+}' k_{2-}'} \right]^{\frac{1}{2}},$$

$$M(-,-;+,-,-,-) = \frac{2e^4 A_3'^* q_{-\perp} Z_{+-}^*}{q_{--} [q_{++} q_{-+}^3 k_{1+}' k_{1-}' k_{2+}' k_{2-}']^{\frac{1}{2}}}$$

$$\doteq \frac{e^4 |A_3'|}{q_{-+}} \left[\frac{8(q_+ \cdot q_-)}{k_{1+}' k_{1-}' k_{2+}' k_{2-}'} \right]^{\frac{1}{2}},$$

$$M(-,-;-,-,+,-) = \frac{2e^4 A_2'(2,1) q_{-\perp} Z_{+-}^*}{q_{--} [q_{++} q_{-+}^3 k_{1+}' k_{1-}' k_{2+}' k_{2-}']^{\frac{1}{2}}}$$

$$\doteq \frac{e^4 |A_2'(2,1)|}{q_{-+}} \left[\frac{8(q_+ \cdot q_-)}{k_{1+}' k_{1-}' k_{2+}' k_{2-}'} \right]^{\frac{1}{2}},$$

$$M(-,-;-,-,-,+) = \frac{2e^4 A_2'(1,2) q_{-\perp} Z_{+-}^*}{q_{--} [q_{++} q_{-+}^3 k_{1+}' k_{1-}' k_{2+}' k_{2-}']^{\frac{1}{2}}}$$

$$\doteq \frac{e^4 |A_2'(1,2)|}{q_{-+}} \left[\frac{8(q_+ \cdot q_-)}{k_{1+}' k_{1-}' k_{2+}' k_{2-}'} \right]^{\frac{1}{2}},$$

9.19. $e^+e^- \to e^+e^-\gamma\gamma$ (NO Z-EXCHANGE; $m \neq 0$)

$$M(-,-;-,-,-,-) = \frac{2e^4 A_1'^* q_{-\perp} Z_{+-}^*}{[q_{++}^3 q_{-+}^3 k_{1+}' k_{1-}' k_{2+}' k_{2-}']^{\frac{1}{2}}}$$

$$\doteq \frac{e^4 |A_1'|}{q_{+-}} \left[\frac{8(q_+ \cdot q_-)}{k_{1+}' k_{1-}' k_{2+}' k_{2-}'} \right]^{\frac{1}{2}}. \quad (9.270)$$

Unpolarized squared matrix element:

$$\overline{|M|^2} = \frac{e^8 \left[4(q_+ \cdot q_-)^2 + q_{++}^2 q_{--}^2 + q_{+-}^2 q_{-+}^2 \right]}{2E^2 k_{1+}' k_{1-}' k_{2+}' k_{2-}' q_{+-} q_{-+}}$$

$$\times \left\{ \left[1 + \left(\frac{2q_{+0} + k_{1+}' + k_{2+}'}{2q_{+0}} \right)^2 \right] |A_1'|^2 + |A_2'(1,2)|^2 \right.$$

$$\left. + |A_2'(2,1)|^2 + |A_3'|^2 + |A_4'(1,2)|^2 + |A_4'(2,1)|^2 \right\}. \quad (9.271)$$

Unpolarized cross section:

$$d\sigma = \frac{\alpha^4 \left[4(q_+ \cdot q_-)^2 + q_{++}^2 q_{--}^2 + q_{+-}^2 q_{-+}^2 \right]}{256\pi^2 E^4 k_{1+}' k_{1-}' k_{2+}' k_{2-}' q_{+-} q_{-+}}$$

$$\times \left\{ \left[1 + \left(\frac{2q_{+0} + k_{1+}' + k_{2+}'}{2q_{+0}} \right)^2 \right] |A_1'|^2 \right.$$

$$\left. + |A_2'(1,2)|^2 + |A_2'(2,1)|^2 + |A_3'|^2 + |A_4'(1,2)|^2 + |A_4'(2,1)|^2 \right\}$$

$$\times \delta^4(p_+ + p_- - q_+ - q_- - k_1 - k_2) \frac{d^3\vec{q}_+ \, d^3\vec{q}_- \, d^3\vec{k}_1 \, d^3\vec{k}_2}{q_{+0} \, q_{-0} \, k_{10} \, k_{20}}. \quad (9.272)$$

9.19.14 $\vec{k}_1$ and $\vec{k}_2$ nearly parallel to $\vec{q}_-$

The doubly primed quantities $k_{i\pm}''$ and $k_{i\perp}''$, $i = 1, 2$, are evaluated in the rotated frame where $\vec{q}_-$ determines the positive z-axis (see also Section 7.4.3). The four-vector q in eqns (9.274) is obtained by applying a space reflection to q_-.

Definitions:

$$Z_{+-} = q_{++} q_{--} - q_{+\perp}^* q_{-\perp},$$

$$Z_{ij}'' = k_{i+}'' k_{j-}'' - k_{i\perp}''^* k_{j\perp}'', \qquad i, j = 1, 2,$$

$$\Delta_{12}'' = 2(q_- \cdot k_1) + 2(q_- \cdot k_2) + 2(k_1 \cdot k_2),$$

$$A_1'' = \frac{2q_{-0}^2}{(q_- \cdot k_1)(q_- \cdot k_2)} \left[k_{1-}'' k_{2-}'' + \frac{m^2 Z_{12}'' Z_{21}''}{4q_{-0}^2 \Delta_{12}''} \right],$$

$$A_2''(1,2) = \frac{1}{\Delta_{12}''} \left[\frac{k_{1-}''(2q_{-0} + k_{1+}'')(2q_{-0}k_{2-}'' + Z_{12}'')}{(q_- \cdot k_1)} \right.$$

$$\left. + \frac{2q_{-0}k_{1-}''k_{2-}''(2q_{-0} + k_{1+}'' + k_{2+}'') + m^2 k_{2+}'' k_{1\perp}''^* k_{2\perp}''/2q_{-0}}{(q_- \cdot k_2)} \right],$$

$$A_3'' = \frac{m}{(q_- \cdot k_1)(q_- \cdot k_2)} \left[k_{1+}'' k_{2-}'' k_{1\perp}'' + k_{1-}'' k_{2+}'' k_{2\perp}'' \right.$$

$$\left. - \frac{(k_{1\perp}'' + k_{2\perp}'')Z_{12}''Z_{21}''}{\Delta_{12}''} \right],$$

$$A_4''(1,2) = \frac{m k_{1+}'' k_{1\perp}''}{2q_{-0}\Delta_{12}''} \left[\frac{2q_{-0}k_{2-}'' + Z_{12}''^*}{(q_- \cdot k_1)} + \frac{2q_{-0}k_{2-}''}{(q_- \cdot k_2)} \right]. \quad (9.273)$$

Photon polarizations:

$$\not{\ell}_i^{\pm} = N_i[\not{k}_i \not{q}_- \not{q}(1 \pm \gamma_5) + \not{q} \not{q}_- \not{k}_i(1 \mp \gamma_5)],$$

$$N_i^{-1} = q_{-0}^2 [32 k_{i+}'' k_{i-}'']^{\frac{1}{2}}, \qquad i = 1, 2. \quad (9.274)$$

Nonvanishing helicity amplitudes:

$$M(+,+;+,+,+,+) = \frac{2e^4 A_1'' q_{-\perp}^* Z_{+-}}{q_{--}[q_{++}q_{-+}^3 k_{1+}'' k_{1-}'' k_{2+}'' k_{2-}'']^{\frac{1}{2}}}$$

$$\doteq \frac{e^4 |A_1''|}{q_{-+}} \left[\frac{8(q_+ \cdot q_-)}{k_{1+}'' k_{1-}'' k_{2+}'' k_{2-}''} \right]^{\frac{1}{2}},$$

$$M(+,+;+,+,+,-) = \frac{2e^4 A_2''^*(1,2) q_{-\perp}^* Z_{+-}}{[q_{++}^3 q_{-+}^3 k_{1+}'' k_{1-}'' k_{2+}'' k_{2-}'']^{\frac{1}{2}}}$$

$$\doteq \frac{e^4 |A_2''(1,2)|}{q_{+-}} \left[\frac{8(q_+ \cdot q_-)}{k_{1+}'' k_{1-}'' k_{2+}'' k_{2-}''} \right]^{\frac{1}{2}},$$

$$M(+,+;+,+,-,+) = \frac{2e^4 A_2''^*(2,1) q_{-\perp}^* Z_{+-}}{[q_{++}^3 q_{-+}^3 k_{1+}'' k_{1-}'' k_{2+}'' k_{2-}'']^{\frac{1}{2}}}$$

$$\doteq \frac{e^4 |A_2''(2,1)|}{q_{+-}} \left[\frac{8(q_+ \cdot q_-)}{k_{1+}'' k_{1-}'' k_{2+}'' k_{2-}''} \right]^{\frac{1}{2}},$$

9.19. $e^+e^- \to e^+e^- \gamma\gamma$ (NO Z-EXCHANGE; $m \neq 0$)

$$M(+,+;+,-,+,+) = -\frac{2e^4 A_3'' q_{-\perp}^* Z_{+-}}{[q_{++}^3 q_{-+}^3 k_{1+}'' k_{1-}'' k_{2+}'' k_{2-}'']^{\frac{1}{2}}}$$

$$\doteq \frac{e^4 |A_3''|}{q_{+-}} \left[\frac{8(q_+ \cdot q_-)}{k_{1+}'' k_{1-}'' k_{2+}'' k_{2-}''}\right]^{\frac{1}{2}},$$

$$M(+,+;+,+,-,-) = \frac{2e^4 A_1''^* q_{-\perp}^* Z_{+-}}{[q_{++}^3 q_{-+}^3 k_{1+}'' k_{1-}'' k_{2+}'' k_{2-}'']^{\frac{1}{2}}}$$

$$\doteq \frac{e^4 |A_1''|}{q_{+-}} \left[\frac{8(q_+ \cdot q_-)}{k_{1+}'' k_{1-}'' k_{2+}'' k_{2-}''}\right]^{\frac{1}{2}},$$

$$M(+,+;+,-,+,-) = -\frac{2e^4 A_4''(1,2) q_{-\perp}^* Z_{+-}}{[q_{++}^3 q_{-+}^3 k_{1+}'' k_{1-}'' k_{2+}'' k_{2-}'']^{\frac{1}{2}}}$$

$$\doteq \frac{e^4 |A_4''(1,2)|}{q_{+-}} \left[\frac{8(q_+ \cdot q_-)}{k_{1+}'' k_{1-}'' k_{2+}'' k_{2-}''}\right]^{\frac{1}{2}},$$

$$M(+,+;+,-,-,+) = -\frac{2e^4 A_4''(2,1) q_{-\perp}^* Z_{+-}}{[q_{++}^3 q_{-+}^3 k_{1+}'' k_{1-}'' k_{2+}'' k_{2-}'']^{\frac{1}{2}}}$$

$$\doteq \frac{e^4 |A_4''(2,1)|}{q_{+-}} \left[\frac{8(q_+ \cdot q_-)}{k_{1+}'' k_{1-}'' k_{2+}'' k_{2-}''}\right]^{\frac{1}{2}},$$

$$M(+,-;+,+,+,-) = \frac{e^4 A_4''^*(2,1) q_{-\perp}}{E q_{+-}} \left[\frac{q_{++}^3}{q_{-+} k_{1+}'' k_{1-}'' k_{2+}'' k_{2-}''}\right]^{\frac{1}{2}}$$

$$\doteq \frac{e^4 |A_4''(2,1)| q_{++} q_{--}}{E [q_{+-} q_{-+} k_{1+}'' k_{1-}'' k_{2+}'' k_{2-}'']^{\frac{1}{2}}},$$

$$M(+,-;+,+,-,+) = \frac{e^4 A_4''^*(1,2) q_{-\perp}}{E q_{+-}} \left[\frac{q_{++}^3}{q_{-+} k_{1+}'' k_{1-}'' k_{2+}'' k_{2-}''}\right]^{\frac{1}{2}}$$

$$\doteq \frac{e^4 |A_4''(1,2)| q_{++} q_{--}}{E [q_{+-} q_{-+} k_{1+}'' k_{1-}'' k_{2+}'' k_{2-}'']^{\frac{1}{2}}},$$

$$M(+,-;+,-,+,+) = \frac{e^4 A_1'' q_{-\perp}}{E q_{+-}} \left[\frac{q_{++}^3}{q_{-+} k_{1+}'' k_{1-}'' k_{2+}'' k_{2-}''}\right]^{\frac{1}{2}}$$

$$\doteq \frac{e^4 |A_1''| q_{++} q_{--}}{E [q_{+-} q_{-+} k_{1+}'' k_{1-}'' k_{2+}'' k_{2-}'']^{\frac{1}{2}}},$$

$$M(+,-;-,+,+,+) = -\frac{e^4 A_1'' q_{+\perp}}{E} \left[\frac{q_{+-}}{q_{--} k_{1+}'' k_{1-}'' k_{2+}'' k_{2-}''} \right]^{\frac{1}{2}}$$

$$\doteq \frac{e^4 q_{+-} |A_1''|}{E} \left[\frac{q_{++}}{q_{--} k_{1+}'' k_{1-}'' k_{2+}'' k_{2-}''} \right]^{\frac{1}{2}},$$

$$M(-,+;+,+,+,-) = -\frac{e^4 A_4''^*(2,1) q_{+\perp}^*}{E} \left[\frac{q_{-+}}{q_{++} k_{1+}'' k_{1-}'' k_{2+}'' k_{2-}''} \right]^{\frac{1}{2}}$$

$$\doteq \frac{e^4 |A_4''(2,1)|}{E} \left[\frac{q_{+-} q_{-+}}{k_{1+}'' k_{1-}'' k_{2+}'' k_{2-}''} \right]^{\frac{1}{2}},$$

$$M(-,+;+,+,-,+) = -\frac{e^4 A_4''^*(1,2) q_{+\perp}^*}{E} \left[\frac{q_{-+}}{q_{++} k_{1+}'' k_{1-}'' k_{2+}'' k_{2-}''} \right]^{\frac{1}{2}}$$

$$\doteq \frac{e^4 |A_4''(1,2)|}{E} \left[\frac{q_{+-} q_{-+}}{k_{1+}'' k_{1-}'' k_{2+}'' k_{2-}''} \right]^{\frac{1}{2}},$$

$$M(-,+;+,-,+,+) = -\frac{e^4 A_1'' q_{+\perp}^*}{E} \left[\frac{q_{-+}}{q_{++} k_{1+}'' k_{1-}'' k_{2+}'' k_{2-}''} \right]^{\frac{1}{2}}$$

$$\doteq \frac{e^4 |A_1''|}{E} \left[\frac{q_{+-} q_{-+}}{k_{1+}'' k_{1-}'' k_{2+}'' k_{2-}''} \right]^{\frac{1}{2}},$$

$$M(-,+;-,+,+,+) = \frac{e^4 A_1'' q_{-\perp}^*}{E} \left[\frac{q_{++}^3}{q_{-+}^3 k_{1+}'' k_{1-}'' k_{2+}'' k_{2-}''} \right]^{\frac{1}{2}}$$

$$\doteq \frac{e^4 |A_1''|}{E q_{-+}} \left[\frac{q_{++}^3 q_{--}}{k_{1+}'' k_{1-}'' k_{2+}'' k_{2-}''} \right]^{\frac{1}{2}},$$

$$M(+,-;+,+,-,-) = \frac{e^4 A_3''^* q_{-\perp}}{E q_{+-}} \left[\frac{q_{++}^3}{q_{-+} k_{1+}'' k_{1-}'' k_{2+}'' k_{2-}''} \right]^{\frac{1}{2}}$$

$$\doteq \frac{e^4 |A_3''| q_{++} q_{--}}{E [q_{+-} q_{-+} k_{1+}'' k_{1-}'' k_{2+}'' k_{2-}'']^{\frac{1}{2}}},$$

$$M(+,-;+,-,+,-) = \frac{e^4 A_2''(2,1) q_{-\perp}}{E q_{+-}} \left[\frac{q_{++}^3}{q_{-+} k_{1+}'' k_{1-}'' k_{2+}'' k_{2-}''} \right]^{\frac{1}{2}}$$

$$\doteq \frac{e^4 |A_2''(2,1)| q_{++} q_{--}}{E [q_{+-} q_{-+} k_{1+}'' k_{1-}'' k_{2+}'' k_{2-}'']^{\frac{1}{2}}},$$

9.19. $e^+e^- \to e^+e^-\gamma\gamma$ (NO Z-EXCHANGE; $m \neq 0$)

$$M(+,-;+,-,-,+) = \frac{e^4 A_2''(1,2)q_{-\perp}}{Eq_{+-}} \left[\frac{q_{++}^3}{q_{-+}k_{1+}''k_{1-}''k_{2+}''k_{2-}''}\right]^{\frac{1}{2}}$$

$$\doteq \frac{e^4|A_2''(1,2)|q_{++}q_{--}}{E\left[q_{+-}q_{-+}k_{1+}''k_{1-}''k_{2+}''k_{2-}''\right]^{\frac{1}{2}}},$$

$$M(+,-;-,+,+,-) = -\frac{e^4 A_2''^*(1,2)q_{+\perp}}{E} \left[\frac{q_{-+}}{q_{++}k_{1+}''k_{1-}''k_{2+}''k_{2-}''}\right]^{\frac{1}{2}}$$

$$\doteq \frac{e^4|A_2''(1,2)|}{E} \left[\frac{q_{+-}q_{-+}}{k_{1+}''k_{1-}''k_{2+}''k_{2-}''}\right]^{\frac{1}{2}},$$

$$M(+,-;-,+,-,+) = -\frac{e^4 A_2''^*(2,1)q_{+\perp}}{E} \left[\frac{q_{-+}}{q_{++}k_{1+}''k_{1-}''k_{2+}''k_{2-}''}\right]^{\frac{1}{2}}$$

$$\doteq \frac{e^4|A_2''(2,1)|}{E} \left[\frac{q_{+-}q_{-+}}{k_{1+}''k_{1-}''k_{2+}''k_{2-}''}\right]^{\frac{1}{2}},$$

$$M(+,-;-,-,+,+) = \frac{e^4 A_3'' q_{+\perp}}{E} \left[\frac{q_{-+}}{q_{++}k_{1+}''k_{1-}''k_{2+}''k_{2-}''}\right]^{\frac{1}{2}}$$

$$\doteq \frac{e^4|A_3''|}{E} \left[\frac{q_{+-}q_{-+}}{k_{1+}''k_{1-}''k_{2+}''k_{2-}''}\right]^{\frac{1}{2}},$$

$$M(-,+;+,+,-,-) = -\frac{e^4 A_3''^* q_{+\perp}^*}{E} \left[\frac{q_{-+}}{q_{++}k_{1+}''k_{1-}''k_{2+}''k_{2-}''}\right]^{\frac{1}{2}}$$

$$\doteq \frac{e^4|A_3''|}{E} \left[\frac{q_{+-}q_{-+}}{k_{1+}''k_{1-}''k_{2+}''k_{2-}''}\right]^{\frac{1}{2}},$$

$$M(-,+;+,-,+,-) = -\frac{e^4 A_2''(2,1)q_{+\perp}^*}{E} \left[\frac{q_{-+}}{q_{++}k_{1+}''k_{1-}''k_{2+}''k_{2-}''}\right]^{\frac{1}{2}}$$

$$\doteq \frac{e^4|A_2''(2,1)|}{E} \left[\frac{q_{+-}q_{-+}}{k_{1+}''k_{1-}''k_{2+}''k_{2-}''}\right]^{\frac{1}{2}},$$

$$M(-,+;+,-,-,+) = -\frac{e^4 A_2''(1,2)q_{+\perp}^*}{E} \left[\frac{q_{-+}}{q_{++}k_{1+}''k_{1-}''k_{2+}''k_{2-}''}\right]^{\frac{1}{2}}$$

$$\doteq \frac{e^4|A_2''(1,2)|}{E} \left[\frac{q_{+-}q_{-+}}{k_{1+}''k_{1-}''k_{2+}''k_{2-}''}\right]^{\frac{1}{2}},$$

$$M(-,+;-,+,+,-) = \frac{e^4 A_2''^*(1,2) q_{-\perp}^*}{E q_{+-}} \left[\frac{q_{++}^3}{q_{-+} k_{1+}'' k_{1-}'' k_{2+}'' k_{2-}''} \right]^{\frac{1}{2}}$$

$$\doteq \frac{e^4 |A_2''(1,2)| q_{++} q_{--}}{E [q_{+-} q_{-+} k_{1+}'' k_{1-}'' k_{2+}'' k_{2-}'']^{\frac{1}{2}}},$$

$$M(-,+;-,+,-,+) = \frac{e^4 A_2''^*(2,1) q_{-\perp}^*}{E q_{+-}} \left[\frac{q_{++}^3}{q_{-+} k_{1+}'' k_{1-}'' k_{2+}'' k_{2-}''} \right]^{\frac{1}{2}}$$

$$\doteq \frac{e^4 |A_2''(2,1)| q_{++} q_{--}}{E [q_{+-} q_{-+} k_{1+}'' k_{1-}'' k_{2+}'' k_{2-}'']^{\frac{1}{2}}},$$

$$M(-,+;-,-,+,+) = -\frac{e^4 A_3'' q_{-\perp}^*}{E q_{+-}} \left[\frac{q_{++}^3}{q_{-+} k_{1+}'' k_{1-}'' k_{2+}'' k_{2-}''} \right]^{\frac{1}{2}}$$

$$\doteq \frac{e^4 |A_3''| q_{++} q_{--}}{E [q_{+-} q_{-+} k_{1+}'' k_{1-}'' k_{2+}'' k_{2-}'']^{\frac{1}{2}}},$$

$$M(+,-;+,-,-,-) = \frac{e^4 A_1''^* q_{-\perp}}{E} \left[\frac{q_{++}^3}{q_{-+}^3 k_{1+}'' k_{1-}'' k_{2+}'' k_{2-}''} \right]^{\frac{1}{2}}$$

$$\doteq \frac{e^4 |A_1''|}{E q_{-+}} \left[\frac{q_{++}^3 q_{--}}{k_{1+}'' k_{1-}'' k_{2+}'' k_{2-}''} \right]^{\frac{1}{2}},$$

$$M(+,-;-,+,-,-) = -\frac{e^4 A_1''^* q_{+\perp}}{E} \left[\frac{q_{-+}}{q_{++} k_{1+}'' k_{1-}'' k_{2+}'' k_{2-}''} \right]^{\frac{1}{2}}$$

$$\doteq \frac{e^4 |A_1''|}{E} \left[\frac{q_{+-} q_{-+}}{k_{1+}'' k_{1-}'' k_{2+}'' k_{2-}''} \right]^{\frac{1}{2}},$$

$$M(+,-;-,-,+,-) = \frac{e^4 A_4''(1,2) q_{+\perp}}{E} \left[\frac{q_{-+}}{q_{++} k_{1+}'' k_{1-}'' k_{2+}'' k_{2-}''} \right]^{\frac{1}{2}}$$

$$\doteq \frac{e^4 |A_4''(1,2)|}{E} \left[\frac{q_{+-} q_{-+}}{k_{1+}'' k_{1-}'' k_{2+}'' k_{2-}''} \right]^{\frac{1}{2}},$$

$$M(+,-;-,-,-,+) = \frac{e^4 A_4''(2,1) q_{+\perp}}{E} \left[\frac{q_{-+}}{q_{++} k_{1+}'' k_{1-}'' k_{2+}'' k_{2-}''} \right]^{\frac{1}{2}}$$

$$\doteq \frac{e^4 |A_4''(2,1)|}{E} \left[\frac{q_{+-} q_{-+}}{k_{1+}'' k_{1-}'' k_{2+}'' k_{2-}''} \right]^{\frac{1}{2}},$$

9.19. $e^+e^- \to e^+e^- \gamma\gamma$ (NO Z-EXCHANGE; $m \neq 0$)

$$M(-,+;+,-,-,-) = -\frac{e^4 A_1''^* q_{+\perp}^*}{E}\left[\frac{q_{+-}}{q_{--}k_{1+}''k_{1-}''k_{2+}''k_{2-}''}\right]^{\frac{1}{2}}$$

$$\doteq \frac{e^4 q_{+-}|A_1''|}{E}\left[\frac{q_{++}}{q_{--}k_{1+}''k_{1-}''k_{2+}''k_{2-}''}\right]^{\frac{1}{2}},$$

$$M(-,+;-,+,-,-) = \frac{e^4 A_1''^* q_{-\perp}^*}{Eq_{+-}}\left[\frac{q_{++}^3}{q_{-+}k_{1+}''k_{1-}''k_{2+}''k_{2-}''}\right]^{\frac{1}{2}}$$

$$\doteq \frac{e^4 |A_1''|q_{++}q_{--}}{E\,[q_{+-}q_{-+}k_{1+}''k_{1-}''k_{2+}''k_{2-}'']^{\frac{1}{2}}},$$

$$M(-,+;-,-,+,-) = -\frac{e^4 A_4''(1,2)q_{-\perp}^*}{Eq_{+-}}\left[\frac{q_{++}^3}{q_{-+}k_{1+}''k_{1-}''k_{2+}''k_{2-}''}\right]^{\frac{1}{2}}$$

$$\doteq \frac{e^4 |A_4''(1,2)|q_{++}q_{--}}{E\,[q_{+-}q_{-+}k_{1+}''k_{1-}''k_{2+}''k_{2-}'']^{\frac{1}{2}}},$$

$$M(-,+;-,-,-,+) = -\frac{e^4 A_4''(2,1)q_{-\perp}^*}{Eq_{+-}}\left[\frac{q_{++}^3}{q_{-+}k_{1+}''k_{1-}''k_{2+}''k_{2-}''}\right]^{\frac{1}{2}}$$

$$\doteq \frac{e^4 |A_4''(2,1)|q_{++}q_{--}}{E\,[q_{+-}q_{-+}k_{1+}''k_{1-}''k_{2+}''k_{2-}'']^{\frac{1}{2}}},$$

$$M(-,-;-,+,+,-) = \frac{2e^4 A_4''^*(2,1)q_{+\perp}Z_{+-}^*}{[q_{++}^3 q_{-+}^3 k_{1+}''k_{1-}''k_{2+}''k_{2-}'']^{\frac{1}{2}}}$$

$$\doteq \frac{e^4 |A_4''(2,1)|}{q_{+-}}\left[\frac{8(q_+ \cdot q_-)}{k_{1+}''k_{1-}''k_{2+}''k_{2-}''}\right]^{\frac{1}{2}},$$

$$M(-,-;-,+,-,+) = \frac{2e^4 A_4''^*(1,2)q_{-\perp}Z_{+-}^*}{[q_{++}^3 q_{-+}^3 k_{1+}''k_{1-}''k_{2+}''k_{2-}'']^{\frac{1}{2}}}$$

$$\doteq \frac{e^4 |A_4''(1,2)|}{q_{+-}}\left[\frac{8(q_+ \cdot q_-)}{k_{1+}''k_{1-}''k_{2+}''k_{2-}''}\right]^{\frac{1}{2}},$$

$$M(-,-;-,-,+,+) = \frac{2e^4 A_1'' q_{-\perp}Z_{+-}^*}{[q_{+-}^3 q_{-+}^3 k_{1+}''k_{1-}''k_{2+}''k_{2-}'']^{\frac{1}{2}}}$$

$$\doteq \frac{e^4 |A_1''|}{q_{+-}}\left[\frac{8(q_+ \cdot q_-)}{k_{1+}''k_{1-}''k_{2+}''k_{2-}''}\right]^{\frac{1}{2}},$$

$$M(-,-;-,+,-,-) = \frac{2e^4 A_3''^* q_{-\perp} Z_{+-}^*}{[q_{++}^3 q_{-+}^3 k_{1+}'' k_{1-}'' k_{2+}'' k_{2-}'']^{\frac{1}{2}}}$$

$$\doteq \frac{e^4 |A_3''|}{q_{+-}} \left[\frac{8(q_+ \cdot q_-)}{k_{1+}'' k_{1-}'' k_{2+}'' k_{2-}''} \right]^{\frac{1}{2}},$$

$$M(-,-;-,-,+,-) = \frac{2e^4 A_2''(2,1) q_{-\perp} Z_{+-}^*}{[q_{++}^3 q_{-+}^3 k_{1+}'' k_{1-}'' k_{2+}'' k_{2-}'']^{\frac{1}{2}}}$$

$$\doteq \frac{e^4 |A_2''(2,1)|}{q_{+-}} \left[\frac{8(q_+ \cdot q_-)}{k_{1+}'' k_{1-}'' k_{2+}'' k_{2-}''} \right]^{\frac{1}{2}},$$

$$M(-,-;-,-,-,+) = \frac{2e^4 A_2''(1,2) q_{-\perp} Z_{+-}^*}{[q_{++}^3 q_{-+}^3 k_{1+}'' k_{1-}'' k_{2+}'' k_{2-}'']^{\frac{1}{2}}}$$

$$\doteq \frac{e^4 |A_2''(1,2)|}{q_{+-}} \left[\frac{8(q_+ \cdot q_-)}{k_{1+}'' k_{1-}'' k_{2+}'' k_{2-}''} \right]^{\frac{1}{2}},$$

$$M(-,-;-,-,-,-) = \frac{2e^4 A_1''^* q_{-\perp} Z_{+-}^*}{q_{--}[q_{++} q_{-+}^3 k_{1+}'' k_{1-}'' k_{2+}'' k_{2-}'']^{\frac{1}{2}}}$$

$$\doteq \frac{e^4 |A_1''|}{q_{-+}} \left[\frac{8(q_+ \cdot q_-)}{k_{1+}'' k_{1-}'' k_{2+}'' k_{2-}''} \right]^{\frac{1}{2}}. \quad (9.275)$$

Unpolarized squared matrix element:

$$\overline{|M|^2} = \frac{e^8 \left[4(q_+ \cdot q_-)^2 + q_{++}^2 q_{--}^2 + q_{+-}^2 q_{-+}^2 \right]}{2 E^2 k_{1+}'' k_{1-}'' k_{2+}'' k_{2-}'' q_{+-} q_{-+}}$$

$$\times \left\{ \left[1 + \left(\frac{2q_{-0} + k_{1+}'' + k_{2+}''}{2q_{-0}} \right)^2 \right] |A_1''|^2 + |A_2''(1,2)|^2 \right.$$

$$\left. + |A_2''(2,1)|^2 + |A_3''|^2 + |A_4''(1,2)|^2 + |A_4''(2,1)|^2 \right\}. \quad (9.276)$$

Unpolarized cross section:

$$d\sigma = \frac{\alpha^4 \left[4(q_+ \cdot q_-)^2 + q_{++}^2 q_{--}^2 + q_{+-}^2 q_{-+}^2 \right]}{256 \pi^2 E^4 k_{1+}'' k_{1-}'' k_{2+}'' k_{2-}'' q_{+-} q_{-+}} \left\{ \left[1 + \left(\frac{2q_{-0} + k_{1+}'' + k_{2+}''}{2q_{-0}} \right)^2 \right] \right.$$

$$\left. \times |A_1''|^2 + |A_2''(1,2)|^2 + |A_2''(2,1)|^2 + |A_3''|^2 + |A_4''(1,2)|^2 + |A_4''(2,1)|^2 \right\}$$

$$\times \delta^4(p_+ + p_- - q_+ - q_- - k_1 - k_2) \frac{d^3 \vec{q}_+ \, d^3 \vec{q}_- \, d^3 \vec{k}_1 \, d^3 \vec{k}_2}{q_{+0} \, q_{-0} \, k_{10} \, k_{20}}. \quad (9.277)$$

10
Summary of QCD formulae

All the following formulae are presented in the centre-of-mass system, with total energy denoted by $2E$. Unless otherwise stated, we take the positive z-axis along $\vec{p}_+$. For an arbitrary four-vector k, we frequently use the notation [see eqns (6.8)]

$$k_\pm = k_0 \pm k_z, \qquad k_\perp = k_x + i k_y. \tag{10.1}$$

Throughout this chapter, we neglect the effects of a finite electron and/or quark mass, except in the Sections 10.57 through 10.77, where we present the formulae for the production of heavy quarkonia. Consequently, the formulae are not valid in certain kinematical configurations. For a discussion of the ranges of validity of the approximation, we refer to Section 3.4. We also mention that, for the QCD processes in this chapter, the possible contributions due to Z-exchange are not taken into account.

The labels of the listed helicity amplitudes always refer to the particles in the order in which they appear in the process. For example, for $e^+e^- \to \bar{q}qg$, the helicity amplitude $M(+,-;-,+,+)$ describes the annihilation of a positive helicity e^+ and a negative helicity e^- into a negative helicity $\bar{q}$, a positive helicity q and a positive helicity g. As usual, the symbol '$\doteq$' is to be read as 'an equality sign modulo a phase factor'. In this chapter, the symbol $\overline{|M|^2}$ is used to denote the square of the absolute value of M summed over the final state degrees of freedom (polarization and color) and averaged over the initial state degrees of freedom. The summation over repeated color indices is always implied.

10.1 $e^+e^- \to \bar{q}q$ (no Z-exchange)

Process:
$$e^+(p_+) + e^-(p_-) \to \bar{q}(q_+,j) + q(q_-,i). \tag{10.2}$$

Invariants:
$$s = 2(p_+ \cdot p_-), \qquad t = -2(p_+ \cdot q_+), \qquad u = -2(p_+ \cdot q_-). \tag{10.3}$$

Nonvanishing helicity amplitudes:

$$M(+,-;+,-) = ee_q \delta_{ij} \frac{q_{+\perp}}{E} \left(\frac{q_{++}}{q_{-+}}\right)^{\frac{1}{2}} \doteq ee_q \delta_{ij} \frac{q_{++}}{E},$$

$$M(+,-;-,+) = ee_q \delta_{ij} \frac{q_{-\perp}}{E} \left(\frac{q_{-+}}{q_{++}}\right)^{\frac{1}{2}} \doteq ee_q \delta_{ij} \frac{q_{-+}}{E},$$

$$M(-,+;+,-) = ee_q\delta_{ij}\frac{q^*_{-\perp}}{E}\left(\frac{q_{-+}}{q_{++}}\right)^{\frac{1}{2}} \doteq ee_q\delta_{ij}\frac{q_{-+}}{E},$$

$$M(-,+;-,+) = ee_q\delta_{ij}\frac{q^*_{+\perp}}{E}\left(\frac{q_{++}}{q_{-+}}\right)^{\frac{1}{2}} \doteq ee_q\delta_{ij}\frac{q_{++}}{E}. \quad (10.4)$$

Unpolarized squared matrix element:

$$\overline{|M|^2} = 6e^2e_q^2\frac{t^2+u^2}{s^2}. \quad (10.5)$$

Unpolarized cross section:

$$\frac{d\sigma}{dt} = 6\pi\alpha^2 Q_f^2\frac{t^2+u^2}{s^4}. \quad (10.6)$$

10.2 $e^+e^- \to \bar{q}qg$ (no Z-exchange)

Process:
$$e^+(p_+) + e^-(p_-) \to \bar{q}(q_+,j) + q(q_-,i) + g(k,a). \quad (10.7)$$

Invariants and definitions:

$$s = 2(p_+ \cdot p_-), \quad t = -2(p_+ \cdot q_+), \quad u = -2(p_+ \cdot q_-),$$

$$s' = 2(q_+ \cdot q_-), \quad t' = -2(p_- \cdot q_-), \quad u' = -2(p_- \cdot q_+),$$

$$Z_{+-} = q_{++}q_{--} - q^*_{+\perp}q_{-\perp}, \quad Z_{-+} = q_{-+}q_{+-} - q^*_{-\perp}q_{+\perp}. \quad (10.8)$$

Gluon polarizations:

$$\not{k}^\pm = N[\not{q}_- \not{q}_+ \not{k}(1\mp\gamma_5) + \not{k}\not{q}_+\not{q}_-(1\pm\gamma_5)],$$

$$N^{-1} = 4[(q_+ \cdot q_-)(q_+ \cdot k)(q_- \cdot k)]^{\frac{1}{2}}. \quad (10.9)$$

Nonvanishing helicity amplitudes:

$$M(+,-;+,-,+) = -\frac{ee_q g T^a_{ij} q_{-\perp} Z_{+-}}{[q_{++}q_{-+}(q_+ \cdot q_-)(q_+ \cdot k)(q_- \cdot k)]^{\frac{1}{2}}},$$

$$M(+,-;-,+,+) = -\frac{ee_q g T^a_{ij} q_{+\perp} Z_{-+}}{[q_{++}q_{-+}(q_+ \cdot q_-)(q_+ \cdot k)(q_- \cdot k)]^{\frac{1}{2}}},$$

$$M(-,+;+,-,+) = -\frac{ee_q g T^a_{ij} q^*_{+\perp} Z_{-+}}{q_{+-}}\left[\frac{q_{-+}}{q_{++}(q_+ \cdot q_-)(q_+ \cdot k)(q_- \cdot k)}\right]^{\frac{1}{2}},$$

$$M(-,+;-,+,+) = -\frac{ee_q g T^a_{ij} q^*_{-\perp} Z_{+-}}{q_{--}}\left[\frac{q_{++}}{q_{-+}(q_+ \cdot q_-)(q_+ \cdot k)(q_- \cdot k)}\right]^{\frac{1}{2}},$$

10.3. $e^+e^- \to \bar{q}q\gamma$ (NO Z-EXCHANGE)

$$M(+,-;+,-,-) = -\frac{ee_q g T^a_{ij} q_{-\perp} Z^*_{+-}}{q_{--}} \left[\frac{q_{++}}{q_{-+}(q_+ \cdot q_-)(q_+ \cdot k)(q_- \cdot k)}\right]^{\frac{1}{2}},$$

$$M(+,-;-,+,-) = -\frac{ee_q g T^a_{ij} q_{+\perp} Z^*_{-+}}{q_{+-}} \left[\frac{q_{-+}}{q_{++}(q_+ \cdot q_-)(q_+ \cdot k)(q_- \cdot k)}\right]^{\frac{1}{2}},$$

$$M(-,+;+,-,-) = -\frac{ee_q g T^a_{ij} q^*_{+\perp} Z^*_{-+}}{[q_{++}q_{-+}(q_+ \cdot q_-)(q_+ \cdot k)(q_- \cdot k)]^{\frac{1}{2}}},$$

$$M(-,+;-,+,-) = -\frac{ee_q g T^a_{ij} q^*_{-\perp} Z^*_{+-}}{[q_{++}q_{-+}(q_+ \cdot q_-)(q_+ \cdot k)(q_- \cdot k)]^{\frac{1}{2}}}. \quad (10.10)$$

Squared absolute values of the nonvanishing helicity amplitudes, summed over the final state color degrees of freedom:

$$|M(+,-;+,-,+)|^2 = |M(-,+;-,+,-)|^2 = \frac{8(ee_q g)^2 q^2_{--}}{(q_+ \cdot k)(q_- \cdot k)},$$

$$|M(+,-;-,+,+)|^2 = |M(-,+;+,-,-)|^2 = \frac{8(ee_q g)^2 q^2_{+-}}{(q_+ \cdot k)(q_- \cdot k)},$$

$$|M(-,+;+,-,+)|^2 = |M(+,-;-,+,-)|^2 = \frac{8(ee_q g)^2 q^2_{-+}}{(q_+ \cdot k)(q_- \cdot k)},$$

$$|M(-,+;-,+,+)|^2 = |M(+,-;+,-,-)|^2 = \frac{8(ee_q g)^2 q^2_{++}}{(q_+ \cdot k)(q_- \cdot k)}. \quad (10.11)$$

Unpolarized squared matrix element:

$$\overline{|M|^2} = 4(ee_q g)^2 \frac{t^2 + t'^2 + u^2 + u'^2}{s(q_+ \cdot k)(q_- \cdot k)}. \quad (10.12)$$

Unpolarized cross section:

$$d\sigma = \frac{\alpha^2 \alpha_s Q_f^2}{2\pi^2} \frac{t^2 + t'^2 + u^2 + u'^2}{s^2(q_+ \cdot k)(q_- \cdot k)} \delta^4(p_+ + p_- - q_+ - q_- - k) \frac{d^3\vec{q}_+ \, d^3\vec{q}_- \, d^3\vec{k}}{q_{+0} \, q_{-0} \, k_0}. \quad (10.13)$$

10.3 $e^+e^- \to \bar{q}q\gamma$ (no Z-exchange)

Process:
$$e^+(p_+) + e^-(p_-) \to \bar{q}(q_+,j) + q(q_-,i) + \gamma(k). \quad (10.14)$$

Invariants and definitions:

$$s = 2(p_+ \cdot p_-), \quad t = -2(p_+ \cdot q_+), \quad u = -2(p_+ \cdot q_-),$$

$$s' = 2(q_+ \cdot q_-), \quad t' = -2(p_- \cdot q_-), \quad u' = -2(p_- \cdot q_+),$$

10. SUMMARY OF QCD FORMULAE

$$B = Q_f + \frac{2E}{s'}\left[2q_{-z} + \frac{k_\perp q^*_{-\perp}}{k_+} - \frac{k^*_\perp q_{-\perp}}{k_-}\right],$$

$$Z_{+-} = q_{++}q_{--} - q^*_{+\perp}q_{-\perp}, \qquad Z_{-+} = q_{-+}q_{+-} - q^*_{-\perp}q_{+\perp}.$$

(10.15)

Photon polarizations:

$$\not{k}^\pm = N[\not{p}_+ \not{p}_- \not{k}(1\mp\gamma_5) - \not{k}\,\not{p}_+\,\not{p}_-(1\pm\gamma_5)],$$

$$N^{-1} = E^2[32k_+k_-]^{\frac{1}{2}}.$$

(10.16)

Nonvanishing helicity amplitudes:

$$M(+,-;+,-,+) = -\frac{e^3\delta_{ij}q_{-\perp}Z_{+-}B}{[q_{++}q_{-+}(q_+\cdot q_-)(q_+\cdot k)(q_-\cdot k)]^{\frac{1}{2}}},$$

$$M(+,-;-,+,+) = -\frac{e^3\delta_{ij}q_{+\perp}Z_{-+}B}{[q_{++}q_{-+}(q_+\cdot q_-)(q_+\cdot k)(q_-\cdot k)]^{\frac{1}{2}}},$$

$$M(-,+;+,-,+) = -\frac{e^3\delta_{ij}q^*_{+\perp}Z_{-+}B}{q_{+-}}\left[\frac{q_{-+}}{q_{++}(q_+\cdot q_-)(q_+\cdot k)(q_-\cdot k)}\right]^{\frac{1}{2}},$$

$$M(-,+;-,+,+) = -\frac{e^3\delta_{ij}q^*_{-\perp}Z_{+-}B}{q_{--}}\left[\frac{q_{++}}{q_{-+}(q_+\cdot q_-)(q_+\cdot k)(q_-\cdot k)}\right]^{\frac{1}{2}},$$

$$M(+,-;+,-,-) = -\frac{e^3\delta_{ij}q_{-\perp}Z^*_{+-}B^*}{q_{--}}\left[\frac{q_{++}}{q_{-+}(q_+\cdot q_-)(q_+\cdot k)(q_-\cdot k)}\right]^{\frac{1}{2}},$$

$$M(+,-;-,+,-) = -\frac{e^3\delta_{ij}q_{+\perp}Z^*_{-+}B^*}{q_{+-}}\left[\frac{q_{-+}}{q_{++}(q_+\cdot q_-)(q_+\cdot k)(q_-\cdot k)}\right]^{\frac{1}{2}},$$

$$M(-,+;+,-,-) = -\frac{e^3\delta_{ij}q^*_{+\perp}Z^*_{-+}B^*}{[q_{++}q_{-+}(q_+\cdot q_-)(q_+\cdot k)(q_-\cdot k)]^{\frac{1}{2}}},$$

$$M(-,+;-,+,-) = -\frac{e^3\delta_{ij}q^*_{-\perp}Z^*_{+-}B^*}{[q_{++}q_{-+}(q_+\cdot q_-)(q_+\cdot k)(q_-\cdot k)]^{\frac{1}{2}}}.$$

(10.17)

Squared absolute values of the nonvanishing helicity amplitudes, summed over the final state color degrees of freedom:

$$|M(+,-;+,-,+)|^2 = |M(-,+;-,+,-)|^2 = \frac{6e^6|B|^2q^2_{--}}{(q_+\cdot k)(q_-\cdot k)},$$

$$|M(+,-;-,+,+)|^2 = |M(-,+;+,-,-)|^2 = \frac{6e^6|B|^2q^2_{+-}}{(q_+\cdot k)(q_-\cdot k)},$$

10.4. $e^+e^- \to \bar{q}qgg$ (NO Z-EXCHANGE)

$$|M(-,+;+,-,+)|^2 = |M(+,-;-,+,-)|^2 = \frac{6e^6|B|^2 q_{-+}^2}{(q_+ \cdot k)(q_- \cdot k)},$$

$$|M(-,+;-,+,+)|^2 = |M(+,-;+,-,-)|^2 = \frac{6e^6|B|^2 q_{++}^2}{(q_+ \cdot k)(q_- \cdot k)}. \quad (10.18)$$

Unpolarized squared matrix element:

$$\overline{|M|^2} = -3e^6(v_p - Q_f v_q)^2 \frac{t^2 + t'^2 + u^2 + u'^2}{ss'}, \quad (10.19)$$

where [see eqns (4.27)]

$$v_p = \frac{p_+}{(p_+ \cdot k)} - \frac{p_-}{(p_- \cdot k)}, \qquad v_q = \frac{q_+}{(q_+ \cdot k)} - \frac{q_-}{(q_- \cdot k)}. \quad (10.20)$$

Unpolarized cross section:

$$d\sigma = -\frac{3\alpha^3}{8\pi^2}(v_p - Q_f v_q)^2 \frac{t^2 + t'^2 + u^2 + u'^2}{s^2 s'}$$

$$\times \delta^4(p_+ + p_- - q_+ - q_- - k) \frac{d^3\vec{q}_+ \, d^3\vec{q}_- \, d^3\vec{k}}{q_{+0}\, q_{-0}\, k_0}. \quad (10.21)$$

10.4 $e^+e^- \to \bar{q}qgg$ (no Z-exchange)

Process:

$$e^+(p_+) + e^-(p_-) \to \bar{q}(k_3, j) + q(k_4, i) + g(k_1, a) + g(k_2, b). \quad (10.22)$$

Definitions:

$$w_i = k_{i\perp}/k_{i+}, \qquad i = 1,\ldots,4,$$

$$Z_{ij} = k_{i+}k_{j-} - k_{i\perp}^* k_{j\perp}, \qquad i,j = 1,\ldots,4,$$

$$C = \frac{1}{|w_1 - w_2|^2 \, |w_1 - w_3| \, |w_1 - w_4| \, |w_2 - w_3| \, |w_2 - w_4|},$$

$$D = \frac{1}{|w_1 - w_2|^2 \, |w_1 - w_3|^2 \, |w_1 - w_4|^2 \, |w_2 - w_3|^2 \, |w_2 - w_4|^2},$$

$$c_1 = |Z_{12}Z_{34} - Z_{14}Z_{32}|^2 + \frac{(k_2 \cdot k_4) Z_{32}^*(Z_{12}^* + Z_{14}^*)(Z_{12}Z_{34} - Z_{14}Z_{32})}{2E(E - k_{30})}$$

$$+ \frac{(k_1 \cdot k_3) Z_{14}(Z_{12} + Z_{32})(Z_{12}^* Z_{34}^* - Z_{14}^* Z_{32}^*)}{2E(E - k_{40})},$$

$$d_1 = Z_{14}Z_{32}^*(Z_{12}+Z_{32})(Z_{12}^*+Z_{14}^*)$$

$$+\frac{(k_1\cdot k_4)Z_{32}^*(Z_{12}^*+Z_{14}^*)(Z_{12}Z_{34}-Z_{14}Z_{32})}{2E(E-k_{30})}$$

$$+\frac{(k_2\cdot k_3)Z_{14}(Z_{12}+Z_{32})(Z_{12}^*Z_{34}^*-Z_{14}^*Z_{32}^*)}{2E(E-k_{40})},$$

$$c_2 = |Z_{12}Z_{43}-Z_{13}Z_{42}|^2$$

$$+\frac{(k_2\cdot k_3)Z_{42}^*(Z_{12}^*+Z_{13}^*)(Z_{12}Z_{43}-Z_{13}Z_{42})}{2E(E-k_{40})}$$

$$+\frac{(k_1\cdot k_4)Z_{13}(Z_{12}+Z_{42})(Z_{12}^*Z_{43}^*-Z_{13}^*Z_{42}^*)}{2E(E-k_{30})},$$

$$d_2 = Z_{13}Z_{42}^*(Z_{12}+Z_{42})(Z_{12}^*+Z_{13}^*)$$

$$+\frac{(k_1\cdot k_3)Z_{42}^*(Z_{12}^*+Z_{13}^*)(Z_{12}Z_{43}-Z_{13}Z_{42})}{2E(E-k_{40})}$$

$$+\frac{(k_2\cdot k_4)Z_{13}(Z_{12}+Z_{42})(Z_{12}^*Z_{43}^*-Z_{13}^*Z_{42}^*)}{2E(E-k_{30})},$$

$$c_3 = |Z_{21}Z_{34}-Z_{24}Z_{31}|^2$$

$$+\frac{(k_1\cdot k_4)Z_{31}(Z_{21}+Z_{24})(Z_{21}^*Z_{34}^*-Z_{24}^*Z_{31}^*)}{2E(E-k_{30})}$$

$$+\frac{(k_2\cdot k_3)Z_{24}^*(Z_{21}^*+Z_{31}^*)(Z_{21}Z_{34}-Z_{24}Z_{31})}{2E(E-k_{40})},$$

$$d_3 = Z_{24}^*Z_{31}(Z_{21}^*+Z_{31}^*)(Z_{21}+Z_{24})$$

$$+\frac{(k_2\cdot k_4)Z_{31}(Z_{21}+Z_{24})(Z_{21}^*Z_{34}^*-Z_{24}^*Z_{31}^*)}{2E(E-k_{30})}$$

$$+\frac{(k_1\cdot k_3)Z_{24}^*(Z_{21}^*+Z_{31}^*)(Z_{21}Z_{34}-Z_{24}Z_{31})}{2E(E-k_{40})},$$

$$c_4 = |Z_{21}Z_{43}-Z_{23}Z_{41}|^2+\frac{(k_1\cdot k_3)Z_{41}(Z_{21}+Z_{23})(Z_{21}^*Z_{43}^*-Z_{23}^*Z_{41}^*)}{2E(E-k_{40})}$$

$$+\frac{(k_2\cdot k_4)Z_{23}^*(Z_{21}^*+Z_{41}^*)(Z_{21}Z_{43}-Z_{23}Z_{41})}{2E(E-k_{30})},$$

10.4. $e^+e^- \to \bar{q}qgg$ (NO Z-EXCHANGE)

$$d_4 = Z_{23}^* Z_{41}(Z_{21}^* + Z_{41}^*)(Z_{21} + Z_{23})$$

$$+ \frac{(k_2 \cdot k_3) Z_{41}(Z_{21} + Z_{23})(Z_{21}^* Z_{43}^* - Z_{23}^* Z_{41}^*)}{2E(E - k_{40})}$$

$$+ \frac{(k_1 \cdot k_4) Z_{23}^* (Z_{21}^* + Z_{41}^*)(Z_{21} Z_{43} - Z_{23} Z_{41})}{2E(E - k_{30})}. \quad (10.23)$$

Gluon polarizations:

$$\not{\epsilon}_i^{\pm} = N_i[\not{k}_4 \not{k}_3 \not{k}_i(1 \mp \gamma_5) + \not{k}_i \not{k}_3 \not{k}_4(1 \pm \gamma_5)],$$

$$N_i^{-1} = \sqrt{2} k_{i+} k_{3+} k_{4+} |w_i - w_3| |w_i - w_4| |w_3 - w_4|, \qquad i = 1, 2. \quad (10.24)$$

Nonvanishing helicity amplitudes:

$$M(+,-;+,-,+,+) = \frac{4ee_q g^2 C}{k_{1+} k_{2+}} \left(\frac{k_{4+}}{k_{3+}}\right)^{\frac{1}{2}} \frac{w_4^2 (w_1 - w_2)^* (w_3 - w_4)^*}{w_3 - w_4}$$

$$\times [(T^a T^b)_{ij}(w_1 - w_3)(w_2 - w_4) - (T^b T^a)_{ij}(w_1 - w_4)(w_2 - w_3)],$$

$$M(+,-;-,+,+,+) = \frac{4ee_q g^2 C}{k_{1+} k_{2+}} \left(\frac{k_{3+}}{k_{4+}}\right)^{\frac{1}{2}} \frac{w_4^2 (w_1 - w_2)^* (w_3 - w_4)^*}{w_3 - w_4}$$

$$\times [(T^a T^b)_{ij}(w_1 - w_4)(w_2 - w_3) - (T^b T^a)_{ij}(w_1 - w_3)(w_2 - w_4)],$$

$$M(-,+;+,-,+,+) = -\frac{4ee_q g^2 C}{k_{1+} k_{2+}} \left(\frac{k_{4+}}{k_{3+}}\right)^{\frac{1}{2}} \frac{(w_1 - w_2)^* (w_3 - w_4)^*}{w_3 - w_4}$$

$$\times [(T^a T^b)_{ij}(w_1 - w_3)(w_2 - w_4) - (T^b T^a)_{ij}(w_1 - w_4)(w_2 - w_3)],$$

$$M(-,+;-,+,+,+) = -\frac{4ee_q g^2 C}{k_{1+} k_{2+}} \left(\frac{k_{3+}}{k_{4+}}\right)^{\frac{1}{2}} \frac{(w_1 - w_2)^* (w_3 - w_4)^*}{(w_3 - w_4)}$$

$$\times [(T^a T^b)_{ij}(w_1 - w_4)(w_2 - w_3) - (T^b T^a)_{ij}(w_1 - w_3)(w_2 - w_4)^*],$$

$$M(+,-;+,-,+,-) = -\frac{2ee_q g^2 C [c_1 (T^a T^b)_{ij} + d_1 (T^b T^a)_{ij}]}{E(k_{1+} k_{2+})^3 (k_{3+} k_{4+})^{\frac{3}{2}} |w_2|^2 w_4^*},$$

$$M(+,-;+,-,-,+) = -\frac{2ee_q g^2 C [c_3^* (T^a T^b)_{ij} + d_3^* (T^b T^a)_{ij}]}{E(k_{1+} k_{2+})^3 (k_{3+} k_{4+})^{\frac{3}{2}} |w_2|^2 w_4^*},$$

$$M(+,-;-,+,+,-) = -\frac{2ee_q g^2 C [c_2 (T^a T^b)_{ij} + d_2 (T^b T^a)_{ij}]}{E(k_{1+} k_{2+})^3 (k_{3+} k_{4+})^{\frac{3}{2}} |w_2|^2 w_3^*},$$

$$M(+,-;-,+,-,+) = -\frac{2ee_qg^2C[c_4^*(T^aT^b)_{ij} + d_4^*(T^bT^a)_{ij}]}{E(k_{1+}k_{2+})^3(k_{3+}k_{4+})^{\frac{3}{2}}|w_2|^2w_3^*},$$

$$M(-,+;+,-,+,-) = -\frac{2ee_qg^2C[c_4(T^aT^b)_{ij} + d_4(T^bT^a)_{ij}]}{E(k_{1+}k_{2+})^3(k_{3+}k_{4+})^{\frac{3}{2}}|w_2|^2w_3},$$

$$M(-,+;+,-,-,+) = -\frac{2ee_qg^2C[c_2^*(T^aT^b)_{ij} + d_2^*(T^bT^a)_{ij}]}{E(k_{1+}k_{2+})^3(k_{3+}k_{4+})^{\frac{3}{2}}|w_2|^2w_3},$$

$$M(-,+;-,+,+,-) = -\frac{2ee_qg^2C[c_3(T^aT^b)_{ij} + d_3(T^bT^a)_{ij}]}{E(k_{1+}k_{2+})^3(k_{3+}k_{4+})^{\frac{3}{2}}|w_2|^2w_4},$$

$$M(-,+;-,+,-,+) = -\frac{2ee_qg^2C[c_1^*(T^aT^b)_{ij} + d_1^*(T^bT^a)_{ij}]}{E(k_{1+}k_{2+})^3(k_{3+}k_{4+})^{\frac{3}{2}}|w_2|^2w_4},$$

$$M(+,-;+,-,-,-) = -\frac{4ee_qg^2C^*}{k_{1+}k_{2+}}\left(\frac{k_{3+}}{k_{4+}}\right)^{\frac{1}{2}}\frac{(w_1-w_2)(w_3-w_4)}{(w_3-w_4)^*}$$
$$\times[(T^aT^b)_{ij}(w_1-w_4)^*(w_2-w_3)^* - (T^bT^a)_{ij}(w_1-w_3)^*(w_2-w_4)^*],$$

$$M(+,-;-,+,-,-) = -\frac{4ee_qg^2C^*}{k_{1+}k_{2+}}\left(\frac{k_{4+}}{k_{3+}}\right)^{\frac{1}{2}}\frac{(w_1-w_2)(w_3-w_4)}{(w_3-w_4)^*}$$
$$\times[(T^aT^b)_{ij}(w_1-w_3)^*(w_2-w_4)^* - (T^bT^a)_{ij}(w_1-w_4)^*(w_2-w_3)^*],$$

$$M(-,+;+,-,-,-) = \frac{4ee_qg^2C^*}{k_{1+}k_{2+}}\left(\frac{k_{3+}}{k_{4+}}\right)^{\frac{1}{2}}\frac{w_4^{*2}(w_1-w_2)(w_3-w_4)}{(w_3-w_4)^*}$$
$$\times[(T^aT^b)_{ij}(w_1-w_4)^*(w_2-w_3)^* - (T^bT^a)_{ij}(w_1-w_3)^*(w_2-w_4)^*],$$

$$M(-,+;-,+,-,-) = \frac{4ee_qg^2C^*}{k_{1+}k_{2+}}\left(\frac{k_{4+}}{k_{3+}}\right)^{\frac{1}{2}}\frac{w_4^{*2}(w_1-w_2)(w_3-w_4)}{(w_3-w_4)^*}$$
$$\times[(T^aT^b)_{ij}(w_1-w_3)^*(w_2-w_4)^* - (T^bT^a)_{ij}(w_1-w_4)^*(w_2-w_3)^*].$$
(10.25)

Squared absolute values of the nonvanishing helicity amplitudes, summed over the final state color degrees of freedom:

$$|M(+,-;+,-,+,+)|^2 = |M(-,+;-,+,-,-)|^2$$

10.4. $e^+e^- \to \bar{q}qgg$ (NO Z-EXCHANGE)

$$= \frac{32e^2 e_q^2 g^4 k_{4+}|w_4|^4 D}{3k_{1+}^2 k_{2+}^2 k_{3+}} \Big[9|w_1 - w_3|^2|w_2 - w_4|^2$$

$$+ 9|w_1 - w_4|^2|w_2 - w_3|^2 - |w_1 - w_2|^2|w_3 - w_4|^2\Big],$$

$$|M(+,-;-,+,+,+)|^2 = |M(-,+;+,-,-,-)|^2$$

$$= \frac{32e^2 e_q^2 g^4 k_{3+}|w_4|^4 D}{3k_{1+}^2 k_{2+}^2 k_{4+}} \Big[9|w_1 - w_3|^2|w_2 - w_4|^2$$

$$+ 9|w_1 - w_4|^2|w_2 - w_3|^2 - |w_1 - w_2|^2|w_3 - w_4|^2\Big],$$

$$|M(-,+;+,-,+,+)|^2 = |M(+,-;-,+,-,-)|^2$$

$$= \frac{32e^2 e_q^2 g^4 k_{4+} D}{3k_{1+}^2 k_{2+}^2 k_{3+}} \Big[9|w_1 - w_3|^2|w_2 - w_4|^2$$

$$+ 9|w_1 - w_4|^2|w_2 - w_3|^2 - |w_1 - w_2|^2|w_3 - w_4|^2\Big],$$

$$|M(-,+;-,+,+,+)|^2 = |M(+,-;+,-,-,-)|^2$$

$$= \frac{32e^2 e_q^2 g^4 k_{3+} D}{3k_{1+}^2 k_{2+}^2 k_{4+}} \Big[9|w_1 - w_3|^2|w_2 - w_4|^2$$

$$+ 9|w_1 - w_4|^2|w_2 - w_3|^2 - |w_1 - w_2|^2|w_3 - w_4|^2\Big],$$

$$|M(+,-;+,-,+,-)|^2 = |M(-,+;-,+,-,+)|^2$$

$$= \frac{4e^2 e_q^2 g^4 D[7|c_1 + d_1|^2 + 9|c_1 - d_1|^2]}{3E^2(k_{1+}k_{2+})^6(k_{3+}k_{4+})^3|w_1 - w_2|^2|w_2|^4|w_4|^2},$$

$$|M(+,-;+,-,-,+)|^2 = |M(-,+;-,+,+,-)|^2$$

$$= \frac{4e^2 e_q^2 g^4 D[7|c_3 + d_3|^2 + 9|c_3 - d_3|^2]}{3E^2(k_{1+}k_{2+})^6(k_{3+}k_{4+})^3|w_1 - w_2|^2|w_1|^4|w_4|^2},$$

$$|M(+,-;-,+,+,-)|^2 = |M(-,+;+,-,-,+)|^2$$

$$= \frac{4e^2 e_q^2 g^4 D[7|c_2 + d_2|^2 + 9|c_2 - d_2|^2]}{3E^2(k_{1+}k_{2+})^6(k_{3+}k_{4+})^3|w_1 - w_2|^2|w_2|^4|w_3|^2},$$

$$|M(+,-;-,+,-,+)|^2 = |M(-,+;+,-,+,-)|^2$$

$$= \frac{4e^2 e_q^2 g^4 D[7|c_4 + d_4|^2 + 9|c_4 - d_4|^2]}{3E^2(k_{1+}k_{2+})^6(k_{3+}k_{4+})^3|w_1 - w_2|^2|w_1|^4|w_3|^2}.$$
(10.26)

Unpolarized squared matrix element:

$$\overline{|M|^2} = \tfrac{2}{3} e^2 e_q^2 g^4 D \left\{ \frac{8(k_{3+}^2 + k_{3-}^2 + k_{4+}^2 + k_{4-}^2)}{k_{1+}^2 k_{2+}^2 k_{3+} k_{4+}} \left[9|w_1 - w_3|^2|w_2 - w_4|^2 \right. \right.$$

$$\left. + 9|w_1 - w_4|^2|w_2 - w_3|^2 - |w_1 - w_2|^2|w_3 - w_4|^2 \right]$$

$$+ \frac{1}{3E^2(k_{1+}k_{2+})^6(k_{3+}k_{4+})^3|w_1 - w_2|^2}$$

$$\times \left[\frac{7|c_1 + d_1|^2 + 9|c_1 - d_1|^2}{|w_2|^4|w_4|^2} + \frac{7|c_2 + d_2|^2 + 9|c_2 - d_2|^2}{|w_2|^4|w_3|^2} \right.$$

$$\left. \left. + \frac{7|c_3 + d_3|^2 + 9|c_3 - d_3|^2}{|w_1|^4|w_4|^2} + \frac{7|c_4 + d_4|^2 + 9|c_4 - d_4|^2}{|w_1|^4|w_3|^2} \right] \right\}.$$
(10.27)

Unpolarized cross section:

$$d\sigma = \frac{\alpha^2 \alpha_S^2 Q_f^2 D^2}{192\pi^4 E^2} \left\{ \frac{8(k_{3+}^2 + k_{3-}^2 + k_{4+}^2 + k_{4-}^2)}{k_{1+}^2 k_{2+}^2 k_{3+} k_{4+}} \left[9|w_1 - w_3|^2|w_2 - w_4|^2 \right. \right.$$

$$\left. + 9|w_1 - w_4|^2|w_2 - w_3|^2 - |w_1 - w_2|^2|w_3 - w_4|^2 \right]$$

$$+ \frac{1}{3E^2(k_{1+}k_{2+})^6(k_{3+}k_{4+})^3|w_1 - w_2|^2}$$

$$\times \left[\frac{7|c_1 + d_1|^2 + 9|c_1 - d_1|^2}{|w_2|^4|w_4|^2} + \frac{7|c_2 + d_2|^2 + 9|c_2 - d_2|^2}{|w_2|^4|w_3|^2} \right.$$

$$\left. \left. + \frac{7|c_3 + d_3|^2 + 9|c_3 - d_3|^2}{|w_1|^4|w_4|^2} + \frac{7|c_4 + d_4|^2 + 9|c_4 - d_4|^2}{|w_1|^4|w_3|^2} \right] \right\}$$

$$\times \delta^4(p_+ + p_- - k_1 - k_2 - k_3 - k_4) \frac{d^3\vec{k}_1 \, d^3\vec{k}_2 \, d^3\vec{k}_3 \, d^3\vec{k}_4}{k_{10} k_{20} k_{30} k_{40}}.$$
(10.28)

10.5 $e^+e^- \to q\bar{q}q'\bar{q}'$ (no Z-exchange)

Process (different quark flavors):

$$e^+(p_+) + e^-(p_-) \to q(k_1,i) + \bar{q}(k_2,j) + q'(k_3,m) + \bar{q}'(k_4,n). \quad (10.29)$$

Definitions:

$$Z_{ij} = k_{i+}k_{j-} - k^*_{i\perp}k_{j\perp}, \qquad i,j = 1,\ldots,4,$$

$$B = \frac{T^a_{ij}T^a_{mn}}{4E^2(k_{1+}k_{2+}k_{3+}k_{4+})^{\frac{1}{2}}},$$

$$F(1,2,3,4) = \frac{k_{4\perp}}{(k_3 \cdot k_4)k_{4-}} \left[\frac{k_{2+}Z^*_{14}(Z_{31} + Z_{34})}{E - k_{20}} + \frac{k_{1\perp}k^*_{3\perp}Z_{23}(Z^*_{24} + Z^*_{34})}{k_{3-}(E - k_{10})} \right]. \quad (10.30)$$

Nonvanishing helicity amplitudes:

$$M(+,-;+,-,+,-) = -eg^2 B[e_q F(1,2,4,3) + e'_q F(3,4,2,1)],$$

$$M(+,-;+,-,-,+) = -eg^2 B[e_q F(1,2,3,4) - e'_q F(4,3,2,1)],$$

$$M(+,-;-,+,+,-) = -eg^2 B[-e_q F(2,1,4,3) + e'_q F(3,4,1,2)],$$

$$M(+,-;-,+,-,+) = -eg^2 B[-e_q F(2,1,3,4) - e'_q F(4,3,1,2)],$$

$$M(-,+;+,-,+,-) = -eg^2 B[-e_q F^*(2,1,3,4) - e'_q F^*(4,3,1,2)],$$

$$M(-,+;+,-,-,+) = -eg^2 B[-e_q F^*(2,1,4,3) + e'_q F^*(3,4,1,2)],$$

$$M(-,+;-,+,+,-) = -eg^2 B[e_q F^*(1,2,3,4) - e'_q F^*(4,3,2,1)],$$

$$M(-,+;-,+,-,+) = -eg^2 B[e_q F^*(1,2,4,3) + e'_q F^*(3,4,2,1)]. \quad (10.31)$$

Squared absolute values of the nonvanishing helicity amplitudes, summed over the final state color degrees of freedom:

$$|M(+,-;+,-,+,-)|^2 = |M(-,+;-,+,-,+)|^2$$
$$= 2e^2g^4B^2\Big[e_q^2|F(1,2,4,3)|^2 + e_q'^2|F(3,4,2,1)|^2$$
$$+2e_qe_q'\text{Re}\Big(F(1,2,4,3)F^*(3,4,2,1)\Big)\Big],$$

$$|M(+,-;+,-,-,+)|^2 = |M(-,+;-,+,+,-)|^2$$
$$= 2e^2g^4B^2\Big[e_q^2|F(1,2,3,4)|^2 + e_q'^2|F(4,3,2,1)|^2$$
$$-2e_qe_q'\text{Re}\Big(F(1,2,3,4)F^*(4,3,2,1)\Big)\Big],$$

$$|M(+,-;-,+,+,-)|^2 = |M(-,+;+,-,-,+)|^2$$
$$= 2e^2g^4B^2\Big[e_q^2|F(2,1,4,3)|^2 + e_q'^2|F(3,4,1,2)|^2$$
$$-2e_qe_q'\text{Re}\Big(F(2,1,4,3)F^*(3,4,1,2)\Big)\Big],$$

$$|M(+,-;-,+,-,+)|^2 = |M(-,+;+,-,+,-)|^2$$
$$= 2e^2g^4B^2\Big[e_q^2|F(2,1,3,4)|^2 + e_q'^2|F(4,3,1,2)|^2$$
$$+2e_qe_q'\text{Re}\Big(F(2,1,3,4)F^*(4,3,1,2)\Big)\Big]. \quad (10.32)$$

Unpolarized squared matrix element:

$$\overline{|M|^2} = e^2g^4B^2$$
$$\times\Big\{e_q^2\Big[|F(1,2,3,4)|^2 + |F(1,2,4,3)|^2 + |F(2,1,3,4)|^2 + |F(2,1,4,3)|^2\Big]$$
$$+e_q'^2\Big[|F(3,4,1,2)|^2 + |F(4,3,1,2)|^2 + |F(3,4,2,1)|^2 + |F(4,3,2,1)|^2\Big]$$
$$+2e_qe_q'\text{Re}\Big[F(1,2,4,3)F^*(3,4,2,1) - F(1,2,3,4)F^*(4,3,2,1)$$
$$+F(2,1,3,4)F^*(4,3,1,2) - F(2,1,4,3)F^*(3,4,1,2)\Big]\Big\}. \quad (10.33)$$

10.6. $e^+e^- \to q\bar{q}q\bar{q}$ (NO Z-EXCHANGE)

Unpolarized cross section:

$$d\sigma = \frac{\alpha^2 \alpha_S^2}{2048\pi^4 E^6 k_{1+}k_{2+}k_{3+}k_{4+}}$$

$$\times \Big\{ Q_f^2 \Big[|F(1,2,3,4)|^2 + |F(1,2,4,3)|^2 + |F(2,1,3,4)|^2 + |F(2,1,4,3)|^2 \Big]$$

$$+ Q_f'^2 \Big[|F(3,4,1,2)|^2 + |F(4,3,1,2)|^2 + |F(3,4,2,1)|^2 + |F(4,3,2,1)|^2 \Big]$$

$$+ 2Q_f Q_f' \text{Re}\Big[F(1,2,4,3)F^*(3,4,2,1) - F(1,2,3,4)F^*(4,3,2,1)$$

$$+ F(2,1,3,4)F^*(4,3,1,2) - F(2,1,4,3)F^*(3,4,1,2) \Big] \Big\}$$

$$\times \delta^4(p_+ + p_- - k_1 - k_2 - k_3 - k_4) \frac{d^3\vec{k}_1 \, d^3\vec{k}_2 \, d^3\vec{k}_3 \, d^3\vec{k}_4}{k_{10} k_{20} k_{30} k_{40}}. \qquad (10.34)$$

10.6 $e^+e^- \to q\bar{q}q\bar{q}$ (no Z-exchange)

Process (identical quark flavors):

$$e^+(p_+) + e^-(p_-) \to q(k_1,i) + \bar{q}(k_2,j) + q(k_3,m) + \bar{q}(k_4,n). \qquad (10.35)$$

Definitions:

$$Z_{ij} = k_{i+}k_{j-} - k_{i\perp}^* k_{j\perp}, \qquad i,j = 1,\ldots,4,$$

$$B = \frac{1}{4E^2(k_{1+}k_{2+}k_{3+}k_{4+})^{\frac{1}{2}}},$$

$$F(1,2,3,4) = \frac{k_{4\perp}}{(k_3 \cdot k_4)k_{4-}} \left[\frac{k_{2+} Z_{14}^*(Z_{31} + Z_{34})}{E - k_{20}} + \frac{k_{1\perp} k_{3\perp}^* Z_{23}(Z_{24}^* + Z_{34}^*)}{k_{3-}(E - k_{10})} \right]. \qquad (10.36)$$

Nonvanishing helicity amplitudes:

$$M(+,-;+,+,-,-) = -ee_g^2 B T_{mj}^a T_{in}^a [F(2,3,4,1) - F(1,4,3,2)],$$

$$M(+,-;+,-,+,-) = -ee_q g^2 B \Big\{ T_{ij}^a T_{mn}^a [F(1,2,4,3) + F(3,4,2,1)]$$

$$- T_{mj}^a T_{in}^b [F(3,2,4,1) + F(1,4,2,3)] \Big\},$$

$$M(+,-;+,-,-,+) = -ee_q g^2 B T_{ij}^a T_{mn}^a [F(1,2,3,4) - (4,3,2,1)],$$

$$M(+,-;-,+,+,-) = ee_q g^2 B T_{ij}^a T_{mn}^a [F(2,1,4,3) - F(3,4,1,2)],$$

$$M(+,-;-,+,-,+) = ee_q g^2 B\{T_{ij}^a T_{mn}^a [F(2,1,3,4) + F(4,3,1,2)]$$
$$- T_{mj}^a T_{in}^a [F(2,3,1,4) + F(4,1,3,2)]\},$$

$$M(+,-;-,-,+,+) = ee_q g^2 B T_{mj}^a T_{in}^a [F(3,2,1,4) - F(4,1,2,3)],$$

$$M(-,+;+,+,-,-) = ee_q g^2 B T_{mj}^a T_{in}^a [F^*(3,2,1,4) - F^*(4,1,2,3)],$$

$$M(-,+;+,-,+,-) = ee_g^2 B\{T_{ij}^a T_{mn}^a [F^*(2,1,3,4) + F^*(4,3,1,2)]$$
$$- T_{mj}^a T_{in}^a [F^*(2,3,1,4) + F^*(4,1,3,2)]\},$$

$$M(-,+;+,-,-,+) = ee_q g^2 B T_{ij}^a T_{mn}^a [F^*(2,1,4,3) - F^*(3,4,1,2)],$$

$$M(-,+;-,+,+,-) = ee_q g^2 B T_{ij}^a T_{mn}^a [F^*(4,3,2,1) - F^*(1,2,3,4)],$$

$$M(-,+;-,+,-,+) = -ee_q g^2 B\{T_{ij}^a T_{mn}^a [F^*(1,2,4,3) + F^*(3,4,2,1)]$$
$$- T_{mj}^a T_{in}^b [F^*(3,2,4,1) + F^*(1,4,2,3)]\},$$

$$M(-,+;-,-,+,+) = ee_q g^2 B T_{mj}^a T_{in}^a [F^*(1,4,3,2) - F^*(2,3,4,1)]. \quad (10.37)$$

Squared absolute values of the nonvanishing helicity amplitudes, summed over the final state color degrees of freedom:

$$|M(+,-;+,+,-,-)|^2 = |M(-,+;-,-,+,+)|^2$$
$$= 2e^2 e_q^2 g^4 B^2 \{|F(2,3,4,1)|^2 + |F(1,4,3,2)|^2$$
$$- 2\text{Re}[F(2,3,4,1)F^*(1,4,3,2)]\},$$

10.6. $e^+e^- \to q\bar{q}q\bar{q}$ (NO Z-EXCHANGE)

$|M(+,-;+,-,+,-)|^2 = |M(-,+;-,+,-,+)|^2$

$= 2e^2 e_q^2 g^4 B^2$

$\times \Big\{ |F(1,2,4,3)|^2 + |F(3,4,2,1)|^2 + |F(3,2,4,1)|^2 + |F(1,4,2,3)|^2$

$\quad + 2\text{Re}\big[F(1,2,4,3)F^*(3,4,2,1) + F(3,4,2,1)F^*(1,4,2,3)\big]$

$\quad + \dfrac{2}{3}\text{Re}\big[\big(F(1,2,4,3) + F(3,4,2,1)\big)\big(F^*(3,4,2,1) + F^*(1,4,2,3)\big)\big]\Big\}$,

$|M(+,-;+,-,-,+)|^2 = |M(-,+;-,+,+,-)|^2$

$\qquad = 2e^2 e_q^2 g^4 B^2 \Big\{ |F(1,2,3,4)|^2 + |F(4,3,2,1)|^2$

$\qquad\qquad - 2\text{Re}\big[F(1,2,3,4)F^*(4,3,2,1)\big]\Big\}$,

$|M(+,-;-,+,+,-)|^2 = |M(-,+;+,-,-,+)|^2$

$\qquad = 2e^2 e_q^2 g^4 B^2 \Big\{ |F(2,1,4,3)|^2 + |F(3,4,1,2)|^2$

$\qquad\qquad - 2\text{Re}\big[F(2,1,4,3)F^*(3,4,1,2)\big]\Big\}$,

$|M(+,-;-,+,-,+)|^2 = |M(-,+;+,-,+,-)|^2$

$= 2e^2 e_q^2 g^4 B^2$

$\times \Big\{ |F(2,1,3,4)|^2 + |F(4,3,1,2)|^2 + |F(2,3,1,4)|^2 + |F(4,1,3,2)|^2$

$\quad + 2\text{Re}\big[F(2,1,3,4)F^*(4,3,1,2) + F(2,3,1,4)F^*(4,1,3,2)\big]$

$\quad + \dfrac{2}{3}\text{Re}\big[\big(F(2,1,3,4) + F(4,3,1,2)\big)\big(F^*(2,3,1,4) + F^*(4,1,3,2)\big)\big]\Big\}$,

$$|M(+,-;-,-,+,+)|^2 = |M(-,+;+,+,-,-)|^2$$
$$= 2e^2 e_q^2 g^4 B^2 \Big\{ |F(3,2,1,4)|^2 + |F(4,1,2,3)|^2$$
$$- 2\mathrm{Re}\Big[F(3,2,1,4)F^*(4,1,2,3)\Big] \Big\}. \quad (10.38)$$

Unpolarized squared matrix element:

$$\overline{|M|^2} = e^2 e_q^2 g^4 B^2$$
$$\times \Big\{ |F(1,2,3,4)|^2 + |F(1,2,4,3)|^2 + |F(2,1,3,4)|^2 + |F(2,1,4,3)|^2$$
$$+|F(4,3,2,1)|^2 + |F(3,4,1,2)|^2 + |F(3,4,2,1)|^2 + |F(3,2,4,1)|^2$$
$$+|F(1,4,2,3)|^2 + |F(4,3,1,2)|^2 + |F(2,3,1,4)|^2 + |F(4,1,3,2)|^2$$
$$+|F(2,3,4,1)|^2 + |F(1,4,3,2)|^2 + |F(3,2,1,4)|^2 + |F(4,1,2,3)|^2$$
$$+2\mathrm{Re}\Big[F(1,2,4,3)F^*(3,4,2,1) - F(1,2,3,4)F^*(4,3,2,1)$$
$$+F(2,1,3,4)F^*(4,3,1,2) - F(2,1,4,3)F^*(3,4,1,2)$$
$$+F(3,2,4,1)F^*(1,4,2,3) - F(2,3,4,1)F^*(1,4,3,2)$$
$$+F(2,3,1,4)F^*(4,1,3,2) - F(3,2,1,4)F^*(4,1,2,3)$$
$$+\frac{1}{3}[F(2,1,3,4) + F(4,3,1,2)][F^*(2,3,1,4) + F^*(4,1,3,2)]$$
$$+\frac{1}{3}[F(1,2,4,3) + F(3,4,2,1)][F^*(3,2,4,1) + F^*(1,4,2,3)]\Big] \Big\}.$$
$$(10.39)$$

Unpolarized cross section:

$$d\sigma = \frac{\alpha^2 \alpha_S^2 Q_f^2}{2048\pi^4 E^6 k_{1+} k_{2+} k_{3+} k_{4+}}$$

$$\times \Big\{ |F(1,2,3,4)|^2 + |F(1,2,4,3)|^2 + |F(2,1,3,4)|^2 + |F(2,1,4,3)|^2$$

$$+ |F(4,3,2,1)|^2 + |F(3,4,1,2)|^2 + |F(3,4,2,1)|^2 + |F(3,2,4,1)|^2$$

$$+ |F(1,4,2,3)|^2 + |F(4,3,1,2)|^2 + |F(2,3,1,4)|^2 + |F(4,1,3,2)|^2$$

$$+ |F(2,3,4,1)|^2 + |F(1,4,3,2)|^2 + |F(3,2,1,4)|^2 + |F(4,1,2,3)|^2$$

$$+ 2\text{Re}\Big[F(1,2,4,3)F^*(3,4,2,1) - F(1,2,3,4)F^*(4,3,2,1)$$

$$+ F(2,1,3,4)F^*(4,3,1,2) - F(2,1,4,3)F^*(3,4,1,2)$$

$$+ F(3,2,4,1)F^*(1,4,2,3) - F(2,3,4,1)F^*(1,4,3,2)$$

$$+ F(2,3,1,4)F^*(4,1,3,2) - F(3,2,1,4)F^*(4,1,2,3)$$

$$+ \frac{1}{3}[F(2,1,3,4) + F(4,3,1,2)][F^*(2,3,1,4) + F^*(4,1,3,2)]$$

$$+ \frac{1}{3}[F(1,2,4,3) + F(3,4,2,1)][F^*(3,2,4,1) + F^*(1,4,2,3)]\Big]\Big\}$$

$$\times \delta^4(p_+ + p_- - k_1 - k_2 - k_3 - k_4) \frac{d^3\vec{k}_1 \, d^3\vec{k}_2 \, d^3\vec{k}_3 \, d^3\vec{k}_4}{k_{10} k_{20} k_{30} k_{40}}. \quad (10.40)$$

10.7 $qq' \to qq'$

Process (different quark flavors):

$$q(p_+, i) + q'(p_-, j) \to q(q_+, m) + q'(q_-, n). \quad (10.41)$$

Definitions:

$$s = 2(p_+ \cdot p_-), \qquad t = -2(p_+ \cdot q_+), \qquad u = -2(p_+ \cdot q_-). \quad (10.42)$$

Nonvanishing helicity amplitudes:

$$M(+,+;+,+) = 4g^2 T^a_{mi} T^a_{nj} \frac{E q^*_{+\perp}}{(q_{++} q^3_{+-})^{\frac{1}{2}}},$$

$$M(+,-;+,-) = -2g^2 T^a_{mi} T^a_{nj} q_{-\perp} \left(\frac{q_{++}}{q^3_{+-}}\right)^{\frac{1}{2}},$$

$$M(-,+;-,+) = -2g^2 T^a_{mi} T^a_{nj} q^*_{-\perp} \left(\frac{q_{++}}{q^3_{+-}}\right)^{\frac{1}{2}},$$

$$M(-,-;-,-) = 4g^2 T^a_{mi} T^a_{nj} \frac{E q_{+\perp}}{(q_{++} q^3_{+-})^{\frac{1}{2}}}. \tag{10.43}$$

Squared absolute values of the nonvanishing helicity amplitudes, summed over the initial and final state color degrees of freedom:

$$|M(+,+;+,+)|^2 = |M(-,-;-,-)|^2 = 32 g^4 \frac{E^2}{q^2_{+-}},$$

$$|M(+,-;+,-)|^2 = |M(-,+;-,+)|^2 = 8 g^4 \frac{q^2_{++}}{q^2_{+-}}. \tag{10.44}$$

Unpolarized squared matrix element:

$$\overline{|M|^2} = \frac{4}{9} g^4 \frac{s^2 + u^2}{t^2}. \tag{10.45}$$

Unpolarized cross section:

$$\frac{d\sigma}{dt} = \frac{4\pi \alpha_S^2}{9} \frac{s^2 + u^2}{s^2 t^2}. \tag{10.46}$$

10.8 $\bar{q} q \to \bar{q}' q'$

Process (different quark flavors):

$$\bar{q}(p_+, i) + q(p_-, j) \to \bar{q}'(q_+, m) + q'(q_-, n). \tag{10.47}$$

Invariants:

$$s = 2(p_+ \cdot p_-), \quad t = -2(p_+ \cdot q_+), \quad u = -2(p_+ \cdot q_-). \tag{10.48}$$

Nonvanishing helicity amplitudes:

$$M(+,-;+,-) = g^2 T^a_{ij} T^a_{nm} \frac{q_{+\perp}}{E} \left(\frac{q_{++}}{q_{+-}}\right)^{\frac{1}{2}},$$

$$M(+,-;-,+) = g^2 T^a_{ij} T^a_{nm} \frac{q_{-\perp}}{E} \left(\frac{q_{+-}}{q_{++}}\right)^{\frac{1}{2}},$$

$$M(-,+;+,-) = g^2 T^a_{ij} T^a_{nm} \frac{q^*_{-\perp}}{E} \left(\frac{q_{+-}}{q_{++}}\right)^{\frac{1}{2}},$$

$$M(-,+;-,+) = g^2 T^a_{ij} T^a_{nm} \frac{q^*_{+\perp}}{E} \left(\frac{q_{++}}{q_{+-}}\right)^{\frac{1}{2}}. \tag{10.49}$$

Squared absolute values of the nonvanishing helicity amplitudes, summed over the color degrees of freedom of the initial state and the final state:

$$|M(+,-;+,-)|^2 = |M(-,+;-,+)|^2 = 2g^4 \frac{q_{++}^2}{E^2},$$

$$|M(+,-;-,+)|^2 = |M(-,+;+,-)|^2 = 2g^4 \frac{q_{+-}^2}{E^2}. \qquad (10.50)$$

Unpolarized squared matrix element:

$$\overline{|M|^2} = \frac{4}{9} g^4 \frac{t^2 + u^2}{s^2}. \qquad (10.51)$$

Unpolarized cross section:

$$\frac{d\sigma}{dt} = \frac{4\pi\alpha_S^2}{9} \frac{t^2 + u^2}{s^4}. \qquad (10.52)$$

10.9 $qq \to qq$

Process (identical quark flavors):

$$q(p_+, i) + q(p_-, j) \to q(q_+, m) + q(q_-, n). \qquad (10.53)$$

Invariants:

$$s = 2(p_+ \cdot p_-), \qquad t = -2(p_+ \cdot q_+), \qquad u = -2(p_+ \cdot q_-). \qquad (10.54)$$

Nonvanishing helicity amplitudes:

$$M(+,+;+,+) = 4g^2 \frac{Eq_{+\perp}^*}{\sqrt{q_{++}q_{+-}}} \left[\frac{T_{mi}^a T_{nj}^a}{q_{+-}} + \frac{T_{ni}^a T_{mj}^a}{q_{++}} \right],$$

$$M(+,-;+,-) = -2g^2 T_{mi}^a T_{nj}^a q_{-\perp} \left(\frac{q_{++}}{q_{+-}^3} \right)^{\frac{1}{2}},$$

$$M(+,-;-,+) = 2g^2 T_{ni}^a T_{mj}^a q_{+\perp} \left(\frac{q_{+-}}{q_{++}^3} \right)^{\frac{1}{2}},$$

$$M(-,+;+,-) = 2g^2 T_{ni}^a T_{mj}^a q_{+\perp}^* \left(\frac{q_{+-}}{q_{++}^3} \right)^{\frac{1}{2}},$$

$$M(-,+;-,+) = -2g^2 T_{mi}^a T_{nj}^a q_{-\perp}^* \left(\frac{q_{++}}{q_{+-}^3} \right)^{\frac{1}{2}},$$

$$M(-,-;-,-) = 4g^2 \frac{Eq_{+\perp}}{\sqrt{q_{++}q_{+-}}} \left[\frac{T_{mi}^a T_{nj}^a}{q_{+-}} + \frac{T_{ni}^a T_{mj}^a}{q_{++}} \right]. \qquad (10.55)$$

Squared absolute values of the nonvanishing helicity amplitudes, summed over the color degrees of freedom of the initial state and the final state:

$$|M(+,+;+,+)|^2 = |M(-,-;-,-)|^2 = 8g^4 \frac{s^2}{t^2 u^2}\left(s^2 - \frac{8}{3}tu\right),$$

$$|M(+,-;+,-)|^2 = |M(-,+;-,+)|^2 = 8g^4 \frac{u^2}{t^2},$$

$$|M(+,-;-,+)|^2 = |M(-,+;+,-)|^2 = 8g^4 \frac{t^2}{u^2}. \tag{10.56}$$

Unpolarized squared matrix element:

$$\overline{|M|^2} = \frac{4}{9}g^4\left[\frac{s^4 + t^4 + u^4}{t^2 u^2} - \frac{8s^2}{3tu}\right]. \tag{10.57}$$

Unpolarized cross section:

$$\frac{d\sigma}{dt} = \frac{4\pi\alpha_s^2}{9}\left[\frac{s^4 + t^4 + u^4}{s^2 t^2 u^2} - \frac{8}{3tu}\right]. \tag{10.58}$$

10.10 $\bar{q}q \to \bar{q}q$

Process (identical quark flavors):

$$\bar{q}(p_+,i) + q(p_-,j) \to \bar{q}(q_+,m) + q(q_-,n). \tag{10.59}$$

Invariants:

$$s = 2(p_+ \cdot p_-), \quad t = -2(p_+ \cdot q_+), \quad u = -2(p_+ \cdot q_-). \tag{10.60}$$

Nonvanishing helicity amplitudes:

$$M(+,+;+,+) = 4g^2 T^a_{nj} T^a_{im} \frac{E q^*_{-\perp}}{(q_{++}q^3_{+-})^{\frac{1}{2}}},$$

$$M(+,-;+,-) = g^2 q_{+\perp}\left(\frac{q_{++}}{q_{+-}}\right)^{\frac{1}{2}}\left[\frac{T^a_{ij}T^a_{nm}}{E} - \frac{2T^a_{nj}T^a_{im}}{q_{+-}}\right],$$

$$M(+,-;-,+) = g^2 T^a_{ij} T^a_{nm} \frac{q_{-\perp}}{E}\left(\frac{q_{+-}}{q_{++}}\right)^{\frac{1}{2}},$$

$$M(-,+;+,-) = g^2 T^a_{ij} T^a_{nm} \frac{q^*_{-\perp}}{E}\left(\frac{q_{+-}}{q_{++}}\right)^{\frac{1}{2}},$$

$$M(-,+;-,+) = g^2 q^*_{+\perp}\left(\frac{q_{++}}{q_{+-}}\right)^{\frac{1}{2}}\left[\frac{T^a_{ij}T^a_{nm}}{E} - \frac{2T^a_{nj}T^a_{im}}{q_{+-}}\right],$$

$$M(-,-;-,-) = 4g^2 T^a_{nj} T^a_{im} \frac{E q_{-\perp}}{(q_{++}q^3_{+-})^{\frac{1}{2}}}. \tag{10.61}$$

Squared absolute values of the nonvanishing helicity amplitudes, summed over the color degrees of freedom of the initial state and the final state:

$$|M(+,+;+,+)|^2 = |M(-,-;-,-)|^2 = 8g^4 \frac{s^2}{t^2},$$

$$|M(+,-;+,-)|^2 = |M(-,+;-,+)|^2 = 8g^4 \frac{u^2}{s^2 t^2}\left[u^2 - \frac{8}{3}st\right],$$

$$|M(+,-;-,+)|^2 = |M(-,+;+,-)|^2 = 8g^4 \frac{t^2}{s^2}. \tag{10.62}$$

Unpolarized squared matrix element:

$$\overline{|M|^2} = \frac{4g^4}{9s^2 t^2}\left[s^4 + t^4 + u^4 - \frac{8}{3}stu^2\right]. \tag{10.63}$$

Unpolarized cross section:

$$\frac{d\sigma}{dt} = \frac{4\pi\alpha_S^2}{9s^4 t^2}\left[s^4 + t^4 + u^4 - \frac{8}{3}stu^2\right]. \tag{10.64}$$

10.11 $\gamma q \to \gamma q$

Process:
$$\gamma(k) + q(p,i) \to \gamma(k') + q(p',j). \tag{10.65}$$

Positive z-axis: along $\vec{k}$.
Invariants:

$$s = 2(k\cdot p), \qquad t = -2(k\cdot k'), \qquad u = -2(k\cdot p'). \tag{10.66}$$

Photon polarizations:

$$\not{\epsilon}^{*\pm}(k) = N[\not{p}'\not{p}\not{k}(1\pm\gamma_5) - \not{k}\not{p}'\not{p}(1\mp\gamma_5)],$$

$$\not{\epsilon}^{\pm}(k') = N[\not{p}'\not{p}\not{k}'(1\mp\gamma_5) - \not{k}'\not{p}'\not{p}(1\pm\gamma_5)],$$

$$N^{-1} = (2stu)^{\frac{1}{2}}. \tag{10.67}$$

Nonvanishing helicity amplitudes:

$$M(+,+;+,+) = -e_q^2 \delta_{ij} \frac{p'^*_\perp}{p'_-}\left(\frac{8E}{p'_+}\right)^{\frac{1}{2}},$$

$$M(+,-;+,-) = -e_q^2 \delta_{ij} p'_\perp \left(\frac{2}{Ep'_+}\right)^{\frac{1}{2}},$$

$$M(-,+;-,+) = -e_q^2 \delta_{ij} p'^*_\perp \left(\frac{2}{Ep'_+}\right)^{\frac{1}{2}},$$

$$M(-,-;-,-) = -e_q^2 \delta_{ij} \frac{p'_\perp}{p'_-} \left(\frac{8E}{p'_+}\right)^{\frac{1}{2}}. \tag{10.68}$$

Squared absolute values of the nonvanishing helicity amplitudes, summed over the color degrees of freedom of the initial state and the final state:

$$|M(+,+;+,+)|^2 = |M(-,-;-,-)|^2 = -12 e_q^4 \frac{s}{u},$$

$$|M(+,-;+,-)|^2 = |M(-,+;-,+)|^2 = -12 e_q^4 \frac{u}{s}. \tag{10.69}$$

Unpolarized squared matrix element:

$$\overline{|M|^2} = -2 e_q^4 \frac{s^2 + u^2}{su}. \tag{10.70}$$

Unpolarized cross section:

$$\frac{d\sigma}{dt} = -2\pi \alpha^2 Q_f^4 \frac{s^2 + u^2}{s^3 u}. \tag{10.71}$$

10.12 $\gamma q \to g q$

Process:
$$\gamma(k) + q(p,i) \to g(k',a) + q(p',j). \tag{10.72}$$

Positive z-axis: along $\vec{k}$.
Invariants:
$$s = 2(k \cdot p), \qquad t = -2(k \cdot k'), \qquad u = -2(k \cdot p'). \tag{10.73}$$

Photon and gluon polarizations:

$$\rlap{/}{\epsilon}^{*\pm}(k) = N[\rlap{/}{p}' \rlap{/}{p} \rlap{/}{k}(1 \pm \gamma_5) - \rlap{/}{k} \rlap{/}{p}' \rlap{/}{p}(1 \mp \gamma_5)],$$

$$\rlap{/}{\epsilon}^{\pm}(k') = N[\rlap{/}{k}' \rlap{/}{p}' \rlap{/}{p}(1 \pm \gamma_5) + \rlap{/}{p} \rlap{/}{p}' \rlap{/}{k}'(1 \mp \gamma_5)],$$

$$N^{-1} = (2stu)^{\frac{1}{2}}. \tag{10.74}$$

Nonvanishing helicity amplitudes:

$$M(+,+;+,+) = e_q g T^a_{ji} \frac{p'^*_\perp}{p'_-} \left(\frac{8E}{p'_+}\right)^{\frac{1}{2}},$$

$$M(+,-;+,-) = e_q g T^a_{ji} p'_\perp \left(\frac{2}{Ep'_+}\right)^{\frac{1}{2}},$$

$$M(-,+;-,+) = e_q g T^a_{ji} p'^*_\perp \left(\frac{2}{Ep'_+}\right)^{\frac{1}{2}},$$

$$M(-,-;-,-) = e_q g T^a_{ji} \frac{p'_\perp}{p'_-} \left(\frac{8E}{p'_+}\right)^{\frac{1}{2}}. \tag{10.75}$$

Squared absolute values of the nonvanishing helicity amplitudes, summed over the color degrees of freedom of the initial state and the final state:

$$|M(+,+;+,+)|^2 = |M(-,-;-,-)|^2 = -16 e_q^2 g^2 \frac{s}{u},$$

$$|M(+,-;+,-)|^2 = |M(-,+;-,+)|^2 = -16 e_q^2 g^2 \frac{u}{s}. \tag{10.76}$$

Unpolarized squared matrix element:

$$\overline{|M|^2} = -\frac{8}{3} e_q^2 g^2 \frac{s^2 + u^2}{su}. \tag{10.77}$$

Unpolarized cross section:

$$\frac{d\sigma}{dt} = -\frac{8\pi \alpha \alpha_S Q_f^2}{3} \frac{s^2 + u^2}{s^3 u}. \tag{10.78}$$

10.13 $gq \to \gamma q$

Process:
$$g(k,a) + q(p,i) \to \gamma(k') + q(p',j). \tag{10.79}$$

Positive z-axis: along $\vec{k}$.
Invariants:

$$s = 2(k \cdot p), \qquad t = -2(k \cdot k'), \qquad u = -2(k \cdot p'). \tag{10.80}$$

Gluon and photon polarizations:

$$\not{\epsilon}^{*\pm}(k) = N[\not{p} \not{p}' \not{k}(1 \pm \gamma_5) + \not{k} \not{p}' \not{p}(1 \mp \gamma_5)],$$

$$\not{\epsilon}^{\pm}(k') = N[\not{p}' \not{p} \not{k}'(1 \mp \gamma_5) - \not{k}' \not{p}' \not{p}(1 \pm \gamma_5)],$$

$$N^{-1} = (2stu)^{\frac{1}{2}}. \tag{10.81}$$

Nonvanishing helicity amplitudes:

$$M(+,+;+,+) = e_q g T^a_{ji} \frac{p'^*_\perp}{p'_-} \left(\frac{8E}{p'_+}\right)^{\frac{1}{2}},$$

$$M(+,-;+,-) = e_q g T^a_{ji} p'_\perp \left(\frac{2}{Ep'_+}\right)^{\frac{1}{2}},$$

$$M(-,+;-,+) = e_q g T^a_{ji} p'^*_\perp \left(\frac{2}{Ep'_+}\right)^{\frac{1}{2}},$$

$$M(-,-;-,-) = e_q g T^a_{ji} \frac{p'_\perp}{p'_-} \left(\frac{8E}{p'_+}\right)^{\frac{1}{2}}. \qquad (10.82)$$

Squared absolute values of the nonvanishing helicity amplitudes, summed over the color degrees of freedom of the initial state and the final state:

$$|M(+,+;+,+)|^2 = |M(-,-;-,-)|^2 = -16 e_q^2 g^2 \frac{s}{u},$$

$$|M(+,-;+,-)|^2 = |M(-,+;-,+)|^2 = -16 e_q^2 g^2 \frac{u}{s}. \qquad (10.83)$$

Unpolarized squared matrix element:

$$\overline{|M|^2} = -\frac{1}{3} e_q^2 g^2 \frac{s^2+u^2}{su}. \qquad (10.84)$$

Unpolarized cross section:

$$\frac{d\sigma}{dt} = -\frac{\pi \alpha \alpha_s Q_f^2}{3} \frac{s^2+u^2}{s^3 u}. \qquad (10.85)$$

10.14 $gq \to gq$

Process:
$$g(k,a) + q(p,i) \to g(k',b) + q(p',j). \qquad (10.86)$$

Positive z-axis: along $\vec{k}$.
Invariants:

$$s = 2(k \cdot p), \qquad t = -2(k \cdot k'), \qquad u = -2(k \cdot p'). \qquad (10.87)$$

Gluon polarizations:

$$\rlap{/}{\epsilon}^{*\pm}(k) = N[\rlap{/}{p}\,\rlap{/}{p}'\,\rlap{/}{k}(1 \pm \gamma_5) + \rlap{/}{k}\,\rlap{/}{p}'\,\rlap{/}{p}(1 \mp \gamma_5)],$$

$$\rlap{/}{\epsilon}^{\pm}(k') = N[\rlap{/}{k}'\,\rlap{/}{p}'\,\rlap{/}{p}(1 \pm \gamma_5) + \rlap{/}{p}\,\rlap{/}{p}'\,\rlap{/}{k}'(1 \mp \gamma_5)],$$

$$N^{-1} = (2stu)^{\frac{1}{2}}. \qquad (10.88)$$

10.15. $\gamma \bar{q} \to \gamma \bar{q}$

Nonvanishing helicity amplitudes:

$$M(+,+;+,+) = -2g^2 \frac{p'^*_\perp}{p'_-} \left(\frac{2E}{p'^3_+}\right)^{\frac{1}{2}} \left[2E(T^aT^b)_{ji} - p'_-(T^bT^a)_{ji}\right],$$

$$M(+,-;+,-) = -g^2 p'_\perp \left(\frac{2}{Ep'^3_+}\right)^{\frac{1}{2}} \left[2E(T^aT^b)_{ji} - p'_-(T^bT^a)_{ji}\right],$$

$$M(-,+;-,+) = -g^2 p'^*_\perp \left(\frac{2}{Ep'^3_+}\right)^{\frac{1}{2}} \left[2E(T^aT^b)_{ji} - p'_-(T^bT^a)_{ji}\right],$$

$$M(-,-;-,-) = -2g^2 \frac{p'_\perp}{p'_-} \left(\frac{2E}{p'^3_+}\right)^{\frac{1}{2}} \left[2E(T^aT^b)_{ji} - p'_-(T^bT^a)_{ji}\right]. \tag{10.89}$$

Squared absolute values of the nonvanishing helicity amplitudes, summed over the color degrees of freedom of the initial state and the final state:

$$|M(+,+;+,+)|^2 = |M(-,-;-,-)|^2 = 48g^4 \left[\frac{s^2}{t^2} - \frac{4s}{9u}\right],$$

$$|M(+,-;+,-)|^2 = |M(-,+;-,+)|^2 = 48g^4 \left[\frac{u^2}{t^2} - \frac{4u}{9s}\right]. \tag{10.90}$$

Unpolarized squared matrix element:

$$\overline{|M|^2} = g^4(s^2 + u^2)\left[\frac{1}{t^2} - \frac{4}{9su}\right]. \tag{10.91}$$

Unpolarized cross section:

$$\frac{d\sigma}{dt} = \pi \alpha_s^2 \frac{s^2 + u^2}{s^2} \left[\frac{1}{t^2} - \frac{4}{9su}\right]. \tag{10.92}$$

10.15 $\gamma \bar{q} \to \gamma \bar{q}$

Process:
$$\gamma(k) + \bar{q}(p,i) \to \gamma(k') + \bar{q}(p',j). \tag{10.93}$$

Positive z-axis: along $\vec{k}$.

Invariants:
$$s = 2(k \cdot p), \qquad t = -2(k \cdot k'), \qquad u = -2(k \cdot p'). \tag{10.94}$$

Photon polarizations:

$$\not{\epsilon}^{*\pm}(k) = N[\not{k}\not{p}\not{p}'(1\mp\gamma_5) - \not{p}\not{p}'\not{k}(1\pm\gamma_5)],$$

$$\not{\epsilon}^{\pm}(k') = N[\not{k}'\not{p}\not{p}'(1\pm\gamma_5) - \not{p}\not{p}'\not{k}'(1\mp\gamma_5)],$$

$$N^{-1} = (2stu)^{\frac{1}{2}}. \tag{10.95}$$

Nonvanishing helicity amplitudes:

$$M(+,+;+,+) = e_q^2 \delta_{ij} \frac{p'^*_\perp}{p'_-} \left(\frac{8E}{p'_+}\right)^{\frac{1}{2}},$$

$$M(+,-;+,-) = e_q^2 \delta_{ij} p'_\perp \left(\frac{2}{Ep'_+}\right)^{\frac{1}{2}},$$

$$M(-,+;-,+) = e_q^2 \delta_{ij} p'^*_\perp \left(\frac{2}{Ep'_+}\right)^{\frac{1}{2}},$$

$$M(-,-;-,-) = e_q^2 \delta_{ij} \frac{p'_\perp}{p'_-} \left(\frac{8E}{p'_+}\right)^{\frac{1}{2}}. \qquad (10.96)$$

Squared absolute values of the nonvanishing helicity amplitudes, summed over the color degrees of freedom of the initial state and the final state:

$$|M(+,+;+,+)|^2 = |M(-,-;-,-)|^2 = -12e_q^4 \frac{s}{u},$$

$$|M(+,-;+,-)|^2 = |M(-,+;-,+)|^2 = -12e_q^4 \frac{u}{s}. \qquad (10.97)$$

Unpolarized squared matrix element:

$$\overline{|M|^2} = -8e_q^4 \frac{s^2+u^2}{su}. \qquad (10.98)$$

Unpolarized cross section:

$$\frac{d\sigma}{dt} = -8\pi\alpha^2 Q_f^4 \frac{s^2+u^2}{s^3 u}. \qquad (10.99)$$

10.16 $\gamma \bar{q} \to g \bar{q}$

Process:

$$\gamma(k) + \bar{q}(p,i) \to g(k',a) + \bar{q}(p',j). \qquad (10.100)$$

Positive z-axis: along $\vec{k}$.
Invariants:

$$s = 2(k \cdot p), \qquad t = -2(k \cdot k'), \qquad u = -2(k \cdot p'). \qquad (10.101)$$

Photon and gluon polarizations:

$$\not{\epsilon}^{*\pm}(k) = N[\not{k}\not{p}\not{p}'(1\mp\gamma_5) - \not{p}\not{p}'\not{k}(1\pm\gamma_5)],$$

$$\not{\epsilon}^{\pm}(k') = N[\not{k}'\not{p}'\not{p}(1\pm\gamma_5) + \not{p}\not{p}'\not{k}'(1\mp\gamma_5)],$$

$$N^{-1} = (2stu)^{\frac{1}{2}}. \qquad (10.102)$$

10.17. $g\bar{q} \to \gamma\bar{q}$

Nonvanishing helicity amplitudes:

$$M(+,+;+,+) = -e_q g T_{ij}^a \frac{p_\perp'^*}{p_-'} \left(\frac{8E}{p_+'}\right)^{\frac{1}{2}},$$

$$M(+,-;+,-) = -e_q g T_{ij}^a p_\perp' \left(\frac{2}{Ep_+'}\right)^{\frac{1}{2}},$$

$$M(-,+;-,+) = -e_q g T_{ij}^a p_\perp'^* \left(\frac{2}{Ep_+'}\right)^{\frac{1}{2}},$$

$$M(-,-;-,-) = -e_q g T_{ij}^a \frac{p_\perp'}{p_-'} \left(\frac{8E}{p_+'}\right)^{\frac{1}{2}}. \qquad (10.103)$$

Squared absolute values of the nonvanishing helicity amplitudes, summed over the color degrees of freedom of the initial state and the final state:

$$|M(+,+;+,+)|^2 = |M(-,-;-,-)|^2 = -16 e_q^2 g^2 \frac{s}{u},$$

$$|M(+,-;+,-)|^2 = |M(-,+;-,+)|^2 = -16 e_q^2 g^2 \frac{u}{s}. \qquad (10.104)$$

Unpolarized squared matrix element:

$$\overline{|M|^2} = -\frac{8}{3} e_q^2 g^2 \frac{s^2 + u^2}{su}. \qquad (10.105)$$

Unpolarized cross section:

$$\frac{d\sigma}{dt} = -\frac{8\pi\alpha\alpha_S Q_f^2}{3} \frac{s^2 + u^2}{s^3 u}. \qquad (10.106)$$

10.17 $g\bar{q} \to \gamma\bar{q}$

Process:
$$g(k,a) + \bar{q}(p,i) \to \gamma(k') + \bar{q}(p',j). \qquad (10.107)$$

Positive z-axis: along $\vec{k}$.

Invariants:
$$s = 2(k \cdot p), \qquad t = -2(k \cdot k'), \qquad u = -2(k \cdot p'). \qquad (10.108)$$

Gluon polarizations:

$$\not{\epsilon}^{*\pm}(k) = N[\not{p}\,\not{p}'\,\not{k}(1 \pm \gamma_5) + \not{k}\,\not{p}'\,\not{p}(1 \mp \gamma_5)],$$

$$\not{\epsilon}^{\pm}(k') = N[\not{k}'\,\not{p}\,\not{p}'(1 \pm \gamma_5) - \not{p}\,\not{p}'\,\not{k}'(1 \mp \gamma_5)],$$

$$N^{-1} = (2stu)^{\frac{1}{2}}. \qquad (10.109)$$

Nonvanishing helicity amplitudes:

$$M(+,+;+,+) = -e_q g T^a_{ij} \frac{p'^*_\perp}{p'_-} \left(\frac{8E}{p'_+}\right)^{\frac{1}{2}},$$

$$M(+,-;+,-) = -e_q g T^a_{ij} p'_\perp \left(\frac{2}{E p'_+}\right)^{\frac{1}{2}},$$

$$M(-,+;-,+) = -e_q g T^a_{ij} p'^*_\perp \left(\frac{2}{E p'_+}\right)^{\frac{1}{2}},$$

$$M(-,-;-,-) = -e_q g T^a_{ij} \frac{p'_\perp}{p'_-} \left(\frac{8E}{p'_+}\right)^{\frac{1}{2}}. \qquad (10.110)$$

Squared absolute values of the nonvanishing helicity amplitudes, summed over the color degrees of freedom of the initial state and the final state:

$$|M(+,+;+,+)|^2 = |M(-,-;-,-)|^2 = -16 e_q^2 g^2 \frac{s}{u},$$

$$|M(+,-;+,-)|^2 = |M(-,+;-,+)|^2 = -16 e_q^2 g^2 \frac{u}{s}. \qquad (10.111)$$

Unpolarized squared matrix element:

$$\overline{|M|^2} = -\tfrac{1}{3} e_q^2 g^2 \frac{s^2 + u^2}{su}. \qquad (10.112)$$

Unpolarized cross section:

$$\frac{d\sigma}{dt} = -\tfrac{1}{3} \pi \alpha \alpha_S Q_f^2 \frac{s^2 + u^2}{s^3 u}. \qquad (10.113)$$

10.18 $g\bar{q} \to g\bar{q}$

Process:
$$g(k,a) + \bar{q}(p,i) \to g(k',b) + \bar{q}(p',j). \qquad (10.114)$$

Positive z-axis: along $\vec{k}$.
Invariants:
$$s = 2(k \cdot p), \qquad t = -2(k \cdot k'), \qquad u = -2(k \cdot p'). \qquad (10.115)$$

Gluon polarizations:

$$\not{\epsilon}^{*\pm}(k) = N[\not{p}\,\not{p}'\,\not{k}(1 \pm \gamma_5) + \not{k}\,\not{p}'\,\not{p}(1 \mp \gamma_5)],$$

$$\not{\epsilon}^{\pm}(k') = N[\not{k}'\,\not{p}'\,\not{p}(1 \pm \gamma_5) + \not{p}\,\not{p}'\,\not{k}'(1 \mp \gamma_5)],$$

$$N^{-1} = (2stu)^{\frac{1}{2}}. \qquad (10.116)$$

10.19. $\bar{q}q \to \gamma\gamma$

Nonvanishing helicity amplitudes:

$$M(+,+;+,+) = -2g^2 \frac{p'^*_\perp}{p'_-} \left(\frac{2E}{p'^3_+}\right)^{\frac{1}{2}} \left[p'_-(T^aT^b)_{ij} - 2E(T^bT^a)_{ij}\right],$$

$$M(+,-;+,-) = -g^2 p'_\perp \left(\frac{2}{Ep'^3_+}\right)^{\frac{1}{2}} \left[p'_-(T^aT^b)_{ij} - 2E(T^bT^a)_{ij}\right],$$

$$M(-,+;-,+) = -g^2 p'^*_\perp \left(\frac{2}{Ep'^3_+}\right)^{\frac{1}{2}} \left[p'_-(T^aT^b)_{ij} - 2E(T^bT^a)_{ij}\right],$$

$$M(-,-;-,-) = -2g^2 \frac{p'_\perp}{p'_-} \left(\frac{2E}{p'^3_+}\right)^{\frac{1}{2}} \left[p'_-(T^aT^b)_{ij} - 2E(T^bT^a)_{ij}\right].$$

(10.117)

Squared absolute values of the nonvanishing helicity amplitudes, summed over the color degrees of freedom of the initial state and the final state:

$$|M(+,+;+,+)|^2 = |M(-,-;-,-)|^2 = 48g^4 \left[\frac{s^2}{t^2} - \frac{4s}{9u}\right],$$

$$|M(+,-;+,-)|^2 = |M(-,+;-,+)|^2 = 48g^4 \left[\frac{u^2}{t^2} - \frac{4u}{9s}\right]. \quad (10.118)$$

Unpolarized squared matrix element:

$$\overline{|M|^2} = g^4(s^2 + u^2)\left[\frac{1}{t^2} - \frac{4}{9su}\right]. \quad (10.119)$$

Unpolarized cross section:

$$\frac{d\sigma}{dt} = \pi\alpha_s^2 \frac{s^2 + u^2}{s^2}\left[\frac{1}{t^2} - \frac{4}{9su}\right]. \quad (10.120)$$

10.19 $\bar{q}q \to \gamma\gamma$

Process:
$$\bar{q}(p_+, i) + q(p_-, j) \to \gamma(k_1) + \gamma(k_2). \quad (10.121)$$

Invariants:
$$s = 2(p_+ \cdot p_-), \qquad t = -2(p_+ \cdot k_1), \qquad u = -2(p_+ \cdot k_2). \quad (10.122)$$

Photon polarizations:

$$\not{\epsilon}^\pm(k_i) = N[\not{k}_i \not{p}_+ \not{p}_-(1 \pm \gamma_5) - \not{p}_+ \not{p}_- \not{k}_i(1 \mp \gamma_5)], \qquad i = 1, 2,$$

$$N = (2stu)^{-\frac{1}{2}}. \quad (10.123)$$

Nonvanishing helicity amplitudes:

$$M(+,-;+,-) = 2e_q^2 \delta_{ij} \frac{k_{1\perp}}{k_{1-}},$$

$$M(+,-;-,+) = 2e_q^2 \delta_{ij} \frac{k_{2\perp}}{k_{2-}},$$

$$M(-,+;+,-) = 2e_q^2 \delta_{ij} \frac{k_{2\perp}^*}{k_{2-}},$$

$$M(-,+;-,+) = 2e_q^2 \delta_{ij} \frac{k_{1\perp}^*}{k_{1-}}. \quad (10.124)$$

Squared absolute values of the nonvanishing helicity amplitudes, summed over the color degrees of freedom of the initial state:

$$|M(+,-;+,-)|^2 = |M(-,+;-,+)|^2 = 12e_q^4 \frac{u}{t},$$

$$|M(+,-;-,+)|^2 = |M(-,+;+,-)|^2 = 12e_q^4 \frac{t}{u}. \quad (10.125)$$

Unpolarized squared matrix element:

$$\overline{|M|^2} = \frac{2}{3} e_q^4 \frac{t^2 + u^2}{tu}. \quad (10.126)$$

Unpolarized cross section:

$$\frac{d\sigma}{dt} = \frac{2\pi \alpha^2 Q_f^4}{3s^2} \frac{t^2 + u^2}{tu}. \quad (10.127)$$

10.20 $\bar{q}q \to g\gamma$

Process:
$$\bar{q}(p_+, i) + q(p_-, j) \to g(k_1, a) + \gamma(k_2). \quad (10.128)$$

Invariants:
$$s = 2(p_+ \cdot p_-), \quad t = -2(p_+ \cdot k_1), \quad u = -2(p_+ \cdot k_2). \quad (10.129)$$

Gluon and photon polarizations:

$$\not{\epsilon}^\pm(k_1) = N[\not{k}_1 \not{p}_- \not{p}_+(1 \pm \gamma_5) + \not{p}_+ \not{p}_- \not{k}_1(1 \mp \gamma_5)],$$

$$\not{\epsilon}^\pm(k_2) = N[\not{k}_2 \not{p}_+ \not{p}_-(1 \pm \gamma_5) - \not{p}_+ \not{p}_- \not{k}_2(1 \mp \gamma_5)],$$

$$N = (2stu)^{-\frac{1}{2}}. \quad (10.130)$$

10.21. $\bar{q}q \to gg$

Nonvanishing helicity amplitudes:

$$M(+,-;+,-) = 2e_q g T^a_{ij} \frac{k_{2\perp}}{k_{2+}},$$

$$M(+,-;-,+) = 2e_q g T^a_{ij} \frac{k_{1\perp}}{k_{1+}},$$

$$M(-,+;+,-) = 2e_q g T^a_{ij} \frac{k^*_{1\perp}}{k_{1+}},$$

$$M(-,+;-,+) = 2e_q g T^a_{ij} \frac{k^*_{2\perp}}{k_{2+}}. \tag{10.131}$$

Squared absolute values of the nonvanishing helicity amplitudes, summed over the color degrees of freedom of the initial state and the final state:

$$|M(+,-;+,-)|^2 = |M(-,+;-,+)|^2 = 16 e_q^2 g^2 \frac{u}{t},$$

$$|M(+,-;-,+)|^2 = |M(-,+;+,-)|^2 = 16 e_q^2 g^2 \frac{t}{u}. \tag{10.132}$$

Unpolarized squared matrix element:

$$\overline{|M|^2} = \frac{8}{9} e_q^2 g^2 \frac{t^2 + u^2}{tu}. \tag{10.133}$$

Unpolarized cross section:

$$\frac{d\sigma}{dt} = \frac{8\pi \alpha \alpha_S Q_f^2}{9 s^2} \frac{t^2 + u^2}{tu}. \tag{10.134}$$

10.21 $\bar{q}q \to gg$

Process:

$$\bar{q}(p_+, i) + q(p_-, j) \to g(k_1, a) + g(k_2, b). \tag{10.135}$$

Invariants:

$$s = 2(p_+ \cdot p_-), \quad t = -2(p_+ \cdot k_1), \quad u = -2(p_+ \cdot k_2). \tag{10.136}$$

Gluon polarizations:

$$\not{\epsilon}^\pm(k_i) = N[\not{k}_i \not{p}_- \not{p}_+(1 \pm \gamma_5) + \not{p}_+ \not{p}_- \not{k}_i(1 \mp \gamma_5)], \quad i = 1, 2,$$

$$N = (2stu)^{-\frac{1}{2}}. \tag{10.137}$$

Nonvanishing helicity amplitudes:

$$M(+,-;+,-) = -g^2 \frac{k_{2\perp}}{Ek_{2+}} \left[k_{1+}(T^aT^b)_{ij} + k_{2+}(T^bT^a)_{ij} \right],$$

$$M(+,-;-,+) = -g^2 \frac{k_{1\perp}}{Ek_{1+}} \left[k_{1+}(T^aT^b)_{ij} + k_{2+}(T^bT^a)_{ij} \right],$$

$$M(-,+;+,-) = -g^2 \frac{k_{1\perp}^*}{Ek_{1+}} \left[k_{1+}(T^aT^b)_{ij} + k_{2+}(T^bT^a)_{ij} \right],$$

$$M(-,+;-,+) = -g^2 \frac{k_{2\perp}^*}{Ek_{2+}} \left[k_{1+}(T^aT^b)_{ij} + k_{2+}(T^bT^a)_{ij} \right]. \quad (10$$

Squared absolute values of the nonvanishing helicity amplitudes, sum over the color degrees of freedom of the initial state and the final state:

$$|M(+,-;+,-)|^2 = |M(-,+;-,+)|^2 = 48g^4 \left[\frac{4t}{9u} - \frac{t^2}{s^2} \right],$$

$$|M(+,-;-,+)|^2 = |M(-,+;+,-)|^2 = 48g^4 \left[\frac{4u}{9t} - \frac{u^2}{s^2} \right]. \quad (10$$

Unpolarized squared matrix element:

$$\overline{|M|^2} = \frac{8}{3}g^4(t^2+u^2)\left[\frac{4}{9tu} - \frac{1}{s^2}\right]. \quad (10$$

Unpolarized cross section:

$$\frac{d\sigma}{dt} = \frac{8\pi\alpha_S^2}{3}\frac{t^2+u^2}{s^2}\left[\frac{4}{9tu} - \frac{1}{s^2}\right]. \quad (10.$$

10.22 $\gamma\gamma \to \bar{q}q$

Process:
$$\gamma(k_1) + \gamma(k_2) \to \bar{q}(q',i) + q(q,j). \quad (10.$$

Positive z-axis: along $\vec{k}_1$.
Invariants:

$$s = 2(k_1 \cdot k_2), \qquad t = -2(k_1 \cdot q'), \qquad u = -2(k_1 \cdot q). \quad (10.$$

Photon polarizations:

$$\not{\epsilon}^\pm(k_i) = N[\not{A}\not{A}'\not{k}_i(1\pm\gamma_5) - \not{k}_i\not{A}\not{A}'(1\mp\gamma_5)], \qquad i=1,2.$$

$$N = (2stu)^{-\frac{1}{2}}. \quad (10.$$

Nonvanishing helicity amplitudes:

$$M(+,-;+,-) = 2e_q^2\delta_{ij}\left(\frac{q_-}{q_+}\right)^{\frac{1}{2}},$$

$$M(+,-;-,+) = -2e_q^2\delta_{ij}\left(\frac{q_+}{q_-}\right)^{\frac{1}{2}},$$

$$M(-,+;+,-) = -2e_q^2\delta_{ij}\left(\frac{q_+}{q_-}\right)^{\frac{1}{2}},$$

$$M(-,+;-,+) = 2e_q^2\delta_{ij}\left(\frac{q_-}{q_+}\right)^{\frac{1}{2}}. \qquad (10.145)$$

Squared absolute values of the nonvanishing helicity amplitudes, summed over the color degrees of freedom of the final state:

$$|M(+,-;+,-)|^2 = |M(-,+;-,+)|^2 = 12e_q^4\frac{u}{t},$$

$$|M(+,-;-,+)|^2 = |M(-,+;+,-)|^2 = 12e_q^4\frac{t}{u}. \qquad (10.146)$$

Unpolarized squared matrix element:

$$\overline{|M|^2} = 6e_q^4\frac{t^2+u^2}{tu}. \qquad (10.147)$$

Unpolarized cross section:

$$\frac{d\sigma}{dt} = \frac{6\pi\alpha^2 Q_f^4}{s^2}\frac{t^2+u^2}{tu}. \qquad (10.148)$$

10.23 $g\gamma \to \bar{q}q$

Process:
$$g(k_1,a) + \gamma(k_2) \to \bar{q}(q',i) + q(q,j). \qquad (10.149)$$

Positive z-axis: along $\vec{k}_1$.
Invariants:
$$s = 2(k_1 \cdot k_2), \qquad t = -2(k_1 \cdot q'), \qquad u = -2(k_1 \cdot q). \qquad (10.150)$$

Gluon and photon polarizations:

$$\rlap{/}{\epsilon}^{*\pm}(k_1) = N[\rlap{/}{A}\,\rlap{/}{A}'\,\rlap{/}{k}_1(1\pm\gamma_5) + \rlap{/}{k}_1\,\rlap{/}{A}'\,\rlap{/}{A}(1\mp\gamma_5)],$$

$$\rlap{/}{\epsilon}^{*\pm}(k_2) = N[\rlap{/}{A}\,\rlap{/}{A}'\,\rlap{/}{k}_2(1\pm\gamma_5) - \rlap{/}{k}_2\,\rlap{/}{A}\,\rlap{/}{A}'(1\mp\gamma_5)],$$

$$N = (2stu)^{-\frac{1}{2}}. \qquad (10.151)$$

Nonvanishing helicity amplitudes:

$$M(+,-;+,-) = 2e_q g T^a_{ji} \left(\frac{q_-}{q_+}\right)^{\frac{1}{2}},$$

$$M(+,-;-,+) = -2e_q g T^a_{ji} \left(\frac{q_+}{q_-}\right)^{\frac{1}{2}},$$

$$M(-,+;+,-) = -2e_q g T^a_{ji} \left(\frac{q_+}{q_-}\right)^{\frac{1}{2}},$$

$$M(-,+;-,+) = 2e_q g T^a_{ji} \left(\frac{q_-}{q_+}\right)^{\frac{1}{2}}. \tag{10.152}$$

Squared absolute values of the nonvanishing helicity amplitudes, summed over the color degrees of freedom of the initial state and the final state:

$$|M(+,-;+,-)|^2 = |M(-,+;-,+)|^2 = 16 e_q^2 g^2 \frac{u}{t},$$

$$|M(+,-;-,+)|^2 = |M(-,+;+,-)|^2 = 16 e_q^2 g^2 \frac{t}{u}. \tag{10.153}$$

Unpolarized squared matrix element:

$$\overline{|M|^2} = e_q^2 g^2 \frac{t^2 + u^2}{tu}. \tag{10.154}$$

Unpolarized cross section:

$$\frac{d\sigma}{dt} = \frac{\pi \alpha \alpha_s Q_f^2}{s^2} \frac{t^2 + u^2}{tu}. \tag{10.155}$$

10.24 $gg \to \bar{q}q$

Process:
$$g(k_1, a) + g(k_2, b) \to \bar{q}(q', i) + q(q, j). \tag{10.156}$$

Positive z-axis: along $\vec{k}_1$.
Invariants:

$$s = 2(k_1 \cdot k_2), \qquad t = -2(k_1 \cdot q'), \qquad u = -2(k_1 \cdot q). \tag{10.157}$$

Gluon polarizations:

$$\not{\epsilon}^{*\pm}(k_i) = N[\not{q}\not{q}'\not{k}_i(1 \pm \gamma_5) + \not{k}_i \not{q}' \not{q}(1 \mp \gamma_5)], \qquad i = 1, 2,$$

$$N = (2stu)^{-\frac{1}{2}}. \tag{10.158}$$

Nonvanishing helicity amplitudes:

$$M(+,-;+,-) = \frac{g^2}{E}\left(\frac{q_-}{q_+}\right)^{\frac{1}{2}}\left[q_+(T^aT^b)_{ji} + q_-(T^bT^a)_{ji}\right],$$

$$M(+,-;-,+) = -\frac{g^2}{E}\left(\frac{q_+}{q_-}\right)^{\frac{1}{2}}\left[q_+(T^aT^b)_{ji} + q_-(T^bT^a)_{ji}\right],$$

$$M(-,+;+,-) = -\frac{g^2}{E}\left(\frac{q_+}{q_-}\right)^{\frac{1}{2}}\left[q_+(T^aT^b)_{ji} + q_-(T^bT^a)_{ji}\right],$$

$$M(-,+;-,+) = \frac{g^2}{E}\left(\frac{q_-}{q_+}\right)^{\frac{1}{2}}\left[q_+(T^aT^b)_{ji} + q_-(T^bT^a)_{ji}\right]. \quad (10.159)$$

Squared absolute values of the nonvanishing helicity amplitudes, summed over the color degrees of freedom of the initial state and the final state:

$$|M(+,-;+,-)|^2 = |M(-,+;-,+)|^2 = 48g^4\left[\frac{4u}{9t} - \frac{u^2}{s^2}\right],$$

$$|M(+,-;-,+)|^2 = |M(-,+;+,-)|^2 = 48g^4\left[\frac{4t}{9u} - \frac{t^2}{s^2}\right]. \quad (10.160)$$

Unpolarized squared matrix element:

$$\overline{|M|^2} = \frac{3}{8}g^4(t^2+u^2)\left[\frac{4}{9tu} - \frac{1}{s^2}\right]. \quad (10.161)$$

Unpolarized cross section:

$$\frac{d\sigma}{dt} = \frac{3\pi\alpha_S^2}{8}\frac{t^2+u^2}{s^2}\left[\frac{4}{9tu} - \frac{1}{s^2}\right]. \quad (10.162)$$

10.25 $gg \to gg$

Process:
$$g(k_1,a) + g(k_2,b) \to g(k_3,c) + g(k_4,d). \quad (10.163)$$

Positive z-axis: along $\vec{k}_1$.
Invariants:
$$s = 2(k_1 \cdot k_2), \qquad t = -2(k_1 \cdot k_3), \qquad u = -2(k_1 \cdot k_4). \quad (10.164)$$

Gluon polarizations:
$$\not{\epsilon}^{*\pm}(k_1) = N[\not{k}_3 \not{k}_2 \not{k}_1(1 \pm \gamma_5) + \not{k}_1 \not{k}_2 \not{k}_3(1 \mp \gamma_5)],$$

$$\epsilon^{*\pm}(k_2) = N[\rlap{/}{k}_4\, \rlap{/}{k}_3\, \rlap{/}{k}_2(1\pm\gamma_5) + \rlap{/}{k}_1\, \rlap{/}{k}_2\, \rlap{/}{k}_3(1\mp\gamma_5)],$$

$$\epsilon^{\pm}(k_3) = N[\rlap{/}{k}_3\, \rlap{/}{k}_2\, \rlap{/}{k}_1(1\pm\gamma_5) + \rlap{/}{k}_1\, \rlap{/}{k}_2\, \rlap{/}{k}_3(1\mp\gamma_5)],$$

$$\epsilon^{\pm}(k_4) = N[\rlap{/}{k}_4\, \rlap{/}{k}_3\, \rlap{/}{k}_2(1\pm\gamma_5) + \rlap{/}{k}_2\, \rlap{/}{k}_3\, \rlap{/}{k}_4(1\mp\gamma_5)],$$

$$N = (2stu)^{-\frac{1}{2}}. \qquad (10.165)$$

Nonvanishing helicity amplitudes:

$$M(+,+;+,+) = 2g^2 \frac{s}{tu}\left(-u f^{abm} f^{cdm} + s f^{adm} f^{bcm}\right),$$

$$M(+,-;+,-) = 2g^2 \frac{u}{st}\left(-u f^{abm} f^{cdm} + s f^{adm} f^{bcm}\right),$$

$$M(+,-;-,+) = 2g^2 \frac{t}{su}\left(t f^{abm} f^{cdm} + s f^{acm} f^{bdm}\right),$$

$$M(-,+;+,-) = 2g^2 \frac{t}{su}\left(t f^{abm} f^{cdm} + s f^{acm} f^{bdm}\right),$$

$$M(-,+;-,+) = 2g^2 \frac{u}{st}\left(-u f^{abm} f^{cdm} + s f^{adm} f^{bcm}\right),$$

$$M(-,-;-,-) = 2g^2 \frac{s}{tu}\left(-u f^{abm} f^{cdm} + s f^{adm} f^{bcm}\right). \qquad (10.166)$$

Squared absolute values of the nonvanishing helicity amplitudes, summed over the color degrees of freedom of the initial state and the final state:

$$|M(+,+;+,+)|^2 = |M(-,-;-,-)|^2 = 144 g^4 \frac{s^2(s^2+t^2+u^2)}{t^2 u^2},$$

$$|M(+,-;+,-)|^2 = |M(-,+;-,+)|^2 = 144 g^4 \frac{u^2(s^2+t^2+u^2)}{s^2 t^2},$$

$$|M(+,-;-,+)|^2 = |M(-,+;+,-)|^2 = 144 g^4 \frac{t^2(s^2+t^2+u^2)}{s^2 u^2}.$$
$$(10.167)$$

Unpolarized squared matrix element:

$$\overline{|M|^2} = \frac{9}{8} g^4 \frac{(s^4+t^4+u^4)(s^2+t^2+u^2)}{s^2 t^2 u^2}. \qquad (10.168)$$

Unpolarized cross section:

$$\frac{d\sigma}{dt} = \frac{9\pi\alpha_s^2}{8} \frac{(s^4+t^4+u^4)(s^2+t^2+u^2)}{s^4 t^2 u^2}. \qquad (10.169)$$

Note that
$$(s^2 + t^2 + u^2)^2 = 2(s^4 + t^4 + u^4) \tag{10.170}$$
in this case.

10.26 $qq' \to qq'\gamma$

Process (different quark flavors):
$$q(p_+,i) + q'(p_-,j) \to q(q_+,m) + q'(q_-,n) + \gamma(k). \tag{10.171}$$

Invariants and definitions:

$$s = 2(p_+ \cdot p_-), \quad t = -2(p_+ \cdot q_+), \quad u = -2(p_+ \cdot q_-),$$
$$s' = 2(q_+ \cdot q_-), \quad t' = -2(p_- \cdot q_-), \quad u' = -2(p_- \cdot q_+),$$

$$B = Q_f - Q'_f \left[1 - \frac{k_\perp q^*_{+\perp}}{k_+ q_{+-}}\right] \frac{q_{-+}k_- - q^*_{-\perp}k_\perp}{2(q_- \cdot k)},$$

$$Z_{-+} = q_{-+}q_{+-} - q^*_{-\perp}q_{+\perp}. \tag{10.172}$$

Photon polarizations:

$$\not{\epsilon}^\pm = N[\not{q}_+ \not{p}_+ \not{k}(1 \mp \gamma_5) - \not{k} \not{q}_+ \not{p}_+(1 \pm \gamma_5)],$$

$$N^{-1} = 4E[k_- q_{+-}(q_+ \cdot k)]^{\frac{1}{2}}. \tag{10.173}$$

Nonvanishing helicity amplitudes:

$$M(+,+;+,+,+) = 4g^2 e T^a_{mi} T^a_{nj} B \frac{Eq^*_{+\perp}}{[k_- q_{++} q_{+-} q_{-+}(q_+ \cdot k)]^{\frac{1}{2}}},$$

$$M(+,+;+,+,-) = g^2 e T^a_{mi} T^a_{nj} B^* \frac{q^*_{+\perp} Z^2_{-+}}{E[k_- q_{++} q^3_{+-} q^3_{-+}(q_+ \cdot k)]^{\frac{1}{2}}},$$

$$M(+,-;+,-,+) = 2g^2 e T^a_{mi} T^a_{nj} B \frac{q^*_{+\perp} q^2_{-\perp}}{[k_- q_{++} q_{+-} q^3_{-+}(q_+ \cdot k)]^{\frac{1}{2}}},$$

$$M(-,+;-,+,+) = 2g^2 e T^a_{mi} T^a_{nj} B q^*_{+\perp} \left[\frac{q_{++}}{k_- q_{+-} q_{-+}(q_+ \cdot k)}\right]^{\frac{1}{2}},$$

$$M(+,-;+,-,-) = 2g^2 e T^a_{mi} T^a_{nj} B^* q_{+\perp} \left[\frac{q_{++}}{k_- q_{+-} q_{-+}(q_+ \cdot k)}\right]^{\frac{1}{2}},$$

$$M(-,+;-,+,-) = 2g^2 e T^a_{mi} T^a_{nj} B^* \frac{q_{+\perp} q^{*2}_{-\perp}}{[k_- q_{++} q_{+-} q^3_{-+}(q_+ \cdot k)]^{\frac{1}{2}}},$$

$$M(-,-;-,-,+) = g^2 e T^a_{mi} T^a_{nj} B \frac{q_{+\perp} Z^{*2}_{-+}}{E[k_- q_{++} q^3_{+-} q^3_{-+}(q_+ \cdot k)]^{\frac{1}{2}}},$$

$$M(-,-;-,-,-) = 4g^2 e T^a_{mi} T^a_{nj} B^* \frac{E q_{+\perp}}{[k_- q_{++} q_{+-} q_{-+}(q_+ \cdot k)]^{\frac{1}{2}}}. \quad (1$$

Squared absolute values of the nonvanishing helicity amplitudes, su: over the color degrees of freedom of the initial state and the final stat<

$$|M(+,+;+,+,+)|^2 = |M(-,-;-,-,-)|^2 = \frac{32 g^4 e^2 |B|^2 E^2}{k_- q_{-+}(q_+ \cdot k)},$$

$$|M(+,+;+,+,-)|^2 = |M(-,-;-,-,+)|^2 = \frac{8 g^4 e^2 |B|^2 (q_+ \cdot q_-}{E^2 k_- q_{-+}(q_+ \cdot k)}$$

$$|M(+,-;+,-,+)|^2 = |M(-,+;-,+,-)|^2 = \frac{8 g^4 e^2 |B|^2 q^2_{--}}{k_- q_{-+}(q_+ \cdot k)},$$

$$|M(-,+;-,+,+)|^2 = |M(+,-;+,-,-)|^2 = \frac{8 g^4 e^2 |B|^2 q^2_{++}}{k_- q_{-+}(q_+ \cdot k)}.$$

$$(1$$

Unpolarized squared matrix element:

$$\overline{|M|^2} = -\frac{2}{9} g^4 e^2 (Q_f v_+ + Q'_f v_-)^2 \frac{s^2 + s'^2 + u^2 + u'^2}{tt'}, \quad (1$$

where

$$v_+ = \frac{p_+}{(p_+ \cdot k)} - \frac{q_+}{(q_+ \cdot k)}, \qquad v_- = \frac{p_-}{(p_- \cdot k)} - \frac{q_-}{(q_- \cdot k)}. \quad (1$$

Unpolarized cross section:

$$d\sigma = -\frac{\alpha_s^2 \alpha}{36 \pi^2} (Q_f v_+ + Q'_f v_-)^2 \frac{s^2 + s'^2 + u^2 + u'^2}{stt'}$$

$$\times \delta^4(p_+ + p_- - q_+ - q_- - k) \frac{d^3\vec{q}_+ \, d^3\vec{q}_- \, d^3\vec{k}}{q_{+0} \, q_{-0} \, k_0}. \quad (1$$

10.27 $qq' \to qq'g$

Process (different quark flavors):

$$q(p_+, i) + q'(p_-, j) \to q(q_+, m) + q'(q_-, n) + g(k, a). \quad (1$$

Invariants and definitions:

$$s = 2(p_+ \cdot p_-), \qquad t = -2(p_+ \cdot q_+), \qquad u = -2(p_+ \cdot q_-),$$

$$s' = 2(q_+ \cdot q_-), \qquad t' = -2(p_- \cdot q_-), \qquad u' = -2(p_- \cdot q_+),$$

10.27. $qq' \to qq'g$

$$Z(k, q_+) = k_+ q_{+-} - k_\perp^* q_{+\perp}, \qquad Z(k, q_-) = k_+ q_{--} - k_\perp^* q_{-\perp},$$

$$B = \frac{1}{k_+[k_- q_{+-} - (q_+ \cdot k)]^{\frac{1}{2}}} \left[k_+ q_{+-} (T^a T^b)_{mi} T^b_{nj} \right.$$
$$\left. - Z^*(k, q_+) T^b_{mi} (T^b T^a)_{nj} + \frac{Z^*(k, q_+) Z(k, q_-)}{2(q_- \cdot k)} T^b_{mi} (T^a T^b)_{nj} \right],$$

$$B' = \frac{1}{k_+[k_- q_{+-} - (q_+ \cdot k)]^{\frac{1}{2}}} \left[k_+ q_{+-} (T^a T^b)_{mi} T^b_{nj} \right.$$
$$\left. - Z(k, q_+) T^b_{mi} (T^b T^a)_{nj} + \frac{Z(k, q_+) Z^*(k, q_-)}{2(q_- \cdot k)} T^b_{mi} (T^a T^b)_{nj} \right],$$

$$C = \frac{1}{3} \left[\frac{4}{k_+ k_-} + \frac{2(q_+ \cdot q_-)}{(q_+ \cdot k)(q_- \cdot k)} - \frac{q_{+-}}{k_-(q_+ \cdot k)} \right.$$
$$\left. - \frac{q_{-+}}{k_+(q_- \cdot k)} + \frac{7 q_{++}}{k_+(q_+ \cdot k)} + \frac{7 q_{--}}{k_-(q_- \cdot k)} \right]. \qquad (10.180)$$

Gluon polarizations:

$$\rlap{/}{e}^\pm = N[\rlap{/}{k}\, \rlap{/}{q}_+\, \rlap{/}{p}_+ (1 \pm \gamma_5) + \rlap{/}{p}_+\, \rlap{/}{q}_+\, \rlap{/}{k} (1 \mp \gamma_5)],$$

$$N^{-1} = 4E[k_- q_{+-} - (q_+ \cdot k)]^{\frac{1}{2}}. \qquad (10.181)$$

Nonvanishing helicity amplitudes:

$$M(+, +; +, +, +) = -\frac{4g^3 B E q_{+\perp}^*}{q_{+-} \sqrt{q_{++} q_{-+}}},$$

$$M(+, +; +, +, -) = -\frac{2g^3 B'(q_+ \cdot q_-) q_{+\perp}^*}{E q_{+-} \sqrt{q_{++} q_{-+}}},$$

$$M(+, -; +, -, +) = -\frac{2g^3 B q_{+\perp}^* q_{-\perp}^2}{q_{+-} \sqrt{q_{++} q_{-+}^3}},$$

$$M(-, +; -, +, +) = -\frac{2g^3 B q_{+\perp}^*}{q_{+-}} \left(\frac{q_{++}}{q_{-+}}\right)^{\frac{1}{2}},$$

$$M(+, -; +, -, -) = -\frac{2g^3 B' q_{+\perp}}{q_{+-}} \left(\frac{q_{++}}{q_{-+}}\right)^{\frac{1}{2}},$$

$$M(-, +; -, +, -) = -\frac{2g^3 B' q_{+\perp} q_{-\perp}^{*2}}{q_{+-} \sqrt{q_{++} q_{-+}^3}},$$

$$M(-,-;-,-,+) = -\frac{2g^3 B(q_+ \cdot q_-) q_{+\perp}}{E q_{+-} \sqrt{q_{++} q_{-+}}},$$

$$M(-,-;-,-,-) = -\frac{4g^3 B' E q_{+\perp}}{q_{+-} \sqrt{q_{++} q_{-+}}}. \quad (10.182)$$

Squared absolute values of the nonvanishing helicity amplitudes, summed over the color degrees of freedom of the initial state and the final state:

$$|M(+,+;+,+,+)|^2 = |M(-,-;-,-,-)|^2 = \frac{16 g^6 C E^2}{q_{+-} q_{-+}},$$

$$|M(+,+;+,+,-)|^2 = |M(-,-;-,-,+)|^2 = \frac{4 g^6 C (q_+ \cdot q_-)^2}{E^2 q_{+-} q_{-+}},$$

$$|M(+,-;+,-,+)|^2 = |M(-,+;-,+,-)|^2 = \frac{4 g^6 C q_{--}^2}{q_{+-} q_{-+}},$$

$$|M(-,+;-,+,+)|^2 = |M(+,-;+,-,-)|^2 = \frac{4 g^6 C q_{++}^2}{q_{+-} q_{-+}}. \quad (10.183)$$

Unpolarized squared matrix element:

$$\overline{|M|^2} = \frac{2g^6}{27} \left[\frac{7(p_+ \cdot q_-)}{(p_+ \cdot k)(q_- \cdot k)} + \frac{7(p_- \cdot q_+)}{(p_- \cdot k)(q_+ \cdot k)} - \frac{(p_+ \cdot q_+)}{(p_+ \cdot k)(q_+ \cdot k)} \right.$$
$$\left. - \frac{(p_- \cdot q_-)}{(p_- \cdot k)(q_- \cdot k)} + \frac{2(p_+ \cdot p_-)}{(p_+ \cdot k)(p_- \cdot k)} + \frac{2(q_+ \cdot q_-)}{(q_+ \cdot k)(q_- \cdot k)} \right]$$
$$\times \frac{s^2 + s'^2 + u^2 + u'^2}{tt'}. \quad (10.184)$$

Unpolarized cross section:

$$d\sigma = \frac{\alpha_S^3}{108\pi^2} \left[\frac{7(p_+ \cdot q_-)}{(p_+ \cdot k)(q_- \cdot k)} + \frac{7(p_- \cdot q_+)}{(p_- \cdot k)(q_+ \cdot k)} - \frac{(p_+ \cdot q_+)}{(p_+ \cdot k)(q_+ \cdot k)} \right.$$
$$\left. - \frac{(p_- \cdot q_-)}{(p_- \cdot k)(q_- \cdot k)} + \frac{2(p_+ \cdot p_-)}{(p_+ \cdot k)(p_- \cdot k)} + \frac{2(q_+ \cdot q_-)}{(q_+ \cdot k)(q_- \cdot k)} \right]$$
$$\times \frac{s^2 + s'^2 + u^2 + u'^2}{stt'} \delta^4(p_+ + p_- - q_+ - q_- - k) \frac{d^3 \vec{q}_+ \, d^3 \vec{q}_- \, d^3 \vec{k}}{q_{+0} \, q_{-0} \, k_0}.$$
$$(10.185)$$

10.28 $\bar{q}q \to \bar{q}'q'\gamma$

Process (different quark flavors):

$$\bar{q}(p_+, i) + q(p_-, j) \to \bar{q}'(q_+, m) + q'(q_-, n) + \gamma(k). \tag{10.186}$$

Invariants and definitions:

$$s = 2(p_+ \cdot p_-), \quad t = -2(p_+ \cdot q_+), \quad u = -2(p_+ \cdot q_-),$$

$$s' = 2(q_+ \cdot q_-), \quad t' = -2(p_- \cdot q_-), \quad u' = -2(p_- \cdot q_+),$$

$$Z_{+-} = q_{++}q_{--} - q^*_{+\perp}q_{-\perp}, \qquad Z_{-+} = q_{-+}q_{+-} - q^*_{-\perp}q_{+\perp},$$

$$B = Q_f + Q'_f \frac{E(2k_+k_-q_{-z} + k_-k^*_\perp q_{-\perp} - k_+k_\perp q^*_{-\perp})}{2(q_+ \cdot k)(q_- \cdot k)}. \tag{10.187}$$

Photon polarizations:

$$\not{e}^\pm = N[\not{p}_+ \not{p}_- \not{k}(1 \mp \gamma_5) - \not{k}\not{p}_+ \not{p}_-(1 \pm \gamma_5)],$$

$$N^{-1} = E^2[32k_+k_-]^{\frac{1}{2}}. \tag{10.188}$$

Nonvanishing helicity amplitudes:

$$M(+,-;+,-,+) = -\frac{g^2 e T^a_{ij} T^a_{nm} q_{-\perp} Z_{+-} B}{(q_+ \cdot q_-)} \left[\frac{2}{k_+k_- q_{++} q_{-+}}\right]^{\frac{1}{2}},$$

$$M(+,-;-,+,+) = -\frac{g^2 e T^a_{ij} T^a_{nm} q_{+\perp} Z_{-+} B}{(q_+ \cdot q_-)} \left[\frac{2}{k_+k_- q_{++} q_{-+}}\right]^{\frac{1}{2}},$$

$$M(-,+;+,-,+) = -\frac{g^2 e T^a_{ij} T^a_{nm} q^*_{+\perp} Z_{-+} B}{q_{+-}(q_+ \cdot q_-)} \left[\frac{2q_{-+}}{k_+k_- q_{++}}\right]^{\frac{1}{2}},$$

$$M(-,+;-,+,+) = -\frac{g^2 e T^a_{ij} T^a_{nm} q^*_{-\perp} Z_{+-} B}{q_{--}(q_+ \cdot q_-)} \left[\frac{2q_{++}}{k_+k_- q_{-+}}\right]^{\frac{1}{2}},$$

$$M(+,-;+,-,-) = -\frac{g^2 e T^a_{ij} T^a_{nm} q_{-\perp} Z^*_{+-} B^*}{q_{--}(q_+ \cdot q_-)} \left[\frac{2q_{++}}{k_+k_- q_{-+}}\right]^{\frac{1}{2}},$$

$$M(+,-;-,+,-) = -\frac{g^2 e T^a_{ij} T^a_{nm} q_{+\perp} Z^*_{-+} B^*}{q_{+-}(q_+ \cdot q_-)} \left[\frac{2q_{-+}}{k_+k_- q_{++}}\right]^{\frac{1}{2}},$$

$$M(-,+;+,-,-) = -\frac{g^2 e T^a_{ij} T^a_{nm} q^*_{+\perp} Z^*_{-+} B^*}{(q_+ \cdot q_-)} \left[\frac{2}{k_+k_- q_{++} q_{-+}}\right]^{\frac{1}{2}},$$

$$M(-,+;-,+,-) = -\frac{g^2 e T^a_{ij} T^a_{nm} q^*_{-\perp} Z^*_{+-} B^*}{(q_+ \cdot q_-)} \left[\frac{2}{k_+k_- q_{++} q_{-+}}\right]^{\frac{1}{2}}. \tag{10.189}$$

Squared absolute values of the nonvanishing helicity amplitudes, summed over the color degrees of freedom of the initial state and the final state:

$$|M(+,-;+,-,+)|^2 = |M(-,+;-,+,-)|^2 = 8g^4 e^2 \frac{q_{--}^2 |B|^2}{k_+ k_- (q_+ \cdot q_-)},$$

$$|M(+,-;-,+,+)|^2 = |M(-,+;+,-,-)|^2 = 8g^4 e^2 \frac{q_{+-}^2 |B|^2}{k_+ k_- (q_+ \cdot q_-)},$$

$$|M(-,+;+,-,+)|^2 = |M(+,-;-,+,-)|^2 = 8g^4 e^2 \frac{q_{-+}^2 |B|^2}{k_+ k_- (q_+ \cdot q_-)},$$

$$|M(-,+;-,+,+)|^2 = |M(+,-;+,-,-)|^2 = 8g^4 e^2 \frac{q_{++}^2 |B|^2}{k_+ k_- (q_+ \cdot q_-)}. \tag{10.190}$$

Unpolarized squared matrix element:

$$\overline{|M|^2} = -\frac{2}{9} g^4 e^2 (Q_f v_p - Q'_f v_q)^2 \frac{t^2 + t'^2 + u^2 + u'^2}{ss'}, \tag{10.191}$$

where [see eqns (4.27)]

$$v_p = \frac{p_+}{(p_+ \cdot k)} - \frac{p_-}{(p_- \cdot k)}, \qquad v_q = \frac{q_+}{(q_+ \cdot k)} - \frac{q_-}{(q_- \cdot k)}. \tag{10.192}$$

Unpolarized cross section:

$$d\sigma = -\frac{\alpha_S^2 \alpha}{36\pi^2} (Q_f v_p - Q'_f v_q)^2 \frac{t^2 + t'^2 + u^2 + u'^2}{s^2 s'}$$

$$\times \delta^4(p_+ + p_- - q_+ - q_- - k) \frac{d^3 \vec{q}_+ \, d^3 \vec{q}_- \, d^3 \vec{k}}{q_{+0} \, q_{-0} \, k_0}. \tag{10.193}$$

10.29 $\bar{q} q \to \bar{q}' q' g$

Process (different quark flavors):

$$\bar{q}(p_+, i) + q(p_-, j) \to \bar{q}'(q_+, m) + q'(q_-, n) + g(k, a). \tag{10.194}$$

Invariants and definitions:

$$s = 2(p_+ \cdot p_-), \qquad t = -2(p_+ \cdot q_+), \qquad u = -2(p_+ \cdot q_-),$$

$$s' = 2(q_+ \cdot q_-), \qquad t' = -2(p_- \cdot q_-), \qquad u' = -2(p_- \cdot q_+),$$

10.29. $\bar{q}q \to \bar{q}'q'g$ 411

$$Z_{+-} = q_{++}q_{--} - q^*_{+\perp}q_{-\perp}, \qquad Z_{-+} = q_{-+}q_{+-} - q^*_{-\perp}q_{+\perp},$$

$$B = \frac{1}{(2k_+k_-)^{\frac{1}{2}}}\left[2(T^aT^b)_{ij}T^b_{nm} + \frac{q_{-+}k_- - q^*_{-\perp}k_\perp}{(q_-\cdot k)}T^b_{ij}(T^aT^b)_{nm}\right.$$
$$\left. - \frac{q_{++}k_- - q^*_{+\perp}k_\perp}{(q_+\cdot k)}T^b_{ij}(T^bT^a)_{nm}\right],$$

$$B' = \frac{1}{(2k_+k_-)^{\frac{1}{2}}}\left[2(T^aT^b)_{ij}T^b_{nm} + \frac{q_{-+}k_- - q_{-\perp}k^*_\perp}{(q_-\cdot k)}T^b_{ij}(T^aT^b)_{nm}\right.$$
$$\left. - \frac{q_{++}k_- - q_{+\perp}k^*_\perp}{(q_+\cdot k)}T^b_{ij}(T^bT^a)_{nm}\right],$$

$$C = \frac{1}{3}\left[-\frac{2}{k_+k_-} - \frac{(q_+\cdot q_-)}{(q_+\cdot k)(q_-\cdot k)} + \frac{7q_{+-}}{k_-(q_+\cdot k)}\right.$$
$$\left. + \frac{7q_{-+}}{k_+(q_-\cdot k)} + \frac{2q_{++}}{k_+(q_+\cdot k)} + \frac{2q_{--}}{k_-(q_-\cdot k)}\right].$$
(10.195)

Gluon polarizations:

$$\not{\epsilon}^\pm = N[\not{k}\not{p}_+\not{p}_-(1\pm\gamma_5) + \not{p}_-\not{p}_+\not{k}(1\mp\gamma_5)],$$

$$N^{-1} = E^2[32k_+k_-]^{\frac{1}{2}}. \qquad (10.196)$$

Nonvanishing helicity amplitudes:

$$M(+,-;+,-,+) = \frac{g^3 B q_{-\perp} Z_{+-}}{(q_+\cdot q_-)\sqrt{q_{++}q_{-+}}},$$

$$M(+,-;-,+,+) = \frac{g^3 B q_{+\perp} Z_{-+}}{(q_+\cdot q_-)\sqrt{q_{++}q_{-+}}},$$

$$M(-,+;+,-,+) = \frac{g^3 B q^*_{+\perp} Z_{-+}}{q_{+-}(q_+\cdot q_-)}\left(\frac{q_{-+}}{q_{++}}\right)^{\frac{1}{2}},$$

$$M(-,+;-,+,+) = \frac{g^3 B q^*_{-\perp} Z_{+-}}{q_{--}(q_+\cdot q_-)}\left(\frac{q_{++}}{q_{-+}}\right)^{\frac{1}{2}},$$

$$M(+,-;+,-,-) = \frac{g^3 B' q_{-\perp} Z^*_{+-}}{q_{--}(q_+\cdot q_-)}\left(\frac{q_{++}}{q_{-+}}\right)^{\frac{1}{2}},$$

$$M(+,-;-,+,-) = \frac{g^3 B' q_{+\perp} Z^*_{-+}}{q_{+-}(q_+\cdot q_-)}\left(\frac{q_{-+}}{q_{++}}\right)^{\frac{1}{2}},$$

$$M(-,+;+,-,-) = \frac{g^3 B' q^*_{+\perp} Z^*_{-+}}{(q_+ \cdot q_-)\sqrt{q_{++}q_{-+}}},$$

$$M(-,+;-,+,-) = \frac{g^3 B' q^*_{-\perp} Z^*_{+-}}{(q_+ \cdot q_-)\sqrt{q_{++}q_{-+}}}. \qquad (10.197)$$

Squared absolute values of the nonvanishing helicity amplitudes, summed over the color degrees of freedom of the initial state and the final state:

$$|M(+,-;+,-,+)|^2 = |M(-,+;-,+,-)|^2 = \frac{2g^6 C q^2_{--}}{(q_+ \cdot q_-)},$$

$$|M(+,-;-,+,+)|^2 = |M(-,+;+,-,-)|^2 = \frac{2g^6 C q^2_{+-}}{(q_+ \cdot q_-)},$$

$$|M(-,+;+,-,+)|^2 = |M(+,-;-,+,-)|^2 = \frac{2g^6 C q^2_{-+}}{(q_+ \cdot q_-)},$$

$$|M(-,+;-,+,+)|^2 = |M(+,-;+,-,-)|^2 = \frac{2g^6 C q^2_{++}}{(q_+ \cdot q_-)}. \qquad (10.198)$$

Unpolarized squared matrix element:

$$\overline{|M|^2} = \frac{2g^6}{27} \left[\frac{7(p_+ \cdot q_+)}{(p_+ \cdot k)(q_+ \cdot k)} + \frac{7(p_- \cdot q_-)}{(p_- \cdot k)(q_- \cdot k)} - \frac{(p_+ \cdot p_+)}{(p_+ \cdot k)(p_+ \cdot k)} \right.$$

$$\left. - \frac{(q_+ \cdot q_-)}{(q_+ \cdot k)(q_- \cdot k)} + \frac{2(p_+ \cdot q_-)}{(p_+ \cdot k)(q_- \cdot k)} + \frac{2(p_- \cdot q_+)}{(p_- \cdot k)(q_+ \cdot k)} \right]$$

$$\times \frac{t^2 + t'^2 + u^2 + u'^2}{ss'}. \qquad (10.199)$$

Unpolarized cross section:

$$d\sigma = \frac{\alpha_S^3}{108\pi^2} \left[\frac{7(p_+ \cdot q_+)}{(p_+ \cdot k)(q_+ \cdot k)} + \frac{7(p_- \cdot q_-)}{(p_- \cdot k)(q_- \cdot k)} - \frac{(p_+ \cdot p_-)}{(p_+ \cdot k)(p_- \cdot k)} \right.$$

$$\left. - \frac{(q_+ \cdot q_-)}{(q_+ \cdot k)(q_- \cdot k)} + \frac{2(p_+ \cdot q_-)}{(p_+ \cdot k)(q_- \cdot k)} + \frac{2(p_- \cdot q_+)}{(p_- \cdot k)(q_+ \cdot k)} \right]$$

$$\times \frac{t^2 + t'^2 + u^2 + u'^2}{s^2 s'} \delta^4(p_+ + p_- - q_+ - q_- - k) \frac{d^3\vec{q}_+ \, d^3\vec{q}_- \, d^3\vec{k}}{q_{+0} \, q_{-0} \, k_0}. \qquad (10.200)$$

10.30 $qq \to qq\gamma$

Process (identical quark flavors):

$$q(p_+, i) + q(p_-, j) \to q(q_+, m) + q(q_-, n) + \gamma(k). \qquad (10.201)$$

Invariants and definitions:

$$s = 2(p_+ \cdot p_-), \quad t = -2(p_+ \cdot q_+), \quad u = -2(p_+ \cdot q_-),$$

$$s' = 2(q_+ \cdot q_-), \quad t' = -2(p_- \cdot q_-), \quad u' = -2(p_- \cdot q_+),$$

$$Z_{+-} = q_{++}q_{--} - q^*_{+\perp}q_{-\perp}, \qquad Z_{-+} = q_{-+}q_{+-} - q^*_{-\perp}q_{+\perp},$$

$$B = 1 - \left[1 - \frac{k_\perp q^*_{+\perp}}{k_+ q_{+-}}\right] \frac{q_{-+}k_- - q^*_{-\perp}k_\perp}{2(q_- \cdot k)}. \qquad (10.202)$$

Photon polarizations:

$$\not{\epsilon}^\pm = N[\not{q}_+ \not{p}_+ \not{k}(1 \mp \gamma_5) - \not{k} \not{q}_+ \not{p}_+(1 \pm \gamma_5)],$$

$$N^{-1} = 4E[k_- q_{+-}(q_+ \cdot k)]^{\frac{1}{2}}. \qquad (10.203)$$

Nonvanishing helicity amplitudes:

$$M(+,+;+,+,+) = 4g^2 e_q BE \left[\frac{q_{+-}}{k_- q_{++} q_{-+}(q_+ \cdot k)}\right]^{\frac{1}{2}}$$
$$\times \left[\frac{q^*_{+\perp} T^a_{mi} T^a_{nj}}{q_{+-}} - \frac{q^*_{-\perp} T^a_{ni} T^a_{mj}}{q_{--}}\right],$$

$$M(+,+;+,+,-) = -g^2 e_q B^* \frac{Z_{+-} Z_{-+}}{E q_{--} [k_- q_{++} q_{+-} q_{-+}(q_+ \cdot k)]^{\frac{1}{2}}}$$
$$\times \left[\frac{q^*_{-\perp} T^a_{mi} T^a_{nj}}{q_{-+}} - \frac{q^*_{+\perp} T^a_{ni} T^a_{mj}}{q_{++}}\right],$$

$$M(+,-;+,-,+) = 2g^2 e_q B T^a_{mi} T^a_{nj} \frac{q^*_{+\perp} q^2_{-\perp}}{[k_- q_{++} q_{+-} q^3_{-+}(q_+ \cdot k)]^{\frac{1}{2}}},$$

$$M(+,-;-,+,+) = -2g^2 e_q B T^a_{ni} T^a_{mj} \frac{q^*_{-\perp} q^2_{+\perp}}{q_{--}} \left[\frac{q_{+-}}{k_- q^3_{++} q_{-+}(q_+ \cdot k)}\right]^{\frac{1}{2}},$$

$$M(-,+;+,-,+) = -2g^2 e_q B T^a_{ni} T^a_{mj} \frac{q^*_{-\perp}}{q_{--}} \left[\frac{q_{+-} q_{-+}}{k_- q_{++}(q_+ \cdot k)}\right]^{\frac{1}{2}},$$

$$M(-,+;-,+,+) = 2g^2 e_q B T^a_{mi} T^a_{nj} q^*_{+\perp} \left[\frac{q_{++}}{k_- q_{+-} q_{-+}(q_+ \cdot k)}\right]^{\frac{1}{2}},$$

$$M(+,-;+,-,-) = 2g^2 e_q B T^a_{mi} T^a_{nj} q_{+\perp} \left[\frac{q_{++}}{k_- q_{+-} q_{-+}(q_+ \cdot k)}\right]^{\frac{1}{2}},$$

$$M(+,-;-,+,-) = -2g^2 e_q B^* T^a_{ni} T^a_{mj} \frac{q_{-\perp}}{q_{--}} \left[\frac{q_{+-} q_{-+}}{k_- q_{++}(q_+ \cdot k)}\right]^{\frac{1}{2}},$$

$$M(-,+;+,-,-) = -2g^2 e_q B^* T^a_{ni} T^a_{mj} \frac{q_{-\perp} q^{*2}_{+\perp}}{q_{--}} \left[\frac{q_{+-}}{k_- q^3_{++} q_{-+}(q_+ \cdot k)}\right]^{\frac{1}{2}},$$

$$M(-,+;-,+,-) = 2g^2 e_q B^* T^a_{mi} T^a_{nj} \frac{q_{+\perp} q^{*2}_{-\perp}}{[k_- q_{++} q_{+-} q^3_{-+}(q_+ \cdot k)]^{\frac{1}{2}}},$$

$$M(-,-;-,-,+) = -g^2 e_q B \frac{Z^*_{+-} Z^*_{-+}}{E q_{--}[k_- q_{++} q_{+-} q_{-+}(q_+ \cdot k)]^{\frac{1}{2}}}$$

$$\times \left[\frac{q_{-\perp} T^a_{mi} T^a_{nj}}{q_{-+}} - \frac{q_{+\perp} T^a_{ni} T^a_{mj}}{q_{++}}\right],$$

$$M(-,-;-,-,-) = 4g^2 e_q B^* E \left[\frac{q_{+-}}{k_- q_{++} q_{-+}(q_+ \cdot k)}\right]^{\frac{1}{2}}$$

$$\times \left[\frac{q_{+\perp} T^a_{mi} T^a_{nj}}{q_{+-}} - \frac{q_{-\perp} T^a_{ni} T^a_{mj}}{q_{--}}\right]. \quad (10.204)$$

Squared absolute values of the nonvanishing helicity amplitudes, summed over the color degrees of freedom of the initial state and the final state:

$$|M(+,+;+,+,+)|^2 = |M(-,-;-,-,-)|^2$$

$$= 32g^4 e_q^2 |B|^2 \frac{E^2 q_{+-}}{k_- q_{++} q_{-+}(q_+ \cdot k)} \left[\frac{q_{++}}{q_{+-}} + \frac{q_{-+}}{q_{--}} + \frac{q_{+\perp} q^*_{-\perp} + q^*_{+\perp} q_{-\perp}}{3 q_{+-} q_{--}}\right],$$

$$|M(+,+;+,+,-)|^2 = |M(-,-;-,-,+)|^2$$

$$= 8g^4 e_q^2 |B|^2 \frac{(q_+ \cdot q_-)^2}{E^2 k_- q_{--}(q_+ \cdot k)} \left[\frac{q_{--}}{q_{-+}} + \frac{q_{+-}}{q_{++}} + \frac{q_{+\perp} q^*_{-\perp} + q^*_{+\perp} q_{-\perp}}{q_{++} q_{-+}}\right],$$

$$|M(+,-;+,-,+)|^2 = |M(-,+;-,+,-)|^2$$

$$= 8g^4 e_q^2 |B|^2 \frac{q^2_{--}}{k_- q_{-+}(q_+ \cdot k)},$$

$$|M(+,-;-,+,+)|^2 = |M(-,+;+,-,-)|^2$$

$$= 8g^4 e_q^2 |B|^2 \frac{q_{+-}^3}{k_- q_{++} q_{--} (q_+ \cdot k)},$$

$$|M(-,+;+,-,+)|^2 = |M(+,-;-,+,-)|^2$$

$$= 8g^4 e_q^2 |B|^2 \frac{q_{+-} q_{-+}^2}{k_- q_{++} q_{--} (q_+ \cdot k)},$$

$$|M(-,+;-,+,+)|^2 = |M(+,-;+,-,-)|^2$$

$$= 8g^4 e_q^2 |B|^2 \frac{q_{++}^2}{k_- q_{-+} (q_+ \cdot k)}. \qquad (10.205)$$

Unpolarized squared matrix element:

$$\overline{|M|^2} = -\frac{2}{9} g^4 e_q^2 \frac{(v_p - v_q)^2}{tt'uu'} \left[ss'(s^2 + s'^2) + tt'(t^2 + t'^2) \right.$$

$$\left. + uu'(u^2 + u'^2) - \frac{4}{3}(s^2 + s'^2)(ss' - tt' - uu') \right], \qquad (10.206)$$

where [see eqns (4.27)]

$$v_p = \frac{p_+}{(p_+ \cdot k)} - \frac{p_-}{(p_- \cdot k)}, \qquad v_q = \frac{q_+}{(q_+ \cdot k)} - \frac{q_-}{(q_- \cdot k)}. \qquad (10.207)$$

Unpolarized cross section:

$$d\sigma = -\frac{\alpha_S^2 \alpha Q_f^2}{36\pi^2} \frac{(v_p - v_q)^2}{stt'uu'} \left[ss'(s^2 + s'^2) + tt'(t^2 + t'^2) \right.$$

$$\left. + uu'(u^2 + u'^2) - \frac{4}{3}(s^2 + s'^2)(ss' - tt' - uu') \right]$$

$$\times \delta^4(p_+ + p_- - q_+ - q_- - k) \frac{d^3\vec{q}_+ \, d^3\vec{q}_- \, d^3\vec{k}}{q_{+0} \, q_{-0} \, k_0}. \qquad (10.208)$$

10.31 $qq \to qqg$

Process (identical quark flavors):

$$q(p_+, i) + q(p_-, j) \to q(q_+, m) + q(q_-, n) + g(k, a). \qquad (10.209)$$

Invariants and definitions:

$$s = 2(p_+ \cdot p_-), \qquad t = -2(p_+ \cdot q_+), \qquad u = -2(p_+ \cdot q_-),$$

$$s' = 2(q_+ \cdot q_-), \qquad t' = -2(p_- \cdot q_-), \qquad u' = -2(p_- \cdot q_+),$$

$$Z(k, q_+) = k_+ q_{+-} - k_\perp^* q_{+\perp}, \qquad Z(k, q_-) = k_+ q_{--} - k_\perp^* q_{-\perp},$$

$$B_1 = \frac{1}{k_+[k_- q_{+-} - (q_+ \cdot k)]^{\frac{1}{2}}} \Bigg[k_+ q_{+-} (T^a T^b)_{mi} T^b_{nj}$$
$$- Z^*(k, q_+) T^b_{mi}(T^b T^a)_{nj} + \frac{Z^*(k, q_+) Z(k, q_-)}{2(q_- \cdot k)} T^b_{mi}(T^a T^b)_{nj} \Bigg],$$

$$B_1' = \frac{1}{k_+[k_- q_{+-} - (q_+ \cdot k)]^{\frac{1}{2}}} \Bigg[k_+ q_{+-} (T^a T^b)_{mi} T^b_{nj}$$
$$- Z(k, q_+) T^b_{mi}(T^b T^a)_{nj} + \frac{Z(k, q_+) Z^*(k, q_-)}{2(q_- \cdot k)} T^b_{mi}(T^a T^b)_{nj} \Bigg],$$

$$B_2 = \frac{1}{k_+[k_- q_{+-} - (q_+ \cdot k)]^{\frac{1}{2}}} \Bigg[\frac{Z^*(k, q_+) Z(k, q_-)}{2(q_- \cdot k)} (T^a T^b)_{ni} T^b_{mj}$$
$$- Z^*(k, q_+) T^b_{ni}(T^b T^a)_{mj} + k_+ q_{+-} T^b_{ni}(T^a T^b)_{mj} \Bigg],$$

$$B_2' = \frac{1}{k_+[k_- q_{+-} - (q_+ \cdot k)]^{\frac{1}{2}}} \Bigg[\frac{Z(k, q_+) Z^*(k, q_-)}{2(q_- \cdot k)} (T^a T^b)_{ni} T^b_{mj}$$
$$- Z(k, q_+) T^b_{ni}(T^b T^a)_{mj} + k_+ q_{+-} T^b_{ni}(T^a T^b)_{mj} \Bigg],$$

$$C_1 = \frac{1}{3} \Bigg[\frac{4}{k_+ k_-} + \frac{2(q_+ \cdot q_-)}{(q_+ \cdot k)(q_- \cdot k)} - \frac{q_{+-}}{k_-(q_+ \cdot k)}$$
$$- \frac{q_{-+}}{k_+(q_- \cdot k)} + \frac{7 q_{++}}{k_-(q_+ \cdot k)} + \frac{7 q_{--}}{k_-(q_- \cdot k)} \Bigg],$$

$$C_2 = \frac{1}{3} \Bigg[\frac{4}{k_+ k_-} + \frac{2(q_+ \cdot q_-)}{(q_+ \cdot k)(q_- \cdot k)} - \frac{q_{--}}{k_-(q_- \cdot k)}$$
$$- \frac{q_{++}}{k_+(q_+ \cdot k)} + \frac{7 q_{-+}}{k_+(q_- \cdot k)} + \frac{7 q_{+-}}{k_-(q_+ \cdot k)} \Bigg],$$

$$C_3 = \frac{1}{9} \Bigg[-\frac{20}{k_+ k_-} - \frac{10(q_+ \cdot q_-)}{(q_+ \cdot k)(q_- \cdot k)} + \frac{q_{--}}{k_-(q_- \cdot k)}$$
$$+ \frac{q_{++}}{k_+(q_+ \cdot k)} + \frac{q_{-+}}{k_+(q_- \cdot k)} + \frac{q_{+-}}{k_-(q_+ \cdot k)} \Bigg]. \qquad (10.210)$$

10.31. $qq \to qqg$ 417

Gluon polarizations:

$$k^{\pm} = N[\not{k}\,\not{A}_+\,\not{p}_+(1\pm\gamma_5)+ \not{p}_+\,\not{A}_+\,\not{k}(1\mp\gamma_5)],$$

$$N^{-1} = 4E[k_-q_{+-}(q_+\cdot k)]^{\frac{1}{2}}. \qquad (10.211)$$

Nonvanishing helicity amplitudes:

$$M(+,+;+,+,+) = -\frac{4g^3 E}{\sqrt{q_{++}q_{-+}}}\left[\frac{q^*_{+\perp}B_1}{q_{+-}} - \frac{q^*_{-\perp}B_2}{q_{--}}\right],$$

$$M(+,+;+,+,-) = -\frac{2g^3(q_+\cdot q_-)}{E\sqrt{q_{++}q_{-+}}}\left[\frac{q^*_{+\perp}B'_1}{q_{+-}} - \frac{q^*_{-\perp}B'_2}{q_{--}}\right],$$

$$M(+,-;+,-,+) = -\frac{2g^3 B_1 q^*_{+\perp}q^2_{-\perp}}{q_{+-}\sqrt{q_{++}q^3_{-+}}},$$

$$M(+,-;-,+,+) = -\frac{2g^3 B_2 q^*_{-\perp}q^2_{+\perp}}{q_{--}\sqrt{q^3_{++}q_{-+}}},$$

$$M(-,+;+,-,-) = -\frac{2g^3 B_2 q^*_{-\perp}}{q_{--}}\left(\frac{q_{-+}}{q_{++}}\right)^{\frac{1}{2}},$$

$$M(-,+;-,+,+) = -\frac{2g^3 B_1 q^*_{+\perp}}{q_{+-}}\left(\frac{q_{++}}{q_{-+}}\right)^{\frac{1}{2}},$$

$$M(+,-;+,-,-) = -\frac{2g^3 B'_1 q_{+\perp}}{q_{+-}}\left(\frac{q_{++}}{q_{-+}}\right)^{\frac{1}{2}},$$

$$M(+,-;-,+,-) = -\frac{2g^3 B'_2 q_{-\perp}}{q_{--}}\left(\frac{q_{-+}}{q_{++}}\right)^{\frac{1}{2}},$$

$$M(-,+;+,-,-) = -\frac{2g^3 B'_2 q_{-\perp}q^2_{+\perp}}{q_{--}\sqrt{q^3_{++}q_{-+}}},$$

$$M(-,+;-,+,-) = -\frac{2g^3 B'_1 q_{+\perp}q^{*2}_{-\perp}}{q_{+-}\sqrt{q_{++}q^3_{-+}}},$$

$$M(-,-;-,-,+) = -\frac{2g^3(q_+\cdot q_-)}{E\sqrt{q_{++}q_{-+}}}\left[\frac{q_{+\perp}B_1}{q_{+-}} - \frac{q_{-\perp}B_2}{q_{--}}\right],$$

$$M(-,-;-,-,-) = -\frac{4g^3 E}{\sqrt{q_{++}q_{-+}}}\left[\frac{q_{+\perp}B'_1}{q_{+-}} - \frac{q_{-\perp}B'_2}{q_{--}}\right]. \quad (10.212)$$

Squared absolute values of the nonvanishing helicity amplitudes, summed over the color degrees of freedom of the initial state and the final state:

$$|M(+,+;+,+,+)|^2 = |M(-,-;-,-,-)|^2$$

$$= 16g^6 E^2 \left[\frac{C_1}{q_{+-}q_{-+}} + \frac{C_2}{q_{++}q_{--}} - \frac{C_3[q_{++}q_{--} + q_{+-}q_{-+} - 2(q_+ \cdot q_-)]}{q_{++}q_{+-}q_{-+}q_{--}} \right],$$

$$|M(+,+;+,+,-)|^2 = |M(-,-;-,-,+)|^2$$

$$= \frac{4g^6(q_+ \cdot q_-)^2}{E^2} \left[\frac{C_1}{q_{+-}q_{-+}} + \frac{C_2}{q_{++}q_{--}} - \frac{C_3[q_{++}q_{--} + q_{+-}q_{-+} - 2(q_+ \cdot q_-)]}{q_{++}q_{+-}q_{-+}q_{--}} \right],$$

$$|M(+,-;+,-,+)|^2 = |M(-,+;-,+,-)|^2 = \frac{4g^6 C_1 q_{--}^2}{q_{+-}q_{-+}},$$

$$|M(+,-;-,+,+)|^2 = |M(-,+;+,-,-)|^2 = \frac{4g^6 C_2 q_{+-}^2}{q_{++}q_{--}},$$

$$|M(-,+;+,-,+)|^2 = |M(+,-;-,+,-)|^2 = \frac{4g^6 C_2 q_{-+}^2}{q_{++}q_{--}},$$

$$|M(-,+;-,+,+)|^2 = |M(+,-;+,-,-)|^2 = \frac{4g^6 C_1 q_{++}^2}{q_{+-}q_{-+}}. \qquad (10.213)$$

Unpolarized squared matrix element:

$$\overline{|M|^2} = \frac{2g^6}{27} \Biggl\{ \left[\frac{7(p_+ \cdot q_-)}{(p_+ \cdot k)(q_- \cdot k)} + \frac{7(p_- \cdot q_+)}{(p_- \cdot k)(q_+ \cdot k)} - \frac{(p_+ \cdot q_+)}{(p_+ \cdot k)(q_+ \cdot k)} \right. $$

$$\left. - \frac{(p_- \cdot q_-)}{(p_- \cdot k)(q_- \cdot k)} + \frac{2(p_+ \cdot p_-)}{(p_+ \cdot k)(p_- \cdot k)} + \frac{2(q_+ \cdot q_-)}{(q_+ \cdot k)(q_- \cdot k)} \right] \frac{s^2 + s'^2 + u^2 + u'^2}{tt'}$$

$$+ \left[\frac{7(p_+ \cdot q_+)}{(p_+ \cdot k)(q_+ \cdot k)} + \frac{7(p_- \cdot q_-)}{(p_- \cdot k)(q_- \cdot k)} - \frac{(p_+ \cdot q_-)}{(p_+ \cdot k)(q_- \cdot k)} - \frac{(p_- \cdot q_+)}{(p_- \cdot k)(q_+ \cdot k)} \right.$$

$$\left. + \frac{2(p_+ \cdot p_-)}{(p_+ \cdot k)(p_- \cdot k)} + \frac{2(q_+ \cdot q_-)}{(q_+ \cdot k)(q_- \cdot k)} \right] \frac{s^2 + s'^2 + t^2 + t'^2}{uu'}$$

$$- \frac{1}{3} \left[\frac{(p_- \cdot q_+)}{(p_- \cdot k)(q_+ \cdot k)} + \frac{(p_+ \cdot q_-)}{(p_+ \cdot k)(q_- \cdot k)} + \frac{(p_+ \cdot q_+)}{(p_+ \cdot k)(q_+ \cdot k)} + \frac{(p_- \cdot q_-)}{(p_- \cdot k)(q_- \cdot k)} \right.$$

$$\left. - \frac{10(p_+ \cdot p_-)}{(p_+ \cdot k)(p_- \cdot k)} - \frac{10(q_+ \cdot q_-)}{(q_+ \cdot k)(q_- \cdot k)} \right] \frac{(s^2 + s'^2)(tt' + uu' - ss')}{tt'uu'} \Biggr\}.$$

$$(10.214)$$

10.32. $\bar{q}q \to \bar{q}q\gamma$

Unpolarized cross section:

$$d\sigma = \frac{\alpha_S^3}{108\pi^2 s} \left\{ \left[\frac{7(p_+ \cdot q_-)}{(p_+ \cdot k)(q_- \cdot k)} + \frac{7(p_- \cdot q_+)}{(p_- \cdot k)(q_+ \cdot k)} - \frac{(p_+ \cdot q_+)}{(p_+ \cdot k)(q_+ \cdot k)} \right. \right.$$

$$- \frac{(p_- \cdot q_-)}{(p_- \cdot k)(q_- \cdot k)} + \frac{2(p_+ \cdot p_-)}{(p_+ \cdot k)(p_- \cdot k)} + \frac{2(q_+ \cdot q_-)}{(q_+ \cdot k)(q_- \cdot k)} \Bigg] \frac{s^2 + s'^2 + u^2 + u'^2}{tt'}$$

$$+ \left[\frac{7(p_+ \cdot q_+)}{(p_+ \cdot k)(q_+ \cdot k)} + \frac{7(p_- \cdot q_-)}{(p_- \cdot k)(q_- \cdot k)} - \frac{(p_+ \cdot q_-)}{(p_+ \cdot k)(q_- \cdot k)} - \frac{(p_- \cdot q_+)}{(p_- \cdot k)(q_+ \cdot k)} \right.$$

$$\left. + \frac{2(p_+ \cdot p_-)}{(p_+ \cdot k)(p_- \cdot k)} + \frac{2(q_+ \cdot q_-)}{(q_+ \cdot k)(q_- \cdot k)} \right] \frac{s^2 + s'^2 + t^2 + t'^2}{uu'}$$

$$- \frac{1}{3} \left[\frac{(p_- \cdot q_+)}{(p_- \cdot k)(q_+ \cdot k)} + \frac{(p_+ \cdot q_-)}{(p_+ \cdot k)(q_- \cdot k)} + \frac{(p_+ \cdot q_+)}{(p_+ \cdot k)(q_+ \cdot k)} + \frac{(p_- \cdot q_-)}{(p_- \cdot k)(q_- \cdot k)} \right.$$

$$\left. - \frac{10(p_+ \cdot p_-)}{(p_+ \cdot k)(p_- \cdot k)} - \frac{10(q_+ \cdot q_-)}{(q_+ \cdot k)(q_- \cdot k)} \right] \frac{(s^2 + s'^2)(tt' + uu' - ss')}{tt'uu'} \Bigg\}$$

$$\times \delta^4(p_+ + p_- - q_+ - q_- - k) \frac{d^3\vec{q}_+ \, d^3\vec{q}_- \, d^3\vec{k}}{q_{+0} \, q_{-0} \, k_0}. \tag{10.215}$$

10.32 $\bar{q}q \to \bar{q}q\gamma$

Process (identical quark flavors):

$$\bar{q}(p_+, i) + q(p_-, j) \to \bar{q}(q_+, m) + q(q_-, n) + \gamma(k). \tag{10.216}$$

Invariants and definitions:

$$s = 2(p_+ \cdot p_-), \quad t = -2(p_+ \cdot q_+), \quad u = -2(p_+ \cdot q_-),$$

$$s' = 2(q_+ \cdot q_-), \quad t' = -2(p_- \cdot q_-), \quad u' = -2(p_- \cdot q_+),$$

$$B = 1 + \frac{E(2k_+ k_- q_{-z} + k_- k_\perp^* q_{-\perp} - k_+ k_\perp q_{-\perp}^*)}{2(q_+ \cdot k)(q_- \cdot k)},$$

$$Z_{-+} = q_{-+} q_{+-} - q_{-\perp}^* q_{+\perp}. \tag{10.217}$$

Photon polarizations:

$$\not{\epsilon}^\pm = N[\not{p}_+ \not{p}_- \not{k}(1 \mp \gamma_5) - \not{k} \not{p}_+ \not{p}_-(1 \pm \gamma_5)],$$

$$N^{-1} = E^2 [32 k_+ k_-]^{\frac{1}{2}}. \tag{10.218}$$

Nonvanishing helicity amplitudes:

$$M(+,+;+,+,+) = -\frac{8g^2 e_q T^a_{nj} T^a_{im} E q^*_{+\perp} B}{q_{+-}[2k_+ k_- q_{++} q_{-+}]^{\frac{1}{2}}},$$

$$M(+,+;+,+,-) = -\frac{4g^2 e_q T^a_{nj} T^a_{im} q^*_{+\perp}(q_+ \cdot q_-) B^*}{E q_{+-}[2k_+ k_- q_{++} q_{-+}]^{\frac{1}{2}}},$$

$$M(+,-;+,-,+) = \frac{\sqrt{2} g^2 e_q q^2_{-\perp} q^*_{+\perp} B}{q_{+-}[k_+ k_- q_{++} q^3_{-+}]^{\frac{1}{2}}} \left[\frac{Z_{-+} T^a_{ij} T^a_{nm}}{(q_+ \cdot q_-)} - 2T^a_{nj} T^a_{im} \right],$$

$$M(+,-;-,+,+) = -\frac{\sqrt{2} g^2 e_q T^a_{ij} T^a_{nm} q_{+\perp} Z_{-+} B}{(q_+ \cdot q_-)[k_+ k_- q_{++} q_{-+}]^{\frac{1}{2}}},$$

$$M(-,+;+,-,+) = -\frac{g^2 e_q T^a_{ij} T^a_{nm} q^*_{+\perp} Z_{-+} B}{q_{+-}(q_+ \cdot q_-)} \left[\frac{2q_{-+}}{k_+ k_- q_{++}} \right]^{\frac{1}{2}},$$

$$M(-,+;-,+,+) = \frac{g^2 e_q q^*_{+\perp} B}{q_{+-}} \left[\frac{2q_{++}}{k_+ k_- q_{-+}} \right]^{\frac{1}{2}} \left[\frac{Z_{-+} T^a_{ij} T^a_{nm}}{(q_+ \cdot q_-)} - 2T^a_{nj} T^a_{im} \right],$$

$$M(+,-;+,-,-) = \frac{g^2 e_q q_{+\perp} B^*}{q_{+-}} \left[\frac{2q_{++}}{k_+ k_- q_{-+}} \right]^{\frac{1}{2}} \left[\frac{Z^*_{-+} T^a_{ij} T^a_{nm}}{(q_+ \cdot q_-)} - 2T^a_{nj} T^a_{im} \right],$$

$$M(+,-;-,+,-) = -\frac{g^2 e_q T^a_{ij} T^a_{nm} q_{+\perp} Z^*_{-+} B^*}{q_{+-}(q_+ \cdot q_-)} \left[\frac{2q_{-+}}{k_+ k_- q_{++}} \right]^{\frac{1}{2}},$$

$$M(-,+;+,-,-) = -\frac{\sqrt{2} g^2 e_q T^a_{ij} T^a_{nm} q^*_{+\perp} Z^*_{-+} B^*}{(q_+ \cdot q_-)[k_+ k_- q_{++} q_{-+}]^{\frac{1}{2}}},$$

$$M(-,+;-,+,-) = \frac{\sqrt{2} g^2 e_q q^{*2}_{-\perp} q_{+\perp} B^*}{q_{+-}[k_+ k_- q_{++} q^3_{-+}]^{\frac{1}{2}}} \left[\frac{Z^*_{-+} T^a_{ij} T^a_{nm}}{(q_+ \cdot q_-)} - 2T^a_{nj} T^a_{im} \right],$$

$$M(-,-;-,-,+) = -\frac{4g^2 e_q T^a_{nj} T^a_{im} q_{+\perp}(q_+ \cdot q_-) B}{E q_{+-}[2k_+ k_- q_{++} q_{-+}]^{\frac{1}{2}}},$$

$$M(-,-;-,-,-) = -\frac{8g^2 e_q T^a_{nj} T^a_{im} E q_{+\perp} B^*}{q_{+-}[2k_+ k_- q_{++} q_{-+}]^{\frac{1}{2}}}. \qquad (10.219)$$

Squared absolute values of the nonvanishing helicity amplitudes, summed over the color degrees of freedom of the initial state and the final state:

$$|M(+,+;+,+,+)|^2 = |M(-,-;-,-,-)|^2 = 64g^4 e_q^2 \frac{E^2 |B|^2}{k_+ k_- q_{+-} q_{-+}},$$

$$|M(+,+;+,+,-)|^2 = |M(-,-;-,-,+)|^2 = 16g^4 e_q^2 \frac{(q_+ \cdot q_-)^2 |B|^2}{E^2 k_+ k_- q_{+-} q_{-+}},$$

10.33. $\bar{q}q \to \bar{q}qg$

$$|M(+,-;+,-,+)|^2 = |M(-,+;-,+,-)|^2$$

$$= 8g^4 e_q^2 \frac{q_{--}^2 |B|^2}{k_+ k_-} \left[\frac{1}{(q_+ \cdot q_-)} + \frac{2}{q_{+-}q_{-+}} + \frac{Z_{-+} + Z_{-+}^*}{3q_{+-}q_{-+}(q_+ \cdot q_-)} \right],$$

$$|M(+,-;-,+,+)|^2 = |M(-,+;+,-,-)|^2 = 8g^4 e_q^2 \frac{q_{+-}^2 |B|^2}{k_+ k_-(q_+ \cdot q_-)},$$

$$|M(-,+;+,-,+)|^2 = |M(+,-;-,+,-)|^2 = 8g^4 e_q^2 \frac{q_{-+}^2 |B|^2}{k_+ k_-(q_+ \cdot q_-)},$$

$$|M(-,+;-,+,+)|^2 = |M(+,-;+,-,-)|^2$$

$$= 8g^4 e_q^2 \frac{q_{++}^2 |B|^2}{k_+ k_-} \left[\frac{1}{(q_+ \cdot q_-)} + \frac{2}{q_{+-}q_{-+}} + \frac{Z_{-+} + Z_{-+}^*}{3q_{+-}q_{-+}(q_+ \cdot q_-)} \right]. \quad (10.220)$$

Unpolarized squared matrix element:

$$\overline{|M|^2} = -\frac{2}{9} g^4 e_q^2 \frac{(v_p - v_q)^2}{ss'tt'} \left[ss'(s^2 + s'^2) + tt'(t^2 + t'^2) \right.$$

$$\left. + uu'(u^2 + u'^2) + \frac{4}{3}(u^2 + u'^2)(ss' + tt' - uu') \right], \quad (10.221)$$

where [see eqns (4.27)]

$$v_p = \frac{p_+}{(p_+ \cdot k)} - \frac{p_-}{(p_- \cdot k)}, \qquad v_q = \frac{q_+}{(q_+ \cdot k)} - \frac{q_-}{(q_- \cdot k)}. \quad (10.222)$$

Unpolarized cross section:

$$d\sigma = -\frac{\alpha_S^2 \alpha Q_f^2}{36\pi^2} \frac{(v_p - v_q)^2}{s^2 s'tt'} \left[ss'(s^2 + s'^2) + tt'(t^2 + t'^2) \right.$$

$$\left. + uu'(u^2 + u'^2) + \frac{4}{3}(u^2 + u'^2)(ss' + tt' - uu') \right]$$

$$\times \delta^4(p_+ + p_- - q_+ - q_- - k) \frac{d^3 \vec{q}_+ \, d^3 \vec{q}_- \, d^3 \vec{k}}{q_{+0} \, q_{-0} \, k_0}. \quad (10.223)$$

10.33 $\bar{q}q \to \bar{q}qg$

Process (identical quark flavors):

$$\bar{q}(p_+, i) + q(p_-, j) \to \bar{q}(q_+, m) + q(q_-, n) + g(k, a). \quad (10.224)$$

Invariants and definitions:

$$s = 2(p_+ \cdot p_-), \quad t = -2(p_+ \cdot q_+), \quad u = -2(p_+ \cdot q_-),$$

$$s' = 2(q_+ \cdot q_-), \quad t' = -2(p_- \cdot q_-), \quad u' = -2(p_- \cdot q_+),$$

$$Z_{+-} = q_{++}q_{--} - q_{+\perp}^* q_{-\perp}, \qquad Z_{-+} = q_{-+}q_{+-} - q_{-\perp}^* q_{+\perp},$$

$$B_1 = \frac{1}{(2k_+k_-)^{\frac{1}{2}}}\left[2(T^aT^b)_{ij}T^b_{nm} + \frac{q_{-+}k_- - q_{-\perp}^*k_\perp}{(q_-\cdot k)}T^b_{ij}(T^aT^b)_{nm}\right.$$
$$\left. - \frac{q_{++}k_- - q_{+\perp}^*k_\perp}{(q_+\cdot k)}T^b_{ij}(T^bT^a)_{nm}\right],$$

$$B_1' = \frac{1}{(2k_+k_-)^{\frac{1}{2}}}\left[2(T^aT^b)_{ij}T^b_{nm} + \frac{q_{-+}k_- - q_{-\perp}k_\perp^*}{(q_-\cdot k)}T^b_{ij}(T^aT^b)_{nm}\right.$$
$$\left. - \frac{q_{++}k_- - q_{+\perp}k_\perp^*}{(q_+\cdot k)}T^b_{ij}(T^bT^a)_{nm}\right],$$

$$B_2 = \frac{1}{(2k_+k_-)^{\frac{1}{2}}}\left[2T^b_{nj}(T^aT^b)_{im} + \frac{q_{-+}k_- - q_{-\perp}^*k_\perp}{(q_-\cdot k)}(T^aT^b)_{nj}T^b_{im}\right.$$
$$\left. - \frac{q_{++}k_- - q_{+\perp}^*k_\perp}{(q_+\cdot k)}T^b_{nj}(T^bT^a)_{im}\right],$$

$$B_2' = \frac{1}{(2k_+k_-)^{\frac{1}{2}}}\left[2T^b_{nj}(T^aT^b)_{im} + \frac{q_{-+}k_- - q_{-\perp}k_\perp^*}{(q_-\cdot k)}(T^aT^b)_{nj}T^b_{im}\right.$$
$$\left. - \frac{q_{++}k_- - q_{+\perp}k_\perp^*}{(q_+\cdot k)}T^b_{nj}(T^bT^a)_{im}\right],$$

$$C_1 = \frac{1}{3}\left[-\frac{2}{k_+k_-} - \frac{(q_+\cdot q_-)}{(q_+\cdot k)(q_-\cdot k)} + \frac{7q_{+-}}{k_-(q_+\cdot k)}\right.$$
$$\left. + \frac{7q_{-+}}{k_+(q_-\cdot k)} + \frac{2q_{++}}{k_+(q_+\cdot k)} + \frac{2q_{--}}{k_-(q_-\cdot k)}\right],$$

$$C_2 = \frac{1}{3}\left[\frac{14}{k_+k_-} + \frac{7(q_+\cdot q_-)}{(q_+\cdot k)(q_-\cdot k)} - \frac{q_{+-}}{k_-(q_+\cdot k)}\right.$$
$$\left. - \frac{q_{-+}}{k_+(q_-\cdot k)} + \frac{2q_{++}}{k_+(q_+\cdot k)} + \frac{2q_{--}}{k_-(q_-\cdot k)}\right],$$

$$C_3 = \frac{1}{9}\left[\frac{2}{k_+k_-} + \frac{(q_+\cdot q_-)}{(q_+\cdot k)(q_-\cdot k)} + \frac{q_{+-}}{k_-(q_+\cdot k)}\right.$$
$$\left. + \frac{q_{-+}}{k_+(q_-\cdot k)} - \frac{10q_{++}}{k_+(q_+\cdot k)} - \frac{10q_{--}}{k_-(q_-\cdot k)}\right]. \qquad (10.225)$$

10.33. $\bar{q}q \to \bar{q}qg$

Gluon polarizations:

$$\slashed{t}^\pm = N[\slashed{k}\,\slashed{p}_+\,\slashed{p}_-(1\pm\gamma_5) + \slashed{p}_-\,\slashed{p}_+\,\slashed{k}(1\mp\gamma_5)],$$

$$N^{-1} = E^2[32k_+k_-]^{\frac{1}{2}}. \tag{10.226}$$

Nonvanishing helicity amplitudes:

$$M(+,+;+,+,+) = \frac{4g^3 B_2 E q_{+\perp}^*}{q_{+-}\sqrt{q_{++}q_{-+}}},$$

$$M(+,+;+,+,-) = -\frac{g^3 B_2' q_{-\perp}^* Z_{+-} Z_{-+}}{E q_{+-} q_{--}\sqrt{q_{++}q_{-+}^3}},$$

$$M(+,-;+,-,+) = \frac{g^3 q_{-\perp}}{\sqrt{q_{++}q_{-+}}}\left[\frac{B_1 Z_{+-}}{(q_+\cdot q_-)} + \frac{2B_2 q_{-\perp} q_{+\perp}^*}{q_{+-}q_{-+}}\right],$$

$$M(+,-;-,+,+) = \frac{g^3 B_1 q_{+\perp} Z_{-+}}{(q_+\cdot q_-)\sqrt{q_{++}q_{-+}}},$$

$$M(-,+;+,-,+) = \frac{g^3 B_1 q_{+\perp}^* Z_{-+}}{q_{+-}(q_+\cdot q_-)}\left(\frac{q_{-+}}{q_{++}}\right)^{\frac{1}{2}},$$

$$M(-,+;-,+,+) = g^3\left(\frac{q_{++}}{q_{-+}}\right)^{\frac{1}{2}}\left[\frac{B_1 q_{-\perp}^* Z_{+-}}{q_{--}(q_+\cdot q_-)} + \frac{2B_2 q_{+\perp}^*}{q_{+-}}\right],$$

$$M(+,-;+,-,-) = g^3\left(\frac{q_{++}}{q_{-+}}\right)^{\frac{1}{2}}\left[\frac{B_1' q_{-\perp} Z_{+-}^*}{q_{--}(q_+\cdot q_-)} + \frac{2B_2' q_{+\perp}}{q_{+-}}\right],$$

$$M(+,-;-,+,-) = \frac{g^3 B' q_{+\perp} Z_{-+}^*}{q_{+-}(q_+\cdot q_-)}\left(\frac{q_{-+}}{q_{++}}\right)^{\frac{1}{2}},$$

$$M(-,+;+,-,-) = \frac{g^3 B' q_{+\perp}^* Z_{-+}^*}{(q_+\cdot q_-)\sqrt{q_{++}q_{-+}}},$$

$$M(-,+;-,+,-) = \frac{g^3 q_{-\perp}^*}{\sqrt{q_{++}q_{-+}}}\left[\frac{B_1' Z_{+-}^*}{(q_+\cdot q_-)} + \frac{2B_2' q_{-\perp}^* q_{+\perp}}{q_{+-}q_{-+}}\right],$$

$$M(-,-;-,-,+) = -\frac{g^3 B_2 q_{-\perp} Z_{+-}^* Z_{-+}^*}{E q_{+-} q_{--}\sqrt{q_{++}q_{-+}^3}},$$

$$M(-,-;-,-,-) = \frac{4g^3 B_2' E q_{+\perp}}{q_{+-}\sqrt{q_{++}q_{-+}}}. \tag{10.227}$$

Squared absolute values of the nonvanishing helicity amplitudes, summed over the color degrees of freedom of the initial state and the final state:

$$|M(+,+;+,+,+)|^2 = |M(-,-;-,-,-)|^2 = \frac{16g^6 C_1 E^2}{q_{+-}q_{-+}},$$

$$|M(+,+;+,+,-)|^2 = |M(-,-;-,-,+)|^2 = \frac{4g^6 C_2 (q_+ \cdot q_-)^2}{E^2 q_{+-} q_{-+}},$$

$$|M(+,-;+,-,+)|^2 = |M(-,+;-,+,-)|^2$$

$$= 4g^6 q_{--}^2 \left[\frac{C_1}{2(q_+ \cdot q_-)} + \frac{C_2}{q_{+-} q_{-+}} - \frac{C_3 [q_{+-} q_{-+} - q_{++} q_{--} + 2(q_+ \cdot q_-)]}{2 q_{+-} q_{-+} (q_+ \cdot q_-)} \right],$$

$$|M(+,-;-,+,+)|^2 = |M(-,+;+,-,-)|^2 = \frac{2g^6 C_1 q_{+-}^2}{(q_+ \cdot q_-)},$$

$$|M(-,+;+,-,+)|^2 = |M(+,-;-,+,-)|^2 = \frac{2g^6 C_1 q_{-+}^2}{(q_+ \cdot q_-)},$$

$$|M(-,+;-,+,+)|^2 = |M(+,-;+,-,-)|^2$$

$$= 4g^6 q_{++}^2 \left[\frac{C_1}{2(q_+ \cdot q_-)} + \frac{C_2}{q_{+-} q_{-+}} - \frac{C_3 [q_{+-} q_{-+} - q_{++} q_{--} + 2(q_+ \cdot q_-)]}{2 q_{+-} q_{-+} (q_+ \cdot q_-)} \right].$$

(10.228)

Unpolarized squared matrix element:

$$\overline{|M|^2} = \frac{2g^6}{27} \left\{ \left[\frac{7(p_- \cdot q_-)}{(p_- \cdot k)(q_- \cdot k)} + \frac{7(p_+ \cdot q_+)}{(p_+ \cdot k)(q_+ \cdot k)} - \frac{(p_+ \cdot p_-)}{(p_+ \cdot k)(p_- \cdot k)} \right. \right.$$

$$\left. - \frac{(q_+ \cdot q_-)}{(q_+ \cdot k)(q_- \cdot k)} + \frac{2(p_- \cdot q_+)}{(p_- \cdot k)(q_+ \cdot k)} + \frac{2(p_+ \cdot q_-)}{(p_+ \cdot k)(q_- \cdot k)} \right] \frac{t^2 + t'^2 + u^2 + u'^2}{ss'}$$

$$+ \left[\frac{7(p_+ \cdot p_-)}{(p_+ \cdot k)(p_- \cdot k)} + \frac{7(q_+ \cdot q_-)}{(q_+ \cdot k)(q_- \cdot k)} - \frac{(p_- \cdot q_-)}{(p_- \cdot k)(q_- \cdot k)} - \frac{(p_+ \cdot q_+)}{(p_+ \cdot k)(q_+ \cdot k)} \right.$$

$$\left. + \frac{2(p_- \cdot q_+)}{(p_- \cdot k)(q_+ \cdot k)} + \frac{2(p_+ \cdot q_-)}{(p_+ \cdot k)(q_- \cdot k)} \right] \frac{s^2 + s'^2 + u^2 + u'^2}{tt'}$$

$$- \frac{1}{3} \left[\frac{(p_+ \cdot q_+)}{(p_+ \cdot k)(q_+ \cdot k)} + \frac{(p_- \cdot q_-)}{(p_- \cdot k)(q_- \cdot k)} + \frac{(p_+ \cdot p_-)}{(p_+ \cdot k)(p_- \cdot k)} + \frac{(q_+ \cdot q_-)}{(q_+ \cdot k)(q_- \cdot k)} \right.$$

$$\left. \left. - \frac{10(p_- \cdot q_+)}{(p_- \cdot k)(q_+ \cdot k)} - \frac{10(p_+ \cdot q_-)}{(p_+ \cdot k)(q_- \cdot k)} \right] \frac{(u^2 + u'^2)(ss' + tt' - uu')}{ss'tt'} \right\}.$$

(10.229)

Unpolarized cross section:

$$d\sigma = \frac{\alpha_S^3}{108\pi^2 s} \left\{ \left[\frac{7(p_- \cdot q_-)}{(p_- \cdot k)(q_- \cdot k)} + \frac{7(p_+ \cdot q_+)}{(p_+ \cdot k)(q_+ \cdot k)} - \frac{(p_+ \cdot p_-)}{(p_+ \cdot k)(p_- \cdot k)} \right. \right.$$

$$\left. - \frac{(q_+ \cdot q_-)}{(q_+ \cdot k)(q_- \cdot k)} + \frac{2(p_- \cdot q_+)}{(p_- \cdot k)(q_+ \cdot k)} + \frac{2(p_+ \cdot q_-)}{(p_+ \cdot k)(q_- \cdot k)} \right] \frac{t^2 + t'^2 + u^2 + u'^2}{ss'}$$

$$+ \left[\frac{7(p_+ \cdot p_-)}{(p_+ \cdot k)(p_- \cdot k)} + \frac{7(q_+ \cdot q_-)}{(q_+ \cdot k)(q_- \cdot k)} - \frac{(p_- \cdot q_-)}{(p_- \cdot k)(q_- \cdot k)} - \frac{(p_+ \cdot q_+)}{(p_+ \cdot k)(q_+ \cdot k)} \right.$$

$$\left. + \frac{2(p_- \cdot q_+)}{(p_- \cdot k)(q_+ \cdot k)} + \frac{2(p_+ \cdot q_-)}{(p_+ \cdot k)(q_- \cdot k)} \right] \frac{s^2 + s'^2 + u^2 + u'^2}{tt'}$$

$$- \frac{1}{3} \left[\frac{(p_+ \cdot q_+)}{(p_+ \cdot k)(q_+ \cdot k)} + \frac{(p_- \cdot q_-)}{(p_- \cdot k)(q_- \cdot k)} + \frac{(p_+ \cdot p_-)}{(p_+ \cdot k)(p_- \cdot k)} + \frac{(q_+ \cdot q_-)}{(q_+ \cdot k)(q_- \cdot k)} \right.$$

$$\left. \left. - \frac{10(p_- \cdot q_+)}{(p_- \cdot k)(q_+ \cdot k)} - \frac{10(p_+ \cdot q_-)}{(p_+ \cdot k)(q_- \cdot k)} \right] \frac{(u^2 + u'^2)(ss' + tt' - uu')}{ss'tt'} \right\}$$

$$\times \delta^4(p_+ + p_- - q_+ - q_- - k) \frac{d^3\vec{q}_+ \, d^3\vec{q}_- \, d^3\vec{k}}{q_{+0} \, q_{-0} \, k_0}. \tag{10.230}$$

10.34 $\gamma q \to \gamma \gamma q$

Process:
$$\gamma(k_1) + q(p,i) \to \gamma(k_2) + \gamma(k_3) + q(p',j). \tag{10.231}$$

Positive z-axis: along $\vec{k}_1$.
Definitions:

$$Z(p',k_i) = p'_+ k_{i-} - p'^*_\perp k_{i\perp},$$

$$Z(k_i, p') = k_{i+} p'_- - k^*_{i\perp} p'_\perp, \qquad i = 2,3. \tag{10.232}$$

Photon polarizations:

$$\slashed{\epsilon}^{*\pm}(k_1) = N_1[\slashed{p}' \slashed{p} \slashed{k}_1(1 \pm \gamma_5) - \slashed{k}_1 \slashed{p}' \slashed{p}(1 \mp \gamma_5)],$$

$$\slashed{\epsilon}^{\pm}(k_i) = N_i[\slashed{p}' \slashed{p} \slashed{k}_i(1 \mp \gamma_5) - \slashed{k}_i \slashed{p}' \slashed{p}(1 \pm \gamma_5)], \qquad i = 2,3,$$

$$N_1^{-1} = E^2[32 p'_+ p'_-]^{\frac{1}{2}},$$

$$N_i^{-1} = 4E[p'_+ k_{i+}(p' \cdot k_i)]^{\frac{1}{2}}, \qquad i = 2,3. \tag{10.233}$$

Nonvanishing helicity amplitudes:

$$M(+,+;+,+,+) = -\frac{4e_q^3 \delta_{ij} E^2 p_\perp'^*}{[Ep_-' k_{2+} k_{3+}(p'\cdot k_2)(p'\cdot k_3)]^{\frac{1}{2}}},$$

$$M(+,+;+,-,+) = -e_q^3 \delta_{ij} p_\perp'^* Z(k_2, p') \left[\frac{(p'\cdot k_2)}{E^3 p_-'^3 k_{2+} k_{3+}(p'\cdot k_3)}\right]^{\frac{1}{2}},$$

$$M(+,+;-,+,+) = -e_q^3 \delta_{ij} p_\perp'^* Z(k_3, p') \left[\frac{(p'\cdot k_3)}{E^3 p_-'^3 k_{2+} k_{3+}(p'\cdot k_2)}\right]^{\frac{1}{2}},$$

$$M(+,-;+,+,-) = -2e_q^3 \delta_{ij} p_\perp' \left[\frac{Ep_-'}{k_{2+} k_{3+}(p'\cdot k_2)(p'\cdot k_3)}\right]^{\frac{1}{2}},$$

$$M(-,+;+,-,+) = -e_q^3 \delta_{ij} p_\perp'^* Z(k_3, p') \left[\frac{k_{3+}}{Ep'^3 k_{2+}(p'\cdot k_2)(p'\cdot k_3)}\right]^{\frac{1}{2}},$$

$$M(-,+;-,+,+) = \frac{e_q^3 \delta_{ij} k_{2\perp}^* Z(p', k_2)}{k_{2-}} \left[\frac{k_{2+}}{Ep_-' k_{3+}(p'\cdot k_2)(p'\cdot k_3)}\right]^{\frac{1}{2}},$$

$$M(+,-;+,-,-) = \frac{e_q^3 \delta_{ij} k_{2\perp} Z^*(p', k_2)}{k_{2-}} \left[\frac{k_{2+}}{Ep_-' k_{3+}(p'\cdot k_2)(p'\cdot k_3)}\right]^{\frac{1}{2}},$$

$$M(+,-;-,+,-) = -e_q^3 \delta_{ij} p_\perp' Z^*(k_3, p') \left[\frac{k_{3+}}{Ep'^3 k_{2+}(p'\cdot k_2)(p'\cdot k_3)}\right]^{\frac{1}{2}},$$

$$M(-,+;-,-,+) = -2e_q^3 \delta_{ij} p_\perp'^* \left[\frac{Ep_-'}{k_{2+} k_{3+}(p'\cdot k_2)(p'\cdot k_3)}\right]^{\frac{1}{2}},$$

$$M(-,-;+,-,-) = -e_q^3 \delta_{ij} p_\perp' Z^*(k_3, p') \left[\frac{(p'\cdot k_3)}{E^3 p'^3 k_{2+} k_{3+}(p'\cdot k_2)}\right]^{\frac{1}{2}},$$

$$M(-,-;-,+,-) = -e_q^3 \delta_{ij} p_\perp' Z^*(k_2, p') \left[\frac{(p'\cdot k_2)}{E^3 p'^3 k_{2+} k_{3+}(p'\cdot k_3)}\right]^{\frac{1}{2}},$$

$$M(-,-;-,-,-) = -\frac{4e_q^3 \delta_{ij} E^2 p_\perp'}{[Ep_-' k_{2+} k_{3+}(p'\cdot k_2)(p'\cdot k_3)]^{\frac{1}{2}}}. \quad (10.234)$$

Squared absolute values of the nonvanishing helicity amplitudes, summed over the color degrees of freedom of the initial state and the final state:

$$|M(+,+;+,+,+)|^2 = |M(-,-;-,-,-)|^2 = \frac{48 e_q^6 E^3 p_+'}{k_{2+} k_{3+}(p'\cdot k_2)(p'\cdot k_3)},$$

$$|M(+,+;+,-,+)|^2 = |M(-,-;-,+,-)|^2 = \frac{6 e_q^6 p_+' (p'\cdot k_2)^2}{E^3 p_-' k_{3+}(p'\cdot k_3)},$$

10.35. $\gamma q \to g \gamma q$

$$|M(+,+;-,+,+)|^2 = |M(-,-;+,-,-)|^2 = \frac{6e_q^6 p'_+ (p' \cdot k_3)^2}{E^3 p'_- k_{2+}(p' \cdot k_2)},$$

$$|M(+,-;+,+,-)|^2 = |M(-,+;-,-,+)|^2 = \frac{12e_q^6 E p'_+ p'^2_-}{k_{2+}k_{3+}(p' \cdot k_2)(p' \cdot k_3)},$$

$$|M(-,+;+,-,+)|^2 = |M(+,-;-,+,-)|^2 = \frac{6e_q^6 p'_+ k_{3+}^2}{E p'_- k_{2+}(p' \cdot k_2)},$$

$$|M(-,+;-,+,+)|^2 = |M(+,-;+,-,-)|^2 = \frac{6e_q^6 p'_+ k_{2+}^2}{E p'_- k_{3+}(p' \cdot k_3)}. \tag{10.235}$$

Unpolarized squared matrix element:

$$\overline{|M|^2} = 2e_q^6 (p \cdot p') \frac{\sum_{i=1}^{3}(p \cdot k_i)(p' \cdot k_i)[(p \cdot k_i)^2 + (p' \cdot k_i)^2]}{\prod_{i=1}^{3}(p \cdot k_i)(p' \cdot k_i)}. \tag{10.236}$$

Unpolarized cross section:

$$d\sigma = \frac{\alpha^3 Q_f^6 (p \cdot p')}{16\pi^2 E^2} \frac{\sum_{i=1}^{3}(p \cdot k_i)(p' \cdot k_i)[(p \cdot k_i)^2 + (p' \cdot k_i)^2]}{\prod_{i=1}^{3}(p \cdot k_i)(p' \cdot k_i)}$$

$$\times \delta^4(p + k_1 - p' - k_2 - k_3) \frac{d^3\vec{p'}\, d^3\vec{k_2}\, d^3\vec{k_3}}{p'_0\, k_{20}\, k_{30}}. \tag{10.237}$$

10.35 $\gamma q \to g \gamma q$

Process:

$$\gamma(k_1) + q(p,i) \to g(k_2,b) + \gamma(k_3) + q(p',j). \tag{10.238}$$

Positive z-axis: along $\vec{k}_1$.
Definitions:

$$Z(p', k_i) = p'_+ k_{i-} - p'^{*}_\perp k_{i\perp},$$

$$Z(k_i, p') = k_{i+} p'_- - k^{*}_{i\perp} p'_\perp, \qquad i = 2,3. \tag{10.239}$$

Gluon and photon polarizations:

$$\rlap{/}{\epsilon}^{*\pm}(k_1) = N_1[\rlap{/}{p}'\rlap{/}{p}\rlap{/}{k}_1(1\pm\gamma_5) - \rlap{/}{k}_1\rlap{/}{p}'\rlap{/}{p}(1\mp\gamma_5)],$$

$$\rlap{/}{\epsilon}^{\pm}(k_2) = N_2[\rlap{/}{k}_2\rlap{/}{p}'\rlap{/}{p}(1\pm\gamma_5) + \rlap{/}{p}\rlap{/}{p}'\rlap{/}{k}_2(1\mp\gamma_5)],$$

$$\rlap{/}{\epsilon}^{\pm}(k_3) = N_3[\rlap{/}{p}'\rlap{/}{p}\rlap{/}{k}_3(1\mp\gamma_5) - \rlap{/}{k}_3\rlap{/}{p}'\rlap{/}{p}(1\pm\gamma_5)],$$

$$N_1^{-1} = E^2[32 p'_+ p'_-]^{\frac{1}{2}},$$

$$N_2^{-1} = 4E[p'_+ k_{2+}(p'\cdot k_2)]^{\frac{1}{2}},$$

$$N_3^{-1} = 4E[p'_+ k_{3+}(p'\cdot k_3)]^{\frac{1}{2}}. \qquad (10.240)$$

Nonvanishing helicity amplitudes:

$$M(+,+;+,+,+) = \frac{4ge_q^2 T_{ji}^b E^2 p'^*_\perp}{[Ep'_- k_{2+} k_{3+}(p'\cdot k_2)(p'\cdot k_3)]^{\frac{1}{2}}},$$

$$M(+,+;+,-,+) = ge_q^2 T_{ji}^b p'^*_\perp Z(k_2,p')\left[\frac{(p'\cdot k_2)}{E^3 p'^3_- k_{2+} k_{3+}(p'\cdot k_3)}\right]^{\frac{1}{2}},$$

$$M(+,+;-,+,+) = ge_q^2 T_{ji}^b p'^*_\perp Z(k_3,p')\left[\frac{(p'\cdot k_3)}{E^3 p'^3_- k_{2+} k_{3+}(p'\cdot k_2)}\right]^{\frac{1}{2}},$$

$$M(+,-;+,+,-) = 2ge_q^2 T_{ji}^b p'_-\left[\frac{Ep'_-}{k_{2+} k_{3+}(p'\cdot k_2)(p'\cdot k_3)}\right]^{\frac{1}{2}},$$

$$M(-,+;+,-,+) = ge_q^2 T_{ji}^b p'^*_\perp Z(k_3,p')\left[\frac{k_{3+}}{Ep'^3_- k_{2+}(p'\cdot k_2)(p'\cdot k_3)}\right]^{\frac{1}{2}},$$

$$M(-,+;-,+,+) = -\frac{ge_q^2 T_{ji}^b k^*_{2\perp} Z(p',k_2)}{k_{2-}}\left[\frac{k_{2+}}{Ep'_- k_{3+}(p'\cdot k_2)(p'\cdot k_3)}\right]^{\frac{1}{2}},$$

$$M(+,-;+,-,-) = -\frac{ge_q^2 T_{ji}^b k_{2\perp} Z^*(p',k_2)}{k_{2-}}\left[\frac{k_{2+}}{Ep'_- k_{3+}(p'\cdot k_2)(p'\cdot k_3)}\right]^{\frac{1}{2}},$$

$$M(+,-;-,+,-) = ge_q^2 T_{ji}^b p'_\perp Z^*(k_3,p')\left[\frac{k_{3+}}{Ep'^3_- k_{2+}(p'\cdot k_2)(p'\cdot k_3)}\right]^{\frac{1}{2}},$$

$$M(-,+;-,-,+) = 2ge_q^2 T_{ji}^b p'^*_\perp\left[\frac{Ep'_-}{k_{2+} k_{3+}(p'\cdot k_2)(p'\cdot k_3)}\right]^{\frac{1}{2}},$$

10.35. $\gamma q \to g \gamma q$

$$M(-,-;+,-,-) = g e_q^2 T_{ji}^b p'_\perp Z^*(k_3,p') \left[\frac{(p' \cdot k_3)}{E^3 p'^3_- k_{2+} k_{3+} (p' \cdot k_2)}\right]^{\frac{1}{2}},$$

$$M(-,-;-,+,-) = g e_q^2 T_{ji}^b p'_\perp Z^*(k_2,p') \left[\frac{(p' \cdot k_2)}{E^3 p'^3_- k_{2+} k_{3+} (p' \cdot k_3)}\right]^{\frac{1}{2}},$$

$$M(-,-;-,-,-) = \frac{4 g e_q^2 T_{ji}^b E^2 p'_\perp}{[E p'_- k_{2+} k_{3+} (p' \cdot k_2)(p' \cdot k_3)]^{\frac{1}{2}}}. \quad (10.241)$$

Squared absolute values of the nonvanishing helicity amplitudes, summed over the color degrees of freedom of the initial state and the final state:

$$|M(+,+;+,+,+)|^2 = |M(-,-;-,-,-)|^2 = \frac{64 g^2 e_q^4 E^3 p'_+}{k_{2+} k_{3+} (p' \cdot k_2)(p' \cdot k_3)},$$

$$|M(+,+;+,-,+)|^2 = |M(-,-;-,+,-)|^2 = \frac{8 g^2 e_q^4 p'_+ (p' \cdot k_2)^2}{E^3 p'_- k_{3+} (p' \cdot k_3)},$$

$$|M(+,+;-,+,+)|^2 = |M(-,-;+,-,-)|^2 = \frac{8 g^2 e_q^4 p'_+ (p' \cdot k_3)^2}{E^3 p'_- k_{2+} (p' \cdot k_2)},$$

$$|M(+,-;+,+,-)|^2 = |M(-,+;-,-,+)|^2 = \frac{16 g^2 e_q^4 E p'_+ p'^2_-}{k_{2+} k_{3+} (p' \cdot k_2)(p' \cdot k_3)},$$

$$|M(-,+;+,-,+)|^2 = |M(+,-;-,+,-)|^2 = \frac{8 g^2 e_q^4 p'_+ k_{3+}^2}{E p'_- k_{2+} (p' \cdot k_2)},$$

$$|M(-,+;-,+,+)|^2 = |M(+,-;+,-,-)|^2 = \frac{8 g^2 e_q^4 p'_+ k_{2+}^2}{E p'_- k_{3+} (p' \cdot k_3)}. \quad (10.242)$$

Unpolarized squared matrix element:

$$\overline{|M|^2} = \frac{8}{3} g^2 e_q^4 (p \cdot p') \frac{\sum_{i=1}^{3}(p \cdot k_i)(p' \cdot k_i)[(p \cdot k_i)^2 + (p' \cdot k_i)^2]}{\prod_{i=1}^{3}(p \cdot k_i)(p' \cdot k_i)}. \quad (10.243)$$

Unpolarized cross section:

$$d\sigma = \frac{\alpha_S \alpha^2 Q_f^4 (p \cdot p')}{12 \pi^2 E^2} \frac{\sum_{i=1}^{3}(p \cdot k_i)(p' \cdot k_i)[(p \cdot k_i)^2 + (p' \cdot k_i)^2]}{\prod_{i=1}^{3}(p \cdot k_i)(p' \cdot k_i)}$$

$$\times \delta^4(p + k_1 - p' - k_2 - k_3) \frac{d^3\vec{p}'\, d^3\vec{k}_2\, d^3\vec{k}_3}{p'_0\, k_{20}\, k_{30}}. \quad (10.244)$$

10.36 $gq \to \gamma\gamma q$

Process:
$$g(k_1,a) + q(p,i) \to \gamma(k_2) + \gamma(k_3) + q(p',j). \quad (10.245)$$

Positive z-axis: along $\vec{k}_1$.

Definitions:
$$Z(p',k_i) = p'_+ k_{i-} - p'^*_\perp k_{i\perp},$$

$$Z(k_i,p') = k_{i+} p'_- - k^*_{i\perp} p'_\perp, \qquad i=2,3. \quad (10.246)$$

Gluon and photon polarizations:

$$\not{\epsilon}^{*\pm}(k_1) = N_1[\not{p}\,\not{p}'\,\not{k}_1(1\pm\gamma_5) + \not{k}_1\,\not{p}'\,\not{p}(1\mp\gamma_5)],$$

$$\not{\epsilon}^\pm(k_2) = N_2[\not{k}_2\,\not{p}'\,\not{p}(1\pm\gamma_5) - \not{p}'\,\not{p}\,\not{k}_2(1\mp\gamma_5)],$$

$$\not{\epsilon}^\pm(k_3) = N_3[\not{k}_3\,\not{p}'\,\not{p}(1\pm\gamma_5) - \not{p}'\,\not{p}\,\not{k}_3(1\mp\gamma_5)],$$

$$N_1^{-1} = E^2[32p'_+ p'_-]^{\frac{1}{2}},$$

$$N_2^{-1} = 4E[p'_+ k_{2+}(p'\cdot k_2)]^{\frac{1}{2}},$$

$$N_3^{-1} = 4E[p'_+ k_{3+}(p'\cdot k_3)]^{\frac{1}{2}}. \quad (10.247)$$

Nonvanishing helicity amplitudes:

$$M(+,+;+,+,+) = \frac{4ge_q^2 T^a_{ji} E^2 p'^*_\perp}{[Ep'_- k_{2+} k_{3+}(p'\cdot k_2)(p'\cdot k_3)]^{\frac{1}{2}}},$$

$$M(+,+;+,-,+) = ge_q^2 T^a_{ji} p'^*_\perp Z(k_2,p') \left[\frac{(p'\cdot k_2)}{E^3 p'^3_- k_{2+} k_{3+}(p'\cdot k_3)}\right]^{\frac{1}{2}},$$

$$M(+,+;-,+,+) = ge_q^2 T^a_{ji} p'^*_\perp Z(k_3,p') \left[\frac{(p'\cdot k_3)}{E^3 p'^3_- k_{2+} k_{3+}(p'\cdot k_2)}\right]^{\frac{1}{2}},$$

$$M(+,-;+,+,-) = 2ge_q^2 T^a_{ji} p'_\perp \left[\frac{Ep'_-}{k_{2+} k_{3+}(p'\cdot k_2)(p'\cdot k_3)}\right]^{\frac{1}{2}},$$

$$M(-,+;+,-,+) = ge_q^2 T^a_{ji} p'^*_\perp Z(k_3,p') \left[\frac{k_{3+}}{Ep'^3_- k_{2+}(p'\cdot k_2)(p'\cdot k_3)}\right]^{\frac{1}{2}},$$

$$M(-,+;-,+,+) = -\frac{ge_q^2 T^a_{ji} k^*_{2\perp} Z(p',k_2)}{k_{2-}} \left[\frac{k_{2+}}{Ep'_- k_{3+}(p'\cdot k_2)(p'\cdot k_3)}\right]^{\frac{1}{2}},$$

10.36. $gq \to \gamma\gamma q$

$$M(+,-;+,-,-) = -\frac{ge_q^2 T_{ji}^a k_{2\perp} Z^*(p',k_2)}{k_{2-}} \left[\frac{k_{2+}}{Ep'_- k_{3+}(p'\cdot k_2)(p'\cdot k_3)}\right]^{\frac{1}{2}},$$

$$M(+,-;-,+,-) = ge_q^2 T_{ji}^a p'_\perp Z^*(k_3,p') \left[\frac{k_{3+}}{Ep'^3_- k_{2+}(p'\cdot k_2)(p'\cdot k_3)}\right]^{\frac{1}{2}},$$

$$M(-,+;-,-,+) = 2ge_q^2 T_{ji}^a p'^*_\perp \left[\frac{Ep'_-}{k_{2+}k_{3+}(p'\cdot k_2)(p'\cdot k_3)}\right]^{\frac{1}{2}},$$

$$M(-,-;+,-,-) = ge_q^2 T_{ji}^a p'_\perp Z^*(k_3,p') \left[\frac{(p'\cdot k_3)}{E^3 p'^3_- k_{2+}k_{3+}(p'\cdot k_2)}\right]^{\frac{1}{2}},$$

$$M(-,-;-,+,-) = ge_q^2 T_{ji}^a p'_\perp Z^*(k_2,p') \left[\frac{(p'\cdot k_2)}{E^3 p'^3_- k_{2+}k_{3+}(p'\cdot k_3)}\right]^{\frac{1}{2}},$$

$$M(-,-;-,-,-) = \frac{4ge_q^2 T_{ji}^a E^2 p'_\perp}{[Ep'_- k_{2+}k_{3+}(p'\cdot k_2)(p'\cdot k_3)]^{\frac{1}{2}}}. \qquad (10.248)$$

Squared absolute values of the nonvanishing helicity amplitudes, summed over the color degrees of freedom of the initial state and the final state:

$$|M(+,+;+,+,+)|^2 = |M(-,-;-,-,-)|^2 = \frac{64g^2 e_q^4 E^3 p'_+}{k_{2+}k_{3+}(p'\cdot k_2)(p'\cdot k_3)},$$

$$|M(+,+;+,-,+)|^2 = |M(-,-;-,+,-)|^2 = \frac{8g^2 e_q^4 p'_+ (p'\cdot k_2)^2}{E^3 p'_- k_{3+}(p'\cdot k_3)},$$

$$|M(+,+;-,+,+)|^2 = |M(-,-;+,-,-)|^2 = \frac{8g^2 e_q^4 p'_+ (p'\cdot k_3)^2}{E^3 p'_- k_{2+}(p'\cdot k_2)},$$

$$|M(+,-;+,+,-)|^2 = |M(-,+;-,-,+)|^2 = \frac{16g^2 e_q^4 E p'_+ p'^2_-}{k_{2+}k_{3+}(p'\cdot k_2)(p'\cdot k_3)},$$

$$|M(-,+;+,-,+)|^2 = |M(+,-;-,+,-)|^2 = \frac{8g^2 e_q^4 p'_+ k_{3+}^2}{Ep'_- k_{2+}(p'\cdot k_2)},$$

$$|M(-,+;-,+,+)|^2 = |M(+,-;+,-,-)|^2 = \frac{8g^2 e_q^4 p'_+ k_{2+}^2}{Ep'_- k_{3+}(p'\cdot k_3)}. \qquad (10.249)$$

Unpolarized squared matrix element:

$$\overline{|M|^2} = \frac{8}{3} g^2 e_q^4 (p\cdot p') \frac{\sum_{i=1}^{3}(p\cdot k_i)(p'\cdot k_i)[(p\cdot k_i)^2 + (p'\cdot k_i)^2]}{\prod_{i=1}^{3}(p\cdot k_i)(p'\cdot k_i)}. \qquad (10.250)$$

Unpolarized cross section:

$$d\sigma = \frac{\alpha_S \alpha^2 Q_f^4 (p \cdot p')}{12\pi^2 E^2} \frac{\sum_{i=1}^{3}(p \cdot k_i)(p' \cdot k_i)[(p \cdot k_i)^2 + (p' \cdot k_i)^2]}{\prod_{i=1}^{3}(p \cdot k_i)(p' \cdot k_i)}$$

$$\times \delta^4(p + k_1 - p' - k_2 - k_3)\frac{d^3\vec{p'}\,d^3\vec{k_2}\,d^3\vec{k_3}}{p'_0\,k_{20}\,k_{30}}. \qquad (10.251)$$

10.37 $\gamma q \to g g q$

Process:

$$\gamma(k_1) + q(p,i) \to g(k_2, b) + g(k_3, c) + q(p', j). \qquad (10.252)$$

Positive z-axis: along $\vec{k}_1$.
Definitions:

$$Z(k_i, k_j) = k_{i+}k_{j-} - k_{i\perp}^* k_{j\perp}, \qquad i, j = 2, 3,$$

$$Z(p', k_i) = p'_+ k_{i-} - p'^*_\perp k_{i\perp}, \qquad i = 2, 3,$$

$$Z(k_i, p') = k_{i+}p'_- - k_{i\perp}^* p'_\perp, \qquad i = 2, 3. \qquad (10.253)$$

Photon and gluon polarizations:

$$\not{k}_1^{*\pm} = N_1[\not{k}_1 \not{p}' \not{p}(1 \mp \gamma_5) - \not{p}' \not{p} \not{k}_1(1 \pm \gamma_5)],$$

$$N_1^{-1} = E^2[32p'_+ p'_-]^{\frac{1}{2}},$$

$$\not{k}_i^\pm = N_i[\not{k}_i \not{p} \not{p}'(1 \pm \gamma_5) + \not{p}' \not{p} \not{k}_i(1 \mp \gamma_5)],$$

$$N_i^{-1} = 4E[p'_+ k_{i+}(p' \cdot k_i)]^{\frac{1}{2}}, \qquad i = 2, 3. \qquad (10.254)$$

Nonvanishing helicity amplitudes:

$$M(+,+;+,+,+) = \frac{2g^2 e_q p'^*_\perp}{p'_+ k_{2-} k_{3-}(k_2 \cdot k_3)} \left[\frac{E^3}{p'_- k_{2+} k_{3+}(p' \cdot k_2)(p' \cdot k_3)}\right]^{\frac{1}{2}}$$

$$\times \left[k_{2-}Z(k_2,k_3)Z^*(p',k_3)(T^b T^c)_{ji} + k_{3-}Z(k_3,k_2)Z^*(p',k_2)(T^c T^b)_{ji}\right],$$

10.37. $\gamma q \to g g q$

$$M(+,+;+,-,+) = \frac{g^2 e_q p'^*_\perp Z(k_2,p')}{2p'_+ k_{2-} k_{3-}(k_2 \cdot k_3)} \left[\frac{(p' \cdot k_2)}{E^3 p'^3_- k_{2+} k_{3+}(p' \cdot k_3)}\right]^{\frac{1}{2}}$$

$$\times \left[k_{2-} Z^*(k_2,k_3) Z(p',k_3)(T^b T^c)_{ji} + k_{3-} Z^*(k_3,k_2) Z(p',k_2)(T^c T^b)_{ji}\right],$$

$$M(+,+;-,+,+) = \frac{g^2 e_q p'^*_\perp Z(k_3,p')}{2p'_+ k_{2-} k_{3-}(k_2 \cdot k_3)} \left[\frac{(p' \cdot k_3)}{E^3 p'^3_- k_{2+} k_{3+}(p' \cdot k_2)}\right]^{\frac{1}{2}}$$

$$\times \left[k_{2-} Z^*(k_2,k_3) Z(p',k_3)(T^b T^c)_{ji} + k_{3-} Z^*(k_3,k_2) Z(p',k_2)(T^c T^b)_{ji}\right],$$

$$M(+,-;+,+,-) = \frac{g^2 e_q p'_\perp}{p'_+ k_{2-} k_{3-}(k_2 \cdot k_3)} \left[\frac{E p'_-}{k_{2+} k_{3+}(p' \cdot k_2)(p' \cdot k_3)}\right]^{\frac{1}{2}}$$

$$\times \left[k_{2-} Z(k_2,k_3) Z^*(p',k_3)(T^b T^c)_{ji} + k_{3-} Z(k_3,k_2) Z^*(p',k_2)(T^c T^b)_{ji}\right],$$

$$M(-,+;+,-,+) = \frac{g^2 e_q p'^*_\perp Z(k_3,p')}{2p'_+ k_{2-} k_{3-}(k_2 \cdot k_3)} \left[\frac{k_{3+}}{E p'^3_- k_{2+}(p' \cdot k_2)(p' \cdot k_3)}\right]^{\frac{1}{2}}$$

$$\times \left[k_{2-} Z(k_2,k_3) Z^*(p',k_3)(T^b T^c)_{ji} + k_{3-} Z(k_3,k_2) Z^*(p',k_2)(T^c T^b)_{ji}\right],$$

$$M(-,+;-,+,+) = \frac{g^2 e_q p'^*_\perp Z(k_2,p')}{2p'_+ k_{2-} k_{3-}(k_2 \cdot k_3)} \left[\frac{k_{2+}}{E p'^3_- k_{3+}(p' \cdot k_2)(p' \cdot k_3)}\right]^{\frac{1}{2}}$$

$$\times \left[k_{2-} Z(k_2,k_3) Z^*(p',k_3)(T^b T^c)_{ji} + k_{3-} Z(k_3,k_2) Z^*(p',k_2)(T^c T^b)_{ji}\right],$$

$$M(+,-;+,-,-) = \frac{g^2 e_q p'_\perp Z^*(k_2,p')}{2p'_+ k_{2-} k_{3-}(k_2 \cdot k_3)} \left[\frac{k_{2+}}{E p'^3_- k_{3+}(p' \cdot k_2)(p' \cdot k_3)}\right]^{\frac{1}{2}}$$

$$\times \left[k_{2-} Z^*(k_2,k_3) Z(p',k_3)(T^b T^c)_{ji} + k_{3-} Z^*(k_3,k_2) Z(p',k_2)(T^c T^b)_{ji}\right],$$

$$M(+,-;-,+,-) = \frac{g^2 e_q p'_\perp Z^*(k_3,p')}{2p'_+ k_{2-} k_{3-}(k_2 \cdot k_3)} \left[\frac{k_{3+}}{E p'^3_- k_{2+}(p' \cdot k_2)(p' \cdot k_3)}\right]^{\frac{1}{2}}$$

$$\times \left[k_{2-} Z^*(k_2,k_3) Z(p',k_3)(T^b T^c)_{ji} + k_{3-} Z^*(k_3,k_2) Z(p',k_2)(T^c T^b)_{ji}\right],$$

$$M(-,+;-,-,+) = \frac{g^2 e_q p'^*_\perp}{p'_+ k_{2-} k_{3-}(k_2 \cdot k_3)} \left[\frac{E p'_-}{k_{2+} k_{3+}(p' \cdot k_2)(p' \cdot k_3)}\right]^{\frac{1}{2}}$$

$$\times \left[k_{2-} Z^*(k_2,k_3) Z(p',k_3)(T^b T^c)_{ji} + k_{3-} Z^*(k_3,k_2) Z(p',k_2)(T^c T^b)_{ji}\right],$$

$$M(-,-;+,-,-) = \frac{g^2 e_q p'_\perp Z^*(k_3,p')}{2p'_+ k_{2-} k_{3-}(k_2 \cdot k_3)} \left[\frac{(p' \cdot k_3)}{E^3 p'^3_- k_{2+} k_{3+}(p' \cdot k_2)}\right]^{\frac{1}{2}}$$

$$\times \left[k_{2-} Z(k_2,k_3) Z^*(p',k_3)(T^b T^c)_{ji} + k_{3-} Z(k_3,k_2) Z^*(p',k_2)(T^c T^b)_{ji}\right],$$

$$M(-,-;-,+,-) = \frac{g^2 e_q p'_\perp Z^*(k_2,p')}{2p'_+ k_{2-} k_{3-}(k_2 \cdot k_3)} \left[\frac{(p' \cdot k_2)}{E^3 p'^3_- k_{2+} k_{3+}(p' \cdot k_3)}\right]^{\frac{1}{2}}$$

$$\times \left[k_{2-} Z(k_2,k_3) Z^*(p',k_3)(T^b T^c)_{ji} + k_{3-} Z(k_3,k_2) Z^*(p',k_2)(T^c T^b)_{ji}\right],$$

$$M(-,-;-,-,-) = \frac{2g^2 e_q p'_\perp}{p'_+ k_{2-} k_{3-}(k_2 \cdot k_3)} \left[\frac{E^3}{p'_- k_{2+} k_{3+}(p' \cdot k_2)(p' \cdot k_3)}\right]^{\frac{1}{2}}$$

$$\times \left[k_{2-} Z^*(k_2,k_3) Z(p',k_3)(T^b T^c)_{ji} + k_{3-} Z^*(k_3,k_2) Z(p',k_2)(T^c T^b)_{ji}\right].$$

(10.255)

Squared absolute values of the nonvanishing helicity amplitudes, summed over the color degrees of freedom of the initial state and the final state:

$$|M(+,+;+,+,+)|^2 = |M(-,-;-,-,-)|^2$$

$$= \frac{32 g^4 e_q^2 E^3 \left[9 k_{2+}(p' \cdot k_3) + 9 k_{3+}(p' \cdot k_2) - p'_+(k_2 \cdot k_3)\right]}{3 k_{2+} k_{3+}(k_2 \cdot k_3)(p' \cdot k_2)(p' \cdot k_3)},$$

$$|M(+,+;+,-,+)|^2 = |M(-,-;-,+,-)|^2$$

$$= \frac{4 g^4 e_q^2 (p' \cdot k_2)^2 \left[9 k_{2+}(p' \cdot k_3) + 9 k_{3+}(p' \cdot k_2) - p'_+(k_2 \cdot k_3)\right]}{3 E^3 p'_- k_{3+}(k_2 \cdot k_3)(p' \cdot k_3)},$$

$$|M(+,+;-,+,+)|^2 = |M(-,-;+,-,-)|^2$$

$$= \frac{4 g^4 e_q^2 (p' \cdot k_3)^2 \left[9 k_{2+}(p' \cdot k_3) + 9 k_{3+}(p' \cdot k_2) - p'_+(k_2 \cdot k_3)\right]}{3 E^3 p'_- k_{2+}(k_2 \cdot k_3)(p' \cdot k_2)},$$

$$|M(+,-;+,+,-)|^2 = |M(-,+;-,-,+)|^2$$

$$= \frac{8 g^4 e_q^2 E p'^2_- \left[9 k_{2+}(p' \cdot k_3) + 9 k_{3+}(p' \cdot k_2) - p'_+(k_2 \cdot k_3)\right]}{3 k_{2+} k_{3+}(k_2 \cdot k_3)(p' \cdot k_2)(p' \cdot k_3)},$$

10.38. $gq \to g\gamma q$

$$|M(-,+;+,-,+)|^2 = |M(+,-;-,+,-)|^2$$

$$= \frac{4g^4 e_q^2 k_{3+}^2 \left[9k_{2+}(p' \cdot k_3) + 9k_{3+}(p' \cdot k_2) - p'_+(k_2 \cdot k_3)\right]}{3Ep'_- k_{2+}(k_2 \cdot k_3)(p' \cdot k_2)},$$

$$|M(-,+;-,+,+)|^2 = |M(+,-;+,-,-)|^2$$

$$= \frac{4g^4 e_q^2 k_{2+}^2 \left[9k_{2+}(p' \cdot k_3) + 9k_{3+}(p' \cdot k_2) - p'_+(k_2 \cdot k_3)\right]}{3Ep'_- k_{3+}(k_2 \cdot k_3)(p' \cdot k_3)}. \quad (10.256)$$

Unpolarized squared matrix element:

$$\overline{|M|^2} = \frac{4}{9}g^4 e_q^2 \frac{9(p \cdot k_2)(p' \cdot k_3) + 9(p \cdot k_3)(p' \cdot k_2) - (p \cdot p')(k_2 \cdot k_3)}{(k_2 \cdot k_3)}$$

$$\times \frac{\sum_{i=1}^{3}(p \cdot k_i)(p' \cdot k_i)[(p \cdot k_i)^2 + (p' \cdot k_i)^2]}{\prod_{i=1}^{3}(p \cdot k_i)(p' \cdot k_i)}. \quad (10.257)$$

Unpolarized cross section:

$$d\sigma = \frac{\alpha_S^2 \alpha Q_f^2}{72\pi^2 E^2} \frac{9(p \cdot k_2)(p' \cdot k_3) + 9(p \cdot k_3)(p' \cdot k_2) - (p \cdot p')(k_2 \cdot k_3)}{(k_2 \cdot k_3)}$$

$$\times \frac{\sum_{i=1}^{3}(p \cdot k_i)(p' \cdot k_i)[(p \cdot k_i)^2 + (p' \cdot k_i)^2]}{\prod_{i=1}^{3}(p \cdot k_i)(p' \cdot k_i)}$$

$$\times \delta^4(k_1 + p - k_2 - k_3 - p') \frac{d^3\vec{k}_2 \, d^3\vec{k}_3 \, d^3\vec{p}'}{k_{20} \, k_{30} \, p'_0}. \quad (10.258)$$

10.38 $gq \to g\gamma q$

Process:

$$g(k_1, a) + q(p, i) \to g(k_2, b) + \gamma(k_3) + q(p', j). \quad (10.259)$$

Positive z-axis: along $\vec{k}_1$.
Invariants and definitions:

$$s = 2(k_1 \cdot p), \qquad t = -2(k_1 \cdot k_2), \qquad u = -2(k_1 \cdot p'),$$

$$s' = 2(p' \cdot k_2), \qquad t' = -2(p \cdot p'), \qquad u' = -2(k_2 \cdot p).$$

$$Z(p', k_i) = p'_+ k_{i-} - p'^*_\perp k_{i\perp},$$

$$Z(k_i, p') = k_{i+} p'_- - k^*_{i\perp} p'_\perp, \qquad i = 2, 3. \qquad (10.260)$$

Gluon and photon polarizations:

$$\not{\epsilon}^\pm(k_1) = N_1[\not{p} \not{p}' \not{k}_1(1 \pm \gamma_5) + \not{k}_1 \not{p}' \not{p}(1 \mp \gamma_5)],$$

$$\not{\epsilon}^\pm(k_2) = N_2[\not{k}_2 \not{p}' \not{p}(1 \pm \gamma_5) + \not{p} \not{p}' \not{k}_2(1 \mp \gamma_5)],$$

$$\not{\epsilon}^\pm(k_3) = N_3[\not{p}' \not{p} \not{k}_3(1 \mp \gamma_5) - \not{k}_3 \not{p}' \not{p}(1 \pm \gamma_5)],$$

$$N_1^{-1} = E^2 [32 p'_+ p'_-]^{\frac{1}{2}},$$

$$N_2^{-1} = 4E[p'_+ k_{2+}(p' \cdot k_2)]^{\frac{1}{2}},$$

$$N_3^{-1} = 4E[p'_+ k_{3+}(p' \cdot k_3)]^{\frac{1}{2}}. \qquad (10.261)$$

Nonvanishing helicity amplitudes:

$$M(+,+;+,+,+) = -\frac{4g^2 e_q E^2 p'^*_\perp}{p'_+ k_{2-} [E p'_- k_{2+} k_{3+}(p' \cdot k_2)(p' \cdot k_3)]^{\frac{1}{2}}}$$
$$\times \left[Z^*(p', k_2)(T^a T^b)_{ji} + p'_\perp k^*_{2\perp}(T^b T^a)_{ji} \right],$$

$$M(+,+;+,-,+) = -\frac{g^2 e_q p'^*_\perp Z(p', k_2)}{p'_+ k_{2-}} \left[\frac{(p' \cdot k_2)}{E^3 p'^3_- k_{2+} k_{3+}(p' \cdot k_3)} \right]^{\frac{1}{2}}$$
$$\times \left[Z(k_2, p')(T^a T^b)_{ji} - p'_- k_{2+}(T^b T^a)_{ji} \right],$$

$$M(+,+;-,+,+) = -\frac{g^2 e_q p'^*_\perp Z(k_3, p')}{p'_+ k_{2-}} \left[\frac{(p' \cdot k_3)}{E^3 p'^3_- k_{2+} k_{3+}(p' \cdot k_2)} \right]^{\frac{1}{2}}$$
$$\times \left[Z(p', k_2)(T^a T^b)_{ji} + p'^*_\perp k_{2\perp}(T^b T^a)_{ji} \right],$$

$$M(+,-;+,+,-) = -\frac{2g^2 e_q p'_\perp}{p'_+ k_{2-}} \left[\frac{E p'_-}{k_{2+} k_{3+}(p' \cdot k_2)(p' \cdot k_3)} \right]^{\frac{1}{2}}$$
$$\times \left[Z^*(p', k_2)(T^a T^b)_{ji} + p'_\perp k^*_{2\perp}(T^b T^a)_{ji} \right],$$

10.38. $gq \to g\gamma q$

$$M(-,+;+,-,+) = -\frac{g^2 e_q p'^*_\perp Z(k_3,p')}{p'_+ k_{2-}} \left[\frac{k_{3+}}{Ep'^3_- k_{2+}(p'\cdot k_2)(p'\cdot k_3)}\right]^{\frac{1}{2}}$$
$$\times \left[Z^*(p',k_2)(T^aT^b)_{ji} + p'_\perp k^*_{2\perp}(T^bT^a)_{ji}\right],$$

$$M(-,+;-,+,+) = \frac{g^2 e_q k^*_{2\perp} Z(k_2,p')}{k_{2-}[Ep'^3_- k_{2+}k_{3+}(p'\cdot k_2)(p'\cdot k_3)]^{\frac{1}{2}}}$$
$$\times \left[Z^*(k_2,p')(T^aT^b)_{ji} - p'_- k_{2+}(T^bT^a)_{ji}\right],$$

$$M(+,-;+,-,-) = \frac{g^2 e_q k_{2\perp} Z^*(k_2,p')}{k_{2-}[Ep'^3_- k_{2+}k_{3+}(p'\cdot k_2)(p'\cdot k_3)]^{\frac{1}{2}}}$$
$$\times \left[Z(k_2,p')(T^aT^b)_{ji} - p'_- k_{2+}(T^bT^a)_{ji}\right],$$

$$M(+,-;-,+,-) = -\frac{g^2 e_q p'_\perp Z^*(k_3,p')}{p'_+ k_{2-}} \left[\frac{k_{3+}}{Ep'^3_- k_{2+}(p'\cdot k_2)(p'\cdot k_3)}\right]^{\frac{1}{2}}$$
$$\times \left[Z(p',k_2)(T^aT^b)_{ji} + p'^*_\perp k_{2\perp}(T^bT^a)_{ji}\right],$$

$$M(-,+;-,-,+) = -\frac{2g^2 e_q p'^*_\perp}{p'_+ k_{2-}} \left[\frac{Ep'_-}{k_{2+}k_{3+}(p'\cdot k_2)(p'\cdot k_3)}\right]^{\frac{1}{2}}$$
$$\times \left[Z(p',k_2)(T^aT^b)_{ji} + p'^*_\perp k_{2\perp}(T^bT^a)_{ji}\right],$$

$$M(-,-;+,-,-) = -\frac{g^2 e_q p'_\perp Z^*(k_3,p')}{p'_+ k_{2-}} \left[\frac{(p'\cdot k_3)}{E^3 p'^3_- k_{2+}k_{3+}(p'\cdot k_2)}\right]^{\frac{1}{2}}$$
$$\times \left[Z^*(p',k_2)(T^aT^b)_{ji} + p'_\perp k^*_{2\perp}(T^bT^a)_{ji}\right],$$

$$M(-,-;-,+,-) = -\frac{g^2 e_q p'_\perp Z^*(p',k_2)}{p'_+ k_{2-}} \left[\frac{(p'\cdot k_2)}{E^3 p'^3_- k_{2+}k_{3+}(p'\cdot k_3)}\right]^{\frac{1}{2}}$$
$$\times \left[Z^*(k_2,p')(T^aT^b)_{ji} - p'_- k_{2+}(T^bT^a)_{ji}\right],$$

$$M(-,-;-,-,-) = -\frac{4g^2 e_q E^2 p'_\perp}{p'_+ k_{2-}[Ep'_- k_{2+}k_{3+}(p'\cdot k_2)(p'\cdot k_3)]^{\frac{1}{2}}}$$
$$\times \left[Z(p',k_2)(T^aT^b)_{ji} + p'^*_\perp k_{2\perp}(T^bT^a)_{ji}\right].$$

(10.262)

Squared absolute values of the nonvanishing helicity amplitudes, summed over the color degrees of freedom of the initial state and the final state:

$$|M(+,+;+,+,+)|^2 = |M(-,-;-,-,-)|^2$$

$$= \frac{32g^4 e_q^2 E^3[18(p' \cdot k_2) + 9p'_- k_{2+} - p'_+ k_{2-}]}{3k_{2+}k_{2-}k_{3+}(p' \cdot k_2)(p' \cdot k_3)},$$

$$|M(+,+;+,-,+)|^2 = |M(-,-;-,+,-)|^2$$

$$= \frac{4g^4 e_q^2 (p' \cdot k_2)^2 [18(p' \cdot k_2) + 9p'_- k_{2+} - p'_+ k_{2-}]}{3E^3 p'_- k_{2-} k_{3+}(p' \cdot k_3)},$$

$$|M(+,+;-,+,+)|^2 = |M(-,-;+,-,-)|^2$$

$$= \frac{4g^4 e_q^2 (p' \cdot k_3)^2 [18(p' \cdot k_2) + 9p'_- k_{2+} - p'_+ k_{2-}]}{3E^3 p'_- k_{2+} k_{2-}(p' \cdot k_2)},$$

$$|M(+,-;+,+,-)|^2 = |M(-,+;-,-,+)|^2$$

$$= \frac{8g^4 e_q^2 E p'^2_-[18(p' \cdot k_2) + 9p'_- k_{2+} - p'_+ k_{2-}]}{3k_{2+}k_{2-}k_{3+}(p' \cdot k_2)(p' \cdot k_3)},$$

$$|M(-,+;+,-,+)|^2 = |M(+,-;-,+,-)|^2$$

$$= \frac{4g^4 e_q^2 k_{3+}^2[18(p' \cdot k_2) + 9p'_- k_{2+} - p'_+ k_{2-}]}{3E p'_- k_{2+} k_{2-}(p' \cdot k_2)},$$

$$|M(-,+;-,+,+)|^2 = |M(+,-;+,-,-)|^2$$

$$= \frac{4g^4 e_q^2 k_{2+}^2[18(p' \cdot k_2) + 9p'_- k_{2+} - p'_+ k_{2-}]}{3E p'_- k_{2-} k_{3+}(p' \cdot k_3)}.$$

(10.263)

Unpolarized squared matrix element:

$$\overline{|M|^2} = \frac{g^4 e_q^2 (9ss' + 9uu' - tt')}{72(k_1 \cdot k_2)} \frac{\sum_{i=1}^{3}(p \cdot k_i)(p' \cdot k_i)[(p \cdot k_i)^2 + (p' \cdot k_i)^2]}{\prod_{i=1}^{3}(p \cdot k_i)(p' \cdot k_i)}.$$

(10.264)

Unpolarized cross section:

$$d\sigma = \frac{\alpha_S^2 \alpha Q_f^2}{576\pi^2} \frac{9ss' + 9uu' - tt'}{s(k_1 \cdot k_2)} \frac{\sum_{i=1}^{3}(p \cdot k_i)(p' \cdot k_i)[(p \cdot k_i)^2 + (p' \cdot k_i)^2]}{\prod_{i=1}^{3}(p \cdot k_i)(p' \cdot k_i)}$$

$$\times \delta^4(p + k_1 - p' - k_2 - k_3) \frac{d^3\vec{p}' \, d^3\vec{k}_2 \, d^3\vec{k}_3}{p'_0 \, k_{20} \, k_{30}}. \qquad (10.265)$$

10.39 $gq \to ggq$

Process:

$$g(k_1, a) + q(p, i) \to g(k_2, b) + g(k_3, c) + q(p', j). \qquad (10.266)$$

Positive z-axis: along $\vec{k}_1$.
Definitions:

$$Z(k_i, k_j) = k_{i+} k_{j-} - k_{i\perp}^* k_{j\perp}, \qquad i,j = 2,3,$$

$$Z(p', k_i) = p'_+ k_{i-} - p'^*_\perp k_{i\perp}, \qquad i = 2,3,$$

$$Z(k_i, p') = k_{i+} p'_- - k_{i\perp}^* p'_\perp, \qquad i = 2,3,$$

$$F = \frac{40 p'_+}{9} - 4\left[\frac{p'_- k_{2+} + 2(p' \cdot k_2)}{k_{2-}} + \frac{p'_- k_{3+} + 2(p' \cdot k_3)}{k_{3-}}\right.$$

$$+ \frac{k_{2+}(p' \cdot k_3) + k_{3+}(p' \cdot k_2)}{(k_2 \cdot k_3)}\right] + \frac{36}{p'_+}\left[\frac{k_{2+}(p' \cdot k_2)[p'_- k_{3+} + 2(p' \cdot k_3)]}{k_{2-}(k_2 \cdot k_3)}\right.$$

$$+ \frac{k_{3+}(p' \cdot k_3)[p'_- k_{2+} + 2(p' \cdot k_2)]}{k_{3-}(k_2 \cdot k_3)} + \frac{2 p'_-[k_{2+}(p' \cdot k_3) + k_{3+}(p' \cdot k_2)]}{k_{2-} k_{3-}}\right].$$

$$(10.267)$$

Gluon polarizations:

$$\not{\epsilon}_1^{*\pm} = N_1[\not{p} \not{p}' \not{k}_1(1 \pm \gamma_5) + \not{k}_1 \not{p}' \not{p}(1 \mp \gamma_5)],$$

$$N_1^{-1} = E^2[32 p'_+ p'_-]^{\frac{1}{2}},$$

$$\not{\epsilon}_i^{\pm} = N_i[\not{k}_i \not{p} \not{p}'(1 \pm \gamma_5) + \not{p}' \not{p} \not{k}_i(1 \mp \gamma_5)],$$

$$N_i^{-1} = 4E[p'_+ k_{i+}(p' \cdot k_i)]^{\frac{1}{2}}, \qquad i = 2,3. \qquad (10.268)$$

Nonvanishing helicity amplitudes:

$$M(+,+;+,+,+) = \frac{2g^3 p'^*_\perp}{p'^2_+ k_{2-} k_{3-}} \left[\frac{E^3}{p'_- k_{2+} k_{3+} (p' \cdot k_2)(p' \cdot k_3)}\right]^{\frac{1}{2}}$$

$$\times \Biggl\{ (T^a T^b T^c)_{ji} \frac{Z(k_2, k_3) Z^*(p', k_2) Z^*(p', k_3)}{(k_2 \cdot k_3)}$$

$$+ (T^a T^c T^b)_{ji} \frac{Z(k_3, k_2) Z^*(p', k_2) Z^*(p', k_3)}{(k_2 \cdot k_3)}$$

$$+ (T^b T^a T^c)_{ji} 2 p'_\perp k^*_{2\perp} Z^*(p', k_3) - (T^b T^c T^a)_{ji} \frac{p'_\perp k^*_{2\perp} Z(k_3, k_2) Z^*(p', k_3)}{(k_2 \cdot k_3)}$$

$$+ (T^c T^a T^b)_{ji} 2 p'_\perp k^*_{3\perp} Z^*(p', k_2) - (T^c T^b T^a)_{ji} \frac{p'_\perp k^*_{3\perp} Z(k_2, k_3) Z^*(p', k_2)}{(k_2 \cdot k_3)} \Biggr\},$$

$$M(+,+;+,-,+) = \frac{g^3 p'^*_\perp Z(k_2, p')}{2 p'^2_+ k_{2-} k_{3-}} \left[\frac{(p' \cdot k_2)}{E^3 p'^3_- k_{2+} k_{3+} (p' \cdot k_3)}\right]^{\frac{1}{2}}$$

$$\times \Biggl\{ (T^a T^b T^c)_{ji} \frac{Z^*(k_2, k_3) Z(p', k_2) Z(p', k_3)}{(k_2 \cdot k_3)}$$

$$+ (T^a T^c T^b)_{ji} \frac{Z^*(k_3, k_2) Z(p', k_2) Z(p', k_3)}{(k_2 \cdot k_3)}$$

$$+ (T^b T^a T^c)_{ji} 2 p'^*_\perp k_{2\perp} Z(p', k_3) - (T^b T^c T^a)_{ji} \frac{p'^*_\perp k_{2\perp} Z^*(k_3, k_2) Z(p', k_3)}{(k_2 \cdot k_3)}$$

$$+ (T^c T^a T^b)_{ji} 2 p'^*_\perp k_{3\perp} Z(p', k_2) - (T^c T^b T^a)_{ji} \frac{p'^*_\perp k_{3\perp} Z^*(k_2, k_3) Z(p', k_2)}{(k_2 \cdot k_3)} \Biggr\},$$

$$M(+,+;-,+,+) = \frac{g^3 p'^*_\perp Z(k_3, p')}{2 p'^2_+ k_{2-} k_{3-}} \left[\frac{(p' \cdot k_3)}{E^3 p'^3_- k_{2+} k_{3+} (p' \cdot k_2)}\right]^{\frac{1}{2}}$$

$$\times \Biggl\{ (T^a T^b T^c)_{ji} \frac{Z^*(k_2, k_3) Z(p', k_2) Z(p', k_3)}{(k_2 \cdot k_3)}$$

$$+ (T^a T^c T^b)_{ji} \frac{Z^*(k_3, k_2) Z(p', k_2) Z(p', k_3)}{(k_2 \cdot k_3)}$$

$$+ (T^b T^a T^c)_{ji} 2 p'^*_\perp k_{2\perp} Z(p', k_3) - (T^b T^c T^a)_{ji} \frac{p'^*_\perp k_{2\perp} Z^*(k_3, k_2) Z(p', k_3)}{(k_2 \cdot k_3)}$$

$$+ (T^c T^a T^b)_{ji} 2 p'^*_\perp k_{3\perp} Z(p', k_2) - (T^c T^b T^a)_{ji} \frac{p'^*_\perp k_{3\perp} Z^*(k_2, k_3) Z(p', k_2)}{(k_2 \cdot k_3)} \Biggr\},$$

10.39. $gq \to ggq$

$$M(+,-;+,+,-) = \frac{g^3 p'_\perp}{p'^2_+ k_{2-} k_{3-}} \left[\frac{E p'_-}{k_{2+} k_{3+} (p' \cdot k_2)(p' \cdot k_3)} \right]^{\frac{1}{2}}$$

$$\times \Bigg\{ (T^a T^b T^c)_{ji} \frac{Z(k_2,k_3) Z^*(p',k_2) Z^*(p',k_3)}{(k_2 \cdot k_3)}$$

$$+ (T^a T^c T^b)_{ji} \frac{Z(k_3,k_2) Z^*(p',k_2) Z^*(p',k_3)}{(k_2 \cdot k_3)}$$

$$+ (T^b T^a T^c)_{ji} 2 p'_\perp k^*_{2\perp} Z^*(p',k_3) - (T^b T^c T^a)_{ji} \frac{p'_\perp k^*_{2\perp} Z(k_3,k_2) Z^*(p',k_3)}{(k_2 \cdot k_3)}$$

$$+ (T^c T^a T^b)_{ji} 2 p'_\perp k^*_{3\perp} Z^*(p',k_2) - (T^c T^b T^a)_{ji} \frac{p'_\perp k^*_{3\perp} Z(k_2,k_3) Z^*(p',k_2)}{(k_2 \cdot k_3)} \Bigg\},$$

$$M(-,+;+,-,+) = \frac{g^3 p'^*_\perp Z(k_3,p')}{2 p'^2_+ k_{2-} k_{3-}} \left[\frac{k_{3+}}{E p'^3_- k_{2+} (p' \cdot k_2)(p' \cdot k_3)} \right]^{\frac{1}{2}}$$

$$\times \Bigg\{ (T^a T^b T^c)_{ji} \frac{Z(k_2,k_3) Z^*(p',k_2) Z^*(p',k_3)}{(k_2 \cdot k_3)}$$

$$+ (T^a T^c T^b)_{ji} \frac{Z(k_3,k_2) Z^*(p',k_2) Z^*(p',k_3)}{(k_2 \cdot k_3)}$$

$$+ (T^b T^a T^c)_{ji} 2 p'_\perp k^*_{2\perp} Z^*(p',k_3) - (T^b T^c T^a)_{ji} \frac{p'_\perp k^*_{2\perp} Z(k_3,k_2) Z^*(p',k_3)}{(k_2 \cdot k_3)}$$

$$+ (T^c T^a T^b)_{ji} 2 p'_\perp k^*_{3\perp} Z^*(p',k_2) - (T^c T^b T^a)_{ji} \frac{p'_\perp k^*_{3\perp} Z(k_2,k_3) Z^*(p',k_2)}{(k_2 \cdot k_3)} \Bigg\},$$

$$M(-,+;-,+,+) = \frac{g^3 p'^*_\perp Z(k_2,p')}{2 p'^2_+ k_{2-} k_{3-}} \left[\frac{k_{2+}}{E p'^3_- k_{3+} (p' \cdot k_2)(p' \cdot k_3)} \right]^{\frac{1}{2}}$$

$$\times \Bigg\{ (T^a T^b T^c)_{ji} \frac{Z(k_2,k_3) Z^*(p',k_2) Z^*(p',k_3)}{(k_2 \cdot k_3)}$$

$$+ (T^a T^c T^b)_{ji} \frac{Z(k_3,k_2) Z^*(p',k_2) Z^*(p',k_3)}{(k_2 \cdot k_3)}$$

$$+ (T^b T^a T^c)_{ji} 2 p'_\perp k^*_{2\perp} Z^*(p',k_3) - (T^b T^c T^a)_{ji} \frac{p'_\perp k^*_{2\perp} Z(k_3,k_2) Z^*(p',k_3)}{(k_2 \cdot k_3)}$$

$$+ (T^c T^a T^b)_{ji} 2 p'_\perp k^*_{3\perp} Z^*(p',k_2) - (T^c T^b T^a)_{ji} \frac{p'_\perp k^*_{3\perp} Z(k_2,k_3) Z^*(p',k_2)}{(k_2 \cdot k_3)} \Bigg\},$$

$$M(+,-;+,-,-) = \frac{g^3 p'_\perp Z^*(k_2,p')}{2p'^2_+ k_{2-} k_{3-}} \left[\frac{k_{2+}}{E p'^3_- k_{3+}(p' \cdot k_2)(p' \cdot k_3)} \right]^{\frac{1}{2}}$$

$$\times \Bigg\{ (T^a T^b T^c)_{ji} \frac{Z^*(k_2,k_3)Z(p',k_2)Z(p',k_3)}{(k_2 \cdot k_3)}$$

$$+ (T^a T^c T^b)_{ji} \frac{Z^*(k_3,k_2)Z(p',k_2)Z(p',k_3)}{(k_2 \cdot k_3)}$$

$$+ (T^b T^a T^c)_{ji} 2 p'^*_\perp k_{2\perp} Z(p',k_3) - (T^b T^c T^a)_{ji} \frac{p'^*_\perp k_{2\perp} Z^*(k_3,k_2) Z(p',k_3)}{(k_2 \cdot k_3)}$$

$$+ (T^c T^a T^b)_{ji} 2 p'^*_\perp k_{3\perp} Z(p',k_2) - (T^c T^b T^a)_{ji} \frac{p'^*_\perp k_{3\perp} Z^*(k_2,k_3) Z(p',k_2)}{(k_2 \cdot k_3)} \Bigg\},$$

$$M(+,-;-,+,-) = \frac{g^3 p'_\perp Z^*(k_3,p')}{2p'^2_+ k_{2-} k_{3-}} \left[\frac{k_{3+}}{E p'^3_- k_{2+}(p' \cdot k_2)(p' \cdot k_3)} \right]^{\frac{1}{2}}$$

$$\times \Bigg\{ (T^a T^b T^c)_{ji} \frac{Z^*(k_2,k_3)Z(p',k_2)Z(p',k_3)}{(k_2 \cdot k_3)}$$

$$+ (T^a T^c T^b)_{ji} \frac{Z^*(k_3,k_2)Z(p',k_2)Z(p',k_3)}{(k_2 \cdot k_3)}$$

$$+ (T^b T^a T^c)_{ji} 2 p'^*_\perp k_{2\perp} Z(p',k_3) - (T^b T^c T^a)_{ji} \frac{p^*_\perp k_{2\perp} Z^*(k_3,k_2) Z(p',k_3)}{(k_2 \cdot k_3)}$$

$$+ (T^c T^a T^b)_{ji} 2 p'^*_\perp k_{3\perp} Z(p',k_2) - (T^c T^b T^a)_{ji} \frac{p'^*_\perp k_{3\perp} Z^*(k_2,k_3) Z(p',k_2)}{(k_2 \cdot k_3)} \Bigg\},$$

$$M(-,+;-,-,+) = \frac{g^3 p'^*_\perp}{p'^2_+ k_{2-} k_{3-}} \left[\frac{E p'_-}{k_{2+} k_{3+}(p' \cdot k_2)(p' \cdot k_3)} \right]^{\frac{1}{2}}$$

$$\times \Bigg\{ (T^a T^b T^c)_{ji} \frac{Z^*(k_2,k_3)Z(p',k_2)Z(p',k_3)}{(k_2 \cdot k_3)}$$

$$+ (T^a T^c T^b)_{ji} \frac{Z^*(k_3,k_2)Z(p',k_2)Z(p',k_3)}{(k_2 \cdot k_3)}$$

$$+ (T^b T^a T^c)_{ji} 2 p'^*_\perp k_{2\perp} Z(p',k_3) - (T^b T^c T^a)_{ji} \frac{p'^*_\perp k_{2\perp} Z^*(k_3,k_2) Z(p',k_3)}{(k_2 \cdot k_3)}$$

$$+ (T^c T^a T^b)_{ji} 2 p'^*_\perp k_{3\perp} Z(p',k_2) - (T^c T^b T^a)_{ji} \frac{p'^*_\perp k_{3\perp} Z^*(k_2,k_3) Z(p',k_2)}{(k_2 \cdot k_3)} \Bigg\},$$

10.39. $gq \to ggq$

$$M(-,-;+,-,-) = \frac{g^3 p'_\perp Z^*(k_3, p')}{2p'^2_+ k_{2-} k_{3-}} \left[\frac{(p' \cdot k_3)}{E^3 p'^3_- k_{2+} k_{3+} (p' \cdot k_2)} \right]^{\frac{1}{2}}$$

$$\times \left\{ (T^a T^b T^c)_{ji} \frac{Z(k_2, k_3) Z^*(p', k_2) Z^*(p', k_3)}{(k_2 \cdot k_3)} \right.$$

$$+ (T^a T^c T^b)_{ji} \frac{Z(k_3, k_2) Z^*(p', k_2) Z^*(p', k_3)}{(k_2 \cdot k_3)}$$

$$+ (T^b T^a T^c)_{ji} 2 p'_\perp k^*_{2\perp} Z^*(p', k_3) - (T^b T^c T^a)_{ji} \frac{p'_\perp k^*_{2\perp} Z(k_3, k_2) Z^*(p', k_3)}{(k_2 \cdot k_3)}$$

$$+ \left. (T^c T^a T^b)_{ji} 2 p'_\perp k^*_{3\perp} Z^*(p', k_2) - (T^c T^b T^a)_{ji} \frac{p'_\perp k^*_{3\perp} Z(k_2, k_3) Z^*(p', k_2)}{(k_2 \cdot k_3)} \right\},$$

$$M(-,-;-,+,-) = \frac{g^3 p'_\perp Z^*(k_2, p')}{2p'^2_+ k_{2-} k_{3-}} \left[\frac{(p' \cdot k_2)}{E^3 p'^3_- k_{2+} k_{3+} (p' \cdot k_3)} \right]^{\frac{1}{2}}$$

$$\times \left\{ (T^a T^b T^c)_{ji} \frac{Z(k_2, k_3) Z^*(p', k_2) Z^*(p', k_3)}{(k_2 \cdot k_3)} \right.$$

$$+ (T^a T^c T^b)_{ji} \frac{Z(k_3, k_2) Z^*(p', k_2) Z^*(p', k_3)}{(k_2 \cdot k_3)}$$

$$+ (T^b T^a T^c)_{ji} 2 p'_\perp k^*_{2\perp} Z^*(p', k_3) - (T^b T^c T^a)_{ji} \frac{p'_\perp k^*_{2\perp} Z(k_3, k_2) Z^*(p', k_3)}{(k_2 \cdot k_3)}$$

$$+ \left. (T^c T^a T^b)_{ji} 2 p'_\perp k^*_{3\perp} Z^*(p', k_2) - (T^c T^b T^a)_{ji} \frac{p'_\perp k^*_{3\perp} Z(k_2, k_3) Z^*(p', k_2)}{(k_2 \cdot k_3)} \right\},$$

$$M(-,-;-,-,-) = \frac{2 g^3 p'_\perp}{p'^2_+ k_{2-} k_{3-}} \left[\frac{E^3}{p'_- k_{2+} k_{3+} (p' \cdot k_2)(p' \cdot k_3)} \right]^{\frac{1}{2}}$$

$$\times \left\{ (T^a T^b T^c)_{ji} \frac{Z^*(k_2, k_3) Z(p', k_2) Z(p', k_3)}{(k_2 \cdot k_3)} \right.$$

$$+ (T^a T^c T^b)_{ji} \frac{Z^*(k_3, k_2) Z(p', k_2) Z(p', k_3)}{(k_2 \cdot k_3)}$$

$$+ (T^b T^a T^c)_{ji} 2 p'^*_\perp k_{2\perp} Z(p', k_3) - (T^b T^c T^a)_{ji} \frac{p'^*_\perp k_{2\perp} Z(k_3, k_2) Z(p', k_3)}{(k_2 \cdot k_3)}$$

$$+ \left. (T^c T^a T^b)_{ji} 2 p'^*_\perp k_{3\perp} Z(p', k_2) - (T^c T^b T^a)_{ji} \frac{p'^*_\perp k_{3\perp} Z^*(k_2, k_3) Z(p', k_2)}{(k_2 \cdot k_3)} \right\}.$$

(10.269)

Squared absolute values of the nonvanishing helicity amplitudes, summed over the color degrees of freedom of the initial state and the final state:

$$|M(+,+;+,+,+)|^2 = |M(-,-;-,-,-)|^2 = \frac{4g^6 E^3 F}{k_{2+}k_{3+}(p'\cdot k_2)(p'\cdot k_3)},$$

$$|M(+,+;+,-,+)|^2 = |M(-,-;-,+,-)|^2 = \frac{g^6 (p'\cdot k_2)^2 F}{2E^3 p'_- k_{3+}(p'\cdot k_3)},$$

$$|M(+,+;-,+,+)|^2 = |M(-,-;+,-,-)|^2 = \frac{g^6 (p'\cdot k_3)^2 F}{2E^3 p'_- k_{2+}(p'\cdot k_2)},$$

$$|M(+,-;+,+,-)|^2 = |M(-,+;-,-,+)|^2 = \frac{g^6 E p'^2_- F}{k_{2+}k_{3+}(p'\cdot k_2)(p'\cdot k_3)},$$

$$|M(-,+;+,-,+)|^2 = |M(+,-;-,+,-)|^2 = \frac{g^6 k^2_{3+} F}{2E p'_- k_{2+}(p'\cdot k_2)},$$

$$|M(-,+;-,+,+)|^2 = |M(+,-;+,-,-)|^2 = \frac{g^6 k^2_{2+} F}{2E p'_- k_{3+}(p'\cdot k_3)}.$$

(10.270)

Unpolarized squared matrix element:

$$\overline{|M|^2} = \frac{g^6}{108} \frac{\sum_{i=1}^{3}(p\cdot k_i)(p'\cdot k_i)[(p\cdot k_i)^2 + (p'\cdot k_i)^2]}{\prod_{i=1}^{3}(p\cdot k_i)(p'\cdot k_i)}$$

$$\times \left\{ 10(p\cdot p') - 9\left[\frac{(p\cdot k_1)(p'\cdot k_2) + (p\cdot k_2)(p'\cdot k_1)}{(k_1\cdot k_2)} \right.\right.$$

$$+ \frac{(p\cdot k_2)(p'\cdot k_3) + (p\cdot k_3)(p'\cdot k_2)}{(k_2\cdot k_3)}$$

$$\left.+ \frac{(p\cdot k_3)(p'\cdot k_1) + (p\cdot k_1)(p'\cdot k_3)}{(k_1\cdot k_3)} \right] + \frac{81}{(p\cdot p')}$$

$$\times \left[\frac{(p\cdot k_1)(p'\cdot k_1)[(p\cdot k_2)(p'\cdot k_3) + (p\cdot k_3)(p'\cdot k_2)]}{(k_1\cdot k_2)(k_1\cdot k_3)} \right.$$

$$+ \frac{(p\cdot k_2)(p'\cdot k_2)[(p\cdot k_3)(p'\cdot k_1) + (p\cdot k_1)(p'\cdot k_3)]}{(k_1\cdot k_2)(k_2\cdot k_3)}$$

$$\left.\left.+ \frac{(p\cdot k_3)(p'\cdot k_3)[(p\cdot k_1)(p'\cdot k_2) + (p\cdot k_2)(p'\cdot k_1)]}{(k_1\cdot k_3)(k_2\cdot k_3)} \right] \right\}.$$

(10.271)

10.40. $\gamma\bar{q} \to \gamma\gamma\bar{q}$

Unpolarized cross section:

$$d\sigma = \frac{\alpha_S^3}{1728\pi^2} \frac{\sum_{i=1}^{3}(p\cdot k_i)(p'\cdot k_i)[(p\cdot k_i)^2 + (p'\cdot k_i)^2]}{(p\cdot k_1)\prod_{i=1}^{3}(p\cdot k_i)(p'\cdot k_i)} \Bigg\{ 10(p\cdot p')$$

$$-9\Bigg[\frac{(p\cdot k_1)(p'\cdot k_2) + (p\cdot k_2)(p'\cdot k_1)}{(k_1\cdot k_2)} + \frac{(p\cdot k_2)(p'\cdot k_3) + (p\cdot k_3)(p'\cdot k_2)}{(k_2\cdot k_3)}$$

$$+\frac{(p\cdot k_3)(p'\cdot k_1) + (p\cdot k_1)(p'\cdot k_3)}{(k_1\cdot k_3)}\Bigg]$$

$$+\frac{81}{(p\cdot p')}\Bigg[\frac{(p\cdot k_1)(p'\cdot k_1)[(p\cdot k_2)(p'\cdot k_3) + (p\cdot k_3)(p'\cdot k_2)]}{(k_1\cdot k_2)(k_1\cdot k_3)}$$

$$+\frac{(p\cdot k_2)(p'\cdot k_2)[(p\cdot k_3)(p'\cdot k_1) + (p\cdot k_1)(p'\cdot k_3)]}{(k_1\cdot k_2)(k_2\cdot k_3)}$$

$$+\frac{(p\cdot k_3)(p'\cdot k_3)[(p\cdot k_1)(p'\cdot k_2) + (p\cdot k_2)(p'\cdot k_1)]}{(k_1\cdot k_3)(k_2\cdot k_3)}\Bigg]\Bigg\}$$

$$\times \delta^4(k_1 + p - k_2 - k_3 - p') \frac{d^3\vec{k}_2\, d^3\vec{k}_3\, d^3\vec{p'}}{k_{20}\, k_{30}\, p'_0}. \tag{10.272}$$

10.40 $\gamma\bar{q} \to \gamma\gamma\bar{q}$

Process:
$$\gamma(k_1) + \bar{q}(p,i) \to \gamma(k_2) + \gamma(k_3) + \bar{q}(p',j). \tag{10.273}$$

Positive z-axis: along $\vec{k}_1$.
Definitions:
$$Z(p',k_i) = p'_+ k_{i-} - p'^*_\perp k_{i\perp},$$

$$Z(k_i,p') = k_{i+} p'_- - k^*_{i\perp} p'_\perp, \qquad i=2,3. \tag{10.274}$$

Photon polarizations:

$$\displaystyle{\not}\epsilon^{*\pm}(k_1) = N_1[\displaystyle{\not}p\,\displaystyle{\not}p'\,\displaystyle{\not}k_1(1\pm\gamma_5) - \displaystyle{\not}k_1\,\displaystyle{\not}p\,\displaystyle{\not}p'(1\mp\gamma_5)],$$

$$\displaystyle{\not}\epsilon^{\pm}(k_i) = N_i[\displaystyle{\not}p\,\displaystyle{\not}p'\,\displaystyle{\not}k_i(1\mp\gamma_5) - \displaystyle{\not}k_i\,\displaystyle{\not}p\,\displaystyle{\not}p'(1\pm\gamma_5)], \qquad i=2,3,$$

$$N_1^{-1} = E^2[32 p'_+ p'_-]^{\frac{1}{2}},$$

$$N_i^{-1} = 4E[p'_+ k_{i+}(p'\cdot k_i)]^{\frac{1}{2}}, \qquad i=2,3. \tag{10.275}$$

Nonvanishing helicity amplitudes:

$$M(+,+;+,+,+) = \frac{4e_q^3 \delta_{ij} E^2 p'^*_\perp}{[Ep'_- k_{2+} k_{3+}(p' \cdot k_2)(p' \cdot k_3)]^{\frac{1}{2}}},$$

$$M(+,+;+,-,+) = e_q^3 \delta_{ij} p'^*_\perp Z(k_2,p') \left[\frac{(p' \cdot k_2)}{E^3 p'^3_- k_{2+} k_{3+}(p' \cdot k_3)}\right]^{\frac{1}{2}},$$

$$M(+,+;-,+,+) = e_q^3 \delta_{ij} p'^*_\perp Z(k_3,p') \left[\frac{(p' \cdot k_3)}{E^3 p'^3_- k_{2+} k_{3+}(p' \cdot k_2)}\right]^{\frac{1}{2}},$$

$$M(+,-;+,+,-) = 2e_q^3 \delta_{ij} p'_\perp \left[\frac{Ep'_-}{k_{2+} k_{3+}(p' \cdot k_2)(p' \cdot k_3)}\right]^{\frac{1}{2}},$$

$$M(-,+;+,-,+) = e_q^3 \delta_{ij} p'^*_\perp Z(k_3,p') \left[\frac{k_{3+}}{Ep'^3_- k_{2+}(p' \cdot k_2)(p' \cdot k_3)}\right]^{\frac{1}{2}},$$

$$M(-,+;-,+,+) = -\frac{e_q^3 \delta_{ij} k^*_{2\perp} Z(p',k_2)}{k_{2-}} \left[\frac{k_{2+}}{Ep'_- k_{3+}(p' \cdot k_2)(p' \cdot k_3)}\right]^{\frac{1}{2}},$$

$$M(+,-;+,-,-) = -\frac{e_q^3 \delta_{ij} k_{2\perp} Z^*(p',k_2)}{k_{2-}} \left[\frac{k_{2+}}{Ep'_- k_{3+}(p' \cdot k_2)(p' \cdot k_3)}\right]^{\frac{1}{2}},$$

$$M(+,-;-,+,-) = e_q^3 \delta_{ij} p'_\perp Z^*(k_3,p') \left[\frac{k_{3+}}{Ep'^3_- k_{2+}(p' \cdot k_2)(p' \cdot k_3)}\right]^{\frac{1}{2}},$$

$$M(-,+;-,-,+) = 2e_q^3 \delta_{ij} p'^*_\perp \left[\frac{Ep'_-}{k_{2+} k_{3+}(p' \cdot k_2)(p' \cdot k_3)}\right]^{\frac{1}{2}},$$

$$M(-,-;+,-,-) = e_q^3 \delta_{ij} p'_\perp Z^*(k_3,p') \left[\frac{(p' \cdot k_3)}{E^3 p'^3_- k_{2+} k_{3+}(p' \cdot k_2)}\right]^{\frac{1}{2}},$$

$$M(-,-;-,+,-) = e_q^3 \delta_{ij} p'_\perp Z^*(k_2,p') \left[\frac{(p' \cdot k_2)}{E^3 p'^3_- k_{2+} k_{3+}(p' \cdot k_3)}\right]^{\frac{1}{2}},$$

$$M(-,-;-,-,-) = \frac{4e_q^3 \delta_{ij} E^2 p'_\perp}{[Ep'_- k_{2+} k_{3+}(p' \cdot k_2)(p' \cdot k_3)]^{\frac{1}{2}}}. \quad (10.276)$$

Squared absolute values of the nonvanishing helicity amplitudes, summed over the color degrees of freedom of the initial state and the final state:

$$|M(+,+;+,+,+)|^2 = |M(-,-;-,-,-)|^2 = \frac{48 e_q^6 E^3 p'_+}{k_{2+} k_{3+}(p' \cdot k_2)(p' \cdot k_3)},$$

$$|M(+,+;+,-,+)|^2 = |M(-,-;-,+,-)|^2 = \frac{6 e_q^6 p'_+ (p' \cdot k_2)^2}{E^3 p'_- k_{3+}(p' \cdot k_3)},$$

10.41. $\gamma\overline{q} \to g\gamma\overline{q}$

$$|M(+,+;-,+,+)|^2 = |M(-,-;+,-,-)|^2 = \frac{6e_q^6 p'_+ (p' \cdot k_3)^2}{E^3 p'_- k_{2+}(p' \cdot k_2)},$$

$$|M(+,-;+,+,-)|^2 = |M(-,+;-,-,+)|^2 = \frac{12 e_q^6 E p'_+ p'^2_-}{k_{2+} k_{3+}(p' \cdot k_2)(p' \cdot k_3)},$$

$$|M(-,+;+,-,+)|^2 = |M(+,-;-,+,-)|^2 = \frac{6 e_q^6 p'_+ k_{3+}^2}{E p'_- k_{2+}(p' \cdot k_2)},$$

$$|M(-,+;-,+,+)|^2 = |M(+,-;+,-,-)|^2 = \frac{6 e_q^6 p'_+ k_{2+}^2}{E p'_- k_{3+}(p' \cdot k_3)}.$$

(10.277)

Unpolarized squared matrix element:

$$\overline{|M|^2} = 2 e_q^6 (p \cdot p') \frac{\sum_{i=1}^{3}(p \cdot k_i)(p' \cdot k_i)[(p \cdot k_i)^2 + (p' \cdot k_i)^2]}{\prod_{i=1}^{3}(p \cdot k_i)(p' \cdot k_i)}. \qquad (10.278)$$

Unpolarized cross section:

$$d\sigma = \frac{\alpha^3 Q_f^6 (p \cdot p')}{16\pi^2 E^2} \frac{\sum_{i=1}^{3}(p \cdot k_i)(p' \cdot k_i)[(p \cdot k_i)^2 + (p' \cdot k_i)^2]}{\prod_{i=1}^{3}(p \cdot k_i)(p' \cdot k_i)}$$

$$\times \delta^4(p + k_1 - p' - k_2 - k_3) \frac{d^3\vec{p'}\, d^3\vec{k_2}\, d^3\vec{k_3}}{p'_0\, k_{20}\, k_{30}}. \qquad (10.279)$$

10.41 $\gamma\overline{q} \to g\gamma\overline{q}$

Process:
$$\gamma(k_1) + \overline{q}(p,i) \to g(k_2,b) + \gamma(k_3) + \overline{q}(p',j). \qquad (10.280)$$

Positive z-axis: along $\vec{k}_1$.
Definitions:

$$Z(p', k_i) = p'_+ k_{i-} - p'^*_\perp k_{i\perp},$$

$$Z(k_i, p') = k_{i+} p'_- - k^*_{i\perp} p'_\perp, \qquad i = 2, 3. \qquad (10.281)$$

Gluon and photon polarizations:

$$\rlap{/}{\epsilon}^{*\pm}(k_1) = N_1 [\rlap{/}{p}\, \rlap{/}{p}'\, \rlap{/}{k}_1 (1 \pm \gamma_5) - \rlap{/}{k}_1\, \rlap{/}{p}\, \rlap{/}{p}' (1 \mp \gamma_5)],$$

$$\ell^{\pm}(k_2) = N_2[\not{k}_2 \not{p}' \not{p}(1 \pm \gamma_5) + \not{p} \not{p}' \not{k}_2(1 \mp \gamma_5)],$$

$$\ell^{\pm}(k_3) = N_3[\not{p} \not{p}' \not{k}_3(1 \mp \gamma_5) - \not{k}_3 \not{p} \not{p}'(1 \pm \gamma_5)],$$

$$N_1^{-1} = E^2[32 p'_+ p'_-]^{\frac{1}{2}},$$

$$N_2^{-1} = 4E[p'_+ k_{2+}(p' \cdot k_2)]^{\frac{1}{2}},$$

$$N_3^{-1} = 4E[p'_+ k_{3+}(p' \cdot k_3)]^{\frac{1}{2}}. \tag{10.282}$$

Nonvanishing helicity amplitudes:

$$M(+,+;+,+,+) = \frac{4g e_q^2 T_{ij}^b E^2 p'^*_\perp}{[E p'_- k_{2+} k_{3+}(p' \cdot k_2)(p' \cdot k_3)]^{\frac{1}{2}}},$$

$$M(+,+;+,-,+) = g e_q^2 T_{ij}^b p'^*_\perp Z(k_2, p') \left[\frac{(p' \cdot k_2)}{E^3 p'^3_- k_{2+} k_{3+}(p' \cdot k_3)}\right]^{\frac{1}{2}},$$

$$M(+,+;-,+,+) = g e_q^2 T_{ij}^b p'^*_\perp Z(k_3, p') \left[\frac{(p' \cdot k_3)}{E^3 p'^3_- k_{2+} k_{3+}(p' \cdot k_2)}\right]^{\frac{1}{2}},$$

$$M(+,-;+,+,-) = 2g e_q^2 T_{ij}^b p'_\perp \left[\frac{E p'_-}{k_{2+} k_{3+}(p' \cdot k_2)(p' \cdot k_3)}\right]^{\frac{1}{2}},$$

$$M(-,+;+,-,+) = g e_q^2 T_{ij}^b p'^*_\perp Z(k_3, p') \left[\frac{k_{3+}}{E p'^3_- k_{2+}(p' \cdot k_2)(p' \cdot k_3)}\right]^{\frac{1}{2}},$$

$$M(-,+;-,+,+) = -\frac{g e_q^2 T_{ij}^b k^*_{2\perp} Z(p', k_2)}{k_{2-}} \left[\frac{k_{2+}}{E p'_- k_{3+}(p' \cdot k_2)(p' \cdot k_3)}\right]^{\frac{1}{2}},$$

$$M(+,-;+,-,-) = -\frac{g e_q^2 T_{ij}^b k_{2\perp} Z^*(p', k_2)}{k_{2-}} \left[\frac{k_{2+}}{E p'_- k_{3+}(p' \cdot k_2)(p' \cdot k_3)}\right]^{\frac{1}{2}},$$

$$M(+,-;-,+,-) = g e_q^2 T_{ij}^b p'_\perp Z^*(k_3, p') \left[\frac{k_{3+}}{E p'^3_- k_{2+}(p' \cdot k_2)(p' \cdot k_3)}\right]^{\frac{1}{2}},$$

$$M(-,+;-,-,+) = 2g e_q^2 T_{ij}^b p'^*_\perp \left[\frac{E p'_-}{k_{2+} k_{3+}(p' \cdot k_2)(p' \cdot k_3)}\right]^{\frac{1}{2}},$$

$$M(-,-;+,-,-) = g e_q^2 T_{ij}^b p'_\perp Z^*(k_3, p') \left[\frac{(p' \cdot k_3)}{E^3 p'^3_- k_{2+} k_{3+}(p' \cdot k_2)}\right]^{\frac{1}{2}},$$

$$M(-,-;-,+,-) = g e_q^2 T_{ij}^b p'_\perp Z^*(k_2, p') \left[\frac{(p' \cdot k_2)}{E^3 p'^3_- k_{2+} k_{3+}(p' \cdot k_3)}\right]^{\frac{1}{2}},$$

10.42. $g\bar{q} \to \gamma\gamma\bar{q}$

$$M(-,-;-,-,-) = \frac{4ge_q^2 T_{ij}^b E^2 p'_\perp}{[Ep'_- k_{2+} k_{3+}(p' \cdot k_2)(p' \cdot k_3)]^{\frac{1}{2}}}. \tag{10.283}$$

Squared absolute values of the nonvanishing helicity amplitudes, summed over the color degrees of freedom of the initial state and the final state:

$$|M(+,+;+,+,+)|^2 = |M(-,-;-,-,-)|^2 = \frac{64g^2 e_q^4 E^3 p'_+}{k_{2+} k_{3+}(p' \cdot k_2)(p' \cdot k_3)},$$

$$|M(+,+;+,-,+)|^2 = |M(-,-;-,+,-)|^2 = \frac{8g^2 e_q^4 p'_+ (p' \cdot k_2)^2}{E^3 p'_- k_{3+}(p' \cdot k_3)},$$

$$|M(+,+;-,+,+)|^2 = |M(-,-;+,-,-)|^2 = \frac{8g^2 e_q^4 p'_+ (p' \cdot k_3)^2}{E^3 p'_- k_{2+}(p' \cdot k_2)},$$

$$|M(+,-;+,+,-)|^2 = |M(-,+;-,-,+)|^2 = \frac{16g^2 e_q^4 E p'_+ p'^2_-}{k_{2+} k_{3+}(p' \cdot k_2)(p' \cdot k_3)},$$

$$|M(-,+;+,-,+)|^2 = |M(+,-;-,+,-)|^2 = \frac{8g^2 e_q^4 p'_+ k_{3+}^2}{E p'_- k_{2+}(p' \cdot k_2)},$$

$$|M(-,+;-,+,+)|^2 = |M(+,-;+,-,-)|^2 = \frac{8g^2 e_q^4 p'_+ k_{2+}^2}{E p'_- k_{3+}(p' \cdot k_3)}. \tag{10.284}$$

Unpolarized squared matrix element:

$$\overline{|M|^2} = \frac{8}{3} g^2 e_q^4 (p \cdot p') \frac{\sum_{i=1}^{3}(p \cdot k_i)(p' \cdot k_i)[(p \cdot k_i)^2 + (p' \cdot k_i)^2]}{\prod_{i=1}^{3}(p \cdot k_i)(p' \cdot k_i)}. \tag{10.285}$$

Unpolarized cross section:

$$d\sigma = \frac{\alpha_S \alpha^2 Q_f^4 (p \cdot p')}{12\pi^2 E^2} \frac{\sum_{i=1}^{3}(p \cdot k_i)(p' \cdot k_i)[(p \cdot k_i)^2 + (p' \cdot k_i)^2]}{\prod_{i=1}^{3}(p \cdot k_i)(p' \cdot k_i)}$$

$$\times \delta^4(p + k_1 - p' - k_2 - k_3) \frac{d^3\vec{p}'\, d^3\vec{k}_2\, d^3\vec{k}_3}{p'_0\, k_{20}\, k_{30}}. \tag{10.286}$$

10.42 $g\bar{q} \to \gamma\gamma\bar{q}$

Process:
$$g(k_1, a) + \bar{q}(p, i) \to \gamma(k_2) + \gamma(k_3) + \bar{q}(p', j). \tag{10.287}$$

Positive z-axis: along $\vec{k}_1$.
Definitions:

$$Z(p', k_i) = p'_+ k_{i-} - p'^*_\perp k_{i\perp},$$

$$Z(k_i, p') = k_{i+} p'_- - k^*_{i\perp} p'_\perp, \qquad i = 2, 3. \qquad (10.288)$$

Gluon and photon polarizations:

$$\not{\epsilon}^{*\pm}(k_1) = N_1 [\not{p} \not{p}' \not{k}_1 (1 \pm \gamma_5) + \not{k}_1 \not{p}' \not{p} (1 \mp \gamma_5)],$$

$$N_1^{-1} = E^2 [32 p'_+ p'_-]^{\frac{1}{2}},$$

$$\not{\epsilon}^{\pm}(k_i) = N_i [\not{k}_i \not{p} \not{p}' (1 \pm \gamma_5) - \not{p} \not{p}' \not{k}_i (1 \mp \gamma_5)],$$

$$N_i^{-1} = 4 E [p'_+ k_{i+} (p' \cdot k_i)]^{\frac{1}{2}}, \qquad i = 2, 3. \qquad (10.289)$$

Nonvanishing helicity amplitudes:

$$M(+,+;+,+,+) = \frac{4 g e_q^2 T_{ij}^a E^2 p'^*_\perp}{[E p'_- k_{2+} k_{3+} (p' \cdot k_2)(p' \cdot k_3)]^{\frac{1}{2}}},$$

$$M(+,+;+,-,+) = g e_q^2 T_{ij}^a p'^*_\perp Z(k_2, p') \left[\frac{(p' \cdot k_2)}{E^3 p'^3_- k_{2+} k_{3+} (p' \cdot k_3)} \right]^{\frac{1}{2}},$$

$$M(+,+;-,+,+) = g e_q^2 T_{ij}^a p'^*_\perp Z(k_3, p') \left[\frac{(p' \cdot k_3)}{E^3 p'^3_- k_{2+} k_{3+} (p' \cdot k_2)} \right]^{\frac{1}{2}},$$

$$M(+,-;+,+,-) = 2 g e_q^2 T_{ij}^a p'_\perp \left[\frac{E p'_-}{k_{2+} k_{3+} (p' \cdot k_2)(p' \cdot k_3)} \right]^{\frac{1}{2}},$$

$$M(-,+;+,-,+) = g e_q^2 T_{ij}^a p'^*_\perp Z(k_3, p') \left[\frac{k_{3+}}{E p'^3_- k_{2+} (p' \cdot k_2)(p' \cdot k_3)} \right]^{\frac{1}{2}},$$

$$M(-,+;-,+,+) = -\frac{g e_q^2 T_{ij}^a k^*_{2\perp} Z(p', k_2)}{k_{2-}} \left[\frac{k_{2+}}{E p'_- k_{3+} (p' \cdot k_2)(p' \cdot k_3)} \right]^{\frac{1}{2}},$$

$$M(+,-;+,-,-) = -\frac{g e_q^2 T_{ij}^a k_{2\perp} Z^*(p', k_2)}{k_{2-}} \left[\frac{k_{2+}}{E p'_- k_{3+} (p' \cdot k_2)(p' \cdot k_3)} \right]^{\frac{1}{2}},$$

10.42. $g\bar{q} \to \gamma\gamma\bar{q}$

$$M(+,-;-,+,-) = ge_q^2 T_{ij}^a p'_\perp Z^*(k_3, p') \left[\frac{k_{3+}}{Ep'^3_- k_{2+}(p' \cdot k_2)(p' \cdot k_3)}\right]^{\frac{1}{2}},$$

$$M(-,+;-,-,+) = 2ge_q^2 T_{ij}^a p'^*_\perp \left[\frac{Ep'_-}{k_{2+} k_{3+}(p' \cdot k_2)(p' \cdot k_3)}\right]^{\frac{1}{2}},$$

$$M(-,-;+,-,-) = ge_q^2 T_{ij}^a p'_\perp Z^*(k_3, p') \left[\frac{(p' \cdot k_3)}{E^3 p'^3_- k_{2+} k_{3+}(p' \cdot k_2)}\right]^{\frac{1}{2}},$$

$$M(-,-;-,+,-) = ge_q^2 T_{ij}^a p'_\perp Z^*(k_2, p') \left[\frac{(p' \cdot k_2)}{E^3 p'^3_- k_{2+} k_{3+}(p' \cdot k_3)}\right]^{\frac{1}{2}},$$

$$M(-,-;-,-,-) = \frac{4ge_q^2 T_{ij}^a E^2 p'_\perp}{[Ep'_- k_{2+} k_{3+}(p' \cdot k_2)(p' \cdot k_3)]^{\frac{1}{2}}}. \quad (10.290)$$

Squared absolute values of the nonvanishing helicity amplitudes, summed over the color degrees of freedom of the initial state and the final state:

$$|M(+,+;+,+,+)|^2 = |M(-,-;-,-,-)|^2 = \frac{64g^2 e_q^4 E^3 p'_+}{k_{2+} k_{3+}(p' \cdot k_2)(p' \cdot k_3)},$$

$$|M(+,+;+,-,+)|^2 = |M(-,-;-,+,-)|^2 = \frac{8g^2 e_q^4 p'_+ (p' \cdot k_2)^2}{E^3 p'_- k_{3+}(p' \cdot k_3)},$$

$$|M(+,+;-,+,+)|^2 = |M(-,-;+,-,-)|^2 = \frac{8g^2 e_q^4 p'_+ (p' \cdot k_3)^2}{E^3 p'_- k_{2+}(p' \cdot k_2)},$$

$$|M(+,-;+,+,-)|^2 = |M(-,+;-,-,+)|^2 = \frac{16g^2 e_q^4 E p'_+ p'^2_-}{k_{2+} k_{3+}(p' \cdot k_2)(p' \cdot k_3)},$$

$$|M(-,+;+,-,+)|^2 = |M(+,-;-,+,-)|^2 = \frac{8g^2 e_q^4 p'_+ k_{3+}^2}{Ep'_- k_{2+}(p' \cdot k_2)},$$

$$|M(-,+;-,+,+)|^2 = |M(+,-;+,-,-)|^2 = \frac{8g^2 e_q^4 p'_+ k_{2+}^2}{Ep'_- k_{3+}(p' \cdot k_3)}. \quad (10.291)$$

Unpolarized squared matrix element:

$$\overline{|M|^2} = \frac{8}{3} g^2 e_q^4 (p \cdot p') \frac{\sum_{i=1}^{3}(p \cdot k_i)(p' \cdot k_i)[(p \cdot k_i)^2 + (p' \cdot k_i)^2]}{\prod_{i=1}^{3}(p \cdot k_i)(p' \cdot k_i)}. \quad (10.292)$$

Unpolarized cross section:

$$d\sigma = \frac{\alpha_S \alpha^2 Q_f^4 (p \cdot p')}{12\pi^2 E^2} \frac{\sum_{i=1}^{3}(p \cdot k_i)(p' \cdot k_i)[(p \cdot k_i)^2 + (p' \cdot k_i)^2]}{\prod_{i=1}^{3}(p \cdot k_i)(p' \cdot k_i)}$$

$$\times \delta^4(p + k_1 - p' - k_2 - k_3) \frac{d^3\vec{p'}\, d^3\vec{k_2}\, d^3\vec{k_3}}{p'_0\, k_{20}\, k_{30}}. \qquad (10.293)$$

10.43 $\gamma\bar{q} \to gg\bar{q}$

Process:

$$\gamma(k_1) + \bar{q}(p,i) \to g(k_2,b) + g(k_3,c) + \bar{q}(p',j). \qquad (10.294)$$

Positive z-axis: along $\vec{k}_1$.
Definitions:

$$\begin{aligned}
Z(k_i, k_j) &= k_{i+}k_{j-} - k_{i\perp}^* k_{j\perp}, & i,j &= 2,3, \\
Z(p', k_i) &= p'_+ k_{i-} - p'^*_\perp k_{i\perp}, & i &= 2,3, \\
Z(k_i, p') &= k_{i+}p'_- - k_{i\perp}^* p'_\perp, & i &= 2,3. \qquad (10.295)
\end{aligned}$$

Photon and gluon polarizations:

$$\rlap{/}{\epsilon}_1^{*\pm} = N_1[\rlap{/}{k}_1 \rlap{/}{p} \rlap{/}{p}'(1 \mp \gamma_5) - \rlap{/}{p} \rlap{/}{p}' \rlap{/}{k}_1 (1 \pm \gamma_5)],$$

$$N_1^{-1} = E^2 [32 p'_+ p'_-]^{\frac{1}{2}},$$

$$\rlap{/}{\epsilon}_i^{\pm} = N_i[\rlap{/}{k}_i \rlap{/}{p} \rlap{/}{p}'(1 \pm \gamma_5) + \rlap{/}{p}' \rlap{/}{p} \rlap{/}{k}_i (1 \mp \gamma_5)],$$

$$N_i^{-1} = 4E[p'_+ k_{i+}(p' \cdot k_i)]^{\frac{1}{2}}, \qquad i = 2,3. \qquad (10.296)$$

Nonvanishing helicity amplitudes:

$$M(+,+;+,+,+) = \frac{2g^2 e_q p'^*_\perp}{p'_+ k_{2-} k_{3-} (k_2 \cdot k_3)} \left[\frac{E^3}{p'_- k_{2+} k_{3+} (p' \cdot k_2)(p' \cdot k_3)}\right]^{\frac{1}{2}}$$

$$\times \left[k_{3-} Z(k_3, k_2) Z^*(p', k_2)(T^b T^c)_{ij} + k_{2-} Z(k_2, k_3) Z^*(p', k_3)(T^c T^b)_{ij}\right],$$

10.43. $\gamma\bar{q} \to gg\bar{q}$

$$M(+,+;+,-,+) = \frac{g^2 e_q p'^*_\perp Z(k_2,p')}{2p'_+ k_{2-} k_{3-}(k_2 \cdot k_3)} \left[\frac{(p' \cdot k_2)}{E^3 p'^3_- k_{2+} k_{3+}(p' \cdot k_3)}\right]^{\frac{1}{2}}$$
$$\times \left[k_{3-} Z^*(k_3,k_2) Z(p',k_2)(T^b T^c)_{ij} + k_{2-} Z^*(k_2,k_3) Z(p',k_3)(T^c T^b)_{ij}\right],$$

$$M(+,+;-,+,+) = \frac{g^2 e_q p'^*_\perp Z(k_3,p')}{2p'_+ k_{2-} k_{3-}(k_2 \cdot k_3)} \left[\frac{(p' \cdot k_3)}{E^3 p'^3_- k_{2+} k_{3+}(p' \cdot k_2)}\right]^{\frac{1}{2}}$$
$$\times \left[k_{3-} Z^*(k_3,k_2) Z(p',k_2)(T^b T^c)_{ij} + k_{2-} Z^*(k_2,k_3) Z(p',k_3)(T^c T^b)_{ij}\right],$$

$$M(+,-;+,+,-) = \frac{g^2 e_q p'_\perp}{p'_+ k_{2-} k_{3-}(k_2 \cdot k_3)} \left[\frac{E p'_-}{k_{2+} k_{3+}(p' \cdot k_2)(p' \cdot k_3)}\right]^{\frac{1}{2}}$$
$$\times \left[k_{3-} Z(k_3,k_2) Z^*(p',k_2)(T^b T^c)_{ij} + k_{2-} Z(k_2,k_3) Z^*(p',k_3)(T^c T^b)_{ij}\right],$$

$$M(-,+;+,-,+) = \frac{g^2 e_q p'^*_\perp Z(k_3,p')}{2p'_+ k_{2-} k_{3-}(k_2 \cdot k_3)} \left[\frac{k_{3+}}{E p'^3_- k_{2+}(p' \cdot k_2)(p' \cdot k_3)}\right]^{\frac{1}{2}}$$
$$\times \left[k_{3-} Z(k_3,k_2) Z^*(p',k_2)(T^b T^c)_{ij} + k_{2-} Z(k_2,k_3) Z^*(p',k_3)(T^c T^b)_{ij}\right],$$

$$M(-,+;-,+,+) = \frac{g^2 e_q p'^*_\perp Z(k_2,p')}{2p'_+ k_{2-} k_{3-}(k_2 \cdot k_3)} \left[\frac{k_{2+}}{E p'^3_- k_{3+}(p' \cdot k_2)(p' \cdot k_3)}\right]^{\frac{1}{2}}$$
$$\times \left[k_{3-} Z(k_3,k_2) Z^*(p',k_2)(T^b T^c)_{ij} + k_{2-} Z(k_2,k_3) Z^*(p',k_3)(T^c T^b)_{ij}\right],$$

$$M(+,-;+,-,-) = \frac{g^2 e_q p'_\perp Z^*(k_2,p')}{2p'_+ k_{2-} k_{3-}(k_2 \cdot k_3)} \left[\frac{k_{2+}}{E p'^3_- k_{3+}(p' \cdot k_2)(p' \cdot k_3)}\right]^{\frac{1}{2}}$$
$$\times \left[k_{3-} Z^*(k_3,k_2) Z(p',k_2)(T^b T^c)_{ij} + k_{2-} Z^*(k_2,k_3) Z(p',k_3)(T^c T^b)_{ij}\right],$$

$$M(+,-;-,+,-) = \frac{g^2 e_q p'_\perp Z^*(k_3,p')}{2p'_+ k_{2-} k_{3-}(k_2 \cdot k_3)} \left[\frac{k_{3+}}{E p'^3_- k_{2+}(p' \cdot k_2)(p' \cdot k_3)}\right]^{\frac{1}{2}}$$
$$\times \left[k_{3-} Z^*(k_3,k_2) Z(p',k_2)(T^b T^c)_{ij} + k_{2-} Z^*(k_2,k_3) Z(p',k_3)(T^c T^b)_{ij}\right],$$

$$M(-,+;-,-,+) = \frac{g^2 e_q p'^*_\perp}{p'_+ k_{2-} k_{3-}(k_2 \cdot k_3)} \left[\frac{E p'_-}{k_{2+} k_{3+}(p' \cdot k_2)(p' \cdot k_3)}\right]^{\frac{1}{2}}$$
$$\times \left[k_{3-} Z^*(k_3,k_2) Z(p',k_2)(T^b T^c)_{ij} + k_{2-} Z^*(k_2,k_3) Z(p',k_3)(T^c T^b)_{ij}\right],$$

$$M(-,-;+,-,-) = \frac{g^2 e_q p'_\perp Z^*(k_3,p')}{2p'_+ k_{2-} k_{3-}(k_2 \cdot k_3)} \left[\frac{(p' \cdot k_3)}{E^3 p'^3_- k_{2+} k_{3+}(p' \cdot k_2)}\right]^{\frac{1}{2}}$$

$$\times \left[k_{3-} Z(k_3,k_2) Z^*(p',k_2)(T^b T^c)_{ij} + k_{2-} Z(k_2,k_3) Z^*(p',k_3)(T^c T^b)_{ij}\right],$$

$$M(-,-;-,+,-) = \frac{g^2 e_q p'_\perp Z^*(k_2,p')}{2p'_+ k_{2-} k_{3-}(k_2 \cdot k_3)} \left[\frac{(p' \cdot k_2)}{E^3 p'^3_- k_{2+} k_{3+}(p' \cdot k_3)}\right]^{\frac{1}{2}}$$

$$\times \left[k_{3-} Z(k_3,k_2) Z^*(p',k_2)(T^b T^c)_{ij} + k_{2-} Z(k_2,k_3) Z^*(p',k_3)(T^c T^b)_{ij}\right],$$

$$M(-,-;-,-,-) = \frac{2g^2 e_q p'_\perp}{p'_+ k_{2-} k_{3-}(k_2 \cdot k_3)} \left[\frac{E^3}{p'_- k_{2+} k_{3+}(p' \cdot k_2)(p' \cdot k_3)}\right]^{\frac{1}{2}}$$

$$\times \left[k_{3-} Z^*(k_3,k_2) Z(p',k_2)(T^b T^c)_{ij} + k_{2-} Z^*(k_2,k_3) Z(p',k_3)(T^c T^b)_{ij}\right].$$
(10.297)

Squared absolute values of the nonvanishing helicity amplitudes, summed over the color degrees of freedom of the initial state and the final state:

$$|M(+,+;+,+,+)|^2 = |M(-,-;-,-,-)|^2$$

$$= \frac{32 g^4 e_q^2 E^3 \left[9 k_{2+}(p' \cdot k_3) + 9 k_{3+}(p' \cdot k_2) - p'_+(k_2 \cdot k_3)\right]}{3 k_{2+} k_{3+}(k_2 \cdot k_3)(p' \cdot k_2)(p' \cdot k_3)},$$

$$|M(+,+;+,-,+)|^2 = |M(-,-;-,+,-)|^2$$

$$= \frac{4 g^4 e_q^2 (p' \cdot k_2)^2 \left[9 k_{2+}(p' \cdot k_3) + 9 k_{3+}(p' \cdot k_2) - p'_+(k_2 \cdot k_3)\right]}{3 E^3 p'_- k_{3+}(k_2 \cdot k_3)(p' \cdot k_3)},$$

$$|M(+,+;-,+,+)|^2 = |M(-,-;+,-,-)|^2$$

$$= \frac{4 g^4 e_q^2 (p' \cdot k_3)^2 \left[9 k_{2+}(p' \cdot k_3) + 9 k_{3+}(p' \cdot k_2) - p'_+(k_2 \cdot k_3)\right]}{3 E^3 p'_- k_{2+}(k_2 \cdot k_3)(p' \cdot k_2)},$$

$$|M(+,-;+,+,-)|^2 = |M(-,+;-,-,+)|^2$$

$$= \frac{8 g^4 e_q^2 E p'^2_- \left[9 k_{2+}(p' \cdot k_3) + 9 k_{3+}(p' \cdot k_2) - p'_+(k_2 \cdot k_3)\right]}{3 k_{2+} k_{3+}(k_2 \cdot k_3)(p' \cdot k_2)(p' \cdot k_3)},$$

10.44. $g\bar{q} \to g\gamma\bar{q}$

$$|M(-,+;+,-,+)|^2 = |M(+,-;-,+,-)|^2$$

$$= \frac{4g^4 e_q^2 k_{3+}^2 \left[9k_{2+}(p' \cdot k_3) + 9k_{3+}(p' \cdot k_2) - p'_+(k_2 \cdot k_3)\right]}{3Ep'_- k_{2+}(k_2 \cdot k_3)(p' \cdot k_2)},$$

$$|M(-,+;-,+,+)|^2 = |M(+,-;+,-,-)|^2$$

$$= \frac{4g^4 e_q^2 k_{2+}^2 \left[9k_{2+}(p' \cdot k_3) + 9k_{3+}(p' \cdot k_2) - p'_+(k_2 \cdot k_3)\right]}{3Ep'_- k_{3+}(k_2 \cdot k_3)(p' \cdot k_3)}.$$

(10.298)

Unpolarized squared matrix element:

$$\overline{|M|^2} = \frac{4}{9}g^4 e_q^2 \frac{9(p \cdot k_2)(p' \cdot k_3) + 9(p \cdot k_3)(p' \cdot k_2) - (p \cdot p')(k_2 \cdot k_3)}{(k_2 \cdot k_3)}$$

$$\times \frac{\sum_{i=1}^{3}(p \cdot k_i)(p' \cdot k_i)[(p \cdot k_i)^2 + (p' \cdot k_i)^2]}{\prod_{i=1}^{3}(p \cdot k_i)(p' \cdot k_i)}.$$

(10.299)

Unpolarized cross section:

$$d\sigma = \frac{\alpha_S^2 \alpha Q_f^2}{72\pi^2 E^2} \frac{9(p \cdot k_2)(p' \cdot k_3) + 9(p \cdot k_3)(p' \cdot k_2) - (p \cdot p')(k_2 \cdot k_3)}{(k_2 \cdot k_3)}$$

$$\times \frac{\sum_{i=1}^{3}(p \cdot k_i)(p' \cdot k_i)[(p \cdot k_i)^2 + (p' \cdot k_i)^2]}{\prod_{i=1}^{3}(p \cdot k_i)(p' \cdot k_i)}$$

$$\times \delta^4(k_1 + p - k_2 - k_3 - p') \frac{d^3 \vec{k}_2 \, d^3 \vec{k}_3 \, d^3 \vec{p}'}{k_{20} \, k_{30} \, p'_0}.$$

(10.300)

10.44 $g\bar{q} \to g\gamma\bar{q}$

Process:

$$g(k_1, a) + \bar{q}(p, i) \to g(k_2, b) + \gamma(k_3) + \bar{q}(p', j).$$ (10.301)

Positive z-axis: along $\vec{k}_1$.
Invariants and definitions:

$$s = 2(k_1 \cdot p), \quad t = -2(k_1 \cdot k_2), \quad u = -2(k_1 \cdot p'),$$

$$s' = 2(p' \cdot k_2), \quad t' = -2(p \cdot p'), \quad u' = -2(k_2 \cdot p),$$

$$Z(p', k_i) = p'_+ k_{i-} - p'^*_\perp k_{i\perp},$$

$$Z(k_i, p') = k_{i+} p'_- - k^*_{i\perp} p'_\perp, \qquad i = 2,3. \qquad (10.302)$$

Gluon and photon polarizations:

$$\not{\epsilon}^{*\pm}(k_1) = N_1[\not{p}\,\not{p}'\,\not{k}_1(1 \pm \gamma_5) + \not{k}_1\,\not{p}'\,\not{p}(1 \mp \gamma_5)],$$

$$\not{\epsilon}^{\pm}(k_2) = N_2[\not{k}_2\,\not{p}'\,\not{p}(1 \pm \gamma_5) + \not{p}\,\not{p}'\,\not{k}_2(1 \mp \gamma_5)],$$

$$\not{\epsilon}^{\pm}(k_3) = N_3[\not{p}\,\not{p}'\,\not{k}_3(1 \mp \gamma_5) - \not{k}_3\,\not{p}\,\not{p}'(1 \pm \gamma_5)],$$

$$N_1^{-1} = E^2[32 p'_+ p'_-]^{\frac{1}{2}},$$

$$N_2^{-1} = 4E[p'_+ k_{2+}(p' \cdot k_2)]^{\frac{1}{2}},$$

$$N_3^{-1} = 4E[p'_+ k_{3+}(p' \cdot k_3)]^{\frac{1}{2}}. \qquad (10.303)$$

Nonvanishing helicity amplitudes:

$$M(+,+;+,+,+) = \frac{4g^2 e_q E^2 p'^*_\perp}{p'_+ k_{2-} [E p'_- k_{2+} k_{3+} (p' \cdot k_2)(p' \cdot k_3)]^{\frac{1}{2}}}$$
$$\times \left[p'_\perp k^*_{2\perp} (T^a T^b)_{ij} + Z^*(p', k_2)(T^b T^a)_{ij} \right],$$

$$M(+,+;+,-,+) = -\frac{g^2 e_q p'^*_\perp Z(p', k_2)}{p'_+ k_{2-}} \left[\frac{(p' \cdot k_2)}{E^3 p'^3_- k_{2+} k_{3+}(p' \cdot k_3)} \right]^{\frac{1}{2}}$$
$$\times \left[p'_- k_{2+}(T^a T^b)_{ij} - Z(k_2, p')(T^b T^a)_{ij} \right],$$

$$M(+,+;-,+,+) = \frac{g^2 e_q p'^*_\perp Z(k_3, p')}{p'_+ k_{2-}} \left[\frac{(p' \cdot k_3)}{E^3 p'^3_- k_{2+} k_{3+}(p' \cdot k_2)} \right]^{\frac{1}{2}}$$
$$\times \left[p'^*_\perp k_{2\perp}(T^a T^b)_{ij} + Z(p', k_2)(T^b T^a)_{ij} \right],$$

$$M(+,-;+,+,-) = \frac{2g^2 e_q p'_\perp}{p'_+ k_{2-}} \left[\frac{E p'_-}{k_{2+} k_{3+}(p' \cdot k_2)(p' \cdot k_3)} \right]^{\frac{1}{2}}$$
$$\times \left[p'_\perp k^*_{2\perp}(T^a T^b)_{ij} + Z^*(p', k_2)(T^b T^a)_{ij} \right],$$

10.44. $g\bar{q} \to g\gamma\bar{q}$

$$M(-,+;+,-,+) = \frac{g^2 e_q p'^*_\perp Z(k_3,p')}{p'_+ k_{2-}} \left[\frac{k_{3+}}{Ep'^3_- k_{2+}(p'\cdot k_2)(p'\cdot k_3)}\right]^{\frac{1}{2}}$$
$$\times \left[p'_\perp k^*_{2\perp}(T^a T^b)_{ij} + Z^*(p',k_2)(T^b T^a)_{ij}\right],$$

$$M(-,+;-,+,+) = \frac{g^2 e_q k^*_{2\perp} Z(k_2,p')}{k_{2-}[Ep'^3_- k_{2+}k_{3+}(p'\cdot k_2)(p'\cdot k_3)]^{\frac{1}{2}}}$$
$$\times \left[p'_- k_{2+}(T^a T^b)_{ij} - Z^*(k_2,p')(T^b T^a)_{ij}\right],$$

$$M(+,-;+,-,-) = \frac{g^2 e_q k_{2\perp} Z^*(k_2,p')}{k_{2-}[Ep'^3_- k_{2+}k_{3+}(p'\cdot k_2)(p'\cdot k_3)]^{\frac{1}{2}}}$$
$$\times \left[p'_- k_{2+}(T^a T^b)_{ij} - Z(k_2,p')(T^b T^a)_{ij}\right],$$

$$M(+,-;-,+,-) = \frac{g^2 e_q p'_\perp Z^*(k_3,p')}{p'_+ k_{2-}} \left[\frac{k_{3+}}{Ep'^3_- k_{2+}(p'\cdot k_2)(p'\cdot k_3)}\right]^{\frac{1}{2}}$$
$$\times \left[p'^*_\perp k_{2\perp}(T^a T^b)_{ij} + Z(p',k_2)(T^b T^a)_{ij}\right],$$

$$M(-,+;-,-,+) = \frac{2g^2 e_q p'^*_\perp}{p'_+ k_{2-}} \left[\frac{Ep'_-}{k_{2+}k_{3+}(p'\cdot k_2)(p'\cdot k_3)}\right]^{\frac{1}{2}}$$
$$\times \left[p'^*_\perp k_{2\perp}(T^a T^b)_{ij} + Z(p',k_2)(T^b T^a)_{ij}\right],$$

$$M(-,-;+,-,-) = \frac{g^2 e_q p'_\perp Z^*(k_3,p')}{p'_+ k_{2-}} \left[\frac{(p'\cdot k_3)}{E^3 p'^3_- k_{2+} k_{3+}(p'\cdot k_2)}\right]^{\frac{1}{2}}$$
$$\times \left[p'_\perp k^*_{2\perp}(T^a T^b)_{ij} + Z^*(p',k_2)(T^b T^a)_{ij}\right],$$

$$M(-,-;-,+,-) = -\frac{g^2 e_q p'_\perp Z^*(p',k_2)}{p'_+ k_{2-}} \left[\frac{(p'\cdot k_2)}{E^3 p'^3_- k_{2+} k_{3+}(p'\cdot k_3)}\right]^{\frac{1}{2}}$$
$$\times \left[p'_- k_{2+}(T^a T^b)_{ij} - Z^*(k_2,p')(T^b T^a)_{ij}\right],$$

$$M(-,-;-,-,-) = \frac{4g^2 e_q E^2 p'_\perp}{p'_+ k_{2-}[Ep'_- k_{2+} k_{3+}(p'\cdot k_2)(p'\cdot k_3)]^{\frac{1}{2}}}$$
$$\times \left[p'^*_\perp k_{2\perp}(T^a T^b)_{ij} + Z(p',k_2)(T^b T^a)_{ij}\right].$$

(10.304)

Squared absolute values of the nonvanishing helicity amplitudes, summed over the color degrees of freedom of the initial state and the final state:

$$|M(+,+;+,+,+)|^2 = |M(-,-;-,-,-)|^2$$
$$= \frac{32g^4 e_q^2 E^3 [18(p' \cdot k_2) + 9p'_- k_{2+} - p'_+ k_{2-}]}{3k_{2+} k_{2-} k_{3+} (p' \cdot k_2)(p' \cdot k_3)},$$

$$|M(+,+;+,-,+)|^2 = |M(-,-;-,+,-)|^2$$
$$= \frac{4g^4 e_q^2 (p' \cdot k_2)^2 [18(p' \cdot k_2) + 9p'_- k_{2+} - p'_+ k_{2-}]}{3E^3 p'_- k_{2-} k_{3+} (p' \cdot k_3)},$$

$$|M(+,+;-,+,+)|^2 = |M(-,-;+,-,-)|^2$$
$$= \frac{4g^4 e_q^2 (p' \cdot k_3)^2 [18(p' \cdot k_2) + 9p'_- k_{2+} - p'_+ k_{2-}]}{3E^3 p'_- k_{2+} k_{2-} (p' \cdot k_2)},$$

$$|M(+,-;+,+,-)|^2 = |M(-,+;-,-,+)|^2$$
$$= \frac{8g^4 e_q^2 E p'^2_- [18(p' \cdot k_2) + 9p'_- k_{2+} - p'_+ k_{2-}]}{3k_{2+} k_{2-} k_{3+} (p' \cdot k_2)(p' \cdot k_3)},$$

$$|M(-,+;+,-,+)|^2 = |M(+,-;-,+,-)|^2$$
$$= \frac{4g^4 e_q^2 k_{3+}^2 [18(p' \cdot k_2) + 9p'_- k_{2+} - p'_+ k_{2-}]}{3E p'_- k_{2+} k_{2-} (p' \cdot k_2)},$$

$$|M(-,+;-,+,+)|^2 = |M(+,-;+,-,-)|^2$$
$$= \frac{4g^4 e_q^2 k_{2+}^2 [18(p' \cdot k_2) + 9p'_- k_{2+} - p'_+ k_{2-}]}{3E p'_- k_{2-} k_{3+} (p' \cdot k_3)}.$$

(10.305)

Unpolarized squared matrix element:

$$\overline{|M|^2} = \frac{g^4 e_q^2 (9ss' + 9uu' - tt')}{72(k_1 \cdot k_2)} \frac{\sum_{i=1}^{3} (p \cdot k_i)(p' \cdot k_i)[(p \cdot k_i)^2 + (p' \cdot k_i)^2]}{\prod_{i=1}^{3} (p \cdot k_i)(p' \cdot k_i)}.$$

(10.306)

10.45. $g\bar{q} \to gg\bar{q}$

Unpolarized cross section:

$$d\sigma = \frac{\alpha_s^2 \alpha Q_f^2}{576\pi^2} \frac{9ss' + 9uu' - tt'}{s(k_1 \cdot k_2)} \frac{\sum_{i=1}^{3}(p \cdot k_i)(p' \cdot k_i)[(p \cdot k_i)^2 + (p' \cdot k_i)^2]}{\prod_{i=1}^{3}(p \cdot k_i)(p' \cdot k_i)}$$

$$\times \delta^4(p + k_1 - p' - k_2 - k_3) \frac{d^3\vec{p}' \, d^3\vec{k}_2 \, d^3\vec{k}_3}{p'_0 \, k_{20} \, k_{30}}. \qquad (10.307)$$

10.45 $g\bar{q} \to gg\bar{q}$

Process:

$$g(k_1, a) + \bar{q}(p, i) \to g(k_2, b) + g(k_3, c) + \bar{q}(p', j). \qquad (10.308)$$

Positive z-axis: along $\vec{k}_1$.
Definitions:

$$F = \frac{40 p'_+}{9} - 4 \left[\frac{p'_- k_{2+} + 2(p' \cdot k_2)}{k_{2-}} + \frac{p'_- k_{3+} + 2(p' \cdot k_3)}{k_{3-}} \right.$$

$$+ \frac{k_{2+}(p' \cdot k_3) + k_{3+}(p' \cdot k_2)}{(k_2 \cdot k_3)} \right] + \frac{36}{p'_+} \left[\frac{k_{2+}(p' \cdot k_2)[p'_- k_{3+} + 2(p' \cdot k_3)]}{k_{2-}(k_2 \cdot k_3)} \right.$$

$$+ \frac{k_{3+}(p' \cdot k_3)[p'_- k_{2+} + 2(p' \cdot k_2)]}{k_{3-}(k_2 \cdot k_3)} + \frac{2p'_-[k_{2+}(p' \cdot k_3) + k_{3+}(p' \cdot k_2)]}{k_{2-} k_{3-}} \right],$$

$$Z(k_i, k_j) = k_{i+} k_{j-} - k^*_{i\perp} k_{j\perp}, \qquad i, j = 2, 3,$$

$$Z(p', k_i) = p'_+ k_{i-} - p'^*_\perp k_{i\perp}, \qquad i = 2, 3,$$

$$Z(k_i, p') = k_{i+} p'_- - k^*_{i\perp} p'_\perp, \qquad i = 2, 3. \qquad (10.309)$$

Gluon polarizations:

$$\not{\epsilon}_1^{*\pm} = N_1[\not{p}' \not{p} \not{k}_1(1 \pm \gamma_5) + \not{k}_1 \not{p} \not{p}'(1 \mp \gamma_5)],$$

$$N_1^{-1} = E^2[32 p'_+ p'_-]^{\frac{1}{2}},$$

$$\not{\epsilon}_i^{\pm} = N_i[\not{k}_i \not{p}' \not{p}(1 \pm \gamma_5) + \not{p} \not{p}' \not{k}_i(1 \mp \gamma_5)],$$

$$N_i^{-1} = 4E[p'_+ k_{i+}(p' \cdot k_i)]^{\frac{1}{2}}, \qquad i = 2, 3. \qquad (10.310)$$

Nonvanishing helicity amplitudes:

$$M(+,+;+,+,+) = \frac{2g^3 p'^*_\perp}{p'^2_+ k_{2-} k_{3-}} \left[\frac{E^3}{p'_- k_{2+} k_{3+}(p' \cdot k_2)(p' \cdot k_3)} \right]^{\frac{1}{2}}$$

$$\times \left\{ -(T^a T^b T^c)_{ij} \frac{p'_\perp k^*_{3\perp} Z(k_2,k_3) Z^*(p',k_2)}{(k_2 \cdot k_3)} \right.$$

$$-(T^a T^c T^b)_{ij} \frac{p'_\perp k^*_{2\perp} Z(k_3,k_2) Z^*(p',k_3)}{(k_2 \cdot k_3)}$$

$$+(T^b T^a T^c)_{ij} 2 p'_\perp k^*_{3\perp} Z^*(p',k_2) + (T^b T^c T^a)_{ij} \frac{Z(k_3,k_2) Z^*(p',k_2) Z^*(p',k_3)}{(k_2 \cdot k_3)}$$

$$\left. +(T^c T^a T^b)_{ij} 2 p'_\perp k^*_{2\perp} Z^*(p',k_3) + (T^c T^b T^a)_{ij} \frac{Z(k_2,k_3) Z^*(p',k_2) Z^*(p',k_3)}{(k_2 \cdot k_3)} \right\},$$

$$M(+,+;+,-,+) = \frac{g^3 p'^*_\perp Z(k_2,p')}{2 p'^2_+ k_{2-} k_{3-}} \left[\frac{(p' \cdot k_2)}{E^3 p'^3_- k_{2+} k_{3+} (p' \cdot k_3)} \right]^{\frac{1}{2}}$$

$$\times \left\{ -(T^a T^b T^c)_{ij} \frac{p'^*_\perp k_{3\perp} Z^*(k_2,k_3) Z(p',k_2)}{(k_2 \cdot k_3)} \right.$$

$$-(T^a T^c T^b)_{ij} \frac{p'^*_\perp k_{2\perp} Z^*(k_3,k_2) Z(p',k_3)}{(k_2 \cdot k_3)}$$

$$+(T^b T^a T^c)_{ij} 2 p'^*_\perp k_{3\perp} Z(p',k_2) + (T^b T^c T^a)_{ij} \frac{Z^*(k_3,k_2) Z(p',k_2) Z(p',k_3)}{(k_2 \cdot k_3)}$$

$$\left. +(T^c T^a T^b)_{ij} 2 p'^*_\perp k_{2\perp} Z(p',k_3) + (T^c T^b T^a)_{ij} \frac{Z^*(k_2,k_3) Z(p',k_2) Z(p',k_3)}{(k_2 \cdot k_3)} \right\},$$

$$M(+,+;-,+,+) = \frac{g^3 p'^*_\perp Z(k_3,p')}{2 p'^2_+ k_{2-} k_{3-}} \left[\frac{(p' \cdot k_3)}{E^3 p'^3_- k_{2+} k_{3+} (p' \cdot k_2)} \right]^{\frac{1}{2}}$$

$$\times \left\{ -(T^a T^b T^c)_{ij} \frac{p'^*_\perp k_{3\perp} Z^*(k_2,k_3) Z(p',k_2)}{(k_2 \cdot k_3)} \right.$$

$$-(T^a T^c T^b)_{ij} \frac{p'^*_\perp k_{2\perp} Z^*(k_3,k_2) Z(p',k_3)}{(k_2 \cdot k_3)}$$

$$+(T^b T^a T^c)_{ij} 2 p'^*_\perp k_{3\perp} Z(p',k_2) + (T^b T^c T^a)_{ij} \frac{Z^*(k_3,k_2) Z(p',k_2) Z(p',k_3)}{(k_2 \cdot k_3)}$$

$$\left. +(T^c T^a T^b)_{ij} 2 p'^*_\perp k_{2\perp} Z(p',k_3) + (T^c T^b T^a)_{ij} \frac{Z^*(k_2,k_3) Z(p',k_2) Z(p',k_3)}{(k_2 \cdot k_3)} \right\},$$

10.45. $g\,\overline{q} \to g\,g\,\overline{q}$

$$M(+,-;+,+,-) = \frac{g^3 p'_\perp}{p'^2_+ k_{2-} k_{3-}} \left[\frac{E p'_-}{k_{2+} k_{3+}(p' \cdot k_2)(p' \cdot k_3)}\right]^{\frac{1}{2}}$$

$$\times \Bigg\{ -(T^a T^b T^c)_{ij} \frac{p'_\perp k^*_{3\perp} Z(k_2, k_3) Z^*(p', k_2)}{(k_2 \cdot k_3)}$$

$$-(T^a T^c T^b)_{ij} \frac{p'_\perp k^*_{2\perp} Z(k_3, k_2) Z^*(p', k_3)}{(k_2 \cdot k_3)}$$

$$+(T^b T^a T^c)_{ij} 2 p'_\perp k^*_{3\perp} Z^*(p', k_2) + (T^b T^c T^a)_{ij} \frac{Z(k_3, k_2) Z^*(p', k_2) Z^*(p', k_3)}{(k_2 \cdot k_3)}$$

$$+(T^c T^a T^b)_{ij} 2 p'_\perp k^*_{2\perp} Z^*(p', k_3) + (T^c T^b T^a)_{ij} \frac{Z(k_2, k_3) Z^*(p', k_2) Z^*(p', k_3)}{(k_2 \cdot k_3)} \Bigg\},$$

$$M(-,+;+,-,+) = \frac{g^3 p'^*_\perp Z(k_3, p')}{2 p'^2_+ k_{2-} k_{3-}} \left[\frac{k_{3+}}{E p'^3_- k_{2+}(p' \cdot k_2)(p' \cdot k_3)}\right]^{\frac{1}{2}}$$

$$\times \Bigg\{ -(T^a T^b T^c)_{ij} \frac{p'_\perp k^*_{3\perp} Z(k_2, k_3) Z^*(p', k_2)}{(k_2 \cdot k_3)}$$

$$-(T^a T^c T^b)_{ij} \frac{p'_\perp k^*_{2\perp} Z(k_3, k_2) Z^*(p', k_3)}{(k_2 \cdot k_3)}$$

$$+(T^b T^a T^c)_{ij} 2 p'_\perp k^*_{3\perp} Z^*(p', k_2) + (T^b T^c T^a)_{ij} \frac{Z(k_3, k_2) Z^*(p', k_2) Z^*(p', k_3)}{(k_2 \cdot k_3)}$$

$$+(T^c T^a T^b)_{ij} 2 p'_\perp k^*_{2\perp} Z^*(p', k_3) + (T^c T^b T^a)_{ij} \frac{Z(k_2, k_3) Z^*(p', k_2) Z^*(p', k_3)}{(k_2 \cdot k_3)} \Bigg\},$$

$$M(-,+;-,+,+) = \frac{g^3 p'^*_\perp Z(k_2, p')}{2 p'^2_+ k_{2-} k_{3-}} \left[\frac{k_{2+}}{E p'^3_- k_{3+}(p' \cdot k_2)(p' \cdot k_3)}\right]^{\frac{1}{2}}$$

$$\times \Bigg\{ -(T^a T^b T^c)_{ij} \frac{p'_\perp k^*_{3\perp} Z(k_2, k_3) Z^*(p', k_2)}{(k_2 \cdot k_3)}$$

$$-(T^a T^c T^b)_{ij} \frac{p'_\perp k^*_{2\perp} Z(k_3, k_2) Z^*(p', k_3)}{(k_2 \cdot k_3)}$$

$$+(T^b T^a T^c)_{ij} 2 p'_\perp k^*_{3\perp} Z^*(p', k_2) + (T^b T^c T^a)_{ij} \frac{Z(k_3, k_2) Z^*(p', k_2) Z^*(p', k_3)}{(k_2 \cdot k_3)}$$

$$+(T^c T^a T^b)_{ij} 2 p'_\perp k^*_{2\perp} Z^*(p', k_3) + (T^c T^b T^a)_{ij} \frac{Z(k_2, k_3) Z^*(p', k_2) Z^*(p', k_3)}{(k_2 \cdot k_3)} \Bigg\},$$

$$M(+,-;+,-,-) = \frac{g^3 p'_\perp Z^*(k_2,p')}{2p'^2_+ k_{2-}k_{3-}} \left[\frac{k_{2+}}{Ep'^3_- k_{3+}(p'\cdot k_2)(p'\cdot k_3)}\right]^{\frac{1}{2}}$$

$$\times \left\{ -(T^a T^b T^c)_{ij} \frac{p'^*_\perp k_{3\perp} Z^*(k_2,k_3) Z(p',k_2)}{(k_2 \cdot k_3)} \right.$$

$$-(T^a T^c T^b)_{ij} \frac{p'^*_\perp k_{2\perp} Z^*(k_3,k_2) Z(p',k_3)}{(k_2 \cdot k_3)}$$

$$+(T^b T^a T^c)_{ij} 2p'^*_\perp k_{3\perp} Z(p',k_2) + (T^b T^c T^a)_{ij} \frac{Z^*(k_3,k_2) Z(p',k_2) Z(p',k_3)}{(k_2 \cdot k_3)}$$

$$\left. +(T^c T^a T^b)_{ij} 2p'^*_\perp k_{2\perp} Z(p',k_3) + (T^c T^b T^a)_{ij} \frac{Z^*(k_2,k_3) Z(p',k_2) Z(p',k_3)}{(k_2 \cdot k_3)} \right\},$$

$$M(+,-;-,+,-) = \frac{g^3 p'_\perp Z^*(k_3,p')}{2p'^2_+ k_{2-}k_{3-}} \left[\frac{k_{3+}}{Ep'^3_- k_{2+}(p'\cdot k_2)(p'\cdot k_3)}\right]^{\frac{1}{2}}$$

$$\times \left\{ -(T^a T^b T^c)_{ij} \frac{p'^*_\perp k_{3\perp} Z^*(k_2,k_3) Z(p',k_2)}{(k_2 \cdot k_3)} \right.$$

$$-(T^a T^c T^b)_{ij} \frac{p'^*_\perp k_{2\perp} Z^*(k_3,k_2) Z(p',k_3)}{(k_2 \cdot k_3)}$$

$$+(T^b T^a T^c)_{ij} 2p'^*_\perp k_{3\perp} Z(p',k_2) + (T^b T^c T^a)_{ij} \frac{Z^*(k_3,k_2) Z(p',k_2) Z(p',k_3)}{(k_2 \cdot k_3)}$$

$$\left. +(T^c T^a T^b)_{ij} 2p'^*_\perp k_{2\perp} Z(p',k_3) + (T^c T^b T^a)_{ij} \frac{Z^*(k_2,k_3) Z(p',k_2) Z(p',k_3)}{(k_2 \cdot k_3)} \right\},$$

$$M(-,+;-,-,+) = \frac{g^3 p'^*_\perp}{p'^2_+ k_{2-}k_{3-}} \left[\frac{Ep'_-}{k_{2+}k_{3+}(p'\cdot k_2)(p'\cdot k_3)}\right]^{\frac{1}{2}}$$

$$\times \left\{ -(T^a T^b T^c)_{ij} \frac{p'^*_\perp k_{3\perp} Z^*(k_2,k_3) Z(p',k_2)}{(k_2 \cdot k_3)} \right.$$

$$-(T^a T^c T^b)_{ij} \frac{p'^*_\perp k_{2\perp} Z^*(k_3,k_2) Z(p',k_3)}{(k_2 \cdot k_3)}$$

$$+(T^b T^a T^c)_{ij} 2p'^*_\perp k_{3\perp} Z(p',k_2) + (T^b T^c T^a)_{ij} \frac{Z^*(k_3,k_2) Z(p',k_2) Z(p',k_3)}{(k_2 \cdot k_3)}$$

$$\left. +(T^c T^a T^b)_{ij} 2p'^*_\perp k_{2\perp} Z(p',k_3) + (T^c T^b T^a)_{ij} \frac{Z^*(k_2,k_3) Z(p',k_2) Z(p',k_3)}{(k_2 \cdot k_3)} \right\},$$

10.45. $g\bar{q} \to gg\bar{q}$

$$M(-,-;+,-,-) = \frac{g^3 p'_\perp Z^*(k_3,p')}{2p'^2_+ k_{2-} k_{3-}} \left[\frac{(p'\cdot k_3)}{E^3 p'^3_- k_{2+} k_{3+}(p'\cdot k_2)}\right]^{\frac{1}{2}}$$

$$\times \Bigg\{-(T^a T^b T^c)_{ij} \frac{p'_\perp k^*_{3\perp} Z(k_2,k_3) Z^*(p',k_2)}{(k_2\cdot k_3)}$$

$$-(T^a T^c T^b)_{ij} \frac{p'_\perp k^*_{2\perp} Z(k_3,k_2) Z^*(p',k_3)}{(k_2\cdot k_3)}$$

$$+(T^b T^a T^c)_{ij} 2p'_\perp k^*_{3\perp} Z^*(p',k_2) + (T^b T^c T^a)_{ij} \frac{Z(k_3,k_2) Z^*(p',k_2) Z^*(p',k_3)}{(k_2\cdot k_3)}$$

$$+(T^c T^a T^b)_{ij} 2p'_\perp k^*_{2\perp} Z^*(p',k_3) + (T^c T^b T^a)_{ij} \frac{Z(k_2,k_3) Z^*(p',k_2) Z^*(p',k_3)}{(k_2\cdot k_3)}\Bigg\},$$

$$M(-,-;-,+,-) = \frac{g^3 p'_\perp Z^*(k_2,p')}{2p'^2_+ k_{2-} k_{3-}} \left[\frac{(p'\cdot k_2)}{E^3 p'^3_- k_{2+} k_{3+}(p'\cdot k_3)}\right]^{\frac{1}{2}}$$

$$\times \Bigg\{-(T^a T^b T^c)_{ij} \frac{p'_\perp k^*_{3\perp} Z(k_2,k_3) Z^*(p',k_2)}{(k_2\cdot k_3)}$$

$$-(T^a T^c T^b)_{ij} \frac{p'_\perp k^*_{2\perp} Z(k_3,k_2) Z^*(p',k_3)}{(k_2\cdot k_3)}$$

$$+(T^b T^a T^c)_{ij} 2p'_\perp k^*_{3\perp} Z^*(p',k_2) + (T^b T^c T^a)_{ij} \frac{Z(k_3,k_2) Z^*(p',k_2) Z^*(p',k_3)}{(k_2\cdot k_3)}$$

$$+(T^c T^a T^b)_{ij} 2p'_\perp k^*_{2\perp} Z^*(p',k_3) + (T^c T^b T^a)_{ij} \frac{Z(k_2,k_3) Z^*(p',k_2) Z^*(p',k_3)}{(k_2\cdot k_3)}\Bigg\},$$

$$M(-,-;-,-,-) = \frac{2g^3 p'_\perp}{p'^2_+ k_{2-} k_{3-}} \left[\frac{E^3}{p'_- k_{2+} k_{3+}(p'\cdot k_2)(p'\cdot k_3)}\right]^{\frac{1}{2}}$$

$$\times \Bigg\{-(T^a T^b T^c)_{ij} \frac{p'^*_\perp k_{3\perp} Z^*(k_2,k_3) Z(p',k_2)}{(k_2\cdot k_3)}$$

$$-(T^a T^c T^b)_{ij} \frac{p'^*_\perp k_{2\perp} Z^*(k_3,k_2) Z(p',k_3)}{(k_2\cdot k_3)}$$

$$+(T^b T^a T^c)_{ij} 2p'^*_\perp k_{3\perp} Z(p',k_2) + (T^b T^c T^a)_{ij} \frac{Z^*(k_3,k_2) Z(p',k_2) Z(p',k_3)}{(k_2\cdot k_3)}$$

$$+(T^c T^a T^b)_{ij} 2p'^*_\perp k_{2\perp} Z(p',k_3) + (T^c T^b T^a)_{ij} \frac{Z^*(k_2,k_3) Z(p',k_2) Z(p',k_3)}{(k_2\cdot k_3)}\Bigg\}.$$

(10.311)

Squared absolute values of the nonvanishing helicity amplitudes, summed over the color degrees of freedom of the initial state and the final state:

$$|M(+,+;+,+,+)|^2 = |M(-,-;-,-,-)|^2 = \frac{4g^6 E^3 F}{k_{2+}k_{3+}(p'\cdot k_2)(p'\cdot k_3)},$$

$$|M(+,+;+,-,+)|^2 = |M(-,-;-,+,-)|^2 = \frac{g^6 (p'\cdot k_2)^2 F}{2E^3 p'_- k_{3+}(p'\cdot k_3)},$$

$$|M(+,+;-,+,+)|^2 = |M(-,-;+,-,-)|^2 = \frac{g^6 (p'\cdot k_3)^2 F}{2E^3 p'_- k_{2+}(p'\cdot k_2)},$$

$$|M(+,-;+,+,-)|^2 = |M(-,+;-,-,+)|^2 = \frac{g^6 E p'^2_- F}{k_{2+}k_{3+}(p'\cdot k_2)(p'\cdot k_3)},$$

$$|M(-,+;+,-,+)|^2 = |M(+,-;-,+,-)|^2 = \frac{g^6 k_{3+}^2 F}{2E p'_- k_{2+}(p'\cdot k_2)},$$

$$|M(-,+;-,+,+)|^2 = |M(+,-;+,-,-)|^2 = \frac{g^6 k_{2+}^2 F}{2E p'_- k_{3+}(p'\cdot k_3)}.$$

(10.312)

Unpolarized squared matrix element:

$$\overline{|M|^2} = \frac{g^6}{108} \frac{\sum_{i=1}^{3}(p\cdot k_i)(p'\cdot k_i)[(p\cdot k_i)^2 + (p'\cdot k_i)^2]}{\prod_{i=1}^{3}(p\cdot k_i)(p'\cdot k_i)}$$

$$\times \left\{ 10(p\cdot p') - 9\left[\frac{(p\cdot k_1)(p'\cdot k_2) + (p\cdot k_2)(p'\cdot k_1)}{(k_1\cdot k_2)} \right.\right.$$

$$+ \frac{(p\cdot k_2)(p'\cdot k_3) + (p\cdot k_3)(p'\cdot k_2)}{(k_2\cdot k_3)}$$

$$\left. + \frac{(p\cdot k_3)(p'\cdot k_1) + (p\cdot k_1)(p'\cdot k_3)}{(k_1\cdot k_3)}\right] + \frac{81}{(p\cdot p')}$$

$$\times \left[\frac{(p\cdot k_1)(p'\cdot k_1)[(p\cdot k_2)(p'\cdot k_3) + (p\cdot k_3)(p'\cdot k_2)]}{(k_1\cdot k_2)(k_1\cdot k_3)}\right.$$

$$+ \frac{(p\cdot k_2)(p'\cdot k_2)[(p\cdot k_3)(p'\cdot k_1) + (p\cdot k_1)(p'\cdot k_3)]}{(k_1\cdot k_2)(k_2\cdot k_3)}$$

$$\left.\left. + \frac{(p\cdot k_3)(p'\cdot k_3)[(p\cdot k_1)(p'\cdot k_2) + (p\cdot k_2)(p'\cdot k_1)]}{(k_1\cdot k_3)(k_2\cdot k_3)}\right]\right\}.$$

(10.313)

10.46. $\bar{q}q \to \gamma\gamma\gamma$

Unpolarized cross section:

$$d\sigma = \frac{\alpha_S^3}{1728\pi^2} \frac{\sum_{i=1}^{3}(p\cdot k_i)(p'\cdot k_i)[(p\cdot k_i)^2 + (p'\cdot k_i)^2]}{(p\cdot k_1)\prod_{i=1}^{3}(p\cdot k_i)(p'\cdot k_i)}$$

$$\times \left\{ 10(p\cdot p') - 9\left[\frac{(p\cdot k_1)(p'\cdot k_2) + (p\cdot k_2)(p'\cdot k_1)}{(k_1\cdot k_2)}\right.\right.$$

$$+ \frac{(p\cdot k_2)(p'\cdot k_3) + (p\cdot k_3)(p'\cdot k_2)}{(k_2\cdot k_3)}$$

$$\left.+ \frac{(p\cdot k_3)(p'\cdot k_1) + (p\cdot k_1)(p'\cdot k_3)}{(k_1\cdot k_3)}\right] + \frac{81}{(p\cdot p')}$$

$$\times \left[\frac{(p\cdot k_1)(p'\cdot k_1)[(p\cdot k_2)(p'\cdot k_3) + (p\cdot k_3)(p'\cdot k_2)]}{(k_1\cdot k_2)(k_1\cdot k_3)}\right.$$

$$+ \frac{(p\cdot k_2)(p'\cdot k_2)[(p\cdot k_3)(p'\cdot k_1) + (p\cdot k_1)(p'\cdot k_3)]}{(k_1\cdot k_2)(k_2\cdot k_3)}$$

$$\left.\left.+ \frac{(p\cdot k_3)(p'\cdot k_3)[(p\cdot k_1)(p'\cdot k_2) + (p\cdot k_2)(p'\cdot k_1)]}{(k_1\cdot k_3)(k_2\cdot k_3)}\right]\right\}$$

$$\times \delta^4(k_1 + p - k_2 - k_3 - p')\frac{d^3\vec{k}_2\, d^3\vec{k}_3\, d^3\vec{p}'}{k_{20}\, k_{30}\, p'_0}. \tag{10.314}$$

10.46 $\bar{q}q \to \gamma\gamma\gamma$

Process:

$$\bar{q}(p_+, i) + q(p_-, j) \to \gamma(k_1) + \gamma(k_2) + \gamma(k_3). \tag{10.315}$$

Photon polarizations:

$$\rlap{/}{e}^{\pm}(k_i) = N_i[\rlap{/}{k}_i\, \rlap{/}{p}_+\, \rlap{/}{p}_-(1\pm\gamma_5) - \rlap{/}{p}_+\, \rlap{/}{p}_-\, \rlap{/}{k}_i(1\mp\gamma_5)],$$

$$N_i^{-1} = E^2[32 k_{i+} k_{i-}]^{\frac{1}{2}}, \qquad i = 1, 2, 3. \tag{10.316}$$

Nonvanishing helicity amplitudes:

$$M(+,-;+,+,-) = -\frac{\sqrt{8}\, e_q^3\, \delta_{ij}\, k_{3-} k_{3\perp}}{[k_{1+} k_{1-} k_{2+} k_{2-} k_{3+} k_{3-}]^{\frac{1}{2}}},$$

$$M(+,-;+,-,+) = -\frac{\sqrt{8}\, e_q^3\, \delta_{ij}\, k_{2-} k_{2\perp}}{[k_{1+} k_{1-} k_{2+} k_{2-} k_{3+} k_{3-}]^{\frac{1}{2}}},$$

$$M(+,-;-,+,+) = -\frac{\sqrt{8}e_q^3 \delta_{ij} k_{1-} k_{1\perp}}{[k_{1+}k_{1-}k_{2+}k_{2-}k_{3+}k_{3-}]^{\frac{1}{2}}},$$

$$M(-,+;+,+,-) = \frac{\sqrt{8}e_q^3 \delta_{ij} k_{3+} k_{3\perp}^*}{[k_{1+}k_{1-}k_{2+}k_{2-}k_{3+}k_{3-}]^{\frac{1}{2}}},$$

$$M(-,+;+,-,+) = \frac{\sqrt{8}e_q^3 \delta_{ij} k_{2+} k_{2\perp}^*}{[k_{1+}k_{1-}k_{2+}k_{2-}k_{3+}k_{3-}]^{\frac{1}{2}}},$$

$$M(-,+;-,+,+) = \frac{\sqrt{8}e_q^3 \delta_{ij} k_{1+} k_{1\perp}^*}{[k_{1+}k_{1-}k_{2+}k_{2-}k_{3+}k_{3-}]^{\frac{1}{2}}},$$

$$M(+,-;+,-,-) = \frac{\sqrt{8}e_q^3 \delta_{ij} k_{1+} k_{1\perp}}{[k_{1+}k_{1-}k_{2+}k_{2-}k_{3+}k_{3-}]^{\frac{1}{2}}},$$

$$M(+,-;-,+,-) = \frac{\sqrt{8}e_q^3 \delta_{ij} k_{2+} k_{2\perp}}{[k_{1+}k_{1-}k_{2+}k_{2-}k_{3+}k_{3-}]^{\frac{1}{2}}},$$

$$M(+,-;-,-,+) = \frac{\sqrt{8}e_q^3 \delta_{ij} k_{3+} k_{3\perp}}{[k_{1+}k_{1-}k_{2+}k_{2-}k_{3+}k_{3-}]^{\frac{1}{2}}},$$

$$M(-,+;+,-,-) = -\frac{\sqrt{8}e_q^3 \delta_{ij} k_{1-} k_{1\perp}^*}{[k_{1+}k_{1-}k_{2+}k_{2-}k_{3+}k_{3-}]^{\frac{1}{2}}},$$

$$M(-,+;-,+,-) = -\frac{\sqrt{8}e_q^3 \delta_{ij} k_{2-} k_{2\perp}^*}{[k_{1+}k_{1-}k_{2+}k_{2-}k_{3+}k_{3-}]^{\frac{1}{2}}},$$

$$M(-,+;-,-,+) = -\frac{\sqrt{8}e_q^3 \delta_{ij} k_{3-} k_{3\perp}^*}{[k_{1+}k_{1-}k_{2+}k_{2-}k_{3+}k_{3-}]^{\frac{1}{2}}}. \quad (10.317)$$

Squared absolute values of the nonvanishing helicity amplitudes, summed over the color degrees of freedom of the initial state:

$$|M(+,-;+,+,-)|^2 = |M(-,+;-,-,+)|^2 = \frac{24e_q^6 k_{3-}^2}{k_{1+}k_{1-}k_{2+}k_{2-}},$$

$$|M(+,-;+,-,+)|^2 = |M(-,+;-,+,-)|^2 = \frac{24e_q^6 k_{2-}^2}{k_{1+}k_{1-}k_{3+}k_{3-}},$$

$$|M(+,-;-,+,+)|^2 = |M(-,+;+,-,-)|^2 = \frac{24e_q^6 k_{1-}^2}{k_{2+}k_{2-}k_{3+}k_{3-}},$$

$$|M(-,+;+,+,-)|^2 = |M(+,-;-,-,+)|^2 = \frac{24e_q^6 k_{3+}^2}{k_{1+}k_{1-}k_{2+}k_{2-}},$$

$$|M(-,+;+,-,+)|^2 = |M(+,-;-,+,-)|^2 = \frac{24e_q^6 k_{2+}^2}{k_{1+}k_{1-}k_{3+}k_{3-}},$$

10.47. $\bar{q}q \to g\gamma\gamma$

$$|M(-,+;-,+,+)|^2 = |M(+,-;+,-,-)|^2 = \frac{24e_q^6 k_{1+}^2}{k_{2+}k_{2-}k_{3+}k_{3-}}. \tag{10.318}$$

Unpolarized squared matrix element:

$$\overline{|M|^2} = \frac{4e_q^6 E^2}{3} \frac{\sum_{i=1}^{3}(p_+ \cdot k_i)(p_- \cdot k_i)[(p_+ \cdot k_i)^2 + (p_- \cdot k_i)^2]}{\prod_{i=1}^{3}(p_+ \cdot k_i)(p_- \cdot k_i)}. \tag{10.319}$$

Unpolarized cross section:

$$d\sigma = \frac{\alpha^3 Q_f^6}{24\pi^2} \frac{\sum_{i=1}^{3}(p_+ \cdot k_i)(p_- \cdot k_i)[(p_+ \cdot k_i)^2 + (p_- \cdot k_i)^2]}{\prod_{i=1}^{3}(p_+ \cdot k_i)(p_- \cdot k_i)}$$

$$\times \delta^4(p_+ + p_- - k_1 - k_2 - k_3) \frac{d^3\vec{k}_1 \, d^3\vec{k}_2 \, d^3\vec{k}_3}{k_{10} \, k_{20} \, k_{30}}. \tag{10.320}$$

10.47 $\bar{q}q \to g\gamma\gamma$

Process:

$$\bar{q}(p_+, i) + q(p_-, j) \to g(k_1, a) + \gamma(k_2) + \gamma(k_3). \tag{10.321}$$

Photon polarizations:

$$\not{\epsilon}^{\pm}(k_1) = N_1[\not{k}_1 \not{p}_- \not{p}_+ (1 \pm \gamma_5) + \not{p}_+ \not{p}_- \not{k}_1 (1 \mp \gamma_5)],$$

$$\not{\epsilon}^{\pm}(k_i) = N_i[\not{k}_i \not{p}_+ \not{p}_- (1 \pm \gamma_5) - \not{p}_+ \not{p}_- \not{k}_i (1 \mp \gamma_5)], \qquad i = 2,3,$$

$$N_i^{-1} = E^2[32k_{i+}k_{i-}]^{\frac{1}{2}}, \qquad i = 1,2,3. \tag{10.322}$$

Nonvanishing helicity amplitudes:

$$M(+,-;+,+,-) = -\frac{\sqrt{8}ge_q^2 T_{ij}^a k_{3-}k_{3\perp}}{[k_{1+}k_{1-}k_{2+}k_{2-}k_{3+}k_{3-}]^{\frac{1}{2}}},$$

$$M(+,-;+,-,+) = -\frac{\sqrt{8}ge_q^2 T_{ij}^a k_{2-}k_{2\perp}}{[k_{1+}k_{1-}k_{2+}k_{2-}k_{3+}k_{3-}]^{\frac{1}{2}}},$$

$$M(+,-;-,+,+) = -\frac{\sqrt{8}ge_q^2 T_{ij}^a k_{1-}k_{1\perp}}{[k_{1+}k_{1-}k_{2+}k_{2-}k_{3+}k_{3-}]^{\frac{1}{2}}},$$

$$M(-,+;+,+,-) = \frac{\sqrt{8}ge_q^2 T_{ij}^a k_{3+} k_{3\perp}^*}{[k_{1+}k_{1-}k_{2+}k_{2-}k_{3+}k_{3-}]^{\frac{1}{2}}},$$

$$M(-,+;+,-,+) = \frac{\sqrt{8}ge_q^2 T_{ij}^a k_{2+} k_{2\perp}^*}{[k_{1+}k_{1-}k_{2+}k_{2-}k_{3+}k_{3-}]^{\frac{1}{2}}},$$

$$M(-,+;-,+,+) = \frac{\sqrt{8}ge_q^2 T_{ij}^a k_{1+} k_{1\perp}^*}{[k_{1+}k_{1-}k_{2+}k_{2-}k_{3+}k_{3-}]^{\frac{1}{2}}},$$

$$M(+,-;+,-,-) = \frac{\sqrt{8}ge_q^2 T_{ij}^a k_{1+} k_{1\perp}}{[k_{1+}k_{1-}k_{2+}k_{2-}k_{3+}k_{3-}]^{\frac{1}{2}}},$$

$$M(+,-;-,+,-) = \frac{\sqrt{8}ge_q^2 T_{ij}^a k_{2+} k_{2\perp}}{[k_{1+}k_{1-}k_{2+}k_{2-}k_{3+}k_{3-}]^{\frac{1}{2}}},$$

$$M(+,-;-,-,+) = \frac{\sqrt{8}ge_q^2 T_{ij}^a k_{3+} k_{3\perp}}{[k_{1+}k_{1-}k_{2+}k_{2-}k_{3+}k_{3-}]^{\frac{1}{2}}},$$

$$M(-,+;+,-,-) = -\frac{\sqrt{8}ge_q^2 T_{ij}^a k_{1-} k_{1\perp}^*}{[k_{1+}k_{1-}k_{2+}k_{2-}k_{3+}k_{3-}]^{\frac{1}{2}}},$$

$$M(-,+;-,+,-) = -\frac{\sqrt{8}ge_q^2 T_{ij}^a k_{2-} k_{2\perp}^*}{[k_{1+}k_{1-}k_{2+}k_{2-}k_{3+}k_{3-}]^{\frac{1}{2}}},$$

$$M(-,+;-,-,+) = -\frac{\sqrt{8}ge_q^2 T_{ij}^a k_{3-} k_{3\perp}^*}{[k_{1+}k_{1-}k_{2+}k_{2-}k_{3+}k_{3-}]^{\frac{1}{2}}}. \quad (10.323)$$

Squared absolute values of the nonvanishing helicity amplitudes, summed over the color degrees of freedom of the initial state and the final state:

$$|M(+,-;+,+,-)|^2 = |M(-,+;-,-,+)|^2 = \frac{32g^2 e_q^4 k_{3-}^2}{k_{1+}k_{1-}k_{2+}k_{2-}},$$

$$|M(+,-;+,-,+)|^2 = |M(-,+;-,+,-)|^2 = \frac{32g^2 e_q^4 k_{2-}^2}{k_{1+}k_{1-}k_{3+}k_{3-}},$$

$$|M(+,-;-,+,+)|^2 = |M(-,+;+,-,-)|^2 = \frac{32g^2 e_q^4 k_{1-}^2}{k_{2+}k_{2-}k_{3+}k_{3-}},$$

$$|M(-,+;+,+,-)|^2 = |M(+,-;-,-,+)|^2 = \frac{32g^2 e_q^4 k_{3+}^2}{k_{1+}k_{1-}k_{2+}k_{2-}},$$

$$|M(-,+;+,-,+)|^2 = |M(+,-;-,+,-)|^2 = \frac{32g^2 e_q^4 k_{2+}^2}{k_{1+}k_{1-}k_{3+}k_{3-}},$$

10.48. $\bar{q}q \to gg\gamma$

$$|M(-,+;-,+,+)|^2 = |M(+,-;+,-,-)|^2 = \frac{32g^2 e_q^4 k_{1+}^2}{k_{2+}k_{2-}k_{3+}k_{3-}}. \qquad (10.324)$$

Unpolarized squared matrix element:

$$\overline{|M|^2} = \frac{16g^2 e_q^4 E^2}{9} \frac{\sum_{i=1}^{3}(p_+ \cdot k_i)(p_- \cdot k_i)[(p_+ \cdot k_i)^2 + (p_- \cdot k_i)^2]}{\prod_{i=1}^{3}(p_+ \cdot k_i)(p_- \cdot k_i)}. \qquad (10.325)$$

Unpolarized cross section:

$$d\sigma = \frac{\alpha_S \alpha^2 Q_f^4}{18\pi^2} \frac{\sum_{i=1}^{3}(p_+ \cdot k_i)(p_- \cdot k_i)[(p_+ \cdot k_i)^2 + (p_- \cdot k_i)^2]}{\prod_{i=1}^{3}(p_+ \cdot k_i)(p_- \cdot k_i)}$$

$$\times \delta^4(p_+ + p_- - k_1 - k_2 - k_3) \frac{d^3\vec{k_1}\, d^3\vec{k_2}\, d^3\vec{k_3}}{k_{10}\, k_{20}\, k_{30}}. \qquad (10.326)$$

10.48 $\bar{q}q \to gg\gamma$

Process:

$$\bar{q}(p_+,i) + q(p,j) \to g(k_1,a) + g(k_2,b) + \gamma(k_3). \qquad (10.327)$$

Invariants and definitions:

$$\begin{aligned}
s &= 2(p_+ \cdot p_-), & t &= -2(p_+ \cdot k_1), & u &= -2(p_+ \cdot k_2), \\
s' &= 2(k_1 \cdot k_2), & t' &= -2(p_- \cdot k_2), & u' &= -2(p_- \cdot k_1), \\
Z_{12} &= k_{1+}k_{2-} - k_{1\perp}^* k_{2\perp}, & & & Z_{21} &= k_{2+}k_{1-} - k_{2\perp}^* k_{1\perp},
\end{aligned}$$

$$A = [k_{1+}k_{1-}k_{2+}k_{2-}k_{3+}k_{3-}]^{\frac{1}{2}}. \qquad (10.328)$$

Gluon and photon polarizations:

$$\begin{aligned}
\not{\epsilon}^{\pm}(k_1) &= N_1[\not{k}_1\, \not{p}_-\, \not{p}_+(1 \pm \gamma_5) + \not{p}_+\, \not{p}_-\, \not{k}_1(1 \mp \gamma_5)], \\
\not{\epsilon}^{\pm}(k_2) &= N_2[\not{k}_2\, \not{p}_-\, \not{p}_+(1 \pm \gamma_5) + \not{p}_+\, \not{p}_-\, \not{k}_2(1 \mp \gamma_5)], \\
\not{\epsilon}^{\pm}(k_3) &= N_3[\not{p}_+\, \not{p}_-\, \not{k}_3(1 \mp \gamma_5) - \not{k}_3\, \not{p}_+\, \not{p}_-(1 \pm \gamma_5)], \\
N_i^{-1} &= E^2[32 k_{i+} k_{i-}]^{\frac{1}{2}}, & i &= 1,2,3. \qquad (10.329)
\end{aligned}$$

Nonvanishing helicity amplitudes:

$$M(+,-;+,+,-) = -\frac{\sqrt{2}g^2 e_q k_{3-} k_{3\perp}}{A(k_1 \cdot k_2)} \left[Z_{12}(T^a T^b)_{ij} + Z_{21}(T^b T^a)_{ij} \right],$$

$$M(+,-;+,-,+) = -\frac{\sqrt{2}g^2 e_q k_{2-} k_{2\perp}}{A(k_1 \cdot k_2)} \left[Z_{12}(T^a T^b)_{ij} + Z_{21}(T^b T^a)_{ij} \right],$$

$$M(+,-;-,+,+) = -\frac{\sqrt{2}g^2 e_q k_{1-} k_{1\perp}}{A(k_1 \cdot k_2)} \left[Z_{12}(T^a T^b)_{ij} + Z_{21}(T^b T^a)_{ij} \right],$$

$$M(-,+;+,+,-) = \frac{\sqrt{2}g^2 e_q k_{3+} k_{3\perp}^*}{A(k_1 \cdot k_2)} \left[Z_{12}(T^a T^b)_{ij} + Z_{21}(T^b T^a)_{ij} \right],$$

$$M(-,+;+,-,+) = \frac{\sqrt{2}g^2 e_q k_{2+} k_{2\perp}^*}{A(k_1 \cdot k_2)} \left[Z_{12}(T^a T^b)_{ij} + Z_{21}(T^b T^a)_{ij} \right],$$

$$M(-,+;-,+,+) = \frac{\sqrt{2}g^2 e_q k_{1+} k_{1\perp}^*}{A(k_1 \cdot k_2)} \left[Z_{12}(T^a T^b)_{ij} + Z_{21}(T^b T^a)_{ij} \right],$$

$$M(+,-;+,-,-) = \frac{\sqrt{2}g^2 e_q k_{1+} k_{1\perp}}{A(k_1 \cdot k_2)} \left[Z_{12}^*(T^a T^b)_{ij} + Z_{21}^*(T^b T^a)_{ij} \right],$$

$$M(+,-;-,+,-) = \frac{\sqrt{2}g^2 e_q k_{2+} k_{2\perp}}{A(k_1 \cdot k_2)} \left[Z_{12}^*(T^a T^b)_{ij} + Z_{21}^*(T^b T^a)_{ij} \right],$$

$$M(+,-;-,-,+) = \frac{\sqrt{2}g^2 e_q k_{3+} k_{3\perp}}{A(k_1 \cdot k_2)} \left[Z_{12}^*(T^a T^b)_{ij} + Z_{21}^*(T^b T^a)_{ij} \right],$$

$$M(-,+;+,-,-) = -\frac{\sqrt{2}g^2 e_q k_{1-} k_{1\perp}^*}{A(k_1 \cdot k_2)} \left[Z_{12}^*(T^a T^b)_{ij} + Z_{21}^*(T^b T^a)_{ij} \right],$$

$$M(-,+;-,+,-) = -\frac{\sqrt{2}g^2 e_q k_{2-} k_{2\perp}^*}{A(k_1 \cdot k_2)} \left[Z_{12}^*(T^a T^b)_{ij} + Z_{21}^*(T^b T^a)_{ij} \right],$$

$$M(-,+;-,-,+) = -\frac{\sqrt{2}g^2 e_q k_{3-} k_{3\perp}^*}{A(k_1 \cdot k_2)} \left[Z_{12}^*(T^a T^b)_{ij} + Z_{21}^*(T^b T^a)_{ij} \right].$$

(10.330)

Squared absolute values of the nonvanishing helicity amplitudes, summed over the color degrees of freedom of the initial state and the final state:

$$|M(+,-;+,+,-)|^2 = |M(-,+;-,-,+)|^2$$

$$= \frac{8g^4 e_q^2 k_{3+} k_{3-}^3}{3A^2(k_1 \cdot k_2)} \left[9k_{1+} k_{2-} + 9k_{1-} k_{2+} - 2(k_1 \cdot k_2) \right],$$

10.48. $\bar{q}q \to gg\gamma$

$$|M(+,-;+,-,+)|^2 = |M(-,+;-,+,-)|^2$$
$$= \frac{8g^4 e_q^2 k_{2+} k_{2-}^3}{3A^2(k_1 \cdot k_2)}[9k_{1+}k_{2-} + 9k_{1-}k_{2+} - 2(k_1 \cdot k_2)],$$

$$|M(+,-;-,+,+)|^2 = |M(-,+;+,-,-)|^2$$
$$= \frac{8g^4 e_q^2 k_{1+} k_{1-}^3}{3A^2(k_1 \cdot k_2)}[9k_{1+}k_{2-} + 9k_{1-}k_{2+} - 2(k_1 \cdot k_2)],$$

$$|M(-,+;+,+,-)|^2 = |M(+,-;-,-,+)|^2$$
$$= \frac{8g^4 e_q^2 k_{3+}^3 k_{3-}}{3A^2(k_1 \cdot k_2)}[9k_{1+}k_{2-} + 9k_{1-}k_{2+} - 2(k_1 \cdot k_2)],$$

$$|M(-,+;+,-,+)|^2 = |M(+,-;-,+,-)|^2$$
$$= \frac{8g^4 e_q^2 k_{2+}^3 k_{2-}}{3A^2(k_1 \cdot k_2)}[9k_{1+}k_{2-} + 9k_{1-}k_{2+} - 2(k_1 \cdot k_2)],$$

$$|M(-,+;-,+,+)|^2 = |M(+,-;+,-,-)|^2$$
$$= \frac{8g^4 e_q^2 k_{1+}^3 k_{1-}}{3A^2(k_1 \cdot k_2)}[9k_{1+}k_{2-} + 9k_{1-}k_{2+} - 2(k_1 \cdot k_2)].$$
(10.331)

Unpolarized squared matrix element:

$$\overline{|M|^2} = \frac{2g^4 e_q^2}{27s'}(9tt' + 9uu' - ss')\frac{\sum_{i=1}^{3}(p_+ \cdot k_i)(p_- \cdot k_i)[(p_+ \cdot k_i)^2 + (p_- \cdot k_i)^2]}{\prod_{i=1}^{3}(p_+ \cdot k_i)(p_- \cdot k_i)}.$$
(10.332)

Unpolarized cross section:

$$d\sigma = \frac{\alpha_S^2 \alpha Q_f^2}{108\pi^2}\frac{9tt' + 9uu' - ss'}{ss'}\frac{\sum_{i=1}^{3}(p_+ \cdot k_i)(p_- \cdot k_i)[(p_+ \cdot k_i)^2 + (p_- \cdot k_i)^2]}{\prod_{i=1}^{3}(p_+ \cdot k_i)(p_- \cdot k_i)}$$

$$\times \delta^4(p_+ + p_- - k_1 - k_2 - k_3)\frac{d^3\vec{k}_1\, d^3\vec{k}_2\, d^3\vec{k}_3}{k_{10}\, k_{20}\, k_{30}}.$$
(10.333)

10.49 $\bar{q}q \to ggg$

Process:

$$\bar{q}(p_+, i) + q(p_-, j) \to g(k_1, a) + g(k_2, b) + g(k_3, c). \qquad (10.334)$$

Definitions:

$$C = (k_{1+}k_{1-}k_{2+}k_{2-}k_{3+}k_{3-})^{-\frac{1}{2}},$$

$$Z_{ij} = k_{i+}k_{j-} - k_{i\perp}^* k_{j\perp}, \qquad i,j = 1, 2, 3,$$

$$\begin{aligned}F = \frac{80}{9} &- 4\left[\frac{k_{1+}k_{2-} + k_{1-}k_{2+}}{(k_1 \cdot k_2)} + \frac{k_{1+}k_{3-} + k_{1-}k_{3+}}{(k_1 \cdot k_3)}\right. \\ &\left. + \frac{k_{2+}k_{3-} + k_{2-}k_{3+}}{(k_2 \cdot k_3)}\right] + 18\left[\frac{k_{1+}k_{1-}(k_{2+}k_{3-} + k_{2-}k_{3+})}{(k_1 \cdot k_2)(k_1 \cdot k_3)}\right. \\ &\left. + \frac{k_{2+}k_{2-}(k_{3+}k_{1-} + k_{3-}k_{1+})}{(k_1 \cdot k_2)(k_2 \cdot k_3)} + \frac{k_{3+}k_{3-}(k_{1+}k_{2-} + k_{1-}k_{2+})}{(k_1 \cdot k_3)(k_2 \cdot k_3)}\right].\end{aligned}$$
$$(10.335)$$

Gluon polarizations:

$$\not{\epsilon}_i^\pm = N_i[\not{k}_i \not{p}_- \not{p}_+(1 \pm \gamma_5) + \not{p}_+ \not{p}_- \not{k}_i(1 \mp \gamma_5)],$$

$$N_i^{-1} = E^2[32k_{i+}k_{i-}]^{\frac{1}{2}}, \qquad i = 1, 2, 3. \qquad (10.336)$$

Nonvanishing helicity amplitudes:

$$\begin{aligned}M(+, -; +, +, -) = &\frac{g^3 C k_{3-} k_{3\perp} Z_{13} Z_{32} Z_{21}}{2\sqrt{2}(k_1 \cdot k_2)(k_1 \cdot k_3)(k_2 \cdot k_3)} \\ &\times \left[(T^a T^b T^c)_{ij} \frac{Z_{31}^*}{k_{3+} k_{1-}} - (T^a T^c T^b)_{ij} \frac{Z_{21}^*}{k_{2+} k_{1-}}\right. \\ &\left. - (T^b T^a T^c)_{ij} \frac{Z_{32}^*}{k_{3+} k_{2-}} + (T^b T^c T^a)_{ij} \frac{Z_{12}^*}{k_{1+} k_{2-}}\right. \\ &\left. + (T^c T^a T^b)_{ij} \frac{Z_{23}^*}{k_{2+} k_{3-}} - (T^c T^b T^a)_{ij} \frac{Z_{13}^*}{k_{1+} k_{3-}}\right],\end{aligned}$$

10.49. $\bar{q}q \to ggg$

$$M(+,-;+,-,+) = \frac{g^3 C k_{2-} k_{2\perp} Z_{13} Z_{32} Z_{21}}{2\sqrt{2}(k_1 \cdot k_2)(k_1 \cdot k_3)(k_2 \cdot k_3)}$$

$$\times \left[(T^a T^b T^c)_{ij} \frac{Z^*_{31}}{k_{3+} k_{1-}} - (T^a T^c T^b)_{ij} \frac{Z^*_{21}}{k_{2+} k_{1-}} \right.$$

$$-(T^b T^a T^c)_{ij} \frac{Z^*_{32}}{k_{3+} k_{2-}} + (T^b T^c T^a)_{ij} \frac{Z^*_{12}}{k_{1+} k_{2-}}$$

$$\left. +(T^c T^a T^b)_{ij} \frac{Z^*_{23}}{k_{2+} k_{3-}} - (T^c T^b T^a)_{ij} \frac{Z^*_{13}}{k_{1+} k_{3-}} \right],$$

$$M(+,-;-,+,+) = \frac{g^3 C k_{1-} k_{1\perp} Z_{13} Z_{32} Z_{21}}{2\sqrt{2}(k_1 \cdot k_2)(k_1 \cdot k_3)(k_2 \cdot k_3)}$$

$$\times \left[(T^a T^b T^c)_{ij} \frac{Z^*_{31}}{k_{3+} k_{1-}} - (T^a T^c T^b)_{ij} \frac{Z^*_{21}}{k_{2+} k_{1-}} \right.$$

$$-(T^b T^a T^c)_{ij} \frac{Z^*_{32}}{k_{3+} k_{2-}} + (T^b T^c T^a)_{ij} \frac{Z^*_{12}}{k_{1+} k_{2-}}$$

$$\left. +(T^c T^a T^b)_{ij} \frac{Z^*_{23}}{k_{2+} k_{3-}} - (T^c T^b T^a)_{ij} \frac{Z^*_{13}}{k_{1+} k_{3-}} \right],$$

$$M(-,+;+,+,-) = -\frac{g^3 C k_{3+} k^*_{3\perp} Z_{13} Z_{32} Z_{21}}{2\sqrt{2}(k_1 \cdot k_2)(k_1 \cdot k_3)(k_2 \cdot k_3)}$$

$$\times \left[(T^a T^b T^c)_{ij} \frac{Z^*_{31}}{k_{3+} k_{1-}} - (T^a T^c T^b)_{ij} \frac{Z^*_{21}}{k_{2+} k_{1-}} \right.$$

$$-(T^b T^a T^c)_{ij} \frac{Z^*_{32}}{k_{3+} k_{2-}} + (T^b T^c T^a)_{ij} \frac{Z^*_{12}}{k_{1+} k_{2-}}$$

$$\left. +(T^c T^a T^b)_{ij} \frac{Z^*_{23}}{k_{2+} k_{3-}} - (T^c T^b T^a)_{ij} \frac{Z^*_{13}}{k_{1+} k_{3-}} \right],$$

$$M(-,+;+,-,+) = -\frac{g^3 C k_{2+} k^*_{2\perp} Z_{13} Z_{32} Z_{21}}{2\sqrt{2}(k_1 \cdot k_2)(k_1 \cdot k_3)(k_2 \cdot k_3)}$$

$$\times \left[(T^a T^b T^c)_{ij} \frac{Z^*_{31}}{k_{3+} k_{1-}} - (T^a T^c T^b)_{ij} \frac{Z^*_{21}}{k_{2+} k_{1-}} \right.$$

$$-(T^b T^a T^c)_{ij} \frac{Z^*_{32}}{k_{3+} k_{2-}} + (T^b T^c T^a)_{ij} \frac{Z^*_{12}}{k_{1+} k_{2-}}$$

$$\left. +(T^c T^a T^b)_{ij} \frac{Z^*_{23}}{k_{2+} k_{3-}} - (T^c T^b T^a)_{ij} \frac{Z^*_{13}}{k_{1+} k_{3-}} \right],$$

$$M(-,+;-,+,+) = -\frac{g^3 C k_{1+} k_{1\perp}^* Z_{13} Z_{32} Z_{21}}{2\sqrt{2}(k_1 \cdot k_2)(k_1 \cdot k_3)(k_2 \cdot k_3)}$$

$$\times \left[(T^a T^b T^c)_{ij} \frac{Z_{31}^*}{k_{3+}k_{1-}} - (T^a T^c T^b)_{ij} \frac{Z_{21}^*}{k_{2+}k_{1-}} \right.$$

$$- (T^b T^a T^c)_{ij} \frac{Z_{32}^*}{k_{3+}k_{2-}} + (T^b T^c T^a)_{ij} \frac{Z_{12}^*}{k_{1+}k_{2-}}$$

$$\left. + (T^c T^a T^b)_{ij} \frac{Z_{23}^*}{k_{2+}k_{3-}} - (T^c T^b T^a)_{ij} \frac{Z_{13}^*}{k_{1+}k_{3-}} \right],$$

$$M(+,-;+,-,-) = -\frac{g^3 C k_{1+} k_{1\perp} Z_{13}^* Z_{32}^* Z_{21}^*}{2\sqrt{2}(k_1 \cdot k_2)(k_1 \cdot k_3)(k_2 \cdot k_3)}$$

$$\times \left[(T^a T^b T^c)_{ij} \frac{Z_{31}}{k_{3+}k_{1-}} - (T^a T^c T^b)_{ij} \frac{Z_{21}}{k_{2+}k_{1-}} \right.$$

$$- (T^b T^a T^c)_{ij} \frac{Z_{32}}{k_{3+}k_{2-}} + (T^b T^c T^a)_{ij} \frac{Z_{12}}{k_{1+}k_{2-}}$$

$$\left. + (T^c T^a T^b)_{ij} \frac{Z_{23}}{k_{2+}k_{3-}} - (T^c T^b T^a)_{ij} \frac{Z_{13}}{k_{1+}k_{3-}} \right],$$

$$M(+,-;-,+,-) = -\frac{g^3 C k_{2+} k_{2\perp} Z_{13}^* Z_{32}^* Z_{21}^*}{2\sqrt{2}(k_1 \cdot k_2)(k_1 \cdot k_3)(k_2 \cdot k_3)}$$

$$\times \left[(T^a T^b T^c)_{ij} \frac{Z_{31}}{k_{3+}k_{1-}} - (T^a T^c T^b)_{ij} \frac{Z_{21}}{k_{2+}k_{1-}} \right.$$

$$- (T^b T^a T^c)_{ij} \frac{Z_{32}}{k_{3+}k_{2-}} + (T^b T^c T^a)_{ij} \frac{Z_{12}}{k_{1+}k_{2-}}$$

$$\left. + (T^c T^a T^b)_{ij} \frac{Z_{23}}{k_{2+}k_{3-}} - (T^c T^b T^a)_{ij} \frac{Z_{13}}{k_{1+}k_{3-}} \right],$$

$$M(+,-;-,-,+) = -\frac{g^3 C k_{3+} k_{3\perp} Z_{13}^* Z_{32}^* Z_{21}^*}{2\sqrt{2}(k_1 \cdot k_2)(k_1 \cdot k_3)(k_2 \cdot k_3)}$$

$$\times \left[(T^a T^b T^c)_{ij} \frac{Z_{31}}{k_{3+}k_{1-}} - (T^a T^c T^b)_{ij} \frac{Z_{21}}{k_{2+}k_{1-}} \right.$$

$$- (T^b T^a T^c)_{ij} \frac{Z_{32}}{k_{3+}k_{2-}} + (T^b T^c T^a)_{ij} \frac{Z_{12}}{k_{1+}k_{2-}}$$

$$\left. + (T^c T^a T^b)_{ij} \frac{Z_{23}}{k_{2+}k_{3-}} - (T^c T^b T^a)_{ij} \frac{Z_{13}}{k_{1+}k_{3-}} \right],$$

10.49. $\bar{q}q \to ggg$

$$M(-,+;+,-,-) = \frac{g^3 C k_{1-} k_{1\perp}^* Z_{13}^* Z_{32}^* Z_{21}^*}{2\sqrt{2}(k_1 \cdot k_2)(k_1 \cdot k_3)(k_2 \cdot k_3)}$$

$$\times \left[(T^a T^b T^c)_{ij} \frac{Z_{31}}{k_{3+} k_{1-}} - (T^a T^c T^b)_{ij} \frac{Z_{21}}{k_{2+} k_{1-}} \right.$$

$$- (T^b T^a T^c)_{ij} \frac{Z_{32}}{k_{3+} k_{2-}} + (T^b T^c T^a)_{ij} \frac{Z_{12}}{k_{1+} k_{2-}}$$

$$\left. + (T^c T^a T^b)_{ij} \frac{Z_{23}}{k_{2+} k_{3-}} - (T^c T^b T^a)_{ij} \frac{Z_{13}}{k_{1+} k_{3-}} \right],$$

$$M(-,+;-,+,-) = \frac{g^3 C k_{2-} k_{2\perp}^* Z_{13}^* Z_{32}^* Z_{21}^*}{2\sqrt{2}(k_1 \cdot k_2)(k_1 \cdot k_3)(k_2 \cdot k_3)}$$

$$\times \left[(T^a T^b T^c)_{ij} \frac{Z_{31}}{k_{3+} k_{1-}} - (T^a T^c T^b)_{ij} \frac{Z_{21}}{k_{2+} k_{1-}} \right.$$

$$- (T^b T^a T^c)_{ij} \frac{Z_{32}}{k_{3+} k_{2-}} + (T^b T^c T^a)_{ij} \frac{Z_{12}}{k_{1+} k_{2-}}$$

$$\left. + (T^c T^a T^b)_{ij} \frac{Z_{23}}{k_{2+} k_{3-}} - (T^c T^b T^a)_{ij} \frac{Z_{13}}{k_{1+} k_{3-}} \right],$$

$$M(-,+;-,-,+) = \frac{g^3 C k_{3-} k_{3\perp}^* Z_{13}^* Z_{32}^* Z_{21}^*}{2\sqrt{2}(k_1 \cdot k_2)(k_1 \cdot k_3)(k_2 \cdot k_3)}$$

$$\times \left[(T^a T^b T^c)_{ij} \frac{Z_{31}}{k_{3+} k_{1-}} - (T^a T^c T^b)_{ij} \frac{Z_{21}}{k_{2+} k_{1-}} \right.$$

$$- (T^b T^a T^c)_{ij} \frac{Z_{32}}{k_{3+} k_{2-}} + (T^b T^c T^a)_{ij} \frac{Z_{12}}{k_{1+} k_{2-}}$$

$$\left. + (T^c T^a T^b)_{ij} \frac{Z_{23}}{k_{2+} k_{3-}} - (T^c T^b T^a)_{ij} \frac{Z_{13}}{k_{1+} k_{3-}} \right].$$

(10.337)

Squared absolute values of the nonvanishing helicity amplitudes, summed over the color degrees of freedom of the initial state and the final state:

$$|M(+,-;+,+,-)|^2 = |M(-,+;-,-,+)|^2 = \frac{g^6 k_{3-}^2 F}{k_{1+} k_{1-} k_{2+} k_{2-}},$$

$$|M(+,-;+,-,+)|^2 = |M(-,+;-,+,-)|^2 = \frac{g^6 k_{2-}^2 F}{k_{1+} k_{1-} k_{3+} k_{3-}},$$

$$|M(+,-;-,+,+)|^2 = |M(-,+;+,-,-)|^2 = \frac{g^6 k_{1-}^2 F}{k_{2+} k_{2-} k_{3+} k_{3-}},$$

$$|M(-,+;+,+,-)|^2 = |M(+,-;-,-,+)|^2 = \frac{g^6 k_{3+}^2 F}{k_{1+} k_{1-} k_{2+} k_{2-}},$$

$$|M(-,+;+,-,+)|^2 = |M(+,-;-,+,-)|^2 = \frac{g^6 k_{2+}^2 F}{k_{1+} k_{1-} k_{3+} k_{3-}},$$

$$|M(-,+;-,+,+)|^2 = |M(+,-;+,-,-)|^2 = \frac{g^6 k_{1+}^2 F}{k_{2+} k_{2-} k_{3+} k_{3-}}.$$

(10.338)

Unpolarized squared matrix element:

$$\overline{|M|^2} = \frac{2g^6}{81} \frac{\sum_{i=1}^{3}(p_+ \cdot k_i)(p_- \cdot k_i)[(p_+ \cdot k_i)^2 + (p_- \cdot k_i)^2]}{\prod_{i=1}^{3}(p_+ \cdot k_i)(p_- \cdot k_i)}$$

$$\times \left\{ 10(p_+ \cdot p_-) - 9 \left[\frac{(p_+ \cdot k_1)(p_- \cdot k_2) + (p_+ \cdot k_2)(p_- \cdot k_1)}{(k_1 \cdot k_2)} \right. \right.$$

$$+ \frac{(p_+ \cdot k_2)(p_- \cdot k_3) + (p_+ \cdot k_3)(p_- \cdot k_2)}{(k_2 \cdot k_3)}$$

$$\left. + \frac{(p_+ \cdot k_3)(p_- \cdot k_1) + (p_+ \cdot k_1)(p_- \cdot k_3)}{(k_1 \cdot k_3)} \right] + \frac{81}{(p_+ \cdot p_-)}$$

$$\times \left[\frac{(p_+ \cdot k_1)(p_- \cdot k_1)[(p_+ \cdot k_2)(p_- \cdot k_3) + (p_+ \cdot k_3)(p_- \cdot k_2)]}{(k_1 \cdot k_2)(k_1 \cdot k_3)} \right.$$

$$+ \frac{(p_+ \cdot k_2)(p_- \cdot k_2)[(p_+ \cdot k_3)(p_- \cdot k_1) + (p_+ \cdot k_1)(p_- \cdot k_3)]}{(k_1 \cdot k_2)(k_2 \cdot k_3)}$$

$$\left. \left. + \frac{(p_+ \cdot k_3)(p_- \cdot k_3)[(p_+ \cdot k_1)(p_- \cdot k_2) + (p_+ \cdot k_2)(p_- \cdot k_1)]}{(k_1 \cdot k_3)(k_2 \cdot k_3)} \right] \right\}.$$

(10.339)

Unpolarized cross section:

$$d\sigma = \frac{\alpha_S^3}{648\pi^2} \frac{\sum_{i=1}^{3}(p_+ \cdot k_i)(p_- \cdot k_i)[(p_+ \cdot k_i)^2 + (p_- \cdot k_i)^2]}{(p_+ \cdot p_-)\prod_{i=1}^{3}(p_+ \cdot k_i)(p_- \cdot k_i)}$$

$$\times \left\{ 10(p_+ \cdot p_-) - 9 \left[\frac{(p_+ \cdot k_1)(p_- \cdot k_2) + (p_+ \cdot k_2)(p_- \cdot k_1)}{(k_1 \cdot k_2)} \right. \right.$$

$$+ \frac{(p_+ \cdot k_2)(p_- \cdot k_3) + (p_+ \cdot k_3)(p_- \cdot k_2)}{(k_2 \cdot k_3)}$$

$$+\frac{(p_+ \cdot k_3)(p_- \cdot k_1) + (p_+ \cdot k_1)(p_- \cdot k_3)}{(k_1 \cdot k_3)}\Bigg] + \frac{81}{(p_+ \cdot p_-)}$$

$$\times \Bigg[\frac{(p_+ \cdot k_1)(p_- \cdot k_1)[(p_+ \cdot k_2)(p_- \cdot k_3) + (p_+ \cdot k_3)(p_- \cdot k_2)]}{(k_1 \cdot k_2)(k_1 \cdot k_3)}$$

$$+\frac{(p_+ \cdot k_2)(p_- \cdot k_2)[(p_+ \cdot k_3)(p_- \cdot k_1) + (p_+ \cdot k_1)(p_- \cdot k_3)]}{(k_1 \cdot k_2)(k_2 \cdot k_3)}$$

$$+\frac{(p_+ \cdot k_3)(p_- \cdot k_3)[(p_+ \cdot k_1)(p_- \cdot k_2) + (p_+ \cdot k_2)(p_- \cdot k_1)]}{(k_1 \cdot k_3)(k_2 \cdot k_3)}\Bigg]\Bigg\}$$

$$\times \delta^4(p_+ + p_- - k_1 - k_2 - k_3)\frac{d^3\vec{k}_1\, d^3\vec{k}_2\, d^3\vec{k}_3}{k_{10}\, k_{20}\, k_{30}}. \qquad (10.340)$$

10.50 $\gamma\gamma \to \bar{q}q\gamma$

Process:
$$\gamma(k_1) + \gamma(k_2) \to \bar{q}(q',i) + q(q,j) + \gamma(k_3). \qquad (10.341)$$

Positive z-axis: along $\vec{k}_1$.

Definitions:
$$Z(q,q') = q_+ q'_- - q^*_\perp q'_\perp, \qquad Z(q',q) = q'_+ q_- - q'^*_\perp q_\perp. \qquad (10.342)$$

Photon polarizations:

$$\slashed{\epsilon}^{*\pm}(k_1) = N_1[\slashed{A}\,\slashed{A}'\,\slashed{k}_1(1\pm\gamma_5) - \slashed{k}_1\,\slashed{A}\,\slashed{A}'(1\mp\gamma_5)],$$

$$\slashed{\epsilon}^{*\pm}(k_2) = N_2[\slashed{A}\,\slashed{A}'\,\slashed{k}_2(1\pm\gamma_5) - \slashed{k}_2\,\slashed{A}\,\slashed{A}'(1\mp\gamma_5)],$$

$$\slashed{\epsilon}^{\pm}(k_3) = N_3[\slashed{A}\,\slashed{A}'\,\slashed{k}_3(1\mp\gamma_5) - \slashed{k}_3\,\slashed{A}\,\slashed{A}'(1\pm\gamma_5)],$$

$$N_1^{-1} = 4E[q_- q'_-(q\cdot q')]^{\frac{1}{2}},$$

$$N_2^{-1} = 4E[q_+ q'_+(q\cdot q')]^{\frac{1}{2}},$$

$$N_3^{-1} = 4[(q\cdot q')(q\cdot k_3)(q'\cdot k_3)]^{\frac{1}{2}}. \qquad (10.343)$$

Nonvanishing helicity amplitudes:

$$M(+,+;+,-,+) = \frac{2e_q^3 \delta_{ij}(q_+ q'_+ + q_\perp q'^*_\perp)}{E q_+ q'_+}\left[\frac{(q\cdot q')(q'\cdot k_3)}{q_- q'_-(q\cdot k_3)}\right]^{\frac{1}{2}},$$

$$M(+,+;-,+,+) = \frac{2e_q^3\delta_{ij}(q_+q'_+ + q^*_\perp q'_\perp)}{Eq_+q'_+}\left[\frac{(q\cdot q')(q\cdot k_3)}{q_-q'_-(q'\cdot k_3)}\right]^{\frac{1}{2}},$$

$$M(+,-;+,-,+) = -\frac{2e_q^3\delta_{ij}q_\perp q'^*_\perp}{q_+q'_+}\left[\frac{q_-(q\cdot q')}{q'_-(q\cdot k_3)(q'\cdot k_3)}\right]^{\frac{1}{2}},$$

$$M(+,-;-,+,+) = \frac{2e_q^3\delta_{ij}q'_\perp q^*_\perp}{q_+q'_+}\left[\frac{q'_-(q\cdot q')}{q_-(q\cdot k_3)(q'\cdot k_3)}\right]^{\frac{1}{2}},$$

$$M(-,+;+,-,+) = -2e_q^3\delta_{ij}q_+\left[\frac{(q\cdot q')}{q_-q'_-(q\cdot k_3)(q'\cdot k_3)}\right]^{\frac{1}{2}},$$

$$M(-,+;-,+,+) = 2e_q^3\delta_{ij}q'_+\left[\frac{(q\cdot q')}{q_-q'_-(q\cdot k_3)(q'\cdot k_3)}\right]^{\frac{1}{2}},$$

$$M(+,-;+,-,-) = 2e_q^3\delta_{ij}q'_+\left[\frac{(q\cdot q')}{q_-q'_-(q\cdot k_3)(q'\cdot k_3)}\right]^{\frac{1}{2}},$$

$$M(+,-;-,+,-) = -2e_q^3\delta_{ij}q_+\left[\frac{(q\cdot q')}{q_-q'_-(q\cdot k_3)(q'\cdot k_3)}\right]^{\frac{1}{2}},$$

$$M(-,+;+,-,-) = \frac{2e_q^3\delta_{ij}q'^*_\perp q_\perp}{q_+q'_+}\left[\frac{q'_-(q\cdot q')}{q_-(q\cdot k_3)(q'\cdot k_3)}\right]^{\frac{1}{2}},$$

$$M(-,+;-,+,-) = -\frac{2e_q^3\delta_{ij}q^*_\perp q'_\perp}{q_+q'_+}\left[\frac{q_-(q\cdot q')}{q'_-(q\cdot k_3)(q'\cdot k_3)}\right]^{\frac{1}{2}},$$

$$M(-,-;+,-,-) = \frac{2e_q^3\delta_{ij}(q_+q'_+ + q_\perp q'^*_\perp)}{Eq_+q'_+}\left[\frac{(q\cdot q')(q\cdot k_3)}{q_-q'_-(q'\cdot k_3)}\right]^{\frac{1}{2}},$$

$$M(-,-;-,+,-) = \frac{2e_q^3\delta_{ij}(q_+q'_+ + q^*_\perp q'_\perp)}{Eq_+q'_+}\left[\frac{(q\cdot q')(q'\cdot k_3)}{q_-q'_-(q\cdot k_3)}\right]^{\frac{1}{2}}.$$

(10.3⋯)

Squared absolute values of the nonvanishing helicity amplitudes, summed over the color degrees of freedom of the final state:

$$|M(+,+;+,-,+)|^2 = |M(-,-;-,+,-)|^2 = \frac{12e_q^6(q\cdot q')(q'\cdot k_3)^2}{E^4 q_+ q_- q'_+ q'_-},$$

$$|M(+,+;-,+,+)|^2 = |M(-,-;+,-,-)|^2 = \frac{12e_q^6(q\cdot q')(q\cdot k_3)^2}{E^4 q_+ q_- q'_+ q'_-},$$

$$|M(+,-;+,-,+)|^2 = |M(-,+;-,+,-)|^2 = \frac{12e_q^6 q_\perp^2(q\cdot q')}{q_+q'_+(q\cdot k_3)(q'\cdot k_3)},$$

$$|M(+,-;-,+,+)|^2 = |M(-,+;+,-,-)|^2 = \frac{12e_q^6 q_-^{\prime 2}(q \cdot q')}{q_+ q_+'(q \cdot k_3)(q' \cdot k_3)},$$

$$|M(-,+;+,-,+)|^2 = |M(+,-;-,+,-)|^2 = \frac{12e_q^6 q_+^2(q \cdot q')}{q_- q_-'(q \cdot k_3)(q' \cdot k_3)},$$

$$|M(-,+;-,+,+)|^2 = |M(+,-;+,-,-)|^2 = \frac{12e_q^6 q_+^{\prime 2}(q \cdot q')}{q_- q_-'(q \cdot k_3)(q' \cdot k_3)}.$$
(10.345)

Unpolarized squared matrix element:

$$\overline{|M|^2} = 6e_q^6(q \cdot q') \frac{\sum_{i=1}^{3}(q \cdot k_i)(q' \cdot k_i)[(q \cdot k_i)^2 + (q' \cdot k_i)^2]}{\prod_{i=1}^{3}(q \cdot k_i)(q' \cdot k_i)}. \quad (10.346)$$

Unpolarized cross section:

$$d\sigma = \frac{3\alpha^3 Q_f^6 (q \cdot q')}{16\pi^2 E^2} \frac{\sum_{i=1}^{3}(q \cdot k_i)(q' \cdot k_i)[(q \cdot k_i)^2 + (q' \cdot k_i)^2]}{\prod_{i=1}^{3}(q \cdot k_i)(q' \cdot k_i)}$$

$$\times \delta^4(k_1 + k_2 - q - q' - k_3) \frac{d^3\vec{q}\, d^3\vec{q}'\, d^3\vec{k}_3}{q_0\, q_0'\, k_{30}}. \quad (10.347)$$

10.51 $\gamma\gamma \to \bar{q}qg$

Process:
$$\gamma(k_1) + \gamma(k_2) \to \bar{q}(q',i) + q(q,j) + g(k_3,c). \quad (10.348)$$

Positive z-axis: along $\vec{k}_1$.
Definitions:

$$Z(q,q') = q_+ q_-' - q_\perp^* q_\perp', \qquad Z(q',q) = q_+' q_- - q_\perp'^* q_\perp. \quad (10.349)$$

Photon and gluon polarizations:

$$\not{\epsilon}^{*\pm}(k_1) = N_1[\not{A}\, \not{A}'\, \not{k}_1(1 \pm \gamma_5) - \not{k}_1\, \not{A}\, \not{A}'(1 \mp \gamma_5)],$$

$$\not{\epsilon}^{*\pm}(k_2) = N_2[\not{A}\, \not{A}'\, \not{k}_2(1 \pm \gamma_5) - \not{k}_2\, \not{A}\, \not{A}'(1 \mp \gamma_5)],$$

$$\not{\epsilon}^{\pm}(k_3) = N_3[\not{A}'\, \not{A}\, \not{k}_3(1 \mp \gamma_5) + \not{k}_3\, \not{A}\, \not{A}'(1 \pm \gamma_5)],$$

$$N_1^{-1} = 4E[q_-q'_-(q\cdot q')]^{\frac{1}{2}},$$

$$N_2^{-1} = 4E[q_+q'_+(q\cdot q')]^{\frac{1}{2}},$$

$$N_3^{-1} = 4[(q\cdot q')(q\cdot k_3)(q'\cdot k_3)]^{\frac{1}{2}}. \qquad (10.35)$$

Nonvanishing helicity amplitudes:

$$M(+,+;+,-,+) = \frac{2ge_q^2 T_{ji}^c(q_+q'_+ + q_\perp q'^*_\perp)}{Eq_+q'_+}\left[\frac{(q\cdot q')(q'\cdot k_3)}{q_-q'_-(q\cdot k_3)}\right]^{\frac{1}{2}},$$

$$M(+,+;-,+,+) = \frac{2ge_q^2 T_{ji}^c(q_+q'_+ + q^*_\perp q'_\perp)}{Eq_+q'_+}\left[\frac{(q\cdot q')(q\cdot k_3)}{q_-q'_-(q'\cdot k_3)}\right]^{\frac{1}{2}},$$

$$M(+,-;+,-,+) = -\frac{2ge_q^2 T_{ji}^c q_\perp q'^*_\perp}{q_+q'_+}\left[\frac{q_-(q\cdot q')}{q'_-(q\cdot k_3)(q'\cdot k_3)}\right]^{\frac{1}{2}},$$

$$M(+,-;-,+,+) = \frac{2ge_q^2 T_{ji}^c q'_\perp q^*_\perp}{q_+q'_+}\left[\frac{q'_-(q\cdot q')}{q_-(q\cdot k_3)(q'\cdot k_3)}\right]^{\frac{1}{2}},$$

$$M(-,+;+,-,+) = -2ge_q^2 T_{ji}^c q_+\left[\frac{(q\cdot q')}{q_-q'_-(q\cdot k_3)(q'\cdot k_3)}\right]^{\frac{1}{2}},$$

$$M(-,+;-,+,+) = 2ge_q^2 T_{ji}^c q'_+\left[\frac{(q\cdot q')}{q_-q'_-(q\cdot k_3)(q'\cdot k_3)}\right]^{\frac{1}{2}},$$

$$M(+,-;+,-,-) = 2ge_q^2 T_{ji}^c q'_+\left[\frac{(q\cdot q')}{q_-q'_-(q\cdot k_3)(q'\cdot k_3)}\right]^{\frac{1}{2}},$$

$$M(+,-;-,+,-) = -2ge_q^2 T_{ji}^c q_+\left[\frac{(q\cdot q')}{q_-q'_-(q\cdot k_3)(q'\cdot k_3)}\right]^{\frac{1}{2}},$$

$$M(-,+;+,-,-) = \frac{2ge_q^2 T_{ji}^c q'^*_\perp q_\perp}{q_+q'_+}\left[\frac{q'_-(q\cdot q')}{q_-(q\cdot k_3)(q'\cdot k_3)}\right]^{\frac{1}{2}},$$

$$M(-,+;-,+,-) = -\frac{2ge_q^2 T_{ji}^c q^*_\perp q'_\perp}{q_+q'_+}\left[\frac{q_-(q\cdot q')}{q'_-(q\cdot k_3)(q'\cdot k_3)}\right]^{\frac{1}{2}},$$

$$M(-,-;+,-,-) = \frac{2ge_q^2 T_{ji}^c(q_+q'_+ + q_\perp q'^*_\perp)}{Eq_+q'_+}\left[\frac{(q\cdot q')(q\cdot k_3)}{q_-q'_-(q'\cdot k_3)}\right]^{\frac{1}{2}},$$

$$M(-,-;-,+,-) = \frac{2ge_q^2 T_{ji}^c(q_+q'_+ + q^*_\perp q'_\perp)}{Eq_+q'_+}\left[\frac{(q\cdot q')(q'\cdot k_3)}{q_-q'_-(q\cdot k_3)}\right]^{\frac{1}{2}}.$$

$$(10.35)$$

Squared absolute values of the nonvanishing helicity amplitudes, summed over the color degrees of freedom of the final state:

$$|M(+,+;+,-,+)|^2 = |M(-,-;-,+,-)|^2 = \frac{16g^2 e_q^4 (q \cdot q')(q' \cdot k_3)^2}{E^4 q_+ q_- q'_+ q'_-},$$

$$|M(+,+;-,+,+)|^2 = |M(-,-;+,-,-)|^2 = \frac{16g^2 e_q^4 (q \cdot q')(q \cdot k_3)^2}{E^4 q_+ q_- q'_+ q'_-},$$

$$|M(+,-;+,-,+)|^2 = |M(-,+;-,+,-)|^2 = \frac{16g^2 e_q^4 q_-^2 (q \cdot q')}{q_+ q'_+ (q \cdot k_3)(q' \cdot k_3)},$$

$$|M(+,-;-,+,+)|^2 = |M(-,+;+,-,-)|^2 = \frac{16g^2 e_q^4 q'^2_- (q \cdot q')}{q_+ q'_+ (q \cdot k_3)(q' \cdot k_3)},$$

$$|M(-,+;+,-,+)|^2 = |M(+,-;-,+,-)|^2 = \frac{16g^2 e_q^4 q_+^2 (q \cdot q')}{q_- q'_- (q \cdot k_3)(q' \cdot k_3)},$$

$$|M(-,+;-,+,+)|^2 = |M(+,-;+,-,-)|^2 = \frac{16g^2 e_q^4 q'^2_+ (q \cdot q')}{q_- q'_- (q \cdot k_3)(q' \cdot k_3)}.$$
(10.352)

Unpolarized squared matrix element:

$$\overline{|M|^2} = 8g^2 e_q^4 (q \cdot q') \frac{\sum_{i=1}^{3}(q \cdot k_i)(q' \cdot k_i)[(q \cdot k_i)^2 + (q' \cdot k_i)^2]}{\prod_{i=1}^{3}(q \cdot k_i)(q' \cdot k_i)}. \quad (10.353)$$

Unpolarized cross section:

$$d\sigma = \frac{\alpha_S \alpha^2 Q_f^4 (q \cdot q')}{4\pi^2 E^2} \frac{\sum_{i=1}^{3}(q \cdot k_i)(q' \cdot k_i)[(q \cdot k_i)^2 + (q' \cdot k_i)^2]}{\prod_{i=1}^{3}(q \cdot k_i)(q' \cdot k_i)}$$

$$\times \delta^4(k_1 + k_2 - q - q' - k_3) \frac{d^3\vec{q}\, d^3\vec{q}'\, d^3\vec{k}_3}{q_0\, q'_0\, k_{30}}. \quad (10.354)$$

10.52 $\gamma g \to \bar{q} q \gamma$

Process:
$$\gamma(k_1) + g(k_2, b) \to \bar{q}(q', i) + q(q, j) + \gamma(k_3). \quad (10.355)$$

Positive z-axis: along $\vec{k}_1$.
Definitions:

$$Z(q, q') = q_+ q'_- - q^*_\perp q'_\perp, \qquad Z(q', q) = q'_+ q_- - q'^*_\perp q_\perp. \quad (10.356)$$

Photon and gluon polarizations:

$$\epsilon^{*\pm}(k_1) = N_1[\not{q}\,\not{q}'\,\not{k}_1(1\pm\gamma_5) - \not{k}_1\,\not{q}\,\not{q}'(1\mp\gamma_5)],$$

$$\epsilon^{*\pm}(k_2) = N_2[\not{q}\,\not{q}'\,\not{k}_2(1\pm\gamma_5) + \not{k}_2\,\not{q}'\,\not{q}(1\mp\gamma_5)],$$

$$\epsilon^{\pm}(k_3) = N_3[\not{q}\,\not{q}'\,\not{k}_3(1\mp\gamma_5) - \not{k}_3\,\not{q}\,\not{q}'(1\pm\gamma_5)],$$

$$N_1^{-1} = 4E[q_- q'_-(q\cdot q')]^{\frac{1}{2}},$$

$$N_2^{-1} = 4E[q_+ q'_+(q\cdot q')]^{\frac{1}{2}},$$

$$N_3^{-1} = 4[(q\cdot q')(q\cdot k_3)(q'\cdot k_3)]^{\frac{1}{2}}. \qquad (10.357)$$

Nonvanishing helicity amplitudes:

$$M(+,+;+,-,+) = \frac{2ge_q^2 T_{ji}^b(q_+ q'_+ + q_\perp q'^*_\perp)}{E q_+ q'_+}\left[\frac{(q\cdot q')(q'\cdot k_3)}{q_- q'_-(q\cdot k_3)}\right]^{\frac{1}{2}},$$

$$M(+,+;-,+,+) = \frac{2ge_q^2 T_{ji}^b(q_+ q'_+ + q^*_\perp q'_\perp)}{E q_+ q'_+}\left[\frac{(q\cdot q')(q\cdot k_3)}{q_- q'_-(q'\cdot k_3)}\right]^{\frac{1}{2}},$$

$$M(+,-;+,-,+) = -\frac{2ge_q^2 T_{ji}^b q_\perp q'^*_\perp}{q_+ q'_+}\left[\frac{q_-(q\cdot q')}{q'_-(q\cdot k_3)(q'\cdot k_3)}\right]^{\frac{1}{2}},$$

$$M(+,-;-,+,+) = \frac{2ge_q^2 T_{ji}^b q'_\perp q^*_\perp}{q_+ q'_+}\left[\frac{q'_-(q\cdot q')}{q_-(q\cdot k_3)(q'\cdot k_3)}\right]^{\frac{1}{2}},$$

$$M(-,+;+,-,+) = -2ge_q^2 T_{ji}^b q_+ \left[\frac{(q\cdot q')}{q_- q'_-(q\cdot k_3)(q'\cdot k_3)}\right]^{\frac{1}{2}},$$

$$M(-,+;-,+,+) = 2ge_q^2 T_{ji}^b q'_+ \left[\frac{(q\cdot q')}{q_- q'_-(q\cdot k_3)(q'\cdot k_3)}\right]^{\frac{1}{2}},$$

$$M(+,-;+,-,-) = 2ge_q^2 T_{ji}^b q'_+ \left[\frac{(q\cdot q')}{q_- q'_-(q\cdot k_3)(q'\cdot k_3)}\right]^{\frac{1}{2}},$$

$$M(+,-;-,+,-) = -2ge_q^2 T_{ji}^b q_+ \left[\frac{(q\cdot q')}{q_- q'_-(q\cdot k_3)(q'\cdot k_3)}\right]^{\frac{1}{2}},$$

$$M(-,+;+,-,-) = \frac{2ge_q^2 T_{ji}^b q'^*_\perp q_\perp}{q_+ q'_+}\left[\frac{q'_-(q\cdot q')}{q_-(q\cdot k_3)(q'\cdot k_3)}\right]^{\frac{1}{2}},$$

10.52. $\gamma g \to \bar{q} q \gamma$

$$M(-,+;-,+,-) = -\frac{2ge_q^2 T_{ji}^b q_\perp^* q_\perp'}{q_+ q_+'} \left[\frac{q_-(q \cdot q')}{q_-'(q \cdot k_3)(q' \cdot k_3)}\right]^{\frac{1}{2}},$$

$$M(-,-;+,-,-) = \frac{2ge_q^2 T_{ji}^b (q_+ q_+' + q_\perp q_\perp'^*)}{Eq_+ q_+'} \left[\frac{(q \cdot q')(q \cdot k_3)}{q_- q_-'(q' \cdot k_3)}\right]^{\frac{1}{2}},$$

$$M(-,-;-,+,-) = \frac{2ge_q^2 T_{ji}^b (q_+ q_+' + q_\perp^* q_\perp')}{Eq_+ q_+'} \left[\frac{(q \cdot q')(q' \cdot k_3)}{q_- q_-'(q \cdot k_3)}\right]^{\frac{1}{2}}.$$

(10.358)

Squared absolute values of the nonvanishing helicity amplitudes, summed over the color degrees of freedom of the final state:

$$|M(+,+;+,-,+)|^2 = |M(-,-;-,+,-)|^2 = \frac{16g^2 e_q^4 (q \cdot q')(q' \cdot k_3)^2}{E^4 q_+ q_- q_+' q_-'},$$

$$|M(+,+;-,+,+)|^2 = |M(-,-;+,-,-)|^2 = \frac{16g^2 e_q^4 (q \cdot q')(q \cdot k_3)^2}{E^4 q_+ q_- q_+' q_-'},$$

$$|M(+,-;+,-,+)|^2 = |M(-,+;-,+,-)|^2 = \frac{16g^2 e_q^4 q_-^2 (q \cdot q')}{q_+ q_+'(q \cdot k_3)(q' \cdot k_3)},$$

$$|M(+,-;-,+,+)|^2 = |M(-,+;+,-,-)|^2 = \frac{16g^2 e_q^4 q_-'^2 (q \cdot q')}{q_+ q_+'(q \cdot k_3)(q' \cdot k_3)},$$

$$|M(-,+;+,-,+)|^2 = |M(+,-;-,+,-)|^2 = \frac{16g^2 e_q^4 q_+^2 (q \cdot q')}{q_- q_-'(q \cdot k_3)(q' \cdot k_3)},$$

$$|M(-,+;-,+,+)|^2 = |M(+,-;+,-,-)|^2 = \frac{16g^2 e_q^4 q_+'^2 (q \cdot q')}{q_- q_-'(q \cdot k_3)(q' \cdot k_3)}.$$

(10.359)

Unpolarized squared matrix element:

$$\overline{|M|^2} = 8g^2 e_q^4 (q \cdot q') \frac{\sum_{i=1}^{3}(q \cdot k_i)(q' \cdot k_i)[(q \cdot k_i)^2 + (q' \cdot k_i)^2]}{\prod_{i=1}^{3}(q \cdot k_i)(q' \cdot k_i)}.$$

(10.360)

Unpolarized cross section:

$$d\sigma = \frac{\alpha_S \alpha^2 Q_f^4 (q \cdot q')}{4\pi^2 E^2} \frac{\sum_{i=1}^{3}(q \cdot k_i)(q' \cdot k_i)[(q \cdot k_i)^2 + (q' \cdot k_i)^2]}{\prod_{i=1}^{3}(q \cdot k_i)(q' \cdot k_i)}$$

$$\times \delta^4(k_1 + k_2 - q - q' - k_3) \frac{d^3\vec{q}\, d^3\vec{q}'\, d^3\vec{k}_3}{q_0 q_0' k_{30}}.$$

(10.361)

10.53 $\gamma g \to \bar{q} q g$

Process:
$$\gamma(k_1) + g(k_2, b) \to \bar{q}(q', i) + q(q, j) + g(k_3, c). \tag{10.362}$$

Positive z-axis: along $\vec{k}_1$.

Definitions:
$$Z(q, q') = q_+ q'_- - q^*_\perp q'_\perp, \qquad Z(q', q) = q'_+ q_- - q'^*_\perp q_\perp,$$
$$Z(k_3, q) = k_{3+} q_- - k^*_{3\perp} q_\perp, \qquad Z(k_3, q') = k_{3+} q'_- - k^*_{3\perp} q'_\perp. \tag{10.363}$$

Gluon and photon polarizations:
$$\not{\epsilon}^{*\pm}(k_1) = N_1[\not{k}_1 \not{q} \not{q}'(1 \mp \gamma_5) - \not{q} \not{q}' \not{k}_1(1 \pm \gamma_5)],$$

$$\not{\epsilon}^{*\pm}(k_2) = N_2[\not{q}' \not{q} \not{k}_2(1 \pm \gamma_5) + \not{k}_2 \not{q} \not{q}'(1 \mp \gamma_5)],$$

$$\not{\epsilon}^{\pm}(k_3) = N_3[\not{k}_3 \not{q} \not{q}'(1 \pm \gamma_5) + \not{q}' \not{q} \not{k}_3(1 \mp \gamma_5)],$$

$$N_1^{-1} = 4E[q_- q'_-(q \cdot q')]^{\frac{1}{2}},$$

$$N_2^{-1} = 4E[q_+ q'_+(q \cdot q')]^{\frac{1}{2}},$$

$$N_3^{-1} = 4[(q \cdot q')(q \cdot k_3)(q' \cdot k_3)]^{\frac{1}{2}}. \tag{10.364}$$

Nonvanishing helicity amplitudes:
$$M(+, +; +, -, +) = -\frac{g^2 e_q (q_+ q'_+ + q_\perp q'^*_\perp)}{E q_+ q'_+ k_{3+}} \left[\frac{(q' \cdot k_3)}{q_-^3 q'^3_-(q \cdot q')(q \cdot k_3)}\right]^{\frac{1}{2}}$$
$$\times \left[q'_- Z^*(q', q) Z(k_3, q)(T^b T^c)_{ji} + q_- Z^*(q, q') Z(k_3, q')(T^c T^b)_{ji}\right],$$

$$M(+, +; -, +, +) = -\frac{g^2 e_q (q_+ q'_+ + q^*_\perp q'_\perp)}{E q_+ q'_+ k_{3+}} \left[\frac{(q \cdot k_3)}{q_-^3 q'^3_-(q \cdot q')(q' \cdot k_3)}\right]^{\frac{1}{2}}$$
$$\times \left[q'_- Z^*(q', q) Z(k_3, q)(T^b T^c)_{ji} + q_- Z^*(q, q') Z(k_3, q')(T^c T^b)_{ji}\right],$$

$$M(+, -; +, -, +) = \frac{g^2 e_q q_\perp Z(q', q)}{q_+ q'_+ k_{3+} [q_- q'_-(q \cdot q')(q \cdot k_3)(q' \cdot k_3)]^{\frac{1}{2}}}$$
$$\times \left[q'^*_\perp Z^*(k_3, q)(T^b T^c)_{ji} - q^*_\perp Z^*(k_3, q')(T^c T^b)_{ji}\right],$$

10.53. $\gamma g \to \overline{q} q g$

$$M(+,-;-,+,+) = \frac{g^2 e_q q'_\perp Z(q,q')}{q_+ q'_+ k_{3+} [q_- q'_- (q \cdot q')(q \cdot k_3)(q' \cdot k_3)]^{\frac{1}{2}}}$$
$$\times \left[q'^*_\perp Z^*(k_3,q)(T^b T^c)_{ji} - q^*_\perp Z^*(k_3,q')(T^c T^b)_{ji} \right],$$

$$M(-,+;+,-,+) = -\frac{g^2 e_q q_+}{k_{3+} [q^3_- q'^3_- (q \cdot q')(q \cdot k_3)(q' \cdot k_3)]^{\frac{1}{2}}}$$
$$\times \left[q'_- Z(q',q) Z^*(k_3,q)(T^b T^c)_{ji} + q_- Z(q,q') Z^*(k_3,q')(T^c T^b)_{ji} \right],$$

$$M(-,+;-,+,+) = -\frac{g^2 e_q q'_+}{k_{3+} [q^3_- q'^3_- (q \cdot q')(q \cdot k_3)(q' \cdot k_3)]^{\frac{1}{2}}}$$
$$\times \left[q'_- Z(q',q) Z^*(k_3,q)(T^b T^c)_{ji} + q_- Z(q,q') Z^*(k_3,q')(T^c T^b)_{ji} \right],$$

$$M(+,-;+,-,-) = -\frac{g^2 e_q q'_+}{k_{3+} [q^3_- q'^3_- (q \cdot q')(q \cdot k_3)(q' \cdot k_3)]^{\frac{1}{2}}}$$
$$\times \left[q'_- Z^*(q',q) Z(k_3,q)(T^b T^c)_{ji} + q_- Z^*(q,q') Z(k_3,q')(T^c T^b)_{ji} \right],$$

$$M(+,-;-,+,-) = -\frac{g^2 e_q q_+}{k_{3+} [q^3_- q'^3_- (q \cdot q')(q \cdot k_3)(q' \cdot k_3)]^{\frac{1}{2}}}$$
$$\times \left[q'_- Z^*(q',q) Z(k_3,q)(T^b T^c)_{ji} + q_- Z^*(q,q') Z(k_3,q')(T^c T^b)_{ji} \right],$$

$$M(-,+;+,-,-) = \frac{g^2 e_q q'^*_\perp Z^*(q,q')}{q_+ q'_+ k_{3+} [q_- q'_- (q \cdot q')(q \cdot k_3)(q' \cdot k_3)]^{\frac{1}{2}}}$$
$$\times \left[q'_\perp Z(k_3,q)(T^b T^c)_{ji} - q_\perp Z(k_3,q')(T^c T^b)_{ji} \right],$$

$$M(-,+;-,+,-) = \frac{g^2 e_q q^*_\perp Z^*(q',q)}{q_+ q'_+ k_{3+} [q_- q'_- (q \cdot q')(q \cdot k_3)(q' \cdot k_3)]^{\frac{1}{2}}}$$
$$\times \left[q'_\perp Z(k_3,q)(T^b T^c)_{ji} - q_\perp Z(k_3,q')(T^c T^b)_{ji} \right],$$

$$M(-,-;+,-,-) = -\frac{g^2 e_q (q_+ q'_+ + q_\perp q'^*_\perp)}{E q_+ q'_+ k_{3+}} \left[\frac{(q \cdot k_3)}{q^3_- q'^3_- (q \cdot q')(q' \cdot k_3)} \right]^{\frac{1}{2}}$$
$$\times \left[q'_- Z(q',q) Z^*(k_3,q)(T^b T^c)_{ji} + q_- Z(q,q') Z^*(k_3,q')(T^c T^b)_{ji} \right],$$

$$M(-,-;-,+,-) = -\frac{g^2 e_q(q_+ q'_+ + q^*_\perp q'_\perp)}{E q_+ q'_+ k_{3+}} \left[\frac{(q' \cdot k_3)}{q^3_- q'^3_-(q \cdot q')(q \cdot k_3)} \right]^{\frac{1}{2}}$$

$$\times \left[q'_- Z(q',q) Z^*(k_3,q) (T^b T^c)_{ji} + q_- Z(q,q') Z^*(k_3,q') (T^c T^b)_{ji} \right].$$

(10.365)

Squared absolute values of the nonvanishing helicity amplitudes, summed over the color degrees of freedom of the initial state and the final state:

$$|M(+,+;+,-,+)|^2 = |M(-,-;-,+,-)|^2$$

$$= \frac{8g^4 e_q^2 (q' \cdot k_3)^2 \left[9q'_+(q \cdot k_3) + 9q_+(q' \cdot k_3) - k_{3+}(q \cdot q') \right]}{3 E^4 q_+ q_- q'_+ q'_- k_{3+}},$$

$$|M(+,+;-,+,+)|^2 = |M(-,-;+,-,-)|^2$$

$$= \frac{8g^4 e_q^2 (q \cdot k_3)^2 \left[9q'_+(q \cdot k_3) + 9q_+(q' \cdot k_3) - k_{3+}(q \cdot q') \right]}{3 E^4 q_+ q_- q'_+ q'_- k_{3+}},$$

$$|M(+,-;+,-,+)|^2 = |M(-,+;-,+,-)|^2$$

$$= \frac{4g^4 e_q^2 q_-^2 \left[9q'_+(q \cdot k_3) + 9q_+(q' \cdot k_3) - k_{3+}(q \cdot q') \right]}{3 q_+ q'_+ k_{3+}(q \cdot k_3)(q' \cdot k_3)},$$

$$|M(+,-;-,+,+)|^2 = |M(-,+;+,-,-)|^2$$

$$= \frac{8g^4 e_q^2 q'^2_- \left[9q'_+(q \cdot k_3) + 9q_+(q' \cdot k_3) - k_{3+}(q \cdot q') \right]}{3 q_+ q'_+ k_{3+}(q \cdot k_3)(q' \cdot k_3)},$$

$$|M(-,+;+,-,+)|^2 = |M(+,-;-,+,-)|^2$$

$$= \frac{8g^4 e_q^2 q_+^2 \left[9q'_+(q \cdot k_3) + 9q_+(q' \cdot k_3) - k_{3+}(q \cdot q') \right]}{3 q_- q'_-(q \cdot k_3)(q' \cdot k_3)},$$

$$|M(-,+;-,+,+)|^2 = |M(+,-;+,-,-)|^2$$

$$= \frac{8g^4 e_q^2 q'^2_+ \left[9q'_+(q \cdot k_3) + 9q_+(q' \cdot k_3) - k_{3+}(q \cdot q') \right]}{3 q_- q'_- k_{3+}(q \cdot k_3)(q' \cdot k_3)}.$$

(10.366)

Unpolarized squared matrix element:

$$\overline{|M|^2} = \frac{g^4 e_q^2}{6(k_2 \cdot k_3)} \frac{\sum_{i=1}^{3}(q \cdot k_i)(q' \cdot k_i)[(q \cdot k_i)^2 + (q' \cdot k_i)^2]}{\prod_{i=1}^{3}(q \cdot k_i)(q' \cdot k_i)}$$

$$\times [9(q \cdot k_2)(q' \cdot k_3) + 9(q \cdot k_3)(q' \cdot k_2) - (q \cdot q')(k_2 \cdot k_3)]. \quad (10.367)$$

Unpolarized cross section:

$$d\sigma = \frac{\alpha_S^2 \alpha Q_f^2}{192\pi^2 E^2 (k_2 \cdot k_3)} \frac{\sum_{i=1}^{3}(q \cdot k_i)(q' \cdot k_i)[(q \cdot k_i)^2 + (q' \cdot k_i)^2]}{\prod_{i=1}^{3}(q \cdot k_i)(q' \cdot k_i)}$$

$$\times [9(q \cdot k_2)(q' \cdot k_3) + 9(q \cdot k_3)(q' \cdot k_2) - (q \cdot q')(k_2 \cdot k_3)]$$

$$\times \delta^4(k_1 + k_2 - q - q' - k_3) \frac{d^3\vec{q}\, d^3\vec{q}\,' d^3\vec{k}_3}{q_0 \, q_0' \, k_{30}}. \quad (10.368)$$

10.54 $gg \to \bar{q} q \gamma$

Process:

$$g(k_1, a) + g(k_2, b) \to \bar{q}(q', i) + q(q, j) + \gamma(k_3). \quad (10.369)$$

Positive z-axis: along $\vec{k}_1$.
Invariants and definitions:

$$s = 2(k_1 \cdot k_2), \quad t = -2(k_1 \cdot q'), \quad u = -2(k_1 \cdot q),$$
$$s' = 2(q \cdot q'), \quad t' = -2(k_2 \cdot q), \quad u' = -2(k_2 \cdot q'),$$
$$Z(q, q') = q_+ q'_- - q_\perp^* q'_\perp, \qquad Z(q', q) = q'_+ q_- - q'^*_\perp q_\perp. \quad (10.370)$$

Gluon and photon polarizations:

$$\not{\epsilon}^{*\pm}(k_1) = N_1[\not{A}\not{A}'\not{k}_1(1 \pm \gamma_5) + \not{k}_1 \not{A}' \not{A}(1 \mp \gamma_5)],$$

$$\not{\epsilon}^{*\pm}(k_2) = N_2[\not{A}\not{A}'\not{k}_2(1 \pm \gamma_5) + \not{k}_2 \not{A}' \not{A}(1 \mp \gamma_5)],$$

$$\not{\epsilon}^{\pm}(k_3) = N_3[\not{A}\not{A}'\not{k}_3(1 \mp \gamma_5) - \not{k}_3 \not{A} \not{A}'(1 \pm \gamma_5)],$$

$$N_1^{-1} = 4E[q_-q'_-(q\cdot q')]^{\frac{1}{2}},$$

$$N_2^{-1} = 4E[q_+q'_+(q\cdot q')]^{\frac{1}{2}},$$

$$N_3^{-1} = 4[(q\cdot q')(q\cdot k_3)(q'\cdot k_3)]^{\frac{1}{2}}. \tag{10.371}$$

Nonvanishing helicity amplitudes:

$$M(+,+;+,-,+) = \frac{g^2 e_q(q_+q'_+ + q_\perp q'^*_\perp)}{Eq_+q'_+}\left[\frac{(q'\cdot k_3)}{q_-q'_-(q\cdot q')(q\cdot k_3)}\right]^{\frac{1}{2}}$$
$$\times \left[Z^*(q,q')(T^aT^b)_{ji} + Z^*(q',q)(T^bT^a)_{ji}\right],$$

$$M(+,+;-,+,+) = \frac{g^2 e_q(q_+q'_+ + q^*_\perp q'_\perp)}{Eq_+q'_+}\left[\frac{(q\cdot k_3)}{q_-q'_-(q\cdot q')(q'\cdot k_3)}\right]^{\frac{1}{2}}$$
$$\times \left[Z^*(q,q')(T^aT^b)_{ji} + Z^*(q',q)(T^bT^a)_{ji}\right],$$

$$M(+,-;+,-,+) = \frac{g^2 e_q Z(q',q)}{q_+q'_+}\left[\frac{q_-}{q'_-(q\cdot q')(q\cdot k_3)(q'\cdot k_3)}\right]^{\frac{1}{2}}$$
$$\times \left[q_+q'_-(T^aT^b)_{ji} - q_\perp q'^*_\perp(T^bT^a)_{ji}\right],$$

$$M(+,-;-,+,+) = \frac{g^2 e_q Z(q,q')}{q_+q'_+}\left[\frac{q'_-}{q_-(q\cdot q')(q\cdot k_3)(q'\cdot k_3)}\right]^{\frac{1}{2}}$$
$$\times \left[q^*_\perp q'_\perp(T^aT^b)_{ji} - q_-q'_+(T^bT^a)_{ji}\right],$$

$$M(-,+;+,-,+) = -\frac{g^2 e_q q_+}{[q_-q'_-(q\cdot q')(q\cdot k_3)(q'\cdot k_3)]^{\frac{1}{2}}}$$
$$\times \left[Z(q,q')(T^aT^b)_{ji} + Z(q',q)(T^bT^a)_{ji}\right],$$

$$M(-,+;-,+,+) = \frac{g^2 e_q q'_+}{[q_-q'_-(q\cdot q')(q\cdot k_3)(q'\cdot k_3)]^{\frac{1}{2}}}$$
$$\times \left[Z(q,q')(T^aT^b)_{ji} + Z(q',q)(T^bT^a)_{ji}\right],$$

$$M(+,-;+,-,-) = \frac{g^2 e_q q'_+}{[q_-q'_-(q\cdot q')(q\cdot k_3)(q'\cdot k_3)]^{\frac{1}{2}}}$$
$$\times \left[Z^*(q,q')(T^aT^b)_{ji} + Z^*(q',q)(T^bT^a)_{ji}\right],$$

10.54. $gg \to \bar{q}q\gamma$

$$M(+,-;-,+,-) = -\frac{g^2 e_q q_+}{[q_- q'_-(q \cdot q')(q \cdot k_3)(q' \cdot k_3)]^{\frac{1}{2}}}$$
$$\times \left[Z^*(q,q')(T^a T^b)_{ji} + Z^*(q',q)(T^b T^a)_{ji}\right],$$

$$M(-,+;+,-,-) = \frac{g^2 e_q Z^*(q,q')}{q_+ q'_+} \left[\frac{q'_-}{q_-(q \cdot q')(q \cdot k_3)(q' \cdot k_3)}\right]^{\frac{1}{2}}$$
$$\times \left[q_\perp q'^*_\perp (T^a T^b)_{ji} - q_- q'_+(T^b T^a)_{ji}\right],$$

$$M(-,+;-,+,-) = \frac{g^2 e_q Z^*(q',q)}{q_+ q'_+} \left[\frac{q_-}{q'_-(q \cdot q')(q \cdot k_3)(q' \cdot k_3)}\right]^{\frac{1}{2}}$$
$$\times \left[q_+ q'_-(T^a T^b)_{ji} - q^*_\perp q'_\perp (T^b T^a)_{ji}\right],$$

$$M(-,-;+,-,-) = \frac{g^2 e_q (q_+ q'_+ + q_\perp q'^*_\perp)}{E q_+ q'_+} \left[\frac{(q \cdot k_3)}{q_- q'_-(q \cdot q')(q' \cdot k_3)}\right]^{\frac{1}{2}}$$
$$\times \left[Z(q,q')(T^a T^b)_{ji} + Z(q',q)(T^b T^a)_{ji}\right],$$

$$M(-,-;-,+,-) = \frac{g^2 e_q (q_+ q'_+ + q^*_\perp q'_\perp)}{E q_+ q'_+} \left[\frac{(q' \cdot k_3)}{q_- q'_-(q \cdot q')(q \cdot k_3)}\right]^{\frac{1}{2}}$$
$$\times \left[Z(q,q')(T^a T^b)_{ji} + Z(q',q)(T^b T^a)_{ji}\right]. \qquad (10.372)$$

Squared absolute values of the nonvanishing helicity amplitudes, summed over the color degrees of freedom of the initial state and the final state:

$$|M(+,+;+,-,+)|^2 = |M(-,-;-,+,-)|^2$$
$$= \frac{8 g^4 e_q^2 (q' \cdot k_3)^2}{3 E^4 q_+ q_- q'_+ q'_-} \left[9 q_+ q'_- + 9 q_- q'_+ - 2(q \cdot q')\right],$$

$$|M(+,+;-,+,+)|^2 = |M(-,-;+,-,-)|^2$$
$$= \frac{4 g^4 e_q^2 (q \cdot k_3)^2}{3 E^4 q_+ q_- q'_+ q'_-} \left[9 q_+ q'_- + 9 q_- q'_+ - 2(q \cdot q')\right],$$

$$|M(+,-;+,-,+)|^2 = |M(-,+;-,+,-)|^2$$
$$= \frac{4 g^4 e_q^2 q_-^2}{3 q_+ q'_+ (q \cdot k_3)(q' \cdot k_3)} \left[9 q_+ q'_- + 9 q_- q'_+ - 2(q \cdot q')\right],$$

$$|M(+,-;-,+,+)|^2 = |M(-,+;+,-,-)|^2$$

$$= \frac{4g^4 e_q^2 q_-'^2}{3q_+ q_+' (q \cdot k_3)(q' \cdot k_3)} \left[9q_+ q_-' + 9q_- q_+' - 2(q \cdot q')\right],$$

$$|M(-,+;+,-,+)|^2 = |M(+,-;-,+,-)|^2$$

$$= \frac{4g^4 e_q^2 q_+^2}{3q_- q_-' (q \cdot k_3)(q' \cdot k_3)} \left[9q_+ q_-' + 9q_- q_+' - 2(q \cdot q')\right],$$

$$|M(-,+;-,+,+)|^2 = |M(+,-;+,-,-)|^2$$

$$= \frac{4g^4 e_q^2 q_+'^2}{3q_- q_-' (q \cdot k_3)(q' \cdot k_3)} \left[9q_+ q_-' + 9q_- q_+' - 2(q \cdot q')\right]. \tag{10.373}$$

Unpolarized squared matrix element:

$$\overline{|M|^2} = \frac{g^4 e_q^2}{96s}(9tt'+9uu'-ss') \frac{\sum_{i=1}^{3}(q \cdot k_i)(q' \cdot k_i)[(q \cdot k_i)^2 + (q' \cdot k_i)^2]}{\prod_{i=1}^{3}(q \cdot k_i)(q' \cdot k_i)}. \tag{10.374}$$

Unpolarized cross section:

$$d\sigma = \frac{\alpha_s^2 \alpha Q_f^2}{768\pi^2} \frac{9tt' + 9uu' - ss'}{s^2} \frac{\sum_{i=1}^{3}(q \cdot k_i)(q' \cdot k_i)[(q \cdot k_i)^2 + (q' \cdot k_i)^2]}{\prod_{i=1}^{3}(q \cdot k_i)(q' \cdot k_i)}$$

$$\times \delta^4(k_1 + k_2 - q - q' - k_3) \frac{d^3\vec{q}\, d^3\vec{q}'\, d^3\vec{k}_3}{q_0\, q_0'\, k_{30}}. \tag{10.375}$$

10.55 $gg \to \bar{q}qg$

Process:

$$g(k_1, a) + g(k_2, b) \to \bar{q}(q', i) + q(q, j) + g(k_3, c). \tag{10.376}$$

Positive z-axis: along $\vec{k}_1$.
Definitions:

$$Z(q, q') = q_+ q_-' - q_\perp^* q_\perp', \qquad Z(q', q) = q_+' q_- - q_\perp'^* q_\perp,$$

$$Z(q, k_3) = q_+ k_{3-} - q_\perp^* k_{3\perp}, \qquad Z(k_3, q) = k_{3+} q_- - k_{3\perp}^* q_\perp,$$

$$Z(q', k_3) = q_+' k_{3-} - q_\perp'^* k_{3\perp}, \qquad Z(k_3, q') = k_{3+} q_-' - k_{3\perp}^* q_\perp',$$

10.55. $gg \to \bar{q}qg$

$$F = \frac{40}{9}(q \cdot q') - 4\left[\frac{q_-(q' \cdot k_3) + q'_-(q \cdot k_3)}{k_{3-}} + \frac{q_+(q' \cdot k_3) + q'_+(q \cdot k_3)}{k_{3+}}\right.$$

$$\left.+\frac{q_+q'_- + q_-q'_+}{2}\right] + \frac{36}{(q \cdot q')}\left[\frac{q_+q'_+[q_-(q' \cdot k_3) + q'_-(q \cdot k_3)]}{2k_{3+}}\right.$$

$$\left.+\frac{q_-q'_-[q_+(q' \cdot k_3) + q'_+(q \cdot k_3)]}{2k_{3-}} + \frac{(q \cdot k_3)(q' \cdot k_3)(q_+q'_+ + q_-q'_-)}{k_{3+}k_{3-}}\right].$$

(10.377)

Gluon polarizations:

$$\not{\epsilon}_i^{*\pm} = N_i[\not{q}' \not{q} \not{k}_i(1 \pm \gamma_5) + \not{k}_i \not{q} \not{q}'(1 \mp \gamma_5)], \qquad i = 1, 2,$$

$$N_1^{-1} = 4E[q_-q'_-(q \cdot q')]^{\frac{1}{2}},$$

$$N_2^{-1} = 4E[q_+q'_+(q \cdot q')]^{\frac{1}{2}},$$

$$\not{\epsilon}_3^{\pm} = N_3[\not{k}_3 \not{q} \not{q}'(1 \pm \gamma_5) + \not{q}' \not{q} \not{k}_3(1 \mp \gamma_5)],$$

$$N_3^{-1} = 4[(q \cdot q')(q \cdot k_3)(q' \cdot k_3)]^{\frac{1}{2}}. \tag{10.378}$$

Nonvanishing helicity amplitudes:

$$M(+,+;+,-,+) = -\frac{g^3(q_+q'_+ + q_\perp q'^*_\perp)Z^*(q,q')Z^*(q',q)(q' \cdot k_3)^{\frac{1}{2}}}{2Eq_+q'_+[q_-q'_-(q \cdot q')^3(q \cdot k_3)]^{\frac{1}{2}}}$$

$$\times \left\{(T^aT^bT^c)_{ji}\frac{Z(k_3,q)}{k_{3+}q_-} - (T^aT^cT^b)_{ji}\frac{Z(k_3,q)Z(q',k_3)}{k_{3+}q_-q'_+k_{3-}}\right.$$

$$+(T^bT^aT^c)_{ji}\frac{Z(q,k_3)}{q_+k_{3-}} - (T^bT^cT^a)_{ji}\frac{Z(k_3,q')Z(q,k_3)}{k_{3+}q'_-q_+k_{3-}}$$

$$\left.+(T^cT^aT^b)_{ji}\frac{Z(q',k_3)}{q'_+k_{3-}} + (T^cT^bT^a)_{ji}\frac{Z(k_3,q')}{k_{3+}q'_-}\right\},$$

$$M(+,+;-,+,+) = -\frac{g^3(q_+q'_+ + q^*_\perp q'_\perp)Z^*(q,q')Z^*(q',q)(q\cdot k_3)^{\frac{1}{2}}}{2Eq_+q'_+[q_-q'_-(q\cdot q')^3(q'\cdot k_3)]^{\frac{1}{2}}}$$

$$\times \left\{ (T^aT^bT^c)_{ji}\frac{Z(k_3,q)}{k_{3+}q_-} - (T^aT^cT^b)_{ji}\frac{Z(k_3,q)Z(q',k_3)}{k_{3+}q_- q'_+ k_{3-}} \right.$$

$$+(T^bT^aT^c)_{ji}\frac{Z(q,k_3)}{q_+k_{3-}} - (T^bT^cT^a)_{ji}\frac{Z(k_3,q')Z(q,k_3)}{k_{3+}q'_- q_+ k_{3-}}$$

$$\left. +(T^cT^aT^b)_{ji}\frac{Z(q',k_3)}{q'_+ k_{3-}} + (T^cT^bT^a)_{ji}\frac{Z(k_3,q')}{k_{3+}q'_-} \right\},$$

$$M(+,-;+,-,+) = -\frac{g^3Z^2(q',q)}{2q'_+}\left[\frac{q_-q'_-}{(q\cdot q')^3(q\cdot k_3)(q'\cdot k_3)}\right]^{\frac{1}{2}}$$

$$\times \left\{ (T^aT^bT^c)_{ji}\frac{Z^*(k_3,q)}{k_{3+}q_-} - (T^aT^cT^b)_{ji}\frac{Z^*(k_3,q)Z^*(q',k_3)}{k_{3+}q_- q'_+ k_{3-}} \right.$$

$$+(T^bT^aT^c)_{ji}\frac{Z^*(q,k_3)}{q_+k_{3-}} - (T^bT^cT^a)_{ji}\frac{Z^*(k_3,q')Z^*(q,k_3)}{k_{3+}q'_- q_+ k_{3-}}$$

$$\left. +(T^cT^aT^b)_{ji}\frac{Z^*(q',k_3)}{q'_+ k_{3-}} + (T^cT^bT^a)_{ji}\frac{Z^*(k_3,q')}{k_{3+}q'_-} \right\},$$

$$M(+,-;-,+,+) = \frac{g^3Z^2(q,q')}{2q_+}\left[\frac{q_-q'_-}{(q\cdot q')^3(q\cdot k_3)(q'\cdot k_3)}\right]^{\frac{1}{2}}$$

$$\times \left\{ (T^aT^bT^c)_{ji}\frac{Z^*(k_3,q)}{k_{3+}q_-} - (T^aT^cT^b)_{ji}\frac{Z^*(k_3,q)Z^*(q',k_3)}{k_{3+}q_- q'_+ k_{3-}} \right.$$

$$+(T^bT^aT^c)_{ji}\frac{Z^*(q,k_3)}{q_+k_{3-}} - (T^bT^cT^a)_{ji}\frac{Z^*(k_3,q')Z^*(q,k_3)}{k_{3+}q'_- q_+ k_{3-}}$$

$$\left. +(T^cT^aT^b)_{ji}\frac{Z^*(q',k_3)}{q'_+ k_{3-}} + (T^cT^bT^a)_{ji}\frac{Z^*(k_3,q')}{k_{3+}q'_-} \right\},$$

$$M(-,+;+,-,+) = \frac{g^3q_+Z(q,q')Z(q',q)}{2[q_-q'_-(q\cdot q')^3(q\cdot k_3)(q'\cdot k_3)]^{\frac{1}{2}}}$$

$$\times \left\{ (T^aT^bT^c)_{ji}\frac{Z^*(k_3,q)}{k_{3+}q_-} - (T^aT^cT^b)_{ji}\frac{Z^*(k_3,q)Z^*(q',k_3)}{k_{3+}q_- q'_+ k_{3-}} \right.$$

$$+(T^bT^aT^c)_{ji}\frac{Z^*(q,k_3)}{q_+k_{3-}} - (T^bT^cT^a)_{ji}\frac{Z^*(k_3,q')Z^*(q,k_3)}{k_{3+}q'_- q_+ k_{3-}}$$

$$\left. +(T^cT^aT^b)_{ji}\frac{Z^*(q',k_3)}{q'_+ k_{3-}} + (T^cT^bT^a)_{ji}\frac{Z^*(k_3,q')}{k_{3+}q'_-} \right\},$$

10.55. $gg \to \bar{q}qg$

$$M(-,+;-,+,+) = -\frac{g^3 q'_+ Z(q,q')Z(q',q)}{2[q_-q'_-(q\cdot q')^3(q\cdot k_3)(q'\cdot k_3)]^{\frac{1}{2}}}$$

$$\times \left\{ (T^aT^bT^c)_{ji}\frac{Z^*(k_3,q)}{k_{3+}q_-} - (T^aT^cT^b)_{ji}\frac{Z^*(k_3,q)Z^*(q',k_3)}{k_{3+}q_-q'_+k_{3-}} \right.$$

$$+ (T^bT^aT^c)_{ji}\frac{Z^*(q,k_3)}{q_+k_{3-}} - (T^bT^cT^a)_{ji}\frac{Z^*(k_3,q')Z^*(q,k_3)}{k_{3+}q'_-q_+k_{3-}}$$

$$\left. + (T^cT^aT^b)_{ji}\frac{Z^*(q',k_3)}{q'_+k_{3-}} + (T^cT^bT^a)_{ji}\frac{Z^*(k_3,q')}{k_{3+}q'_-} \right\},$$

$$M(+,-;+,-,-) = -\frac{g^3 q'_+ Z(q,q')Z^*(q',q)}{2[q_-q'_-(q\cdot q')^3(q\cdot k_3)(q'\cdot k_3)]^{\frac{1}{2}}}$$

$$\times \left\{ (T^aT^bT^c)_{ji}\frac{Z(k_3,q)}{k_{3+}q_-} - (T^aT^cT^b)_{ji}\frac{Z(k_3,q)Z(q',k_3)}{k_{3+}q_-q'_+k_{3-}} \right.$$

$$+ (T^bT^aT^c)_{ji}\frac{Z(q,k_3)}{q_+k_{3-}} - (T^bT^cT^a)_{ji}\frac{Z(k_3,q')Z(q,k_3)}{k_{3+}q'_-q_+k_{3-}}$$

$$\left. + (T^cT^aT^b)_{ji}\frac{Z(q',k_3)}{q'_+k_{3-}} + (T^cT^bT^a)_{ji}\frac{Z(k_3,q')}{k_{3+}q'_-} \right\},$$

$$M(+,-;-,+,-) = \frac{g^3 q_+ Z^*(q,q')Z^*(q',q)}{2[q_-q'_-(q\cdot q')^3(q\cdot k_3)(q'\cdot k_3)]^{\frac{1}{2}}}$$

$$\times \left\{ (T^aT^bT^c)_{ji}\frac{Z(k_3,q)}{k_{3+}q_-} - (T^aT^cT^b)_{ji}\frac{Z(k_3,q)Z(q',k_3)}{k_{3+}q_-q'_+k_{3-}} \right.$$

$$+ (T^bT^aT^c)_{ji}\frac{Z(q,k_3)}{q_+k_{3-}} - (T^bT^cT^a)_{ji}\frac{Z(k_3,q')Z(q,k_3)}{k_{3+}q'_-q_+k_{3-}}$$

$$\left. + (T^cT^aT^b)_{ji}\frac{Z(q',k_3)}{q'_+k_{3-}} + (T^cT^bT^a)_{ji}\frac{Z(k_3,q')}{k_{3+}q'_-} \right\},$$

$$M(-,+;+,-,-) = \frac{g^3 Z^{*2}(q,q')}{2q_+}\left[\frac{q_-q'_-}{(q\cdot q')^3(q\cdot k_3)(q'\cdot k_3)}\right]^{\frac{1}{2}}$$

$$\times \left\{ (T^aT^bT^c)_{ji}\frac{Z(k_3,q)}{k_{3+}q_-} - (T^aT^cT^b)_{ji}\frac{Z(k_3,q)Z(q',k_3)}{k_{3+}q_-q'_+k_{3-}} \right.$$

$$+ (T^bT^aT^c)_{ji}\frac{Z(q,k_3)}{q_+k_{3-}} - (T^bT^cT^a)_{ji}\frac{Z(k_3,q')Z(q,k_3)}{k_{3+}q'_-q_+k_{3-}}$$

$$\left. + (T^cT^aT^b)_{ji}\frac{Z(q',k_3)}{q'_+k_{3-}} + (T^cT^bT^a)_{ji}\frac{Z(k_3,q')}{k_{3+}q'_-} \right\},$$

$$M(-,+;-,+,-) = -\frac{g^3 Z^{*2}(q',q)}{2q'_+}\left[\frac{q_-q'_-}{(q\cdot q')^3(q\cdot k_3)(q'\cdot k_3)}\right]^{\frac{1}{2}}$$

$$\times\left\{(T^aT^bT^c)_{ji}\frac{Z(k_3,q)}{k_{3+}q_-}-(T^aT^cT^b)_{ji}\frac{Z(k_3,q)Z(q',k_3)}{k_{3+}q_-q'_+k_{3-}}\right.$$

$$+(T^bT^aT^c)_{ji}\frac{Z(q,k_3)}{q_+k_{3-}}-(T^bT^cT^a)_{ji}\frac{Z(k_3,q')Z(q,k_3)}{k_{3+}q'_-q_+k_{3-}}$$

$$\left.+(T^cT^aT^b)_{ji}\frac{Z(q',k_3)}{q'_+k_{3-}}+(T^cT^bT^a)_{ji}\frac{Z(k_3,q')}{k_{3+}q'_-}\right\},$$

$$M(-,-;+,-,-) = -\frac{g^3(q_+q'_+q_\perp q'^*_\perp)Z(q,q')Z(q',q)(q\cdot k_3)^{\frac{1}{2}}}{2Eq_+q'_+[q_-q'_-(q\cdot q')^3(q'\cdot k_3)]^{\frac{1}{2}}}$$

$$\times\left\{(T^aT^bT^c)_{ji}\frac{Z^*(k_3,q)}{k_{3+}q_-}-(T^aT^cT^b)_{ji}\frac{Z^*(k_3,q)Z^*(q',k_3)}{k_{3+}q_-q'_+k_{3-}}\right.$$

$$+(T^bT^aT^c)_{ji}\frac{Z^*(q,k_3)}{q_+k_{3-}}-(T^bT^cT^a)_{ji}\frac{Z^*(k_3,q')Z^*(q,k_3)}{k_{3+}q'_-q_+k_{3-}}$$

$$\left.+(T^cT^aT^b)_{ji}\frac{Z^*(q',k_3)}{q'_+k_{3-}}+(T^cT^bT^a)_{ji}\frac{Z^*(k_3,q')}{k_{3+}q'_-}\right\},$$

$$M(-,-;-,+,-) = -\frac{g^3(q_+q'_+ + q^*_\perp q'_\perp)Z(q,q')Z(q',q)(q'\cdot k_3)^{\frac{1}{2}}}{2Eq_+q'_+[q_-q'_-(q\cdot q')^3(q\cdot k_3)]^{\frac{1}{2}}}$$

$$\times\left\{(T^aT^bT^c)_{ji}\frac{Z^*(k_3,q)}{k_{3+}q_-}-(T^aT^cT^b)_{ji}\frac{Z^*(k_3,q)Z^*(q',k_3)}{k_{3+}q_-q'_+k_{3-}}\right.$$

$$+(T^bT^aT^c)_{ji}\frac{Z^*(q,k_3)}{q_+k_{3-}}-(T^bT^cT^a)_{ji}\frac{Z^*(k_3,q')Z^*(q,k_3)}{k_{3+}q'_-q_+k_{3-}}$$

$$\left.+(T^cT^aT^b)_{ji}\frac{Z^*(q',k_3)}{q'_+k_{3-}}+(T^cT^bT^a)_{ji}\frac{Z^*(k_3,q')}{k_{3+}q'_-}\right\}.$$

(10.379)

Squared absolute values of the nonvanishing helicity amplitudes, summed over the color degrees of freedom of the initial state and the final state:

$$|M(+,+;+,+,-)|^2 = |M(-,-;-,-,+)|^2 = \frac{4g^6(q\cdot k_3)^2 F}{E^4 q_+q_-q'_+q'_-},$$

$$|M(+,+;+,-,+)|^2 = |M(-,-;-,+,-)|^2 = \frac{g^6(q'\cdot k_3)^2 F}{E^4 q_+q_-q'_+q'_-},$$

$$|M(+,-;+,+,-)|^2 = |M(-,+;-,-,+)|^2 = \frac{g^6 q'^2_- F}{q_+q'_+(\cdot k_3)(q'\cdot k_3)},$$

10.55. $gg \to \bar{q}qg$

$$|M(+,-;+,-,+)|^2 = |M(-,+;-,+,-)|^2 = \frac{g^6 q_-^2 F}{q_+ q'_+ (q \cdot k_3)(q' \cdot k_3)},$$

$$|M(-,+;+,+,-)|^2 = |M(+,-;-,-,+)|^2 = \frac{g^6 q'^2_+ F}{q_- q'_- (q \cdot k_3)(q' \cdot k_3)},$$

$$|M(-,+;+,-,+)|^2 = |M(+,-;-,+,-)|^2 = \frac{g^6 q_+^2 F}{q_- q'_- (q \cdot k_3)(q' \cdot k_3)}.$$

(10.380)

Unpolarized squared matrix element:

$$\overline{|M|^2} = \frac{g^6}{288} \frac{\sum_{i=1}^{3}(q \cdot k_i)(q' \cdot k_i)[(q \cdot k_i)^2 + (q' \cdot k_i)^2]}{\prod_{i=1}^{3}(q \cdot k_i)(q' \cdot k_i)}$$

$$\times \left\{ 10(q \cdot q') - 9 \left[\frac{(q \cdot k_1)(q' \cdot k_2) + (q \cdot k_2)(q' \cdot k_1)}{(k_1 \cdot k_2)} \right. \right.$$

$$+ \frac{(q \cdot k_2)(q' \cdot k_3) + (q \cdot k_3)(q' \cdot k_2)}{(k_2 \cdot k_3)}$$

$$\left. + \frac{(q \cdot k_3)(q' \cdot k_1) + (q \cdot k_1)(q' \cdot k_3)}{(k_1 \cdot k_3)} \right] + \frac{81}{(q \cdot q')}$$

$$\times \left[\frac{(q \cdot k_1)(q' \cdot k_1)[(q \cdot k_2)(q' \cdot k_3) + (q \cdot k_3)(q' \cdot k_2)]}{(k_1 \cdot k_2)(k_1 \cdot k_3)} \right.$$

$$+ \frac{(q \cdot k_2)(q' \cdot k_2)[(q \cdot k_3)(q' \cdot k_1) + (q \cdot k_1)(q' \cdot k_3)]}{(k_1 \cdot k_2)(k_2 \cdot k_3)}$$

$$\left. \left. + \frac{(q \cdot k_3)(q' \cdot k_3)[(q \cdot k_1)(q' \cdot k_2) + (q \cdot k_2)(q' \cdot k_1)]}{(k_1 \cdot k_3)(k_2 \cdot k_3)} \right] \right\}.$$

(10.381)

Unpolarized cross section:

$$d\sigma = \frac{\alpha_S^3}{4608\pi^2} \frac{\sum_{i=1}^{3}(q \cdot k_i)(q' \cdot k_i)[(q \cdot k_i)^2 + (q' \cdot k_i)^2]}{\prod_{i=1}^{3}(q \cdot k_i)(q' \cdot k_i)}$$

$$\times \left\{ 10(q \cdot q') - 9\left[\frac{(q \cdot k_1)(q' \cdot k_2) + (q \cdot k_2)(q' \cdot k_1)}{(k_1 \cdot k_2)} \right.\right.$$

$$+ \frac{(q \cdot k_2)(q' \cdot k_3) + (q \cdot k_3)(q' \cdot k_2)}{(k_2 \cdot k_3)}$$

$$\left.+ \frac{(q \cdot k_3)(q' \cdot k_1) + (q \cdot k_1)(q' \cdot k_3)}{(k_1 \cdot k_3)} \right] + \frac{81}{(q \cdot q')}$$

$$\times \left[\frac{(q \cdot k_1)(q' \cdot k_1)[(q \cdot k_2)(q' \cdot k_3) + (q \cdot k_3)(q' \cdot k_2)]}{(k_1 \cdot k_2)(k_1 \cdot k_3)} \right.$$

$$+ \frac{(q \cdot k_2)(q' \cdot k_2)[(q \cdot k_3)(q' \cdot k_1) + (q \cdot k_1)(q' \cdot k_3)]}{(k_1 \cdot k_2)(k_2 \cdot k_3)}$$

$$\left.\left.+ \frac{(q \cdot k_3)(q' \cdot k_3)[(q \cdot k_1)(q' \cdot k_2) + (q \cdot k_2)(q' \cdot k_1)]}{(k_1 \cdot k_3)(k_2 \cdot k_3)} \right] \right\}$$

$$\times \delta^4(k_1 + k_2 - k_3 - q - q') \frac{d^3\vec{k}_3 \, d^3\vec{q} \, d^3\vec{q'}}{k_{30} \, q_0 \, q'_0} \,. \qquad (10.382)$$

10.56 $gg \to ggg$

Process:

$$g(k_1, a) + g(k_2, b) \to g(k_3, c) + g(k_4, d) + g(k_5, e) \,. \qquad (10.383)$$

Positive z-axis: along $\vec{k}_1$.
Definitions:

$$Z_{ij} = k_{i+}k_{j-} - k_{i\perp}^* k_{j\perp}, \qquad\qquad i, j = 3, 4, 5 \,,$$

$$F = 108\left[\frac{2(k_3 \cdot k_4)(k_3 \cdot k_5)}{k_{3+}k_{3-}k_{4+}k_{5-}} + \frac{2(k_3 \cdot k_4)(k_3 \cdot k_5)}{k_{3+}k_{3-}k_{4-}k_{5+}} + \frac{2(k_3 \cdot k_4)(k_4 \cdot k_5)}{k_{3+}k_{4+}k_{4-}k_{5-}} \right.$$

$$+ \frac{2(k_3 \cdot k_4)(k_4 \cdot k_5)}{k_{3-}k_{4+}k_{4-}k_{5+}} + \frac{2(k_3 \cdot k_5)(k_4 \cdot k_5)}{k_{3+}k_{4-}k_{5+}k_{5-}} + \frac{2(k_3 \cdot k_5)(k_4 \cdot k_5)}{k_{3-}k_{4+}k_{5+}k_{5-}}$$

$$\left.+ \frac{(k_3 \cdot k_4)}{k_{3+}k_{4-}} + \frac{(k_3 \cdot k_4)}{k_{3-}k_{4+}} + \frac{(k_3 \cdot k_5)}{k_{3+}k_{5-}} + \frac{(k_3 \cdot k_5)}{k_{3-}k_{5+}} + \frac{(k_4 \cdot k_5)}{k_{4+}k_{5-}} + \frac{(k_4 \cdot k_5)}{k_{4-}k_{5+}} \right],$$

10.56. $gg \to ggg$

$$\begin{aligned}B = {} & \frac{Z_{35}}{k_{3+}k_{5-}}\mathrm{Im}[\mathrm{Tr}\,(T^aT^bT^cT^dT^e)] - \frac{Z_{34}}{k_{3+}k_{4-}}\mathrm{Im}[\mathrm{Tr}\,(T^aT^bT^cT^eT^d)] \\ & - \frac{Z_{45}}{k_{4+}k_{5-}}\mathrm{Im}[\mathrm{Tr}\,(T^aT^bT^dT^cT^e)] + \frac{Z_{43}}{k_{4+}k_{3-}}\mathrm{Im}[\mathrm{Tr}\,(T^aT^bT^dT^eT^c)] \\ & + \frac{Z_{54}}{k_{5+}k_{4-}}\mathrm{Im}[\mathrm{Tr}\,(T^aT^bT^eT^cT^d)] - \frac{Z_{53}}{k_{5+}k_{3-}}\mathrm{Im}[\mathrm{Tr}\,(T^aT^bT^eT^dT^c)] \\ & - \frac{Z_{35}Z_{43}\mathrm{Im}[\mathrm{Tr}\,(T^aT^cT^bT^dT^e)]}{k_{3+}k_{5-}k_{4+}k_{3-}} + \frac{Z_{34}Z_{53}\mathrm{Im}[\mathrm{Tr}\,(T^aT^cT^bT^eT^d)]}{k_{3+}k_{4-}k_{5+}k_{3-}} \\ & + \frac{Z_{34}Z_{45}\mathrm{Im}[\mathrm{Tr}\,(T^aT^dT^bT^cT^e)]}{k_{3+}k_{4-}k_{4+}k_{5-}} - \frac{Z_{43}Z_{54}\mathrm{Im}[\mathrm{Tr}\,(T^aT^dT^bT^eT^c)]}{k_{4+}k_{3-}k_{5+}k_{4-}} \\ & - \frac{Z_{35}Z_{54}\mathrm{Im}[\mathrm{Tr}\,(T^aT^eT^bT^cT^d)]}{k_{3+}k_{5-}k_{5+}k_{4-}} + \frac{Z_{45}Z_{53}\mathrm{Im}[\mathrm{Tr}\,(T^aT^eT^bT^dT^c)]}{k_{4+}k_{5-}k_{5+}k_{3-}}.\end{aligned}$$
(10.384)

Gluon polarizations:

$$\begin{aligned}\not{\epsilon}_1^{*\pm} & = N_1[\not{k}_4\,\not{k}_5\,\not{k}_1(1\pm\gamma_5)+\not{k}_1\,\not{k}_5\,\not{k}_4(1\mp\gamma_5)],\\ N_1^{-1} & = 4E[k_{4-}k_{5-}(k_4\cdot k_5)]^{\frac{1}{2}},\\ \not{\epsilon}_2^{*\pm} & = N_2[\not{k}_4\,\not{k}_5\,\not{k}_2(1\pm\gamma_5)+\not{k}_2\,\not{k}_5\,\not{k}_4(1\mp\gamma_5)],\\ N_2^{-1} & = 4E[k_{4+}k_{5+}(k_4\cdot k_5)]^{\frac{1}{2}},\\ \not{\epsilon}_3^{\pm} & = N_3[\not{k}_3\,\not{k}_5\,\not{k}_4(1\pm\gamma_5)+\not{k}_4\,\not{k}_5\,\not{k}_3(1\mp\gamma_5)],\\ N_3^{-1} & = 4[(k_3\cdot k_4)(k_3\cdot k_5)(k_4\cdot k_5)]^{\frac{1}{2}},\\ \not{\epsilon}_4^{\pm} & = N_4[\not{k}_4\,\not{k}_1\,\not{k}_3(1\pm\gamma_5)+\not{k}_3\,\not{k}_1\,\not{k}_4(1\mp\gamma_5)],\\ N_4^{-1} & = 4E[k_{3-}k_{4-}(k_3\cdot k_4)]^{\frac{1}{2}},\\ \not{\epsilon}_5^{\pm} & = N_5[\not{k}_5\,\not{k}_1\,\not{k}_3(1\pm\gamma_5)+\not{k}_3\,\not{k}_1\,\not{k}_5(1\mp\gamma_5)],\\ N_5^{-1} & = 4E[k_{3-}k_{5-}(k_3\cdot k_5)]^{\frac{1}{2}}.\end{aligned}$$
(10.385)

Nonvanishing helicity amplitudes:

$$M(+,+;+,+,+) = \frac{8g^3 B^* E^2 Z_{43}^* Z_{53}^*}{k_{3-}(k_3 \cdot k_4)(k_3 \cdot k_5)\left[k_{4+}k_{5+}(k_4 \cdot k_5)\right]^{\frac{1}{2}}},$$

$$M(+,+;+,+,-) = \frac{g^3 B Z_{43} Z_{53}^* Z_{45}^{*2}(k_3 \cdot k_4)}{E^2 k_{3-}k_{5-}(k_3 \cdot k_5)\left[k_{4+}^3 k_{5+}(k_4 \cdot k_5)\right]^{\frac{1}{2}}},$$

$$M(+,+;+,-,+) = \frac{g^3 B Z_{43}^* Z_{53} Z_{54}^{*2}(k_3 \cdot k_5)}{E^2 k_{3-}k_{4-}(k_3 \cdot k_4)\left[k_{4+}k_{5+}^3(k_4 \cdot k_5)\right]^{\frac{1}{2}}},$$

$$M(+,+;-,+,+) = \frac{2g^3 B Z_{43}^* Z_{53}^*}{E^2 k_{3-}(k_3 \cdot k_4)(k_3 \cdot k_5)}\left[\frac{(k_4 \cdot k_5)^3}{k_{4+}k_{5+}}\right]^{\frac{1}{2}},$$

$$M(+,-;+,+,-) = \frac{g^3 B^* k_{5-} Z_{43}^* Z_{53} Z_{45}^2}{k_{3-}(k_3 \cdot k_4)(k_3 \cdot k_5)\left[k_{4+}^3 k_{5+}(k_4 \cdot k_5)^3\right]^{\frac{1}{2}}},$$

$$M(+,-;+,-,+) = \frac{g^3 B^* k_{4-} Z_{43} Z_{53}^* Z_{54}^2}{k_{3-}(k_3 \cdot k_4)(k_3 \cdot k_5)\left[k_{4+}k_{5+}^3(k_4 \cdot k_5)^3\right]^{\frac{1}{2}}},$$

$$M(+,-;-,+,+) = \frac{2g^3 B^* k_{3-} Z_{43} Z_{53}}{(k_3 \cdot k_4)(k_3 \cdot k_5)\left[k_{4+}k_{5+}(k_4 \cdot k_5)\right]^{\frac{1}{2}}},$$

$$M(-,+;+,+,-) = \frac{g^3 B^* Z_{43}^* Z_{53} Z_{54}^2}{k_{3-}k_{4-}(k_3 \cdot k_4)(k_3 \cdot k_5)}\left[\frac{k_{5+}}{k_{4+}(k_4 \cdot k_5)^3}\right]^{\frac{1}{2}},$$

$$M(-,+;+,-,+) = \frac{g^3 B^* Z_{43} Z_{53}^* Z_{45}^2}{k_{3-}k_{5-}(k_3 \cdot k_4)(k_3 \cdot k_5)}\left[\frac{k_{4+}}{k_{5+}(k_4 \cdot k_5)^3}\right]^{\frac{1}{2}},$$

$$M(-,+;-,+,+) = \frac{2g^3 B^* k_{3\perp}^{*2} k_{4\perp} k_{5\perp} Z_{34} Z_{35}}{k_{3-}k_{4-}k_{5-}(k_3 \cdot k_4)(k_3 \cdot k_5)\left[k_{4+}k_{5+}(k_4 \cdot k_5)\right]^{\frac{1}{2}}},$$

$$M(+,-;+,-,-) = \frac{2g^3 B k_{3\perp}^2 k_{4\perp}^* k_{5\perp}^* Z_{34}^* Z_{35}^*}{k_{3-}k_{4-}k_{5-}(k_3 \cdot k_4)(k_3 \cdot k_5)\left[k_{4+}k_{5+}(k_4 \cdot k_5)\right]^{\frac{1}{2}}},$$

$$M(+,-;-,+,-) = \frac{g^3 B Z_{43}^* Z_{53} Z_{45}^{*2}}{k_{3-}k_{5-}(k_3 \cdot k_4)(k_3 \cdot k_5)}\left[\frac{k_{4+}}{k_{5+}(k_4 \cdot k_5)^3}\right]^{\frac{1}{2}},$$

$$M(+,-;-,-,+) = \frac{g^3 B Z_{43} Z_{53}^* Z_{54}^{*2}}{k_{3-}k_{4-}(k_3 \cdot k_4)(k_3 \cdot k_5)}\left[\frac{k_{5+}}{k_{4+}(k_4 \cdot k_5)^3}\right]^{\frac{1}{2}},$$

$$M(-,+;+,-,-) = \frac{2g^3 B k_{3-} Z_{43}^* Z_{53}^*}{(k_3 \cdot k_4)(k_3 \cdot k_5)\left[k_{4+}k_{5+}(k_4 \cdot k_5)\right]^{\frac{1}{2}}},$$

$$M(-,+;-,+,-) = \frac{g^3 B k_{4-} Z_{43}^* Z_{53} Z_{54}^{*2}}{k_{3-}(k_3 \cdot k_4)(k_3 \cdot k_5)\left[k_{4+}k_{5+}^3(k_4 \cdot k_5)^3\right]^{\frac{1}{2}}},$$

10.56. $gg \to ggg$

$$M(-,+;-,-,+) = \frac{g^3 B k_{5-} Z_{43} Z^*_{53} Z^{*2}_{45}}{k_{3-}(k_3 \cdot k_4)(k_3 \cdot k_5)[k^3_{4+} k_{5+}(k_4 \cdot k_5)^3]^{\frac{1}{2}}},$$

$$M(-,-;+,-,-) = \frac{2g^3 B^* Z_{43} Z_{53}}{E^2 k_{3-}(k_3 \cdot k_4)(k_3 \cdot k_5)} \left[\frac{(k_4 \cdot k_5)^3}{k_{4+} k_{5+}} \right]^{\frac{1}{2}},$$

$$M(-,-;-,+,-) = \frac{g^3 B^* Z_{43} Z^*_{53} Z^2_{54}(k_3 \cdot k_5)}{E^2 k_{3-} k_{4-}(k_3 \cdot k_4)[k_{4+} k^3_{5+}(k_4 \cdot k_5)]^{\frac{1}{2}}},$$

$$M(-,-;-,-,+) = \frac{g^3 B^* Z^*_{43} Z_{53} Z^2_{45}(k_3 \cdot k_4)}{E^2 k_{3-} k_{5-}(k_3 \cdot k_5)[k^3_{4+} k_{5+}(k_4 \cdot k_5)]^{\frac{1}{2}}},$$

$$M(-,-;-,-,-) = \frac{8 g^3 B E^2 Z_{43} Z_{53}}{k_{3-}(k_3 \cdot k_4)(k_3 \cdot k_5)[k_{4+} k_{5+}(k_4 \cdot k_5)]^{\frac{1}{2}}}. \quad (10.386)$$

Squared absolute values of the nonvanishing helicity amplitudes, summed over the color degrees of freedom of the initial state and the final state:

$$|M(+,+;+,+,+)|^2 = |M(-,-;-,-,-)|^2 = \frac{16 g^6 F E^4}{(k_3 \cdot k_4)(k_3 \cdot k_5)(k_4 \cdot k_5)},$$

$$|M(+,+;+,+,-)|^2 = |M(-,-;-,-,+)|^2 = \frac{g^6 F (k_3 \cdot k_4)^3}{E^4 (k_3 \cdot k_5)(k_4 \cdot k_5)},$$

$$|M(+,+;+,-,+)|^2 = |M(-,-;-,+,-)|^2 = \frac{g^6 F (k_3 \cdot k_5)^3}{E^4 (k_3 \cdot k_4)(k_4 \cdot k_5)},$$

$$|M(+,+;-,+,+)|^2 = |M(-,-;+,-,-)|^2 = \frac{g^6 F (k_4 \cdot k_5)^3}{E^4 (k_3 \cdot k_4)(k_3 \cdot k_5)},$$

$$|M(+,-;+,+,-)|^2 = |M(-,+;-,-,+)|^2 = \frac{g^6 F k^4_{5-}}{(k_3 \cdot k_4)(k_3 \cdot k_5)(k_4 \cdot k_5)},$$

$$|M(+,-;+,-,+)|^2 = |M(-,+;-,+,-)|^2 = \frac{g^6 F k^4_{4-}}{(k_3 \cdot k_4)(k_3 \cdot k_5)(k_4 \cdot k_5)},$$

$$|M(+,-;-,+,+)|^2 = |M(-,+;+,-,-)|^2 = \frac{g^6 F k^4_{3-}}{(k_3 \cdot k_4)(k_3 \cdot k_5)(k_4 \cdot k_5)},$$

$$|M(-,+;+,+,-)|^2 = |M(+,-;-,-,+)|^2 = \frac{g^6 F k^4_{5+}}{(k_3 \cdot k_4)(k_3 \cdot k_5)(k_4 \cdot k_5)},$$

$$|M(-,+;+,-,+)|^2 = |M(+,-;-,+,-)|^2 = \frac{g^6 F k^4_{4+}}{(k_3 \cdot k_4)(k_3 \cdot k_5)(k_4 \cdot k_5)},$$

$$|M(-,+;-,+,+)|^2 = |M(+,-;+,-,-)|^2 = \frac{g^6 F k^4_{3+}}{(k_3 \cdot k_4)(k_3 \cdot k_5)(k_4 \cdot k_5)}. \quad (10.387)$$

Unpolarized squared matrix element:

$$\overline{|M|^2} = \frac{27g^6}{16} \frac{\sum\limits_{i<j}(k_i \cdot k_j)^4}{\prod\limits_{i<j}(k_i \cdot k_j)}$$

$$\times \Big[(k_1 \cdot k_2)(k_1 \cdot k_3)(k_2 \cdot k_4)(k_3 \cdot k_5)(k_4 \cdot k_5)$$

$$+ (k_1 \cdot k_2)(k_1 \cdot k_3)(k_2 \cdot k_5)(k_3 \cdot k_4)(k_4 \cdot k_5)$$

$$+ (k_1 \cdot k_2)(k_1 \cdot k_4)(k_2 \cdot k_3)(k_3 \cdot k_5)(k_4 \cdot k_5)$$

$$+ (k_1 \cdot k_2)(k_1 \cdot k_4)(k_2 \cdot k_5)(k_3 \cdot k_4)(k_3 \cdot k_5)$$

$$+ (k_1 \cdot k_2)(k_1 \cdot k_5)(k_2 \cdot k_3)(k_3 \cdot k_4)(k_4 \cdot k_5)$$

$$+ (k_1 \cdot k_2)(k_1 \cdot k_5)(k_2 \cdot k_4)(k_3 \cdot k_4)(k_3 \cdot k_5)$$

$$+ (k_1 \cdot k_3)(k_1 \cdot k_4)(k_2 \cdot k_3)(k_2 \cdot k_5)(k_4 \cdot k_5)$$

$$+ (k_1 \cdot k_3)(k_1 \cdot k_4)(k_2 \cdot k_4)(k_2 \cdot k_5)(k_3 \cdot k_5)$$

$$+ (k_1 \cdot k_3)(k_1 \cdot k_5)(k_2 \cdot k_3)(k_2 \cdot k_4)(k_4 \cdot k_5)$$

$$+ (k_1 \cdot k_3)(k_1 \cdot k_5)(k_2 \cdot k_4)(k_2 \cdot k_5)(k_3 \cdot k_4)$$

$$+ (k_1 \cdot k_4)(k_1 \cdot k_5)(k_2 \cdot k_3)(k_2 \cdot k_4)(k_3 \cdot k_5)$$

$$+ (k_1 \cdot k_4)(k_1 \cdot k_5)(k_2 \cdot k_3)(k_2 \cdot k_5)(k_3 \cdot k_4) \Big] .$$

(10.388)

Unpolarized cross section:

$$d\sigma = \frac{27\alpha_S^3}{256\pi^2} \frac{\sum_{i<j}(k_i \cdot k_j)^4}{(k_1 \cdot k_2)\prod_{i<j}(k_i \cdot k_j)} \Big[(k_1 \cdot k_2)(k_1 \cdot k_3)(k_2 \cdot k_4)(k_3 \cdot k_5)(k_4 \cdot k_5)$$

$$+(k_1 \cdot k_2)(k_1 \cdot k_3)(k_2 \cdot k_5)(k_3 \cdot k_4)(k_4 \cdot k_5)$$

$$+(k_1 \cdot k_2)(k_1 \cdot k_4)(k_2 \cdot k_3)(k_3 \cdot k_5)(k_4 \cdot k_5)$$

$$+(k_1 \cdot k_2)(k_1 \cdot k_4)(k_2 \cdot k_5)(k_3 \cdot k_4)(k_3 \cdot k_5)$$

$$+(k_1 \cdot k_2)(k_1 \cdot k_5)(k_2 \cdot k_3)(k_3 \cdot k_4)(k_4 \cdot k_5)$$

$$+(k_1 \cdot k_2)(k_1 \cdot k_5)(k_2 \cdot k_4)(k_3 \cdot k_4)(k_3 \cdot k_5)$$

$$+(k_1 \cdot k_3)(k_1 \cdot k_4)(k_2 \cdot k_3)(k_2 \cdot k_5)(k_4 \cdot k_5)$$

$$+(k_1 \cdot k_3)(k_1 \cdot k_4)(k_2 \cdot k_4)(k_2 \cdot k_5)(k_3 \cdot k_5)$$

$$+(k_1 \cdot k_3)(k_1 \cdot k_5)(k_2 \cdot k_3)(k_2 \cdot k_4)(k_4 \cdot k_5)$$

$$+(k_1 \cdot k_3)(k_1 \cdot k_5)(k_2 \cdot k_4)(k_2 \cdot k_5)(k_3 \cdot k_4)$$

$$+(k_1 \cdot k_4)(k_1 \cdot k_5)(k_2 \cdot k_3)(k_2 \cdot k_4)(k_3 \cdot k_5)$$

$$+(k_1 \cdot k_4)(k_1 \cdot k_5)(k_2 \cdot k_3)(k_2 \cdot k_5)(k_3 \cdot k_4)\Big]$$

$$\times \delta^4(k_1 + k_2 - k_3 - k_4 - k_5) \frac{d^3\vec{k}_3 \, d^3\vec{k}_4 \, d^3\vec{k}_5}{k_{30} \, k_{40} \, k_{50}}. \quad (10.389)$$

10.57 $gg \to {}^1S_0$

Process:
$$g(k_1, a) + g(k_2, b) \to {}^1S_0(p). \quad (10.390)$$

Invariants and definitions:
$$s = 2(k_1 \cdot k_2) = p^2 = M^2. \quad (10.391)$$

Gluon polarizations (the four-vector k is a generally positioned four-vector with $k^2 = 0$):

$$\epsilon^{*\pm}(k_1) = N[\slashed{k}_1 \slashed{k} \slashed{k}_2(1 \mp \gamma_5) + \slashed{k}_2 \slashed{k} \slashed{k}_1(1 \pm \gamma_5)],$$

$$\epsilon^{*\pm}(k_2) = N[\slashed{k}_2 \slashed{k} \slashed{k}_1(1 \mp \gamma_5) + \slashed{k}_1 \slashed{k} \slashed{k}_2(1 \pm \gamma_5)],$$

$$N^{-1} = 4[(k_1 \cdot k_2)(k_1 \cdot k)(k_2 \cdot k)]^{\frac{1}{2}}. \tag{10.392}$$

Nonvanishing helicity amplitudes [the definition of the quantity R_0 is given in eqn (8.11)]:

$$M(+,+) = -\frac{g^2 R_0 \delta_{ab}}{\sqrt{3\pi M}},$$

$$M(-,-) = \frac{g^2 R_0 \delta_{ab}}{\sqrt{3\pi M}}. \tag{10.393}$$

Squared absolute values of the nonvanishing helicity amplitudes, summed over the color degrees of freedom of the initial state:

$$|M(+,+)|^2 = |M(-,-)|^2 = \frac{8g^4 R_0^2}{3\pi M}. \tag{10.394}$$

Unpolarized squared matrix element:

$$\overline{|M|^2} = \frac{g^4 R_0^2}{48\pi M}. \tag{10.395}$$

Unpolarized cross section:

$$\sigma = \frac{\pi^2 \alpha_S^2 R_0^2}{3M^3} \delta(s - M^2). \tag{10.396}$$

10.58 $gg \to {}^3P_0$

Process:
$$g(k_1, a) + g(k_2, b) \to {}^3P_0(p). \tag{10.397}$$

Invariants and definitions:
$$s = 2(k_1 \cdot k_2) = p^2 = M^2. \tag{10.398}$$

Gluon polarizations (the four-vector k is a generally positioned four-vector with $k^2 = 0$):

$$\epsilon^{*\pm}(k_1) = N[\slashed{k}_1 \slashed{k} \slashed{k}_2(1 \mp \gamma_5) + \slashed{k}_2 \slashed{k} \slashed{k}_1(1 \pm \gamma_5)],$$

$$\epsilon^{*\pm}(k_2) = N[\slashed{k}_2 \slashed{k} \slashed{k}_1(1 \mp \gamma_5) + \slashed{k}_1 \slashed{k} \slashed{k}_2(1 \pm \gamma_5)],$$

$$N^{-1} = 4[(k_1 \cdot k_2)(k_1 \cdot k)(k_2 \cdot k)]^{\frac{1}{2}}. \tag{10.399}$$

Nonvanishing helicity amplitudes [the definition of the quantity R'_1 is given in eqn (8.12)]:

$$M(+,+) = M(-,-) = -\frac{6ig^2 R'_1 \delta_{ab}}{\sqrt{3\pi M^3}}. \qquad (10.400)$$

Squared absolute values of the nonvanishing helicity amplitudes, summed over the color degrees of freedom of the initial state:

$$|M(+,+)|^2 = |M(-,-)|^2 = \frac{96g^4 R'^2_1}{\pi M^3}. \qquad (10.401)$$

Unpolarized squared matrix element:

$$\overline{|M|^2} = \frac{3g^4 R'^2_1}{4\pi M^3}. \qquad (10.402)$$

Unpolarized cross section:

$$\sigma = \frac{12\pi^2 \alpha_s^2 R'^2_1}{M^5} \delta(s - M^2). \qquad (10.403)$$

10.59 $gg \to {}^3P_2$

Process:
$$g(k_1, a) + g(k_2, b) \to {}^3P_2(p). \qquad (10.404)$$

Invariants and definitions:
$$s = 2(k_1 \cdot k_2) = p^2 = M^2. \qquad (10.405)$$

Gluon polarizations (the four-vector k is a generally positioned four-vector with $k^2 = 0$):

$$\slashed{\epsilon}^{*\pm}(k_1) = N[\slashed{k}_1 \slashed{k} \slashed{k}_2(1 \mp \gamma_5) + \slashed{k}_2 \slashed{k} \slashed{k}_1(1 \pm \gamma_5)],$$

$$\slashed{\epsilon}^{*\pm}(k_2) = N[\slashed{k}_2 \slashed{k} \slashed{k}_1(1 \mp \gamma_5) + \slashed{k}_1 \slashed{k} \slashed{k}_2(1 \pm \gamma_5)],$$

$$N^{-1} = 4[(k_1 \cdot k_2)(k_1 \cdot k)(k_2 \cdot k)]^{\frac{1}{2}}. \qquad (10.406)$$

Nonvanishing helicity amplitudes [the definition of the quantity R'_1 is given in eqn (8.12) and the quantity $\epsilon^{\alpha\beta}$ denotes the polarization tensor of the ${}^3P_2$]:

$$M(+,-) = \frac{16ig^2 R'_1 N^2 \delta_{ab} \epsilon^{\alpha\beta}}{\sqrt{\pi M^3}} \Big\{ 2k_{1\alpha} k_{1\beta} \left[(k_1 \cdot k)^2 + (k_2 \cdot k)^2\right] + \tfrac{1}{2} M^4 k_\alpha k_\beta$$

$$+ 2M^2 k_{1\alpha} k_\beta (k_1 - k_2 \cdot k) - i\epsilon_{\alpha\mu\nu\rho} k^{1\mu} k^{2\nu} k^\rho \left[2k_{1\beta}(k_1 - k_2 \cdot k) + M^2 k_\beta\right] \Big\},$$

$$M(-,+) = \frac{16ig^2 R_1' N^2 \delta_{ab}\epsilon^{\alpha\beta}}{\sqrt{\pi}M^3}\left\{2k_{1\alpha}k_{1\beta}\left[(k_1\cdot k)^2 + (k_2\cdot k)^2\right] + \tfrac{1}{2}M^4 k_\alpha k_\beta\right.$$
$$\left. + 2M^2 k_{1\alpha}k_\beta(k_1 - k_2\cdot k) + i\epsilon_{\alpha\mu\nu\rho}k^{1\mu}k^{2\nu}k^\rho\left[2k_{1\beta}(k_1 - k_2\cdot k) + M^2 k_\beta\right]\right\}. \tag{10.407}$$

Squared absolute values of the nonvanishing helicity amplitudes, summed over the color degrees of freedom of the initial state and summed over the $^3P_2$ polarizations:

$$|M(+,-)|^2 = |M(-,+)|^2 = \frac{128g^4 R_1'^2}{\pi M^3}. \tag{10.408}$$

Unpolarized squared matrix element:

$$\overline{|M|^2} = \frac{g^4 R_1'^2}{\pi M^3}. \tag{10.409}$$

Unpolarized cross section:

$$\sigma = \frac{16\pi^2 \alpha_s^2 R_1'^2}{M^5}\delta(s - M^2). \tag{10.410}$$

10.60 $qg \to q\,{}^1S_0$

Process:
$$q(p_+, i) + g(k, a) \to q(q_+, j) + {}^1S_0(p). \tag{10.411}$$

Invariants and definitions:
$$s = 2(p_+\cdot k), \qquad t = -2(p_+\cdot q_+), \qquad u = -2(k\cdot q_+),$$
$$s + t + u = p^2 = M^2. \tag{10.412}$$

Gluon polarizations:
$$\not{\epsilon}^\pm = N[\not{k}\,\not{q}_+\,\not{p}_+(1\mp\gamma_5) + \not{p}_+\,\not{q}_+\,\not{k}(1\pm\gamma_5)],$$
$$N^{-1} = E^2[32 q_{++}q_{+-}]^{\frac{1}{2}}. \tag{10.413}$$

Nonvanishing helicity amplitudes [the definition of the quantity R_0 is given in eqn (8.11)]:

$$M(+,+;+) = \frac{4g^3 R_0 T^a_{ji}}{t - M^2}\left[\frac{E^3}{3\pi M q_{+-}}\right]^{\frac{1}{2}},$$

$$M(+,-;+) = -\frac{2g^3 R_0 T^a_{ji} q_{++}}{t - M^2} \left[\frac{E}{3\pi M q_{+-}}\right]^{\frac{1}{2}},$$

$$M(-,+;-) = \frac{2g^3 R_0 T^a_{ji} q_{++}}{t - M^2} \left[\frac{E}{3\pi M q_{+-}}\right]^{\frac{1}{2}},$$

$$M(-,-;-) = -\frac{4g^3 R_0 T^a_{ji}}{t - M^2} \left[\frac{E^3}{3\pi M q_{+-}}\right]^{\frac{1}{2}}. \tag{10.414}$$

Squared absolute values of the nonvanishing helicity amplitudes, summed over the color degrees of freedom of the initial state and the final state:

$$|M(+,+;+)|^2 = |M(-,-;-)|^2 = -\frac{8g^6 R_0^2 s^2}{3\pi M t (t-M^2)^2},$$

$$|M(+,-;+)|^2 = |M(-,+;-)|^2 = -\frac{8g^6 R_0^2 u^2}{3\pi M t (t-M^2)^2}. \tag{10.415}$$

Unpolarized squared matrix element:

$$\overline{|M|^2} = -\frac{g^6 R_0^2 (s^2 + u^2)}{18\pi M t (t-M^2)^2}. \tag{10.416}$$

Unpolarized cross section:

$$\frac{d\sigma}{dt} = -\frac{2\pi \alpha_S^3 R_0^2 (s^2 + u^2)}{9 M s^2 t (t-M^2)^2}. \tag{10.417}$$

10.61 $qg \to q\,^3P_0$

Process:
$$q(p_+, i) + g(k, a) \to q(q_+, j) + {}^3P_0(p). \tag{10.418}$$

Invariants and definitions:

$$s = 2(p_+ \cdot k), \qquad t = -2(p_+ \cdot q_+), \qquad u = -2(k \cdot q_+),$$

$$s + t + u = p^2 = M^2. \tag{10.419}$$

Gluon polarizations:

$$\not{k}^{*\pm} = N[\not{k}\not{q}_+ \not{p}_+(1 \mp \gamma_5) + \not{p}_+ \not{q}_+ \not{k}(1 \pm \gamma_5)],$$

$$N^{-1} = E^2 [32 q_{++} q_{+-}]^{\frac{1}{2}}. \tag{10.420}$$

Nonvanishing helicity amplitudes [the definition of the quantity R'_1 is given in eqn (8.12)]:

$$M(+,+;+) = \frac{2ig^3 R'_1 T^a_{ji} s(t-3M^2)}{(t-M^2)^2 [3\pi M^3 E q_{+-}]^{\frac{1}{2}}},$$

$$M(+,-;+) = \frac{4ig^3 R'_1 T^a_{ji}(t-3M^2) q_{++}}{(t-M^2)^2} \left[\frac{E}{3\pi M^3 q_{+-}}\right]^{\frac{1}{2}},$$

$$M(-,+;-) = \frac{4ig^3 R'_1 T^a_{ji}(t-3M^2) q_{++}}{(t-M^2)^2} \left[\frac{E}{3\pi M^3 q_{+-}}\right]^{\frac{1}{2}},$$

$$M(-,-;-) = \frac{2ig^3 R'_1 T^a_{ji} s(t-3M^2)}{(t-M^2)^2 [3\pi M^3 E q_{+-}]^{\frac{1}{2}}}. \quad (10.421)$$

Squared absolute values of the nonvanishing helicity amplitudes, summed over the color degrees of freedom of the initial state and the final state:

$$|M(+,+;+)|^2 = |M(-,-;-)|^2 = -\frac{32 g^6 R'^2_1 s^2 (t-3M^2)^2}{3\pi M^3 t (t-M^2)^4},$$

$$|M(+,-;+)|^2 = |M(-,+;-)|^2 = -\frac{32 g^6 R'^2_1 u^2 (t-3M^2)^2}{3\pi M^3 t (t-M^2)^4}. \quad (10.422)$$

Unpolarized squared matrix element:

$$\overline{|M|^2} = -\frac{2 g^6 R'^2_1 (s^2+u^2)(t-3M^2)^2}{9\pi M^3 t (t-M^2)^4}. \quad (10.423)$$

Unpolarized cross section:

$$\frac{d\sigma}{dt} = -\frac{8\pi \alpha_s^3 R'^2_1 (s^2+u^2)(t-3M^2)^2}{9 M^3 s^2 t (t-M^2)^4}. \quad (10.424)$$

10.62 $qg \to q\,^3P_1$

Process:
$$q(p_+,i) + g(k,a) \to q(q_+,j) + {}^3P_1(p). \quad (10.425)$$

Invariants and definitions:
$$s = 2(p_+ \cdot k), \qquad t = -2(p_+ \cdot q_+), \qquad u = -2(k \cdot q_+),$$

$$s+t+u = p^2 = M^2. \quad (10.426)$$

Gluon polarizations:

$$\not{k}^{*\pm} = N[\not{k}\,\not{A}_+\,\not{p}_+(1\mp\gamma_5) + \not{p}_+\,\not{A}_+\,\not{k}(1\pm\gamma_5)],$$

$$N^{-1} = E^2[32 q_{++} q_{+-}]^{\frac{1}{2}}. \quad (10.427)$$

10.62. $qg \to q\,{}^3P_1$

Nonvanishing helicity amplitudes [the definition of the quantity R'_1 is given in eqn (8.12) and the four-vector ϵ denotes the polarization of the ${}^3P_1$]:

$$M(+,+;+) = \frac{i\sqrt{8}g^3 R'_1 T^a_{ji}}{(t-M^2)^2\sqrt{\pi M E q_{+-}}}$$
$$\times \left[(t+u)(p_+ \cdot \epsilon) + (s-t)(q_+ \cdot \epsilon) + 2i\epsilon^{\alpha\beta\gamma\delta}k_\alpha p_{+\beta} q_{+\gamma} \epsilon_\delta\right],$$

$$M(+,-;+) = \frac{i\sqrt{8}g^3 R'_1 T^a_{ji}}{(t-M^2)^2\sqrt{\pi M E q_{+-}}}$$
$$\times \left[(t-u)(p_+ \cdot \epsilon) - (s+t)(q_+ \cdot \epsilon) + 2i\epsilon^{\alpha\beta\gamma\delta}k_\alpha p_{+\beta} q_{+\gamma} \epsilon_\delta\right],$$

$$M(-,+;-) = -\frac{i\sqrt{8}g^3 R'_1 T^a_{ji}}{(t-M^2)^2\sqrt{\pi M E q_{+-}}}$$
$$\times \left[(t-u)(p_+ \cdot \epsilon) - (s+t)(q_+ \cdot \epsilon) - 2i\epsilon^{\alpha\beta\gamma\delta}k_\alpha p_{+\beta} q_{+\gamma} \epsilon_\delta\right],$$

$$M(-,-;-) = -\frac{i\sqrt{8}g^3 R'_1 T^a_{ji}}{(t-M^2)^2\sqrt{\pi M E q_{+-}}}$$
$$\times \left[(t+u)(p_+ \cdot \epsilon) + (s-t)(q_+ \cdot \epsilon) - 2i\epsilon^{\alpha\beta\gamma\delta}k_\alpha p_{+\beta} q_{+\gamma} \epsilon_\delta\right].$$
(10.428)

Squared absolute values of the nonvanishing helicity amplitudes, summed over the color degrees of freedom of the initial state and the final state and over the polarization states of the ${}^3P_1$:

$$|M(+,+;+)|^2 = |M(-,-;-)|^2 = -\frac{64g^6 R'^2_1 (s^2 t + 2suM^2)}{\pi M^3 (t-M^2)^4},$$

$$|M(+,-;+)|^2 = |M(-,+;-)|^2 = -\frac{64g^6 R'^2_1 (u^2 t + 2suM^2)}{\pi M^3 (t-M^2)^4}.$$
(10.429)

Unpolarized squared matrix element:

$$\overline{|M|^2} = -\frac{4g^6 R'^2_1}{3\pi M^3 (t-M^2)^4}\left[(s^2+u^2)t + 4suM^2\right]. \tag{10.430}$$

Unpolarized cross section:

$$\frac{d\sigma}{dt} = -\frac{16\pi\alpha_s^3 R'^2_1}{3M^3 s^2 (t-M^2)^4}\left[(s^2+u^2)t + 4suM^2\right]. \tag{10.431}$$

10.63 $qg \to q\,^3P_2$

Process:
$$q(p_+, i) + g(k, a) \to q(q_+, j) + {}^3P_2(p). \tag{10.432}$$

Invariants and definitions:
$$s = 2(p_+ \cdot k), \qquad t = -2(p_+ \cdot q_+), \qquad u = -2(k \cdot q_+),$$
$$s + t + u = p^2 = M^2. \tag{10.433}$$

Gluon polarizations:
$$\not{k}^{*\pm} = N[\not{k}\,\not{q}_+\,\not{p}_+(1 \mp \gamma_5) + \not{p}_+\,\not{q}_+\,\not{k}(1 \pm \gamma_5)],$$
$$N^{-1} = E^2[32 q_{++} q_{+-}]^{\frac{1}{2}}. \tag{10.434}$$

Nonvanishing helicity amplitudes [the definition of the quantity R'_1 is given in eqn (8.12) and the quantity $\epsilon^{\alpha\beta}$ denotes the polarization tensor of the $^3P_2$]:

$$M(+,+;+) = -\frac{4ig^3 R'_1 T^a_{ji} \epsilon^{\alpha\beta}}{s(t-M^2)^2} \left[\frac{-2M}{\pi t^3}\right]^{\frac{1}{2}} \Big\{(t+u)^2 p_{+\alpha} p_{+\beta}$$
$$+ 2(st + su - tM^2) p_{+\alpha} q_{+\beta} + (s^2 + t^2) q_{+\alpha} q_{+\beta}$$
$$+ 2i[(t+u)p_{+\alpha} + (s-t)q_{+\alpha}] \epsilon_{\mu\nu\rho\beta} k^\mu p^\nu_+ q^\rho_+ \Big\},$$

$$M(+,-;+) = \frac{4ig^3 R'_1 T^a_{ji} \epsilon^{\alpha\beta}}{u(t-M^2)^2} \left[\frac{-2M}{\pi t^3}\right]^{\frac{1}{2}} \Big\{(t^2 + u^2) p_{+\alpha} p_{+\beta}$$
$$+ 2(su + tu - tM^2) p_{+\alpha} q_{+\beta} + (s+t)^2 q_{+\alpha} q_{+\beta}$$
$$+ 2i[(t-u)p_{+\alpha} - (s+t)q_{+\alpha}] \epsilon_{\mu\nu\rho\beta} k^\mu p^\nu_+ q^\rho_+ \Big\},$$

$$M(-,+;-) = \frac{4ig^3 R'_1 T^a_{ji} \epsilon^{\alpha\beta}}{u(t-M^2)^2} \left[\frac{-2M}{\pi t^3}\right]^{\frac{1}{2}} \Big\{(t^2 + u^2) p_{+\alpha} p_{+\beta}$$
$$+ 2(su + tu - tM^2) p_{+\alpha} q_{+\beta} + (s+t)^2 q_{+\alpha} q_{+\beta}$$
$$- 2i[(t-u)p_{+\alpha} - (s+t)q_{+\alpha}] \epsilon_{\mu\nu\rho\beta} k^\mu p^\nu_+ q^\rho_+ \Big\},$$

$$M(-,-;-) = -\frac{4ig^3 R_1' T_{ji}^a \epsilon^{\alpha\beta}}{s(t-M^2)^2}\left[\frac{-2M}{\pi t^3}\right]^{\frac{1}{2}}\{(t+u)^2 p_{+\alpha}p_{+\beta}$$

$$+2(st+su-tM^2)p_{+\alpha}q_{+\beta}+(s^2+t^2)q_{+\alpha}q_{+\beta}$$

$$-2i[(t+u)p_{+\alpha}+(s-t)q_{+\alpha}]\epsilon_{\mu\nu\rho\beta}k^\mu p_+^\nu q_+^\rho\}. \quad (10.435)$$

Squared absolute values of the nonvanishing helicity amplitudes, summed over the color degrees of freedom of the initial state and the final state and over the polarization states of the $^3P_2$:

$$|M(+,+;+)|^2 = |M(-,-;-)|^2$$

$$= -\frac{64g^6 R_1'^2}{3\pi M^3 t(t-M^2)^4}(s^2t^2+6stuM^2+6u^2M^4),$$

$$|M(+,-;+)|^2 = |M(-,+;-)|^2$$

$$= -\frac{64g^6 R_1'^2}{3\pi M^3 t(t-M^2)^4}(u^2t^2+6stuM^2+6s^2M^4). \quad (10.436)$$

Unpolarized squared matrix element:

$$\overline{|M|^2} = -\frac{4g^6 R_1'^2}{9\pi M^3 t(t-M^2)^4}\left[(s^2+u^2)(t^2+6M^4)+12stuM^2\right]. \quad (10.437)$$

Unpolarized cross section:

$$\frac{d\sigma}{dt} = -\frac{16\pi\alpha_S^3 R_1'^2}{9M^3 s^2 t(t-M^2)^4}\left[(s^2+u^2)(t^2+6M^4)+12stuM^2\right]. \quad (10.438)$$

10.64 $\bar{q}g \to \bar{q}\,^1S_0$

Process:
$$\bar{q}(p_+,i)+g(k,a) \to \bar{q}(q_+,j)+{}^1S_0(p). \quad (10.439)$$

Invariants and definitions:
$$s=2(p_+\cdot k), \quad t=-2(p_+\cdot q_+), \quad u=-2(k\cdot q_+),$$

$$s+t+u=p^2=M^2. \quad (10.440)$$

Gluon polarizations:
$$\not{\epsilon}^{*\pm} = N[\not{k}\not{q}_+\not{p}_+(1\mp\gamma_5)+\not{p}_+\not{q}_+\not{k}(1\pm\gamma_5)],$$

$$N^{-1} = E^2[32q_{++}q_{+-}]^{\frac{1}{2}}. \quad (10.441)$$

Nonvanishing helicity amplitudes [the definition of the quantity R_0 is given in eqn (8.11)]:

$$M(+,+;+) = \frac{4g^3 R_0 T^a_{ij}}{t - M^2} \left[\frac{E^3}{3\pi M q_{+-}}\right]^{\frac{1}{2}},$$

$$M(+,-;+) = -\frac{2g^3 R_0 T^a_{ij} q_{++}}{t - M^2} \left[\frac{E}{3\pi M q_{+-}}\right]^{\frac{1}{2}},$$

$$M(-,+;-) = \frac{2g^3 R_0 T^a_{ij} q_{++}}{t - M^2} \left[\frac{E}{3\pi M q_{+-}}\right]^{\frac{1}{2}},$$

$$M(-,-;-) = -\frac{4g^3 R_0 T^a_{ij}}{t - M^2} \left[\frac{E^3}{3\pi M q_{+-}}\right]^{\frac{1}{2}}. \tag{10.442}$$

Squared absolute values of the nonvanishing helicity amplitudes, summed over the color degrees of freedom of the initial state and the final state:

$$|M(+,+;+)|^2 = |M(-,-;-)|^2 = -\frac{8g^6 R_0^2 s^2}{3\pi M t(t - M^2)^2},$$

$$|M(+,-;+)|^2 = |M(-,+;-)|^2 = -\frac{8g^6 R_0^2 u^2}{3\pi M t(t - M^2)^2}. \tag{10.443}$$

Unpolarized squared matrix element:

$$\overline{|M|^2} = -\frac{g^6 R_0^2}{18\pi M t(t - M^2)^2}(s^2 + u^2). \tag{10.444}$$

Unpolarized cross section:

$$\frac{d\sigma}{dt} = -\frac{2\pi \alpha_S^3 R_0^2 (s^2 + u^2)}{9M s^2 t(t - M^2)^2}. \tag{10.445}$$

10.65 $\bar{q}g \to \bar{q}\,^3P_0$

Process:
$$\bar{q}(p_+, i) + g(k, a) \to \bar{q}(q_+, j) + {}^3P_0(p). \tag{10.446}$$

Invariants and definitions:
$$s = 2(p_+ \cdot k), \quad t = -2(p_+ \cdot q_+), \quad u = -2(k \cdot q_+),$$

$$s + t + u = p^2 = M^2. \tag{10.447}$$

Gluon polarizations:

$$\rlap{/}{k}^{*\pm} = N[\rlap{/}{k}\,\rlap{/}{A}_+ \rlap{/}{p}_+(1 \mp \gamma_5) + \rlap{/}{p}_+ \rlap{/}{A}_+ \rlap{/}{k}(1 \pm \gamma_5)],$$

$$N^{-1} = E^2 [32 q_{++} q_{+-}]^{\frac{1}{2}}. \tag{10.448}$$

Nonvanishing helicity amplitudes [the definition of the quantity R_1' is given in eqn (8.12)]:

$$M(+,+;+) = \frac{2ig^3 R_1' T_{ij}^a s(t - 3M^2)}{(t - M^2)^2 [3\pi M^3 E q_{+-}]^{\frac{1}{2}}},$$

$$M(+,-;+) = \frac{4ig^3 R_1' T_{ij}^a (t - 3M^2) q_{++}}{(t - M^2)^2} \left[\frac{E}{3\pi M^3 q_{+-}}\right]^{\frac{1}{2}},$$

$$M(-,+;-) = \frac{4ig^3 R_1' T_{ij}^a (t - 3M^2) q_{++}}{(t - M^2)^2} \left[\frac{E}{3\pi M^3 q_{+-}}\right]^{\frac{1}{2}},$$

$$M(-,-;-) = \frac{2ig^3 R_1' T_{ij}^a s(t - 3M^2)}{(t - M^2)^2 [3\pi M^3 E q_{+-}]^{\frac{1}{2}}}. \quad (10.449)$$

Squared absolute values of the nonvanishing helicity amplitudes, summed over the color degrees of freedom of the initial state and the final state:

$$|M(+,+;+)|^2 = |M(-,-;-)|^2 = -\frac{32g^6 R_1'^2 s^2 (t - 3M^2)^2}{3\pi M^3 t (t - M^2)^4},$$

$$|M(+,-;+)|^2 = |M(-,+;-)|^2 = -\frac{32g^6 R_1'^2 u^2 (t - 3M^2)^2}{3\pi M^3 t (t - M^2)^4}. \quad (10.450)$$

Unpolarized squared matrix element:

$$\overline{|M|^2} = -\frac{2g^6 R_1'^2 (s^2 + u^2)(t - 3M^2)^2}{9\pi M^3 t (t - M^2)^4}. \quad (10.451)$$

Unpolarized cross section:

$$\frac{d\sigma}{dt} = -\frac{8\pi \alpha_s^3 R_1'^2 (s^2 + u^2)(t - 3M^2)^2}{9 M^3 s^2 t (t - M^2)^4}. \quad (10.452)$$

10.66 $\bar{q} g \to \bar{q} \, ^3P_1$

Process:
$$\bar{q}(p_+, i) + g(k, a) \to \bar{q}(q_+, j) + {}^3P_1(p). \quad (10.453)$$

Invariants and definitions:
$$s = 2(p_+ \cdot k), \quad t = -2(p_+ \cdot q_+), \quad u = -2(k \cdot q_+),$$

$$s + t + u = p^2 = M^2. \quad (10.454)$$

Gluon polarizations:
$$\rlap{/}{\epsilon}^{*\pm} = N[\rlap{/}{k} \rlap{/}{q}_+ \rlap{/}{p}_+ (1 \mp \gamma_5) + \rlap{/}{p}_+ \rlap{/}{q}_+ \rlap{/}{k}(1 \pm \gamma_5)],$$

$$N^{-1} = E^2 [32 q_{++} q_{+-}]^{\frac{1}{2}}. \quad (10.455)$$

Nonvanishing helicity amplitudes [the definition of the quantity R'_1 is given in eqn (8.12) and the four-vector ϵ denotes the polarization of the $^3P_1$]:

$$M(+,+;+) = \frac{i\sqrt{8}g^3 R'_1 T^a_{ij}}{(t-M^2)^2\sqrt{\pi M E}q_{+-}}\Big[(t+u)(p_+\cdot\epsilon)$$
$$+(s-t)(q_+\cdot\epsilon)+2i\epsilon^{\alpha\beta\gamma\delta}k_\alpha p_{+\beta}q_{+\gamma}\epsilon_\delta\Big],$$

$$M(+,-;+) = \frac{i\sqrt{8}g^3 R'_1 T^a_{ij}}{(t-M^2)^2\sqrt{\pi M E}q_{+-}}\Big[(t-u)(p_+\cdot\epsilon)$$
$$-(s+t)(q_+\cdot\epsilon)+2i\epsilon^{\alpha\beta\gamma\delta}k_\alpha p_{+\beta}q_{+\gamma}\epsilon_\delta\Big],$$

$$M(-,+;-) = -\frac{i\sqrt{8}g^3 R'_1 T^a_{ij}}{(t-M^2)^2\sqrt{\pi M E}q_{+-}}\Big[(t-u)(p_+\cdot\epsilon)$$
$$-(s+t)(q_+\cdot\epsilon)-2i\epsilon^{\alpha\beta\gamma\delta}k_\alpha p_{+\beta}q_{+\gamma}\epsilon_\delta\Big],$$

$$M(-,-;-) = -\frac{i\sqrt{8}g^3 R'_1 T^a_{ij}}{(t-M^2)^2\sqrt{\pi M E}q_{+-}}\Big[(t+u)(p_+\cdot\epsilon)$$
$$+(s-t)(q_+\cdot\epsilon)-2i\epsilon^{\alpha\beta\gamma\delta}k_\alpha p_{+\beta}q_{+\gamma}\epsilon_\delta\Big]. \quad (10.456)$$

Squared absolute values of the nonvanishing helicity amplitudes, summed over the color degrees of freedom of the initial state and the final state and over the polarization states of the $^3P_1$:

$$|M(+,+;+)|^2 = |M(-,-;-)|^2 = -\frac{64g^6 R'^2_1(s^2 t + 2suM^2)}{\pi M^3(t-M^2)^4},$$

$$|M(+,-;+)|^2 = |M(-,+;-)|^2 = -\frac{64g^6 R'^2_1(u^2 t + 2suM^2)}{\pi M^3(t-M^2)^4}. \quad (10.457)$$

Unpolarized squared matrix element:

$$\overline{|M|^2} = -\frac{4g^6 R'^2_1}{3\pi M^3(t-M^2)^4}\Big[(s^2+u^2)t + 4suM^2\Big]. \quad (10.458)$$

Unpolarized cross section:

$$\frac{d\sigma}{dt} = -\frac{16\pi\alpha_s^3 R'^2_1}{3M^3 s^2(t-M^2)^4}\Big[(s^2+u^2)t + 4suM^2\Big]. \quad (10.459)$$

10.67 $\bar{q}g \to \bar{q}\,^3P_2$

Process:
$$\bar{q}(p_+,i) + g(k,a) \to \bar{q}(q_+,j) + {}^3P_2(p). \tag{10.460}$$

Invariants and definitions:
$$s = 2(p_+ \cdot k), \qquad t = -2(p_+ \cdot q_+), \qquad u = -2(k \cdot q_+),$$
$$s + t + u = p^2 = M^2. \tag{10.461}$$

Gluon polarizations:
$$\not{\epsilon}^{*\pm} = N[\not{k}\,\not{A}_+\,\not{p}_+(1\mp\gamma_5) + \not{p}_+\,\not{A}_+\,\not{k}(1\pm\gamma_5)],$$
$$N^{-1} = E^2[32q_{++}q_{+-}]^{\frac{1}{2}}. \tag{10.462}$$

Nonvanishing helicity amplitudes [the definition of the quantity R'_1 is given in eqn (8.12) and the quantity $\epsilon^{\alpha\beta}$ denotes the polarization tensor of the ${}^3P_2$]:

$$M(+,+;+) = -\frac{4ig^3 R'_1 T^a_{ij}\epsilon^{\alpha\beta}}{s(t-M^2)^2}\left[\frac{-2M}{\pi t^3}\right]^{\frac{1}{2}}\Big\{(t+u)^2 p_{+\alpha}p_{+\beta}$$
$$+2(st+su-tM^2)p_{+\alpha}q_{+\beta} + (s^2+t^2)q_{+\alpha}q_{+\beta}$$
$$+2i[(t+u)p_{+\alpha} + (s-t)q_{+\alpha}]\epsilon_{\mu\nu\rho\beta}k^\mu p_+^\nu q_+^\rho\Big\},$$

$$M(+,-;+) = \frac{4ig^3 R'_1 T^a_{ij}\epsilon^{\alpha\beta}}{u(t-M^2)^2}\left[\frac{-2M}{\pi t^3}\right]^{\frac{1}{2}}\Big\{(t^2+u^2)p_{+\alpha}p_{+\beta}$$
$$+2(su+tu-tM^2)p_{+\alpha}q_{+\beta} + (s+t)^2 q_{+\alpha}q_{+\beta}$$
$$+2i[(t-u)p_{+\alpha} - (s+t)q_{+\alpha}]\epsilon_{\mu\nu\rho\beta}k^\mu p_+^\nu q_+^\rho\Big\},$$

$$M(-,+;-) = \frac{4ig^3 R'_1 T^a_{ij}\epsilon^{\alpha\beta}}{u(t-M^2)^2}\left[\frac{-2M}{\pi t^3}\right]^{\frac{1}{2}}\Big\{(t^2+u^2)p_{+\alpha}p_{+\beta}$$
$$+2(su+tu-tM^2)p_{+\alpha}q_{+\beta} + (s+t)^2 q_{+\alpha}q_{+\beta}$$
$$-2i[(t-u)p_{+\alpha} - (s+t)q_{+\alpha}]\epsilon_{\mu\nu\rho\beta}k^\mu p_+^\nu q_+^\rho\Big\},$$

$$M(-,-;-) = -\frac{4ig^3 R_1' T_{ij}^a \epsilon^{\alpha\beta}}{s(t-M^2)^2} \left[\frac{-2M}{\pi t^3}\right]^{\frac{1}{2}} \{(t+u)^2 p_{+\alpha} p_{+\beta}$$

$$+ 2(st+su-tM^2) p_{+\alpha} q_{+\beta} + (s^2+t^2) q_{+\alpha} q_{+\beta}$$

$$- 2i[(t+u) p_{+\alpha} + (s-t) q_{+\alpha}] \epsilon_{\mu\nu\rho\beta} k^\mu p_+^\nu q_+^\rho \} \,.$$
(10.4

Squared absolute values of the nonvanishing helicity amplitudes, summ over the color degrees of freedom of the initial state and the final state over the polarization states of the $^3P_2$:

$$|M(+,+;+)|^2 = |M(-,-;-)|^2$$

$$= -\frac{64 g^6 R_1'^2}{3\pi M^3 t(t-M^2)^4} (s^2 t^2 + 6stuM^2 + 6u^2 M^4) \,,$$

$$|M(+,-;+)|^2 = |M(-,+;-)|^2$$

$$= -\frac{64 g^6 R_1'^2}{3\pi M^3 t(t-M^2)^4} (u^2 t^2 + 6stuM^2 + 6s^2 M^4) \,.$$
(10.4

Unpolarized squared matrix element:

$$\overline{|M|^2} = -\frac{4 g^6 R_1'^2}{9\pi t(t-M^2)^4} \left[(s^2+u^2)(t^2+6M^4) + 12stuM^2\right] \,. \quad (10.4$$

Unpolarized cross section:

$$\frac{d\sigma}{dt} = -\frac{16\pi \alpha_s^3 R_1'^2}{9 s^2 t(t-M^2)^4} \left[(s^2+u^2)(t^2+6M^4) + 12stuM^2\right] \,. \quad (10.4$$

10.68 $\bar{q}q \to g\,^1 S_0$

Process:
$$\bar{q}(p_+, i) + q(p_-, j) \to g(k,a) + {}^1S_0(p) \,. \quad (10.4$$

Invariants and definitions:

$$s = 2(p_+ \cdot p_-), \qquad t = -2(p_+ \cdot k), \qquad u = -2(p_- \cdot k) \,,$$

$$s + t + u = p^2 = M^2 \,. \tag{10.4}$$

10.69. $\bar{q}q \to g\,{}^3P_0$

Gluon polarizations:

$$\not{\epsilon}^{\pm} = N[\not{k}\,\not{p}_-\,\not{p}_+(1 \pm \gamma_5) + \not{p}_+\,\not{p}_-\,\not{k}(1 \mp \gamma_5)],$$

$$N^{-1} = E^2[32k_+k_-]^{\frac{1}{2}}. \tag{10.469}$$

Nonvanishing helicity amplitudes [the definition of the quantity R_0 is given in eqn (8.11)]:

$$M(+,-;+) = -\frac{g^3 R_0 T^a_{ij} k_\perp}{s - M^2}\left[\frac{2k_+}{3\pi M k_-}\right]^{\frac{1}{2}},$$

$$M(-,+;+) = \frac{g^3 R_0 T^a_{ij} k^*_\perp}{s - M^2}\left[\frac{2k_-}{3\pi M k_+}\right]^{\frac{1}{2}},$$

$$M(+,-;-) = -\frac{g^3 R_0 T^a_{ij} k_\perp}{s - M^2}\left[\frac{2k_-}{3\pi M k_+}\right]^{\frac{1}{2}},$$

$$M(-,+;-) = \frac{g^3 R_0 T^a_{ij} k^*_\perp}{s - M^2}\left[\frac{2k_+}{3\pi M k_-}\right]^{\frac{1}{2}}. \tag{10.470}$$

Squared absolute values of the nonvanishing helicity amplitudes, summed over the color degrees of freedom of the initial state and the final state:

$$|M(+,-;+)|^2 = |M(-,+;-)|^2 = \frac{8g^6 R_0^2 u^2}{3\pi M s(s - M^2)^2},$$

$$|M(-,+;+)|^2 = |M(+,-;-)|^2 = \frac{8g^6 R_0^2 t^2}{3\pi M s(s - M^2)^2}. \tag{10.471}$$

Unpolarized squared matrix element:

$$\overline{|M|^2} = \frac{4g^6 R_0^2(t^2 + u^2)}{27\pi M s(s - M^2)^2}. \tag{10.472}$$

Unpolarized cross section:

$$\frac{d\sigma}{dt} = \frac{16\pi \alpha_S^3 R_0^2(t^2 + u^2)}{27 M s^3(s - M^2)^2}. \tag{10.473}$$

10.69 $\bar{q}q \to g\,{}^3P_0$

Process:

$$\bar{q}(p_+,i) + q(p_-,j) \to g(k,a) + {}^3P_0(p). \tag{10.474}$$

10. SUMMARY OF QCD FORMULAE

Invariants and definitions:

$$s = 2(p_+ \cdot p_-), \quad t = -2(p_+ \cdot k), \quad u = -2(p_- \cdot k),$$

$$s + t + u = p^2 = M^2. \tag{10.475}$$

Gluon polarizations:

$$\rlap{/}{\epsilon}^{\pm} = N[\rlap{/}{k} \rlap{/}{p}_- \rlap{/}{p}_+(1 \pm \gamma_5) + \rlap{/}{p}_+ \rlap{/}{p}_- \rlap{/}{k}(1 \mp \gamma_5)],$$

$$N^{-1} = E^2[32 k_+ k_-]^{\frac{1}{2}}. \tag{10.476}$$

Nonvanishing helicity amplitudes [the definition of the quantity R_1' is given in eqn (8.12)]:

$$M(+,-;+) = \frac{4ig^3 R_1' T_{ij}^a (s - 3M^2) k_\perp}{(s - M^2)^2} \left[\frac{k_+}{6\pi M^3 k_-} \right]^{\frac{1}{2}},$$

$$M(-,+;+) = -\frac{4ig^3 R_1' T_{ij}^a (s - 3M^2) k_\perp^*}{(s - M^2)^2} \left[\frac{k_+}{6\pi M^3 k_-} \right]^{\frac{1}{2}},$$

$$M(+,-;-) = -\frac{4ig^3 R_1' T_{ij}^a (s - 3M^2) k_\perp}{(s - M^2)^2} \left[\frac{k_+}{6\pi M^3 k_-} \right]^{\frac{1}{2}},$$

$$M(-,+;-) = \frac{4ig^3 R_1' T_{ij}^a (s - 3M^2) k_\perp^*}{(s - M^2)^2} \left[\frac{k_+}{6\pi M^3 k_-} \right]^{\frac{1}{2}}. \tag{10.477}$$

Squared absolute values of the nonvanishing helicity amplitudes, summed over the color degrees of freedom of the initial state and the final state:

$$|M(+,-;+)|^2 = |M(-,+;-)|^2 = \frac{32 g^6 R_1'^2 (s - 3M^2)^2 u^2}{3\pi M^3 s(s - M^2)^4},$$

$$|M(+,-;-)|^2 = |M(-,+;+)|^2 = \frac{32 g^6 R_1'^2 (s - 3M^2)^2 t^2}{3\pi M^3 s(s - M^2)^4}. \tag{10.478}$$

Unpolarized squared matrix element:

$$\overline{|M|^2} = \frac{16 g^6 R_1'^2 (s - 3M^2)^2 (t^2 + u^2)}{27 \pi M^3 s (s - M^2)^4}. \tag{10.479}$$

Unpolarized cross section:

$$\frac{d\sigma}{dt} = \frac{64\pi \alpha_s^3 R_1'^2 (s - 3M^2)^2 (t^2 + u^2)}{27 M^3 s^3 (s - M^2)^4}. \tag{10.480}$$

10.70 $\bar{q}q \to g\,{}^3P_1$

Process:
$$\bar{q}(p_+, i) + q(p_-, j) \to g(k, a) + {}^3P_1(p). \qquad (10.481)$$

Invariants and definitions:
$$s = 2(p_+ \cdot p_-), \qquad t = -2(p_+ \cdot k), \qquad u = -2(p_- \cdot k),$$

$$s + t + u = p^2 = M^2. \qquad (10.482)$$

Gluon polarizations:
$$\not{\epsilon}^{\pm} = N[\not{k}\,\not{p}_-\,\not{p}_+(1 \pm \gamma_5) + \not{p}_+\,\not{p}_-\,\not{k}(1 \mp \gamma_5)],$$

$$N^{-1} = E^2[32k_+k_-]^{\frac{1}{2}}. \qquad (10.483)$$

Nonvanishing helicity amplitudes [the definition of the quantity R'_1 is given in eqn (8.12) and the four-vector ϵ denotes the polarization vector of the ${}^3P_1$]:

$$M(+,-;+) = -\frac{2ig^3 R'_1 T^a_{ij} k_\perp}{E(s-M^2)^2 \sqrt{\pi M k_+ k_-}} \Big[(u-s)(p_+ \cdot \epsilon)$$
$$-(s+t)(p_- \cdot \epsilon) - 2i\epsilon^{\alpha\beta\gamma\delta}p_{+\alpha}p_{-\beta}k_\gamma\epsilon_\delta\Big],$$

$$M(-,+;+) = -\frac{2ig^3 R'_1 T^a_{ij} k^*_\perp}{E(s-M^2)^2 \sqrt{\pi M k_+ k_-}} \Big[(s+u)(p_+ \cdot \epsilon)$$
$$+(s-t)(p_- \cdot \epsilon) - 2i\epsilon^{\alpha\beta\gamma\delta}p_{+\alpha}p_{-\beta}k_\gamma\epsilon_\delta\Big],$$

$$M(+,-;-) = \frac{2ig^3 R'_1 T^a_{ij} k_\perp}{E(s-M^2)^2 \sqrt{\pi M k_+ k_-}} \Big[(s+u)(p_+ \cdot \epsilon)$$
$$+(s-t)(p_- \cdot \epsilon) + 2i\epsilon^{\alpha\beta\gamma\delta}p_{+\alpha}p_{-\beta}k_\gamma\epsilon_\delta\Big],$$

$$M(-,+;-) = \frac{2ig^3 R'_1 T^a_{ij} k^*_\perp}{E(s-M^2)^2 \sqrt{\pi M k_+ k_-}} \Big[(u-s)(p_+ \cdot \epsilon)$$
$$-(s+t)(p_- \cdot \epsilon) + 2i\epsilon^{\alpha\beta\gamma\delta}p_{+\alpha}p_{-\beta}k_\gamma\epsilon_\delta\Big]. \qquad (10.484)$$

Squared absolute values of the nonvanishing helicity amplitudes, summed over the color degrees of freedom of the initial state and the final state and over the polarization states of the ${}^3P_1$:

$$|M(+,-;+)|^2 = |M(-,+;-)|^2 = \frac{64g^6 R'^2_1(su^2 + 2tuM^2)}{\pi M^3 (s-M^2)^4},$$

$$|M(+,-;-)|^2 = |M(-,+;+)|^2 = \frac{64g^6 R'^2_1(st^2 + 2tuM^2)}{\pi M^3 (s-M^2)^4}. \qquad (10.485)$$

Unpolarized squared matrix element:

$$\overline{|M|^2} = \frac{32g^6 R_1'^2}{9\pi M^3(s-M^2)^4}[s(t^2+u^2)+4tuM^2].\qquad(10.486)$$

Unpolarized cross section:

$$\frac{d\sigma}{dt} = \frac{128\pi\alpha_S^3 R_1'^2}{9M^3 s^2(s-M^2)^4}[s(t^2+u^2)+4tuM^2].\qquad(10.487)$$

10.71 $\bar{q}q \to g\,^3P_2$

Process:
$$\bar{q}(p_+,i) + q(p_-,j) \to g(k,a) + {}^3P_2(p).\qquad(10.488)$$

Invariants and definitions:

$$s = 2(p_+ \cdot p_-), \qquad t = -2(p_+ \cdot k), \qquad u = -2(p_- \cdot k),$$

$$s+t+u = p^2 = M^2.\qquad(10.489)$$

Gluon polarizations:

$$\not{k}^\pm = N[\not{k}\,\not{p}_-\,\not{p}_+(1\pm\gamma_5) + \not{p}_+\,\not{p}_-\,\not{k}(1\mp\gamma_5)],$$

$$N^{-1} = E^2[32k_+k_-]^{\frac{1}{2}}.\qquad(10.490)$$

Nonvanishing helicity amplitudes [the definition of the quantity R_1' is given in eqn (8.12) and the quantity $\epsilon^{\alpha\beta}$ denotes the polarization tensor of the $^3P_2$]:

$$M(+,-;+) = \frac{ig^3 R_1' T_{ij}^a k_\perp \epsilon^{\alpha\beta}}{s^2(s-M^2)^2}\left[\frac{32M}{\pi k_+^3 k_-}\right]^{\frac{1}{2}}\Big\{(s^2+u^2)p_{+\alpha}p_{+\beta}$$

$$-2(su+tu-sM^2)p_{+\alpha}p_{-\beta} + (s+t)^2 p_{-\alpha}p_{-\beta}$$

$$+2i[(s-u)p_{+\alpha} + (s+t)p_{-\alpha}]\epsilon_{\mu\nu\rho\beta}k^\mu p_+^\nu p_-^\rho\Big\},$$

$$M(-,+;+) = \frac{ig^3 R_1' T_{ij}^a k_\perp^* \epsilon^{\alpha\beta}}{s^2(s-M^2)^2}\left[\frac{32M}{\pi k_+ k_-^3}\right]^{\frac{1}{2}}\Big\{(s+u)^2 p_{+\alpha}p_{+\beta}$$

$$-2(st+tu-sM^2)p_{+\alpha}p_{-\beta} + (s^2+t^2)p_{-\alpha}p_{-\beta}$$

$$-2i[(s+u)p_{+\alpha} + (s-t)p_{-\alpha}]\epsilon_{\mu\nu\rho\beta}k^\mu p_+^\nu p_-^\rho\Big\},$$

10.71. $\bar{q}q \to g\,^3P_2$

$$M(+,-;-) = \frac{ig^3 R'_1 T^a_{ij} k_\perp \epsilon^{\alpha\beta}}{s^2(s-M^2)^2} \left[\frac{32M}{\pi k_+^3 k_-}\right]^{\frac{1}{2}} \{(s^2+u^2)p_{+\alpha}p_{+\beta}$$

$$-2(su+tu-sM^2)p_{+\alpha}p_{-\beta} + (s+t)^2 p_{-\alpha}p_{-\beta}$$

$$-2i[(s-u)p_{+\alpha} + (s+t)p_{-\alpha}]\epsilon_{\mu\nu\rho\beta}k^\mu p^\nu_+ p^\rho_-\},$$

$$M(-,+;-) = \frac{ig^3 R'_1 T^a_{ij} k^*_\perp \epsilon^{\alpha\beta}}{s^2(s-M^2)^2} \left[\frac{32M}{\pi k_+ k_-^3}\right]^{\frac{1}{2}} \{(s+u)^2 p_{+\alpha}p_{+\beta}$$

$$-2(st+tu-sM^2)p_{+\alpha}p_{-\beta} + (s^2+t^2)p_{-\alpha}p_{-\beta}$$

$$+2i[(s+u)p_{+\alpha} + (s-t)p_{-\alpha}]\epsilon_{\mu\nu\rho\beta}k^\mu p^\nu_+ p^\rho_-\}. \quad (10.491)$$

Squared absolute values of the nonvanishing helicity amplitudes, summed over the color degrees of freedom of the initial state and the final state and over the polarization states of the $^3P_2$:

$$|M(+,-;+)|^2 = |M(-,+;-)|^2$$

$$= \frac{64g^6 R_1'^2}{3\pi M^3 s(s-M^2)^4} (s^2 u^2 + 6stuM^2 + 6t^2 M^4),$$

$$|M(-,+;+)|^2 = |M(+,-;-)|^2$$

$$= \frac{64g^6 R_1'^2}{3\pi M^3 s(s-M^2)^4} (s^2 t^2 + 6stuM^2 + 6u^2 M^4).$$

$$(10.492)$$

Unpolarized squared matrix element:

$$\overline{|M|^2} = \frac{32g^6 R_1'^2}{27\pi M^3 s(s-M^2)^4} \left[(t^2+u^2)(s^2+6M^4) + 12stuM^2\right]. \quad (10.493)$$

Unpolarized cross section:

$$\frac{d\sigma}{dt} = \frac{128\pi \alpha_S^3 R_1'^2}{27 M^3 s^3 (s-M^2)^4} \left[(t^2+u^2)(s^2+6M^4) + 12stuM^2\right]. \quad (10.494)$$

10.72 $gg \to g\,^1S_0$

Process:

$$g(k_1,a) + g(k_2,b) \to g(k_3,c) + {}^1S_0(p). \quad (10.495)$$

Positive z-axis: along $\vec{k}_1$.
Invariants and definitions:

$$s = 2(k_1 \cdot k_2), \qquad t = -2(k_2 \cdot k_3), \qquad u = -2(k_1 \cdot k_3),$$

$$s + t + u = p^2 = M^2. \tag{10.496}$$

Gluon polarizations:

$$\begin{aligned}
\slashed{\epsilon}^{*\pm}(k_1) &= N[\slashed{k}_1 \slashed{k}_2 \slashed{k}_3(1 \mp \gamma_5) + \slashed{k}_3 \slashed{k}_2 \slashed{k}_1(1 \pm \gamma_5)], \\
\slashed{\epsilon}^{*\pm}(k_2) &= N[\slashed{k}_2 \slashed{k}_3 \slashed{k}_1(1 \mp \gamma_5) + \slashed{k}_1 \slashed{k}_3 \slashed{k}_2(1 \pm \gamma_5)], \\
\slashed{\epsilon}^{\pm}(k_3) &= N[\slashed{k}_3 \slashed{k}_1 \slashed{k}_2(1 \pm \gamma_5) + \slashed{k}_2 \slashed{k}_1 \slashed{k}_3(1 \mp \gamma_5)], \\
N^{-1} &= E^2[32k_{3+}k_{3-}]^{\frac{1}{2}}.
\end{aligned} \tag{10.497}$$

Nonvanishing helicity amplitudes [the definition of the quantity R_0 is given in eqn (8.11)]:

$$\begin{aligned}
M(+,+;+) &= -\frac{ig^3 R_0 f^{abc}}{E^2[24\pi M k_{3+}k_{3-}]^{\frac{1}{2}}} \frac{s^2(st+tu+us)}{(s-M^2)(t-M^2)(u-M^2)}, \\
M(+,+;-) &= -\frac{ig^3 R_0 f^{abc}}{E^2[24\pi M k_{3+}k_{3-}]^{\frac{1}{2}}} \frac{M^4(st+tu+us)}{(s-M^2)(t-M^2)(u-M^2)}, \\
M(+,-;+) &= \frac{ig^3 R_0 f^{abc}}{E^2[24\pi M k_{3+}k_{3-}]^{\frac{1}{2}}} \frac{u^2(st+tu+us)}{(s-M^2)(t-M^2)(u-M^2)}, \\
M(-,+;+) &= \frac{ig^3 R_0 f^{abc}}{E^2[24\pi M k_{3+}k_{3-}]^{\frac{1}{2}}} \frac{t^2(st+tu+us)}{(s-M^2)(t-M^2)(u-M^2)}, \\
M(+,-;-) &= -\frac{ig^3 R_0 f^{abc}}{E^2[24\pi M k_{3+}k_{3-}]^{\frac{1}{2}}} \frac{t^2(st+tu+us)}{(s-M^2)(t-M^2)(u-M^2)}, \\
M(-,+;-) &= -\frac{ig^3 R_0 f^{abc}}{E^2[24\pi M k_{3+}k_{3-}]^{\frac{1}{2}}} \frac{u^2(st+tu+us)}{(s-M^2)(t-M^2)(u-M^2)}, \\
M(-,-;+) &= \frac{ig^3 R_0 f^{abc}}{E^2[24\pi M k_{3+}k_{3-}]^{\frac{1}{2}}} \frac{M^4(st+tu+us)}{(s-M^2)(t-M^2)(u-M^2)}, \\
M(-,-;-) &= \frac{ig^3 R_0 f^{abc}}{E^2[24\pi M k_{3+}k_{3-}]^{\frac{1}{2}}} \frac{s^2(st+tu+us)}{(s-M^2)(t-M^2)(u-M^2)}.
\end{aligned} \tag{10.498}$$

Squared absolute values of the nonvanishing helicity amplitudes, summed over the color degrees of freedom of the initial state and the final state:

$$|M(+,+;+)|^2 = |M(-,-;-)|^2$$

$$= \frac{16g^6 R_0^2 s^4 (st+tu+us)^2}{\pi M stu[(s-M^2)(t-M^2)(u-M^2)]^2},$$

$$|M(+,+;-)|^2 = |M(-,-;+)|^2$$

$$= \frac{16g^6 R_0^2 M^8 (st+tu+us)^2}{\pi M stu[(s-M^2)(t-M^2)(u-M^2)]^2},$$

$$|M(+,-;+)|^2 = |M(-,+;-)|^2$$

$$= \frac{16g^6 R_0^2 u^4 (st+tu+us)^2}{\pi M stu[(s-M^2)(t-M^2)(u-M^2)]^2},$$

$$|M(-,+;+)|^2 = |M(+,-;-)|^2$$

$$= \frac{16g^6 R_0^2 t^4 (st+tu+us)^2}{\pi M stu[(s-M^2)(t-M^2)(u-M^2)]^2}. \quad (10.499)$$

Unpolarized squared matrix element:

$$\overline{|M|^2} = \frac{g^6 R_0^2}{8\pi M stu} \frac{(M^8+s^4+t^4+u^4)(st+tu+us)^2}{[(s-M^2)(t-M^2)(u-M^2)]^2}. \quad (10.500)$$

Unpolarized cross section:

$$\frac{d\sigma}{dt} = \frac{\pi \alpha_S^3 R_0^2}{2 M s^3 tu} \frac{(M^8+s^4+t^4+u^4)(st+tu+us)^2}{[(s-M^2)(t-M^2)(u-M^2)]^2}. \quad (10.501)$$

10.73 $gg \to g\,^3S_1$

Process:
$$g(k_1,a) + g(k_2,b) \to g(k_3,c) + \,^3S_1(p). \quad (10.502)$$

Positive z-axis: along $\vec{k}_1$.
Invariants and definitions:

$$s = 2(k_1 \cdot k_2), \qquad t = -2(k_2 \cdot k_3), \qquad u = -2(k_1 \cdot k_3),$$

$$s+t+u = p^2 = M^2. \quad (10.503)$$

Gluon polarizations:

$$\not{\epsilon}^{*\pm}(k_1) = N[\not{k}_1 \not{k}_2 \not{k}_3(1 \mp \gamma_5) + \not{k}_3 \not{k}_2 \not{k}_1(1 \pm \gamma_5)],$$

$$\not{\epsilon}^{*\pm}(k_2) = N[\not{k}_2 \not{k}_1 \not{k}_3(1 \mp \gamma_5) + \not{k}_3 \not{k}_1 \not{k}_2(1 \pm \gamma_5)],$$

$$\not{\epsilon}^{\pm}(k_3) = N[\not{k}_3 \not{k}_1 \not{k}_2(1 \pm \gamma_5) + \not{k}_2 \not{k}_1 \not{k}_3(1 \mp \gamma_5)],$$

$$N^{-1} = E^2[32 k_{3+} k_{3-}]^{\frac{1}{2}}. \tag{10.504}$$

Nonvanishing helicity amplitudes [the definition of the quantity R_0 is given in eqn (8.11) and the four-vector ϵ denotes the polarization vector of the $^3S_1$]:

$$M(+,+;+) = -\frac{g^3 R_0 d^{abc} \sqrt{M}}{E^2 [24\pi k_{3+} k_{3-}]^{\frac{1}{2}}} \frac{s}{(s - M^2)(t - M^2)(u - M^2)}$$

$$\times \left[t(t+u)(k_1 \cdot \epsilon) - u(t+u)(k_2 \cdot \epsilon) \right.$$

$$\left. + s(t-u)(k_3 \cdot \epsilon) - 2i(t+u)\epsilon^{\alpha\beta\gamma\delta} k_{1\alpha} k_{2\beta} k_{3\gamma} \epsilon_\delta \right],$$

$$M(+,-;+) = -\frac{g^3 R_0 d^{abc} \sqrt{M}}{E^2 [24\pi k_{3+} k_{3-}]^{\frac{1}{2}}} \frac{t}{(s - M^2)(t - M^2)(u - M^2)}$$

$$\times \left[t(s-u)(k_1 \cdot \epsilon) + u(s+u)(k_2 \cdot \epsilon) \right.$$

$$\left. + s(s+u)(k_3 \cdot \epsilon) + 2i(s+u)\epsilon^{\alpha\beta\gamma\delta} k_{1\alpha} k_{2\beta} k_{3\gamma} \epsilon_\delta \right],$$

$$M(-,+;+) = \frac{g^3 R_0 d^{abc} \sqrt{M}}{E^2 [24\pi k_{3+} k_{3-}]^{\frac{1}{2}}} \frac{u}{(s - M^2)(t - M^2)(u - M^2)}$$

$$\times \left[t(s+t)(k_1 \cdot \epsilon) + u(s-t)(k_2 \cdot \epsilon) \right.$$

$$\left. + s(s+t)(k_3 \cdot \epsilon) - 2i(s+t)\epsilon^{\alpha\beta\gamma\delta} k_{1\alpha} k_{2\beta} k_{3\gamma} \epsilon_\delta \right],$$

$$M(+,-;-) = \frac{g^3 R_0 d^{abc} \sqrt{M}}{E^2 [24\pi k_{3+} k_{3-}]^{\frac{1}{2}}} \frac{u}{(s - M^2)(t - M^2)(u - M^2)}$$

$$\times \left[t(s+t)(k_1 \cdot \epsilon) + u(s-t)(k_2 \cdot \epsilon) \right.$$

$$\left. + s(s+t)(k_3 \cdot \epsilon) + 2i(s+t)\epsilon^{\alpha\beta\gamma\delta} k_{1\alpha} k_{2\beta} k_{3\gamma} \epsilon_\delta \right],$$

10.73. $gg \to g\,^3S_1$

$$M(-,+;-) = -\frac{g^3 R_0 d^{abc}\sqrt{M}}{E^2[24\pi k_{3+}k_{3-}]^{\frac{1}{2}}} \frac{t}{(s-M^2)(t-M^2)(u-M^2)}$$

$$\times [t(s-u)(k_1\cdot\epsilon) + u(s+u)(k_2\cdot\epsilon)$$

$$+s(s+u)(k_3\cdot\epsilon) - 2i(s+u)\epsilon^{\alpha\beta\gamma\delta}k_{1\alpha}k_{2\beta}k_{3\gamma}\epsilon_\delta]\,,$$

$$M(-,-;-) = -\frac{g^3 R_0 d^{abc}\sqrt{M}}{E^2[24\pi k_{3+}k_{3-}]^{\frac{1}{2}}} \frac{s}{(s-M^2)(t-M^2)(u-M^2)}$$

$$\times [t(t+u)(k_1\cdot\epsilon) - u(t+u)(k_2\cdot\epsilon)$$

$$+s(t-u)(k_3\cdot\epsilon) + 2i(t+u)\epsilon^{\alpha\beta\gamma\delta}k_{1\alpha}k_{2\beta}k_{3\gamma}\epsilon_\delta]\,.$$
(10.505)

Squared absolute values of the nonvanishing helicity amplitudes, summed over the color degrees of freedom of the initial state and the final state and over the polarization states of the $^3S_1$:

$$|M(+,+;+)|^2 = |M(-,-;-)|^2 = \frac{160g^6 R_0^2 M s^2}{9\pi[(t-M^2)(u-M^2)]^2},$$

$$|M(+,-;+)|^2 = |M(-,+;-)|^2 = \frac{160g^6 R_0^2 M t^2}{9\pi[(s-M^2)(u-M^2)]^2},$$

$$|M(-,+;+)|^2 = |M(+,-;-)|^2 = \frac{160g^6 R_0^2 M u^2}{9\pi[(s-M^2)(t-M^2)]^2}.$$
(10.506)

Unpolarized squared matrix element:

$$\overline{|M|^2} = \frac{5g^6 R_0^2 M}{36\pi}\left[\frac{s^2}{[(t-M^2)(u-M^2)]^2}\right.$$

$$\left.+\frac{t^2}{[(s-M^2)(u-M^2)]^2} + \frac{u^2}{[(s-M^2)(t-M^2)]^2}\right]. \quad (10.507)$$

Unpolarized cross section:

$$\frac{d\sigma}{dt} = \frac{5\pi\alpha_s^3 R_0^2 M}{9s^2}\left[\frac{s^2}{[(t-M^2)(u-M^2)]^2}\right.$$

$$\left.+\frac{t^2}{[(s-M^2)(u-M^2)]^2} + \frac{u^2}{[(s-M^2)(t-M^2)]^2}\right]. \quad (10.508)$$

10.74 $gg \to g\,{}^1P_1$

Process:
$$g(k_1, a) + g(k_2, b) \to g(k_3, c) + {}^1P_1(p). \qquad (10.509)$$

Positive z-axis: along $\vec{k}_1$.
Invariants and definitions:
$$s = 2(k_1 \cdot k_2), \qquad t = -2(k_2 \cdot k_3), \qquad u = -2(k_1 \cdot k_3),$$

$$s + t + u = p^2 = M^2, \qquad P = st + tu + us, \qquad Q = stu. \qquad (10.510)$$

Gluon polarizations:
$$\not{\epsilon}^{*\pm}(k_1) = N[\not{k}_1 \not{k}_2 \not{k}_3(1 \mp \gamma_5) + \not{k}_3 \not{k}_2 \not{k}_1(1 \pm \gamma_5)],$$

$$\not{\epsilon}^{*\pm}(k_2) = N[\not{k}_2 \not{k}_3 \not{k}_1(1 \mp \gamma_5) + \not{k}_1 \not{k}_3 \not{k}_2(1 \pm \gamma_5)],$$

$$\not{\epsilon}^{\pm}(k_3) = N[\not{k}_3 \not{k}_1 \not{k}_2(1 \pm \gamma_5) + \not{k}_2 \not{k}_1 \not{k}_3(1 \mp \gamma_5)],$$

$$N^{-1} = E^2 [32 k_{3+} k_{3-}]^{\frac{1}{2}}. \qquad (10.511)$$

Nonvanishing helicity amplitudes [the definition of the quantity R'_1 is given in eqn (8.12) and the four-vector ϵ denotes the polarization vector of the ${}^1P_1$]:

$$M(+, +; +) = -\frac{ig^3 R'_1 d^{abc}}{E^2 [2\pi M k_{3+} k_{3-}]^{\frac{1}{2}}} \frac{s}{(s - M^2)^2 (t - M^2)(u - M^2)}$$

$$\times \left\{ \frac{(k_1 \cdot \epsilon)}{t - M^2} \left[tM^6 - (su + t^2)M^4 - stuM^2 + 2s^2 tu \right] \right.$$

$$- \frac{(k_2 \cdot \epsilon)}{u - M^2} \left[uM^6 - (st + u^2)M^4 - stuM^2 + 2s^2 tu \right]$$

$$\left. - 2i(s - M^2)M^2 \epsilon^{\alpha\beta\gamma\delta} k_{1\alpha} k_{2\beta} k_{3\gamma} \epsilon_\delta \right\},$$

$$M(+, -; +) = -\frac{ig^3 R'_1 d^{abc}}{E^2 [2\pi M k_{3+} k_{3-}]^{\frac{1}{2}}} \frac{t}{(s - M^2)(t - M^2)^2 (u - M^2)}$$

$$\times \left\{ \frac{(k_2 \cdot \epsilon)}{u - M^2} \left[uM^6 - (st + u^2)M^4 - stuM^2 + 2st^2 u \right] \right.$$

$$+ \frac{(k_3 \cdot \epsilon)}{s - M^2} \left[sM^6 - (tu + s^2)M^4 - stuM^2 + 2st^2 u \right]$$

$$\left. + 2i(t - M^2)M^2 \epsilon^{\alpha\beta\gamma\delta} k_{1\alpha} k_{2\beta} k_{3\gamma} \epsilon_\delta \right\},$$

10.74. $gg \to g\,^1P_1$

$$M(-,+;+) = \frac{ig^3 R_1' d^{abc}}{E^2[2\pi M k_{3+}k_{3-}]^{\frac{1}{2}}} \frac{u}{(s-M^2)(t-M^2)(u-M^2)^2}$$

$$\times \left\{ \frac{(k_1 \cdot \epsilon)}{t-M^2} \left[tM^6 - (su+t^2)M^4 - stuM^2 + 2stu^2\right] \right.$$

$$+ \frac{(k_3 \cdot \epsilon)}{s-M^2} \left[sM^6 - (tu+s^2)M^4 - stuM^2 + 2stu^2\right]$$

$$\left. -2i(u-M^2)M^2 \epsilon^{\alpha\beta\gamma\delta} k_{1\alpha}k_{2\beta}k_{3\gamma}\epsilon_\delta \right\},$$

$$M(+,-;-) = \frac{ig^3 R_1' d^{abc}}{E^2[2\pi M k_{3+}k_{3-}]^{\frac{1}{2}}} \frac{u}{(s-M^2)(t-M^2)(u-M^2)^2}$$

$$\times \left\{ \frac{(k_1 \cdot \epsilon)}{t-M^2} \left[tM^6 - (su+t^2)M^4 - stuM^2 + 2stu^2\right] \right.$$

$$+ \frac{(k_3 \cdot \epsilon)}{s-M^2} \left[sM^6 - (tu+s^2)M^4 - stuM^2 + 2stu^2\right]$$

$$\left. +2i(u-M^2)M^2 \epsilon^{\alpha\beta\gamma\delta} k_{1\alpha}k_{2\beta}k_{3\gamma}\epsilon_\delta \right\},$$

$$M(-,+;-) = -\frac{ig^3 R_1' d^{abc}}{E^2[2\pi M k_{3+}k_{3-}]^{\frac{1}{2}}} \frac{t}{(s-M^2)(t-M^2)^2(u-M^2)}$$

$$\times \left\{ \frac{(k_2 \cdot \epsilon)}{u-M^2} \left[uM^6 - (st+u^2)M^4 - stuM^2 + 2st^2u\right] \right.$$

$$+ \frac{(k_3 \cdot \epsilon)}{s-M^2} \left[sM^6 - (tu+s^2)M^4 - stuM^2 + 2st^2u\right]$$

$$\left. -2i(t-M^2)M^2 \epsilon^{\alpha\beta\gamma\delta} k_{1\alpha}k_{2\beta}k_{3\gamma}\epsilon_\delta \right\},$$

$$M(-,-;-) = -\frac{ig^3 R_1' d^{abc}}{E^2[2\pi M k_{3+}k_{3-}]^{\frac{1}{2}}} \frac{s}{(s-M^2)^2(t-M^2)(u-M^2)}$$

$$\times \left\{ \frac{(k_1 \cdot \epsilon)}{t-M^2} \left[tM^6 - (su+t^2)M^4 - stuM^2 + 2s^2tu\right] \right.$$

$$- \frac{(k_2 \cdot \epsilon)}{u-M^2} \left[uM^6 - (st+u^2)M^4 - stuM^2 + 2s^2tu\right]$$

$$\left. +2i(s-M^2)M^2 \epsilon^{\alpha\beta\gamma\delta} k_{1\alpha}k_{2\beta}k_{3\gamma}\epsilon_\delta \right\},$$

$$M(+,+;-) = \frac{ig^3 R'_1 d^{abc}}{E^2[32\pi M k_{3+} k_{3-}]^{\frac{1}{2}}} \frac{1}{(s-M^2)(t-M^2)(u-M^2)}$$

$$\times \left\{ \frac{2(k_1 \cdot \epsilon)(s-u)}{t-M^2} \left[3tM^4 - (su+3t^2)M^2 + t(su+2t^2) \right] \right.$$

$$+ \frac{2(k_2 \cdot \epsilon)(t-s)}{u-M^2} \left[3uM^4 - (st+3u^2)M^2 + u(st+2u^2) \right]$$

$$- \frac{2(k_3 \cdot \epsilon)(u-t)}{s-M^2} \left[3sM^4 - (tu+3s^2)M^2 + s(tu+2s^2) \right]$$

$$\left. -8iM^4 \epsilon^{\alpha\beta\gamma\delta} k_{1\alpha} k_{2\beta} k_{3\gamma} \epsilon_\delta \right\},$$

$$M(-,-;+) = \frac{ig^3 R'_1 d^{abc}}{E^2[32\pi M k_{3+} k_{3-}]^{\frac{1}{2}}} \frac{1}{(s-M^2)(t-M^2)(u-M^2)}$$

$$\times \left\{ \frac{2(k_1 \cdot \epsilon)(s-u)}{t-M^2} \left[3tM^4 - (su+3t^2)M^2 + t(su+2t^2) \right] \right.$$

$$+ \frac{2(k_2 \cdot \epsilon)(t-s)}{u-M^2} \left[3uM^4 - (st+3u^2)M^2 + u(st+2u^2) \right]$$

$$- \frac{2(k_3 \cdot \epsilon)(u-t)}{s-M^2} \left[3sM^4 - (tu+3s^2)M^2 + s(tu+2s^2) \right]$$

$$\left. +8iM^4 \epsilon^{\alpha\beta\gamma\delta} k_{1\alpha} k_{2\beta} k_{3\gamma} \epsilon_\delta \right\}. \tag{10.512}$$

Squared absolute values of the nonvanishing helicity amplitudes, summed over the color degrees of freedom of the initial state and the final state and over the polarization states of the $^1P_1$:

$$|M(+,+;+)|^2 = |M(-,-;-)|^2$$

$$= \frac{640 g^6 R_1'^2 s^2}{3\pi M[(s-M^2)(t-M^2)(u-M^2)]^2}$$

$$\times \left[M^4 + \frac{2stu(s^2+M^4)}{(s-M^2)(t-M^2)(u-M^2)} \right],$$

$$|M(+,+;-)|^2 = |M(-,-;+)|^2$$

$$= \frac{640 g^6 R_1'^2 M^7 [-M^2(st+tu+us)+5stu]}{3\pi[(s-M^2)(t-M^2)(u-M^2)]^3},$$

10.75. $gg \to g\,^3P_0$

$$|M(+,-;+)|^2 = |M(-,+;-)|^2$$

$$= \frac{640g^6 R_1'^2 t^2}{3\pi M[(s-M^2)(t-M^2)(u-M^2)]^2}$$

$$\times \left[M^4 + \frac{2stu(t^2+M^4)}{(s-M^2)(t-M^2)(u-M^2)}\right],$$

$$|M(-,+;+)|^2 = |M(+,-;-)|^2$$

$$= \frac{640g^6 R_1'^2 u^2}{3\pi M[(s-M^2)(t-M^2)(u-M^2)]^2}$$

$$\times \left[M^4 + \frac{2stu(u^2+M^4)}{(s-M^2)(t-M^2)(u-M^2)}\right]. \qquad (10.513)$$

Unpolarized squared matrix element:

$$\overline{|M|^2} = \frac{5g^6 R_1'^2}{3\pi M} \frac{1}{[(s-M^2)(t-M^2)(u-M^2)]^2}$$

$$\times \left\{ M^4(s^2+t^2+u^2+M^4) \right.$$

$$\left. + \frac{2stu[s^4+t^4+u^4+M^4(s^2+t^2+u^2)+2M^8]}{(s-M^2)(t-M^2)(u-M^2)} \right\}. \qquad (10.514)$$

Unpolarized cross section:

$$\frac{d\sigma}{dt} = \frac{40\pi\alpha_S^3 R_1'^2}{3Ms^2} \frac{-M^{10}P + M^6 P^2 + Q(5M^8 - 7M^4 P + 2P^2) + 4M^2 Q^2}{(Q - M^2 P)^3}. \qquad (10.515)$$

10.75 $gg \to g\,^3P_0$

Process:
$$g(k_1, a) + g(k_2, b) \to g(k_3, c) + {}^3P_0(p). \qquad (10.516)$$

Positive z-axis: along $\vec{k}_1$.
Invariants and definitions:

$$s = 2(k_1 \cdot k_2), \qquad t = -2(k_2 \cdot k_3), \qquad u = -2(k_1 \cdot k_3),$$

$$s + t + u = p^2 = M^2,$$

$$P = st + tu + us, \qquad Q = stu. \qquad (10.517)$$

Gluon polarizations:

$$\not{\epsilon}^{*\pm}(k_1) = N[\not{k}_1 \not{k}_2 \not{k}_3(1 \mp \gamma_5) + \not{k}_3 \not{k}_2 \not{k}_1(1 \pm \gamma_5)],$$

$$\not{\epsilon}^{*\pm}(k_2) = N[\not{k}_2 \not{k}_3 \not{k}_1(1 \mp \gamma_5) + \not{k}_1 \not{k}_3 \not{k}_2(1 \pm \gamma_5)],$$

$$\not{\epsilon}^{\pm}(k_3) = N[\not{k}_3 \not{k}_1 \not{k}_2(1 \pm \gamma_5) + \not{k}_2 \not{k}_1 \not{k}_3(1 \mp \gamma_5)],$$

$$N^{-1} = E^2[32 k_{3+} k_{3-}]^{\frac{1}{2}}. \tag{10.518}$$

Nonvanishing helicity amplitudes [the definition of the quantity R'_1 is given in eqn (8.12)]:

$$M(+,+;+) = M(-,-;-)$$

$$= \frac{ig^3 R'_1 f^{abc}}{E^2 [6\pi M^3 k_{3+} k_{3-}]^{\frac{1}{2}}} \frac{s^2[t^2 u^2 - stu(s - 6M^2) - 3s^2 M^2(s - M^2)]}{(s - M^2)(t - M^2)^2(u - M^2)^2},$$

$$M(+,+;-) = M(-,-;+)$$

$$= \frac{ig^3 R'_1 f^{abc}}{E^2} \left[\frac{3M^5}{2\pi k_{3+} k_{3-}}\right]^{\frac{1}{2}} \frac{st + tu + us}{(s - M^2)(t - M^2)(u - M^2)},$$

$$M(+,-;+) = M(-,+;-)$$

$$= \frac{ig^3 R'_1 f^{abc}}{E^2 [6\pi M^3 k_{3+} k_{3-}]^{\frac{1}{2}}} \frac{t^2[s^2 u^2 - stu(t - 6M^2) - 3t^2 M^2(t - M^2)]}{(s - M^2)^2(t - M^2)(u - M^2)^2},$$

$$M(-,+;+) = M(+,-;-)$$

$$= \frac{ig^3 R'_1 f^{abc}}{E^2 [6\pi M^3 k_{3+} k_{3-}]^{\frac{1}{2}}} \frac{u^2[s^2 t^2 - stu(u - 6M^2) - 3u^2 M^2(u - M^2)]}{(s - M^2)^2(t - M^2)^2(u - M^2)}. \tag{10.519}$$

Squared absolute values of the nonvanishing helicity amplitudes, summed over the color degrees of freedom of the initial state and the final state:

$$|M(+,+;+)|^2 = |M(-,-;-)|^2$$

$$= \frac{64 g^6 R'^2_1 s}{\pi M^3 tu} \frac{[PQ + 3s^2 M^2 P + 2sM^2 Q]^2}{(s - M^2)^2(t - M^2)^4(u - M^2)^4},$$

10.76. $gg \to g\,^3P_1$

$$|M(+,+;-)|^2 = |M(-,-;+)|^2$$

$$= \frac{576g^6 R_1'^2 M^5 P^2}{\pi Q[(s-M^2)(t-M^2)(u-M^2)]},$$

$$|M(+,-;+)|^2 = |M(-,+;-)|^2$$

$$= \frac{64g^6 R_1'^2 t}{\pi M^3 su} \frac{[PQ + 3t^2 M^2 P + 2tM^2 Q]^2}{(s-M^2)^4(t-M^2)^2(u-M^2)^4},$$

$$|M(-,+;+)|^2 = |M(+,-;-)|^2$$

$$= \frac{64g^6 R_1'^2 u}{\pi M^3 st} \frac{[PQ + 3u^2 M^2 P + 2uM^2 Q]^2}{(s-M^2)^4(t-M^2)^4(u-M^2)^2}. \quad (10.520)$$

Unpolarized squared matrix element:

$$\overline{|M|^2} = \frac{g^6 R_1'^2}{\pi M^3 Q(Q-M^2 P)^4} \left[9M^4 P^4(M^8 - 2M^4 P + P^2)\right.$$

$$-6M^2 P^3 Q(2M^8 - 5M^4 P + P^2) - P^2 Q^2(M^8 + 2M^4 P - P^2)$$

$$\left. + 2M^2 PQ^3(M^4 - P) + 6M^4 Q^4\right]. \quad (10.521)$$

Unpolarized cross section:

$$\frac{d\sigma}{dt} = \frac{4\pi \alpha_S^3 R_1'^2}{s^2 M^3 Q(Q-M^2 P)^4} \left[9M^4 P^4(M^8 - 2M^4 P + P^2)\right.$$

$$-6M^2 P^3 Q(2M^8 - 5M^4 P + P^2) - P^2 Q^2(M^8 + 2M^4 P - P^2)$$

$$\left. + 2M^2 PQ^3(M^4 - P) + 6M^4 Q^4\right]. \quad (10.522)$$

10.76 $gg \to g\,^3P_1$

Process:
$$g(k_1, a) + g(k_2, b) \to g(k_3, c) + {}^3P_1(p). \quad (10.523)$$

Positive z-axis: along $\vec{k}_1$.
Invariants and definitions:

$$s = 2(k_1 \cdot k_2), \qquad t = -2(k_2 \cdot k_3), \qquad u = -2(k_1 \cdot k_3),$$

$$s + t + u = p^2 = M^2,$$

$$P = st + tu + us, \qquad\qquad Q = stu. \quad (10.524)$$

Gluon polarizations:

$$\rlap{/}\epsilon^{*\pm}(k_1) = N[\rlap{/}k_1 \rlap{/}k_2 \rlap{/}k_3(1\mp\gamma_5) + \rlap{/}k_3 \rlap{/}k_2 \rlap{/}k_1(1\pm\gamma_5)],$$

$$\rlap{/}\epsilon^{*\pm}(k_2) = N[\rlap{/}k_2 \rlap{/}k_1 \rlap{/}k_3(1\mp\gamma_5) + \rlap{/}k_3 \rlap{/}k_1 \rlap{/}k_2(1\pm\gamma_5)],$$

$$\rlap{/}\epsilon^{\pm}(k_3) = N[\rlap{/}k_3 \rlap{/}k_1 \rlap{/}k_2(1\pm\gamma_5) + \rlap{/}k_2 \rlap{/}k_1 \rlap{/}k_3(1\mp\gamma_5)],$$

$$N^{-1} = E^2[32k_{3+}k_{3-}]^{\frac{1}{2}}. \quad (10.525)$$

Nonvanishing helicity amplitudes [the definition of the quantity R_1' is given in eqn (8.12) and the four-vector ϵ denotes the polarization vector of the $^3P_1$]:

$$M(+,+;+) = -\frac{ig^3 R_1' f^{abc}}{2E^2\sqrt{\pi M}} \frac{s}{(s-M^2)(t-M^2)^2(u-M^2)^2}$$

$$\times\Big\{(k_1\cdot\epsilon)(s+t)[st(t-s)+tu(t-u)+su(u-s)]$$

$$+(k_2\cdot\epsilon)(s+u)[su(u-s)+tu(u-t)+st(t-s)]$$

$$-2i\epsilon^{\alpha\beta\gamma\delta}k_{1\alpha}k_{2\beta}k_{3\gamma}\epsilon_\delta(t-u)(st+tu+us-s^2)\Big\},$$

$$M(+,-;+) = -\frac{ig^3 R_1' f^{abc}}{2E^2\sqrt{\pi M}} \frac{t}{(s-M^2)^2(t-M^2)(u-M^2)^2}$$

$$\times\Big\{(k_2\cdot\epsilon)(t+u)[st(s-t)+tu(u-t)+su(u-s)]$$

$$-(k_3\cdot\epsilon)(s+t)[st(s-t)+tu(u-t)+su(s-u)]$$

$$-2i\epsilon^{\alpha\beta\gamma\delta}k_{1\alpha}k_{2\beta}k_{3\gamma}\epsilon_\delta(s-u)(st+tu+us-t^2)\Big\},$$

$$M(-,+;+) = -\frac{ig^3 R_1' f^{abc}}{2E^2\sqrt{\pi M}} \frac{u}{(s-M^2)^2(t-M^2)^2(u-M^2)}$$

$$\times\Big\{(k_1\cdot\epsilon)(t+u)[st(t-s)+tu(t-u)+su(s-u)]$$

$$-(k_3\cdot\epsilon)(s+u)[st(s-t)+tu(t-u)+su(s-u)]$$

$$-2i\epsilon^{\alpha\beta\gamma\delta}k_{1\alpha}k_{2\beta}k_{3\gamma}\epsilon_\delta(t-s)(st+tu+us-u^2)\Big\},$$

10.76. $gg \to g\,^3P_1$

$$M(+,-;-) = \frac{ig^3 R'_1 f^{abc}}{2E^2\sqrt{\pi M}} \frac{u}{(s-M^2)^2(t-M^2)^2(u-M^2)}$$
$$\times \Big\{(k_1 \cdot \epsilon)(t+u)[st(t-s)+tu(t-u)+su(s-u)]$$
$$-(k_3 \cdot \epsilon)(s+u)[st(s-t)+tu(t-u)+su(s-u)]$$
$$+2i\epsilon^{\alpha\beta\gamma\delta}k_{1\alpha}k_{2\beta}k_{3\gamma}\epsilon_\delta(t-s)(st+tu+us-u^2)\Big\},$$

$$M(-,+;-) = \frac{ig^3 R'_1 f^{abc}}{2E^2\sqrt{\pi M}} \frac{t}{(s-M^2)^2(t-M^2)(u-M^2)^2}$$
$$\times \Big\{(k_2 \cdot \epsilon)(t+u)[st(s-t)+tu(u-t)+su(u-s)]$$
$$-(k_3 \cdot \epsilon)(s+t)[st(s-t)+tu(u-t)+su(s-u)]$$
$$+2i\epsilon^{\alpha\beta\gamma\delta}k_{1\alpha}k_{2\beta}k_{3\gamma}\epsilon_\delta(s-u)(st+tu+us-t^2)\Big\},$$

$$M(-,-;-) = \frac{ig^3 R'_1 f^{abc}}{2E^2\sqrt{\pi M}} \frac{s}{(s-M^2)(t-M^2)^2(u-M^2)^2}$$
$$\times \Big\{(k_1 \cdot \epsilon)(s+t)[st(t-s)+tu(t-u)+su(u-s)]$$
$$+(k_2 \cdot \epsilon)(s+u)[su(u-s)+tu(u-t)+st(t-s)]$$
$$+2i\epsilon^{\alpha\beta\gamma\delta}k_{1\alpha}k_{2\beta}k_{3\gamma}\epsilon_\delta(t-u)(st+tu+us-s^2)\Big\}.$$
(10.526)

Squared absolute values of the nonvanishing helicity amplitudes, summed over the color degrees of freedom of the initial state and the final state and over the polarization states of the $^3P_1$:

$$|M(+,+;+)|^2 = |M(-,-;-)|^2$$
$$= \frac{96 g^6 R'^2_1 s^2}{\pi M^3 (s-M^2)^2 (t-M^2)^4 (u-M^2)^4}$$
$$\times \Big\{(s+t)(t+u)(u+s)[st(s-t)^2+tu(t-u)^2+su(s-u)^2]$$
$$+M^2(t-u)^2(st+tu+us-s^2)^2\Big\},$$

$$|M(+,-;+)|^2 = |M(-,+;-)|^2$$

$$= \frac{96g^6 R_1'^2 t^2}{\pi M^3 (s-M^2)^4 (t-M^2)^2 (u-M^2)^4}$$

$$\times \left\{(s+t)(t+u)(u+s)[st(s-t)^2 + tu(t-u)^2 + su(s-u)^2]\right.$$

$$\left. + M^2 (s-u)^2 (st+tu+us-t^2)^2\right\},$$

$$|M(-,+;+)|^2 = |M(+,-;-)|^2$$

$$= \frac{96g^6 R_1'^2 u^2}{\pi M^3 (s-M^2)^4 (t-M^2)^4 (u-M^2)^2}$$

$$\times \left\{(s+t)(t+u)(u+s)[st(s-t)^2 + tu(t-u)^2 + su(s-u)^2]\right.$$

$$\left. + M^2 (s-t)^2 (st+tu+us-u^2)^2\right\}. \qquad (10.527)$$

Unpolarized squared matrix element:

$$\overline{|M|^2} = \frac{3g^6 R_1'^2}{\pi M^3} \frac{P^2 [M^2 P^2 (M^4 - 4P) - 2Q(M^8 - 5M^4 P - P^2) - 15M^2 Q^2]}{(Q-M^2 P)^4}. \qquad (10.528)$$

Unpolarized cross section:

$$\frac{d\sigma}{dt} = \frac{12\pi \alpha_s^3 R_1'^2}{M^3 s^2} \frac{P^2 [M^2 P^2 (M^4 - 4P) - 2Q(M^8 - 5M^4 P - P^2) - 15M^2 Q^2]}{(Q-M^2 P)^4}. \qquad (10.529)$$

10.77 $gg \to g\,^3P_2$

Process:
$$g(k_1,a) + g(k_2,b) \to g(k_3,c) + {}^3P_2(p). \qquad (10.530)$$

Positive z-axis: along $\vec{k}_1$.
Invariants and definitions:

$$s = 2(k_1 \cdot k_2), \qquad t = -2(k_2 \cdot k_3), \qquad u = -2(k_1 \cdot k_3),$$

$$s + t + u = p^2 = M^2, \qquad P = st + tu + us, \qquad Q = stu. \qquad (10.531)$$

Gluon polarizations:

$$\not{\epsilon}^{*\pm}(k_1) = N[\not{k}_1 \not{k}_2 \not{k}_3 (1 \mp \gamma_5) + \not{k}_3 \not{k}_2 \not{k}_1 (1 \pm \gamma_5)],$$

10.77. $gg \to g\,^3P_2$

$$\not{\epsilon}^{*\pm}(k_2) = N[\not{k}_2\not{k}_1\not{k}_3(1\mp\gamma_5) + \not{k}_3\not{k}_1\not{k}_2(1\pm\gamma_5)],$$

$$\not{\epsilon}^{\pm}(k_3) = N[\not{k}_3\not{k}_1\not{k}_2(1\pm\gamma_5) + \not{k}_2\not{k}_1\not{k}_3(1\mp\gamma_5)],$$

$$N^{-1} = E^2[32k_{3+}k_{3-}]^{\frac{1}{2}}. \tag{10.532}$$

Nonvanishing helicity amplitudes [the definition of the quantity R'_1 is given in eqn (8.12) and the quantity $\epsilon^{\alpha\beta}$ denotes the polarization tensor of the $^3P_2$]:

$$M(+,+;+) = -\frac{ig^3 R'_1 f^{abc}\epsilon^{\alpha\beta}}{E^2(s-M^2)(t-M^2)(u-M^2)}\left[\frac{2M}{\pi k_{3+}k_{3-}}\right]^{\frac{1}{2}}$$

$$\times\left\{s^2\left[\frac{k_{1\alpha}}{t(t-M^2)} - \frac{k_{2\alpha}}{u(u-M^2)}\right]\left[k_{1\beta}(tM^2-su) - k_{2\beta}(uM^2-st)\right]\right.$$

$$+\frac{st+tu+us}{stu}\Bigg[(s-t-u)(tk_{1\alpha}-uk_{2\alpha})(tk_{1\beta}-uk_{2\beta})$$

$$-sk_{3\alpha}\Big(t(2t-s)k_{1\beta}+u(2u-s)k_{2\beta}\Big) + \frac{s^2(t^2+u^2)}{s-M^2}k_{3\alpha}k_{3\beta}\Bigg]$$

$$-2i\epsilon_{\mu\nu\rho\alpha}k_1^\mu k_2^\nu k_3^\rho\Bigg[k_{1\beta}\left(\frac{u^2(t-s)}{t(t-M^2)} - t+u - \frac{tu}{s} - \frac{t(st+tu+us)}{su}\right)$$

$$\left.-k_{2\beta}\left(\frac{t^2(u-s)}{u(u-M^2)} - u+t - \frac{tu}{s} - \frac{u(st+tu+us)}{st}\right)\Bigg]\right\},$$

$$M(+,-;+) = -\frac{ig^3 R'_1 f^{abc}\epsilon^{\alpha\beta}}{E^2(s-M^2)(t-M^2)(u-M^2)}\left[\frac{2M}{\pi k_{3+}k_{3-}}\right]^{\frac{1}{2}}$$

$$\times\left\{t^2\left[\frac{k_{2\alpha}}{u(u-M^2)} + \frac{k_{3\alpha}}{s(s-M^2)}\right]\left[k_{2\beta}(uM^2-st) + k_{3\beta}(sM^2-tu)\right]\right.$$

$$+\frac{st+tu+us}{stu}\Bigg[(t-s-u)(uk_{2\alpha}+sk_{3\alpha})(uk_{2\beta}+sk_{3\beta})$$

$$+tk_{1\alpha}\Big(u(t-2u)k_{2\beta}+s(2s-t)k_{3\beta}\Big) + \frac{t^2(s^2+u^2)}{t-M^2}k_{1\alpha}k_{1\beta}\Bigg]$$

$$-2i\epsilon_{\mu\nu\rho\alpha}k_1^\mu k_2^\nu k_3^\rho\Bigg[k_{2\beta}\left(\frac{s^2(u-t)}{u(u-M^2)} - u+s - \frac{su}{t} - \frac{u(st+tu+us)}{st}\right)$$

$$\left.+k_{3\beta}\left(\frac{u^2(s-t)}{s(s-M^2)} - s+u - \frac{su}{t} - \frac{s(st+tu+us)}{tu}\right)\Bigg]\right\},$$

$$M(-,+;+) = -\frac{ig^3 R_1' f^{abc} \epsilon^{\alpha\beta}}{E^2(s-M^2)(t-M^2)(u-M^2)} \left[\frac{2M}{\pi k_{3+} k_{3-}}\right]^{\frac{1}{2}}$$

$$\times \left\{ u^2 \left[\frac{k_{1\alpha}}{t(t-M^2)} + \frac{k_{3\alpha}}{s(s-M^2)} \right] \left[k_{1\beta}(tM^2 - su) + k_{3\beta}(sM^2 - tu) \right] \right.$$

$$+ \frac{st + tu + us}{stu} \left[(u - t - s)(tk_{1\alpha} + sk_{3\alpha})(tk_{1\beta} + sk_{3\beta}) \right.$$

$$\left. - uk_{2\alpha} \left(t(2t-u)k_{1\beta} - s(2s-u)k_{3\beta} \right) + \frac{u^2(s^2+t^2)}{u-M^2} k_{2\alpha} k_{2\beta} \right]$$

$$+ 2i\epsilon_{\mu\nu\rho\alpha} k_1^\mu k_2^\nu k_3^\rho \left[k_{1\beta} \left(\frac{s^2(t-u)}{t(t-M^2)} - t + s - \frac{st}{u} - \frac{t(st+tu+us)}{su} \right) \right.$$

$$\left. \left. + k_{3\beta} \left(\frac{t^2(s-u)}{s(s-M^2)} - s + t - \frac{st}{u} - \frac{s(st+tu+us)}{tu} \right) \right] \right\},$$

$$M(+,-;-) = -\frac{ig^3 R_1' f^{abc} \epsilon^{\alpha\beta}}{E^2(s-M^2)(t-M^2)(u-M^2)} \left[\frac{2M}{\pi k_{3+} k_{3-}}\right]^{\frac{1}{2}}$$

$$\times \left\{ u^2 \left[\frac{k_{1\alpha}}{t(t-M^2)} + \frac{k_{3\alpha}}{s(s-M^2)} \right] \left[k_{1\beta}(tM^2 - su) + k_{3\beta}(sM^2 - tu) \right] \right.$$

$$+ \frac{st + tu + us}{stu} \left[(u - t - s)(tk_{1\alpha} + sk_{3\alpha})(tk_{1\beta} + sk_{3\beta}) \right.$$

$$\left. - uk_{2\alpha} \left(t(2t-u)k_{1\beta} - s(2s-u)k_{3\beta} \right) + \frac{u^2(s^2+t^2)}{u-M^2} k_{2\alpha} k_{2\beta} \right]$$

$$- 2i\epsilon_{\mu\nu\rho\alpha} k_1^\mu k_2^\nu k_3^\rho \left[k_{1\beta} \left(\frac{s^2(t-u)}{t(t-M^2)} - t + s - \frac{st}{u} - \frac{t(st+tu+us)}{su} \right) \right.$$

$$\left. \left. + k_{3\beta} \left(\frac{t^2(s-u)}{s(s-M^2)} - s + t - \frac{st}{u} - \frac{s(st+tu+us)}{tu} \right) \right] \right\},$$

$$M(-,+;-) = -\frac{ig^3 R_1' f^{abc} \epsilon^{\alpha\beta}}{E^2(s-M^2)(t-M^2)(u-M^2)} \left[\frac{2M}{\pi k_{3+} k_{3-}}\right]^{\frac{1}{2}}$$

$$\times \left\{ t^2 \left[\frac{k_{2\alpha}}{u(u-M^2)} + \frac{k_{3\alpha}}{s(s-M^2)} \right] \left[k_{2\beta}(uM^2 - st) + k_{3\beta}(sM^2 - tu) \right] \right.$$

$$+ \frac{st + tu + us}{stu} \left[(t - s - u)(uk_{2\alpha} + sk_{3\alpha})(uk_{2\beta} + sk_{3\beta}) \right.$$

$$+ tk_{1\alpha}\Big(u(t-2u)k_{2\beta} + s(2s-t)k_{3\beta}\Big) + \frac{t^2(s^2+u^2)}{t-M^2}k_{1\alpha}k_{1\beta}\Big]$$

$$+2i\epsilon_{\mu\nu\rho\alpha}k_1^\mu k_2^\nu k_3^\rho \Bigg[k_{2\beta}\left(\frac{s^2(u-t)}{u(u-M^2)} - u + s - \frac{su}{t} - \frac{u(st+tu+us)}{st}\right)$$

$$+k_{3\beta}\left(\frac{u^2(s-t)}{s(s-M^2)} - s + u - \frac{su}{t} - \frac{s(st+tu+us)}{tu}\right)\Bigg]\Bigg\},$$

$$M(-,-;-) = -\frac{ig^3 R_1' f^{abc}\epsilon^{\alpha\beta}}{E^2(s-M^2)(t-M^2)(u-M^2)}\left[\frac{2M}{\pi k_{3+}k_{3-}}\right]^{\frac{1}{2}}$$

$$\times\Bigg\{s^2\left[\frac{k_{1\alpha}}{t(t-M^2)} - \frac{k_{2\alpha}}{u(u-M^2)}\right]\Big[k_{1\beta}(tM^2 - su) - k_{2\beta}(uM^2 - st)\Big]$$

$$+\frac{st+tu+us}{stu}\Bigg[(s-t-u)(tk_{1\alpha} - uk_{2\alpha})(tk_{1\beta} - uk_{2\beta})$$

$$-sk_{3\alpha}\Big(t(2t-s)k_{1\beta} + u(2u-s)k_{2\beta}\Big) + \frac{s^2(t^2+u^2)}{s-M^2}k_{3\alpha}k_{3\beta}\Bigg]$$

$$+2i\epsilon_{\mu\nu\rho\alpha}k_1^\mu k_2^\nu k_3^\rho \Bigg[k_{1\beta}\left(\frac{u^2(t-s)}{t(t-M^2)} - t + u - \frac{tu}{s} - \frac{t(st+tu+us)}{su}\right)$$

$$-k_{2\beta}\left(\frac{t^2(u-s)}{u(u-M^2)} - u + t - \frac{tu}{s} - \frac{u(st+tu+us)}{st}\right)\Bigg]\Bigg\}.$$

(10.533)

Squared absolute values of the nonvanishing helicity amplitudes, summed over the color degrees of freedom of the initial state and the final state and over the polarization states of the $^3P_2$:

$$|M(+,+;+)|^2 = |M(-,-;-)|^2$$

$$= \frac{64g^6 R_1'^2}{\pi M^3 Q(Q-M^2P)^4}\Big[12M^8P^4(3s^2 - 4sM^2 + M^4)$$

$$-12s(s-3M^2)M^4P^5 - 3(25s^2 - 33sM^2 + 8M^4)M^6P^3Q$$

$$+12(s^2 - 4sM^2 - 3M^4)M^2P^4Q + (8s^2 + 9sM^2 - 15M^4)M^4P^2Q^2$$

$$-2(s^2 - 5sM^2 - 30M^4)P^3Q^2 + (29s^2 - 51sM^2 + 18M^4)M^2PQ^3$$

$$+2(s-11M^2)P^2Q^3 - (9s-11M^2)M^2Q^4\Big],$$

$$|M(+,-;+)|^2 = |M(-,+;-)|^2$$

$$= \frac{64g^6 R_1'^2}{\pi M^3 Q(Q-M^2P)^4} \Big[12M^8 P^4 (3t^2 - 4tM^2 + M^4)$$

$$-12t(t-3M^2)M^4P^5 - 3(25t^2 - 33tM^2 + 8M^4)M^6P^3Q$$

$$+12(t^2 - 4tM^2 - 3M^4)M^2P^4Q + (8t^2 + 9tM^2 - 15M^4)M^4P^2Q^2$$

$$-2(t^2 - 5tM^2 - 30M^4)P^3Q^2 + (29t^2 - 51tM^2 + 18M^4)M^2PQ^3$$

$$+2(t-11M^2)P^2Q^3 - (9t-11M^2)M^2Q^4 \Big],$$

$$|M(-,+;+)|^2 = |M(+,-;-)|^2$$

$$= \frac{64g^6 R_1'^2}{\pi M^3 Q(Q-M^2P)^4} \Big[12M^8 P^4 (3u^2 - 4uM^2 + M^4)$$

$$-12u(u-3M^2)M^4P^5 - 3(25u^2 - 33uM^2 + 8M^4)M^6P^3Q$$

$$+12(u^2 - 4uM^2 - 3M^4)M^2P^4Q + (8u^2 + 9uM^2 - 15M^4)M^4P^2Q^2$$

$$-2(u^2 - 5uM^2 - 30M^4)P^3Q^2 + (29u^2 - 51uM^2 + 18M^4)M^2PQ^3$$

$$+2(u-11M^2)P^2Q^3 - (9u-11M^2)M^2Q^4 \Big]. \quad (10.534)$$

Unpolarized squared matrix element:

$$\overline{|M|^2} = \frac{g^6 R_1'^2}{\pi M^3 Q(Q-M^2P)^4} \Big[12M^4 P^4 (M^8 - 2M^4 P + P^2)$$

$$-3M^2 P^3 Q(8M^8 - M^4 P + 4P^2) - 2P^2 Q^2 (7M^8 - 43M^4 P - P^2)$$

$$+M^2 PQ^3 (16M^4 - 61P) + 12M^4 Q^4 \Big]. \quad (10.535)$$

Unpolarized cross section:

$$\frac{d\sigma}{dt} = \frac{4\pi \alpha_S^3 R_1'^2}{M^3 s^2 Q(Q-M^2P)^4} \Big[12M^4 P^4 (M^8 - 2M^4 P + P^2)$$

$$-3M^2 P^3 Q(8M^8 - M^4 P + 4P^2) - 2P^2 Q^2 (7M^8 - 43M^4 P - P^2)$$

$$+M^2 PQ^3 (16M^4 - 61P) + 12M^4 Q^4 \Big]. \quad (10.536)$$

11
Polarization

> Beam polarization
> gives information
> on the electron's inclination
> to behave with aberration.

In the preceding chapters, we expounded a convenient way to calculate helicity amplitudes for various processes at high energies. In this chapter, we show how those helicity amplitudes must be combined for the description of processes in which arbitrary spin polarizations occur. We first treat the case of fermion polarization in detail and then analyse the polarization effects for photons and gluons. Throughout this chapter, we again neglect the finite mass effects for simplicity.

11.1 Fermions

The positive helicity states for fermions, introduced in Chapter 3 [eqns (3.1)], describe particles with spin polarization along the direction of motion, whereas negative helicity states have their spin polarization opposite to the direction of motion. This can be seen as follows.

A fermion at rest, with mass m, is described by the spinor $u(\vec{0})$ which satisfies the Dirac equation

$$(\gamma^0 - 1)\, u(\vec{0}) = 0. \tag{11.1}$$

With our representation of the γ-matrices, eqns (2.6), we have two independent solutions,

$$u_+(\vec{0}) = \frac{1}{\sqrt{2}} \begin{pmatrix} 1 \\ 0 \\ 1 \\ 0 \end{pmatrix}, \tag{11.2}$$

and

$$u_-(\vec{0}) = \frac{1}{\sqrt{2}} \begin{pmatrix} 0 \\ 1 \\ 0 \\ 1 \end{pmatrix}. \tag{11.3}$$

Both spinors are eigenstates of Σ_z, the spin-operator in the z-direction, where $\vec{\Sigma}$ is given by

$$\vec{\Sigma} = \begin{pmatrix} \vec{\sigma} & 0 \\ 0 & \vec{\sigma} \end{pmatrix}, \tag{11.4}$$

i.e.,
$$\Sigma_z u_+(\vec{0}) = +u_+(\vec{0}),$$
$$\Sigma_z u_-(\vec{0}) = -u_-(\vec{0}). \tag{11.5}$$

Similarly, one finds that $(u_+ \pm u_-)/\sqrt{2}$ are eigenstates of Σ_x, while $(u_+ \pm iu_-)/\sqrt{2}$ are eigenstates of Σ_y:

$$\Sigma_x \frac{u_+ \pm u_-}{\sqrt{2}} = \pm \frac{u_+ \pm u_-}{\sqrt{2}}, \qquad \Sigma_y \frac{u_+ \pm iu_-}{\sqrt{2}} = \pm \frac{u_+ \pm iu_-}{\sqrt{2}}. \tag{11.6}$$

How does one describe fermions moving along the z-direction with, for example, spin polarization along the positive x-axis? It suffices to boost the states (11.2) and (11.3) in the z-direction, which transforms the spinors $u_\pm$ into helicity eigenstates once the fermion mass in neglected. Such a boost does not affect the spin in the x-direction. Hence, the amplitude for spin in the positive x-direction ($s_x = +1$) is simply given by the following linear combination of helicity amplitudes $M(\pm)$:

$$M(s_x = +1) = \frac{1}{\sqrt{2}} [M(+) + M(-)]. \tag{11.7}$$

Obviously, one also finds that

$$M(s_x = -1) = \frac{1}{\sqrt{2}} [M(+) - M(-)],$$

$$M(s_y = +1) = \frac{1}{\sqrt{2}} [M(+) + iM(-)],$$

$$M(s_y = -1) = \frac{1}{\sqrt{2}} [M(+) - iM(-)]. \tag{11.8}$$

The same analysis can be repeated for the antifermions, described by the v-type spinors. Now,

$$v_+(\vec{0}) = \frac{1}{\sqrt{2}} \begin{pmatrix} 0 \\ -1 \\ 0 \\ 1 \end{pmatrix}, \tag{11.9}$$

and

$$v_-(\vec{0}) = \frac{1}{\sqrt{2}} \begin{pmatrix} 1 \\ 0 \\ -1 \\ 0 \end{pmatrix}. \tag{11.10}$$

are the two independent solutions of the Dirac equation

$$(\gamma^0 + 1) v(\vec{0}) = 0. \tag{11.11}$$

11.2. AN EXAMPLE: $e^+e^- \to \mu^+\mu^-$

Through the same boost, they are transformed into helicity eigenstates. But now,

$$\Sigma_z v_\pm = \mp v_\pm,$$

$$\Sigma_x \frac{v_- \mp v_+}{\sqrt{2}} = \pm \frac{v_- \mp v_+}{\sqrt{2}},$$

$$\Sigma_y \frac{v_- \mp iv_+}{\sqrt{2}} = \pm \frac{v_- \mp iv_+}{\sqrt{2}}. \tag{11.12}$$

To describe transversely polarized antifermions, we thus have to use the following combinations of helicity amplitudes:

$$M(s_x = +1) = \frac{1}{\sqrt{2}}[M(-) + M(+)],$$

$$M(s_x = -1) = \frac{1}{\sqrt{2}}[M(-) - M(+)],$$

$$M(s_y = +1) = \frac{1}{\sqrt{2}}[M(-) + iM(+)],$$

$$M(s_y = -1) = \frac{1}{\sqrt{2}}[M(-) - iM(+)]. \tag{11.13}$$

Since the expressions $\bar{u}u$ and $\bar{v}v$ are invariant under rotations, it follows that the states $(\bar{u}_+ - \bar{u}_-)/\sqrt{2}$ and $(\bar{v}_- - \bar{v}_+)/\sqrt{2}$ describe the spin states with $s_x = +1$. In the same way, one can construct the other transverse spin states involving the spinors $\bar{u}$ or $\bar{v}$. The results of this section are summarized in Table 11.1.

11.2 An example: $e^+e^- \to \mu^+\mu^-$

When e^+ and e^- beams are kept circulating in storage rings for long periods of time, the particles tend to align their spins along the direction of the magnetic field which keeps them in circular orbits [42]. This magnetic field is perpendicular to the plane of motion and can be taken to point in the x-direction. (Although very desirable, it is by no means simple to achieve large polarizations in high energy accelerators, because of machine resonances, beam-beam interactions, solenoid fields of the detectors, etc.)

In this section, we analyse the process

$$e^+(p_+) + e^-(p_-) \to \mu^+(q_+) + \mu^-(q_-), \tag{11.14}$$

for the case where the e^+ end e^- spin polarizations are thus directed along the x-direction. As usual, we take the positive z-direction along $\vec{p}_+$.

Table 11.1: Spinor combinations for transverse polarization.

$s_x = +1$	$s_x = -1$	$s_y = +1$	$s_y = -1$
$\dfrac{u_+ + u_-}{\sqrt{2}}$	$\dfrac{u_+ - u_-}{\sqrt{2}}$	$\dfrac{u_+ + iu_-}{\sqrt{2}}$	$\dfrac{u_+ - iu_-}{\sqrt{2}}$
$\dfrac{\bar{u}_+ - \bar{u}_-}{\sqrt{2}}$	$\dfrac{\bar{u}_+ + \bar{u}_-}{\sqrt{2}}$	$\dfrac{\bar{u}_+ + i\bar{u}_-}{\sqrt{2}}$	$\dfrac{\bar{u}_+ - i\bar{u}_-}{\sqrt{2}}$
$\dfrac{v_- + v_+}{\sqrt{2}}$	$\dfrac{v_- - v_+}{\sqrt{2}}$	$\dfrac{v_- + iv_+}{\sqrt{2}}$	$\dfrac{v_- - iv_+}{\sqrt{2}}$
$\dfrac{\bar{v}_- - \bar{v}_+}{\sqrt{2}}$	$\dfrac{\bar{v}_- + \bar{v}_+}{\sqrt{2}}$	$\dfrac{\bar{v}_- + i\bar{v}_+}{\sqrt{2}}$	$\dfrac{\bar{v}_- - i\bar{v}_+}{\sqrt{2}}$

Let $M(\lambda_1, \lambda_2; \lambda_3, \lambda_4)$ be the helicity amplitudes for this process, with λ_1 (λ_2) the helicity of the e^+ (e^-) and λ_3 (λ_4) that of the μ^+ (μ^-). The nonvanishing helicity amplitudes are given by [see eqns (9.10)]:

$$M(+,-;+,-) = e^2 \frac{q_{+\perp}}{E} \left(\frac{q_{++}}{q_{-+}}\right)^{\frac{1}{2}},$$

$$M(+,-;-,+) = e^2 \frac{q_{-\perp}}{E} \left(\frac{q_{-+}}{q_{++}}\right)^{\frac{1}{2}},$$

$$M(-,+;-,+) = e^2 \frac{q^*_{+\perp}}{E} \left(\frac{q_{++}}{q_{-+}}\right)^{\frac{1}{2}},$$

$$M(-,+;+,-) = e^2 \frac{q^*_{-\perp}}{E} \left(\frac{q_{-+}}{q_{++}}\right)^{\frac{1}{2}}. \quad (11.15)$$

Denoting $s_x = +1$ by $\uparrow$ and $s_x = -1$ by $\downarrow$, we know from the preceding section that the amplitude for $s_x(e^+) = +1$ is given by

$$M(\uparrow, \lambda_2; \lambda_3, \lambda_4) = \frac{1}{\sqrt{2}}[M(-,\lambda_2;\lambda_3\lambda_4) - M(+,\lambda_2;\lambda_3\lambda_4)]. \quad (11.16)$$

It then follows that, for $s_x(e^+) = +1$ and $s_x(e^-) = +1$,

$$M(\uparrow,\uparrow;\lambda_3,\lambda_4) = \tfrac{1}{2}[M(-,+;\lambda_3,\lambda_4) - M(+,-;\lambda_3,\lambda_4)], \quad (11.17)$$

11.2. AN EXAMPLE: $e^+e^- \to \mu^+\mu^-$

as $M(+,+;\lambda_3,\lambda_4) = M(-,-;\lambda_3,\lambda_4) = 0$ in the high energy limit.

The remaining combinations of spin polarization can be worked out similarly and the result reads

$$M(\uparrow,\uparrow;\lambda_3,\lambda_4) = M(\downarrow,\downarrow;\lambda_3,\lambda_4)$$

$$= \tfrac{1}{2}[M(-,+;\lambda_3,\lambda_4) - M(+,-;\lambda_3,\lambda_4)],$$

$$M(\uparrow,\downarrow;\lambda_3,\lambda_4) = M(\downarrow,\uparrow;\lambda_3,\lambda_4)$$

$$= \tfrac{1}{2}[M(-,+;\lambda_3,\lambda_4) + M(+,-;\lambda_3,\lambda_4)]. \quad (11.18)$$

Let us concentrate on the important case for which $s_x(e^+) = +1$ and $s_x(e^-) = -1$. There are only two amplitudes, i.e.,

$$M(\uparrow,\downarrow;+,-) = \frac{e^2}{2E}\left[q_{+\perp}\left(\frac{q_{++}}{q_{-+}}\right)^{\frac{1}{2}} + q_{-\perp}^*\left(\frac{q_{-+}}{q_{++}}\right)^{\frac{1}{2}}\right],$$

$$M(\uparrow,\downarrow;-,+) = \frac{e^2}{2E}\left[q_{+\perp}^*\left(\frac{q_{++}}{q_{-+}}\right)^{\frac{1}{2}} + q_{-\perp}\left(\frac{q_{-+}}{q_{++}}\right)^{\frac{1}{2}}\right], \quad (11.19)$$

and

$$|M(\uparrow,\downarrow;+,-)|^2 = |M(\uparrow,\downarrow;-,+)|^2 = \frac{e^4}{E^2}(q_{+y}^2 + q_{+z}^2). \quad (11.20)$$

Summing over the helicities of the outgoing μ^+ and μ^-, we have

$$\frac{d\sigma(\uparrow\downarrow)}{dt} = \frac{\pi\alpha^2}{8E^6}(q_{+y}^2 + q_{+z}^2), \quad (11.21)$$

with

$$t = -2(p_+ \cdot q_+). \quad (11.22)$$

This formula supposes complete spin polarization along the x-axis. More realistic, however, is the case of partial polarization. If P denotes the degree of polarization, assumed to be equal in magnitude for the e^+ and e^- beams, the cross section becomes

$$\frac{d\sigma}{dt} = P^2\frac{d\sigma(\uparrow\downarrow)}{dt} + P(1-P)\left[\frac{d\sigma(\uparrow\uparrow)}{dt} + \frac{d\sigma(\downarrow\downarrow)}{dt}\right] + (1-P)^2\frac{d\sigma(\downarrow\uparrow)}{dt}$$

$$= \frac{\pi\alpha^2}{16E^6}\left[q_{+0}^2 + q_{+z}^2 - (1-2P)^2(q_{+x}^2 - q_{+y}^2)\right], \quad (11.23)$$

with

$$\frac{d\sigma(\downarrow\uparrow)}{dt} = \frac{d\sigma(\uparrow\downarrow)}{dt} = \frac{\pi\alpha^2}{8E^6}(q_{+y}^2 + q_{+z}^2),$$

$$\frac{d\sigma(\uparrow\uparrow)}{dt} = \frac{d\sigma(\downarrow\downarrow)}{dt} = \frac{\pi\alpha^2}{8E^6}(q_{+x}^2 + q_{+z}^2). \quad (11.24)$$

Clearly, when $P = 1/2$, i.e., no polarization, we recover the unpolarized cross section of eqn (9.12).

11.3 Photons and gluons

Throughout this book, we used polarization vectors for incoming photons and gluons of the form

$$\not{\epsilon}^{*\pm}(k) = N[\not{k}_3\,\not{k}_2\,\not{k}(1\pm\gamma_5)+ \not{k}\,\not{k}_2\,\not{k}_3(1\mp\gamma_5)], \qquad (11.25)$$

where k is the four-momentum of the particle and k_2 and k_3 are taken to be light-like. Let us first convince ourselves that eqn (11.25) corresponds to a positive helicity state for ϵ^{*+} and a negative helicity state for ϵ^{*-}.

The four-vector potential of a photon or gluon with momentum k^μ is given by

$$A^\mu(x;k) = \frac{1}{[(2\pi)^3 2k_0]^{\frac{1}{2}}} \left[\epsilon^{*\mu}(k) e^{-i(k\cdot x)} + \epsilon^\mu(k) e^{i(k\cdot x)} \right]. \qquad (11.26)$$

Let us take $k^\mu = (E, 0, 0, E)$, i.e., motion along the z-axis. To the expression (11.26) then correspond the following components of the electric field:

$$E_x(x;k) = \left[\frac{2E}{(2\pi)^3}\right]^{\frac{1}{2}} \left[\mathrm{Re}\!\left(\epsilon_x^*(k)\right) \sin(k\cdot x) - \mathrm{Im}\!\left(\epsilon_x^*(k)\right) \cos(k\cdot x) \right],$$

$$E_y(x;k) = \left[\frac{2E}{(2\pi)^3}\right]^{\frac{1}{2}} \left[\mathrm{Re}\!\left(\epsilon_y^*(k)\right) \sin(k\cdot x) - \mathrm{Im}\!\left(\epsilon_y^*(k)\right) \cos(k\cdot x) \right].$$

$$(11.27)$$

From eqn (11.25), it follows that

$$\mathrm{Re}\!\left(\epsilon_x^{*+}(k)\right) = \mathrm{Im}\!\left(\epsilon_y^{*+}(k)\right) = 2NE(k_{3x}k_{2-} - k_{2x}k_{3-}),$$

$$\mathrm{Im}\!\left(\epsilon_x^{*+}(k)\right) = -\mathrm{Re}\!\left(\epsilon_y^{*+}(k)\right) = 2NE(k_{2y}k_{3-} - k_{3y}k_{2-}).$$

$$(11.28)$$

A simple calculation shows that

$$4N^2 E^2 \left[(k_{3x}k_{2-} - k_{2x}k_{3-})^2 + (k_{2y}k_{3-} - k_{3y}k_{2-})^2 \right] = \tfrac{1}{2}, \qquad (11.29)$$

as

$$N^{-1} = 4E[k_{2-}k_{3-}(k_2\cdot k_3)]^{\frac{1}{2}}. \qquad (11.30)$$

Hence, writing

$$\mathrm{Re}\!\left(\epsilon_x^{*+}(k)\right) = -\frac{1}{\sqrt{2}}\sin\delta,$$

$$\mathrm{Im}\!\left(\epsilon_x^{*+}(k)\right) = -\frac{1}{\sqrt{2}}\cos\delta, \qquad (11.31)$$

11.3. PHOTONS AND GLUONS

we have

$$E_x(x;k) = \left[\frac{E}{(2\pi)^3}\right]^{\frac{1}{2}} \cos\bigl((k \cdot x) + \delta\bigr),$$

$$E_y(x;k) = \left[\frac{E}{(2\pi)^3}\right]^{\frac{1}{2}} \cos\bigl((k \cdot x) + \delta - \frac{\pi}{2}\bigr). \quad (11.32)$$

The amplitudes of E_x and E_y are found to be equal and the phase difference between E_y and E_x is seen to be $-\pi/2$. This is precisely the definition of positive helicity, also called right-handed circular polarization. (In optics, the convention is, for some reason, just the other way around: the case of (11.32) is called left-handed circular polarization [43].)

The same manipulations lead to the conclusion that ϵ^{*-} denotes photons or gluons with negative helicities or right-handed circular polarizations.

How do we describe the case of linear polarization? Consider the combination

$$\epsilon^{*\|} = \frac{\epsilon_q^{*+} + \epsilon_q^{*-}}{\sqrt{2}}, \quad (11.33)$$

where

$$\slashed{\epsilon}_q^{*\pm} = N_q[\slashed{p}\,\slashed{q}\,\slashed{k}(1\pm\gamma_5) + \slashed{k}\,\slashed{q}\,\slashed{p}(1\mp\gamma_5)], \quad (11.34)$$

with

$$p^\mu = (1,0,0,-1), \quad q^\mu = (1,1,0,0), \quad N_q^{-1} = E\sqrt{32}. \quad (11.35)$$

It follows that

$$\text{Re}\left(\epsilon_x^{*\|}\right) = -1,$$

$$\text{Im}\left(\epsilon_x^{*\|}\right) = \text{Re}\left(\epsilon_y^{*\|}\right) = \text{Im}\left(\epsilon_y^{*\|}\right) = 0, \quad (11.36)$$

and

$$E_x(x;k) = -\left[\frac{2E}{(2\pi)^3}\right]^{\frac{1}{2}} \sin(k \cdot x),$$

$$E_y(x;k) = 0. \quad (11.37)$$

Thus, $\epsilon^{*\|}$ is found to describe the case of linear polarization along the x-axis.

In practice, we want to perform the calculations with the polarization vectors $\epsilon^{*\pm}$ given by eqn (11.25). To go from $\epsilon^{*\pm}$ to $\epsilon_q^{*\pm}$, we use the relation [see eqn (4.13)]

$$\epsilon_q^{*\pm} = e^{\pm i\alpha}\epsilon^{*\pm} + \beta_\pm k, \quad (11.38)$$

with
$$e^{\pm i\alpha} = -(\epsilon_q^{*\pm} \cdot \epsilon^{*\mp}), \tag{11.39}$$

and the quantities $\beta_\pm$ are irrelevant because of gauge invariance. The phase factor $e^{i\alpha}$ is easily evaluated:

$$\begin{aligned} e^{i\alpha} &= -NN_q \text{Tr}[\not{p}\,\not{A}\,\not{k}\,\not{k}_3\,\not{k}_2\,\not{k}(1-\gamma_5)] \\ &= -\frac{k_{2\perp} Z_{23}}{k_{2+}[2k_{2-}k_{3-}(k_2\cdot k_3)]^{\frac{1}{2}}}, \end{aligned} \tag{11.40}$$

with
$$Z_{23} = k_{2+} k_{3-} - k_{2\perp}^* k_{3\perp}. \tag{11.41}$$

Let $M(+)$ and $M(-)$ be the helicity amplitudes for a given process calculated with the polarization vectors $\epsilon^{*\pm}$. The amplitude for the process involving a photon or a gluon with linear polarization along the x-axis is then given by

$$M(x) = -\frac{k_{2\perp} Z_{23} M(+) + k_{2\perp}^* Z_{23}^* M(-)}{2k_{2+}[k_{2-} k_{3-}(k_2 \cdot k_3)]^{\frac{1}{2}}}. \tag{11.42}$$

Replacing q^μ by $q'^\mu = (1,0,1,0)$ in the above formulae, we obtain the description of linear polarization along the y-axis. In that case, the phase factor becomes

$$e^{i\alpha'} = -\frac{ik_{2\perp} Z_{23}}{k_{2+}[2k_{2-} k_{3-}(k_2 \cdot k_3)]^{\frac{1}{2}}}, \tag{11.43}$$

and
$$M(y) = -\frac{i[k_{2\perp} Z_{23} M(+) + k_{2\perp}^* Z_{23}^* M(-)]}{2k_{2+}[k_{2-} k_{3-}(k_2 \cdot k_3)]^{\frac{1}{2}}} \tag{11.44}$$

is the corresponding amplitude.

When the motion of the photon or gluon takes place along the negative z-axis, we choose $\epsilon_q^{*\pm}$ as in eqn (11.34), but with $p^\mu = (1,0,0,1)$. Analogous to the above α, the phase factor for linear polarization along the x-axis then becomes

$$e^{i\beta} = \frac{k_{2\perp}^* Z_{32}}{k_{2-}[2k_{2+} k_{3+}(k_2 \cdot k_3)]^{\frac{1}{2}}}, \tag{11.45}$$

with
$$Z_{32} = k_{3+} k_{2-} - k_{3\perp}^* k_{2\perp}. \tag{11.46}$$

Taking $q^\mu = (1,0,1,0)$, we obtain photons or gluons with linear polarization along the y-axis, and the relevant phase factor is

$$e^{i\beta'} = \frac{ik_{2\perp}^* Z_{32}}{k_{2-}[2k_{2+} k_{3+}(k_2 \cdot k_3)]^{\frac{1}{2}}}. \tag{11.47}$$

The results of this section are summarized in Table 11.2.

11.4. AN EXAMPLE: $gg \to g\, ^1S_0$

Table 11.2: Helicity combinations for linear polarization. The quantities α, α', β and β' are given by eqns (11.40), (11.43), (11.45) and (11.47).

	$M(x)$	$M(y)$
$k_z = +E$	$\dfrac{e^{i\alpha} M(+) + e^{-i\alpha} M(-)}{\sqrt{2}}$	$\dfrac{e^{i\alpha'} M(+) + e^{-i\alpha'} M(-)}{\sqrt{2}}$
$k_z = -E$	$\dfrac{e^{i\beta} M(+) + e^{-i\beta} M(-)}{\sqrt{2}}$	$\dfrac{e^{i\beta'} M(+) + e^{-i\beta'} M(-)}{\sqrt{2}}$

11.4 An example: $gg \to g\, ^1S_0$

In Section 10.72, we presented the helicity amplitudes for the process

$$g(k_1, a) + g(k_2, b) \to g(k_3, c) + {}^1S_0(p), \qquad (11.48)$$

with the following choice of polarization vectors for the gluons:

$$\begin{aligned}
\slashed{\epsilon}^{*\pm}(k_1) &= N[\slashed{k}_1\, \slashed{k}_2\, \slashed{k}_3(1 \mp \gamma_5) + \slashed{k}_3\, \slashed{k}_2\, \slashed{k}_1(1 \pm \gamma_5)], \\
\slashed{\epsilon}^{*\pm}(k_2) &= N[\slashed{k}_2\, \slashed{k}_3\, \slashed{k}_1(1 \mp \gamma_5) + \slashed{k}_1\, \slashed{k}_3\, \slashed{k}_2(1 \pm \gamma_5)], \\
\slashed{\epsilon}^{\pm}(k_3) &= N[\slashed{k}_3\, \slashed{k}_1\, \slashed{k}_2(1 \pm \gamma_5) + \slashed{k}_2\, \slashed{k}_1\, \slashed{k}_3(1 \mp \gamma_5)], \\
N^{-1} &= E^2[32 k_{3+} k_{3-}]^{\frac{1}{2}}.
\end{aligned} \qquad (11.49)$$

By taking the positive z-axis along $\vec{k}_1$, we obtained the nonvanishing helicity amplitudes in eqns (10.498).

Suppose we want to calculate the cross section for the case of incoming gluons which are both linearly polarized in directions perpendicular to the z-axis. For example, gluon 1 is linearly polarized in the x-direction and gluon 2 is linearly polarized in the y-direction. For gluon 1, we then have to combine helicity amplitudes with the phase factor $e^{i\alpha}$, given by eqn (11.40), and for gluon 2 we have to use $e^{i\beta'}$, given by eqn (11.47). They are simply

$$e^{i\alpha} = \frac{k_{3\perp}}{[k_{3+}k_{3-}]^{\frac{1}{2}}} = e^{i\phi},$$

$$e^{i\beta'} = \frac{ik_{3\perp}^*}{[k_{3+}k_{3-}]^{\frac{1}{2}}} = ie^{-i\phi},$$

(11.50)

with ϕ the azimuthal angle of $\vec{k}_3$.

From eqns (10.498), it then follows that

$$M(x,y;+) = \tfrac{i}{2}\Big[M(+,+;+) - e^{2i\phi}M(+,-;+)$$

$$+ e^{-2i\phi}M(-,+;+) - M(-,-;+)\Big]$$

$$= \frac{g^3 R_0 f^{abc}(st + tu + us)\big[M^4 + s^2 + u^2 e^{2i\phi} - t^2 e^{-2i\phi}\big]}{2E^2[24\pi M k_{3+}k_{3-}]^{\frac{1}{2}}(s-M^2)(t-M^2)(u-M^2)},$$

$$M(x,y;-) = \tfrac{i}{2}\Big[M(+,+;-) - e^{2i\phi}M(+,-;-)$$

$$+ e^{-2i\phi}M(-,+;-) - M(-,-;-)\Big]$$

$$= \frac{g^3 R_0 f^{abc}(st + tu + us)\big[M^4 + s^2 - t^2 e^{2i\phi} + u^2 e^{-2i\phi}\big]}{2E^2[24\pi M k_{3+}k_{3-}]^{\frac{1}{2}}(s-M^2)(t-M^2)(u-M^2)}.$$

(11.51)

Summing over the color degrees of freedom of the initial state and the final state, we find

$$|M(x,y;+)|^2 = |M(x,y;-)|^2$$

$$= \frac{4g^6 R_0^2 (st + tu + us)^2}{\pi M stu[(s-M^2)(t-M^2)(u-M^2)]^2}$$

$$\times \Big\{\big[M^4 + s^2 + (u^2 - t^2)\cos(2\phi)\big]^2 + (u^2 + t^2)^2 \sin^2(2\phi)\Big\},$$

(11.52)

and the polarized cross section, averaged over the initial state color degrees of freedom and summed over the final state degrees of freedom, reads

$$\frac{d\sigma}{dt} = \frac{\pi \alpha_s^3 R_0^2}{2Ms^3 tu} \frac{(st + tu + us)^2}{[(s-M^2)(t-M^2)(u-M^2)]^2}$$

$$\times \Big\{\big[M^4 + s^2 + (u^2 - t^2)\cos(2\phi)\big]^2 + (u^2 + t^2)^2 \sin^2(2\phi)\Big\}.$$

(11.53)

This example was particularly simple, as the different helicity amplitudes could be nicely combined. This was due in part to the fact that the color structure for each helicity amplitude was given by an overall factor f^{abc}. In applications to other processes, one may encounter more complicated situations where the helicity amplitudes are linear combinations of different color matrices. In that case, it is probably best to resort to complex arithmetic. One first evaluates all the coefficients of the color matrices in each amplitude as complex numbers depending on the components of the four-momenta in the process. Multiplying the coefficients with the appropriate phase factors then yields the amplitudes for the polarized scattering process, and the corresponding cross section is obtained by adding the squared absolute values of the polarization amplitudes, summed and/or averaged over the color degrees of freedom.

12
Beamstrahlung

> Photons mediate,
> but electrons radiate:
> The loss of counting rate
> makes people fulminate.

12.1 Electron-positron linear colliders

In this book, the helicity method is applied to a variety of processes. These processes are conveniently described in terms of Feynman diagrams, and as a result all formulae obtained so far, including those of Chapters 9 and 10, are each proportional to powers of the relevant coupling constants. Such applications of the helicity method are by now completely understood, although the calculations remain to be worked out for processes involving more particles, and further simplifications can be achieved in some cases. As an example, the results (10.512) for the production of the $^1P_1$ heavy quarkonium state through $g\,g \to g\,^1P_1$ have been obtained only recently, the previous version being significantly more complicated.

These are not the only known applications of the helicity method. However, these other applications have not yet reached a state of completion comparable to those already described. In this chapter, which is essentially the last chapter of the book, we attempt to present one of these other applications, namely that to beamstrahlung. Beamstrahlung refers to the bremsstrahlung due to the interpenetration of two bunches of electrons and positrons. Its importance to physics comes about as follows.

As already mentioned in Chapter 1, the maximum centre-of-mass energy for e^+e^- physics at present is about 100 GeV, provided by LEP at CERN, Geneva, Switzerland. In a few years, this maximum energy of LEP will be doubled to 200 GeV, using superconducting accelerating cavities. Since the energy loss due to synchrotron radiation is proportional to the fourth power of the energy of the circulating electron and positron bunches, it is unlikely for the centre-of-mass energy at LEP to exceed significantly this 200 GeV. In order to reach the TeV regime, with circular colliding machines, a larger ring is necessary. Such a project has not been planned, and is quite unlikely.

Since the limiting factor is synchrotron radiation, a possible alternative is the electron-positron linear collider, which consists of two linear accelerators, one for electrons and the other for positrons, in opposite directions. Such colliders were first proposed by Tigner [44] a quarter of a century ago. One of the major difficulties of building such colliders is due to the fact that the electron bunch and the positron bunch interpenetrate each other only once. This is to be contrasted with circular colliders, such as LEP, PETRA at DESY and PEP and SPEAR at SLAC, where the electron and positron bunches cross each other many times, even in one millisecond.

The single-pass nature of e^+e^- linear colliders imposes strong constraints on the design of such accelerators. After one collision, the energies contained in the electron and the positron bunches are completely lost, since no practical way of recovering them has been found. Accordingly, the number of particles in each bunch is limited by power requirement, which is typically taken to be 100 or 200 megawatts, comparable to that of LEP. On the other hand, the potentially interesting annihilation cross sections decrease with increasing energies as the square of the centre-of-mass energy. Therefore, with the power limitation, a useful luminosity, which must compensate the decrease of the annihilation cross sections, can be reached only by making the radius of the bunches exceedingly small: typically less than one micron, often much less [45]. During the last few years, there are discussions of three TeV e^+e^- linear colliders:

1. CLIC (CERN Linear Collider) at CERN, with a centre-of-mass energy of 2 TeV;

2. TLC (TeV Linear Collider) at SLAC, with the same centre-of-mass energy; and

3. Super, with a centre-of-mass energy of 10 TeV.

As knowledge about linear colliders increases, the parameters for CLIC and TLC change. On the other hand, those for Super, which is an accelerator for the far future, remain at the values given by Richter [46]. The main parameters are

$$f = 100 \text{ Hz},$$

$$R(\text{radius of bunches}) = 5 \text{ Å},$$

$$N(\text{number of particles in each bunch}) = 3 \times 10^8,$$

$$\overline{L}_b(\text{length of bunch}) = 0.3 \text{ μm}. \tag{12.1}$$

With these parameters, the number density of the electron and positron bunches is about $10^{27}/\text{cm}^3$, i.e., the density is roughly that of water.

Such very high densities have the following consequence. Although synchrotron radiation is avoided by using linear accelerators, there is nevertheless significant photon radiation due to the interpenetration of such very dense bunches. This photon radiation is called beamstrahlung [47], and is studied in this chapter in the high energy limit for some simple cases. The presentation follows that of Jacob and Wu [48].

12.2 Nature of approximations

Similar to many other problems in field theory, beamstrahlung is best studied using Feynman diagrams as described in Chapter 2.

It is however not feasible to apply the the method of Feynman diagrams directly. As given for example by (12.1), there are always many millions of electrons and positrons in each bunch. No method has ever been developed to study the behaviour of a Feynman diagram involving millions of incoming electron lines and millions of incoming positron lines. Therefore, by the nature of the problem of beamstrahlung, approximations must be introduced at the beginning. In this section, these approximations are discussed. In the future, if a decision is ever reached to construct an electron-positron linear collider in the TeV regime, these approximations will probably be removed or at least replaced by better ones. Where feasible, we shall attempt here to give a discussion of how these improvements may be carried out.

12.2.1 Single-particle approximation

Consider the interpenetration of an electron bunch and a positron bunch. If the two bunches are accurately centred with respect to each other, then there is a mutual focusing effect that tends to reduce the transverse sizes of the bunches. This effect is useful because it increases the luminosity of the electron-positron collider.

The mutual focusing is necessarily a collective effect and hence very difficult to extract from the Feynman diagrams. Thus, the first assumption to be imposed is that the mutual focusing effect is small.

In order to put this effect of mutual focusing on a more quantitative basis, consider an electron penetrating through a positron bunch (or equivalently a positron through an electron bunch). While a single electron does not disturb the positron bunch in any significant way, the positron bunch acts as a focusing lens for the electron. Thus, there is a focal length, taken at the distance R, the mean radius of the positron bunch. In accelerator physics, the disruption factor D is defined [49] essentially as the ratio of the bunch length $\overline{L}_b$ to its focal length:

$$D = \frac{r_e \overline{L}_b N}{R^2 \gamma}, \qquad (12.2)$$

where r_e is the classical radius of the electron, which is equal to 2.818×10^{-15} m, and γ is the ratio of the beam energy to the electron mass. For the Super machine of Richter, $\gamma = 10^7$ and, hence,

$$D = 0.1. \qquad (12.3)$$

Our first assumption that the mutual focusing effect is small can be restated in the form that the focal length is much longer than the bunch length, i.e.,

$$D \ll 1. \tag{12.4}$$

This is well satisfied by the Super machine. When the relation (12.4) holds, the electron and the positron bunches are not significantly distorted by each other. Under this circumstance, it is justified to apply the single-particle approximation. More precisely, the problem of beamstrahlung can be replaced by the problem of a single electron penetrating and being deflected by the collective effect of a positron bunch (or equivalently, a positron by that of an electron bunch). For the most important case of the radiation of a single photon, the process under consideration is

$$e + \text{bunch} \to e + \gamma + \text{bunch}. \tag{12.5}$$

Clearly, this single-particle approximation already simplifies the problem greatly.

Let us discuss briefly how this condition (12.4) of small disruption factor may be removed. The issue is essentially the determination of the charge distribution of a bunch as seen by an electron or positron of the other bunch. It is feasible to carry out this determination classically, through the electromagnetic interaction between the bunches, and such a classical approximation is likely to be quite accurate for the purpose of calculating beamstrahlung. With the resulting charge distribution found in this way, the single-particle approximation can again be applied. The main novel feature is that different electrons (positrons) encounter different charge distributions of the positron (electron) bunch.

12.2.2 External field approximation

For the process (12.5), it is natural to use the external field approximation, i.e., the approximation of replacing the bunch by an external field. This is most easily accomplished by a Lorentz transform to the coordinate system where the bunch is at rest, and then replace the bunch by an external electrostatic potential. Relative motion of the positrons (or electrons) in the bunch is neglected.

With this external field approximation, the Feynman diagram for the process (12.5) is now the one shown in Fig. 12.1, where each cross denotes an interaction with the external electrostatic potential. There are of course many additional Feynman diagrams with internal photon lines describing radiative corrections, but these additional diagrams will not be studied here.

There is a fundamental difference between the Feynman diagram depicted in Fig. 12.1 and those of the earlier chapters of this book, namely, that of Fig. 12.1 stands for an *infinite* set of Feynman diagrams. Let n_i be the number of crosses, i.e., interactions with the external field, to the left of the

12.2. NATURE OF APPROXIMATIONS

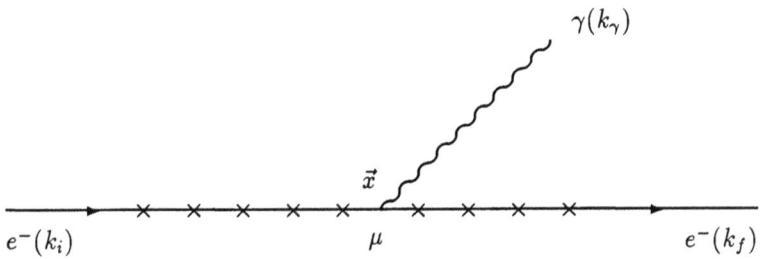

Fig. 12.1: Feynman diagram for $e^- +$ bunch $\to e^- + \gamma +$ bunch.

$ee\gamma$ vertex of Fig. 12.1, and n_f be the corresponding number to the right. [In the figure, $n_i = 5$ and $n_f = 4$.] Then, both n_i and n_f need to be summed over all integers 0, 1, 2, 3..., with the exception of $n_i = n_f = 0$. Because of this sum, the beamstrahlung cross section is not simply proportional to some power of the fine structure constant, contrary to, for example, all the cases given in Chapters 9 and 10. Indeed, it is not a polynomial in the fine structure constant either.

Consider the exceptional term $n_i = n_f = 0$ just mentioned. Since it represents the process $e^- \to e^-\gamma$ without any external field, and, hence, must be zero due to energy-momentum conservation, it is also correct to include this term. With this additional term that contributes nothing, n_i and n_f are summed independently over all non-negative integers. These sums have very simple interpretations. The sum over n_i means that the incoming electron plane wave should be replaced by the electron wave function in the presence of the external electrostatic potential, and similarly the sum over n_f means the replacement of the outgoing electron plane wave by a similar wave function with the external potential. Fortunately, the determination of the necessary wave functions at high energies in the presence of the external potential was completely solved in the fifties [50,51] and are in a form suitable for the study of beamstrahlung.

The external potential cannot be translationally invariant, since it is due to an electron or positron bunch of finite extent. Therefore, it is more natural and simpler to express these wave functions using coordinate space rather than momentum space. In other words, the matrix element for the Feynman diagram, or rather the infinite set of Feynman diagrams, of Fig. 12.1 is most simply expressed as an integral over the product of the wave functions in coordinate space:

$$M_\mu(\vec{k}_i, \vec{k}_f, \vec{k}_\gamma) = -ie \int d^3\vec{x}\, e^{-i(\vec{k}_\gamma \cdot \vec{x})}\, \overline{\psi}_f(\vec{x})\gamma_\mu \psi_i(\vec{x}), \qquad (12.6)$$

where, as mentioned above, ψ_i and ψ_f are the initial and final electron wave functions in the presence of the external potential due to the positron bunch.

12.2.3 Important length scales

Before attempting to evaluate this matrix element, it is necessary to understand the various important parameters associated with the process (12.1), and to express the regime of interest to be studied in terms of these parameters.

Three characteristic lengths are particularly relevant. The first one is the correlation length $\overline{L}_c$. The notation here is to use a bar to denote a length in the laboratory system (which is also the c.m. system) where the incoming electron and positron have opposite momenta, while the corresponding quantity without a bar means the length in the system where the positron (electron) bunch is at rest. Thus, $\overline{L}_c$ and the bunch length $\overline{L}_b$ are both taken in the laboratory system, while

$$L_c = \gamma \overline{L}_c, \qquad L_b = \gamma \overline{L}_b \quad \text{and} \quad L_e = \gamma \overline{L}_e, \qquad (12.7)$$

where the virtual electron length $L_e = \gamma \overline{L}_e$ will be defined in the next paragraph. As already mentioned, when an electron (positron) passes through the oppositely moving positron (electron) bunch, it is deflected and focused by its electromagnetic interaction with the bunch as a whole. The correlation length $\overline{L}_c$ is defined as the distance it travels for a deflection in angle of $1/\gamma$, computed on the basis of a transverse distance R and an average charge density over $\overline{L}_b$.

The second characteristic length is the bunch length $\overline{L}_b$. The third one is the virtual electron length $\overline{L}_e$, that is the distance over which coherent radiation by the impinging electron is allowed before it enters the bunch. It is simply

$$\overline{L}_e = \gamma/m = \gamma r_e/\alpha, \qquad (12.8)$$

where m is the electron mass.

It may be instructive to give the approximate values of these characteristic lengths for the Super machine as described by (12.1). They are

$$\begin{aligned} \overline{L}_c &= 2 \times 10^{-10} \text{ m}, & L_c &= 2 \times 10^{-3} \text{ m}, \\ \overline{L}_b &= 3 \times 10^{-7} \text{ m}, & L_b &= 3 \text{ m}, \\ \overline{L}_e &= 4 \times 10^{-6} \text{ m}, & L_e &= 40 \text{ m}. \end{aligned} \qquad (12.9)$$

It is observed in this case that

$$\overline{L}_c \ll \overline{L}_b \ll \overline{L}_e. \qquad (12.10)$$

For the study of beamstrahlung at high energies, $\overline{L}_c$ is not really an appropriate characteristic length, because it is actually not a relevant correlation length. The relevant correlation length is instead [52]

$$\overline{\ell}_c = (\overline{L}_c^2 \overline{L}_e)^{\frac{1}{3}}, \qquad (12.11)$$

12.2. NATURE OF APPROXIMATIONS

or, equivalently,
$$\ell_c = (L_c^2 L_e)^{\frac{1}{3}}. \tag{12.12}$$

This relation can be seen most readily as follows. During the transverse deflection of the electron as seen in the laboratory system, the mass of the electron is the beam energy $\overline{E}$, not the rest mass m. Since $\overline{E} \gg m$, this correlation length ℓ_c is expected to be independent of m. In other words, since $\overline{L}_b$ is clearly unrelated to any correlation length, $\overline{\ell}_c$ must be of the form $\overline{L}_c^{1-a} \overline{L}_e^a$, where the power a is determined so that the power dependences of $\overline{L}_c$ and $\overline{L}_e$ on m cancel out. Since $\overline{L}_c$ is defined to be the distance travelled for a deflection in angle of $1/\gamma = m/\overline{E}$, $\overline{L}_c$ is proportional to m. On the other hand, it follows from eqn (12.8) that

$$\overline{L}_e = \overline{E}/m^2. \tag{12.13}$$

The required cancellation of the dependence on m yields immediately the result of eqn (12.12).

For the Super machine, it follows from the numerical values of eqns (12.9) that
$$\overline{\ell}_c = 5 \times 10^{-9} \text{ m}, \qquad \ell_c = 0.05 \text{ m}. \tag{12.14}$$

Thus, in this case,
$$\overline{\ell}_c \ll \overline{L}_b \ll \overline{L}_e. \tag{12.15}$$

Note that (12.15) implies (12.10), but (12.10) does not necessarily imply the relations (12.15).

Throughout the remainder of this chapter, the inequalities (12.15) will be assumed to hold.

12.2.4 Shape of bunch

It remains to choose the class of charge distributions for the bunch. The actual charge distributions are very complicated and depend on the detailed design of the accelerating structures of the linear collider. Since we do not wish to work with only the most idealized charge distribution, we reach the following compromise.

The transverse distribution of the charge is taken to be uniform and rotationally symmetric. In other words, the charge distribution in the transverse xy-plane is taken to be a constant (which, however, depends on the longitudinal variable z) when $x^2 + y^2 < R^2$, where the radius R is independent of z. For $x^2 + y^2 > R^2$, charges are absent.

The longitudinal distribution of the charge, however, is allowed to vary in a fairly arbitrary way. To avoid unnecessary complications, this longitudinal charge distribution is not allowed to be zero for any finite value of z. This innocent condition is amply satisfied for realistic bunches. For large values of $|z|$, the charge distribution is assumed to decrease exponentially. This is also quite realistic. For technical reasons, longitudinal distributions that approach zero for large $|z|$ as a Gaussian, or more generally faster than any exponential, are excluded from consideration. In any case, such rapid approaches to zero are believed not to occur for realistic bunches.

At the time when this chapter is being written, the case of the transversally nonuniform bunch has not been solved. This case is expected to have interesting features. If the bunch is rotationally symmetric, then an electron that goes through the bunch at the axis of symmetry is not deflected. At the other extreme, an electron that travels at a large distance from the axis experiences very little electromagnetic force and is thus deflected by only a small angle. There is therefore a distance from the axis where the deflection angle is maximal. At this distance, the matrix element, suitably defined, for the process (12.1) is especially large, and the method of treatment to be discussed here is not adequate. We can look forward to many interesting developments in the theory of beamstrahlung in the next few years.

12.3 Electron wave functions

The matrix element for the Feynman diagram of Fig. 12.1, after summation over all possible numbers of interactions with the external field, is given by eqn (12.6). In order to evaluate this expression for the matrix element, we need a knowledge of $\psi_i(\vec{x})$ and $\psi_f(\vec{x})$, the initial and final electron wave functions which are solutions of the Dirac equation in the presence of the external field.

Even though the external potentials are relatively simple for the bunches described in Section 12.2.4, it is nevertheless not possible to solve the Dirac equation explicitly. Fortunately, because of the very high energy of the incoming electron, especially in the frame where the bunch is at rest, the so-called high energy approximation [50,51] developed in the fifties can be used for beamstrahlung.

In applying this high energy approximation, the following point is especially important. As specified by the condition (12.15), the bunch length $\overline{L}_b$ is much longer than the correlation length $\overline{\ell}_c$. This implies that the angle of deflection of the electron by the positron bunch must be taken into account and cannot be neglected. In terms of the high energy approximation, this means that it is not sufficient to use the leading approximation [50]; rather the next-order approximation [51] must be employed.

The high energy approximation is slightly more complicated for the Dirac equation of spin-$\frac{1}{2}$ than the Klein-Gordon case of spin-0. Accordingly, before considering the realistic case of the Dirac equation in the next section, we first carry out the formulation for the spin-0 case.

The formalism is actually quite simple. Let the spin-0 wave function $\psi(\vec{x})$, which satisfies the Klein-Gordon equation [53]

$$\{[E - eV(\vec{x})]^2 + \nabla^2 - m^2\}\psi(\vec{x}) = 0, \qquad (12.16)$$

12.3. ELECTRON WAVE FUNCTIONS

be expressed in the form

$$\psi(\vec{x}) = e^{i\eta(\vec{x})} A(\vec{x}). \tag{12.17}$$

Substitution into eqn (12.16) then gives

$$\{[E - eV(\vec{x})]^2 - |\nabla \eta(\vec{x})|^2 + 2i\nabla \eta(\vec{x}) \cdot \nabla + i[\nabla^2 \eta(\vec{x})] + \nabla^2 - m^2\} A(\vec{x}) = 0. \tag{12.18}$$

In as much as the writing of $\psi(\vec{x})$ in terms of $A(\vec{x})$ and $\eta(\vec{x})$ is not unique, it is permissible to split eqn (12.18) into two equations. For the present purpose, it is especially convenient to choose the following splitting:

$$[E - eV(\vec{x})]^2 - |\nabla \eta(\vec{x})|^2 - m^2 = 0 \tag{12.19}$$

and

$$\{2\nabla \eta(\vec{x}) \cdot \nabla + [\nabla^2 \eta(\vec{x})] - i\nabla^2\} A(\vec{x}) = 0. \tag{12.20}$$

This choice has the advantage that the phase $\eta(\vec{x})$ is completely determined by eqn (12.19), since the amplitude $A(\vec{x})$ does not appear there. The equations (12.19) and (12.20) are exact. For application to beamstrahlung where the energy E is very large, both the phase $\eta(\vec{x})$ and the amplitude $A(\vec{x})$ can be expanded in powers of the momentum $k = (E^2 - m^2)^{\frac{1}{2}}$. This will be carried out later on in this section for the phase.

Once this high energy approximation is understood for the Klein-Gordon equation, it can also be used for the Dirac equation. Here, the electron wave function satisfies

$$[E - eV(\vec{x}) + i\vec{\alpha} \cdot \nabla - m\beta] \psi(\vec{x}) = 0, \tag{12.21}$$

where β is simply γ^0 and $\vec{\alpha}$ is as usual $\gamma^0 \vec{\gamma}$. Once again, $\psi(\vec{x})$ is expressed in the form of eqn (12.17)

$$\psi(\vec{x}) = e^{i\eta(\vec{x})} A(\vec{x}), \tag{12.22}$$

except that here $A(\vec{x})$ is a four-component spinor. It remains to choose convenient equations for the phase $\eta(\vec{x})$ and the amplitude $A(\vec{x})$.

The choice is: *use the same phase $\eta(\vec{x})$ for the Dirac equation as for the Klein-Gordon equation.* In other words, the choice is for the quantity $\eta(\vec{x})$ here also to satisfy eqn (12.19).

The basic reason for this choice is that this phase $\eta(\vec{x})$ is closely related to the classical path of the particle under consideration. Because the classical path, and hence this phase, is determined by classical optics, they are independent of the spin of the particle.

The substitution of the form (12.22) into the Dirac equation (12.21) gives simply

$$[E - eV(\vec{x}) - \vec{\alpha} \cdot \nabla \eta(\vec{x}) + i\vec{\alpha} \cdot \nabla - m\beta] A(\vec{x}) = 0. \tag{12.23}$$

With $\eta(\vec{x})$ known, this determines the amplitude $A(\vec{x})$. We proceed to determine approximately from eqn (12.19) the phase $\eta(\vec{x})$, common to the Klein-Gordon equation and the Dirac equation. Let $\vec{k}$ be the momentum of the electron, so that the phase of the incident wave function is $\vec{k}\cdot\vec{x}$. Without loss of generality, let this momentum $\vec{k}$ be in the xz-plane. For application to beamstrahlung, where z is the direction of motion of the electrons in the bunch, the case of interest is $k_z \gg k_x$.

Expand $\eta(\vec{x})$ to second order in the sense that [50,51]

$$\eta(\vec{x}) = \vec{k}\cdot\vec{x} + \eta_0(\vec{x}) + \frac{1}{k}\eta_1(\vec{x}) + \mathcal{O}(\frac{1}{k^2}) \qquad (12.24)$$

in the limit of large k. Then, as $z \to -\infty$, the boundary conditions are

$$\eta_0(\vec{x}) \to 0 \quad \text{and} \quad \eta_1(\vec{x}) \to 0. \qquad (12.25)$$

The gradient of $\eta(\vec{x})$, as given by eqn (12.24) is

$$\nabla\eta = \vec{k} + \nabla\eta_0 + \frac{1}{k}\nabla\eta_1, \qquad (12.26)$$

and hence

$$|\nabla\eta|^2 = k^2 + 2\vec{k}\cdot\nabla\eta_0 + 2\hat{k}\cdot\eta_1 + |\nabla\eta_0|^2$$

$$= k^2 + 2k\frac{\partial\eta_0}{\partial z} + \left(2\frac{\partial\eta_1}{\partial z} + |\nabla\eta_0|^2 + 2k_x\frac{\partial\eta_0}{\partial x}\right). \qquad (12.27)$$

Substitution into eqn (12.19) then gives

$$-2EeV + e^2V^2 - 2k\frac{\partial\eta_0}{\partial z} - \left(2\frac{\partial\eta_1}{\partial z} + |\nabla\eta_0|^2 + 2k_x\frac{\partial\eta_0}{\partial x}\right) = 0. \qquad (12.28)$$

Since $E = k + \mathcal{O}(k^{-1})$, the terms of order k and order 1 give respectively

$$\frac{\partial\eta_0}{\partial z} = -eV \qquad (12.29)$$

and

$$\frac{\partial\eta_1}{\partial z} = -k_x\frac{\partial\eta_0}{\partial x} - \frac{1}{2}\left[\left(\frac{\partial\eta_0}{\partial x}\right)^2 + \left(\frac{\partial\eta_0}{\partial y}\right)^2\right]. \qquad (12.30)$$

It is now merely a matter of integrating over z to get $\eta(\vec{x})$ to the desired accuracy. With the boundary conditions (12.25), the result is

$$\eta(\vec{x}) = \vec{k}\cdot\vec{x} - S(x,y,z) + \frac{1}{2k}\int_{-\infty}^{z} dz'\left\{2k_x(z-z')\frac{\partial eV(x,y,z')}{\partial z'}\right.$$

$$\left. - \left[\frac{\partial S(x,y,z')}{\partial x}\right]^2 - \left[\frac{\partial S(x,y,z')}{\partial y}\right]^2\right\}, \qquad (12.31)$$

where
$$S(x,y,z) = \int_{-\infty}^{z} dz' \, eV(x,y,z'). \tag{12.32}$$

The terms in eqn (12.31) have simple interpretations. The first one, $\vec{k} \cdot \vec{x}$, is that of the incident wave function. The second one, $-S(x,y,z)$, is the additional phase shift for a classical trajectory parallel to the z-axis. The last term, proportional to k^{-1}, is the correction to the phase shift due to the bending of the classical trajectory in the external field. Since the bunch length $\bar{L}_b$ is much larger than the correlation length $\bar{\ell}_c$, this $1/k$ term must be kept; otherwise the coherence is much overestimated. This explicit solution (12.31) shows clearly that the high energy approximation is simpler than the WKB approximation [54] or physical optics.

It only remains to determine the amplitude $A(\vec{x})$. For the Klein–Gordon case, the leading-order approximation to eqn (12.20) is simply

$$\frac{\partial A}{\partial z} = 0, \tag{12.33}$$

because $\nabla \eta(\vec{x}) \sim \vec{k}$ gives the only piece proportional to $\vec{k}$. A comparison with the incident plane wave $\exp(i\vec{k} \cdot \vec{x})$ leads to the leading-order approximation $A(\vec{x}) = 1$. Similarly, for the Dirac case, it follows from the helicity method that, with m neglected and the incident plane wave in a helicity state, the leading-order approximation is

$$A(\vec{x}) = \text{helicity state along the classical trajectory.} \tag{12.34}$$

In other words, the deviation of the classical trajectory from the line parallel to the z-axis needs to be taken into account, just as it is in eqn (12.31). Explicitly, if the incident plane wave is

$$\psi^{\text{inc}}(\vec{x}) = e^{i\vec{k} \cdot \vec{x}} u_0, \tag{12.35}$$

where u_0 is a spinor describing the electron, then eqn (12.34) is

$$A(x,y,z) = \left\{ 1 + \frac{1}{2k} \vec{\alpha}_\perp \cdot [\nabla_\perp S(x,y,z)] \right\} u_0, \tag{12.36}$$

where the subscript $\perp$ refers to the transverse x-y directions. A first-order perturbative calculation gives a slightly more complicated expression

$$A(x,y,z) = \left\{ 1 + \frac{1}{2k} \left[\vec{\alpha}_\perp \cdot \nabla_\perp S(x,y,z) \right.\right.$$
$$\left.\left. + \int_{-\infty}^{z} dz' \left(\frac{\partial^2}{\partial x^2} + \frac{\partial^2}{\partial y^2} \right) S(x,y,z') \right] \right\} u_0. \tag{12.37}$$

Beamstrahlung calculations have mostly been based on the simpler approximation (12.36), because the additional term in (12.37) is spin independent and hence contributes little in the case where the disruption factor is small. As already discussed in Section 12.2.1, this condition of small disruption factor is needed in order to apply the single-particle approximation that leads to the Feynman diagram of Fig. 12.1.

12.4 Cross section for $e + \text{bunch} \to e + \gamma + \text{bunch}$

It is the purpose of this section to obtain an explicit expression for the e
distribution of the beamstrahlung photon on the basis of the electron
functions of Section 12.3. In the next section, this explicit expression
be evaluated asymptotically under the condition (12.15). The bunch
is the one described in Section 12.2.4, namely it is longitudinally nonun
but transversely uniform. The simplest example of such a charge distrib
is

$$\text{sech}^2(2z/L_b). \tag{1}$$

Such a distribution, with exponential decrease for large $|z|$, is actually
a good approximation for realistic bunches in electron-positron linear c
ers when complications such as those due to head-tail interaction are 1
control. In fact, the distribution (12.38) is often a better approximation
a Gaussian.

12.4.1 Electrostatic potential

When the relative motion of the positrons in the bunch is neglected, the
external potential experienced by the incident electron is electrostatic i
frame where the bunch is at rest. Thus the Dirac equation for the ele
indeed takes the form of eqn (12.21).

In this frame where the bunch is at rest, let L_b be the nominal b
length. The average charge density is then

$$\frac{Ne}{\pi L_b R^2}. \tag{1}$$

Let this bunch length L_b be used as the scale along the bunch, taken to
the z-direction, then we define a normalized charge density ρ by expre
the actual charge density ρ_0 as

$$\rho_0(z) = \frac{Ne}{\pi L_b R^2} \rho\left(\frac{z}{L_b}\right). \tag{1}$$

The electrostatic potential for such a charge distribution is

$$V(x,y,z) = \frac{Ne}{\pi L_b R^2} \int_{-\infty}^{\infty} dz' \rho\left(\frac{z'}{L_b}\right)$$

$$\times \int dx' \, dy' \, [(x-x')^2 + (y-y')^2 + (z-z')^2]^{-\frac{1}{2}}, \tag{1}$$

where the x' and y' integrations are over a circle $x'^2 + y'^2 < R^2$.
R is very much smaller than L_b (by nearly ten orders of magnitude i
case of the Super machine for example), the charge distribution $\rho(z'/L_b)$
be replaced very accurately by $\rho(z/L_b)$ in the expression for the diffe

12.4. CROSS SECTION FOR $e + bunch \to c + \gamma + bunch$

$V(x, y, z) - V(0, 0, z)$. With this approximation, $V(x, y, z)$ is given by, for $x^2 + y^2 \leq R^2$,

$$eV(x, y, z) = eV(0, 0, z) + \frac{N\alpha}{L_b} \frac{x^2 + y^2}{R^2} \rho\left(\frac{z}{L_b}\right), \quad (12.42)$$

where, by eqn (12.41),

$$V(0, 0, z) = \frac{2Ne}{L_b R^2} \int_{-\infty}^{\infty} dz' \rho\left(\frac{z'}{L_b}\right) \left\{ [R^2 + (z - z')^2]^{\frac{1}{2}} - |z - z'| \right\}. \quad (12.43)$$

Even though the two terms on the right-hand side of eqn (12.42) are comparable in magnitude, the first term is independent of x and y and, hence, leads to a constant phase shift in the matrix element. Such a constant phase shift is of no consequence and, therefore, it is sufficient to use

$$eV(r, z) = \frac{N\alpha}{L_b} \frac{r^2}{R^2} \rho\left(\frac{z}{L_b}\right), \quad (12.44)$$

where of course $r^2 = x^2 + y^2$.

12.4.2 Initial and final phases

With eqn (12.22) applied to initial and final wave functions,

$$\psi_i(\vec{x}) = e^{i\phi_i(\vec{x})} A_i(\vec{x}),$$

$$\psi_f(\vec{x}) = e^{i\phi_f(\vec{x})} A_f(\vec{x}), \quad (12.45)$$

the matrix element of eqn (12.6) is

$$M_\mu(\vec{k}_i, \vec{k}_f, \vec{k}_\gamma) = -ie \int_{-\infty}^{\infty} dz \int dx\, dy\, e^{i\phi(\vec{x})} \overline{A}_f(\vec{x}) \gamma_\mu A_i(\vec{x}), \quad (12.46)$$

where

$$\phi(\vec{x}) = \phi_i(\vec{x}) - \phi_f(\vec{x}) - \vec{k}_\gamma \cdot \vec{x}, \quad (12.47)$$

and the x-y integration is again over the circle $x^2 + y^2 < R^2$. The reason why it is unnecessary to consider $r > R$ is that the transverse size of the electron beam is also taken to be R and, hence, there is no incident electron where $r > R$.

In order to make use of eqn (12.31) to write down the phases $\phi_i(\vec{x})$ and $\phi_f(\vec{x})$ for the electrostatic potential (12.31), it is convenient to define

$$\tau(\xi) = \int_{-\infty}^{\xi} d\xi' \rho(\xi'),$$

$$T(\xi) = \int_{-\infty}^{\xi} d\xi' [\tau(\xi')]^2,$$

$$U(\xi) = \int_{-\infty}^{\xi} d\xi'\, \tau(\xi'),$$

$$\tau_f(\xi) = \int_{\xi}^{\infty} d\xi'\, \rho(\xi'),$$

$$T_f(\xi) = \int_{\xi}^{\infty} d\xi'\, [\tau_f(\xi')]^2,$$

$$U_f(\xi) = \int_{\xi}^{\infty} d\xi'\, \tau_f(\xi'), \qquad (12.48)$$

where

$$\xi = z/L_b. \qquad (12.49)$$

Note that because of the way $\rho(\xi)$ is normalized, $\tau(\xi)$ has the property that $\tau(\xi) \to 1$ as $\xi \to \infty$. This implies a simple relation between $\tau(\xi)$ and $\tau_f(\xi)$, namely

$$\tau(\xi) + \tau_f(\xi) = 1. \qquad (12.50)$$

Since the incoming electron momentum is along the z-direction, it follows from eqn (12.31) that $\phi_i(\vec{x})$ is simply

$$\phi_i(\vec{x}) = k_i z - N\alpha \frac{r^2}{R^2}\left[\tau\left(\frac{z}{L_b}\right) + \frac{2}{k_i}\frac{N\alpha L_b}{R^2} T\left(\frac{z}{L_b}\right)\right]. \qquad (12.51)$$

For the phase $\phi_f(\vec{x})$, it is necessary to reverse the z-axis in eqn (12.31), leading to

$$\phi_f(\vec{x}) = \vec{k}_f \cdot \vec{x} + S_f(x,y,z) + \frac{1}{2k_f}\int_z^{\infty} dz'\bigg\{2k_{fx}(z'-z)\frac{\partial eV(x,y,z')}{\partial x}$$

$$- \left[\frac{\partial S_f(x,y,z')}{\partial x}\right]^2 - \left[\frac{\partial S_f(x,y,z')}{\partial y}\right]^2\bigg\}, \qquad (12.52)$$

where

$$S_f(x,y,z) = \int_z^{\infty} dz'\, eV(x,y,z'), \qquad (12.53)$$

and $\vec{k}_f$ has been taken to be in the xz-plane. By eqn (12.44) and the definitions (12.48), here

$$S_f(x,y,z) = N\alpha \frac{r^2}{R^2} \tau_f\left(\frac{z}{L_b}\right), \qquad (12.54)$$

and

$$\phi_f(\vec{x}) = \vec{k}_f \cdot \vec{x} + N\alpha \frac{r^2}{R^2}\tau_f\left(\frac{z}{L_b}\right) + \frac{2N\alpha L_b}{k_f R^2}\left[k_{fx} x U_f\left(\frac{z}{L_b}\right) + N\alpha \frac{r^2}{R^2} T_f\left(\frac{z}{L_b}\right)\right]. \qquad (12.55)$$

12.4. CROSS SECTION FOR $e + bunch \to e + \gamma + bunch$

Substitution of eqns (12.51) and (12.55) into the definition (12.47) gives

$$\phi(\vec{x}) = k_i z - (\vec{k}_f + \vec{k}_\gamma) \cdot \vec{x} - N\alpha \frac{r^2}{R^2} - \frac{2}{k_i}(N\alpha)^2 \frac{r^2}{R^4} L_b T\left(\frac{z}{L_b}\right)$$

$$- \frac{2N\alpha L_b}{k_f R^2}\left[k_{fx} x U_f\left(\frac{z}{L_b}\right) + N\alpha \frac{r^2}{R^2} T_f\left(\frac{z}{L_b}\right)\right]. \qquad (12.56)$$

The first three terms on the right-hand side of eqn (12.56), namely

$$(k_i - k_{fz} - k_{\gamma z})z - (k_{fx} + k_{\gamma x})x - k_{\gamma y} y - \frac{N\alpha}{R^2}(x^2 + y^2) \qquad (12.57)$$

reach a maximum value at

$$x = x_0 = -\frac{R^2}{2N\alpha}(k_{fx} + k_{\gamma x}),$$

$$y = y_0 = -\frac{R^2}{2N\alpha}k_{\gamma y}. \qquad (12.58)$$

This point lies within the range of the x-y integration when

$$(k_{fx} + k_{\gamma x})^2 + k_{\gamma y}^2 < \left(\frac{2N\alpha}{R}\right)^2. \qquad (12.59)$$

Under this circumstance, these x-y integrations can be carried out with the method of stationary phase to give for the quantity M_μ of eqn (12.46)

$$M_\mu(\vec{k}_i, \vec{k}_f, \vec{k}_\gamma) = -\frac{e\pi R^2}{N\alpha} \int_{-\infty}^{\infty} dz\, e^{i\phi_0(z)} \overline{A}_f(\vec{x})\gamma_\mu A_i(\vec{x})\bigg|_{\substack{x=x_0 \\ y=y_0}}. \qquad (12.60)$$

where

$$\phi_0(z) = \phi(x_0, y_0, z). \qquad (12.61)$$

In order to get an explicit formula for this $\phi_0(z)$, it is useful to express U_f and T_f, which appear in eqn (12.56), in terms of U and T. This is most easily accomplished by differentiating the definitions (12.48):

$$T'(\xi) = -[\tau_f(\xi)]^2 = -\tau^2(\xi) + 2\tau(\xi) - 1$$

$$= [-T(\xi) + 2U(\xi) - \xi]' \qquad (12.62)$$

and

$$U'_f(\xi) = -\tau_f(\xi) = -1 + \tau(\xi) = [-\xi + U(\xi)]'. \qquad (12.63)$$

Therefore,

$$T_f(\xi) = -T(\xi) + 2U(\xi) - \xi + C_T \qquad (12.64)$$

and
$$U_f(\xi) = U(\xi) - \xi + C_U, \tag{12.65}$$

where C_T and C_U are two constants. These constants are given by

$$C_U = \int_{-\infty}^{\infty} d\xi\, \xi\, \rho(\xi) \tag{12.66}$$

and

$$C_T = 2\int_{-\infty}^{\infty} d\xi\, \xi\, \rho(\xi)\, \tau_f(\xi), \tag{12.67}$$

but these explicit expressions are not needed.

Let X be the fraction of incident energy carried by the outgoing photon. Thus,

$$k_\gamma = X k_i \quad \text{and} \quad k_f = (1-X)k_i. \tag{12.68}$$

In terms of this quantity X, the longitudinal momentum transfer is

$$k_i - k_{fz} - k_{\gamma z} = \frac{1}{2X(1-X)k_i}[X k_{fx}^2 + (1-X)(k_{\gamma x}^2 + k_{\gamma y}^2) + X^2 m^2]. \tag{12.69}$$

Here, the last term proportional to m^2 is kept only for the purpose of avoiding a logarithmic divergence, which becomes $\log L_e$ as we shall see later. The substitution of eqns (12.58) and (12.64)–(12.69) into eqn (12.56) gives

$$\phi_0(z) = \frac{L_b}{2X(1-X)k_i}\Big\{[(k_{fx}+k_{\gamma x})^2 + k_{\gamma y}^2] X^2\, T\Big(\frac{z}{L_b}\Big)$$
$$-2[k_{\gamma x}(k_{fx}+k_{\gamma x}) + k_{\gamma y}^2] X\, U\Big(\frac{z}{L_b}\Big) + (k_{\gamma x}^2 + k_{\gamma y}^2 + X^2 m^2)\frac{z}{L_b}\Big\} + \phi_{00}, \tag{12.70}$$

where ϕ_{00} is an irrelevant constant given by

$$\phi_{00} = -\frac{L_b}{(1-X)k_i}\Big\{k_{fx}(k_{fx}+k_{\gamma x})C_U - [(k_{fx}+k_{\gamma x})^2 + k_{\gamma y}^2]\Big[C_T - \frac{R^2(1-X)k_i}{4N\alpha L_b}\Big]\Big\}. \tag{12.71}$$

12.4.3 Integration over transverse momenta

Beginning with the expression (12.60) for the matrix element for the beamstrahlung process (12.5), we calculate the square of the matrix element and sum over spin. It is assumed that the electron and the positron bunches are both rotationally symmetrical and uniform in the transverse directions up to a radius R, and, furthermore, that they are perfectly aligned so that the axes of the bunches coincide. Further integrations over the transverse momenta of the outgoing electron and outgoing photon then give the energy distribution

12.4. CROSS SECTION FOR $e + bunch \to e + \gamma + bunch$

of the beamstrahlung photon as

$$I(X) = \frac{\alpha}{2\pi^2 k_i^2} \left[\frac{R}{2N\alpha}\right]^2 s(X) X^3 (1-X)$$
$$\times \int_{-\infty}^{\infty} dz \int_{-\infty}^{\infty} dz' F(z,z') \exp\left[i\frac{Xm^2(z-z')}{2(1-X)k_i}\right], \quad (12.72)$$

where $s(X)$ is the spin factor

$$s(X) = \frac{2 - 2X + X^2}{2(1-X)}, \quad (12.73)$$

and $F(z, z')$ involves the integration over transverse momenta:

$$F(z,z') = \int k_{fx}\, dk_{fx}\, dk_{\gamma x}\, dk_{\gamma y}\, e^{i\phi(z,z')}$$
$$\times \left\{ \left[(k_{fx} + k_{\gamma x})X\,\tau\!\left(\frac{z}{L_b}\right) - k_{\gamma x}\right]\left[(k_{fx} + k_{\gamma x})X\,\tau\!\left(\frac{z'}{L_b}\right) - k_{\gamma x}\right]\right.$$
$$\left. + k_{\gamma y}^2 \left[X\,\tau\!\left(\frac{z}{L_b}\right) - 1\right]\left[X\,\tau\!\left(\frac{z'}{L_b}\right) - 1\right]\right\}, \quad (12.74)$$

with

$$\phi(z,z') = [\phi_0(z) - \phi_0(z')]\big|_{m=0}. \quad (12.75)$$

If the spin of the electron were zero, then $s(X)$ would simply be 1. In eqn (12.72), the two z and z' integrations come from the matrix element and its complex conjugate. The region of integration for eqn (12.74) is given by (12.59).

It is the purpose of this section to carry out the integration on the right-hand side of eqn (12.74) in order to get a simple expression for $F(z, z')$.

By eqn (12.70) for ϕ_0, the exponent in eqn (12.74) is

$$\phi(z,z') = [(k_{fx}+k_{\gamma x})^2 + k_{\gamma y}^2]A + [k_{\gamma x}(k_{fx}+k_{\gamma x}) + k_{\gamma y}^2]B + (k_{\gamma x}^2 + k_{\gamma y}^2)C, \quad (12.76)$$

where

$$A = \frac{XL_b}{2(1-X)k_i}\left[T\!\left(\frac{z}{L_b}\right) - T\!\left(\frac{z'}{L_b}\right)\right],$$
$$B = -\frac{L_b}{(1-X)k_i}\left[U\!\left(\frac{z}{L_b}\right) - U\!\left(\frac{z'}{L_b}\right)\right],$$
$$C = \frac{1}{2X(1-X)k_i}(z - z'). \quad (12.77)$$

With this notation, the quantity $F(z, z')$ takes the form

$$F(z, z') = -i\left\{X^2\tau\left(\frac{z}{L_b}\right)\tau\left(\frac{z'}{L_b}\right)\frac{\partial}{\partial A}\right.$$
$$\left. -X\left[\tau\left(\frac{z}{L_b}\right)+\tau\left(\frac{z'}{L_b}\right)\right]\frac{\partial}{\partial B}+\frac{\partial}{\partial C}\right\}F_0(z,z'), \quad (12.78)$$

where

$$F_0(z, z') = \int k_{fx}\, dk_{fx}\, dk_{\gamma x}\, dk_{\gamma y}\, e^{i\phi(z,z')}, \quad (12.79)$$

still with the region (12.59) of integration. By eqn (12.76), this F_0 is

$$F_0(z, z') = (2\pi)^{-1}\int dk_{fx}\, dk_{fy}\, dk_{\gamma x}\, dk_{\gamma y}$$

$$\times \exp i\{[(k_{fx}+k_{\gamma x})^2 + (k_{fy}+k_{\gamma y})^2]A$$

$$+[k_{\gamma x}(k_{fx}+k_{\gamma x})+k_{\gamma y}(k_{fy}+k_{\gamma y})]B + (k_{\gamma x}^2+k_{\gamma y}^2)C\}. \quad (12.80)$$

It is now clear that a more natural variable is

$$\vec{k}_{T\perp} = \vec{k}_{f\perp} + \vec{k}_{\gamma\perp}. \quad (12.81)$$

With this variable, the quantity $F_0(z, z')$ of eqn (12.80) is

$$F_0(z, z') = (2\pi)^{-1}\int dk_{Tx}\, dk_{Ty}\, dk_{\gamma x}\, dk_{\gamma y}$$

$$\times \exp i[(k_{Tx}+k_{Ty})^2 A + (k_{Tx}k_{\gamma x}+k_{Ty}k_{\gamma y})B + (k_{\gamma x}^2+k_{\gamma y}^2)C], \quad (12.82)$$

where the region of integration (12.59) is simply

$$k_{Tx}^2 + k_{Ty}^2 < \left(\frac{2N\alpha}{R}\right)^2, \quad (12.83)$$

independent of $\vec{k}_{\gamma\perp}$. Therefore,

$$F_0(z, z') = -\frac{2\pi}{4AC-B^2}\left\{1 - \exp\left[i\left(\frac{2N\alpha}{R}\right)^2 \frac{4AC-B^2}{4C}\right]\right\}. \quad (12.84)$$

It remains to apply the differential operator on the right-hand side of eqn (12.78) to this expression for $F_0(z, z')$. Only after differentiation, the

12.4. CROSS SECTION FOR $e + bunch \to e + \gamma + bunch$

explicit values of A, B, C as given by eqns (12.77) can be used. But, eqns (12.77) imply that

$$\frac{\partial}{\partial z} = \frac{1}{2X(1-X)k_i}\left[X^2\tau\left(\frac{z}{L_b}\right)^2\frac{\partial}{\partial A} - 2X\tau\left(\frac{z}{L_b}\right)\frac{\partial}{\partial B} + \frac{\partial}{\partial C}\right],$$

$$\frac{\partial}{\partial z'} = \frac{-1}{2X(1-X)k_i}\left[X^2\tau\left(\frac{z'}{L_b}\right)^2\frac{\partial}{\partial A} - 2X\tau\left(\frac{z'}{L_b}\right)\frac{\partial}{\partial B} + \frac{\partial}{\partial C}\right]. \quad (12.85)$$

Therefore,

$$X^2\tau\left(\frac{z}{L_b}\right)\tau\left(\frac{z'}{L_b}\right)\frac{\partial}{\partial A} - X\left[\tau\left(\frac{z}{L_b}\right) + \tau\left(\frac{z'}{L_b}\right)\right]\frac{\partial}{\partial B} + \frac{\partial}{\partial C}$$

$$= -\tfrac{1}{2}X^2\left[\tau\left(\frac{z}{L_b}\right) - \tau\left(\frac{z'}{L_b}\right)\right]^2\frac{\partial}{\partial A} + X(1-X)k_i\left(\frac{\partial}{\partial z} - \frac{\partial}{\partial z'}\right). \quad (12.86)$$

This operator is to be applied to the quantity $F_0(z, z')$ of eqn (12.84) to give $F(z, z')$, which is then substituted into eqn (12.72) to yield the energy distribution of the beamstrahlung photon. For this purpose, integration by parts can be used to deal with the last term on the right-hand side of eqn (12.86), once with respect to z and once with respect to z'. Since the ranges of integration for both variables are the entire real line, the integrated terms vanish. Moreover, because of the structure of the integrand in eqn (12.72), the result of the integrations by parts is proportional to m^2 and is hence negligible. With this approximation, $F(z, z')$ is given by

$$F(z, z') = -\pi i X^2\left[\tau\left(\frac{z}{L_b}\right) - \tau\left(\frac{z'}{L_b}\right)\right]^2$$

$$\times \frac{\partial}{\partial A}\frac{1}{4AC - B^2}\left\{1 - \exp\left[i\left(\frac{2N\alpha}{R}\right)^2\frac{4AC - B^2}{4C}\right]\right\}. \quad (12.87)$$

A more convenient form is

$$F(z, z') = -\frac{\pi i X^2}{4C}\left[\tau\left(\frac{z}{L_b}\right) - \tau\left(\frac{z'}{L_b}\right)\right]^2 \int_0^\omega d\eta\, \eta\, \exp\left[i\eta\frac{4AC - B^2}{4C}\right], \quad (12.88)$$

with

$$\omega = \left(\frac{2N\alpha}{R}\right)^2. \quad (12.89)$$

If we define

$$W(\xi, \xi') = T(\xi) - T(\xi') - (\xi - \xi')^{-1}[U(\xi) - U(\xi')]^2, \quad (12.90)$$

then it is readily verified that this function has an interesting symmetry property in that it can be expressed equally well in terms of the functions $T_f(\xi)$ and $U_f(\xi)$ of eqns (12.48):

$$-W(\xi,\xi') = T_f(\xi) - T_f(\xi') - (\xi - \xi')^{-1}[U_f(\xi) - U_f(\xi')]^2, \quad (12.91)$$

as a consequence of eqns (12.64) and (12.65). With this quantity W, the substitution of eqns (12.77) and (12.88) into eqn (12.72) gives the energy distribution of the beamstrahlung photon as

$$I(X) = \frac{-i\alpha}{2(2\pi)k_i}\left[\frac{R}{2N\alpha}\right]^2 s(X) \int_{-\infty}^{\infty} dz \int_{-\infty}^{\infty} dz' (z-z')^{-1}\left[\tau\left(\frac{z}{L_b}\right) - \tau\left(\frac{z'}{L_b}\right)\right]^2$$

$$\times \int_0^\omega d\eta\, \eta \exp\left\{i\frac{X}{2(1-X)k_i}\left[m^2(z-z') + \eta L_b W\left(\frac{z}{L_b},\frac{z'}{L_b}\right)\right]\right\}, \quad (12.92)$$

where ω is given by eqn (12.89), k_i is the energy of the electron or positron under consideration, X is the fractional energy of the radiated photon (i.e., the photon energy divided by k_i) and $s(X)$ is given by eqn (12.73).

In the remainder of this chapter, this quantity $I(X)$ is to be evaluated under the assumption (12.15). In terms of this $I(X)$, the average fractional beamstrahlung energy loss is

$$\delta = \int_0^1 dX\, X\, I(X). \quad (12.93)$$

12.5 Energy distribution of beamstrahlung photon

The expression (12.92) actually depends on the quantities k_i and L_b only through the ratio L_b/k_i. It is therefore the same in the laboratory as in the frame where the bunch is at rest. To make this explicit, it is only necessary to scale z by L_b. If η is scaled by $(2N\alpha/R)^2$, then

$$I(X) = \frac{-i\alpha}{\pi}\Lambda^3 s(X) \int_{-\infty}^{\infty} d\xi \int_{-\infty}^{\infty} d\xi'\, (\xi-\xi')^{-1}[\tau(\xi)-\tau(\xi')]^2 e^{i\beta(\xi-\xi')}$$

$$\times \int_0^1 d\eta\, \eta \exp\left[i\eta\frac{2X}{1-X}\Lambda^3 W(\xi,\xi')\right], \quad (12.94)$$

where

$$\beta = \frac{Xm^2 L_b}{2(1-X)k_i} = \frac{X}{1-X}\frac{L_b}{L_e} \quad (12.95)$$

and

$$\Lambda = \left[\frac{N^2\alpha^2 L_b}{k_i R^2}\right]^{\frac{1}{3}} = \frac{L_b}{\ell_c}. \quad (12.96)$$

12.5. ENERGY DISTRIBUTION OF BEAMSTRAHLUNG PHOTON

In order to evaluate the right-hand side of eqn (12.94) asymptotically for large values of Λ, it is necessary to know the region where $W(\xi, \xi')$ is positive. It follows from the Schwarz inequality

$$\left[\int_{\xi'}^{\xi} d\xi''\, \tau(\xi'')\right]^2 \leq \left[\int_{\xi'}^{\xi} d\xi''\, [\tau(\xi'')]^2\right]\left[\int_{\xi'}^{\xi} d\xi''\right] \qquad (12.97)$$

that

$$[U(\xi) - U(\xi')]^2 \leq (\xi - \xi')[T(\xi) - T(\xi')]. \qquad (12.98)$$

If the observation is made that the longitudinal charge distribution is nowhere zero inside any realistic bunch, then $\tau(\xi)$ is a strictly increasing function of ξ inside the bunch, and for eqn (12.98) the equality sign holds only for $\xi = \xi'$. Therefore,

$$\text{sign of } W(\xi, \xi') = \text{sign of } (\xi - \xi'). \qquad (12.99)$$

This property (12.99) is of central importance in the development of this section.

12.5.1 Mellin transform

We follow the procedure of Jacob and Wu [48].

In order to determine the asymptotic behaviour of $I(X)$ with the conditions (12.15), we employ the method of Mellin transforms, which has been used to analyse the high energy behaviour of Feynman diagrams [55,56]. The Mellin transform of (12.94) is to be carried out with respect to the variable Λ, defined by eqn (12.96). Thus, we consider the photon spectrum as a function of Λ with the parameter X, rather than a function of X with the parameter Λ. With this in mind, we write

$$I(X) = \frac{\alpha}{\pi} s(X) K(\Lambda), \qquad (12.100)$$

which defines $K(\Lambda)$. Of course, $I(X)$ and $K(\Lambda)$ are actually functions of both X and Λ.

With this notation, define the Mellin transform

$$\overline{K}(\zeta) = \int_0^\infty d\Lambda\, \Lambda^{-1-\zeta} K(\Lambda). \qquad (12.101)$$

Since, as seen from eqn (12.94),

$$K(\Lambda) = \mathcal{O}(\Lambda) \qquad (12.102)$$

as $\Lambda \to \infty$, and

$$K(\Lambda) = \mathcal{O}(\Lambda^3) \qquad (12.103)$$

as $\Lambda \to 0$, the $\overline{K}(\zeta)$ as defined by eqn (12.101) is analytic for

$$3 > \text{Re}\,\zeta > 1. \qquad (12.104)$$

The residue of $\overline{K}(\zeta)$ at $\zeta = 1$ gives the leading behaviour of $K(\Lambda)$, and hence of $I(X)$, in the limit $\ell_c \ll L_b$. This residue is computed in the next section. Higher-order terms, including the end and nonuniformity effects, are to be extracted from the behaviour of the analytic continuation of $\overline{K}(\zeta)$ at $\zeta = 0$ and $\zeta = -1$ — see Sections 12.5.4 and 12.5.6, respectively. [The behaviour of $\overline{K}(\zeta)$ near $\zeta = 3$, on the contrary, is of no interest. It gives merely the behaviour of $I(X)$ for the opposite limit $\ell_c \gg L_b$.]

Remaining in the region (12.104), we can substitute (12.94) into (12.101) to get an explicit expression for $\overline{K}(\zeta)$. The Λ integral is

$$\int_0^\infty d\Lambda\, \Lambda^{-1-\zeta}\, \Lambda^3 \exp\left[i\eta \frac{2X}{1-X}\Lambda^3 W(\xi,\xi')\right]$$

$$= \begin{cases} \frac{1}{3}ie^{-i\pi\zeta/6}\Gamma(1-\zeta/3)\left[\eta\dfrac{2X}{1-X}W(\xi,\xi')\right]^{-1+\zeta/3} & \text{for } W(\xi,\xi') > 0, \\[1em] -\frac{1}{3}ie^{i\pi\zeta/6}\Gamma(1-\zeta/3)\left[-\eta\dfrac{2X}{1-X}W(\xi,\xi')\right]^{-1+\zeta/3} & \text{for } W(\xi,\xi') < 0. \end{cases}$$

(12.105)

Because of (12.99), this can be rewritten as

$$\int_0^\infty d\Lambda\, \Lambda^{-1-\zeta}\, \Lambda^3 \exp\left[i\eta \frac{2X}{1-X}\Lambda^3 W(\xi,\xi')\right]$$

$$= \tfrac{1}{3}i\,\mathrm{sg}(\xi-\xi')e^{-i\pi(\zeta/6)\mathrm{sg}(\xi-\xi')}\Gamma(1-\zeta/3)\left[\eta\frac{2X}{1-X}|W(\xi,\xi')|\right]^{-1+\zeta/3}.$$

(12.106)

The substitution into eqn (12.101) then gives, using eqns (12.100) and (12.94),

$$\overline{K}(\zeta) = \tfrac{1}{3}\Gamma(1-\zeta/3)\left[\frac{2X}{1-X}\right]^{-1+\zeta/3}$$

$$\times \int_{-\infty}^\infty d\xi \int_{-\infty}^\infty d\xi'\, |\xi-\xi'|^{-1}[\tau(\xi)-\tau(\xi')]^2$$

$$\times e^{i\beta(\xi-\xi')} e^{-i\pi(\zeta/6)\mathrm{sg}(\xi-\xi')} |W(\xi,\xi')|^{-1+\zeta/3} \int_0^1 d\eta\, \eta^{\zeta/3}$$

$$= (3+\zeta)^{-1}\Gamma(1-\zeta/3)\left[\frac{2X}{1-X}\right]^{-1+\zeta/3} \int_{-\infty}^\infty d\xi \int_{-\infty}^\infty d\xi'\, |\xi-\xi'|^{-1}$$

$$\times [\tau(\xi)-\tau(\xi')]^2\, e^{i\beta(\xi-\xi')}\, e^{-i\pi(\zeta/6)\mathrm{sg}(\xi-\xi')}\, |W(\xi,\xi')|^{-1+\zeta/3}.$$

(12.107)

Its behaviour near $\zeta = 1, 0, -1$ is to be determined in succession.

12.5.2 Residue of $\overline{K}(\zeta)$ at $\zeta = 1$

At $\zeta = 1$, $\overline{K}(\zeta)$ has a simple pole. Let R_1 be its residue, i.e.,

$$R_1 = \lim_{\zeta \to 1} (\zeta - 1) \overline{K}(\zeta). \tag{12.108}$$

The contribution to this residue comes from the vicinity of $\xi = \xi'$. Let

$$\bar{\xi} = \tfrac{1}{2}(\xi + \xi') \tag{12.109}$$

and

$$\mu = \xi - \xi'. \tag{12.110}$$

For μ small, it follows from eqns (12.48) that

$$\tau(\xi) - \tau(\xi') \sim \mu \rho(\bar{\xi}),$$

$$T(\xi) - T(\xi') \sim \mu \tau(\bar{\xi})^2 + \tfrac{1}{4}\mu^3[\tau'(\bar{\xi})^2 + \tau(\bar{\xi})\tau''(\bar{\xi})],$$

$$U(\xi) - U(\xi') \sim \mu \tau(\bar{\xi}) + \tfrac{1}{8}\mu^3 \tau''(\bar{\xi}), \tag{12.111}$$

and hence, from eqn (12.90),

$$W(\xi, \xi') \sim \tfrac{1}{12} \mu^3 \rho(\bar{\xi})^2. \tag{12.112}$$

Therefore, in the vicinity of $\zeta = 1$, $\overline{K}(\zeta)$ is approximately given by

$$\overline{K}(\zeta) \sim \tfrac{1}{4}\Gamma(\tfrac{2}{3}) \left[\frac{2X}{1-X}\right]^{-\tfrac{2}{3}} \int_{-\infty}^{\infty} d\bar{\xi} \int_{-1}^{1} d\mu \, |\mu| \, \rho(\bar{\xi})^2 e^{-i\pi(\text{sg}\,\mu)/6}$$

$$\times \left[\tfrac{1}{12}|\mu|^3 \rho(\bar{\xi})^2\right]^{-1+\zeta/3}$$

$$\sim \Gamma(\tfrac{2}{3}) \left[\frac{4X}{1-X}\right]^{-\tfrac{2}{3}} 3^{7/6} \frac{1}{\zeta - 1} \int_{-\infty}^{\infty} d\bar{\xi} \, [\rho(\bar{\xi})]^{2/3}, \tag{12.113}$$

and the residue R_1 is

$$R_1 = \Gamma(\tfrac{2}{3}) \left[\frac{4X}{1-X}\right]^{-\tfrac{2}{3}} 3^{7/6} \int_{-\infty}^{\infty} d\bar{\xi} \, [\rho(\bar{\xi})]^{2/3}. \tag{12.114}$$

One recognizes here several of the factors entering the expression for radiation 'during bunch crossing' [57,58].

12.5.3 Analytic continuation into the region $1 > \text{Re}\,\zeta > 0$

The considerations of Section 12.5.2 indicate that $\bar{\xi}$ and μ of eqns (12.109) and (12.110) are more appropriate variables. In terms of these variables, the quantity $\overline{K}(\zeta)$ of (12.107) is

$$\overline{K}(\zeta) = (3+\zeta)^{-1}\Gamma(1-\zeta/3)\left[\frac{2X}{1-X}\right]^{-1+\zeta/3} \int_{-\infty}^{\infty} d\bar{\xi} \int_{-\infty}^{\infty} d\mu\, |\mu|^{-1}$$
$$\times [\tau(\bar{\xi}+\mu/2) - \tau(\bar{\xi}-\mu/2)]^2\, e^{i\beta\mu}\, e^{-i\pi(\zeta/6)\text{sg}\,\mu}\, |\tilde{W}(\bar{\xi},\mu)|^{-1+\zeta/3}, \tag{12.115}$$

where

$$\tilde{W}(\bar{\xi},\mu) = W(\xi,\xi') \tag{12.116}$$

is an odd function of μ and has the property from (12.112) that

$$\tilde{W}(\bar{\xi},\mu) \sim \tfrac{1}{12}\mu^3\, \rho(\bar{\xi})^2, \tag{12.117}$$

for small μ.

Before we can study the behaviour of $\overline{K}(\zeta)$ in the vicinity of $\zeta = 0$, we must rewrite the right-hand side of (12.115) in such a way that it makes sense in that vicinity. This can be accomplished in a variety of ways; we choose one that facilitates further analytic continuation to near $\zeta = -1$. Our procedure involves rewriting the μ integral using suitable contours. We assume that $\rho(\xi)$ is analytic so that analytic continuation to complex values of μ is possible, at least for small values of $|\text{Im}\,\mu|$.

Consider the following factor in the integrand of (12.115):

$$L(\mu) = |\mu|^{-1} e^{-i\pi(\zeta/6)\text{sg}\,\mu}\, |\tilde{W}(\bar{\xi},\mu)|^{-1+\zeta/3}$$
$$= \begin{cases} \mu^{-1} e^{-i\pi\zeta/6} [\tilde{W}(\bar{\xi},\mu)]^{-1+\zeta/3} & \text{for } \mu > 0, \\ -\mu^{-1} e^{i\pi\zeta/6} [-\tilde{W}(\bar{\xi},\mu)]^{-1+\zeta/3} & \text{for } \mu < 0. \end{cases} \tag{12.118}$$

Starting with the above expression for $\mu > 0$, we continue analytically to $\mu < 0$. The results are, using (12.117),

$$L(\mu e^{i\pi}) = \mu^{-1} e^{5i\pi\zeta/6} [\tilde{W}(\bar{\xi},\mu)]^{-1+\zeta/3}, \tag{12.119}$$

and

$$L(\mu e^{-i\pi}) = \mu^{-1} e^{-7i\pi\zeta/6} [\tilde{W}(\bar{\xi},\mu)]^{-1+\zeta/3}. \tag{12.120}$$

We attempt to write the function $L(-\mu)$, as defined by eqn (12.118), as a linear combination of the two different contributions given by (12.119) and (12.120):

$$L(-\mu) = C_1\, L(\mu e^{i\pi}) + C_2\, L(\mu e^{-i\pi}), \tag{12.121}$$

where the sum of C_1 and C_2 should be 1, i.e.,

$$C_1 + C_2 = 1. \tag{12.122}$$

Since $\tilde{W}(\bar{\xi},\mu)$ is an odd function of μ, it follows from (12.118)–(12.121) that

$$C_1 e^{2i\pi\zeta/3} + C_2 e^{-i\pi\zeta/3} = 1. \tag{12.123}$$

Eqns (12.122) and (12.123) give

$$C_1 = e^{-i\pi\zeta/3} \frac{\sin(2\pi\zeta/3)}{\sin \pi \zeta} \tag{12.124}$$

and

$$C_2 = e^{2i\pi\zeta/3} \frac{\sin(\pi\zeta/3)}{\sin \pi \zeta}. \tag{12.125}$$

Note that C_1 and C_2 are finite as $\zeta \to 0$, but not for $\zeta \to -1$. This property turns out to be important.

Using eqn (12.121) together with eqns (12.124) and (12.125), we can rewrite (12.115) in the form

$$\begin{aligned}\overline{K}(\zeta) &= (3+\zeta)^{-1}\Gamma(1-\zeta/3)\left[\frac{2X}{1-X}\right]^{-1+\zeta/3} \int_{-\infty}^{\infty} d\bar{\xi} \int_{C} d\mu\, \mu^{-1} \\ &\quad \times [\tau(\bar{\xi}+\mu/2) - \tau(\bar{\xi}-\mu/2)]^2\, e^{i\beta\mu}\, e^{-i\pi\zeta/6}\, [\tilde{W}(\bar{\xi},\mu)]^{-1+\zeta/3}, \end{aligned} \tag{12.126}$$

where, for the μ integration, $\int_C$ is defined by the weighted average of two contour integrations:

$$\int_C = e^{-i\pi\zeta/3} \frac{\sin(2\pi\zeta/3)}{\sin \pi \zeta} \int_{C_+} + e^{2i\pi\zeta/3} \frac{\sin(\pi\zeta/3)}{\sin \pi \zeta} \int_{C_-}. \tag{12.127}$$

In (12.127), the contours of integration C_+ and C_- are along the real axis except for an indentation at the origin into the upper and lower half planes respectively, as shown on Fig. 12.2.

Eqn (12.126) gives the desired analytic continuation of $\overline{K}(\zeta)$ into the enlarged region

$$3 > \text{Re}\,\zeta > 0 \tag{12.128}$$

except for the pole at $\zeta = 1$ already studied in Section 12.5.2.

We can now proceed to the study of the singularity at $\zeta = 0$.

12.5.4 Residue of $\overline{K}(\zeta)$ at $\zeta = 0$

While the residue of $\overline{K}(\zeta)$ at $\zeta = 1$ comes from the region of integration in the vicinity of $\xi = \xi'$, that at $\zeta = 0$ comes from the region of large $|\bar{\xi}|$.

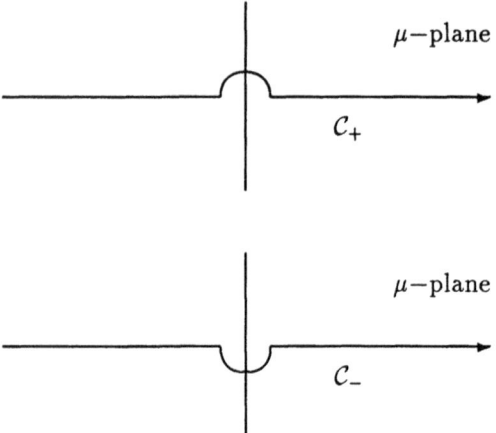

Fig. 12.2: The contours of integration $\mathcal{C}_+$ and $\mathcal{C}_-$ in the μ-plane.

Having in mind the hyperbolic secant distribution (12.38), we assume more generally that $\rho(\xi)$ decreases exponentially as $\xi \to \infty$ and $\xi \to -\infty$. More explicitly, let

$$\rho(\xi) \sim \begin{cases} c_1 e^{a_1 \xi}, & \text{as } \xi \to -\infty, \\ c_2 e^{-a_2 \xi}, & \text{as } \xi \to \infty, \end{cases} \qquad (12.129)$$

where a_1, a_2, c_1 and c_2 are four positive constants. By a translation on ξ, the values of c_1 and c_2 can be made equal; however, such a choice has no particular advantage.

Similar to (12.108), let R_0 be the residue of $\overline{K}(\zeta)$ at $\zeta = 0$, i.e.,

$$R_0 = \lim_{\zeta \to 0} \zeta \, \overline{K}(\zeta). \qquad (12.130)$$

In this section, we calculate R_0.

Let τ_1 and $\tilde{W}_1$ be the τ and $\tilde{W}$ corresponding to

$$\rho_1(\xi) = c_1 \, e^{a_1 \xi}, \qquad (12.131)$$

while τ_2 and $\tilde{W}_2$ correspond to

$$\rho_2(\xi) = c_2 \, e^{-a_2 \xi}. \qquad (12.132)$$

Then from (12.126), near $\zeta = 0$,

$$\overline{K}(\zeta) \sim \overline{K}_1(\zeta) + \overline{K}_2(\zeta), \qquad (12.133)$$

12.5. ENERGY DISTRIBUTION OF BEAMSTRAHLUNG PHOTON

where

$$\overline{K}_1(\zeta) = \frac{1-X}{6X} \int_{-\infty}^{0} d\overline{\xi} \left(\frac{2}{3}\int_{C_+} + \frac{1}{3}\int_{C_-}\right) d\mu\, \mu^{-1}$$
$$\times [\tau_1(\overline{\xi}+\mu/2) - \tau_1(\overline{\xi}-\mu/2)]^2 e^{i\beta\mu} [\tilde{W}_1(\overline{\xi},\mu)]^{-1+\zeta/3} \quad (12.134)$$

and

$$\overline{K}_2(\zeta) = \frac{1-X}{6X} \int_{0}^{\infty} d\overline{\xi} \left(\frac{2}{3}\int_{C_+} + \frac{1}{3}\int_{C_-}\right) d\mu\, \mu^{-1}$$
$$\times [\tau_2(\overline{\xi}+\mu/2) - \tau_2(\overline{\xi}-\mu/2)]^2 e^{i\beta\mu} [\tilde{W}_2(\overline{\xi},\mu)]^{-1+\zeta/3}. \quad (12.135)$$

Since $\overline{K}_2(\zeta)$ can be obtained from $\overline{K}_1(\zeta)$ by the replacements $a_1 \to a_2$ and $c_1 \to c_2$, it is sufficient to concentrate on $\overline{K}_1(\zeta)$.

From eqns (12.48) and (12.90), the ρ_1 of eqn (12.131) leads to

$$\tau_1(\xi) = (c_1/a_1) e^{a_1\xi},$$

$$T_1(\xi) = \tfrac{1}{2}(c_1^2/a_1^3) e^{2a_1\xi},$$

$$U_1(\xi) = (c_1/a_1^2) e^{a_1\xi}, \quad (12.136)$$

and

$$W_1(\xi,\xi') = \frac{c_1^2}{2a_1^3}\left[e^{2a_1\xi} - e^{2a_1\xi'} - \frac{2}{a_1(\xi-\xi')}\left(e^{a_1\xi} - e^{a_1\xi'}\right)^2\right]. \quad (12.137)$$

The important point here is that the right-hand side of eqn (12.137) is of the form of a product $e^{2a\overline{\xi}}$ times a function of μ. Therefore, the $\overline{\xi}$ integration in eqn (12.134) can be carried out easily, leading to

$$\overline{K}_1(\zeta) \sim \zeta^{-1}\frac{1-X}{2X}\left(\frac{2}{3}\int_{C_+} + \frac{1}{3}\int_{C_-}\right) d\mu\, e^{2i\beta\mu/a_1} \frac{\sinh\mu}{\mu\cosh\mu - \sinh\mu} \quad (12.138)$$

for ζ small. This exhibits explicitly the pole structure at $\zeta = 0$. Since the value of c_1 can be changed by shifting ξ, it cannot appear in the residue at $\zeta = 0$; indeed, it does not.

Let

$$E(\beta) = \left(\frac{2}{3}\int_{C_+} + \frac{1}{3}\int_{C_-}\right) d\mu\, e^{i\beta\mu}\frac{\sinh\mu}{\mu\cosh\mu - \sinh\mu}, \quad (12.139)$$

then the residue of $\overline{K}(\zeta)$ at $\zeta = 0$ is

$$R_0 = \frac{1-X}{2X}\left[E\left(\frac{2\beta}{a_1}\right) + E\left(\frac{2\beta}{a_2}\right)\right]. \quad (12.140)$$

It remains to evaluate this function $E(\beta)$ for small β. Since β is positive, we can close the contour of integration in the upper half plane to get

$$E(\beta) = -2\pi \left[\beta - \sum_{n=1}^{\infty} \frac{\exp(-\beta y_n)}{y_n}\right], \qquad (12.14)$$

where y_n is the nth positive zero of

$$y - \tan y = 0. \qquad (12.14)$$

The first few zeros are at [59]

$$y_1 = 4.493\,409\,458,$$

$$y_2 = 7.725\,251\,838,$$

$$y_3 = 10.904\,121\,66, \qquad (12.14)$$

etc. From (12.141), it is clear that $E(\beta)$ is a decreasing function of β for positive β. For large n, y_n is asymptotically given by

$$y_n \sim (n+\tfrac{1}{2})\pi - [(n+\tfrac{1}{2})\pi]^{-1}. \qquad (12.14)$$

We know that [60]

$$\sum_{n=0}^{\infty} (n+\tfrac{1}{2})^{-1} e^{-\beta \pi n} = \Phi(e^{-\pi\beta}, 1, \tfrac{1}{2})$$

$$= 2\,_2F_1(1, \tfrac{1}{2}; \tfrac{3}{2}; e^{-\beta\pi})$$

$$= h_0'' - \log(1 - e^{-\beta\pi}) + \mathcal{O}(\beta)$$

$$= 2\log 2 - \log(\beta\pi) + \mathcal{O}(\beta), \qquad (12.14)$$

where Φ is defined on p. 27 [eqn (1)], $_2F_1$ is the hypergeometric function [p. 30, eqn (10)], h_0'' is defined on p. 110 [eqns (12) and (13)], all of the Bateman Manuscript Project [60]. Therefore,

$$E(\beta) = 2(-\log\beta + K_e) + \mathcal{O}(\beta) \qquad (12.14)$$

as $\beta \to 0^+$, where K_e is defined by

$$K_e = -\log\pi + 2\log 2 - 2 + \sum_{n=1}^{\infty} \left(\frac{\pi}{y_n} - \frac{2}{2n+1}\right)$$

$$= -1.713\,150\,32. \qquad (12.14)$$

Finally, the substitution into eqn (12.140) gives

$$R_0 = -\frac{1-X}{X}\left(\log\frac{4\beta^2}{a_1 a_2} - 2K_e\right). \qquad (12.14)$$

12.5.5 Analytic continuation into the region $0 > \text{Re}\,\zeta > -1$

In order to solve the problem of the effect of the density gradient for a distribution such as (12.38), we need to study the behaviour of $\overline{K}(\zeta)$ near $\zeta = -1$. For this purpose, we must rewrite the right-hand side of eqn (12.126) in such a way that it makes sense in that vicinity. Similar to the development of Section 12.5.3, the method of deforming the contour of integration is to be used. Unlike that of Section 12.5.3, this deformation is applied to the $\overline{\xi}$ integration.

First, we assume that the domain of analyticity in $\overline{\xi}$ is large enough to permit the deformation of the contour. However, the result to be obtained in Section 12.5.6 can be written in a way that this analyticity is not essential. Since the singularity of $\overline{K}(\zeta)$ at $\zeta = 0$ is a pole, not a branch point, and the result (12.140) on R_0 is of the form of two separate contributions from the two ends of the bunch, the analytic continuation can be carried out via either the upper half plane or the lower half plane.

The required analytic continuation is performed by merely rewriting $\overline{K}(\zeta)$, eqn (12.126), in the form

$$\overline{K}(\zeta) = (3+\zeta)^{-1}\Gamma(1-\zeta/3)\left[\frac{2X}{1-X}\right]^{-1+\zeta/3}\int_{\overline{c}}d\overline{\xi}\int_{C}d\mu\,\mu^{-1}$$

$$\times [\tau(\overline{\xi}+\mu/2) - \tau(\overline{\xi}-\mu/2)]^2\,e^{i\beta\mu}\,e^{-i\pi\zeta/6}\,[\tilde{W}(\overline{\xi},\mu)]^{-1+\zeta/3}, \quad (12.149)$$

where the contour $\overline{C}$ can be, for example, any of the four shown in Fig. 12.3. Except for the simple poles at $\zeta = 1$ and $\zeta = 0$, eqn (12.149) is valid in the further enlarged region

$$3 > \text{Re}\,\zeta > -1. \quad (12.150)$$

12.5.6 Residue of $\overline{K}(\zeta)$ at $\zeta = -1$

Once again, $\overline{K}(\zeta)$ has a simple pole at $\zeta = -1$. Similar to eqns (12.108) and (12.130), let R_{-1} be the residue

$$R_{-1} = \lim_{\zeta \to -1}(\zeta + 1)\overline{K}(\zeta). \quad (12.151)$$

The origin of this pole is exceptionally simple: for ζ near -1, the μ integration is approximately, by (12.127),

$$\int_{C} \sim -\frac{\sqrt{3}}{2\pi(\zeta+1)}e^{i\pi/3}\left[\int_{C_-}-\int_{C_+}\right]$$

$$= -\frac{\sqrt{3}}{2\pi(\zeta+1)}e^{i\pi/3}\int_{C_0}, \quad (12.152)$$

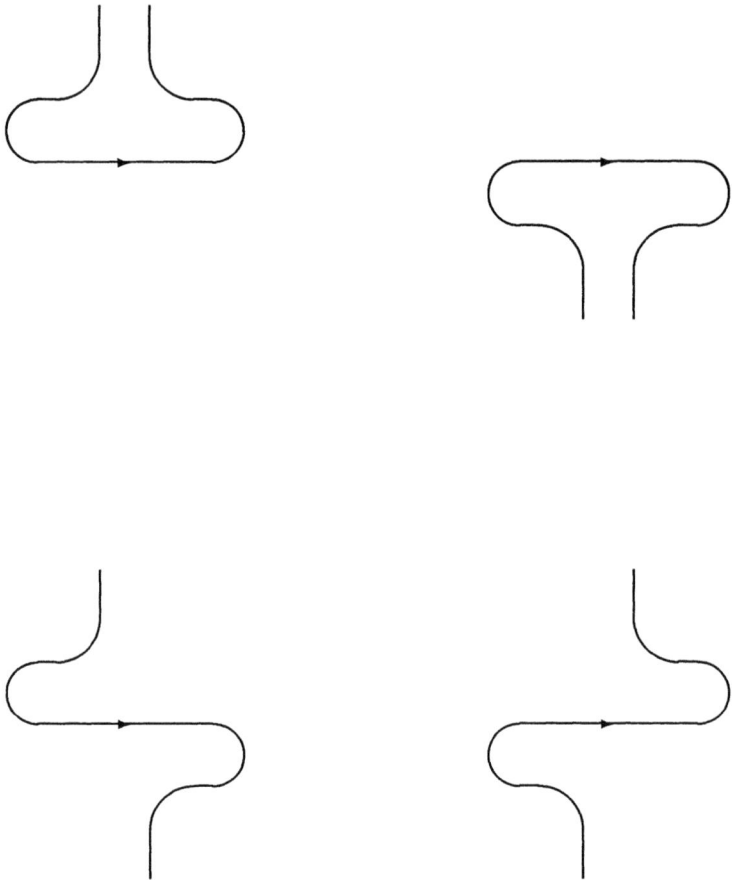

Fig. 12.3: Four possible choices of the contour $\overline{C}$ of integration. The horizontal straight portions of the four contours are on the real axis.

12.5. ENERGY DISTRIBUTION OF BEAMSTRAHLUNG PHOTON

where $C_0 = C_- - C_+$ is a small counterclockwise circle around the origin. Therefore, by eqn (12.149), the residue is

$$R_{-1} = \frac{-i\sqrt{3}}{4\pi} \Gamma(\tfrac{4}{3}) \left[\frac{2X}{1-X}\right]^{-4/3} \int_{\bar{C}} d\bar{\xi} \int_{C_0} d\mu\, \mu^{-1}$$

$$\times [\tau(\bar{\xi}+\mu/2) - \tau(\bar{\xi}-\mu/2)]^2\, e^{i\beta\mu}\, [\tilde{W}(\bar{\xi},\mu)]^{-4/3}. \quad (12.153)$$

In order to evaluate the integration over μ, it is necessary to expand the integrand for small μ [57]. Since

$$\tau(\bar{\xi}+\mu/2) = \tau + \frac{1}{2}\tau'\mu + \frac{1}{8}\tau''\mu^2 + \frac{1}{48}\tau'''\mu^3 + \frac{1}{384}\tau''''\mu^4 + \mathcal{O}(\mu^5), \quad (12.154)$$

with

$$\tau = \tau(\bar{\xi}), \qquad \tau' = \tau'(\bar{\xi}), \quad (12.155)$$

etc., it follows that

$$\tau(\bar{\xi}+\mu/2) - \tau(\bar{\xi}-\mu/2) = \tau'\mu + \frac{1}{24}\tau'''\mu^3 + \mathcal{O}(\mu^5), \quad (12.156)$$

$$U(\bar{\xi}+\mu/2) - U(\bar{\xi}-\mu/2) = \frac{1}{2}\int_{-\mu}^{\mu} d\nu\, \tau(\bar{\xi}+\nu/2)$$

$$= \tau\mu + \frac{1}{24}\tau''\mu^3 + \frac{1}{1920}\tau''''\mu^5 + \mathcal{O}(\mu^7), \quad (12.157)$$

and

$$T(\bar{\xi}+\mu/2) - T(\bar{\xi}-\mu/2) = \frac{1}{2}\int_{-\mu}^{\mu} d\nu\, [\tau(\bar{\xi}+\nu/2)]^2$$

$$= \tau^2\mu + \frac{1}{12}(\tau\tau'' + \tau'^2)\mu^3 + \frac{1}{960}(\tau\tau'''' + 4\tau'\tau''' + 3\tau''^2)\mu^5 + \mathcal{O}(\mu^7). \quad (12.158)$$

The substitution of both eqns (12.157) and (12.158) into the definition of $\tilde{W}(\bar{\xi},\mu)$, eqn (12.116), with (12.90) then gives

$$\tilde{W}(\bar{\xi},\mu) = T(\bar{\xi}+\mu/2) - T(\bar{\xi}-\mu/2) - \mu^{-1}[U(\bar{\xi}+\mu/2) - U(\bar{\xi}-\mu/2)]^2$$

$$= \frac{1}{12}\tau'^2\mu^3 + \frac{1}{720}(3\tau'\tau''' + \tau''^2)\mu^5 + \mathcal{O}(\mu^7), \quad (12.159)$$

and hence, from eqns (12.153) and (12.156),

$$R_{-1} = \frac{-i\sqrt{3}}{4\pi} \Gamma(\tfrac{4}{3}) \left[\frac{2X}{1-X}\right]^{-4/3} \int_{\bar{C}} d\bar{\xi} \int_{C_0} d\mu\, \mu[\tau'^2 + \frac{1}{12}\tau'\tau'''\mu^2 + \mathcal{O}(\mu^4)]$$

$$\times \mu^{-4}\left[\frac{\tau'^2}{12}\right]^{-4/3}\left[1 - \frac{1}{45}\frac{3\tau'\tau''' + \tau''^2}{\tau'^2}\mu^2 + \mathcal{O}(\mu^4)\right]. \quad (12.160)$$

Since $\tau' = \rho(\bar{\xi})$, it is now straightforward to pick out the residue at $\mu = 0$ to get the result

$$R_{-1} = \frac{2}{15}\Gamma(\tfrac{4}{3})\left[\frac{4X}{1-X}\right]^{-4/3} 3^{5/6}\int_{\bar{C}} d\bar{\xi}\,\rho(\bar{\xi})^{-8/3}[3\rho(\bar{\xi})\rho''(\bar{\xi}) - 4\rho'(\bar{\xi})^2]. \tag{12.161}$$

This differs from the previously obtained results [57,61,62] in an essential way, namely the contour of integration is not the real axis, but, for example, any of the paths shown in Fig. 12.3. With such a path of integration, the integral is well defined, and *no cutoff is needed*.

In some circumstances, it is possible to deform the contour $\bar{C}$ back to the real axis at the expense of subtracting and adding some terms. For example, if the condition (12.129) is strengthened to be

$$\rho(\xi) \sim \begin{cases} c_1 e^{a_1\xi} + o(e^{5a_1\xi/3}), & \text{as } \xi \to -\infty, \\ c_2 e^{-a_2\xi} + o(e^{-5a_2\xi/3}), & \text{as } \xi \to \infty, \end{cases} \tag{12.162}$$

then the subtraction of the integrand calculated on the basis of the leading behaviour makes it possible to use the real axis as the contour of integration. Such a procedure makes it unnecessary to assume analyticity in $\bar{\xi}$. However, the resulting formula for R_{-1} is not elegant and also not especially informative.

Eqn (12.161) can be simplified slightly by an integration by parts. Because of the contour $\bar{C}$, the integrated terms are zero and the result is

$$R_{-1} = \frac{2}{15}\Gamma(\tfrac{4}{3})\left[\frac{4X}{1-X}\right]^{-4/3} 3^{5/6}\int_{\bar{C}} d\xi\,\rho(\xi)^{-8/3}\,\rho'(\xi)^2. \tag{12.163}$$

Although the integrand is positive on the real axis, R_{-1} may be positive or negative.

12.5.7 Hyperbolic secant distribution

As an example, we apply our results to the hyperbolic secant distribution (12.38):

$$\rho(\xi) = \text{sech}^2(2\xi) = \frac{4e^{4\xi}}{(1+e^{4\xi})^2}. \tag{12.164}$$

Comparison with (12.129) shows that

$$a_1 = a_2 = 4. \tag{12.165}$$

It follows from (12.114), (12.140) and (12.163) that, for the present case,

$$R_1 = \frac{1}{2\pi}\left[\frac{81(1-X)}{8X}\right]^{2/3}[\Gamma(\tfrac{2}{3})]^4,$$

12.5. ENERGY DISTRIBUTION OF BEAMSTRAHLUNG PHOTON

$$R_0 = \frac{1-X}{X} E\left(\frac{X}{1-X} \frac{L_b}{2L_e}\right),$$

$$R_{-1} = -\frac{1}{5\pi} \left[\frac{1-X}{4X}\right]^{4/3} 12^{1/3} [\Gamma(\tfrac{1}{3})]^4. \qquad (12.166)$$

Since R_{-1} is negative, contrary to previous conjecture [61], it may be worthwhile to give a derivation of (12.166). For the hyperbolic secant charge distribution (12.164), the R_{-1} of (12.163) is

$$R_{-1} = \frac{2}{15} \Gamma(\tfrac{4}{3}) \left[\frac{1-X}{4X}\right]^{4/3} 2^{2/3} 3^{5/6} I_{-1}, \qquad (12.167)$$

where

$$I_{-1} = \int_{\bar{C}} d\xi \left[\frac{e^\xi}{(1+e^\xi)^2}\right]^{-8/3} \left[\frac{(1-e^\xi)e^\xi}{(1+e^\xi)^3}\right]^2. \qquad (12.168)$$

Define a function of a complex variable η by

$$I_{-1}(\eta) = \int_{\bar{C}} d\xi \left[\frac{e^\xi}{(1+e^\xi)^2}\right]^{-2+\eta} \left[\frac{(1-e^\xi)e^\xi}{(1+e^\xi)^3}\right]^2. \qquad (12.169)$$

Then, $I_{-1}(\eta)$ has a simple pole at $\eta = 0$, and the desired I_{-1} is

$$I_{-1} = I_{-1}(-\tfrac{2}{3}). \qquad (12.170)$$

For $\mathrm{Re}\,\eta > 0$, the contour $\bar{C}$, as shown for example in Fig. 12.3, can be replaced by the real axis, and thus the evaluation of I_{-1} is straightforward:

$$I_{-1}(\eta) = \frac{2}{\eta} \frac{[\Gamma(1+\eta)]^2}{\Gamma(2+2\eta)}. \qquad (12.171)$$

This shows explicitly the pole structure at $\eta = 0$, and also that it is positive for $\eta > 0$. However, analytic continuation to negative values of η gives

$$I_{-1} = I_{-1}(-\tfrac{2}{3}) = -3 \frac{[\Gamma(\tfrac{1}{3})]^2}{\Gamma(\tfrac{2}{3})}, \qquad (12.172)$$

which is negative. The lesson is that we have to be very careful in providing reliable arguments about even the sign of higher-order corrections.

12.5.8 Summary

When the correlation length ℓ_c is much shorter than the nominal bunch length L_b, the energy spectrum of the beamstrahlung photon is approximately given by

$$I(X) \sim \frac{\alpha}{2\pi} \frac{2-2X+X^2}{1-X} \left[\frac{L_b}{\ell_c} R_1 + R_0 + \frac{\ell_c}{L_b} R_{-1}\right], \qquad (12.173)$$

where R_1, R_0 and R_{-1} are given by eqns (12.114), (12.140) and (12.163), respectively. The extra factor present in the Dirac case has been explicitly written down.

It is instructive to rewrite this result in terms of the actual charge density $\rho_0(z)$ rather than the normalized charge density $\rho(z/L_b)$. They are related by eqn (12.40). The result is

$$I(X) \sim \frac{\alpha}{2\pi} \frac{2 - 2X + X^2}{1 - X}$$

$$\times \left\{ \Gamma(\tfrac{2}{3}) \left[\frac{1-X}{4X}\right]^{2/3} 3^{7/6} \left[\frac{\pi^2 R^2}{k_i}\right]^{1/3} \int_{-\infty}^{\infty} dz \, [e\rho_0(z)]^{2/3} \right.$$

$$+ \frac{1-X}{2X} \left[E\left(\frac{2X}{1-X} \frac{1}{a_{10} L_e}\right) + E\left(\frac{2X}{1-X} \frac{1}{a_{20} L_e}\right) \right]$$

$$\left. + \tfrac{2}{3}\Gamma(\tfrac{4}{3}) \left[\frac{1-X}{4X}\right]^{4/3} 3^{-1/6} \left[\frac{k_i}{\pi^2 R^2}\right]^{1/3} \int_{\mathcal{C}} dz \, [e\rho_0(z)]^{-8/3} [e\rho_0'(z)]^2 \right\},$$

(12.174)

where, by eqn (12.129),

$$a_{10} = \lim_{z \to -\infty} \rho_0'(z)/\rho_0(z),$$

$$a_{20} = -\lim_{z \to \infty} \rho_0'(z)/\rho_0(z). \tag{12.175}$$

Examples of the contour $\overline{\mathcal{C}}$ of integration are shown in Fig. 12.3. The important point to be noticed in eqn (12.174) is that the nominal bunch length L_b does not appear anywhere on the right-hand side. This is the way it should be, because there is nothing to prevent us from using $2L_b$ or $\tfrac{1}{2}L_b$, for example, instead of L_b in all the intermediate steps.

The result (12.173), or equivalently (12.174), takes the form of three terms. Roughly speaking, the first term gives the beamstrahlung during beam crossing, while the second term gives the effects of the bunch ends. Both terms can be sizeable for very high energies.

The value of the third term depends on the variation of charge densities $\rho_0'(z)$ and $\rho_0''(z)$ along the bunch. Contrary to the claims of Chen and Yokoya [61], it is not large. At least for the special case of the hyperbolic secant bunch, this term is negative, reducing slightly the beamstrahlung energy loss.

In obtaining the present result, we have assumed an exponential decrease in the charge density far away from the centre of the bunch. This is as good a description as, if not better than, a Gaussian bunch for realistic cases. Nevertheless, it may be asked what the corresponding result looks like for a Gaussian bunch. Unfortunately, the answer is that it is very much more complicated, and hence, probably of little practical use.

12.6 Discussions

So far as methodology is concerned, the main lesson that one learns from studying beamstrahlung in TeV electron-positron linear colliders is the same one as from the other chapters of this book, namely, Feynman diagrams can be used with great advantage. This point is worth emphasizing because, while everybody agrees that Feynman diagrams are useful for the calculation of radiative processes such as those listed in Chapters 9 and 10, there has been great resistance in the use of Feynman diagrams for beamstrahlung and related processes. Indeed, until very recently, Feynman diagrams have been used only in the work of Jacob and Wu. Fortunately, this situation is changing, and others are now in the process of adopting such methods.

The study of the radiation from a charged particle in a magnetic field has a long history [63]. Much emphasis has been placed on the case of an infinite uniform magnetic field. In this case, it is very difficult, if not impossible, to make efficient use of Feynman diagrams. In view of the large and excellent literature on this case, it was natural to make use of this knowledge to deal with the case of beamstrahlung in high-energy electron-positron colliders [64]. This is perhaps the original reason why Feynman diagrams were not used at the beginning [65]. By now, however, there is no reason whatsoever not to use the powerful tool provided by Feynman.

The study of radiative processes for high-energy electron-positron linear colliders is still in its infancy. In this chapter, we have considered only the process (12.5), which is analogous to $e^+ + e^- \rightarrow e^+ + e^- + \gamma$. Similarly, in analogy with

$$e^+ + e^- \rightarrow e^+ + e^- + \gamma + \gamma, \qquad (12.176)$$

$$\gamma + e^{\pm} \rightarrow e^+ + e^- + e^{\pm}, \qquad (12.177)$$

and

$$e^+ + e^- \rightarrow e^+ + e^- + e^+ + e^-, \qquad (12.178)$$

we can list the following processes which involve electron and/or positron bunches:

$$e^{\pm} + \text{bunch} \rightarrow e^{\pm} + \gamma + \gamma + \text{bunch}, \qquad (12.179)$$

$$\gamma + \text{bunch} \rightarrow e^+ + e^- + \text{bunch}, \qquad (12.180)$$

and

$$e^{\pm} + \text{bunch} \rightarrow e^+ + e^- + e^{\pm} + \text{bunch}. \qquad (12.181)$$

There are of course many, many other similar processes of interest. The Feynman diagrams for (12.180) and (12.181) are shown in Fig. 12.4 and Fig. 12.5. Of these examples, so far only (12.180) has been studied [66]. For actual applications, the process (12.181) is especially important, but, at the time of writing this chapter, it still has not been studied in detail. We can look forward in the near future to a rapid increase in our knowledge of such processes, and we hope that the inclusion of this chapter in this book may

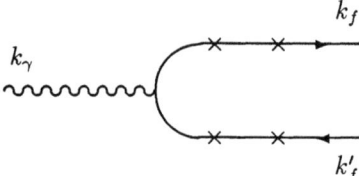

Fig. 12.4: Feynman diagram for pair production by an photon in a bunch electrons (positrons).

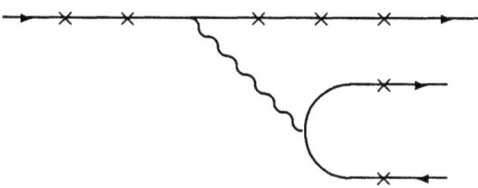

Fig. 12.5: Feynman diagram for pair production by an electron in a bunch of positrons.

stimulate the interest in these and other processes involving electron and positron bunches.

We conclude this chapter with a discussion of the implication of the studies of beamstrahlung and pair production in the design and construction of future electron-positron linear colliders with energies of 1 TeV on 1 TeV or higher. Such a discussion is necessarily preliminary – it is conceivable that somebody may come up with a totally new and unexpected idea that changes our thinking on such accelerators. Nevertheless, a discussion at this time may still be of interest.

As already mentioned in Section 12.1, a power limitation of 100 or 200 megawatts together with the necessity of useful luminosity make it essential to have very small bunch radii. This in turn leads to significant beamstrahlung energy loss. Fortunately, this energy loss, although varying from event to event, does not cause too much complication in the design of the detector and in the analysis of the experimental data. The main effect is rather that a higher-energy accelerator is needed for a given desirable centre-of-mass e

12.6. DISCUSSIONS

ergy. For example, if the average beamstrahlung energy loss is 17 %, which is a reasonable value, then, to achieve 2 TeV in the centre-of-mass energy, the energies of the electron and positron beams should be about 1.2 TeV.

The creation of electron-positron pairs [66] by the beamstrahlung photon seems to cause much more trouble. Because both electrons and positrons are created, there is no simple way to deflect them away from the focusing devices. At present, the implication of the presence of such pairs on the design of the final focus has not been completely understood.

Although the complications from both beamstrahlung and pair production are by themselves not fatal, it must be remembered that the design of the multi-TeV electron-positron linear colliders such as CLIC and TLC is marginal even without these extra difficulties. It therefore seems that, if the power limitation of about 200 megawatts remains imposed, it is rather unlikely that, for the electron-positron system, an effective centre-of-mass energy of 2 TeV with a useful luminosity can be reached before the beginning of the next century.

13
Outlook

The use of helicity
is not an eccentricity:
It is pure simplicity
which brings felicity.

13.1 The ubiquity of the photon

The ubiquity of the photon is a well-known fact in particle physics. Photons are necessarily present whenever charged particles take part in a process. The same is true for gluons when strongly interacting particles collide or decay. Photons and gluons being the carriers of the electromagnetic force and the strong force, they play a fundamental role in particle physics and their ubiquity only underscores this fact. It is therefore essential to study carefully the processes in which they take part. To understand their behaviour is to understand two fundamental forces in nature and the helicity method is a key ingredient in this exploration.

We hope that, by now, the reader is convinced that the helicity method is an efficient way for obtaining cross section formulae in the high-energy limit. Nevertheless, we learn from the examples studied in the previous chapters that the final result of the calculation is often much simpler than the intermediate formulae which had to be worked out. This is because the starting point of the calculation usually consists of many Feynman diagrams. As a result, we strongly suspect that there is still room for improvement in the helicity method.

In this chapter, we present a series of recent developments of the helicity method, which can be regarded either as extensions of the method to physical situations not previously covered in this book, or as attempts to streamline the calculations even further. We also discuss some unsolved problems and present them as challenges to the reader.

13.2 Further developments

13.2.1 Supersymmetry

Supersymmetry [67] is a symmetry which relates bosons and fermions. In most supersymmetric theories, the number of fermionic degrees of freedom is equal to the number of bosonic degrees of freedom. The simplest supersymmetric model, the Wess-Zumino model, describes a massive Majorana fermion and its superpartners are a scalar field and a pseudoscalar field. We thus have two fermionic degrees of freedom, the two spin states of the fermion, and two bosonic degrees of freedom. In the case of unbroken supersymmetry, the bosons and the fermions have the same mass.

Supersymmetry can provide relations among scattering amplitudes involving one kind of particle and scattering amplitudes involving its superpartners. This idea was exploited by Parke and Taylor [68], who embedded QCD in a SO(2) extended supersymmetric version of QCD. This larger theory encompasses, besides still more particles, the usual gluons (g), massless spin-$\frac{1}{2}$ gluinos (λ) and massless complex scalar gluons (ϕ). Because of supersymmetry Ward identities [69], they derived that the helicity amplitudes for $gg \to gg$ are simply related to gluino amplitudes, which, in turn, are related to the massless scalar amplitudes. More specifically, they find that

$$M(g_+, g_+; g_+, g_+) \doteq M(\lambda_+, \lambda_+; \lambda_+, \lambda_+) \doteq M(\phi_+, \phi_+; \phi_+, \phi_+), \qquad (13.1)$$

where the subscripts on g and λ denote the helicities of the particles and where ϕ_+ is some combination of different components of the ϕ-field. As processes involving external scalars are often easier to calculate than those involving external spin-1 particles, Parke and Taylor propose to calculate the scalar amplitude first. The relation (13.1) then yields the $M(g_+, g_+; g_+, g_+)$ amplitude and the remaining helicity amplitudes for $gg \to gg$ are obtained through crossing. In this way, the cross section for the gluonic process could be calculated from the scalar amplitudes only.

For processes like $gg \to ggg$ and $gg \to gggg$, other relations of the type (13.1) also exist. For example,

$$M(g_+, g_+; g_+, g_+, g_+) \doteq \frac{(k_1 \cdot k_2)}{(k_4 \cdot k_5)} M(\phi_+, \phi_+; g_+\phi_+, \phi_+), \qquad (13.2)$$

where $k_i, i = 1, \ldots, 5$, denotes the four-momentum of gluon i. Again, the purely gluonic process is expressed in terms of an associated process involving scalars, which is simpler to calculate. For more details of this method, we refer to Parke and Taylor [68], where also the cross sections for $gg \to ggg$ and $gg \to gggg$ are explicitly listed.

Additional supersymmetry relations for the $gg \to gggg$ helicity amplitudes were also derived by Kunszt [70].

13.2.2 Phase choice of polarization vectors

Throughout this book, we made extensive use of the photon polarization vectors of the form

$$\displaystyle{\not}{\epsilon}^{\pm}(k) = N\left[\displaystyle{\not}{k}\,\displaystyle{\not}{p}\,\displaystyle{\not}{q}(1 \pm \gamma_5) - \displaystyle{\not}{p}\,\displaystyle{\not}{q}\,\displaystyle{\not}{k}(1 \mp \gamma_5)\right], \qquad (13.3)$$

which introduced two light-like reference four-vectors p and q. Xu, Zhang and Chang [71] noted that one such reference four-vector is sufficient. They

13.2. FURTHER DEVELOPMENTS

observe that the four-vectors $\epsilon^{\pm}$, defined by

$$\epsilon_\mu^+(k,q) = \frac{\bar{u}(q)\gamma_\mu(1-\gamma_5)u(k)}{\sqrt{2}\,\bar{u}(q)(1+\gamma_5)u(k)},$$

$$\epsilon_\mu^-(k,q) = \frac{\bar{u}(q)\gamma_\mu(1+\gamma_5)u(k)}{\sqrt{2}\,\bar{u}(k)(1-\gamma_5)u(q)}, \qquad (13.4)$$

satisfy all the requirements

$$(k \cdot \epsilon^{\pm}) = (\epsilon^+ \cdot \epsilon^+) = (\epsilon^- \cdot \epsilon^-) = 0,$$

$$(\epsilon^+ \cdot \epsilon^-) = -1. \qquad (13.5)$$

Hence, the expressions (1.4) can also be used in the calculation of helicity amplitudes. To this end, we simply observe that

$$\not{\epsilon}^+(k,q) = \frac{(1-\gamma_5)u(k)\bar{u}(q)(1+\gamma_5) + (1+\gamma_5)u(q)\bar{u}(k)(1-\gamma_5)}{\sqrt{2}\,\bar{u}(q)(1+\gamma_5)u(k)},$$

$$\not{\epsilon}^-(k,q) = \frac{(1+\gamma_5)u(k)\bar{u}(q)(1-\gamma_5) + (1-\gamma_5)u(q)\bar{u}(k)(1+\gamma_5)}{\sqrt{2}\,\bar{u}(k)(1-\gamma_5)u(q)}.$$

$$(13.6)$$

If a photon is radiated from a fermion line, one should take one of the external four-vectors of the fermions to play the role of the reference four-vector q. In this way, one again encounters the usual simplifications in the spinor algebra, similar to what happened in Chapters 4–6.

When radiation from two different fermion lines occurs, such as $e^+e^- \to \mu^+\mu^-\gamma$, one would like to use different reference four-vectors. This can again be done by noting that, for the expressions (1.1),

$$\epsilon_\mu^{\pm}(k,q) = \epsilon_\mu^{\pm}(k,p) + \beta_\pm k_\mu. \qquad (13.7)$$

This relation is to be compared with eqn (4.13): the main difference is the absence of the a phase factor relating $\epsilon^{\pm}(k,q)$ to $\epsilon^{\pm}(k,p)$. As a consequence, the bookkeeping of the phase factors for the different contributions to a specific helicity amplitude is somewhat simplified: it is now done automatically through the normalization factor of $\epsilon^{\pm}$ in eqns (1.4), which is complex.

This technique can also be used for QCD calculations, and Xu, Zhang and Chang [71] applied it to the evaluation of the helicity amplitudes for $q\bar{q} \to q'\bar{q}'gg$, which is a double bremsstrahlung process.

Other applications of this technique have been presented by Gunion and Kunszt, who calculated the helicity amplitudes for $gg \to q\bar{q}q'\bar{q}'$ and for $gg \to q\bar{q}\ell\bar{\ell}$, where ℓ denotes any light lepton [72].

In the same way, Kleiss and Stirling [73] calculated the subprocesses for $p\bar{p} \to W^{\pm}/Z^0 + n$ jets, $n = 0, 1, 2$, such as, for example, $u\bar{u} \to Z^0$, $u\bar{u} \to Z^0 g$, $u\bar{u} \to Z^0 u\bar{u}$ and $u\bar{u} \to Z^0 gg$. Here, u denotes the u-quark.

Finally, we also mention the work of Gunion and Kalinowski [74], who study the process $gg \to gggg$ and the analysis by Kunszt [70] of the process $gg \to gg\bar{q}q$.

13.2.3 Weyl-van der Waerden formalism

The helicity states for fermions are in fact described by two-component spinors because of the helicity projection operators $(1 \pm \gamma_5)/2$. For this reason, one can nicely reformulate the helicity method using the Weyl-van der Waerden formalism [75], which explicitly refers to the nonvanishing components of the spinors only.

Thus, the helicity spinors $u_+(p) = v_-(p)$ of eqn (6.10) and $u_-(p) = v_+(p)$ of eqn (6.11) are replaced by

$$u_+(p) = v_-(p) \;\to\; p_A,$$

$$u_-(p) = v_+(p) \;\to\; p^{\dot{A}}, \qquad A, \dot{A} = 1, 2, \qquad (13.8)$$

where

$$p_A = \frac{1}{\sqrt{p_+}} \begin{pmatrix} p_+ \\ p_\perp \end{pmatrix}, \qquad p^{\dot{A}} = \frac{1}{\sqrt{p_+}} \begin{pmatrix} -p_\perp^* \\ p_+ \end{pmatrix}. \qquad (13.9)$$

The undotted indices refer to the upper components of the Dirac spinor, whereas the dotted indices refer to the lower components.

The raising and lowering of indices is done with the antisymmetric tensor

$$\epsilon_{\dot{A}\dot{B}} = \epsilon^{AB} = \begin{pmatrix} 0 & 1 \\ -1 & 0 \end{pmatrix}, \qquad (13.10)$$

i.e.,

$$p^A = p_B \epsilon^{BA} = \frac{1}{\sqrt{p_+}} \begin{pmatrix} -p_\perp \\ p_+ \end{pmatrix},$$

$$p_{\dot{A}} = \epsilon_{\dot{A}\dot{B}} p^{\dot{B}} = \frac{1}{\sqrt{p_+}} \begin{pmatrix} p_+ \\ p_\perp^* \end{pmatrix}. \qquad (13.11)$$

It then follows from eqns (1.9) and (1.11) that

$$\bar{u}_+(p) = \bar{v}_-(p) \;\to\; p_{\dot{A}},$$

$$\bar{u}_-(p) = \bar{v}_+(p) \;\to\; p^A, \qquad (13.12)$$

in this formalism.

One also defines a spinor product

$$<pq> = p_A q^A = \frac{p_\perp q_+ - p_+ q_\perp}{\sqrt{p_+ q_+}}, \qquad (13.13)$$

13.2. FURTHER DEVELOPMENTS

which satisfies the the properties

$$< pq > + < qp > = 0,$$

$$|< pq >|^2 = 2(p \cdot q). \tag{13.14}$$

As a result,

$$p_{\dot{A}} q^{\dot{A}} = < pq >^* . \tag{13.15}$$

Similarly, the matrix

$$\not{k} = \begin{pmatrix} 0 & 0 & k_+ & k_\perp^* \\ 0 & 0 & k_\perp & k_- \\ k_- & -k_\perp^* & 0 & 0 \\ -k_\perp & k_+ & 0 & 0 \end{pmatrix}, \tag{13.16}$$

with $k^2 = 0$, has the following nonvanishing matrix elements

$$(\not{k})^{\dot{A}B} = k^{\dot{A}} k^B, \qquad (\not{k})_{A\dot{B}} = k_A k_{\dot{B}}. \tag{13.17}$$

One then readily finds that

$$\begin{aligned}
\bar{u}_+(p)\, u_-(q) &= p_{\dot{A}} q^{\dot{A}} = < pq >^*, \\
\bar{u}_-(p)\, u_+(q) &= p^A q_A = - < pq >, \\
\bar{u}_+(p)\, \not{k} u_+(q) &= p_{\dot{A}} k^{\dot{A}} k^B q_B = - < pk >^* < kq >, \\
\bar{u}_-(p)\, \not{k} u_-(q) &= p^A k_A k_{\dot{B}} q^{\dot{B}} = - < pk >< kq >^*, \tag{13.18}
\end{aligned}$$

etc.

Also, for the photon polarization vectors (1.6), we have that

$$\begin{aligned}
\left(\not{\epsilon}^+(k,q) \right)^{\dot{A}B} &= \frac{\sqrt{2}}{< kq >} k^{\dot{A}} q^B, \\
\left(\not{\epsilon}^+(k,q) \right)_{A\dot{B}} &= \frac{\sqrt{2}}{< kq >} q_A k_{\dot{B}}, \\
\left(\not{\epsilon}^-(k,q) \right)^{\dot{A}B} &= \frac{\sqrt{2}}{< kq >^*} q^{\dot{A}} k^B, \\
\left(\not{\epsilon}^-(k,q) \right)_{A\dot{B}} &= \frac{\sqrt{2}}{< kq >^*} k_A q_{\dot{B}}, \tag{13.19}
\end{aligned}$$

while all other components vanish.

With this formalism, one can now reformulate the entire helicity method. All helicity amplitudes are ultimately expressed in terms of spinor products (13.13), which are closely related to our quantities Z_{ij}, defined in eqn (6.13).

This program has been carried out by Berends and Giele [76], where the translation of four-component Dirac expressions in terms of Weyl-van der Waerden expressions is extensively discussed. Applications of this technique to the purely gluonic processes $gg \to gg$, $gg \to ggg$ and $gg \to gggg$ and to the neutrino counting processes $e^+e^- \to \nu\bar{\nu}\gamma$ and $e^+e^- \to \nu\bar{\nu}\gamma\gamma$ can be found in Berends et al. [76].

13.2.4 Quantum gravity

Quantum gravity is the gauge theory of gravitational interactions. Being a gauge theory, its perturbative treatment is quite similar to that of QED or QCD. The main difference, from the point of view of calculations, is the fact that the graviton is a gauge particle with spin 2. It is described by a polarization tensor $\epsilon^{\mu\nu}(k)$, which satisfies the following conditions:

$$\epsilon^{\mu\nu}(k) = \epsilon^{\nu\mu}(k), \qquad \epsilon^\mu_\mu = 0, \qquad k_\mu \epsilon^{\mu\nu} = 0. \qquad (13.20)$$

Using the polarization vectors $\epsilon^\pm_\mu(k)$, which we introduced for the photons or the gluons, it is possible to construct explicit representations for the graviton polarization tensor. It suffices to note that

$$\epsilon^\pm_{\mu\nu}(k) = \epsilon^\pm_\mu(k)\epsilon^\pm_\nu(k) \qquad (13.21)$$

satisfies all the requirements of eqns (13.20). The quantity $\epsilon^+_{\mu\nu}$ thus describes a helicity $+2$ graviton, and $\epsilon^-_{\mu\nu}$ a helicity -2 graviton.

These polarization tensors can now be introduced in the helicity amplitudes of a process involving a graviton with four-momentum k and the evaluation of the helicity amplitudes is found to proceed in complete analogy with the QED or QCD case.

This was done by Su [77], who examined several quantum gravity processes in this way. It was seen that, for single graviton bremsstrahlung, the cross section again factorized, this time into a 'graviton infrared factor' and a lower order, nonradiative cross section. Consider, for example, the cross section for

$$e^+(p_+) + e^-(p_-) \to \mu^+(q_+) + \mu^-(q_-) + G(k), \qquad (13.22)$$

in lowest order mediated through the exchange of a virtual photon. Here, $G(k)$ denotes the graviton with momentum k. The calculation then shows that the cross section, in the high-energy limit, is proportional to

13.2. FURTHER DEVELOPMENTS

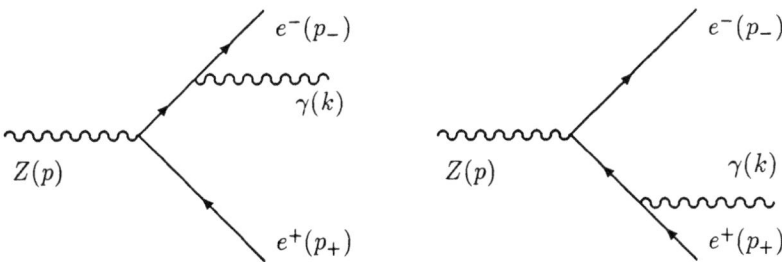

Fig. 13.1: Feynman diagrams for $Z \to e^+ e^- \gamma$.

$$\overline{|M|^2} = \frac{e^4 \kappa^2}{32} \frac{t^2 + t'^2 + u^2 + u'^2}{ss'}$$
$$\times \left[8 - \frac{s^2}{(p_+ \cdot k)(p_- \cdot k)} - \frac{s'^2}{(q_+ \cdot k)(q_- \cdot k)} + \frac{t^2}{(p_+ \cdot k)(q_+ \cdot k)} \right.$$
$$\left. + \frac{t'^2}{(p_- \cdot k)(q_- \cdot k)} + \frac{u^2}{(p_+ \cdot k)(q_- \cdot k)} + \frac{u'^2}{(p_- \cdot k)(q_+ \cdot k)} \right]. \quad (13.23)$$

In this formula, the quantity κ is related to G, Newton's gravitational constant, through $\kappa^2 = 32\pi G$, and the invariants s, s', t, t', u and u' are defined as usual by

$$s = 2(p_+ \cdot p_-), \quad t = -2(p_+ \cdot q_+), \quad u = -2(p_+ \cdot q_-),$$
$$s' = 2(q_+ \cdot q_-), \quad t' = -2(p_- \cdot q_-), \quad u' = -2(p_- \cdot q_+). \quad (13.24)$$

13.2.5 Polarization vectors for massive spin-1 particles

It is possible to adapt the helicity method to the description of processes involving massive spin-1 particles. However, as we have not worked out a general method for this purpose, we shall merely illustrate a possible way of dealing with massive spin-1 particles. To this end, we choose the process

$$Z(p) \to e^+(p_+) + e^-(p_-) + \gamma(k), \quad (13.25)$$

with the Feynman diagrams depicted in Fig. 13.1.

The corresponding Feynman amplitudes are

$$M_1 = -e^2\bar{u}(p_-)\,\rlap{/}{\epsilon}(k)\frac{\rlap{/}{p}_-+\rlap{/}{k}}{2(p_-\cdot k)}\,\rlap{/}{\epsilon}^*(p)[a(1-\gamma_5)+b(1+\gamma_5)]v(p_+),$$

$$M_2 = -e^2\bar{u}(p_-)\,\rlap{/}{\epsilon}^*(p)[a(1-\gamma_5)+b(1+\gamma_5)]\frac{-\rlap{/}{p}_+-\rlap{/}{k}}{2(p_+\cdot k)}\,\rlap{/}{\epsilon}(k)v(p_+),$$

(13.26)

with $\epsilon^*_\mu(p)$ the polarization vector of the Z-boson. The relation of the quantities a and b with the parameters of the standard model are given in eqns (2.24).

First, we choose the following polarization vectors for the photon:

$$\rlap{/}{\epsilon}^\pm(k) = N[\rlap{/}{p}_-\rlap{/}{p}_+\rlap{/}{k}(1\mp\gamma_5) - \rlap{/}{k}\,\rlap{/}{p}_-\rlap{/}{p}_+(1\pm\gamma_5)],$$

$$N^{-1} = 4[(p_+\cdot p_-)(p_+\cdot k)(p_-\cdot k)]^{\frac{1}{2}}.$$

(13.27)

It follows that the helicity amplitudes are

$$M(+,-,+) = 2e^2 aN\,\bar{u}(p_-)\,\rlap{/}{\epsilon}^*(p)(\rlap{/}{p}_++\rlap{/}{k})\,\rlap{/}{p}_-(1-\gamma_5)v(p_+),$$

$$M(-,+,+) = 2e^2 bN\,\bar{u}(p_-)\,\rlap{/}{\epsilon}^*(p)(\rlap{/}{p}_++\rlap{/}{k})\,\rlap{/}{p}_-(1+\gamma_5)v(p_+),$$

$$M(+,-,-) = 2e^2 aN\,\bar{u}(p_-)\,\rlap{/}{p}_+(\rlap{/}{p}_-+\rlap{/}{k})\,\rlap{/}{\epsilon}^*(p)(1-\gamma_5)v(p_+),$$

$$M(-,+,-) = 2e^2 bN\,\bar{u}(p_-)\,\rlap{/}{\epsilon}^*(p)(\rlap{/}{p}_++\rlap{/}{k})\,\rlap{/}{p}_-(1+\gamma_5)v(p_+).$$

(13.28)

A massive spin-1 particle has three polarization states. Consider the following three expressions

$$\rlap{/}{\epsilon}^*_1(p) = N[\rlap{/}{p}'\,\rlap{/}{p}_+\,\rlap{/}{p}_-(1+\gamma_5) + \rlap{/}{p}_-\,\rlap{/}{p}_+\,\rlap{/}{p}'(1-\gamma_5)],$$

$$\rlap{/}{\epsilon}^*_2(p) = N[\rlap{/}{p}'\,\rlap{/}{p}_+\,\rlap{/}{p}_-(1-\gamma_5) + \rlap{/}{p}_-\,\rlap{/}{p}_+\,\rlap{/}{p}'(1+\gamma_5)],$$

$$\rlap{/}{\epsilon}^*_3(p) = \frac{1}{M_Z}\left[\rlap{/}{p} - \frac{M_Z^2}{(p\cdot p_-)}\,\rlap{/}{p}_-\right],$$

(13.29)

with

$$p' = p_+ + k - \frac{(p_+\cdot k)}{(p\cdot p_-)}p_-.$$

(13.30)

The expressions $\rlap{/}{\epsilon}^*_i(p)$, $i=1,2,3$, satisfy the relations $(p\cdot\epsilon^*_i)=0$. Furthermore, they are orthogonal to one another, i.e.,

$$(\epsilon^*_i\cdot\epsilon_j) = 0, \qquad i\neq j,$$

(13.31)

13.2. FURTHER DEVELOPMENTS

and they are normalized:

$$(\epsilon_i^* \cdot \epsilon_i) = -1, \qquad i = 1, 2, 3. \tag{13.32}$$

These expressions are thus possible polarization states for the Z-particle. It then suffices to evaluate the helicity amplitudes, eqns (13.28), for each $\not{\epsilon}_i^*(p)$, $i = 1, 2, 3$, eqns (13.29), and to square the absolute values of the resulting expressions to obtain the cross section.

A simple calculation shows that the helicity amplitude $M(+, -, +)$ yields the following contributions upon insertion of $\not{\epsilon}_i^*(p)$:

$$M_1(+, -, +) = 0,$$

$$M_2(+, -, +) = \frac{e^2 a(p \cdot p_-)}{(p_+ \cdot k)(p_- \cdot k)} \bar{u}(p_-) \not{k}(1 - \gamma_5) v(p_+),$$

$$M_3(+, -, +) = 0. \tag{13.33}$$

Consequently,

$$\sum_{\text{pol}} |M(+, -, +)|^2 = \frac{16 e^4 a^2 (p \cdot p_-)^2}{(p_+ \cdot k)(p_- \cdot k)}. \tag{13.34}$$

The remaining helicity amplitudes are readily evaluated in the same way. The unpolarized squared matrix element, summed over the final state polarization degrees of freedom and averaged over the polarizations of the Z-boson, then becomes [78]

$$\overline{|M|^2} = \frac{16 e^4 (a^2 + b^2)}{3} \frac{(p \cdot p_+)^2 + (p \cdot p_-)^2}{(p_+ \cdot k)(p_- \cdot k)}. \tag{13.35}$$

Very similar polarization vectors describing the spin states of the Z were used by Böhm and Sack [79], who study the related process $e^+ e^- \to \gamma Z$.

Of course, this example is so simple that one could equally well use the standard techniques of summing over the polarization states of the Z. However, for more complicated situations, the method we presented here could prove to be more convenient. For example, if the polarization vector of the Z appears in a Feynman diagram in a closed fermion loop, our method allows one to calculate the associated trace and to express the result in terms of four-momenta only. Of course, the price to be paid is that one must calculate the trace thrice.

The way of introducing polarization vectors for massive spin-1 particles, like we did in eqns (13.29), is by no means unique. An alternative method, using massive spinors to define ϵ_i, $i = 1, 2, 3$, has been presented by Passarino [80].

13.3 Unsolved problems

Regarding beamstrahlung and associated topics such as pair production, there are numerous unsolved problems as already discussed in Chapter 12. Many of these unsolved problems are critical for the actual design of a future multi-TeV electron-positron collider. In this section, we list a number of other important and unsolved problems; they are not directly related to beamstrahlung.

13.3.1 Collinear fermions and gluons

In Chapters 3–6, we developed the helicity method in the high-energy limit, where the fermion masses were neglected. In Section 3.4, we stated the ranges of validity for this approximation: the energy of the incoming particles must be sufficiently large and no collinear particles should be present.

In Chapter 7, we showed how the corrections due to a finite fermion mass could be taken into account at the level of the spin amplitudes. This allowed us to treat the case of photon or gluon emission in directions nearly collinear to a fermion direction to leading order in m/E. Here, m denotes the fermion mass and E is the incoming energy.

The helicity method developed there does not allow us to describe the cases of nearly collinear fermions or nearly collinear gluons. In Section 3.4, we gave some examples of configurations where difficulties of this kind arise. It is not unlikely that a more or less simple modification of the helicity method can be devised which would permit the treatment of these special configurations, but it remains to be worked out.

To treat the case of collinear fermions, Berends, Daverveldt and Kleiss [81] propose to evaluate the spin amplitudes for a given process exactly, i.e., without neglecting the fermion masses. To this end, they first rewrite the massive spinors as

$$u_+(p) = \frac{1}{\sqrt{2(p \cdot k_0)}} (\slashed{p} + m) u_-(k_0),$$

$$u_-(p) = \frac{1}{\sqrt{2(p \cdot k_0)}} (\slashed{p} + m) \slashed{k}_1 u_-(k_0), \qquad (13.36)$$

where $u_-(k_0)$ is a negative helicity massless spinor. The four-vectors k_0 and k_1 must be generally positioned and satisfy the relations

$$k_0^2 = 0, \qquad k_1^2 = -1, \qquad (k_0 \cdot k_1) = 0. \qquad (13.37)$$

The spin amplitudes can then be expressed in terms of the components of the different four-momenta in the usual way, and the cross section is obtained by adding the squared absolute values of the spin amplitudes.

13.3. UNSOLVED PROBLEMS

This technique has been applied to the calculation of the cross section for $e^+e^- \to e^+e^-\ell^+\ell^-$ ($\ell = e, \mu, \tau$) by these authors and by Maña and Martinez [82] to $e^+e^- \to e^+e^-\gamma$ for outgoing fermions near the beam direction.

It is clear, however, that, with this method, the formulae are not simple as many mass terms are retained which could in fact be neglected.

13.3.2 Relation to quaternions

In the helicity method, one frequently deals with expressions of the type

$$A_\pm = \ldots \not{k}_a \not{k}_b \not{k}_c \not{k}_d \, u_\pm(k_e). \tag{13.38}$$

Although we use four-dimensional γ-matrices, the problem is essentially a two-dimensional one. Indeed, we have that [see eqns (6.10),(6.11)]

$$u_+(p) = \begin{pmatrix} \chi_+ \\ 0 \end{pmatrix}, \qquad u_-(p) = \begin{pmatrix} 0 \\ \chi_- \end{pmatrix}, \tag{13.39}$$

with

$$\chi_+ = \frac{1}{\sqrt{p_+}} (p_0 \mathbb{1} + \vec{p} \cdot \vec{\sigma}) \begin{pmatrix} 1 \\ 0 \end{pmatrix},$$

$$\chi_- = \frac{1}{\sqrt{p_+}} (p_0 \mathbb{1} - \vec{p} \cdot \vec{\sigma}) \begin{pmatrix} 0 \\ 1 \end{pmatrix}. \tag{13.40}$$

The σ-matrices are the familiar Pauli matrices, given in eqns (2.7). With our representation (2.6) of the γ-matrices, we have

$$\not{k} = \begin{pmatrix} 0 & k_0 + \vec{k} \cdot \vec{\sigma} \\ k_0 - \vec{k} \cdot \vec{\sigma} & 0 \end{pmatrix}. \tag{13.41}$$

Using eqn (13.41) in the evaluation of A_+, eqn (13.38), we find that the four-dimensional matrix multiplication reduces to a two-dimensional one because of the vanishing of the lower components of u_+, eqn (13.39). We are thus left with problem of evaluating

$$\ldots (k_{a0} + \vec{k}_a \cdot \vec{\sigma})(k_{b0} - \vec{k}_b \cdot \vec{\sigma})(k_{c0} + \vec{k}_c \cdot \vec{\sigma})(k_{d0} - \vec{k}_d \cdot \vec{\sigma})(k_{e0} + \vec{k}_e \cdot \vec{\sigma}). \tag{13.42}$$

This problem can be formulated in a different way by introducing quaternions. A quaternion q [83] is a linear combination of the type

$$q = q_0 + q_1 e_1 + q_2 e_2 + q_3 e_3, \tag{13.43}$$

where the quantities q_i, $i = 0, \ldots, 3$, are ordinary (complex) numbers and where the imaginary units e_i, $i = 1, \ldots, 3$, satisfy the following multiplication rules

$$e_i e_j = \epsilon_{ijk} e_k - \delta_{ij}, \tag{13.44}$$

with ϵ_{ijk} the totally antisymmetric tensor in 3 dimensions ($\epsilon_{123} = +1$).

It is immediately seen that the matrices $-i\vec{\sigma}$ obey the same multiplication rules as the units e_i. We can thus represent the expression $k_0 \mathbb{1} + \vec{k} \cdot \vec{\sigma}$ by the quaternion q, if we identify

$$q_0 = k_0, \qquad q_1 = ik_x, \qquad q_2 = ik_y, \qquad q_3 = ik_z. \qquad (13.45)$$

Similarly, the expression $k_0 \mathbb{1} - \vec{k} \cdot \vec{\sigma}$ can be represented by $\bar{q}$, where $\bar{q}$ is the (quaternionic) conjugate quaternion:

$$\bar{q} = k_0 - ik_x e_1 - ik_y e_2 - ik_z e_3. \qquad (13.46)$$

The problem of the evaluation of the matrix product (13.42) is thus reduced to a multiplication of quaternions of the type

$$\ldots q_a \bar{q}_b q_c \bar{q}_d q_e. \qquad (13.47)$$

It could very well be that, for the evaluation of long expressions of the type (13.42), a computer would need less time to perform the associated quaternionic multiplications than for the straightforward two-dimensional matrix multiplications.

13.3.3 Loops in Feynman amplitudes

The helicity method produced some remarkable simplifications in the calculation of bremsstrahlung processes in the high-energy limit. This was achieved trough the introduction, in a covariant way, of explicit polarization vectors for the radiated gluons and/or photons.

It seems natural to suppose that similar ideas could be fruitful also for the calculation of loop corrections at high energies and large momentum transfers. The external field approximation used in Chapter 12 is in fact derived from multiloop Feynman diagrams [84]. However, this is a case where the transverse momentum transfer is small and, hence, qualitatively very different from our concerns here.

In Appendix B, we present the calculation of the process $\gamma\gamma \to \gamma\gamma$, which involves a loop integral in the lowest order Feynman amplitudes. Through a good choice for the photon polarization vectors, we could circumvent the problem of ultraviolet divergences which are normally encountered with the standard procedures. From this example, it appears that some amount of simplification occurs and this is presumably also true for other loop calculations.

Although the helicity method reduced the number of terms for this example, we did not succeed in developing a useful technique based on the helicity method for the Feynman integrals themselves. We tried out several ideas on the matter, but we must admit that we could not come up with anything we dare to present: all our efforts produced more complications than simplifications...

13.4 Epilogue

'When a macroscopic object is illuminated by an electromagnetic wave, the resulting field is, in general, fairly complicated and not easily calculated. If the wavelength of the illuminating electromagnetic wave is small compared with the dimensions of the object, however, geometrical optics takes over, and reflection and refraction become a suitable and sufficient description for many situations. As a concrete example, let us imagine that we are in a room equipped with a red light bulb at one instant and with a blue light bulb at a later time. It is a common experience that, as the light bulb is changed, the objects in the room change their colors but *not* their shapes and sizes. This implies, in particular, that the scattering cross section does not depend on the light frequency. That the shapes and sizes are independent of the wavelength simplifies the physics—and our daily life—greatly. Just imagine trying to sit in a chair that has different shapes for different color rays in the sunbeam!' [85]

Compared with the classical case, the quantum situation is much more subtle and interesting. Indeed, both classically and quantum mechanically, simplifications can be expected and do occur when the wavelength of the incident particle is much smaller than the sizes of the interacting objects. In the classical case, geometrical and scaling rules apply, albeit with due complications from caustics including foci and shadow boundaries.

The situation which is both physically important and well understood is the elastic and diffractive scattering of hadrons at very high energies. Here, the behaviour of the total cross sections (obtained through the optical theorem) and elastic differential cross sections was predicted theoretically [86] and later verified experimentally in the case of proton-antiproton scattering [87]. In particular, both the total cross section and the ratio of the integrated elastic cross section to the total cross section increase with increasing energy.

The situation treated in the present book is a second case where great simplifications occur when the centre-of-mass energy is much larger than the rest masses of the scattering and produced particles. This simplification was first found by direct calculation, as mentioned in the Preface, and later understood through the helicity method, which is the central subject here.

It is natural to speculate that there are other situations in particle physics where similar simplifications occur. This speculation must be true. Nevertheless, at the moment, it is by no means clear where to look. Let us conclude this book with a possible suggestion. It is the best one we can think of; whether it will work remains to be seen.

In the first instance given above, the total energy is very large, but the transverse momentum transfers are not large compared with the rest masses of the particles. In the second instance, the subject of this book, both the total energy and the transverse momentum transfers are very large compared with the rest masses and are of the same order of magnitude in most cases. An intermediate instance, between these two, suggests itself, where it is perhaps not unreasonable to also expect simplifications.

In this intermediate case, the total energy is very large compared with the transverse momentum transfers, which are in turn much larger than the rest masses of the particles. For particle physics, this case can perhaps be expected to be important at the SSC (Superconducting Super Collider) in Texas and the LHC (Large Hadron Collider) at CERN, where the centre-of-mass energies are respectively 17 and 40 GeV, while the transverse momentum transfers are very roughly of the order of 100 GeV/c. Since this case is intermediate, features of both the first and the second instances are expected to appear. A very preliminary step in this direction was taken a few years ago [88]. If our speculation is right, we can look forward to new fascinating developments in particle physics. *Felix qui potuit rerum cognoscere causas!*

Appendix A
Traces: cut-and-paste

Manipulations of spinor expressions and traces of γ-matrices are key ingredients for an efficient application of the helicity amplitude method. Many formulae are known in the literature [89] for rewriting products of traces of the type $\text{Tr}[...\gamma_\mu...]\,\text{Tr}[...\gamma^\mu...]$ or similarly for spinor expressions. Unfortunately, it is not always easy to remember these formulae as they often differ from case to case.

In the helicity amplitude method, one deals with expressions involving the projection operators $(1 \pm \gamma_5)/2$ and contractions of the type $\not{k}$ with $k^2 = 0$. For these situations, it is often possible to factorize a given expression or conversely to combine a product of expressions into a single one. It is the purpose of this appendix to show how this 'cut-and-paste' technique can easily be implemented without the need of remembering more or less complicated formulae.

The starting point in this game is the observation that, for a massless spinor $u(p)$,

$$(1 \pm \gamma_5)u(p)\bar{u}(p)(1 \mp \gamma_5) = (1 \pm \gamma_5)\Big(\sum_{\text{pol}} u(p)\bar{u}(p)\Big)(1 \mp \gamma_5)$$

$$= 2(1 \pm \gamma_5)\,\not{p}. \tag{A.1}$$

In this way, a product of spinor expressions can be combined in a single spinor expression, or, by using the formula backwards, one can factorize it. The same is true for traces involving $(1 \pm \gamma_5)$. For example, for $p^2 = q^2 = 0$,

$$\text{Tr}[\not{a}\,\not{b}\,\not{p}\,\not{c}\,\not{d}\,\not{q}(1+\gamma_5)]$$

$$= \tfrac{1}{4}\text{Tr}[\not{a}\,\not{b}(1+\gamma_5)u(p)\,\bar{u}(p)\,\not{c}\,\not{d}(1-\gamma_5)u(q)\,\bar{u}(q)(1+\gamma_5)]$$

$$= \tfrac{1}{2}\bar{u}(p)\,\not{c}\,\not{d}(1-\gamma_5)u(q)\,\bar{u}(q)\,\not{a}\,\not{b}(1+\gamma_5)u(p). \tag{A.2}$$

This property is useful for generating overall factors consisting of scalar products. If, in the previous example, $b = c$ with $b^2 = 0$, the l.h.s. is equal to

$$2(b \cdot p)\text{Tr}[\not{a}\,\not{b}\,\not{d}\,\not{q}(1+\gamma_5)], \tag{A.3}$$

which contains a factor $(b \cdot p)$, whereas in the r.h.s. none of the two factors do. Of course, it is hidden in some way in the r.h.s.

This technique is also very useful for simplifying expressions which contain repeated indices, for example,

$$M = \bar{u}(k_1)\gamma_\mu(1-\gamma_5)u(k_2)\bar{u}(k_3)\gamma^\mu(1+\gamma_5)u(k_4), \tag{A.4}$$

where k_2 and k_3 are light-like vectors. Obviously,

$$\begin{aligned} M &= \frac{\bar{u}(k_1)\gamma_\mu(1-\gamma_5)u(k_2)\bar{u}(k_2)(1+\gamma_5)u(k_3)\bar{u}(k_3)\gamma^\mu(1+\gamma_5)u(k_4)}{\bar{u}(k_2)(1+\gamma_5)u(k_3)} \\ &= 4\frac{\bar{u}(k_1)\gamma_\mu \not{k}_2 \not{k}_3 \gamma^\mu u(k_4)}{\bar{u}(k_2)(1+\gamma_5)u(k_3)} \\ &= 16(k_2 \cdot k_3)\frac{\bar{u}(k_1)(1+\gamma_5)u(k_4)}{\bar{u}(k_2)(1+\gamma_5)u(k_3)} \\ &= 16(k_2 \cdot k_3)\frac{\bar{u}(k_1)(1+\gamma_5)u(k_4)\bar{u}(k_3)(1-\gamma_5)u(k_2)}{\text{Tr}[\not{k}_2(1+\gamma_5)\not{k}_3(1-\gamma_5)]} \\ &= 2\bar{u}(k_1)(1+\gamma_5)u(k_4)\bar{u}(k_3)(1-\gamma_5)u(k_2). \end{aligned} \tag{A.5}$$

This expression no longer contains repeated indices.

In this example, things worked out nicely because the combination $1-\gamma_5$ in the first spinor expression matched the $1+\gamma_5$ combination in the second one. What if the $1+\gamma_5$ factor is changed into $1-\gamma_5$? Take

$$M' = \bar{u}(k_1)\gamma_\mu(1-\gamma_5)u(k_2)\bar{u}(k_3)\gamma^\mu(1-\gamma_5)u(k_4), \tag{A.6}$$

and let us perform similar manipulations, but with an extra four-vector, a, as follows

$$M' = \frac{\bar{u}(k_1)\gamma_\mu(1-\gamma_5)u(k_2)\bar{u}(k_2)\not{a}(1-\gamma_5)u(k_3)\bar{u}(k_3)\gamma^\mu(1-\gamma_5)u(k_4)}{\bar{u}(k_2)\not{a}(1-\gamma_5)u(k_3)}. \tag{A.7}$$

Clearly, we must choose $a \neq k_2, k_3$. Then,

$$\begin{aligned} M' &= 4\frac{\bar{u}(k_1)\gamma_\mu \not{k}_2 \not{a} \not{k}_3 \gamma^\mu(1-\gamma_5)u(k_4)}{\bar{u}(k_2)\not{a}(1-\gamma_5)u(k_3)} \\ &= -8\frac{\bar{u}(k_1)\not{k}_3 \not{a} \not{k}_2(1-\gamma_5)u(k_4)}{\bar{u}(k_2)\not{a}(1-\gamma_5)u(k_3)}. \end{aligned} \tag{A.8}$$

If k_1 is also light-like, we can choose $a = k_1$, and rewrite M'

$$\begin{aligned} M' &= -16(k_1 \cdot k_3)\frac{\bar{u}(k_1)\not{k}_2(1-\gamma_5)u(k_4)}{\bar{u}(k_2)\not{k}_1(1-\gamma_5)u(k_3)} \\ &= -16(k_1 \cdot k_3)\frac{\bar{u}(k_1)\not{k}_2(1-\gamma_5)u(k_4)\bar{u}(k_3)\not{k}_1(1-\gamma_5)u(k_2)}{\text{Tr}[\not{k}_2 \not{k}_1(1-\gamma_5)\not{k}_3(1+\gamma_5)\not{k}_1]} \\ &= -\frac{1}{(k_1 \cdot k_2)}\bar{u}(k_1)\not{k}_2(1-\gamma_5)u(k_4)\bar{u}(k_3)\not{k}_1(1-\gamma_5)u(k_2). \end{aligned} \tag{A.9}$$

APPENDIX A. TRACES: CUT-AND-PASTE

One sees that it is again possible to eliminate the repeated indices, but only at the expense of introducing a factor $(k_1 \cdot k_2)$ in the denominator. One could also choose $a = k_4$, with $k_4^2 = 0$, in which case one obtains

$$M' = -\frac{1}{(k_3 \cdot k_4)} \bar{u}(k_1) \not{k}_3 (1-\gamma_5) u(k_4) \bar{u}(k_3) \not{k}_4 (1-\gamma_5) u(k_2), \quad (A.10)$$

but now, $(k_3 \cdot k_4)$ is present in the denominator. It thus appears that there are several ways of rewriting M', and one is free to choose the most convenient one.

Another useful formula is

$$\gamma_\mu \text{Tr}[\gamma^\mu \not{a} \not{b} \not{c}(1 \pm \gamma_5)] = 2[\not{a} \not{b} \not{c}(1 \pm \gamma_5) + \not{c} \not{b} \not{a}(1 \mp \gamma_5)]. \quad (A.11)$$

Its proof goes as follows. Consider

$$\not{a} \not{b} \not{c}(1 \pm \gamma_5) + \not{c} \not{b} \not{a}(1 \mp \gamma_5), \quad (A.12)$$

which is a 4×4 matrix. It can be decomposed in the 16 independent combinations of γ-matrices: $\mathbb{1}$, γ_μ, $[\gamma_\mu, \gamma_\nu]$, $\gamma_\mu \gamma_5$ and γ_5:

$$\not{a} \not{b} \not{c}(1 \pm \gamma_5) + \not{c} \not{b} \not{a}(1 \mp \gamma_5)$$

$$= S\mathbb{1} + V^\mu \gamma_\mu + T^{\mu\nu}[\gamma_\mu, \gamma_\nu] + A^\mu \gamma_\mu \gamma_5 + P\gamma_5. \quad (A.13)$$

Multiplying both sides of this equation with the 16 independent γ-matrices and taking the trace, one finds

$$S = 0, \qquad T^{\mu\nu} = 0, \qquad A^\mu = 0, \qquad P = 0,$$

$$V^\mu = \tfrac{1}{4} \text{Tr}[\gamma^\mu \not{a} \not{b} \not{c}(1 \pm \gamma_5) + \gamma^\mu \not{c} \not{b} \not{a}(1 \mp \gamma_5)]$$

$$= \tfrac{1}{2} \text{Tr}[\gamma^\mu \not{a} \not{b} \not{c}(1 \pm \gamma_5)]. \quad (A.14)$$

Hence,

$$\gamma_\mu \text{Tr}[\gamma^\mu \not{a} \not{b} \not{c}(1 \pm \gamma_5)] = 2\gamma_\mu V^\mu, \quad (A.15)$$

which completes the proof.

Appendix B
The process $\gamma\gamma \to \gamma\gamma$

The usefulness of the helicity method is by no means restricted to calculations involving tree diagrams. In this appendix, we illustrate its applicability to a simple process described, in lowest order, by a set of one-loop Feynman diagrams:

$$\gamma(k_1) + \gamma(k_2) \to \gamma(k_3) + \gamma(k_4). \tag{B.1}$$

There are six one-loop Feynman diagrams, which are given in Fig. B.1. They were first evaluated by Karplus and Neuman [90], using the standard techniques. It is easily seen that the last three diagrams are equal to the first three. We therefore only calculate the first three and give them an extra factor 2. If the photons are all acollinear, then the electron mass can be neglected and the Feynman amplitudes are readily written down:

$$M = 2(M_1 + M_2 + M_3),$$

$$M_1 = ie^4 \int \frac{d^4p}{(2\pi)^4} \text{Tr}[\not{\epsilon}_1^* \not{p} \not{\epsilon}_4(\not{p}+\not{k}_4) \not{\epsilon}_3(\not{p}+\not{k}_1+\not{k}_2) \not{\epsilon}_2^*(\not{p}+\not{k}_1)]$$

$$\times \frac{1}{[p^2+i\epsilon][(p+k_4)^2+i\epsilon][(p+k_1+k_2)^2+i\epsilon][(p+k_1)^2+i\epsilon]},$$

$$M_2 = ie^4 \int \frac{d^4p}{(2\pi)^4} \text{Tr}[\not{\epsilon}_1^* \not{p} \not{\epsilon}_3(\not{p}+\not{k}_3) \not{\epsilon}_2^*(\not{p}+\not{k}_1-\not{k}_4) \not{\epsilon}_4(\not{p}+\not{k}_1)]$$

$$\times \frac{1}{[p^2+i\epsilon][(p+k_3)^2+i\epsilon][(p+k_1-k_4)^2+i\epsilon][(p+k_1)^2+i\epsilon]},$$

$$M_3 = ie^4 \int \frac{d^4p}{(2\pi)^4} \text{Tr}[\not{\epsilon}_1^* \not{p} \not{\epsilon}_3(\not{p}+\not{k}_3) \not{\epsilon}_4(\not{p}+\not{k}_1+\not{k}_2) \not{\epsilon}_2^*(\not{p}+\not{k}_1)]$$

$$\times \frac{1}{[p^2+i\epsilon][(p+k_3)^2+i\epsilon][(p+k_1+k_2)^2+i\epsilon][(p+k_1)^2+i\epsilon]}. \tag{B.2}$$

The standard procedure for calculating one-loop Feynman diagrams consists of first combining the denominators into a single denominator. This is done by repeatedly using the trick of introducing extra integration variables, called Feynman parameters [91]:

$$\frac{1}{A^\alpha B^\beta} = \frac{\Gamma(\alpha+\beta)}{\Gamma(\alpha)\Gamma(\beta)} \int_0^1 dx \frac{x^{\alpha-1}(1-x)^{\beta-1}}{[Ax+B(1-x)]^{\alpha+\beta}}. \tag{B.3}$$

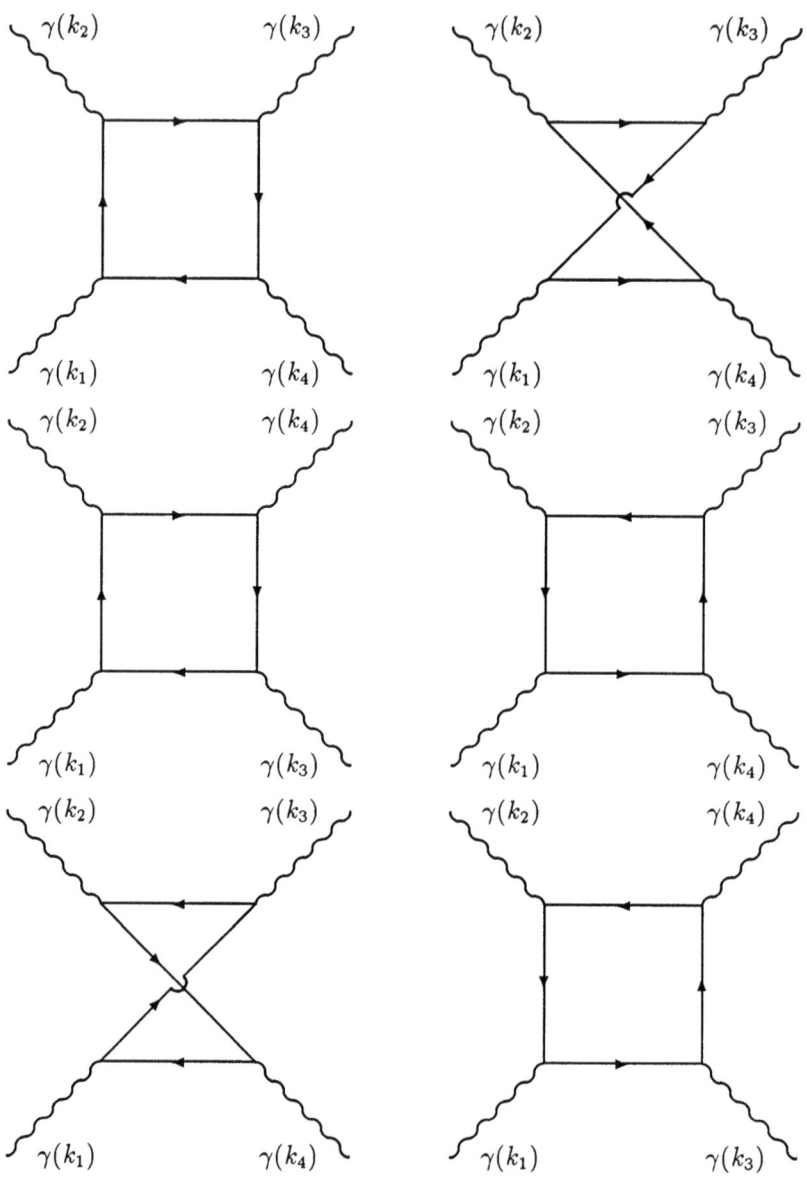

Fig. B.2: Feynman diagrams for $\gamma + \gamma \to \gamma + \gamma$.

APPENDIX B. THE PROCESS $\gamma\gamma \to \gamma\gamma$

In this way, the denominator is cast in the form

$$p^2 + 2(p \cdot K) + L + i\epsilon, \tag{B.4}$$

where, in general, the four-vector K is a linear combination of the four-vectors in the process. The quantity L is then a scalar quantity depending on the invariants. Both K and L depend on the Feynman parameters.

The next step consists in shifting the integration variable

$$p \to p - K, \tag{B.5}$$

which makes the denominator a function of p^2 only, i.e.,

$$p^2 + L - K^2 + i\epsilon. \tag{B.6}$$

In this case, the shift (B.5) can be performed without paying attention to eventual surface terms as the integrals (B.2) are only logarithmically divergent.

After these manipulations, the expression M_1, for example, reads

$$M_1 = ie^4 \int \frac{d^4p}{(2\pi)^4} \int_0^1 dx \int_0^1 dy \int_0^1 dz \, \text{Tr}[\not{\epsilon}_1^*(\not{p} - \not{K}) \not{\epsilon}_4(\not{p} - \not{K} + \not{k}_4)$$

$$\not{\epsilon}_3(\not{p} - \not{K} + \not{k}_1 + \not{k}_2) \not{\epsilon}_2^*(\not{p} - \not{K} + \not{k}_1)]$$

$$\times \frac{6z(1-z)}{\{p^2 + z(1-z)[sy(1-x) + ux(1-y)] + i\epsilon\}^4}, \tag{B.7}$$

with

$$s = (k_1 + k_2)^2, \qquad t = (k_1 - k_3)^2, \qquad u = (k_1 - k_4)^2, \tag{B.8}$$

and

$$K = (1-z)k_1 + y(1-z)k_2 + xzk_4. \tag{B.9}$$

We now proceed to calculate the various helicity amplitudes for this process denoted by $M(\lambda_1, \lambda_2; \lambda_3, \lambda_4)$. Suppose we first evaluate $M(+,+;-,+)$. To this end, we choose

$$\not{\epsilon}_1^{+*} = N[\not{k}_4 \not{k}_3 \not{k}_1(1+\gamma_5) + \not{k}_1 \not{k}_3 \not{k}_4(1-\gamma_5)],$$

$$\not{\epsilon}_2^{+*} = N[\not{k}_4 \not{k}_3 \not{k}_2(1+\gamma_5) + \not{k}_2 \not{k}_3 \not{k}_4(1-\gamma_5)],$$

$$\not{\epsilon}_3^{-} = N[\not{k}_4 \not{k}_1 \not{k}_3(1+\gamma_5) + \not{k}_3 \not{k}_1 \not{k}_4(1-\gamma_5)],$$

$$\not{\epsilon}_4^{+} = N[\not{k}_4 \not{k}_1 \not{k}_3(1+\gamma_5) + \not{k}_3 \not{k}_1 \not{k}_4(1-\gamma_5)], \tag{B.10}$$

with
$$N = (2stu)^{-\frac{1}{2}}.\qquad\text{(B.1)}$$

Note that photons 3 and 4 have the same polarization vectors, which possible because of their opposite helicities.

Inserting eqns (B.10) and (B.11) into the expression (B.7) for M_1, we fi that its contribution to $M(+,+;-,+)$ reads

$$M_1(+,+;-,+) = ie^4 \int \frac{d^4p}{(2\pi)^4} \int_0^1 dx \int_0^1 dy \int_0^1 dz\, 6z(1-z)$$

$$\times \frac{[T_1^{(+)} + T_1^{(-)}]}{\left[p^2 + z(1-z)[sy(1-x) + ux(1-y)] + i\epsilon\right]^4},\qquad\text{(B.1)}$$

with

$$T_1^{(+)} = \frac{2}{(stu)^2}\mathrm{Tr}\Big[\slashed{k}_1\,\slashed{k}_3\,\slashed{k}_4(\slashed{p}-\slashed{K})\,\slashed{k}_3\,\slashed{k}_1\,\slashed{k}_4(\slashed{p}-\slashed{K}+\slashed{k}_4)$$

$$\times\,\slashed{k}_3\,\slashed{k}_1\,\slashed{k}_4(\slashed{p}-\slashed{K}+\slashed{k}_1+\slashed{k}_2)\,\slashed{k}_2\,\slashed{k}_3\,\slashed{k}_4(\slashed{p}-\slashed{K}+\slashed{k}_1)(1+\gamma_5)$$

$$T_1^{(-)} = \frac{2}{(stu)^2}\mathrm{Tr}\Big[\slashed{k}_4\,\slashed{k}_3\,\slashed{k}_1(\slashed{p}-\slashed{K})\,\slashed{k}_4\,\slashed{k}_1\,\slashed{k}_3(\slashed{p}-\slashed{K}+\slashed{k}_4)$$

$$\times\,\slashed{k}_4\,\slashed{k}_1\,\slashed{k}_3(\slashed{p}-\slashed{K}+\slashed{k}_1+\slashed{k}_2)\,\slashed{k}_4\,\slashed{k}_3\,\slashed{k}_2(\slashed{p}-\slashed{K}+\slashed{k}_1)(1-\gamma_5)$$
(B.1)

We now want to perform the p-integral. As the numerator of expre sion (B.12) contains four powers of p and the denominator only eight powe of p, we can expect a logarithmic divergence in the integral. We thus have regularize the expression (B.12). If we take a regularization procedure whi respects Lorentz covariance, we can replace

$$p_\alpha p_\beta p_\gamma p_\delta \to A(p^2)^2(g_{\alpha\beta}g_{\gamma\delta} + g_{\alpha\gamma}g_{\beta\delta} + g_{\alpha\delta}g_{\beta\gamma}),$$

$$p_\alpha p_\beta \to \tfrac{1}{4}p^2 g_{\alpha\beta},\qquad\text{(B.1)}$$

where the coefficient A depends on the chosen regularization scheme. I symmetric integration, we can omit all the terms with odd powers in p. T replacement (B.14) in eqns (B.13) then shows that, with our choice (B.1 of polarization vectors, the $(p^2)^2$ and p^2 terms vanish. To see this, it suffic to use the formulae

$$\gamma_\alpha\,\slashed{A}\,\slashed{B}\,\slashed{C}\,\gamma^\alpha = -2\,\slashed{C}\,\slashed{B}\,\slashed{A},$$

$$\gamma_\alpha\,\slashed{A}\,\slashed{B}\,\slashed{C}\,\slashed{D}\,\slashed{E}\,\gamma^\alpha = -2\,\slashed{E}\,\slashed{D}\,\slashed{C}\,\slashed{B}\,\slashed{A}.\qquad\text{(B.1)}$$

APPENDIX B. THE PROCESS $\gamma\gamma \to \gamma\gamma$

Moreover,

$$T_1^{(+)} = \frac{-2}{(stu)^2} y(1-y)^2 z(1-z)^3$$

$$\times \text{Tr}[\not{k}_1 \not{k}_3 \not{k}_4 \not{k}_2 \not{k}_3 \not{k}_1 \not{k}_4 \not{k}_2 \not{k}_3 \not{k}_1 \not{k}_4 \not{k}_1 \not{k}_2 \not{k}_3 \not{k}_4 \not{k}_2]$$

$$= -4uty(1-y)^2 z(1-z)^3,$$

$$T_1^{(-)} = \frac{-2}{(stu)^2} y(1-y)^2 z(1-z)^3$$

$$\times \text{Tr}[\not{k}_4 \not{k}_3 \not{k}_1 \not{k}_2 \not{k}_4 \not{k}_1 \not{k}_3 \not{k}_2 \not{k}_4 \not{k}_1 \not{k}_3 \not{k}_2 \not{k}_4 \not{k}_3 \not{k}_2 \not{k}_1]$$

$$= -4uty(1-y)^2 z(1-z)^3. \tag{B.16}$$

With the formula

$$i \int \frac{d^4p}{(2\pi)^4} \frac{1}{[p^2 - C + i\epsilon]^\alpha} = \frac{(-1)^{\alpha+1}}{16\pi^2} \frac{\Gamma(\alpha-2)}{\Gamma(\alpha)} [C - i\epsilon]^{2-\alpha}, \qquad (\alpha > 2), \tag{B.17}$$

we obtain

$$M_1(+,+;-,+) = \frac{e^4}{16\pi^2} \int_0^1 dx \int_0^1 dy \int_0^1 dz \frac{8uty(1-y)^2(1-z)^2}{[sy(1-x) + ux(1-y) + i\epsilon]^2}$$

$$= \frac{e^4}{12\pi^2} \frac{t}{s}. \tag{B.18}$$

Similar manipulations on M_2 and M_3 yield

$$M_2(+,+;-,+) = -\frac{e^4}{6\pi^2},$$

$$M_3(+,+;-,+) = \frac{e^4}{12\pi^2} \frac{u}{s}. \tag{B.19}$$

Adding these three contributions and taking into account the factor 2 of eqn (B.2), then leads to

$$M(+,+;-,+) = -\frac{e^4}{2\pi^2}. \tag{B.20}$$

For the calculation of the helicity amplitude $M(+,+;+,+)$, we choose

$$\not{\epsilon}_1^{+*} = N[\not{k}_2 \not{k}_3 \not{k}_1(1+\gamma_5) + \not{k}_1 \not{k}_3 \not{k}_2(1-\gamma_5)],$$

$$\not{\epsilon}_2^{+*} = N[\not{k}_1 \not{k}_3 \not{k}_2(1+\gamma_5) + \not{k}_2 \not{k}_3 \not{k}_1(1-\gamma_5)],$$

$$\slashed{\epsilon}_3^+ = N[\slashed{k}_3\,\slashed{k}_1\,\slashed{k}_2(1+\gamma_5) + \slashed{k}_2\,\slashed{k}_1\,\slashed{k}_3(1-\gamma_5)],$$

$$\slashed{\epsilon}_4^+ = N[\slashed{k}_4\,\slashed{k}_1\,\slashed{k}_2(1+\gamma_5) + \slashed{k}_2\,\slashed{k}_1\,\slashed{k}_4(1-\gamma_5)], \quad \text{(B.21)}$$

with

$$N = (2stu)^{-\frac{1}{2}}. \quad \text{(B.22)}$$

Again, the divergent $(p^2)^2$ terms are found to vanish, but, this time, the p^2 terms contribute. They are readily evaluated using

$$i\int \frac{d^4p}{(2\pi)^4} \frac{p^2}{[p^2 - C + i\epsilon]^\alpha} = \frac{(-1)^\alpha}{16\pi^2} \frac{2\Gamma(\alpha - 3)}{\Gamma(\alpha)} [C - i\epsilon]^{3-\alpha}, \quad (\alpha > 3). \quad \text{(B.23)}$$

One is then left with the integrals over the Feynman parameters, which are slightly more complicated for this helicity amplitude. The result is

$$M(+,+;+,+) = -\frac{e^4}{2\pi^2}\left\{\frac{t^2 + u^2}{2s^2}\left[\ln^2\left(\frac{u}{t}\right) + \pi^2\right] + \frac{u-t}{s}\ln\left(\frac{u}{t}\right) + 1\right\}. \quad \text{(B.24)}$$

A convenient choice of polarization vectors for $M(+,+;-,-)$ is

$$\slashed{\epsilon}_1^{+*} = N[\slashed{k}_4\,\slashed{k}_3\,\slashed{k}_1(1+\gamma_5) + \slashed{k}_1\,\slashed{k}_3\,\slashed{k}_4(1-\gamma_5)],$$

$$\slashed{\epsilon}_2^{+*} = N[\slashed{k}_4\,\slashed{k}_3\,\slashed{k}_2(1+\gamma_5) + \slashed{k}_2\,\slashed{k}_3\,\slashed{k}_4(1-\gamma_5)],$$

$$\slashed{\epsilon}_3^- = N[\slashed{k}_4\,\slashed{k}_1\,\slashed{k}_3(1+\gamma_5) + \slashed{k}_3\,\slashed{k}_1\,\slashed{k}_4(1-\gamma_5)],$$

$$\slashed{\epsilon}_4^- = N[\slashed{k}_3\,\slashed{k}_1\,\slashed{k}_4(1+\gamma_5) + \slashed{k}_4\,\slashed{k}_1\,\slashed{k}_3(1-\gamma_5)], \quad \text{(B.25)}$$

where the normalization factor N is given by eqn (B.22). This choice of ϵ's again eliminates the $(p^2)^2$ terms, and one obtains the simple result

$$M(+,+;-,-) = \frac{e^4}{2\pi^2}. \quad \text{(B.26)}$$

All the remaining helicity amplitudes can be obtained from eqns (B.20), (B.24) and (B.26) through crossing. For example, $M(+,-;+,-)$ is obtained from $M(+,+;+,+)$ by replacing $k_2 \leftrightarrow -k_4$. As a result, $s \leftrightarrow u$ [see eqns (B.8)] and the logarithms in eqn (B.24) become complex. We find

$$M(+,-;+,-) \doteq \frac{e^4}{2\pi^2}\left\{\frac{s^2 + t^2}{2u^2}\ln^2\left(\frac{-t}{s}\right) - \frac{s-t}{u}\ln\left(\frac{-t}{s}\right) + 1\right.$$

$$\left. + i\pi\left[\frac{s^2 + t^2}{u^2}\ln\left(\frac{-t}{s}\right) - \frac{s-t}{u}\right]\right\}. \quad \text{(B.27)}$$

APPENDIX B. THE PROCESS $\gamma\gamma \to \gamma\gamma$

A complete list of helicity amplitudes for this process, including the finite fermion mass effects, can be found in Costantini, De Tollis and Pistoni [92]. They were obtained by introducing photon polarization vectors in a noncovariant way.

To summarize the results of this appendix, we can state that the helicity method is also applicable to processes involving closed-loop Feynman diagrams. For the process $\gamma + \gamma \to \gamma + \gamma$, we found that the divergence problem, associated with the integrals over the virtual momenta, could be nicely circumvented by a clever choice of polarization vectors for the photons.

Appendix C
Color traces

In this appendix, we list the color traces which are needed for the evaluation of the squared helicity amplitudes summed over the color degrees of freedom. We first briefly outline the method.

The SU(3) color matrices T^a, $a = 1, \ldots, 8$, are Hermitian and satisfy

$$\text{Tr}\,(T^a) = 0, \qquad \text{Tr}\,(T^a T^b) = \tfrac{1}{2} \delta_{ab}. \qquad (C.1)$$

Together with the 3×3 unit matrix $\mathbb{1}$, they constitute a basis for the 3×3 matrices. Any 3×3 matrix M can be written in the form

$$M = 2 \sum_a T^a \,\text{Tr}\,(MT^a) + \tfrac{1}{3} \text{Tr}\,(M)\, \mathbb{1}. \qquad (C.2)$$

Applying this relation to the matrix M, whose elements are

$$M_{ij} = \delta_{pi}\, \delta_{qj}, \qquad (C.3)$$

with p and q fixed numbers from the set $(1, 2, 3)$, we find

$$\delta_{pi}\, \delta_{qj} = 2 \sum_a T^a_{ij} \sum_{k,\ell} (\delta_{pk} \delta_{q\ell} T^a_{\ell k}) + \tfrac{1}{3} \delta_{pq} \delta_{ij}, \qquad (C.4)$$

or

$$\sum_a T^a_{ij} T^a_{qp} = \tfrac{1}{2} \delta_{pi} \delta_{qj} - \tfrac{1}{6} \delta_{pq} \delta_{ij}. \qquad (C.5)$$

This relation is very useful for the evaluation of color traces involving a summation over a color index. For example, let O_1 and O_2 be any combination of the color matrices. Then,

$$\sum_a \text{Tr}\,(T^a O_1 T^a O_2) = \sum_{i,j,p,q} \sum_a T^a_{ij}(O_1)_{jq} T^a_{qp}(O_2)_{pi}$$
$$= \tfrac{1}{2} \text{Tr}\,(O_1)\,\text{Tr}\,(O_2) - \tfrac{1}{6} \text{Tr}\,(O_1 O_2). \qquad (C.6)$$

Relation (C.5) can also be used when the matrices T^a appear in different traces. For example,

$$\sum_a \text{Tr}\,(T^a O_1)\,\text{Tr}\,(T^a O_2) = \tfrac{1}{2} \text{Tr}\,(O_1 O_2) - \tfrac{1}{6} \text{Tr}\,(O_1)\,\text{Tr}\,(O_2). \qquad (C.7)$$

In both cases, (C.6) and (C.7), the resulting expressions contain fewer color matrices and are thus easier to evaluate.

We now list the color traces which are needed for the summation over the color degrees of freedom in the squared helicity amplitudes. One can also derive many other equations by exploiting the invariance of the traces under cyclic permutation of the color matrices. In what follows, the summation over the repeated color indices $a, b, \ldots, e$ is always understood.

C.1 2 color matrices

$$\operatorname{Tr}(T^aT^a) = 4. \qquad (\text{C.8})$$

C.2 4 color matrices

$$\operatorname{Tr}(T^aT^b)\operatorname{Tr}(T^aT^b) = 2,$$

$$\operatorname{Tr}(T^aT^bT^aT^b) = -\frac{2}{3},$$

$$\operatorname{Tr}(T^aT^bT^bT^a) = \frac{16}{3}. \qquad (\text{C.9})$$

C.3 6 color matrices

$$\operatorname{Tr}(T^aT^bT^c)\operatorname{Tr}(T^aT^bT^c) = -\frac{2}{3},$$

$$\operatorname{Tr}(T^aT^bT^c)\operatorname{Tr}(T^aT^cT^b) = \frac{7}{3},$$

$$\operatorname{Tr}(T^aT^aT^bT^c)\operatorname{Tr}(T^bT^c) = \frac{8}{3},$$

$$\operatorname{Tr}(T^aT^aT^cT^b)\operatorname{Tr}(T^bT^c) = \frac{8}{3},$$

$$\operatorname{Tr}(T^aT^bT^aT^c)\operatorname{Tr}(T^bT^c) = -\frac{1}{3},$$

$$\operatorname{Tr}(T^aT^bT^cT^a)\operatorname{Tr}(T^bT^c) = \frac{8}{3},$$

$$\operatorname{Tr}(T^aT^cT^aT^b)\operatorname{Tr}(T^bT^c) = -\frac{1}{3},$$

$$\operatorname{Tr}(T^aT^cT^bT^a)\operatorname{Tr}(T^bT^c) = \frac{8}{3},$$

$$\operatorname{Tr}(T^aT^bT^cT^aT^bT^c) = \frac{10}{9},$$

$$\operatorname{Tr}(T^aT^bT^cT^aT^cT^b) = \frac{1}{9},$$

APPENDIX C. COLOR TRACES

$$\text{Tr}\,(T^a T^b T^c T^b T^a T^c) = \frac{1}{9},$$

$$\text{Tr}\,(T^a T^b T^c T^b T^c T^a) = -\frac{8}{9},$$

$$\text{Tr}\,(T^a T^b T^c T^c T^a T^b) = -\frac{8}{9},$$

$$\text{Tr}\,(T^a T^b T^c T^c T^b T^a) = \frac{64}{9}. \tag{C.10}$$

C.4   8 color matrices

$$\text{Tr}\,(T^a T^b T^c T^d)\,\text{Tr}\,(T^a T^b T^c T^d) = \frac{2}{3},$$

$$\text{Tr}\,(T^a T^b T^c T^d)\,\text{Tr}\,(T^a T^b T^d T^c) = -\frac{1}{3},$$

$$\text{Tr}\,(T^a T^b T^c T^d)\,\text{Tr}\,(T^a T^c T^b T^d) = -\frac{1}{3},$$

$$\text{Tr}\,(T^a T^b T^c T^d)\,\text{Tr}\,(T^a T^c T^d T^b) = -\frac{1}{3},$$

$$\text{Tr}\,(T^a T^b T^c T^d)\,\text{Tr}\,(T^a T^d T^b T^c) = -\frac{1}{3},$$

$$\text{Tr}\,(T^a T^b T^c T^d)\,\text{Tr}\,(T^a T^d T^c T^b) = \frac{19}{6},$$

$$\text{Re}[\text{Tr}\,(T^a T^b T^c T^d)]\,\text{Re}[\text{Tr}\,(T^a T^b T^c T^d)] = \frac{23}{12},$$

$$\text{Re}[\text{Tr}\,(T^a T^b T^c T^d)]\,\text{Re}[\text{Tr}\,(T^a T^b T^d T^c)] = -\frac{1}{3},$$

$$\text{Re}[\text{Tr}\,(T^a T^b T^c T^d)]\,\text{Re}[\text{Tr}\,(T^a T^c T^b T^d)] = -\frac{1}{3},$$

$$\text{Re}[\text{Tr}\,(T^a T^b T^c T^d)]\,\text{Re}[\text{Tr}\,(T^a T^c T^d T^b)] = -\frac{1}{3},$$

$$\text{Re}[\text{Tr}\,(T^a T^b T^c T^d)]\,\text{Re}[\text{Tr}\,(T^a T^d T^b T^c)] = -\frac{1}{3},$$

$$\text{Re}[\text{Tr}\,(T^a T^b T^c T^d)]\,\text{Re}[\text{Tr}\,(T^a T^d T^c T^b)] = \frac{23}{12}. \tag{C.11}$$

C.5   10 color matrices

$$\text{Im}[\text{Tr}\,(T^a T^b T^c T^d T^e)]\,\text{Im}[\text{Tr}\,(T^a T^b T^c T^d T^e)] = \frac{55}{24},$$

$$\text{Im}[\text{Tr}\,(T^a T^b T^c T^d T^e)]\,\text{Im}[\text{Tr}\,(T^a T^b T^c T^e T^d)] = -\frac{7}{12},$$

$$\text{Im}[\text{Tr}\,(T^a T^b T^c T^d T^e)]\,\text{Im}[\text{Tr}\,(T^a T^b T^d T^c T^e)] = -\frac{7}{12},$$

$$\text{Im}[\text{Tr}\,(T^a T^b T^c T^d T^e)]\,\text{Im}[\text{Tr}\,(T^a T^b T^d T^e T^c)] = \frac{1}{12},$$

$$\text{Im}[\text{Tr}\,(T^a T^b T^c T^d T^e)]\,\text{Im}[\text{Tr}\,(T^a T^b T^e T^c T^d)] = \frac{1}{12},$$

$$\text{Im}[\text{Tr}\,(T^a T^b T^c T^d T^e)]\,\text{Im}[\text{Tr}\,(T^a T^b T^e T^d T^c)] = \frac{7}{12},$$

$$\text{Im}[\text{Tr}\,(T^a T^b T^c T^d T^e)]\,\text{Im}[\text{Tr}\,(T^a T^c T^b T^d T^e)] = -\frac{7}{12},$$

$$\text{Im}[\text{Tr}\,(T^a T^b T^c T^d T^e)]\,\text{Im}[\text{Tr}\,(T^a T^c T^b T^e T^d)] = -\frac{1}{12},$$

$$\text{Im}[\text{Tr}\,(T^a T^b T^c T^d T^e)]\,\text{Im}[\text{Tr}\,(T^a T^c T^d T^b T^e)] = \frac{1}{12},$$

$$\text{Im}[\text{Tr}\,(T^a T^b T^c T^d T^e)]\,\text{Im}[\text{Tr}\,(T^a T^c T^d T^e T^b)] = -\frac{7}{12},$$

$$\text{Im}[\text{Tr}\,(T^a T^b T^c T^d T^e)]\,\text{Im}[\text{Tr}\,(T^a T^c T^e T^b T^d)] = 0,$$

$$\text{Im}[\text{Tr}\,(T^a T^b T^c T^d T^e)]\,\text{Im}[\text{Tr}\,(T^a T^c T^e T^d T^b)] = -\frac{1}{12},$$

$$\text{Im}[\text{Tr}\,(T^a T^b T^c T^d T^e)]\,\text{Im}[\text{Tr}\,(T^a T^d T^b T^c T^e)] = \frac{1}{12},$$

$$\text{Im}[\text{Tr}\,(T^a T^b T^c T^d T^e)]\,\text{Im}[\text{Tr}\,(T^a T^d T^b T^e T^c)] = 0,$$

$$\text{Im}[\text{Tr}\,(T^a T^b T^c T^d T^e)]\,\text{Im}[\text{Tr}\,(T^a T^d T^c T^b T^e)] = \frac{7}{12},$$

$$\text{Im}[\text{Tr}\,(T^a T^b T^c T^d T^e)]\,\text{Im}[\text{Tr}\,(T^a T^d T^c T^e T^b)] = -\frac{1}{12},$$

$$\text{Im}[\text{Tr}\,(T^a T^b T^c T^d T^e)]\,\text{Im}[\text{Tr}\,(T^a T^d T^e T^b T^c)] = \frac{1}{12},$$

APPENDIX C. COLOR TRACES

$$\text{Im}[\text{Tr}\,(T^aT^bT^cT^dT^e)]\,\text{Im}[\text{Tr}\,(T^aT^dT^eT^cT^b)] = \frac{7}{12},$$

$$\text{Im}[\text{Tr}\,(T^aT^bT^cT^dT^e)]\,\text{Im}[\text{Tr}\,(T^aT^eT^bT^cT^d)] = -\frac{7}{12},$$

$$\text{Im}[\text{Tr}\,(T^aT^bT^cT^dT^e)]\,\text{Im}[\text{Tr}\,(T^aT^eT^bT^dT^c)] = -\frac{1}{12},$$

$$\text{Im}[\text{Tr}\,(T^aT^bT^cT^dT^e)]\,\text{Im}[\text{Tr}\,(T^aT^eT^cT^bT^d)] = -\frac{1}{12},$$

$$\text{Im}[\text{Tr}\,(T^aT^bT^cT^dT^e)]\,\text{Im}[\text{Tr}\,(T^aT^eT^cT^dT^b)] = \frac{7}{12},$$

$$\text{Im}[\text{Tr}\,(T^aT^bT^cT^dT^e)]\,\text{Im}[\text{Tr}\,(T^aT^eT^dT^bT^c)] = \frac{7}{12},$$

$$\text{Im}[\text{Tr}\,(T^aT^bT^cT^dT^e)]\,\text{Im}[\text{Tr}\,(T^aT^eT^dT^cT^b)] = -\frac{55}{24}. \qquad (\text{C.12})$$

Bibliography

[1] S. Tomonaga, *Prog. Theor. Phys.* **1** (1946), 27;
J. Schwinger, *Phys. Rev.* **74** (1948), 1439;
R.P. Feynman, *Phys. Rev.* **76** (1949), 749; *ibid.*, 769;
F.J. Dyson, *Phys. Rev.* **75** (1949), 486; *ibid.*, 1736.

[2] W.E. Lamb, Jr. and R.C. Retherford, *Phys. Rev.* **72** (1947), 241; *ibid.* **75** (1949), 1325; *ibid.* **79** (1950), 549; *ibid.* **81** (1951), 222; *ibid.* **85** (1952), 259; *ibid.* **86** (1952), 1014.

[3] J. Schwinger, *Phys. Rev.* **73** (1948), 416; *ibid.* **76** (1949), 790.

[4] C.N. Yang and R.L. Mills, *Phys. Rev.* **96** (1954), 191.

[5] M. Gell-Mann, *Phys. Lett.* **8** (1964), 214;
G. Zweig, CERN preprints TH-401 and TH-412 (1964), unpublished.

[6] S.L. Glashow, *Nucl. Phys.* **22** (1961), 579;
S. Weinberg, *Phys. Rev. Lett.* **19** (1967), 1264;
A. Salam, in *Elementary Particle Theory: Relativistic Groups and Analiticity*, Nobel Symposium No. 8, ed. N. Svartholm (Almqvist and Wiksells, Stockholm, 1968), p. 367.

[7] F. Englert and R. Brout, *Phys. Rev. Lett.* **13** (1964), 321;
P.W. Higgs, *Phys. Lett.* **12** (1964), 132; *Phys. Rev. Lett.* **13** (1964), 508; *Phys. Rev.* **145** (1966), 1156;
G.S. Guralnik, C.R. Hagen and T.W.B. Kibble, *Phys. Rev. Lett.* **13** (1964), 585;
T.W.B. Kibble, *Phys. Rev.* **155** (1967), 1554.

[8] H. Fritzsch and M. Gell-Mann, in *Broken Scale Invariance and the Light Cone*, eds. M. Dal Cin, G. Iverson and A. Perlmutter (Gordon and Breach, New York, 1971), p. 1;
H. Fritzsch and M. Gell-Mann, in *Proceedings XVI International Conference on High Energy Physics*, Chicago 1972, eds. J.D. Jackson and A. Roberts, vol. II, p. 135;
H. Fritzsch, M. Gell-Mann and H. Leutwyler, *Phys. Lett.* **47B** (1973), 365.

[9] Sau Lan Wu and G. Zobernig, *Z. Phys.* **C2** (1979), 107; TASSO Note No. 84, June 1979;
B.H. Wiik, in *Proceedings of the International Conference on Neutrinos, Weak Interactions and Cosmology*, Bergen 1979, eds. A. Haatuft and C. Jarlskog, vol. I, p. 113;

P. Söding, in *Proceedings of the European Physical Society International Conference on High-Energy Physics*, Geneva 1979, vol. I, p. 271;
TASSO Collaboration, R. Brandelik, W. Braunschweig, K. Gather, V. Kadansky, K. Lübelsmeyer, P. Mättig, H.-U. Martyn, G. Peise, J. Rimkus, H.G. Sander, D. Schmitz, A. Schultz von Dratzig, D. Trines, W. Wallraff, H. Boerner, H.M. Fischer, H. Hartmann, E. Hilger, W. Hillen, G. Knop, W. Korbach, P. Leu, B. Löhr, F. Roth, W. Rühmer, R. Wedemeyer, N. Wermes, M. Wollstadt, R. Bühring, R. Fohrmann, D. Heyland, H. Hultschig, P. Joos, W. Koch, U. Kötz, H. Kowalski, A. Ladage, D. Lüke, H.L. Lynch, G. Mikenberg, D. Notz, J. Pyrlik, R. Riethmüller, M. Schliwa, P. Söding, B.H. Wiik, G. Wolf, M. Holder, G. Poelz, J. Ringel, O. Römer, R. Rüsch, P. Schmüser, D.M. Binnie, P.J. Dornan, N.A. Downie, D.A. Garbutt, W.G. Jones, S.L. Lloyd, D. Pandoulas, A. Pevsner, J. Sedgebeer, S. Yarker, C. Youngman, R.J. Barlow, R.J. Cashmore, J. Illingworth, M. Ogg, G.L. Salmon, K.W. Bell, W. Chinowsky, B. Foster, J.C. Hart, J. Proudfoot, D.R. Quarrie, D.H. Saxon, P.L. Woodworth, Y. Eisenberg, U. Karshon, E. Kogan, D. Revel, E. Ronat, A. Shapira, J. Freeman, P. Lecomte, T. Meyer, S.L. Wu and G. Zobernig, *Phys. Lett.* **86B** (1979), 243;
MARK-J Collaboration, D.P. Barber, U. Becker, H. Benda, A. Boehm, J.G. Branson, J. Bron, D. Buikman, J. Burger, C.C. Chang, H.S. Chen, M. Chen, C.P. Cheng, Y.S. Chu, R. Clare, P. Duinker, G.Y. Fang, H. Fesefeldt, D. Fong, M. Fukushima, J.C. Guo, A. Hariri, G. Herten, M.C. Ho, H.K. Hsu, T.T. Hsu, R.W. Kadel, W. Krenz, J. Li, Q.Z. Li, M. Lu, D. Luckey, D.A. Ma, C.M. Ma, G.G.G. Massaro, T. Matsuda, H. Newman, J. Paradiso, F.P. Poschmann, J.P. Revol, M. Rohde, H. Rykaczewski, K. Sinram, H.W. Tang, L.G. Tang, S.C.C. Ting, K.L. Tung, F. Vannucci, X.R. Wang, P.S. Wei, M. White, G.H. Wu, T.W. Wu, J.P. Xi, P.C. Yang, X.H. Yu, N.L. Zhang and R.Y. Zhu, *Phys. Rev. Lett.* **43** (1979), 830;
PLUTO Collaboration, Ch. Berger, H. Genzel, R. Grigull, W. Lackas, F. Raupach, A. Klovning, E. Lillestöl, E. Lillethun, J.A. Skard, H. Ackermann, G. Alexander, F. Barreiro, J. Bürger, L. Criegee, H.C. Dehne, R. Devenish, A. Eskreys, G. Flügge, G. Franke, W. Gabriel, Ch. Gerke, G. Knies, E. Lehmann, H.D. Mertiens, K.H. Pape, H.D. Reich, B. Stella, T.N. Ranga Swamy, U. Timm, W. Wagner, P. Waloschek, G.G. Winter, W. Zimmermann, O. Achterberg, V. Blobel, L. Boesten, H. Kapitza, B. Koppitz, W. Lührsen, R. Maschuw, R. van Staa, H. Spitzer, C.Y. Chang, R.G. Glasser, R.G. Kellogg, K.H. Lau, B. Sechi-Zorn, A. Skuja, G. Welch, G.T. Zorn, A. Bäcker, S. Brandt, K. Derikum, A. Diekmann, C. Grupen, H.J. Meyer, B. Neumann, M. Rost, G. Zech, T. Azemoon, H.J. Daum, H. Meyer, O. Meyer, M. Rössler, D. Schmidt and K. Wacker, *Phys. Lett.* **86B** (1979), 418;

JADE Collaboration, W. Bartel, T. Canzler, D. Cords, P. Dittmann, R. Eichler, R. Felst, D. Haidt, S. Kawabata, H. Krehbiel, B. Naroska, L.H. O'Neill, J. Olsson, P. Steffen, W.L. Yen, E. Elsen, M. Helm, A. Petersen, P. Warming, G. Weber, H. Drumm, J. Heintze, G. Heinzelmann, R.D. Heuer, J. von Krogh, P. Lennert, H. Matsumura, T. Nozaki, H. Rieseberg, A. Wagner, D.C. Darvill, F. Foster, G. Hughes, H. Wriedt, J. Allison, J. Armitage, I. Duerdoth, J. Hassard, F. Loebinger, B. King, A. Macbeth, H. Mills, P.G. Murphy, H. Prosper, K. Stephens, D. Clarke, M.C. Goddard, R. Hedgecock, R. Marshall, G.F. Pearce, M. Imori, T. Kobayashi, S. Komamiya, M. Koshiba, M. Minowa, S. Orito, A. Sato, T. Suda, H. Takeda, Y. Totsuka, Y. Watanabe, S. Yamada and C. Yanagisawa, *Phys. Lett.* **91B** (1980), 142.

[10] UA1 Collaboration, G. Arnison, A. Astbury, B. Aubert, C. Bacci, G. Bauer, A. Bézaguet, R.K. Böck, T.J.V. Bowcock, M. Calvetti, T. Carroll, P. Catz, P. Cennini, S. Centro, F. Ceradini, S. Cittolin, D. Cline, C. Cochet, J. Colas, M. Corden, D. Dallman, D. Dau, M. DeBeer, M. Della Negra, M. Demoulin, D. Denegri, A. Di Ciaccio, D. DiBitonto, L. Dobrzynski, J.D. Dowell, M. Edwards, K. Eggert, E. Eisenhandler, N. Ellis, P. Erhard, H. Faissner, M. Fincke, G. Fontaine, R. Frey, R. Frühwirth, J. Garvey, S. Geer, C. Ghesquière, P. Ghez, K.L. Giboni, W.R. Gibson, Y. Giraud-Héraud, A. Givernaud, A. Godinec, G. Grayer, P. Gutierrez, T. Hansl-Kozanecka, W.J. Haynes, L.O. Hertzberger, C. Hodges, D. Hoffmann, H. Hoffmann, D.J. Holthuizen, R.J. Homer, A. Honma, W. Jank, G. Jorat, P.I.P. Kalmus, V. Karimäki, R. Keeler, I. Kenyon, A. Kernan, R. Kinnunen, H. Kowalski, W. Kozanecki, D. Kryn, F. Lacava, J.-P. Laugier, J.-P. Lees, H. Lehmann, K. Leuchs, R. Leuchs, A. Lévêque, D. Linglin, E. Locci, M. Loret, J.-J. Malosse, T. Markiewicz, G. Maurin, T. McMahon, J.-P. Mendiburu, M.-N. Minard, M. Mohammadi, K. Morgan, M. Moricca, H. Muirhead, F. Muller, A.K. Nandi, L. Naumann, A. Norton, A. Orkin-Lecourtois, L. Paoluzi, F. Pauss, G. Petrucci, G. Piano Mortari, E. Pietarinen, M. Pimiä, A. Placci, J.P. Porte, E. Radermacher, J. Ransdell, H. Reithler, J.-P. Revol, J. Rich, M. Rijssenbeek, C. Roberts, J. Rohlf, P. Rossi, C. Rubbia, B. Sadoulet, G. Sajot, G. Salvi, G. Salvini, J. Sass, J. Saudraix, A. Savoy-Navarro, D. Schinzel, W. Scott, T.P. Shah, D. Smith, M. Spiro, J. Strauss, J. Streets, K. Sumorok, F. Szoncso, C. Tao, G. Thompson, J. Timmer, E. Tscheslog, J. Tuominiemi, S. Van der Meer, B. Van Eijk, J.-P. Vialle, J. Vrana, V. Vuillemin, H.D. Wahl, P. Watkins, J. Wilson, R. Wilson, C.-E. Wulz, Y.G. Xie, M. Yvert and E. Zurfluh, *Phys. Lett.* **122B** (1983), 103; *ibid.* **126B** (1983), 398; *ibid.* **129B** (1983), 273; *ibid.* **134B** (1984), 469; *ibid.* **135B** (1984), 250;
UA2 Collaboration, P. Bagnaia, M. Banner, R. Battiston, Ph. Bloch, F. Bonaudi, K. Borer, M. Borghini, J. Bürger, P. Cenci, J.-C. Chol-

let, A.G. Clark, C. Conta, P. Darriulat, L. Di Lella, J. Dines-Hansen, P.-A. Dorsaz, R. Engelmann, L. Fayard, M. Fraternali, D. Froidevaux, G. Fumagalli, J.-M. Gaillard, O. Gildemeister, V.G. Goggi, C. Gössling, H. Grote, B. Hahn, H. Hänni, J.R. Hansen, P. Hansen, N. Harnew, T. Himel, V. Hungerbühler, P. Jenni, O. Kofoed-Hansen, E. Lançon, M. Livan, S. Loucatos, B. Madsen, P. Mani, B. Mansoulié, G.C. Mantovani, L. Mapelli, B. Merkel, M. Mermikides, R. Møllerud, B. Nilsson, C. Onions, G. Parrour, F. Pastore, H. Plothow-Besch, M. Polverel, J.-P. Repellin, A. Rimoldi, A. Rothenberg, A. Roussarie, G. Sauvage, J. Schacher, J.L. Siegrist, H.M. Steiner, G. Stimpfl, F. Stocker, M. Swartz, J. Teiger, S. Tovey, V. Vercesi, A.R. Weidberg, H. Zaccone, J.A. Zakrzewski and W. Zeller, Phys. Lett. **122B** (1983), 476; ibid. **129B** (1983), 130; ibid. **139B** (1984), 105; Z. Phys. **C24** (1984), 1.

[11] G. Altarelli and G. Martinelli, in Physics at LEP, eds. J. Ellis and R. Peccei, CERN 86-02 (1986), p. 56.

[12] H.J. Bhabha, Proc. Roy. Soc., Ser. A **154** (1935), 195.

[13] P.A.M. Dirac, Proc. Roy. Soc., Ser. A **117** (1928), 610; ibid. **118** (1928), 351.

[14] Y. Ne'eman, Nucl. Phys. **26** (1961), 222;
M. Gell-Mann, Phys. Rev. **125** (1962), 1067;
D.R. Speiser and J. Tarski, J. Math. Phys. **4** (1963), 588.

[15] Tai Tsun Wu, Phys. Rev. **125** (1962), 1436.

[16] N. Bohr, R. Peierls and G. Placzek, Nature (London) **144** (1939), 200.

[17] L.D. Faddeev and V.N. Popov, Phys. Lett. **25B** (1967), 29.

[18] Particle Data Group, G.P. Yost, R.M. Barnett, I. Hinchliffe, G.R. Lynch, A. Rittenberg, R.R. Ross, M. Suzuki, T.G. Trippe, C.G. Wohl, B. Armstrong, G.S. Wagman, F.C. Porter, L. Montanet, M. Aguilar-Benitez, J.J. Hernandez, G. Conforti, R.L. Crawford, K.R. Schubert, M. Roos, N.A. Törnqvist, G. Höhler, K. Hagiwara, S. Kawabata, D.M. Manley, K.A. Olive, K.G. Hayes, R.H. Schindler, B. Cabrera, R.E. Scrock, R.A. Eichler, L.D. Roper and W.P. Trower, Phys. Lett. **204B** (1988), 105.

[19] P.A.M. Dirac, Proc. Cambridge Phil. Soc. **26** (1930), 261.

[20] S. Mandelstam, Phys. Rev. **112** (1958), 1344.

[21] P. De Causmaecker, R. Gastmans, W. Troost and Tai Tsun Wu, Phys. Lett. **105B** (1981), 215; Nucl. Phys. **B206** (1982), 53.

[22] N.A. Voronov, *Zh. Eksp. Teor. Fiz.* **64** (1973), 1889 [English translation: *Sov. Phys.-JETP* **37** (1973), 953].

[23] F.A. Berends, R. Kleiss, P. De Causmaecker, R. Gastmans and Tai Tsun Wu, *Phys. Lett.* **103B** (1981), 124.

[24] F.A. Berends, R. Kleiss, P. De Causmaecker, R. Gastmans, W. Troost and Tai Tsun Wu, *Nucl. Phys.* **B206** (1982), 61.

[25] D.R. Yennie, S.C. Frautschi and H. Suura, *Ann. Phys. (New York)* **13** (1961), 379.

[26] JADE Collaboration, W. Bartel, D. Cords, P. Dittmann, R. Eichler, R. Felst, D. Haidt, H. Krehbiel, K. Meier, B. Naroska, L.H. O'Neill, P. Steffen, H. Wenniger, Y. Zhang, E. Elsen, A. Petersen, P. Warming, G. Weber, S. Bethke, H. Drumm, J. Heintze, G. Heinzelmann, K.H. Hellenbrand, R.D. Heuer, J. von Krogh, P. Lennert, S. Kawabata, H. Matsumura, T. Nozaki, J. Olsson, H. Rieseberg, A. Wagner, A. Bell, F. Foster, G. Hughes, H. Wriedt, J. Allison, A.H. Ball, G. Bamford, R. Barlow, C. Bowdery, I.P. Duerdoth, J.F. Hassard, B.T. King, F.K. Loebinger, A.A. Macbeth, H. McCann, H.E. Mills, P.G. Murphy, D. Clarke, M.C. Goddard, R. Marshall, G.F. Pearce, J. Kanzaki, T. Kobayashi, S. Komamiya, M. Koshiba, M. Minowa, M. Nozaki, S. Odaka, S. Orito, A. Sato, H. Takeda, Y. Totsuka, Y. Watanabe, S. Yamada and C. Yanagisawa, *Phys. Lett.* **108B** (1982), 140;
MARK-J Collaboration, B. Adeva, D.P. Barber, U. Becker, J. Berdugo, G. Berghoff, A. Böhm, J.G. Branson, J.D. Burger, M. Capell, M. Cerrada, C.C. Chang, H.S. Chen, M. Chen, M.L. Chen, M.Y. Chen, C.P. Cheng, R. Clare, E. Deffur, P. Duinker, Z.Y. Feng, H.S. Fesefeldt, D. Fong, M. Fukushima, D. Harting, T. Hebbeker, G. Herten, M.C. Ho, M.M. Ilyas, D.Z. Jiang, D. Kooijman, W. Krenz, Q.Z. Li, D. Luckey, E.J. Luit, C. Maña, G.G.G. Massaro, T. Matsuda, H. Newman, M. Pohl, F.P. Poschmann, J.P. Revol, M. Rohde, H. Rykaczewski, J.A. Rubio, J. Salicio, I. Schulz, K. Sinram, M. Steuer, G.M. Swider, H.W. Tang, D. Teuchert, S.C.C. Ting, K.L. Tung, F. Vannucci, M.Q. Wang, M. White, S.X. Wu, R.Y. Zhu and Y.C. Zhu, *Phys. Rev. Lett.* **48** (1982), 1701;
CELLO Collaboration, H.J. Behrend, Ch. Chen, H. Fenner, J.H. Field, V. Schröder, H. Sindt, G. d'Agostini, W.D. Apel, S. Banerjee, J. Bodenkamp, D. Chrobaczek, J. Engler, G. Flügge, D.C. Fries, W. Fues, K. Gamerdinger, G. Hopp, H. Küster, H. Müller, H. Randoll, G. Schmidt, H. Schneider, W. de Boer, G. Buschhorn, G. Grindhammer, P. Grosse-Wiesmann, B. Gunderson, C. Kiesling, R. Kotthaus, U. Kruse, H. Lierl, D. Lüers, T. Meyer, H. Oberlack, P. Schacht, M.J. Schachter, G. Carnesecchi, P. Colas, A. Cordier, M. Davier, D. Fournier, J.F. Grivaz, J. Haissinski, V. Journé, A. Klarsfeld, F. Laplanche, F. Le Diberder,

U. Mallik, J.J. Veillet, R. George, M. Goldberg, B. Grossetête, O. Hamon, F. Kapusta, F. Kovacs, G. London, L. Poggioli, M. Rivoal, R. Aleksan, J. Bouchez, G. Cozzika, Y. Ducros, A. Gaidot, S. Jadach, Y. Lavagne, J. Pamela, J.P. Pansart and F. Pierre, Z. Phys. C14 (1982), 283;
PLUTO Collaboration, Ch. Berger, H. Genzel, W. Lackas, J. Pielorz, F. Raupach, W. Wagner, H. Flølo, A. Klovning, E. Lillestöl, J.M. Olsen, J. Bürger, L. Criegee, Ch. Dehne, A. Deuter, A. Eskreys, G. Franke, M. Gaspero, Ch. Gerke, U. Jacobs, G. Knies, B. Lewendel, U. Maurus, J. Meyer, U. Michelsen, K.H. Pape, B. Stella, U. Timm, P. Waloschek, G.G. Winter, S.T. Xue, M. Zachara, W. Zimmermann, P.J. Bussey, S.L. Cartwright, J.B. Daiton, B.T. King, C. Raine, J.M. Scarr, I.O. Skillicorn, K.M. Smith, J.C. Thomson, O. Achterberg, V. Blobel, D. Burkart, K. Diehlmann, H. Kapitza, B. Koppitz, M. Krüger, W. Lührsen, M. Poppe, H. Spitzer, R. van Staa, C.Y. Chang, R.G. Glasser, R.G. Kellogg, S.J. Maxfield, R.O. Polvado, B. Sechi-Zorn, J.A. Skard, A. Skuja, A.J. Tylka, G.E. Welch, G.T. Zorn, F. Almeida, A. Bäcker, F. Barreiro, S. Brandt, K. Derikum, C. Grupen, H.J. Meyer, H. Müller, B. Neumann, M. Rost, K. Stupperich, G. Zech, G. Alexander, G. Bella, Y. Gnat, J. Grunhaus, H.J. Daum, H. Junge, K. Kraski, C. Maxeiner, H. Meyer and D. Schmidt, Z. Phys. C21 (1983), 53;
TASSO Collaboration, M. Althoff, W. Braunschweig, F.J. Kirschfink, K. Lübelsmeyer, H.-U. Martyn, G. Peise, J. Rimkus, P. Rosskamp, H.G. Sander, D. Schmitz, H. Siebke, W. Wallraff, H.M. Fischer, H. Hartmann, W. Hillen, A. Jocksch, G. Knop, L. Köpke, H. Kolanoski, H. Kück, R. Wedemeyer, N. Wermes, M. Wollstadt, H. Burkhardt, Y. Eisenberg, K. Gather, H. Hultschig, P. Joos, W. Koch, U. Kötz, H. Kowalski, A. Ladage, B. Löhr, D. Lüke, P. Mättig, D. Notz, J. Pyrlik, D.R. Quarrie, M. Rushton, W. Schütte, D. Trines, G. Wolf, Ch. Xiao, R. Fohrmann, E. Hilger, T. Kracht, H.L. Krasemann, P. Leu, E. Lohrmann, D. Pandoulas, G. Poelz, B.H. Wiik, R. Beuselinck, D.M. Binnie, A.J. Campbell, P.J. Dornan, B. Foster, D.A. Garbutt and C. Jenkins, Z. Phys. C22 (1984), 13.

[27] P. De Causmaecker, doctoral thesis, University of Leuven (1983).

[28] B. Naroska, Phys. Rep. 148 (1987), 67.

[29] W.T. Ford, A.L. Read, Jr., J.G. Smith, A. Marini, I. Peruzzi, M. Piccolo, F. Ronga, L.A. Baksay, H.R. Band, W.L. Faissler, M.W. Gettner, G.P. Goderre, B. Gottschalk, R.B. Hurst, O.A. Meyer, J.H. Moromisato, W.D. Shambroom, E. von Goeler, R. Weinstein, J.V. Allaby, W.W. Ash, G.B. Chadwick, S.H. Clearwater, R.W. Coombes, Y. Goldschmidt-Clermont, H.S. Kaye, K.H. Lau, R.E. Leedy, R.L. Messner, S.J. Michalowski, K. Rich, D.M. Ritson,

L.J. Rosenberg, D.E. Wiser, R.W. Zdarko, D.E. Groom, H.Y. Lee, E.C. Loh, M.C. Delfino, B.K. Heltsley, J.R. Johnson, T.L. Lavine, T. Maruyama and R. Prepost, *Phys. Rev. Lett.* **51** (1983), 257.

[30] B. Adeva, D.P. Barber, U. Becker, G.D. Bei, J. Berdugo, G. Berghoff, A. Böhm, J.G. Branson, D. Buikman, J.D. Burger, M. Cerrada, C.C. Chang, G.F. Chen, H.S. Chen, M. Chen, M.L. Chen, M.Y. Chen, C.P. Cheng, R. Clare, E. Deffur, P. Duinker, Z.Y. Feng, H. Fesefeldt, D. Fong, M. Fukushima, J.C. Guo, D. Harting, T. Hebbeker, G. Herten, M.C. Ho, M.M. Ilyas, D.Z. Jiang, D. Kooijman, W. Krenz, Q.Z. Li, D. Luckey, E.J. Luit, C. Maña, G.G.G. Massaro, T. Matsuda, H. Newman, M. Pohl, F.P. Poschmann, J.-P. Revol, M. Rohde, H. Rykaczewski, A. Rubio, J. Salicio, I. Schulz, K. Sinram, M. Steuer, G.M. Swider, H.W. Tang, D. Teuchert, S.C.C. Ting, K.L. Tung, F. Vannucci, Y.X. Wang, M. White, S.X. Wu, T.W. Wu, C.C. Yu, Y.Q. Zeng, N.L. Zhang and R.Y. Zhu, *Phys. Rev. Lett.* **48** (1982), 967;
JADE Collaboration, W. Bartel, L. Becker, C. Bowdery, D. Cords, R. Felst, D. Haidt, J. Huttunen, H. Junge, G. Knies, H. Krehbiel, P. Laurikainen, R. Meinke, B. Naroska, J. Olsson, E. Pietarinen, D. Schmidt, P. Steffen, P. Warming, M. Zachara, G. Dietrich, E. Elsen, G. Heinzelmann, H. Kado, K. Meier, A. Petersen, U. Schneekloth, G. Weber, K. Ambrus, S. Bethke, A. Dieckmann, J. Heintze, K.H. Hellenbrand, R.D. Heuer, S. Komamiya, J.v. Krogh, P. Lennert, H. Matsumura, H. Rieseberg, J. Spitzer, A. Wagner, A. Finch, F. Foster, G. Hughes, T. Nozaki, J. Nye, J. Allison, J. Baines, A.H. Ball, R.J. Barlow, J. Chrin, I.P. Duerdoth, F.K. Loebinger, A.A. Macbeth, H. McCann, H.E. Mills, P.G. Murphy, K. Stephens, R.G. Glasser, B. Sechi-Zorn, J.A.J. Skard, S. Wagner, G.T. Zorn, S.L. Cartwright, D. Clarke, R. Marshall, J.B. Whittaker, J. Kanzaki, T. Kawamoto, T. Kobayashi, M. Koshiba, M. Minowa, M. Nozaki, S. Orito, A. Sato, H. Takeda, T. Takeshita, Y. Totsuka and S. Yamada, *Z. Phys.* **C24** (1984), 223.

[31] P. De Causmaecker, R. Gastmans, W. Troost and Tai Tsun Wu, *Phys. Lett.* **105B** (1981), 215;
CALKUL Collaboration, F.A. Berends, P. De Causmaecker, R. Gastmans, R. Kleiss, W. Troost and Tai Tsun Wu, *Nucl. Phys.* **B239** (1984), 395.

[32] D. Danckaert, P. De Causmaecker, R. Gastmans, W. Troost and Tai Tsun Wu, *Phys. Lett.* **114B** (1982), 203.

[33] Ch. Berger, H. Genzel, R. Grigull, W. Lackas, F. Raupach, A. Klovning, E. Lillestöl, J.A. Skard, H. Ackermann, J. Bürger, L. Criegee, H.C. Dehne, A. Eskreys, G. Franke, W. Gabriel, Ch. Gerke, G. Knies, E. Lehmann, H.D. Mertiens, U. Michelsen, K.H. Pape, H.D. Reich,

M. Scarr, B. Stella, U. Timm, W. Wagner, P. Waloschek, G.G. Winter, W. Zimmermann, O. Achterberg, V. Blobel, L. Boesten, V. Hepp, H. Kapitza, B. Koppitz, B. Lewendel, W. Lührsen, R. van Staa, H. Spitzer, C.Y. Chang, R.G. Glasser, R.G. Kellogg, K.H. Lau, R.O. Polvado, B. Sechi-Zorn, A. Skuja, G. Welch, G.T. Zorn, A. Bäcker, F. Barreiro, S. Brandt, K. Derikum, C. Grupen, H.J. Meyer, B. Neumann, M. Rost, G. Zech, H.J. Daum, H. Meyer, O. Meyer, M. Rössler and D. Schmidt, Phys. Lett. **97B** (1980), 459;
Sau Lan Wu, Z. Phys. **C9** (1981), 329; Phys. Scr. **25** (1982), 221;
JADE Collaboration, W. Bartel, D. Cords, P. Dittmann, R. Eichler, R. Felst, D. Haidt, H. Krehbiel, K. Meier, B. Naroska, L.H. O'Neill, P. Steffen, H. Wenninger, E. Elsen, A. Petersen, P. Warming, G. Weber, S. Bethke, H. Drumm, J. Heintze, G. Heinzelmann, K.H. Hellenbrand, R.D. Heuer, J. von Krogh, P. Lennert, S. Kawabata, S. Komamiya, H. Matsumura, T. Nozaki, J. Olsson, H. Rieseberg, A. Wagner, A. Bell, F. Foster, G. Hughes, H. Wriedt, J. Allison, A.H. Ball, G. Bamford, R. Barlow, C. Bowdery, I.P. Duerdoth, J.F. Hassard, B.T. King, F.K. Loebinger, A.A. Macbeth, H. McCann, H.E. Mills, P.G. Murphy, K. Stephens, D. Clarke, M.C. Goddard, R. Marshall, G.F. Pearce, J. Kanzaki, T. Kobayashi, M. Koshiba, M. Minowa, M. Nozaki, S. Odaka, S. Orito, A. Sato, H. Takeda, Y. Totsuka, Y. Watanabe, S. Yamada and C. Yanagisawa, Phys. Lett. **115B** (1982), 338.

[34] R.K. Ellis, D.A. Ross and A.E. Terrano, Phys. Rev. Lett. **45** (1980), 1226; Nucl. Phys. **B178** (1981), 421;
A. Ali, J.G. Körner, Z. Kunszt, E. Pietarinen, G. Kramer, G. Schierholz and J. Willrodt, Phys. Lett. **82B** (1979), 285; Nucl. Phys. **B167** (1980), 454;
K. Fabricius, I. Schmitt, G. Schierholz and G. Kramer, Phys. Lett. **97B** (1980), 431;
K.J.F. Gaemers and J.A.M. Vermaseren, Z. Phys. **C7** (1980), 81;
J.G. Körner, G. Schierholz and J. Willrodt, Nucl. Phys. **B185** (1981), 365;
J.A.M. Vermaseren, K.J.F. Gaemers and S.J. Oldham, Nucl. Phys. **B187** (1981), 301;
K. Fabricius, G. Kramer, G. Schierholz and I. Schmitt, Z. Phys. **C11** (1982), 315;
T.D. Gottschalk and D. Sivers, Phys. Rev. **D21** (1980), 102;
F. Gutbrod, G. Kramer and G. Schierholz, Z. Phys. **C21** (1984), 235.

[35] CALKUL Collaboration, F.A. Berends, P. De Causmaecker, R. Gastmans, R. Kleiss, W. Troost and Tai Tsun Wu, Nucl. Phys. **B239** (1984), 382.

[36] See Particle Data Group [18], pp. 329–332.

[37] Chang Chao-Hsi, *Nucl. Phys.* **B172** (1980), 425;
R. Baier and R. Rückl, *Phys. Lett.* **102B** (1981), 364; *Nucl. Phys.* **B208** (1982), 381; *Z. Phys.* **C19** (1983), 251;
B. Humpert, *Phys. Lett.* **184B** (1987), 105.

[38] E.L. Berger and D. Jones, *Phys. Rev.* **D23** (1983), 1521;
Z. Kunszt, *Phys. Lett.* **207B** (1988), 103.

[39] G. Guberina, J.H. Kühn, R.D. Peccei and R. Rückl, *Nucl. Phys.* **B174** (1980), 317.

[40] K. Hagiwara, A.D. Martin and A.W. Peacock, *Z. Phys.* **C33** (1986), 135.

[41] R. Gastmans, W. Troost and Tai Tsun Wu, *Phys. Lett.* **184B** (1987), 257; *Nucl. Phys.* **B291** (1987), 731.

[42] A.A. Sokolov and I.M. Ternov, *Dokl. Akad. Nauk SSSR* **153** (1963), 1052 [English translation: *Sov. Phys.-Dokl.* **8** (1964), 1203].

[43] M. Born and E. Wolf, *Principles of Optics* (Pergamon Press, Oxford, 1964), pp. 28–32.

[44] M. Tigner, *Nuovo Cimento* **37** (1965), 1228.

[45] B. Richter, *Nucl. Instrum. Methods* **136** (1976), 47;
U. Amaldi, *Phys. Lett.* **61B** (1976), 313;
U. Amaldi, in *Frontiers of Particle Beams*, eds. M. Month and S. Turner (Lecture Notes in Physics, No. 296) (Springer, Berlin, 1988), p. 341;
W. Schnell, *ibid.*, p. 461.

[46] B. Richter, *IEEE Trans. Nucl. Sci.* **NS–32** (1985), 3828.

[47] T. Himel and J. Siegrist, in *Laser Acceleration of Particles* (Malibu, California, 1985), eds. C. Joshi and T. Katsouleas (AIP Conf. Proc. No.130) (AIP, New York, 1985), p. 602;
R.J. Noble, *Nucl. Instrum. Methods Phys. Res.* **A256** (1987), 427.

[48] M. Jacob and Tai Tsun Wu, *Nucl. Phys.* **B318** (1989), 53.

[49] R. Hollebeek, *Nucl. Instrum. Methods* **184** (1981), 333.

[50] L.I. Schiff, *Phys. Rev.* **103** (1956), 443;
D.S. Saxon, *Phys. Rev.* **107** (1957), 871;
R. Glauber, in *Lectures in Theoretical Physics*, lectures delivered at the Summer Institute for Theoretical Physics, University of Colorado, 1958, eds. W.E. Brittin and L.G. Dunham (Interscience Publ., New York,

1959), vol. I, p. 315;
Hung Cheng and Tai Tsun Wu, in *High Energy Collisions*, ed. C.N. Yang (Gordon and Breach, New York, 1969), p. 329.

[51] Tai Tsun Wu, *Phys. Rev.* **108** (1957), 466;
D.S. Saxon and L.I. Schiff, *Nuovo Cimento* **6** (1957), 614;
V.P. Maslov and M.V. Fedoriuk, *Semi-Classical Approximation in Quantum Mechanics* (Reidel, Dordrecht, 1981), ch. 14.

[52] M. Jacob and Tai Tsun Wu, *Phys. Lett.* **197B** (1987), 253.

[53] W. Gordon, *Z. Phys.* **40** (1926), 117;
O. Klein, *Z. Phys.* **41** (1927), 407.

[54] G. Wentzel, *Z. Phys.* **38** (1926), 518;
H.A. Kramers, *Z. Phys.* **39** (1926), 828;
L. Brillouin, *C. R. H. Acad. Sci.* **183** (1926), 24; *J. Phys. Radium* **7** (1926), 353.

[55] J.D. Bjorken and Tai Tsun Wu, *Phys. Rev.* **130** (1963) 2566.

[56] Hung Cheng and Tai Tsun Wu, *Expanding Protons: Scattering at High Energies*, (MIT Press, Cambridge, 1987).

[57] M. Jacob and Tai Tsun Wu, *Nucl. Phys.* **B314** (1989), 334.

[58] R. Blankenbecler and S.D. Drell, *Phys. Rev.* **D37** (1988), 3308.

[59] National Bureau of Standards, *Handbook of Mathematical Functions*, eds. M. Abramowitz and I.A. Stegun (1964), p. 224.

[60] Bateman Manuscript Project, *Higher Transcendental Functions*, ed. A. Erdélyi (McGraw-Hill, New York, 1953), vol. I.

[61] P. Chen, in *Frontiers of Particle Beams*, eds. M. Month and S. Turner (Lecture Notes in Physics, No. 296) (Springer, Berlin, 1988), p. 495;
P. Chen and K. Yokoya, *Phys. Rev.* **D38** (1988), 987.

[62] M. Bell and J.S. Bell, *Nucl. Instrum. Methods Phys. Res.* **A275** (1989), 258.

[63] A.A. Sokolov, N.P. Klepikov and I.M. Ternov, *Zh. Eksp. Teor. Fiz.* **24** (1953), 249;
J. Schwinger, *Proc. Nat. Acad. Sci. USA* **40** (1954), 132;
A.N. Matveev, *Zh. Eksp. Teor. Fiz.* **31** (1956), 479 [English translation: *Sov. Phys.-JETP* **4** (1957), 409];
A.I. Nikishov and V.I. Ritus, *Zh. Eksp. Teor. Fiz.* **52** (1967), 1707 [English translation: *Sov. Phys.-JETP* **25** (1967), 1135];

V.N. Baĭer and V.M. Katkov, *Zh. Eksp. Teor. Fiz.* **53** (1967), 1478 [English translation: *Sov. Phys.-JETP* **26** (1968), 854]; *Zh. Eksp. Teor. Fiz.* **55** (1968), 1542 [English translation: *Sov. Phys.-JETP* **28** (1969), 807]; A.A. Sokolov and I.M. Ternov, *Synchrotron Radiation* (Pergamon Press, Oxford, 1968).

[64] M. Jacob and Tai Tsun Wu, *Nucl. Phys.* **B303** (1988), 389.

[65] R. Blankenbecler and S.D. Drell, *Phys. Rev.* **D36** (1987), 277; M. Bell and J.S. Bell, *Part. Accel.* **22** (1988), 301.

[66] M. Jacob and Tai Tsun Wu, *Phys. Lett.* **221B** (1989), 203; *Nucl. Phys.* **B327** (1989), 285.

[67] Yu.A. Gol'fand and E.P. Likhtman, *Pis'ma Zh. Eksp. Teor. Fiz.* **13** (1971), 452 [English translation: *JETP Lett.* **13** (1971), 323]; D.V. Volkov and V.P. Akulov, *Phys. Lett.* **46B** (1973), 109; J. Wess and B. Zumino, *Phys. Lett.* **49B** (1974), 52; *Nucl. Phys.* **B70** (1974), 39.

[68] S.J. Parke and T.R. Taylor, *Phys. Lett.* **157B** (1985), 81; *Nucl. Phys.* **B269** (1986), 410.

[69] J.C. Ward, *Phys. Rev.* **78** (1950), 182.

[70] Z. Kunszt, *Nucl. Phys.* **B271** (1986), 333.

[71] Z. Xu, D.-H. Zhang and L. Chang, *Nucl. Phys.* **B291** (1987), 392.

[72] J.F. Gunion and Z. Kunszt, *Phys. Lett.* **159B** (1985), 167; *ibid.* **161B** (1985), 333.

[73] R. Kleiss and W.J. Stirling, *Nucl. Phys.* **B262** (1985), 235.

[74] J.F. Gunion and J. Kalinowski, *Phys. Rev.* **D34** (1986), 2119.

[75] H. Weyl, *Gruppentheorie und Quantummechanik* (Leipzig, 1928); B.L. van der Waerden, *Goettinger Nachrichten* (1929), 100.

[76] F.A. Berends and W. Giele, *Nucl. Phys.* **B294** (1987), 700; F.A. Berends, G.J.H. Burgers, C. Maña, M. Martinez and W.L. van Neerven, *Nucl. Phys.* **B301** (1988), 583.

[77] S.-Q. Su, doctoral thesis, University of Leuven (1982).

[78] D. Albert, W.J. Marciano, D. Wyler and Z. Parsa, *Nucl. Phys.* **B166** (1980), 460.

[79] M. Böhm and Th. Sack, *Z. Phys.* **C35** (1987), 119.

[80] G. Passarino, *Nucl. Phys.* **B237** (1984), 249.

[81] F.A. Berends, P.H. Daverveldt and R. Kleiss, *Phys. Lett.* **148B** (1984), 489; *Nucl. Phys.* **B253** (1985), 441.

[82] C. Maña and M. Martinez, *Nucl. Phys.* **B287** (1987), 601.

[83] W.R. Hamilton, Brougham Bridge over the Royal Canal at Cabra (1843).

[84] Hung Cheng and Tai Tsun Wu, *Phys. Rev.* **182** (1969), 1873; *ibid.* **186** (1969), 1611.

[85] See Cheng and Wu [56], p. 4.

[86] Hung Cheng and Tai Tsun Wu, *Phys. Rev. Lett.* **24** (1970), 1456; C. Bourrely, J. Soffer and Tai Tsun Wu, *Phys. Lett.* **121B** (1983), 284; *Nucl. Phys.* **B247** (1984), 15; *Phys. Rev. Lett.* **54** (1985), 757; *Z. Phys.* **C37** (1988), 369 [note that the vertical scale of Fig. 1a is mislabeled in this paper].

[87] UA4 Collaboration, R. Battiston, M. Bozzo, P.L. Braccini, F. Carbonara, R. Carrara, R. Castaldi, F. Cervelli, G. Chiefari, E. Drago, M. Haguenauer, B. Koene, L. Linssen, G. Matthiae, L. Merola, M. Napolitano, V. Palladino, G. Sanguinetti, G. Sciacca, G. Sette, R. van Swol, J. Timmermans, C. Vannini, J. Velasco and F. Visco, *Phys. Lett.* **115B** (1982), 333; *ibid.* **117B** (1982), 126; *ibid.* **127B** (1983), 472; UA1 Collaboration, G. Arnison, A. Astbury, B. Aubert, C. Bacci, R. Bernabei, A. Bézaguet, R. Böck, T.J.V. Bowcock, M. Calvetti, T. Carroll, P. Catz, S. Centro, F. Ceradini, S. Cittolin, A.M. Cnops, C. Cochet, J. Colas, M. Corden, D. Dallman, S. D'Angelo, M. DeBeer, M. Della Negra, M. Demoulin, D. Denegri, R. Desalvo, D. DiBitonto, L. Dobrzynski, J.D. Dowell, M. Edwards, K. Eggert, E. Eisenhandler, N. Ellis, P. Erhard, H. Faissner, G. Fontaine, J.P. Fournier, R. Frey, R. Frühwirth, J. Garvey, S. Geer, C. Ghesquière, P. Ghez, K.L. Giboni, W.R. Gibson, Y. Giraud-Héraud, A. Givernaud, A. Godinec, G. Grayer, P. Gutierrez, R. Haidan, T. Hansl-Kozanecka, W.J. Haynes, L.O. Hertzberger, C. Hodges, D. Hoffmann, H. Hoffmann, G. von Holtey, D.J. Holthuizen, R.J. Homer, A. Honma, W. Jank, P.I.P. Kalmus, V. Karimäki, R. Keeler, I. Kenyon, A. Kernan, R. Kinnunen, H. Kowalski, W. Kozanecki, D. Kryn, F. Lacava, J.-P. Laugier, J.-P. Lees, H. Lehmann, R. Leuchs, A. Lévêque, D. Linglin, E. Locci, T. Markiewicz, G. Maurin, T. McMahon, J.-P. Mendiburu, M.-N. Minard, K. Morgan, M. Moricca, F. Muller, A.K. Nandi, L. Naumann, A. Norton, A. Orkin-Lecourtois, L. Paoluzi, G. Piano Mortari, M. Pimiä, A. Placci, M. Rabany, E. Radermacher, J. Ransdell, H. Reithler, J. Rich,

M. Rijssenbeek, C. Roberts, C. Rubbia, B. Sadoulet, G. Sajot, G. Salvi, G. Salvini, J. Sass, J. Saudraix, A. Savoy-Navarro, D. Schinzel, W. Scott, T.P. Shah, M. Spiro, J. Strauss, K. Sumorok, F. Szoncso, C. Tao, G. Thompson, J. Timmer, E. Tscheslog, J. Tuominiemi, J.-P. Vialle, G. Vismara, J. Vrana, V. Vuillemin, H.D. Wahl, P. Watkins, J. Wilson, M. Yvert and E. Zurfluh, *Phys. Lett.* **121B** (1983), 77.

[88] P. Osland and Tai Tsun Wu, *Nucl. Phys.* **B288** (1987), 77; *ibid.*, 95.

[89] J.S.R. Chisholm, *Nuovo Cimento* **30** (1963), 426;
J. Kahane, *J. Math. Phys.* **9** (1968), 1732;
A. Sirlin, *Nucl. Phys.* **B192** (1981), 93.

[90] R. Karplus and M. Neuman, *Phys. Rev.* **80** (1950), 380.

[91] R.P. Feynman, *Phys. Rev.* **76** (1949), 769 [eqn (14a)];
J. Schwinger, *Phys. Rev.* **76** (1949), 790 [eqn (2.82)].

[92] V. Costantini, B. De Tollis and G. Pistoni, *Nuovo Cimento* **2A** (1971), 733.

Author Index

Achterberg, O. 2, 44, 70, *620*, *624*, *626*
Ackermann, H. 2, 70, *620*, *625*
Adeva, B. 44, 65, *623*, *625*
Aguilar-Benitez, M. 17, 101, *622*, *627*
Akulov, V.P. 587, *629*
Albert, D. 595, *629*
Aleksan, R. 44, *624*
Alexander, G. 2, 44, *620*, *624*
Ali, A. 70, *626*
Allaby, J.V. 65, *624*
Allison, J. 2, 44, 65, 70, *621*, *623*, *625-6*
Almeida, F. 44, *624*
Altarelli, G. 2, *622*
Althoff, M. 44, *624*
Amaldi, U. 550, *627*
Ambrus, K. 65, *625*
Apel, W.D. 44, *623*
Armitage, J. 2, *621*
Armstrong, B. 17, 101, *622*, *627*
Arnison, G. 2, 16, 599, *621*, *630*
Ash, W.W. 65, *624*
Astbury, A. 2, 16, 599, *621*, *630*
Aubert, B. 2, 16, 599, *621*, *630*
Azemoon, T. 2, *620*

Bacci, C. 2, 16, 599, *621*, *630*
Bäcker, A. 2, 44, 70, *620*, *624*, *626*
Bagnaia, P. 2, 16, *621*
Baier, R. 101, *627*
Baĭer, V.N. 583, *628*
Baines, J. 65, *625*
Baksay, L.A. 65, *624*
Ball, A.H. 44, 65, 70, *623*, *625-6*
Bamford, G. 44, 70, *623*, *626*
Band, H.R. 65, *624*
Banerjee, S. 44, *623*

Banner, M. 2, 16, *621*
Barber, D.P. 2, 44, 65, *620*, *623*, *625*
Barlow, R.J. 2, 44, 65, 70, *620*, *623*, *626*
Barnett, R.M. 17, 101, *622*, *627*
Barreiro, F. 2, 44, 70, *620*, *624*, *626*
Bartel, W. 2, 44, 65, 70, *621*, *623*, *625-6*
Battiston, R. 2, 16, 599, *621*, *630*
Bauer, G. 2, 16, *621*
Becker, L. 65, *625*
Becker, U. 2, 44, 65, *620*, *623*, *625*
Behrend, H.J. 44, *623*
Bei, G.D. 65, *625*
Bell, A. 44, 70, *623*, *626*
Bell, J.S. 580, 583, *628-9*
Bell, K.W. 2, *620*
Bell, M. 580, 583, *628-9*
Bella, G. 44, *624*
Benda, H. 2, *620*
Berdugo, J. 44, 65, *623*, *625*
Berends, F.A. 23, 31, 41, 65, 68, 70, 80, 84, 592, 596, *623*, *625-6*, *629-30*
Berger, Ch. 2, 44, 70, *620*, *624-5*
Berger, E.L. 102, *627*
Berghoff, G. 44, 65, *623*, *625*
Bernabei, R. 599, *630*
Bethke, S. 44, 65, 70, *623*, *625-6*
Beuselinck, R. 44, *624*
Bézaguet, A. 2, 16, 599, *621*, *630*
Bhabha, H.J. 3, 39, 65, *622*
Binnie, D.M. 2, 44, *620*, *624*
Bjorken, J.D. 569, *628*
Blankenbecler, R. 571, 583, *628-9*
Blobel, V. 2, 44, 70, *620*, *624*, *626*
Bloch, Ph. 2, 16, *621*
Böck, R.K. 2, 16, 599, *621*, *630*

Bodenkamp, J. 44, *623*
Boerner, H. 2, *620*
Boesten, L. 2, 70, *620*, *626*
Böhm, A. 2, 44, 65, *620*, *623*, *625*
Böhm, M. 595, *629*
Bohr, N. 13, *622*
Bonaudi, F. 2, 16, *621*
Borer, K. 2, 16, *621*
Borghini, M. 2, 16, *621*
Born, M. 543, *627*
Bouchez, J. 44, *624*
Bourrely, C. 599, *630*
Bowcock, T.J.V. 2, 16, 599, *621*, *630*
Bowdery, C. 44, 65, 70, *623*, *625-6*
Bozzo, M. 599, *630*
Braccini, P.L. 599, *630*
Brandelik, R. 2, *620*
Brandt, S. 2, 44, 70, *620*, *624*, *626*
Branson, J.G. 2, 44, 65, *620*, *623*, *625*
Braunschweig, W. 2, 44, *620*, *624*
Brillouin, L. 559, *628*
Bron, J. 2, *620*
Brout, R. 1-2, 9, *619*
Bühring, R. 2, *620*
Buikman, D. 2, 65, *620*, *625*
Bürger, J. 2, 16, 44, 70, *620-1*, *624-25*
Burger, J.D. 44, 65, *623*, *625*
Burgers, G.J.H. 592, *629*
Burkart, D. 44, *624*
Burkhardt, H. 44, *624*
Buschhorn, G. 44, *623*
Bussey, P.J. 44, *624*

Cabrera, B. 17, 101, *622*, *627*
Calvetti, M. 2, 16, 599, *621*, *630*
Campbell, A.J. 44, *624*
Canzler, T. 2, *621*
Capell, M. 44, *623*
Carbonara, F. 599, *630*
Carnesecchi, G. 44, *623*
Carrara, R. 599, *630*

Carroll, T. 2, 16, 599, *621*, *630*
Cartwright, S.L. 44, 65, *624-5*
Cashmore, R.J. 2, *620*
Castaldi, R. 599, *630*
Catz, P. 2, 16, 599, *621*, *630*
Cenci, P. 2, 16, *621*
Cennini, P. 2, 16, *621*
Centro, S. 2, 16, 599, *621*, *630*
Ceradini, F. 2, 16, 599, *621*, *630*
Cerrada, M. 44, 65, *623*, *625*
Cervelli, F. 599, *630*
Chadwick, G.B. 65, *624*
Chang, C.C. 2, 44, 65, *620*, *623*, *625*
Chang, C.-H. 101, *627*
Chang, C.Y. 2, 44, 70, *620*, *624*, *626*
Chang, L. 588-9, *629*
Chen, Ch. 44, *623*
Chen, G.F. 65, *625*
Chen, H.S. 2, 44, 65, *620*, *623*, *625*
Chen, M. 2, 44, 65, *620*, *623*, *625*
Chen, M.L. 44, 65, *623*, *625*
Chen, M.Y. 44, 65, *623*, *625*
Chen, P. 580-2, *628*
Cheng, C.P. 2, 44, 65, *620*, *623*, *625*
Cheng, H. 553, 556, 569, 598-9, *628-30*
Chiefari, G. 599, *630*
Chinowsky, W. 2, *620*
Chisholm, J.S.R. 601, *631*
Chollet, J.-C. 2, 16, *621-622*
Chrin, J. 65, *625*
Chrobaczek, D. 44, *623*
Chu, Y.S. 2, *620*
Cittolin, S. 2, 16, 599, *621*, *630*
Clare, R. 2, 44, 65, *620*, *623*, *625*
Clark, A.G. 2, 16, *622*
Clarke, D. 2, 44, 65, 70, *621*, *623*, *625-6*
Clearwater, S.H. 65, *624*
Cline, D. 2, 16, *621*
Cnops, A.M. 599, *630*
Cochet, C. 2, 16, 599, *621*, *630*

Colas, J. 2, 16, 599, *621*, *630*
Colas, P. 44, *623*
Conforti, G. 17, 101, *622*, *627*
Conta, C. 2, 16, *622*
Coombes, R.W. 65, *624*
Corden, M. 2, 16, 599, *621*, *630*
Cordier, A. 44, *623*
Cords, D. 2, 44, 65, 70, *621*, *623*, *625–6*
Costantini, V. 611, *631*
Cozzika, G. 44, *624*
Crawford, R.L. 17, 101, *622*, *627*
Criegee, L. 2, 44, 70, *620*, *624–5*

d'Agostini, G. 44, *623*
Daiton, J.B. 44, *624*
Dallman, D. 2, 16, 599, *621*, *630*
Danckaert, D. 68, 70, *625*
D'Angelo, S. 599, *630*
Darriulat, P. 2, 16, *622*
Darvill, D.C. 2, *621*
Dau, D. 2, 16, *621*
Daum, H.J. 2, 44, 70, *620*, *624*, *626*
Daverveldt, P.H. 596, *630*
Davier, M. 44, *623*
DeBeer, M. 2, 16, 599, *621*, *630*
de Boer, W. 44, *623*
De Causmaecker, P. 22–3, 31, 41, 55, 65, 68, 70, 80–1, 84, *622–6*
Deffur, E. 44, 65, *623*, *625*
Dehne, Ch. 44, *624*
Dehne, H.C. 2, 70, *620*, *625*
Delfino, M.C. 65, *625*
Della Negra, M. 2, 16, 599, *621*, *630*
Demoulin, M. 2, 16, 599, *621*, *630*
Denegri, D. 2, 16, 599, *621*, *630*
Derikum, K. 2, 44, 70, *620*, *624*, *626*
Desalvo, R. 599, *630*
De Tollis, B. 611, *631*
Deuter, A. 44, *624*
Devenish, R. 2, *620*

DiBitonto, D. 2, 16, 599, *621*, *630*
Di Ciaccio, A. 2, 16, *621*
Dieckmann, A. 65, *625*
Diehlmann, K. 44, *624*
Diekmann, A. 2, *620*
Dietrich, G. 65, *625*
Di Lella, L. 2, 16, *622*
Dines-Hansen, J. 2, 16, *622*
Dirac, P.A.M. 7, 18, *622*
Dittmann, P. 2, 44, 70, *621*, *623*, *626*
Dobrzynski, L. 2, 16, 599, *621*, *630*
Dornan, P.J. 2, 44, *620*, *624*
Dorsaz, P.-A. 2, 16, *622*
Dowell, J.D. 2, 16, 599, *621*, *630*
Downie, N.A. 2, *620*
Drago, E. 599, *630*
Drell, S.D. 571, 583, *628–9*
Drumm, H. 2, 44, 70, *621*, *623*, *626*
Ducros, Y. 44, *624*
Duerdoth, I. 2, *621*
Duerdoth, I.P. 44, 65, 70, *623*, *625–6*
Duinker, P. 2, 44, 65, *620*, *623*, *625*
Dyson, F.J. 1, 7, *619*

Edwards, M. 2, 16, 599, *621*, *630*
Eggert, K. 2, 16, 599, *621*, *630*
Eichler, R. 2, 44, 70, *621*, *623*, *626*
Eichler, R.A. 17, 101, *622*, *627*
Eisenberg, Y. 2, 44, *620*, *624*
Eisenhandler, E. 2, 16, 599, *621*, *630*
Ellis, N. 2, 16, 599, *621*, *630*
Ellis, R.K. 70, *626*
Elsen, E. 2, 44, 65, 70, *621*, *623*, *625–6*
Engelmann, R. 2, 16, *622*
Engler, J. 44, *623*
Englert, F. 1–2, 9, *619*
Erhard, P. 2, 16, 599, *621*, *630*
Eskreys, A. 2, 44, 70, *620*, *624–5*

Fabricius, K. 70, *626*
Faddeev, L.D. 13, *622*

AUTHOR INDEX

Faissler, W.L. 65, *624*
Faissner, H. 2, 16, 599, *621*, *630*
Fang, G.Y. 2, *620*
Fayard, L. 2, 16, *622*
Fedoriuk, M.V. 553, 556, *628*
Felst, R. 2, 44, 65, 70, *621*, *623*, *625-6*
Feng, Z.Y. 44, 65, *623*, *625*
Fenner, H. 44, *623*
Fesefeldt, H. 2, 65, *620*, *625*
Fesefeldt, H.S. 44, *623*
Feynman, R.P. 1, 7, 605, *619*, *631*
Field, J.H. 44, *623*
Finch, A. 65, *625*
Fincke, M. 2, 16, *621*
Fischer, H.M. 2, 44, *620*, *624*
Fløio, H. 44, *624*
Flügge, G. 2, 44, *620*, *623*
Fohrmann, R. 2, 44, *620*, *624*
Fong, D. 2, 44, 65, *620*, *623*, *625*
Fontaine, G. 2, 16, 599, *621*, *630*
Ford, W.T. 65, *624*
Foster, B. 2, 44, *620*, *624*
Foster, F. 2, 44, 65, 70, *621*, *623*, *625-6*
Fournier, D. 44, *623*
Fournier, J.P. 599, *630*
Franke, G. 2, 44, 70, *620*, *624-5*
Fraternali, M. 2, 16, *622*
Frautschi, S.C. 39, *623*
Freeman, J. 2, *620*
Frey, R. 2, 16, 599, *621*, *630*
Fries, D.C. 44, *623*
Fritzsch, H. 1, 11, *619*
Froidevaux, D. 2, 16, *622*
Frühwirth, R. 2, 16, 599, *621*, *630*
Fues, W. 44, *623*
Fukushima, M. 2, 44, 65, *620*, *623*, *625*
Fumagalli, G. 2, 16, *622*

Gabriel, W. 2, 70, *620*, *625*
Gaemers, K.J.F. 70, *626*
Gaidot, A. 44, *624*
Gaillard, J.-M. 2, 16, *622*
Gamerdinger, K. 44, *623*
Garbutt, D.A. 2, 44, *620*, *624*
Garvey, J. 2, 16, 599, *621*, *630*
Gaspero, M. 44, *624*
Gastmans, R. 22-3, 31, 41, 65, 68, 70, 80, 84, 105, *622-3*, *625-7*
Gather, K. 2, 44, *620*, *624*
Geer, S. 2, 16, 599, *621*, *630*
Gell-Mann, M. 1, 11, *619*, *622569*
Genzel, H. 2, 44, 70, *620*, *624-5*
George, R. 44, *624*
Gerke, Ch. 2, 44, 70, *620*, *624-5*
Gettner, M.W. 65, *624*
Ghesquière, C. 2, 16, 599, *621*, *630*
Ghez, P. 2, 16, 599, *621*, *630*
Giboni, K.L. 2, 16, 599, *621*, *630*
Gibson, W.R. 2, 16, 599, *621*, *630*
Giele, W. 592, *629*
Gildemeister, O. 2, 16, *622*
Giraud-Héraud, Y. 2, 16, 599, *621*, *630*
Givernaud, A. 2, 16, 599, *621*, *630*
Glashow, S. 1, 16, *619*
Glasser, R.G. 2, 44, 65, 70, *620*, *624-6*
Glauber, R. 553, 556, *627*
Gnat, Y. 44, *624*
Goddard, M.C. 2, 44, 70, *621*, *623*, *626*
Goderre, G.P. 65, *624*
Godinec, A. 2, 16, 599, *621*, *630*
Goggi, V.G. 2, 16, *622*
Goldberg, M. 44, *624*
Goldschmidt-Clermont, Y. 65, *624*
Gol'fand, Yu.A. 587, *629*
Gordon, W. 556, *628*
Gössling, C. 2, 16, *622*
Gottschalk, B. 65, *624*
Gottschalk, T.D. 70, *626*
Grayer, G. 2, 16, 599, *621*, *630*
Grigull, R. 2, 70, *620*, *625*
Grindhammer, G. 44, *623*

Grivaz, J.F. 44, *623*
Groom, D.E. 65, *625*
Grossetête, B. 44, *624*
Grosse-Wiesmann, P. 44, *623*
Grote, H. 2, 16, *622*
Grunhaus, J. 44, *624*
Grupen, C. 2, 44, 70, *620*, *624*, *626*
Guberina, G. 104, *627*
Gunderson, B. 44, *623*
Gunion, J.F. 589–90, *629*
Guo, J.C. 2, 65, *620*, *625*
Guralnik, G.S. 1–2, 9, *619*
Gutbrod, F. 70, *626*
Gutierrez, P. 2, 16, 599, *621*, *630*

Hagen, C.R. 1–2, 9, *619*
Hagiwara, K. 17, 101, 105, *622*, *627*
Haguenauer, M. 599, *630*
Hahn, B. 2, 16, *622*
Haidan, R. 599, *630*
Haidt, D. 2, 44, 65, 70, *621*, *623*, *625–6*
Haissinski, J. 44, *623*
Hamilton, W.R. 597, *630*
Hamon, O. 44, *624*
Hänni, H. 2, 16, *622*
Hansen, J.R. 2, 16, *622*
Hansen, P. 2, 16, *622*
Hansl-Kozanecka, T. 2, 16, 599, *621*, *630*
Hariri, A. 2, *620*
Harnew, N. 2, 16, *622*
Hart, J.C. 2, *620*
Harting, D. 44, 65, *623*, *625*
Hartmann, H. 2, 44, *620*, *624*
Hassard, J. 2, *621*
Hassard, J.F. 44, 70, *623*, *626*
Hayes, K.G. 17, 101, *622*, *627*
Haynes, W.J. 2, 16, 599, *621*, *630*
Hebbeker, T. 44, 65, *623*, *625*
Hedgecock, R. 2, *621*
Heintze, J. 2, 44, 65, 70, *621*, *623*, *625–6*
Heinzelmann, G. 2, 44, 65, 70, *621*, *623*, *625–6*
Hellenbrand, K.H. 44, 65, 70, *623*, *625–6*
Helm, M. 2, *621*
Heltsley, B.K. 65, *625*
Hepp, V. 70, *626*
Hernandez, J.J. 17, 101, *622*, *627*
Herten, G. 2, 44, 65, *620*, *623*, *625*
Hertzberger, L.O. 2, 16, 599, *621*, *630*
Heuer, R.D. 2, 44, 65, 70, *621*, *623*, *625–6*
Heyland, D. 2, *620*
Higgs, P.W. 1–2, 9, *619*
Hilger, E. 2, 44, *620*, *624*
Hillen, W. 2, 44, *620*, *624*
Himel, T. 2, 16, 550, *622*, *627*
Hinchliffe, I. 17, 101, *622*, *627*
Ho, M.C. 2, 44, 65, *620*, *625*
Hodges, C. 2, 16, 599, *621*, *630*
Hoffmann, D. 2, 16, 599, *621*, *630*
Hoffmann, H. 2, 16, 599, *621*, *630*
Höhler, G. 17, 101, *622*, *627*
Holder, M. 2, *620*
Hollebeek, R. 551, *627*
Holthuizen, D.J. 2, 16, 599, *621*, *630*
Homer, R.J. 2, 16, 599, *621*, *630*
Honma, A. 2, 16, 599, *621*, *630*
Hopp, G. 44, *623*
Hsu, H.K. 2, *620*
Hsu, T.T. 2, *620*
Hughes, G. 2, 44, 65, 70, *621*, *623*, *625–6*
Hultschig, H. 2, 44, *620*, *624*
Humpert, B. 101, *627*
Hungerbühler, V. 2, 16, *622*
Hurst, R.B. 65, *624*
Huttunen, J. 65, *625*

Illingworth, J. 2, *620*
Ilyas, M.M. 44, 65, *623*, *625*
Imori, M. 2, *621*

Jacob, M. 551, 554, 569, 571, 579–80, 583, 585, *627–9*

Jacobs, U. 44, *624*
Jadach, S. 44, *624*
Jank, W. 2, 16, 599, *621*, *630*
Jenkins, C. 44, *624*
Jenni, P. 2, 16, *622*
Jiang, D.Z. 44, 65, *623*, *625*
Jocksch, A. 44, *624*
Johnson, J.R. 65, *625*
Jones, D. 102, *627*
Jones, W.G. 2, *620*
Joos, P. 2, 44, *620*, *624*
Jorat, G. 2, 16, *621*
Journé, V. 44, *623*
Junge, H. 44, 65, *624-5*

Kadansky, V. 2, *620*
Kadel, R.W. 2, *620*
Kado, H. 65, *625*
Kahane, J. 601, *630*
Kalmus, P.I.P. 2, 16, 599, *621*, *630*
Kalinowski, J. 590, *629*
Kanzaki, J. 44, 65, 70, *623*, *625-6*
Kapitza, H. 2, 44, 70, *620*, *624*, *626*
Kapusta, F. 44, *624*
Karimäki, V. 2, 16, 599, *621*, *630*
Karplus, R. 605, *631*
Karshon, U. 2, *620*
Katkov, V.M. 583, *628*
Kawabata, S. 2, 17, 44, 70, 101, *621-3*, *626-7*
Kawamoto, T. 65, *625*
Kaye, H.S. 65, *624*
Keeler, R. 2, 16, 599, *621*, *630*
Kellog, R.G. 2, 44, 70, *620*, *624*, *626*
Kenyon, I. 2, 16, 599, *621*, *630*
Kernan, A. 2, 16, 599, *621*, *630*
Kibble, T.W.B. 1-2, 9, *619*
Kiesling, C. 44, *623*
King, B. 2, *621*
King, B.T. 44, 70, *623-4*, *626*
Kinnunen, R. 2, 16, 599, *621*, *630*
Kirschfink, F.J. 44, *624*
Klarsfeld, A. 44, *623*

Klein, O. 556, *628*
Kleiss, R. 23, 31, 41, 65, 68, 70, 80, 84, 589, 596, *623*, *625-6*, *629-30*
Klepikov, N.P. 583, *628*
Klovning, A. 2, 44, 70, *620*, *624-5*
Knies, G. 2, 44, 65, 70, *620*, *624-5*
Knop, G. 2, 44, *620*, *624*
Kobayashi, T. 2, 44, 65, 70, *621*, *623*, *626*
Koch, W. 2, 44, *620*, *624*
Koene, B. 599, *630*
Kofoed-Hansen, O. 2, 16, *622*
Kogan, E. 2, *620*
Kolanoski, H. 44, *624*
Komamiya, S. 2, 44, 65, 70, *621*, *623*, *625-6*
Kooijman, D. 44, 65, *623*, *625*
Köpke, L. 44, *624*
Koppitz, B. 2, 44, 70, *620*, *624*, *626*
Korbach, W. 2, *620*
Körner, J.G. 70, *626*
Koshiba, M. 2, 44, 65, 70, *621*, *623*, *625-6*
Kotthaus, R. 44, *623*
Kötz, U. 2, 44, *620*, *624*
Kovacs, F. 44, *624*
Kowalski, H. 2, 16, 44, 599, *620-1*, *624*, *630*
Kozanecki, W. 2, 16, 599, *621*, *630*
Kracht, T. 44, *624*
Kramer, G. 70, *626*
Kramers, H.A. 559, *628*
Krasemann, H.L. 44, *624*
Kraski, K. 44, *624*
Krehbiel, H, 2, 44, 65, 70, *621*, *623*, *625-6*
Krenz, W. 2, 44, 65, *620*, *623*, *625*
Krüger, M. 44, *624*
Kruse, U. 44, *623*
Kryn, D. 2, 16, 599, *621*, *630*
Kück, H. 44, *624*
Kühn, J.H. 104, *627*
Kunszt, Z. 70, 102, 588-90, *626-7*,

Kunszt, Z. (cont'd) *629*
Küster, H. 44, *623*

Lacava, F. 2, 16, 599, *621*, *630*
Lackas, W. 2, 44, 70, *620*, *624*, *625*
Ladage, A. 2, 44, *620*, *624*
Lamb, Jr. W.E. 1, *619*
Lançon, E. 2, 16, *622*
Laplanche, F. 44, *623*
Lau, K.H. 2, 65, 70, *620*, *624*, *626*
Laugier, J.-P. 2, 16, 599, *621*, *630*
Laurikainen, P. 65, *625*
Lavagne, Y. 44, *624*
Lavine, T.L. 65, *625*
Lecomte, P. 2, *620*
Le Diberder, F. 44, *623*
Lee, H.Y. 65, *625*
Leedy, R.E. 65, *624*
Lees, J.-P. 2, 16, 599, *621*, *630*
Lehmann, E. 2, 70, *620*, *625*
Lehmann, H. 2, 16, 599, *621*, *630*
Lennert, P. 2, 44, 65, 70, *621*, *623*, *625-6*
Leu, P. 2, 44, *620*, *624*
Leuchs, K. 2, 16, *621*
Leuchs, R. 2, 16, 599, *621*, *630*
Leutwyler, H. 1, 11, *619*
Lévêque, A. 2, 16, 599, *621*, *630*
Lewendel, B. 44, 70, *624*, *626*
Li, J. 2, *620*
Li, Q.Z. 2, 44, 65, *620*, *623*, *625*
Lierl, H. 44, *623*
Likhtman, E.P. 587, *629*
Lillestöl, E. 2, 44, 70, *620*, *624-5*
Lillethun, E. 2, *620*
Linglin, D. 2, 16, 599, *621*, *630*
Linssen, L. 599, *630*
Livan, M. 2, 16, *622*
Lloyd, S.L. 2, *620*
Locci, E. 2, 16, 599, *621*, *630*
Loebinger, F. 2, *621*
Loebinger, F.K. 44, 65, 70, *623*, *625-6*
Loh, E.C. 65, *625*

Löhr, B. 2, 44, *620*, *624*
Lohrmann, E. 44, *624*
London, G. 44, *624*
Loret, M. 2, 16, *621*
Loucatos, S. 2, 16, *622*
Lu, M. 2, *620*
Lübelsmeyer, K. 2, 44, *620*, *624*
Luckey, D. 2, 44, 65, *620*, *623*, *625*
Lüers, D. 44, *623*
Lührsen, W. 2, 44, 70, *620*, *624*, *626*
Luit, E.J. 44, 65, *623*, *625*
Lüke, D. 2, 44, *620*, *624*
Lynch, G.R. 17, 101, *622*, *627*
Lynch, H.L. 2, *620*

Ma, C.M. 2, *620*
Ma, D.A. 2, *620*
Macbeth, A. 2, *621*
Macbeth, A.A. 44, 65, 70, *623*, *625-6*
Madsen, B. 2, 16, *622*
Mallik, U. 44, *624*
Malosse, J.-J. 2, 16, *621*
Maña, C. 44, 65, 592, 597, *623*, *625*, *629-630*
Mandelstam, S. 18, *622*
Mani, P. 2, 16, *622*
Manley, D.M. 17, 101, *622*, *627*
Mansoulié, B. 2, 16, *622*
Mantovani, G.C. 2, 16, *622*
Mapelli, L. 2, 16, *622*
Marciano, W.J. 595, *629*
Marini, A. 65, *624*
Markiewicz, T.W. 2, 16, 599, *621*, *630*
Marshall, R. 2, 44, 65, 70, *621*, *623*, *625-6*
Martin, A.D. 105, 120, *627*
Martinelli, G. 2, *622*
Martinez, M. 592, 597, *629-30*
Martyn, H.-U. 2, 44, *620*, *624*
Maruyama, T. 65, *625*
Maschuw, R. 2, *620*

Maslov, V.P. 553, 556, *628*
Massaro, G.G.G. 2, 44, 65, *620, 623, 625*
Matsuda, T. 2, 44, 65, *620, 623, 625*
Matsumura, H. 2, 44, 65, 70, *621, 623, 626*
Matthiae, G. 599, *630*
Mättig, P. 2, 44, *620, 624*
Matveev, A.N. 583, *628*
Maurin, G. 2, 16, 599, *621, 630*
Maurus, U. 44, *624*
Maxeiner, C. 44, *624*
Maxfield, S.J. 44, *624*
McCann, H. 44, 65, 70, *623, 625-6*
McMahon, T. 2, 16, 599, *621, 630*
Mendiburu, J.-P. 2, 16, 599, *621, 630*
Meier, K. 44, 65, 70, *623, 625-6*
Meinke, R. 65, *625*
Merkel, B. 2, 16, *622*
Mermikides, M. 2, 16, *622*
Merola, L. 599, *630*
Mertiens, H.D. 2, 70, *620, 625*
Messner, R.L. 65, *624*
Meyer, H. 2, 44, 70, *620, 624, 626*
Meyer, H.J. 2, 44, 70, *620, 624, 626*
Meyer, J. 44, *624*
Meyer, O. 2, 70, *620, 626*
Meyer, O.A. 65, *624*
Meyer, T. 2, 44, *620, 623*
Michalowski, S.J. 65, *624*
Michelsen, U. 44, 70, *624-5*
Mikenberg, G. 2, *620*
Mills, H. 2, *621*
Mills, H.E. 44, 65, 70, *623, 625-6*
Mills, R.L. 1, 11, *619*
Minard, M.-N. 2, 16, 599, *621, 630*
Minowa, M. 2, 44, 65, 70, *621, 623, 626*
Mohammadi, M. 2, 16, *621*
Møllerud, R. 2, 16, *622*
Montanet, L. 17, 101, *622, 627*
Morgan, K. 2, 16, 599, *621, 630*

Moricca, M. 2, 16, 599, *621, 630*
Moromisato, J.H. 65, *624*
Muirhead, H. 2, 16, *621*
Muller, F. 2, 16, 599, *621, 630*
Müller, H. 44, *623*
Murphy, P.G. 2, 44, 65, 70, *621, 623, 626*

Nandi, A.K. 2, 16, 599, *621, 630*
Napolitano, M. 599, *630*
Naroska, B. 2, 44, 65, 70, *621, 623, 625-6*
Naumann, L. 2, 16, 599, *621, 630*
Ne'eman, Y. 11, *622*
Neuman, M. 605, *630*
Neumann, B. 2, 44, 70, *620, 624, 626*
Newman, H. 2, 44, 65, *620, 623, 625*
Nikishov, A.I. 583, *628*
Nilsson, B. 2, 16, *622*
Noble, R.J. 550, *627*
Norton, A. 2, 16, 599, *621, 630*
Notz, D. 2, 44, *620, 624*
Nozaki, M. 44, 65, 70, *623, 625-6*
Nozaki, T. 2, 44, 65, 70, *621, 623, 625-6*
Nye, J. 65, *625*

Oberlack, H. 44, *623*
Odaka, S. 44, 70, *623, 626*
Ogg, M. 2, *620*
Oldham, S.J. 70, *626*
Olive, K.A. 17, 101, *622, 627*
Olsen, J.M. 44, *624*
Olsson, J. 2, 44, 65, 70, *621, 623, 625-6*
O'Neill, L.H. 2, 44, 70, *621, 623, 626*
Onions, C. 2, 16, *622*
Orito, S. 2, 44, 65, 70, *621, 623, 626*
Orkin-Lecourtois, A. 2, 16, 599, *621, 630*
Osland, P. 600, *631*

Palladino, V. 599, *630*
Pamela, J. 44, *624*
Pandoulas, D. 2, 44, *620*, *624*
Pansart, J.P. 44, *624*
Paoluzi, L. 2, 16, 599, *621*, *630*
Pape, K.H. 2, 44, 70, *620*, *624*
Paradiso, J. 2, *620*
Parke, S.J. 588, *629*
Parrour, G. 2, 16, *622*
Parsa, Z. 595, *629*
Passarino, G. 595, *630*
Pastore, F. 2, 16, *622*
Pauss, F. 2, 16, *621*
Peacock, A.W. 105, *627*
Pearce, G.F. 2, 44, 70, *621*, *623*, *626*
Peccei, R.D. 104, *627*
Peierls, R. 13, *622*
Peise, G. 2, 44, *620*, *624*
Peruzzi, I. 65, *624*
Petersen, A. 2, 44, 65, 70, *621*, *623*, *625-6*
Petrucci, G. 2, 16, *621*
Pevsner, A. 2, *620*
Piano Mortari, G. 2, 16, 599, *621*, *630*
Piccolo, M. 65, *624*
Pielorz, J. 44, *624*
Pierre, F. 44, *624*
Pietarinen, E. 2, 16, 65, 70, *621*, *625-6*
Pimiä, M. 2, 16, 599, *621*, *630*
Pistoni, G. 611, *631*
Placci, A. 2, 16, 599, *621*, *630*
Placzek, G. 13, *622*
Plothow-Besch, H. 2, 16, *622*
Poelz, G. 2, 44, *620*, *624*
Pohl, M. 44, 65, *623*, *625*
Poggioli, L. 44, *624*
Polvado, R.O. 44, 70, *624*, *626*
Polverel, M. 2, 16, *622*
Poppe, M. 44, *624*
Popov, V.N. 13, *622*
Porte, J.P. 2, 16, *621*

Porter, F.C. 17, 101, *622*, *627*
Poschmann, F.P. 2, 44, 65, *620*, *625*
Prepost, R. 65, *625*
Prosper, H. 2, *621*
Proudfoot, J. 2, *620*
Pyrlik, J. 2, 44, *620*, *624*

Quarrie, D.R. 2, 44, *620*, *624*

Rabany, M. 599, *630*
Radermacher, E. 2, 16, 599, *621*, *630*
Raine, C. 44, *624*
Randoll, H. 44, *623*
Ranga Swamy, T.N. 2, *620*
Ransdell, J. 2, 16, 599, *621*, *630*
Raupach, F. 2, 44, 70, *620*, *624-5*
Read, Jr. A.L. 65, *624*
Reich, H.D. 2, 70, *620*, *625*
Reithler, H. 2, 16, 599, *621*, *630*
Repellin, J.-P. 2, 16, *622*
Retherford, R.C. 1, *619*
Revel, D. 2, *620*
Revol, J.-P. 2, 16, 44, 65, *620-1*, *623*, *625*
Rich, J. 2, 16, 599, *621*, *630*
Rich, K. 65, *624*
Richter, B. 550, *627*
Rieseberg, H. 2, 44, 65, 70, *621*, *623*, *626*
Riethmüller, R. 2, *620*
Rijssenbeek, M. 2, 16, 599, *621*, *631*
Rimkus, J. 2, 44, *620*, *624*
Rimoldi, A. 2, 16, *622*
Ringel, J. 2, *620*
Ritson, D.M. 65, *624*
Rittenberg, A. 17, 101, *622*, *627*
Ritus, V.I. 583, *628*
Rivoal, M. 44, *624*
Roberts, C. 2, 16, 599, *621*, *631*
Rohde, M. 2, 44, 65, *620*, *625*
Rohlf, J. 2, 16, *621*
Römer, O. 2, *620*

Ronat, E. 2, *620*
Ronga, F. 65, *624*
Roos, M. 17, 101, *622*, *627*
Roper, L.D. 17, 101, *622*, *627*
Rosenberg, L.J. 65, *625*
Ross, D.A. 70, *626*
Ross, R.R. 17, 101, *622*, *627*
Rossi, P. 2, 16, *621*
Rosskamp, P. 44, *624*
Rössler, M. 2, 70, *620*, *626*
Rost, M. 2, 44, 70, *620*, *624*, *626*
Roth, F. 2, *620*
Rothenberg, A. 2, 16, *622*
Roussarie, A. 2, 16, *622*
Rubbia, C. 2, 16, 599, *621*, *631*
Rubio, A. 65, *625*
Rubio, J.A. 44, *623*
Rückl, R. 101, 104, *627*
Rühmer, W. 2, *620*
Rüsch, R. 2, *620*
Rushton, M. 44, *624*
Rykaczewski, H. 2, 44, 65, *620*, *623*

Sack, Th. 595, *629*
Sadoulet, B. 2, 16, 599, *621*, *631*
Sajot, G. 2, 16, 599, *621*, *631*
Salam, A. 1, 16, *619*
Salicio, J. 44, 65, *623*, *625*
Salmon, G.L. 2, *620*
Salvi, G. 2, 16, 599, *621*, *631*
Salvini, G. 2, 16, 599, *621*, *631*
Sander, H.G. 2, 44, *620*, *624*
Sanguinetti, G. 599, *630*
Sass, J. 2, 16, 599, *621*, *631*
Sato, A. 2, 44, 65, 70, *621*, *623*, *626*
Saudraix, J. 2, 16, 599, *621*, *631*
Sauvage, G. 2, 16, *622*
Savoy-Navarro, A. 2, 16, 599, *621*, *631*
Saxon, D.H. 2, *620*
Saxon, D.S. 553, 556, *627-8*
Scarr, J.M. 44, *624*
Scarr, M. 70, *626*
Schacher, J. 2, 16, *622569*

Schacht, P. 44, *6234*
Schachter, M.J. 44, *6234*
Schierholz, G. 70, *626*
Schiff, L.I. 553, 556, *627-8*
Schindler, R.H. 17, 101, *622*, *627*
Schinzel, D. 2, 16, 599, *621*, *631*
Schliwa, M. 2, *620*
Schmidt, D. 2, 44, 65, 70, *620*, *624-6*
Schmidt, G. 44, *623*
Schmitt, I. 70, *626*
Schmitz, D. 2, 44, *620*, *624*
Schmüser, P. 2, *620*
Schneekloth, U. 65, *625*
Schneider, H. 44, *623*
Schnell, W. 550, *627*
Schröder, V. 44, *623*
Schubert, K.R. 17, 101, *622*, *627*
Schultz von Dratzig, A. 2, *620*
Schulz, I. 44, 65, *623*, *625*
Schütte, W. 44, *624*
Schwinger, J. 1, 7, 583, 605, *619*, *628*, *631*
Sciacca, G. 599, *630*
Scott, W. 2, 16, 599, *621*, *631*
Scrock, R.E. 17, 101, *622*, *627*
Sechi-Zorn, B. 2, 44, 65, 70, *620*, *624*, *626*
Sedgebeer, J. 2, *620*
Sette, G. 599, *630*
Shah, T.P. 2, 16, 599, *621*, *631*
Shambroom, W.D. 65, *624*
Shapira, A. 2, *620*
Siebke, H. 44, *624*
Siegrist, J. 550, *627*
Siegrist, J.L. 2, 16, *622*
Sindt, H. 44, *623*
Sinram, K. 2, 44, 65, *620*, *623*, *625*
Sirlin, A. 601, *631*
Sivers, D. 70, *626*
Shapira, A. *620*
Skard, J.A. 2, 44, 70, *620*, *624-5*
Skard, J.A.J. 65, *625*
Skillicorn, I.O. 44, *624*

Skuja, A. 2, 44, 70, *620*, *624*, *626*
Smith, D. 2, 16, *621*
Smith, J.G. 65, *624*
Smith, K.M. 44, *624*
Söding, P. 2, *620*
Soffer, J. 599, *630*
Sokolov, A.A. 539, 583, *627–9*
Speiser, D.R. 11, *622*
Spiro, M. 2, 16, 599, *621*, *631*
Spitzer, H. 2, 44, 70, *620*, *624*, *626*
Spitzer, J. 65, *626*
Steffen, P. 2, 44, 65, 70, *621*, *623*, *625–6*
Steiner, H.M. 2, 16, *622*
Stella, B. 2, 44, 70, *620*, *624*, *626*
Stephens, K. 2, 65, 70, *621*, *625–6*
Steuer, M. 44, 65, *623*, *625*
Stimpfl, G. 2, 16, *622*
Stirling, W.J. 589, *629*
Stocker, F. 2, 16, *622*
Strauss, J. 2, 16, 599, *621*, *631*
Streets, J. 2, 16, *621*
Strupperich, K. 44, *624*
Su, S.-Q. 592, *629*
Suda, T. 2, *621*
Sumorok, K. 2, 16, 599, *621*, *631*
Suura, H. 39, *623*
Suzuki, M. 17, 101, *622*, *627*
Swartz, M. 2, 16, *622*
Swider, G.M. 44, 65, *623*, *625*
Szonsco, F. 2, 16, 599, *621*, *631*

Takeda, H. 2, 44, 65, 70, *621*, *623*, *626*
Takeshita, T. 65, *625*
Tang, H.W. 2, 44, 65, *620*, *623*, *625*
Tang, L.G. 2, *620*
Tao, C. 2, 16, 599, *621*, *631*
Tarski, J. 11, *622*
Taylor, T.R. 588, *629*
Teiger, J. 2, 16, *622*
Ternov, I.M. 539, 583, *627–9*
Terrano, A.E. 70, *626*
Teuchert, D. 44, 65, *623*, *625*

Thompson, G. 2, 16, 599, *621*, *6*
Thomson, J.C. 44, *624*
Tigner, M. 549, *627*
Timm, U. 2, 44, 70, *620*, *624*, *62*
Timmer, J. 2, 16, 599, *621*, *631*
Timmermans, J. 599, *630*
Ting, S.C.C. 2, 44, 65, *620*, *62*, *625*
Tomonaga, S. 1, 7, *619*
Törnqvist, N.A. 17, 101, *622*, *62*
Totsuka, Y. 2, 44, 65, 70, *621*, *62*, *625–6*
Tovey, S. 2, 16, *622*
Trines, D. 2, 44, *620*, *624*
Trippe, T.G. 17, 101, *622*, *627*
Troost, W. 22–3, 31, 41, 65, 68, 7(80, 84, 105, *622–3*, *625–7*
Trower, W.P. 17, 101, *622*, *627*
Tscheslog, E. 2, 16, 599, *621*, *631*
Tung, K.L. 2, 44, 65, *620*, *623*, *62*
Tuominiemi, J. 2, 16, 599, *621*, *6*
Tylka, A.J. 44, *624*

Van der Meer, S. 2, 16, *621*
van der Waerden, B.L. 590, *629*
Van Eijk, B. 2, 16, *621*
van Neerven, W.L. 592, *629*
Vannini, C. 599, *630*
Vannucci, F. 2, 44, 65, *620*, *625*
van Staa, R. 2, 44, 70, *620*, *62*, *626*
van Swol, R. 599, *630*
Veillet, J.J. 44, *624*
Velasco, J. 599, *630*
Vercesi, V. 2, 16, *622*
Vermaseren, J.A.M. 70, *626*
Vialle, J.-P. 2, 16, 599, *621*, *631*
Visco, F. 599, *630*
Vismara, G. 599, *631*
Volkov, D.V. 587, *629*
von Goeler, E. 65, *624*
von Holtey, G. 599, *630*
von Krogh, J. 2, 44, 65, 70, *62*, *623*, *625–6*

Voronov, N.A. 22, *623*
Vrana, J. 2, 16, 599, *621*, *631*
Vuillemin, V. 2, 16, 599, *621*, *631*

Wacker, K. 2, *620*
Wagman, G.S. 17, 101, *622*, *627*
Wagner, A. 2, 44, 65, 70, *621*, *623*, *625-6*
Wagner, S. 65, *625*
Wagner, W. 2, 44, 70, *620*, *624*, *626*
Wahl, H.D. 2, 16, 599, *621*, *631*
Wallraff, W. 2, 44, *620*, *624*
Waloschek, P. 2, 44, 70, *620*, *624*, *626*
Wang, M.Q. 44, *623*
Wang, X.R. 2, *620*
Wang, Y.X. 65, *625*
Ward, J.C. 588, *629*
Warming, P. 2, 44, 65, 70, *621*, *623*, *625-6*
Watanabe, Y. 2, 44, 70, *621*, *623*, *626*
Watkins, P. 2, 16, 599, *621*, *631*
Weber, G. 2, 44, 65, 70, *621*, *623*, *625-6*
Wedemeyer, R. 2, 44, *620*, *624*
Wei, P.S. 2, *620*
Weidberg, A.R. 2, 16, *622*
Weinberg, S. 1, 16, *619*
Weinstein, R. 65, *624*
Welch, G. 2, 70, *620*, *626*
Welch, G.E. 44, *624*
Wenniger, H. 44, 70, *623*, *626*
Wentzel, G. 559, *628*
Wermes, N. 2, 44, *620*, *624*
Wess, J. 587, *629*
Weyl, H. 590, *629*
White, M. 2, 44, 65, *620*, *623*, *625*
Whittaker, J.B. 65, *625*
Wiik, B.H. 2, 44, *619*, *624*
Willrodt, J. 70, *626*
Wilson, J. 2, 16, 599, *621*, *631*
Wilson, R. 2, 16, *621*
Winter, G.G. 2, 44, 70, *620*, *624*, *626*

Wiser, D.E. 65, *625*
Wohl, C.G. 17, 101, *622*, *627*
Wolf, E. 543, *627*
Wolf, G. 2, 44, *620*, *624*
Wollstadt, M. 2, 44, *620*, *624*
Woodworth, P.L. 2, *620*
Wriedt, H. 2, 44, 70, *621*, *623*, *626*
Wu, G.H. 2, *620*
Wu, S.L. 2, 70, *619-20*, *626*
Wu, S.X. 44, 65, *623*, *625*
Wu, T.T. 12, 22-3, 31, 41, 65, 68, 70, 80, 84, 105, 551, 553-4, 556, 558, 569, 571, 579-80, 583, 585, 598-600, *622-3*, *625-31*
Wu, T.W. 2, 65, *620*, *625*
Wulz, C.-E. 2, 16, *621*
Wyler, D. 595, *629*

Xi, J.P. 2, *620*
Xiao, Ch. 44, *624*
Xie, Y.G. 2, 16, *621*
Xu, Z. 588-9, *629*
Xue, S.T. 44, *624*

Yamada, S. 2, 44, 65, 70, *621*, *623*, *625-6*
Yanagisawa, C. 2, 44, 70, *621*, *623*, *626*
Yang, C.N. 1, 11, *619*
Yang, P.C. 2, *620*
Yarker, S. 2, *620*
Yen, W.L. 2, *621*
Yennie, D.R. 39, *623*
Yokoya, K. 580-2, *628*
Yost, G.P. 17, 101, *622*, *627*
Youngman, C. 2, *620*
Yu, C.C. 65, *625*
Yu, X.H. 2, *620*
Yvert, M. 2, 16, 599, *621*, *631*

Zaccone, H. 2, ·16, *622*
Zachara, M. 44, 65, *624-5*
Zakrzewski, J.A. 2, 16, *622*
Zdarko, R.W. 65, *625*

Zech, G. 2, 44, 70, *620*, *624*, *626*
Zeller, W. 2, 16, *622*
Zeng, Y.Q. 65, *625*
Zhang, D.-H. 588-9, *629*
Zhang, N.L. 2, 65, *620*, *625*
Zhang, Y. 44, *623*
Zhu, R.Y. 2, 44, 65, *620*, *623*, *625*
Zhu, Y.C. 44, *623*
Zimmerman, W. 2, 44, 70, *620*, *624*, *626*
Zobernig, G. 2, *619-20*
Zorn, G.T. 2, 44, 65, 70, *620*, *624-6*
Zumino, B. 587, *628*
Zurfluh, E. 2, 16, 599, *621*, *631*
Zweig, G. 1, *619*

Subject Index

Abelian gauge symmetry 1
accelerator
  linear 549-50, 583-5

Beam luminosity 3, 65, 550, 584-5
beamstrahlung 549-85
  photon energy 564-5, 567-8, 581-2
Bhabha scattering 3, 28, 44, 79, 122-4
bremsstrahlung
  correction 2-3
  double 3, 65-78
    in QCD 70-8
    in QED 65-9
  gluon 3, 23-4
  mass effects in 79-100
  multiple 79
  process 2, 29
  single
    in QCD 47-64
    in QED 31-46
bunch
  charge distribution 555, 560, 580-2
  correlation length 554-5, 559, 581
  crossing 571, 582
  length 550-2, 554-5, 559-60, 581
  mean radius 551
  parameters 550
  shape 555-6

Charge asymmetry in $e^+e^- \to \mu^+\mu^-$ 3, 44, 65
CLIC 550, 585
collinearity factor 89-98
collinear particles 28-9, 78-9, 596-7

color
  degree of freedom 1-2, 9, 62
  matrices 11, 613-17
    traces of 11, 51, 62, 614-17
cross sections 14-16

Decay rates 16
Dirac equation 8, 23, 537-8, 557-60
disruption factor 551

Electron
  classical radius 551, 554
  virtual - length 554
electron-positron annihilation
  into e^+e^- 3, 28, 44, 79
  into $e^+e^-\ell^+\ell^-$ 597
  into $e^+e^-\gamma$ 28, 39-44, 46, 597
  into $e^+e^-\gamma\gamma$ 21, 65, 69, 100
  into $\gamma\gamma$ 14, 18-19, 24-7, 35, 43
  into $\gamma\gamma\gamma$ 31-5
  into $\gamma\gamma\gamma\gamma$ 65-9, 100
  into $\mu^+\mu^-$ 2-3, 44, 82-3, 539-42
  into $\mu^+\mu^-\gamma$ 28, 35-9, 45-6, 55, 79, 82-4, 91-5
  into $\mu^+\mu^-\gamma\gamma$ 65, 69, 97-100
  into $\nu\bar{\nu}\gamma$ 592
  into $\nu\bar{\nu}\gamma\gamma$ 592
  into $q\bar{q}$ 3, 14-15
  into $q\bar{q}g$ 3, 29, 52-5
  into $q\bar{q}gg$ 3, 29, 70-4
  into $q\bar{q}q\bar{q}$ 3, 70, 77-8
  into $q\bar{q}q'\bar{q}'$ 3, 74-7
  into $Z\gamma$ 595
  into 2 jets 3
  into 3 jets 3, 48, 52-5
  into 4 jets 3, 70-8
electroweak theory 1-2

Fermi-Dirac statistics 1
Feynman diagrams 7–19, 583
    for e^-+bunch→ $e^-\gamma$+bunch 553
    for $e^\pm$+bunch→ $e^+e^-e^\pm$+bunch 583–4
    for $e^+e^- \to e^+e^-\gamma$ 39–41
    for $e^+e^- \to \gamma\gamma$ 14, 18, 24–5
    for $e^+e^- \to \gamma\gamma\gamma$ 31–2
    for $e^+e^- \to \gamma\gamma\gamma\gamma$ 65–6
    for $e^+e^- \to \mu^+\mu^-$ 82
    for $e^+e^- \to \mu^+\mu^-\gamma$ 36, 45–6
    for $e^+e^- \to q\bar{q}$ 14–15
    for $e^+e^- \to q\bar{q}g$ 52–3
    for $e^+e^- \to q\bar{q}gg$ 70–2
    for $e^+e^- \to q\bar{q}q'\bar{q}'$ 75–6
    for $ggg \to {}^{2S+1}L_J$ 102–3
    for γ+bunch→ e^+e^-+bunch 583–4
    for $\gamma\gamma \to \gamma\gamma$ 605–6
    for $q\bar{q} \to gg$ 48–9
    for $qq' \to qq'g$ 56–7
    for $Z \to e^+e^-$ 16–17
    for $Z \to e^+e^-\gamma$ 593–4
    loops in 12, 598, 605–11
Feynman parameters 605
Feynman rules 7–14

Gamma-matrices 8–9, 16, 68
    traces of 601–3
gauge invariance 15
gauge particle 1–2
geometrical optics 599
Glashow-Weinberg-Salam model 1–2, 16
gluino 588
gluon 1–2
    four - vertex 10–11, 47
    three - vertex 10–11, 47, 56, 58
graviton 592–3
gravity 592–3

Helicity
    amplitudes 4, 21
        for $e^+e^- \to e^+e^-$ 122–4
        for $e^+e^- \to e^+e^-\ell^+\ell^-$ 597

for $e^+e^- \to e^+e^-\gamma$ 39–44, 130–6, 166–78, 597
for $e^+e^- \to e^+e^-\gamma\gamma$ 145–53, 265–368
for $e^+e^- \to \gamma\gamma$ 25–7, 119–20
for $e^+e^- \to \gamma\gamma\gamma$ 31–4, 124–5, 154–7
for $e^+e^- \to \gamma\gamma\gamma\gamma$ 65–9, 137–9, 179–93
for $e^+e^- \to \mu^+\mu^-$ 82, 92, 120–1, 540
for $e^+e^- \to \mu^+\mu^-\gamma$ 37–9, 91–3, 125–30, 157–66
for $e^+e^- \to \mu^+\mu^-\gamma\gamma$ 97–100, 139–45, 193–265
for $e^+e^- \to \nu\bar{\nu}\gamma$ 592
for $e^+e^- \to \nu\bar{\nu}\gamma\gamma$ 592
for $e^+e^- \to q\bar{q}$ 369–70
for $e^+e^- \to q\bar{q}g$ 52–5, 370–1
for $e^+e^- \to q\bar{q}\gamma$ 371–3
for $e^+e^- \to q\bar{q}gg$ 70–4, 373–8
for $e^+e^- \to q\bar{q}q\bar{q}$ 77, 381–5
for $e^+e^- \to q\bar{q}q'\bar{q}'$ 76–7, 379–81
for $e^+e^- \to Z\gamma$ 595
for $gg \to gg$ 403–4, 588, 592
for $gg \to ggg$ 496–501, 588, 592
for $gg \to gggg$ 588, 590, 592
for $gg \to g\,{}^1S_0$ 111, 519–21
for $gg \to g\,{}^3S_1$ 521–3
for $gg \to g\,{}^1P_1$ 524–7
for $gg \to g\,{}^3P_0$ 527–9
for $gg \to g\,{}^3P_1$ 529–32
for $gg \to g\,{}^3P_2$ 532–6
for $gg \to {}^1S_0$ 501–2
for $gg \to {}^3P_0$ 502–3
for $gg \to {}^3P_2$ 503–4
for $gg \to \bar{q}q$ 402–3
for $gg \to \bar{q}qg$ 490–6
for $gg \to \bar{q}qgg$ 590
for $gg \to \bar{q}q\gamma$ 487–90
for $gg \to \bar{q}q\bar{\ell}\ell$ 589

SUBJECT INDEX

Helicity (cont'd)
  amplitudes
    for $gg \to \bar{q}q\bar{q}'q'$ 589
    for $g\gamma \to \bar{q}q$ 401–2
    for $g\gamma \to \bar{q}qg$ 484–7
    for $g\gamma \to \bar{q}q\gamma$ 481–3
    for $\gamma\gamma \to \gamma\gamma$ 607–11
    for $\gamma\gamma \to \bar{q}q$ 400–1
    for $\gamma\gamma \to \bar{q}qg$ 479–81
    for $\gamma\gamma \to \bar{q}q\gamma$ 477–9
    for $gq \to qg$ 392–3
    for $gq \to q\gamma$ 391–2
    for $gq \to qgg$ 439–45
    for $gq \to qg\gamma$ 435–9
    for $gq \to q\gamma\gamma$ 430–2
    for $gq \to q\,^1S_0$ 504–5
    for $gq \to q\,^3P_0$ 505–6
    for $gq \to q\,^3P_1$ 506–7
    for $gq \to q\,^3P_2$ 508–9
    for $g\bar{q} \to \bar{q}g$ 396–7
    for $g\bar{q} \to \bar{q}\gamma$ 395–6
    for $g\bar{q} \to \bar{q}gg$ 459–65
    for $g\bar{q} \to \bar{q}g\gamma$ 455–9
    for $g\bar{q} \to \bar{q}\gamma\gamma$ 449–52
    for $g\bar{q} \to \bar{q}\,^1S_0$ 509–10
    for $g\bar{q} \to \bar{q}\,^3P_0$ 510–1
    for $g\bar{q} \to \bar{q}\,^3P_1$ 511–12
    for $g\bar{q} \to \bar{q}\,^3P_2$ 513–14
    for $\gamma q \to qg$ 390–1
    for $\gamma q \to q\gamma$ 389–90
    for $\gamma q \to qgg$ 432–5
    for $\gamma q \to qg\gamma$ 427–9
    for $\gamma q \to q\gamma\gamma$ 425–7
    for $\gamma\bar{q} \to \bar{q}g$ 394–5
    for $\gamma\bar{q} \to \bar{q}\gamma$ 393–4
    for $\gamma\bar{q} \to \bar{q}gg$ 452–5
    for $\gamma\bar{q} \to \bar{q}g\gamma$ 447–9
    for $\gamma\bar{q} \to \bar{q}\gamma\gamma$ 445–7
    for $qq \to qq$ 387–8
    for $qq \to qqg$ 415–19
    for $qq \to qq\gamma$ 413–15
    for $q\bar{q} \to gg$ 48–52, 399–400
    for $q\bar{q} \to g\gamma$, 398–9
    for $q\bar{q} \to ggg$ 472–7
    for $q\bar{q} \to gg\gamma$ 469–71
    for $q\bar{q} \to g\gamma\gamma$ 467–9
    for $q\bar{q} \to ggZ$ 589
    for $q\bar{q} \to g\,^1S_0$ 514–15
    for $q\bar{q} \to g\,^3P_0$ 515–16
    for $q\bar{q} \to g\,^3P_1$ 517–18
    for $q\bar{q} \to g\,^3P_2$ 518–19
    for $q\bar{q} \to gZ$ 589
    for $q\bar{q} \to \gamma\gamma$ 397–8
    for $q\bar{q} \to \gamma\gamma\gamma$ 465–7
    for $q\bar{q} \to q\bar{q}$ 388–9
    for $q\bar{q} \to q\bar{q}g$ 421–5
    for $q\bar{q} \to q\bar{q}\gamma$ 419–21
    for $q\bar{q} \to q\bar{q}Z$ 589
    for $q\bar{q} \to q'\bar{q}'$ 386–7
    for $q\bar{q} \to q'\bar{q}'g$ 410–2
    for $q\bar{q} \to q'\bar{q}'gg$ 589
    for $q\bar{q} \to q'\bar{q}'\gamma$ 409–10
    for $q\bar{q} \to Z$ 589
    for $qq' \to qq'$ 385–6
    for $qq' \to qq'g$ 55–63, 406–8
    for $qq' \to qq'\gamma$ 405–6
    for $Z \to e^+e^-\gamma$ 593–5
    for $ggg \to {}^1S_0$ 106–10
    for $ggg \to {}^3S_1$ 112
    for $ggg \to {}^1P_1$ 112–14
    for $ggg \to {}^3P_0$ 114–15
    for $ggg \to {}^3P_1$ 115–16
    for $ggg \to {}^3P_2$ 116–18
  states 4, 21–9
    for fermions 21–2, 68
    for gluons 23–4
    for photons 22–3
    for quarks 55
Higgs boson 2, 9
Higgs mechanism 1
high energy limit 4
  validity of 19, 27–9

Infrared factor 39, 62, 592
intermediate vector boson 2

Jet 2–3

Klein-Gordon equation 556–9

LEP 3, 549–50
lepton 2
LHC 600

Mandelstam variables 18
mass effects 19, 28–9, 79–100
   for amplitudes 84–100
   for double bremsstrahlung 95–7
   for $e^+e^- \to \mu^+\mu^-\gamma$ 82–4, 91–5
   for $e^+e^- \to \mu^+\mu^-\gamma\gamma$ 97–100
   for single bremsstrahlung 80–2, 88–91
Mellin transform 569–70
metric 7

Neutrino 2, 592
non-Abelian gauge symmetry 1, 11

Optical theorem 13, 599

Pauli matrices 8, 597–8
PEP 43, 65, 550
PETRA 2, 44, 65, 550
photon 1
   helicity *see* helicity
photon-photon scattering 605–11
$p\bar{p}$ collider 2–3
polarization 537–47
   degrees of freedom 7
   for fermions 537–9
   for gluons 542–5
   for gravitons 592
   for massive spin-1 particles 593–5
   for photons 542–5, 588–9, 591
   in $e^+e^- \to \mu^+\mu^-$ 539–42
   in $gg \to g\,{}^1S_0$ 545–7
   linear 543–5
   sum 15, 112, 117
   vector 4, 22–4, 35, 47–8, 64
      phase choice of 35, 588–90

QCD 1–3, 9–11, 101
   structure constants 11
QED 1–3, 9

quark 1–3
quarkonia 3, 101–18, 549
   production 101–18
quaternions 597–8

Regularization 608

Schwarz inequality 569
SPEAR 550
spin
   factor 565, 582
   operator 537
spinor 8, 21, 68, 537–9
   product 590
   two-component 590–1
SSC 600
standard model 1–2
Super 550–1, 554–5, 560
supersymmetry 587–8
symmetry
   local 1
   factor 12–13, 16
synchrotron radiation 549–50

Tevatron 3
TLC 550, 585

Wess-Zumino model 587–8
Weyl-van der Waerden formalism 590–2
WKB-approximation 559
W-particle 1–2

Yang-Mills theory 1

Z-decay 16–17, 593–5
Z-exchange 44–6, 65, 121–4, 127–30, 133–6
Z-particle 1–2, 16
Z-propagator 45
Z-width 45

www.ingramcontent.com/pod-product-compliance
Ingram Content Group UK Ltd.
Pitfield, Milton Keynes, MK11 3LW, UK
UKHW022152230426
12049UKWH00003BA/53